The Role of Chemistry in the Evolution of Molecular Medicine

The Role of Chemistry in the Evolution of Molecular Medicine

27-29 June, 2003
Szeged, Hungary

A Tribute to Professor Albert Szent-Györgyi

Proceedings of the Symposium held at the University of Szeged
to celebrate the 110th Birthday of Professor Albert Szent-Györgyi

Edited by

Imre G. Csizmadia
Botond Penke
Gábor Tóth

Assistants to the Editors

Katalin Bicsérdy
Suzanne K. Lau
Michelle A. Sahai

2004

ELSEVIER

Amsterdam • Boston • Heidelberg • London • New York • Oxford • Paris
San Diego • San Francisco • Singapore • Sydney • Tokyo

ELSEVIER B.V.
Radarweg 29
P.O. Box 211, 1000 AE
Amsterdam, The Netherlands

ELSEVIER Inc.
525 B Street, Suite 1900
San Diego, CA 92101-4495
USA

ELSEVIER Ltd
The Boulevard, Langford Lane
Kidlington, Oxford OX5 1GB
UK

ELSEVIER Ltd
84 Theobalds Road
London WC1X 8RR
UK

First edition 2004

Library of Congress Cataloging in Publication Data
A catalog record is available from the Library of Congress.

British Library Cataloguing in Publication Data
A catalogue record is available from the British Library.

Reprinted from the Journal of Molecular Structure (Theochem), Vol. 666-667.

Printed and bound in the United Kingdom
Transferred to Digital Print 2010

ISBN: 9780444515889

FSC
Mixed Sources
Product group from well-managed
forests and other controlled sources

Cert no. SGS-COC-2953
www.fsc.org
© 1996 Forest Stewardship Council

SPONSORS

The following organizations are greatly acknowledged for their help and support: The Hungarian Academy of Sciences, the University of Szeged and last but not least the pharmaceutical company of Gedeon Richter.

GEDEON RICHTER LTD.

PREFACE

This Volume contains the Proceedings of the Symposium held at the University of Szeged (27–29 June 2003) under the title:

"THE ROLE OF CHEMISTRY IN THE EVOLUTION OF MOLECULAR MEDICINE"
A tribute to Professor Albert Szent-Györgyi

The Symposium we are reporting here was groundbreaking in more than one way, laying down the foundation for the future of the field of molecular medicine. In fact, the effervescent atmosphere with all its excitement appeared to be indicative of a New Renaissance.

One of the novelties of this Symposium was that it represented the large age gap spanning almost 60 years, as some of the participants were 20 years old, while others were perhaps nearly 80 years young. The inexperienced came to *learn from* and be *inspired by* those who had already acquired experience.

The other novelty was the extent of the participants' interdisciplinary backgrounds. In fact, participants represented a wide range of disciplines in Molecular Science, including Mathematics, Physics, Computer Science, Chemistry, Biochemistry, Biology and Medicine, covering the whole territory in agreement with the legacy of Professor Albert Szent-Györgyi.

The Symposium as well as the present volume was particularly inspired by the booklet published in 1960 by Albert Szent-Györgyi under the title:

Introduction to Submolecular Biology

The booklet is reprinted here in its entirety in the Appendix of this volume.

The research of Professor Albert Szent-Györgyi, which led to the Nobel Prize in 1937 already foreshadowed the conclusion that the human body is nothing else than an automatic molecular machine, in which not wheels but molecules are turning and are ultimately responsible for all bodily functions.

However, the ultimate legacy of Professor Albert Szent-Györgyi is best summarized, by himself, in the mid 20th Century as printed in the above booklet (see page 856 in this volume).

> *"The distance between those abstruse quantum mechanical calculations*
> *and the patient bed may not be as great as believed"*

We know that his prophetic vision has not been fulfilled as yet; the time is yet to come, perhaps in the 21st Century, when his legacy will be practiced on a daily basis.

Imre G. Csizmadia   Botond Penke   Gábor Tóth

Contents

ELSEVIER

Journal of Molecular Structure (Theochem) 666–667 (2003) 1–9

www.elsevier.com/locate/theochem

Solid state physics of biological macromolecules: the legacy of Albert Szent-Györgyi

János Ladik*

Chair for Theoretical Chemistry and Laboratory of the National Foundation for Cancer Research, Friedrich-Alexander University, Erlangen-Nuremberg, Egerland Str. 3, D-91058 Erlangen-Nurnberg, Germany

Abstract

A. Szent-Györgyi contributing so much to the citric acid (oxygen metabolism) cycle in which subsequent electron transfers occur in eight proteins has come to the idea already in 1941 that proteins have to be conductors. This hypothesis was first not accepted because of the too large gap and non-periodic nature of a protein chain.

With the development of the theory of disordered systems and the occurrence of high-speed computers it was possible to show that both proteins and nucleotide base stacks are good hopping semiconductors, proving the correctness of Szent-Györgyi's original idea.

Ab initio Hartree–Fock correlation corrected band structures were computed both for homopolypeptides and periodic nucleotide base stacks. On the basis of these band structures applying the intermediate exciton theory the UV spectra of different biopolymers were calculated. The results for a cytosine stack, polyglycine and polyalanine are in good agreement with experiments.

Finally, the expected further development of the quantum theory of biopolymers are discussed.
© 2003 Elsevier B.V. All rights reserved.

Keywords: Hopping conductivity of disordered proteins; a.c. conductivity of aperiodic nucleotide base stacks; The frequency-, water content-, temperature-dependence of the conductivity of DNA

1. Introduction

In 1937, Albert Szent-Györgyi obtained the Nobel Prize in Medicine and Physiology besides for discovering vitamin C for his seminal contributions to the establishment of the oxygen metabolism cycle (citric acid cycle, Szent-Györgyi–Krebs cycle) [1]. In his works he has demonstrated that cellular respiration is strongly accelerated by catalytic amounts of succinate, fumerate, malate or oxoloacetate and that these compounds were interconverted according to the reaction sequence succinate → fumerate → malate → oxoloacetate [1]. One should mention that Szent-Györgyi was the only Hungarian who was resident of the country (here in Szeged) when he has obtained the Prize.

In the citric acid cycle, the different reactions are catalyzed by different (altogether eight) enzymes. This means that in running around the cycle electrons have to be transferred through the eight-enzymatic proteins. This observation has led Szent-Györgyi to the hypothesis that proteins are conductors, which make possible these electron transfers [2,3].

* Tel.: +49-9131-85-28831; fax: +49-9131-85-277-36.
E-mail address: janos.ladik@chemie.uni-erlangen.de (J. Ladik).

0166-1280/$ - see front matter © 2003 Elsevier B.V. All rights reserved.
doi:10.1016/j.theochem.2003.08.050

Szent-Györgyi's hypothesis was for many years not accepted by the scientific community. This has had two reasons.

1. The periodic part of the proteins, the peptide group (see polyglycin) has a gap value of about 7.5 eV, which is far too large for the existence of thermal semiconduction.
2. Real proteins are non-periodic due to their different side chains and different conformations of them.

In 1941 (and also for many years after) the theory of aperiodic (disordered) chains was not worked out and no high-speed electronic computers were not available to make a calculation of the frequency-dependent hopping conductivity of disordered polymeric chains as proteins. As it will be shown in this paper the situation in the meantime has been changed completely. One can calculate the frequency-dependent hopping conductivity of proteins (and also of DNA) if one knows the sequence of the amino acids or nucleotide bases, respectively, together with the conformation of the chain. In both cases, it has turned out that at frequencies above 10^{10}–10^{11} s^{-1} the a.c. hopping conductivity is quite large. It should be mentioned that the time scale of the elementary steps in enzymatic reactions is $\approx 10^{13}$ s, that is, the results have a large biological relevance.

In Section 2, it will be qualitatively outlined how the electronic density of states (DOS), hopping frequencies and a.c. conductivity were calculated in the case of aperiodic protein chains or nucleotide stacks. In the case of DNA (where experimental results are available) the a.c. conductivity as the function of frequency, water content and temperature is in complete agreement with the measurements.

In a subsequent point, the treatment of periodic base stacks and polypeptides will be outlined in the HF + Moeller–Plesset2 [5] level and the resulting band structures will be discussed [6]. The UV (excitonic spectrum) of biopolymers will be discussed afterwards. Finally, an outlook will be given for the developments, which can be expected in the near future.

2. The hopping conductivity of proteins and aperiodic nucleotide base stacks

The calculation of hopping conductivities was performed in four steps. First, one calculates the total electronic DOS of the disordered chains. For this, the so-called negative factor counting (NFC) method [7] was applied in its most general form [8] (taking into account also S–S-bridges in cystin residues [9]). Here and subsequently no mathematical formulations are given (because it is quite complicated and therefore they would not fit in the general scope of this Symposium Proceedings). The interested reader can find out the technical details from the ample references. Of course, the results will be given in the different figures.

Having calculated DOS, as next step one can apply the standard inverse iteration method [10] to determine the localized wave functions (Anderson localization) belonging to the physically interesting levels (the levels belonging to the upper part of the filled valence band and those belonging to the lower part of the unfilled conduction band). In this way one can treat both hole and electron hopping conduction.

As third step, the so-called hopping frequencies (primary jump rates) were computed using the simplified expression for the electron–acoustic phonon interaction given in the book of Mott and Davis [11]. In the case of proteins, the vibrations of the side chains with respect to each other are the acoustic phonons. In DNA, the motion of the stacked base pairs in the direction of the main axis of the double helix can be described by acoustic phonons. There are, of course, many other vibrations in proteins or DNA, but the above-mentioned motions correspond to the direction of the conduction in both systems.

Using a generalized form of a random walk theory developed by Lax and co-workers [12], one can calculate the frequency-dependent diffusion constant $D(\omega)$ of the electrons and holes, respectively. Substituting these into Einstein's relation

$$\sigma(\omega) = \frac{ne^2}{k_B T} D(\omega) \tag{1}$$

where n is the number of charge carriers per unit volume, k_B is the Boltzmann's constant and T is the absolute temperature. One finally obtains the frequency-dependent complex conductivity. $\sigma(\omega)$ is also T-dependent, as one can see from Eq. (1), but $D(\omega)$ also shows a strong temperature-dependence [12]. The above sketched four-step procedure is described in detail in Ref. [13].

This procedure was first applied to native pig insulin (in its active form) at room temperature [14] (which has 51 amino acid residues and 3 disulphur bridges), taking into account its first 50 unfilled levels. Its amino acid sequence and detailed conformations are known. In Fig. 1 the $|\sigma(\omega)|$ versus ω curve is shown.

One has found that $\sigma_R(\omega)$ (σ_R is the real part of σ) and $|\sigma(\omega)|$ show a very similar ω-dependence. (The reason of this is that $\lim_{\omega \to \infty} \sigma_I(\omega) \to 0$, as it is easy to show [15].) The $\sigma(\omega)$ curve has a saturation value of $5 \times 10^{-3} \, \Omega^{-1} \, cm^{-1}$ at $\omega = 10^{11} \, s^{-1}$ (it should be pointed out that conductivity measurements still can be performed at such large frequencies as it was done in the case of DNA with the help of microwave techniques.)

Since there are no a.c. measurements on proteins, one can compare these results only with the hopping conductivity of amorphous glasses like chalcenogides (see for instance Fig. 15 of Ref. [11] in the frequency range of 10^4–$10^8 \, s^{-1}$). One finds that $|\sigma(\omega)|$ lies between those of Te_2AsSi and As_2Se_3. (The same is true for the $\sigma(\omega)$ values of hen egg white lysozyme.)

The same kind of calculation has been performed also for the inactive form of pig insulin [16]. In this case, the saturation value at $\omega = 10^{10} \, s^{-1}$ is by two orders of magnitude smaller ($|\sigma(\omega)| = 10^{-4} \, \Omega^{-1} \times cm^{-1}$). Since in the inactive form of pig insulin the sequence is the same, only its conformation is different, this shows that the conductivity is strongly dependent on the conformation of a protein molecule.

As mentioned before $|\sigma(\omega)|$ was also computed for hen egg white lysozyme [15], which is an enzyme with 129 amino acid residues and again 3 S–S bridges. In the active form of this protein, the saturation value of $|\sigma(\omega)|$ is $10^{-4} \, \Omega^{-1} \, cm^{-1}$ at $\omega = 10^{11} \, s^{-1}$. In the inactive form it has a saturation value already at $\omega = 10^8 \, s^{-1}$ [17]. This shows again the important role of the conformation of a protein in its conduction properties. It is especially important in the case of the active sites of enzymes, which in most cases change their comformation (especially in the presence of reactants) quite easily [17]. These considerations show that in the case of enzymatic reactions charge transport hopping conductivity is very probably quite important. The investigations were extended also to the native and inhibited form of subtilisin [17]. In the latter case at $\omega = 10^{10} \, s^{-1}$ $|\sigma(\omega)|$ has a saturation value of $10^{-3} \, \Omega^{-1} \, cm^{-1}$, one order of magnitude larger than in its original native form ($10^{-4} \, \Omega^{-1} \, cm^{-1}$). In Table 1, $|\sigma(\omega)|$ values of the calculated proteins at $\omega = 10^{11} \, s^{-1}$ and room temperature are summarized.

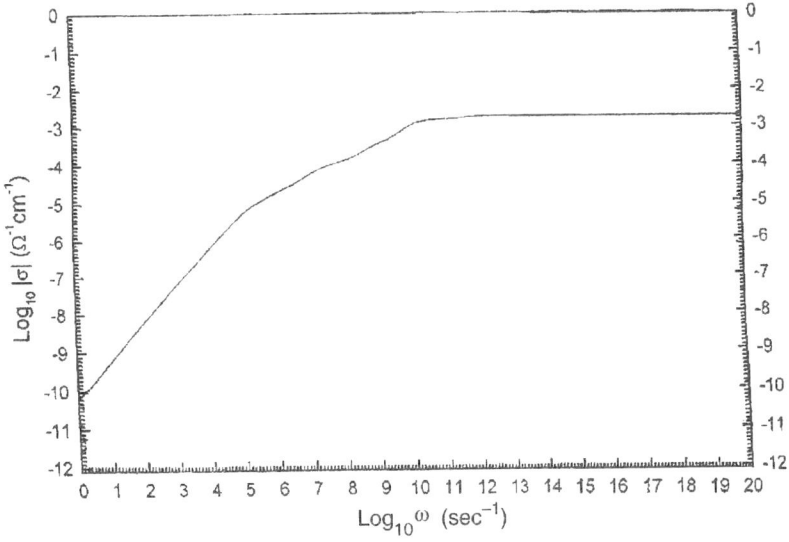

Fig. 1. $|\sigma(\omega)|$ (in $\Omega^{-1} \, cm^{-1}$) $-$ ω (in s^{-1}) curve of native pig insulin (in its active form).

Table 1
The calculated hopping conductivities of different proteins (in $\Omega^{-1}\,cm^{-1}$) at $\omega = 10^{11}\,s^{-1}$ and $T = 298$ K

Native pig insulin	Inactive pig insulin	Active egg white lysozyme	Inactive egg white lysozyme	Native subtilisin	Inhibited subtilisin
10^{-2}	10^{-4}	10^{-4}	10^{-5}	10^{-4}	10^{-3}

All these results show that proteins have a significant hopping conductivity (in a DNA-nucleohistone complex in the nucleosomes, the negatively charged DNA molecule gives over electrons to the protein that possess some positively charged side chains) if a segment of a folded protein closes a small angle with the local effective electric field. To find out the role of electron transfer through a protein molecule one also has to take into account also the other electron transfer channels as: (1) the electronic motion coupled to the motion of protons in hydrated proteins; (2) hopping between different segments of folded proteins and (3) multi-channel tunnelling between more distant segments of the protein. Since all these mechanisms occur in many channels, to find out the overall electron transfer through a protein an appropriately modified form of the Feynman path integral formalism could be applied. This is, of course, not an easy undertaking. We can conclude anyway that Szent-Györgyi's hypothesis from 1941 [2,3] is right, proteins under certain circumstances can be quite good semiconductors. The hopping conductivity values of different proteins, which we have computed are only lower bounds because a minimal basis set was used and no correlation corrections of the electronic structure were taken into account. Namely, it can be shown that both effects increase the hopping conductivity values [18].

In a subsequent series of calculations, the hopping conductivity of aperiodic nucleotide base (pair) stacks were performed. For these investigations, the same four-step procedure was applied as in the case of proteins. The sequence of the bases and the conformation from a part of a human oncogene [19] was taken. One should mention here that according to previous calculations [20–22] on periodic polynucleotides (base stacks with super-phosphate backbone of DNA) have shown that the highest filled (n^*) and the lowest unfilled ($n^* + 1$) bands come from the base stacks and only the $n^* - 1$ and $n^* + 2$ bands, respectively, bands originate from the backbone.

Therefore, the bound structure of the periodic stacks and the distributions of the physically interesting levels of the aperiodic stacks describe quite well the corresponding strands of DNA. Further, the alternating positive and negative charges on the different constituents (net negative charges on the bases, positive ones on the sugar groups, again negative one on the phosphate and positive K^+ counter-ions [20–22]) screen each other out. Therefore, the DOS histograms obtained for the aperiodic base stacks have to be quite similar to those which one would have obtained for native DNA in its B conformation [23].

In Fig. 2, we have shown the DOS [24] of a 100 base and base pairs long segment of the aperiodic stack corresponding to Ref. [19].

It is apparent that the fundamental gaps are larger than they were in the periodic cases (see for instance Refs. [4, p. 77, 25]). In Fig. 3, $\sigma(\omega)$ as a function of ω is given for a double strand with 200 levels and for a single strand applying in both cases the oncogene sequence.

At room temperature and $\omega = 10^{10}\,s^{-1}$ $|\sigma(\omega)| = 5 \times 10^{-1}\,\Omega^{-1}\,cm^{-1}$ (which is one and a half orders of magnitude larger than in the case of native pig insulin) for the base pair sequence with 200 levels in the valence bands regions and about $10^{-5}\,\Omega^{-1}\,cm^{-1}$ with only 100 levels in the valence bands regions. For both cases Clementi's minimal basis [26] was used.

In further computations besides Clementi's minimal basis, his double ζ one [27] was applied for a single stack of the oncogene sequence taking into account 100 levels in the valence bands regions. As one can see in Fig. 4 the saturation value of $|\sigma(\omega)|$ is $\approx 7 \times 10^{-7}\,\Omega^{-1}\,cm^{-1}$ at $\omega = 10^{5}\,s^{-1}$ and $\approx 7 \times 10^{-6}\,\Omega^{-1}\,cm^{-1}$ at $\omega = 10^{12}\,s^{-1}$ (Fig. 4).

We have corrected the double $|\sigma(\omega)| - \omega$ curves for correlation at the MP2 level [18]. For this purpose, we have introduced into the matrix elements of the different units (bases or base pairs) these corrections before the DOS calculation with the help of the NFC procedure (for the mathematical details see Ref. [28]).

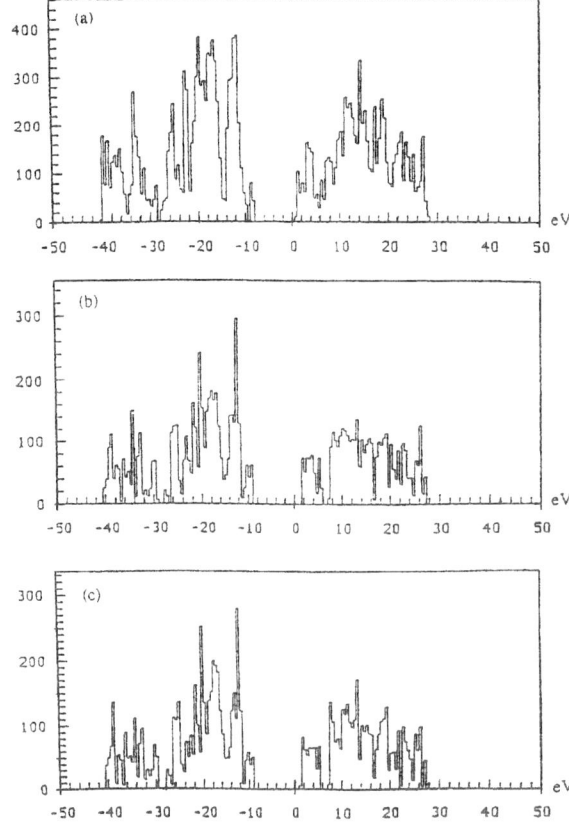

Fig. 2. The electronic density of states of a 100-base and base pair segment of the human oncogene described in Ref. [19]. (a) STO-3G basis, (b) 6-31G basis, (c) 6-31 + MP2.

As one can see from Fig. 4, the introduction of correlation has increased $|\sigma(\omega)|$ by two and a half orders at magnitude. ($|\sigma(\omega)|$ is $\approx 4 \times 10^{-3}\,\Omega^{-1}\,cm^{-1}$ at $\omega = 5 \times 10^{10}\,s^{-1}$.) To summarize, the better basis increased $|\sigma(\omega)|$ by one and the MP2 correlation by additional 2.5 orders of magnitudes. For this reason, the former (minimal basis) $|\sigma(\omega)|$ value for a base pair stack can be considered only as a loose lower bound. (It should be around $10\,\Omega^{-1}\,cm^{-1}$ if no other effects would interfere.) It should be pointed out that our results are in good agreement with a measurement of Gruener at al. [29], who have investigated the frequency-dependence of the conductivity of lambda phage DNA at room temperature by electron energy loss in a cavity resonator. They have obtained for $|\sigma(\omega)|$ $24\,\Omega^{-1}cm^{-1}$ in wet DNA. They have also found that if DNA is dried, its conductivity decreases by half an order of magnitude. This result is also in

qualitative agreement with our result [30], according to which for a single stack $|\sigma(\omega)| = 10^{-4}\,\Omega^{-1}\,cm^{-1}$ in the presence and $10^{-6}\,\Omega^{-1}\,cm^{-1}$ in the absence of water [30]. This large effect of water can be explained first of all by the fact that dry DNA is not anymore in a B form and therefore the overlap of the stacked bases is smaller. (In the case of Gruener et al. their purification procedure has left still 15% water in the sample, which is enough to keep the DNA molecules in their B form.)

It should be mentioned that most recently there is a measurement on wet spun macroscopically oriented calf thymus Li-DNA in an a.c. field and at low frequencies between 20 and $10^6 cm^{-1}$ [31]. Combining the data of Ref. [31] with those of Ref. [29], one obtains (with some interruptions) the $|\sigma(\omega)| - \omega$ curve between 10^{-3} and 10^{15} Hz. Comparing this curve with our theoretical one ($10^{-3} - 10^{20}\,s^{-1}$), one finds a very good agreement in all parts of the curve where experimental results are available (with the exception of optical frequencies where the measured $\sigma(\omega)$ values refer to excited states of the base pairs). It is also very interesting to point out that our estimated $|\sigma(\omega)|$ value of $10\,\Omega^{-1}\,cm^{-1}$, if one takes into account correlation, agrees also quite well with the value of Kutjnjak et al. [31]. Finally, in a further paper we have calculated the temperature-dependence of the hopping conductivity [32]. We have found that $|\sigma(\omega)|$ first decreases exponentially with T (the range $60 \times K \leq T \leq 360$ K was investigated) for a base pair stack (oncogene sequence) until about 60 K (at $\omega = 10^{11}\,s^{-1}$). A 60 K its value is not T-dependent anymore. This behaviour is again in good agreement with the experimental findings [29,31].

The activation energy of the conductivity (ΔE) calculated from these T-dependence shows (see Fig. 5 and Ref. [33]) that at $T = 60$ K $\Delta E = 0.030$ eV, increases exponentially with T between 70 and 120 K. (ΔE has its maximum at 120 K, where its value is 0.159 eV.) By further increase of T it decreases not very steeply. $\Delta E = 0.114$ eV at 310 K. This shows that at larger temperatures the dominant mechanism of conductivity is variable range hopping and below 60 K [32,33] multi-channel tunnelling (Fig. 5). Of course, further experimental and theoretical investigations are needed to find out in detail the mechanisms of a.c. conductivity in DNA.

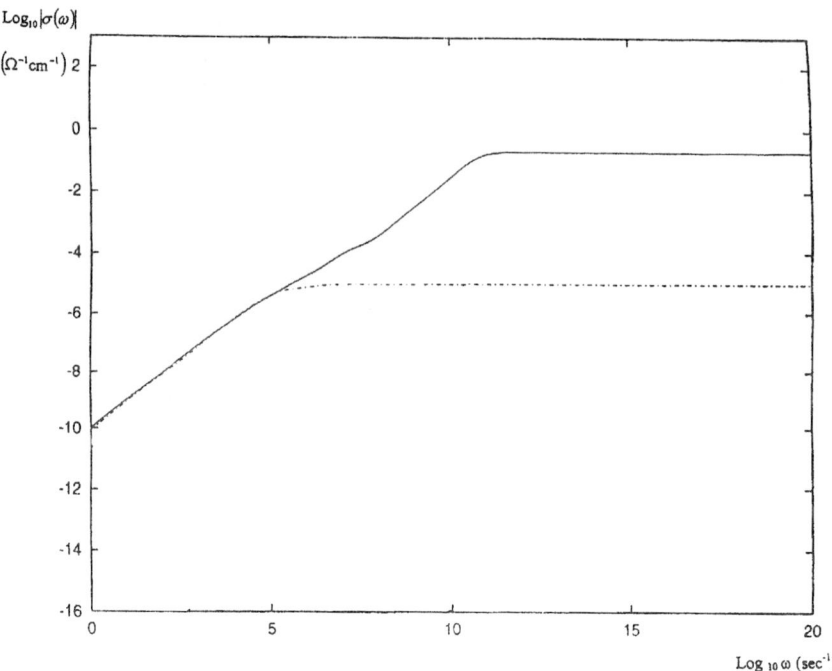

Fig. 3. $|\sigma(\omega)|$ as a function of ω ($T = 298$ K). $\omega_{sat} = 10^{10}$ s^{-1} if 200 levels were taken into account in the valence bands region (—) and $\omega_{sat} = 10^{5}$ s^{-1} if only 100 levels were considered. ($-\cdot-\cdot-$).

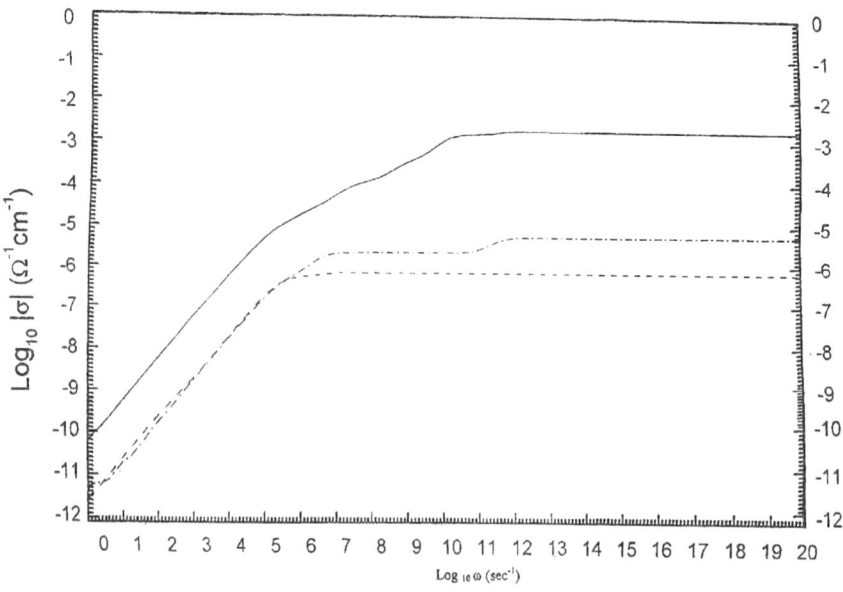

Fig. 4. The $|\sigma(\omega)| - \omega$ for a single stranded base stack (oncogene sequence) with 100 levels in the valence bands region with a minimal basis ($-\,-\,-$), with a double ζ one ($-\cdot-\cdot-$) and with a double ζ one and correlation (MP2) (—).

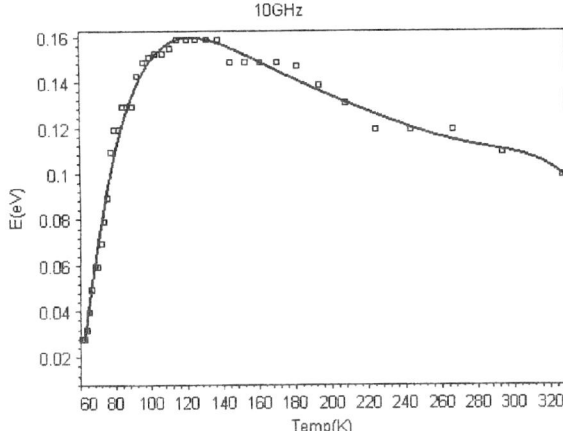

Fig. 5. The mechanism of charge transfer at different temperatures in aperiodic nucleotide base stacks [33].

3. Calculation of the band structures and spectra of periodic nucleotide base stacks and homopolypeptides

Though native proteins and DNA have a non-periodic sequence of their constituents, one can obtain more insight into their electronic structure and properties from their periodic counterparts.

If one has a 1D periodic system one can formulate the calculations of their band structures and wave functions, the crystal orbitals (CO-s), which are a linear combination of Bloch orbitals, if one takes into account their periodic symmetry. In the case of simple translation, this was done first by Löwdin [34] and in more detail (taking into account an arbitrary number of orbitals in the unit cell) by Del Re et al. and André et al., respectively, [35]. The method was extended for the case of general periodic symmetry (like helix or zip-zag operation) [36]. The latter development has made it possible to determine the band structure of homopolypeptides and of periodic nucleotide base (pair) stacks.

As next step, the band structure calculation method was worked out to take into account the major part of correlation using either many body perturbation theory in its Moeller–Plesset (MP) form [5,37] or applying the coupled cluster theory to polymers [38,39].

Both the HF and the HF + MP2 band structures were calculated for a larger number of periodic polymers. For the ones which do not play a role in biology, see Chapter 2 of Ref. [4].

For detailed calculation of the periodic nucleotide base stacks, see Ref. [25]. In this paper, the basis set-, k- and number of neighbours' dependence were investigated in detail. It was found that with a double ζ basis supplemented by a set of p functions centered at the middle plane of the stacking distance (at 3.36/2 = 1.68 Å distance from the stacked base) with 21k-points in the first Brillouin zone and with second neighbours' interactions the HF + MP2 calculation provides a rather realistic band structure with a gap value of 6.5 eV, which is close to the value estimated on the basis of the intermediate exciton spectrum of a cytosine stack [40].

There are a larger number of correlation corrected [MP2] calculations on homopolypeptides. The HF + MP2 gaps turned out to be too large [41,42], which means that one has to use a better basis set than the double ζ and probably has to apply for correlation the higher MP2 + MP3 + MP4 approximation.

In subsequent density functional theoretical calculations (DFT), 12 homopolypeptides were calculated in the LDA approximation [43] and 19 ones in the β pleated sheet form in the BLYP one [44]). In all cases the fundamental gaps were far too small. The density gradient containing BLYP calculation changed the gap only in a small amount. We expect that the B3LYP DF investigations (which are in progress) will increase the gap but probably not in the desired amount, because they contain only partly the HF exchange. Most probably one can obtain still more correct gap values if one applies a suitable form of the optimized effective potential approximation to polymers. Despite this situation, it was worthwhile to apply the different forms of the DFT to homopolypeptides because

1. it has given some insight of its electronic structure, and

2. it will be easy to change the band structure programs for calculations of the total energy per unit cell for which the BLYP and B3LYP DFT methods work quite well.

Using their correlation corrected band structures with the aid of the intermediate exciton theory [45], the UV spectra of a cytosine (C) stack [46] and that of polyglycine and polyalanine [47] were calculated applying the double ζ basis. In the case of the C stack, the first singlet exciton band has its minimum at 4.7 eV and its width is 0.7 eV. These results agree quite well with experiments [48].

The results also show the importance to take into account charge transfer-type excitations as compared to the results of Frenkel's exciton theory (where the excitations occur within the same cell). The intensity (oscillator strength) value of the first singlet exciton band of the C stack is by 26% smaller than the corresponding monomer value. This so-called hypochromicity is well known in the spectrum in DNA. Due to the $\sum_i f_i = 1$ rule, the second theoretical exciton band shows a hyperchromicity of 12%.

In the case of a polyglycine and a polyalanine, the calculations applying a 6-31G* basis, assuming an α-helix conformation and taking into account fifth neighbours' interactions resulted in the case of a polyglycine as the lower edge of the first singlet exciton band 6.2 eV and for polyalanine the result is 6.1 eV. In both cases the band widths are ≈ 0.4 eV. One finds again hypochromicity in the first singlet exciton band, a hyperchromicity in the second one for both homopolypeptides. Again, one has to take into account (like in the case of the stacks) charge transfer-type excitations until 4–5th neighbours to obtain saturated results. Finally, one should point out that the small differences in the spectra of these two homopolypeptides indicate that the side chains of the amino acid residues play a minor role in the spectra of proteins (at a given conformation) which is dominated by the $\pi \rightarrow \pi^*$ transitions of the $-N-\overset{O}{\underset{H}{C}}-$ parts of the peptide groups.

4. Outlook

With the very fast development of the power of computers and that of the numerical techniques one can expect that in the next years it will be possible to perform more accurate calculations on biopolymers and their properties. As next step, we will be able to treat much better the interactions between different biopolymers that is very important for their biological functions.

One expects that one can treat simultaneously the different channels of charge transfer in biopolymer taking simultaneously into account: (1) coherent (Bloch-type conduction); (2) hopping conductivity; (3) multi-channel tunnelling (using a suitably modified Feynmann path integral or WKB formalism) and (4) electron transfer coupled to protons or ions.

One expects that before long the very important protein folding problems will be solved and in this way one will be able to better understand the mechanism of action of enzymes and other functions of proteins.

Further, one hopes that one will be able to treat with the help of quantum theory the concerted action of proteins (proteomics) and in this way one can handle theoretically the electron transfer through a chain of proteins as it happens for instance in the Szent-Györgyi-Krebs cycle.

Acknowledgements

I would like to express my deep gratitude to late Professor Albert Szent-Györgyi, whose way of thinking has inspired me very much. I am indebted very much to Professors I.G. Csizmadia and B. Penke for inviting me to the International Symposium, which has given the impetus to write this small review.

References

[1] See for instance:D. Voet, J.G. Voet, Biochemistry, Wiley, New York, 1995, p. 540.
[2] A. Szent-Györgyi, Science 93 (1941) 609.
[3] A. Szent-Györgyi, Nature 148 (1941) 157.
[4] J.J. Ladik, Quantum Theory of Polymers as Solids, Plenum Press, New York, 1988, p. 289.
[5] C. Moeller, M.S. Plesset, Phys. Rev. 46 (1934) 618.
[6] J.J. Ladik, locit., p. 281. S.Suhai, Int. J. Quantum Chem.111984223. S.SuhaiJ. Mol. Struct. (Theochem.)123198597.
[7] P. Dean, Proc. R. Soc., Lond., Ser. A 245 (1960) 507. P. Dean, Proc. R. Soc., Lond., Ser. A A260 (1961) 263. P. Dean, Rev. Mod. Phys. 44 (1972) 127.
[8] R.S. Day, F. Martino, Chem. Phys. Lett. 84 (1981) 86.
[9] Y.-J. Ye, J. Math. Chem. 14 (1993) 121.
[10] J.H. Wilkilson. The Algebraic Eigenvalue Problem, Clarendon Press, Oxford, 1965, p. 633.
[11] M.F. Mott, E.A. Davis, Electronic Processes in Non-crystalline Materials, Clarendon Press, Oxford, 1971.
[12] T. Scher, M. Lax, Phys. Rev. B7 (1973) 4491. T. Odagaki, M. Lax, Phys. Rev. B24 (1981) 5284. T. Odagaki, M. Lax, Phys. Rev. B26 (1982) 6480.

[13] J.J. Ladik, Phys. Rep. 313 (1999) 171.
[14] Y.-J. Ye, J. Ladik, Phys. Rev. B48 (1993) 5120.
[15] Y.-J. Ye, J. Ladik, Int. J. Quantum Chem. 52 (1994) 52.
[16] Y.-J. Ye, J. Ladik, Phys. Rev. B51 (1995) 13091.
[17] Y.-J. Ye, J. Ladik, Phys. Chem. Phys. Med. NMR 28 (1986) 123.
[18] Y.-J. Ye, R.S. Chen, A. Martinez, P. Otto, J. Ladik, Solid State Commun. 112 (1999) 139.
[19] G.I. Bell, R. Picklet, J. Writter, Nucl. Acid Res. 8 (1980) 409.
[20] J. Ladik, S. Suhai, Phys. Lett. A77 (1980) 25.
[21] P. Otto, E. Clementi, J. Ladik, J. Chem. Phys. 78 (1983) 4547.
[22] E. Clementi, G. Corongiu, Int. J. Quantum Chem. QBS9 (1982) 213.
[23] R.A. Dickerson, H.R. Drew, B.N. Couner, R.M. Wing, A.V. Fratini, M.L. Kopka, Science 216 (1982) 475.
[24] Y.-J. Ye, R.S. Chen, A. Martinez, P. Otto, J. Ladik, Physica B279 (2000) 246.
[25] F. Bogar, J. Ladik, Chem. Phys. 237 (1998) 273.
[26] L. Gianolo, R. Pavoni, E. Clementi, Gazz. Chim. Ital. 108 (1978) 108.
[27] L. Gianolo, E. Clementi, Gazz. Chim. Ital. 110 (1980) 179.
[28] C.M. Liegener, Chem. Phys. 133 (1989) 173.
[29] P. Tran, B. Alavi, G. Gruener, Phys. Rev. Lett. 85 (2000) 1564.
[30] Y.-J. Ye, R.S. Chen, F. Chen, J. Sun, J. Ladik, Solid State Commun. 119 (2001) 175.
[31] Z. Kutjnjak, C. Filipic, R. Podgornik, L. Nordenskjöld, N. Korolev, Phys. Rev. Lett. 90 (2003) 98101.
[32] L. Shen, Y.-J. Ye, J. Ladik, Solid State Commun. 121 (2003) 35.
[33] J. Čížek, A. Martinez, J. Ladik, J. Mol. Struct. (Theochem) 626 (2003) 77.

[34] P.-O. Löwdin, Adv. Phys. 5 (1956) 1. J.J. Ladik, Quantum Theory of Polymers as SolidsPlenum PressNew York (1988) p. 289, Chapter 1.
[35] G. Del Re, J. Ladik, G. Biczó, Phys. Rev. 155 (1967) 997. J.-M. Anré, L. Guverneur, G. Leroy, Int. J. Quantum Chem. 1 (1967) 427 see also p. 451.
[36] A. Blumen, C. Merkel, Phys. Status Solidi B83 (1977) 4259. C.A. Nicolaides, D. Beck (Eds.), Excited States in Quantum Chemistry, Reidel, Dordrecht (1979) 495.
[37] S. Suhai, Phys. Rev. B27 (1983) 3506 see also. J.J. Ladik, Quantum Theory of Polymers as Solids, Plenum Press (1988) Chapter 5.
[38] J. Čížek, J. Chem. Phys. 45 (1966) 4256. J. Čížek, Adv. Quantum Chem. 3 (1969) 35. J. Čížek, J. Paldus, Int. J. Quantum Chem. 5 (1971) 339.
[39] W. Förner, R. Knab, J. Čížek, J. Ladik, J. Chem. Phys. 106 (1997) 10248.
[40] J. Ladik, A. Sujianto, P. Otto, J. Mol. Struct. (Theochem) 228 (1991) 271.
[41] F. Bogar, V. Van Doren, J. Ladik, Phys. Chem. Chem. Phys. 3 (2001) 5426.
[42] F. Bogar, J. Ladik, Phys. Chem. Chem. Phys. 5 (2003) 953.
[43] F. Bogar, J. Ladik, Acta Phys. Chem. Debr. 34–35 (2003) 51.
[44] F. Bogar, J. Ladik, Int. J. Quant. Chem. (submitted).
[45] S. Suhai, Phys. Rev. B29 (1984) 4570 see also. J.J. Ladik, Quantum Theory of Polymers as Solids, Plenum Press, New York (1988) p. 289, Chapter 8.
[46] S. Suhai, Int. J. Quantum Chem. 11 (1984) 223.
[47] S. Suhai, J. Mol. Struct. 123 (1985) 97.
[48] I. Tinoco Jr, J. Am. Chem. Soc. 82 (1960) 4785. H. Devoe, I. Tinoco Jr, J. Mol. Biol. 4 (1968) 51881.

ELSEVIER

Journal of Molecular Structure (Theochem) 666–667 (2003) 11–24

THEO
CHEM

www.elsevier.com/locate/theochem

From submolecular biology to submolecular medicine
The legacy of Albert Szent-Györgyi

Imre G. Csizmadia[a,b,*]

[a]*Department of Chemistry, University of Toronto, 80 St George Street, Toronto, Ont., Canada M5S 3H6*
[b]*Department of Medical Chemistry, University of Szeged, Dóm tér 8, Szeged 6720, Hungary*

Abstract

The legacy of Professor Albert Szent-Györgyi is best summarized, by himself, in the mid 20th Century.

"The distance between those abstruse quantum mechanical calculations and the patient bed may not be as great as believed"

We know that his prophetic vision has not been fulfilled as yet; the time is yet to come, perhaps in the 21st Century, when his legacy will be practiced on a daily basis. It may take several generations before we can walk that proverbial short distance from our research computer to the patient bed. It will probably require a 'Grande Armée' of researchers covering the whole spectrum of Mathematics, Physics, Computer Science, Chemistry, Biochemistry, Biology and Medicine, fighting a constant uphill battle in order convert Szent-Györgyi's legacy to reality. In contrast to our present daily practice with our future eschatological hope, the current situation is reviewed and the future direction of the overall process is outlined that shows the progression from Classical Medicine through Molecular Medicine all the way to Submolecular or Quantum Medicine.
© 2003 Elsevier B.V. All rights reserved.

Keywords: Albert Szent-Györgyi; Molecular medicine; Citric acid cycle

1. A tribute to Albert Szent-Györgyi

The title of this paper refers explicitly to the legacy of Professor Albert Szent-Györgyi, (Fig. 1), whom we are celebrating with the present *Special Issue*, on his 110th birthday. Some legacies are short in duration and easy to unfold, but this one is different.

His legacy is more than just a legacy. It is, in fact, a scientific prophecy and it will take most, if not all, of the 21st Century to fulfill it as he formulated in the 1950s and published in 1960 [1]:

"The distance between those abstruse quantum mechanical calculations and the patient bed may not be as great as believed"

This quote summarizes the essence of the prophetic legacy of Albert Szent-Györgyi in the most profound way.

It is hard to speak after the Prophet has spoken! Our basic problem with this prophetic legacy of Szent-Györgyi is that it takes several generations to walk that proverbial short distance from the research computer to the patient bed.

* Address: Department of Chemistry, University of Toronto, 80 St George Street, Toronto, Ont., Canada M5S 3H6. Tel.: +1-416-978-3598; fax: +1-416-978-8775.
E-mail address: icsizmad@alchemy.chem.utoronto.ca (I.G. Csizmadia).

0166-1280/$ - see front matter © 2003 Elsevier B.V. All rights reserved.
doi:10.1016/j.theochem.2003.08.052

Fig. 1. Nobel prize and portrait of Szent Györgyi.

Consequently, my task is not easy. I am expected to cover the territory not only for our own generation but for the next generation as well, to say the least. In other words, I need to contrast our present daily practice with our future eschatological hope. Thus, while I need to talk about our current battles at the research front, I also need to encourage our younger colleagues to pick up the flag when we cannot carry it any further.

The Nobel Prize was awarded to Professor Albert Szent-Györgyi in 1937 for deciphering some of the intricacies of the Citric Acid Cycle (Fig. 2). That cycle is also referred to as the Szent-Györgyi–Krebs cycle in Europe.

This was the first of such biochemical cycles discovered and it has laid the foundation of *Molecular Medicine*. As the result of his pioneering work the conclusion becomes unavoidable, namely, the human body is nothing else than an automatic molecular machine, in which not wheels but molecules are turning and are ultimately responsible for all bodily functions. However, one can go beyond the molecular level to the submolecular or electronic level. From that famous quotation presented above, it was clear that this was Szent-Györgyi's Quest at the middle of the 20th Century.

He already foresaw, at that time, the emerging new field of Submolecular or Quantum Medicine. According to lingering stories he was seeking out the help of physicists at the time when neither computer hardware nor software was available. After listening carefully to the Nobel Laureate, the Physicists started to laugh when they discovered that those bioactive molecules contained more than one electron. Cleary, this is the fate of everybody who dares to walk ahead of his time.

Today, a half a Century later, we see the situation a bit more clearly. It has become self-evident that while the thesis of Molecular Medicine is that:

All diseases start at the molecular level, thus, ultimately, all cures must be achieved at the molecular level

we need to go a level further to reach the submolecular level.

Accordingly, the thesis of *Submolecular Medicine* is that:

All diseases are the result of some unfavourable electron distribution within the human body, thus ultimately all cures must represent a favourable perturbation on the ill-distributed electron density.

By studying life, at the submolecular or electronic level, utilizing the techniques of Quantum Mechanics, surprising new results may emerge. The practical application of such new knowledge could be far reaching for the Pharmaceutical Industry.

2. Basic principles of submolecular medicine

From the composition of the human body (Table 1) it is possible to estimate its content, using only the major components.

For a person, who weighs 100 kg, it is estimated that his body consists of 0.98×10^{28} atomic nuclei and 2.81×10^{28} electrons. While most of this person's weight (over 99.9%) resides within the atomic nuclei, the total volume of these nuclei is much less than a micro micro-litre (μl). Thus, the human body is virtually an empty bag. In contrast to this, the 2.81×10^{28} electrons carry hardly any weight (less than 0.1%). According to the final results of Table 1 (2.81×10^{28}) the total mass for the electrons is calculated to be about 25 g after taking into account Avogadro's number (6.03×10^{23}) and the relative mass of the electron (1/1840):

$$(2.81 \times 10^{28}/6.03 \times 10^{23}) \times (1/1840) = 25.35 \text{ g}$$

Hence, if we consider the human body as a large bag, that is 100 l in volume, it would be filled up by 25 g of electrons behaving as a uniform 'electron gas'. Of course, due to the presence of the atomic nuclei the electron density will not be completely uniform. Rather, it will assume some particular distribution in which it will be relatively high in the vicinity of the atomic nuclei. Also the various organs, like the heart and lungs would exhibit different electron density as may be appropriate for denser and less dense parts.

Virtually an infinite number of possible electron distributions may occur from such a collection of atomic nuclei and electrons. Since only one such, electron distribution can be regarded to represent 'Perfect Health', very similar, or only slightly

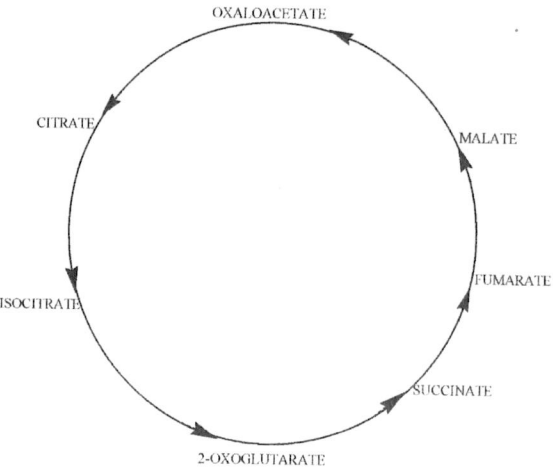

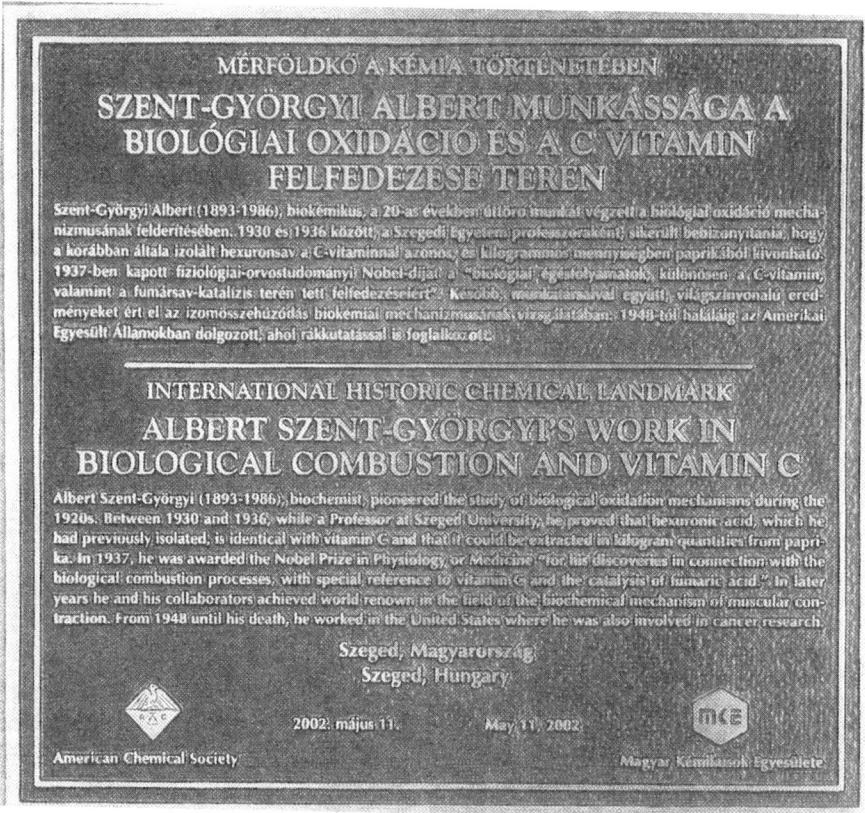

Fig. 2. Szent Györgyi–Krebs biochemical pathway and brass plaque at the University of Szeged.

different, electron distributions must represent very good, but not perfectly healthy, states. This is case for a well-defined margin. As the electron distribution gets further and further away from the ideal or perfect distribution we reach a large collection of ill-distributed electron densities, which are exhibited by people suffering from various diseases. In many cases, when these deviations are located close to the ideal

Table 1
The major components of the human body

	Weight (%)	Gram/100 kg	Mol/100 kg	No. of atoms/100 kg	No. of electrons/100
Oxygen	65	65,000	4062	24497×10^{23}	147717×10^{23}
Carbon	18	18,000	1500	9045×10^{23}	54270×10^{23}
Hydrogen	10	10,000	10000	60300×10^{23}	60300×10^{23}
Nitrogen	3	3000	214	1290×10^{23}	9030×10^{23}
Calcium	1.5	1500	357	215×10^{23}	4300×10^{23}
Phosphorous	1.0	1000	32.3	194×10^{23}	2910×10^{23}
Potassium	0.35	350	875	53×10^{23}	1007×10^{23}
Sulfur	0.25	250	781	47×10^{23}	752×10^{23}
Sodium	0.15	150	652	41×10^{23}	451×10^{23}
Chlorine	0.15	150	395	24×10^{23}	408×10^{23}
Magnesium	0.05	50	200	12×10^{23}	144×10^{23}
Iron	0.0004	–	–	–	–
Iodine	0.00004	–	–	–	–
Total	–	–	–	95721×10^{23}	281289×10^{23}
Rounded Total	–	–	–	0.96×10^{28}	2.81×10^{28}

distribution, the ill-distributed electron density may be forced to move back toward the ideal distribution. This is the reason why the Pharmaceutical Industry is producing 'drugs to cure' for people whose electron distribution has deviated only a little from the ideal distribution. Once the electron distribution can no longer be changed back to normal, i.e. the illness becomes irreversible, death results.

The above information can also be translated to the language of thermodynamics.

The healthy human body is the most organized molecular machine, so its entropy is at its lowest state. When illness strikes, the human body will become less organized internally. So with increasing disorganization, its entropy is on its rise ($\Delta S > 0$). Eventually, at death, entropy reaches its highest value. The opposite trend can be said about the change in Gibbs free energy (ΔG) on going from the healthy body down to the sick body all the way to the dead body ($\Delta G < 0$), as illustrated in Fig. 3.

3. The question of steady-state or virtual stationary electron distribution

The human body is a very complicated molecular machine. A good portion of its activity lays in energy processing and converting the chemical energy of food to biological energy (ATP) in its metabolic machinery. The mitochondria, present practically in all cells, carry out this job in the form of a 'biological fuel cell' (Fig. 4).

In a biological fuel cell, the reductive and oxidative compartments are separated. This is analogous to the cathodic and anodic channels of a regular fuel cell.

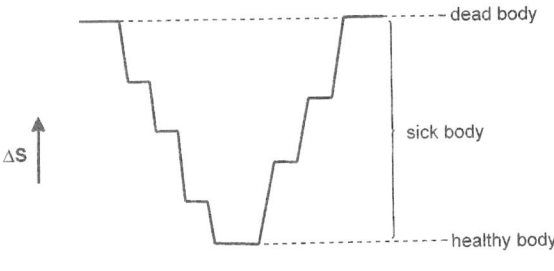

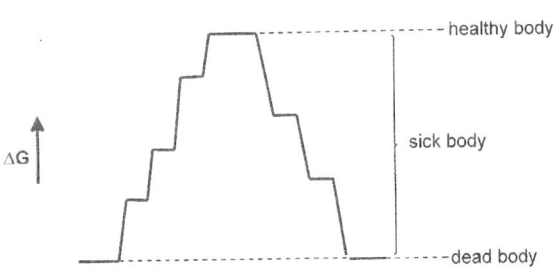

Fig. 3. A schematic illustrations of the key thermodynamic functions (ΔS and ΔG) for the human body in different states.

I.G. Csizmadia / Journal of Molecular Structure (Theochem) 666–667 (2003) 11–24

Of course in the biological fuel cell (i.e. the mitochondria) the electrons are not carried through a piece of wire but are transmitted through hydride ions $[:H^{(-)}]$ which are carried by a redox coenzyme such as $NAD^{(+)}$ in the form of NADH. All of these are illustrated in Fig. 4.

Thus the whole human body may be considered as a power-generator consisting of a very large number of biological fuel cells.

The human body not only produces energy for its physical and mental work but must refine its own fuel through its digestive system. The body also has to renew itself implying that after the programmed cell-death, the dead cells need to be replaced. It is also necessary to produce offspring for the survival of the species. Both of these processes require genetic information (DNA) and a reproduction mechanism.

The essence of the foregoing is that the human body is a dynamic molecular machine, which is never stationary. Substances enter in the form of food, and exit in the form of excretion. Oxygen is inhaled for the wet-combustion of the refined fuel. Exhaust, in the form of CO_2 and H_2O is exhaled as illustrated in Fig. 5.

Clearly, this molecular machinery is in constant move, perhaps not at constant speed because at night time it is slower and at daytime it is faster. Nevertheless, with all its cycles, including daily as well as monthly (menstrual) ones, one may still consider the human body (even allowing for variations from adulthood to old ages) to be in some steady-state of

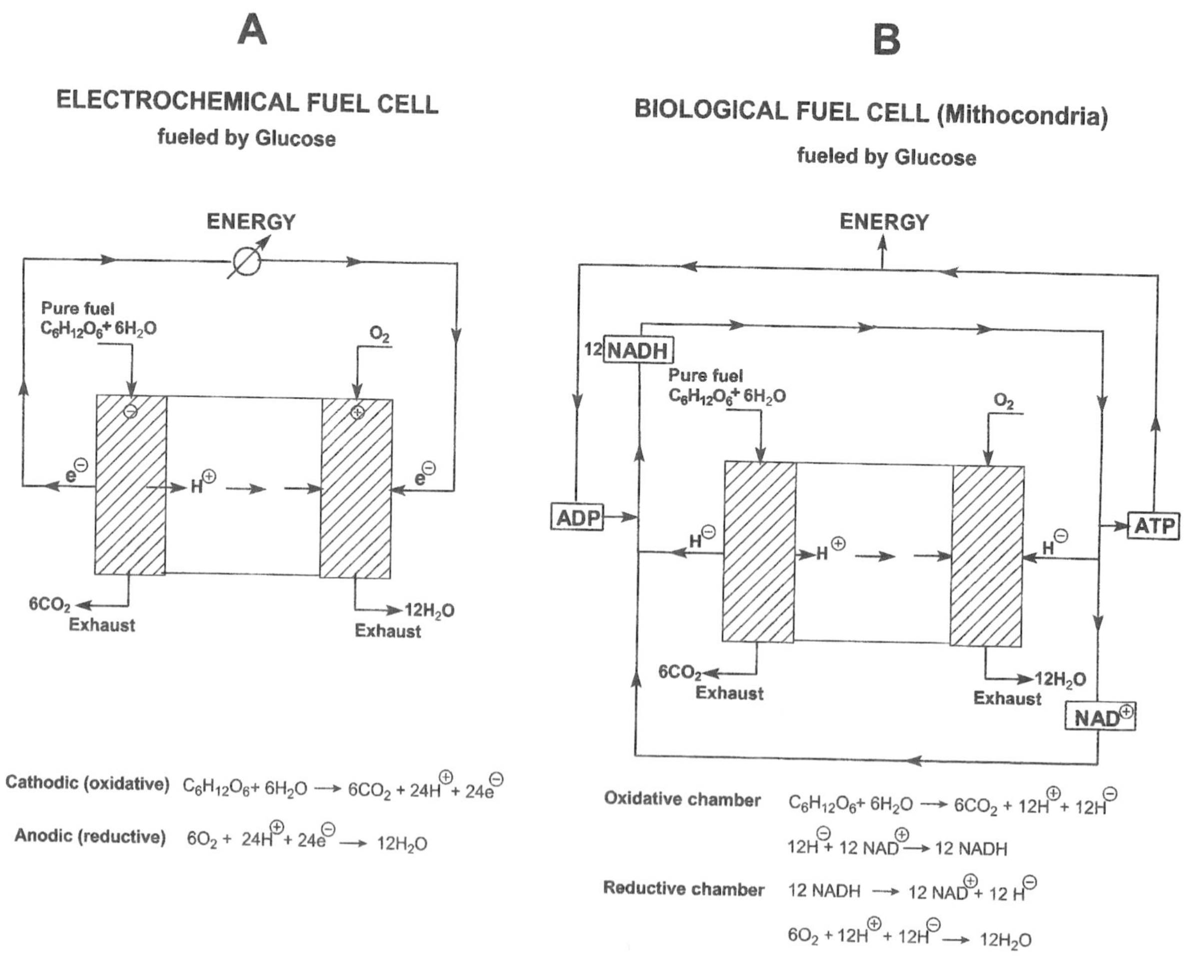

Fig. 4. Fuel cells: electrical (left) and biological (right).

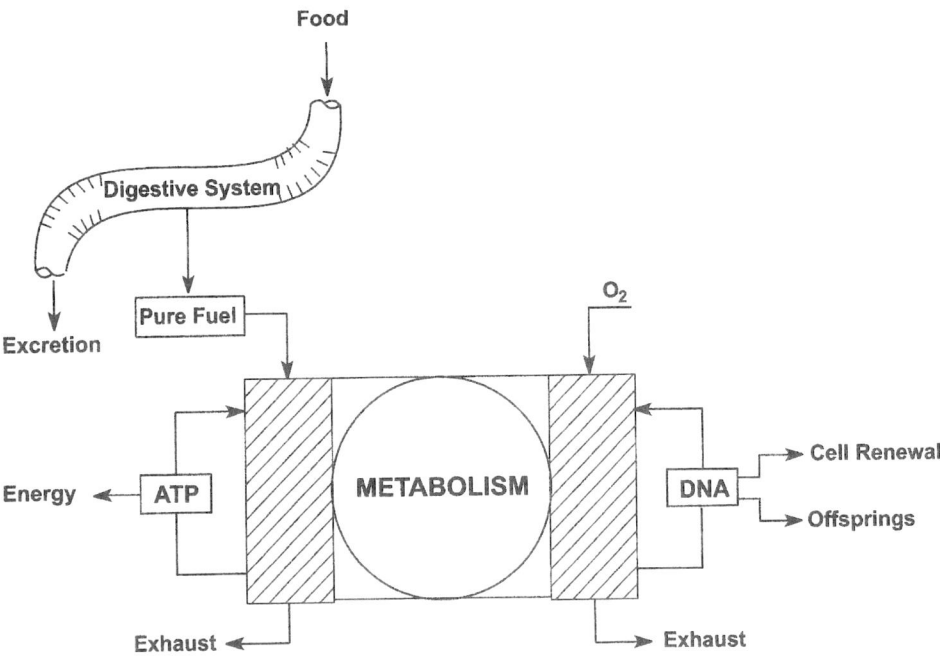

Fig. 5. Human body as a self reproducing machine.

electron distribution. Such steady state of otherwise time dependent electron distribution of the human body is distinctly different when it is healthy from the distribution that may exist when it is sick. Thus, healing at the electronic level may well be considered as the next step for Medicine and the associated Pharmaceutical Industry.

4. The current state of research

Recognizing that the yardstick for human beings is metre and the yardstick for atoms and small molecules is angstrom (1 Å = 10^{-10} m) it becomes self evident the nine order of magnitude (i.e. 10 Å × 10^9 = 1 m) shift in scale must cover a rather complicated hierarchy. It is not trivial therefore to build the human body from as set of small molecules of the dimension of 10 Å. This biological organization is illustrated in Fig. 6.

The demarcation line, showing computational limit at the dawn of the 21st Century illustrates that we can compute only at the level of Chemistry and Biochemistry but not as yet at the level of Biology.

Speaking of the chemical level, our primary goal is configurations and conformations for molecular

structures and mechanism for reactions (Fig. 7). Later, a schematic illustration for one of each of these classes will be given.

As for as Bioactive molecules are concerned some important classes are listed (Fig. 8). The actual number of molecules has no final count.

Since practically the whole periodic table is present in the human body, admittedly some of the elements in rather low concentration, the number of actual molecules must be very large. Computations for many of these bioactive molecules are reported in this volume.

There are, in this volume an unusually large number of papers authored by young researchers. However, if we consider the magnitude of the problems the human body poses even the proverbial Grande Armée would not be large enough to do all the research required. The magnitude of the project ahead of us supersedes the magnitude of the Human Genome project. This means that we need to change policy in our approach to compile a database as large as required for the deciphering the secrets of the human body. The project would get a boost if we were to rely on the millions of undergraduate students of the planet. In the same time those who

BIOLOGICAL ORGANIZATION

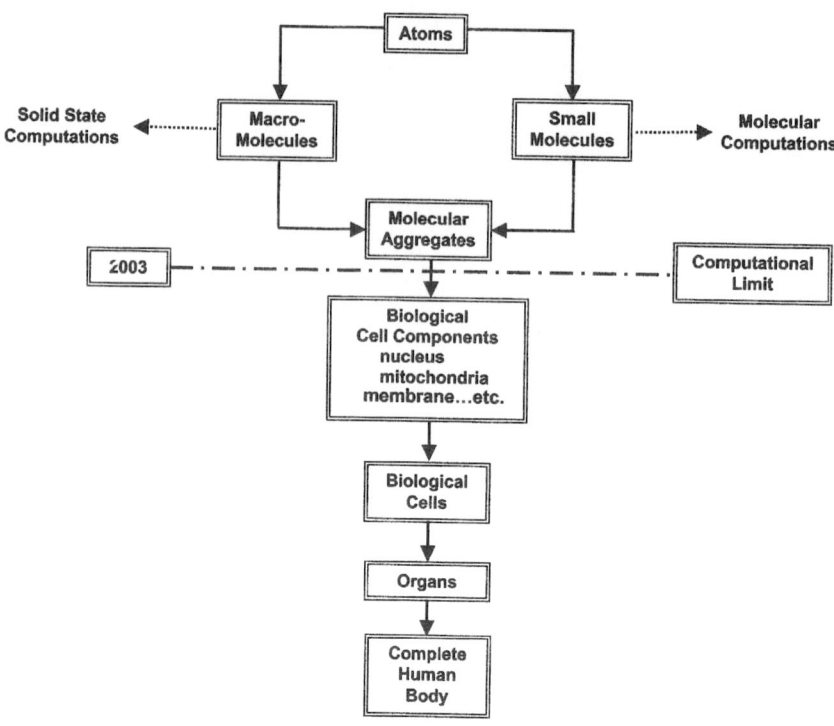

Fig. 6. Biological organization.

were to participate in such research projects would become better graduate students. Professor Kenneth Bartlett wrote a paper in this volume in which he speaks about the integration process of Teaching and Research.

In order to illustrate some of the complexities involved let us briefly consider two topics:

1. Peptide folding
2. Biological methylation

4.1. Peptide folding

The detailed knowledge of peptide folding may be considered to be a prelude to the understanding of

MOLECULAR COMPUTATIONS AT THE DAWN OF THE 21th CENTURY

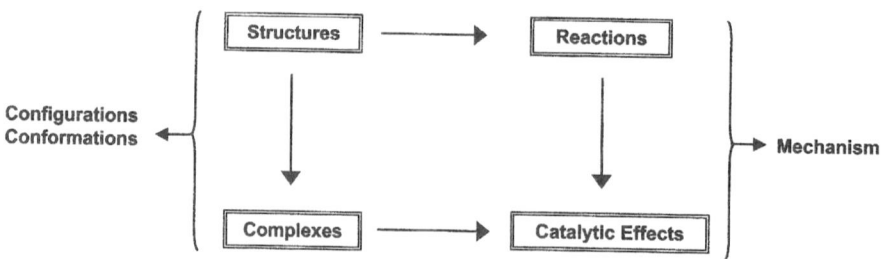

Fig. 7. Computations in the 21st century.

BIO-ACTIVE MOLECULES

Intrinsic stabilities and solvent effect

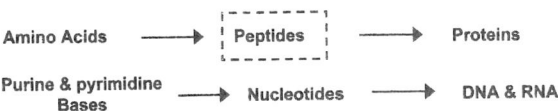

Carbohydrates (pentoses, hexoses)
Lypids (phospholypids and cell membranes)
Vitamines (B$_1$,B$_2$, B$_6$, B$_{12}$, etc.)
Co-Enzymes (NADH, FAD etc.)
Anti-Oxidants (Lycopenes, Flavones, Vitamin C and E
Neurotransmitters (Seratonine, Cathecolamines)
Drugs(Carvedilol, Seleginin, D-amino acids, β-amino acids

Fig. 8. Bioactive molecules.

protein folding [1,7]. However, even as small a unit such as tetrapeptides represent an astronomical problem (Fig. 9).

Numerous publications testify concerning the complexity of peptide folding [2,3] which is a prelude to the understanding of protein folding [1,3]. The establishment of a conformational database seems to be essential to a successful approach. There are hopes to automate the input generation as well as the tabulation of output results and management of the generated database in the foreseeable future [1,5].

Working with Professor Emil Pai and Michelle Sahai some advances were made in the case of a tetrapeptide project Pro-Pro-Thr-Pro (Fig. 10) [4] by

starting with the central dipeptide moiety: Pro-Thr [6]. This will form the basis of the design of an anti-bacterial drug which protects immunoglobulin A (IgA) against bacterial proteolytic enzymes.

Multivariable Fourier analysis of potential energy surfaces might eventually lead to analytic functions which could represent Ramachandran type potential energy hypersurfaces describing peptide folding [7].

Finally, it will be inevitable to analyze the entropy change (ΔS) along the Ramachandran map [8] to determine the relative amount of information present in the various conformations associated with the entropy contribution ($-T\Delta S$) to the free energy (ΔG) of peptide folding.

Even though only four small areas are singled out, nevertheless it is abundantly clear that peptide folding in a mega-project. Numerous papers present in this volume will illustrate, more clearly, the depth of the problem at hand.

4.2. Biological methylation

Even though the methyl group is perhaps the smallest function group its biological transfer is very important. There is scattered evidence that in depression methyl transfer does not occur at the level it is required.

MULTITUDES OF PEPTIDE CONFORMATIONS FOR FULLY TRANSPEPTIDES

Number of Amino acid	Sequences	Backbone (BB)		Side chain	
		Number of BB Rotors	Max No. of BB Conformers	Number of SC Rotors	Max No. of SC Conformers
1	20=20	2	$3^2 = 9$	1	$3^1 = 3$
2	20^2=400	4	$3^4 = 81$	2	$3^2 = 9$
3	20^3=8000	6	$3^6 = 729$	3	$3^3 = 27$
4	20^4=160000	8	$3^8 = 6561$	4	$3^4 = 81$

One sequence out of the 160000 sequences:

Thr　　　Val　　　Ser　　　Leu

BB(1)[SC(1)] x BB(2)[SC(2)] x BB(3)[SC(3)] x BB(4)[SC(4)]

9 [9]　　x　　9 [3]　　x　　9 [9]　　x　　9 [9]　　= 9^4[9^{3}x3] = 6 561 x [2 187] = 14 348 907 conformers

Fig. 9. Peptide conformations.

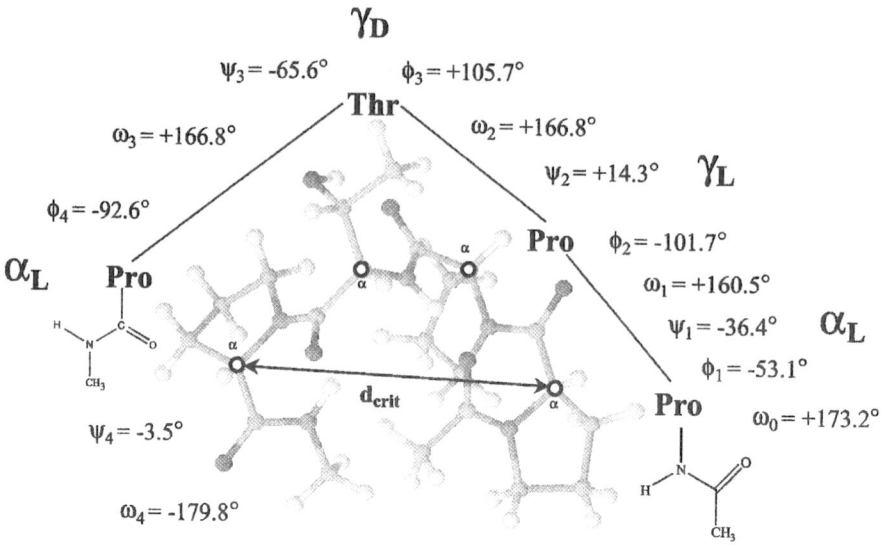

MeCO-Pro-Pro-Thr-Pro-NHMe

Fig. 10. The IgA tetrapeptide sequence.

Neuro-transmitters are methylated by SAM as illustrated in Fig. 11. In this process, methionine (Met) is converted to homocysteine (Hcy).

The remethylation of homocysteine [9,10] to methionine [11] requires tetrahydrofolate (derivative) which is also shown in Fig. 11. The tetrahydrofolate derivatives [12] involved in C_1 transfer is summarized in Fig. 12.

Papers of numerous young researches are included in this volume which illustrate that although there is a courageous start we have hardly scratched the surface. Thus the road ahead of us is not easy, yet it is very long. We only get comfort from the ancient proverb that 'even the longest journey starts with the first step'.

Before closing, I wish to provide some moral conclusion for the encouragement for our younger colleagues who will carry the burden of this multi-generation project to fulfill Szent-Györgyi's prophetic legacy.

5. Moral conclusions

To turn to the more philosophical aspect of the implications of our research I would like to admit, that most of us like fairy tales. We do like them, because there is more truth in a fairy tale than in reality. Reality tells us how things actually are and fairy tales tell us how things really ought to be. In fairy tales, irrespectively whether it is old or new, it is always the good that wins and the evil looses at the end. Countless examples can be drawn from the literature of the world from the classical 'Snow White and the Seven Dwarfs' to the contemporary 'Star Wars'.

When we examine the great religions of the world we have the impression that the Universe is a battle ground and the war is between good and evil. Such a concept has appeared many times, during the History of Art throughout the centuries.

Many of the undergraduates I have been teaching in Toronto during the years were and are preparing to enter Medical Schools. Most of the professors, from the older generations, tempted to view this with skepticism saying that they want to become Medical Doctors only because of the money. When I asked them why they choose that field most of them told me that they wanted to do something that they consider to be 'doing good'.

Teaching for 40 years I became accustomed to see that most of my students, are in fact, telling the truth. Thus, I am convinced, that in addition to their realism toward money they have a deep-seated conviction that they have to be warriors on the side of good rather

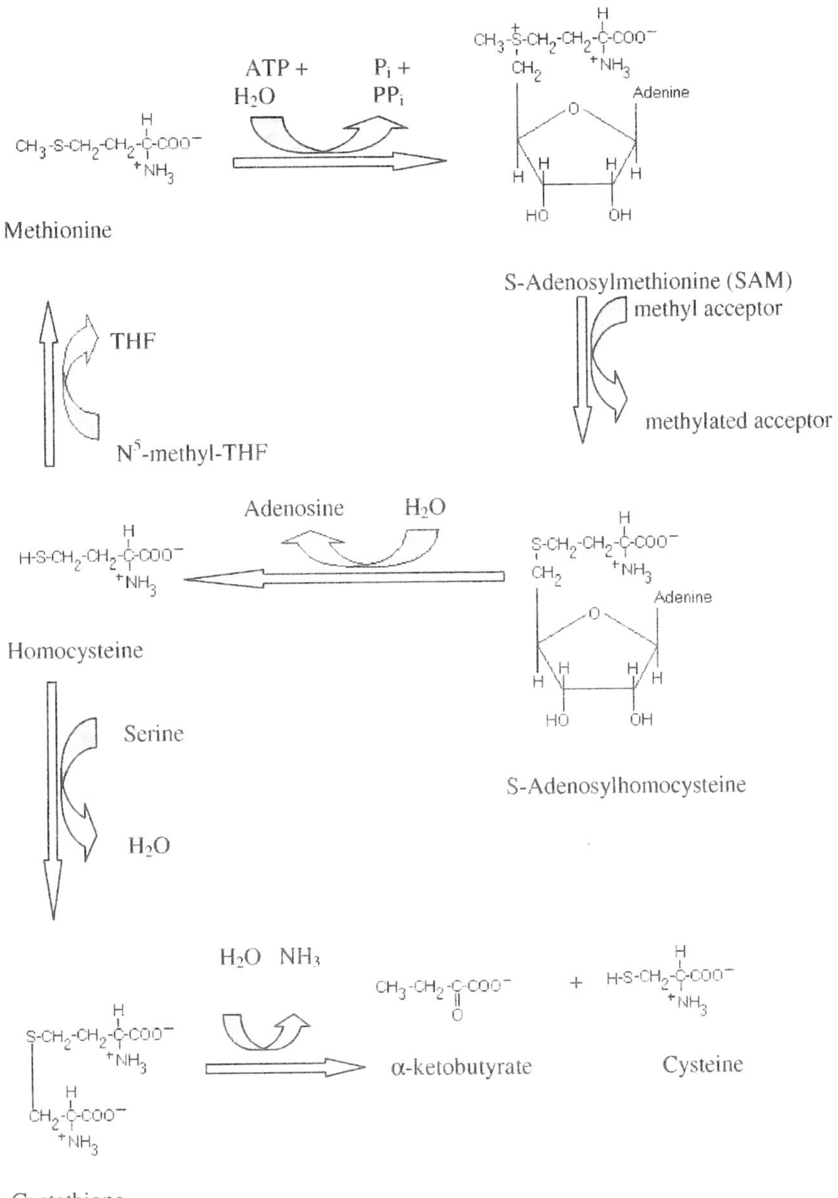

Fig. 11. Methionine cycle.

than on the side of evil while they may not become Dr Schweitzer nevertheless they wish to help in making the world a better place.

When we consider one of the most basis premises of Quantum Mechanics, the 'Uncertainty Principle', we learn that the only way to observe a small object, like an electron, is by carrying out a certain interaction with another particle, like a photon, which leads to some perturbation (Fig. 13).

Thus, by simply observing an electron, we altered the state of the electron. Because the electron is in the Universe we also altered the state of the Universe by virtue of the fact that we altered the state of the electron which is in the Universe. Thus, our action

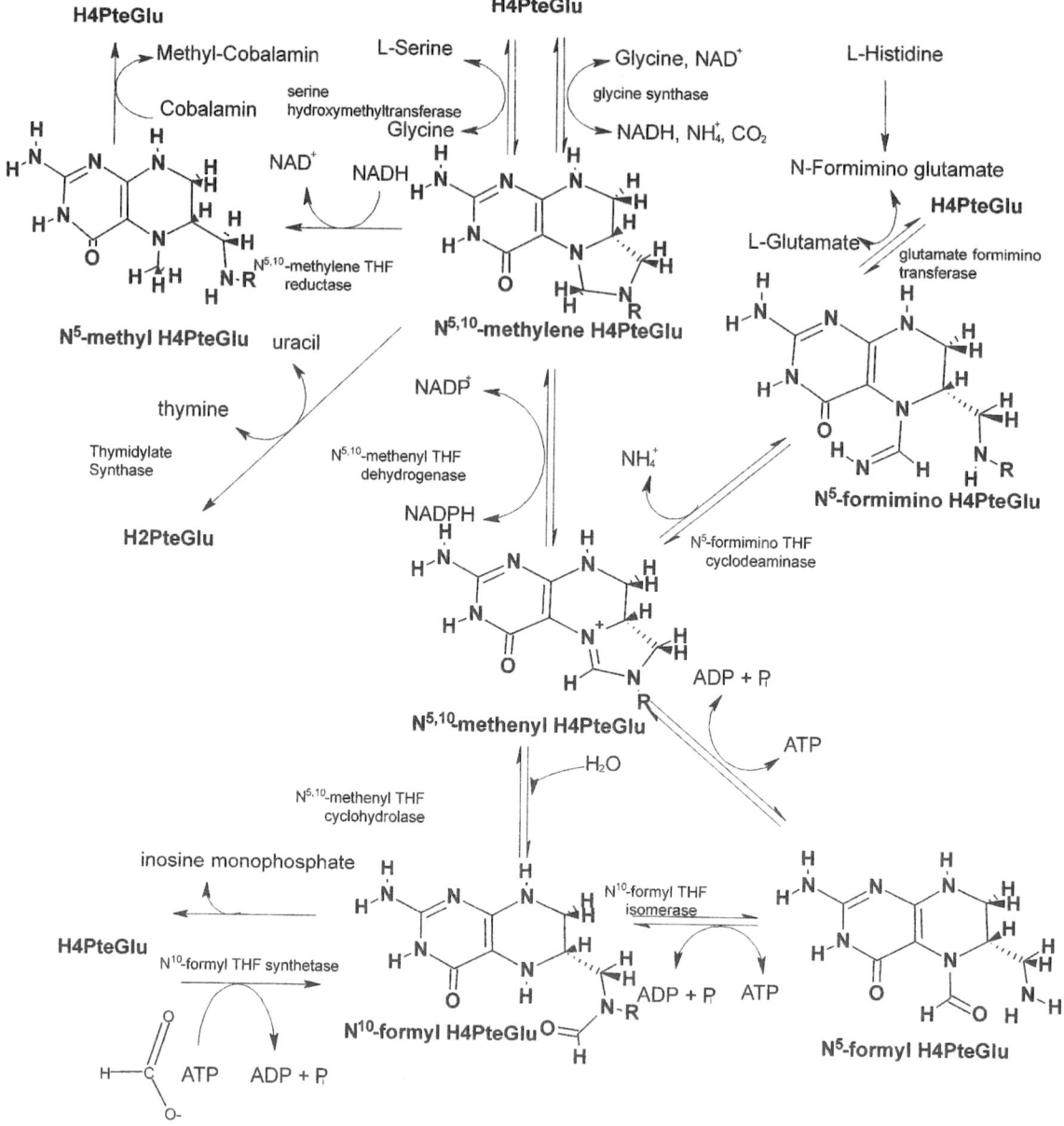

Fig. 12. Tetrahydrofolate derivatives.

does not go unnoticed. This is also true in a general sense. Medical healing, by any therapy, is also a perturbation on the ill-distributed electron density within the human body. As we altered the state of electron distribution in the body of a single human being we also altered the state of the Universe by

virtue of the fact that single human being is also within the Universe.

Finally information theory tells us that the increase of information leads to negative entropy change ($\Delta S < 0$). Structures, irrespectively whether it is the structure of galaxies or those of molecules also

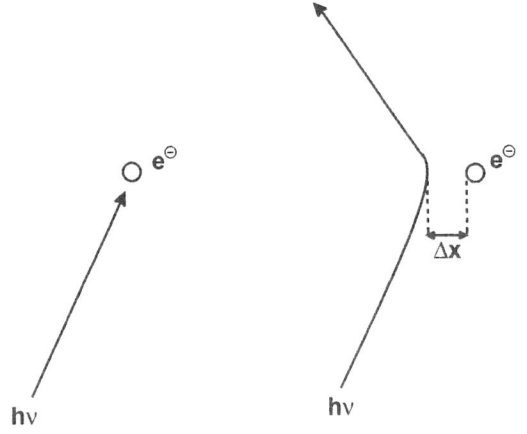

Original State
Before Observation   **Perturbed State**
                                After Observation

Fig. 13. Observation and uncertainty.

represent information (Fig. 14). Consequently, certain conformational change could in fact reduce or increase the entropy of a molecule and thereby the entropy of the entire Universe. Also, when we administer a drug to form a drug–receptor complex the new structure contains new information and therefore the entropy is expected to be reduced.

So, irrespectively if we start from the higher aspirations of mankind or from the most fundamental science of the human race the conclusion is

inescapable; with our actions, we do make a difference in the world because we live in a

"Participatory Universe"

We should have been warned at birth that we are entering into a *Participatory Universe* but perhaps the present warning comes just in time for our young colleagues. At least from now on, they shall be keenly aware that any act that they commit, let it be good or evil, will, in fact, be written into the evolutionary direction of the Universe because we do live in a

"Participatory Universe"

It remains to be seen if at some future date philosophers and theologians will discuss good an evil deeds in terms of negative and positive entropy changes. Nevertheless, today I am highly appreciative of the motivation of my colleagues, who wish to do something good when they plan to enter in to any one of the seven fields that represents a full spectrum of Molecular Science from Mathematics to Medicine (Fig. 15).

I sincerely hope that their own efforts as well as their students' effort will shift the focus of Physicians from Molecular Medicine to Submolecular Medicine in accordance with the legacy of Professor Albert Szent-Györgyi.

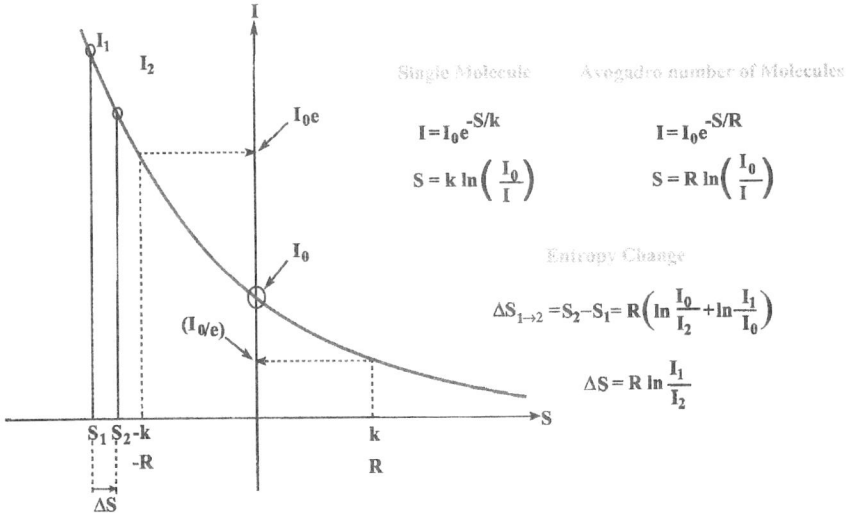

Fig. 14. Entropy and information.

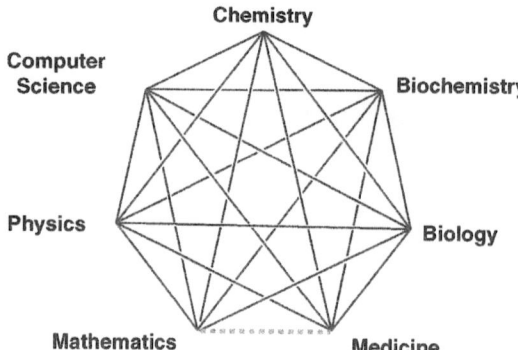

Fig. 15. The seven fields of molecular science.

In closing, I would like to leave for them, for the long road ahead, two encouraging quotations and a warning.

Quotations

Nearly everyone takes the limits of his vision for the limits of the world. A few do not. Join them—Arthur Schopenhauer

We must not be afraid of dreaming the impossible if we want the seemingly impossible to become a reality—Vaclav Havel

Warning

The future usually arrives a little bit sooner than we are ready to give up the present.

Acknowledgements

The author wishes to thank the Ministry of Education for a Szent Györgyi Visiting Professorship.

References

[1] G.A. Chasse, A.M. Rodriguez, M.L. Mak, E. Deretey, A. Perczel, C.P. Sosa, R.D. Enriz, I.G. Csizmadia, Peptide and protein folding, J. Mol. Struct. (THEOCHEM) 537 (2001) 319–361.

[2] A. Perczel, I.G. Csizmadia, Ab initio conformational analysis of protein subunits, in: A. Greenberg, C.M. Breneman, J.F. Liebman (Eds.), The Amide Linkage: Structural Significance in Chemistry, Biochemistry and Materials Science, Wiley/Interscience, New York, 2000, pp. 409–461, ISBN 0-471-35893-2.

[3] M. Sahai, G.A. Chass, B. Penke, I.G. Csizmadia, 2003. An ab initio study on selected conformational features of MeCo-L-Ala-L-Ala(βL)-L-Ala-NHMe as XxxAlaZzz tripeptide motif within a protein structure. THEOCHEM, in this issue.

[4] M. Berg, G.A. Chasse, E. Deretey, A.K. Füzéry, B.M. Fung, D.Y.K. Fung, H. Henry-Riyad, A.C. Lin, M.L. Mak, A. Mantas, M. Patel, I.V. Repyakh, M. Staikova, S.J. Salpietro, T.-H. Tang, J.C. Vank, A. Perczel, Ö. Farkas, L.L. Torday, Z. Székely, I.G. Csizmadia, Prospects in computational molecular medicine. A millennial mega-project on peptide folding, J. Mol. Struct. 500 (2000) 5–58. Millennium Volume.

[5] G.A. Chass, THEOCHEM, in this issue.

[6] M. Sahai, G.A. Chass, D.H. Setiadi, B. Penke, E.F. Pai, I.G. Csizmadia, 2003. A model study of the IgA hinge region. The systematic exploratory study of the backbone conformation of MeCo-L-Pro-L-Thr-NHMe. THEOCHEM, in this issue.

[7] T.A.K. Kehoe, M.R. Peterson, G.A. Chass, B. Viskolcz, L. Stacho, I.G. Csizmadia, 2003. The fitting and functional analysis of a double rotor potential energy surface for the *R* and *S* enantiomers of 1-chloro-3-flouro-isobutane. THEOCHEM, in this issue.

[8] S.J. Salpietro, B. Viskolcz, I.G. Csizmadia, 2003. An ab initio study on the entropy of various backbone conformers for the HCO-Gly-Gly-Gly-NH_2 tripeptide motif. THEOCHEM, in this issue.

[9] A.R. Sheraly, R.V. Chang, G.A. Chass, Multidimensional conformational analysis of the sidechain conformers of the fully extended backbone (β_L) of N-Ac-Homocysteine-NHMe; an ab initio exploratory study, J. Mol. Struct. (THEOCHEM) 619 (2002) 21.

[10] A.R. Sheraly, G.A. Chass, I.G. Csizmadia, 2003. The multidimensional conformational analysis for the backbone across the disrotatory axis at selected side-chain conformers of N-Ac-Homocysteine-NHMe—an ab initio exploratory study. THEOCHEM, in this issue.

[11] A. Láng, A. Perczel, 2003. Exploration of the conformational space of a sulphur-containing amino acid. THEOCHEM, in this issue.

[12] J.H. Keller, G.A. Chass, I.G. Csizmadia, 2003. An isodesmic comparison of the C_1 modified reduced pteridine ring as a folic acid model. THEOCHEM, in this issue.

Journal of Molecular Structure (Theochem) 666–667 (2003) 25–30

www.elsevier.com/locate/theochem

Gaussian-based first-principles calculations on large systems using the Fourier Transform Coulomb method

L. Füsti-Molnár*, P. Pulay

Department of Chemistry and Biochemistry, University of Arkansas, Fayetteville, AR 72701, USA

Abstract

The Fourier Transform Coulomb method is described and discussed. The method reduces the quadratic scaling of the calculation of the Coulomb energy and operator matrix elements with the molecular size to an essentially linear scaling. It also reduces the quartic scaling with respect to basis set quality with constant molecular size to quadratic or lower. A major computational expense in our initial implementation, the calculation of contributions from integrals with one core-type basis function is greatly improved in the current program. This is particularly advantageous in analytical derivative calculations. The method is illustrated with large calculations on alpha helical polyalanines. The largest calculations involve over 3600 basis functions for Ala_{15}. Current work on further improvements of the efficiency of the method is discussed.
© 2003 Elsevier B.V. All rights reserved.

Keywords: Density functional theory; Coulomb operator; Gaussian; Plane wave; Linear scaling

1. Introduction

The computational modeling of biological systems requires the calculation of the energies and properties of large molecules. Ideally, one would like to carry out these calculations at a high level of accuracy. In practice, the best feasible level at the current state of technology is density functional theory (DFT) with a reasonably good basis set. Such calculations, if carried out using traditional quantum chemistry techniques (e.g. Gaussian basis sets, direct SCF), are expensive. For large calculations, the computing effort is determined mainly by the scaling of the algorithm. Practically useful algorithms must have polynomial scaling, i.e. a suitable measure of the computing effort (e.g. flop count or execution time)

should behave approximately as a polynomial function of the characteristic size of the calculation (e.g. the number of contracted basis functions or atoms): $t = a_n s^n + a_{n-1} s^{n-1} + \dots$. The order of the scaling is the leading term of the polynomial. Methods with steeper (exponential or factorial) scaling are considered computationally unfeasible for large systems. We must distinguish formal and asymptotic scaling. Formal scaling does not take into account the sparsity of the problem, for instance the fact that some of the numbers become negligibly small in large molecules. Asymptotic scaling assumes that all such quantities (e.g. overlap between distant orbitals) are zero. The scaling of real calculations lies between the formal and asymptotic limits, approaching the latter as the system size increases.

If we restrict ourselves to DFT, then the theoretical level is determined by two parameters: the basis set and the exchange-correlation functional. Computing

---

* Corresponding author.
 E-mail address: fusti@uark.edu (L. Füsti-Molnár).

0166-1280/$ - see front matter © 2003 Elsevier B.V. All rights reserved.
doi:10.1016/j.theochem.2003.08.114

effort usually has only a mild dependence on the latter. The most widely used model for scaling assumes a constant theoretical level, and considers the dependence of the computational effort on the molecular size. An alternative measure, scaling with increasing basis set size at constant molecular size is a less frequently used but also important quantity.

Scaling is only one aspect of computational behavior; the prefactor, i.e. the coefficient of the leading term in the polynomial is also important. Elaborate algorithms often scale with a lower order than simpler algorithms but have a higher prefactor, with the result that, for smaller systems, a simple algorithm outperform the elaborate one. This behavior can be characterized by the crossover point, i.e. the system size where the latter becomes equivalent to the former. In our opinion, however, a more meaningful measure is the onset point where the elaborate algorithm becomes twice as efficient as the simple one.

The main computational steps in a DFT calculation are the following:

1. The calculation of the Coulomb energy, and matrix elements of the Coulomb operator. The formal scaling of this step is quartic, becoming asymptotically quadratic for large molecules with constant basis set. Scaling with the basis set size at constant molecular size is quartic, both formally and asymptotically.

2. Calculation of the exact (Hartree–Fock) exchange energy and its matrix elements. This step is present only in hybrid DFT calculations, and in Hartree–Fock theory. There is intensive effort underway to develop exchange-correlation functionals that match the performance of hybrid functionals without the inclusion of exact exchange. This effort has been largely successful for organic molecules [1]. However, for transition metals, hybrid functionals still outperform pure DFT [2]. The formal scaling of exact exchange with the molecular size is quartic, going asymptotically to linear in large insulating molecules [3]. However, the transition to linear scaling is rather late in three-dimensional systems. A simple way of proving the asymptotic linear scaling of the exact exchange is to use localized occupied orbitals, which is always possible in insulators. An exchange matrix element

is given by

$$K_{\mu\lambda} = (\mu\nu|\lambda\sigma)D_{\nu\sigma}$$

where $(\mu\nu|\lambda\sigma)$ is a two-electron integral in the Mulliken notation, and the density matrix is defined as $D_{\nu\sigma} = \Sigma_i C_{\nu i} C_{\sigma i}$. Greek letters denote atomic orbitals (AOs) and i runs over the occupied molecular orbitals (MOs); $C_{\nu i}$ is the coefficient of the νth AO in MO i. The localization of the MOs requires that the AOs ν and σ are spatially close. The fast exponential decrease of the AOs requires that AOs μ and ν, and also λ and σ be adjacent. Taken together, all four AOs must be located in the same region of space for the exchange contribution being non-negligible, leading to linear scaling.

3. Calculation of the exchange-correlation energy and its matrix elements. This operation scales formally as the cube of the system size. Asymptotically, it becomes linear in a proper formulation.

4. Matrix manipulations, in particular matrix diagonalization. These steps scale as the cube of the number of (contracted) basis functions. Fock matrix diagonalization can be replaced by alternative procedures that exhibit linear scaling asymptotically but the transition to linear scaling is rather late, particularly for larger basis sets.

5. Calculation of the one-electron Coulomb energy (nuclear attraction energy) and its matrix elements show cubic formal scaling, flattening out to quadratic but with a low prefactor. These terms can be included with the electronic Coulomb terms and will not be discussed separately.

As the above list shows, the worst scaling, both formally and asymptotically, is shown by the electron–electron Coulomb repulsion. Considerable effort has been spent on improving the efficiency of this term in the last 15 years. Most progress has been achieved in the application of the fast multipole algorithm to continuous charge distributions [4–6]. If properly implemented, this method is very accurate. However, its onset point is rather late, and therefore the savings for intermediate sized molecules are often modest, particularly if the system has three-dimensional topology.

Another popular method is density fitting. In this method, the electron density is expanded in an intermediate density basis set [7]. This can be regarded formally as inserting a "resolution of identity" (RI) operator, $\Sigma_{\mu\nu}|\mu > (S^{-1})_{\mu\nu} < \nu|$ in the two-electron integrals; μ and ν are the density basis functions. If the density basis is complete, the above operator is the identity operator, and allows the factorization of the four-index integrals to products of three-index quantities. For this reason, the method is often called the RI method, although the more descriptive Density Fitting nomenclature is increasingly used. Recently, Ahlrichs and co-workers [8,9] have completed an outstanding implementation of this method. It speeds up the calculation of the Coulomb terms by about an order of magnitude. However, its accuracy is somewhat limited, and its asymptotic scaling is not better than that of the traditional method (quadratic).

In the following, we will describe the Fourier Transform Coulomb (FTC) method. In this method, we retain the usual Gaussian basis sets. However, the Coulomb energy and operator matrix elements are calculated using an intermediate plane-wave expansion, like in density fitting. The special properties of plane waves allow a very fast and low-scaling evaluation of the Coulomb operator. Moreover, the orthogonal nature of the plane waves eliminates the difficulties that arise for very large, nearly overcomplete density basis sets. Therefore, the precision of the method is high. It yields total energies to a precision better than a microhartree (1 μE_h) for small molecules and better than 0.1 μE_h/atom for large ones. This high precision can be relaxed where not needed, e.g. in preliminary calculations, early stages of geometry optimization or molecular dynamics runs. However, we consider it important to maintain compatibility with existing Gaussian basis programs.

The plane-wave based calculation of the Coulomb energy in Gaussian basis was pioneered by Parrinello and co-workers under the acronym GAPW [10,11]. Ref. [9] is particularly noteworthy because it describes an all-electron (no pseudopotential) method, and, uniquely in the physics literature, it lists calculated total energies which can be directly compared to other programs.

Our goals in the FTC program are the same as those of the Parrinello group. However, the implementations are very different. In particular, we aim at much higher precision than Ref. [11] (sub-μE_h for medium-sized molecules).

2. The Fourier Transform Coulomb method

The initial implementation of the FTC method was described earlier [12]. We start by constructing a box, a parallelepiped, which contains essentially all the electronic charge of the molecule. For full accuracy, the box should be rather large, even for small molecules (see the examples later). We also define d, the maximum density of the plane waves. This is related to the more commonly used kinetic energy cutoff by $E_{cutoff} = d^2\pi^2/2\ E_h = d^2\pi^2$ Ry. A typical value is $d = 3.5\ a_0^{-1}$, corresponding to $E_{cutoff} \approx$ 120.9 Ry = 60.45 E_h. The plane wave basis necessarily describes an infinite periodic system. For molecules, the error arising from the repeated periodic images of the molecule is eliminated by the technique described in Ref. [13]. This is the analytical version of the method of Hockney [14] and it is exact. It is based on truncating the Coulomb operator. The truncated operator is equal to the usual $1/|\mathbf{r}|$ Coulomb operator up to a maximum distance D and is zero beyond this distance. The Fourier transform of the truncated Coulomb operator is given by a simple formula [13]. We also introduce a real-space rectangular grid with spacing $h = d^{-1}$. Functions can be represented either on the real-space grid or in the plane wave (momentum) basis. For functions which can be represented exactly in the plane wave space the two representations are isomorphic. The efficacy of the method is largely dependent on the fact that transformation from real space grid to momentum space representation can be accomplished efficiently by the Fast Fourier (FFT) algorithm.

Following Krack and Parrinello [11], we partition the Gaussian basis set in two subsets: core and valence functions. Note that this distinction is purely mathematical and has nothing to do with the physical role of these functions. Basis functions in the valence category can be described to essentially full accuracy by the plane wave basis. Coulomb integrals over the valence functions (and also some mixed integrals involving both valence and core functions) are evaluated in the plane wave basis; the remaining integrals are evaluated by alternative techniques, e.g. by the traditional electron repulsion integral (ERI) method. The partitioning is done mainly by the value of the orbital exponent but depends weakly also on the angular momentum. For example using a grid density of $3.5\,a_0^{-1}$, an s-type function can be represented virtually exactly if its exponent is below $2.45\,a_0^{-2}$. For a p-type function, the exponent must be less than $2.37\,a_0^{-2}$, etc. Contractions interfere with this scheme because they mix high and low exponent basis functions. The correct way of handling contractions is to decompose them to a core and a valence component, evaluate the energy or Fock matrix contributions separately, and add the two contributions. This has not yet implemented, and therefore we either treat contractions, which include both low and high exponents as core functions, or slightly decontract the basis. Both of these measures somewhat diminish the efficiency of the code.

Two-electron integrals containing only valence-type basis functions are treated by somewhat modified versions of techniques widely used in solid-state physics [15,16]. The basic idea is that the Coulomb operator is diagonal in momentum space:

$$|\mathbf{r} - \mathbf{r}'| = (1/2\pi^2) \int |\mathbf{k}|^{-2} \exp\{i\mathbf{k}(\mathbf{r} - \mathbf{r}')\}d\mathbf{k}$$

The molecular electronic density is first evaluated on the real-space grid. Because of the separability of Gaussians into x, y, z components, and the rectangular nature of the grid, this can be done very efficiently. The real-space electron density is transformed to momentum space using the FFT algorithm. Division by the value of the squared momentum yields the momentum space representation of the electrostatic potential of the electrons. This is transformed back to real space, using again FFT. The matrix elements of the Coulomb operator are obtained by numerical quadrature; note that this quadrature is exact. By introducing boxes which fully contain a given basis function, both the calculation of the density and the numerical quadrature can be implemented in a linear scaling manner. The FFT steps scale slightly higher, like $O(N \log N)$. However, the absolute time needed for FFT is minute. For example the two FFT steps needed for the calculation of the Coulomb part of the Fock matrix for taxol, $C_{47}H_{51}NO_{14}$, using a slightly decontracted 6-311G(2df, 2pd) basis (2860 contracted basis functions), takes only 15 s CPU time on an AMD XP 1800 + processor [12]. The number of plane waves used to describe the valence electron density is 5.8×10^6. It is clear that the slightly superlinear $N \log N$ formal scaling of the FFT will not affect the overall scaling until much larger calculations.

The FTC method can also be used to evaluate the contributions of integrals of the form $(cv|v'v'')$ (in the Mulliken notation), where c denotes a core-type (compact) basis function and v, v' and v'' denote valence-type (diffuse) basis functions, even though c cannot be expressed by the limited plane-wave expansion. This surprising fact is easy to understand: although the product cv has momentum components beyond those contained in the plane wave basis, these do not interact with $v'v''$ which has only low momentum components [13]. However, the numerical quadrature cannot be performed accurately with the original core-type function c. For accurate quadrature, c has to be subjected to a low-pass filter operation. This can be accomplished by transforming it to the plane-wave (momentum) space, removing all components above the limiting momentum k_{max}, and transforming it back to real space. Unfortunately, this operation generates long, slowly decaying oscillatory tails in the filtered basis functions. These delocalized tails are analogous to Fresnel diffraction in optics. In our initial implementation, the computationally most expensive part of the plane-wave part of the program involved these delocalized filtered core-type functions.

In the current implementation, we have greatly improved this part of the code. It can be easily proven that the filtered core-like basis functions, which become delocalized as a result of the low-pass filtering operation, can be restricted to the (small) range of the original (unfiltered) basis functions, provided that this operation does not introduce high momentum components. We introduce an extended range for the core-like basis functions, which is larger than their original range. The filtered basis functions are smoothly reduced to zero in the interval between the original and the extended range, using a polynomial attenuation factor which goes from 1 at the outer limit of the original range to 0 at the boundary of the extended range, with its n first derivatives vanishing at both the outer and the inner limit. To satisfy these conditions, the degree of the attenuation polynomial must be $2n + 1$. For example the eight coefficients of a seventh-degree polynomial can be chosen to satisfy the eight conditions $f(r_{in}) = 1, f(r_{out}) = 0, f^{(k)}(r_{in}) = f^{(k)}(r_{out}) = 0, \quad k = 1, 2, 3$.

Outside of the extended range, the filtered core basis functions are identically zero. Restricting their range results in a significant speed-up of the code, particularly in the calculation of the gradients (forces). The details of this will be published separately [17]

The major remaining computational task in the calculation of the Coulomb energy and Coulomb matrix elements is the evaluation of the contribution of integrals with two or more core-type functions. In the current code, this is done by using traditional ERI techniques. However, this limits the ultimate speedup of the method. Although the overwhelming majority of the integrals is handled by the FTC part of the code, traditional ERI techniques are inherently slow and constitute a significant bottleneck.

Integrals with two core-like functions can have the form of $(cv|c'v')$ or $(cc'|vv')$. The latter are relatively few in number because they require that the overlap between the two core-type functions c and c' be non-zero. They can be treated in principle by the FTC code, although this is not done in the current program.

Integrals of the form $(cv|c'v')$ constitute currently the bulk of the computational work associated with core-type functions. Fortunately, they lend themselves to an efficient multipole-type approximation if the two charge densities, cv and $c'v'$ are not overlapping and sufficiently separated. This part of the code is being currently implemented. Preliminary tests show that it will reduce the effort for these integrals by more than an order of magnitude. Other integrals, e.g. $(cc'|c''v)$ can be calculated together with this type but they are computationally not important because they require the overlap of two core-type basis functions.

3. Examples and discussion

Fig. 1 shows a comparison of FTC timings with a traditional ERI calculation for polyalanines, $H(NHCH(CH_3)CO)_nOH$, $n = 1, 2, 5, 10, 15$. The molecules are in the α-helical conformation, and the basis is a slightly decontracted version of the 6-311G(2df,2pd) set. Note that the FTC timings have been multiplied by 10 for display in the same graph as the traditional ERI timings.

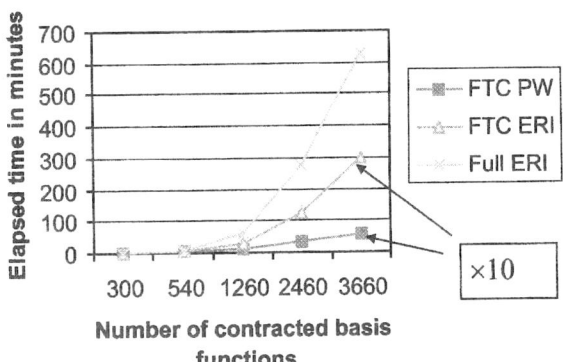

Fig. 1. Elapsed times (in minutes, on a 2.4 GHz Pentium 4 processor) for the construction of the Coulomb operator matrix elements for polyalanines, $(Ala_n,$ alpha helix conformation), using a slightly decontracted version of the 6-311G(2df,2pd) basis. Upper curve: traditional ERI code. Middle curve: the ERI part of the FTC code. Lower curve: the plane-wave part of the FTC code. Note that the latter two numbers have been multiplied by 10 to enable better comparison.

Fig. 2 shows the scaling for alanine$_5$, $H(NHCH(CH_3)CO)_5OH$, with various basis set. It illustrates the point that the FTC method is particularly appropriate for large basis set calculations, because of its low scaling with the basis set quality. It is particularly promising for NMR chemical shift calculations, as these require large basis sets.

In the current program, three major bottlenecks remain after eliminating the valence part of

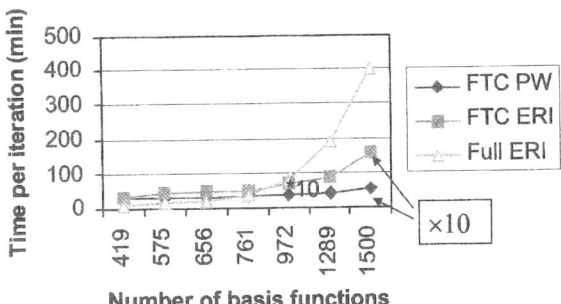

Fig. 2. Elapsed times (in minutes, on a 2.4 GHz Pentium 4 processor) for the construction of the Coulomb operator matrix elements for the alanine pentamer, $(Ala)_n$, using a slightly decontracted version of the 6-311G(2df,2pd) basis. Upper curve: traditional ERI code. Middle curve: the ERI part of the FTC code. Lower curve: the plane-wave part of the FTC code. Note that the latter two numbers have been multiplied by 10 to enable better comparison.

the Coulomb repulsion: calculation contributions from the remaining ERIs, the calculation of the exchange-correlation matrix elements, and matrix manipulations, mainly diagonalization. The first can be treated by an atom-centered multipole expansion. This converges rapidly, as the charge distributions are all compact. To accelerate the calculation of the exchange-correlation energy and its matrix elements significantly will probably require a fundamental change of the usual Becke numerical quadrature scheme [18]. We are planning to implement a scheme which combines a Cartesian integration using the plane-wave gridpoints with a small spherical grid in the vicinity of the nuclei.

As these examples show, the FTC method is an efficient way of extending the range of ab initio DFT calculations to large molecular systems important for biological systems. In order to exhaust its full potential, other computational steps, which are currently not important, will have to be accelerated. We expect that the method will ultimately result in a speed-up by two orders of magnitude for a good-quality calculation on a moderately large (~ 100 atoms) molecule. It is particularly advantageous for analytical derivatives (only the gradient has been implemented so far), and for calculations requiring diffuse basis functions (e.g. for negative ions).

References

[1] J. Baker, P. Pulay, J. Chem. Phys. 117 (2002) 1441.
[2] J. Baker, P. Pulay, J. Comput. Chem. 24 (2003) 1184.
[3] E. Schwegler, M. Callacombe, J. Chem. Phys. 105 (1996) 2726.
[4] C.A. White, B.G. Johnson, P.W.M. Gill, M. Head-Gordon, Chem. Phys. Lett. 230 (1994) 8.
[5] C.A. White, M. Head-Gordon, J. Chem. Phys. 101 (1994) 6593.
[6] M.C. Strain, G.E. Scuseria, M.J. Frisch, Science 271 (1996) 51.
[7] B.I. Dunlap, Phys. Rev. A 42 (1990) 1127.
[8] K. Eichkorn, O. Treutler, H. Öhm, M. Häser, R. Ahlrichs, Chem. Phys. Lett. 240 (1995) 283.
[9] K. Eichkorn, F. Weigend, O. Trutler, R. Ahlrichs, Theor. Chim. Acta 97 (1997) 119.
[10] G. Lippert, J. Hutter, M. Parrinello, Theor. Chem. Acc. 103 (1999) 124.
[11] M. Krack, M. Parrinello, Phys. Chem. Chem. Phys. 2 (2000) 2105.
[12] L. Füsti-Molnár, P. Pulay, J. Chem. Phys. 117 (2002) 7827.
[13] L. Füsti-Molnár, P. Pulay, J. Chem. Phys. 116 (2002) 7795.
[14] R.W. Hockney, in: B. Alder, S. Fernbach, M. Rothenberg (Eds.), Methods of Computational Physics, vol. 9, Academic Press, New York, 1970, p. 136.
[15] M.C. Payne, M.P. Teter, D.C. Allan, J.D. Joannopoulos, Rev. Mod. Phys. 64 (1992) 1045.
[16] G. Galli, M. Parrinello, in: M. Meyer, V. Pontikis (Eds.), Computer Simulation in Material Science, Kluwer Academic Publishers, Dordrecht, 1991, pp. 32, Chapter 3.2.
[17] L. Füsti-Molnár, J. Chem. Phys. (2003) in press.
[18] A.D. Becke, J. Chem. Phys. 88 (1988) 2547.

ELSEVIER

Journal of Molecular Structure (Theochem) 666–667 (2003) 31–39

THEO
CHEM

www.elsevier.com/locate/theochem

Geometry optimization methods for modeling large molecules

Ödön Farkas[a,*], H. Bernhard Schlegel[b]

[a]Department of Organic Chemistry, Eötvös Loránd University, Budapest, Hungary
[b]Department of Chemistry, Wayne State University, Detroit, USA

Abstract

Geometry optimization is an essential part of quantum chemical applications. The diversity of the scaling of different methods from linear to exponential implies that there are different requirements for a chosen optimization method. The proposed method aims to meet two requirements, good scaling with size and reliability, which would be a good match for redundant internal coordinate system-based optimization techniques with linear scaling coordinate transformation. The new optimization algorithm uses screened Cholesky decomposition for coordinate transformations and an iterative subspace optimization method. The iterative subspace appears in the course of any optimization. However, few methods are available for using such information efficiently. The Geometry Optimization using Direct Inversion in the Iterative Subspace method is known to have good scaling and efficiency, but poor reliability. Building a Hessian-like matrix in the iterative subspace allows one to take advantage of the reliability offered by Rational Function Optimization, Eigenvector Following and Trust Radius Method (TRM), but still avoid a consequent computational penalty. Also, the new approach steps away from the regular quadratic approximation related to the Newton methods by assuming a simple linear connection between gradient and coordinate changes.
© 2003 Elsevier B.V. All rights reserved.

Keywords: Geometry optimization using direct inversion in the iterative subspace; Rational function optimization; Trust radius method; Sparse Cholesky decomposition

1. Introduction

Quantum chemical geometry optimization methods evolved rapidly over the last three decades. A major developmental milestone included the analytic gradients of the potential energy and the methods based on them, such as the quasi-Newton methods and their modifications. Hessian update techniques allowed information to be collected for the potential energy surface (PES), which accelerated the optimization process. The quadratic line search (QLS) [1–5], rational function optimization

(RFO) [6], trust radius model (TRM) [1–5,7,8] and trust radius update [1–5,7,8] made such methods more reliable. Originally, the geometry optimization was performed in Cartesian or in Z-matrix internal coordinates. Redundant internal coordinates [9,10] took precedence only in the last decade of the 20th century. Other improvements, such as the Geometry Optimization using Direct Inversion in the Iterative Subspace (GDIIS) [11,12] or using natural [13] or delocalized [14,15] internal coordinates has also been considered. It is now commonly agreed that an efficient optimization method [7,8] for quantum chemical applications should use the Hessian update, line search or GDIIS, and RFO in the framework of redundant internal coordinates. The general use of

---

* Corresponding author.
    E-mail address: farkas@para.chem.elte.hu (Ö. Farkas).

0166-1280/$ - see front matter © 2003 Elsevier B.V. All rights reserved.
doi:10.1016/j.theochem.2003.08.010

such optimization techniques is, however, prevented by the quadratic ($O(N^2)$) scaling of their memory usage and the regularly cubic, $O(N^3)$, scaling of their computational demand with the number of variables to optimize. Current advances in the scaling of some frequently used quantum chemical methods, like Hartree–Fock, density functional theory or semi-empiricals, necessitate the development of new optimization techniques to match the required scaling with size, without compromising traditional efficiency and reliability. Our previous studies [16,17] revealed that the overall computational bottleneck can be reduced to an asymptotic quadratic, $O(N^2)$, scaling using an updated inverse technique for solving the systems of linear equations in question for both, the coordinate transformations and the RFO or TRM optimization step. The updated inverse technique provides similar efficiency and reliability as regular quantum chemical optimization methods but it still needs large, $O(N^2)$, storage of full matrices in memory. Other efforts have been focused on achieving an overall linear, $O(N)$, scaling for the coordinate transformations [18–21] to make the use of redundant internal coordinates affordable for much larger systems. The present paper describes an alternative way, a linear scaling equivalent of singular value decomposition (SVD) or generalized inverse, for solving linear equations with non-definite sparse matrixes for the coordinate transformations necessary in redundant internal coordinate-based optimizations. We also outline a new optimization technique similar to GDIIS, which can employ a step size control in the spirit of RFO and TRM through a generalized Hessian built in the iterative subspace.

The regular optimization process, such as the 'Berny' optimization algorithm of the GAUSSIAN [22] program contains two practically equivalent computational bottlenecks; one is the transformation of the forces into internal coordinates and then the transformation of the optimization step back to Cartesian coordinates. The coordinate transformations are based on the use of the Wilson B-matrix [23], which collects the partial derivatives of the internal coordinates with respect to Cartesians

$$B_{i,j} = \frac{\partial q_i}{\partial x_j} \Rightarrow dq = B\,dx \text{ and } \mathbf{f}_x = \mathbf{B}^t\mathbf{f}_q \qquad (1)$$

where q and x denote internal and Cartesian coordinates, respectively. The optimization process needs the transformation of forces ($\mathbf{f}_x$) given in Cartesian coordinates, and also the transformation of the internal coordinate step ($\Delta\mathbf{q}$). Due to the curvilinear nature of the internal coordinates, the finite internal coordinate step can be transformed by a few iterations of solving the corresponding equations ($\Delta\mathbf{q} \approx \mathbf{B}\Delta\mathbf{x}$). For detailed remarks on solving such equations, see Appendix A. In the case of constrained optimizations, an extra projection of the forces and the optimization step is required.

The other bottleneck arises from the computation of the RFO or TRM step, which is in fact a Newton–Raphson step with a modified, shifted Hessian matrix

$$-\Delta\mathbf{f} = (\mathbf{H} + \lambda\mathbf{I})\Delta\mathbf{x} \qquad (2)$$

where $\mathbf{H}$ is an approximate Hessian (force constant matrix) in the case of quasi-Newton methods. The appropriate diagonal shift to the Hessian is defined by the method of choice (RFO or TRM), and is usually carried out by solving Eq. (2) for $\Delta\mathbf{x}$ each time during the course of determining λ iteratively. A practical way of solving Eq. (2) implies the diagonalization of the Hessian, which is a cubically scaling computational bottleneck.

2. Method

2.1. Coordinate transformations

Paizs et al. pointed out that any set of redundant internal coordinates could be constructed from their complete, but non-redundant, subsets as linear combinations [20]. They also concluded that this also applies to matrices $\mathbf{B}^t\mathbf{B}$ and $\mathbf{BB}^t$, and their rows. The full Cholesky factorization of positive semi-definite matrices results in zero diagonal values. Becausse of consequent divisions by zero, the full Cholesky factorization (regardless of the sparsity of the matrix in question) can only be applied to positive definite matrices. The zero (or in practice very small) diagonal values, however, indicate rows that can be produced as a linear combination of previously processed rows. Removing such rows from further examination results in the screened Cholesky factorization. This creates a non-redundant set of linear combinations. In practice, we use a small positive threshold value, such as 2×10^{-5}, for diagonal elements to select the rows for removal. We also noted that the rank (number of non-zero diagonal values) of the decomposition might change by varying the threshold value. With proper ordering, the screened Cholesky factors can reproduce the original symmetric matrix accurately, although in general, the multiplication of a vector by the 'screened Cholesky inverse' of the matrix via forward–backward substitution using the screened Cholesky factors does not give the same results as multiplying by the generalized inverse. Nevertheless, linear equations of the form $\mathbf{y} = \mathbf{A}\mathbf{x}$ can be solved using the screened Cholesky factors of positive semi-definite matrices $\mathbf{A}\mathbf{A}^t$ and $\mathbf{A}^t\mathbf{A}$. The screened Cholesky factorization-based equation solving method can be directly applied to the equations arising from the coordinate transformations (see Eq. (1)). The screened Cholesky decomposition has been tested in practice and give identical results identical to the regular generalized inverse-based transformations; however, its thorough mathematical proof should be given later. For details and the constrained optimization formulae, see Appendix B. The main advantage of the screened Cholesky factorization over the approximate or shifted Cholesky decomposition is that it can substantially reduce the required storage and computational demand for highly redundant systems.

2.2. Iterative subspace optimization

The overall efficiency of the optimization method also depends on the choiced algorithm of choice. Quasi-Newton related algorithms could greatly benefit from using redundant internal coordinates, since a good approximation to the PES can be used for a much wider range than in Cartesian coordinates. The problem is that the approximation to the PES regularly stored in a full Hessian matrix results in at least an $O(N^2)$ scaling computational bottleneck. A starting guess Hessian matrix can be stored in sparse, or even diagonal form and limited memory Hessian update techniques have been formulated to achieve linear scaling [24–27]. These techniques proved to be more efficient than available alternatives, such as pure conjugate gradient (CG)-based methods. Limited memory update techniques, have some disadvantages,

which include the lack of convergence acceleration via RFO/TRM type shifting to the Hessian. In addition, they are not advised to use for transition state optimizations. The GDIIS method is also a possible choice to consider, since its memory and computational demand can be easily controlled by the number of vectors stored and used in the iterative subspace. The reliability of the GDIIS method is unfortunately very poor, especially for large-scale non-linear problems, such as for the geometry optimization of biomolecules. Our previously described method for controlling GDIIS [12] appears to be satisfactory, but requires an expensive computation for a reference step, which is currently RFO. It is, however, an instructive task to find the reason for the otherwise surprisingly good performance of the GDIIS technique closed to convergence. The first thing to note is that GDIIS assumes a simple linear connection between coordinate and force (or gradient) changes. This condition seems to be similar to the quadratic approximation for the PES. The quadratic approximation, however, presumes that the linear connection is related to the second derivative, Hessian matrix of the PES, in the form of a symmetric real matrix:

$$-\Delta\mathbf{f} = \mathbf{H}\Delta\mathbf{x} \qquad (3)$$

The GDIIS method only assumes that the force change related to any linear combination of the collected optimization steps (coordinate changes) can be formed using the same combination of the corresponding force changes:

$$\sum\gamma_i\Delta\mathbf{x}_i = \Delta\mathbf{x} \Leftrightarrow \sum\gamma_i\Delta\mathbf{f}_i = \Delta\mathbf{f} \qquad (4)$$

In fact, Eq. (4) states a more generic linear connection between coordinate and force changes, which is valid for the whole set of stored coordinate and force changes when no redundancy occurs:

$$-\mathbf{F} = \mathbf{H}_G\mathbf{X} \qquad (5)$$

Matrices $\mathbf{F}$ and $\mathbf{X}$ collect the coordinate and force changes, respectively. Matrix $\mathbf{H}_G$, expressing the linear connection, is the generalized Hessian that is not necessarily symmetric, not even for quadratic PESs. For the purpose of optimization, it may serve a similar purpose as the Hessian in the quasi-Newton methods. Based on Eq. (5), it is also possible to find a suitable way of obtaining $\mathbf{H}_G$.

Orthonormalizing the column vectors of matrix $\mathbf{X}$ via Schmidt orthogonalization results in a unitary matrix on the right side that leads us to the expression for obtaining $\mathbf{H}_G$:

$$-\mathbf{FV} = \mathbf{H}_G\mathbf{XV} = \mathbf{H}_G\mathbf{U} \Rightarrow -\mathbf{FVU}^t = \mathbf{H}_G \tag{6}$$

where matrix $\mathbf{V}$ orthonormalizes $\mathbf{X}$. Matrix $\mathbf{VU}^t$ is the generalized inverse of $\mathbf{X}$ if no redundancy occurs. The purpose of using Schmidt orthogonalization instead of a generalized inverse provides higher priority for the latest coordinate and force changes. $\mathbf{H}_G$ is usually not symmetric; therefore, its direct diagonalization is not feasible. The tools of SVD can provide the singular values and two sets of eigenvectors:

$$\mathbf{H}_G = \mathbf{L}\Lambda\mathbf{R}^t \tag{7}$$

The singular values are positive numbers by definition; however, if we change the sign of one of the eigenvector pairs and the corresponding singular value in order to constrain the scalar product of the corresponding eigenvectors to be non-negative

$$\mathbf{l}_i^t\mathbf{r}_i \geq 0 \tag{8}$$

they may gain signs. The resulting diagonal values of Λ can serve as pseudo-eigenvalues, allowing RFO/TRM type step size control of the optimization step. It is important to note that the latest force usually cannot be represented in the subspace of collected force changes. Therefore, only a part of the optimization step can be computed using $\mathbf{H}_G$. The resulting step corresponds to minimizing the force in the iterative subspace, such as in GDIIS. The residual force then can be used to compute the residual optimization step. Examination of the difference between GDIIS and a line search reveals that since the residual force is perpendicular to the subspace of force (or error vector) changes, the quadratic line search finds an energy extreme and its residual force is perpendicular to the latest step, or if it is generalized to higher dimensions to the subspace of coordinate changes, while GDIIS ends to minimize the force (or an arbitrary error vector). This difference suggests that a projected generalized Hessian for the iterative subspace should be constructed, to serve the purpose of multidimensional search (MDS) using a generalized quadratic line search to higher dimensions:

$$-\mathbf{UU}^t\mathbf{FVU}^t = \mathbf{H}_{MDS} \tag{9}$$

The advantage of using the projection of the force changes into the coordinate change subspace is that the left and right side eigenvectors of $\mathbf{H}_{MDS}$ span the same subspace. In addition, the generalized Hessian for MDS can more sensitively detect the proximity of higher order critical points with negative pseudo-eigenvalues. RFO/TRM style step size controls can also be applied to $\mathbf{H}_{MDS}$, but unfortunately, these kinds of corrections are not capable of ensuring an energy lowering step direction, since $\mathbf{H}_{MDS}$ is generally non-symmetric for more than one dimension. The construction of an efficient, but symmetric Hessian for the iterative subspace is a key target for further studies in that field.

The presented iterative subspace optimization (ISSO) scheme produces dense Hessian matrixes in the full space, resulting in the same computational demand than the regular $O(N^3)$ scaling techniques. The rank of these matrixes, however, is not larger than the number of coordinate changes used. The common subspace of coordinate and force changes can be constructed reducing the required storage and computational cost significantly. If the maximum number of used vectors remains constant, then the ISSO method requires linearly scaling storage and computational effort. The details about the construction of the iterative subspace efficiently can be found in Appendix C.

3. Results and discussion

The new methods presented here have been implemented in the development version of Gaussian [28]. The preliminary test results on storage and CPU requirements of the coordinate transformations (namely the computation of the Cholesky factors) can be found in Figs. 1 and 2. Proper reordering is essential for the efficiency of the factorization. Thus, a divide-and-conquer-based approach will be developed for this purpose, as suggested by Nemeth et al. [21]. Note that the factorization of $\mathbf{G}_q = \mathbf{BB}^t$ is less demanding than the factorization of $\mathbf{G}_x = \mathbf{B}^t\mathbf{B}$ in the screened Cholesky formalism. It is not easy to find examples to compare the performance of different optimization algorithms on large flexible molecules because they tend to converge to different local minima. However, the optimization of taxol

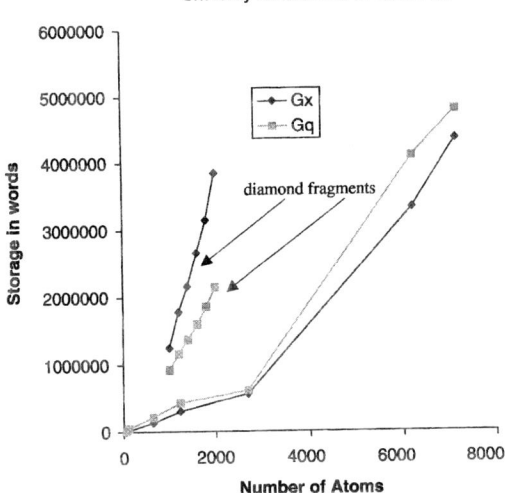

Fig. 1. Diagram for required storage (including indexing) vs. number of atoms. The diamond fragments (3D network examples) are clearly separated from the others, which are mostly proteins. For the definition of G_q and G_x, see Appendix B.

(113 atoms) in the UFF [29] forcefield results the same minimum for the three selected optimization methods used for large molecules. The results are summarized in Table 1. The linear scaling ISSO method uses only a diagonal updated guess Hessian

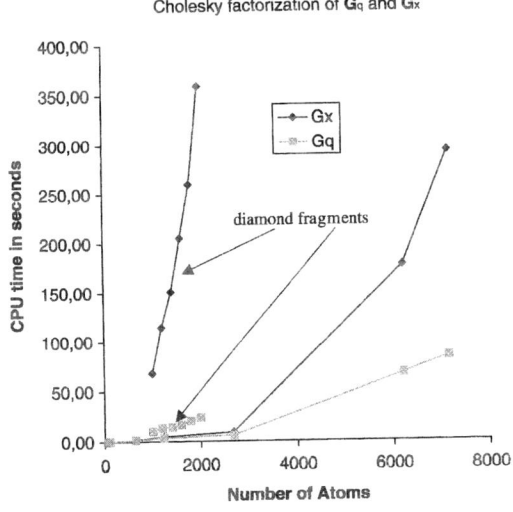

Fig. 2. Diagram for required CPU time on 500 MHz Alpha processor vs. number of atoms. The diamond fragments (3D network examples) are clearly separated from the others, which are mostly proteins. For the definition of G_q and G_x, see Appendix B.

Table 1
The number of optimization steps required optimizing taxol in the UFF [29] molecular mechanics forcefield

Optimization method	Number of optimization steps
Controlled GDIIS, full Hessian [12]	40
ISSO, diagonal full space Hessian	127
CG, Cartesian coordinates	1153

for the full space and therefore, needs substantially more steps to converge than the controlled GDIIS. However, it is still only a small fraction of the steps used by the CG optimizer in Cartesian coordinates. Further improvements are expected using a sparse updated Hessian in the full space and implementing a symmetric MDS Hessian for guiding the optimization to the quadratic region of a proper critical point more efficiently.

4. Summary

The aim of the new methods presented is to provide efficient linear scaling optimization tools for the emerging linear scaling quantum chemical methods, including QM/MM and MM calculations. The use of redundant internal coordinates allows more accurate calculations for large molecules with the aid of ISSO and MDS, which also opens the opportunity for the development of transition state optimization algorithms for large molecules. In general, the screened Cholesky decomposition can be considered for solving large, redundant systems of linear equations.

Acknowledgements

This work has been supported by Gaussian, Inc. (http://www.gaussian.com) and the Hungarian Research Foundation (OTKA D29446). The authors would also like to thank NSF, MTA and OTKA for an international collaboration grant. The computations have been performed in part using the High Performance Computing Center of the Eötvös Loránd University.

Appendix A

A.1. Solving linear equations

A system of linear equations can be formulated as $y = Ax$, which can only be solved if the proper Py projection does not alter y:

$$(AA^t)(AA^t)^{-1}y = Py = y' = Ax \qquad (A1)$$

Otherwise, the corresponding equations with y' can be solved using the generalized inverse or SVD formalism:

$$(A^tA)^{-1}A^ty = A^t(AA^t)^{-1}y = x \qquad (A2)$$

If matrix A^tA or AA^t is non-singular, then $AA^t = LL^t$ or $A^tA = KK^t$ Cholesky decomposition is the most efficient way of solving the equations:

$$(A^tA)^{-C}A^ty = x \text{ or } A^t(AA^t)^{-C}y = x \qquad (A3)$$

The $-C$ superscript denotes the Cholesky inverse which is in practice multiplication by the inverse via forward–backward substitution using the corresponding lower/upper triangular Cholesky factors. If A^tA and AA^t are positive semi-definite, then incomplete, shifted or approximate Cholesky factorization and iterative solving is necessary. The removal of columns of the lower and the corresponding rows of the upper triangular factors with diagonals smaller than a small positive ε during the decomposition results in a non-redundant subset:

$$A^tA = L_SL_S^t \text{ and } AA^t = K_SK_S^t \qquad (A4)$$

The number of remaining non-zero diagonals indicates the rank of A. We found that changing ε may change the obtained rank if the original limit is too large. The corresponding Cholesky inverse can be used as a projector

$$A(A^tA)^{-1}A^ty = Py = A(A^tA)^{-S}A^ty = y' \qquad (A5)$$

like the generalized inverse, but solving the equations cannot be done in the usual way, since:

$$(A^tA)^{-S}A^ty \neq x$$

$$A^t(AA^t)^{-S}y \neq x \qquad (A6)$$

$$(A^tA)^{-S}A^ty' \neq x$$

On the other hand,

$$A^t(AA^t)^{-S}y' = x \qquad (A7)$$

leads to the solution in the screened Cholesky formalism:

$$A^t(AA^t)^{-S}A(A^tA)^{-S}A^ty = (A^tA)^{-1}A^t = x \qquad (A8)$$

which is a potential linear scaling equivalent of the generalized inverse. The 'screened Cholesky inverse', denoted by the $-S$ superscript, is not provided explicitly but the multiplications with vectors can be computed via forward–backward substitutions, like in the case of regular Cholesky decomposition.

Appendix B

B.1. Coordinate transformations

Using the first order approximation to the connection between internal and Cartesian coordinates, one can carry out the coordinate transformations for the purpose of redundant internal coordinates

$$B_{i,j} = \frac{\partial q_i}{\partial x_j} \Rightarrow dq = B\,dx \text{ and } f_x = B^t f_q$$

$$G_q = BB^t(N_q \times N_q) \text{ and } G_x = B^tB(N_x \times N_x) \qquad (A9)$$

$$P_q^S = BG_x^{-S}B^t \text{ and } P_x^S = B^tG_q^{-S}B$$

where B is the Wilson B-matrix, q denotes internal, while x denotes Cartesian coordinates and the $-S$ subscipt indicates screened Cholesky inverse. The equations for the internal forces, f_q, and the Cartesian coordinate step, Δx, can be solved using the screened Cholesky decomposition:

$$f_q = BG_x^{-S}P_x^S f_x = G_q^{-1}Bf_x$$
$$\qquad (A10)$$
$$dx = B^tG_q^{-S}P_q^S\,dq = B^tG_q^{-1}\,dq$$

The transformation of the optimization step is valid for infinitesimally small steps only. In practice, we iterate $\Delta x = B^tG_q^{-S}P_q^S\Delta q = B^tG_q^{-1}\Delta q$ until convergence. Constrained optimization necessitates the definition of the constrained B-matrix, which has empty rows for the non-constrained internal coordinates, and then the related matrixes:

$$\mathbf{G}_{q,c} = \mathbf{B}_c\mathbf{B}_c^t \quad \text{and} \quad \mathbf{G}_{x,c} = \mathbf{B}_c^t\mathbf{B}_c \tag{A11}$$

$$\mathbf{P}_{q,c}^S = \mathbf{B}_c\mathbf{G}_{x,c}^{-S}\mathbf{B}_c^t \quad \text{and} \quad \mathbf{P}_{x,c}^S = \mathbf{B}_c^t\mathbf{G}_{q,c}^{-S}\mathbf{B}_c$$

The force transformation is straightforward

$$\mathbf{f}_q = \mathbf{B}\mathbf{G}_x^{-S}(\mathbf{I} - \mathbf{P}_{x,c}^S)\mathbf{P}_x^S\mathbf{f}_x = \mathbf{G}_q^{-1}\mathbf{B}(\mathbf{I} - \mathbf{P}_{x,c})\mathbf{f}_x \tag{A12}$$

but the step transformation has to treat optional non-zero step in the constrained subspace

$$\Delta\mathbf{x} \approx \mathbf{B}^t\mathbf{G}_q^{-S}(\mathbf{P}_{q,c}^S\Delta\mathbf{q}_c + (\mathbf{I} - \mathbf{P}_{q,c}^S)\mathbf{P}_q^S\Delta\mathbf{q}) \tag{A13}$$

which is a little bit more complicated and can be formulated in numerous equivalent ways.

Appendix C

C.1. Iterative subspace

Definition. $\mathbf{W}$ collects M of L dimensional vectors, then a transformation matrix, $\mathbf{V}$ ($N \times M$) is applied to achieve:

$$(\mathbf{WV})^T\mathbf{WV} = \mathbf{V}^T\mathbf{W}^T\mathbf{WV} = \mathbf{I} \tag{A14}$$

The $\mathbf{Q} = \mathbf{WV}$ ($N \times L$) set of vectors forms an orthonormal basis to define a subspace. Any a vector of that subspace (but represented in the original space) can then be represented as $\mathbf{p}$ in the basis of the subspace coordinates ($\mathbf{Q}$) as:

$$\mathbf{p} = \mathbf{Q}^T\mathbf{a} \tag{A15}$$

In general, Eq. (A15) provides the projection of vector $\mathbf{a}$ on to the subspace. The number of dimensions for the original vectors ($\mathbf{W}, \mathbf{a}$) might be large. Therefore, their direct use might be impractical. The necessary values can be obtained using the $\mathbf{O} = \mathbf{W}^T\mathbf{W}$ overlap matrix instead. It is also practical to give any vector $\mathbf{a}$ as a $\mathbf{b}$ linear combination of the original $\mathbf{W}$ vectors

$$\mathbf{a} = \mathbf{Wb} \tag{A16}$$

then the $\mathbf{p}$ representation can be formed as:

$$\mathbf{p}^T = \mathbf{b}^T\mathbf{OV} \tag{A17}$$

Furthermore, all work with the represented vectors in the subspace can be done using only N dimensional vectors and accordingly formed

matrixes with a maximum size of $N \times N$. The resulting vectors in the subspace, e.g. $\mathbf{y}$, then can be transformed back to the original space, denoted as $\mathbf{z}$

$$\mathbf{z} = \mathbf{Qy} \tag{A18}$$

because as it is defined by Eq. (A15):

$$\mathbf{Q}^T\mathbf{z} = \mathbf{Q}^T\mathbf{Qy} = \mathbf{y} \tag{A19}$$

For the purpose of geometry optimization, the $\mathbf{W}$ set of column vectors contains the previously obtained K coordinate and force values as

$$\mathbf{w}_{2i-1} = \mathbf{x}_i, \quad \mathbf{w}_{2i} = \mathbf{f}_i, \qquad i \in \{1, ..., K\} \tag{A20}$$

where the $\mathbf{x}$ and $\mathbf{f}$ vectors denote the coordinate vectors and the corresponding forces (negative of the gradient of the PES), respectively. Later, we assume that $\mathbf{x}_1$ and $\mathbf{f}_1$ belong to the latest point. The iterative subspace can be constructed based on difference vectors from the optimization history. As we noted before, it is practical to use linear combinations of the previously defined vector set (see Eq. (A16)), such that

$$\begin{aligned}\mathbf{x}_{j+1} - \mathbf{x}_j &= \mathbf{w}_{2(j+1)-1} - \mathbf{w}_{2j-1} = \mathbf{Wv}_{2j-1}^S, \\ \mathbf{f}_{j+1} - \mathbf{f}_j &= \mathbf{w}_{2(j+1)} - \mathbf{w}_{2j} = \mathbf{Wv}_{2j}^S,\end{aligned} \quad j \in \{1, ..., K-1\} \tag{A21}$$

for a 'sequential' iterative subspace based on optimization steps or alternatively, a 'central' iterative subspace can be built using differences from the latest point:

$$\begin{aligned}\mathbf{x}_{j+1} - \mathbf{x}_1 &= \mathbf{w}_{2(j+1)-1} - \mathbf{w}_1 = \mathbf{Wv}_{2j-1}^C, \\ \mathbf{f}_{j+1} - \mathbf{f}_1 &= \mathbf{w}_{2(j+1)} - \mathbf{w}_2 = \mathbf{Wv}_{2j}^C,\end{aligned} \quad j \in \{1, ..., K-1\} \tag{A22}$$

Any of the previously defined linear combination sets can then be used to construct an orthonormal basis via Gram–Schmidt orthogonalization, where the necessary scalar products are taken using the $\mathbf{O}$ overlap matrix instead of direct use of the original $\mathbf{W}$ set, as in the following example:

$$(\mathbf{x}_3 - \mathbf{x}_2)^T(\mathbf{x}_4 - \mathbf{x}_3) = (\mathbf{v}_3^S)^T\mathbf{W}^T\mathbf{Wv}_5^S = (\mathbf{v}_3^S)^T\mathbf{Ov}_5^S \tag{A23}$$

The orthogonalization results in a new set of $\mathbf{v}$ vectors, collected in $\mathbf{V}$, which satisfies Eq. (A14), and therefore, the corresponding $\mathbf{Q}$ set of vectors can serve as a basis for the iterative subspace. The iterative subspace is constructed in such a way that any **a** vector needed to calculate an optimization step can be given as a **b** linear combination of the **W** set of vectors and its representation in the iterative subspace can be calculated using Eq. (A17). We note that, the **p** representation (see Eq. (A15)) of a general vector, **a**, in the iterative subspace only represent its projection and the *residuum* vector, **r**, can be calculated as:

$$\mathbf{r} = \mathbf{a} - \mathbf{Qp} = (\mathbf{I} - \mathbf{QQ}^T)\mathbf{a} \qquad (A24)$$

If the *residuum* vector vanishes, then vector **a** can be fully represented in the iterative subspace. In the case of geometry optimization, the iterative subspace Hessian matrix is applied only to the represented part of the actual force vector, while its *residuum* is used to calculate the remaining part of the optimization step calculated in the full space.

C.2. Hessian in the iterative subspace

Further simplifications can be applied for constructing the iterative subspace if we assume, according to GDIIS, a linear relationship between the changes in the coordinate values and the corresponding changes in the forces (or gradients). The optimization history contains stored geometries and corresponding forces. If the assumed linearity applies for them then at any $\mathbf{x}'$ point

$$\mathbf{x}' = \mathbf{x}_1 + \sum_{i=1}^{K} \sum_{j=1}^{i-1} \alpha_{i,j}(\mathbf{x}_i - \mathbf{x}_j) \qquad (A25)$$

of the iterative subspace, the forces appear to be

$$\mathbf{f}' = \mathbf{f}_1 + \sum_{i=1}^{K} \sum_{j=1}^{i-1} \alpha_{i,j}(\mathbf{f}_i - \mathbf{f}_j) \qquad (A26)$$

We note, that the following rearrangement of Eqs. (A25) and (A26)

$$\mathbf{x}' = \sum_{i=1}^{K} \beta_i \mathbf{x}_i$$

$$\mathbf{f}' = \sum_{i=1}^{K} \beta_i \mathbf{f}_i \qquad (A27)$$

leads to the following

$$\sum_{i=1}^{K} \beta_i = 1 \qquad (A28)$$

constraint for the β coefficients, which is a simple result of the fact that the sum of the coefficients for linear combinations of difference vectors is always 0. The linearity condition can be stated in such a way that any linear combination of coordinate changes results in a force change that can be formed with the identical linear combination of the respective force changes. Also, a desired linear combination of force changes implies the same linear combination of coordinate changes, which is the basis of the GDIIS method.

It is also important to note that as a consequence of the assumed linearity, the following relationship holds between changes in coordinates and forces

$$-\Delta\mathbf{f} = \mathbf{H}_G \Delta\mathbf{x} \qquad (A29)$$

where the (otherwise needless) negative sign indicates the similarity with the quadratic approximation to the potential PES, while the G index means 'generic', stating that $\mathbf{H}_G$ is not necessarily symmetric unlike the force constant (or Hessian or second derivative) matrix of the quasi-Newton methods. Further details are discussed in the text.

References

[1] R. Fletcher, Practical Methods of Optimization, Wiley, Chichester, 1987.

[2] P.E. Gill, W. Murray, M.H. Wright, Practical Optimization, Academic Press, London, 1981.

[3] L.E. Scales, Introduction to Non-Linear Optimization, Springer, New York, 1985.

[4] J.E. Dennis, R.B. Schnabel, Numerical Methods for Unconstrained Optimization and Nonlinear Equations, Prentice-Hall, Englewood Cliffs, NJ, 1983.

[5] M.J.D. Powell (Eds.), Nonlinear Optimization, Academic Press, New York, 1982.

[6] A. Banerjee, N. Adams, J. Simons, R. Shepard, Journal of Physical Chemistry 89 (1985) 52–57.

[7] H.B. Schlegel, in: D.R. Yarkony (Ed.), Modern Electronic Structure Theory, World Scientific, Singapore, 1995, pp. 459–500.

[8] H.B. Schlegel, in: P.v.R. Schleyer, N.L. Allinger, T. Clark, J. Gasteiger, P.A. Kollman, H.F. Schaefer III, P.R Schreiner,

(Eds.), Encyclopedia of Computational Chemistry, Wiley, Chichester, 1998, pp. 1136–1142.

[9] P. Pulay, G. Fogarasi, Journal of Chemical Physics 96 (1992) 2856–2860.

[10] C.Y. Peng, P.Y. Ayala, H.B. Schlegel, M.J. Frisch, Journal of Computational Chemistry 17 (1996) 49–56.

[11] P. Csaszar, P. Pulay, Journal of Molecular Structure 114 (1984) 31–34.

[12] Ö. Farkas, H.B. Schlegel, Physical Chemistry Chemical Physics 4 (2002) 11–15.

[13] G. Fogarasi, X.F. Zhou, P.W. Taylor, P. Pulay, Journal of the American Chemical Society 114 (1992) 8191–8201.

[14] J. Baker, A. Kessi, B. Delley, Journal of Chemical Physics 105 (1996) 192–212.

[15] J. Baker, D. Kinghorn, P. Pulay, Journal of Chemical Physics 110 (1999) 4986–4991.

[16] Ö. Farkas, H.B. Schlegel, Journal of Chemical Physics 109 (1998) 7100–7104.

[17] Ö. Farkas, H.B. Schlegel, Journal of Chemical Physics 111 (1999) 10806–10814.

[18] B. Paizs, G. Fogarasi, P. Pulay, Journal of Chemical Physics 109 (1998) 6571–6576.

[19] K. Nemeth, O. Coulaud, G. Monard, J.G. Angyan, Journal of Chemical Physics 113 (2000) 5598–5603.

[20] B. Paizs, J. Baker, S. Suhai, P. Pulay, Journal of Chemical Physics 113 (2000) 6566–6572.

[21] K. Nemeth, O. Coulaud, G. Monard, J.G. Angyan, Journal of Chemical Physics 114 (2001) 9747–9753.

[22] M.J. Frisch, G.W. Trucks, H.B. Schlegel, G.E. Scuseria, M.A. Robb, J.R. Cheeseman, J.A. Montgomery, T. Vreven, K.N. Kudin, J.C. Burant, J.M. Millam, S.S. Iyengar, J. Tomasi, V. Barone, B. Mennucci, M. Cossi, G. Scalmani, N. Rega, G.A. Petersson, H. Nakatsuji, M. Hada, M. Ehara, K. Toyota, R. Fukuda, J. Hasegawa, M. Ishida, T. Nakajima, Y. Honda, O. Kitao, H. Nakai, M. Klene, X. Li, J.E. Knox, H.P. Hratchian, J.B. Cross, C. Adamo, J. Jaramillo, R. Gomperts, R.E. Stratmann, O. Yazyev, A.J. Austin, R. Cammi, C. Pomelli, J.W. Ochterski, P.Y. Ayala, K. Morokuma, G.A. Voth, P. Salvador, J.J. Dannenberg, V.G. Zakrzewski, S. Dapprich, A.D. Daniels, M.C. Strain, O. Farkas, D.K. Malick, A.D.

Rabuck, K. Raghavachari, J.B. Foresman, J.V. Ortiz, Q. Cui, A.G. Baboul, S. Clifford, J. Cioslowski, B.B. Stefanov, G. Liu, A. Liashenko, P. Piskorz, I. Komaromi, R.L. Martin, D.J. Fox, T. Keith, M.A. Al-Laham, C.Y. Peng, A. Nanayakkara, M. Challacombe, P.M.W. Gill, B. Johnson, W. Chen, M.W. Wong, C. Gonzalez, J.A. Pople, Gaussian 03, Revision B.01, Gaussian, Inc., Pittsburgh, PA, 2003.

[23] E.B. Wilson, J.C. Decius, P.C. Cross, Molecular Vibrations: The Theory of Infrared and Raman Vibrational Spectra, Dover, New York, 1980.

[24] D.C. Liu, J. Nocedal, Mathematical Programming 45 (1989) 503–528.

[25] J. Nocedal, Mathematics of Computation 35 (1980) 773–782.

[26] R.H. Byrd, J. Nocedal, R.B. Schnabel, Mathematical Programming 63 (1994) 129–156.

[27] R.H. Byrd, P.H. Lu, J. Nocedal, C.Y. Zhu, Siam Journal on Scientific Computing 16 (1995) 1190–1208.

[28] M.J. Frisch, G.W. Trucks, H.B. Schlegel, G.E. Scuseria, M.A. Robb, J.R. Cheeseman, J.A. Montgomery, T. Vreven, K.N. Kudin, J.C. Burant, J.M. Millam, S.S. Iyengar, J. Tomasi, V. Barone, B. Mennucci, M. Cossi, G. Scalmani, N. Rega, G.A. Petersson, H. Nakatsuji, M. Hada, M. Ehara, K. Toyota, R. Fukuda, J. Hasegawa, M. Ishida, T. Nakajima, Y. Honda, O. Kitao, H. Nakai, M. Klene, X. Li, J.E. Knox, H.P. Hratchian, J.B. Cross, C. Adamo, J. Jaramillo, R. Gomperts, R.E. Stratmann, O. Yazyev, A.J. Austin, R. Cammi, C. Pomelli, J.W. Ochterski, P.Y. Ayala, K. Morokuma, G.A. Voth, P. Salvador, J.J. Dannenberg, V.G. Zakrzewski, S. Dapprich, A.D. Daniels, M.C. Strain, O. Farkas, D.K. Malick, A.D. Rabuck, K. Raghavachari, J.B. Foresman, J.V. Ortiz, Q. Cui, A.G. Baboul, S. Clifford, J. Cioslowski, B.B. Stefanov, G. Liu, A. Liashenko, P. Piskorz, I. Komaromi, R.L. Martin, D.J. Fox, T. Keith, M.A. Al-Laham, C.Y. Peng, A. Nanayakkara, M. Challacombe, P.M.W. Gill, B. Johnson, W. Chen, M.W. Wong, C. Gonzalez, J.A. Pople, Gaussian Development Version, Revision B.02, Gaussian, Inc., Pittsburgh, PA, 2003.

[29] A.K. Rappe, C.J. Casewit, K.S. Colwell, W.A. Goddard, W.M. Skiff, Journal of the American Chemical Society 114 (1992) 10024–10035.

ELSEVIER

Journal of Molecular Structure (Theochem) 666–667 (2003) 41–50

THEO
CHEM

www.elsevier.com/locate/theochem

A MIA enhanced linear scaling approach to the computation of the exchange-correlation terms in DFT/LDA

B. Rousseau[a], C. Van Alsenoy[a,*], A. Peeters[a], F. Bogár[b], G. Paragi[b]

[a]*Department of Chemistry, University of Antwerp, Universiteitsplein 1, Antwerpen B-2610, Belgium*
[b]*Protein Chemistry Research Group, Hungarian Academy of Sciences, University of Szeged, Dóm tér 8, Szeged H-6720, Hungary*

Abstract

The Multiplicative Integral Approximation is applied in the linear scaling local density approximation density functional theory. Our method is a modified version of the algorithm of Stratmann et al. [Chem. Phys. Lett. 257 (1996) 213]. We suggest an alternative shell pair based selection scheme for the identification of non-negligible terms in the expression of charge density and exchange-correlation contribution of Kohn–Sham matrix together with an iterative update procedure for these quantities. These modifications enable us to implement the Multiplicative Integral Approach, which further reduces the computational cost. The linear scaling behaviour as well as the CPU time reduction is demonstrated on a test system containing up to 350 water molecules.
© 2004 Elsevier B.V. All rights reserved.

Keywords: Density functional theory/local density approximation; Exchange-correlation; Linear scaling

1. Introduction

Although the Kohn–Sham formulation of density functional theory (DFT) opened new perspectives in the theoretical treatment of spatially extended systems, the computations for large clusters and biological macromolecules remained unfeasible. This problem induced an intensive development of computer algorithms and resulted in several methods that require a computational effort proportional to the system size (linear scaling methods). Instead of summarizing them, we refer here only to two recent review articles written by Goedecker [2] and Wu and Jayanthi [3].

Besides the treatment of Coulomb interactions and diagonalization of the Kohn–Sham matrix,

the numerical integration needed for the calculation of the exchange part of the Kohn–Sham matrix is the most time-consuming part of a DFT calculation. Stratmann et al. [1] suggested in 1996 a grid point driven linear scaling method which selects the basis functions (significant basis functions) giving non-negligible contribution to the numerical integration during the exchange build-up procedure. Most recently, Challacombe [4] suggested a new hierarchical cubature for the numerical integration of the exchange-correlation matrix. He uses an entirely Cartesian grid and a k-dimensional binary search tree data structure that fits well to the large variability of electron density both in range and in magnitude.

The multiplicative integral approach (MIA) was originally developed for the linear scaling Hartree–Fock treatment of large molecules [7] and was successfully applied in many cases [8]. In this paper,

---

* Corresponding author.

0166-1280/$ - see front matter © 2004 Elsevier B.V. All rights reserved.
doi:10.1016/j.theochem.2003.08.011

we present an application of the MIA in the linear scaling DFT. Our method is based on the algorithm of Stratmann et al. [1] and suggests an alternative shell pair based selection scheme for the identification of non-negligible terms in the expression of charge density and exchange-correlation contribution of Kohn–Sham matrix. This modification enable us to implement the MIA, which further reduces the computational cost. The linear scaling behaviour as well as the CPU time reduction is demonstrated on a test system containing up to 350 water molecules.

2. Exchange-correlation in the direct Kohn–Sham scheme

In the Kohn–Sham theory [9], based on the Hohenberg–Kohn theorems [10], the electron density is determined by the one-electron equation

$$\hat{\mathbf{F}}\psi_k = \varepsilon_k \psi_k$$

in the form of

$$\rho(\mathbf{r}) = \sum_k \psi_k^*(\mathbf{r})\psi_k(\mathbf{r}).$$

The Kohn–Sham operator $\hat{\mathbf{F}}$ is the following

$$\hat{\mathbf{F}} = \hat{\mathbf{t}} + v(\mathbf{r}) + \hat{\mathbf{J}} + \hat{\mathbf{F}}^{xc},$$

where $\hat{\mathbf{t}}$ is the kinetic energy, $v(\mathbf{r})$ is the external potential

$$\hat{\mathbf{J}} = \int \frac{\rho(\mathbf{r}')}{|\mathbf{r} - \mathbf{r}'|}\,d\mathbf{r}'$$

is the Coulomb part of the electron–electron interaction and $\hat{\mathbf{F}}^{xc}$ is the exchange-correlation potential. Expanding the one-electron functions on a M dimensional basis set $\{\chi_\mu\}_{\mu=1}^M$

$$\psi_k(\mathbf{r}) = \sum_{\mu=1}^{M} c_{\mu k}\chi_\mu(\mathbf{r}),$$

the form of the matrix elements of the $\hat{\mathbf{F}}^{xc}$ operator in local density approximation (LDA) is

$$F_{\mu\nu}^{xc} = \int \frac{\partial f(\rho)}{\partial \rho(\mathbf{r})} \chi_\mu(\mathbf{r})\chi_\nu(\mathbf{r})\,d\mathbf{r},$$

where $\int f(\rho(\mathbf{r}))d\mathbf{r}$ is the LDA exchange-correlation energy functional used.

In the DFT treatment of the spatially extended systems a direct approach provides a chance to overcome the storage problem. In this method, the Kohn–Sham matrix is built up in an incremental way. The exchange-correlation part of the Kohn–Sham matrix in the nth iteration is

$$F_{\mu\nu}^{xc(n)} = F_{\mu\nu}^{xc(n-1)} + \int \left(\Delta \frac{\partial f(\rho)}{\partial \rho}\right)^{(n)} \chi_\mu(\mathbf{r})\chi_\nu(\mathbf{r})\,d\mathbf{r}, \quad (1)$$

where

$$\left(\Delta \frac{\partial f(\rho)}{\partial \rho}\right)^{(n)} = \left(\frac{\partial f}{\partial \rho}\right)^{(n)} - \left(\frac{\partial f}{\partial \rho}\right)^{(n-1)}.$$

As the integral in Eq. (1) cannot be evaluated analytically, we have to use a numerical method

$$F_{\mu\nu}^{xc(n)} = F_{\mu\nu}^{xc(n-1)} + \sum_i \Delta f_i^{(n)} \chi_\mu(\mathbf{r}_i)\chi_\nu(\mathbf{r}_i)\,d\mathbf{r}, \quad (2)$$

where

$$\Delta f_i^{(n)} = w_i \left(\Delta \frac{\partial f}{\partial \rho}\right)^{(n)}(\mathbf{r}_i). \quad (3)$$

Here w_i and $\mathbf{r}_i$ are the integration weights and grid point coordinates, respectively.

3. A dominantly linear scaling method for the computation of $\mathbf{F}^{xc}$ matrix

The numerical integration makes the build up of the exchange part of the Kohn–Sham matrix in DFT LDA very time consuming. There are two problematic steps in this procedure. The first being the calculation of the electron density at every grid point and the second being the numerical integration in the expression of the exchange-correlation contribution to the Kohn–Sham matrix elements. Stratmann et al. [1] suggested a modification of the Becke [11] weighting scheme that reduces the computational time of the numerical integration. They also present a strategy to reach a linear scaling algorithm, which contains two basic parts. (1) A non-linear (quadratic) scaling estimation part with low computational cost that identifies the non-negligible terms. To each grid point, those basis functions are selected which have non-negligible value at the grid point. These basis functions are located inside a sphere centered at the given grid point. The number of basis functions inside this sphere is independent of the system size. (2) A linear scaling build up procedure that utilizes the information collected in the first step.

In this section, we present a modification of the first step of Stratmann's scheme that paves our way to the application of MIA in LDA DFT.

3.1. Calculation of the electron density

For the numerical integration in Eq. (2) we need to calculate the values of the function f in every step of the iterative SCF procedure at the grid points which requires the calculation of the density ρ in these points. ρ is built up incrementally, in the nth iteration of the SCF procedure the density is given by

$$\Delta\rho^{(n)}(\mathbf{r}_i) = \rho^{(n)}(\mathbf{r}_i) - \rho^{(n-1)}(\mathbf{r}_i)$$

$$= \sum_{\mu,\nu=1}^{M} \Delta P_{\mu\nu}^{(n)} \chi_\mu(\mathbf{r}_i - \mathbf{R}_\mu)\chi_\nu(\mathbf{r}_i - \mathbf{R}_\nu), \quad (4)$$

where

$$\Delta P_{\mu\nu}^{(n)} = \sum_k \left[c_{\mu k}^{(n)*} c_{\nu k}^{(n)} - c_{\mu k}^{(n-1)*} c_{\nu k}^{(n-1)} \right]$$

is the increment of the charge density-bond order matrix in the nth iteration step and $\mathbf{R}$ denotes the center of the basis function. We mention here that the calculation of density and exchange-correlation potential shifts necessitates the storage of the $\rho^{(n)}(\mathbf{r}_i)$ and $\frac{\partial f}{\partial \rho}(\mathbf{r}_i)$ values at the grid points throughout the calculation. Fortunately, the memory requirements for this step scale linearly with the system size. For a molecule containing 1000 atoms and a grid comprising 5000 points for each atom, requires about twice 38 MB of memory which poses no problem. Formally, the computer time to calculate $\rho(\mathbf{r}_i)$ values in the above form (Eq. (4)) would be proportional to the third power of the system size as the number of grid points (index i) as well as the number of basis functions (indices μ and ν) are proportional to the system size.

The estimation step is based on the shell concept. A contracted Gaussian basis function has the form

$$\chi(\mathbf{r} - \mathbf{R}) = (\mathbf{r} - \mathbf{R})^{\mathbf{l}} \sum_{i=1}^{m} c_i N_i \, e^{-\alpha_i(\mathbf{r}-\mathbf{R})^2},$$

here $\mathbf{R}$ denotes the center of a basis function (in our case an atomic position), c_is are the contraction coefficients, N_i is a normalization factor and finally m is the number of primitive Gaussians in the contraction. We use $(\mathbf{r} - \mathbf{R})^{\mathbf{l}}$ as a compact notation for $\prod_i (r_i - R_i)^{l_i}$, where $\mathbf{l} = (l_1, l_2, l_3)$ and $l = \sum_i l_i$ is the orbital quantum number. As usual, a shell groups all basis functions positioned at the same atomic site that share the same set of exponents. We define the Gaussian representative of the Ith shell in the form $g_I = N_I \, e^{-\alpha_{I\min}(\mathbf{r}-\mathbf{R}_I)^2}$ where $\alpha_{I\min}$ is the minimal exponent of the given shell. The magnitude of the product of two basis functions belonging to the shells I and J is estimated with the overlap of the representative Gaussians of their shells $S_{IJ} = \langle g_I | g_J \rangle$. For the Ith shell, we store the list (Λ_I) of those shells J which have an overlap S_{IJ} larger than a predefined threshold (Condition I). This way, for a given shell, we obtain a restricted number of shells depending only on the neighborhood of the atom the shell is positioned on, but does not depend on the size of the system. We collect the selected pairs together

$$\Gamma = \bigcup_I \Lambda_I.$$

The number of selected shell pairs in Γ will be proportional to the system size. The calculation of density can be written in the form

$$\Delta\rho^{(n)}(\mathbf{r}_i) = \sum_{IJ \in \Gamma} \sum_{\mu_I=1}^{m_I} \sum_{\nu_J=1}^{m_J} \Delta P_{\mu_I \nu_J}^{(n)} \chi_{\mu_I}(\mathbf{r}_i - \mathbf{R}_I)\chi_{\nu_J}(\mathbf{r}_i - \mathbf{R}_J).$$

$$(5)$$

Using the Gaussians representative for shells I and J, the magnitude of the terms in the above sum are estimated. The contribution of a shell pair IJ in Eq. (5) is neglected if the product of their representative Gaussians $g_I(\mathbf{r}_i)g_J(\mathbf{r}_i)$ at a given grid point, is under a predefined threshold (Condition II). Using this estimation a subset of Γ to each grid point $\Gamma(\mathbf{r}_i)$ is selected. The number of pairs in $\Gamma(\mathbf{r}_i)$ depends only on the properties of the shell pairs (their spatial extent) and their distance to the grid point $\mathbf{r}_i$, but are independent of the total system size. As both the number of grid points and the number of shell pairs in Γ, are proportional to the system size, the estimation procedure scales quadratically.

Fortunately, it only requires a small number of operations. Using this low cost estimation procedure we can select a system size independent subset of the shell pairs that make the calculation of ρ in Eq. (5) a linear scaling procedure.

$$\Delta\rho^{(n)}(\mathbf{r}_i) = \sum_{IJ\in\Gamma(\mathbf{r}_i)} \sum_{\mu_I=1}^{m_I} \sum_{\nu_J=1}^{m_J} \Delta P^{(n)}_{\mu_I\nu_J} \chi_{\mu_I}$$
$$(\mathbf{r}_i - \mathbf{R}_I)\chi_{\nu_J}(\mathbf{r}_i - \mathbf{R}_J).$$

The effectiveness of this approach depends on the ratio of the quadratically scaling selection and linear scaling calculation procedure.

3.2. Calculation of $\boldsymbol{F}^{xc}$

The similarity between the expressions for updating, in a given iteration, the densities $\rho(\mathbf{r})$ and the exchange contributions $F^{\mathrm{xc}(n)}_{\mu\nu}$ to the elements of

$$\Delta^{\mathrm{MIA}}_{\mu_I\nu_J} = \int \left(\chi_{\mu_I}\chi_{\nu_J} - \sum_\alpha C^{\mu\nu}_{\alpha_{IJ}}\chi_{\mu_I\alpha} \right)^2 d\mathbf{r}$$

$$A^{(n)}_{\mu_I,\alpha} = \sum_J \sum_{\nu_J=1}^{m} \Delta P^{(n)}_{\mu_I\nu_J} C^{\mu_I\nu_J}_\alpha$$

$$
\begin{aligned}
&\textbf{loop } \mathbf{i} = 1, \textbf{number\_of\_gridpoints} \\
&\quad \textbf{if}(\Delta^{\mathrm{MIA}}_{\mu_I\nu_J}\Delta P^{(n-1)}_{\mu_I\nu_J} \geqq \textbf{THR}) \textbf{ then} \\
&\qquad \Delta\rho^{(n)}(\mathbf{r}_i) = \sum_{IJ\in\Gamma} \sum_{\mu_I=1}^{m_I} \sum_{\nu_J=1}^{m_J} \Delta P^{(n)}_{\mu_I\nu_J}\chi_{\mu_I}(\mathbf{r}_i - \mathbf{R}_I)\chi_{\nu_J}(\mathbf{r}_i - \mathbf{R}_J) \\
\\
&\quad \textbf{else} \\
&\qquad \Delta\rho^{(n)}(r_i) = \sum_{\alpha=1}^{N_{\mathrm{aux}}} \sum_I \sum_{\mu_I=1}^{m_I} A^{(n)}_{\mu_I,\alpha}\chi_{\mu_I,\alpha}(r_i - R_I) \\
&\quad \textbf{endif} \\
&\quad \Delta f_i^{(n)} = w_i \left(\Delta\frac{\partial f}{\partial\rho} \right)^{(n)} (r_i) \\
&\quad \textbf{if}(\Delta^{\mathrm{MIA}}_{\mu_I\nu_J}\Delta P^{(n-1)}_{\mu_I\nu_J} \geqq \textbf{THR}) \textbf{ then} \\
&\qquad F^{\mathrm{xc}(n)}_{\mu\nu} = F^{\mathrm{xc}(n-1)}_{\mu\nu} + \sum_i \Delta f_i^{(n)}\chi_\mu(\mathbf{r}_i)\chi_\nu(\mathbf{r}_i)dr \\
&\quad \textbf{else} \\
&\qquad B^{(n)}_{\mu_I,\alpha} = \sum_i \Delta f_i^{(n)}\chi_{\mu_I,\alpha}(\mathbf{r}_i - \mathbf{R}_I) \\
&\quad \textbf{endif} \\
&\textbf{endloopi}
\end{aligned}
$$

$$\textbf{if}(\Delta^{\mathrm{MIA}}_{\mu_I\nu_J}\Delta P^{(n-1)}_{\mu_I\nu_J} > \textbf{THR})$$

$$\Delta F^{\mathrm{xc}(n)}_{\mu_I\nu_J} = \begin{cases} \sum_{\alpha=1}^{N_{\mathrm{aux}}} C'^{\mu_I\nu_J}_\alpha B^{(n)}_{\mu_I,\alpha} & \text{if } J \in \Lambda_I \\ 0 & \text{otherwise} \end{cases}$$

$$\textbf{endif}$$

Fig. 1. Schematic overview of the calculation of the exchange-correlation contribution to the Kohn–Sham matrix element $F^{\mathrm{xc}(n)}_{\mu\nu}$ in the nth iteration of the SCF procedure.

the Kohn–Sham matrix, is evident from Eqs. (2) and (4). Therefore, also for the exchange-correlation terms, contributing shell pairs will only be those which fulfill Condition I and for a given grid point those which fulfill Condition II. Increments to the F^{xc} matrix elements are calculated as

$$F_{\mu_I \nu_J}^{xc(n)} - F_{\mu_I \nu_J}^{xc(n-1)}$$

$$= \sum_i \begin{cases} \Delta f_i^{(n)} \chi_{\mu_I}(\mathbf{r}_i) \chi_{\nu_J}(\mathbf{r}_i), & \text{if } IJ \in \Gamma(\mathbf{r}_i), \\ 0, & \text{otherwise.} \end{cases} \quad (6)$$

The calculation of this term also scales linearly with system size since only the index (i) runs over the grid points of the whole system while the other two indices (I, J) are restricted to a system size independent subset.

4. The multiplicative integral approximation in DFT

In MIA [7], the product of two basis functions $\chi_{\mu_I}(\mathbf{r} - \mathbf{R}_I)\chi_{\nu_J}(\mathbf{r} - \mathbf{R}_J)$ centered, respectively, at $\mathbf{R}_I$ and $\mathbf{R}_J$ is approximated by

$$\chi_{\mu_I}(\mathbf{r} - \mathbf{R}_I)\chi_{\nu_J}(\mathbf{r} - \mathbf{R}_J) \approx \sum_{\alpha=1}^{N_{aux}} C_\alpha^{\mu_I \nu_J} \chi_{\mu_I, \alpha}(\mathbf{r} - \mathbf{R}_\mu)$$

$$= \sum_{\alpha=1}^{N_{aux}} C_\alpha^{\mu_I \nu_J} \chi_{\mu_I}(\mathbf{r} - \mathbf{R}_I) \sigma_{\alpha_I}(\mathbf{r} - \mathbf{R}_I), \quad (7)$$

where $\{\sigma_{\alpha_I}\}_{\alpha_I = 1, N_{aux}}$ is an auxiliary basis set containing N_{aux} basis functions, centered on the same atom as χ_{μ_I}. The exponents used in this basis set can be chosen to be the same for every shells of a molecule. In practice, an uncontracted P shell with exponent 1.0 and a M shell (S, P and D shells with common

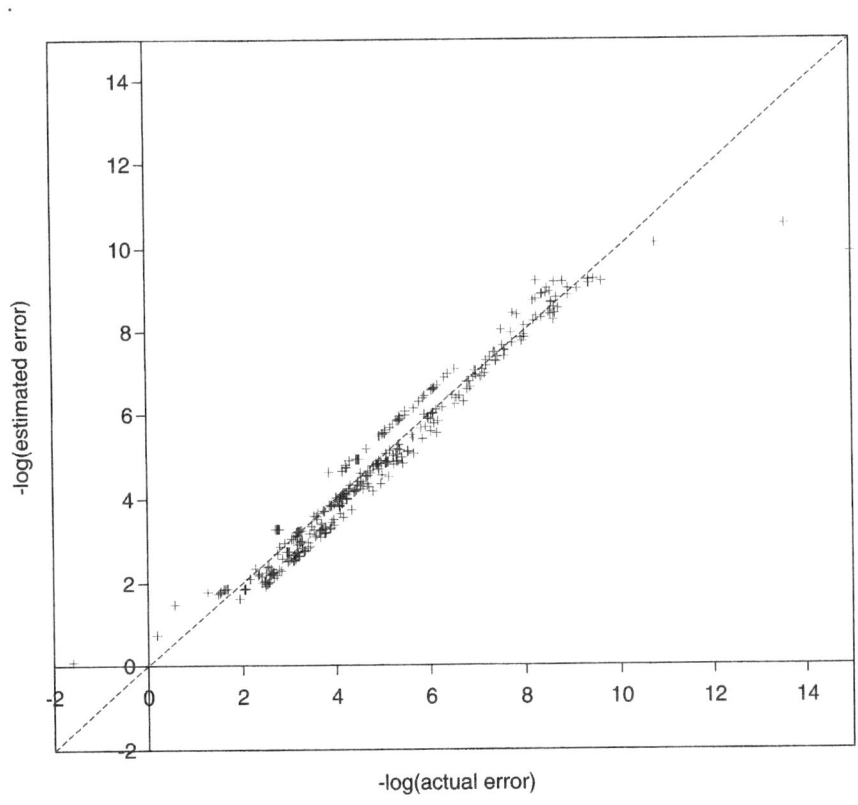

Fig. 2. Comparison of the estimated error with the actual error introduced by the MIA. Each ' + ' represents a shell-combination for a cluster of water molecules.

exponent) with exponent 0.2 were used. The effectiveness of this method was already proven in Hartree–Fock case [8,12]. The accuracy of this approximation depends mainly on the relative position and extent of the basis functions. If we want to keep the accuracy of our calculation over a certain threshold we need an estimation of the error introduced by the MIA approximation. If the estimated error is too large, the contribution is calculated exactly, if not, the MIA value which is easier to calculate, is used.

The error introduced by the MIA-approximation is estimated as

$$\Delta_{\mu_I \nu_J}^{\text{MIA}} = \int \left(\chi_{\mu_I} \chi_{\nu_J} - \sum_\alpha C_\alpha^{\mu_I \nu_J} \chi_{\mu_I \alpha} \right)^2 d\mathbf{r}. \tag{8}$$

The MIA coefficients $C_\alpha^{\mu_I \nu_J}$ are calculated minimising this expression under the condition that the charge of the original distribution is reproduced exactly. This leads to a linear system of N_{aux} equations, which needs

to be solved only for those shell pairs which fulfill Condition I. This way the determination of $C_\alpha^{\mu_I \nu_J}$ coefficients scales linearly with the system size.

4.1. Calculation of ρ and $\mathbf{F}^{xc}$ with MIA

Substituting the MIA form of basis function products into the expression of ρ (Eq. (5)) leads to

$$\Delta \rho^{(n)}(\mathbf{r}_i) = \sum_{\alpha=1}^{N_{\text{aux}}} \sum_I \sum_{\mu_I=1}^{m_I} A_{\mu_I, \alpha}^{(n)} \chi_{\mu_I, \alpha}(\mathbf{r}_i - \mathbf{R}_I), \tag{9}$$

where the auxiliary quantity $A_{\mu_I, \alpha}^{(n)}$ is given by:

$$A_{\mu_I, \alpha}^{(n)} = \sum_J \sum_{\nu_J=1}^m \Delta P_{\mu_I \nu_J}^{(n)} C_\alpha^{\mu_I \nu_J}. \tag{10}$$

This expression is very similar to the one without MIA. The main difference being that the auxiliary basis functions are only dependent upon the functions of the shell I. The value of the auxiliary basis function

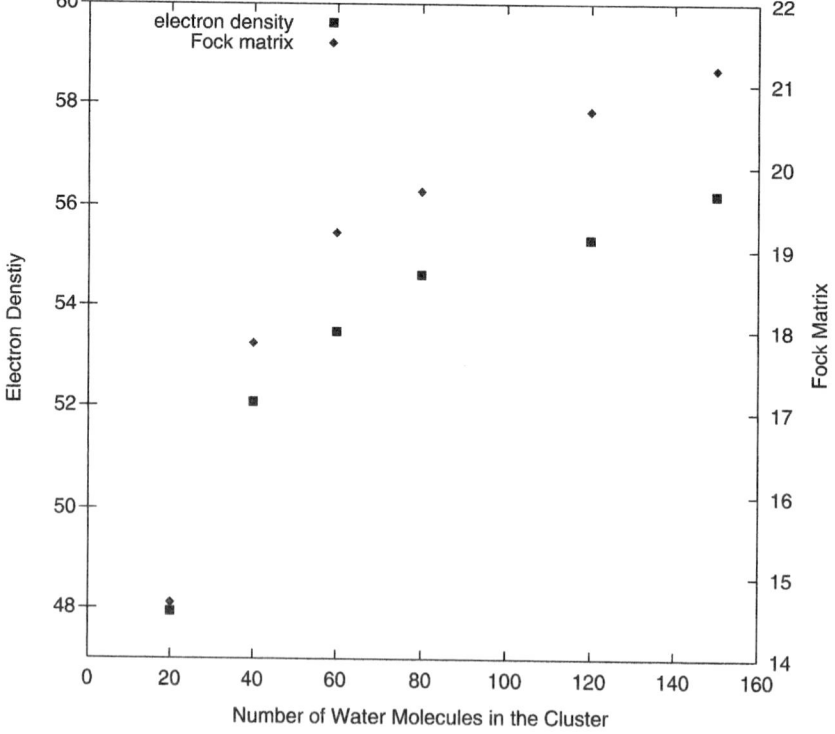

Fig. 3. Percentage of shell-combinations calculated using the MIA as a function of the system size for both the calculation of the electron density (■) and the calculation of the exchange-correlation contribution of the Kohn–Sham matrix (◆), in the first iteration of a water cluster calculation with different number of water molecules.

at a given grid point is the same for all basis functions belonging to a given shell. This way the calculation of the basis function products requires less effort than in non-MIA case.

The significant shells ($\Omega(\mathbf{r}_i)$) for a given grid point can also be selected at this level using the Gaussians representative of the auxiliary and Ith shells. This way the expression for the electron density using MIA is

$$\Delta\rho^{(n)}(\mathbf{r}_i) = \sum_{I\in\Omega(\mathbf{r}_i)} \sum_{\alpha=1}^{N_{\text{aux}}} \sum_{\mu_I=1}^{m_I} A^{(n)}_{\mu_I,\alpha} \chi_{\mu_I,\alpha}(\mathbf{r}_i - \mathbf{R}_I) \quad (11)$$

and

$$A^{(n)}_{\mu_I,\alpha} = \sum_{J\in\Lambda_I} \sum_{\nu_J=1}^{m_J} \Delta P^{(n)}_{\mu_I\nu_J} C^{\mu_I\nu_J}_{\alpha}, \quad (12)$$

where the subset Λ_I of shells is obtained in a way similar to the non-MIA case. In Eq. (12), only those MIA coefficients appear that are calculated for a basis

function pair being in a shell pair which is an element of Γ.

The MIA expression for the matrix elements of the exchange-correlation part of the Kohn–Sham matrix is therefore

$$\Delta F^{\text{xc}(n)}_{\mu_I\nu_J} = \begin{cases} \sum_{\alpha=1}^{N_{\text{aux}}} C^{\mu_I\nu_J}_{\alpha} B^{(n)}_{\mu_I,\alpha} & \text{if } J \in \Lambda_I \\ 0 & \text{otherwise} \end{cases}, \quad (13)$$

where the auxiliary quantity $B_{\mu_I,\alpha}$ is given by

$$B^{(n)}_{\mu_I,\alpha} = \sum_i \Delta f^{(n)}_i \chi_{\mu_I}(\mathbf{r}_i - \mathbf{R}_I)\sigma_\alpha(\mathbf{r}_i - \mathbf{R}_I). \quad (14)$$

In the latest sum only those terms give non-zero contribution for which $I \in \Omega(\mathbf{r}_i)$.

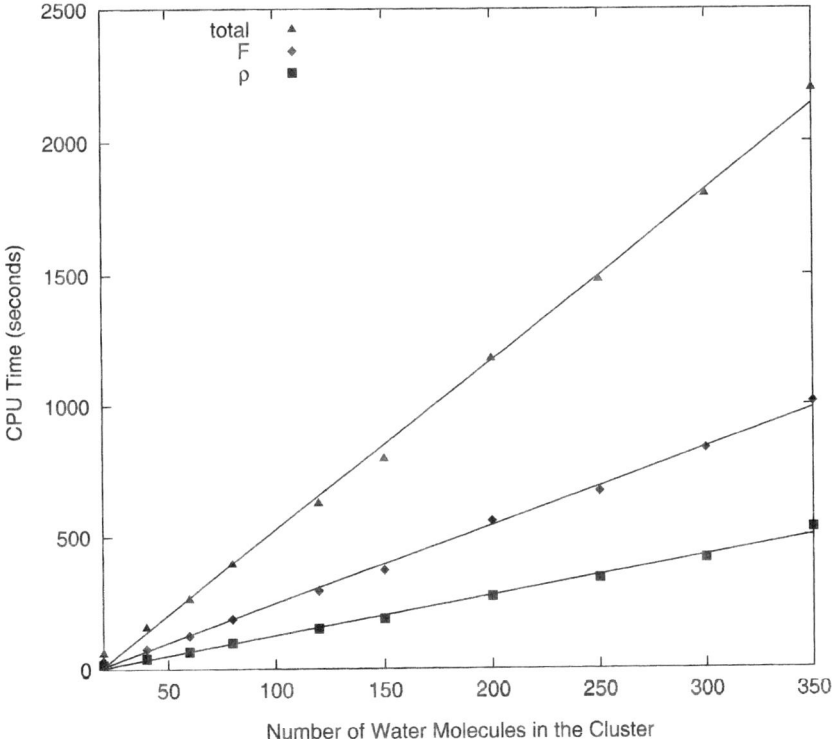

Fig. 4. Scaling behaviour of the DFT/MIA method. The time required for the calculation of electron density (■) and exchange-correlation contribution to the Kohn–Sham matrix (♦) as a function of water molecules included in the test system. '▲' represents the total time for the exchange-correlation contribution at the LDA/321G level at theory using a (64;6,38,86,194,86) grid.

5. Implementation of DFT/MIA

The DFT/MIA method was implemented in program package BRABO [12]. In this section, we describe the algorithm of our method. The accuracy of error estimation in MIA (Eq. (8)) is also discussed. Although the method described above can be applied for any exchange-correlation functional, we have chosen the Slater exchange [5] and VWN correlation [6] for testing purposes. Water clusters [13] were chosen as a test system to demonstrate the scaling properties of the method as well as the CPU time reduction due to the introduction of MIA.

5.1. Algorithm

Here an overview of the calculation of the exchange-correlation contribution in the DFT/MIA method (Fig. 1) is given. Before starting an SCF calculation the MIA error estimates (Eq. (8)) are calculated and the maximum error per shell is stored.

Each SCF iteration starts with the calculation of the Coulomb contribution also using MIA as it is described in Ref. [7]. The necessary MIA coefficients $C\alpha\mu\nu$ are calculated and combined with the increment of the density matrix elements $\Delta P_{\mu\nu}$ using Eq. (12) yielding $A_{\mu,\alpha}$. Then a loop over all atoms in the system is initiated. For a given atom the integration weights and grid points are calculated. Next the electron density in every grid point is updated. Depending upon the magnitude of the error estimate $\Delta_{\mu\nu}^{MIA}$ either Eq. (9) or Eq. (4) (MIA-expansion) is used for this purpose. In practice, the error estimate $\Delta_{\mu\nu}^{MIA}$ is weighted with the largest $\Delta P_{\mu\nu}$ for the given shell-combination. Having calculated the electron density, Δf_i is subsequently calculated using Eq. (3). Again, depending upon the magnitude of the error estimate $\Delta_{\mu\nu}^{MIA}$ (in this case, weighted with the largest value of Δf_i) the Kohn–Sham matrix is updated either using Eq. (6) or Eq. (14) is used to calculate the intermediate quantity $B_{\mu,\alpha}$. Finally, after looping over all atoms in the system,

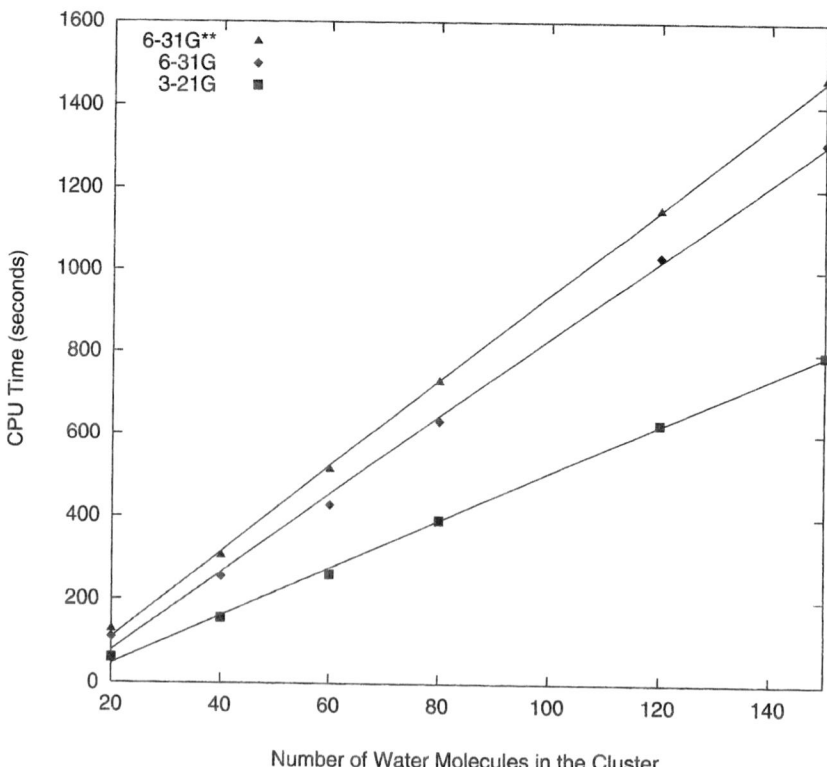

Fig. 5. Scaling behaviour of the CPU time required for the calculation of the exchange-correlation contribution using the DFT/MIA method for the 3-21G (■), 6-31G (♦) and 6-31G** (▲) basis sets. Calculations at the LDA level of theory.

the contributions of the shell-combinations for which the MIA approach was used, are added to the Kohn–Sham matrix using Eq. (13).

5.2. Test calculation on a water cluster

Clusters of water molecules of increasing size from Ref. [13] were chosen as test systems. The 3-21G basis set [14] (unless stated otherwise) and the (64;6,38,86,194,86) grid was used in these calculations and the same auxiliary basis set as in Hartree–Fock case [12] built up from a P shell with an exponent of 1.0 and an M shell with an exponent of 0.2 (13 basis functions), as already mentioned earlier.

5.2.1. Accuracy of error estimation

In Fig. 2, for $(H_2O)_{20}$, the estimated error is plotted versus the actual error which is given by the largest

value of

$$\left| \chi_\mu \chi_\nu - \sum_{\alpha=1}^{N_{aux}} C_\alpha^{\mu\nu} \chi_{\mu,\alpha}(\mathbf{r}_i) \right|, \tag{15}$$

over all grid points. As can be seen from this figure Eq. (2) does indeed give a useful error estimate. During the calculation this error estimate is used to decide whether or not a given shell-combination is calculated using the MIA-approximation.

5.2.2. Ratio of MIA and non-MIA shell pairs

The MIA gains in effectivity with an increasing number of shell pairs involved in the approximation. In Fig. 3, the percentage of shell pairs calculated using MIA as a function of the number of water molecules in a cluster in the first iteration of the SCF procedure is plotted. As can be seen, for the exchange-correlation part of the Kohn–Sham matrix this ratio is 14.5% in

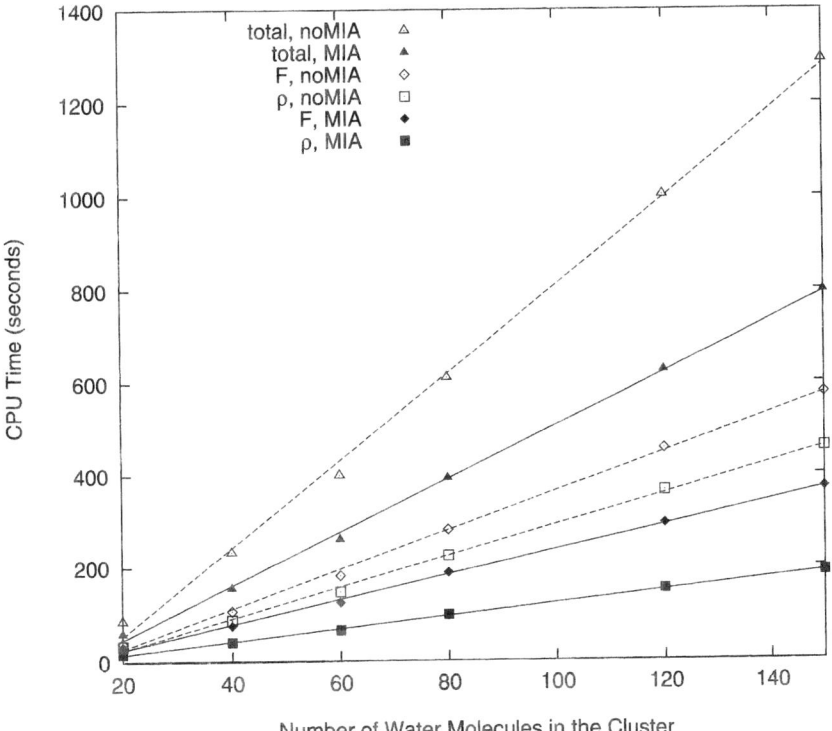

Fig. 6. Comparison of CPU time of LDA (dashed lines) and the LDA/MIA (solid lines) method. The time required for the calculation of electron density (■) and exchange-correlation contribution to the Kohn–Sham matrix (♦) as a function number of water molecules included in the test system. The total time for the exchange-correlation contribution is indicated by '▲'. 3-21G basis set and a (64;6,38,86,194,86) grid were used.

case of 20 water molecules and rises to 21% if our cluster contains 150 water molecules. In the electron density calculation, these percentages are higher still with 48 and 56%, respectively. Percentages for the Kohn–Sham matrix contributions are smaller since, in the first iteration, increase of the functional ($\Delta f_i^{(n)}$) is taken as one.

5.2.3. Scaling behaviour

In Fig. 4, the dependence of CPU time upon the size of the water cluster is presented. Both the calculation of ρ and F^{xc} depend linearly on the cluster size. In case of density, the linear regression yields $1.5n - 25.3$ and for F^{xc} $3.0n - 49.9$, where n stands for the clustersize. Note that, although the scaling behaviour is only illustrated for the LDA functional, a similar scaling behaviour is expected for functionals that include the density-gradient as well as for the force calculation both in LDA and gradient corrected case. This is because, in comparison to the calculation of densities and density-gradients in grid points, the calculation of the functionals themselves is negligible.

In order to check how the scaling behaviour of the DFT/MIA method depends on the basis set used, calculations were performed for clusters up to 150 water molecules comparing the 3-21G, 6-31G and 6-21G** basis sets. As can be seen form the results shown in Fig. 5, our approach scales also linear for more extended basis sets.

5.2.4. MIA versus non-MIA

The CPU time reduction due to the introduction of MIA is shown in Fig. 6. As was described in the theoretical part, the build up of ρ and F^{xc} scales linearly in the non-MIA case too. The CPU time reduction amounts to more than 50% for the calculation of the density while around 30% for F^{xc}.

We expect that in case of a better basis set this reduction will be larger.

Acknowledgements

This work was supported by the Flemish-Hungarian Scientific and Technological Joint Fund (TeT B-2/01 and BIL01/72). One of us (F.B.) is indebted to the National Scientific Research Fund Hungary (OTKA M36803) and the Hungarian Ministry of Education (MU-00094/2002). B.R. acknowledges the Flemish governmental institution IWT for a predoctoral grant. This research was supported by the University of Antwerp under Grant GOA-BOF-UA No. 23.

References

[1] R.E. Stratmann, G.E. Scuseria, M.J. Frisch, Chem. Phys. Lett. 257 (1996) 213.
[2] S. Goedecker, Rev. Mod. Phys. 71 (1999) 1085.
[3] S.Y. Wu, C.S. Jayanthi, Phys. Rep. 358 (2002) 1.
[4] M. Challacombe, J. Chem. Phys. 113 (2000) 10037.
[5] J.C. Slater, Phys. Rev. 81 (1951) 385.
[6] S.H. Vosko, L. Wilk, M. Nusair, Can. J. Phys. 58 (1980) 1200.
[7] C. Van Alsenoy, J. Comput. Chem. 9 (1988) 620.
[8] C. Van Alsenoy, C.H. Yu, A. Peeters, J.M.L. Martin, L. Schafer, J. Phys. Chem. A 113 (2000) 10037. B. Rousseau, R. Keuleers, H.O. Desseyn, H.J. Geise, C. Van Alsenoy, Chem. Phys. Lett. 302 (1999) 55.
[9] W. Kohn, L.J. Sham, Phys. Rev. A 140 (1965) 1133.
[10] P. Hohenberg, W. Kohn, Phys. Rev. B 136 (1964) 864.
[11] A.D. Becke, J. Chem. Phys. 88 (1988) 2547. A.D. Becke, J. Chem. Phys. 286 (1993) 19.
[12] C. Van Alsenoy, A. Peeters, J. Mol. Struct. (Theochem) 286 (1993) 19.
[13] M. Challacombe, E. Schegler, J. Almöf, J. Chem. Phys. 104 (1996) 4685.
[14] J.S. Binkley, J.A. Pople, W.J. Hehre, J. Am. Chem. Soc. 102 (1980) 939.

ELSEVIER

Journal of Molecular Structure (Theochem) 666–667 (2003) 51–59

THEO
CHEM

www.elsevier.com/locate/theochem

Molecules from the Minkowski space: an approach to building 3D molecular structures

G. Imre[a,b], G. Veress[c], A. Volford[d], Ö. Farkas[a,*]

[a]Department of Organic Chemistry, Eötvös Loránd University, 1/A Pázmány Péter St, Budapest H-1117, Hungary
[b]Department of Automation and Applied Informatics, Budapest University of Technology and Economics, Goldman György sq. 3., Budapest H-1521, Hungary
[c]Department of Theoretical Chemistry, Eötvös Loránd University, 1/A Pázmány Péter St, Budapest H-1117, Hungary
[d]ChemAxon Ltd, Máramaros köz 3/a, Budapest H-1037, Hungary

Abstract

In the field of computational chemistry it is usual to have only a partial set of structural information about compounds, like the connectivity or the formula. Individual studies can easily be performed using 'human interfaces' for building input structures. However, automatic, 'batch' processes cannot be applied on a large number of molecules if they imply human intervention. Studies, like QSAR, pharamacophore analysis, reaction prediction might need full, complete 3D information for the compounds of interest. The widespread tools used for structure determination (force-fields or quantum chemical methods) even require a complete set of initial 3D coordinates.

Our approach intends on generating globally valid set of 3D coordinates for small and medium sized molecules, based on local structural criteria. Over against iterative, backtrack based structure predicting algorithms, our method is capable of satisfying partially inconsistent requirements. Such situations are common for structures holding polycyclic, rigid details.

Goals mentioned above can be achieved using coordinates interpreted in a space with a Minkowski metric. Our coordinate assignment process is divided into the following parts: (I) Automatic generation of distance criteria based on chemically relevant local properties, such as bond stretches, bond angles, dihedral angles, etc. (II) Multi-dimensional coordinate assignment which fulfills all the criteria. (III) Geometry optimization using a force field extended to the multi-dimensional Minkowski space. The optimization eliminates the over-3D components and yields the 3D coordinates.
© 2003 Elsevier B.V. All rights reserved.

Keywords: Distance geometry; Minkowski metric; Three-dimensional structure generation; Policycle; Macrocycle; Atomic coordinates

1. Introduction

Distance geometry based methods [1] are important and standard tools for constructing acceptable three-dimensional (3D) coordinates for molecules based on a complete or partial set of interatomic distances. Some other techniques build up 3D structures using a backtracking algorithm [2, 3]. The present work outlines a new method, which after generating interatomic distances can construct temporary coordinates which can completely fulfill the given distance constraints through the use of the Minkowski metric. The temporary coordinates

* Corresponding author.
  E-mail address: farkas@para.chem.elte.hu (O. Farkas).

0166-1280/$ - see front matter © 2003 Elsevier B.V. All rights reserved.
doi:10.1016/j.theochem.2003.08.013

provide the whole set of the distances allowing the algorithm to directly avoid unnecessary proximity of certain atoms. Subsequent geometry optimization or molecular dynamics simulations using a simple, extended molecular mechanics force field to higher dimensions and Minkowski metric may provide a set of acceptable 3D coordinates for the molecule in question. Alternatively, the temporary coordinates can be directly projected into 3D.

2. Method

The purpose of our method is to give valid 3D coordinates for small and medium sized molecules based on partial structural information. The required structural information is practically equivalent to detailed chemical formulae. In this paper, we consider the input parameters being a set of atoms and bonds. Process consists of the following parts: (I) Generating distance criteria based on chemically relevant local properties, such as bond stretches, bond angles, dihedral angles, etc. (II) Assigning multi-dimensional coordinates which fulfills all the criteria in a Minkowski space. (III) Geometry optimization to eliminate the over-3D components and yields the 3D coordinates.

$$INPUT = \langle ATOMS, BONDS \rangle \quad (1)$$

$$ATOMS = \bigcup_{i=1}^{NA} \{ATOM_i\}$$

$$BONDS = \bigcup_{i=1}^{NB} \{BOND_i\}$$

$$ATOM_k = \langle HYB_k \rangle,$$
$$HYB_k \in \{ \text{``S''}, \text{``SP''}, \text{``SP2''}, \text{``SP3''}, \text{``}X\text{''} \}$$

$$BOND_k = \langle BA_{1,k}, BA_{2,k}, BL_k, BO_k \rangle,$$

$$BO_k \in \{ \text{``SINGLE''}, \text{``DOUBLE''}, \text{``TRIPLE''}, \text{``AROMATIC''} \}$$

Interpretation of the variables above:

NA   number of atoms in the examined molecule
NB   number of bonds in the examined molecule

HYB_i hybridization state of atom number i
$BA_{1,i}, BA_{2,i}$ atoms connected by bond number i
BL_i expected length for bond number i
BO_i style of bond number i

The desired output is a set of 3D coordinates for the atoms of the input molecule.

The final coordinates should describe a valid structure for the given molecule. Structures can be characterized by a molecular mechanics force field based energy value as the sum of energy components associated to bonds, dihedrals, etc. The final goal is providing structures with the possibly lowest total energy.

2.1. Interatomic distances

The necessary set of interatomic distances is generated using estimated, desired values for chemically relevant internal coordinates. Some of the internal coordinates can be estimated easily (experimental bond lengths are available, bond angles can be associated to the central atoms hybridization state) or through a selection (dihedrals) in the spirit of the multi-dimensional conformational analysis. The estimation of internal coordinate values then leads to a subset of the interatomic distances. We gave short name 'metrid' to the squares of the distances since they play an important role during the assignment of coordinates in the Minkowski space.

$$DISTANCES = \bigcup_{i=1}^{ND} \{DISTANCE_i\} \quad (2)$$

$$DISTANCE_k = \langle ATOM_{1,k}, ATOM_{2,k}, M_k, P_k \rangle$$

$$M_{req}(ATOM_{1,k}, ATOM_{2,k}) = M_k,$$

$$P_k \in \{ \text{``BOND''}, \text{``ANGLE''}, \text{``TORSION''}, \text{``PROXIMITY''} \}$$

Interpretation of the variables in Eq. (2):

ND   total number of interatomic distance criteria
$ATOM_{1,i}, ATOM_{2,i}$ atoms related to the distance criterion number i

M_i square of the desired distance or metrid number i

P_i priority class of distance number i

$M_{req}(\ )$ in Eq. (2) denotes a distance square, metrid, requirement based on a selected internal coordinate.

Bonds generate the highest priority distance requirements with their desired length.

$$\langle BA_{1,k}, BA_{2,k}, BL_k, BO_k \rangle \in BONDS \qquad (3)$$

$$\Downarrow$$

$$\langle BA_{1,k}, BA_{2,k}, (BL_k)^2, \text{"BOND"} \rangle \in DISTANCES$$

Each atom with more than one ligand generates distance requirements derived from bond angles (Eq. (4)). Fig. 1 shows the interpretation of distances

$$\left. \begin{array}{l} ATOM_k = \langle HYB_k \rangle \in ATOMS \\[4pt] \langle X_1, X_2, BL_l, BO_l \rangle \in BONDS \\[4pt] \langle X_2, X_3, BL_m, BO_m \rangle \in BONDS \end{array} \right\} \Rightarrow$$

$$\langle X_1, X_3, d^2, \text{"ANGLE"} \rangle \in DISTANCES \qquad (4)$$

$$d^2 = (BL_l)^2 + (BL_m)^2 - 2 BL_l BL_m \cos \alpha$$

$$\alpha = \begin{cases} 180° - \dfrac{360°}{n}, & \text{if } BO_l = BO_m = \\ & \text{"AROMATIC", where } n \\ & \text{is the size of aromatic ring} \\[4pt] 180°, & \text{if } HYB_k = \text{"SP"} \\[4pt] 120°, & \text{if } HYB_k = \text{"SP2"} \\[4pt] 109.5°, & \text{if } HYB_k = \text{"SP3"} \\[4pt] \text{if } HYB_k = \text{"X"} & \text{the bond angle is set according} \\ & \text{to an optimal arrangement} \end{cases}$$

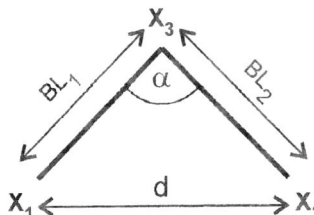

Fig. 1. Interatomic distances in bond angle calculation.

The adjustment of distances related to dihedrals is based on a local heuristic. The configuration of atoms around a bond, the spatial arrangement of the atoms connected to the two end atoms of the bond, is determined by the distances corresponding to all bond lengths, bond angles and one of the dihedrals. The heuristic assignment of dihedrals is based on the following rules:

Let us denote the central bonded atoms with A and B and their neighbors with A_i and B_j. The presence of metrids assigned in previous steps to bond lengths and angles are already available, denote them by $M_{req}(X, Y)$. It is necessary to identify the rings that contain bond A–B (a ring search method is described in Ref. [4]).

If the central bond is labeled as "AROMATIC":

The central bond is part of an aromatic ring. If A_i and B_j atoms are also members of the same aromatic ring then set the dihedral angle φ for A_i–A–B–B_j to 0°

If the central bond is labeled as "DOUBLE":

If CIS/TRANS information is specified for one or more A_x–B_y, atom pairs, then the appropriate dihedral angle, $\varphi = 0°$ for CIS position and $\varphi = 180°$ for TRANS position is used.

If CIS/TRANS position is not specified and A–B bond is an element of a small ring, then specify CIS position for the pair of ring atoms.

If CIS/TRANS position not specified and A–B bond is not an element of a small ring, then specify TRANS position for a connected non-hydrogen atom pair.

If the central bond is labeled as "SINGLE":

If the central bond is element of a chain and both, A and B, has only one non-hydrogen neighbor, then set the dihedral angle to $\varphi = 180°$ for these neighbors.

In all other cases, where A or B has more than one non-hydrogen neighbor or bond A–B is an element of a ring, chose the dihedral angle φ which

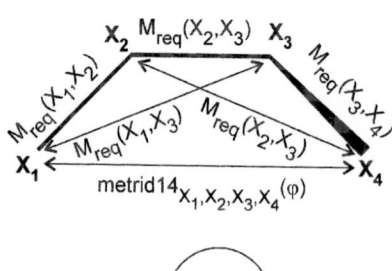

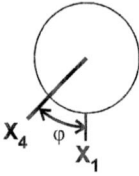

Fig. 2. Interatomic distances in dihedral calculations.

maximizes the following formula:

$$f(\varphi) = \sum_i \sum_j \text{metrid} 14_{A_i, A, B, B_j}(\varphi) \tag{5}$$

Setting dihedral angle φ for atoms $X_1-X_2-X_3-X_4$ (Fig. 2) results in the following metrid (metrid14(φ)):

$$\langle X_1, X_4, \text{metrid} 14_{X_1, X_2, X_3, X_4}(\varphi), \text{``TORSION''}\rangle$$

$$\in \text{DISTA}\pm\text{NCES} \tag{6}$$

$$\text{metrid} 14(\varphi) = (a + (M_{\text{req}}(X_2, X_3))^{1/2} + c)^2$$
$$+ (d \sin \varphi)^2 + (b - d \cos \varphi)^2$$

$$a = \frac{M_{\text{req}}(X_1, X_3) - M_{\text{req}}(X_1, X_2) - M_{\text{req}}(X_2, X_3)}{2(M_{\text{req}}(X_2, X_3))^{1/2}}$$

$$b = (M_{\text{req}}(X_1, X_2) - a^2)^{1/2}$$

$$c = \frac{M_{\text{req}}(X_4, X_2) - M_{\text{req}}(X_4, X_3) - M_{\text{req}}(X_2, X_3)}{2(M_{\text{req}}(X_2, X_3))^{1/2}}$$

$$d = (M_{\text{req}}(X_3, X_4) - c^2)^{1/2}$$

2.2. Assign Minkowski coordinates

The ideal coordinate assignment would satisfy all of the distances generated during previous steps. In most of the cases, they can contain contradictions in 3D, for example, because of a ring closure. By our approach, the assignment is done in two parts: at first, we calculate the coordinates in a multi-dimensional Minkowski space, where all of the distance requirements can be satisfied, then a geometry optimization collapses them into three

(real) dimensions. The assignment minimizes the energy excess originating from the deviation of the actual and the desired internal coordinates values.

First, we illustrate the way of constructing coordinates in the Minkowski space. The square of a distance (metrid) between two points (atoms) is determined in the Minkowski space through a metric tensor (**W** in Eq. (7)):

$$u_{ij} = u(\mathbf{c}_i - \mathbf{c}_j) = u(\mathbf{a}) = d^2(\mathbf{a}) =$$

$$\mathbf{a}^{\text{T}} \begin{bmatrix} \pm 1 & 0 & 0 & 0 \\ 0 & \pm 1 & 0 & 0 \\ & & \ddots & \vdots \\ 0 & 0 & \cdots & \pm 1 \end{bmatrix} \mathbf{a} = \mathbf{a}^{\text{T}} \mathbf{W} \mathbf{a} \tag{7}$$

$$\mathbf{W} = \begin{bmatrix} w_1 & 0 & 0 & 0 \\ 0 & w_2 & 0 & 0 \\ & & \ddots & \vdots \\ 0 & 0 & \cdots & w_n \end{bmatrix}, \qquad w_i \in \{+1, -1\}$$

$$\mathbf{w}^{\text{T}} = \begin{bmatrix} w_1 & w_2 & \cdots & w_n \end{bmatrix}$$

The special nature of this metric is the presence of singular directions which are placed on multi-dimensional cone surfaces (Fig. 3). The metrid along these directions is zero, consequently differing coordinate values do not necessarily induce non-zero distance.

Two Minkowski placement algorithms are given here: the first illustrates the simplicity of the proper multi-dimensional coordinate assignment, whereas the second will provide a more effective way.

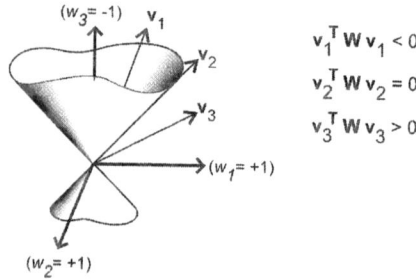

Fig. 3. Singular directions in the Minkowski space along a cone surface.

2.3. Refine-only Minkowsky assignment

We define an iterative algorithm that assigns $(2n - 3)$-dimensional Minkowski coordinates satisfying distance requirements for n points (atoms).

Step 1. Order the atoms to assure that every atom have proper distance requirements, to at least one, previous atom.

Step 2. Assign 1D coordinates with $\mathbf{w} = [1]$ metric to the first two atoms (with number 1 and 2), $\mathbf{c}_i$ denotes the coordinates for atom number i. At the beginning all other atoms have vanishing, [0.0], coordinates (Eq. (8)).

$$\mathbf{c}_1^\mathrm{T} = [0]$$

$$\mathbf{c}_2^\mathrm{T} = \left[(M_\mathrm{req}(0,1)^{1/2} \right] \tag{8}$$

$$\mathbf{c}_i^\mathrm{T} = [0], \qquad i > 2$$

After this step the distance requirements for the first two atoms are satisfied.

Step N. Coordinate assignment for atom number N: Let us denote the dimension of the system before Step N with d and atoms to which metrid defined from atom N by n_i. The defined metrids will be referenced as $M_\mathrm{req}(n_i, N)$.

In this step, we concatenate two new dimensions with metric $[+1 \ -1]$ to the system. The concatenation shown in Eq. (9) aims, the satisfaction of all $M_\mathrm{req}(n_i, N)$ metrid requirements with minimal square sum of new coordinates

$$(\mathbf{w}_c)^\mathrm{T} = [(w^c)_1 \quad (w^c)_2] = [+1 \quad -1]$$

$$(\mathbf{c}_N^c)^\mathrm{T} = [a \quad a]$$

$$(\mathbf{c}_{n_i}^c)^\mathrm{T} = [-x_{n_i} \quad x_{n_i}], \qquad n_i \geq 1 \quad n_i < N$$

$$(\mathbf{c}_j^c)^\mathrm{T} = [0 \quad 0], \qquad \forall i : j \neq n_i$$

$$a = \left(\frac{1}{4} \sum_{i=1} (M_\mathrm{req}(n_i, N) - \mathbf{c}_{n_i}^\mathrm{T} \mathbf{w} \mathbf{c}_{n_i})^2 \right)^{1/4} \tag{9}$$

$$x_{n_i} = \frac{M_\mathrm{req}(n_i, N) - \mathbf{c}_{n_i}^\mathrm{T} \mathbf{w} \mathbf{c}_{n_i}}{4a}$$

$$\mathbf{w}'^\mathrm{T} = \langle \mathbf{w}^\mathrm{T} | (\mathbf{w}^c)^\mathrm{T} \rangle = [w_1 \quad \cdots \quad w_d \quad (w^c)_1 \quad (w^c)_2]$$

$$\mathbf{c}_i'^\mathrm{T} = \langle \mathbf{c}_i^\mathrm{T} | (\mathbf{c}_i^c)^\mathrm{T} \rangle = [c_1 \quad \cdots \quad c_d \quad (c_i^c)_1 \quad (c_i^c)_2]$$

$$d' = d + 2$$

This assignment does not disturb the metrids between previously assigned atoms, but sets all required metrids (and only the required ones) between the currently placed and the previous atoms.

If previously placed atoms with unspecified metrids gets too close to the current atom then a correction of the unspecified metrid by moving the unspecified atom to the singular direction where the previously specified atoms were moved is necessary. The previously placed, unspecified atom is denoted by i and the minimal required metrid is denoted by M_min. Eq. (10) shows the required movement

$$\mathbf{c}_i'' = \mathbf{c}_i' + \frac{M_\mathrm{min} - \mathbf{c}_i^\mathrm{T} \mathbf{w} \mathbf{c}_i}{4a}$$

$$\times [0 \quad 0 \quad \cdots \quad 0 \quad -1 \quad +1]^\mathrm{T}, \tag{10}$$

where $\mathbf{c}_n'^\mathrm{T} = [0 \quad 0 \quad \cdots \quad 0 \quad a \quad a]$

It is, however, possible to contract the proximity corrections with the assignment step by estimating M_min distance requirements for atoms that will be too close to atom N after its placement.

2.4. Placement using triangulation

A more efficient placement algorithm can be constructed using triangulation. The method is based on the concept of decomposing a vector to orthogonal component vectors. The metrid of the vector is the sum of the metrids of its orthogonal components. The orthogonality in the Minkowski space is a metrical orthogonality, since the scalar products should be computed through the metric tensor.

Step 1. Order the atoms to assure that every atom have defined metrid (preferably with "BOND" label) requirement to at least one, previously ordered atom).

Step 2. Assign 1D coordinates with $\mathbf{w}^\mathrm{T} = [1]$ metric to the first two atom (with number 0 and 1):

$$\mathbf{c}_1^\mathrm{T} = [0]$$

$$\mathbf{c}_2^\mathrm{T} = [(M_\mathrm{req}(0,1))^{1/2}] \tag{11}$$

$$\mathbf{c}_i^\mathrm{T} = [0], \ i > 2$$

The other atoms have [0] starting coordinates. After this step the distance requirements for the first two atoms are satisfied.

Step N. Coordinate assignment for atom with number N:

The assignment satisfies metrids by placing and moving atom N in the subspace of previously placed atoms and open new dimension(s) only when it is required. Let n_i denote the index of the connected atoms with defined $M_{req}(n_i, N)$ metrids.

Substep 1. Place atom to its first neighbor: $c_N = c_{n_1}$. Set a scalar value M_{res} to the desired metrid between atom N and its first neighbor: $M_{res} = M_{req}(N, n_1)$. When atom N is moved into any direction with a vector with metric equal with M_{res}, then the distance requirement between it and its first neighbor is satisfied. During the process, we decompose this vector into metrically orthogonal components and at the end of processing each new direction (component) M_{res} become the common residual metrid for the already processed neighbors.

Substep S. In all following steps, we try to satisfy distance requirements between atom N and its neighbors by moving atom N along directions metrically orthogonal to the subspace spanned by the previously processed neighbor atoms [7]. If the currently processed neighbor, n_S, is not lying in the subspace of previously processed neighbors then it can provide a new orthogonal direction to serve as the next component direction. Moving atom N in this direction will decrease the residual metrid M_{res} by the metrid of the displacement vector and also changes the metrid between atom N and current neighbor, n_S.

For the purpose of continuing with Substep 1–Substep $(S - 1)$, a metrically orthonormal basis, **B**, for the subspace spanned by the previously processed neighbor atoms considered in Substep 1–Substep $(S - 1)$ should be present:

$$\forall i < S: \quad \exists v, \quad (c_{n_S} - c_{n_i}) = \mathbf{B}v + c_{n_i} \tag{12}$$

$$\mathbf{b}_i^T \mathbf{w} \mathbf{b}_j = \pm \delta_{i,j}$$

$$\delta_{i,j} = \begin{cases} 1 & \text{if } i = j \\ 0 & \text{if } i \neq j \end{cases}$$

Calculate the actual metrid between N and n_S in the subspace of **B** and see if this vector has a metrically orthogonal component:

$$M_{present} = \sum_i (c_{n_S} - c_N)^T \mathbf{w} \mathbf{b}_i \tag{13}$$

$$\mathbf{v}_{ortho} = (c_{n_S} - c_N) - \sum_i \mathbf{b}_i((c_{n_S} - c_N)^T \mathbf{w} \mathbf{b}_i)$$

If $\mathbf{v}_{ortho}$ is non-singular, based on the residual metrid M_{res}, the required metrid $M(N, n_S)$ and the actual metrid in the subspace of **B**, ($M_{present}$), we can calculate the proper movement for atom N along $\mathbf{v}_{ortho}$ using triangulation to ensure a new common residual metrid M_{res} for all previously processed neighbors, including the current, n_S (Fig. 4). The triangulation is shown in Eq. (14).

$$c_N' = c_N + \lambda \mathbf{v}_{ortho} \tag{14}$$

$$M_{ortho} = u(\mathbf{v}_{ortho}) = \mathbf{v}_{ortho}^T \mathbf{W} \mathbf{v}_{ortho}$$

$$M_{res}' = M_{res} - \lambda^2 M_{ortho}$$

$$M_{res} - \lambda^2 M_{ortho}$$

$$= M_{req}(N, n_S) - M_{present} - (1 - \lambda)^2 M_{ortho}$$

$$\Downarrow$$

$$\lambda = \frac{M_{res} - M_{req}(N, n_S) + M_{present} + M_{ortho}}{2 M_{ortho}}$$

If $\mathbf{v}_{ortho}$ is a singular vector, follow the process with the following neighbors without updating (not moving atom N nor update M_{res}) and correct the metrid for neighbor n_S after all substeps via refinement.

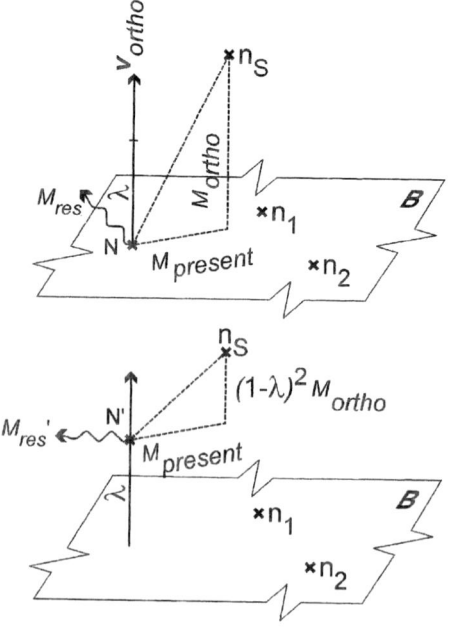

Fig. 4. Triangulation.

Refine. If the triangulation fails for at least one of the neighbors, then we do a refinement (similar to the refine-only placement) with respect to the residual M_{res} metrid to correct all required metrids. Unspecified metrids between atom N and previously placed atoms after the refinement can be calculated by calculating the actual metrid between them after movements and adding M_{res} to it. If any of these metrids will be too low, the distance requirements must be extended with proper M_{min}. requirements. Refinement is done like at the refine-only placement method with the distinction of the respect to M_{res}. Let r_i denote the atoms which need refinement considering $M(N, r_i)$ requirements. The refinement will open and concatenate two new dimensions with metric $[+1 -1]$. Atom N will be moved to satisfy M_{res}, atoms r_i will be moved to a singular direction to satisfy the $M(N, r_i)$ requirements (see Eq. (15) and Fig. 5).

$$(\mathbf{w}^c)^{\mathbf{T}} = [(w^c)_1 \quad (w^c)_2] = [+1 \quad -1]$$

$$(\mathbf{c}_N^c)^{\mathbf{T}} = [a \quad b], \qquad a = p + q, \qquad b = p - q,$$

$$(\mathbf{c}_{r_i}^c)^{\mathbf{T}} = [-x_{r_i} \quad x_{r_i}], \qquad r_i \geq 1, \quad r_i < N$$

$$(\mathbf{c}_j^c)^{\mathbf{T}} = [0 \quad 0], \qquad \forall i : j \neq r_i$$

$$q = \left(\frac{1}{16} \sum_i (M_{req}(r_i, N) - u(\mathbf{c}_{r_i} - \mathbf{c}_N) - M_{res})^2 \right)^{1/4}$$

$$p = \frac{M_{res}}{4q} \tag{15}$$

$$x_{r_i} = \frac{M_{req}(r_i, N) - u(\mathbf{c}_{r_i} - \mathbf{c}_N) - M_{res}}{4d}$$

$$\mathbf{w}'^{\mathbf{T}} = \langle \mathbf{w}^{\mathbf{T}} | (\mathbf{w}^c)^{\mathbf{T}} \rangle = [w_1 \quad \cdots \quad w_d \quad (w^c)_1 \quad (w^c)_2]$$

$$\mathbf{c}_i'^{\mathbf{T}} = \langle \mathbf{c}_i^{\mathbf{T}} | (\mathbf{c}_i^c)^{\mathbf{T}} \rangle = [c_1 \quad \cdots \quad c_d \quad (c_i^c)_1 \quad (c_i^c)_2]$$

$$d' = d + 2$$

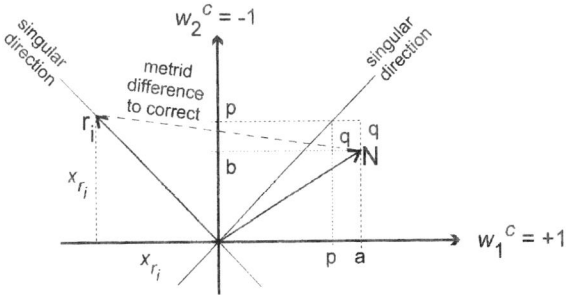

Fig. 5. Refinement.

Place. If atom N has a non-zero residual M_{res} metrid and refinement was not needed, it must be moved toward a direction orthogonal to the subspace spanned by its neighbors (to whom metrid was defined).

In-space placement. With some care taken proper orthogonal direction might be found in the remaining subspace, orthogonal to the subspace spanned by the processed neighbors to avoid opening a new dimension.

Opening one new dimension. If proper direction not found (either no orthogonal direction or placement would lead to undesired proximity), it is required to open a new dimension with the proper metric and using the square root of the absolute value of the residual metrid, M_{res} as coordinate for atom N in that dimension. Before this kind of placement is done it is required to check for proximity, if such a situation occurs, a new distance requirement should be added to the set and process a new substep.

Proximity. After atom N has been placed, all the required distances to the previously placed atoms are set. Some unspecified distances might be too low, so their correction is required. Such a correction can be done with adding a new distance requirement and step back to the previous steps:

$M_{res} = 0$. The distance requirement should be added to the set of distances and process a new substep. (This may, however, lead to a refinement.)

Refinement needed. If refinement was needed, the possible proximity situations could be recognized before refinement, so their correction should be done during the refinement.

New dimension opened $(M_{res} = 0)$. Proximity check should be done before opening a new dimension. If a proximity situation leads to refinement, it will concatenate two new dimensions to the system.

2.5. Geometry optimization

After all the distance requirements fulfilled, the multi-dimensional coordinates are collapsed into three real dimensions using geometry optimization. One possible method is described in Ref. [8]. The energy function which minimum searched is composed from structural and coordinate reduction parts.

The energy function has $n \times d$ parameters, where n is the number of atoms and d is the dimensionality of the system. Before the optimization process, three (real, with $+1$ metric) dimensions are selected, all coordinates in other dimensions will produce energy excess.

In the simplest case, we do the structure optimization by using the weighted distance requirements as energy components:

$$E(c_{1,1}, c_{1,2}, \dots, c_{1,D}, \dots, c_{N,1}, c_{N,2}, \dots, c_{N,D}) =$$

$$\underbrace{\sum_{i,j \text{ distance } i-j \text{ specified}} \gamma(t_{ij})(u_{ij}^{1/2} - M(i,j)^{1/2})^2}_{\text{defined distances}} +$$

$$\underbrace{\sum_{i,j \text{ distance } i-j \text{ not specified}} \alpha \frac{1}{u_{ij}}}_{\text{proximity}} + \underbrace{\sum_{i=1}^{N} \sum_{j=3}^{D} \beta(c_{i,j})^2}_{\text{extra dimension}}$$

$$(16)$$

where

t_{ij}	priority class of $i - j$ distance criteria
$\gamma( )$	coefficient associated to distance criteria classes $\gamma(\text{BOND}) > \gamma(\text{ANGLE}) > \gamma(\text{TORSION})$
α	force constant to avoid non-bonded atoms to get close to each other
β	force constant to ensure the reduction of extra dimensions.

The optimized 3D structure will be valid according to the original, unmodified force field because the modified energy function converges to the original as the extra dimensions disappear.

It is possible to extend other force fields, like Dreiding [5] with coordinate reduction part to be useful in the described method. Such a modification requires a modified transformation of the energy component derivatives into the Minkowski space.

3. Results and discussion

Many polycyclic structures occur where a back-track based coordinate assignment algorithm may fail. Our method respects the priority of the structural requirements, so in cases when tension

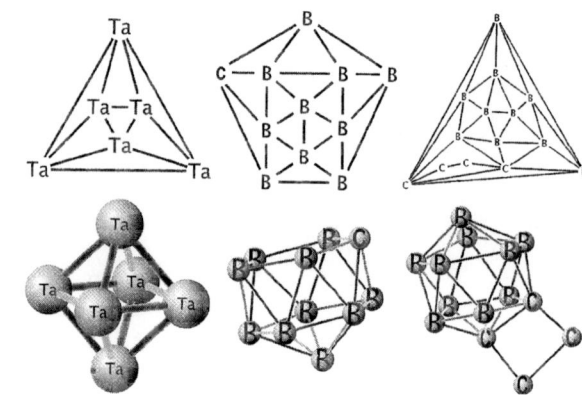

Fig. 6. Three-dimensional systems generated from chemical formulae.

disturbs the lower order requirements (deforms bond angles, torsions) valid structures still can be generated. Fig. 6 shows some polycyclic molecule fragments (see the corresponding smiles strings at [9–11]) where the spatial arrangement is hardly bounded. The method has been implemented in Marvin [6], a Java based chemistry software, which was used for calculating the example 3D structures.

4. Summary

Our method has been tested for months as part of the Marvin program and proved to be capable of providing useful 3D structures even for problematic cases. However, its overall efficiency for larger molecules should be improved. Simple molecular chirality based on chiral centers is treated properly but special cases (e.g. more then three ligands) and proper force constants to force the required chirality when moving from the Minkowski space to 3D should also be included in the extended MM force filed. Further improvements are expected from the introduction of a divide-and-conquer approach.

Acknowledgements

This work was supported by Chemaxon, Ltd and the Hungarian Scientific Fundation (OTKA D-29446).

The authors thank the Chemaxon team for fruitful discussions.

References

[1] L.M. Blumenthal, Theory and Applications of Distance Geometry, American Mathematical Society/Chelsea, Providence, RI/New York, 1970.
[2] J. Sadowski, J. Gasteiger, Chem. Rev. 93 (1993) 2567–2581.
[3] J. Gasteiger, C. Rudolph, J. Sadowski, Tetrahedron Comput. Meth. 3 (6C) (1990) 537–554.
[4] R. Balducci, R.S. Pearlmann, J. Chem. Inf. Comput. Sci. 34 (1994) 822–831.
[5] S.L. Mayo, B.D. Olafson, W.A. Goddard III, J. Phys. Chem. 94 (1990) 8897–8909.
[6] For details on the Marvin program see: http://www.chemaxon.com/marvin.
[7] We refer with 'subspace spanned by atoms' to the subspace which is spanned by all of the vectors between the spanning atoms. Note that the subspace has at most the number of spanning atoms—1 dimensions because the vectors from the origin to each spanning atom are not considered to be in the spanned subspace..
[8] W.H. Press, B.P. Flannery, S.A. Teukolsky, W.T. Vetterling, Numerical Recipies in C, Cambridge University Press, Cambridge, 1988, (Chapter 10).
[9] [Ta]123[Ta]45[Ta]16[Ta]27[Ta]34[Ta]567.
[10] [B@@]123[B@@]45[B@]67[C@]89[B@@]1%10 B2%11%12B34%13B56%14B78%15B9%10%11 B%12%13%14%15.
[11] C1CC2345B678B9%10%11B%12%13%14B69%15 B%12%16%17B%13%18%19B%10%14%20B27%11 B3%18%20C14%16%19B58%15%17.

Journal of Molecular Structure (Theochem) 666–667 (2003) 61–67

www.elsevier.com/locate/theochem

ELSEVIER

Toward a computed structure database: methodology for effective molecular orbital computations

Gregory A. Chass*

Global Institute of Computational Molecular and Materials Sciences (GIOCOMMS), 1422 Edenrose St, Mississauga, Ont., Canada L5V 1H3

Abstract

This work provides a concise description and presentation of selected results emerging from the application of an established modular definition of molecular structure to molecular orbital computations. Perturbations in the relative spatial orientation of the constituent phenyl ring carbon nuclei in benzene, phenol and conformers of Ac-Phenylalanine-NH_2, Ac-Phenylalanine-NHMe and Ac-Tyrosine-NHMe, each geometry optimized at the B3LYP/6-31G(d) level of ab initio theory, were investigated. Several stable conformers were selected for the peptide models. The influence of basis-set refinement was investigated using the Ac-Phenylalanine-NH_2 system, applying the 6-31G(d,p) and 6-31 + G(d) levels of theory. The deviation of the phenyl structural elements from that of benzene was statistically negligible in all cases, with standard deviations of approximately 0.001 Å, 0.15°, 0.20°, for the phenyl ring carbon bond lengths, bond angles, dihedral angles, respectively. Total and relative energies are also presented, showing that the phenyl ring structures of high energy model peptide conformers are also immune to perturbation.
© 2004 Published by Elsevier B.V.

Keywords: Molecular mechanics; Modecular dynamics; B3LYP/6-31G(d,p)

1. Introduction

Computational approaches are now being applied to a great number of problems in chemistry, biochemistry, and biophysics. Insight and solutions to modern challenges are continually evolving due to the continually increasing efficiency of computation. This increase is due in a large part to refined algorithms, increased computational power, building up from and upon past successes as well as a gradually increasing acceptance of theoretical determinations in the scientific community.

The ideal theoretical solution involves using the most accurate approximations and highest levels of theory possible; the true ideal is an infinitely large basis set. In reality, a compromise must be made between the level of theory and the size of the molecular system investigated. Large-scale protein motions and complex inter-molecular reactions, involving a number of species, are typically modeled with coarse approximations, such as molecular mechanics or molecular dynamics. More exacting conformational analysis, and the basis of stability, as well as electrostatic interactions can only be accurately described using ab initio methods, themselves requiring relatively large computational resources.

Each molecular structure that is computed must be computed in its entirety every time a computation

* Present address: Laboratoire Francis Perrin, Service des Photons, Atomes et Molécules (SPAM), CEA-Saclay, 91191, Gif-sur-Yuette, France.

E-mail address: gchass@giocomms.org (G.A. Chass).

0166-1280/$ - see front matter © 2004 Published by Elsevier B.V.
doi:10.1016/j.theochem.2003.08.014

Table 1

Selected phenyl ring bond lengths (R), bond angles (A) and dihedral angles (D), extracted from conformers of benzene, geometry optimized at the B3LYP/6-31G(d) level of theory

R1	R2	R3	R4	R5	A1	A2	A3	A4	D1	D2	D3
1.3966	1.3966	1.3966	1.3966	1.3966	120.00	120.00	120.00	120.00	0.00	0.00	0.00

Table 2

Selected phenyl ring bond lengths (R), bond angles (A) and dihedral angles (D), extracted from conformers of phenol, geometry optimized at the B3LYP/6-31G(d) level of theory

R1	R2	R3	R4	R5	A1	A2	A3	A4	D1	D2	D3
1.3939	1.3977	1.3954	1.3955	1.3981	120.64	119.26	120.45	119.76	−0.80	1.33	−0.86

Table 3

Selected phenyl ring bond lengths (R), bond angles (A) and dihedral angles (D), extracted from conformers of Ac-Phe-NHMe, geometry optimized at the B3LYP/6-31G(d) level of theory

BB	χ^1	χ^2	R1	R2	R3	R4	R5	A1	A2	A3	A4	D1	D2	D3	Rel.E
δ_L	+	+	1.4045	1.3949	1.3971	1.3949	1.3972	120.67	120.42	119.53	119.99	0.13	−0.23	0.12	2.06
δ_L	+	−	1.4033	1.3972	1.3950	1.3971	1.3949	121.07	119.99	119.53	120.42	0.09	0.12	−0.22	2.06
δ_L	−	+	1.4006	1.3968	1.3948	1.3971	1.3941	121.09	120.01	119.53	120.22	0.07	−0.19	0.11	4.11
δ_L	−	−	1.4025	1.3941	1.3971	1.3948	1.3969	120.91	120.22	119.53	120.01	0.09	0.12	−0.19	4.11
α_D	−	−	1.4030	1.3951	1.3969	1.3956	1.3967	121.02	120.12	119.51	120.15	0.10	0.19	−0.28	5.44
α_D	−	+	1.4020	1.3967	1.3956	1.3968	1.3951	120.96	120.15	119.51	120.12	0.09	−0.29	0.19	5.44
α_D	a	+	1.4015	1.3963	1.3951	1.3969	1.3951	121.10	119.99	119.46	120.48	0.07	−0.15	0.09	7.71
γ_L	a	−	1.4018	1.3958	1.3960	1.3959	1.3953	120.52	120.39	119.56	119.98	−0.09	0.02	0.07	0.81
γ_L	a	+	1.4018	1.3952	1.3959	1.3960	1.3958	120.96	119.98	119.56	120.38	−0.10	0.08	0.02	0.82
γ_L	−	+	1.4008	1.3973	1.3944	1.3975	1.3940	121.07	120.04	119.52	120.21	−0.07	−0.15	0.17	1.02
γ_L	−	−	1.4028	1.3941	1.3975	1.3945	1.3973	120.93	120.21	119.52	120.04	0.05	0.18	−0.15	1.13
γ_L	+	−	1.4036	1.3975	1.3946	1.3973	1.3947	121.10	119.94	119.51	120.55	0.05	−0.07	−0.09	0.00
γ_L	+	+	1.4046	1.3947	1.3973	1.3946	1.3974	120.53	120.55	119.51	119.94	0.27	−0.09	−0.06	0.00
δ_D	a	+	1.4023	1.3943	1.3968	1.3950	1.3959	120.76	120.17	119.61	120.05	−0.20	0.01	0.14	8.68
δ_D	a	−	1.4000	1.3959	1.3950	1.3968	1.3943	120.91	120.05	119.60	120.17	−0.10	0.14	0.01	8.69
δ_D	+	+	1.4030	1.3959	1.3963	1.3963	1.3961	120.82	120.25	119.54	120.08	0.12	0.28	−0.31	6.83
δ_D	+	+	1.4032	1.3962	1.3962	1.3964	1.3958	120.96	120.08	119.55	120.25	−0.07	−0.27	0.25	6.82
α_L	a	−	1.4016	1.3979	1.3950	1.3979	1.3941	120.82	120.08	119.62	120.24	−0.10	0.15	−0.07	0.73
ε_D	a	+	1.4042	1.3940	1.3975	1.3943	1.3980	121.06	120.23	119.48	120.10	−0.08	−0.27	0.25	7.69
ε_D	a	−	1.4031	1.3979	1.3942	1.3975	1.3939	121.12	120.09	119.48	120.24	0.13	0.25	−0.27	7.63
γ_D	a	−	1.4008	1.3956	1.3954	1.3962	1.3946	120.82	120.19	119.57	120.04	0.08	0.03	−0.03	4.46
γ_D	a	+	1.3914	1.3838	1.3862	1.3839	1.3864	120.97	120.07	119.53	120.20	−0.10	−0.06	0.06	4.46
γ_D	−	−	1.4021	1.3951	1.3968	1.3957	1.3964	120.89	120.11	119.57	120.15	0.07	0.13	−0.20	2.67
γ_D	−	+	1.4014	1.3964	1.3957	1.3968	1.3951	120.83	120.15	119.57	120.11	0.06	−0.20	0.13	2.67
γ_D	+	−	1.4006	1.3952	1.3956	1.3963	1.3949	120.91	120.05	119.60	120.18	0.26	0.31	−0.26	7.57
γ_D	+	−	1.4002	1.3953	1.3952	1.3964	1.3946	120.96	120.04	119.58	120.23	0.28	0.31	−0.28	7.58
γ_D	+	+	1.4028	1.3944	1.3966	1.3950	1.3955	120.71	120.23	119.58	120.03	−0.30	−0.29	0.29	7.61
β_L	a	−	1.4016	1.3979	1.3950	1.3979	1.3941	120.82	120.08	119.62	120.24	−0.11	0.14	−0.07	0.73
β_L	a	−	1.4016	1.3979	1.3950	1.3979	1.3941	120.82	120.08	119.62	120.24	−0.11	0.14	−0.07	0.73
β_L	a	+	1.4052	1.3941	1.3979	1.3950	1.3980	120.69	120.24	119.62	120.08	−0.03	−0.08	0.14	0.75
β_L	−	−	1.4012	1.3944	1.3962	1.3954	1.3957	120.96	120.07	119.57	120.13	0.09	0.02	−0.05	3.95
β_L	−	+	1.4005	1.3958	1.3954	1.3962	1.3944	120.88	120.13	119.57	120.07	−0.05	−0.04	0.02	3.97
β_L	−	+	1.4006	1.3957	1.3954	1.3962	1.3944	120.88	120.13	119.56	120.07	−0.05	−0.04	0.02	3.97
β_L	+	−	1.4019	1.3953	1.3959	1.3957	1.3952	120.90	120.16	119.54	120.11	0.15	0.16	−0.20	3.02
Mean			1.4018	1.3954	1.3956	1.3958	1.3952	120.89	120.14	119.55	120.16	0.02	0.01	−0.02	
Std. Dev.			0.0023	0.0024	0.0019	0.0023	0.0020	0.15	0.13	0.04	0.14	0.13	0.18	0.17	

is performed. As geometry optimizations are the most prevalent computations performed and stable minima structures are required for more 'exotic' computations, the bulk of computational resources are spent on minimizing input structures energies, through geometric perturbation.

Although every computation must re-optimize the structure when any change is made, to either the structure, method, basis set or local environment (i.e. the inclusion of implicit or explicit solvent or dielectric mediums), pre-optimized structures may be used to reduce the amount of computational steps, or iterations, required. The concept of using

pre-computed structures or structural moieties is in its infancy with only preliminary establishment [1–2]. The existence of a computed structure database would greatly improve the computational efficiency with which geometry optimizations may be carried out.

Using peptide-models as an example, a database containing all possible tri-peptide sequences, in all possible backbone and sidechain orientations, would afford the user with the ability to more effectively characterize the structural and energetic properties of larger protein systems. Pre-computed tri-peptide 'blocks' or 'modules' could be combined to make more accurate hexa- or larger oligo-peptide input

Table 4

Selected phenyl ring bond lengths (R), bond angles (A) and dihedral angles (D), extracted from conformers of Ac-Tyr-NHMe, geometry optimized at the B3LYP/6-31G(d) level of theory

BB	χ^1	χ^2	R1	R2	R3	R4	R5	A1	A2	A3	A4	D1	D2	D3	Rel.E
α_D	a	−	1.4055	1.3913	1.3992	1.3967	1.3963	121.30	120.01	119.63	119.71	0.11	0.02	−0.11	8.12
α_D	+	+	1.4051	1.3895	1.3998	1.3965	1.3974	121.26	120.02	119.71	119.58	0.17	−0.15	−0.07	9.33
β_L	+	a	1.4034	1.3916	1.3985	1.3978	1.3946	121.63	119.72	119.67	119.85	0.13	0.31	−0.37	3.31
β_L	+	−	1.4003	1.3947	1.3980	1.3983	1.3915	121.39	119.90	119.67	119.67	0.16	0.24	−0.29	3.21
β_L	+	−	1.4034	1.3916	1.3985	1.3978	1.3946	121.62	119.72	119.67	119.84	0.13	0.31	−0.37	3.29
β_L	a	−	1.3996	1.3973	1.3973	1.4007	1.3902	121.35	119.83	119.72	119.80	−0.05	0.05	0.08	1.03
β_L	−	+	1.4016	1.3923	1.3978	1.3985	1.3933	121.63	119.65	119.70	119.84	0.08	−0.30	0.29	4.22
β_L	−	−	1.3996	1.3934	1.3985	1.3978	1.3922	121.42	119.83	119.70	119.65	−0.06	0.29	−0.31	4.15
β_L	−	−	1.4027	1.3909	1.3987	1.3977	1.3949	121.70	119.61	119.70	119.89	−0.02	0.16	−0.15	4.27
δ_L	a	−	1.4027	1.3943	1.3978	1.4001	1.3930	121.55	119.66	119.72	120.00	0.02	0.07	−0.01	0.81
δ_L	−	−	1.4053	1.3895	1.4010	1.3963	1.3975	121.57	119.82	119.66	119.75	0.03	0.11	−0.11	4.77
δ_L	+	−	1.4042	1.3938	1.3976	1.3993	1.3938	121.78	119.57	119.64	120.16	0.16	0.02	−0.16	2.10
δ_D	a	a	1.4014	1.3923	1.3977	1.3988	1.3934	121.65	119.62	119.70	119.92	−0.20	0.32	−0.15	8.84
δ_D	a	+	1.4040	1.3904	1.3995	1.3971	1.3955	121.51	119.72	119.72	119.80	−0.15	−0.12	0.30	9.02
δ_D	+	−	1.4047	1.3921	1.3990	1.3982	1.3957	121.54	119.81	119.67	119.81	0.27	0.11	−0.17	7.06
δ_D	+		1.4045	1.3926	1.3989	1.3984	1.3950	121.69	119.64	119.67	119.97	−0.19	−0.13	0.12	6.99
ε_L	−	−	1.4027	1.3908	1.3987	1.3976	1.3950	121.70	119.61	119.70	119.88	−0.01	0.14	−0.15	4.26
ε_D	a	−	1.4045	1.3939	1.3974	1.3998	1.3932	121.85	119.68	119.56	119.98	0.20	0.20	−0.23	7.92
ε_D	a		1.4061	1.3899	1.4007	1.3965	1.3974	121.77	119.82	119.58	119.83	−0.16	−0.11	0.09	7.90
ε_D		−	1.4014	1.3918	1.3977	1.3986	1.3937	121.69	119.59	119.72	119.97	0.15	0.42	−0.38	7.70
γ_L	+	+	1.4071	1.3902	1.4005	1.3966	1.3973	121.24	120.13	119.64	119.66	0.21	−0.01	−0.16	0.31
γ_L	+	−	1.4014	1.3972	1.3967	1.4003	1.3903	121.61	119.66	119.64	120.12	0.12	−0.15	0.00	0.29
γ_L	+	−	1.4046	1.3941	1.3972	1.3997	1.3932	121.83	119.49	119.63	120.31	0.21	−0.12	−0.11	0.00
γ_L	a	+	1.3996	1.3950	1.3977	1.3987	1.3917	121.45	119.72	119.71	119.93	−0.06	0.03	0.03	1.06
γ_L	a	−	1.4006	1.3947	1.3980	1.3983	1.3918	121.01	120.12	119.70	119.53	−0.08	0.02	0.04	0.87
γ_L	−	+	1.3988	1.3971	1.3964	1.4006	1.3899	121.57	119.77	119.63	119.79	−0.11	−0.11	0.13	1.38
γ_L	−	−	1.4048	1.3900	1.4005	1.3964	1.3971	121.65	119.77	119.64	119.78	0.08	0.12	−0.12	1.48
γ_D	+	+	1.4045	1.3907	1.3991	1.3972	1.3948	121.45	119.76	119.73	119.78	−0.23	−0.45	0.49	7.80
γ_D	+	+	1.4045	1.3907	1.3991	1.3972	1.3948	121.45	119.76	119.73	119.78	−0.23	−0.45	0.49	7.80
γ_D	+	−	1.4014	1.3919	1.3977	1.3986	1.3937	121.69	119.58	119.71	119.97	0.14	0.42	−0.38	7.71
γ_D	a	a	1.4025	1.3916	1.3981	1.3982	1.3942	121.56	119.74	119.69	119.78	−0.01	0.18	−0.16	4.74
γ_D	−	+	1.3996	1.3959	1.3976	1.3995	1.3914	121.33	119.88	119.69	119.68	0.04	−0.11	0.04	2.97
Mean			1.4028	1.3927	1.3984	1.3983	1.3939	121.54	119.76	119.68	119.84	0.02	0.04	−0.05	
Std. Dev.			0.0023	0.0023	0.0011	0.0013	0.0022	0.19	0.15	0.04	0.16	0.15	0.22	0.23	

structures. Although the tri-peptide database would initially lack the long-range interactions and other spatial effects, nearest neighboring inductive effects would be well characterized. Subsequent inclusion of long-range interactions, solvation and counter-ions, or the investigation of excited states may be carried out on the optimized tri-peptides. However, the initial exercise itself would prove to be a monumental task at the present time.

Fortunately, structural properties are relatively well conserved for a large number of molecular moieties. Bond lengths, bond angles, and dihedral angles, may be extracted from existing geometry optimized structures, are 're-used' in subsequent computations. Molecular

Table 5

Selected phenyl ring bond lengths (R), bond angles (A) and dihedral angles (D), extracted from conformers of Ac-Phe-NH$_2$, geometry optimized at the B3LYP/6-31G(d) level of theory

BB	χ^1	χ^2	R1	R2	R3	R4	R5	A1	A2	A3	A4	D1	D2	D3	Rel.E
α_L	a	—	1.3992	1.3951	1.3953	1.3962	1.3943	120.75	120.16	119.62	120.03	0.02	0.07	0.06	8.13
α_D	—	+	1.4019	1.3967	1.3956	1.3969	1.3952	120.99	120.14	119.51	120.13	0.06	−0.28	0.21	5.48
α_D	+	—	1.4013	1.3970	1.3948	1.3971	1.3939	121.08	119.87	119.55	120.46	0.26	0.03	−0.20	8.82
α_D	+	+	1.4025	1.3938	1.3971	1.3948	1.3970	120.55	120.46	119.55	119.87	0.09	−0.20	0.03	8.82
α_D	a	—	1.4031	1.3953	1.3966	1.3952	1.3961	120.57	120.48	119.47	119.97	−0.02	0.07	−0.06	7.67
α_D	a	+	1.4019	1.3961	1.3952	1.3966	1.3953	121.12	119.97	119.47	120.48	−0.01	−0.05	0.07	7.67
β_L	—	—	1.4012	1.3945	1.3962	1.3954	1.3956	120.95	120.07	119.58	120.13	0.06	0.04	−0.06	4.25
β_L	—	+	1.4005	1.3956	1.3954	1.3962	1.3944	120.87	120.12	119.58	120.07	−0.02	−0.05	0.03	4.24
β_L	+	—	1.4018	1.3953	1.3959	1.3958	1.3952	120.90	120.15	119.55	120.12	0.14	0.14	−0.16	3.12
β_L	+	+	1.4019	1.3952	1.3958	1.3959	1.3953	120.93	120.13	119.55	120.14	−0.10	−0.15	0.14	3.13
β_L	a	—	1.4018	1.3982	1.3948	1.3979	1.3939	120.82	120.08	119.61	120.26	−0.04	0.16	−0.14	0.70
β_L	a	—	1.4018	1.3982	1.3948	1.3979	1.3939	120.82	120.08	119.61	120.26	−0.04	0.16	−0.14	0.70
β_L	a	+	1.4052	1.3940	1.3979	1.3948	1.3982	120.70	120.26	119.61	120.08	−0.02	−0.13	0.16	0.70
β_L	a	+	1.4052	1.3940	1.3979	1.3948	1.3982	120.70	120.26	119.61	120.08	−0.02	−0.13	0.16	0.70
δ_L	—	—	1.4025	1.3940	1.3971	1.3948	1.3968	120.90	120.21	119.53	120.01	0.09	0.11	−0.20	4.37
δ_L	—	+	1.4006	1.3969	1.3948	1.3971	1.3940	121.07	120.01	119.54	120.21	0.07	−0.20	0.12	4.38
δ_L	+	—	1.4032	1.3971	1.3950	1.3970	1.3949	121.05	120.00	119.53	120.41	0.08	0.10	−0.19	2.40
δ_L	+	+	1.4044	1.3949	1.3971	1.3950	1.3972	120.67	120.41	119.53	120.00	0.11	−0.19	0.10	2.40
δ_L	a	—	1.4018	1.3982	1.3948	1.3979	1.3939	120.82	120.08	119.61	120.26	−0.05	0.16	−0.13	0.70
δ_D	+	—	1.4033	1.3956	1.3965	1.3960	1.3964	120.85	120.24	119.55	120.08	0.14	0.24	−0.29	6.68
δ_D	+	+	1.4034	1.3963	1.3961	1.3965	1.3956	120.98	120.08	119.55	120.24	−0.03	−0.29	0.24	6.68
δ_D	a	—	1.4001	1.3958	1.3951	1.3967	1.3943	120.89	120.04	119.62	120.16	−0.09	0.10	0.04	8.56
δ_D	a	+	1.4022	1.3943	1.3967	1.3951	1.3958	120.75	120.16	119.62	120.04	−0.18	0.04	0.10	8.56
ε_D	+	+	1.4028	1.3945	1.3964	1.3952	1.3952	120.69	120.24	119.58	120.04	−0.38	−0.16	0.24	7.60
ε_D	a	—	1.4031	1.3981	1.3943	1.3975	1.3940	121.10	120.10	119.48	120.23	0.13	0.23	−0.26	7.54
ε_D	a	+	1.4042	1.3940	1.3975	1.3943	1.3981	121.06	120.23	119.49	120.10	−0.06	−0.27	0.24	7.54
γ_L	—	—	1.4029	1.3940	1.3975	1.3943	1.3974	120.91	120.22	119.53	120.03	0.08	0.20	−0.21	1.01
γ_L	—	+	1.4010	1.3974	1.3943	1.3975	1.3940	121.06	120.03	119.53	120.22	−0.05	−0.21	0.20	1.01
γ_L	+	—	1.4037	1.3975	1.3946	1.3973	1.3948	121.09	119.93	119.52	120.54	0.09	−0.04	−0.14	0.00
γ_L	+	+	1.4045	1.3948	1.3973	1.3946	1.3975	120.53	120.54	119.52	119.93	0.27	−0.15	−0.03	0.00
γ_L	a	—	1.4018	1.3957	1.3960	1.3958	1.3953	120.55	120.37	119.55	120.00	−0.05	0.01	0.05	0.88
γ_L	a	+	1.4017	1.3953	1.3959	1.3960	1.3957	120.94	120.00	119.55	120.37	−0.08	0.06	0.00	0.88
γ_D	—	—	1.4022	1.3950	1.3968	1.3956	1.3965	120.90	120.10	119.56	120.17	0.08	0.12	−0.18	2.67
γ_D	—	+	1.4013	1.3964	1.3956	1.3968	1.3950	120.81	120.17	119.56	120.10	0.03	−0.18	0.13	2.67
γ_D	+	—	1.4002	1.3953	1.3951	1.3964	1.3944	120.97	120.01	119.59	120.25	0.27	0.28	−0.25	7.51
γ_D	+	+	1.4028	1.3944	1.3965	1.3951	1.3954	120.68	120.25	119.59	120.01	−0.32	−0.25	0.27	7.51
γ_D	a	—	1.4011	1.3955	1.3955	1.3961	1.3948	120.77	120.21	119.58	120.02	0.09	0.00	−0.03	4.37
γ_D	a	+	1.4017	1.3948	1.3961	1.3955	1.3955	120.97	120.02	119.58	120.21	−0.04	−0.03	0.00	4.37
Mean			1.4023	1.3957	1.3959	1.3960	1.3955	120.86	120.15	119.56	120.15	0.01	−0.02	0.00	
Std. Dev.			0.0014	0.0013	0.0010	0.0011	0.0013	0.17	0.15	0.04	0.15	0.13	0.16	0.16	

fragments such as phenyl rings, peptide bonds, alkyl chains, as well as terminal amines and amides also now show this 'structural conservation,' based on first principles.

This work provides a viewpoint, supported with computed data, of these concepts, making use of selected structural properties based on the relative spatial orientation of the constituent phenyl ring carbon nuclei in selected systems. It is, therefore, proposed that any molecular system containing a phenyl ring could be more effectively geometry optimized, using pre-computed bond lengths, bond angles, and dihedral angles, for benzene. This concept

is illustrative of the ability for any such molecular fragment, displaying conserved structure, to be so used in the construction of input structures for larger computations.

2. Methods

An existing computed peptide structural database was utilized to extract structural results from selected systems, geometry optimized at the B3LYP/6-31G(d) level of theory. The database was constructed from structural data existing in

Table 6
Selected phenyl ring bond lengths (R), bond angles (A) and dihedral angles (D), extracted from conformers of Ac-Phe-NH$_2$, geometry optimized at the B3LYP/6-31 G(d,p) level of theory

BB	χ^1	χ^2	R1	R2	R3	R4	R5	A1	A2	A3	A4	D1	D2	D3	Rel.E
α_L	a	−	1.3990	1.3948	1.3950	1.3959	1.3940	120.74	120.16	119.62	120.03	−0.01	0.08	0.07	8.32
α_D	−	−	1.4028	1.3949	1.3966	1.3953	1.3964	121.00	120.13	119.52	120.14	0.09	0.21	−0.29	5.57
α_D	+	−	1.4011	1.3968	1.3945	1.3969	1.3935	121.06	119.87	119.57	120.46	0.25	0.03	−0.20	8.95
α_D	+	+	1.4023	1.3936	1.3968	1.3945	1.3967	120.54	120.45	119.56	119.88	0.09	−0.22	0.05	8.95
β_L	−	−	1.4010	1.3942	1.3959	1.3951	1.3954	120.93	120.07	119.59	120.13	0.05	0.04	−0.05	4.31
β_L	−	+	1.4004	1.3954	1.3951	1.3959	1.3942	120.85	120.13	119.59	120.07	−0.02	−0.05	0.04	4.31
β_L	+	−	1.4017	1.3951	1.3956	1.3955	1.3949	120.88	120.15	119.56	120.12	0.12	0.13	−0.14	3.16
β_L	+	+	1.4017	1.3949	1.3955	1.3956	1.3950	120.91	120.13	119.56	120.14	−0.09	−0.15	0.14	3.17
β_L	a	−	1.4017	1.3980	1.3945	1.3976	1.3936	120.81	120.09	119.62	120.26	−0.05	0.18	−0.14	0.78
β_L	a	+	1.4051	1.3936	1.3977	1.3945	1.3980	120.68	120.27	119.62	120.08	−0.01	−0.14	0.18	0.78
δ_L	−	−	1.4023	1.3937	1.3968	1.3945	1.3965	120.90	120.21	119.54	120.02	0.08	0.12	−0.20	4.41
δ_L	−	+	1.4005	1.3965	1.3945	1.3968	1.3937	121.05	120.02	119.54	120.21	0.08	−0.20	0.12	4.41
δ_L	+	−	1.4031	1.3969	1.3948	1.3968	1.3946	121.04	120.00	119.54	120.41	0.08	0.09	−0.19	2.43
δ_L	+	+	1.4043	1.3946	1.3968	1.3948	1.3969	120.66	120.41	119.54	120.00	0.10	−0.19	0.09	2.43
δ_D	+	−	1.4032	1.3953	1.3962	1.3958	1.3961	120.83	120.24	119.56	120.08	0.12	0.24	−0.30	6.92
δ_D	+	+	1.4033	1.3961	1.3958	1.3962	1.3953	120.96	120.09	119.56	120.24	−0.02	−0.30	0.25	6.91
δ_D	a	−	1.4000	1.3956	1.3948	1.3965	1.3941	120.88	120.05	119.63	120.16	−0.11	0.10	0.04	8.75
δ_D	a	+	1.4021	1.3941	1.3965	1.3948	1.3956	120.74	120.16	119.63	120.05	−0.18	0.04	0.10	8.75
ε_D	a	−	1.4031	1.3978	1.3940	1.3972	1.3937	121.09	120.11	119.49	120.23	0.12	0.23	−0.26	7.62
ε_D	a	−	1.4031	1.3978	1.3940	1.3972	1.3937	121.09	120.11	119.49	120.23	0.12	0.23	−0.26	7.62
ε_D	a	+	1.4042	1.3937	1.3972	1.3940	1.3978	121.05	120.23	119.49	120.11	−0.08	−0.26	0.23	7.62
γ_L	−	−	1.4028	1.3937	1.3973	1.3941	1.3971	120.90	120.22	119.54	120.03	0.08	0.20	−0.21	1.09
γ_L	−	+	1.4009	1.3972	1.3941	1.3973	1.3937	121.05	120.03	119.54	120.22	−0.05	−0.21	0.20	1.09
γ_L	+	−	1.4036	1.3972	1.3943	1.3971	1.3945	121.07	119.94	119.53	120.54	0.10	−0.03	−0.15	0.00
γ_L	+	+	1.4044	1.3945	1.3971	1.3943	1.3972	120.51	120.54	119.53	119.94	0.26	−0.14	−0.03	0.00
γ_L	a	−	1.4018	1.3954	1.3957	1.3956	1.3950	120.53	120.38	119.56	120.01	−0.05	0.01	0.05	0.94
γ_L	a	+	1.4016	1.3950	1.3956	1.3957	1.3954	120.92	120.01	119.56	120.37	−0.06	0.05	0.01	0.95
γ_D	−	−	1.4020	1.3948	1.3965	1.3953	1.3962	120.89	120.11	119.57	120.17	0.07	0.13	−0.18	2.72
γ_D	−	+	1.4011	1.3962	1.3953	1.3965	1.3947	120.80	120.17	119.57	120.11	0.04	−0.18	0.12	2.72
γ_D	+	+	1.4026	1.3941	1.3962	1.3948	1.3951	120.67	120.25	119.60	120.02	−0.30	−0.27	0.31	7.55
γ_D	a	−	1.4010	1.3952	1.3953	1.3958	1.3946	120.76	120.21	119.59	120.03	0.09	−0.01	−0.03	4.41
γ_D	a	+	1.4015	1.3946	1.3958	1.3953	1.3952	120.95	120.03	119.59	120.21	−0.02	−0.03	−0.01	4.41
Mean			1.4022	1.3953	1.3957	1.3957	1.3953	120.87	120.15	119.56	120.15	0.03	−0.01	−0.02	
Std. Dev.			0.0014	0.0013	0.0010	0.0010	0.0013	0.17	0.15	0.04	0.15	0.11	0.17	0.17	

the literature [3] as well as pre-computed structures in the authors' own personal database. Five bond lengths, four bond angles, and three dihedral angles were used to describe the relative spatial orientation of phenyl ring carbons in benzene (Bz), phenol (Bz-OH), Ac-Phenylalanine-NH$_2$ (Ac-Phe-NH$_2$), Ac-Phenylalanine-NHMe (Ac-Phe-NHMe), Ac-Tyrosine-NHMe (Ac-Tyr-NHMe).

All results used were constructed using an established methodology explicitly defining all $3N - 6$ degrees of intra-molecular freedom explicitly [1].

Subsequent examination of the influence of basis-set extension was performed using the same database for the Ac-Phe-NH$_2$ system. Structural results were similarly extracted from the B3LYP/6-31G(d,p) and B3LYP/6-31 + G(d) geometry optimized structures

and compared with those of the previous B3LYP/6-31G(d) results.

Results were tabulated and relative energies computed for each system. Simple statistical analysis of each structural parameter tabulated was performed, including the determination of the mean and standard deviation for each.

3. Results and discussion

The results of the database extraction are found in Tables 1–4 for the Bz, Bz-OH, Ac-Phe-NHMe and Ac-Tyr-NHMe molecular systems, respectively, computed at the B3LYP/6-31G(d) levels of theory. Results for the Ac-Phe-NH$_2$ peptide model computed at the B3LYP/6-31G(d), B3LYP/6-31G(d,p), and

Table 7

Selected phenyl ring bond lengths (R), bond angles (A) and dihedral angles (D), extracted from conformers of Ac-Phe-NH$_2$, geometry optimized at the B3LYP/6-31 + G(d) level of theory

BB	χ^1	χ^2	R1	R2	R3	R4	R5	A1	A2	A3	A4	D1	D2	D3	Rel.E
α_L	A	–	1.4000	1.3973	1.3965	1.3983	1.3956	120.82	120.11	119.58	120.08	0.07	0.14	−0.01	16.28
α_D	–	–	1.4039	1.3968	1.3986	1.3972	1.3985	120.95	120.10	119.52	120.19	0.08	0.20	−0.24	11.34
α_D	+	–	1.4025	1.3986	1.3964	1.3988	1.3956	121.14	119.87	119.51	120.45	0.28	0.18	−0.28	18.74
α_D	+	+	1.4037	1.3956	1.3988	1.3964	1.3987	120.61	120.45	119.51	119.88	−0.06	−0.26	0.14	18.72
β_L	–	–	1.4043	1.3960	1.3991	1.3961	1.3993	120.96	120.22	119.48	120.05	0.11	0.26	−0.29	10.01
β_L	–	+	1.4021	1.3993	1.3961	1.3991	1.3959	121.10	120.04	119.47	120.24	−0.02	−0.30	0.24	9.99
β_L	+	–	1.4033	1.3972	1.3976	1.3975	1.3970	120.96	120.13	119.52	120.13	0.13	0.20	−0.21	11.97
β_L	+	+	1.4032	1.3970	1.3975	1.3976	1.3972	120.98	120.13	119.51	120.14	−0.10	−0.21	0.17	11.98
β_L	a	+	1.4060	1.3971	1.3989	1.3973	1.3986	120.82	120.21	119.53	120.11	−0.29	−0.09	0.35	9.64
δ_L	–	–	1.4040	1.3957	1.3991	1.3962	1.3991	120.90	120.24	119.49	120.02	0.12	0.13	−0.20	12.56
δ_L	–	+	1.4018	1.3991	1.3962	1.3991	1.3957	121.10	120.01	119.49	120.25	0.04	−0.22	0.12	12.56
δ_D	+	–	1.4050	1.3979	1.3980	1.3980	1.3979	120.97	120.21	119.50	120.10	0.22	0.29	−0.39	15.58
δ_D	+	+	1.4049	1.3979	1.3979	1.3981	1.3978	121.10	120.08	119.50	120.22	0.00	−0.41	0.27	15.61
δ_D	a	–	1.4016	1.3980	1.3965	1.3985	1.3960	120.99	120.03	119.56	120.22	−0.08	0.18	−0.05	16.97
ε_D	a	–	1.4041	1.3995	1.3961	1.3990	1.3960	121.16	120.09	119.43	120.26	0.07	0.36	−0.30	16.43
ε_D	a	+	1.4052	1.3958	1.3988	1.3959	1.3993	121.06	120.24	119.44	120.10	−0.16	−0.33	0.36	0.00
γ_L	–	–	1.4043	1.3960	1.3991	1.3961	1.3993	120.96	120.22	119.48	120.05	0.11	0.26	−0.29	9.99
γ_L	–	+	1.4022	1.3993	1.3961	1.3991	1.3959	121.10	120.04	119.47	120.24	−0.02	−0.31	0.25	9.99
γ_L	+	–	1.4049	1.3992	1.3964	1.3989	1.3969	121.17	119.92	119.49	120.53	0.10	0.15	−0.24	9.84
γ_L	+	+	1.4056	1.3969	1.3988	1.3965	1.3992	120.60	120.53	119.48	119.94	0.12	−0.24	0.12	9.81
γ_L	a	–	1.4026	1.3978	1.3975	1.3979	1.3970	120.63	120.33	119.54	120.01	−0.09	0.04	0.08	10.14
γ_L	a	+	1.4035	1.3970	1.3978	1.3975	1.3977	120.97	120.00	119.54	120.35	−0.12	0.05	0.03	10.15
γ_D	–	–	1.4039	1.3968	1.3987	1.3971	1.3985	120.95	120.10	119.52	120.19	0.08	0.19	−0.24	11.34
γ_D	+	–	1.4016	1.3971	1.3968	1.3981	1.3963	121.04	120.01	119.55	120.23	0.29	0.44	−0.37	16.83
γ_D	+	+	1.4036	1.3962	1.3980	1.3968	1.3970	120.78	120.22	119.54	120.02	−0.43	−0.35	0.39	16.82
γ_D	a	–	1.4019	1.3976	1.3970	1.3980	1.3963	120.84	120.19	119.55	120.04	0.02	0.12	−0.09	13.35
γ_D	a	+	1.4032	1.3964	1.3980	1.3970	1.3976	121.01	120.02	119.55	120.20	−0.04	−0.11	0.10	13.35
Mean			1.4034	1.3974	1.3976	1.3976	1.3974	120.95	120.14	119.51	120.16	0.02	0.01	−0.02	
Std. Dev.			0.0014	0.0012	0.0011	0.0010	0.0013	0.16	0.15	0.04	0.15	0.16	0.25	0.25	

B3LYP/6-31 + G(d) levels of theory are listed in Tables 5–7, respectively. In each table, BB stands for backbone conformation, following the nomenclature well established in the literature [1–4]. χ^1 and χ^2 represent the sidechain (SC) dihedral angles. Phenyl ring carbon bond lengths, bond angles, dihedral angles are represented as R, A, D, respectively. Selected conformers are shown for each.

As is easily apparent, there is almost no structural deviation in the constituent ring carbons, of the phenyl rings for these systems. The standard deviations for all structural results are of no significant value, supporting the claim that a pre-computed phenyl ring may be 'mined' for the relative spatial orientation of constituent carbon nuclei, for use in the geometry optimization of any structure.

Finally, the influence of basis-set extension is negligible and provides confidence that the structural properties are well conserved for differing levels of theory.

4. Conclusions

Although this work deals only with a very ideal and limited example of the transferability of structural properties, it is nonetheless representative of a larger phenomenon. It is with confidence and supportive data that the author encourages the continued construction of a computed structure database, with subsequent use of the 'stored' pre-computed structures.

The use of pre-computed phenyl ring degrees of freedom may not drastically improve the efficiency of computation. The exercise described in this work is nonetheless representative of a methodology that may be extended to other moieties that show structural conservation. The combination of several pre-computed and 'structurally static' fragments would allow for a marked increase in computational efficiency, reducing the number of iterative steps necessary to bring about convergence.

Finally, when any of these fragments does show structural perturbations upon substitution or combination with other fragments, one is afforded an opportunity to quantitatively characterize the molecular basis of these distortions. The formulation of an analytical function to characterize and predict these perturbations is the long term goal of this methodology.

Numerical perturbations in silico may be representative or predictive of macroscopic phenomena in vitro or even in vivo.

Acknowledgements

G. A. Chass wishes to thank Tania A. Pecora, Lucha Alforque, Elena Luzi, Michelle A. Sahai, Jaqueline M. S. Law, Christopher N. J. Marai, David H. Setiadi, Reinard S. Mirasol and Imre G. csizmadia all for helpful discussions and manuscript preparation.

Gac also thanks the Global Institute of Computational Molecular and Materials Sciences for funding and support, allowing this and other related works to have been pursued and completed in all of the following settings: Toronto, Ottawa and Montreal, Canada; Budapest and Szeged, Hungary; Paris and Strasbourg, France; Milano and Napoli, Italy; Girona, Spain; New York City and Omaha, USA. The 'travelling nature' of all of GIOCOMMS works truly attest to the versatility of the power of the internet expecially in the theoretical sciences.

References

[1] G.A. Chass, M.A. Sahai, J.M.S. Law, S. Lovas, O. Farkas, A. Perczel, J.-L. Rivail, I.G. Csizmadia, Toward a computed peptide structure database: the role of a universal atomic numbering system of amino acids in peptides and internal hierarchy of database, Int. J. Quantum Chem. 90 (2) (2002) 933.
[2] M.A. Sahai, S. Lovas, G.A. Chass, A. Varro, J.Gy. Papp, A modular numbering system of selected oligopeptides for molecular computations, J. Mol. Struct. (THEOCHEM) (2004) in press.
[3] G.A. Chass, C.N.J. Marai, I.G. Csizmadia, A.G. Harrison, I.G. Csizmadia, Fragmentation reactions of a$_2$ ions derived from deprotonated dipeptides—a synergy between experiment and theory, J. Phys. Chem. A 106 (2002) 9695.
[4] G.A. Chass, S. Lovas, R.F. Murphy, I.G. Csizmadia, The role of enhanced aromatic π-electron donating aptitude of the tyrosyl sidechain with respect to that of phenylalanine, Eur. Phys. J. D 20 (2002) 481.

ELSEVIER

Journal of Molecular Structure (Theochem) 666–667 (2003) 69–77

THEO
CHEM

www.elsevier.com/locate/theochem

Molecular shape, dimensions, and shape selective catalysis

Gyula Tasi[a,*], István Pálinkó[b], Árpád Molnár[b], István Hannus[a]

[a]Department of Applied and Environmental Chemistry, University of Szeged, Rerrich B. tér 1, H-6720 Szeged, Hungary
[b]Department of Organic Chemistry, University of Szeged, Dóm tér 8, H-6720 Szeged, Hungary

Abstract

Molecular shape is a fundamental quantum mechanical property determined by the spatial distribution of electrons in the molecule. Enzymes, the catalysts of biology, and zeolites, the catalysts of inorganic chemistry, show remarkable shape selective properties. Generally, the active sites are confined within the three-dimensional structures of the catalysts, so the fate of the reactant/intermediate/product molecules depends primarily on their molecular shapes and dimensions. An ab initio method is applied for calculating molecular shapes and dimensions. Selective alkylation of naphthalene over various zeolites is reviewed and discussed in this paper from the viewpoints of these properties.
© 2003 Elsevier B.V. All rights reserved.

Keywords: Molecular shape; Molecular dimensions; Shape selective catalysis; Zeolites; 2,6-Dialkylnaphthalenes

1. Introduction

Molecular shape is determined by the spatial distribution of electrons in the molecule [1]. Since the motion of electrons within a molecule obeys the laws of quantum mechanics, molecular shape is a fundamental quantum mechanical property and must be very different from the shape of a macroscopic body. Molecular shape determined then by the outer regions of the electron distribution plays a fundamental role in molecular recognition, in the interaction of reacting molecules, and, of course, in shape selective catalysis.

Enzymes, the catalysts of biology, and zeolites, the catalysts of inorganic chemistry, show remarkable shape selective properties [2–9]. There are structural similarities between these catalysts: both have channels and cavities [10–15]. Of course, the tertiary

protein structure of an enzyme is much more flexible than the aluminosilicate framework of a zeolite. Generally, the active sites are confined within the three-dimensional (3D) structures of the catalysts in question, thus, the fate of the reactant/intermediate/product molecules depends primarily on their molecular shape and dimensions.

Essentially, there are three types of shape selectivity concerning zeolites [3–8]: reactant selectivity, product selectivity, and (restricted) transition state selectivity. Reactant selectivity occurs when certain molecules in the reactant mixture are too bulky to enter the bulk phase of the zeolite. In this case only those molecules will undergo transformation, which are not excluded from the zeolite micropore system. Product selectivity occurs when certain product molecules formed in the bulk phase of the zeolite are too bulky to diffuse out. These bulky molecules will either be converted to smaller ones and leave the pore system, or they will block the pores deactivating the catalyst. Transition state selectivity occurs when

* Corresponding author. Tel./fax: +36-62-544-619.
E-mail address: tasi@chem.u-szeged.hu (G. Tasi).

certain reactions are prevented, because their transition states require more space than is available in the cavities and/or channels of the zeolite structures.

For zeolites having intersecting channels of different sizes, a further type of shape selectivity was observed: molecular traffic control effect [16]. According to this idea, the small reactant molecules use the smaller channels to enter, while the bulkier product molecules use the larger channels to leave the bulk phase of the zeolite. This could result in the lack of counterdiffusion effects between reactants and products.

Besides the traditional types of shape selectivity mentioned above, there is an inverse shape selectivity concept as well [17–19]. The original interpretation states that in certain cases molecules having an optimal fit within the pore system are formed in the bulk phase of a zeolite [17]. This may explain the formation of branched isomers over linear ones in hydrocracking reactions of n-alkanes over certain zeolites. Recently, this interpretation has been revised [19]: the molecular basis of inverse shape selectivity is related to entropic effects inside the zeolite pore system at almost full saturation. In fact, the zeolite provides an environment in which the length differences between the linear and branched isomers are maximum. It can be seen that the molecular shape and dimensions are key factors in this interpretation too.

2. Computation of molecular shape and dimensions

The ab initio and semiempirical quantum chemical calculations were performed by the GAUSSIAN98 [20], DALTON [21] and PcMol [22] program packages. For determining the conformers of molecules with free torsional angles, methods applied for alkanes were used [23,24].

The following notation is used to designate the ab initio quantum chemical computational results: Hartree-Fock (HF), HF/6-31G*//HF/6-31G*; second order Møller–Plesset (MP2), MP2(full)/6-31G*//MP2(full)/6-31G*, and G2(MP2,SVP) [25]. G2(MP2,SVP) energy calculations on MP2(full)/6-31G* optimized geometry approximate energies at the QCISD(T)/6-311 + G(3df,2p) level. This composite ab initio method affords precise thermochemical data for hydrocarbons. Details concerning

the methods and the basis sets applied are given in the literature [26,27].

Calculation of zero-point vibrational energy (ZPVE) corrections and thermochemical quantities was performed within the rigid rotor-harmonic oscillator (RRHO) approximation [28]. Scaled HF harmonic vibrational frequencies were used in every case, exactly as in the Gaussian methods [29]. For instance, the RRHO approximation was successfully applied for calculating various thermochemical quantities for alkanes [24,30].

Trial geometries for the geometry optimizations were generated with the help of the PcMol package. For the characterization of the stationary points obtained, harmonic vibrational analysis was performed at both HF and MP2 levels.

As far as shape selective catalysis is concerned, molecular shape and size are properties of utmost importance [2–9]. Therefore, a reliable ab initio method is needed for determining them [31].

Generally, mechanistic models are used in theoretical studies concerning shape selective catalysis: possessing a molecular geometry (from molecular mechanical or semiempirical quantum chemical calculations) and atomic van der Waals radii, the molecular surface is obtained as the superposition of rigid atomic spheres (fused-spheres models). The dimensions of the generated molecular shape are then determined by means of molecular graphics [32]. Further ambiguous molecular size concepts without firm quantum mechanical background are also used in the catalytic literature [12–15]: kinetic diameter, critical diameter and critical dimension. These procedures, however, are not reliable quantitative methods [32].

By quantum mechanics, the molecular shape is determined by the spatial distribution of electrons in the molecule. Consequently, it can be generated from 3D electron density maps. For instance, the popular GAUSSIAN program packages allow one to calculate the electron density $\rho(x, y, z)$ at high level of theory at the points of a large cubic grid around the molecule. From these data, points on the ab initio molecular surface (isosurface) can be generated via interpolation using a reliable cutoff value ($\rho_0 =$ constant). To visualize our molecular surfaces on a graphical display, triangulations were performed with the help of the marching cubes algorithm [33].

For determining the ab initio molecular surface and shape, cubic grids consisting of $400 \times 400 \times 400$ points were taken around each molecule. The electron density was calculated at the MP2 level of theory, i.e. including important dynamic electron correlation effects. For each molecule studied, a sequence of molecular shapes and dimensions was generated using a set of cutoff values: {0.0005, 0.001, 0.002, 0.003, 0.004, 0.005}. However, since the differences in shapes and dimensions were consistent on each isodensity level, only the results obtained with $\rho_0 = 0.001$ a.u. are presented here. This 'calibration' was based on zeolite adsorption measurements and the published effective pore sizes of zeolite structures [12–15].

Having obtained the ab initio molecular shapes, the molecular dimensions were calculated via parameter estimation [34]: the smallest parallelepipeds enclosing the molecular shapes in question were determined. To a good approximation, the dimensions of these optimum parallelepipeds can be regarded as the dimensions of the molecules under consideration. Since three parameters determine the spatial orientation of a molecule in the laboratory coordinate system, our parameter space is three-dimensional. The parameters may be the rotational angles around the axes of the laboratory coordinate system. The simplex procedure of Nelder and Mead was used for parameter estimation [35].

Obviously, instead of using a parallelepiped other 3D solid objects (sphere, right circular cylinder, etc.) may be considered. However, the channels of zeolites may be further distorted in the course of diffusion of guest molecules, so the use of a parallelepiped seems to be well justified.

Generally, optimization procedures afford more than one local minimum. The problem can be circumvented by determining all the local minima and selecting the best one, i.e. the global minimum. Since the parameter space is only three-dimensional in our case, this procedure is feasible: it can be performed with relative ease. However, one more criterion has to be taken into account: one of the cross sections of the optimum parallelepiped should be the smallest possible as well.

The procedure described above is fully automatic. Our FORTRAN modules have been linked to the GAUSSIAN98 program package. From Gaussian cube files they generate molecular shapes, triangularized molecular surfaces for graphical visualization, and determine the molecular dimensions by parameter estimation.

Table 1 lists the calculated MP2 molecular dimensions of some small molecules. The n-butane and n-hexane molecules have one and three free C–C–C–C torsional angles, respectively. Table 1 lists all the nonisomorphic conformers (equivalence classes of the conformers) for these molecules [24]. The following notation is used to designate the C–C–C–C torsional angles in the n-alkane molecules: t (trans): $\approx 180°$, g^+ (gauche$^+$): $\approx +60°$, g^-: $\approx -60°$, x^+ (extended gauche$^+$): $\approx +90°$, and x^-: $\approx -90°$ [24]. It is to be seen that the conformers of a given molecule may have very different molecular shapes and dimensions. The last column of Table 1 gives the fractions of the equivalence classes of the conformers of the molecules in the gas phase at 298.15 K and at 1 atm. These fractions were

Table 1

MP2 molecular dimensions of some small molecules and the fractions of their nonisomorphic conformers

Molecule	Dimensions (in Å)	Fraction
O_2	$3.45 \times 3.45 \times 4.55$	–
N_2	$3.65 \times 3.65 \times 4.75$	–
CH_4	$4.14 \times 4.14 \times 4.14$	–
C_2H_4	$4.10 \times 4.20 \times 5.32$	–
Cyclopropane	$4.56 \times 5.26 \times 5.44$	–
CHF_2Cl	$4.76 \times 4.76 \times 5.87$	–
CCl_4	$6.21 \times 6.21 \times 6.21$	–
$CHClCCl_2$	$4.25 \times 6.64 \times 8.15$	–
CCl_2CCl_2	$4.24 \times 6.91 \times 7.34$	–
n-butane		
Trans	$4.58 \times 5.06 \times 8.39$	0.59
Gauche	$5.43 \times 5.54 \times 7.83$	0.41
n-hexane		
$t\,t\,t$	$4.58 \times 5.07 \times 10.94$	0.13772
$g^+\,t\,t$	$5.66 \times 5.70 \times 10.26$	0.41505
$t\,g^+\,t$	$5.43 \times 5.71 \times 10.11$	0.10329
$g^+\,g^+\,t$	$5.54 \times 6.19 \times 9.16$	0.19209
$x^-\,g^+\,t$	$5.80 \times 5.85 \times 9.13$	0.00884
$g^-\,x^+\,t$	$5.86 \times 6.13 \times 9.05$	0.01089
$g^+\,t\,g^+$	$5.53 \times 5.85 \times 9.93$	0.03493
$g^-\,t\,g^+$	$5.41 \times 5.53 \times 9.99$	0.06951
$g^+\,g^+\,g^+$	$5.86 \times 5.95 \times 8.79$	0.02028
$x^-\,g^+\,g^+$	$5.64 \times 6.51 \times 8.04$	0.00386
$g^-\,x^+\,g^+$	$5.79 \times 6.14 \times 8.97$	0.00352
$x^+\,g^-\,x^+$	$5.84 \times 6.57 \times 7.73$	0.00003

calculated at G2(MP2,SVP) level of theory using Boltzmann statistics [36].

The molecular properties of an isolated molecule more or less depend on its vibrational state. So far we have used equilibrium molecular geometries without considering vibrational effects (even zero-point vibrations) for calculating molecular shapes and sizes. By using vibrationally averaged molecular geometries [37], however, it is possible to take the effect of the vibrational motion into consideration at common temperatures on the molecular shape and dimensions. The DALTON program [21] allows one to calculate vibrationally averaged molecular geometries.

3. Zeolites: structure and properties

Zeolites are porous crystalline aluminosilicates with infinite 3D lattice network of TO_4 tetrahedra linked through common oxygen atoms [12–15]. The tetrahedral sites (T atoms) are usually aluminum and silicon atoms, but chemically similar atoms (Ga, Ge, Ti, P, etc.) can be incorporated into the framework. Because of aluminum atoms, or other trivalent T atoms, the framework of a zeolite is generally anionic. The excess negative charges are usually compensated by exchangeable cations (H^+, Na^+, etc.) bound to the framework oxygen atoms close to the aluminum atoms. We are talking about Brønsted acid sites in the case of proton. The catalytic activity of most of the zeolites arises from Brønsted acid sites confined within their 3D structures. Information about the spatial distributions of exchangeable cations in various zeolite structures can be found in the literature [38].

From the TO_4 primary building units more complex secondary building units (4-, 6-, or 8-rings, etc.) can be constructed. Further linkages of these secondary building units result in the 3D frameworks of zeolites. For instance, from sodalite cages (see Fig. 1) we can construct three different kind of zeolites differing only in the linkages of sodalite units: sodalite (SOD), zeolite A (LTA), and faujasite (FAU: X, Y). Fig. 2 displays these structures with their pore openings.

Zeolites can be classified by their pore structure: pores, channels, and cavities of various dimensions are responsible for their molecular sieving and

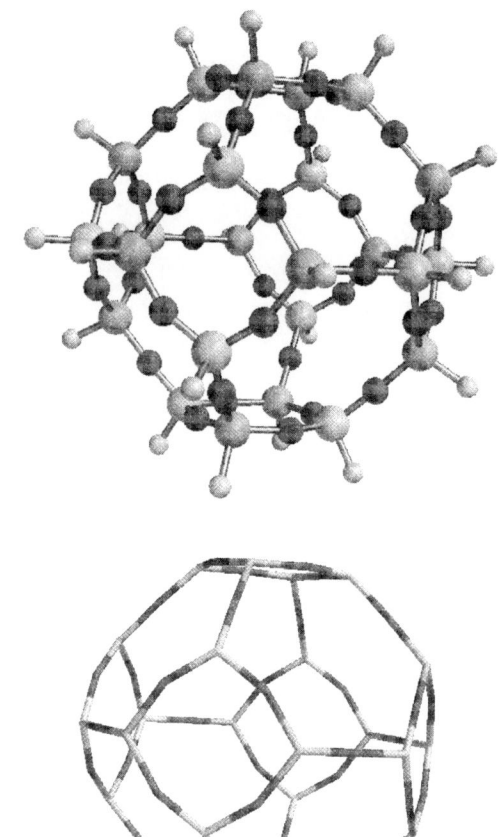

Fig. 1. Ball-and-stick and tube representations of the sodalite unit.

confinement effects. Of the three zeolite structures presented above, SOD and LTA are small-port materials having 6- and 8-ring openings, respectively. Large-port materials like FAU contain 12-ring openings. Table 2 lists the effective pore sizes (estimates of the free dimensions of the pore openings) of some common zeolites. These effective pore sizes are based on crystallographic data assuming a van der Waals radius of 1.35 Å for the framework oxygen atom. However, the free dimensions of the zeolite apertures more or less depend on the temperature, the dehydration degree of the zeolite samples, as well as the exchangeable cations and their spatial distributions. For example, the effective pore size is slightly increasing in the following order K–A < Na–A < Ca–A.

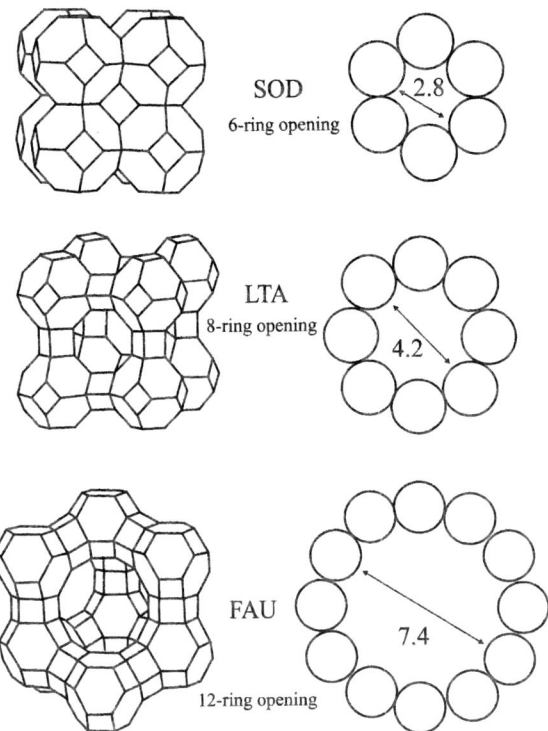

Fig. 2. Various zeolite structures built from sodalite units.

There are several methods for tailoring the molecular sieve effect in zeolites, i.e. for changing the zeolite pore apertures: (i) replacing the exchangeable cations by ion exchange may enlarge or reduce the size of the pore openings; (ii) preadsorption of small polar molecules (water, ammonia, etc.) or formation of strong inorganic complexes in the zeolite

Table 2
The free dimensions of the pore openings of some zeolites

Zeolite	Openings	Effective pore size (in Å)
Sodalite (SOD)	6-rings	2.8
Zeolite A (LTA)	8-rings	4.2
Erionite (ERI)	8-rings	3.6×5.2
Clinoptilolite (HEU)	8-rings	4.1×4.7
		4.0×5.5
	10-rings	4.4×7.2
ZSM-5 (MFI)	10-rings	5.3×5.6
		5.1×5.5
Mordenite (MOR)	8-rings	2.9×5.7
	12-rings	6.7×7.0
Faujasite (FAU)	12-rings	7.4

pore system can significantly reduce the effective pore size; and (iii) hydrolitic pore closure by steaming the zeolite sample.

With the help of the molecular dimensions listed in Table 1, the following shape selective phenomena can be easily rationalized.

At room temperature Na–A zeolite adsorbs methane readily, while the ethylene molecules are adsorbed slowly [12]. The ethylene molecule is somewhat larger than the methane. The bulkier cyclopropane molecule cannot enter the pores of Na–A, therefore its skeletal isomerization reaction does not take place over this zeolite [39].

At temperatures of -183 to $-196\,°C$ Na–A adsorbs oxygen freely, while nitrogen is not adsorbed at all [12]. It can be concluded that the effective pore size of Na–A slightly changes with the temperature and the small difference in the dimensions of the oxygen and nitrogen molecudles allows one to separate them from air. Partially dehydrated Co,Na–A zeolites may also be used for this purpose [39].

Nonburning components (e.g. nitrogen) in the natural gas reduce its thermal value. By controlling the effective pore size of the naturally occurring clinoptilolite, it is possible to separate the nitrogen and methane molecules from the natural gas. It was found that low-sodium clinoptilolite is highly efficient in methane/nitrogen separation [40].

By steaming Na–A at high temperature, the effective pore size can be reduced (hydrolitic pore closure) [12]. For instance, the slow adsorption of CHF_2Cl can be eliminated. More drastic steam treatment even removes all the oxygen capacity at low temperature.

Molecular shape and size of an isolated molecule, however, are not the only factors, which determine the selective adsorption of the molecule in question into zeolites. As a matter of fact, in interacting systems molecular shape and size are not static but dynamic properties: one should consider the interaction between the guest molecule and the host zeolite. This interaction primarily depends on the polarity, polarizability, and degree of saturation of the guest molecule, organophilicity and hydrophobicity of the zeolite structure, and the polarizing power of the exchangeable cations.

For instance, erionite adsorbs n-hexane more readily at room temperature than n-butane [12].

Data listed in Table 1, however, do not support these experimental findings. It can be seen that only the static lengths of the 'all-*trans*' conformers of *n*-hexane and *n*-butane are different: their cross sections are equal at the MP2 level of theory.

4. Shape selective alkylation of naphthalene

The selective synthesis of 2,6-dialkyl-substituted naphthalene derivatives is of considerable interest, which is due to their significant practical importance as valuable starting materials in polymer synthesis. 2,6-Dialkylnaphthalenes are oxidized to naphthalene-2,6-dicarboylic acid, which serves as starting material in the manufacture of specialty polyesters with outstanding properties, mainly poly(ethylene naphthalenedicarboxylate) (PEN). It has wide application in the production of films, fibers, packaging, coatings, and liquid crystal polymers [41,42].

2,6-Dimethylnaphthalene (DMN) can be recovered from kerosene reforming fractions [43] and from fractions of FCC oil [44]. Tedious separation (distillation, absorption, crystallization) and purification steps, however, make these possibilities uneconomical. A process commercialized by Amoco [45] produces 2,6-DMN by alkylating *o*-xylene with butadiene followed by five reaction steps including the isomerization of 1,5-DMN to 2,6-DMN. A more attractive and more economical alternative appears to be the direct alkylation of naphthalene using alkenes, alcohols or alkyl halides. Conventional Friedel-Crafts catalysts and amorphous silica–alumina, however, yield 2,6-dialkylnaphthalenes in low yield. This is due to polyalkylation and the low β,β' selectivity. For example, the 2,6- and 2,7-isomers are formed in equal amounts. Moreover, the desired 2,6-dialkyl isomer is difficult to separate from the 10 possible dialkylnaphthalene isomers.

The shape selective nature and the acidic character of zeolites, however, offer a unique possibility for the selective synthesis of 2,6-dialkylnaphthalenes at high catalyst activity through the alkylation of naphthalene. In fact, similar results can be achieved in the alkylation of 2-alkylnaphthalenes and transalkylation of naphthalene with polyalkylbenzenes or polyalkylnaphthalenes, and in the isomerization of dialkylnaphthalenes.

Initial work focused on the selective preparation of 2,6-DMN. Data disclosed both in the open [46,47] and patent literature [48,49] show that medium-port H-ZSM-5 is capable of producing isomer mixtures with increased 2,6 selectivity. Both the yield and the 2,6/2,7 ratio, however, are low. In turn, H-ZSM-5, when the external acid sites at the surface are neutralized, exhibits higher selectivities in the methylation of methylnaphthalenes [47]. MCM-22 was later found to give results similar to H-ZSM-5 but at higher conversions [50,51]. Recently, better isomeric ratios were achieved in transalkylation over ZSM-12 zeolite [52–54].

Whereas much less attention has been paid to the ethylation of naphthalene [55,56], significant improvements were found in the synthesis of 2,6-diisopropylnaphthalene (DIPN) and 2,6-di-*tert*-butylnaphthalene (DTBN). A further advantage of these processes is the milder reaction conditions, which are due to the involvement of the more stable secondary and tertiary carbocations in the alkylation reaction. In addition, oxidation of a branched benzylic side chain is a more facile process. In the synthesis of 2,6-DIPN higher selectivities were reported over the 12-ring zeolite H-mordenite (H-M) [57–61]. The best results are a 54.2% 2,6-DIPN yield and a 2,6/2,7 ratio of 4.0 [62]. Selectivity values could be further increased by catalyst modifications. Dealumination of mordenite results in an increase in 2,6-selectivity [63–66], which is attributed to the decrease in unit cell dimensions. In addition, the removal of strong acid sites also contributes to selectivity improvements by decreasing the isomerization of 2,6-DIPN on the external surface. Similar effects were observed over catalysts modified by ion exchange [66–68]. The improvement of catalyst performance is ascribed to the selective deactivation of acid sites at the external surface of the zeolite.

The bulky *tert*-butyl group may allow further improvements. Indeed, large-port H-Y and H-beta zeolites showed efficient activities and high selectivities [69–72]. The best values are 84% selectivity with a 2,6/2,7 ratio of 5.9 [72]. Most of these studies attribute the increased β,β' selectivity to the shape-selective effect of the zeolites applied. Isopropylation over H-M was shown to be controlled by steric restriction of the transition state in the zeolite and by the entrance of intermediate product molecules into

Table 3
Molecular dimensions (a, b, and c in Å) and total energies (a.u.) of the conformers of some monoalkyl- and dialkyl-substituted naphthalene molecules at the MP2 level of theory

Molecule	Conformers	Total energy	a	b	c
1-IPN	I (C_I)	−501.96070	5.85	8.62	11.09
	II (C_s)	−501.95783	6.65	8.58	10.31
2-IPN	I (C_s)	−501.96429	6.61	6.61	12.03
	II (C_s)	−501.96363	6.44	7.06	11.99
2,6-DIPN	I (C_{2h})	−619.40526	6.61	6.61	14.23
	II (C_s)	−619.40462	6.69	7.09	14.10
	III (C_{2h})	−619.40395	6.44	7.06	14.12
2,7-DIPN	I (C_{2v})	−619.40518	6.62	7.26	13.76
	II (C_s)	−619.40454	6.70	8.05	13.30
	III (C_{2h})	−619.40388	6.73	8.40	12.57
1-TBN	I (C_s)	−541.10010	6.66	8.56	11.00
	II (C_s)	−541.09266	6.56	8.49	11.01
2-TBN	I (C_s)	−541.11049	6.82	7.04	12.14
	II (C_s)	−541.10930	6.84	7.00	12.24
2,6-DTBN	I (C_{2h})	−697.69769	6.96	6.96	14.26
	II (C_s)	−697.69653	6.97	6.97	14.42
	III (C_{2h})	−697.69531	6.91	6.98	14.52
2,7-DTBN	I (C_{2v})	−697.69761	6.60	8.29	13.38
	II (C_s)	−697.69647	6.70	8.40	14.25
	III (C_{2v})	−697.69524	6.74	8.43	14.57

Total energies at 0 K and include ZPVE corrections.

the pore [73]. A specific shape-selectivity effect occurring at the entrances into the pores of zeolites has recently been suggested [74,75].

How can we explain the high 2,6-DIPN/2,7-DIPN and 2,6-DTBN/2,7-DTBN ratios over H-M and H-Y zeolites, respectively? According to the literature [57, 69,76–78], the molecular dimensions of the 2,6- and 2,7-isomers are the same or very close to each other. We have already shown that the molecular shapes and dimensions of the 2,6- and 2,7-DIPN molecules are very different [31]. And, in fact, this statement is even more valid for the 2,6- and 2,7-DTBN molecules. Table 3 lists the molecular dimensions of all the nonisomorphic conformers of the monoalkyl- and dialkyl-substituted naphthalene molecules in question. Data for the IPN and DIPN molecules are from refs [31,79]. Fig. 3 displays the molecular shapes of the most stable conformers of 1-TBN, 2-TBN, 2,6-DTBN, and 2,7-DTBN.

Data in Table 3 reveal that the cross sections of the most stable conformers of 2-IPN and 2,6-DIPN are the same and they perfectly fit into the elliptical channel of mordenite (6.7 × 7.0 Å, see Table 2). The same is valid for all the conformers of 2-TBN

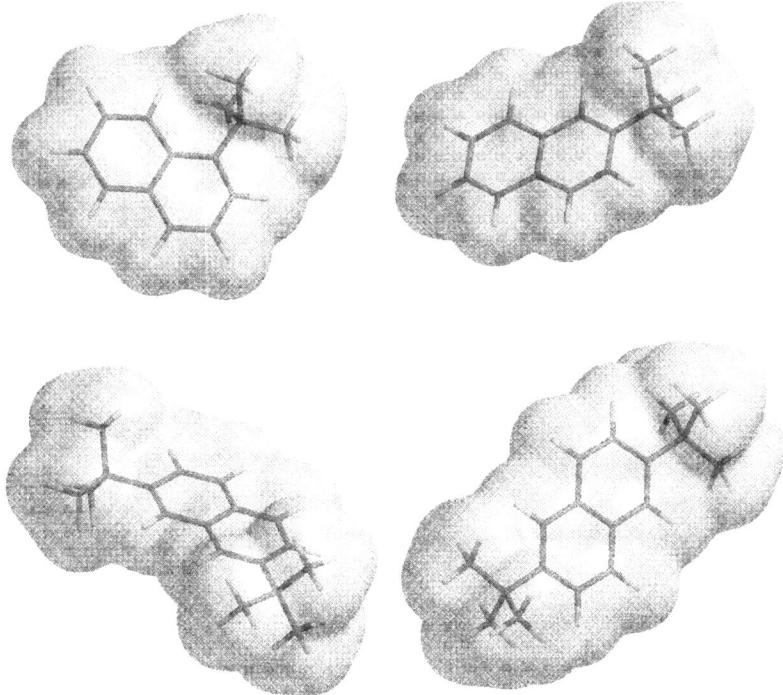

Fig. 3. MP2 molecular shapes of the most stable conformers of 1-TBN, 2-TBN, 2,7-DTBN, and 2,6-DTBN.

and 2,6-DTBN: their cross sections are equal to each other. However, they fit into the larger circular 12-ring pore openings of Y zeolite (7.4 Å, see Table 2) instead. Furthermore, the molecular size difference between 2,6-DTBN and 2,7-DTBN (≈ 1.3 Å) is even larger than that between 2,6-DIPN and 2,7-DIPN (≈ 0.65 Å). This may be the reason why the 2,6/2,7 ratio over Y zeolite (≈ 6) is much greater than that over mordenite (≈ 3).

It can be concluded that the selective formations of 2,6-DIPN over H-M zeolite and 2,6-DTBN over H-Y zeolite are the result of diffusion-controlled shape selective catalysis, i.e. product selectivity [58,72,80] is operative in these cases.

Acknowledgements

This work was sponsored by the National Science Fund of Hungary (OTKA T042825 and T034184) and the János Bolyai Research Fund (G.T.).

References

[1] P.G. Mezey, Shape in Chemistry, VCH Publishers, New York, 1993.
[2] A.L. Lehninger, Biochemistry, Worth, New York, 1975.
[3] P.B. Weisz, Pure Appl. Chem. 52 (1980) 2091.
[4] P.B. Weisz, ACS Symp. Ser. 738 (1999) 18.
[5] S.M. Csicsery, Pure Appl. Chem. 58 (1986) 841.
[6] S.M. Csicsery, Stud. Surf. Sci. Catal. 94 (1995) 1.
[7] C. Song, J.M. Garcés, Y. Sugi, ACS Symp. Ser. 738 (1999) 1.
[8] C.D. Williams, CHIMICA OGGI-Chemistry Today 20 (2002) 74.
[9] P.C.H. Mitchell, Chem. Ind. (London) (1991) 308.
[10] C.A. Tolman, N. Herron, Catal. Today 3 (1988) 235.
[11] M.G.B. Drew, N.J. Jutson, P.C.H. Mitchell, S.A. Wass, Polyhedron 8 (1989) 1817.
[12] D.W. Breck, Zeolite Molecular Sieves, Wiley, New York, 1974.
[13] R.M. Barrer, Zeolite and Clay Minerals as Sorbents and Molecular Sieves, Academic Press, London, 1978.
[14] D.M. Ruthven, Principles of Adsorption and Adsorption Processes, Wiley, New York, 1984.
[15] J.B. Nagy, P. Bodart, I. Hannus, I. Kiricsi, Synthesis, characterization and use of zeolite microporous materials, DecaGen Ltd, Szeged, 1998.
[16] E.G. Derouane, Z. Gabelica, J. Catal. 65 (1980) 486.
[17] D.S. Santilli, T.V. Harris, S.I. Zones, Microporous Mater. 1 (1993) 329.

[18] M. Schenk, B. Smit, T.J.H. Vlugt, T.L.M. Maesen, Angew. Chem. Int. Ed. Engl. 40 (2001) 736.
[19] M. Schenk, S. Calero, T.L.M. Maesen, L.L. Van Benthem, M.G. Verbeck, B. Smit, Angew. Chem. Int. Ed. Engl. 41 (2002) 2500.
[20] M.J. Frisch, G.W. Trucks, H.B. Schlegel, G.E. Scuseria, M.A. Robb, J.R. Cheeseman, V.G. Zakrzewski, J.A. Montgomery Jr., R.E. Stratmann, J.C. Burant, S. Dapprich, J.M. Millam, A.D. Daniels, K.N. Kudin, M.C. Strain, O. Farkas, J. Tomasi, V. Barone, M. Cossi, R. Cammi, B. Mennucci, C. Pomelli, C. Adamo, S. Clifford, J. Ochterski, G.A. Petersson, P.Y. Ayala, Q. Cui, K. Morokuma, D.K. Malick, A.D. Rabuck, K. Raghavachari, J.B. Foresman, J. Cioslowski, J.V. Ortiz, B.B. Stefanov, G. Liu, A. Liashenko, P. Piskorz, I. Komaromi, R. Gomperts, L.R. Martin, D.J. Fox, T. Keith, M.A. Al-Laham, C.Y. Peng, A. Nanayakkara, C. Gonzalez, M. Head-Gordon, E.S. Repogle, J.A. Pople: GAUSSIAN98, Revision A.6, Gaussian, Inc., Pittsburgh, PA, 1998.
[21] "DALTON, a molecular electronic structure program, Release 1.2.1 (2001)", written by T. Helgaker, H.J.Aa. Jensen, P. Jørgensen, J. Olsen, K. Ruud, H. Ågren, A.A. Auer, K.L. Bak, V. Bakken, O. Christiansen, S. Coriani, P. Dahle, E.K. Dalskov, T. Enevoldsen, B. Fernandez, C. Hättig, K. Hald, A. Halkier, H. Heiberg, H. Hettema, D. Jonsson, S. Kirpekar, R. Kobayashi, H. Koch, K.V. Mikkelsen, P. Norman, M.J. Packer, T.B. Pedersen, T.A. Ruden, A. Sanchez, T. Saue, S.P.A. Sauer, B. Schimmelpfenning, K.O. Sylvester-Hvid, P.R. Taylor, and O. Vahtras.
[22] G. Tasi, I. Pálinkó, J. Halász, G. Náray-Szabó, Semiempirical Calculations on Microcomputers, CheMicro Ltd, Budapest, 1992.
[23] G. Tasi, F. Mizukami, J. Chem. Inf. Comput. Sci. 38 (1998) 632.
[24] G. Tasi, F. Mizukami, I. Pálinkó, J. Csontos, W. Győrffy, P. Nair, K. Maeda, M. Toba, S. Niwa, Y. Kiyozumi, I. Kiricsi, J. Phys. Chem. A 102 (1998) 7698.
[25] L.A. Curtiss, P.C. Redfern, B.J. Smith, L. Radom, J. Chem. Phys. 104 (1996) 5148.
[26] W.J. Hehre, L. Radom, P.v.R. Schleyer, J.A. Pople, Ab initio Molecular Orbital Theory, Wiley, New York, 1986.
[27] T. Helgaker, P. Jørgensen, J. Olsen, Molecular Electronic-Structure Theory, Wiley, Chichester, 2000.
[28] D.A. McQuarrie, Statistical Mechanics, Harper & Row, New York, 1976.
[29] L.A. Curtiss, K. Raghavachari, P.C. Redfern, V. Rassolov, J.A. Pople, J. Chem. Phys. 109 (1998) 7764.
[30] D.F. DeTar, J. Phys. Chem. A 102 (1998) 5128.
[31] G. Tasi, F. Mizukami, I. Pálinkó, M. Toba, Á. Kukovecz, J. Phys. Chem. A 105 (2001) 6513.
[32] R.C. Deka, R. Vetrivel, J. Mol. Graphics Mod. 16 (1998) 157.
[33] W.E. Lorensen, H.E. Cline, Comput. Graphics 21 (1987) 163.
[34] L.E. Scales, Introduction to Non-Linear Optimization, Macmillan Publishers, London, 1985.
[35] J.A. Nelder, R. Mead, Comput. J. 7 (1965) 308.
[36] G. Tasi, F. Mizukami, M. Toba, S. Niwa, I. Pálinkó, J. Phys. Chem. A 104 (2000) 1337.

[37] P.-O. Åstrand, K. Ruud, P.R. Taylor, J. Chem. Phys. 112 (2000) 2655.
[38] W. Mortier, Compilation of Extra Framework Sites in Zeolites, Butterworths, Guildford, England, 1982.
[39] G. Tasi, I. Kiricsi, F. Evanics, E. Nagy, P. Fejes, Acta Chim. Hung. 128 (1991) 119.
[40] J.E. Guest, C.D. Williams, Chem. Commun. (2002) 2870.
[41] C. Song, H.H. Schobert, Fuel Process Technol. 34 (1993) 15.
[42] M. Taludger, C.P. Kates, in: J.I. Kroschwitz, M. Howe-Grau (Eds.), Kirkh Othmer Encyclopedia of Chemical Technology, vol. 16, Wiley, New York, 1995, p. 1003.
[43] K. Yano, S. Aizawa, Nippon Mining Co. US Pat., 4,963,248, 1988.
[44] Eur. Chem. News, 1992, Sep. 28, 30.
[45] D.L. Sikkenga, Amoco Co. US Pat., 4,990,717, 1989.
[46] D. Fraenkel, M. Cherniavsky, B. Ittah, M. Levy, J. Catal. 101 (1986) 273.
[47] T. Inui, S.-B. Pu, J. Kugai, Appl. Catal. A 146 (1996) 285.
[48] T. Onodera, T. Sakai, Y. Yamasaki, K. Sumitani, Teijin Petrochem. Ind. US Pat., 4,524,055, 1985.
[49] J. Weitkamp, M. Neuber, W. Holtmann, G. Collin, H. Spengler, Rutgerswerke Aktienges. US Pat., 4,795,847, 1989.
[50] P.J. Angevine, T.F. Degnan, D.O. Marler, Mobil Oil US Pat., 5,001,295, 1991.
[51] M. Motoyuki, K. Yamamoto, A.V. Sapre, J.P. Mc Williams, S.P. Donnelly, Kabushiki Kaisha Kobe Seiko Sho and Mobil Oil US Pat., 6,018,086, 2000; 6,121,501, 2000.
[52] T. Tsutsui, T. Sasaki, Y. Satou, O. Kubota, S. Okada, M. Fujii, Fuji Oil Co. US Pat., 6,204,422, 2001.
[53] G. Pazzuconi, C. Perego, R. Millini, F. Frigerio, R. Mansani, D. Rancati, Enichem SpA US Pat., 6,232,517, 2001.
[54] C. Perego, G. Pazzuconi, R. Mansani, Enichem SpA US Pat., 6,388,158, 2002.
[55] G. Takeuchi, Y. Shimoura, T. Hara, Catal. Lett. 41 (1996) 195.
[56] G. Kamalakar, M.R. Prasad, S.J. Kulkami, S. Narayanan, K.V. Raghavan, Micropor. Mesopor. Mater. 38 (2000) 135.
[57] A. Katayama, M. Toba, G. Takeuchi, F. Mizukami, S. Niwa, S. Mitamura, J. Chem. Soc., Chem. Commun. (1991) 39.
[58] J.A. Horsely, J.D. Fellmann, E.G. Derouane, C.M. Freeman, J. Catal. 147 (1994) 231.
[59] S.-J. Chu, Y.-W. Chen, Appl. Catal. A 123 (1995) 51.
[60] R. Brzozowski, W. Tecza, Appl. Catal. A 166 (1998) 21.
[61] J.D. Fellmann, R.J. Saxton, P.R. Wentrcek, E.G. Derouane, P. Massiani, Catalytica US Pat., 5,026,942, 1991.
[62] P.P.B. Notte, G.M.J.L. Poncelet, M.J.H Remy, P.F.M.G. Lardinois, M.J.M. Van Hoecke, EP Pat., 0 528 096, 1993.
[63] Y. Sugi, T. Matsuzaki, T. Hanaoka, Y. Kubota, J.H. Kim, X. Tu, M. Matsumoto, Stud. Surf. Sci. Catal. 90 (1994) 397.
[64] A.D. Schmitz, C.S. Song, Catal. Today 31 (1996) 19.
[65] C.S. Song, C.R. Acad. Sci. II C 3 (2000) 477.
[66] D. Mravec, J. Chylik, M. Michvocik, M. Hronec, A. Smieskova, P. Hudec, Chem. Pap.-Chem. Zvesti 52 (1998) 218.
[67] A.D. Schmitz, C.S. Song, Catal. Lett. 40 (1996) 59.
[68] I.M. Tseng, J.F. Wu, Y.W. Chen, React. Kinet. Catal. Lett. 63 (1998) 359.
[69] Z. Liu, P. Moreau, F. Fajula, Chem. Commun. (1996) 2653. Z. Liu, P. Moreau, F. Fajula, Appl. Catal. A 159 (1997) 305.
[70] E. Armengol, A. Corma, H. Garcia, J. Proimo, Appl. Catal. A 149 (1997) 411.
[71] K. Smith, S.D. Roberts, Catal. Today 60 (2000) 227.
[72] P. Moreau, Z. Liu, F. Fajula, J. Jofire, Catal. Today 60 (2000) 235.
[73] Y. Sugi, M. Toba, Catal. Today 19 (1994) 187.
[74] R. Brzozowski, W. Skupinski, J. Catal. 210 (2002) 313.
[75] M.G. Cutrufello, I. Ferino, R. Monaci, E. Rombi, V. Solinas, P. Magnoux, M. Guisnet, Appl. Catal. A 241 (2003) 91.
[76] P. Moreau, A. Finiels, P. Geneste, J. Joffre, F. Moreau, J. Solofo, Catal. Today 31 (1996) 11.
[77] C. Song, X. Ma, A.D. Schmitz, H.H. Schobert, Appl. Catal. A 182 (1999) 175.
[78] C. Song, X. Ma, H.H. Schobert, ACS Symp. Ser. 738 (1999) 305.
[79] G. Tasi, I. Pálinkó, F. Mizukami, React. Kinet. Catal. Lett. 74 (2001) 317.
[80] E. Kikuchi, K. Sawada, M. Maeda, T. Matsuda, Stud. Surf. Sci. Catal. 90 (1994) 391.

ELSEVIER

Journal of Molecular Structure (Theochem) 666–667 (2003) 79–87

THEO
CHEM

www.elsevier.com/locate/theochem

The fitting and functional analysis of a double rotor potential energy surface for the *R* and *S* enantiomers of 1-chloro-3-fluoro-isobutane

Tara A.K. Kehoe[a,b,*], Mike R. Peterson[b], Gregory A. Chass[a,b], Bela Viskolcz[c], Laszlo Stacho[d], Imre G. Csizmadia[a,b,e]

[a]*Global Institute of Computational Molecular and Materials Science (GIOCOMMS)@Velocet, 210 Dundas St West, Suite 810, Toronto, Ont., Canada M5G 2E8*
[b]*University of Toronto, 80 St George St, Toronto, Ont., Canada M5S 3H6*
[c]*Department of Chemistry, JGyTFK, University of Szeged, P.O. Box 396, H-6701 Szeged, Hungary*
[d]*Bolyai Institute University of Szeged, Aradi Vertanuk Tere 1, H-6720 Szeged, Hungary*
[e]*Department of Medical Chemistry, University of Szeged, Dom ter 8, H-6720 Szeged, Hungary*

Abstract

A model compound was chosen to see whether it mimics a backbone of an amino acid residue in a peptide structure so that a model Ramachandran potential energy surface could be fitted by a mathematical function. A Fourier series of two independent variables (φ and ψ) has been used to fit a set of grid points representing the surface. To determine the accuracy of the fitted equation vs. the generated data points three grids were examined, $24^2 = 576$ points (15° intervals), $12^2 = 144$ points (30° intervals), and $6^2 = 36$ points (60° intervals). The grid points were generated for the *S* enantiomer and a Fourier expansion was fitted to the grid points along with a functional analysis of each fitted expansion. A series of functions were found for 15, 30, and 60° increments in order to see the lowest limit of resolution of the grid needed for a relatively accurate fit. Ab initio calculations were also carried out for the *R* and *S* enantiomer to fit a 31 term Fourier expansion where a functional analysis determined the location of the critical points from the expansions. Geometry optimizations were preformed to locate more precisely the minima. The optimized minima were then included in a new surface that was fit.
© 2003 Elsevier B.V. All rights reserved.

Keywords: Fourier expansion; Functional analysis; Geometry optimization; Minimum energy conformer; Potential energy surface; Double rotor

1. Introduction

A great impact in medicine will arise when one can determine the rules of protein folding. Since a protein is made up of a sequence of amino acids, one way of attempting to understand the folding process is to individually look at each amino acid residue and locate the potential energy minima of each structure.

* Corresponding author.
E-mail addresses: t.kehoe@utoronto.ca (T.A.K. Kehoe); mikep@onet.on.ca (M.R. Peterson); gchass@giocomms.org (G.A. Chass); viskolcz@jgytf.u-szeged.hu (B. Viskolcz); stacho@math.u-szeged.hu (L. Stacho); icsizmad@chem.utoronto.ca (I.G. Csizmadia).

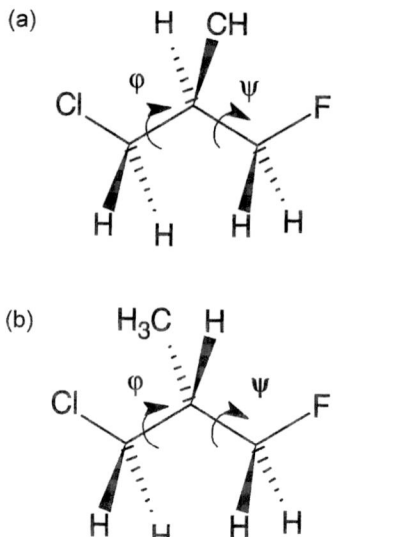

Fig. 1. The fully extended conformation of 1-chloro-3-fluoro-isobutane. (a) *S* absolute configuration. (b) *R* absolute configuration.

The conformation where the potential energy is a minimum would be the approximate conformation each residue would be found in the protein. A gradual process to find the location of potential energy minima would then be preformed to di-peptides, tri-peptides, and so on, to see how each residue is affected by its neighbour. Several questions may be asked. What about the relationships between all of these surfaces? Is there some sort of natural relationship between these surfaces that can be found?

Mathematics has been used for centuries to describe the known and to predict the unknown. Fourier expansions can fit accurately conformational potential energy surfaces and hyper surfaces which describe the surface as a linear combination of trigonometric functions rather than a set of generated data points. This expansion summarizes the surface to a great accuracy, where the location of the minima can be simply found by a functional analysis.

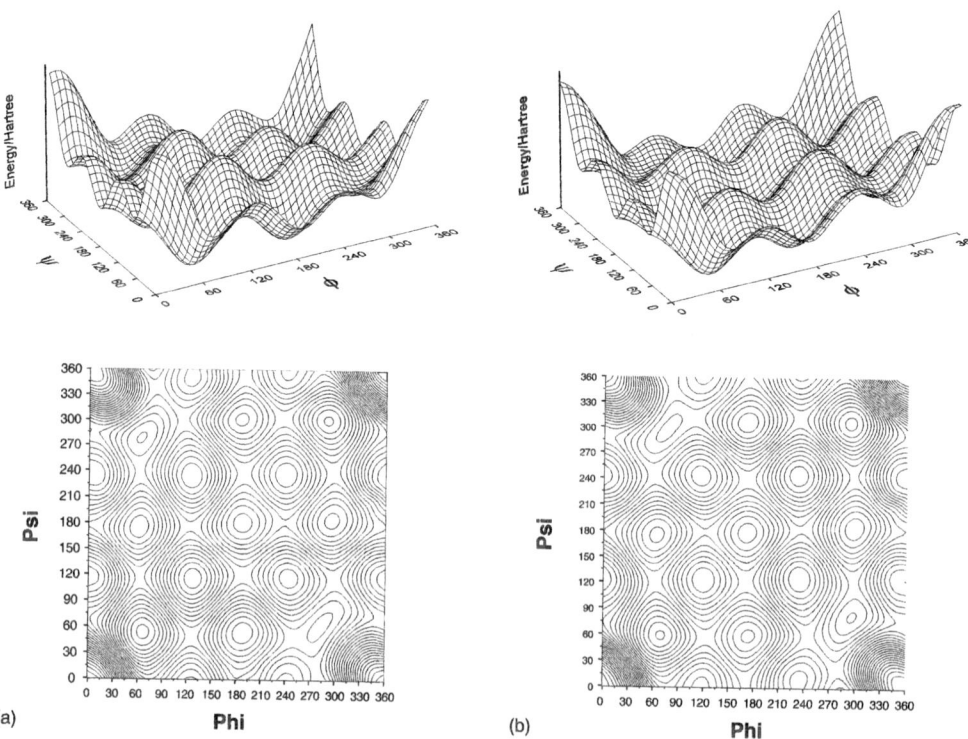

Fig. 2. (a) Ramachandran type PEHS of the *S* configuration model molecule. (Top) Landscape representation. (Bottom) Contour diagram representation. (b) Ramachandran type PEHS of the *R* configuration model molecule. (Top) Landscape representation. (Bottom) Contour diagram representation.

Table 1
Tabulated components of the fitted 28 term Fourier expansion for S-1-chloro-3-fluoro-isobutane

At 15°

−461.30844948	0.64809403 sin ψ	−0.56742157 sin 2ψ	−0.69022701 sin 3ψ	1.32820141 cos ψ	0.35652097 cos 2ψ	4.16135608 cos 3ψ
1.21185436 sin φ	−2.49021582 sin ψ sin φ	−1.40061795 sin 2ψ sin φ	−1.14850552 sin 3ψ sin φ	0 cos ψ sin φ	0 cos 2ψ sin φ	0 cos 3ψ sin φ
−0.87790882 sin 2φ	−1.44918333 sin ψ sin 2φ	−1.25231971 sin 2ψ sin 2φ	−0.82453399 sin 3ψ sin 2φ	0 cos ψ sin 2φ	0 cos 2ψ sin 2φ	0 cos 3ψ sin 2φ
0.64715046 sin 3φ	0 sin ψ sin 3φ	0 sin 2ψ sin 3φ	−0.64472843 sin 3ψ sin 3φ	0 cos ψ sin 3φ	0 cos 2ψ sin 3φ	0 cos 3ψ sin 3φ
2.57596399 cos φ	0 sin ψ cos φ	0 sin 2ψ cos φ	0 sin 3ψ cos φ	4.41334861 cos ψ cos φ	1.80832742 cos 2ψ cos φ	0.53858137 cos 3ψ cos φ
1.74543674 cos 2φ	0 sin ψ cos 2φ	0 sin 2ψ cos 2φ	0 sin 3ψ cos 2φ	1.63413861 cos ψ cos 2φ	1.25253585 cos 2ψ cos 2φ	0.69948369 cos 3ψ cos 2φ
4.52198337 cos 3φ	0 sin ψ cos 3φ	0 sin 2ψ cos 3φ	0 sin 3ψ cos 3φ	0.25352051 cos ψ cos 3φ	0 cos 2ψ cos 3φ	0.47723389 cos 3ψ cos 3φ

At 30°

−461.30986944	0.65080338 sin ψ	−0.56700006 sin 2ψ	−0.68923750 sin 3ψ	1.32753090 cos ψ	0.36438819 cos 2ψ	4.16949583 cos 3ψ
1.20991044 sin φ	−2.49149438 sin ψ sin φ	−1.39762635 sin 2ψ sin φ	−1.13824553 sin 3ψ sin φ	0 cos ψ sin φ	0 cos 2ψ sin φ	0 cos 3ψ sin φ
−0.87788755 sin 2φ	−1.45250963 sin ψ sin 2φ	−1.25563542 sin 2ψ sin 2φ	−0.81786477 sin 3ψ sin 2φ	0 cos ψ sin 2φ	0 cos 2ψ sin 2φ	0 cos 3ψ sin 2φ
0.64848333 sin 3φ	0 sin ψ sin 3φ	0 sin 2ψ sin 3φ	−0.63070556 sin 3ψ sin 3φ	0 cos ψ sin 3φ	0 cos 2ψ sin 3φ	0 cos 3ψ sin 3φ
2.57571291 cos φ	0 sin ψ cos φ	0 sin 2ψ cos φ	0 sin 3ψ cos φ	4.41330725 cos ψ cos φ	1.80465439 cos 2ψ cos φ	0.53186654 cos 3ψ cos φ
1.74622222 cos 2φ	0 sin ψ cos 2φ	0 sin 2ψ cos 2φ	0 sin 3ψ cos 2φ	1.62978494 cos ψ cos 2φ	1.24906458 cos 2ψ cos 2φ	0.70300000 cos 3ψ cos 2φ
4.52506944 cos 3φ	0 sin ψ cos 3φ	0 sin 2ψ cos 3φ	0 sin 3ψ cos 3φ	0.25080956 cos ψ cos 3φ	0 cos 2ψ cos 3φ	0.49203611 cos 3ψ cos 3φ

At 60°

−461.42923611	0.70774964 sin ψ	−0.53799904 sin 2ψ	0 sin 3ψ	1.27475833 cos ψ	0.24865278 cos 2ψ	4.09845833 cos 3ψ
1.10004952 sin φ	−2.59663333 sin ψ sin φ	−1.39825000 sin 2ψ sin φ	0 sin 3ψ sin φ	0 cos ψ sin φ	0 cos 2ψ sin φ	−0.11041343 cos 3ψ sin φ
−0.84856537 sin 2φ	−1.52886667 sin ψ sin 2φ	−1.13351667 sin 2ψ sin 2φ	0 sin 3ψ sin 2φ	0 cos ψ sin 2φ	0 cos 2ψ sin 2φ	−0.05833643 cos 3ψ sin 2φ
0 sin 3φ	0 sin ψ sin 3φ	0 sin 2ψ sin 3φ	0 sin 3ψ sin 3φ	0 cos ψ sin 3φ	0 cos 2ψ sin 3φ	0 cos 3ψ sin 3φ
2.35996389 cos φ	0.05304887 sin ψ cos φ	0 sin 2ψ cos φ	0 sin 3ψ cos φ	4.23794444 cos ψ cos φ	1.63306111 cos 2ψ cos φ	0.56694722 cos 3ψ cos φ
1.47113611 cos 2φ	0 sin ψ cos 2φ	0 sin 2ψ cos 2φ	0 sin 3ψ cos 2φ	1.39283333 cos ψ cos 2φ	1.25217222 cos 2ψ cos 2φ	0.84847500 cos 3ψ cos 2φ
4.48825278 cos 3φ	0 sin ψ cos 3φ	0.06274354 sin 2ψ cos 3φ	0 sin 3ψ cos 3φ	0.27709722 cos ψ cos 3φ	0.23154722 cos 2ψ cos 3φ	0.49203611 cos 3ψ cos 3φ

The objective of this paper is to illustrate a process that can be used to characterize these potential energy surfaces. Thus, rather than looking at the potential energy surface as a set of data points, we now want to look at the surfaces as a mathematical function.

2. Method

2.1. Molecular computations

Grid points at 15° increments were generated for both the S and R enantiomer of 1-chloro-3-fluoro-isobutane (Fig. 1) using Gaussian 98 program [1].

2.2. Multivariable Fourier expansion fitting of potential energy surface

A stepwise regression procedure was performed on grids of 15, 30, and 60° increments to fit a two independent variable (φ, ψ) Fourier expansion to each surface. A functional analysis was then carried out on the three fitted surfaces of 28 function terms (including the constant) to determine the location of the minima from each expansion. Two independent variable (φ, ψ) Fourier expansion of 32 function terms (including the constant) were fit using the same stepwise regression procedure to 30° increment grids for both enantiomers. A functional analysis was performed on both fitted 32 function term expansions.

Table 2
Functional analysis fitted Fourier expansion of the potential energy surface

Conformer	φ	ψ	Energy (Hartree)	ΔE (kcal/mol)
At 15° grid, 28 term expansion for S-1-chloro-3-fluoro-isobutane				
g⁺g⁺	66.05	52.64	−711.4732988	−2.45418471
g⁺a	60.73	173.04	−711.4708271	−0.90322534
g⁺g⁻	60.77	276.25	−711.4655433	2.412464705
a g⁺	189.00	53.40	−711.4719539	−1.61026031
aa	189.07	181.09	−711.4693878	0
ag⁻	185.60	298.65	−711.4715175	−1.33646013
g⁻g⁺	296.19	71.47	−711.4641447	3.29004748
g⁻a	300.63	186.48	−711.4713441	−1.2276204
g⁻g⁻	295.18	299.85	−711.4735714	−2.62525962
At 30° grid, 28 term expansion for S-1-chloro-3-fluoro-isobutane				
g⁺g⁺	66.07	52.68	−711.473291	−2.45082188
g⁺a	60.76	173.08	−711.4708144	−0.89667037
g⁺g⁻	60.86	276.21	−711.4655381	2.414241186
a g⁺	189.06	53.41	−711.471967	−1.61993652
aa	189.09	181.08	−711.4693854	0
ag⁻	185.60	298.62	−711.4715324	−1.34722632
g⁻g⁺	296.07	71.40	−711.4641311	3.297122655
g⁻a	300.61	186.44	−711.4713289	−1.21955815
g⁻g⁻	295.19	299.78	−711.4735709	−2.62642302
At 60° grid, 28 term expansion for S-1-chloro-3-fluoro-isobutane				
g⁺g⁺	62.07	60.98	−711.47239	−1.88591981
g⁺a	58.11	184.91	−711.4709272	0.387429694
g⁺g⁻	62.13	287.75	−711.4630391	3.981891688
a g⁺	186.07	55.00	−711.471737	−1.47612942
aa	185.00	183.24	−711.4693846	0
Ag⁻	182.60	302.19	−711.4715446	−1.35540905
g⁻g⁺	284.17	57.27	−711.463833	3.483676986
g⁻a	297.48	181.97	−711.4715681	−1.37014361
g⁻g⁻	294.64	297.01	−711.4735979	−2.64387157

Table 3
Tabulated components of the fitted 32 term Fourier expansion at 30°

S enantiomer

	sin ψ	sin 2ψ	sin 3ψ	cos ψ	cos 2ψ	cos 3ψ
−461.3098694	−0.65080338 sin ψ	0.56700006 sin 2ψ	0.68923750 sin 3ψ	1.32753090 cos ψ	0.36438819 cos 2ψ	4.16949583 cos 3ψ
−1.20991044 sin φ	−2.49149438 sin ψ sin φ	−1.39762635 sin 2ψ sin φ	−1.13824553 sin 3ψ sin φ	0 cos ψ sin φ	0 cos 2ψ sin φ	0 cos 3ψ sin φ
0.87788755 sin 2φ	−1.45250963 sin ψ sin 2φ	−1.25563542 sin 2ψ sin 2φ	−0.81786477 sin 3ψ sin 2φ	0 cos ψ sin 2φ	0 cos 2ψ sin 2φ	0 cos 3ψ sin 2φ
−0.6484833 sin 3φ	0 sin ψ sin 3φ	−0.20055224 sin 2ψ sin 3φ	−0.63070556 sin 3ψ sin 3φ	0 cos ψ sin 3φ	0 cos 2ψ sin 3φ	−0.16496111 cos 3ψ sin 3φ
2.57571291 cos φ	0 sin ψ cos φ	0 sin 2ψ cos φ	0 sin 3ψ cos φ	4.41330725 cos ψ cos φ	1.80465439 cos 2ψ cos φ	0.53186654 cos 3ψ cos φ
1.74622222 cos 2φ	0 sin ψ cos 2φ	0 sin 2ψ cos 2φ	0 sin 3ψ cos 2φ	1.62978494 cos ψ cos 2φ	1.24906458 cos 2ψ cos 2φ	0.70300000 cos 3ψ cos 2φ
4.52506944 cos 3φ	0 sin ψ cos 3φ	0 sin 2ψ cos 3φ	0.15364722 sin 3ψ cos 3φ	0.25080956 cos ψ cos 3φ	0.20336250 cos 2ψ cos 3φ	0.49203611 cos 3ψ cos 3φ

R enantiomer

	sin ψ	sin 2ψ	sin 3ψ	cos ψ	cos 2ψ	cos 3ψ
−461.3098694	0.65080338 sin ψ	−0.56700006 sin 2ψ	−0.68923750 sin 3ψ	1.32753090 cos ψ	0.36438819 cos 2ψ	4.16949583 cos 3ψ
1.20991044 sin φ	−2.49149438 sin ψ sin φ	−1.39762635 sin 2ψ sin φ	−1.13824553 sin 3ψ sin φ	0 cos ψ sin φ	0 cos 2ψ sin φ	0 cos 3ψ sin φ
−0.87788755 sin 2φ	−1.45250963 sin ψ sin 2φ	−1.25563542 sin 2ψ sin 2φ	−0.81786477 sin 3ψ sin 2φ	0 cos ψ sin 2φ	0 cos 2ψ sin 2φ	0 cos 3ψ sin 2φ
0.64848333 sin 3φ	0 sin ψ sin 3φ	−0.20055224 sin 2ψ sin 3φ	−0.63070556 sin 3ψ sin 3φ	0 cos ψ sin 3φ	0 cos 2ψ sin 3φ	0.16496111 cos 3ψ sin 3φ
2.57571291 cos φ	0 sin ψ cos φ	0 sin 2ψ cos φ	0 sin 3ψ cos φ	4.41330725 cos ψ cos φ	1.80465439 cos 2ψ cos φ	0.53186654 cos 3ψ cos φ
1.74622222 cos 2φ	0 sin ψ cos 2φ	0 sin 2ψ cos 2φ	0 sin 3ψ cos 2φ	1.62978494 cos ψ cos 2φ	1.24906458 cos 2ψ cos 2φ	0.70300000 cos 3ψ cos 2φ
4.52506944 cos 3φ	0 sin ψ cos 3φ	0 sin 2ψ cos 3φ	−0.15364722 sin 3ψ cos 3φ	0.25080956 cos ψ cos 3φ	0.20336250 cos 2ψ cos 3φ	0.49203611 cos 3ψ cos 3φ

2.3. Geometry optimization and subsequence analysis

The nine minima for each enantiomer, S and R, were optimized with the Gaussian 98 program [1] to determine a more precise location of the minima. For each enantiomer, the nine optimized minima were inserted into the corresponding 30° increment surface to obtain a new set of 153 grid points. A Fourier expansion was fitted to each set of 153 grid points followed by a functional analysis. These expansions have 33 function terms (including the constant). All of the stepwise regression procedures and functional analysis have been carried out using software [2,3].

3. Results and discussion

The computed grid points for both the R and S enantiomers are shown graphically in Fig. 2. The following Fourier expansion was the best equation for all the fitted surfaces:

$$E(\varphi, \psi) = c^0 + \sum_{m=0}^{3} \sum_{n=0}^{3} c_{m,n}^1 \cos m\varphi \cos n\psi$$

$$+ c_{m,n}^2 \cos m\varphi \sin n\psi + c_{m,n}^3 \sin m\varphi \cos n\psi$$

$$+ c_{m,n}^4 \sin m\varphi \sin n\psi$$

A systematic stepwise procedure was carried out to fit functions to the potential energy surface of 1-chloro-3-fluoro-isobutane for the S enantiomer at 15, 30, and 60° intervals to see the similarities and differences amongst the equations. The idea is to have the most accurate fit with the least amount of terms. In the regression analysis for the 60° refinement, only the most important 28 terms (including the constant) were kept. In order to compare the 15, 30 and 60° grid refinements 28 term expression had to be used for 15 and 30° refinement to match the limitation of the 60° refinement. The tabulated components of the fitted Fourier expansion for 15, 30, and 60° are found in Table 1. A functional analysis was carried out on the three fitted equations at 15, 30, and 60° increments found in Table 2. The first derivative of the equations was solved for the location of the critical points where nine minima were located for each surface. The Fourier expansion that seemed to be the best choice is found in Table 1.

Table 4
Functional analysis of 32 term fitted Fourier expansion

Conformer	φ	ψ	Energy (Hartree)	ΔE (kcal/mol)
S enantiomer				
g^+g^+	65.06	52.83	−711.4730849	−2.24712084
g^+a	60.34	173.86	−711.4709577	−0.912308551
g^+g^-	61.00	276.78	−711.465195	2.703853994
$a\,g^+$	187.79	53.35	−711.4717184	−1.389646623
aa	187.79	181.31	−711.4695038	0
ag^-	185.94	300.08	−711.4714284	−1.207654918
g^-g^+	291.83	68.10	−711.4641364	3.368130432
g^-a	299.69	186.92	−711.4715779	−1.301497783
g^-g^-	295.05	300.84	−711.4734743	−2.49150712
R enantiomer				
g^+g^+	64.95	59.16	−711.4734743	−2.49150712
g^+a	60.31	173.08	−711.4715779	−1.301497783
g^+g^-	68.17	291.90	−711.4641364	3.368130432
ag^+	174.06	59.92	−711.4714284	−1.207654918
aa	172.21	178.69	−711.4695038	0
ag^-	172.21	306.65	−711.4717184	−1.389646623
g^-g^+	299.00	83.22	−711.465195	2.703853994
g^-a	299.66	186.71	−711.4709577	−0.912308551
g^-g^-	294.94	307.17	−711.4730849	−2.24712084

Table 5
Nine optimized minima

Conformer	φ	ψ	Energy (Hartree)	ΔE (kcal/mol)
S enantiomer				
g^+g^+	66.61	53.69	−711.4729679	−2.160287261
g^+a	61.23	175.43	−711.4709131	−0.870928032
g^+g^-	63.37	278.24	−711.4654592	2.551454405
$a\,g^+$	188.75	53.99	−711.4721491	−1.646524117
aa	187.81	181.41	−711.4695252	0
ag^-	185.88	301.65	−711.4716381	−1.325840151
g^-g^+	283.42	61.97	−711.4645667	3.111510845
g^-a	296.27	184.39	−711.4717223	−1.378718536
g^-g^-	290.98	301.11	−711.4739387	−2.769498442
R enantiomer				
g^+g^+	69.02	58.89	−711.4739387	−2.769498442
g^+a	63.73	175.61	−711.4717223	−1.378718536
g^+g^-	76.58	298.03	−711.4645667	3.111510845
$a\,g^+$	174.12	58.35	−711.4716381	−1.325840151
aa	172.19	178.59	−711.4695252	0
ag^-	171.25	306.01	−711.4721491	−1.646524117
g^-g^+	296.63	81.76	−711.4654592	2.551454405
g^-a	298.77	184.57	−711.4709131	−0.870928032
g^-g^-	293.39	306.31	−711.4729679	−2.160287261

Table 6
Tabulated components of the fitted 33 term Fourier expansion at 30° with the optimized minima included

S enantiomer

-461.3195342	$0.65354330 \sin \psi$	$-0.56688693 \sin 2\psi$	$-0.69250128 \sin 3\psi$	$1.32093506 \cos \psi$	$0.37267006 \cos 2\psi$	$4.18658806 \cos 3\psi$
$1.21766019 \sin \varphi$	-2.48509560 $\sin \psi \sin \varphi$	-1.39506392 $\sin 2\psi \sin \varphi$	-1.14481608 $\sin 3\psi \sin \varphi$	$0 \cos \psi \sin \varphi$	$0 \cos 2\psi \sin \varphi$	-0.11493268 $\cos 3\psi \sin \varphi$
$-0.87603883 \sin 2\varphi$	-1.45079124 $\sin \psi \sin 2\varphi$	-1.25767495 $\sin 2\psi \sin 2\varphi$	-0.82457165 $\sin 3\psi \sin 2\varphi$	$0 \cos \psi \sin 2\varphi$	$0 \cos 2\psi \sin 2\varphi$	$0 \cos 3\psi \sin 2\varphi$
$0.64586732 \sin 3\varphi$	$0 \sin \psi \sin 3\varphi$	-0.19929562 $\sin 2\psi \sin 3\varphi$	-0.62760531 $\sin 3\psi \sin 3\varphi$	$0 \cos \psi \sin 3\varphi$	$0 \cos 2\psi \sin 3\varphi$	0.17101827 $\cos 3\psi \sin 3\varphi$
$2.57723405 \cos \varphi$	$0 \sin \psi \cos \varphi$	$0 \sin 2\psi \cos \varphi$	$0 \sin 3\psi \cos \varphi$	4.41970684 $\cos \psi \cos \varphi$	1.80377986 $\cos 2\psi \cos \varphi$	0.52750387 $\cos 3\psi \cos \varphi$
$1.74986802 \cos 2\varphi$	$0 \sin \psi \cos 2\varphi$	$0 \sin 2\psi \cos 2\varphi$	$0 \sin 3\psi \cos 2\varphi$	1.62744740 $\cos \psi \cos 2\varphi$	1.24472336 $\cos 2\psi \cos 2\varphi$	0.69766075 $\cos 3\psi \cos 2\varphi$
$4.54189872 \cos 3\varphi$	$0 \sin \psi \cos 3\varphi$	$0 \sin 2\psi \cos 3\varphi$	-0.14711437 $\sin 3\psi \cos 3\varphi$	0.26154371 $\cos \psi \cos 3\varphi$	0.18927849 $\cos 2\psi \cos 3\varphi$	0.46272898 $\cos 3\psi \cos 3\varphi$

R enantiomer

-461.3195342	$-0.65354330 \sin \psi$	$0.56688693 \sin 2\psi$	$0.69250128 \sin 3\psi$	$1.32093506 \cos \psi$	$0.37267006 \cos 2\psi$	$4.18658806 \cos 3\psi$
$-1.21766019 \sin \varphi$	-2.48509560 $\sin \psi \sin \varphi$	-1.39506392 $\sin 2\psi \sin \varphi$	-1.14481608 $\sin 3\psi \sin \varphi$	$0 \cos \psi \sin \varphi$	$0 \cos 2\psi \sin \varphi$	0.11493268 $\cos 3\psi \sin \varphi$
$0.87603883 \sin 2\varphi$	-1.45079124 $\sin \psi \sin 2\varphi$	-1.25767495 $\sin 2\psi \sin 2\varphi$	-0.82457165 $\sin 3\psi \sin 2\varphi$	$0 \cos \psi \sin 2\varphi$	$0 \cos 2\psi \sin 2\varphi$	$0 \cos 3\psi \sin 2\varphi$
$-0.64586732 \sin 3\varphi$	$0 \sin \psi \sin 3\varphi$	-0.19929562 $\sin 2\psi \sin 3\varphi$	-0.62760531 $\sin 3\psi \sin 3\varphi$	$0 \cos \psi \sin 3\varphi$	$0 \cos 2\psi \sin 3\varphi$	-0.17101827 $\cos 3\psi \sin 3\varphi$
$2.57723405 \cos \varphi$	$0 \sin \psi \cos \varphi$	$0 \sin 2\psi \cos \varphi$	$0 \sin 3\psi \cos \varphi$	4.41970684 $\cos \psi \cos \varphi$	1.80377986 $\cos 2\psi \cos \varphi$	0.52750387 $\cos 3\psi \cos \varphi$
$1.74986802 \cos 2\varphi$	$0 \sin \psi \cos 2\varphi$	$0 \sin 2\psi \cos 2\varphi$	$0 \sin 3\psi \cos 2\varphi$	1.62744740 $\cos \psi \cos 2\varphi$	1.24472336 $\cos 2\psi \cos 2\varphi$	0.69766075 $\cos 3\psi \cos 2\varphi$
$4.54189872 \cos 3\varphi$	$0 \sin \psi \cos 3\varphi$	$0 \sin 2\psi \cos 3\varphi$	0.14711437 $\sin 3\psi \cos 3\varphi$	0.26154371 $\cos \psi \cos 3\varphi$	0.18927849 $\cos 2\psi \cos 3\varphi$	0.46272898 $\cos 3\psi \cos 3\varphi$

The 30° grid was chosen to be the best comprise.

When the S and R isomers were studied in the 30° refinement it was possible to use a more extensive expansion as was used before. In the present case the regression analysis at the 30° refinement allowed a 32-term expansion (including the constant). The tabulated components for the 32 term Fourier expansion are shown in Table 3. There is a sign change between the R and S for the odd function terms, while there is no sign change for the even function terms. The Functional analysis for the 32 expansions can be found in Table 4.

Gaussian gradient optimizations were performed at the minimum energy points of this function and the results are compared to the minima values found from the fitted function (Table 5). The nine optimized structures were added to the 30° refinement the regression analysis allowed the inclusion of 33 terms (including the constant). The tabulated components of the fitted 33 terms Fourier expansion for both the S and R enantiomer is found in Table 6. A functional analysis was preformed on the 33 term function for both enantiomers found on Table 7.

Table 7
Functional Analysis of 33 term fitted Fourier expansion

Conformer	φ	ψ	Energy (Hartree)	ΔE (kcal/mol)
S enantiomer				
g^+g^+	64.98	52.73	−711.4730625	−2.210913513
g^+a	60.23	173.67	−711.4708962	−0.851530442
g^+g^-	60.90	276.36	−711.4652511	2.690820611
$a\,g^+$	187.86	53.41	−711.4718174	−1.429563162
aa	187.91	181.26	−711.4695392	0
ag^-	185.99	300.01	−711.4715289	−1.248559785
g^-g^+	290.88	66.97	−711.4643434	3.26039073
g^-a	299.52	186.66	−711.4717155	−1.365678878
g^-g^-	294.99	300.79	−711.4736778	−2.596983393
R enantiomer				
g^+g^+	65.01	59.21	−711.4736778	−2.596983393
g^+a	60.48	173.34	−711.4717155	−1.365678878
g^+g^-	69.12	293.03	−711.4643434	3.26039073
$a\,g^+$	174.01	59.99	−711.4715289	−1.248559785
aa	172.09	178.74	−711.4695392	0
ag^-	172.14	306.59	−711.4718174	−1.429563162
g^-g^+	299.11	83.64	−711.4652511	2.690820611
g^-a	299.77	186.33	−711.4708962	−0.851530442
g^-g^-	295.02	307.27	−711.4730625	−2.210913513

The accuracy of the φ and ψ values obtained from the functional analysis of the Fourier expansion can be assessed when compared to the φ and ψ values obtained be direct optimization. The comparison of

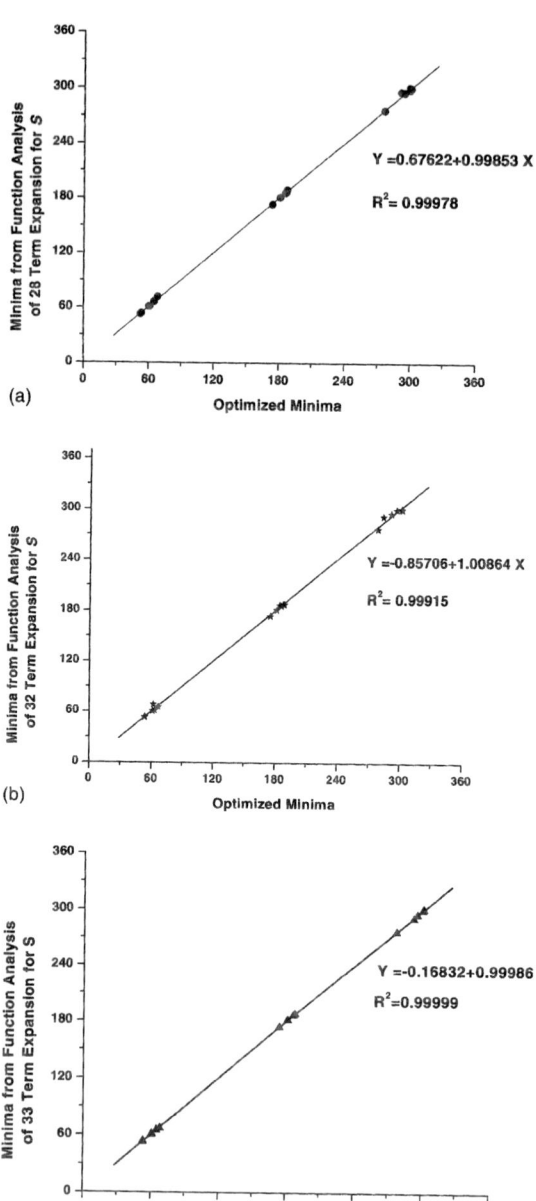

Fig. 3. (a) Minima from Functional Analysis of 27 term expansion for *S* vs. optimized minima. (b) Minima from Functional Analysis of 31 term expansion for *S* vs. optimized minima. (c) Minima from Functional Analysis of 32 term expansion for *S* vs. optimized minima.

the data presented at Table 2 against Table 5 as well on Table 4 against Table 5 and Table 7 against Table 5 are shown graphically in Fig. 3.

4. Conclusions

It appears that 30° resolution of grid is satisfactory to determine the location of the minima (φ, ψ) with respect to the optimized structures with reasonably degree accuracy (i.e. $R^2 = 0.99637582$). It seems that the inclusion of the optimized minima in the generated grid with not improve dramatically the results obtained by the functional analysis.

Acknowledgements

One of us (IGC) would like to thank the Ministry of Education for a Szent Györgyi Visiting Professorship. B.V. is grateful for the Hungarian Scientific Research Found (OTKA F-037648).

References

[1] M.J. Frisch, G.W. Trucks, H.B. Schlegel, G.E. Scuseria, M.A. Robb, J.R. Cheeseman, V.G. Zakrzewski, J.A. Montgomery Jr., R.E. Stratmann, J.C. Burant, S. Dapprich, J.M. Millam, A.D. Daniels, K.N. Kudin, M.C. Strain, O. Farkas, J. Tomasi, V. Barone, M. Cossi, R. Cammi, B. Mennucci, C. Pomelli, C. Adamo, S. Clifford, J. Ochterski, G.A. Petersson, P.Y. Ayala, Q. Cui, K. Morokuma, P. Salvador, J.J. Dannenberg, D.K. Malick, A.D. Rabuck, K. Raghavachari, J.B. Foresman, J. Cioslowski, J.V. Ortiz, A.G. Baboul, B.B. Stefanov, G. Liu, A. Liashenko, P. Piskorz, I. Komaromi, R. Gomperts, R.L. Martin, D.J. Fox, T. Keith, M.A. Al-Laham, C.Y. Peng, A. Nanayakkara, M. Challacombe, P.M.W. Gill, B. Johnson, W. Chen, M.W. Wong, J.L. Andres, C. Gonzalez, M. Head-Gordon, E.S. Replogle, J.A. Pople, Gaussian 98, Revision A.9, Gaussian, Inc., Pittsburgh PA, 2001.
[2] M.R. Peterson, I.G. Csizmadia, J. Am. Chem. Soc. 100 (1978) 6911–6916.
[3] M.R. Peterson, I.G. Csizmadia, in: I.G. Csizmadia (Ed.), Analytic Equations for Conformational Energy Surfaces in Progress in Theoretical Organic Chemistry, vol. 3, Elsevier, New York, 1982.

Journal of Molecular Structure (Theochem) 666–667 (2003) 89–94

THEO
CHEM

www.elsevier.com/locate/theochem

An exploratory ab initio study on the entropy of various backbone conformers for the HCO-Gly-Gly-Gly-NH$_2$ tripeptide motif

Salvatore J. Salpietro[a,*], Béla Viskolcz[b], Imre G. Csizmadia[a,c]

[a]Department of Chemistry, University of Toronto, Toronto, Ont., Canada M5S 3H6
[b]Department of Chemistry, Juhász Gyula Teacher's Training College, University of Szeged, H-6701, P.O. Box 396, Szeged, Hungary
[c]Department of Medical Chemistry, Szeged University, H-6270 Szeged, Dom ter 8, Hungary

Abstract

Ab initio molecular computations were carried out on the HCO-Gly-Gly-Gly-NH$_2$ tripeptide at the RHF/6-311G(d,p) level of theory. The two terminal glycine moieties were kept in the β_L conformations while the conformation of the central glycine was varied. All five minima of the central glycine residue were optimized and in addition to electronic energy (E), the key thermodynamic functions: enthalpy (H), Gibbs-free energy (G) and entropy (S) were obtained. Cross-sections along each of ϕ and ψ passing through the fully extended β_L conformations, were generated from the Ramachandran Potential Energy Surface and the role of entropy change (ΔS) in the folding process is assessed.
© 2003 Elsevier B.V. All rights reserved.

Keywords: Glycine; Tripeptide; Entropy; Gibbs Free Energy; Ab initio; Ramachandran Cross-section; HCO-Gly-Gly-Gly-NH$_2$

As of yet, the process of protein folding is not fully understood. A typical protein has such a multitude of possible conformations that it can occupy through the simple counting argument of Levinthal [1] it was demonstrated that there are too many conformations to permit a random search in the whole conformational space of the protein structure in a timely fashion. Thus, it may be that the conformational search must be restricted in some way as to direct folding to a native conformation which is both stable and kinetically accessible. Furthermore, many purified protein may spontaneously refold in vitro after being denaturized as was shown by experiments performed by Anfinsen [2]. This reversible phenomenon, in the absence of any catalytic biomolecules, suggests that the folding of a protein is governed by the laws of thermodynamics.

In addition to Gibbs free energy change (ΔG), a possible thermodynamic driving force which may direct the unfolded structure along a folding pathway to the bioactive structure is the entropy (S). In 1944 Erwin Schrödinger [3] suggested that living organisms appear to maintain their ordered structure and avoid falling into a state of thermodynamic equilibrium or 'maximum entropy' by feeding upon the negative entropy change ($\Delta S < 0$). It may be the the path that protein molecules take towards their native or bioactive conformation may be the path which minimizes the overall entropy of the system and brings it to an 'ordered' state [4]. Thus, it is the objective of the current study to begin the steps towards determining through ab initio calculation, if

---

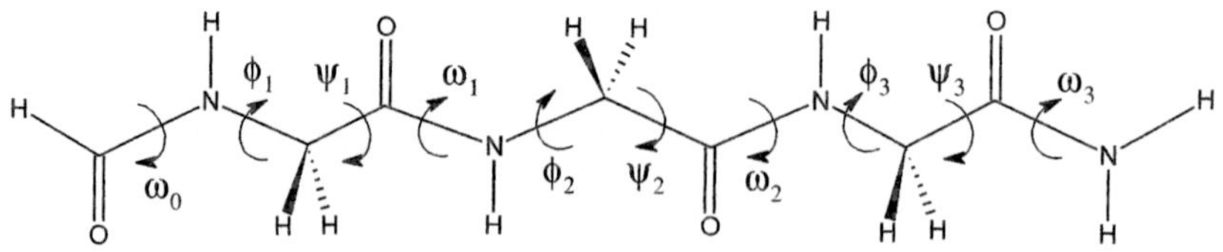

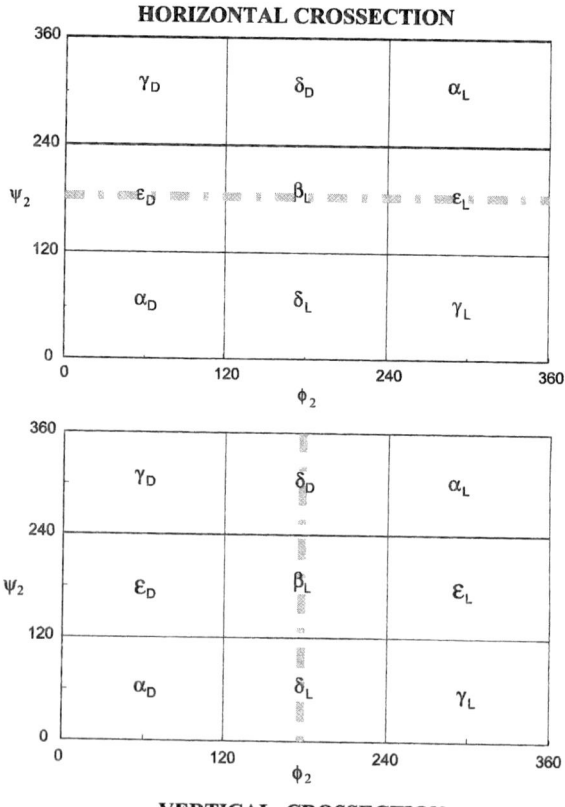

Fig. 1. Schematic diagram with torsional angle definitions for HCO-Gly-Gly-Gly-NH$_2$

entropy may play a role in determining the kinetic pathways towards the native structure of complex bioactive molecules.

The extent of perturbation by neighboring amino acids on the conformations of a given peptide has not been fully studied in all of its completion. For example, one may wonder how different are the Ramachandran Potential Energy Surfaces (PES) for the following glycine moieties in structures **I**, **II**, **III** and **IV**.

HCO-**Gly**-NH$_2$ **I**

MeCO-**Gly**-NHMe   **II**

HCO-Gly-**Gly**-Gly-NH$_2$ **III**

MeCO-Gly-**Gly**-Gly-NHMe   **IV**

Such questions have been explored recently [5] in a study carried out on compound **IV**.

In the current study, ab initio calculations were carried out on the HCO-Gly-Gly-Gly-NH$_2$ [**III**] tripeptide shown in Fig. 1. at the RHF/6-311G(d,p) level of theory using the GAUSSIAN 98 program[6]. For input generation the standard modular numbering systems, developed recently [7,8], has been used.

Minimum energy conformers were obtained using TIGHT optimization where all minima had gradients less than 1.5×10^{-5} a.u. PES scan calculations were carried out for the two cross-sections as shown in Fig. 2 using regular convergence criteria with gradients less than 4.5×10^{-4} a.u. Frequencies were then determined on all constrained geometries from the PES scan as well as on fully optimized geometries from which values for enthalpy (H), Gibbs-free energy (G), and entropy (S) were obtained.

All internal degrees of freedom ($3n - 6$) were calculated by harmonic oscillator approximation. The choice of the number of internal coordinates was the essential question to obtain the correct thermodynamic functions since entropy is changing through internal rotation. We neglected in all

structures those normal vibrations which belonged to the torsional motion studied in this work. The two rotating moieties were systematically changed (along φ and ψ); the corresponding pair of vibrations were therefore neglected.

These low energy vibrations, close to the local minima, have only minor contribution to the ZPE, and enthalpy, but their entropy contributions were the largest components of the total vibrational entropy.

If the rotational reaction path was followed, these vibrations decreased, to zero at the inflexion point,

Fig. 2. One-dimensional PES for HCO-Gly-Gly-Gly-NH$_2$ across the cross section shown. All values are relative to the $\beta_L\beta_L\beta_L$ geometry on the PES.

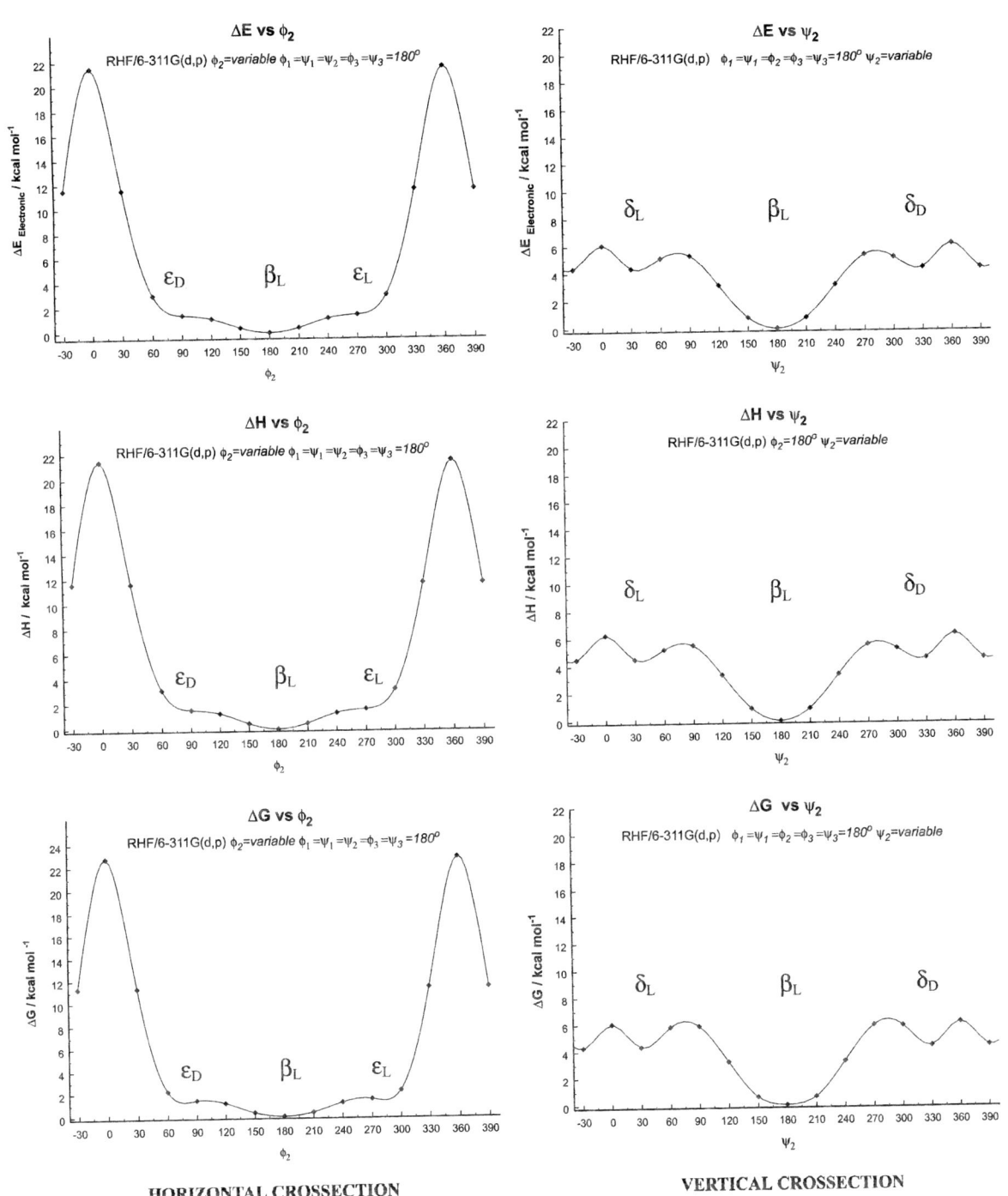

HORIZONTAL CROSSECTION   **VERTICAL CROSSECTION**

Fig. 3. One-dimensional PES for HCO-Gly-Gly-Gly-NH$_2$ across the cross section shown. All values are relative to the $\beta_L\beta_L\beta_L$ geometry on the PES.

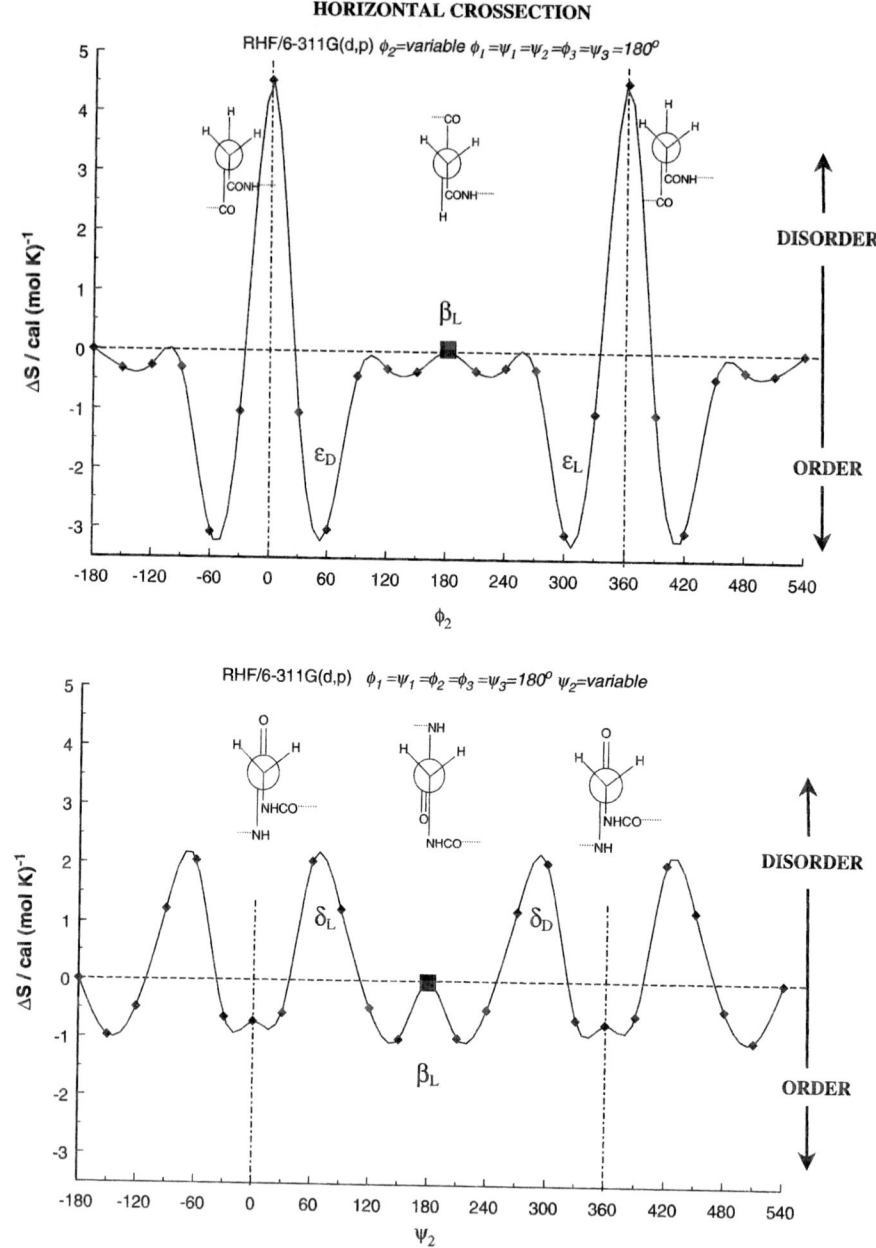

Fig. 4. A schematic illustration of Dipole–Dipole interactions in distinctly different conformations at the central glycine residue in HCO-Gly-Gly-Gly-NH$_2$

after that they become, as usual, imaginary frequencies. Close to the inflection point the entropy function becomes discontinuous. The exact way to get the correct normal coordinate analysis is that this

vibrational mode has to be neglected in frequency calculations.

The location of the two PES scan calculations along φ and ψ are shown in Fig. 2.

Table 1

Optimized geometries for HCO-Gly-Gly-Gly-NH$_2$ at the RHF/6-311g(d,p) level of theory

Backbone conformation	ω_0	Gly1			Gly2			Gly3		
		ϕ_1	ψ_1	ω_1	ϕ_2	ψ_2	ω_2	ϕ_3	ψ_3	ω_3
$\beta_L\alpha_D\beta_L$	Not found									
$\beta_L\epsilon_D\beta_L$	Not found									
$\beta_L\gamma_D\beta_L$	180,061	182,030	174,897	174,829	90,409	-62,723	177,607	186,840	178,604	181,499
$\beta_L\delta_L\beta_L$	181,101	184,836	179,337	174,490	-108,009	7688	181,998	177,667	181,568	177,846
$\beta_L\beta_L\beta_L$	180,000	180,000	180,000	180,000	180,000	180,000	180,000	180,000	180,000	179,986
$\beta_L\delta_D\beta_L$	178,905	175,185	180,658	185,51	108,023	-7697	177,998	182,317	178,426	182,194
$\beta_L\gamma_L\beta_L$	179,940	177,972	185,100	185,172	-90,4093	62,738	182,389	173,177	181,397	178,467
$\beta_L\epsilon_L\beta_L$	Not found									
$\beta_L\alpha_L\beta_L$	Not found									

Table 2

Electronic energy, enthalpy, Gibbs-free energy, and entropy values for the optimized geometries of HCO-Gly-Gly-Gly-NH$_2$ at the RHF/6-311g(d,p) level of theory

Backbone conformation	Electronic energy (HF)	ΔE (kcal/mol)	Enthalpy (HF)	ΔH (kcal/mol)	Gibbs-free energy (HF)	ΔG (kcal/mol)	Entropy (cal/mol K)	ΔS (cal/mol K)
$\beta_L\gamma_D\beta_L$	-789.5893755	2.885	-789.343725	3.049	-789.410607	2.480	140.766	1.910
$\beta_L\delta_L\beta_L$	-789.5905758	2.132	-789.344993	2.253	-789.410742	2.395	138.381	-0.475
$\beta_L\beta_L\beta_L$	-789.5939726	0.000	-789.348584	0.000	-789.414559	0.000	138.856	0.000
$\beta_L\delta_D\beta_L$	-789.5905758	2.132	-789.344992	2.254	-789.410739	2.397	138.375	-0.481
$\beta_L\gamma_L\beta_L$	-789.5893755	2.885	-789.343725	3.049	-789.410607	2.480	140.765	1.909

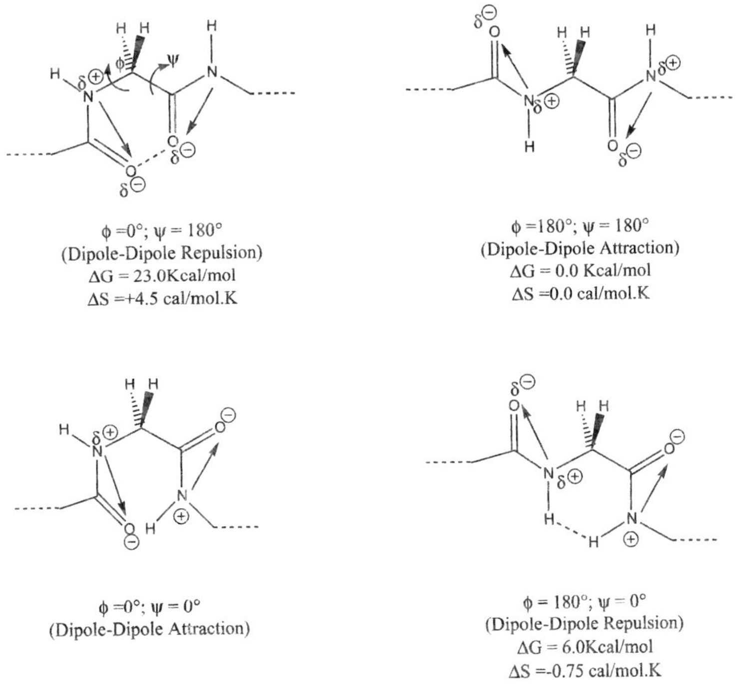

ϕ =0°; ψ = 180°
(Dipole-Dipole Repulsion)
ΔG = 23.0 Kcal/mol
ΔS = +4.5 cal/mol.K

ϕ =180°; ψ = 180°
(Dipole-Dipole Attraction)
ΔG = 0.0 Kcal/mol
ΔS = 0.0 cal/mol.K

ϕ =0°; ψ = 0°
(Dipole-Dipole Attraction)

ϕ = 180°; ψ = 0°
(Dipole-Dipole Repulsion)
ΔG = 6.0 Kcal/mol
ΔS = -0.75 cal/mol.K

Fig. 5. A schematic illustration of Dipole–Dipole Repulsion and Attraction Interactions for certain conformers of amino acid diamides. Gibbs-free energy (G) and entropy (S) are also included for comparison

The resultant thermodynamic curves $(E, H, S$ and $G)$ of the conformational change are shown in Fig. 3.

Generally, the topology of the electronic energy, enthalpy and free energy functions are similar in both cross-sections. However, noticeable difference exist in terms of energetics in the $-60°$ to $+60°$ as well as in the $300–420°$ range.

The two entropy curves (Fig. 4) are completely different from each other, thus they are more characteristic, for each of the cross-sections (Fig. 2). However, both entropy curves have local maxima at $180°$ while at $180°$ the ΔE, ΔH and ΔG functions exhibit global minima.

Optimized dihedral angles and the computed thermodynamic functions are tabulated in Tables 1 and 2, respectively.

The dipole–dipole repulsion plays a role in the destabilization of certain conformers of amino acid diamides as was noted earlier [9]. Such interactions are illustrated in Fig. 5.

It is interesting to note in comparing Figs. 3 and 4 that the ϵ_L and ϵ_D pair behave differently from the δ_L and δ_D pair. The δ pair represents energetic minima on the ΔE, ΔH and ΔG curves while on the ΔS curve it corresponds to a maximum in the 'disorder' range. In contrast the ϵ pair exhibits ΔS minima in the 'order' range yet it is $4–5$ kcal mol^{-1} higher in local minima over the β_L conformation.

Acknowledgements

The authors are grateful for the generous allocation of CPU time granted by the Frederick Biochemical Super Computing Centre at the National Cancer Institute (NCI). One of us (IGC) would like to thank the Ministry of Education for a Szent Györgyi Visiting Professorship. One of us (B.V.) is wishing to acknowledge gratefully the Hungarian Scientific Research Found (OTKA) for a grant, number F-037648.

References

[1] C.J. Levinthal, Chim. Phys. 65 (1968) 44.

[2] C.B. Anfinsen, Science 181 (1973) 223.

[3] E. Schrödinger, What is Life? (1944) 73.

[4] J.N. Onuchic, P.G. Wolynes, Z. Luthey-Schulten, N.D. Socci, Proc. Natl Acad. Sci. 92 (1995) 3626.

[5] A. Mehdizadeh, G.A. Chass, Ö. Farkas, A. Perczel, L.L. Torday, A. Varro, G. Papp, J. Mol. Struct. (THEOCHEM) 588 (2002) 187–200.

[6] M.J. Frisch, G.W. Trucks, H.B.Schlegel, G.E. Scuseria, M.A. Robb, J.R. Cheeseman, V.G. Zakrzewski, J.A. Montgomery, Jr., R.E. Stratmann, J.C. Burant, S. Dapprich, J.M. Millam, A.D. Daniels, K.N. Kudin, M.C. Strain, Ö. Farkas, J. Tomasi, V. Barone, M. Cossi, R. Cammi, B. Mennucci, C. Pomelli, C. Adamo, S. Clifford, J. Ochterski, G.A. Petersson, P.Y. Ayala, Q. Cui, K. Morokuma, P. Salvador, J.J. Dannenberg, D.K. Malick, A.D. Rabuck, K. Raghavachari, J.B. Foresman, J. Cioslowski, J.V. Ortiz, A.G. Baboul, B.B. Stefanov, G. Liu, A. Liashenko, P. Piskorz, I. Komaromi, R. Gomperts, R.L. Martin, D.J. Fox, T. Keith, M.A. Al-Laham, C.Y. Peng, A. Nanayakkara, M. Challacombe, P.M.W. Gill, B. Johnson, W. Chen, M.W. Wong, J.L. Andres, C. Gonzalez, M. Head-Gordon, E.S. Replogle, and J.A. Pople, GAUSSIAN 98, Revision A.9, Gaussian, Inc., Pittsburgh PA, 2001.

[7] G.A. Chass, M.A. Sahai, J.M.S. Law, S. Lovas, Ö. Farkes, A. Perczel, J.-L. Rivail, I.G. Csizmadia, Int. J. Quantum Chem. 90 (2002) 933–968.

[8] M.A. Sahai, S. Lovas, G.A. Chass, B. Penke and I.G. Csizmadia, J. Mol. Struct. (THEOCHEM) (in this issue).

[9] A. Perczel, J.G. Angyan, M. Kajtar, W. Viviani, J.L. Rivail, J.F. Marcoccia, G. Csizmadia, J. Am. Chem. Soc. 113 (1991) 6256–6265.

ELSEVIER

Journal of Molecular Structure (Theochem) 666–667 (2003) 95–97

THEO
CHEM

www.elsevier.com/locate/theochem

Comparison of the extent of hydrogen bonding in H_2O-H_2O and H_2O-CH_4 systems

Cornelia Kozmutza*, Imre Varga, László Udvardi

Elméleti Fizika Tanszék, Fizikai Intézet, Budapesti Műszaki és Gazdaságtudományi Egyetem, Budafoki u. 8, Budapest H-1111, Hungary

Abstract

The displacement of centroids of charge vectors is calculated for localized molecular orbitals for water–water and methane–water complexes. The nature of hydrogen bonding is analyzed in both cases. The centroids of the acceptor-lone pair and the donor-bond pair are all shifted, so they seem to significantly contribute to the H-bond formation in the water–water system. On the other hand, the charge distribution in the acceptor water molecule of the methane–water systems does not contribute substantially to the strengthening of the H-bond, therefore we experience a weaker H-bond in that case.
© 2003 Elsevier B.V. All rights reserved.

PACS: 34.20.Gj; 31.15.Ne; 87.15.Aa

Keywords: Hydrogen bonding; Centroids; Localized orbitals

1. Introduction

Weak interactions, hydrogen bonding in particular, play an important role in determining the structure of biological systems [1]. Directionality is an important property of hydrogen bonds because it influences molecular conformations. It is included in some definitions of the hydrogen bond and has been used to distinguish them from van der Waals interactions [2].

The formation of the hydrogen bond in water is a widely known fact and it has been thoroughly investigated over the past decades [3]. H-bonds result from the approach of proton donor molecule towards an acceptor, forming a bridge of the sort $A-H\cdots B$. The donor atom A is thought to be very

electronegative, e.g. O or N, so is the acceptor atom B, which must also contain at least one lone pair of electrons by which it forms the bridge.

Although carbon is not particularly electronegative, recent experiments and theoretical studies suggest [4,5] that the $C-H\cdots O$ interaction is indeed a hydrogen bond, of course weaker than in the case of water. This interaction bears an exceptional importance in biological systems of nucleic acids, proteins, and carbohydrates [5]. This is the reason why we have chosen also the water–methane system in order to investigate the formation and extent of weak hydrogen bonds. As a comparison we have also used our method to study the strong hydrogen bond in the water–water system.

Methodological details for the treatment of weakly bound systems are explained by Lenthe et al. [6]. Both the different tools accounting for electron correlation and the issue of basis set superposition error are

* Corresponding author.
E-mail address: kozmutza@phy.bme.hu (C. Kozmutza).

0166-1280/$ - see front matter © 2003 Elsevier B.V. All rights reserved.
doi:10.1016/j.theochem.2003.08.017

discussed in that work. The method used in the present work is described in Section 2 where we present our results as well.

2. Method and results

The central diagnostic tool for the existence and the extent of hydrogen bonds is the magnitude of the shift of the centroids of the molecular orbitals. The centroids of the localized molecular orbitals (LMOs), within Boy's localization procedure, are conventionally measured from the heavy nucleus of the lone pair or the bond pair. However, the shift of this quantity for different molecular separations is independent of this choice. The quantity $|\delta\langle\mathbf{r}\rangle|$ has been normalized to the 'length' of the monomer case in the separated molecular orbital scheme (SMO), $|\langle\mathbf{r}\rangle|$. The SMO is an appropriately extended version of the LMO forming LMOs in the supermolecule that resembles the localization properties of the orbitals in the separated monomers [7,8]. In fact these quantities measure the change in the dipole moment of the corresponding molecular orbitals containing electrons of each MO.

For that purpose the software package MONSTERGAUSS with Boy's localization has been used to obtain the molecular orbitals in the SMO scheme. The geometry optimization was performed using GAUSSIAN-98. Two sets of geometries have been used: an optimized structure generated at, e.g. MP2 level of theory and an educated guess, like

a conventional input geometry for Gaussian optimizations. With the aid of these two extremes of accuracy we wished to assess the geometric sensitivity of the method. These geometries for the water–water and the methane–water system are presented in Fig. 1.

Furthermore one may also examine the $\langle x \rangle$, $\langle y \rangle$, $\langle z \rangle$ components of the bond moment vector $\langle\mathbf{r}\rangle$. In constructing the molecular orbitals we have chosen a minimal (MINI) basis.

After normalizing to the case of the monomers we have plotted the lengths of the shifts of the centroids measured from the heavy atoms (O or C, respectively) in Figs. 2 and 3. The decrease of the O–O and the C–O distances perturb some MOs much more than others. Also we may see both increases and decreases of the shifts (red-shifting and blue-shifting).

The two systems were chosen to present the cases of strong hydrogen bond (water–water) and weak hydrogen bond (methane–water). Much larger donor effect is expected for the O–H···O case as compared to the C–H···O case. While the trends are very similar for both cases, in the latter case the shift of the donor-bond pair is only a factor of 2 smaller. On the other hand, the acceptor nature of water within the interaction with methane is much smaller as compared to the interaction with the donor water. As for the case of weak H-bond, however, its centroid is shifted on the cost of other C–H orbitals, meanwhile there is slight but positive shift for the lone pair of the O, too.

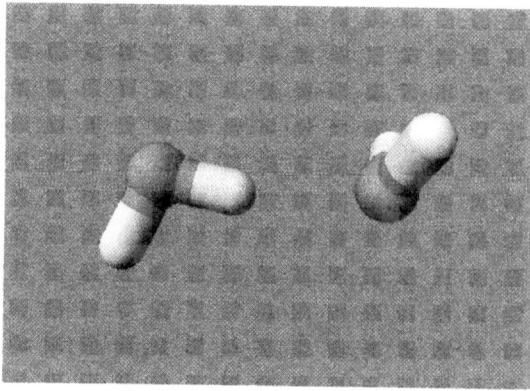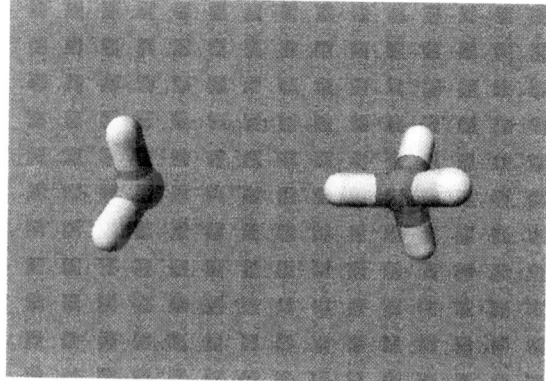

Fig. 1. Two water molecules at an O–O distance of 5.5 a.u. (left panel) and a methane and a water molecule at an O–C distance of 6.8 a.u. (right panel).

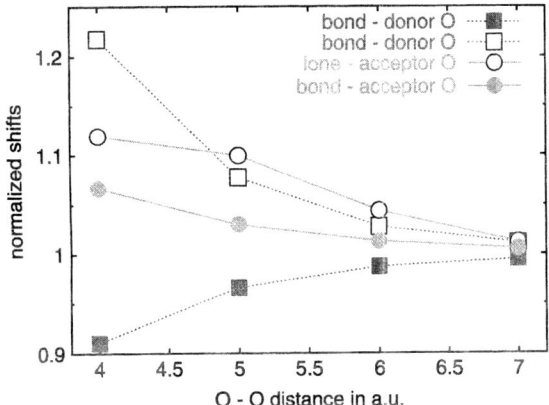

Fig. 2. The distance of the centroids of molecular orbitals measured from the heavy atom (O) in the bond normalized to the value of the monomers. Symbols connected by red (blue) lines (in web version) stand for the molecule on the right (left) hand side depicted on the left panel of Fig. 1.

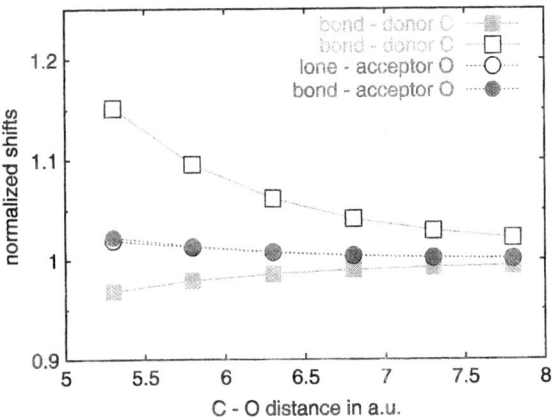

Fig. 3. The distance of the centroids of molecular orbitals measured from the heavy atoms (C and O) in the bond normalized to the value obtained for the monomers. Symbols connected by red (blue) lines (in web version) stand for the methane (water) molecule depicted on the right panel of Fig. 1.

3. Conclusion

In this study, we have shown that under the formation of water–water and methane–water systems, the displacement of the SMOs' centroids for the proton donor molecules significantly move toward the central nucleus (O and C, respectively). However, there appears a remarkable difference in the shift of SMOs' centroids, calculated for the proton acceptors.

The centroids of the two bonds and one of the lone pair are shifted to *promote* the strengthening of the O$\cdots$H bond in the water dimer. On the other hand, the charge distribution in the acceptor molecule of the methane–water system, does not contribute to the elongation of the C$\cdots$H bond. The present results thus suggest—considering the *screening* as the presence of the water molecule as proton acceptor—a shift-dependence on molecular charges of proton donor system. Our method can be very easily extended to molecules with biological importance and apart from the energetic circumstances provides an easy tool to investigate and predict the molecular interactions of nucleic acids, proteins, and carbohydrates attributed to the formation of hydrogen bonds.

We have performed the same calculation for the case of optimized geometries as well as using much more sophisticated basis as well. We found qualitatively no differences, only slight quantitative deviation is detected especially in the case of small separation.

Acknowledgements

This work was partially supported by OTKA (Hungarian Research Fund) under T034832 and T042981 and the Ministry of Education under OM-TST 01327/2002. I.V. is a Bolyai fellow.

References

[1] G.A. Jeffrey, W. Saenger, Hydrogen Bonding in Biological Structures, Springer, Berlin, 1991.
[2] G.R. Desiraju, Th. Steiner, The Weak Hydrogen Bonding in Structural Chemistry and Biology, Oxford University Press, Oxford, 1999.
[3] S. Scheiner, Hydrogen Bonding: A Theoretical Perspective, Oxford University Press, Oxford, 1997, and references therein.
[4] L. Turi, J.J. Dannenberg, J. Phys. Chem. 97 (1993) 7899.
[5] Y. Gu, T. Kar, S. Scheiner, J. Am. Chem. Soc. 121 (1999) 9411 and references therein.
[6] J.H. van Lenthe, J.G.C.M. van Duijneveldt-van de Rijdt, F.B. van Duijneveldt, in: K.P. Lawley (Ed.), Ab Initio Methods in Quantum Chemistry—II, Wiley, New York, 1987, p. 521.
[7] E. Kapuy, C. Kozmutza, Chem. Phys. Lett. 226 (1994) 484.
[8] C. Kozmutza, E. Tfirst, Adv. Quantum Chem. 31 (1998) 231. C. Kozmutza, E. Tfirst, I.G. Csizmadia, Adv. Quantum Chem. 40 (2001) 49 and references therein.

ELSEVIER

Journal of Molecular Structure (Theochem) 666–667 (2003) 99–107

THEO
CHEM

www.elsevier.com/locate/theochem

Intermolecular interactions of small biologically active molecules: acetone, methylamine and water; methyl phosphate, water and divalent ions; phenol and water; N-Ac-L-Gly-NH-Me and water

Maryam Yeganegi[a], Dar'ya Pylypenko[a], Anders Hon[a], Chris Choi[a], Zsolt Zsoldos[b], Gregory A. Chass[a,c,d,e,*], Imre G. Csizmadia[a,c,f]

[a]Department of Chemistry, University of Toronto, 80 St George St, Toronto, Ont., Canada M5S 3H6
[b]Simulated Biomolecular Systems Inc. 135 Queens's Plate Dr, Unit 355, Toronto, Ont., Canada M9W 6V1
[c]Global Institute of Computational Molecular and Materials Science (GIOCOMMS), 1422 Edenrose St., Mississauga, Ont., Canada L5V 1H3
[d]Institut de Science et d'Ingenierie Supramoleculaires 8, allee Gaspard Monge, BP 70028, Strasbourg Cedex 67083, France
[e]Department of Biomedical Sciences, Creighton University, 2500 California plaza, Omaha, NE 68178, USA
[f]Department of Medical Chemistry, Szeged University, H-6720 Szeged, Dom ter 8, Hungary

Abstract

By using quantum mechanical methods, ab initio molecular orbital computations have been carried out on a variety of small biologically active molecules. Since all living processes occur in the aqueous environment, it is necessary to investigate the effects of water solvation when exploring a biological molecule [Comput. Meth. Sci. Technol. 4 (1998) 25, J. Chem. Phys. 117 (2002) 4720]. First, the reactant complex of acetone and methylamine stabilized by water-mediated hydrogen bonding was studied. Then, ab initio characterization of the methyl phosphate interaction with water as well as its Ca^{2+}, Mg^{2+}, and Zn^{2+} complexes [Biochemistry 41 (2002) 3207, Chem. Rev. 96 (1996) 2435, Chem. Rev. 101 (2001) 3] was completed. Finally, studies of phenol and N-Ac-L-Gly-NH-Me with inclusion of one water molecule as starting building blocks for future multi-molecular water solvation of these molecules have been completed. All the data has been generated at the RHF/3-21G level of theory using GAUSSIAN 98 computational program.
© 2003 Published by Elsevier B.V.

Keywords: Intermolecular; Ketone; Glycolysis; Solvation; Energy surface; Hydrogen bonding

1. Introduction

Acetone–methylamine interactions can be viewed as the simplest example of a ketone and an amine

reacting to form an imine (also known as a Schiff base). Schiff base formation is typical of enzyme-substrate (e.g. an aldolase enzyme facilitating the cleavage of fructose-1,6-bisphosphate in the glycolysis pathway [1,6]), drug-receptor (e.g. HIV-1 nuclear import can be inhibited by imine formation with arylene bis(methylketone) [7]), or any other ligand-receptor interactions (e.g. antiphospholipid antibody recognition [8]). Investigating this system will give

* Corresponding author. Address: Department of Chemistry, University of Toronto, 80 St George St, Toronto, Ont., Canada M5S 3H6.
E-mail address: gchass@fixy.org (G.A. Chass).

0166-1280/$ - see front matter © 2003 Published by Elsevier B.V.
doi:10.1016/j.theochem.2003.08.018

a better understanding of how hydrogen bonding affects the mechanism of this reaction.

The methyl phosphate dianion represents the simplest phosphate ester [2]. Since it is highly resistant towards hydrolysis [9], it will serve as a good model for solvation effects. It is used in this study to perform ab initio solvation calculations, and will serve as models for the solvation of more complex phosphate monoesters [10–12]. Solvation of methyl phosphate dianion was calculated with a single water molecule in conjunction with one of the divalent cations Ca^{2+}, Mg^{2+} or Zn^{2+}.

Phenol, has become an important molecule to investigate using molecular orbital calculations at the RHF/3-21G level of theory, especially when it is in the presence of solvents such as water. Despite its toxic effects in living organisms, phenol is present as a substituent in many important biological molecules such as tyrosine [13,14]. Under a water solvation layer, phenol would be represented by a more realistic biological system. Thus, phenol solvation studies can be used as models for tyrosine in biological systems, or a ligand-receptor binding complex between a solute and solvent [3].

Ab initio molecular orbital computations were performed on N-Ac-L-Gly-NH-Me at the RHF/3-21G level of theory in order to investigate the conformational features of this non-essential amino acid. Subsequently, different conformations were solvated in water with variations in the degree of approach of the water molecule, thus enhancing our understanding of its behaviour in an aqueous environment and its biological properties.

2. Methods

2.1. Acetone, methylamine and water interactions

Acetone, water and methylamine molecules were separately optimized (Fig. 2(c)) and the values of their energetic minima were obtained at RHF/3-21G level of theory [4]. The problem of the 180-degree angle and undefined dihedrals appeared when two molecules were connected via a hydrogen bond in the Z-matrix. From a mathematical perspective, if $A11 = 181°$, $D11 = X°$, this would be equivalent to $A11 = 179°$ and $D11 = 180 + X°$. Instead of changing the angle

from 181 to 179°, the program changes the dihedral to the one that corresponds to an angle $> 180°$; thus, the dihedral has lost its physical meaning, although it retains its mathematical significance. Therefore, 'dummy atoms' have been introduced into bimolecular and trimolecular systems in order to avoid problems in computations with 180° degree angles. The best solution found was to locate a 'dummy atom' above each hydrogen bond in the system, forming a right triangle and fixing the bond lengths connecting it to the neighboring atoms in the system. The next step is to fix the rotation of an atom following the 'dummy atom' to that of the 'dummy atom' by freezing its dihedral at 180°. This will enable one to monitor the rotation of the adjacent atom by simply rotating the 'dummy atom' itself. It is important to locate the 'dummy atom' so that it would form an acute angle with the neighboring atoms. Using the above approach, geometry optimized energetic minima, as well as potential energy hypersurfaces for acetone-methylamine and acetone–water–methylamine complexes have been generated.

2.2. Methyl phosphate, water and divalent ions interactions

A standardized numbering system [16] was used to easily accommodate changes of the solvent molecules to the methyl phosphate dianion. Using a standardized numbering system, solvent molecules can be easily changed or supplemented onto the existing solvation complex as additional modules. The orientations about the dihedrals are denoted in relatively the same manner as Tvaroska et al. [16]. For example, the conformation depicted in Fig. 1(a) will be of the conformation $A + c$ (A being alpha $\sim 120°$; and c being zeta $\sim 0°$). Each backbone dihedral angle was calculated at all six possible orientations giving a total of $6 \times 6 = 36$ probable rotamers.

The solvation complex was optimized using the methods implemented by Florian et al. [17] in which the backbone dihedrals alpha [$\alpha = \alpha(C1–O2–P3–O6)$] and zeta [$\xi = \xi(O2–P3–O6–H10)$] were kept constant while other degrees of freedom were relaxed (Fig. 1(a)). The PES was scanned in which dihedrals (α, ξ) were frozen at increments of 30° while other degrees of freedom were relaxed. For systems that GAUSSIAN 98 could not calculate due errors in

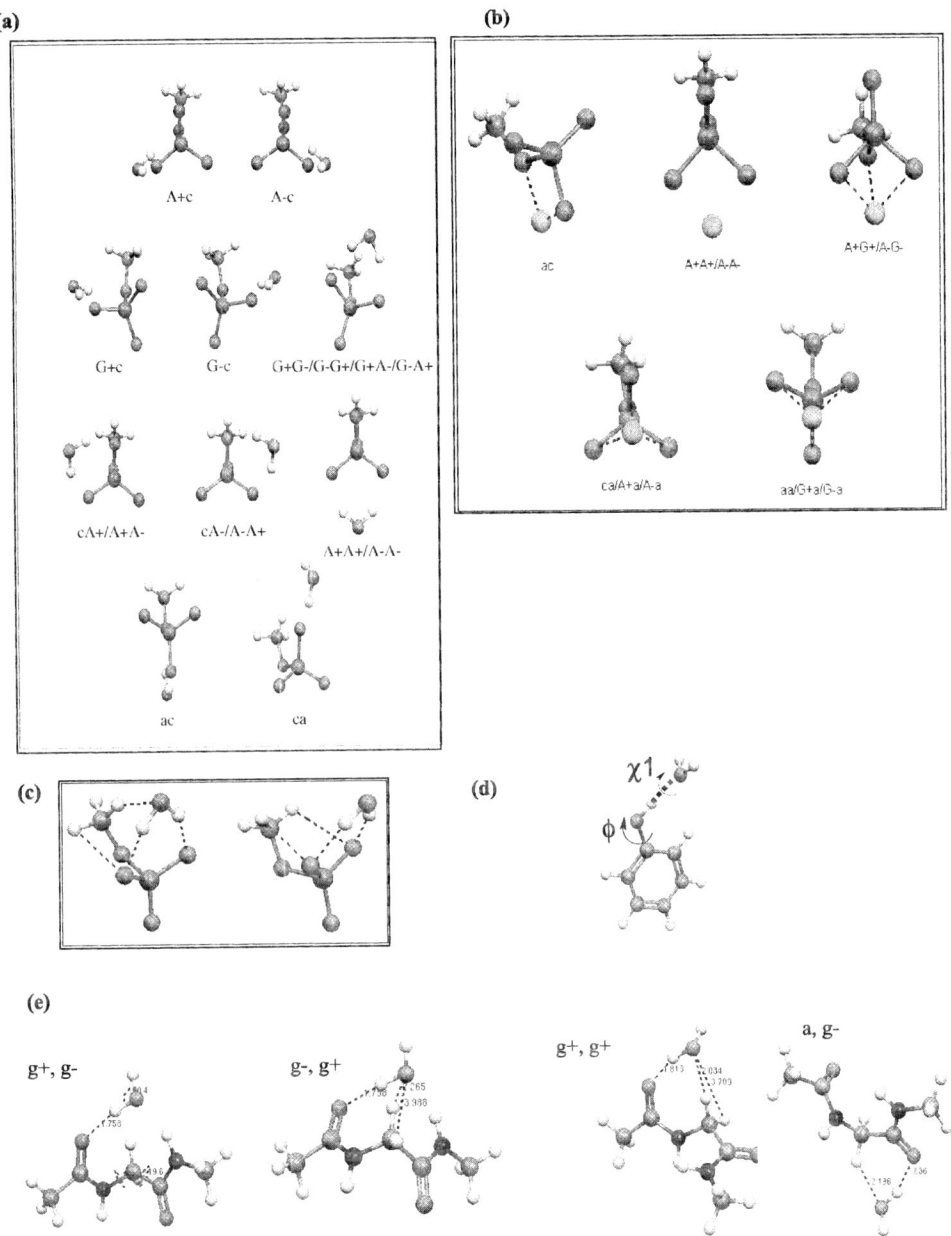

Fig. 1. (a) Found conformations of the methyl phosphate-water solvation complex. Some conformations were found to be equivalent (e.g. A^+A^+ and A^-A^-); (b) existing conformations of the divalent cation solvation complex. Some conformations were equivalent (e.g. A^+A^+ and A^-A^-); (c) the $G^-G^+/G^+G^-/G+A^-/G^-A^+$ conformation showed two different types of hydrogen bonding. The hydrogen bonding between the methyl group and the water (left) is found to be approximately 1.977 kcal mol^{-1} more stable than without the hydrogen bonding between the methyl group and the water (right). The PO-ester bond is also shorter by 0.0039 angstroms when intermolecular hydrogen bonding is present in the complex. (d) Phenol solvated with one water molecule at a head-on hydrogen bonding orientation. Energy a function of two of the three variables studied thus far ϕ and χ_1 rotation, of this solvated model. Hydrogen bond length is also considered as a very important variable of energy. (e) The two most stable solvation complexes of glycine (g^+, g^- and g^-, g^+) are shown. Further formation of hydrogen bonds for complex stabilization (g^-, g^+ and g^+, g^+ and a, g^-) also appear in this figure.

dihedral range, the problematic angle was frozen and scanned at increments of 10° from 100 to 170° along with the frozen dihedrals. Due to the symmetrical nature of methyl phosphate, the alpha torsional angle was scanned from 0 to 180° instead of 0–330°. All calculations were done at the RHF/3-21G level of theory.

2.3. Phenol and water interaction

Ab initio calculations (at the RHF/3-21G level of theory) of energetically optimized confirmations of phenol were studied first (Fig. 2(b)-(i)), followed by phenol solvated with a water molecule (Figs. 1(d) and 2(b)-(ii)) which is the first step to represent phenol under a more real biological system. This complex can serve as a model for compounds containing phenol, such as the amino acid tyrosine (e.g. can influence protein folding with proteins containing tyrosine, under a biologically solvated state), or a ligand-receptor binding model (e.g. drug to a phenol receptor). A scan of ϕ on phenol was done first to generate a potential energy curve, and to observe what geometry the molecule prefers. A double scan of ϕ and $\chi 1$ was followed on the phenol–water molecule complex generating a potential energy surface of the system. A hydrogen bond scan was also performed where energy of the system was a function of hydrogen bond distance, giving a very recognizable type of potential energy curve.

2.4. N-Ac-L-Gly-NH-Me and water complexes

In this study, the molecule *N*-Ac-L-Gly-NH-Me was divided into three modules and numbered accordingly by the standardized numbering system, as shown in Fig. 2 [15]. This numbering and the subsequent Z-matrix served as input for the GAUSSIAN98 program [5,18], which performed ab initio calculations meant to identify all stable conformers of glycine, taking into consideration the stereochemistry of the molecule and its interaction with water.

Glycine consists of one prochiral center (C8). Therefore, the dihedrals of interest are D9 (φ) and D14 (ψ) (Fig. 2), whose variations give us nine different conformations of this amino acid. Each of these conformations was optimized in order to find the most stable dihedrals that glycine tends to take

on. These conformations were then solvated twice, once at O4 and once at O10, with one molecule of water that was individually analyzed through to the RHF/3-21G level of theory and the energies were compared.

3. Results

3.1. Acetone, methylamine and water interactions

Geometry optimizations of acetone-methylamine and acetone–water–methylamine showed that hydrogen bonding stabilizes these systems and the sum of energies of all studied molecules added is higher than the energy of the complexes. Energy of any acetone–water–methylamine complex is lower than that of acetone-methylamine with energy of water molecule added to it, which shows that forming hydrogen bonds stabilize the system. There is not much variability in how many energetic minima can be formed by the studied system (Fig. 3). This suggests that when acetone and methylamine come together, interactions are limited. A potential energy surface generated for the acetone–water–methylamine complex confirms our optimization results, and demonstrates a remarkable trend, where energy of the system is at its lowest when D11 is 0 or 180° (Fig. 4(a)), which allows for the nitrogen of methylamine to interact with methyl-group-hydrogen of acetone forming a stable three-hydrogen-bond-complex. From our results, it is also clear that as nitrogen of methylamine approaches the central carbon atom of the acetone, the energy of the system increases and the hydrogen bonds become longer. This shows that this conformation is close to the reactive state. In a time-dependent, dynamic scenario, this system would be favoured to react under certain conditions, likely due to the increase in energy among other initiating structures located in minima.

3.2. Methyl phosphate, water and divalent ions interactions

Sixteen of the 36 possible conformers of the water solvation complex were found in which 10 were unique conformers (Fig. 1(a)). In the Ca^{2+} solvation, 11 of the 36 possible conformers were found while only

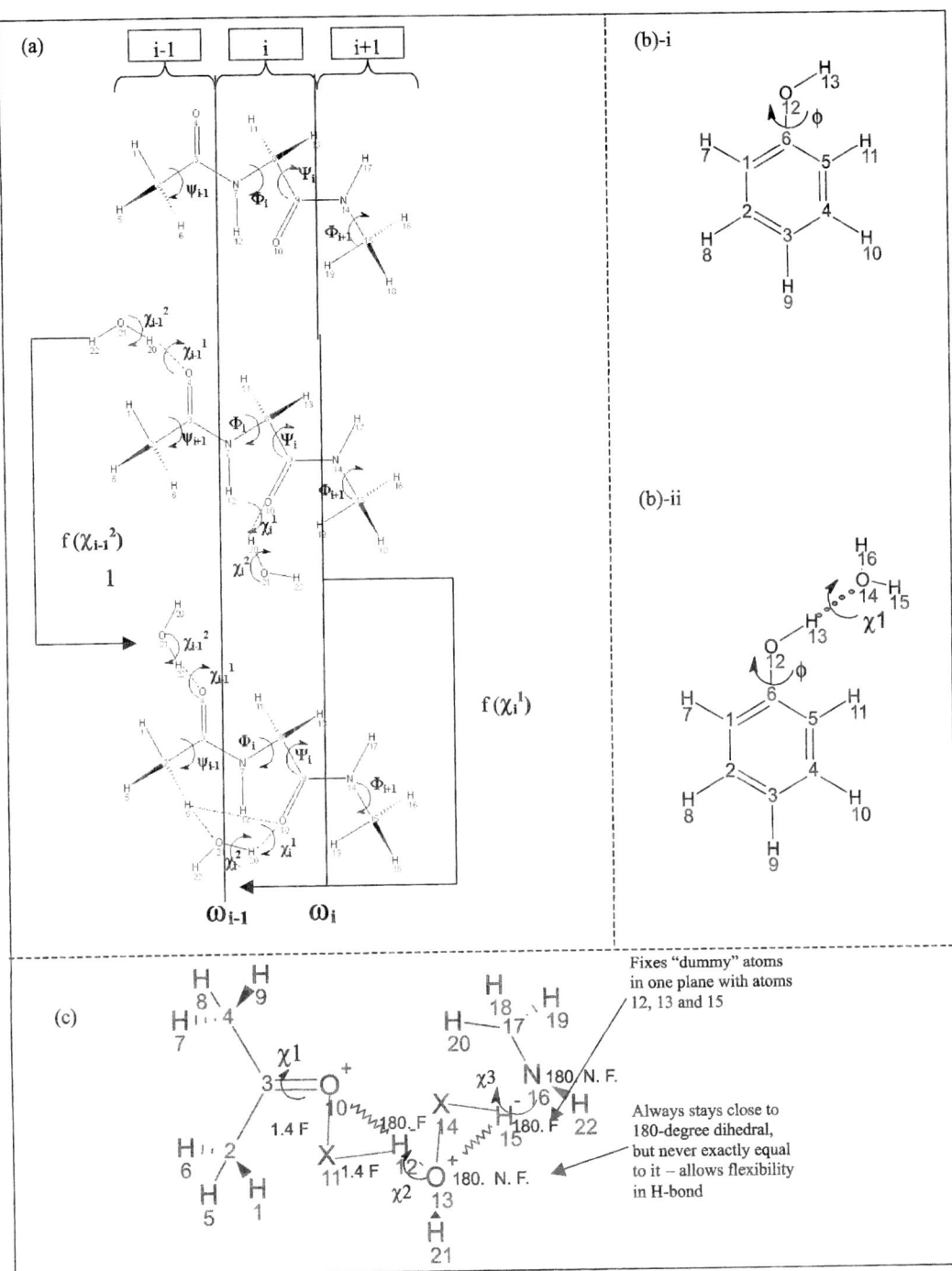

Fig. 2. (a) Modular numbering system of *N*-Ac-L-Gly-NH-Me. Water molecules approach the lone pairs on oxygens 4 and 10. (b) The structure, numbering, and dihedrals studied for (b)-(i): phenol under a non-solvated state (b)-(ii): Phenol solvated with a water molecule (c) Acetone–water–methylamine complex numbered with indicated dihedrals of interest and variables fixed due to usage of 'dummy atoms'. 'F' stands for a fixed variable and NF stands for a not fixed variable. Other variables are not shown on this diagram.

Relative energy (kcal/mol)

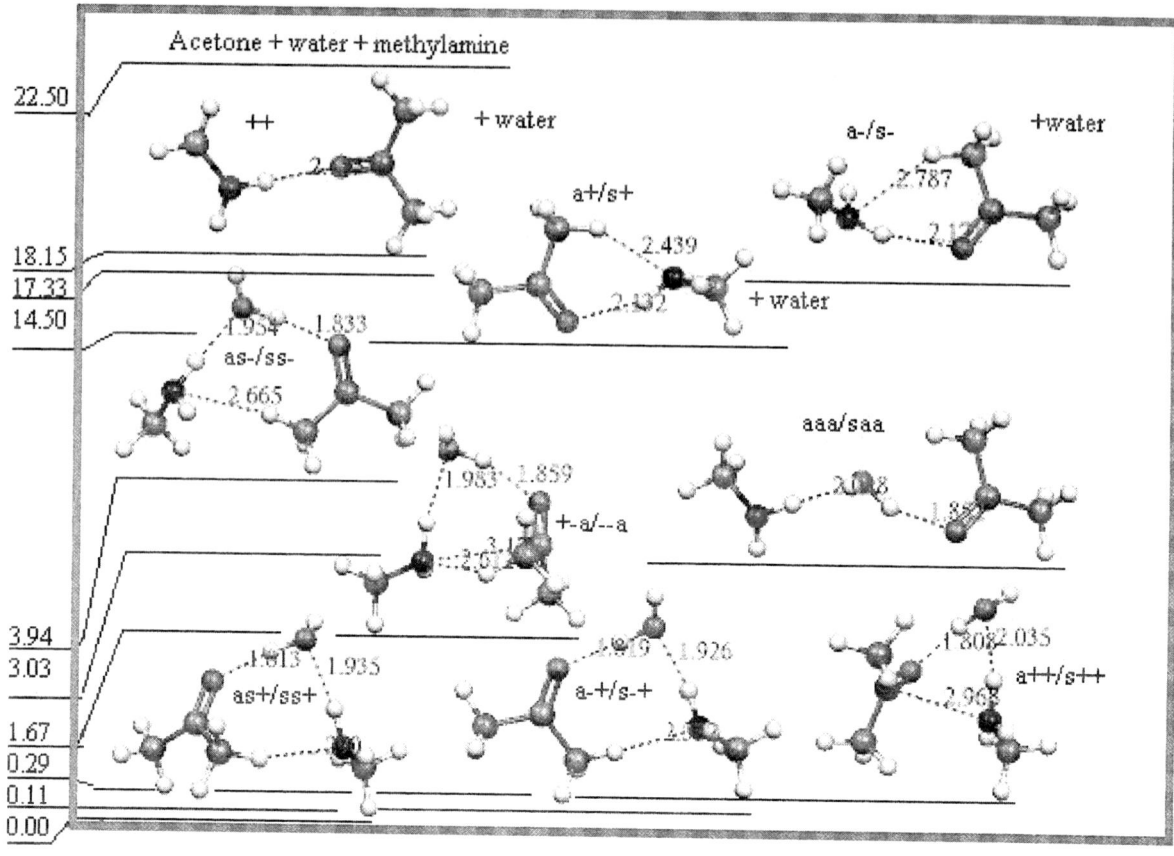

Fig. 3. Energies of different complexes of *N*-methylamine, acetone and water in different conformations.

nine of the 36 possible conformers for the magnesium solvation complex and the zinc solvation complex were found. All the divalent cation solvation complexes yielded five unique conformations (Fig. 1(b)). All the solvation complexes with equivalent conformations exhibited small ranges of geometries and energies indicating the high flexibility of the methyl phosphate dianion under solvation

Hydrogen bonding played a much more significant role in water solvation than in divalent cation solvation. Intermolecular hydrogen bonding was not a requirement for stability in the water solvation complexes; however, its presence increased the stability of the water solvated complex (Fig. 1(c)). Variations in intermolecular hydrogen bonding in the water solvated complex also gave slightly different stabilities in the complex. All divalent cation solvation

complexes showed similar intermolecular hydrogen bonding within the methyl phosphate dianion.

3.3. Phenol and water interactions

The φ rotor is most stable when it is at 0.0 or 180° (in the plane of the ring), whether it is in the non-solvated state or the solvated state. Energy of this solvated system is more stable due to hydrogen bonding (calculations show it is stable by approximately 15.38 kcal mol^{-1}), at the RHF/3-21G level of theory. Rotation of the water molecule about the hydrogen bond is not very significant in forming new energetically minimal conformers of phenol. Hydrogen bond distance appears to play an important role in the energy of the solvated system, and there is only one distinct minimum when D is 1.7–1.8 Å

(a)

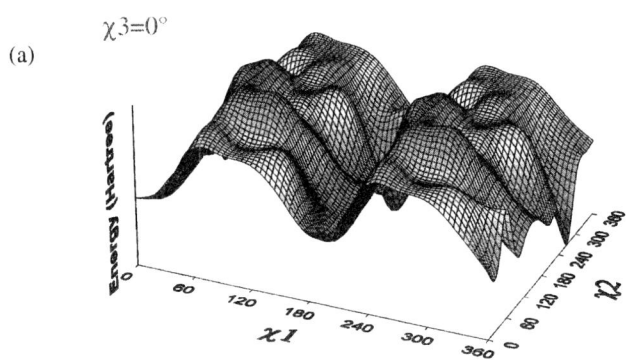

Energy of Phenol as a Function of Hydrogen
Bonding Distance between Water and the Hydroxyl group

(b)

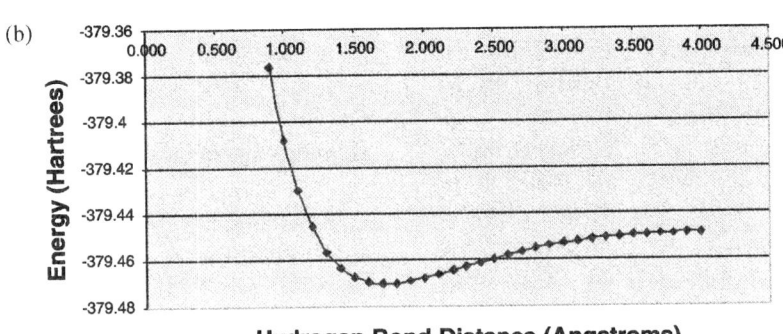

Energy (Hartree) vs. Backbone Conformations
of *N*-Ac-*L*-Gly-NH-Me

(c)

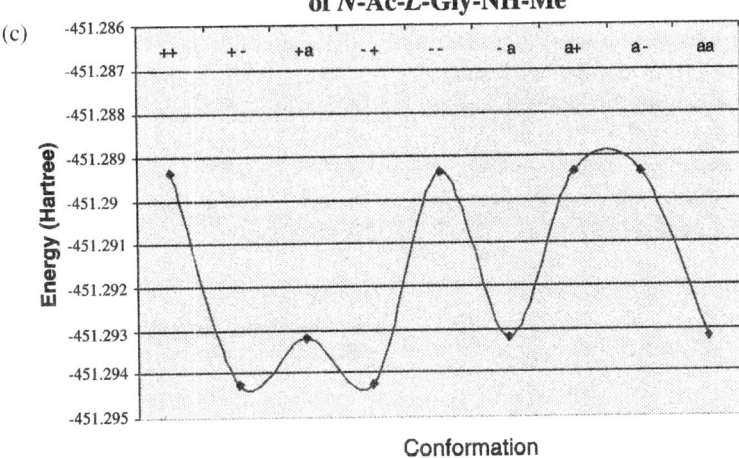

Fig. 4. (a) Potential energy surface of acetone–water–methylamine complex when $\chi3 = 0°$ is shown to demonstrate the shape of all the surfaces generated while scanning three dihedrals of interest. All the other surfaces have similar energy peaks at $\chi1 = 90°$, and similar hollows at $\chi1 = 0°$ and $\chi1 = 180°$, differing in heights and depths; (b) A potential energy curve, $E = f(d)$, where d represents the hydrogen bond distance between the hydroxyl of phenol and water; (c) Relative energies (hartree) of different backbone conformations of Glycine.

(Fig. 4(b)). All hydrogen bond distances, angles, and dihedrals will be used as functions for the energy of this type of solvated system, with more water molecules solvating phenol.

3.4. N-Ac-L-Gly-NH-Me and water

The results of the initial conformational analysis of glycine showed that the most stable backbone conformation was (g^+, g^-), for D9 and D14 respectively (Figs. 1(e) and 4(c)). The results clearly show that solvation increases the molecule's stability by lowering its energy. However, under solvation conditions with the water molecule approaching in the *anti* position to O4 both (g^+, g^-) and (g^-, g^+) backbone conformers are found to be the most stable conformations of glycine (Fig. 1(e)). Solvation of O4 also resulted in formation of hydrogen bonds between the oxygen of water and H11 and H13. However, these hydrogen bonds were not consistently found in all the conformations of glycine. Solvation of O10 also resulted in the formation of a hydrogen bond between the oxygen of water and H12 (Fig. 1(e)). All these hydrogen bonds have contributed to lowering the energy and the stabilization of the complex. Calculation problems arose during the solvation of glycine. For some files, the angle of approach of the water molecule was greater than 180° and the calculations terminated. It was observed that all the files terminated in this manner had the water molecule approaching in the *syn* position. This was overcome by the presence of the dummy atoms between the water and the carbonyl groups and computations were then progressed as expected and results were obtained.

4. Conclusion

Intermolecular interactions are very difficult to study because they may involve flexible 180° angles that cause problems in computations. Starting with small building blocks and solving this problem at a small scale by means of manipulating 'dummy atoms', we can apply it to larger systems to investigate the behaviour of a particular molecule when solvated with a layer of water, or to see how multiple

biologically important molecules change their conformations when interacting, which would allow us to track the mechanisms of interactions.

Starting from one water molecule, a primary solvation layer can eventually be built around phenol and glycine, where many of the variables stated can be studied to determine the most desired solvation layer. Following a primary solvation layer, a secondary solvation layer will be added to determine what its effects would be on the already solvated molecular system. Therefore, it is important to begin with a simple water molecule creating a hydrogen bonding system that will lead to a solvation layer, in the hopes of eventually demonstrating that this explicit solvation model can expand and build upon the results of other implicit SCRF models.

Acknowledgements

This work was supported with grants from the Global Institute Of COmputational Molecular and Materials Sciences (GIOCOMMS), Toronto, Ontario, Canada. We would like to extend our deepest thanks to David H. Setiadi for always being the endless source of information and to Suzanne K. Lau, Jacqueline M.S. Law, Michelle A. Sahai, Christopher N.J. Marai and Jeremy H. Keller for generously devoting their time.

One of the authors (IGC) wishes to thank the Ministry of Education for a Szent-Györgyi Visiting Professorship.

References

[1] M. Jedrzejas, P. Setlow, Chem. Rev. 101 (2001) 3.
[2] D. Wilcox, Chem. Rev. 96 (1996) 2435.
[3] P. O'Brien, D. Herschlag, Biochemistry 41 (2002) 3207.
[4] I. Szareska, J. Rychlewski, U. Rychlewska, Comput. Meth. Sci. Technol. 4 (1998) 25.
[5] GAUSSIAN 98, Revision A.9, M.J. Frisch, G.W. Trucks, H.B. Schlegel, G.E. Scuseria, M.A. Robb, J.R. Cheeseman, V.G. Zakrzewski, J.A. Montgomery, Jr., R.E. Stratmann, J.C. Burant, S. Dapprich, J.M. Millam, A.D. Daniels, K.N. Kudin, M.C. Strain, O. Farkas, J. Tomasi, V. Barone, M. Cossi, R. Cammi, B. Mennucci, C. Pomelli, C. Adamo, S. Clifford, J. Ochterski, G.A. Petersson, P.Y. Ayala, Q. Cui, K. Morokuma, D.K. Malick, A.D. Rabuck, K. Raghavachari,

J.B. Foresman, J. Cioslowski, J.V. Ortiz, A.G. Baboul, B.B. Stefanov, G. Liu, A. Liashenko, P. Piskorz, I. Komaromi, R. Gomperts, R.L. Martin, D.J. Fox, T. Keith, M.A. Al-Laham, C.Y. Peng, A. Nanayakkara, M. Challacombe, P.M.W. Gill, B. Johnson, W. Chen, M.W. Wong, J.L. Andres, C. Gonzalez, M. Head-Gordon, E.S. Replogle, J.A. Pople, Gaussian, Inc., Pittsburgh PA, 1998.

[6] J. McMurry, Organic Chemistry, fifth ed., 2000.

[7] Y. Al-Abed, L. Dubrovsky, B. Ruzsicska, M. Seepersaud, M. Bukrinsky, Biorg. Med. Chem. Lett. 12 (2002) 3117–3119.

[8] P.J. Friedman, et al., Biol. Chem. 277 (2002) 7010–7020.

[9] J. Chin, Acc. Chem. Res. 24 (1991) 145.

[10] C. Alhambra, L. Wu, J. Gao, J. Am. Chem. Soc. 120 (1998) 3858.

[11] M. Bioanciotto, J. Barthelat, A. Vigroux, J. Phys. Chem. A. 106 (2002) 6521.

[12] C. Hu, T. Brinck, J. Phys. Chem. A 103 (1999) 5379.

[13] S. Scheiner, T. Kar, J. Pattanayak, J. Am. Chem. Soc. 124 (2002) 13528.

[14] L. Serrano, M. Bycroft, Fersht, R. Alan, J. Mol. Biol. 218 (1991) 465.

[15] G. Chass, M. Sahai, J. Law, S. Lovas, O. Farkas, A. Perczel, J. Rivail, I. Csizmadia, J. Quantum Chem. 90 (2002) 933.

[16] I. Tvaroska, I. Andre, J. Carver, Theochem 469 (1999) 103.

[17] J. Florian, M. Strajbl, A. Warshel, J. Am. Chem. Soc. 120 (1998) 7959.

[18] M. Iwaoka, M. Okada, S. Tomoda, J. Mol. Struct. (Theochem) 586 (2002) 111–124.

ELSEVIER

Journal of Molecular Structure (Theochem) 666–667 (2003) 109–116

THEO
CHEM

www.elsevier.com/locate/theochem

Conformational study of *N*-alkyl-benzyltetrahydroisoquinolines alkaloid

F.D. Suvire[a], I. Andreu[b], A. Bermejo[b], M.A. Zamora[a], D. Cortes[b], R.D. Enriz[a,*]

[a]*Departamento de Química, Universidad Nacional de San Luis, Chacabuco 915, 5700 San Luis, Argentina*
[b]*Departamento de Farmacología, Farmacognosia y Farmacodinamia, Facultad de Farmacia, Universidad de Valencia, 46100 Burjasot, Valencia, Spain*

Abstract

An exhaustive conformational study on the benzyltetrahydroisoquinolines (BTHIQ) from ab initio (RHF/6-31G(d)) calculations was carried out. The effects of different substituents at chiral C_1 atom were also considered.

Our results indicate that different substituents at C_1 in BTHIQ molecules introduce a significant steric hindrance which, in turn, might be responsible for a conformational restriction favouring or disfavouring the spatial orientation of the lone pairs of N atom allowing or not the electronic attachment with the side chain of Asp residue. These results can serve as an aid for designing suitable structures of BTHIQs for better dopamine D_1-receptor inhibitory activity.
© 2003 Elsevier B.V. All rights reserved.

Keywords: Conformation; Ab initio and DFT calculations; Benzyltetrahydroisoquinolines; Dopamine D_1-receptor

1. Introduction

Some 1-benzyl-1,2,3,4-tetrahydroisoquinoline (BTHIQ) alkaloids have shown high affinity for both D_1 and D_2 dopamine receptors [1–3]. Thus, some BTHIQs bind to dopamine receptors from striatal membranes [3] and in some cases, inhibit dopamine uptake by striatal synaptosomes [2].

The selective affinity for one of these receptors appears to be related to the stereochemistry and the substitution pattern in the A and C rings, as well as N-substitution. Previously, we have reported that in BTHIQs, (1*S*)-enantiomers are 5–15 times more effective at D_1-like and D_2-like dopamine receptors than (1*R*)-enantiomers [4]. The different activities

reported for the structures shown in Table 1 illustrate very well the importance of the chirality at C_1 in BTHIQs and related compounds in order to produce the biological response.

The most active BTHIQs possess a phenyl group as aromatic ring C (compound **f** in Table 1). Interestingly, the replacement of phenyl group by a benzyl group is relatively well tolerated but with a significant loss of activity (compare IC_{50} of compounds **e** and **f**). It should be noted, however, that the spatial ordering adopted by these groups might be, in principle, very different. The selectivity of bis-BTHIQ and tetrahydroprotoberberine alkaloids for D_2 dopamine receptors has been reported previously [7,8]. Moreover, our own binding data [9] suggest that a protoberberine-like conformation of BTHIQs with an unsubstituted benzyl moiety increases their selectivity for D_2 dopamine receptors,

* Corresponding author. Fax: +54-2652-4311301.
E-mail address: denriz@unsl.edu.ar (R.D. Enriz).

0166-1280/$ - see front matter © 2003 Elsevier B.V. All rights reserved.
doi:10.1016/j.theochem.2003.08.019

Table 1
Structures and biological activities (IC_{50}) of representative BTHIQ compounds acting as dopamine D_1-receptor inhibitors

Structure	IC_{50} (µ/mol)	Structure	IC_{50} (µ/mol)
a	9.28 [5]	**d**	0.17 [6]
b	>60 [5]	**e**	5.75×10^{-2} [7]
c	1.8×10^{-4} [6]	**f**	5.25×10^{-3} [7]

while an 'aporfine-like' conformation displays high affinities for both D_1 and D_2 dopamine receptors, with an increase in D_1 dopamine receptor affinity.

In a companion paper, we report the molecular recognition and binding mechanism of BTHIQs to the D_1 dopamine receptor [10]. In that model, we propose a stepwise binding, involving SITE I followed by the rest of the molecule (SITE II) (Fig. 1). It should be noted that on the basis of our results of molecular interaction simulations, it is not prudent to discard the possibility of a third binding site which could be performed by the aromatic ring C (SITE III). It is clear therefore, that in order to obtain a clear profile of the overall recognition process, it is necessary to underline the role of the benzyl group (CH_2-Ph) in the binding mechanism of BTHIQ compounds. The conformational behaviour of BTHIQs is clearly relevant to any consideration of their interactions with the dopamine D_1-receptor. BTHIQ compounds may exist in solution in a number of conformations in equilibrium one another. If it is assumed, as it seems highly probable [10], that the ligand has only a single conformation in its complex with the receptor, then the conformation of this complex must involve a process of conformational selection (or alternatively,

the ability of the ligand to achieve the requisite conformation) which will influence the kinetics and energetics of complex formation.

In principle, at least three different roles might be attributed to the benzyl group of BTHIQs, i.e.:

(i)   conformational restriction
(ii) steric hindrance
(iii) other possibility is that aromatic ring C could contribute to its own interaction through dispersion forces.

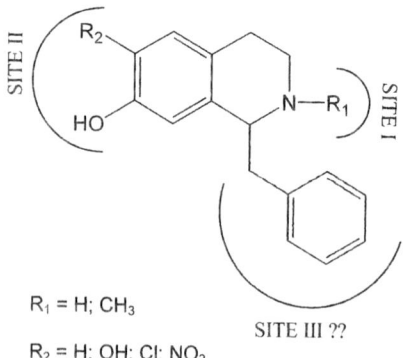

R_1 = H; CH_3

R_2 = H; OH; Cl; NO_2

Fig. 1. Schematic representation of dopamine–D_1-receptor showing putative binding sites.

Although, these concepts are independently formulated, they might be interdependent.

During the course of our study we were intrigued by the conformational preferences of the flexible side chain at C_1 in the BTHIQs. Such a study would ultimately tell us the spatial orientation of ring C, which due to its apparently strong impact on its activity and D_1 or D_2 selectivity, should be very important for the design of new structures. We report here an exhaustive conformational study on the BTHIQs molecules from ab initio calculations. The effects of different substituents at the chiral C_1 atom are also considered in this paper.

2. Computational methods

All the ab initio (RHF/3-21G and RHF/6-31G(d,p)) calculations were performed using the GAUSSIAN 98 program [11]. The optimized geometries of the BTHIQ structures (minima and transition states (TS)) were obtained using the Hartree–Fock methods at the level of the 6-31G(d,p) basis set. All geometry optimizations were made without any geometry constrains. Harmonic frequencies of every state were calculated in order to ensure that the optimized geometry really corresponds to an energy minimum or transition state on the potential energy surface.

The potential energy surface (PES) was calculated with 144 points at 15° increments at RHF/3-21G level of theory. The potential energy curve (PEC) cross-section was calculated by using RHF/6-31G calculations. The energy has been calculated at 15° intervals of the dihedral angle ϕ_5. The rotational energy profile was calculated using frozen parameters (bond angles and lengths) that were previously optimized but all the dihedral angles (except ϕ_5)

were optimized during the scan. The PEC as well as PES were plotted using Axum 5.0. Corresponding minima from the PES were selected and full optimizations were carried out successively at the RHF/3-21G and RHF/6-31G(d,p) level of theory.

3. Results and discussion

The essential conformational problem of BTHIQs (I) (although not the only one) involves two aspects:

(a) conformational behaviour of flexible side chain (torsional angles ϕ_5 and ϕ_6, compound I in Fig. 2) determining the spatial ordering for ring C;
(b) overall shape of ring B which is determined by the orientation of torsion angles $\phi_1 - \phi_4$. In this ring, it is also important to evaluate the conformational influence due to the presence of different substituents at chiral C_1 (compounds II and III in Fig. 2).

3.1. Conformational study of BTHIQ flexible side chain

In order to evaluate the flexible side chain of BTHIQs, we treated it as a double rotor. Torsional angle ϕ_5 was associated with the rotation of the isoquinoline group, while torsional angle ϕ_6 was associated with the rotation of aromatic ring C (Fig. 2). Accumulated experience suggested that when ϕ_5 varies, g^+, a and g^- conformers can be observed. However, when a planar moiety is rotated about a tetrahedral moiety (ϕ_6), only g^+ and g^- conformers may be expected in the vicinity of 90° and −90°,

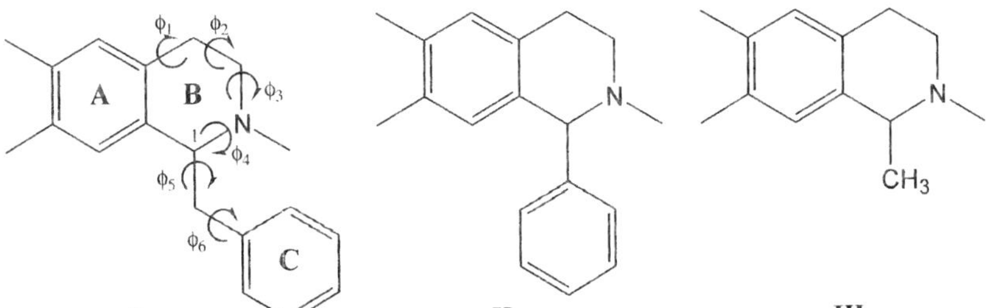

Fig. 2. General structural feature of BTHIQ molecules reported here, showing the different torsional angles (ϕ_1, ϕ_2, ϕ_3, ϕ_4, ϕ_5 and ϕ_6).

respectively. The two dimensional potential energy surface (PES) as described by equation $E = E(\phi_5, \phi_6)$ was computed and plotted for compound **I** and its graphical representation is shown in Fig. 3. Three minima (g^+, a and g^-) can be appreciated in this figure. It should be noted, however, that considering the symmetry of benzene group, only half of the full rotation of torsional angle ϕ_6 is shown here. Therefore, this PES displays six minima as expected. Because the 3-21G basis set is small one, we have attempted a partial verification of the above results by using a more extended basis set (RHF/6-31G). For reasons of economy, we did not recalculate the whole PES but only the essential parts of it. Thus, we have calculated a conformational PEC along the torsional mode ϕ_5 which is shown in Fig. 4. This PEC can be described by equation $E = E(\phi_5)$. Fig. 4 shows the presence of the three minima indicating the frequently expected *anti* effect (i.e. *anti* is more stable than *gauche*). This curve has a global minimum close to 180° ('*anti*') and two local minima at about 70° (g^+) and 280° (g^-). The *cis* barrier at 0° has 15.5 kcal/mol which is twice the height between the *anti* and g^+ and

anti and g^- conformations (7.69 and 9.61 kcal/mol, respectively).

The location of minima exhibited by the PES as well as PEC was used as a guide for geometry optimizations. Low-energy conformations displaying a *trans* arrangement is shown in Fig. 5.

3.2. Conformational study of BTHIQ ring B

Compound **I** in its cationic form possesses two asymmetric atoms (C_1 and N) which are shown in a schematic form in Fig. 6. Therefore, from the theoretical point of view, four conformational isomers (SR, SS, RR and RS) can be drawn for this molecule (Fig. 6). We report here the results obtained for SR and SS forms. Preliminary calculations performed for their respective enantiomeric forms displayed the same energetic profile (results not shown).

In order to determine the spatial ordering adopted by ring B of compound **I**, different possibilities were taken into account. Thus two sofa conformations (also called half-boat), half-chair as well as boat conformations were evaluated. RHF/6-31G(d) calculations

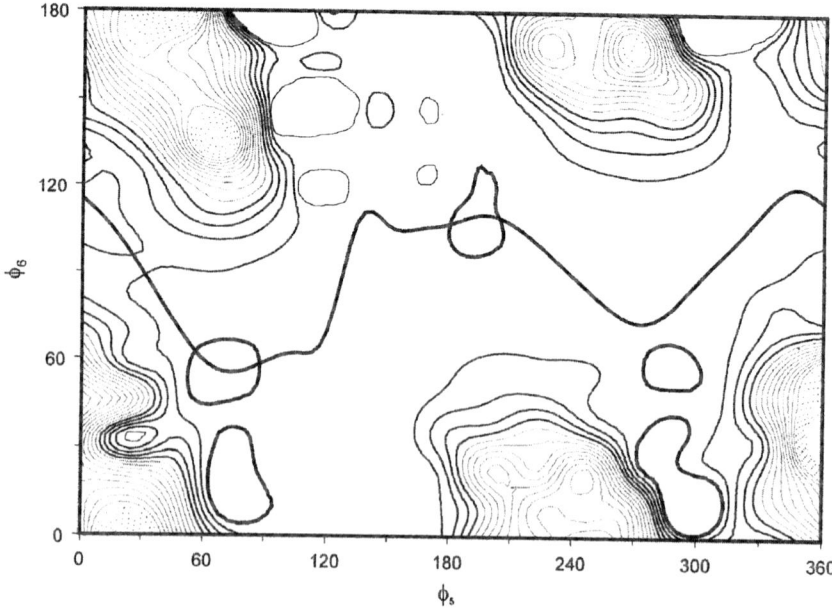

Fig. 3. Conformational PES obtained for compound **I** from RHF/3-21G calculations. Full cycle of rotation (from 0 to 360°) is shown for torsional angle ϕ_5; only half cycle of rotation (from 0 to 180°) is shown for ϕ_6. Potential energy lines are shown at 1 kcal/mol interval, the global minimum being defined at zero (shown in bold). The interconversion pathway among the three minima (followed by ϕ_5) is denoted with a bold line.

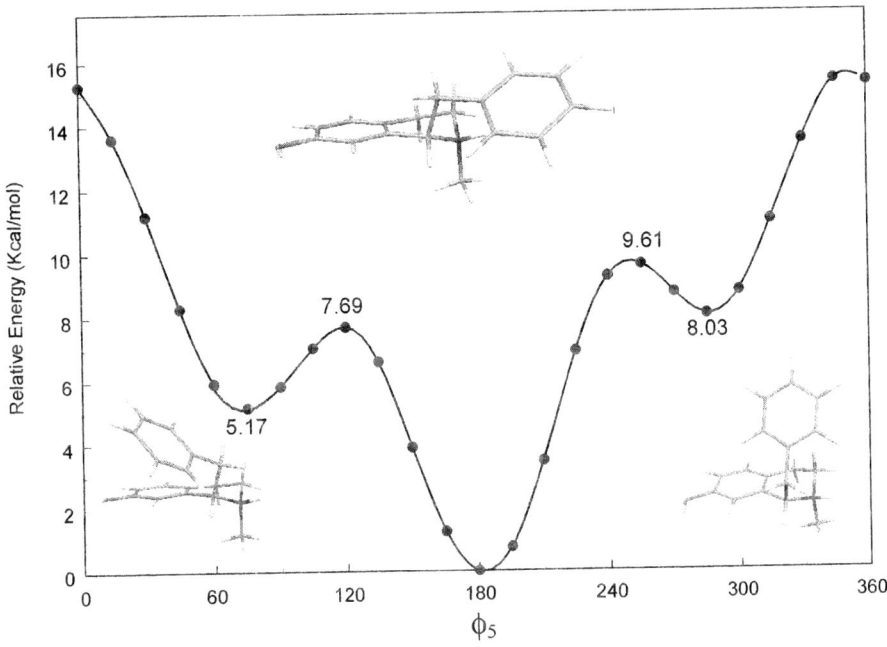

Fig. 4. Conformational PEC obtained for torsional angle ϕ_5 from RHF/6-31G calculations.

predict the half-chair conformation (in fact a somewhat distorted form) as the low energy conformation for ring B in compound **I** (Fig. 5). Low-energy conformation (form **Ib**) displays both substituents, the benzyl and the methyl groups in axial positions and therefore, the hydrogen atom (the putative lone pairs of N atom) is located in an equatorial position (Fig. 5). The second global minimum (form **Ia**) possesses the opposite arrangement with an energy gap of 2 kcal/mol with respect to the global minimum. Conformations **Ic** and **Id** possess 2.5 and 5.8 kcal/mol above the preferred form indicating that these forms are sterically disfavoured with respect to **Ia** and **Ib** (see Fig. 5 for the spatial view and Fig. 7A for the energy gaps). We are also interested in knowing how the interconversion process between these forms

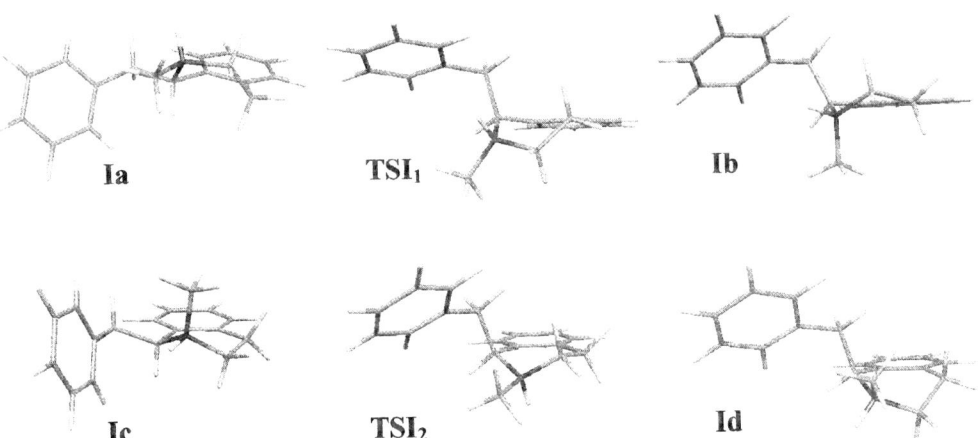

Fig. 5. Spatial view of four low-energy conformations (**Ia, Ib, Ic** and **Id**) of BTHIQ obtained from RHF/6-31G(d,p)) calculations. The transition structures (**TSI₁** and **TSI₂**) connecting these minima are also shown.

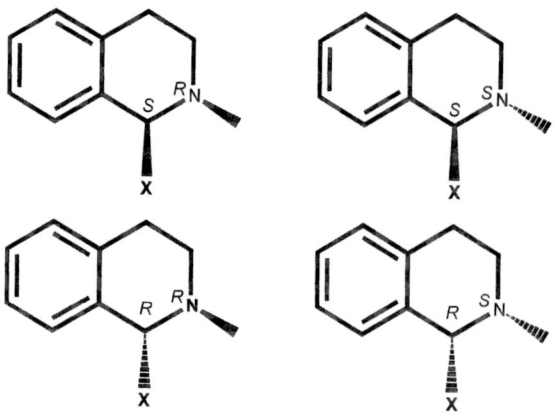

Fig. 6. Schematic representation for the four conformational isomers of compounds reported here.

occurs and therefore information about local and global minima is not enough. We need to know, at least, the molecular shape and also some indication about the dynamic behaviour of its internal degree of freedom. Therefore, we also evaluate the transition state (TS) structures for the interconversion between the conformations. Transition structure TS_1, interconnecting Ia and Ib, is located 10.25 kcal/mol above the global minimum (Fig. 7A), indicating that this conformational interconversion is some what restricted. A transition structure TS_2, connecting conformations Ic and Id, was founded at 13.78 kcal/mol.

At this stage of our work we consider, which might be of interest, to compare the influence of other alternative substituents on ring B resulting from

replacing the benzyl group with a phenyl or methyl group (compounds II and III; Fig. 2). Phenyl group was chosen considering that it is present in the most active BTHIQ compounds (see compound f in Table 1). With respect to methyl group, it is well known that this group is considered a benchmark substituent in the conformational analysis of substituted cyclohexanes [12–15].

Compound II possess two conformations (IIa and IIb) in a clear conformational equilibrium; the energy gap between these forms is only 0.39 kcal/mol and the transition state (TS_1) for this interconversion is located at 4.34 kcal/mol (Fig. 7B). It is interesting to note that the preferred conformation obtained for compound II (form IIa) possesses both phenyl and methyl groups in equatorial position and the hydrogen atom (the putative lone pair of N) in axial position (Fig. 8). In addition, our results indicate that ring inversion in compound II requires substantially lower energy than that required in compound I. This is a striking difference, which could be of great importance from a medicinal chemistry point of view. Molecular interaction simulations [10] indicate that the interaction between the $>$N–CH$_3$ group of BTHIQs and Asp 103 residue (SITE I, Fig. 1) is highly dependent of the spatial orientation of the lone pair of N atom. It is clear therefore that the major molecular flexibility exhibited by phenyl derivatives indicates a major ability of these compounds to achieve the requisite conformation. This fact is in a complete agreement with the higher inhibitory effect of phenyl derivatives in comparison to that of the benzyl

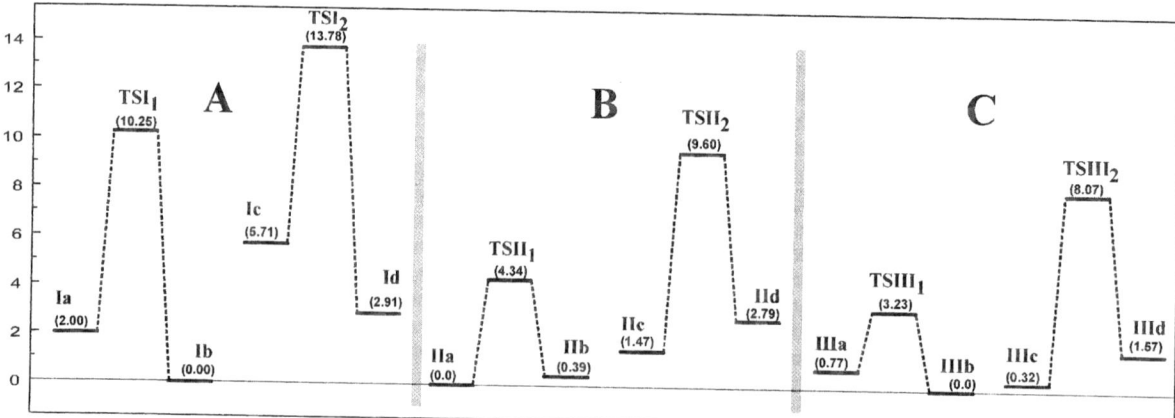

Fig. 7. Energy profiles showing the low-energy conformations and transition structures obtained for compounds I (A), II (B) and III (C).

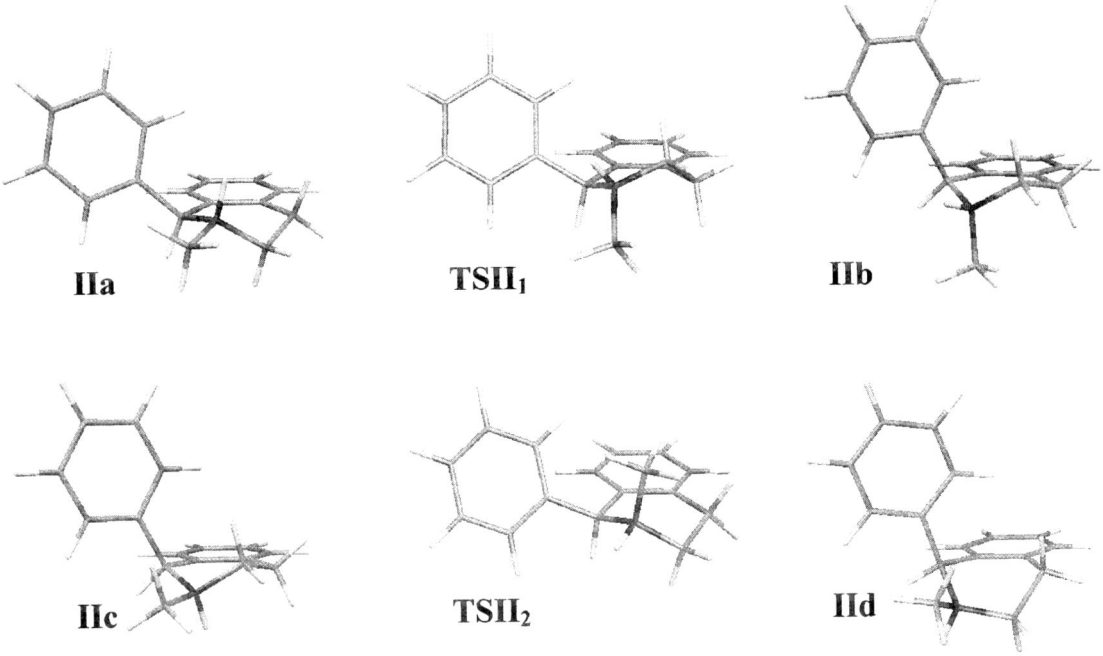

Fig. 8. Spatial view of four low-energy conformations (**IIa**, **IIb**, **IIc** and **IId**) obtained for the phenyl derivative from RHF/6-31G(d,p) calculations. The transition structures (**TSII₁** and **TSII₂**) connecting these minima are also shown.

derivatives. The results obtained for compound **III** are, in general, closely related to those obtained for compound **II** (compare results shown in Fig. 7B and C). Thus, in spite of the different size between phenyl and methyl groups, the conformational behaviour obtained for compounds **II** and **III** are still comparables. Unfortunately, we have no data about the inhibitory effect of methyl derivatives. This information could be crucial to determine if the substituents at C_1 might not interact with a putative third site (SITE III; Fig. 1). At present, we are synthesising these compounds in our lab. In order to obtain experimental support to our proposal.

4. Conclusions

We have presented a detailed study of BTHIQ conformational properties based on ab initio calculations. The PES and PEC obtained for BTHIQ side chain indicates that the low-energy conformation displays a *trans* arrangement. The energy barrier between *trans* and *gauche* conformations are about

8.6 kcal/mol suggesting a moderate molecular flexibility for this moiety. With respect to the conformational behaviour of ring B, energy minima having energies below 2 kcal/mol are separated by a transition barrier of approximately 10.25 kcal/mol and hence, the molecule cannot be expected to move easily from one conformation to the other by gaining this energy from other favourable interactions.

Molecule **II** present a different situation. Energy minima having energies below 0.39 kcal/mol are separated by a transition barrier of 4.34 kcal/mol showing a significant molecular flexibility. The different spatial ordering obtained for ring B in these compounds as well as different molecular flexibility in these moieties can explain, at least in part, the different activities reported for phenyl and benzyl THIQ derivatives.

Theoretical calculations indicate that different substituents at C_1 in BTHIQ molecules introduce a significant steric hindrance which, in turn, might be responsible of a conformational restriction favouring or disfavouring the spatial orientation of the lone pairs of N atom allowing or not the electronic attachment with the side chain of Asp residue (SITE I, Fig. 1).

These results can serve as an aid for designing suitable structures of BTHIQs for better dopamine–D_1-receptor inhibitory activity.

Acknowledgements

This work was supported by grants from Fundación Antorchas and Universidad Nacional de San Luis (U.N.S.L.), Argentina. R.D.E. is member of the Consejo Nacional de Investigaciones Científicas y Técnicas de la República Argentina (CONICET).

References

[1] P. Protais, J. Arbaovi, E.H. Bakkali, A. Bermejo, D. Cortes, J. Nat. Prod. 58 (1995) 1475.

[2] A. Bermejo, P. Protais, M.A. Bázquez, K.S. Rao, M.C. Zafra-Polo, D. Cortes, Nat. Prod. Lett. 6 (1995) 57.

[3] N. Cabedo, P. Protais, B.K. Cassels, D. Cortes, J. Nat. Prod. 61 (1998) 709.

[4] N. Cabedo, I. Andreu, M.C. de Arellano, A. Chagraovi, A. Serrano, A. Bermejo, P. Protais, D. Cortes, J. Med. Chem. 44 (2001) 1794.

[5] P.H. Andersen, F.C. Gronvald, R. Hohlweg, L.B. Hansen, E. Guddal, C. Braestrup, E.B. Nielsen, Eur. J. Pharm. 219 (1992) 45.

[6] B. Hoffman, S.J. Cho, W. Zheng, S. Wyrick, D.E. Nichols, R.B. Mailman, A. Tropsha, J. Med. Chem. 42 (1999) 3217.

[7] D. Cortes, B. Figadere, J. Saez, P. Protais, J. Nat. Prod. 55 (1992) 1281.

[8] D. Cortes, J. Arbaovi, P. Protais, Nat. Prod. Lett. 2 (1993) 233.

[9] I. Andreu, D. Cortes, P. Protais, B.K. Cassels, A. Chagraovi, N. Cabedo, Bioorg. Med. Chem. 8 (2000) 889.

[10] F.D. Suvire, N. Cabedo, A Chagraovi, M.A. Zamora, D. Cortes, R.D. Enriz, J. Mol. Struct. (Theochem), submitted for publication.

[11] M.J. Frisch, G.W. Trucks, H.B. Schlegel, G.E. Scuseria, M.A. Robb, J.R. Cheeseman, V.G. Zakrzewski, J.A. Montgomery Jr., R.E. Stratmann, J.C. Burant, S. Dapprich, J.M. Millam, A.D. Daniels, K.N. Kudin, M.C. Strain, O. Farkas, J. Tomasi, V. Barone, M. Cossi, R. Cammi, B. Mennucci, C. Pomelli, C. Adamo, S. Clifford, J. Ochterski, G.A. Petersson, P.Y. Ayala, Q. Cui, K. Morokuma, D.K. Malick, A.D. Rabuck, K. Raghavachari, J.B. Foresman, J. Cioslowski, J.V. Ortiz, A.G. Baboul, B.B. Stefanov, G. Liu, A. Liashenko, P. Piskorz, I. Komaromi, R. Gomperts, R.L. Martin, D.J. Fox, T. Keith, M.A. Al-Laham, C.Y. Peng, A. Nanayakkara, C. Gonzalez, M. Challacombe, P.M.W. Gill, B. Johnson, W. Chen, M.W. Wong, J.L. Andres, C. Gonzalez, M. Head-Gordon, E.S. Replogle, J.A. Pople, Gaussian 98, Revision A.7, Gaussian, Inc., Pittsburgh PA, 1998.

[12] C.H. Bushweller, Conformational Behavior of Six-membered Rings, VCH, New York, 1995, pp. 25–58.

[13] F. Freeman, M.L. Kasner, Z.M. Tsegai, W.J. Hehre, J. Chem. Ed. 77 (2000) 661.

[14] K.B. Wiberg, J.D. Hammer, H. Castejon, W.F. Bailey, E.L. De-Leon, R.M. Jarret, J. Org. Chem. 64 (1999) 2085.

[15] H. Booth, J.R. Everett, J. Chem. Soc., Perkin Trans. 2 (1980) 255.

ELSEVIER

Journal of Molecular Structure (Theochem) 666–667 (2003) 117–122

THEO CHEM

www.elsevier.com/locate/theochem

Analysis of weakly bound structures: hydrogen bond and the electron density in a water dimer

F. Bartha[a,*], O. Kapuy[b], C. Kozmutza[b], C. Van Alsenoy[c]

[a]*Department of Theoretical Physics, University of Szeged, Tisza Lajos krt. 84-86, Szeged 6720, Hungary*
[b]*Budapest University of Technology and Economics, 1111 Budapest, Budafoki út 8, Hungary*
[c]*Department of Chemistry, University of Antwerp, B-2610 Wilrijk, Universiteitsplein 1, Antwerp, Belgium*

Abstract

The formation of the hydrogen bond is investigated in terms of the charge density. The water dimer is chosen as an example, and the redistribution of the electrons due to the H-bond is discussed. The density difference between interacting and non-interacting monomers is partitioned using localized orbitals and localized partial densities. Although the change in the electronic density is not completely described by the deformation of the two localized orbitals, OH bond and the appropriate lone pair orbital, the binding region is already well characterized with these local densities.
© 2003 Published by Elsevier B.V.

Keywords: Water dimer; Electron density; Localized molecular orbitals

1. Preamble

The isolation and structural analysis of proteins contained from a complex system, whether it be in food or the human body, is of great importance. In this century, it is expected that a detailed analysis of proteins will be made and the accumulated information will be stored in databases. Such accumulation of data will inevitably lead to the generation of healthier foodstuffs. Presumably, the isolation and structural analyses of more and more proteins will also allow early diagnosis of diseases and revolutionize drug design.

Therefore, in order to target a given protein, three tasks must be performed:

(a) isolation of the protein (from food or from the human body),
(b) determination of its structure (sequence and folded 3D structure)
(c) storage of the information using a convenient and searchable database.

The experimental work is very labour intensive. The isolation of proteins is regularly done by gel electrophoresis (GE). Several more refined methods such as one- and two-dimensional, and pH-gradient GE's have also been developed [1] (and references therein). After the GE procedure, proteins are further treated. For example in wheat, one of the storage proteins is glutenin, which can be decomposed to high and low molecular weight subunits (HMW and LMW, respectively); the further separation of these larger subunits is more difficult. Using well-known databases, the composition pattern of proteins can be

* Corresponding author. Tel.: +36-62-54-4807; fax: +36-62-54-4368.
E-mail address: barthaf@physx.u-szeged.hu (F. Bartha).

0166-1280/$ - see front matter © 2003 Published by Elsevier B.V.
doi:10.1016/j.theochem.2003.08.020

identified ([2] and references therein). After the purification of proteins or their constituent fragments, the structural analysis may expand to include other procedures. One of the most effective methods used in this case is mass spectroscopy (MS), which is often combined with the matrix laser desorption ionization (MALDI) technique. Determining the conformation of complex proteins is usually done using nuclear magnetic resonance (NMR) spectroscopy. Current NMR techniques are of the highest level, as attested by the recognition it received when awarded the Nobel Prize in Chemistry, 2002 [3]. Thus, after this brief overview of the experimental methods widely used in protein structural analysis, let us turn to the theoretical field.

2. Introduction

Molecular structures have been studied by theoretical means since the introduction and development of quantum mechanics. The ab initio method originally involved a computational procedure where only the nuclear charges and the position of atoms were considered. Clearly, this method could provide, at least in principle, results that would be in good agreement with experimentally determined data. Later, the ab initio expression was 'altered' and extended to include both experimental and theoretical procedures that are of the highest level within a given series of more or less systematic and consecutive approximations. It is well known that by the ab initio (theoretical) method, even when using its more extended interpretation, large proteins cannot be studied. Only model protein conformation studies can be completed [4]. In this context, large molecules are defined as those having a molecular weight of 10,000 Da or more. Theoretical (ab initio) methods may contribute to the investigation of protein structures larger than this threshold only when local effects are studied. From this, it would follow, that the localized representation must be determined, and the experimental procedure has to determine a well-defined region that should be further theoretically analyzed.

Thus, one may ask whether theoretical methods can contribute to protein analysis. In the investigation

of food by-products, it would be straightforward to use theoretical calculations to help in studying LMW structures. In addition, the interactions between proteins and cofactors, inhibitors, etc. could be treated theoretically after convenient localization. It is also well known that the 'individual character' of NMR signals is highly affected by the position of active nuclei, i.e. by their 'chemical neighborhood'. This could be the gate through which quantum chemistry could enter. The chemical surrounding of relatively small regions can be accurately treated by ab initio methods. The tasks ahead of theoretical researchers are as follows:

- to investigate the largest possible neighbourhood of atoms, whether it be a whole molecule or part of a molecule, using small basis sets at the Hartree-Fock (HF) level,
- to calculate physically well-defined (one-electron) quantities that are less sensitive to geometry and basis set variation, and
- to create well-located parts, preferably localized molecular orbitals (LMOs). The quantities associated to these LMOs are able to describe the fine electronic charge distributions of the given region.

This latter requirement is of special interest, as the procedure can provide information about the effect of a given chemical environment, arrangement of bond and lone pair LMO densities in a molecule [5,6].

Chemical interactions, including the weak interactions of the van der Waals type variety, are extremely important in determining the 3D structure of proteins. Our aim is to investigate using theoretical means, the presence of hydrogen bonds in molecular systems.

The study was performed on the smallest H-bonded structure, the water dimer. The method is based on the calculation of LMO densities and density differences, and making comparisons to the similar quantities in monomer units. The method is under further development and will be soon extended to investigate larger molecules.

3. An overview of the water dimer: structure and methods

The water dimer as prototype of H-bonding has been studied by several theoretical methods. When it was discovered that H-bonds are partly responsible for macromolecular structure, the investigations became still more intensive. Both HF-based many-body perturbation and density functional theory, in addition to other methods, focused on the calculation of the interaction energy in the water dimer and larger water clusters. In spite of the efforts made towards this subject, the interaction energy of a water dimer can only be calculated to a restricted accuracy [7–11]. This is partly due to the sensitivity of the interaction energy to geometric distortions such as red shifting in stretching frequencies. The red shifting hydrogen bond is due to the 'addition' of electron density.

A large number of experimental studies have also been performed related to water in its vapor and/or ice forms, which included free clusters. These measures correlate with theoretical investigations [12–14]. Certain transition structures were calculated within the water dimer, which showed that the barriers of the various tunneling effects do not coincide with previous empirical estimations [15]. The calculations unequivocally suggest a flat nature of the potential energy hypersurface for the water dimer. These results affirm that the water dimer cannot be theoretically described in a useful way by calculating interaction energy. This observation may also hold for H-bonded structures in general. Thus, this suggests that other procedures may be more informative.

Several attempts were made to outline various energy decomposition schemes of intermolecular interactions [16-19]. These procedures provide useful information on the different energy quantities (electrostatic, charge transfer, etc.) in H-bonded systems. The use of localized representation seemed to be advantageous in certain studies where LMOs were obtained by unitary transformations [20,21].

4. Local changes in the electron density

From density functional theory it is known that the ground state one-electron density of a molecule is enough to rebuild the molecular Hamiltonian, and that all the ground state properties of a molecule are, in principle, reflected in the one-electron density. However, we do not have a method to extract the required information directly from this density. When focusing on the weak interaction between parts of a (super-) molecule, it is reasonable to consider changes in the electronic density with the full interaction as allowed or neglected. This difference should give insight into the very nature of the interaction, while irrelevant features that do not contribute to the bond under investigation disappear by subtraction.

For a spatially extended complex, such as a biomolecule with hundreds or thousands of atoms, the interaction is expected to take place at certain active sites with the involvement of only a few nuclei. This means that the interaction could be well characterized by looking at the changes in the electronic density in a relatively small, localized volume surrounding the 'hot spots'. Instead of looking at the electron density of the whole molecule, a local examination is sufficient. The rough implementation of this idea can be carried out using the reduction of the electronic charge density to point charges located at the nuclei and describe the interaction in terms of charge transfer. However, a closer look at the detailed redistribution pattern of the electrons in the H-bond shows that the reduction of point charges must be an oversimplification of the binding process.

Many of our theoretical methods provide the one-electron density as the sum of orbital densities $\rho = \Sigma_i n_i \rho_i$ with $\rho_i(\mathbf{r}) = \phi_i(\mathbf{r}) * \phi_i(\mathbf{r})$. Here the $\phi_i(\mathbf{r})$ denotes the real molecular orbitals and the n_i occupation numbers are the result of a quantum chemical calculation. The molecular orbitals, and, therefore, the ρ_i (orbital) partial densities, are the so-called canonical quantities that 'normally' extend to the whole molecule and are inappropriate for a local description of the confined spatial regions. There are several ways to partition density into the sum of localized charge clouds, one of which is the partitioning based on the standard Boys-localized orbitals. In the calculations presented here, this implies that the single reference closed shell ground state determinant of the quasi-molecule is reproduced with using the unitary transform of the canonical orbitals $\Psi_a = \Sigma U_{ai} \phi_i$ whose orbitals give the same n-electron ground sate, the same one-electron density

ρ, but partial $\rho_a = \Psi_a * \Psi_a$ orbital densities are concentrated best on their centroids.

5. Calculations

The ground state of the water dimer was calculated using the B3LYP hybrid density functional method with the CC-pVTZ (5D, 7F) basis set. For a quantitative approach, larger basis sets and more sophisticated methods are needed, as the qualitative nature of the H-bond is believed to be well described already at this level of the calculations.

The geometry of the $(H_2O)_2$ supermolecule constrained to the C_s symmetry was optimized giving (distances in Å, angles in degrees): O1–H2 = 0.969, O1–H3 = 0.961, H3–O1–O4 = 104.40, O4–H5 = 0.962, O4–H6 = 0.962, H5–O4–H6 = 105.15, O1–O4 = 2.914, (H2–O4 = 1.945) and O1–O4–H7 = 108.54.

In order to study the *intermolecular* interactions in the dimer, the properties of the two non-interacting water molecules were also calculated. When focusing on the changes in the electronic density, it was necessary to use the same optimized dimer geometry in all these calculations. To compensate for the errors due to the incomplete basis set (BSSE), the non-interacting molecules were calculated using the full dimer basis. The 'binding energy' in this context was found to be: $\Delta E = E_{dimer} - E_{monomers} = -4.5$ kcal/mol. This is just a model binding energy, as the monomers are not free space water molecules at infinite separation but the monomers with the internal coordinates fixed at the dimer optimum values.

The valence one-electron density of the dimer is shown in Fig. 1., were the isodensity surface (as in all the figures) is plotted at $\rho = 0.01$(a.u.). To visualize the redistribution of the electrons when forming the hydrogen bond, the $\Delta\rho = \rho_{dimer} - \rho_{monomers}$ equisurface is also plotted in the same figure. Isosurface is determined at $\Delta\rho = \pm0.001$ for density differences used in all the figures. Here, we subtracted the electronic densities of the two non-interacting water molecules from that of the dimer. Dotted surfaces denote negative change, i.e. loss of electrons. Were it a covalent bond, the increase of the electronic density should appear along the bond

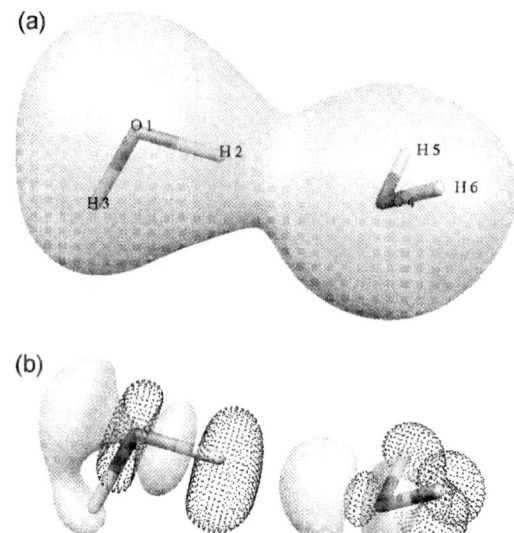

(a)

(b)

Fig. 1. One electron density equisurface ($\rho = 0.01$ a.u.) for a water dimer. (a) Change in the one-electron density due to H-bonding $\Delta\rho = \rho_{dimer} - \rho_{monomers}$. Density difference isosurfaces are at $\Delta\rho = \pm0.001$, dotted surface are volumes with loss of electrons (b).

axes. If it were an ionic bond, an acceptor–donor deformation of the density could have been observed. What is observed instead is that the density difference changes sign several times along the H-bond, which is related to electrons leaving or entering regions along the bond. This peculiar alternation of the density change in the H-bond could have been only an artifact of the calculations. The geometry was optimized for the interacting dimer and the distortions in the internal nuclear conformation of the monomers are also present in the non-interacting monomer calculations where the nuclei are promoted to form the bond. However, test calculations were carried out by placing the free space optimized monomers in a relative position of a H-bonded dimer and the density difference showed the same pattern as with the promoted geometry.

As a next step, the localization of the molecular orbitals was done with the Boys method for the dimer and for the monomers. The orbitals localized on the center of the H-bond region are the OH bond orbital of the left monomer and the lone pair orbital of the right monomer, with both orbitals doubly occupied. See Figs. 2 and 3 for the orbital charge densities in the dimer and the corresponding

(a)

(b)

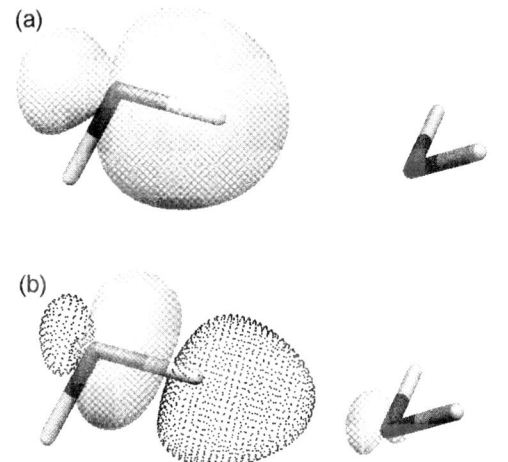

Fig. 2. Localized orbital charge density of the OH bond (a), and its deformation in the H-bond (b).

density difference equisurfaces. These pictures suggest that the electrons in the OH bond contract towards the oxygen atom, whereas the electrons in the lone pair orbitals move towards the partner molecule.

In Fig. 4., the charge density of all the four electrons in question is plotted together with their contribution to the valence electron density difference. One concludes that the qualitative behaviour of the change in the partial densities of the localized electrons well represents the variation of the total

(a)

(b)

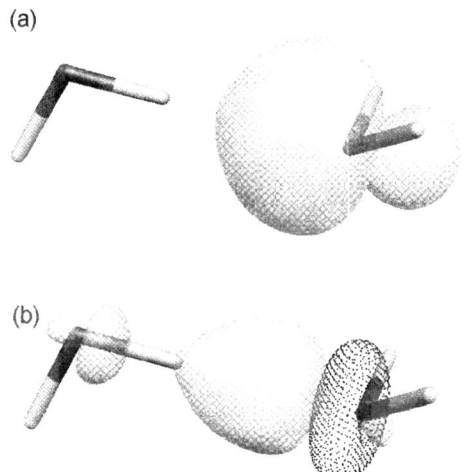

Fig. 3. The charge density of the lone pair electrons participating in the H-bond (a) and the redistribution of the charge in the binding process (b).

(a)

(b)

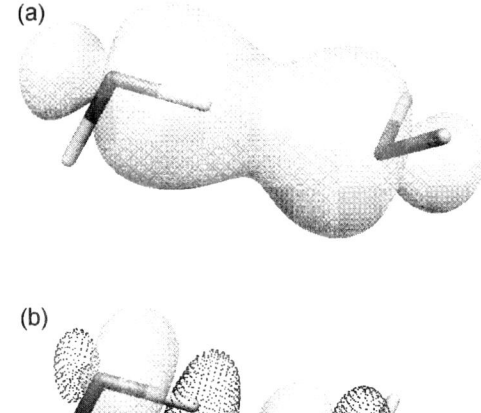

Fig. 4. The partial density of the four electrons mostly involved in forming the H-bond between the two water molecules (a) and the deformation with respect to the non-interacting corresponding monomer orbital densities (b).

charge density in the H-bond region, which is the most important in the binding process.

Acknowledgements

This work was supported by the Ministry of the Flemish Community under contract BIL01/72 in the framework of the bilateral scientific cooperation between Belgium and Hungary.

References

[1] S.A. Forsyth, R.D. Thompson, P.J. Sharp, P.I. Payne, in: R. Lasztity, F. Bekes (Eds.), Proceedings of Third International Workshop on Gluten Proteins, World Science, Budapest, Hungary, 1987, p. 331. M.H. Morel, Cereal Chem., 71, 1994, pp. 238.

[2] E.A. Jackson, M.H. Morel, T. Sontag-Strohm, G. Bramlard, E.V. Metakovsky, R. Redaelli, J. Genet. Breed. 50 (1996) 321.

[3] A. Perczel, Természet Vilga 134/4 (2003) 154.

[4] R.J. Lavrich, D.F. Plusquellic, R.D. Suenram, G.T. Fraser, A.R. Hight Walker, M.J. Tubergen, J. Chem. Phys. 118 (2003) 1253.

[5] E. Kapuy, C. Kozmutza, Chem. Phys. Lett. 226 (1994) 484.

[6] V. Kairys, J.H. Jensen, Chem. Phys. Lett. 315 (1999) 140.

[7] S.J. Chakravorty, E.R. Davidson, J. Phys. Chem. 97 (1993) 6373.

[8] D.R. Hamann, Phys. Rev., B 55 (1997) R10/157.

[9] R. Specchio, A. Famulari, M. Sironi, M. Raimondi, J. Chem. Phys. 111 (1999) 6204.

[10] G.S. Tschumper, M.L. Leininger, B.C. Hoffman, E.F. Valeev, H.F. Schaefer, M. Quack, J. Chem. Phys. 116 (2002) 690.

[11] I. Adamovic, M.A. Freitag, M.S. Gordon, J. Chem. Phys. 118 (2003) 6725.

[12] E. Whalley, Trans. Faraday Soc. 53 (1957) 1578.

[13] O. Björneholm, F. Federmann, S. Kakar, T. Möller, J. Chem. Phys. 111 (1999) 546.

[14] E.D. Isaacs, A. Shukla, P.M. Platzman, D.R. Hamann, B. Barbiellini, C.A. Tulk, J. Phys. Chem. Solids 61 (2000) 403.

[15] B.J. Smith, D.J. Swanton, J.A. Pople, H.F. Schaefer, L. Radom, J. Chem. Phys. 92 (1990) 1240.

[16] C. Kozmutza, E. Tfirst, Adv. Quantum Chem. 31 (1998) 231.

[17] Y. Mo, J. Gao, S.D. Peyerimhoff, J. Chem. Phys. 112 (2000) 5530.

[18] P. Salvador, M. Duran, X. Fradera, J. Chem. Phys. 116 (2002) 6443.

[19] J. Langlet, J. Caillet, J. Berge's, P. Reinhardt, J. Chem. Phys. 118 (2003) 6157.

[20] E. Kapuy, Z. Csépes, C. Kozmutza, Int. J. Quantum Chem. 23 (1983) 981.

[21] E. Kapuy, C. Kozmutza, J. Chem. Phys. 94 (1991) 55.

ELSEVIER

Journal of Molecular Structure (Theochem) 666–667 (2003) 123–129

THEO CHEM

www.elsevier.com/locate/theochem

Ab initio calculations on simple heterocycles and DNA base-pair triplets

G. Paragi[a,*], C. Van Alsenoy[b], B. Penke[a,c], Z. Timár[c]

[a]*Protein Chemistry Research Group of Hungarian Academic of Science, University of Szeged, Dóm tér 8, Szeged H-6720, Hungary*
[b]*Department of Chemistry, University Insstelling Antwerp, B-2610 Wilrijk, Belgium*
[c]*University of Szeged, Department of Medical Chemistry, Dóm tér 8. Szeged, Hungary*

Abstract

Ab initio calculations (HF and B3LYP) were performed on the simplest triplex-forming heterocycles, namely 2-aminopyridine, 2-aminoquinoline, 2-aminopyrrole and 2-aminoindole. Computations show propeller and buckle conformations between purines and triplex-forming heterocycles. Isomorphicity, H-bonding energy and comparison of computational methods are also presented.
© 2004 Elsevier B.V. All rights reserved.

Keywords: DNA; Triplex; Heterocycles; Density functional theory; Hartree–Fock

1. Introduction

Investigation of nucleic acid base triplets runs back over a five-decade. Recent relevance of DNA triple helices resides in the recognition that this phenomenon may be exploited in the war against diseases such as cancer [1,2]. The antigene or triplex strategy targets the working double stranded DNA—that is the cause of the sickness—by a triplex-forming third nucleic acid strand [3]. The specific recognition and binding of the third strand to the doublet DNA is the key of the success. H-bonding is believed to be responsible for the specific binding and partly for stabilization [4].

Watson–Crick H-bonding is responsible for the DNA base-pair formation and Hoogsteen or revers-Hoogsteen type of binding is observed between the major groove triplex-forming structure and the purine DNA base. The availability of the pyrimidine nucleobase from the major groove in the triple helix is restricted by the fact that only large planar molecules with more than two aromatic rings may reach both the purine Hoogsteen sites and the same on the pyrimidine. These big structures will rather intercalate into the double stranded chain thus losing specificity. The idea for the general solution is that, only the purines are targeted in Hoogsteen interaction by one or two-membered planar (aromatic) heterocycles containing the least H-bonding sites possible. We are confident that these structures are 2-aminopyridine, 2-amino-quinoline, 2-aminopyrrole and 2-aminoindole. Semi-empirical pre-calculations were performed by Fenyő et al. [5], but the problem requires higher level of theory calculation (ab initio). Following the line of the hypoxanthine–DNA base-pair investigations, the model systems contain the DNA bases as part of the DNA strand. It should further be noted that

---

* Corresponding author. Tel./fax: +36-62-544-368.
E-mail address: paragi@sol.cc.u-szeged.hu (G. Paragi).

the simplest 'bases' may be modified in order to strengthen the binding or specificity if needed. These are only suggestions towards a possible solution of the problem.

2. Calculation method

The accepted procedure of investigation of DNA base-pairs is ab initio calculation but the results depend on the chosen method. For example, in

a previous paper of hypoxanthine–DNA base-pair investigation [6], we found that Hartree–Fock (HF) calculation sometimes unable resulting in Hoogsteen-bonding geometries that are predicted by Density Functional Theory (DFT) method. By this experience, both types of calculation were performed, namely HF and DFT with Becke's three parameters (B3LYP) exchange-correlation potential. The effects of choice of the basis set were studied. The basis set superposition error was always taken into account in binding energy

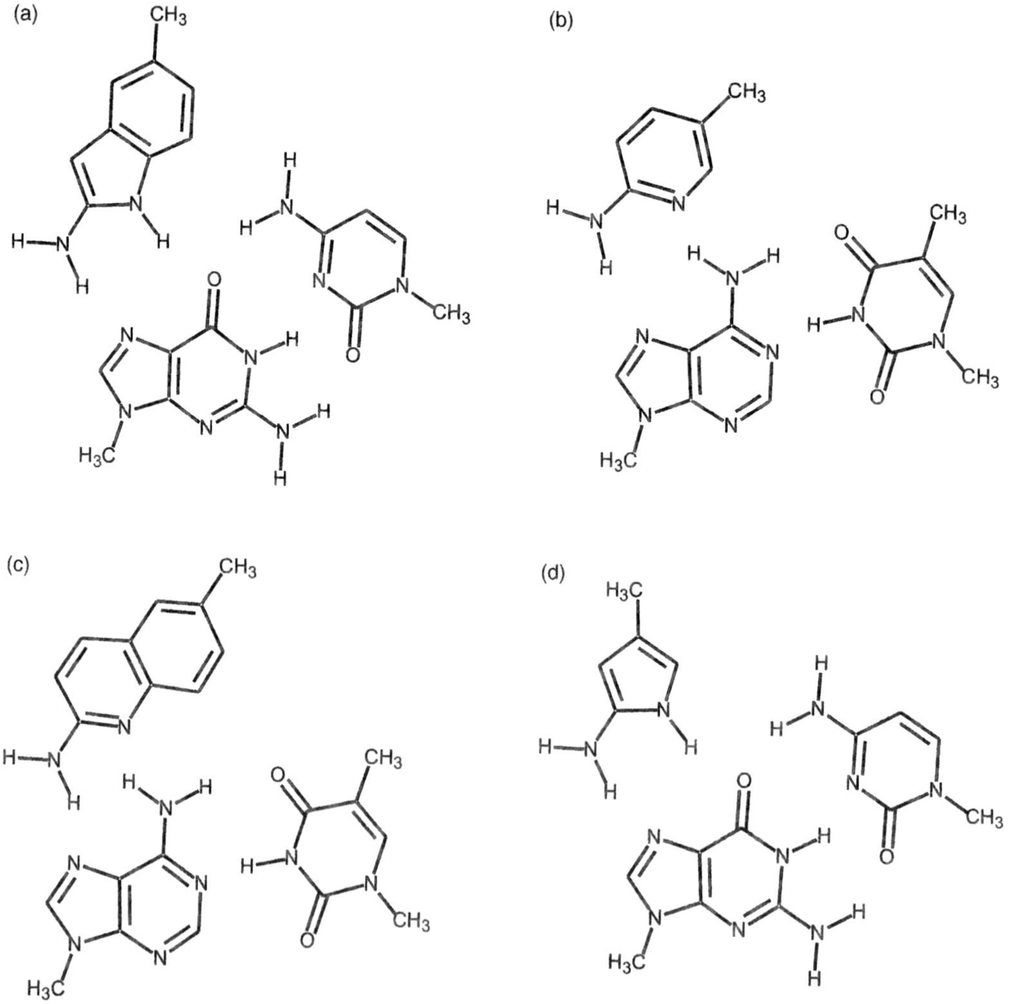

Fig. 1. Starting conformation of the triplex systems. (a) 2-Aminoindole with guanine and cytosine, (b) 2-aminopyridine with adenine and thymine, (c) 2-aminoquinoline with guanine and cytosine, and (d) 2-aminopyrrole with guanine and cytosine. The figures below the boxes show side view aspect of the system.

calculations. GAUSSIAN98 [7] program package were used in all cases.

3. Results and discussion

Second order bonds (mainly Hydrogen bonding) are believed to be responsible for the specific binding and partly for stabilization in biological systems. Therefore, we have investigated the four simplest possible triplex-forming heterocycles

against natural DNA base doublets. Previous studies [4,6] showed that model systems built up from bases of DNA provide enough information for structure analysis and binding energy preference order studies.

The boundary conditions to design the simplest DNA binding third molecule were the following: (1) a triplex-forming heterocycle should only contain one elementary H-bonding moiety, (2) it is an aromatic single ring or condensed aromatic double ring system, (3) all the four designed

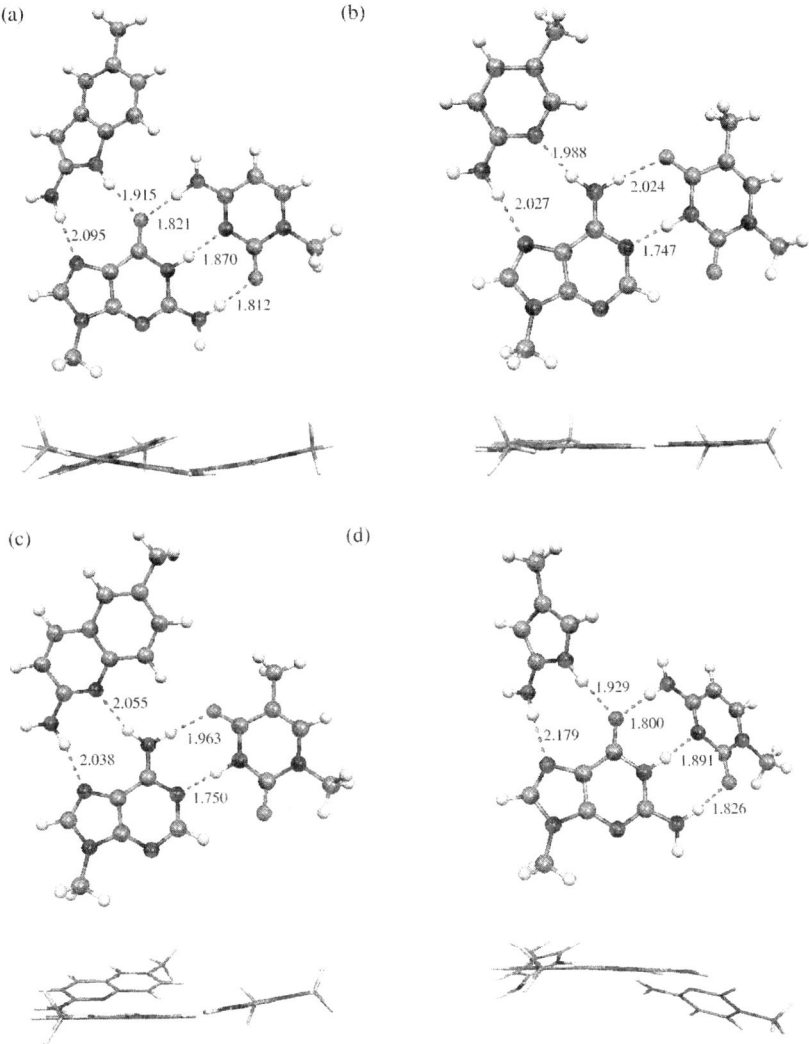

Fig. 2. Optimized triplex form of (a) 2-aminoindole (b) 2-aminopyridine (c) 2-aminoquinoline and (d) 2-aminopyrrole with DNA bases at the HF/3-21G level of theory. The numbers are the length of the hydrogen bonds (in Å). The figures below the boxes show side view aspect of the system.

structures should be isomorphic, (4) oligomers containing the non-natural bases do not support Watson–Crick hybridization, (5) they are chemically and biologically stable.

Four possible starting arrangements of the triplet systems appropriate for the above conditions are shown in Fig. 1.

Optimization were started from planar configuration in all cases. Vibrational frequency analysis was performed after the optimization for the optimized conformations in order to check the type of the geometry. It is a necessary step because the GAUSSIAN program can finish the geometry optimization in a transition state. For example, in case of 2-aminoindole, we got one negative frequency in the spectrum of the optimized geometry at the HF calculation, but the visualization of this negative frequency pointed out that this belongs to the rotation of the methyl group in the 2-aminoindole molecule. It means that this calculation also resulted in stable conformation.

Optimized conformations of HF/3-21G calculations of the four triplex systems are presented in Fig. 2. It shows that in most cases, there are buckle and propeller type twists except the 2-aminopyridine formation (Fig. 2(b)), where the conformation moved just slightly out of the starting plane. The biggest twist can be found at 2-aminopyrrole triplex system (Fig. 2(d)). The lengths of hydrogen bonds are typical. The binding energies of the different components of triplexes are shown in Table 1. Furthermore, Table 1 includes the results of the B3LYP calculations as well. B3LYP method also found stable conformations but the 2-aminoquinoline case. From Fig. 2, in general, it can be drawn that the lengths of the H-bonds of the proposed molecules are longer compared to the H-bonds of the natural DNA bases in the triplex. Therefore, the binding energies for the heterocyclic molecules are less than that of the cytosine or the thymine. Of course, adenine or guanine is bound strongest to the system, because these take part in all the intermolecular interactions in the triplexes. From the calculated energy values, the following consequences can be drawn: the sum of the binding energies of the artificial bases and the pyrimidine base to the purine approximately equals to the binding energy of the purine in the same triplex

Table 1

Dependence of the binding energy (in kcal/mol) of the constituents of the different triplex formations on the calculation method

		HF/3-21G (kcal/mol)	B3LYP/3-21G (kcal/mol)
Triplex I	2-Aminoindole	− 15.153	− 19.116
	Cytosine	− 34.319	− 42.247
	Guanine	− 47.422	− 54.941
Triplex II	2-Aminopyridine	− 9.592	− 15.480
	Thymine	− 13.373	− 18.345
	Adenine	− 23.915	− 35.447
Triplex III	2-Aminoquinoline	− 9.000	−
	Thymine	− 13.584	−
	Adenine	− 22.746	−
Triplex IV	2-Aminopyrrole	− 12.244	− 14.523
	Cytosine	− 33.125	− 38.257
	Guanine	− 43.092	− 50.156

(see Table 1). The little difference is the consequence of the polarization effects induced by the DNA base. The average value for H-bond energy is usually 5–10 kcal/mol. The calculated binding energies are between 9 and 20 kcal/mol, which corresponds well to the above value, because the designed molecules and purines bind to the system by two H-bonds. The dependence of the optimized geometry on the calculation method is summarized in Fig. 3. It can be seen that B3LYP calculations always predict shorter (first column structures) distances, and consequently stronger binding energies (see Tables 1 and 2).

In Table 2, we present the effect of the different basis set and computational method for the 2-aminopyridine system. It is really surprising, that the results of the HF/3-21G and the B3LYP/6-31G** calculations are very close. It should be emphasized that the optimizations were started from the same starting position. In Fig. 4, we present the optimized structure of the 2-aminopyrridine triplex. If we compare the optimized bond lengths in the same calculation method as the function of the basis set, the larger set always provides longer bonds, and consequently weaker binding energies (Table 2). This statement is in good accordance with the results of the hypoxanthine–DNA base-pair investigations [6].

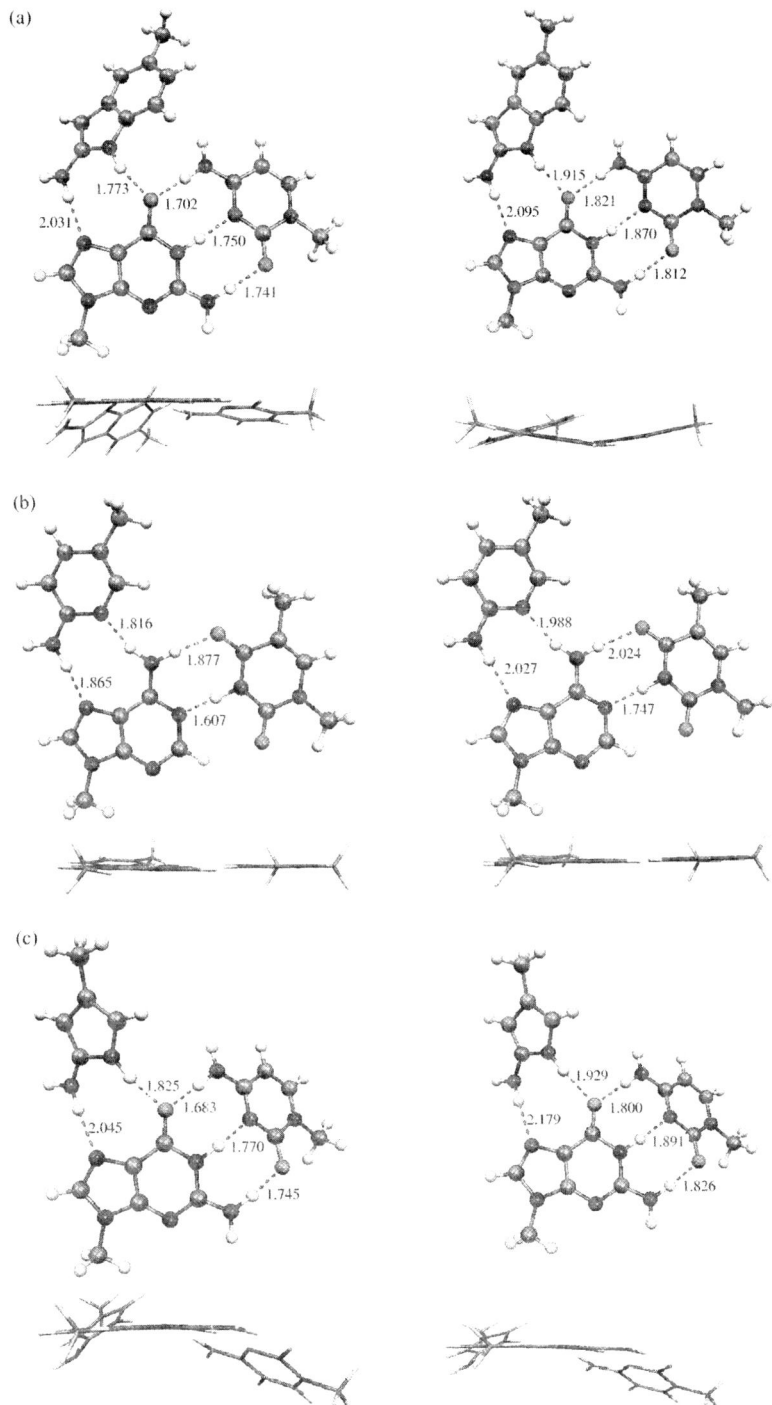

Fig. 3. The dependence of the optimized geometries on the chosen calculation method. In the first column, the results of the B3LYP calculation are placed, in the second column, the Hartree–Fock. The numbers are the H-bond lengths (in Å).The (a), (b) and (c) sign the 2-aminoindole, 2-aminopyridine and the 2-aminopyrrole triplex system, respectively. All the calculations were performed with 3-21G basis set. The figures below the boxes show side view aspect of the system.

Table 2
Dependence of the binding energy (in kcal/mol) of the constituents of the 2-aminopyridine triplex on the calculation method and the choose of the basis set

	HF/3-21G (kcal/mol)	B3LYP/3-21G (kcal/mol)	HF/6-31G** (kcal/mol)	B3LYP/6-31G** (kcal/mol)
2-Aminopyridine	− 9.592	− 15.480	− 6.619	− 10.028
Thymine	− 13.373	− 18.345	− 9.691	− 13.032
Adenine	− 23.915	− 35.447	− 17.054	− 24.388

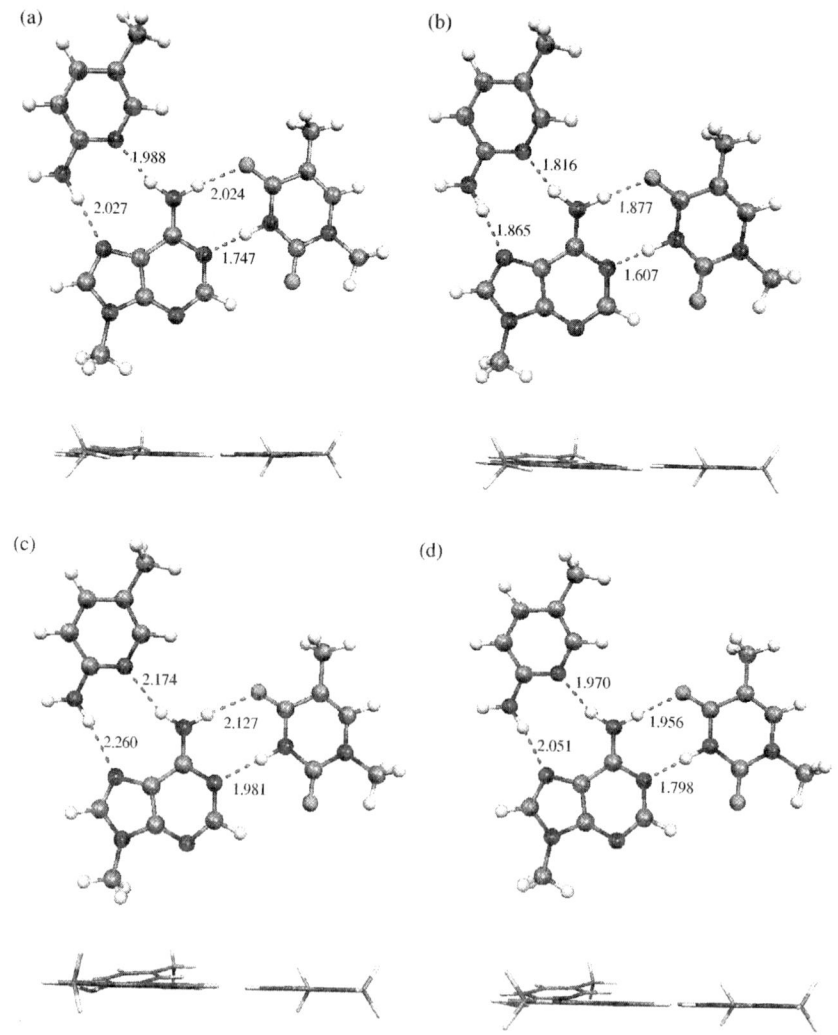

Fig. 4. The dependence of the optimized structures on the calculation method and the choice of the basis set in the 2-aminopyridine case. The first two cases are the HF (a) and the B3LYP (b) results with the smaller (3-21G) basis set, the (c) and the (d) cases are the HF/6-31G** and B3LYP/6-31G** calculations, respectively. The figures below the boxes show side view aspect of the system.

4. Conclusions

The results of calculations of possible triplex formations of DNA bases and four heterocyclic molecules are presented. Stable conformation may be formed in these simplified model systems. The length of H-bonds (and consequently the binding energies of the components) depends on the chosen calculation method and the applied basis set. Comparing Hartree–Fock and DFT (B3LYP) methods, it can be stated that B3LYP calculations always provide shorter bond lengths (Figs. 3 and 4), and consequently stronger binding energies (Tables 1 and 2). According to the basis sets, it can be reported that larger one (6-31G**) resulted in longer bond lengths and thus smaller binding energies. The consequence of these two facts is that cheaper calculations (HF/3-21G) provide very similar results as the more sophisticated (B3LYP/6-31G**) ones. All these statements are in good accordance with the results of previous investigations for DNA base-pairs.

Acknowledgements

This work was partially supported by the Flemish-Hungarian Scientific and Technological Joint Found (TeT B-2/01 and BIL01/72) and the National Scientific Research Fund Hungary (OTKA D38492)

References

[1] J.C. Francois, T. Saisonbehmoaras, C. Helene, Nucleic Acids Res. 16 (1988) 11431.

[2] H.E. Moser, P.B. Dervan, Science 238 (1987) 645.

[3] D. Guianvarc'h, J.L. Fourrey, R. Maurisse, H.S. Sun, R. Benhida, Bioorg. Med. Chem. 11 (2003) 2751.

[4] J. Sponer, P. Hobza, Encyclopedia of Computational Chemistry, Wiley, Chichester, 1998, 777 pp.

[5] R. Fenyő, Z. Timár, I. Pálinkó, B. Penke, J. Mol. Struct. (THEOCHEM) 496 (2000) 101.

[6] G. Paragi, I. Pálinkó, C. Van Alsenoy, I.K. Gyémánt, B. Penke, Z. Timár, New J. Chem. 26 (2002) 1503.

[7] M.J. Frisch, G.W. Trucks, H.B. Schlegel, G.E. Scuseria, M.A. Robb, J.R. Cheeseman, V.G. Zakrzewski, J.A. Montgomery, Jr., R.E. Stratmann, J.C. Burant, S. Dapprich, J.M. Millam, A.D. Daniels, K.N. Kudin, M.C. Strain, O. Farkas, J. Tomasi, V. Barone, M. Cossi, R. Cammi, B. Mennucci, C. Pomelli, C. Adamo, S. Clifford, J. Ochterski, G.A. Petersson, P.Y. Ayala, Q. Cui, K. Morokuma, D.K. Malick, A.D. Rabuck, K. Raghavachari, J.B. Foresman, J. Cioslowski, J.V. Ortiz, A.G. Baboul, B.B. Stefanov, G. Liu, A. Liashenko, P. Piskorz, I. Komaromi, R. Gomperts, R.L. Martin, D.J. Fox, T. Keith, M.A. Al-Laham, C.Y. Peng, A. Nanayakkara, C. Gonzalez, M. Challacombe, P.M.W. Gill, B. Johnson, W. Chen, M.W. Wong, J.L. Andres, C. Gonzalez, M. Head-Gordon, E.S. Replogle, J.A. Pople, GAUSSIAN'98, Revision A.7, Gaussian, Inc., Pittsburgh, PA, 1998.

ELSEVIER

Journal of Molecular Structure (Theochem) 666–667 (2003) 131–134

THEO
CHEM

www.elsevier.com/locate/theochem

Conformational analysis of substituted (E)-4-phenylbut-3-en-2-ones

Timea T. Polgár[a], Gyula Tasi[a],*, Imre G. Csizmadia[b,c,d]

[a]Department of Applied and Environmental Chemistry, University of Szeged, Rerrich B. tér 1, H-6720 Szeged, Hungary
[b]Department of Medical Chemistry, Faculty of General Medicine, University of Szeged, Dóm tér 8, H-6720 Szeged, Hungary
[c]Department of Chemistry. University of Toronto, 80 St. George St., Toronto, Ont., Canada M5S 3H6
[d]Global Institute of COmputational Molecular and Materials Science GIOCOMMS)@1422 Edenrose St., Mississauga, Ont., Canada L5V 1H3

Abstract

The biological role of many medical herbs used in traditional Chinese medicine has already been studied. The antimutagenic activity of BZ's ((E)-4-phenylbut-3-en-2-ones) derivatives isolated from *Scutellaria barbata* was quantitatively analyzed in term of physicochemical parameters. Popelier et al. carried out a quantum topological molecular similarity study on a set of 15 phenylbutenone derivatives. The equilibrium geometries of the molecules in question were determined at the HF/6-31G(d) level of theory. According to their results, all the trial conformations collapsed to the planar configuration with C_s symmetry. This short paper presents that at the second order Møller–Plesset level of theory at least two chiral conformers exist in each case. Of the set of 15 phenylbutenone derivatives, two representative examples were chosen to demonstrate the results of the conformational analysis.
© 2003 Published by Elsevier B.V.

Keywords: Phenylbutenone derivatives; Conformational analysis; Chirality; Quantum molecular similarity; Drug design

1. Introduction

The α,β-unsaturated ketone, (E)-1-(4'-hydroxyphenyl)-but-3-en-2-one, was obtained from the dried plant *Scutellaria barbata* D. Don (*Labiatae*), which has been used as an antitumour agent, antibacterial compound, and diuretic [1–4]. It was found to have moderate antitumour activity and hence further substituted analogues were prepared. Their antitumour activities were analyzed by using an in vitro cell culture system against the K562 human chronic myelogenous leukaemia cell line [5].

* Corresponding author. Tel./fax: +36-62-544-619.
E-mail address: tasi@chem.u-szeged.hu (G. Tasi).

0166-1280/$ - see front matter © 2003 Published by Elsevier B.V.
doi:10.1016/j.theochem.2003.08.022

Popelier et al. determined the equilibrium geometries of the conformers of fifteen derivatives [6]. It was found that all the trial conformations collapsed to the planar configuration with C_s symmetry at Hartree–Fock (HF) level of theory. For the quantum topological molecular similarity (QTMS) analysis [6–8], the most stable conformer was used in each case.

Here we restrict ourselves to present the results of the geometry optimizations performed on two members of the Popelier's set (Nos. 5 and 11, Fig. 1). We determined the equilibrium molecular geometries of the conformers at AM1 [9], HF, and second order Møller–Plesset (MP2) [10] levels of theory. Interestingly, the equilibrium molecular geometries found by Popelier et al. at HF level are transition states at the MP2 level of theory. Quantum molecular similarity

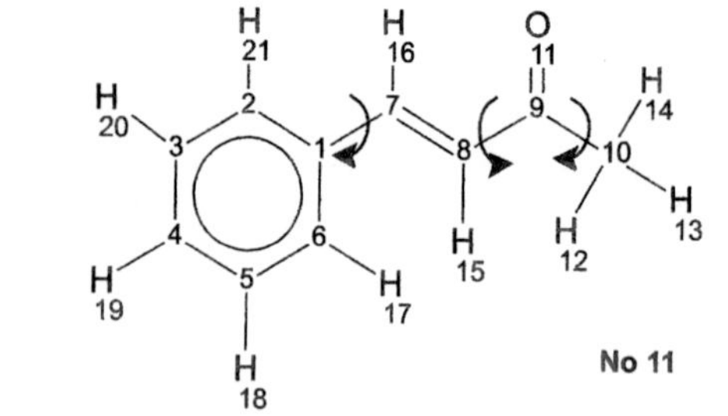

(3E)-4-phenylbut-3-en-2-one

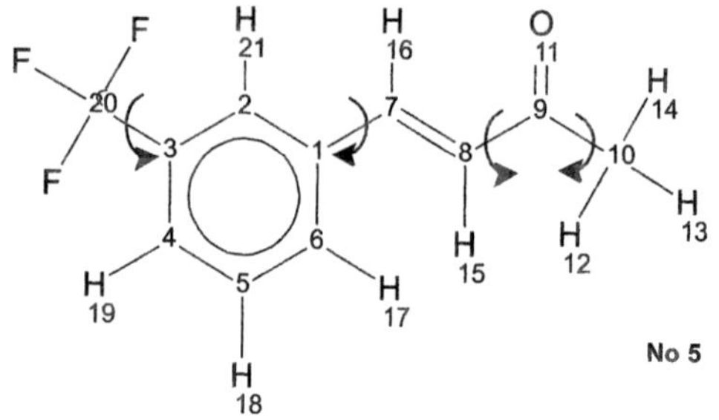

No 11

(3E)-4-[3-(trifluoromethyl)phenyl]but-3-en-2-one

No 5

Fig. 1. Free torsional angles considered in the molecules under consideration.

(QMS) indices [11] were also calculated and we made an effort to point out the part of the molecules that was responsible for the observed biological activity.

2. Methods and software packages

For semiempirical and ab initio quantum chemical calculations, the following program packages were used: PcMol [12], MOPAC 6.0 [13] and GAUSSIAN98 [14]. For drawing and analyzing the molecular structures obtained, gOpenMol [15] and ACD/ChemSketch [16] packages were used.

3. Results and discussion

First, an AM1 geometry optimization using the PcMol program package was performed. With the help of the optimized geometry obtained, 300–500 trial conformations were generated using the Monte-Carlo method (Fig. 1). These initial geometries were then optimized at AM1 level using the MOPAC program. The AM1 geometry optimization

procedure led to numerous identical conformations, which were then omitted. For the rest of the conformers further AM1 geometry optimizations were performed using the GAUSSIAN98 package at very tight level with harmonic vibrational analysis. The identitical conformations obtained were also left out. This procedure provided with the initial conformations for the geometry optimizations at HF/6-31G(d) and MP2(full)/6-31G(d) levels of theory.

For the molecule No. 11, two chiral conformers with C_1 symmetry were found at the MP2 level

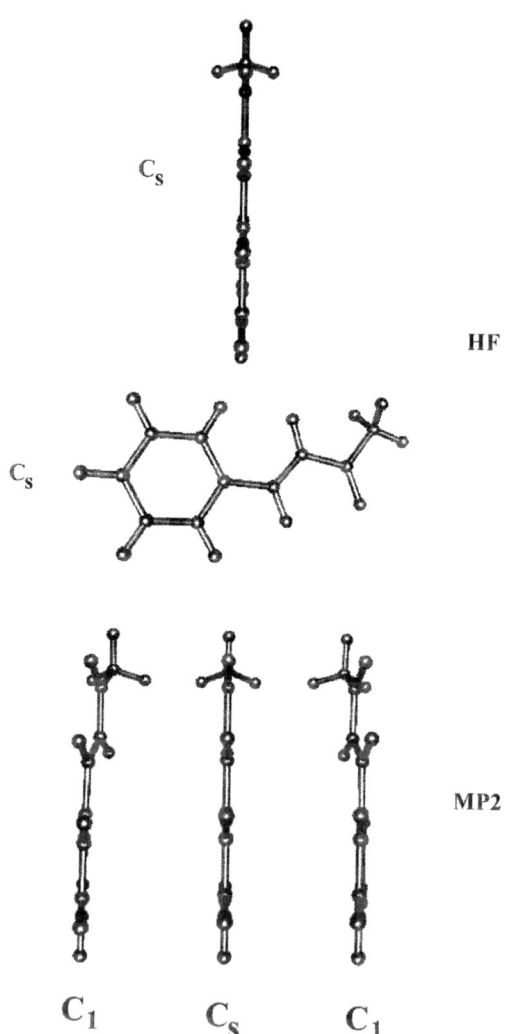

Fig. 2. Two conformers of (3E)-4-phenylbut-3-en-2-one at the MP2 level of theory.

of theory (see Fig. 2). The side chain is moved out of the plane of the ring by approximately 20°. The equilibrium geometry with C_s symmetry at the top of Fig. 2 was obtained at HF level. Below this, the same structure can be seen from an other direction. The lower three geometries are stationary points on the MP2 potential energy hypersurface. It is worth noting that the C_s structure is a transition state at this level of theory. The MP2 conformational activation energy including zero point vibrational energy corrections is 0.31 kJ/mol. This means that both conformers are present at room temperature.

Obviously, the substituents of the phenyl ring more or less increase the number of the conformers. For the molecule No. 5, the CF_3-substituent increases the number of the conformers up to eight (see Fig. 3).

Molecular electrostatic potential (MEP) maps were calculated at the MP2 level of theory. Atomic charges were derived from these MEP maps using the ChelpG method implemented in the GAUSSIAN98 program. These electrostatic potential derived atomic charges (EP atomic charges) were used in QMS studies. These studies let us conclude that in the reaction responsible for the antimutagenic activity of the molecules the segment $C_1–C_7(H_{16})–C_8(H_{15})–C_9$ plays a major role. This result is in accordance with that of Popelier et al. [6] obtained from QTMS analysis.

Acknowledgements

This work was sponsored by the János Bolyai Research Fund (G.T.).

One of the authors (IGC) wishes to thank the Ministry of Education for a Szent-Györgyi Visiting Professorship.

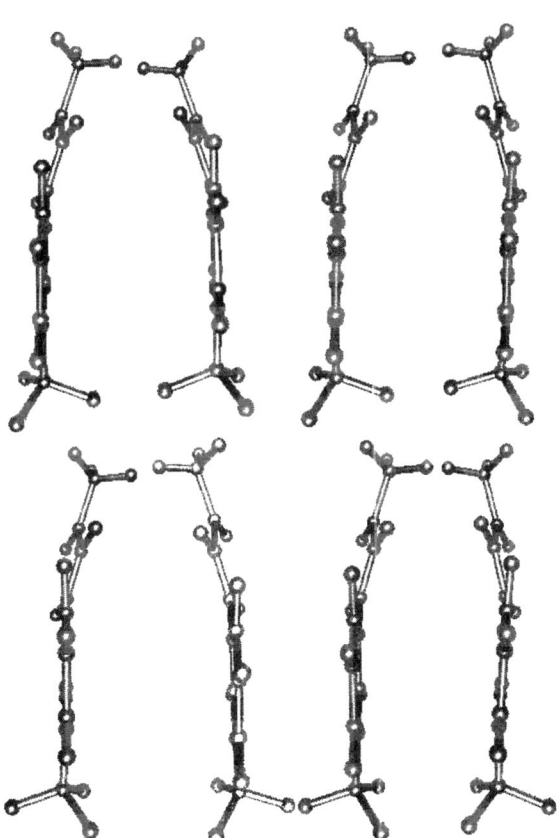

Fig. 3. Eight conformers of (3E)-4-[3-(trifluoromethyl)phenyl]but-3-en-2-one at the MP2 level of theory.

References

[1] N. Motohashi, Y. Ashihara, C. Yamagami, Y. Saito, Mutat. Res. 474 (2001) 113.

[2] C. Yamagami, N. Motohashi, Eur. J. Med. Chem. 37 (2002) 127.

[3] Y. Sato, S. Suzaki, T. Nishikawa, M. Kihara, H. Shibata, T. Higuti, J. Ethnopharmacol. 72 (2000) 483.

[4] C. Yamagami, N. Motohashi, M. Akamatsu, Bioorg. Med. Chem. Lett. 12 (2002) 2281.

[5] S. Ducki, J.A. Hadfield, L.A. Hepworth, N.J. Lawrence, C.-Y. Liu, A.T. McGown, Bioorg. Med. Chem. Lett. 7 (1997) 3091.

[6] S.E. O'Brien, P.L.A. Popelier, J. Chem. Soc., Perkin Trans. 2 (2002) 478.

[7] L.A. Popelier, J. Phys. Chem. A 103 (1999) 2883.

[8] S.E. O'Brien, P.L.A. Popelier, J. Chem. Inf. Comput. Sci. 41 (2001) 764.

[9] M.J.S. Dewar, E.G. Zoebisch, E.F. Healy, J.J.P. Stewart, J. Am. Chem. Soc. 107 (1985) 3902.

[10] C. Møller, M.S. Plesset, Phys. Rev. 46 (1934) 618.

[11] G. Tasi, I. Pálinkó, Top. Curr. Chem. 174 (1995) 45.

[12] G. Tasi, I. Pálinkó, J. Halász, G. Nárai-Szabó, PcMol: Semiempirical Quantum Chemical Calculations on Microcomputers, CheMicro Ltd, Budapest, 1992.

[13] J.J.P. Stewart, J. Comput.-Aided Mol. Des. 4 (1990) 1.

[14] M.J. Frisch, G.W. Trucks, H.B. Schlegel, G.E. Scuseria, M.A. Robb, J.R. Cheeseman, V.G. Zakrzewski, J.A. Montgomery Jr., R.E. Stratmann, J.C. Burant, S. Dapprich, J.M. Millam, A.D. Daniels, K.N. Kudin, M.C. Strain, O. Farkas, J. Tomasi, V. Barone, M. Cossi, R. Cammi, B. Mennucci, C. Pomelli,

C. Adamo, S. Clifford, J. Ochterski, G.A. Petersson, P.Y. Ayala, Q. Cui, K. Morokuma, D.K. Malick, A.D. Rabuck, K. Raghavachari, J.B. Foresman, J. Cioslowski, J.V. Ortiz, B.B. Stefanov, G. Liu, A. Liashenko, P. Piskorz, I. Komaromi, R. Gomperts, L.R. Martin, D.J. Fox, T. Keith, M.A. Al-Laham, C.Y. Peng, A. Nanayakkara, C. Gonzalez, M. Head-Gordon, E.S. Repogle, J.A. Pople, GAUSSIAN98, Revision A.6, Gaussian, Inc., Pittsburgh, PA, 1998.

[15] L. Laaksonen, gOpenMol version 1.30, 1999.

[16] ACD/ChemSketch Freeware Version 5.12, Advanced Chemistry Development Inc., Toronto, Ontario, Canada, M5H 3V9, 2002.

ELSEVIER

Journal of Molecular Structure (Theochem) 666–667 (2003) 135–141

THEO
CHEM

www.elsevier.com/locate/theochem

An exploratory study of the conformational intricacy of selected fluoro-substituted carboxylic acids

Á. Dörnyei[a,*], I.G. Csizmadia[b,c,d]

[a]Department of Inorganic and Analytical Chemistry, University of Szeged, P.O. Box 440, Szeged H-6701, Hungary
[b]Department of Medical Chemistry, University of Szeged, Dóm tér 8, Szeged H-6720, Hungary
[c]Department of Chemistry, University of Toronto, 80 St. George St., Toronto, Ont, Canada M5S 3H6
[d]Global Institute of Computational Molecular and Materials Science (GIOCOMMS), 1422 Edenrose St.,
Mississauga, Ont, Canada L5V 1H3

Abstract

The effects of fluorine and carboxylic acid functional groups was studied on the conformations of $CH_3–CHF–COOH$ and $CH_3–CHF–CHF–COOH$ molecules. The ab initio level of theory investigations were performed using the Hartree–Fock (HF) method with 6-31G** basis set. The preferred conformations were identified and possible intramolecular hydrogen bonds were looked for. Two types of intramolecular hydrogen bonds were found. The first is an $F\cdots HO$ and the other is an $OH\cdots O(carbonyl)$ close contact. The energy contributions of these bondings to the stabilization of the molecules were also estimated.
© 2003 Elsevier B.V. All rights reserved.

Keywords: Ab initio study; Substituted carboxylic acids; Intramolecular hydrogen bond; Conformation; Potential energy surface

1. Introduction

The α- and β-hydroxycarboxylic acids can play important role in complex formation when hard or borderline central ions. There is competition between protons and metal ions for the ligand during the process. Hydrogen bonding can also be envisaged as a donor–acceptor interaction. For substituted carboxylic acids the donor (−OH) or acceptor (=O) nature of a carboxyl group depends on the possible conformations, and on the electronic nature of the interacting substituents.

Large variety of hydroxycarboxylic acids are present in the living organisms, e.g. lactic acid and its derivatives. The intracellular substituted (low molecular mass) carboxylic acids can take part in the transport of metal ions. To study the structure of the metal complexes formed, first, the hydrogen bond system of the possible carrier molecules has to be understood.

The electronic nature of hydroxy-substituted carboxylic acids may be modeled by the possible conformations of fluoro-substituted carboxylic acids, because the fluorine functional group is isoelectronic with the hydroxyl group. 2-Fluoropropionic acid and 2,3-diflourobutyric acid were used to model their respective isoelectronic α- and β-hydroxycarboxylic acids. The structure and the assignation of

* Corresponding author. Tel.: +36-62-544-335; fax: +36-62-420-505.
E-mail address: dagi@petra.hos.u-szeged.hu (Á. Dörnyei).

0166-1280/$ - see front matter © 2003 Elsevier B.V. All rights reserved.
doi:10.1016/j.theochem.2003.08.023

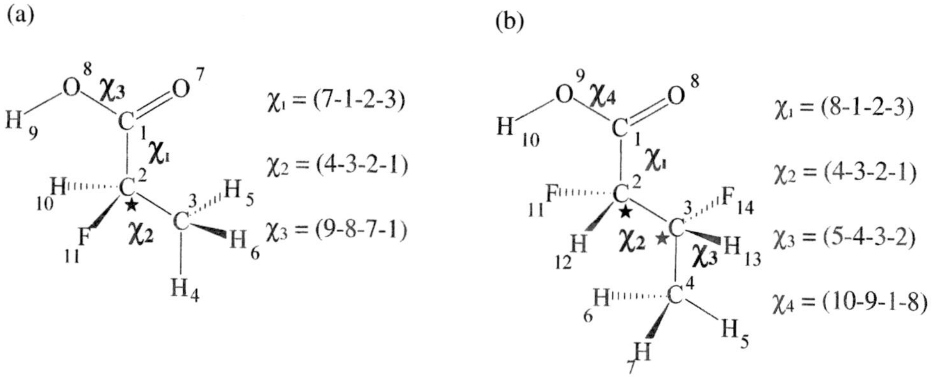

(a)   (b)

Fig. 1. Structure and assignation of the torsional angles of (a) 2-fluoropropionic acid and (b) 2,3-difluorobutyric acid.

the torsional angles of the studied compounds are shown in Fig. 1.

2-Fluoropropionic acid has one chiral center and, thus, two enantiomers (R, S), while the 2,3-difluorobutyric acid has two chiral centers and four diastereoisomers (2R,3R; 2R,3S; 2S,3R; 2S,3S). In the present work the conformational behaviour of (R)-2-fluoropropionic acid and (2R,3R)-2,3-difluorobutyric acid is summarized.

When a planar moiety of two-fold symmetry is connected to a tetrahedral carbon of three-fold symmetry, the torsional conformation potential about that single bond does not necessarily have three minima. In principle it may have six minima, as for methylbenzene. For ethylbenzene, the six degenerate minima are reduced to two unique structures [1]. For propionate ion or propionic acid, the possible six-fold periodicity along the carboxylate ion or carboxylic acid moiety is reduced to a three-fold periodicity. For 3,3-difluoropropionate ion and 3,3-difluoropropionic acid, (g^+,g^+) and (g^-,g^-) minima also disappear from the potential energy surface (PES) [2].

2. Computational methods

GAUSSIAN 98 [3] has been used to perform restricted HF (RHF) calculations on the compounds. All the calculations have been carried out at the RHF/6-31G** level of theory. All geometry optimizations as well as rigid and partially relaxed potential energy curve or surface (PEC/PES) scan calculations were performed for all the model compounds using

the Berny optimization method. All of these calculation were run under normal conditions (Opt = Z-matrix), where all critical points had gradients less than 4.5×10^{-4} a.u. For (R)-2-fluoropropionic acid both rigid and partially relaxed PEC scan calculations were carried out. The partially relaxed calculation means that at any point of the calculated points, the torsional angle variable (χ_1 for PEC) or variables (χ_1, χ_2 for PES) were frozen at certain values while optimizing the rest of the molecule. Both in the rigid and in the partially relaxed PEC scan calculations for the (R)-2-fluoropropionic acid, the χ_1 variable was rotated 15.0° increments (24 points). In the rigid PES scan calculations for the (2R,3R)-2,3-diflourobutyric acid, the χ_1 and χ_2 variables were rotated with 15.0° increments to produce plots consisting of $24 \times 24 = 576$ points. Fully relaxed ab initio calculations were performed to determine the minima on the conformational PEC of (R)-2-fluoropropionic acid and the conformational PES of (2R,3R)-2,3-diflourobutyric acid.

3. Results and discussion

In order to study the rigid PES of (2R,3R)-2,3-difluorobutyric acid, model studies were carried out on the rigid and partially relaxed PEC of (R)-2-fluoropropionic acid. The energy differences between rigid and partially relaxed PEC is displayed in Fig. 2.

The position (χ_1) of the minimum on the rigid PEC is almost identical with that on the partially relaxed PEC if optimized χ_1 torsional angle of an optimized conformation is the starting point of the scan.

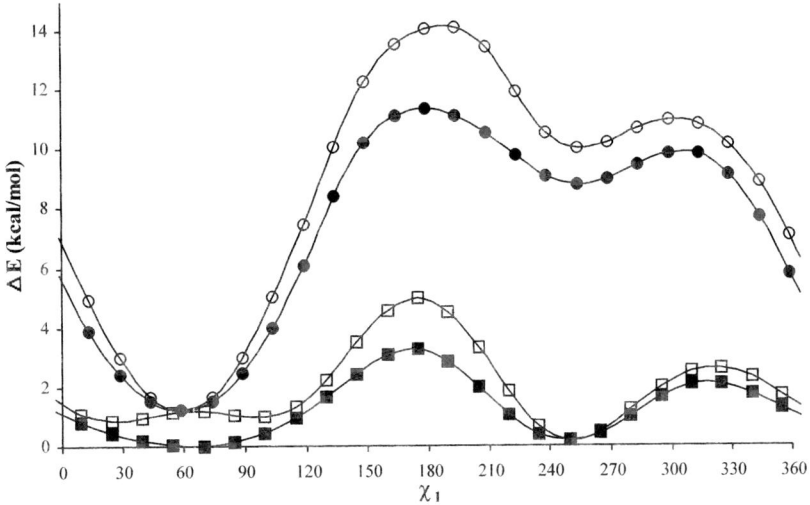

Fig. 2. A comparison of partially relaxed and rigid PEC of (R)-2-fluoropropionic acid (■–rigid and □–partially relaxed PEC in case of *endo* carboxylic –OH: $\chi_3 = 0°$; ●–rigid and ○–partially relaxed PEC in case of *exo* carboxylic –OH: $\chi_3 = 180°$).

Although three minima exist on the rigid PEC of (R)-2-fluoropropionic acid with its carboxylic –OH *exo*, but if we start with the fully relaxed geometry optimization of these minima, the same conformation is found. Thus, rigid PEC or PES scan followed by a fully relaxed geometry optimization can be used for identifying preferred conformations.

Two optimized conformations were found for both the *endo* and the *exo* (R)-2-fluoropropionic acid on PEC. The ball-and-stick representations are shown in Fig. 3 and optimized parameters are tabulated in Table 1. The three-fold periodicity along the carboxylic acid moiety (for propionic acid [2]) is reduced to a two-fold periodicity.

The 2,3-difluoropropionic acid has an additional asymmetric –CHF-group. It has two-fold periodicity across χ_1 and three-fold periodicity across χ_2 torsional angle leading to $2 \times 3 = 6$ minima. The PES and energy contour diagram for the (2R,3R)-2,3-difluorobutyric acid with its carboxylic -OH *endo* are shown in Fig. 4 and 5(a). The contour diagram for the same acid with its carboxylic –OH *exo* is shown in Fig. 5(b). The optimized parameters for the minima on the PES of both the *endo* and the *exo* (2R,3R)-2,3-difluorobutyric acid are summarized in Table 2. The location of the found minima are plotted in the topological diagram (Fig. 6(a) for *endo* carboxylic –OH orientation and Fig. 6(b) for *exo* carboxylic –OH orientation).

The disappearance of the (a, g^+), (a, a) and (a, g^-) minima is observed on both surfaces due to the mountain ridge. The change in energy from the global minimum (ΔE in kcal/mol) for structures with their

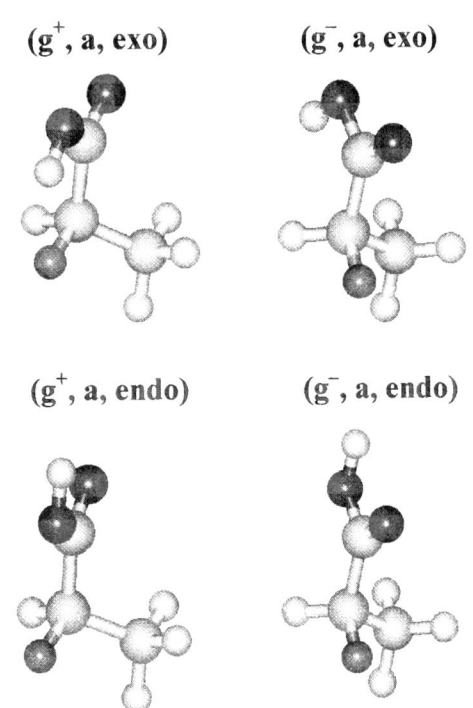

Fig. 3. The optimized structure for the conformations of (R)-2-fluoropropionic acid.

Table 1
Optimized torsional angles (in degree), computed energy values (in hartree), and relative energies (in kcal/mol) for the conformational minima of (R)-2-fluoropropionic acid computed at the RHF/6-31G** level of theory

Initial parameters			Optimized parameters				
χ_1	χ_2	χ_3	$\chi_1(°)$	$\chi_2(°)$	$\chi_3(°)$	E_{min}(hartree)	ΔE (kcal/mol)
g^+	a	endo	65.675	−178.226	0.372	−365.7045148	0.000
a	a		Not found				
g^-	a		−110.346	178.290	−0.260	−365.7042346	0.176
g^+	a	exo	58.746	−178.813	179.847	−365.7025735	1.218
a	a		Not found				
g^-	a		−104.993	177.984	−176.241	−365.6905531	8.761

carboxylic −OH *exo* is higher than for those with their carboxylic −OH *endo*. This can be explained by the fact that when the carboxylic−OH is in *endo* orientation, there can exist hydrogen bond between the −OH hydrogen and the carbonyl oxygen resulting in a more stable structure. However, for fluoro-substituted carboxylic acids, *exo* orientation presents a new possibility for *exo* −OH to establish a hydrogen

bond to the substituent fluorine group. Energy stabilization is also found for the hydrogen bond between the −OH hydrogen and the fluorine as for hydrogen bond between the −OH hydrogen and the carbonyl oxygen, however, the effect was higher for the OH···O hydrogen bond. Distances between atoms where hydrogen bond may exist are tabulated in Table 3.

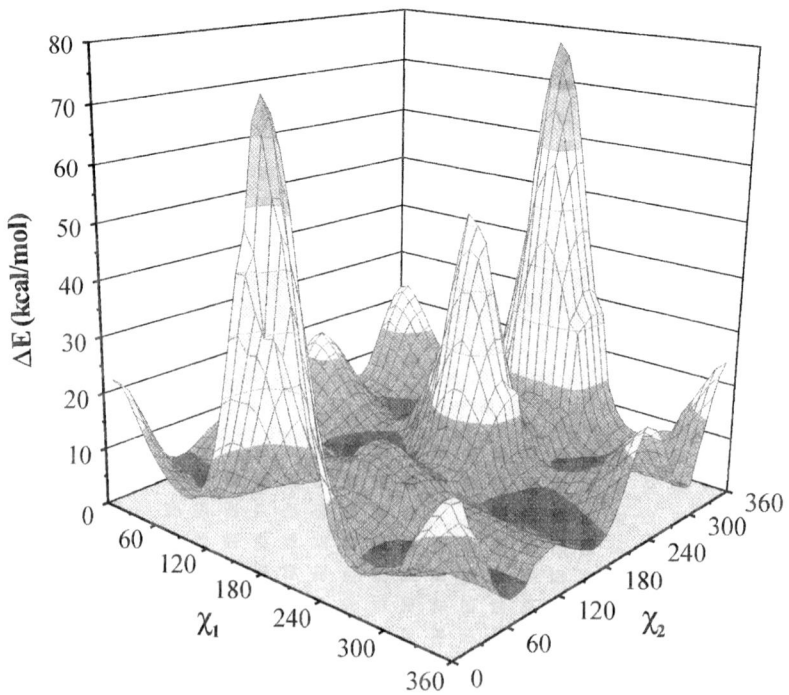

Fig. 4. Potential energy surface landscape of (2R,3R)-2,3-difluorobutyric acid ($\chi_3 = 180°$, $\chi_4 = 0°$ carboxylic −OH in *endo* orientation).

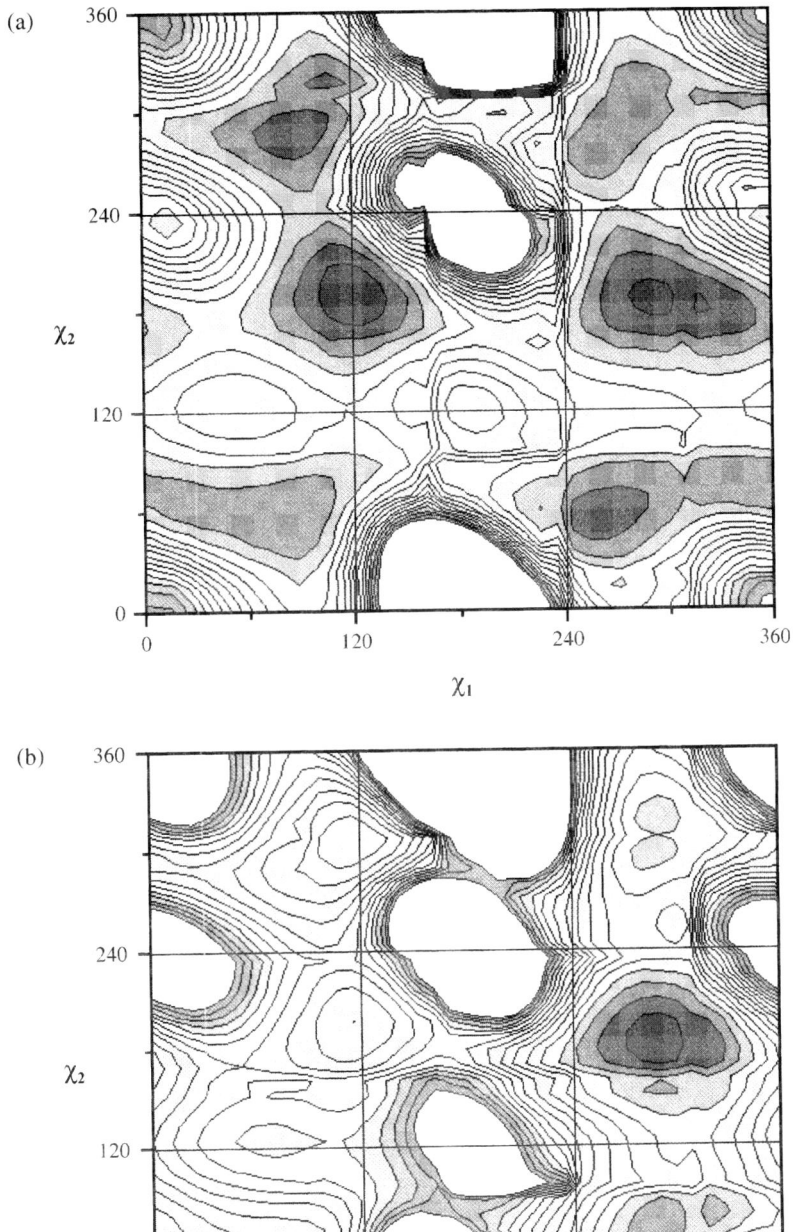

Fig. 5. Potential energy surface contour map of (2R,3R)-2,3-difluorobutyric acid ($\chi_3 = 180°$). (a) $\chi_4 = 0°$ carboxylic –OH in *endo* orientation; (b) $\chi_4 = 180°$ carboxylic –OH in *exo* orientation.

Table 2
Optimized torsional angles (in degree), computed energy values (in hartree), and relative energies (in kcal/mol) for the conformational minima of (2R,3R)-2,3-difluorobutyric acid computed at the RHF/6-31G** level of theory

Initial parameters				Optimized parameters					
χ_1	χ_2	χ_3	χ_4	$\chi_1(°)$	$\chi_2(°)$	$\chi_3(°)$	$\chi_4(°)$	E_{min}(hartree)	ΔE (kcal/mol)
g^+	g^+	a	endo	84.471	58.077	172.668	−0.436	−503.5880028	2.540
g^+	a	a		121.077	−168.872	178.966	−2.119	−503.5911532	0.563
g^+	g^-	a		95.127	−57.259	−177.154	−1.742	−503.5895027	1.599
g^-	g^+	a	endo	−88.578	58.490	173.409	−1.701	−503.5889336	1.956
g^-	a	a		−60.951	−175.116	177.956	1.371	−503.5920507	0.000
g^-	g^-	a		−82.935	−56.393	−175.783	−0.624	−503.5887049	2.100
g^+	g^+	a	exo	78.473	58.246	174.419	174.466	−503.5751745	9.271
g^+	a	a		119.007	−165.751	179.812	178.898	−503.5752513	9.205
g^+	g^-	a		125.577	−53.547	−177.718	−173.255	−503.5843575	3.509
g^-	g^+	a	exo	−63.275	65.127	173.829	−178.407	−503.5868058	1.972
g^-	a	a		−64.359	−176.479	177.425	178.248	−503.5899492	0.000[a]
g^-	g^-	a		−63.337	−47.988	−176.464	−178.192	−503.5848794	3.181

[a] The energy difference between the *endo* and *exo* global minima corresponds to 1.319 kcal/mol.

Further work is in progress for identifying the conformations of the remaining stereoisomers and conformations of the metal complexes formed with already studied molecules.

4. Conclusions

For (2R,3R)-2,3-difluorobutyric acid six minima are found on the PES ($E = E(\chi_1, \chi_2)$ and $\chi_3 = 180°$).

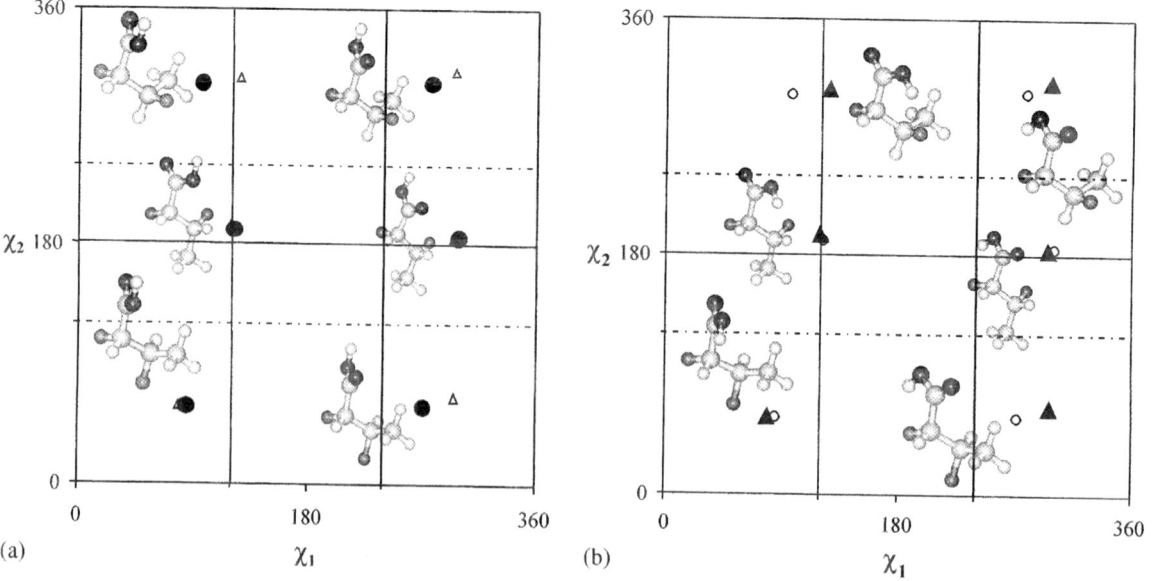

Fig. 6. Potential energy surface topology for the optimized molecular conformations of (2R,3R)-2,3-difluorobutyric acid and their representations. (a) *endo* carboxylic –OH orientation denoted by ●; (b) *exo* carboxylic –OH orientation denoted by ▲.

Table 3
Distances (in Å) between atoms involved in hydrogen bond in the optimized structures found for (R)-2-fluoropropionic acid and (2R,3R)-2,3-difluorobutyric acid

Initial parameters				Optimized parameters					
χ_1	χ_2	χ_3	χ_4	$\chi_1(°)$	$\chi_2(°)$	$\chi_3(°)$	$\chi_4(°)$	Atoms involved in hydrogen bond	Distance (Å)
(R)-2-fluoropropionic acid									
g^+	a	endo		65.675	−178.226	0.372		O_7-H_9	2.277
g^-	a			−110.346	178.290	−0.260		O_7-H_9	2.293
g^+	a	exo		58.746	−178.813	179.847		$F_{11}-H_9$	1.981
(2R,3R)-2,3-difluorobutyric acid									
g^+	g^+	a	endo	84.471	58.077	172.668	−0.436	O_8-H_{10}	2.293
g^+	a	a		121.077	−168.872	178.966	−2.119	O_8-H_{10}	2.296
g^+	g^-	a		95.127	−57.259	−177.154	−1.742	O_8-H_{10}	2.294
g^-	g^+	a	endo	−88.578	58.490	173.409	−1.701	O_8-H_{10}	2.287
g^-	a	a		−60.951	−175.116	177.956	1.371	O_8-H_{10}	2.286
g^-	g^-	a		−82.935	−56.393	−175.783	−0.624	O_8-H_{10}	2.283
g^+	g^-	a	exo	125.577	−53.547	−177.718	−173.255	$F_{14}-H_{10}$	1.975
g^-	g^+	a	exo	−63.275	65.127	173.829	−178.407	$F_{12}-H_{10}$	1.964
g^-	a	a		−64.359	−176.479	177.425	178.248	$F_{12}-H_{10}$	2.013
g^-	g^-	a		−63.337	−47.988	−176.464	−178.192	$F_{12}-H_{10}$	1.986

These conformations are the followings (χ_1, χ_2): (g^+, g^+); (g^+, a); (g^+, g^-); (g^-, g^+); (g^-, a); (g^-, g^-). At $\chi_1 = 180$ maximum exists due to mountain ridge. Two types of intramolecular hydrogen bonds are observed: OH⋯O(carbonyl) with carboxylic −OH *endo* and OH⋯F with carboxylic −OH *exo* (χ_1, χ_2): (g^-, g^+); (g^-, a); (g^-, g^-); (g^+, g^-). The stabilization effect of the hydrogen bonding between OH⋯O is higher than that between OH⋯F.

Acknowledgements

One of the authors (Á. D.) thanks Gábor Paragi and István Pálinkó for the helpful discussions.

One of the authors (IGC) wishes to thank the Ministry of Education for a Szent-Györgyi Visiting Professorship.

References

[1] O. Farkas, S.J. Salpietro, P. Csaszar, I.G. Csizmadia, J. Mol. Struct. (Theochem) 367 (1996) 25.

[2] S.J. Salpietro, A. Perczel, O. Farkas, R.D. Enriz, I.G. Csizmadia, J. Mol. Struct. (Theochem) 497 (2000) 39.

[3] M.J. Frisch, G.W. Trucks, H.B. Schlegel, G.E. Scuseria, M.A. Robb, J.R. Cheeseman, V.G. Zakrzewski, J.A. Montgomery, Jr., R.E. Stratmann, J.C. Burant, S. Dapprich, J.M. Millam, A.D. Daniels, K.N. Kudin, M.C. Strain, O. Farkas, J. Tomasi, V. Barone, M. Cossi, R. Cammi, B. Mennucci, C. Pomelli, C. Adamo, S. Clifford, J. Ochterski, G.A. Petersson, P.Y. Ayala, Q. Cui, K. Morokuma, D.K. Malick, A.D. Rabuck, K. Raghavachari, J.B. Foresman, J. Cioslowski, J.V. Ortiz, A.G. Baboul, B.B. Stefanov, G. Liu, A. Liashenko, P. Piskorz, I. Komaromi, R. Gomperts, R.L. Martin, D.J. Fox, T. Keith, M.A. Al-Laham, C.Y. Peng, A. Nanayakkara, C. Gonzalez, M. Challacombe, P.M.W. Gill, B. Johnson, W. Chen, M.W. Wong, J.L. Andres, C. Gonzalez, M. Head-Gordon, E.S. Replogle, and J.A. Pople, GAUSSIAN 98, Revision A.7, Gaussian, Inc., Pittsburgh PA, 1998.

ELSEVIER

Journal of Molecular Structure (Theochem) 666–667 (2003) 143–152

THEO
CHEM

www.elsevier.com/locate/theochem

A conformational analysis of histamine, and its protonated or deprotonated forms: an ab initio study

I.M. Mandity[a,*], G. Paragi[b], F. Bogár[b], I.G. Csizmadia[c,d,e]

[a]*Institute of Pharmaceutical Chemistry, Szent-Gyorgyi Medical Center, University of Szeged,
Eotvos utca 6, Szeged H-6720, Hungary*
[b]*Protein Chemistry Research Group of Hungarian Academic of Science, University of Szeged,
Dom ter 8, Szeged H-6720, Hungary*
[c]*Department of Chemistry, University of Toronto, 80 St. George St., Toronto, Ont., Canada M5S 3H6*
[d]*Department of Medical Chemistry, Szent-Gyorgyi Medical Center, University of Szeged, Dom ter 8, Szeged H-6720, Hungary*
[e]*Global Institute of Computational Molecular and Materials Science (GIOCOMMS), 1422 Edenrose St.,
Mississauga, Ont., Canada L5V 1H3*

Abstract

Histamine, a neurotransmitter contains an imidazole ring which may be protonated or deprotonated. The present study was dedicated to the two tautomeric and their protonated and deprotonated forms. Throughout the whole study, the side chain nitrogen was kept in its protonated form. Potential energy surface scanning and reaction path calculation between two optimized conformations were carried out using restricted Hartree-Fock method with 3-21G basis set (RHF/3-21G).
© 2004 Published by Elsevier B.V.

Keywords: Histamine; Conformers; RHF/3-21G; DDRP; Conformational analysis

1. Introduction

Histamine is an important biological molecule, which has different activities in animal organisms. (The standard nomenclature for this molecule is 1*H*-imidazole-4(5)-ethanamine. However, (2-aminoethyl)imidazole and following Ganellin, β-(4-imidazolyl)ethylamine, are also used.) This compound also appears in plant tissues. Histamine mediates with at least three different biological receptors: H_1, H_2 and H_3 [1].

The two nitrogens of the imidazole ring are in the 1, 3 position. The H^+ can be localized on the 1 or on the 3 nitrogen of the aromatic system, and the ring can be deprotonated and protonated, as shown in Fig. 1.

This biogenic amine is formed by the metabolization [2] of an important amino acid Histidine (His). We might mention, in passing, that histidine is an essential amino acid. This amino acid must be presented in sufficient quantities in the diet for proper functioning of the human body. Histamine may be oxidized by the monoamine oxidize enzyme and by

* Corresponding author.

E-mail address: mandity@ovrisc.mdche.u-szeged.hu (I.M. Mandity).

0166-1280/$ - see front matter © 2004 Published by Elsevier B.V.
doi:10.1016/j.theochem.2003.08.024

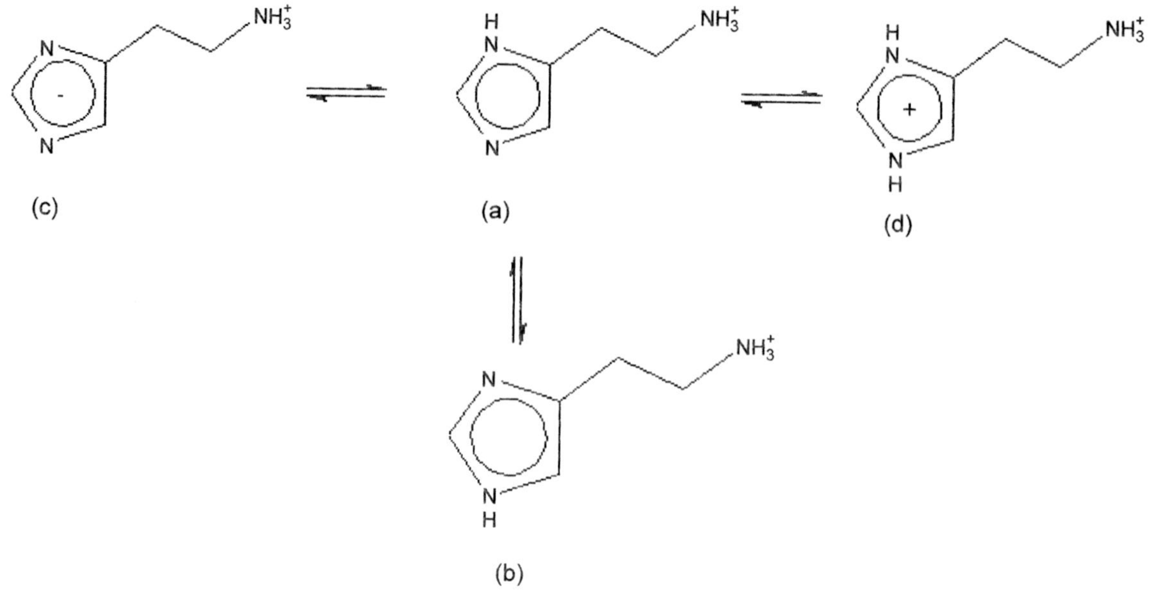

Fig. 1. The four forms of histamine. (a) Side chain protonated histamine, tautomeric form 1. (b) Side chain protonated histamine, tautomeric form 2. (c) Side chain protonated ring deprotonated histamine. (d) Side chain protonated ring protonated histamine.

the diamine oxidize enzyme as well, which is shown in Fig. 2.

When a systematic protonation is made on the imidazole ring, one would expect that such a change would manifest itself in the potential energy surface (PES). It may alter the appearance of the surface, or even its topology as well as the location and the energy values of the optimized minima.

2. Computational method

The molecular structure, stereochemistry, and geometry of protonated, deprotonated and the two tautomeric forms of histamine were exclusively defined in terms of their z-matrix internal co-ordinate system. PES were calculated with 144 points at 30° increments at the RHF/3-21G level of theory. Corresponding minima from the PES were selected and full optimizations were carried out successively at the RHF/3-21G level of theory. Further reaction path analysis were performed between two minima using the Dynamically Defined Reaction Path method (DDRP) [3]. The reaction path investigation was

also performed with RHF/3-21G energy calculation. According to our knowledge, it was the first time when this method was used at this level of theory. We also made semiempirical study with DDRP, but the transition state structure seemed to be irrelevant. The GAUSSIAN98 [4] program was used to carry out all the ab initio computations, and the DDRP calculations were performed with modified Tinker program package [3,5] in combination with the GAUSSIAN98 code.

3. Results and discussions

3.1. Conformational analysis

The four compounds studied [6] conformationally in the present work are shown in Fig. 3. All of them were treated as double rotors. Torsional angle χ_1 was associated with the rotation of the imidazole group, while torsional angle χ_2 was associated with the rotation of the $-CH_2-NH_3^+$ group.

The conformational PES as described by the equation $E = E(\chi_1, \chi_2)$ was computed and plotted

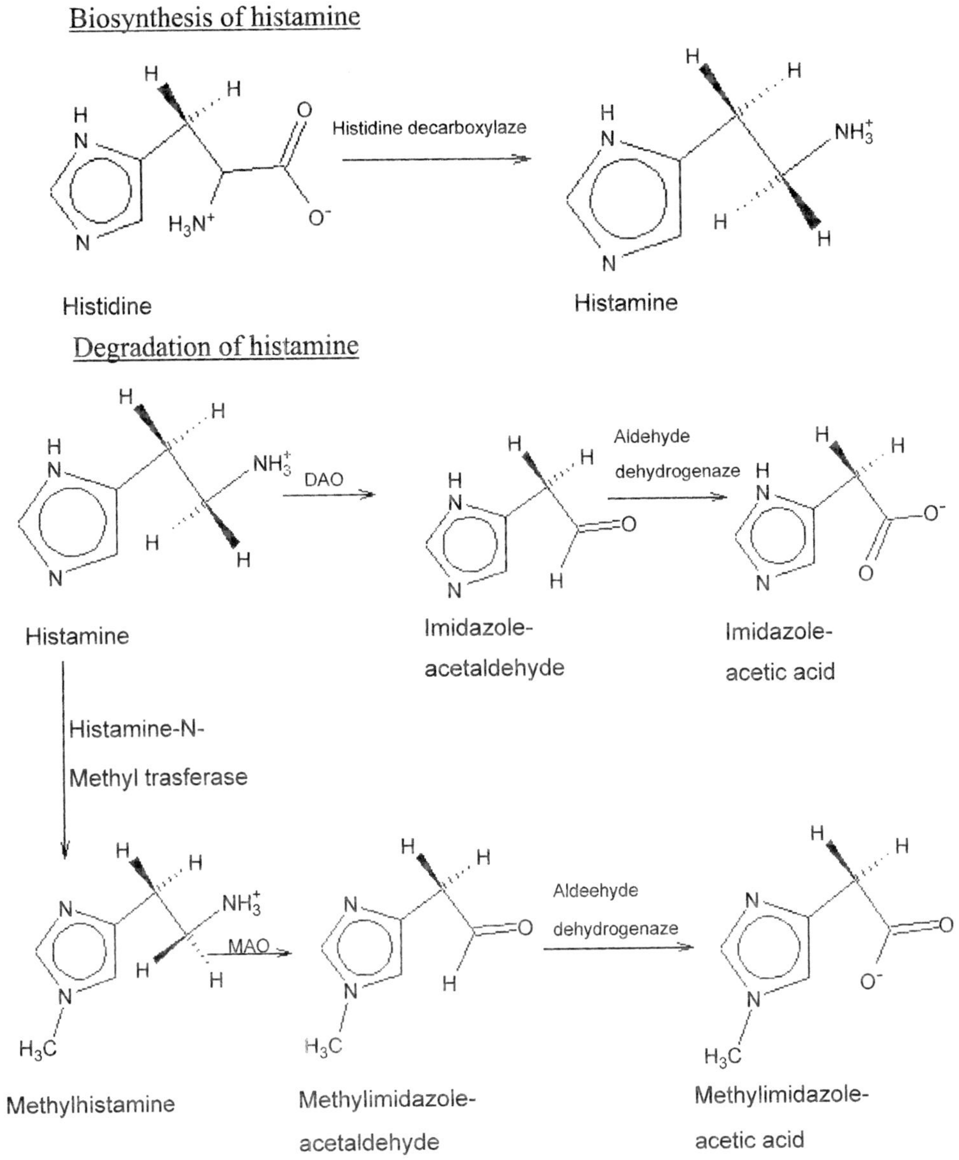

Biosynthesis of histamine

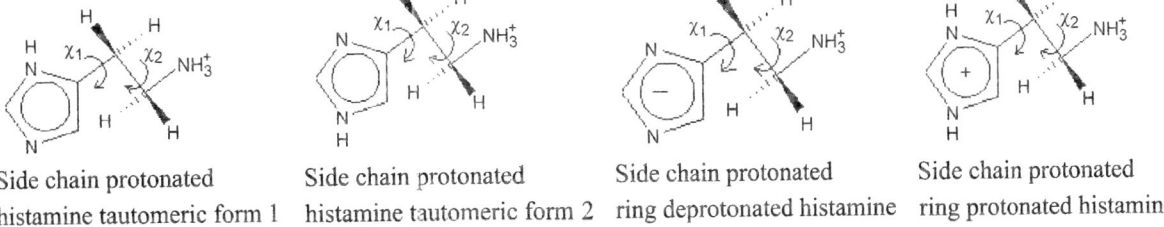

Histidine → (Histidine decarboxylaze) → Histamine

Degradation of histamine

Histamine → (DAO) → Imidazole-acetaldehyde → (Aldehyde dehydrogenaze) → Imidazole-acetic acid

Histamine → (Histamine-N-Methyl trasferase) → Methylhistamine → (MAO) → Methylimidazole-acetaldehyde → (Aldeehyde dehydrogenaze) → Methylimidazole-acetic acid

Fig. 2. A schematic representation of the metabolism of histamine.

Side chain protonated histamine tautomeric form 1

Side chain protonated histamine tautomeric form 2

Side chain protonated ring deprotonated histamine

Side chain protonated ring protonated histamine

Fig. 3. Structure of the four compounds studied.

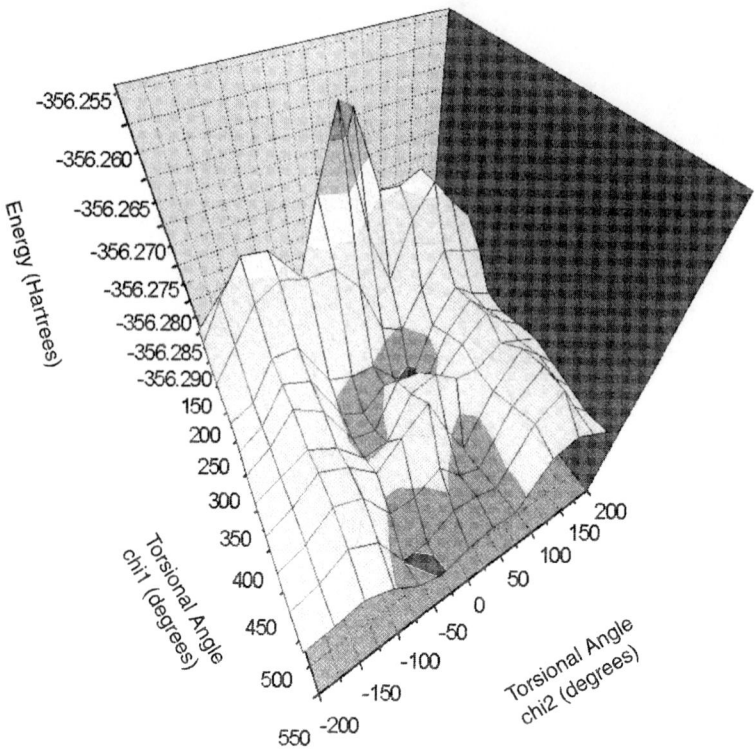

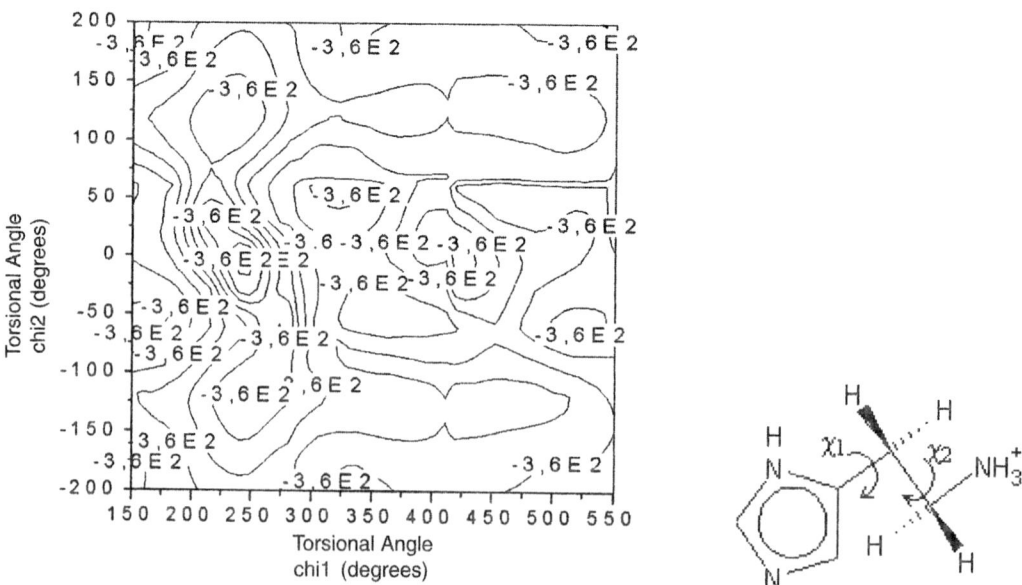

Fig. 4. Side chain protonated histamine tautomeric form 1 conformational surface (up) and contour (bottom).

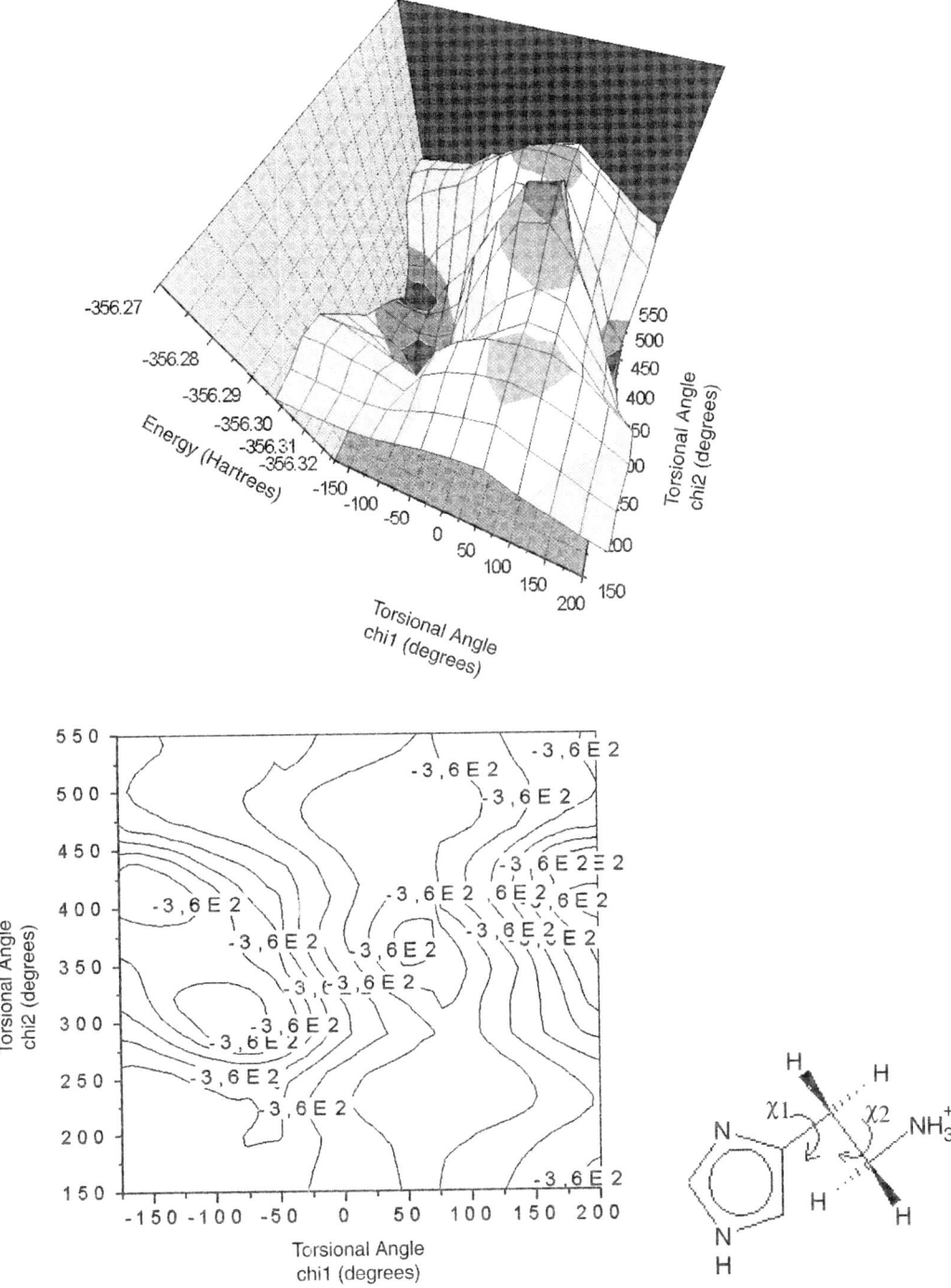

Fig. 5. Side chain protonated histamine tautomeric form 2 conformational surface (up) and contour (bottom).

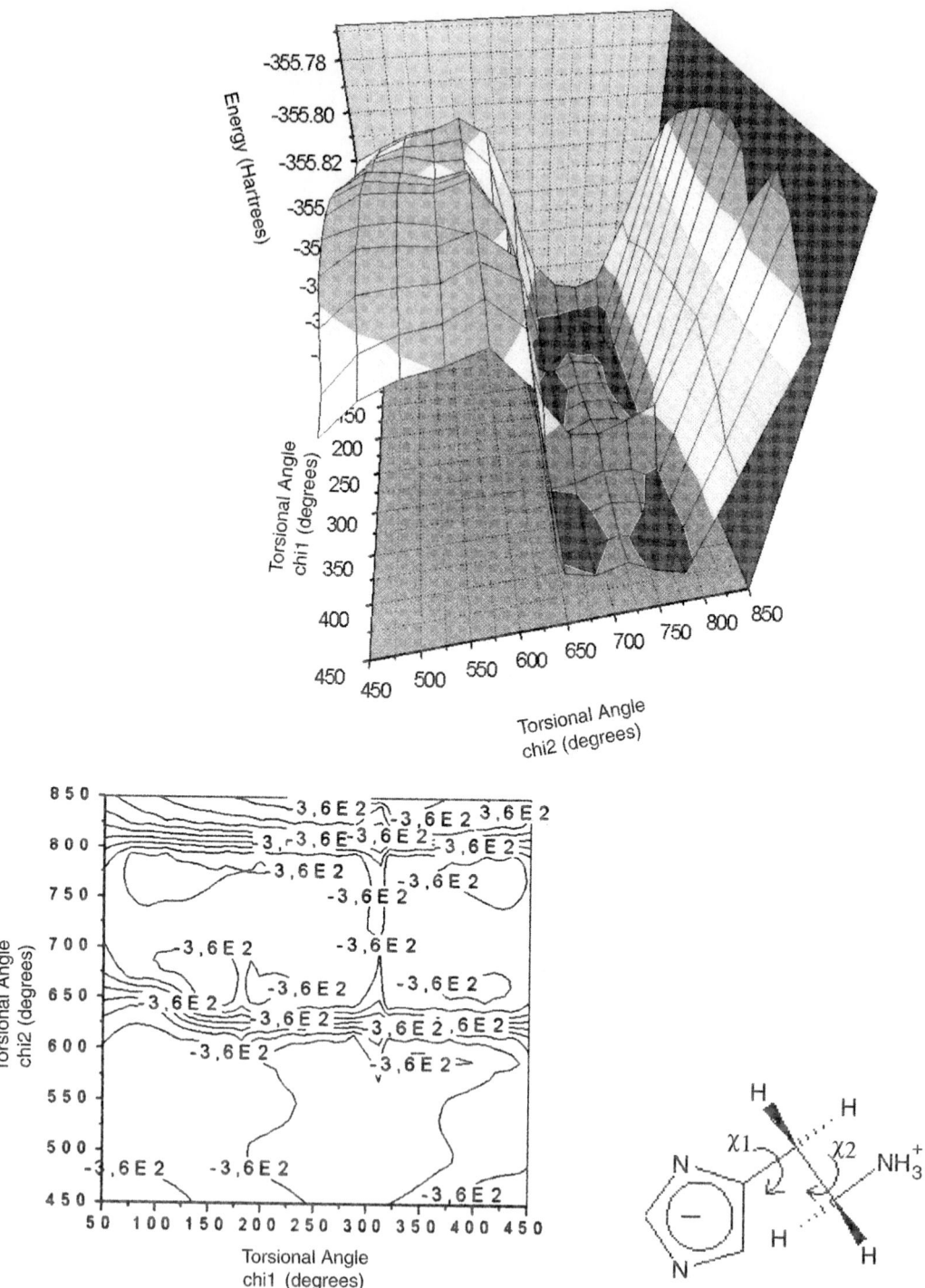

Fig. 6. Side chain protonated ring deprotonated form of histamine conformational surface (up) and contour (bottom).

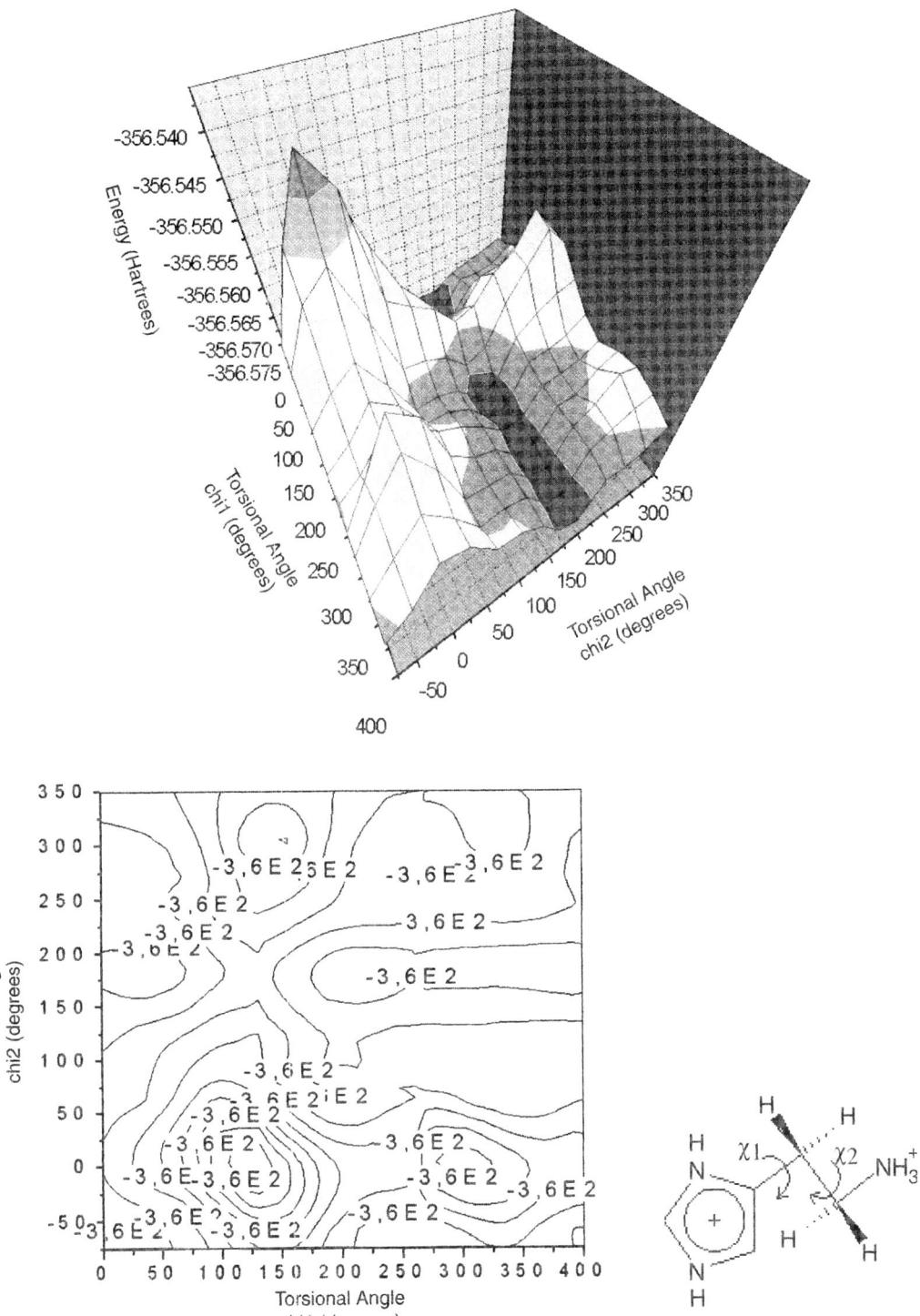

Fig. 7. Side chain protonated ring protonated form of histamine conformational surface (up) and contour (bottom).

Table 1
Optimized energy (in Hartree) and dihedral angles (in degree) of minima for side chain protonated histamine tautomeric form (1)

χ_1 (degree)	χ_2 (degree)	Energy (Hartree)
333.82021	61.84213	−356.28784
513.82021	−58.15787	−356.28874
453.82021	61.84213	−356.28629
333.82021	−58.15787	−356.28673

Table 2
Optimized energy (in Hartree) and dihedral angles (in degree) of minima for side chain protonated histamine tautomeric form (2)

χ_1 (degree)	χ_2 (degree)	Energy (Hartree)
−103.20718	318.45509	−356.32642
−163.20718	408.45509	−356.32774

Table 3
Optimized energy (in Hartree) and dihedral angles (in degree) of minima for side chain protonated ring deprotonated histamine

χ_1 (degree)	χ_2 (degree)	Energy (Hartree)
436.98255	762.27031	−355.90826
406.98255	642.27031	−355.90450
106.98255	792.27031	−355.90946
106.98255	642.27031	−355.91193

for each of the four forms of side chain protonated histamine. The graphical representation of the four PESs are shown in Figs. 4–7, respectively.

The conformational and energetic characteristics of the optimized structures of the four forms of side chain protonated histamine are summarized in Tables 1–4. The conformers which have the lowest total energy are shown in Fig. 8. The conformer of side chain protonated histamine

Table 4
Optimized energy (in Hartree) and dihedral angles (in degree) of minima for side chain protonated ring protonated histamine

χ_1 (degree)	χ_2 (degree)	Energy (Hartree)
35.04327	181.8951	−356.57623
245.04327	181.8951	−356.57514

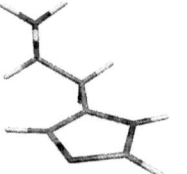

Conformer of side chain protonated histamine tautomeric form 1

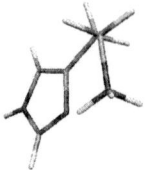

Conformer of side chain protonated histamine tautomeric form 2

Conformer of side chain protonated ring prtonated histamine

Conformer of side chain protonated ring deprotonated histamine

Fig. 8. The four conformers, which have the lowest total energy.

tautomer (1), the conformer of side chain protonated histamine tautomer (2) and the conformer of side chain protonated ring deprotonated histamine have intramolecular hydrogen bound between one of the hydrogens of the NH_3^+ group and one of the nitrogens in the imidazole ring. The conformer of the side chain protonated ring protonated histamine is the *anti* conformer.

3.2. The path way between two minima of side chain protonated histamine tautomeric form (1) (DDRP)

The PES can help us to see how the molecule from an optimized conformation can transform into another one. However, it is always an important question the height of the potential barrier between the two minima, and the transition state of the molecule. Furthermore, the scanning of the whole PES is a very time consuming task, which cannot be performed in many cases. The reaction path methods simplify these tasks, and furthermore, they can help us to scan the 'physically interesting' part of the surface. With the DDRP method, we connected successfully the two minima of the side chain protonated histamine tautomeric form (1). The change of the energy and conformation along the path is shown in Fig. 9. The frequency analysis showed that the conformation belongs to the maximal energy on the path was

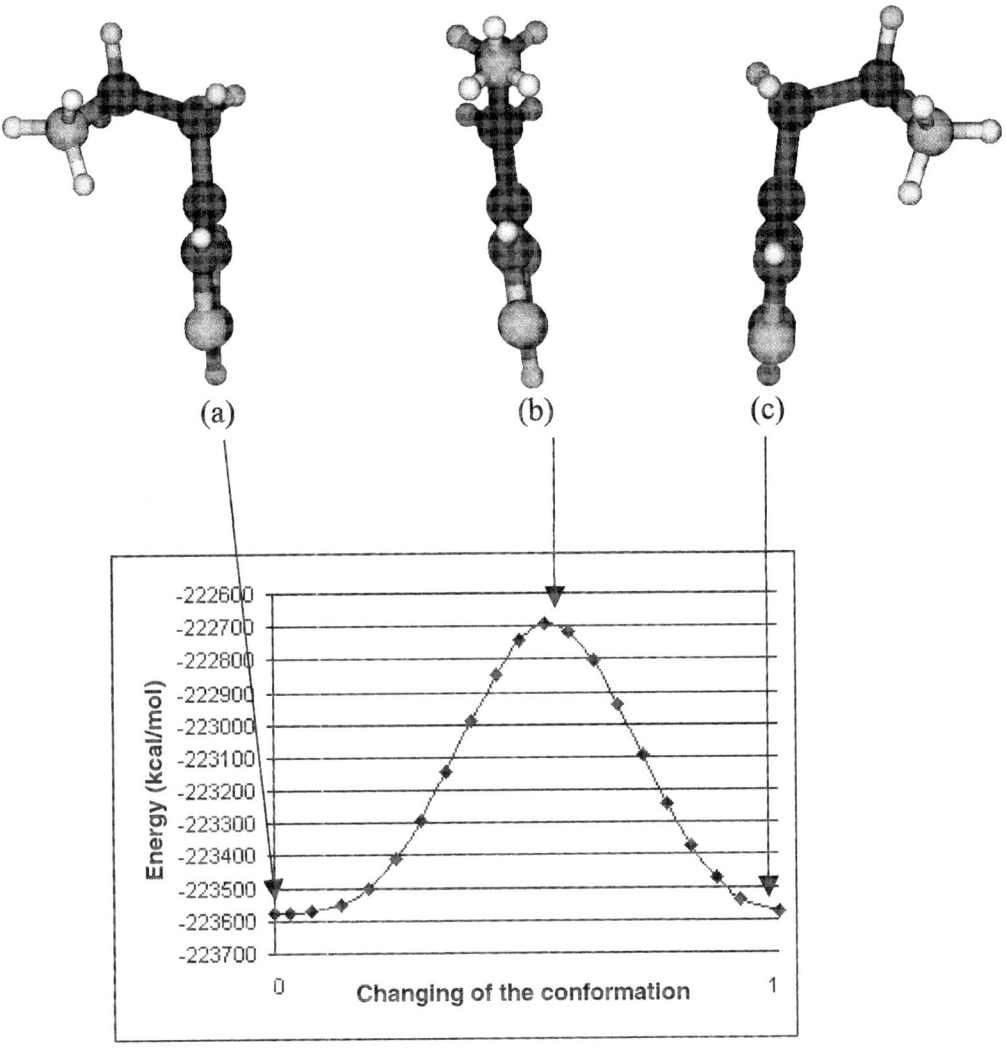

Fig. 9. The changing of the energy and conformation during the path.

a transition state. The conformation of the transition is shown in Fig. 9(b). The height of the potential barrier along the path is 853.15567 kcal/mol.

Acknowledgements

One of us (IGC) would like to thank the Ministry of Education for a Szent-Györgyi Visiting Professorship.

References

[1] A. Hernandez-Laguna, Z. Cruz-Rodriguez, Y.G. Smeyers, G.A. Arteca, J.-L.M. Abboud, O. Tapia, J. Mol. Struct. (Theochem) 335 (1995) 77–87.

[2] B.G. Katzug, Basic and Clinical Pharmacology, eighth ed., Lange Medical Books/McGraw-Hill, Toronto/New York, 2001.

[3] L.L. Stachó, M.I. Bán, Theor. Chim. Acta 83 (1992) 433.

[4] M.J. Frisch, G.W. Trucks, H.B. Schlegel, G.E. Scuseria, M.A. Robb, J.R. Cheeseman, V.G. Zakrzewski, J.A. Montgomery Jr., R.E. Stratmann, J.C. Burant, S. Dapprich, J.M. Millam, A.D. Daniels, K.N. Kudin, M.C. Strain, O. Farkas, J. Tomasi,

V. Barone, M. Cossi, R. Cammi, B. Mennucci, C. Pomelli, C. Adamo, S. Clifford, J. Ochterski, G.A. Petersson, P.Y. Ayala, Q. Cui, K. Morokuma, D.K. Malick, A.D. Rabuck, K. Raghavachari, J.B. Foresman, J. Cioslowski, J.V. Ortiz, A.G. Baboul, B.B. Stefanov, G. Liu, A. Liashenko, P. Piskorz, I. Komaromi, R. Gomperts, R.L. Martin, D.J. Fox, T. Keith, M.A. Al-Laham, C.Y. Peng, A. Nanayakkara, C. Gonzalez, M. Challacombe, P.M.W. Gill, B. Johnson, W. Chen, M.W. Wong, J.L. Andres, C. Gonzalez, M. Head-Gordon, E.S. Replogle, and J.A. Pople, GAUSSIAN'98, Revision A.7, Gaussian, Inc., Pittsburgh, PA, 1998.

[5] P. Ren, J.W. Ponder, J. Comput. Chem. 23 (2002) 1497–1506.

[6] D.M. Gasparo, D.R.P. Almeida, S.M. Dobo, L.L. Torday, A. Varro, J.Gy. Papp, J. Mol. Stuct. (Theochem) 585 (2002) 167–179.

ELSEVIER

Journal of Molecular Structure (Theochem) 666–667 (2003) 153–158

THEO
CHEM

www.elsevier.com/locate/theochem

Conformation analysis of 1,4-pentadien-3-yl radicals by ab initio and DFT methods

Milan Szori, Bela Viskolcz*

Department of Chemistry, JGyTFK, University of Szeged, P.O. Box 396, Szeged H-6701, Hungary

Abstract

1,4-Pentadien-3-yl radical is the simplest acyclic hydrocarbon radical having five conjugated pi-electrons. In this work the 1,4-pentadien-3-yl ($CH_2CHCH^*CHCH_2$) is studied by different level of theoretical methods (UB3LYP, UHF and UMP2 as well) with conformation and energy aspect. All of potential energy surfaces show that three different minima can exist. According to energy calculations with B3LYP/6-31G(d), B3LYP/6-311+G(3df,2p), BH and HLYP/6-311+G(3df,3p) and G3MP2 quantum chemistry models, relative energies of the E,Z and the Z,Z structure of 1,4-pentadien-3-yl radical with respect to the E,E structure of 1,4-pentadien-3-yl are 10.0 ± 0.7 and 26.9 ± 1.9 kJ mol^{-1}, respectively.
© 2003 Elsevier B.V. All rights reserved.

Keywords: 1,4-Pentadien-3-yl radicals; Ab initio; Potential energy surface; Conformation analysis

1. Introduction

1,4-Pentadien-3-yl (C_5H_7) radical is the simplest acyclic hydrocarbon radical having five conjugated electrons so it became an excellent model for analogous large unsaturated delocalized radicals. The nature and size of the resonance or stabilization effects are of broad general interest. We focused our intention to study the conformations of C_5H_7 radical to obtain information on the properties of the delocalization. The C_5H_7 radical is an excellent prototype to test different adequate ab initio methods for larger biologically important radicals. These types of short live species frequently play a central role in many processes including hydrocarbon combustion as well as autoxidation of unsaturated fatty acids [1].

The direct experimental study on the properties of these radicals is only possible in gas phase [2–4]. The earlier experimental results concentrated on the equilibrium of rotational conformers [5,6]. Recently, the electrocycling reactions of C_5H_7 radical were studied by high levels ab initio methods [7].

In addition, 1,4-pentadien-3-yl-type radicals have a longer lifetime than to other radicals found in membranes. These types of radical can be formed in membranes from the ω-3-fatty acid though H-atom abstraction by radicals. The docosahexaenoic acid (DHA) is an ω-3-fatty acid which is essential for the development of our nervous system and visual abilities during the first 6 months of life for the proper functioning of our brains as adults [8,9]. Lack of sufficient DHA may be associated with impaired mental and visual functioning as well as attention-deficit hyperactivity disorder (ADHD) in children [10]. Low levels have also been associated with depression and Alzheimer's disease in adults [11].

* Corresponding author.
E-mail addresses: szorimil@freemail.hu (M. Szori), viskolcz@jgytf.u-szeged.hu (B. Viskolcz).

0166-1280/$ - see front matter © 2003 Elsevier B.V. All rights reserved.
doi:10.1016/j.theochem.2003.08.025

This is the reason why our attention was focused on the 1,4-pentadien-3-yl radical as model molecule to understand the biological behavior and importance of DHA [8].

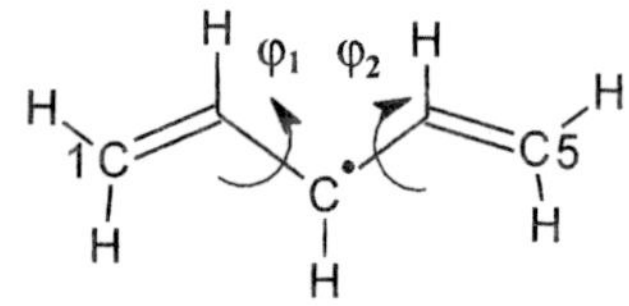

2. Methods

The ab initio and B3LYP calculations were performed with the GAUSSIAN 98 package [12] and we used the B3LYP [13] and BH and HLYP functional form implemented in the Gaussian program. The BH and HLYP method are very much like B3LYP, with the exception that the fraction of HF exchange is 50% (in fact, H and H denotes 'half and half').

For conformational analysis, two dihedral angles were defined (φ_1 and φ_2), as shown by Scheme 1. The step size of the scan was made by 5° between 0° and 360°. In every point, these dihedral angles were fixed and remaining variables of the structure were optimized by B3LYP/6-31G*, HF/6-31G* and MP2/6-31G* quantum chemistry models, respectively. The energy is related to the least energy containing conformer (E,E structure see also Fig. 1).

The accurate relative energies are evaluated by B3LYP/6-31G*, B3LYP/6-311+G(3df,2p), BH and HLYP/6-311+G(3df,3p) and G3MP2 [14,15] levels of theory, based on the MP2/6-31G* geometry and frequency calculations. The frequencies and ZPE energies are scaled by 0.9478 [16].

3. Results and discussion

3.1. Potential energy surfaces

The PES are obtained by three different methods are shown in Figs. 2–4. The PES obtained by HF and B3LYP are similar, magnitude of relative energies are

Fig. 1. The numbering and definition of dihedral angles of 1,4-pentadien-3-yl radical. Definition of dihedral angles: φ_1[H–C3–C2–C1] and φ_2[H–C3–C4–C5].

very close to each other in every point of PES as the Figs. 2 and 3 are shown. Both of them are very symmetric like the 1,4-pentadienyl radical. The relative energy of HF and B3LYP varies to 0 from 138 and 128 kJ mol^{-1}, respectively. On the other hand, the MP2 surface to show disparate characteristic as shown in Fig. 4. PES obtained by MP2/6-31G(d) are more tabulate compared with HF and DFT ones. The relative energy varies to 0 from 90 kJ mol^{-1}, the larger value of relative energy in MP2–PES differences in 48 kJ mol^{-1} from HF–PES.

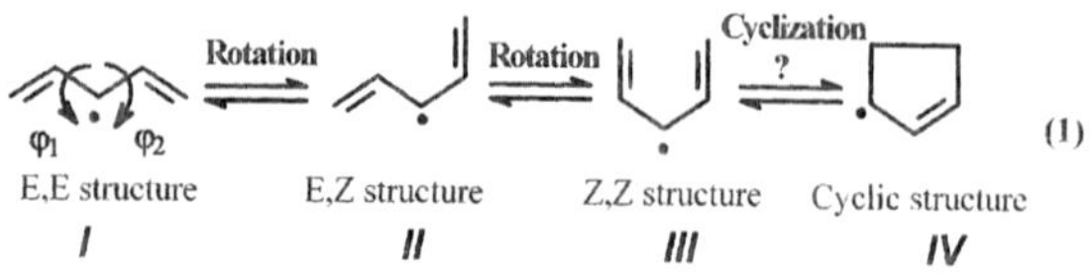

The PES of C_5H_7 radical obtained by HF and DFT methods show similarity, the characteristic of both PES are quasi the same. The larger differences of the relative energies are less than 6 kJ mol^{-1}. The φ_1 and φ_2 dihedral angles of minima agreed within 3°. Both PES have two saddle points. We found two different kind of transition states (TS1, TS2) and three local minima (E,E structure, E,Z structure and Z,Z structure) in these surfaces, the number of local minima are discussed later. The characterization of the correspondent dihedral angles is relative simple (TS1:[$\varphi_1 = 90$; $\varphi_2 = 180$], and TS2: [$\varphi_1 = 0$; $\varphi_2 = 90$] degrees). The relative energies of the saddle point are about 50 kJ mol^{-1}, it is in good agreement with the estimated delocalization energy [2].

The characteristic of full MP2 conformational map differs from map obtained by HF and B3LYP describe above. The differences are listed: (a) the range of relative energy is smaller by 50 kJ mol^{-1}, the MP2 surface is more flat than the other two PES; (b) the shape of the MP2 surface also differs. The cross-sections (Figs. 5 and 6) bring these properties out in strong relief. The MP2 horizontal cross-sections at 180° have local minima 90°, at these dihedral angles the HF and DFT curves have maxima. At 180° the HF and the DFT energy functions have local minima, on the other hand,

Scheme 1.

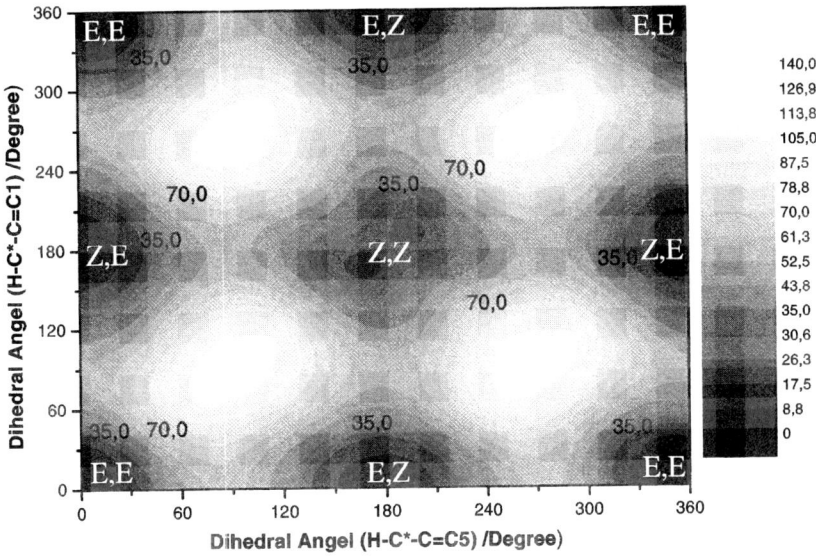

Fig. 2. The rotational conformation map of 1,4-pentadien-3-yl radicals obtained by B3LYP/6-31G*.

the MP2 one has a local maxima. The corresponding minimum of MP2 energy functions can be found at about 160°. It appears from the surface (Figs. 2–4) as well as the cross-sections generated at 180° that the ZZ structure splits up two equivalent minima. The symmetric (both dihedral angles: 180, 180°) structure becomes a transition states. This phenomenon is currently under further investigation. There is a same local minimum in 90, 210 and 270°

(relative energies 28 kJ mol^{-1}). In addition, there are three different magnitudes of energy local maxima about 60, 120 and 180°, adherent values of relative energies following: 32.6, 30.1 and 28.6 kJ mol^{-1}, respectively. This finding points out that taking account of the correct electron correlations are very important to characterize the TS structure of different conformers of C_5H_7 radicals.

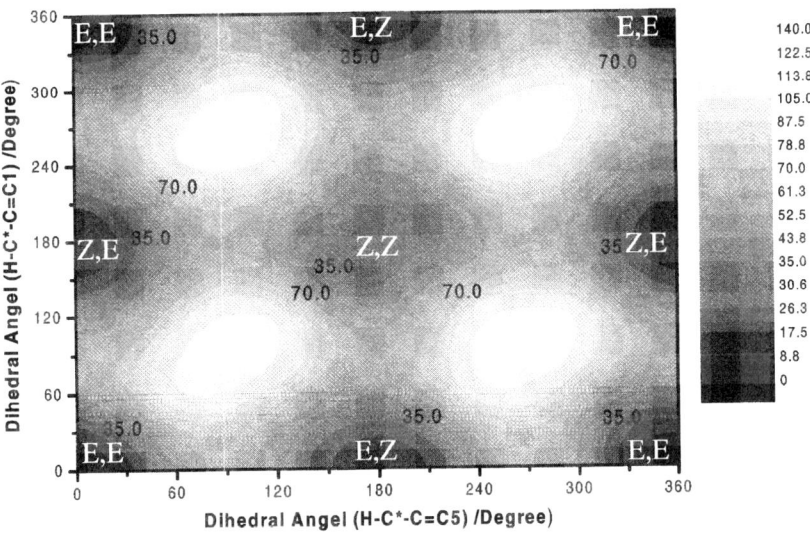

Fig. 3. The rotational conformation map of 1,4-pentadien-3-yl radicals obtained by HF/6-31G*.

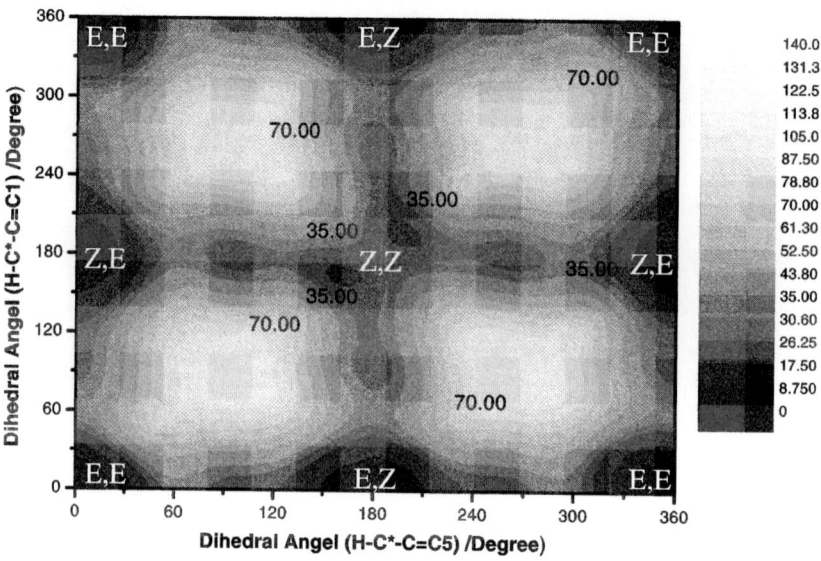

Fig. 4. The rotational conformation map of 1,4-pentadien-3-yl radicals obtained by MP2/6-31G*.

3.2. Energy minima

The relative energies of structure **II**, **III** and also the cyclopentene-3yl (**IV**, c-C_5H_7) radicals are listed in Table 1. The energies of minima **II** and **III** obtained by different levels of theory are in excellent agreement with each other within 2 kJ mol^{-1}. The less stable structure is of **III** by an average energy of 27.7 ± 1.4 kJ mol^{-1}.

The relative energy of minimum **II** was 10.2 ± 0.5 kJ mol^{-1}. The G3MP2 method predicted the lowest relative energy for minima **II** and **III**. The small standard deviation shows how consistent are the theoretical models. The energies of rotational conformers are comparable of the recent CCSD(T) calculations [7], but the latest ones are somewhat lower, 9 and 23 kJ mol^{-1} for **II** and **III**, respectively.

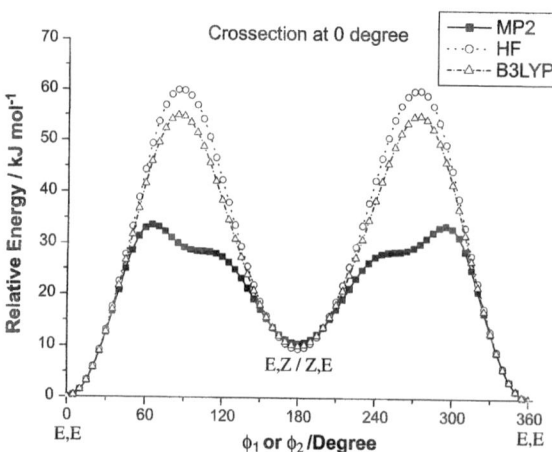

Fig. 5. Cross-section at 0° of PES obtained by HF, DFT and MP2 level of theory.

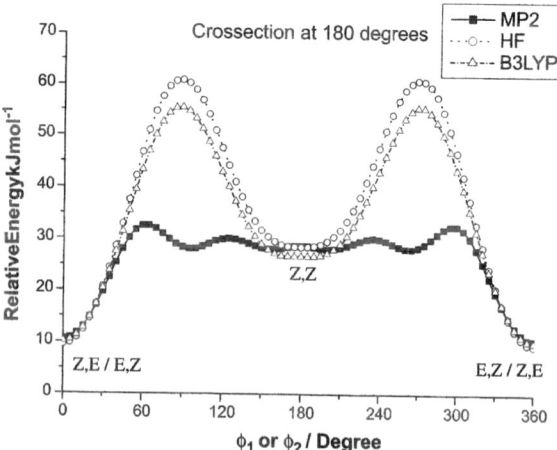

Fig. 6. Cross-section at 180° of potential energy surfaces obtained by HF, DFT and MP2 level of theory.

Table 1
ZPE corrected relative energy with respect to the E,E structure of 1,4-pentadien-3-yl radical in kJ mol^{-1}

	HF/6-311G**	MP2/6-311G**	B3LYP/ 6-311G*	B3LYP/6-311 + G(3df,2p)	BH and HLYP/ 6-311 + G(3df,3p)	G3MP2	CCSD(T)/ cc-pVDZ[a]
E,Z (II)	10.4	10.3	10.5	10.5	10.3	8.6	9
Z,Z (III)	29.7	27.7	27.2	26.3	26.9	23.2	23
IV.	−6.4	−70.1	−24.5	−10.6	−21.3	−32.3	−37

[a] Values taken from Ref. [7].

In the case of stability of cyclopentene-3-yl radicals the applied theoretical methods do not predict a consistence picture. The relative energies of the c-C_5H_7 varied between −6 and −70 kJ mol^{-1}. We expected the energy calculations based on G3MP2 and also CCSD(T) combined with relative small basis set (cc-pVDZ) [17] can lead the hazardous prediction of theory. We would like to carry out MRCI and CASSCF calculation to select a satisfactory theoretical model, which can be used to study for larger delocalized electron systems containing radicals.

4. Conclusion

Three theoretical models (B3LYP/6-31G*, HF/6-31G* and MP2/6-31G* level of quantum chemical theory) were tested to predict the complete conformational maps of 1,4-perntadien-3-yl radical. The surfaces determined by the HF/6-31G* and the B3LYP/6-31G* were quasi the same, and the difference from MP2/6-31G* are remarkable. The relative energies and the shape of MP2 surface varied compared with other two PES. We concluded, there are lots of changes with increasing the level of theory to take in account of the correlation energy. To solve this problem, we need additional probe for determining accuracy values of energy and shape of potential energy surface of 1,4-pentadien-3-yl radical.

Relative energy to E,E structure of 1,4-pentadien-3-yl radical was determined and there are the followings: E,Z structure has 10.0 ± 0.7 kJ mol^{-1} and E,Z structure has 26.9 ± 1.9 kJ mol^{-1}, respectively. To predict the accurate relative energy of

cyclopenten-3-yl radical we need to carry out additional calculations.

Acknowledgements

One of the authors (BV) thanks the Hungarian Scientific Research Found (OTKA F-037648) for financial support.

References

[1] N. Bernoud-Hubac, L.J. Roberts II, Biochemistry 41 (2002) 11466.
[2] K.W. Egger, S.W. Benson, J. Am. Chem. Soc. 88 (1966) 241.
[3] K.B. Clark, P.N. Culshaw, D. Griller, F.P. Lossinog, J.A.M. Simões, J.C. Walton, J. Org. Chem. 56 (1991) 5535.
[4] R. Zils, S. Inomata, T. Imamura, A. Miyoshi, N. Washida, J. Phys. Chem. A. 105 (2001) 1277.
[5] D. Griller, K.U. Ingold, J.C. Walton, J. Am. Chem. Soc. 101 (1979) 758.
[6] P.B. Karadakov, J. Gerratt, G. Raos, D.L. Cooper, M. Raimondi, J. Am. Chem. Soc. 116 (1994) 2075.
[7] C. Martinez, A.L. Cooksy, J. Org. Chem. 67 (2002) 2295.
[8] W.R. Markesbery, J.M. Carney, Brain Pathol. 9 (1999) 133–146.
[9] J.A. Conquer, M.C. Tierney, J. Zecevic, W.J. Bettger, R.H. Fisher, Lipids 35 (12) (2000) 1305–1312.
[10] M.R. Prasad, M.A. Lovell, M. Yatin, H. Dhillon, W.R. Markesbery, J. Biol. Chem. 273 (22) (1998) 13605–13612.
[11] J.D. Morrow, A.R. Tapper, W.E. Zackert, J. Yang, S.C. Sanchez, T.J. Montine, L.J. Roberts, Adv. Exp. Med. Biol. 469 (1999) 343–347.
[12] M.J. Frisch, G.W. Trucks, H.B. Schlegel, G.E. Scuseria, M.A. Robb, J.R. Cheeseman, V.G. Zakrzewski, J.A. Montgomery, Jr., R.E. Stratmann, J.C. Burant, S. Dapprich, J.M. Millam, A.D. Daniels, K.N. Kudin, M.C. Strain, O. Farkas, J. Tomasi, V. Barone, M. Cossi, R. Cammi, B. Mennucci, C. Pomelli, C. Adamo, S. Clifford, J. Ochterski, G.A. Petersson, P.Y. Ayala, Q. Cui, K. Morokuma, D.K. Malick, A.D. Rabuck,

K. Raghavachari, J.B. Foresman, J. Cioslowski, J.V. Ortiz, B.B. Stefanov, G. Liu, A. Liashenko, P. Piskorz, I. Komaromi, R. Gomperts, R.L. Martin, D.J. Fox, T. Keith, M.A. Al-Laham, C.Y. Peng, A. Nanayakkara, C. Gonzalez, M. Challacombe, P.M.W. Gill, B. Johnson, W. Chen, M.W. Wong, J.L. Andres, C. Gonzalez, M. Head-Gordon, E.S. Replogle, J.A. Pople, GAUSSIAN 98, Revision A.6, Gaussian, Inc., Pittsburgh PA, 1998.

[13] A.D. Becke, J. Chem. Phys., 101 (1993) 5648.
[14] L.A. Curtiss, K. Raghavachari, P.C. Redfern, V. Rassolov, J.A. Pople, J. Chem. Phys., 109 (1998) 7764.
[15] A.G. Baboul, L.A. Curtiss, P.C. Redfern, K. Raghavachari, J. Chem. Phys. 110 (1999) 7650.
[16] J.A. Pople, A.P. Scott, M.W. Wong, L. Radom, Israel J. Chem. 33 (1993) 345.
[17] T.H. Dunning, J. Chem. Phys. 90 (1989) 1007.

ELSEVIER

Journal of Molecular Structure (Theochem) 666–667 (2003) 159–162

THEO CHEM

www.elsevier.com/locate/theochem

L-2-Hydroxypropionic acid in aqueous solution— a vibrational spectroscopic and computational study

Z.A. Fekete[a], T. Körtvélyesi[a], J. Andor[a], I. Pálinkó[b,*]

[a]Department of Physical Chemistry, University of Szeged, Szeged, Hungary
[b]Department of Organic Chemistry, University of Szeged, Dom ter 8, Szeged 6720, Hungary

Abstract

The hydrogen-bonded dimers of L-2-hydroxypropionic were studied by FT-IR and Raman spectroscopies and computational methods. The structures of the possible dimers were calculated at the level of ab initio (HF/6-31G(d,p)) quantum chemical method. Calculations on the isolated monomer and dimers revealed a number of possible conformers. Comparison of the calculated and the measured spectra allowed us to choose the dimers most probably present in the aqueous solution of the acid. © 2003 Elsevier B.V. All rights reserved.

Keywords: L-2-Hydroxypropionic acid; FT-IR and Raman spectra; Hydrogen-bonded dimers; HF/6-31G(d,p) ab initio method

1. Introduction

α-Hydroxy carboxylic compounds are theoretically interesting systems, since their two functional groups in close proximity give rise to a variety of possible intra- and intermolecular interactions [1–4]. Recent research interest has also been motivated by their use as monomers for biodegradable polymers [5,6], and by other technological applications in the pharmaceutical industry [7]. The smallest member of this family of compounds with a secondary α-hydroxy functionality is the lactic acid (2-hydroxypropionic acid). It is one of the most important monohydroxy monocarboxylic acid, with additional biological relevance as a metabolic intermediate [8–10] and as a product of fermentation processes.

In this contribution we present a quantum chemical investigation on the monomeric acid and its dimer conformers, in connection with their vibrational spectroscopy. Calculations on the monomer have been published before [1,2,9–14], however, to our knowledge similar treatments on dimers are not available in the literature yet. Our focus is studying the association behavior typical of carboxylic acids computationally as well as experimentally. Therefore, vibrational spectra of this molecule are also registered and compared to those of the computed ones.

2. Methods

2.1. Experimental

Infrared spectra were measured on a Bio-Rad Digilab FTS-65A FT-IR spectrometer equipped with a liquid nitrogen cooled MCT detector. At least 256 scans were collected in the 4000–400 cm^{-1} range, with typical optical resolution of 2 cm^{-1}.

---

* Corresponding author. Tel.: +36-62-544-288; fax: +36-62-544-200.
E-mail address: palinko@chem.u-szeged.hu (I. Pálinkó).

0166-1280/$ - see front matter © 2003 Elsevier B.V. All rights reserved.
doi:10.1016/j.theochem.2003.08.026

Lactic acid (Merck) samples were either analyzed as-received (concentrated aqueous solution of ca. 80%) by casting films on KBr pellet, or from diluted solutions by attenuated total reflection (ATR).

Raman measurements were performed with NIR excitation (1064 nm, Spectra Physics Topaz Series T10-106 C source) on a Bio-Rad Digilab FTS-65R FT-Raman instrument with a liquid nitrogen cooled Ge detector. The laser power at the sample was ca. 500 mW, and typically 512 scans were collected in the $4000-400$ cm^{-1} range with 2 cm^{-1} resolution, at 180° backscattering geometry.

2.2. Calculations

Calculations were performed by ab initio quantum chemical method at the level of HF/6-31G(d,p)//HF/6-31G(d,p) with full geometry optimization [15]. The force matrices of the optimized molecules were positive definite in all cases. The electronic energies were corrected by the zero point vibrational energy (ZPVE), which was scaled for thermochemical data and vibrational analysis applying a scaling factor of 0.8992 [16]. For the calculation of the relative energies, the energy of the most stable monomer corrected with ZPVE was considered. In the evaluation of the hydrogen bonds the criteria of Bondi [17] was used: the distance between the pillar atoms has to be within the sum of their van der Waals radii (O$\cdots$O $\leq$ 304 pm) and the angle O–H$\cdots$O $\geq$ 90°.

3. Results and discussion

The calculated and ZPVE corrected electronic energies of the lactic acid dimers are summarized in Table 1. It also contains the relative energies of the dimers, in parenthesis, compared to the most stable dimer, when dimerisation occurs through the association of the two carboxylic groups. Computations revealed that the stabilities of the associates depend on the conformation of the constituent monomers and the strain in the dimer structures. The most stable structures have two hydrogen bondings. Potentially, the molecule has three hydrogen bond acceptor (two O atoms in the carboxylic group and one in the hydroxy group) and two hydrogen bond donor

Table 1

Electronic energies of lactic acid, monomers and dimers corrected by ZPVE calculated at the level of the HF6-31G(d,p) ab initio method

Structures	E^0 (Hartree) (E_{rel}, kJ/mol)	Structures	E^0 (Hartree) (E_{rel}, kJ/mol)
(a)	-341.62181	(f)	-683.25348 (31.8)
(b)	-341.61839	(g)	-683.25320 (32.5)
(c)	-683.26559 (0)	(h)	-683.24804 (46.1)
(d)	-683.26106 (11.9)	(i)	-683.25385 (30.8)
(e)	-683.25340 (32.0)		

For the structures denoted by (a)–(i), see Fig. 1; relative energies are compared to the most stable dimer.

atoms (the O in the OH part of the carboxylic group and the O in the alcoholic hydroxy group). The main hydrogen bonding possibilities in the dimers considered in the calculations are as follows:

(i)   association through the carboxylic groups by two (carbonyl)O$\cdots$HO hydrogen bonds [structures (c) and (d)]

(ii) bifurcated hydrogen bonds between the carboxylic OH (the oxygen is the donor atom) of one molecule and the carboxylic group of the other (both its carbonyl and alcoholic oxygen play the role of hydrogen bond acceptors) [structure (e)]

(iii) bifurcated hydrogen bonds between the hydroxy OH (the oxygen is the donor atom) of one molecule and the carboxylic group of the other (both its carbonyl and alcoholic oxygen play the role of hydrogen bond acceptors) [structure (f)]

(iv) two (hydroxy)OH$\cdots$O(carbonyl) hydrogen bonds [structure (h)]

(v)   one (hydroxy)OH$\cdots$O(carbonyl) hydrogen bond [structures (g) and (i)].

The monomer acid was geometry optimized too. Two conformations were identified. Structure (b) was found to be more stable due to strong intramolecular (hydroxy)OH$\cdots$(carbonyl)O hydrogen bond.

In the carboxylic–carboxylic dimers [structures (c) and (d)] the O$\cdots$O distances are 276 and 278 pm, respectively, the corresponding (O–H$\cdots$O) angles are

175.3 and 174.7°, respectively, both are well within Bondi's limit of such hydrogen bonds. Out of the two, structure (c) is more stable due to intramolecular (hydroxy)OH···(carbonyl)O hydrogen bonds (O···O distances are 262 and 265 pm, respectively, O–H···O angles are 103.3 and 113.4°, respectively).

In dimers with bifurcated intermolecular hydrogen bonds [structures (e) and (f)], one of the hydrogen bonds is as short as in the carboxylic–carboxylic dimers (O···O distance is typically 278 pm), while

the other is considerably longer (O···O distance is typically 299 pm). These dimers are significantly less stable than any of the carboxylic–carboxylic dimers.

The dimer connected by two (hydroxy)OH···O(-carbonyl) hydrogen bonds [structure (h)] is also considerably less stable than the carboxylic–carboxylic dimers due to the longer hydrogen bonds (O···O distance is 297 pm). Dimers having one intermolecular hydrogen bond [structures (f) and (i)]

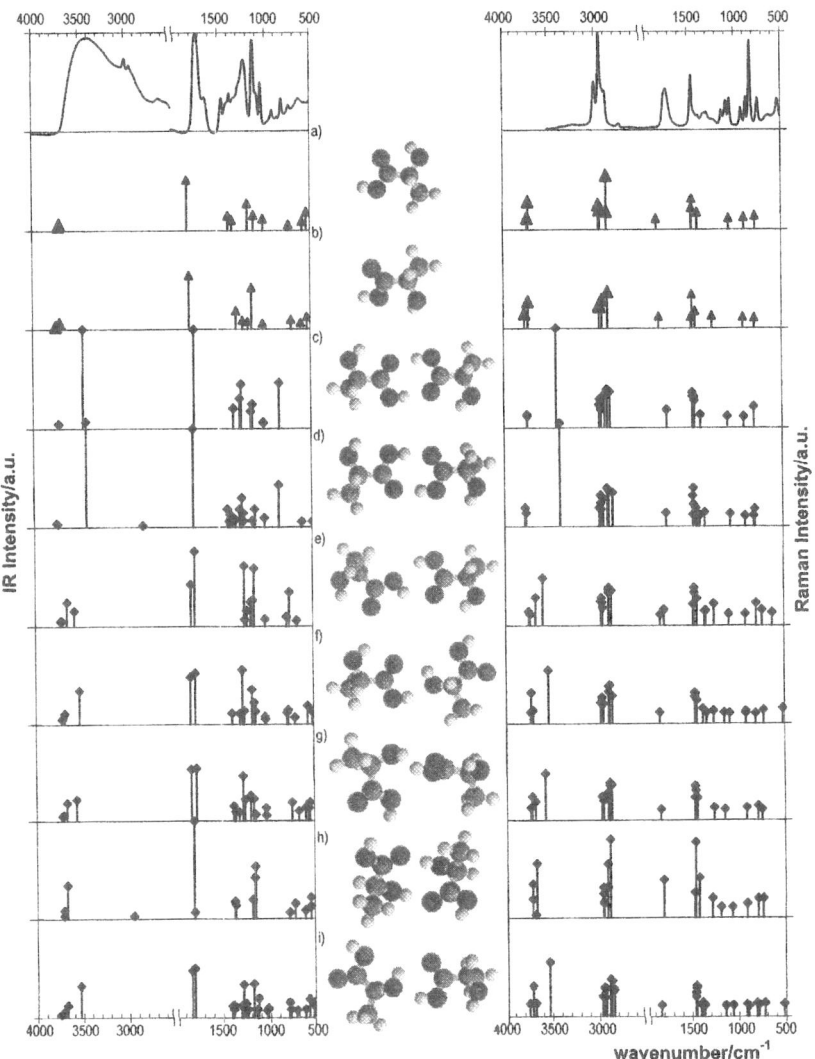

Fig. 1. Measured (aqueous solution; top panel) and calculated vibrational spectra of 2-hydroxypropionic acid conformers (monomers and dimers) (under 2000 cm^{-1} the calculated intensities are shown in a magnified scale: 3 × for the infrared (left) and 6 × for the Raman spectrum).

are not competitive with the carboxylic–carboxylic dimers either as far as stabilities are concerned.

The vibrational spectra of the geometry optimized dimers as well as the monomer were also computed and after appropriate scaling they are depicted in Fig. 1 together with the experimental infrared and Raman spectra.

Obviously, one does not expect complete coincidence between the computed and the measured spectra, because of at least two factors. One of them is the fact that computations were performed on isolated dimers, while the experimental spectra were taken in aqueous solution. Water is a protic solvent capable of participating in hydrogen bonding both as a donor and as an acceptor, therefore hydrogen bonding patterns may be influenced extensively. The other is that various forms of dimers, moreover, hydrogen bonded cyclic or acyclic oligomers larger than dimers may coexist. These structures should have a distribution and one is lucky if one or only few such structures are predominant.

Fortunately, the comparison of the experimental and the computed spectra allowed us to pick the carboxylic–carboxylic dimers as the most abundant formations. The coincidence is excellent between the measured and computed FT-IR spectra and except one band near the high frequency end is very good between those of the Raman spectra. On the basis of energy calculations we may even state that structure (c) is predominant over structure (d).

4. Conclusions

Through assuming various hydrogen bonding arrangements and calculating the geometries and energies of lactic acid dimers (L-2-hydroxypropionic acid) we were successful in determining the predominant species in the aqueous solution of this molecule. The comparison of the computed and the measured vibrational spectra revealed that the most abundant species are the carboxylic–carboxylic dimers. Even the most probable conformer could be identified on the basis of energy calculations. It is the dimer having the maximum number of intramolecular (hydroxy)OH···(carbonyl)O hydrogen bonds.

Acknowledgements

This work was supported by the Ministry of Education of Hungary through grant FKFP 0286/2000. The financial help is highly appreciated. Some of the IR measurements were carried out by Ms B. Kardos. Her assistance is gratefully acknowledged.

References

[1] H. Hollenstein, J. Mol. Struct. (Theochem). 79 (1982) 447.
[2] M.D. Newton, G.A. Jeffrey, J. Am. Chem. Soc. 99 (1977) 2413.
[3] C. Blaquiere, G. Berthon, Inorg. Chim. Acta 135 (1987) 179.
[4] S. Jarmelo, R. Fausto, Phys. Chem. Chem. Phys. 4 (2002) 1555.
[5] G. Cassanas, M. Morssli, E. Fabregue, L. Bardet, J. Raman Spectrosc. 22 (1991) 409.
[6] G. Cassanas, G. Kister, E. Fabregue, M. Morssli, L. Bardet, Spectrochim. Acta A 49 (1991) 271.
[7] S. Jarmelo, T.M.R. Maria, M.L.P. Leitao, R. Fausto, Phys. Chem. Chem. Phys. 3 (2001) 387.
[8] R. Schmidt, J. Gready, J. Mol. Struct. (Theochem). 498 (2000) 101.
[9] K. Norris, J. Gready, J. Mol. Struct. (Theochem). 98 (1993) 99.
[10] K. Norris, J. Gready, J. Mol. Struct. (Theochem). 90 (1992) 109.
[11] H. Hollenstein, R.W. Schar, N. Schwizgebel, G. Grassi, H.H. Gunthard, Spectrochim. Acta A 39 (1983) 193.
[12] L.R. Domingo, J. Andres, V. Moliner, V.S. Safont, J. Am. Chem. Soc. 119 (1997) 6415.
[13] C. Chen, F. Hsu, J. Mol. Struct. (Theochem). 506 (2000) 147.
[14] M. Pecul, A. Rizzo, J. Leszczynski, J. Phys. Chem. A 106 (2002) 11008.
[15] M.J. Frisch, G.W. Trucks, H.B. Schlegel, G.E. Scuseria, M.A. Robb, J.R. Cheeseman, V.G. Zakrzewski, J.J.A. Montgomery, R.E. Stratmann, J.C. Burant, S. Dapprich, J.M. Millam, A.D. Daniels, K.N. Kudin, M.C. Strain, O. Farkas, J. Tomasi, V. Barone, M. Cossi, R. Cammi, B. Mennucci, C. Pomelli, C. Adamo, S. Clifford, J. Ochterski, G.A. Petersson, P.Y. Ayala, Q. Cui, K. Morokuma, D.K. Malick, A.D. Rabuck, K. Raghavachari, J.B. Foresman, J. Clioslowski, J.V. Ortiz, B.B. Stefanov, G. Liu, A. Liashenko, P. Piskorz, I. Komaromi, R. Gomperts, R.L. Martin, D.J. Fox, T. Keith, M.A. Al-Laham, C.Y. Peng, A. Nanayakkara, C. Gonzalez, M. Challacombe, P.M.W. Gill, B. Johnson, W. Chen, M.W. Wong, J.L. Andres, C. Gonzalez, M. Head-Gordon, E.S. Replogle, J.A. Pople, GAUSSIAN 98, Revision A.6, Gaussian, Inc., Pittsburgh, PA, 1998.
[16] A.P. Scott, L. Radom, J. Phys. Chem. 100 (1996) 16502.
[17] A. Bondi, J. Phys. Chem. 68 (1964) 441.

ELSEVIER

Journal of Molecular Structure (Theochem) 666–667 (2003) 163–168

THEO
CHEM

www.elsevier.com/locate/theochem

Intra- and intermolecular hydrogen bondings in steroids— a combined experimental and theoretical study

A. Magyar, Z. Szendi, J.T. Kiss, I. Pálinkó*

Department of Organic Chemistry, University of Szeged, Dóm tér 8, Szeged H-6720, Hungary

Abstract

FTIR measurements on the solutions of previously not studied steroids [(20R)[6′(3′,4′[2′H]-dihydropyranyl)]-pregn-5-ene-3β,20-diol 3(2″-tetrahydropyranyl) ether (1) and (20R)[6′(3′,4′[2′H]-dihydropyranyl)]-pregn-5-ene-3β,20-diol 3-acetate (2)] with varying concentrations revealed, that intra- as well as intermolecular interactions existed for both molecules. The former was more important for compound 1 and the latter for compound 2. Only intermolecular hydrogen bonds were found in the crystalline state, however. Molecular modeling (PM3 code) allowed a fair estimation of intramolecular hydrogen bonding parameters and helped in suggesting possible arrangement for compound 2 in solution.
© 2003 Elsevier B.V. All rights reserved.

Keywords: (20R)[6′(3′,4′[2′H]-dihydropyranyl)]-pregn-5-ene-3β,20-diol 3(2″-tetrahydropyranyl) ether; (20R)[6′(3′,4′[2′H]-dihydropyranyl)]-pregn-5-ene-3β,20-diol 3-acetate; intra- and intermolecular hydrogen bonds; FTIR spectroscopy; Molecular modeling

1. Introduction

Steroids are important compounds from chemical, biological as well as medicinal aspects [1,2]. The steroid skeleton is peculiar, since it is rigid, therefore substituents can sit in different positions providing large stereochemical variety. The substituents and their stereochemistry determine biological activity. When alterations are made in the substituents or in their stereochemistry, alterations are expected in biological activity as well. Thus, modifying the substitution pattern may lead to novel compounds with novel physiological properties [3,4]. Many different chemical ways of modifying the substituents exist, and those are the most valuable, which afford stereoselection. There are methods, which allow building cholesterol like side chains on the D ring of the steroid [5]. One way is to attach a dihydropyranyl group onto the 20-C atom. This ring can be further transformed by epoxidation, then by ring opening, thus providing a route for more complicated side chains [6]. Stereoselection was observed during epoxidation. This was tentatively attributed to an intramolecular hydrogen bond between the proton of the 20-OH group and the pyranyl oxygen. FTIR spectroscopy and molecular modeling were applied for the first time for the model compounds to test the assumption and these results are communicated in this contribution.

2. Experimental

The following compounds were studied:
(20R)[6′(3′,4′[2′H]-dihydropyranyl)]-pregn-5-ene-3β,20-diol 3(2″-tetrahydropyranyl) ether (1) and

---

* Corresponding author. Fax: +36-62-544-200.
E-mail address: palinko@chem.u-szeged.hu (I. Pálinkó).

0166-1280/$ - see front matter © 2003 Elsevier B.V. All rights reserved.
doi:10.1016/j.theochem.2003.08.027

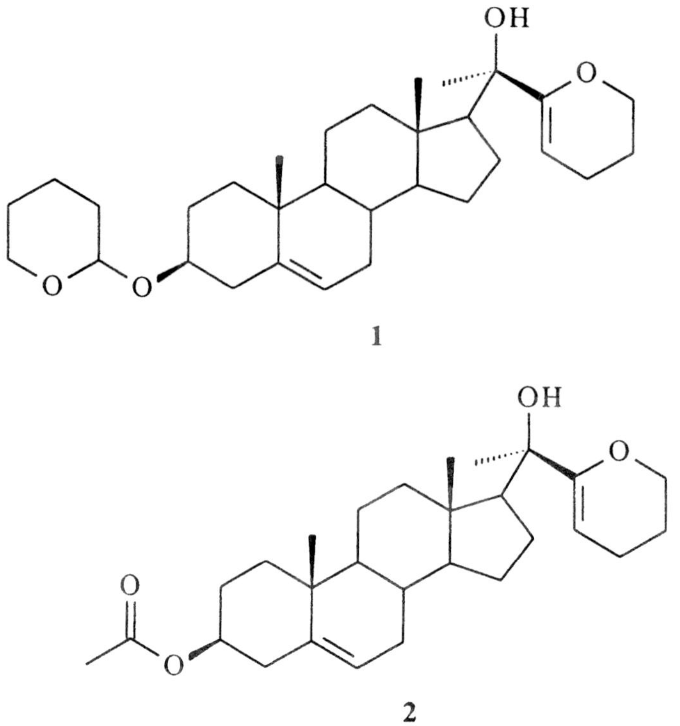

Fig. 1. The steroid molecules studied.

(20R)[6′(3′,4′[2′H]-dihydropyranyl)]-pregn-5-ene-3β,20-diol 3-acetate (2). They were synthesized in our laboratory for the first time and they are depicted in Fig. 1.

The compounds were prepared through a series of reactions [5]. The final products were crystalline and were purified with recrystallisation. Their identity and purity were checked by elemental analysis, FT-IR and NMR spectroscopies and melting point measurements.

Hydrogen bonding interactions were studied by FTIR spectroscopy both by the KBr technique (1.2 mg of the compounds in 200 mg KBr) and in CCl_4 solutions. The concentration of the steroid was varied between 10^{-1} and 10^{-4} mol/dm^3. The FT-IR spectra were taken with a BIORAD FTS-65A/896 spectrometer equipped with a liquid nitrogen cooled MCT detector. Resolution was 2 cm^{-1} and 256 scans were collected for a spectrum. Spectra were evaluated by the WIN-IR software package. The 3800–900 cm^{-1} region was examined. The following assignation was used for identifying the interaction in which the 20-OH was involved (i) non-associated:

above 3600 cm^{-1} (sharp band), (ii) intramolecular hydrogen bond: 3600–3550 cm^{-1} (sharp band) and (iii) intermolecular hydrogen bond(s): 3500–3200 cm^{-1} ([structured]broad band).

Molecular modeling was performed with both compounds applying the PM3 [7] semiempirical quantum chemical code available in the HyperChem 7.0 package [8]. On searching for intermolecular hydrogen bonds two of the optimized molecules were placed in the vicinity of each other enforcing the assumed close contact. On searching for intramolecular hydrogen bonds, the assumed conformation was enforced. Then, optimization started. It was stopped when the gradient norms were less than 0.1 and the force matrices were found to be positive definites verifying that minima were found. After convergence bond lengths were determined. They were accounted as hydrogen bonds when both the distance between the heavy atoms fell within the sum of their van der Waals radii (O···O ≤ 304 pm) compiled by Bondi [9] and the angle defined by the two heavy atoms and hydrogen (donor atom–H–acceptor atom) were larger than 90° [10].

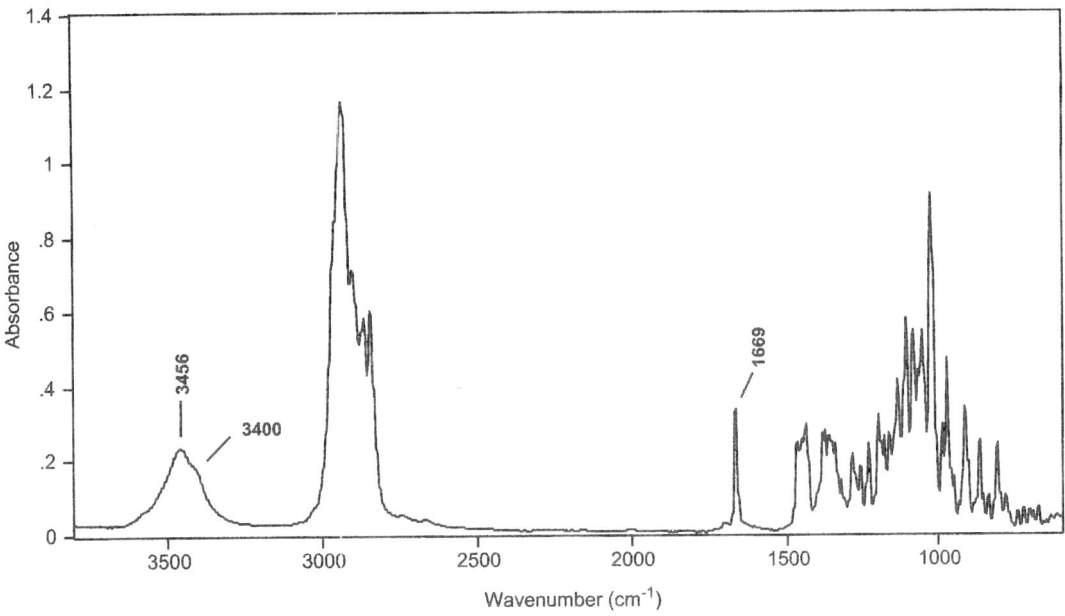

Fig. 2. FTIR spectra of compound **1** in the solid state: KBr pellet.

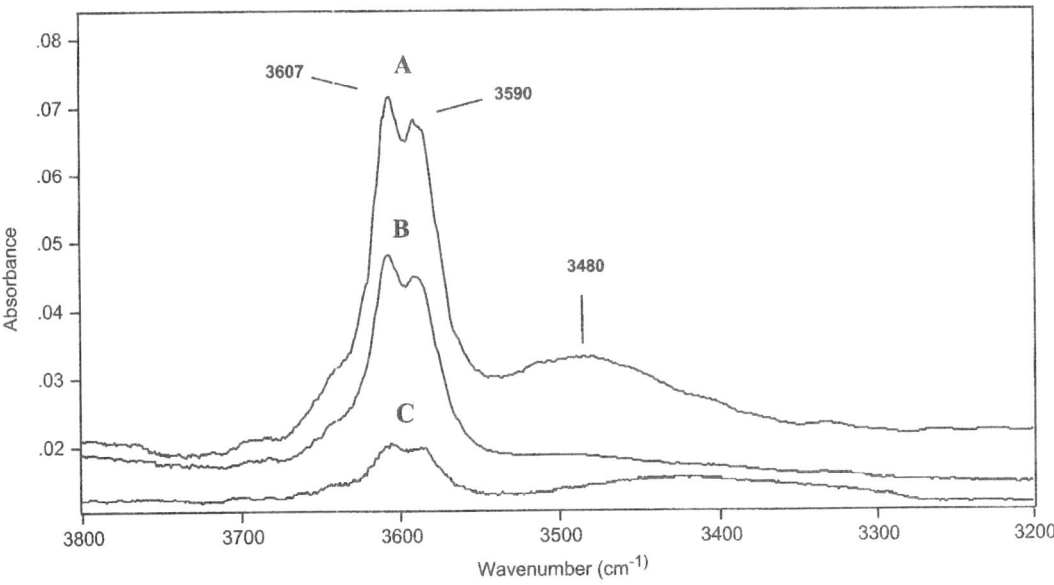

Fig. 3. FTIR spectra of compound **1** in CCl$_4$ solution: (A) 10^{-1} mol/dm^3, (B) 10^{-2} mol/dm^3 and (C) 10^{-3} mol/dm^3 in the 3800 cm^{-1}– 3200 cm^{-1} range.

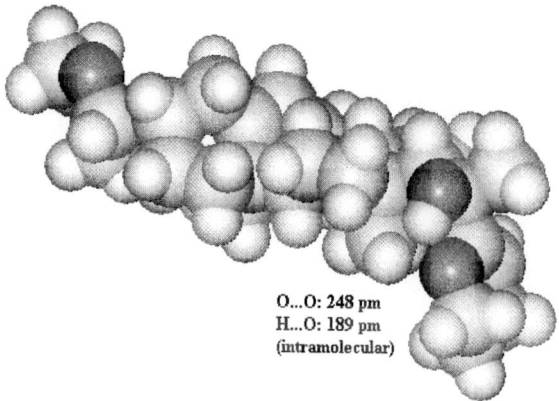

O...O: 248 pm
H...O: 189 pm
(intramolecular)

Fig. 4. Intramolecular hydrogen bonding between the hydrogen of the 20-OH group and the pyranyl oxygen in compound **1**.

3. Results and discussion

3.1. Hydrogen bonding interactions in compound 1

It is to be seen in the FTIR spectrum taken in the solid state (Fig. 2) that there is a relatively broad peak centering at 3456 cm^{-1} in the OH stretching region indicating intermolecular hydrogen bonding interaction.

The band is structured, it has a less intense component at 3400 cm^{-1}. It shows that there is more than one type of feasible intermolecular hydrogen bond. The lack of sharp bands at higher wavenumbers is the sign that intramolecular hydrogen bonding is of low priority.

FTIR measurements in solutions of different steroid concentrations (Fig. 3) revealed that now, intermolecular hydrogen bonding (although it exists) was of low priority here and the most important interaction was the intramolecular one signaled by the intense sharp band at 3590 cm^{-1}.

The third band at 3607 cm^{-1} (not seen in the solid state) is indicative of non-associated OH groups. On dilution the relative intensity of this latter band increased on the expense of the band attributed to intermolecular hydrogen bonding.

In the intramolecular interaction, quite probably, the proton of the 20-OH group gets in close contact with the pyranyl oxygen. Molecular modeling verified that this interaction can exist indeed: a five-membered quasi-ring can be formed (Fig. 4) in the computational experiment.

3.2. Hydrogen bonding interactions in compound 2

Only intermolecular hydrogen bonding was identified for the acetoxy compound in the solid state as well (Fig. 5).

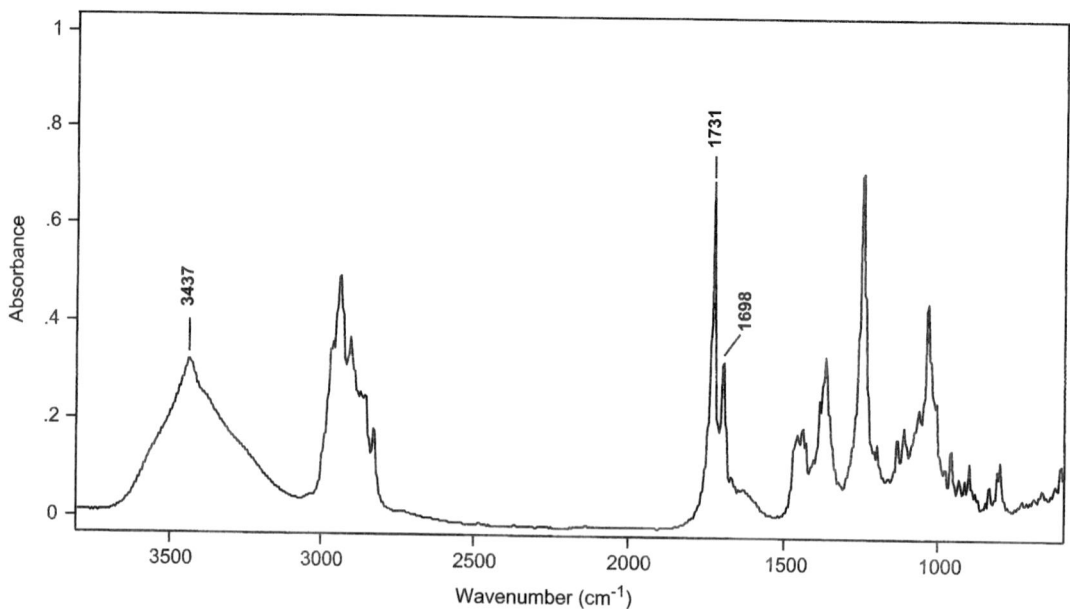

Fig. 5. FTIR spectra of compound **2** in the solid state: KBr pellet.

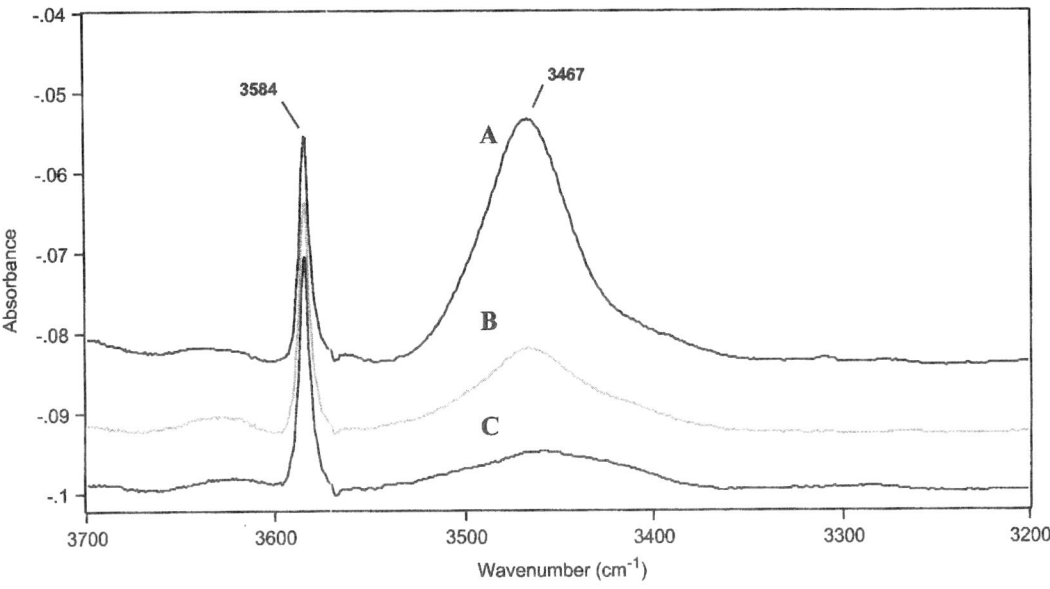

Fig. 6. FTIR spectra of compound **2** in CCl$_4$ solution: (A) 10^{-2} mol/dm^3, (B) 10^{-3} mol/dm^3 and (C) 10^{-4} mol/dm^3 in the 3700 cm^{-1}– 3200 cm^{-1} range.

Although the band centered at 3437 cm^{-1} is structured, but only at the low frequency side. This means again that various types of such interactions may exist. Band typical for non-associated OH groups cannot be seen, however.

Interestingly, the band for non-associated OH groups are not found in the solution spectra either even at extremely low (10^{-4} mol/dm^3) concentration (Fig. 6).

Nevertheless, intramolecular hydrogen bonding (band at 3584 cm^{-1}) is important in solution and its relative significance compared to intermolecular hydrogen bonds is growing on dilution.

The arrangement for the intramolecular hydrogen bonding is the same as for compound **1**, nonetheless, learning about intermolecular interactions becomes essential here. As an example,

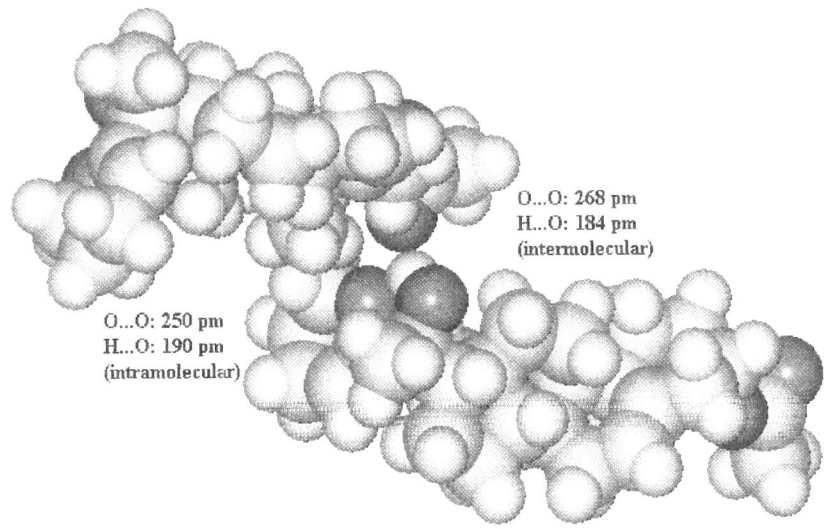

Fig. 7. Intra- and intermolecular hydrogen bonding both involving the hydrogen of the 20-OH group in compound **2**.

we show a dimer obtained from a calculation, where the hydrogen of the OH group is involved in two hydrogen bonds (Fig. 7).

One is the 'usual' intramolecular hydrogen bonding (the acceptor is the pyranyl oxygen) and the other is an intermolecular hydrogen bond with the carbonyl oxygen of another molecule of compound **2**. This minimum structure models well the coexistence of intra- and intermolecular hydrogen bonding interactions revealed in the solution spectra (Fig. 6).

In this compound dilution did not provide with band due to non-associated OH groups. The following explanation may be offered: at lower concentration certain intermolecular hydrogen bonds are broken, but the hydrogen of the OH bond remain in the intramolecular hydrogen bond.

4. Conclusions

IR measurements in the solid state and also in solution with varying steroid concentration indicate that intra- as well as intermolecular hydrogen bonds of the 20-OH group exist and quantum chemical calculations show models for viable steric arrangements. Intramolecular hydrogen bond is formed between the hydrogen of the OH group and the pyranyl oxygen, while a possible intermolecular hydrogen bonding is the close contact of the carbonyl

oxygen in a molecule of compound **2** in solution with the hydrogen in the OH group of another molecule of the same compound. The hydrogen is involved in intramolecular hydrogen bonding at the same time.

Acknowledgements

This work was supported by the Ministry of Education of Hungary through grant FKFP 0286/2000. The financial help is highly appreciated.

References

[1] T.W.G. Solomons, Organic Chemistry, Fifth ed., Wiley, New York, 1992, pp. 432.
[2] B.L. Riggs, S. Khosla, L.J. Melton, Endocr. Rev. 23 (2002) 279.
[3] A. Ullmann, J. Steroid Biochem. 27 (1987) 1009.
[4] A. Ullmann, G. Teutsch, D. Philibert, Sci. Am. 262 (1990) 42.
[5] Z. Szendi, F. Sweet, Steroids 56 (1991) 458.
[6] Z. Szendi, P. Forgó, G. Tasi, Z. Böcskei, L. Nyerges, F. Sweet, Steroids 67 (2002) 31.
[7] J.J.P. Stewart, J. Comput. Chem. 10 (1989) 209 see also page 221.
[8] HyperChem 7.0, Hypercube, Inc., 2001, Gainesville, FL, USA.
[9] A. Bondi, J. Phys. Chem. 68 (1964) 441.
[10] C.B. Aakeröy, T.A. Evans, K.R. Seddon, I. Pálinkó, New J. Chem. 23 (1999) 145.

ELSEVIER

Journal of Molecular Structure (Theochem) 666–667 (2003) 169–218

www.elsevier.com/locate/theochem

A modular numbering system of selected oligopeptides for molecular computations: using pre-computed amino acid building blocks

Michelle A. Sahai[a,b,*], Sándor Lovas[c], Gregory A. Chass[a,b,c,d], Botond Penke[e,f], Imre G. Csizmadia[a,b,f]

[a]Global Institute of COmputational Molecular and Materials Science (GIOCOMMS),
1422 Edenrose St, Mississiauga, Ont., Canada L5V 1H3
[b]Department of Chemistry, University of Toronto, 80 St George St, Toronto, Ont., Canada M5S 3H6
[c]Department of Biomedical Sciences, Creighton University, 2500 California plaza, Omaha, NE 68178, USA
[d]Institut de Science et d'Ingénierie Supramoléculaires, 8 allée Gaspard Monge, BP 70028, 67083 Strasbourg Cedex, France
[e]Department of Medical Chemistry, University of Szeged, Dóm tér 8, H-6720 Szeged, Hungary
[f]Protein Chemistry Research Group, Hungarian Academy of Sciences, University of Szeged, Dóm tér 8, H-6720 Szeged, Hungary

Abstract

A standardized numbering system has been developed to define the relative spatial orientation of all constituent atomic nuclei in model peptide systems. The system is expected to improve efficiency in the formulation of initial structural estimates and analysis of the results emerging from molecular orbital computations on oligopeptides, catalogued in a database. This standard provides each peptide residue with a numeric definition of its own, enabling the numerical differentiation of individual residues in a peptide chain; without the use of visualization tools. The standardized method was used in effective geometry optimizations of selected conformers of selected peptides. Such an arrangement facilitates automation of data generation and proper database management of oligo and polypeptides. The extension from single residues to a variety of oligopeptides is elaborated upon.
© 2003 Elsevier B.V. All rights reserved.

Keywords: Peptide computations; Atomic numbering; Peptide folding; Protein folding; Internal coordinates; Oligopeptide; Ab initio

1. Introduction

The structure, function and properties of all proteins are dependent on their individual amino acid sequences, hence all conformational intricacies

of the constituent amino acids must be fully understood at the molecular level (Fig. 1). Using conformational analysis to determine the most energetically stable and most probable conformations of all amino acids is a monumental task. Research is already taking place to construct a database for all amino acids and their 'found' geometry optimized ab initio backbone and sidechain conformations, at various levels of theory [1]. The goal is to use a systematic procedure with the aid of a standardized numbering system to effectively combine pre-optimized peptide model monomers into di, tri, tetra,

* Corresponding author. Global Institute of COmputational Molecular and Materials Science (GIOCOMMS) 1422 Edenrose St, Mississiauga, Ont., Canada L5V 1H3.

E-mail addresses: msahai@giocomms.org (M.A. Sahai), slovas@bif1.creighton.edu (S. Lovas), gchass@giocomms.org (G.A. Chass), pbotond@mdche.szote.u-szeged.hu (B. Penke), icsizmadia@chem.utoronto.ca (I.G. Csizmadia).

0166-1280/$ - see front matter © 2003 Elsevier B.V. All rights reserved.
doi:10.1016/j.theochem.2003.08.028

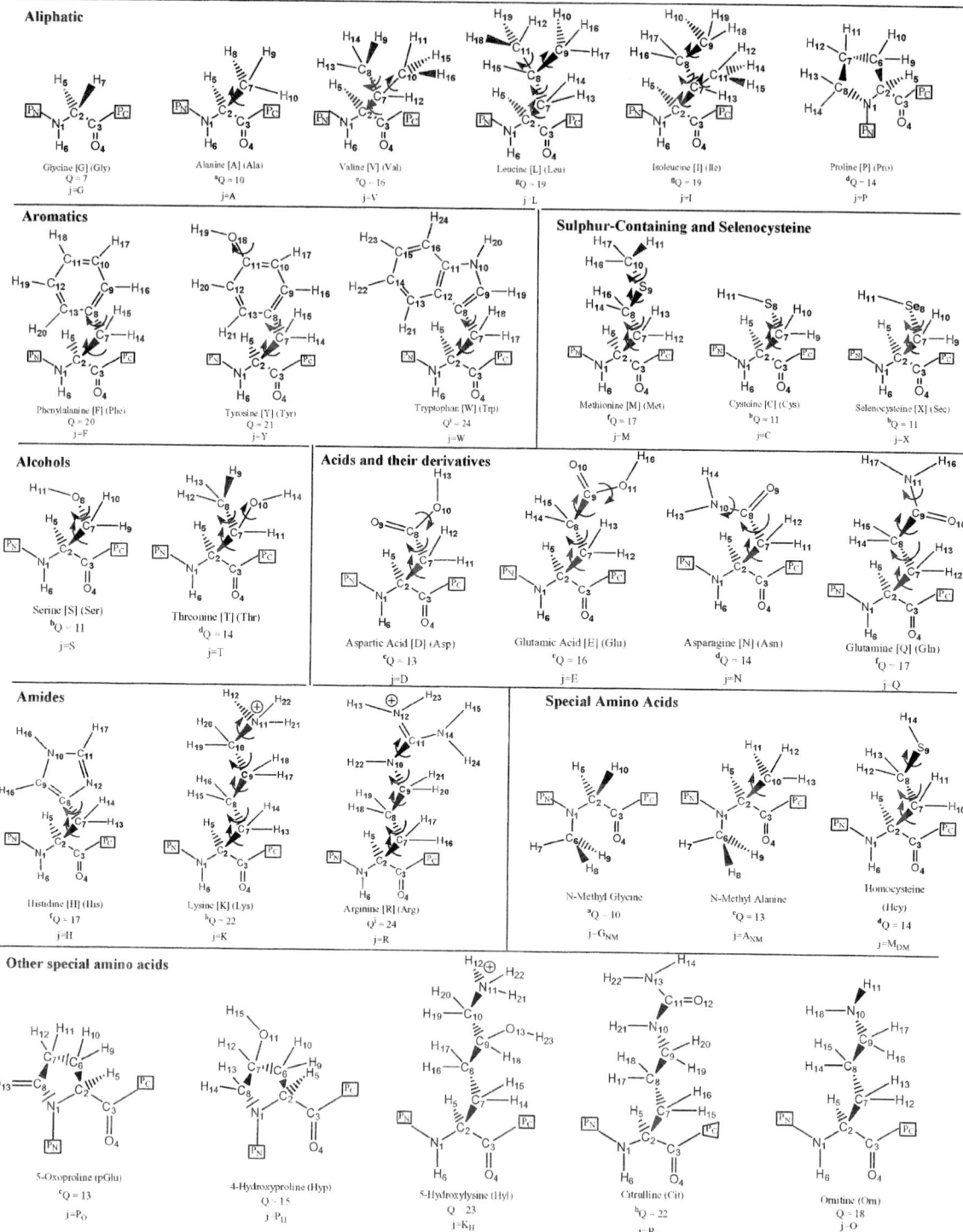

Note : Peptides with similar Q numbers (a, b, c, d, e, f, g, h and i) are differentiated through the examination of their sidechain atomic nuclei

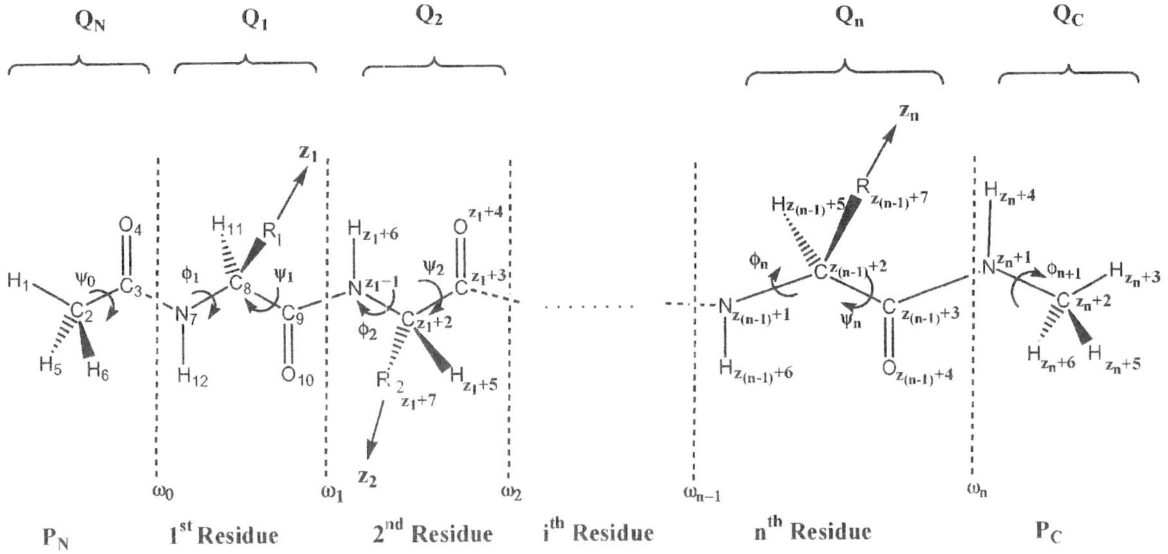

Scheme 1.

penta, and even larger oligo and polypeptides. With these larger peptide structures built up, one may begin to model secondary and eventually tertiary structural elements of proteins, perhaps helping to further the understanding of the folding process in proteins [2]. The backbone conformational characteristics of a protein might be viewed in terms of corresponding conformational potential energy surfaces and hypersurfaces in which long-range interactions are not fully explored.

The simplest amino acids such as glycine (Gly) [3–8] and alanine (Ala) [3–7,9–11] were among the first studied, followed by valine (Val) [5,12], which has one energetically significant sidechain torsional mode of motion (structurally significant dihedral angle). Subsequently, N-formyl-serylamide [13–19] was investigated in considerable detail as was phenylalanine (Phe) [20–23], both of which also have two such significant sidechain dihedral angles; tyrosine (Tyr), with its three sidechain dihedrals, is also under investigation [23]. In addition, conformational studies have been performed on threonine [24], proline (Pro) [25],

aspartate (Asp) [26], aspargine (Asn) [27], cysteine (Cys) [28,29], homocysteine (Hcy) [30] and selenocysteine (Sec) [31]. Ab initio simulations/computations have also been used to characterize the conformers of di [9,32–35], tri [36], tetra [37–40] as well as oligopeptide amides [41].

This work extends the description, through the use of exemplary peptide models, of a proposed standardized numbering system [1] that may be used more efficiently in characterizing the topologically probable (energetically stable) conformer sets of all amino acid and peptide systems.

2. Method

An extensive set of rules has already been outlined on the efficient generation of input data for geometry optimizations [1] which is illustrated in Scheme 1.

All isomeric, enantiomeric, stereochemical and conformational properties are accounted for, numerically defined and then incorporated into this standard. The concept of 'related angles' [1] allows for

Fig. 1. A schematic illustration of amino acid residues. Note that in addition to the 20 naturally occurring amino acids, which have both DNA and RNA codons, the 21st amino acid, selenocysteine (Sec), which has only an RNA codon, is also included. Eight special amino acids, N-methyl glycine (sarcosine), N-methyl alanine, S-demethylated methionine (homocysteine), 5-oxoproline, 4-hydroxyproline (Hyp), 5-hydroxylysine, citrulline and ornitine are also listed.

numerically accurate and highly efficient sampling of the full set of topologically possible conformers as established by Multi-Dimensional-Conformational Analysis [42].

For the automation of input generation and subsequent data extraction, the numeric standard defining all constituent atomic nuclei in the peptide model plays a crucial rule. A standardized method would enable software and automation scripts to immediately recognize structural features necessary for any structure–activity relationships of interest. The standard formulated allows any peptide model to be truncated (at any comprising 'module') or enlarged (with a desired 'module') at any time, without gross perturbation to the remainder of the model.

The use of this modular system is practical but becomes increasingly difficult as the number of residues in a peptide increases. The standard,

therefore, must be extended in such a way that it will function effectively, providing complete numerical clarity to the user, when dealing with simple and even complex peptide chains. The end goal of the standard is to afford the user with the ability to perform and recognize multi-dimensional Quantitative Structure–Activity Relationships in these systems, without always having to resort to the cumbersome use of visualization tools.

The following outlines this standardized system, as presented in Fig. 2, for an amino acid diamide or 'monopeptide'. The formula for any linear oligopeptide is also presented in Fig. 2.

The number of atoms within the amino acid residue or 'module' of a peptide is significant to the standardized numbering method; $Q_{i,j}$ is used to denote the number of atoms in any particular amino acid module j, in position i of a peptide chain (Fig. 2). These two factors are denoted,

For a 'monopeptide':

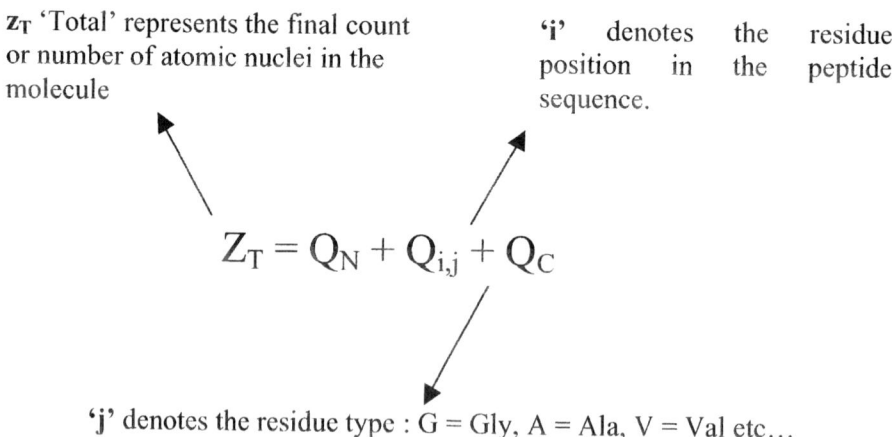

z_T 'Total' represents the final count or number of atomic nuclei in the molecule

'i' denotes the residue position in the peptide sequence.

$$Z_T = Q_N + Q_{i,j} + Q_C$$

'j' denotes the residue type : G = Gly, A = Ala, V = Val etc...

For an oligopeptide:

$$z_T = Q_N + \sum_{i=0}^{n} Q_{i,j} + Q_C$$

Fig. 2. The standardized numbering relationship where z_i is the number of atoms or 'atomic count' up to and including the ith module or residue is defined as $Z_i = Q_N + Q_{i,j}$. The total sum of atoms in a complete 'monopeptide' is $Q_N + Q_{1,j} + Q_C$. For N-acetyl and N-methylamide protective groups, $Q_N = Q_C = 6$. The total sum of atoms in a general oligopeptide is also presented.

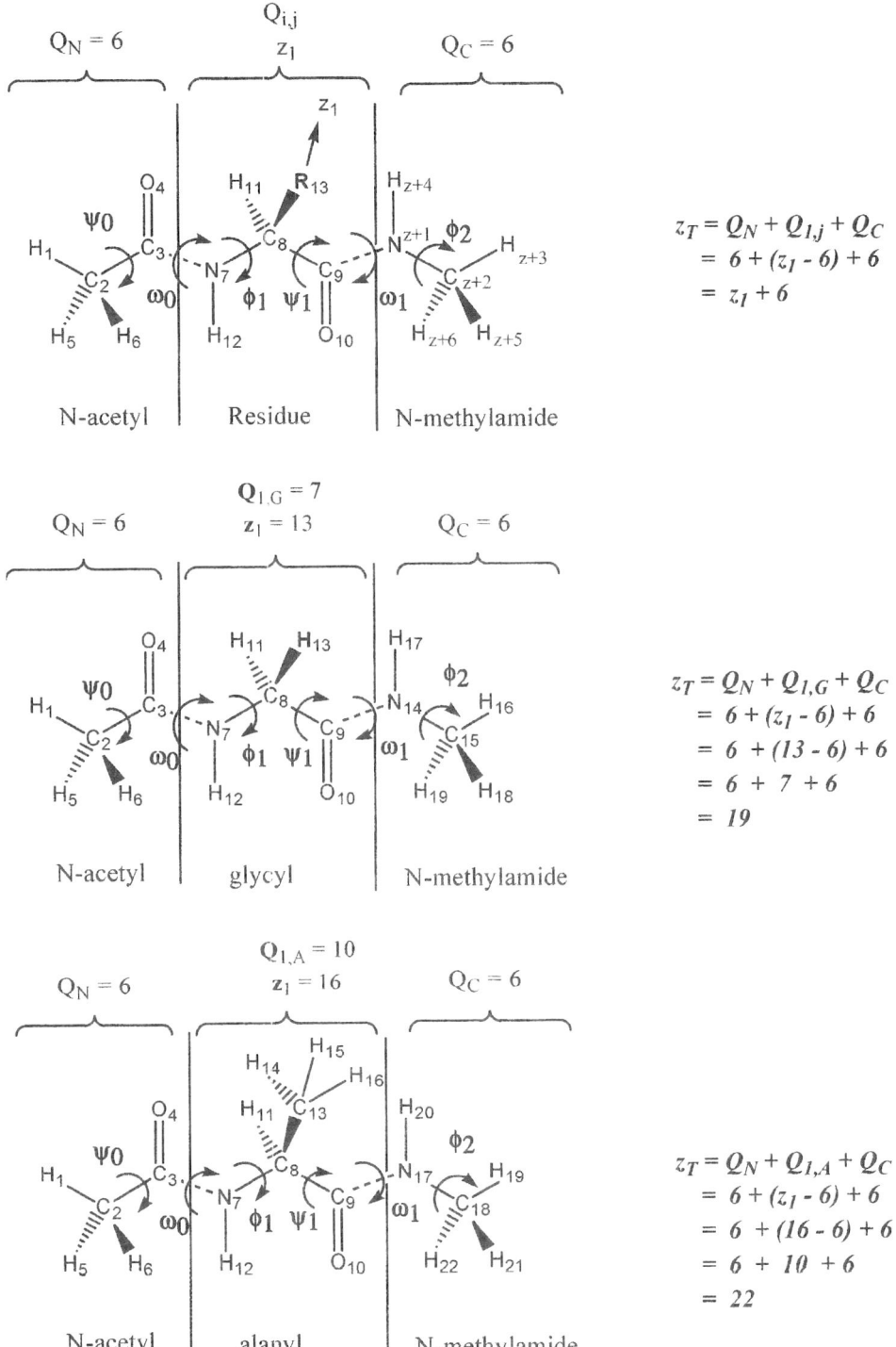

$$z_T = Q_N + Q_{1,j} + Q_C$$
$$= 6 + (z_1 - 6) + 6$$
$$= z_1 + 6$$

$$z_T = Q_N + Q_{1,G} + Q_C$$
$$= 6 + (z_1 - 6) + 6$$
$$= 6 + (13 - 6) + 6$$
$$= 6 + 7 + 6$$
$$= 19$$

$$z_T = Q_N + Q_{1,A} + Q_C$$
$$= 6 + (z_1 - 6) + 6$$
$$= 6 + (16 - 6) + 6$$
$$= 6 + 10 + 6$$
$$= 22$$

Fig. 3. A schematic illustration of the general structure of a 'monopeptide' as compared to N-acetyl-Gly-N-methylamide and N-acetyl-Ala-N-methylamide.

Table 1

Definitions of the dihedral angles for model peptides using the *N*-acetyl and *N*-methylamide protecting groups

$$\left.\begin{array}{l}\psi_0 = D7 \\ \omega_0 = D8\end{array}\right\} \quad Q_N$$

$$\left.\begin{array}{l}\phi_1 = D9 \\ \psi_1 = D[6 + (Q_{1,j} + 1)] \\ \omega_1 = D[6 + (Q_{1,j} + 2)]\end{array}\right\} \quad Q_1$$

$$\left.\begin{array}{l}\phi_2 = D[6 + (Q_{1,j} + 3)] \\ \psi_2 = D[6 + (Q_{1,j} + Q_{2,j} + 1)] \\ \omega_2 = D[6 + (Q_{1,j} + Q_{2,j} + 2)]\end{array}\right\} \quad Q_2$$

$$\vdots$$

$$\left.\begin{array}{l}\phi_i = D[6 + (Q_{1,j} + Q_{2,j} + Q_{3,j} + \cdots + Q_{i-1,j}) + 3] \\ \psi_i = D[6 + (Q_{1,j} + Q_{2,j} + Q_{3,j} + \cdots + Q_{i,j}) + 1] \\ \omega_i = D[6 + (Q_{1,j} + Q_{2,j} + Q_{3,j} + \cdots + Q_{i,j}) + 2]\end{array}\right\} \quad Q_i$$

$$\vdots$$

$$\left.\begin{array}{l}\phi_n = D[6 + (Q_{1,j} + Q_{2,j} + Q_{3,j} + \cdots + Q_{i,j} + \cdots + Q_{n-1,j}) + 3] \\ \psi_n = D[6 + (Q_{1,j} + Q_{2,j} + Q_{3,j} + \cdots + Q_{i,j} + \cdots + Q_{n,j}) + 1] \\ \omega_n = D[6 + (Q_{1,j} + Q_{2,j} + Q_{3,j} + \cdots + Q_{i,j} + \cdots + Q_{n,j}) + 2]\end{array}\right\} \quad Q_n$$

$$\vdots$$

$$Q_C$$

respectively, using subscripted letters: i can be any number since peptide chains can be quite long, but j is unique for each amino acid (Fig. 1), having a specific number of atoms. Therefore, each amino acid should be given a specific identifying variable. This is the most arbitrary portion of the methodology and could very well have its own standard developed in the future. At this point, the one-letter codes are used for each amino acid with DNA and RNA codons. For the other amino acids, under the headings 'special amino acids' and 'other special amino acids' including selenocysteine, some designations have been suggested (Fig. 1). Refinement of this definition is encouraged for future works and extensions of this methodology.

Two other definitions are made, specifically the number of atoms in the N-terminal and C-terminal protecting end groups. Using the general example, we see that this standardized numbering method starts at the *N*-acetyl end, progressing towards the *N*-methylamide at the C-terminus, with numeric priority given to the backbone in eachm module, consecutively along the peptide chain. When using an *N*-acetyl protecting group, there will always be six atoms on the N-terminus of the peptide chain. Therefore, the *N*-acetyl and *N*-methylamide protective groups are represented by $Q_N = 6$ and $Q_C = 6$, respectively. The examples found in this work solely

Table 2

The numeric designations of the dihedral angles for the alanine tripeptide, N-Ac-L-Ala-L-Ala-L-Ala-NH-Me defined using the standardized numbering system for the molecule

Module	ψ	ω	ϕ
N-acetyl	$\psi_0 = D7$	$\omega_0 = D8$	
Alanine 1	$\psi_1 = D[6 + (Q_{1,A} + 1)]$	$\omega_1 = D[6 + (Q_{1,A} + 2)]$	$\phi_1 = D9$
	$\psi_1 = D[6 + (10_{1,A} + 1)] = D17$	$\omega_1 = D[6 + (10_{1,A} + 2)] = D18$	
Alanine 2	$\psi_2 = D[6 + (Q_{1,A} + Q_{2,A} + 1)]$	$\omega_2 = D[6 + (Q_{1,A} + Q_{2,A} + 2)]$	$\phi_2 = D[6 + (Q_{1,A} + 3)]$
	$\psi_2 = D[6 + (10_{1,A} + 10_{2,A} + 1)] = D27$	$\omega_2 = D[6 + (10_{1,A} + 10_{2,A} + 2)] = D28$	$\phi_2 = D[6 + (10_{1,A} + 3)] = D19$
Alanine 3	$\psi_3 = D[6 + (Q_{1,A} + Q_{2,A} + Q_{3,A} + 1)]$	$\omega_3 = D[6 + (Q_{1,A} + Q_{2,A} + Q_{3,A} + 2)]$	$\phi_3 = D[6 + (Q_{1,A} + Q_{2,A} + 3)]$
	$\psi_3 = D[6 + (10_{1,A} + 10_{2,A} + 10_{3,A} + 1)] = D37$	$\omega_3 = D[6 + (10_{1,A} + 10_{2,A} + 10_{3,A} + 2)] = D38$	$\phi_3 = D[6 + (10_{1,A} + 10_{2,A} + 3)] = D29$
N-methylamide			$\phi_4 = D[6 + (Q_{1,A} + Q_{2,A} + Q_{3,A} + 3)]$
			$\phi_4 = D[6 + (10_{1,A} + 10_{2,A} + 10_{3,A}) + 3] = D39$

Table 3
The standardized numbering system applied to tripeptide variations of alanine, valine and arginine

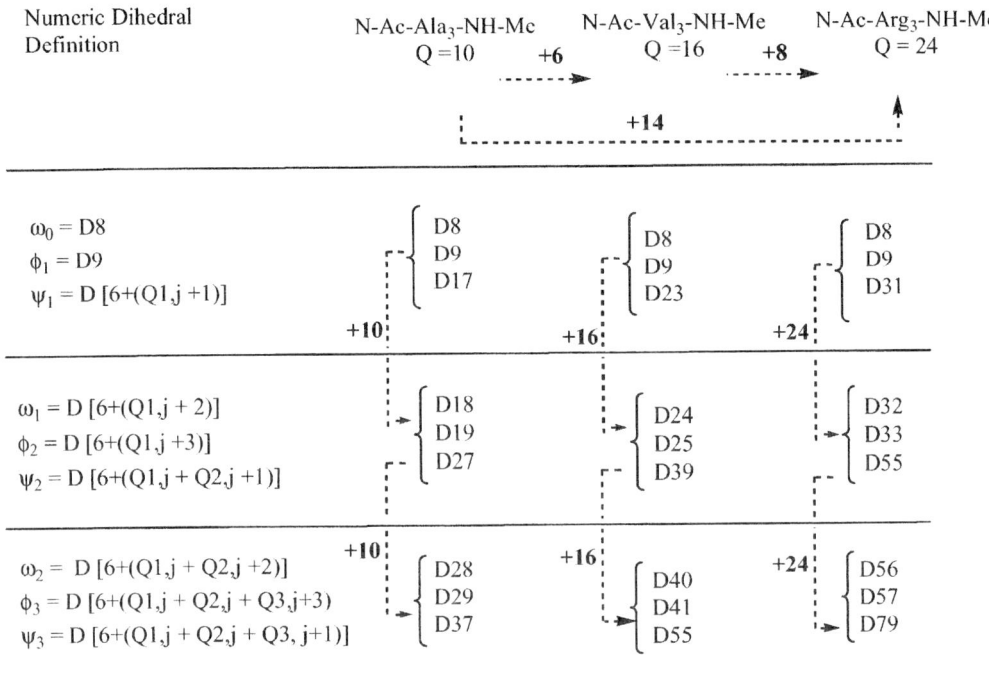

Dihedrals.

Module	'Monopeptide' (n=1)	Dipeptide (n = 2)	Tripeptide (n = 3)	Oligopeptide (n = n)
N-terminus	Q_N	Q_N	Q_N	Q_N
1st peptide module	$Q_{1,j}$	$Q_{1,j}$	$Q_{1,j}$	$Q_{1,j}$
2nd peptide module		$Q_{2,j}$	$Q_{2,j}$	$Q_{2,j}$
3rd peptide module			$Q_{3,j}$	$Q_{3,j}$
4th peptide module				$Q_{4,j}$
. . .				. . .
ith peptide module				$Q_{i,j}$
. . .				. . .
nth peptide module				$Q_{n,j}$
C-terminus	Q_C	Q_C	Q_C	Q_C

Scheme 2.

Table 4

The dependence of the 'lead' atom number of the backbone torsional angles on P_N and Q_i for various forms of amino acids. The resultant number becomes the dihedral number of the particular torsion (e.g. $\phi_1 = D9$ for N-Ac-Val-NH-Me). Shown are four variations of the protecting groups that may be used at either the N- or C-termini

Dihedral	General formula	Neutral (unprotected)	N-Ac- protected form ($P_N = 6$)	NH-Me-protected form ($P_N = 1$)	N-Ac- and NH-Me protected form ($P_N = 6$)
ψ_0	$P_N + 1$	2	7	N/A	7
ω_0	$P_N + 2$	3	8	N/A	8
ϕ_1	$P_N + 3$	4	9	4	9
ψ_1	$P_N + 1 + Q_{1,j}$	$2 + Q_{1,j}$	$7 + Q_{1,j}$	$2 + Q_{1,j}$	$7 + Q_{1,j}$
ω_1	$P_N + 2 + Q_{1,j}$	$3 + Q_{1,j}$	$8 + Q_{1,j}$	$3 + Q_{1,j}$	$8 + Q_{1,j}$
ϕ_2	$P_N + 3 + Q_{1,j}$	$4 + Q_{1,j}$	$9 + Q_{1,j}$	$4 + Q_{1,j}$	$9 + Q_{1,j}$
ψ_2	$P_N + 1 + Q_{1,j} + Q_{2,j}$	$2 + Q_{1,j} + Q_{2,j}$	$7 + Q_{1,j} + Q_{2,j}$	$2 + Q_{1,j} + Q_{2,j}$	$7 + Q_{1,j} + Q_{2,j}$
ω_2	$P_N + 2 + Q_{1,j} + Q_{2,j}$	$3 + Q_{1,j} + Q_{2,j}$	$8 + Q_{1,j} + Q_{2,j}$	$3 + Q_{1,j} + Q_{2,j}$	$8 + Q_{1,j} + Q_{2,j}$
ϕ_3	$P_N + 3 + Q_{1,j} + Q_{2,j}$	$4 + Q_{1,j} + Q_{2,j}$	$9 + Q_{1,j} + Q_{2,j}$	$4 + Q_{1,j} + Q_{2,j}$	$9 + Q_{1,j} + Q_{2,j}$
ψ_3	$P_N + 1 + Q_{1,j} + Q_{2,j} + Q_{3,j}$	$2 + Q_{1,j} + Q_{2,j} + Q_{3,j}$	$7 + Q_{1,j} + Q_{2,j} + Q_{3,j}$	$2 + Q_{1,j} + Q_{2,j} + Q_{3,j}$	$7 + Q_{1,j} + Q_{2,j} + Q_{3,j}$
ω_3	$P_N + 2 + Q_{1,j} + Q_{2,j} + Q_{3,j}$	$3 + Q_{1,j} + Q_{2,j} + Q_{3,j}$	$8 + Q_{1,j} + Q_{2,j} + Q_{3,j}$	$3 + Q_{1,j} + Q_{2,j} + Q_{3,j}$	$8 + Q_{1,j} + Q_{2,j} + Q_{3,j}$
$\phi_4 \ldots$	$P_N + 3 + Q_{1,j} + Q_{2,j} + Q_{3,j}$	$4 + Q_{1,j} + Q_{2,j} + Q_{3,j}$	$9 + Q_{1,j} + Q_{2,j} + Q_{3,j}$	$4 + Q_{1,j} + Q_{2,j} + Q_{3,j}$	$9 + Q_{1,j} + Q_{2,j} + Q_{3,j}$
	$\ldots$	$\ldots$	$\ldots$	$\ldots$	$\ldots$
ϕ_i	$P_N + 3 + Q_{1,j} + Q_{2,j} + \cdots + Q_{i-1,j}$	$4 + Q_{1,j} + Q_{2,j} + \cdots + Q_{i-1,j}$	$9 + Q_{1,j} + Q_{2,j} + \cdots + Q_{i-1,j}$	$4 + Q_{1,j} + Q_{2,j} + \cdots + Q_{i-1,j}$	$9 + Q_{1,j} + Q_{2,j} + \cdots + Q_{i-1,j}$
ψ_i	$P_N + 1 + Q_{1,j} + Q_{2,j} + \cdots + Q_{i,j}$	$2 + Q_{1,j} + Q_{2,j} + \cdots + Q_{i,j}$	$7 + Q_{1,j} + Q_{2,j} + \cdots + Q_{i,j}$	$2 + Q_{1,j} + Q_{2,j} + \cdots + Q_{i,j}$	$7 + Q_{1,j} + Q_{2,j} + \cdots + Q_{i,j}$
$\omega_i \ldots$	$P_N + 2 + Q_{1,j} + Q_{2,j} + \cdots + Q_{i,j}$	$3 + Q_{1,j} + Q_{2,j} + \cdots + Q_{i,j}$	$8 + Q_{1,j} + Q_{2,j} + \cdots + Q_{i,j}$	$3 + Q_{1,j} + Q_{2,j} + \cdots + Q_{i,j}$	$8 + Q_{1,j} + Q_{2,j} + \cdots + Q_{i,j}$
	$\ldots$	$\ldots$	$\ldots$	$\ldots$	$\ldots$
ϕ_n	$P_N + 3 + \sum_{i=1}^{n-1} Q_{i,j}$	$4 + \sum_{i=1}^{n-1} Q_{i,j}$	$9 + \sum_{i=1}^{n-1} Q_{i,j}$	$4 + \sum_{i=1}^{n-1} Q_{i,j}$	$9 + \sum_{i=1}^{n-1} Q_{i,j}$
ψ_n	$P_N + 1 + \sum_{i=1}^{n} Q_{i,j}$	$2 + \sum_{i=1}^{n} Q_{i,j}$	$7 + \sum_{i=1}^{n} Q_{i,j}$	$2 + \sum_{i=1}^{n} Q_{i,j}$	$7 + \sum_{i=1}^{n} Q_{i,j}$
$\omega_n \ldots$	$P_N + 2 + \sum_{i=1}^{n} Q_{i,j}$	$3 + \sum_{i=1}^{n} Q_{i,j}$	$8 + \sum_{i=1}^{n} Q_{i,j}$	$3 + \sum_{i=1}^{n} Q_{i,j}$	$8 + \sum_{i=1}^{n} Q_{i,j}$
	$\ldots$	$\ldots$	$\ldots$	$\ldots$	$\ldots$
Z	$Q_N + \sum_{i=1}^{n} Q_{i,j} + Q_C$	$1 + \sum_{i=1}^{n} Q_{i,j} + 2$	$6 + \sum_{i=1}^{n} Q_{i,j} + 2$	$1 + \sum_{i=1}^{n} Q_{i,j} + 6$	$6 + \sum_{i=1}^{n} Q_{i,j} + 6$

Table 5
Sample internal coordinates for valine with various protecting groups

Neutral (unprotected)	N-Ac-protected form	NH-Me-protected form	N-Ac and NH-Me protected form	
H	H	H	H	P_N (N-terminal protecting group)
	C1 R2		C1 R2	
	C2 R31 A3		C2 R31 A3	
	O3 R4 2A41 D4		O3 R42A41 D4	
	H2 R53A54 D5		H2 R53A54 D5	
	H2 R63A64 D6		H2 R63A64 D6	
N1 R2	N3 R72A71 D7	N1 R2	N3 R72A71 D7	Peptide backbone and Cα–Cβ (Cα-H for Gly)
C2 R31 A3	C7 R83A82 D8	C2 R31 A3	C7 R83A82 D8	
C3 R4 2A41 D4	C8 R97A93 D9	C3 R42A41 D4	C8 R97A93 D9	
O4 R53A52 D5	O9 R108A107 D10	O4 R53A52 D5	O9 R108A107 D10	
H3 R62A61 D6	H8 R117A113 D11	H3 R62A61 D6	H8 R117A113 D11	
H2 R73A74 D7	H7 R123A12 2 D12	H2 R73A74 D7	H7 R123A122 D12	
C3 R82A81 D8	C8 R137A133 D13	C3 R82A81 D8	C8 R137A133 D13	
C13 R98A97 D9	C13 R148A147 D14	C13 R98A97 D9	C13 R148A147 D14	Sidechain
H14 R1013A108 D10	H14 R1513A158 D15	H14 R1013A108 D10	H14 R1513A158 D15	
C13 R118A117 D11	C13 R168A167 D16	C13 R118A117 D11	C13 R168A167 D16	
H16 R1213A128 D12	H16 R1713A178 D17	H16 R1213A128 D12	H16 R1713A178 D17	
H13 R138A137 D13	H13 R188A187 D18	H13 R138A137 D13	H13 R18 8A187 D18	
H14 R1413A148 D14	H14 R1913A198 D19	H14 R1413A148 D14	H14 R1913A198 D19	
H14 R1513A158 D15	H14 R2013A208 D20	H14 R1513A158 D15	H14 R2013A208 D20	
H16 R1613A168 D16	H16 R2113A218 D21	H16 R1613A168 D16	H16 R2113A218 D21	
H16 R1713A178 D17	H16 R2213A238 D22	H16 R1713A178 D17	H16 R2213A238 D22	
O4 R183A182 D18	O9 R238A237 D23	N4 R183A182 D18	N9 R238A237 D23	P_C (C-terminal protecting group)
H18 R194A193 D19	H23 R249A248 D24	C18 R194A193 D19	C23 R249A248 D24	
		H19 R2018A204 D20	H24 R2523A259 D25	
		H18 R214A213 D21	H23 R269A268 D26	
		H19 R2218A224 D22	H24 R2723A279 D27	
		H19 R2318A224 D22	H24 R2823A289 D28	

make use of the N-acetyl and N-methylamide groups to 'protect' the N- and C-terminus, respectively, although other terminal groups have already been explored [1].

An example of this standardized numeric definition is the peptide N-Ac-Gly-NH-Me. The amino acid glycine has been referred to as the simplest of all amino acids with its structure being the deciding factor, not its activity. From Fig. 3, the sidechain (R) that was defined for the general N-acetyl-amino acid-N-methylamide is synonymous to the hydrogen (H) of glycine. If R = CH₃, then N-acetyl-alanine-N-methylamide is defined as shown in the lower part of Fig. 3.

An atomic numbering pattern for a 'mono', di and tri glycine peptide model may follow the exact pattern

as per the generalized definition. The z number or 'atomic count' that defined the number of atoms up to and including the R-group (Fig. 2) will be 13 for glycine and 16 for alanine (Fig. 3).

From Fig. 3, there are seven atoms in the central glycine residue or 'module' in position 1, making that $Q_{1,G} = 7$. Using the formula $z_i = 6 + Q_{i,G}$ (Fig. 2), we end up with $z_1 = 6 + 7 = 13$. Each amino acid or peptide residue has a certain number of atoms making up the structure (Fig. 1). From Fig. 3 it can be seen that the number of atoms in the 'peptide module' of alanine is 10 and therefore, $Q_{1,A} = 10$ for alanine in position 1 of a peptide chain.

The formula $z_i = Q_N + Q_{i,j}$ works effectively for a 'monopeptide', however, this becomes more difficult

Table 6
Symbolic internal coordinates for selected amino acids from Fig. 1 in their *N*-acetyl and *N*-methylated protected forms

	pGlu	Hyp	Hyl	Cit	Orn
P_N	H C1R2 C2R3lA3 O3R42A41D4 H2R53A54D5 H2R63A64D6	H C1R2 C2R3lA3 O3R42A41D4 H2R53A54D5 H2R63A64D6	H C1R2 C2R3lA3 O3R42A41D4 H2R53A54D5 H2R63A64D6	H C1R2 C2R3lA3 O3R42A41D4 H2R53A54D5 H2R63A64D6	H C1R2 C2R3lA3 O3R42A41D4 H2R53A54D5 H2R63A64D6
I	N3R72A71D7 C7R83A82D8 C8R97A93D9 O9R108A107D10 H8R117A113D11 C8R127A123D12	N3R72A71D7 C7R83A82D8 C8R97A93D9 O9R108A107D10 H8R117A113D11 C8R127A123D12	N3R72A71D7 C7R83A82D8 C8R97A93D9 O9R108A107D10 H8R117A113D11 H7R123A122D12 C8R137A133D13	N3R72A71D7 C7R83A82D8 C8R97A93D9 O9R108A107D10 H8R117A113D11 H7R123A122D12 C8R137A133D13	N3R72A71D7 C7R83A82D8 C8R97A93D9 O9R108A107D10 H8R117A113D11 H7R123A122D12 C8R137A133D13
I_{sc}	C12R138A137D13 C13R14l2A148D14 H12R158A157D15 H12R168A167D16 H13R17l2A178D17 H13R18l2A188D18 O14R19l3A19l2D19	C12R138A137D13 C13R14l2A148D14 H12R158A157D15 H12R168A167D16 O13R17l2A178D17 H13R18l2A188D18 H14R19l3A19l2D19 H14R20l3A20l2D20 H17R21l3A21l2D21	C13R148A147D14 C14R15l3A158D15 C15R16l4A16l3D16 N16R17l5A17l4D17 H17R18l6A18l5D18 O15R19l4A19l3D19 H13R208A207D20 H13R218A217D21 H14R22l3A228D22 H14R23l3A238D23 H15R24l4A24l3D24 H16R25l5A25l4D25 H16R26l5A26l4D26 H17R27l6A27l5D27 H17R28l6A28l5D28 H19R29l5A29l4d29	C13R148A147D14 C14R15l3A158D15 N15R16l4A16l3D16 C16R17l5A17l4D17 O17R18l6A18l5D18 N17R19l6A19l5D19 H19R20l7A20l6D20 H13R218A217D21 H13R228A227D22 H14R23l3A238D23 H14R24l3A248D24 H15R25l4A25l3D25 H15R26l4A26l3D26 H16R27l5A27l4D27 H19R28l7A28l6D28	C13R148A147D14 C14R15l3A158D15 N15R16l4A16l3D16 H16R17l5A17l4D17 H13R188A187D18 N13R198A197D19 H14R20l3A208D20 H14R21l3A218D21 H15R22l4A22l3D22 H15R23l4A23l3D23 H16R24l5A24l4D24
P_C	N9R208A207D20 C2lR2120A218D21 H2lR2220A229D22 H20R239A238D23 H2lR2420A249D24 H2lR2520A259D25	N9R228A227D22 C22R239A238D23 H23R242lA249D24 H22R259A258D25 H23R2622A269D26 H23R2722A279D27	N9R308A307D30 C30R319A318D31 H3lR3230A329D32 H30R339A338D33 H3lR3430A349D34 H3lR3530A359D35	N9R298A297D29 C29R309A308D30 H30R3129A319D31 H29R329A328D32 H30R3329A339D33 H30R3429A349D34	N9R258A257D25 C25R269A268D26 H26R2725A279D27 H25R289A288D28 H26R2925A299D29 H26R3025A309D30

The following representations are used in the following symbolic internal coordinate tables: P_N, *N*-acetyl end; P_C, *N*-methylamide end; I, first residue; I_{sc}, first residue's sidechain; II, second residue; II_{sc}, second residue's sidechain. III, third residue; III_{sc}, third residue's sidechain, etc.

to follow when the structure under study becomes a di, tri or even a polypeptide, which is more often the case. This is due to the methodology underlying the use of the Q and z variables.

The resultant numbering pattern depends on the amino acid directly preceding the protecting group at the C-terminus. Hence, the z_i variable will differ for each system so defined. It is therefore, necessary to take the simple numbering system and extend it into a system that works

well with any complex peptide chain sequence of any oligopeptide (Fig. 2). All model peptides (using the *N*-acetyl and *N*-methylamide protecting groups) have specific definitions for their dihedral angles. The definitions presented in Table 1 are an expansion of the ones presented in the previous work [1].

A general rule can be made after examining the concept behind the definitions listed in Table 1.

Table 7
Selected dihedral angles (°) and total energy (hartrees) for various amino acid diamides, 'monopeptides' (N-Ac-amino acid-NH-Me) in their β_L or γ_L backbone conformations geometry optimized at three levels of theory

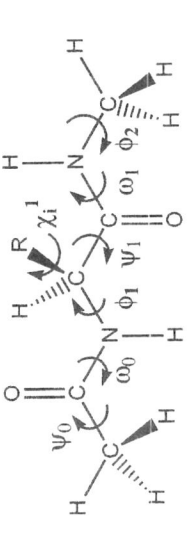

Amino acid	Level of theory	Energy	Dihedral angles												
			ψ_0	ω_0	ϕ_1	ψ_1	ω_1	ϕ_2	χ_1^1	χ_2^1	χ_3^1	χ_4^1	χ_5^1	χ_6^1	η_1
pGlu	RHF/3-21G	−639.8990560	178.14	−173.93	−86.19	73.45	−178.67	−6.48	156.31	33.21	−33.56	21.29	–	–	–
	RHF/6-31G (d)	−643.4831920	178.14	−176.39	−83.67	85.92	−174.92	−170.50	156.92	29.68	−28.96	17.36	–	–	–
	B3LYP/6-31G (d)	−647.3048410	175.98	−174.49	−83.90	75.64	−176.74	−12.03	156.64	30.46	−29.97	18.08	–	–	–
Hyp	RHF/3-21G	−641.0332260	178.19	−173.00	−85.60	72.16	−178.53	−7.98	156.47	34.40	−40.84	30.86	–	–	−77.31
	RHF/6-31G (d)	−644.6142190	−179.75	−172.82	−85.97	78.21	−175.07	−171.54	153.99	31.89	−38.52	29.82	–	–	−76.77
	B3LYP/6-31G (d)	−648.4804250	176.26	−172.59	−84.11	73.07	−176.31	−16.02	155.87	31.97	−38.33	29.38	–	–	−74.89
Hyl	RHF/3-21G	−736.1499370	168.00	178.74	−167.01	166.97	177.64	−3.91	−137.91	77.85	179.10	157.42	−148.32	–	87.01
	RHF/6-31G (d)	−740.2305460	3.87	179.39	−165.06	154.07	−179.39	−166.07	−137.71	78.80	173.47	170.32	−164.75	–	73.07
	B3LYP/6-31G (d)	−744.7461520	0.05	−179.65	−165.01	159.12	178.54	−12.12	−131.97	74.26	173.38	164.82	−151.94	–	83.87
Cit	RHF/3-21G	−789.3402390	−179.60	−174.62	−85.97	70.05	−178.37	−7.10	−170.31	−179.87	−179.69	80.20	−178.76	179.82	–
	RHF/6-31G (d)	−793.7540510	−13.32	−176.76	−85.89	81.08	−173.33	−170.56	−171.03	−179.03	179.69	80.14	−171.24	−167.48	–
	B3LYP/6-31G (d)	−798.5397720	−1.31	−175.67	−83.12	73.92	−175.15	−148.29	−168.23	−179.02	179.51	80.23	−169.51	−166.42	–
Orn	RHF/3-21G	−622.4655660	−179.94	179.77	−163.92	156.31	177.80	−1.10	−173.39	174.95	179.57	−164.06	–	–	–
	RHF/6-31G (d)	−625.9478760	−17.05	−177.51	−154.50	141.54	178.52	−174.34	−173.04	176.05	179.90	−173.16	–	–	–
	B3LYP/6-31G (d)	−629.8191770	−0.84	179.38	−156.49	149.23	176.90	2.90	−170.73	175.30	179.56	−174.91	–	–	–

η represents the hydroxyl (−OH) torsional sidechain angle of a monopeptide.

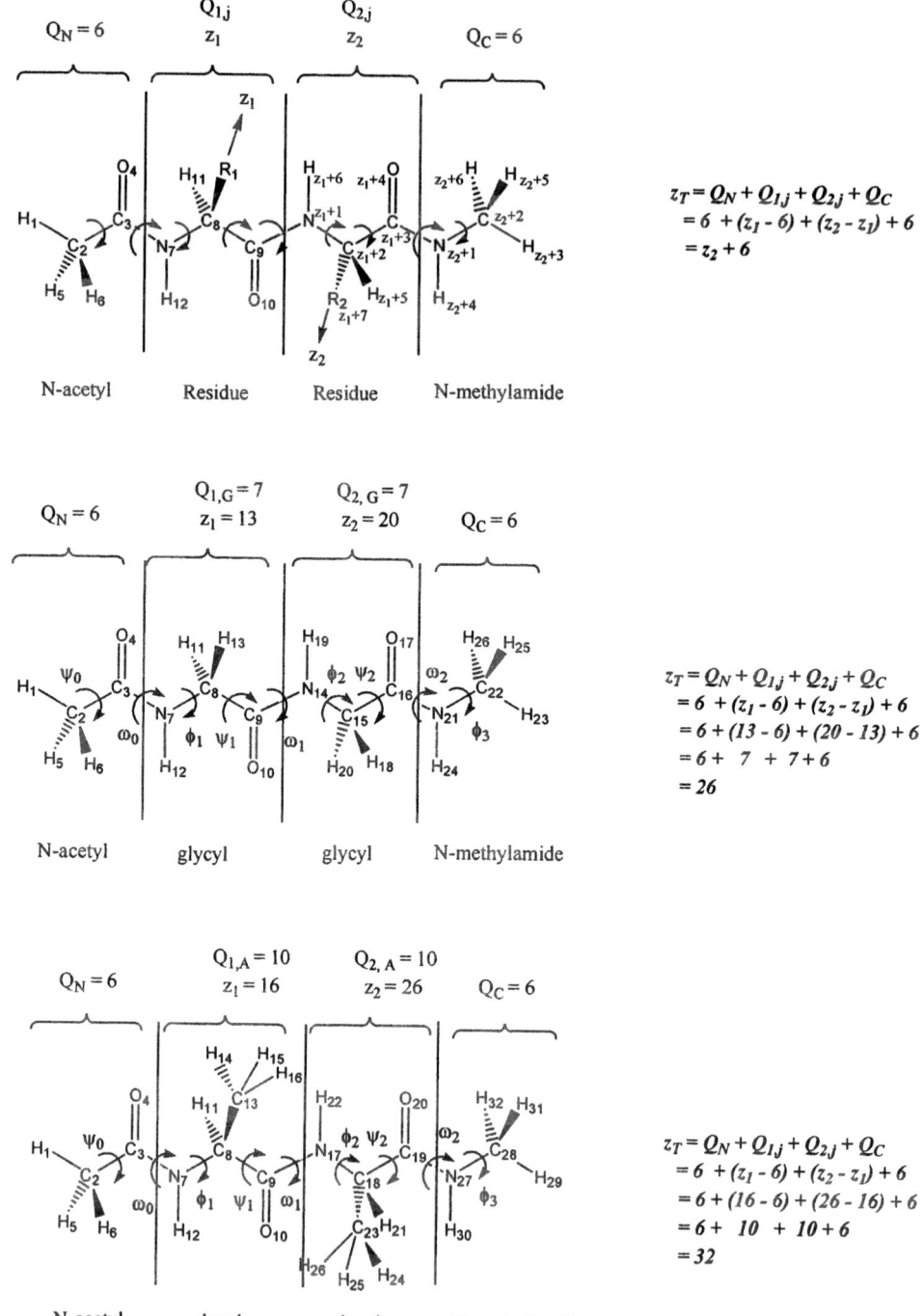

$$z_T = Q_N + Q_{1j} + Q_{2j} + Q_C$$
$$= 6 + (z_1 - 6) + (z_2 - z_1) + 6$$
$$= z_2 + 6$$

$$z_T = Q_N + Q_{1j} + Q_{2j} + Q_C$$
$$= 6 + (z_1 - 6) + (z_2 - z_1) + 6$$
$$= 6 + (13 - 6) + (20 - 13) + 6$$
$$= 6 + 7 + 7 + 6$$
$$= 26$$

$$z_T = Q_N + Q_{1j} + Q_{2j} + Q_C$$
$$= 6 + (z_1 - 6) + (z_2 - z_1) + 6$$
$$= 6 + (16 - 6) + (26 - 16) + 6$$
$$= 6 + 10 + 10 + 6$$
$$= 32$$

Fig. 4. A schematic illustration of the general structure of a dipeptide as compared to N-acetyl-Gly-Gly-N-methylamide and N-acetyl-Ala-Ala-N-methylamide.

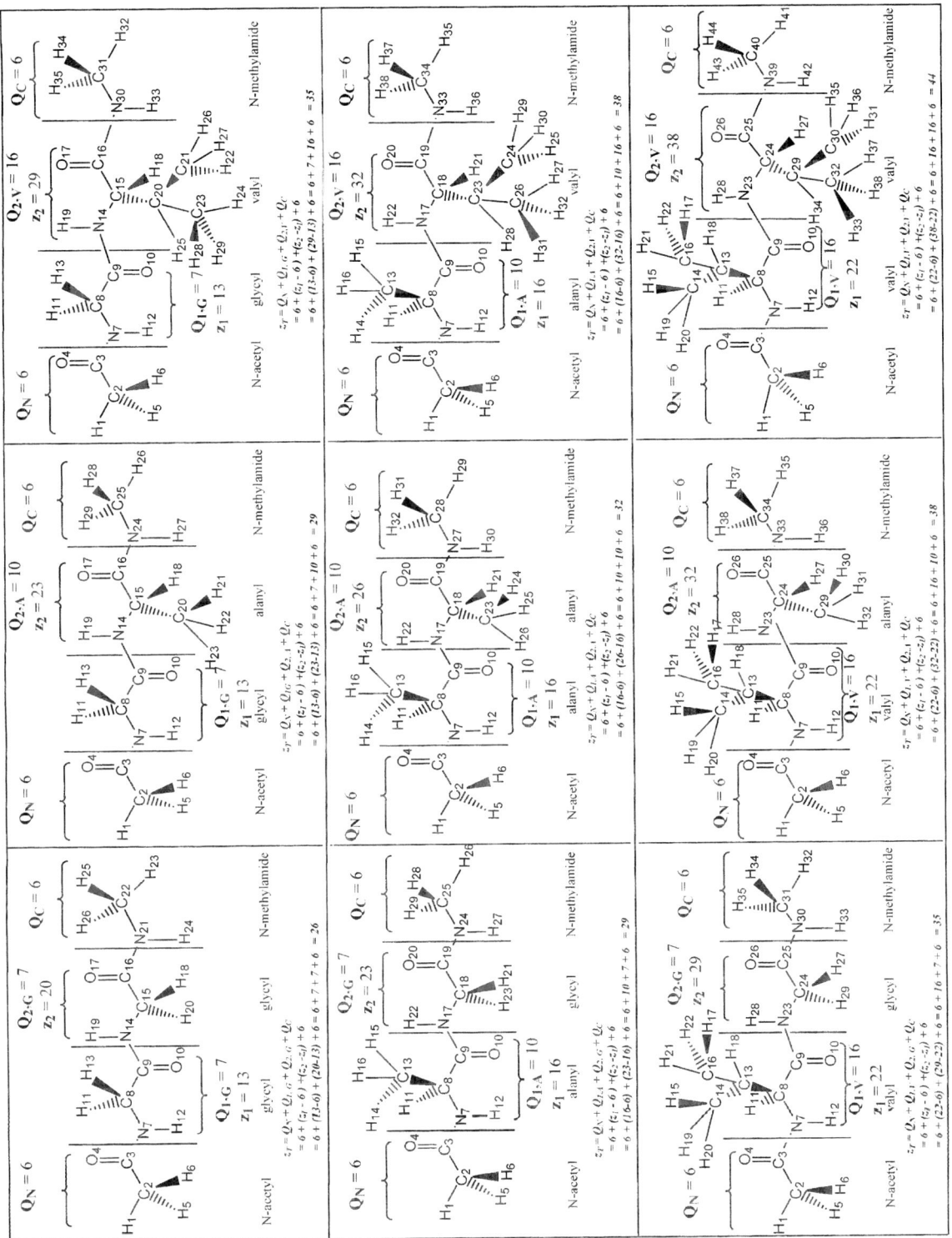

Fig. 5. A schematic illustration of the nine selected dipeptides with the *N*-acetyl and *N*-methylamide protecting groups.

Table 8

Symbolic internal coordinates for nine dipeptides in their N-acetyl and N-methylated protected forms as shown in Fig. 5

	N-acetyl-Ala-Gly-N-methylamide	N-acetyl-Ala-Ala-N-methylamide	N-acetyl-Ala-Val-N-methylamide
P_N	H	H	H
	C2R3*I*A3	C2R3*I*A3	C2R3*I*A3
	O3R42A4*I*D4	O3R42A4*I*D4	O3R42A4*I*D4
	H2R53A54D5	H2R53A54D5	H2R53A54D5
	H2R63A64D6	H2R63A64D6	H2R63A64D6
I	N3R72A7*I*D7	N3R72A7*I*D7	N3R72A7*I*D7
	C7R83A82D8	C7R83A82D8	C7R83A82D8
	C8R97A93D9	C8R97A93D9	C8R97A93D9
	O9R108A107D10	O9R108A107D10	O9R108A107D10
	H8R117A113D11	H8R117A113D11	H8R117A113D11
	H7R123A122D12	H7R123A122D12	H7R123A122D12
	H8R137A133D13	H8R137A133D13	C8R137A133D13
I_{sc}			
II	N9R148A147D14	N9R148A147D14	N9R148A147D14
	C*14*R159A158D15	C*14*R159A158D15	C*14*R159A158D15
	C*15*R16*14*A169D16	C*15*R16*14*A169D16	C*15*R16*14*A169D16
	O*16*R17*15*A17*14*D17	O*16*R17*15*A17*14*D17	O*16*R17*15*A17*14*D17
	H*15*R18*14*A189D18	H*15*R18*14*A189D18	H*15*R18*14*A189D18
	H*14*R199A198D19	H*14*R199A198D19	H*14*R199A198D19
	H*15*R20*14*A209D20	C*15*R20*14*A209D20	C*15*R20*14*A209D20
II_{sc}		H20R21*15*A21*14*D21	C20R21*15*A21*14*D21
		H20R22*15*A22*14*D22	H21R2220A22*15*D22
		H20R23*15*A23*14*D23	C20R23*15*A23*14*D23
			H23R2420A24*15*D24
			H20R25*15*A25*14*D25
			H21R2620A26*15*D26
			H21R2720A27*15*D27
			H23R2820A28*15*D28
			H23R2920A29*15*D29
P_C	N*16*R21*15*A21*14*D21	N*16*R24*15*A24*14*D24	N*16*R30*15*A30*14*D30
	C21R22*16*A22*15*D22	C21R25*16*A25*15*D25	C21R31*16*A31*15*D31
	H22R2321A23*16*D23	H22R2621A26*16*D26	H22R3221A32*16*D32
	H21R24*16*A24*15*D24	H21R27*16*A27*15*D27	H21R33*16*A33*15*D33
	H22R2521A25*16*D25	H22R2821A28*16*D28	H22R3421A34*16*D34
	H22R2621A26*16*D26	H22R2921A29*16*D29	H22R3521A35*16*D35

	N-acetyl-Ala-Gly-N-methylamide	N-acetyl-Ala-Ala-N-methylamide	N-acetyl-Ala-Val-N-methylamide
P_N	H	H	H
	C*I*R2	C*I*R2	C*I*R2
	C2R3*I*A3	C2R3*I*A3	C2R3*I*A3
	O3R42A4*I*D4	O3R42A4*I*D4	O3R42A4*I*D4
	H2R53A54D5	H2R53A54D5	H2R53A54D5
	H2R63A64D6	H2R63A64D6	H2R63A64D6
I	N3R72A7*I*D7	N3R72A7*I*D7	N3R72A7*I*D7
	C7R83A82D8	C7R83A82D8	C7R83A82D8
	C8R97A93D9	C8R97A93D9	C8R97A93D9
	O9R108A107D10	O9R108A107D10	O9R108A107D10
	H8R117A113D11	H8R117A113D11	H8R117A113D11

Table 8 (*continued*)

	N-acetyl-Ala-Gly-*N*-methylamide	*N*-acetyl-Ala-Ala-*N*-methylamide	*N*-acetyl-Ala-Val-*N*-methylamide
	H7R123A122D12	H7R123A122D12	H7R123A122D12
	C8R137A133D13	C8R137A133D13	C8R137A133D13
I_{sc}	H13R148A147D14	H13R148A147D14	H13R148A147D14
	H13R158A157D15	H13R158A157D15	H13R158A157D15
	H13R168A167D16	H13R168A167D16	H13R168A167D16
II	N9R178A177D17	N9R178A177D17	N9R178A177D17
	C17R189A188D18	C17R189A188D18	C17R189A188D18
	C18R1917A199D19	C18R1917A199D19	C18R1917A199D19
	O19R2018A2017D20	O19R2018A2017D20	O19R2018A2017D20
	H18R2117A219D21	H18R2117A219D21	H18R2117A219D21
	H17R229A228D22	H17R229A228D22	H17R229A228D22
	H18R2317A239D23	C18R2317A239D23	C18R2317A239D23
II_{sc}		H23R2418A2417D14	C23R2418A2417D24
		H23R2518A2517D25	H24R2523A2518D25
		H23R2618A2617D26	C23R2618A2617D26
			H26R2723A2718D27
			H23R2818A2817D28
			H24R2923A2918D29
			H24R3023A3018D30
			H26R3123A3118D31
			H26R3223A3218D32
P_C	N19R2418A2417D24	N16R2715A2714D27	N16R3315A3314D33
	C24R2519A2518D25	C21R2816A2815D28	C21R3416A3415D34
	H25R2624A2619D26	H22R2921A2916D29	H22R3521A3516D35
	H24R2719A2718D27	H21R3016A3015D30	H21R3616A3615D36
	H25R2824A2819D28	H22R3121A3116D31	H22R3721A3716D37
	H25R2924A2919D29	H22R3221A3216D32	H22R3821A3816D38

	N-acetyl-Val-Gly-*N*-methylamide	*N*-acetyl-Val-Ala-*N*-methylamide	*N*-acetyl-Val-Val-*N*-methylamide
	H	H	H
P_N	C1R2	C1R2	C1R2
	C2R31A3	C2R31A3	C2R31A3
	O3R42A41D4	O3R42A41D4	O3R42A41D4
	H2R53A54D5	H2R53A54D5	H2R53A54D5
	H2R63A64D6	H2R63A64D6	H2R63A64D6
I	N3R72A71D7	N3R72A71D7	N3R72A71D7
	C7R83A82D8	C7R83A82D8	C7R83A82D8
	C8R97A93D9	C8R97A93D9	C8R97A93D9
	O9R108A107D10	O9R108A107D10	O9R108A107D10
	H8R117A113D11	H8R117A113D11	H8R117A113D11
	H7R123A122D12	H7R123A122D12	H7R123A122D12
	C8R137A133D13	C8R137A133D13	C8R137A133D13
I_{sc}	C13R148A147D14	C13R148A147D14	C13R148A147D14
	H14R1513A158D15	H14R1513A158D15	H14R1513A158D15
	C13R168A167D16	C13R168A167D16	C13R168A167D16
	H16R1713A178D17	H16R1713A178D17	H16R1713A178D17
	H13R188A187D18	H13R188A187D18	H13R188A187D18
	H14R1913A198D19	H14R1913A198D19	H14R1913A198D19

(*continued on next page*)

Table 8 (continued)

	N-acetyl-Val-Gly-N-methylamide	N-acetyl-Val-Ala-N-methylamide	N-acetyl-Val-Val-N-methylamide
	H14R2013A208D20	H14R2013A208D20	H14R2013A208D20
	H16R2113A218D21	H16R2113A218D21	H16R2113A218D21
	H16R2213A238D22	H16R2213A238D22	H16R2213A238D22
II	N9R238A237D23	N9R238A237D23	N9R238A237D23
	C23R249A248D24	C23R249A248D24	C23R249A248D24
	C24R2523A259D25	C24R2523A259D25	C24R2523A259D25
	O25R2624A2623D26	O25R2624A2623D26	O25R2624A2623D26
	H24R2723A279D27	H24R2723A279D27	H24R2723A279D27
	H23R289A288D28	H23R289A288D28	H23R289A288D28
	H24R2923A299D29	H24R2923A299D29	H24R2923A299D29
II$_{sc}$		H29R3024A3023D30	C29R3024A3023D30
		H29R3124A3123D31	H30R3129A3124D31
		H29R3224A3223D32	C29R3224A3223D32
			H32R3329A3324D33
			H29R3424A3423D34
			H30R3529A3524D35
			H30R3629A3624D36
			H32R3729A3724D37
			H32R3829A3824D38
P_C	N25R3024A3023D30	N25R3324A3323D33	N25R3924A3923D39
	C30R3125A3124D31	C30R3425A3424D34	C30R4025A4024D40
	H31R3230A3229D32	H31R3530A3529D35	H31R4130A4129D41
	H30R3325A3324D33	H30R3625A3624D36	H30R4225A4224D42
	H31R3430A3425D34	H31R3730A3725D37	H31R4330A4325D43
	H31R3530A3525D35	H31R3830A3825D38	H31R4430A4425D44

The following general formulas can be used to define all dihedrals except for ψ_0 and ω_1

$$\phi_n = D[6 + (Q_{1,j} + Q_{2,j} + Q_{3,j} + \cdots + Q_{i,j} + \cdots + Q_{n-1,j}) + 3]$$

$$\psi_n = D[6 + (Q_{1,j} + Q_{2,j} + Q_{3,j} + \cdots + Q_{i,j} + \cdots + Q_{n,j}) + 1]$$

$$\omega_n = D[6 + (Q_{1,j} + Q_{2,j} + Q_{3,j} + \cdots + Q_{i,j} + \cdots + Q_{n,j}) + 2]$$

Using these simple definitions allows the system to be extended to define a model tripeptide, N-Ac-Ala-Ala-Ala-NH-Me (Table 2). An extension of this method is exemplified in Table 3, in comparing the tripeptides N-Ac-Val-Val-Val-NH-Me and N-Ac-Arg-Arg-Arg-NH-Me to N-Ac-Ala-Ala-Ala-NH-Me.

There are several advantages of the present method. One can identify the dihedral angles that are changing more significantly in successive iterations of geometry optimizations using molecular orbital computations. One may also extract these parameters very quickly and efficiently at any time during the iterative cycles, in order to determine whether the structure has remained in the desired conformation. Scheme 2 shows the relationships between the number of residues and each Q value.

Of course many different protecting groups may be used at either the N- or C-termini; therefore, the labels P_N and P_C are given to them, respectively, as can be seen in Table 4. In Table 5, the symbolic internal coordinate system for a general amino acid is shown with various N- and C-terminal protecting groups. The first column shows the free amino acid in its neutral form, the second and third columns show the N-only and C-only protected forms, respectively, while in the fourth column, the N- and C-protected forms are presented.

The numbering for the C-terminal protecting group depends on the 'atomic count' of the amino acid that it

Table 9
Selected dihedral angles (°) and total energy (hartrees) for selected dipeptides (N-Ac-X-Y-NH-Me) in their β_L backbone conformations geometry optimized at three levels of theory

Dipeptide	Level of theory	Energy	Dihedral angles												
			ψ_0	ω_0	ϕ_1	ψ_1	ω_1	ϕ_2	ψ_2	ω_2	ϕ_3	χ_1^1	χ_2^1	χ_1^2	χ_2^2
Gly-Gly	RHF/3-21G	−656.9556911	−179.99	179.99	−179.99	−179.95	−179.99	179.93	−179.99	−179.99	−119.36	–	–	–	–
	RHF/6-31G (d)	−660.6410950	−179.98	179.99	−179.99	−179.99	−179.99	179.99	−179.95	−179.99	119.52	–	–	–	–
	B3LYP/6-31G (d)	−664.5480470	−179.99	179.99	−179.99	180.00	180.00	179.98	179.99	180.00	119.28	–	–	–	–
Gly-Ala	RHF/3-21G	−695.7806303	−179.92	179.92	−179.99	179.90	179.08	−167.64	169.59	177.95	119.39	–	–	179.27	–
	RHF/6-31G (d)	−699.6781860	−2.90	−179.66	−179.68	179.66	178.51	−157.25	159.65	179.36	70.75	–	–	−178.61	–
	B3LYP/6-31G (d)	−703.8651780	6.18	177.01	−171.70	179.46	177.21	−158.85	164.50	177.78	119.48	–	–	−178.04	–
Gly-Val	RHF/3-21G	−773.4175734	179.88	179.89	−179.40	−180.02	177.16	−137.68	142.38	178.94	117.30	–	–	−178.74	−174.73
	RHF/6-31G (d)	−777.7446440	3.69	179.46	−179.40	178.84	175.52	−123.84	128.39	−178.42	69.80	–	–	178.83	178.76
	B3LYP/6-31G (d)	−782.4908340	−6.25	−175.61	168.50	178.20	173.11	−129.83	135.49	179.69	113.04	–	–	179.59	179.25
Ala-Gly	RHF/3-21G	−695.7808163	179.52	178.75	−167.54	170.98	178.54	179.22	−179.92	179.99	−119.29	179.47	–	–	–
	RHF/6-31G (d)	−699.6778820	179.12	177.73	−157.16	160.97	177.36	−179.68	−179.72	−179.99	−119.47	−178.31	–	–	–
	B3LYP/6-31G (d)	−703.8652170	−2.69	176.67	−158.76	166.35	178.60	177.54	179.30	179.93	−118.71	−178.24	–	–	–
Ala-Ala	RHF/3-21G	−734.6057508	−179.64	178.91	−167.37	170.68	177.32	−167.41	169.49	177.89	−119.88	179.45	–	179.26	–
	RHF/6-31G (d)	−738.7149770	−11.92	179.67	−157.83	159.93	175.62	−156.42	159.67	179.29	−169.05	−178.35	–	−178.62	–
	B3LYP/6-31G (d)	−743.1819226	−0.12	176.86	−159.09	165.43	174.91	−159.16	165.20	177.59	−118.60	−178.13	–	−178.24	–
Ala-Val	RHF/3-21G	−812.2426800	179.85	178.88	−167.18	169.91	174.90	−137.43	142.48	178.91	−121.30	179.33	–	−178.77	−174.69
	RHF/6-31G (d)	−816.7814880	−13.65	179.88	−157.67	158.40	172.13	−122.80	128.80	−178.51	−170.06	−178.65	–	178.78	178.60
	B3LYP/6-31G (d)	−821.8074050	10.82	175.07	−157.07	164.28	171.18	−128.63	136.78	−179.64	−158.46	−178.28	–	179.74	178.60
Val-Gly	RHF/3-21G	−773.4174802	−179.90	176.47	−137.65	143.92	179.05	179.27	−180.12	179.87	118.93	−178.88	−174.22	–	–
	RHF/6-31G (d)	−777.7441720	−144.81	177.50	−127.50	130.11	179.24	−177.90	−179.74	−178.20	70.22	178.45	178.33	–	–
	B3LYP/6-31G (d)	−782.4905540	9.42	171.46	−128.71	136.43	178.63	−177.90	−179.63	−179.26	115.82	179.29	178.61	–	–
Val-Ala	RHF/3-21G	−812.2424662	−179.60	176.58	−137.62	144.15	177.78	−167.21	169.41	177.85	−0.01	−178.80	−174.17	179.28	–
	RHF/6-31G (d)	−816.7810010	148.64	173.06	−125.47	130.15	178.08	−156.49	159.18	179.20	−49.63	178.25	178.43	−178.70	–
	B3LYP/6-31G (d)	−821.8072480	−7.04	174.12	−130.32	136.94	176.64	−158.69	164.17	177.45	−0.74	179.79	178.78	−178.21	–
Val-Val	RHF/3-21G	−889.8801436	−179.38	176.20	−138.33	148.53	171.44	−112.00	142.65	179.42	−1.90	179.82	−173.83	179.87	179.42
	RHF/6-31G (d)	−894.8477140	147.28	173.13	−123.47	127.98	176.22	−123.44	125.80	−178.03	−49.74	178.27	178.23	179.07	178.94
	B3LYP/6-31G (d)	−900.4330990	−17.02	175.87	−130.85	135.23	174.19	−129.60	133.97	−179.94	−5.30	179.44	178.99	179.49	178.58

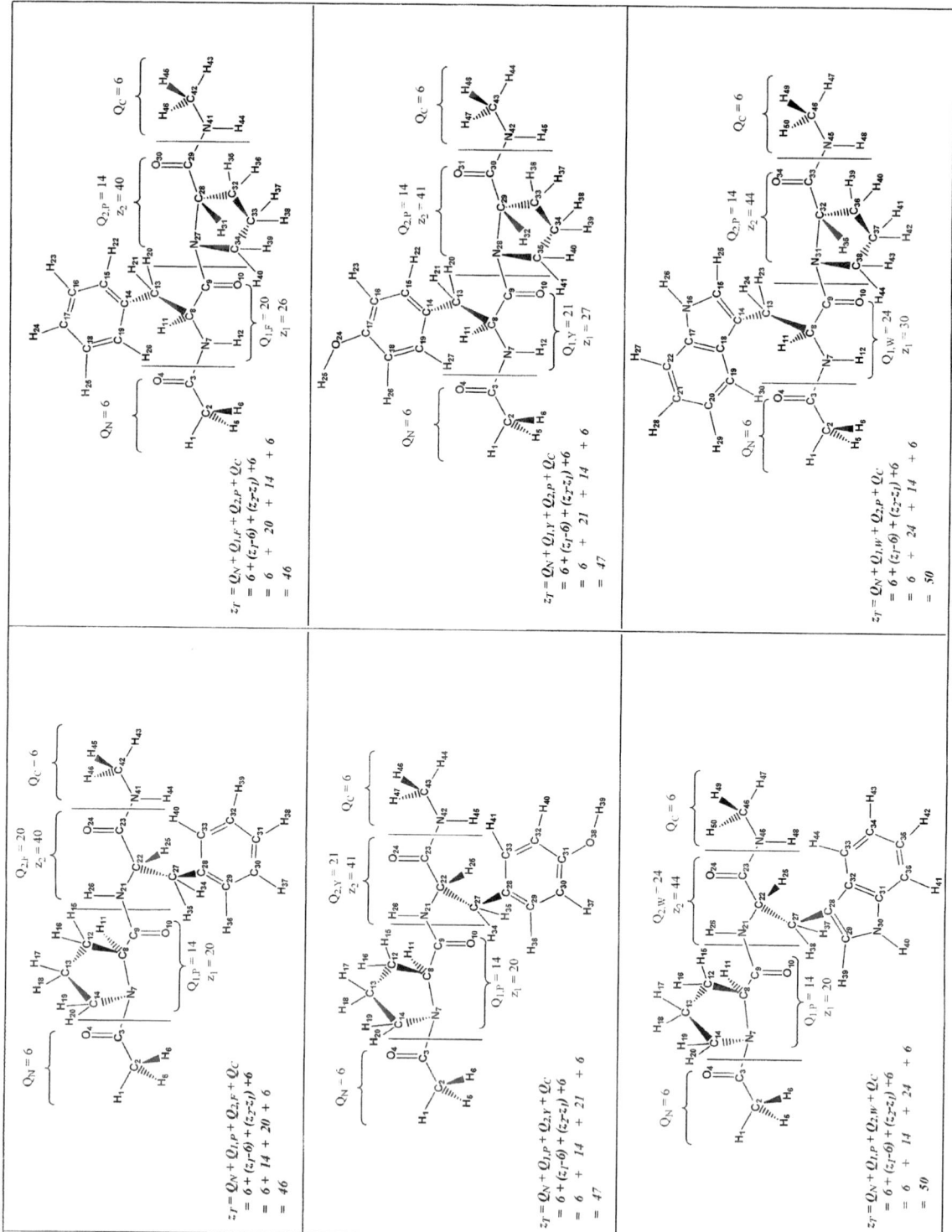

Fig. 6. A schematic illustration of six special dipeptides, Pro-Phe, Phe-Pro, Pro-Tyr, Tyr-Pro, Pro-Trp, and Trp-Pro, from the possible 400 such dipeptides.

Table 10
Symbolic internal coordinates for six special dipeptides from Fig. 6 in their *N*-acetyl and *N*-methylated forms from a possible 400

	N-acetyl-Pro-Phe-*N*-methylamide	*N*-acetyl-Pro-Tyr-*N*-methylamide	*N*-acetyl-Pro-Trp-*N*-methylamide
P_N	H	H	H
	C*1*R2	C*1*R2	C*1*R2
	C2R3*1*A3	C2R3*1*A3	C2R3*1*A3
	O*3*R42A4*1*D4	O*3*R42A4*1*D4	O*3*R42A4*1*D4
	H2R5*3*A5*4*D5	H2R5*3*A5*4*D5	H2R5*3*A5*4*D5
	H2R6*3*A6*4*D6	H2R6*3*A6*4*D6	H2R6*3*A6*4*D6
I	N*3*R72A7*1*D7	N*3*R72A7*1*D7	N*3*R72A7*1*D7
	C7R83A82D8	C7R83A82D8	C7R83A82D8
	C8R97A93D9	C8R97A93D9	C8R97A93D9
	O9R108A107D10	O9R108A107D10	O9R108A107D10
	H*8*R117A11*3*D11	H*8*R117A11*3*D11	H*8*R117A11*3*D11
	C8R127A123D12	C8R127A123D12	C8R127A123D12
I_{sc}	C*12*R138A137D13	C*12*R138A137D13	C*12*R138A137D13
	C*13*R14*12*A148D14	C*13*R14*12*A148D14	C*13*R14*12*A148D14
	H*12*R158A157D15	H*12*R158A157D15	H*12*R158A157D15
	H*12*R168A167D16	H*12*R168A167D16	H*12*R168A167D16
	H*13*R17*12*A178D17	H*13*R17*12*A178D17	H*13*R17*12*A178D17
	H*13*R18*12*A188D18	H*13*R18*12*A188D18	H*13*R18*12*A188D18
	H*14*R19*13*A19*12*D19	H*14*R19*13*A19*12*D19	H*14*R19*13*A19*12*D19
	H*14*R20*13*A20*12*D20	H*14*R20*13*A20*12*D20	H*14*R20*13*A20*12*D20
II	N9R218A217D21	N9R218A217D21	N9R218A217D21
	C2*1*R229A228D22	C2*1*R229A228D22	C2*1*R229A228D22
	C22R23*21*A239D23	C22R23*21*A239D23	C22R23*21*A239D23
	O*23*R2422A242*1*D24	O*23*R2422A242*1*D24	O*23*R2422A242*1*D24
	H22R252*1*A259D25	H22R252*1*A259D25	H22R252*1*A259D25
	H2*1*R269A26*16*D26	H2*1*R269A26*16*D26	H2*1*R269A26*16*D26
	C22R272*1*A279D27	C22R272*1*A279D27	C22R272*1*A279D27
II_{sc}	C27R2822A282*1*D28	C27R2822A282*1*D28	C27R2822A282*1*D28
	C28R2927A2922D29	C28R2927A2922D29	C28R2927A2922D29
	C29R3028A3027D30	C29R3028A3027D30	N29R3028A3027D30
	C*30*R3129A3128D31	C*30*R3129A3128D31	C*30*R3129A3128D31
	C*31*R3230A3229D32	C*31*R3230A3229D32	C*31*R3230A3229D32
	C*32*R333*1*A3330D33	C*32*R333*1*A3330D33	C*32*R333*1*A3330D33
	H27R3422A342*1*D34	H27R3422A342*1*D34	C*33*R3432A343*1*D34
	H27R3522A352*1*D35	H27R3522A352*1*D35	C*34*R35*33*A3532D35
	H29R3628A3627D36	H29R3628A3627D36	C*35*R36*34*A36*33*D36
	H*30*R3729A3728D37	H*30*R3729A3728D37	H27R3722A372*1*D37
	H*31*R3830A3829D38	O*31*R3830A3829D38	H27R3822A382*1*D38
	H*32*R3931A3930D39	H38R393*1*A39*30*D39	H29R3928A3927D39
	H*33*R403*2*A403*1*D40	H*32*R403*1*A4030D40	H*30*R4029A4028D40
		H*33*R413*2*A413*1*D41	H*33*R413*2*A413*1*D41
			H*34*R4233A4232D42
			H*35*R4334A4333D43
			H*36*R44*35*A44*34*D44
P_C	N*23*R4122A412*1*D41	N*23*R4222A422*1*D42	N*23*R4522A452*1*D45
	C4*1*R4223A4222D42	C*42*R4323A4322D43	C*45*R4623A4622D46
	H*42*R434*1*A4323D43	H*43*R4442A4423D44	H*46*R4745A4723D47
	H4*1*R4423A4422D44	H*42*R4523A4522D45	H*45*R4823A4822D48
	H*42*R454*1*A4523D45	H*43*R4642A4623D46	H*46*R494*5*A4923D49

(continued on next page)

Table 10 (continued)

	N-acetyl-Pro-Phe-N-methylamide	N-acetyl-Pro-Tyr-N-methylamide	N-acetyl-Pro-Trp-N-methylamide
P_N	H42R4641A4623D46 N-acetyl-Phe-Pro-Nmethylamide H C1R2 C2R31A3 O3R42A41D4 H2R53A54D5 H2R63A64D6	H43R4742A4723D47 Nacetyl-Tyr-Pro-Nmethylamide H C1R2 C2R31A3 O3R42A41D4 H2R53A54D5 H2R63A64D6	H46R5045A5023D50 Nacetyl-Trp-Pro-Nmethylamide H C1R2 C2R31A3 O3R42A41D4 H2R53A54D5 H2R63A64D6
I	N3R72A71D7 C7R83A82D8 C8R97A93D9 O9R108A107D10 H8R117A113D11 H7R123A122D12 C8R137A133D13	N3R72A71D7 C7R83A82D8 C8R97A93D9 O9R108A107D10 H8R117A113D11 H7R123A122D12 C8R137A133D13	N3R72A71D7 C7R83A82D8 C8R97A93D9 O9R108A107D10 H8R117A113D11 H7R123A122D12 C8R137A133D13
I_{sc}	C13R148A147D14 C14R1513A158D15 C15R1614A1613D16 C16R1715A1714D17 C17R1816A1815D18 C18R1917A1916D19 H13R208A207D20 H13R218A217D21 H15R2214A2213D22 H16R2315A2314D23 H17R2416A2415D24 H18R2517A2516D25 H19R2618A2617D26	C13R148A147D14 C14R1513A158D15 C15R1614A1613D16 C16R1715A1714D17 C17R1816A1815D18 C18R1917A1916D19 H13R208A207D20 H13R218A217D21 H15R2214A2213D22 H16R2315A2314D23 O17R2416A2415D24 H24R2517A2516D25 H18R2617A2616D26 H19R2718A2717D27	C13R148A147D14 C14R1513A158D15 N15R1614A1613D16 C16R1715A1714D17 C17R1816A1815D18 C18R1917A1916D19 C19R2018A2017D20 C20R2119A2118D21 C21R2220A2219D22 H13R238A237D23 H13R248A247D24 H15R2514A2513D25 H16R2615A2614D26 H19R2718A2717D27 H20R2819A2818D28 H21R2920A2919D29 H22R3021A3020D30
II	N9R278A277D27 C27R289A288D28 C28R2927A299D29 O29R3028A3027D30 H28R3127A319D31 C28R3227A329D32 C32R3328A3327D33	N9R288A287D28 C28R299A298D29 C29R3028A309D30 O30R3129A3128D31 H29R3228A329D32 C29R3328A339D33 C33R3429A3428D34	N9R318A317D31 C31R329A328D32 C32R3331A339D33 O33R3432A3431D34 H32R3531A358D35 C32R3631A369D36 C36R3732A3731D37
II_{sc}	C33R3432A3428D34 H32R3528A3527D35 H32R3628A3627D36 H33R3732A3728D37 H33R3832A3828D38 H34R3933A3932D39 H34R4033A4032D40	C34R3533A3529D35 H33R3629A3628D36 H33R3729A3728D37 H34R3833A3829D38 H34R3933A3929D39 H35R4034A4033D40 H35R4134A4133D41	C37R3836A3832D38 H36R3932A3931D39 H36R4032A4031D40 H37R4136A4132D41 H37R4236A4232D42 H38R4337A4336D43 H38R4437A4436D44
P_C	N29R4128A4127D41 C41R4229A4228D42 H42R4341A4329D43 H41R4429A4428D44 H42R4541A4529D45 H42R4641A4629D46	N30R4229A4228D42 C42R4330A4329D43 H43R4442A4430D44 H42R4530A4529D45 H43R4642A4630D46 H43R4742A4730D47	N33R4532A4531D45 C45R4633A4632D46 H46R4745A4733D47 H45R4833A4832D48 H46R4945A4933D49 H46R5045A5033D50

Table 11
Symbolic internal coordinates for four selected tripeptides in their *N*-acetyl and *N*-methylated protected forms as shown in Fig. 7

	N-acetyl-Gly-Gly-Gly-*N*-methylamide	*N*-acetyl-Gly-Ala-Gly-*N*-methylamide	*N*-acetyl-Ala-Gly-Ala-*N*-methylamide	*N*-acetyl-Ala-Ala-Ala-*N*-methylamide
P_N	H	H	H	H
	C*1*R2	C*1*R2	C*1*R2	C*1*R2
	C2R3*1*A3	C2R3*1*A3	C2R3*1*A3	C2R3*1*A3
	O3R42A4*1*D4	O*3*R42A4*1*D4	O*3*R42A4*1*D4	O*3*R42A4*1*D4
	H2R53A54D5	H2R53A54D5	H2R5*3*A54D5	H2R53A54D5
	H2R6*3*A6*4*D6	H2R6*3*A6*4*D6	H2R63A64D6	H2R6*3*A6*4*D6
I	N*3*R72A7*1*D7	N*3*R72A7*1*D7	N*3*R72A7*1*D7	N*3*R72A7*1*D7
	C7R8*3*A82D8	C7R8*3*A82D8	C7R83A82D8	C7R83A82D8
	C8R97A9*3*D9	C8R97A9*3*D9	C8R97A9*3*D9	C8R97A9*3*D9
	O9R108A107D10	O9R108A107D10	O9R108A107D10	O9R108A107D10
	H8R117A11*3*D11	H8R117A11*3*D11	H8R117A11*3*D11	H8R117A11*3*D11
	H7R123A122D12	H7R123A122D12	H7R12*3*A122D12	H7R12*3*A122D12
	H8R137A13*3*D13	C8R137A13*3*D13	C8R137A13*3*D13	C8R137A13*3*D13
I_{sc}			H*13*R148A147D14	H*13*R148A147D14
			H*13*R158A157D15	H*13*R158A157D15
			H*13*R168A167D16	H*13*R168A167D16
II	N9R148A147D14	N9R148A147D14	N9R178A177D17	N9R178A177D17
	C*14*R159A158D15	C*14*R159A158D15	C*17*R189A188D18	C*17*R189A188D18
	C*15*R16*14*A169D16	C*15*R16*14*A169D16	C*18*R19*17*A199D19	C*18*R19*17*A199D19
	O*16*R17*15*A17*14*D17	O*16*R17*15*A17*14*D17	O*19*R20*18*A20*17*D20	O*19*R20*18*A20*17*D20
	H*15*R18*14*A189D18	H*15*R18*14*A189D18	H*18*R21*17*A219D21	H*18*R21*17*A219D21
	H*14*R199A198D19	H*14*R199A198D19	H*17*R229A228D22	H*17*R229A228D22
	H*15*R20*14*A209D20	H*15*R20*14*A209D20	H*18*R23*17*A239D23	C*18*R23*17*A239D23
II_{sc}		H20R21*15*A21*14*D21		H23R24*18*A24*17*D14
		H20R22*15*A22*14*D22		H23R25*18*A25*17*D25
		H20R23*15*A23*14*D23		H23R26*18*A26*17*D26
III	N*16*R21*15*A21*14*D21	N*16*R24*15*A24*14*D24	N*19*R24*18*A24*17*D24	N*19*R27*18*A27*17*D27
	C2*1*R22*16*A22*15*D22	C24R25*16*A25*15*D25	C24R25*19*A25*18*D25	C27R28*19*A28*18*D28
	C22R23*21*A23*16*D23	C25R2624A26*16*D26	C25R2624A26*19*D26	C28R2927A29*19*D29
	O*23*R2422A242*1*D24	O26R2725A2724D27	O26R2725A2724D27	O29R3028A3027D30
	H22R252*1*A25*16*D25	H25R2824A28*16*D28	H25R2824A28*19*D28	H28R3127A31*19*D31
	H2*1*R26*16*A26*15*D26	H24R29*16*A29*15*D29	H24R29*19*A29*18*D29	H27R32*19*A32*18*D32
	H22R272*1*A27*16*D27	H25R3024A30*16*D30	C25R3024A30*19*D21	C28R3327A33*19*D33
III_{sc}			H30R3125A3124D31	H*33*R3428A3427D34
			H30R3225A3224D32	H*33*R3528A3527D35
			H30R3325A3324D33	H*33*R3628A3627D36
P_C	N23R2822A282*1*D28	N26R3125A3124D31	N26R3425A3424D34	N29R3728A3727D37
	C28R2923A2922D29	C*31*R3226A3225D32	C*34*R3526A3525D35	C37R3829A3828D38
	H29R3028A3023D30	H32R333*1*A3326D33	H35R36*34*A3626D36	H38R3937A3929D39
	H28R3123A3122D31	H3*1*R3426A3425D34	H34R3726A3725D37	H37R4029A4028D40
	H29R3228A3223D32	H32R353*1*A3526D35	H35R3834A3826D38	H38R4137A4129D41
	H29R3328A3323D33	H32R363*1*A3626D36	H35R3934A3926D39	H38R4237A4229D42

Table 12

Selected dihedral angles (°) and total energy (hartrees) for four selected tripeptides (N-Ac-X-Y-Z-NH-Me) in their β_L backbone conformations geometry optimized at three levels of theory

Tripeptide	Level of theory	Energy	ψ_0	ω_0	ϕ_1	ψ_1	ω_1	ϕ_2	ψ_2	ω_2	ϕ_3	ψ_3	ω_3	ϕ_4	χ_i^1	χ_i^2	χ_i^3
			Dihedral angles														
Gly-Gly-Gly	RHF/3-21G	−862.6181202	179.99	−179.99	179.94	179.96	179.99	179.97	−179.97	179.99	179.98	−180.00	179.98	179.56	–	–	–
	RHF/6-31G (d)	−867.458843	179.97	−179.99	179.98	179.97	179.99	180.00	180.00	180.00	179.99	−179.98	179.96	179.77	–	–	–
	B3LYP/6-31G (d)	−872.560352	179.92	−179.99	179.98	179.99	180.00	179.99	−179.99	179.99	179.99	−179.96	179.97	179.65	–	–	–
Gly-Ala-Gly	RHF/3-21G	−901.4436328	179.88	−179.98	179.67	179.99	180.00	179.99	−179.99	178.55	179.21	−179.90	179.97	19.25	–	179.27	–
	RHF/6-31G (d)	−906.4957890	0.38	−179.90	179.99	179.60	178.18	−157.66	171.19	177.35	179.56	179.56	177.99	170.16	–	178.48	–
	B3LYP/6-31G (d)	−911.8775770	−147.00	178.96	179.60	179.10	179.01	−157.88	160.79	179.18	179.62	179.56	179.41	1.60	–	178.27	–
Ala-Gly-Ala	RHF/3-21G	−940.2685698	179.63	178.74	−174.09	170.95	173.61	179.45	−179.96	178.50	177.83	169.74	177.99	0.07	179.48	–	179.22
	RHF/6-31G (d)	−945.5325380	−16.14	−179.97	167.57	160.49	177.62	−179.88	−179.91	176.48	167.79	160.33	179.09	−166.84	−178.38	–	178.59
	B3LYP/6-31G (d)	−951.1945330	13.77	173.99	−158.60	167.86	178.97	177.26	−179.42	177.25	157.03	165.48	178.91	−140.13	−177.85	–	178.18
Ala-Ala-Ala	RHF/3-21G	−979.0932909	−179.94	178.84	−157.49	170.77	177.32	−167.81	170.79	175.49	167.72	170.16	178.17	−178.32	179.55	179.14	179.14
	RHF/6-31G (d)	−984.569377	−13.78	179.86	167.37	159.82	175.45	−156.93	160.47	175.40	156.47	160.19	179.20	−169.11	−178.36	178.47	178.66
	B3LYP/6-31G (d)	−990.511373	−8.44	177.82	160.08	166.84	173.93	−158.45	166.44	175.40	160.12	165.48	177.77	0.46	−177.96	178.46	178.31

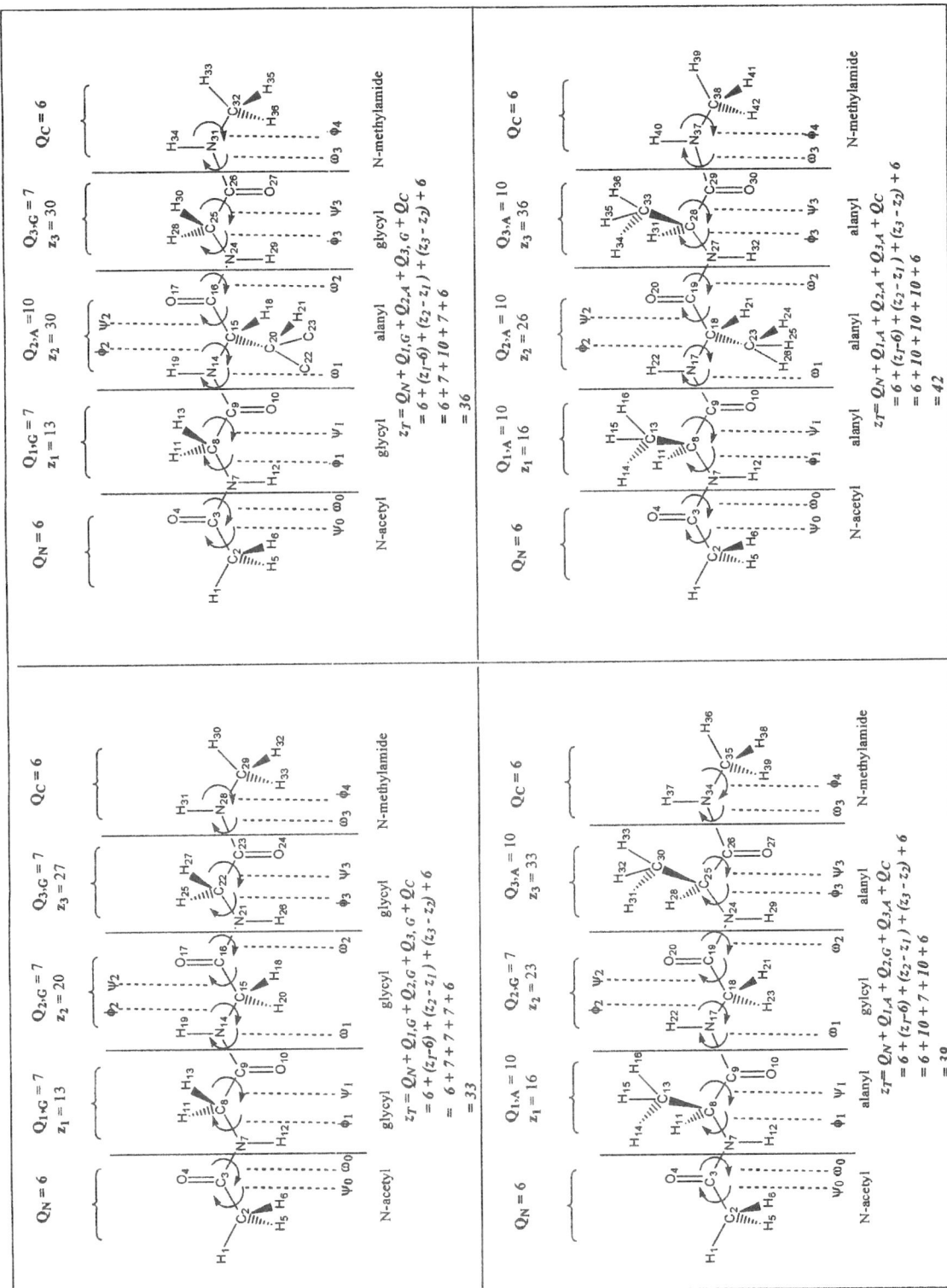

Fig. 7. A schematic illustration of the four selected tripeptides, Gly-Gly-Gly, Gly-Ala-Gly, Ala-Gly-Ala and Ala-Ala-Ala, with the N-acetyl and N-methylamide protecting groups.

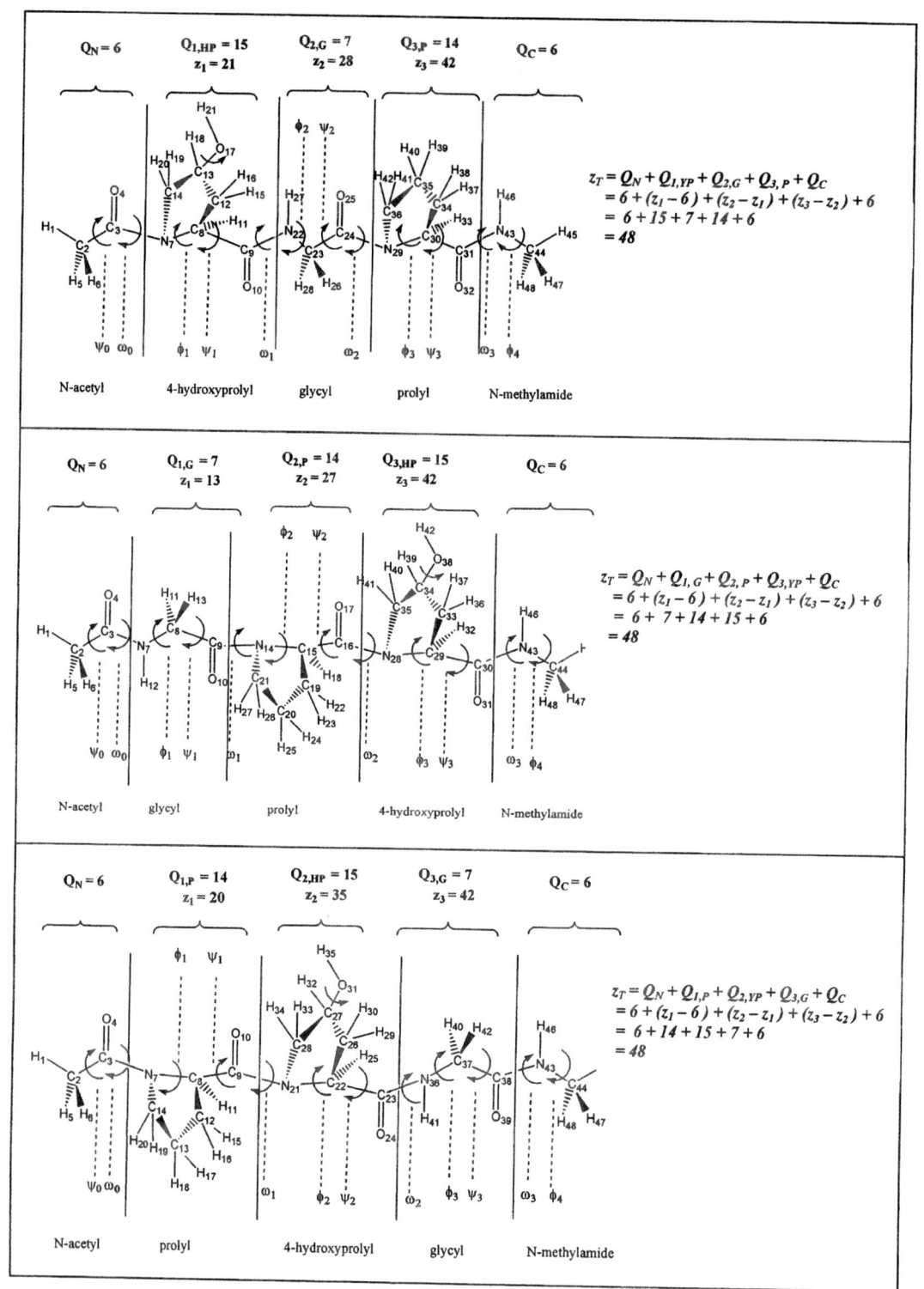

Fig. 8. A schematic illustration of three special tripeptides, Hyp-Pro-Gly, Gly-Pro-Hyp and Pro-Hyp-Gly, from the possible 8000 such tripeptides.

Table 13
Symbolic internal coordinates for three special tripeptides from Fig. 8 in their N-acetyl and N-methylated forms from a possible 8000

	N-acetyl-Hyp-Gly-Pro-N-methylamide	N-acetyl-Gly-Pro-Hyp-N-methylamide	N-acetyl-Pro-Hyp-Gly-N-methylamide
P_N	H	H	H
	C1 R2	C1 R2	C1 R2
	C2 R3 1 A3	C2 R3 1 A3	C2 R3 1 A3
	O3 R42 A41 D4	O3 R42 A41 D4	O3 R42 A41 D4
	H2 R53 A54 D5	H2 R53 A54 D5	H2 R53 A54 D5
	H2 R63 A64 D6	H2 R63 A64 D6	H2 R63 A64 D6
I	N3 R72 A71 D7	N3 R72 A71 D7	N3 R72 A71 D7
	C7 R83 A82 D8	C7 R83 A82 D8	C7 R83 A82 D8
	C8 R97 A93 D9	C8 R97 A93 D9	C8 R97 A93 D9
	O9 R108 A107 D10	O9 R108 A107 D10	O9 R108 A107 D10
	H8 R117 A113 D11	H8 R117 A113 D11	H8 R117 A113 D11
	H7 R123 A122 D12	H7 R123 A122 D12	H7 R123 A122 D12
	H8 R137 A133 D13	H8 R137 A133 D13	C8 R137 A133 D13
I_{sc}	C13 R141 2 A148 D14		C13 R141 2 A148 D14
	H12 R158 A157 D15		H8 R157 A153 D15
	H12 R168 A167 D16		H8 R167 A163 D16
	O13 R171 2 A178 D17		H13 R171 2 A178 D17
	H13 R181 2 A188 D18		H13 R181 2 A188 D18
	H14 R197 A193 D19		H14 R191 3 A191 2 D19
	H14 R207 A203 D20		H14 R201 3 A201 2 D20
	H17 R211 3 A211 2 D21		
II	N9 R228 A227 D22	N9 R148 A147 D14	N9 R218 A217 D21
	C22 R239 A238 D23	C14 R159 A158 D15	C21 R229 A228 D22
	C23 R242 2 A249 D24	C15 R161 4 A169 D16	C22 R232 1 A239 D23
	O24 R2523 A2522 D25	O16 R171 5 A171 4 D17	O23 R242 2 A242 1 D24
	H23 R2622 A269 D26	H15 R181 4 A189 D18	H22 R252 1 A259 D25
	H22 R279 A278 D27	C15 R191 4 A199 D19	C22 R262 1 A269 D26
	H23 R2822 A289 D28	C19 R201 5 A201 4 D20	C26 R272 2 A272 1 D27
II_{sc}		C20 R211 9 A211 5 D21	C27 R2826 A2822 D28
		H19 R221 5 A221 4 D22	H26 R2922 A292 1 D29
		H19 R231 5 A231 4 D23	H26 R3022 A302 1 D30
		H20 R241 9 A241 5 D24	O27 R3126 A3122 D31
		H20 R251 9 A251 5 D25	H27 R3226 A3222 D32
		H21 R2620 A261 9 D26	H28 R3327 A3326 D33
		H21 R2720 A271 9 D27	H28 R3427 A3426 D34
			H34 R3527 A3526 D35
III	N24 R2923 A2922 D29	N16 R281 5 A281 4 D28	N23 R3622 A362 1 D36
	C29 R3024 A3023 D30	C28 R291 6 A291 5 D29	C36 R3723 A3722 D37
	C30 R3129 A3124 D31	C29 R3028 A3027 D30	C37 R3836 A3823 D38
	O31 R3230 A3229 D32	O30 R3129 A3128 D31	O38 R3937 A3936 D39
	H30 R3329 A3324 D33	H29 R3228 A3227 D32	H37 R4036 A4023 D40
	C30 R3429 A3424 D34	C29 R3328 A3327 D33	H36 R4123 A4122 D41
	C34 R3530 A3529 D35	C33 R3429 A3428 D34	H37 R4236 A4223 D42
III_{sc}	C35 R3634 A3630 D36	C34 R3533 A3529 D35	
	H34 R3730 A3729 D37	H33 R3629 A3628 D36	
	H34 R3830 A3829 D38	H33 R3729 A3728 D37	
	H35 R3934 A3930 D39	O34 R3833 A3829 D38	

(continued on next page)

Table 13 (*continued*)

	N-acetyl-Hyp-Gly-Pro-N-methylamide	N-acetyl-Gly-Pro-Hyp-N-methylamide	N-acetyl-Pro-Hyp-Gly-N-methylamide
	H35 R4034 A4030 D40	H34 R3933 A3929 D39	
	H36 R4135 A4134 D41	H35 R4034 A4033 D40	
	H36 R4235 A4234 D42	H35 R4134 A4133 D41	
		H38 R4234 A4233 D42	
P_C	N30 R4329 A4328 D43	N30 R4329 A4328 D43	N38 R4337 A4336 D43
	C43 R4430 A4429 D44	C43 R4430 A4429 D44	C43 R4438 A4437 D44
	H44 R4543 A4530 D45	H44 R4543 A4530 D45	H44 R4543 A4538 D45
	H43 R4630 A4629 D46	H43 R4630 A4629 D46	H43 R4638 A4637 D46
	H44 R4743 A4730 D47	H44 R4743 A4730 D47	H44 R4743 A4738 D47
	H44 R4843 A4830 D48	H44 R4843 A4830 D48	H44 R4843 A4838 D48

is 'protecting' (i.e. which module precedes the C-terminus). Nevertheless for the methylamide protective group, $Q_C = 6$, thereby, defining the equation that explains the complete numbering for any oligopeptide as shown in Fig. 2:

$$z_T = Q_N + \sum_{i=0}^{n} Q_{i,j} + Q_C$$

3. Computational method

The GAUSSIAN 98 program package [43] was used for geometry optimizations. Preliminary structural definitions are often pre-optimized at the AM1 level of theory as was done in the previous work [1]. Such pre-optimizations are usually helpful in making a good 'educated guess' for ab initio input geometries, particularly where bond lengths and bond angles are concerned. However, for this work, optimizations at the AM1 level of theory were excluded for all structures, while geometry optimizations were carried out at the following three split-valence levels of theory, specifically the RHF/3-21G, RHF/6-31G(d) and B3LYP/6-31G(d) levels of theory.

4. Results and discussion

4.1. Monopeptides

The N-acetyl or N-methylamide groups are defined as separate units and can be added or removed, as their internal coordinates do not disturb the other numeric definitions of any other amino acid residue. It is this core amino acid residue that extends to each of the 29 residues, represented in Fig. 1.

The R group defines the sidechain of all amino acids, forming a modular and general structure of the 'monopeptide' that is generated for input into any computation. Sidechains may have the same number of atoms such as leucine (Leu) and isoleucine (Ile), which are isomers, or serine (Ser) and cysteine (Cys), which are congeners. These pairs have identical Q numbers, yet their sidechains need to be differentiated; as introduced in the previous work [1]. As a result, each of the amino acid residues and the N-acetyl and N-methylamide end protecting groups are exclusively defined, using an internal coordinate system to characterize molecular structure, isomeric state, stereochemistry and conformation.

The symbolic internal coordinates of five special cases that are modifications of naturally occurring amino acids are listed in Table 6. These are pGlu, Hyp, Hyl, Cit and Orn. The symbolic internal coordinates for the first 24 amino acids, as presented in Fig. 1, were presented in the previous work [1].

The ab initio calculations carried out at the RHF/3-21G level of theory represent the first phase. Due to a fortuitous cancellation of errors, the RHF/3-21G results are very good representations of the conformational intricacies of peptides. For the five chosen amino acids, selected dihedral angles obtained at the RHF/3-21G, RHF/6-31G(d) and B3LYP/6-31G(d) levels of theory are presented in Table 7.

4.2. Dipeptides

The numbering system defined is easily propagated for the two selected dipeptide (i.e. diamino acid diamide) structures presented in Fig. 4. Seven more

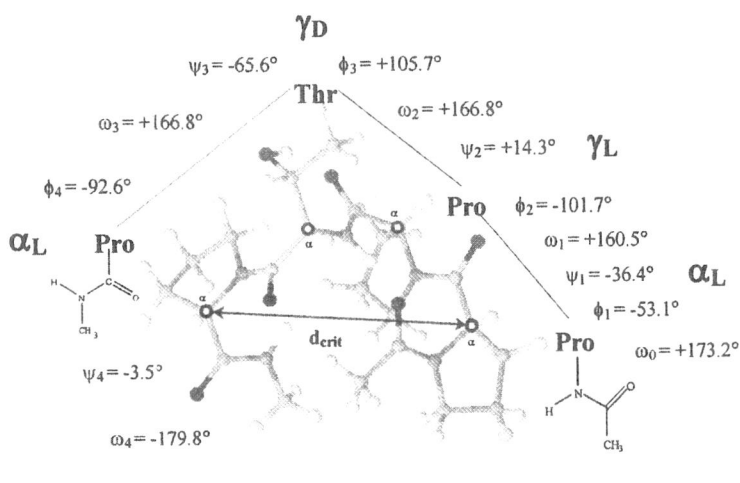

$$z = Q_N + Q_{1,P} + Q_{2,P} + Q_{3,T} + Q_{4,P} + Q_C$$
$$= 6 + (z_1 - 6) + (z_2 - z_1) + (z_3 - z_2) + (z_4 - z_3) + 6$$
$$= 6 + 14 + 14 + 14 + 14 + 6$$
$$= 68$$

(a)

(b)

Fig. 9. A schematic illustration of a special tetrapeptide of Immunoglobulin A hinge region. (a) Numbering system of PPTP, (b) one of the tetrapeptide structures (*N*-acetyl-Pro-Pro-Thr-Pro-*N*-methylamide) optimized at the RHF/3-21G level of theory. This optimized structure corresponds to an $\alpha_L \gamma_L \gamma_D \gamma_L$ conformation. The $\gamma_L \gamma_D$ conformer was identified previously as a β-turn [42], with $d_{crit} < 7$ Å and $-90° \leq \tau_{crit} \leq +90°$. The above structure is only marginally a β-turn, with $d_{crit} = 7.49$ Å and $\tau_{crit} = -97.77°$ (note that τ_{crit} is defined as the dihedral angle involving the four Cα atoms, denoted by the heavy opened circles above].

Table 14

Symbolic internal coordinates of a tetrapeptide in its *N*-acetyl and *N*-methylated in protected form as shown in Fig. 9

P_N	H	
	C*1*R2	
	C2R3*1*A3	
	O3R42A4*1*D4	
	H2R5*3*A54D5	
	H2R6*3*A6*4*D6	
I	N*3*R72A7*1*D7	
	C7R8*3*A82D8	
	C8R97A9*3*D9	
	O9R10*8*A107D10	
	H8R117A11*3*D11	
	C8R127A12*3*D12	
I_{sc}	C*12*R138A137D13	Pro
	C*13*R14*12*A148D14	
	H*12*R158A157D15	
	H*12*R168A167D16	
	H*13*R17*12*A178D17	
	H*13*R18*12*A188D18	
	H*14*R19*13*A19*12*D19	
	H*14*R20*13*A20*12*D20	
II	N9R21*8*A217D21	
	C*21*R229A228D22	
	C22R23*21*A239D23	
	O23R2422A242*1*D24	
	H22R25*21*A259D25	
	C22R26*21*A269D26	
II_{sc}	C26R2722A272*1*D27	Pro
	C27R2826A2822D28	
	H26R2922A292*1*D29	
	H26R3022A302*1*D30	
	H27R3126A3122D31	
	H27R3226A3222D32	
	H28R3327A3326D33	
	H28R3427A3426D34	
III	N23R3522A352*1*D35	
	C35R3623A3622D36	
	C36R3735A3723D37	
	O37R3836A3835D38	
	H36R3935A3923D39	
	H35R4023A4022D40	
	C36R41*3*5A4123D41	
III_{sc}	C*41*R4236A4235D42	Thr
	H42R43*41*A4336D43	
	O*41*R4436A4435D44	
	H*41*R45*36*A4535D45	
	H42R46*41*A4636D46	
	H42R47*41*A4736D47	
	H44R48*41*A4836D48	
IV	N37R4936A4935D49	

Table 14 (*continued*)

	C49R5037A5036D50	
	C50R51*49*A5137D51	
	O*51*R5250A5249D52	
	H50R5349A5337D53	
	C50R5449A5437D54	
IV_{sc}	C54R5550A5549D55	Pro
	C55R5654A5650D56	
	H54R5750A5749D57	
	H54R5850A5849D58	
	H55R5954A5950D59	
	H55R6054A6050D60	
	H56R6155A6154D61	
	H56R6255A6254D62	
P_C	N*51*R6350A6349D63	
	C63R64*51*A6449D64	
	H64R6563A655*1*D65	
	H63R6651A6649D66	
	H64R6763A675*1*D67	
	H64R6863A685*1*D68	

Table 15

Symbolic internal coordinates for two pentapeptides in their *N*-acetyl and *N*-methylated protected forms as shown in Fig. 10

	Leu-enkephalin	Met-enkephalin
P_N	H	H
	C*1*R2	C*1*R2
	C2R3*1*A3	C2R3*1*A3
	O3R42A4*1*D4	O3R42A4*1*D4
	H2R5*3*A54D5	H2R5*3*A54D5
	H2R6*3*A64D6	H2R6*3*A64D6
I	N*3*R72A7*1*D7	N*3*R72A7*1*D7
	C7R8*3*A82D8	C7R8*3*A82D8
	C8R97A9*3*D9	C8R97A9*3*D9
	O9R10*8*A107D10	O9R10*8*A107D10
	H8R117A11*3*D11	H8R117A11*3*D11
	H7R12*3*A122D12	H7R12*3*A122D12
	C8R137A13*3*D13	C8R137A13*3*D13
I_{sc}	C*13*R148A147D14	C*13*R148A147D14
	C*14*R15*13*A158D15	C*14*R15*13*A158D15
	C*15*R16*14*A16*13*D16	C*15*R16*14*A16*13*D16
	C*16*R17*15*A17*14*D17	C*16*R17*15*A17*14*D17
	C*17*R18*16*A18*15*D18	C*17*R18*16*A18*15*D18
	C*18*R19*17*A19*16*D19	C*18*R19*17*A19*16*D19
	H*13*R208A207D20	H*13*R208A207D20
	H*13*R218A217D21	H*13*R218A217D21
	H*15*R22*14*A22*13*D22	H*15*R22*14*A22*13*D22
	H*16*R23*15*A23*14*D23	H*16*R23*15*A23*14*D23
	O*17*R24*16*A24*15*D24	O*17*R24*16*A24*15*D24
	H*18*R25*17*A25*16*D25	H*18*R25*17*A25*16*D25
	H*19*R26*18*A26*17*D26	H*19*R26*18*A26*17*D26
	H24R27*17*A27*16*D27	H24R27*17*A27*16*D27

Table 15 (*continued*)

	Leu-enkephalin	Met-enkephalin
II	N9 R288 A287 D28	N9 R288 A287 D28
	C28 R299 A298 D29	C28 R299 A298 D29
	C29 R3028 A309 D30	C29 R3028 A309 D30
	O30 R3129 A3128 D31	O30 R3129 A3128 D31
	H29 R3228 A329 D32	H29 R3228 A329 D32
	H28 R339 A338 D33	H28 R339 A338 D33
	H29 R3428 A349 D34	H29 R3428 A349 D34
III	N30 R3529 A3528 D35	N30 R3529 A3528 D35
	C35 R3630 A3629 D36	C35 R3630 A3629 D36
	C36 R3735 A3730 D37	C36 R3735 A3730 D37
	O37 R3836 A3835 D38	O37 R3836 A3835 D38
	H36 R3935 A3930 D39	H36 R3935 A3930 D39
	H35 R4030 A4029 D40	H35 R4030 A4029 D40
	H36 R4135 A4130 D41	H36 R4135 A4130 D41
IV	N37 R4236 A4235 D42	N37 R4236 A4235 D42
	C42 R4337 A4336 D43	C42 R4337 A4336 D43
	C43 R4442 A4437 D44	C43 R4442 A4437 D44
	O44 R4543 A4542 D45	O44 R4543 A4542 D45
	H43 R4642 A4637 D46	H43 R4642 A4637 D46
	H42 R4737 A4736 D47	H42 R4737 A4736 D47
	C43 R4842 A4837 D48	C43 R4842 A4837 D48
IV$_{sc}$	C48 R4943 A4942 D49	C48 R4943 A4942 D49
	C49 R5048 A5043 D50	C49 R5048 A5043 D50
	C50 R5149 A5148 D51	C50 R5149 A5148 D51
	C51 R5250 A5249 D52	C51 R5250 A5249 D52
	C52 R5351 A5350 D53	C52 R5351 A5350 D53
	C53 R5452 A5451 D54	C53 R5452 A5451 D54
	H48 R5543 A5542 D55	H48 R5543 A5542 D55
	H48 R5643 A5642 D56	H48 R5643 A5642 D56
	H50 R5749 A5748 D57	H50 R5749 A5748 D57
	H51 R5850 A5849 D58	H51 R5850 A5849 D58
	H52 R5951 A5950 D59	H52 R5951 A5950 D59
	H53 R6052 A6051 D60	H53 R6052 A6051 D60
	H54 R6153 A6152 D61	H54 R6153 A6152 D61
V	N44 R6243 A6242 D62	N44 R6243 A6242 D62
	C62 R6344 A6343 D63	C62 R6344 A6343 D63
	C63 R6462 A6444 D64	C63 R6462 A6444 D64
	O64 R6563 A6562 D65	O64 R6563 A6562 D65
	H63 R6662 A6644 D66	H63 R6662 A6644 D66
	H62 R6744 A6743 D67	H62 R6744 A6743 D67
	C63 R6862 A6844 D68	C63 R6862 A6844 D68
V$_{sc}$	C68 R6963 A6962 D69	C68 R6963 A6962 D69
	C69 R7068 A7063 D70	S69 R7068 A7063 D70
	H70 R7169 A7168 D71	C70 R7169 A7168 D71
	C69 R7268 A7263 D72	H71 R7270 A7269 D72
	H72 R7369 A7368 D73	H68 R7363 A7362 D73
	H68 R7463 A7462 D74	H68 R7463 A7462 D74
	H68 R7563 A7562 D75	H69 R7568 A7563 D75
	H69 R7668 A7663 D76	H69 R7668 A7663 D76
	H70 R7769 A7768 D77	H71 R7770 A7769 D77

Table 15 (*continued*)

	Leu-enkephalin	Met-enkephalin
	H70 R7869 A7868 D78	H71 R7870 A7869 D78
	H72 R7969 A7968 D79	
	H72 R8069 A8068 D80	
P_C	N64 R8163 A8162 D81	N64 R7963 A7962 D79
	C81 R8264 A8263 D82	C79 R8064 A8063 D80
	H82 R8381 A8364 D83	H80 R8179 A8164 D81
	H81 R8464 A8463 D84	H79 R8264 A8263 D82
	H82 R8581 A8564 D85	H80 R8379 A8364 D83
	H82 R8681 A8664 D86	H80 R8479 A8464 D84

structures were drawn to illustrate the continuity of the method and further supply evidence for the efficiency and modularity of the sequential numbering system.

Fig. 5 shows these nine structures. Table 8 presents the symbolic internal coordinates of the nine dipeptides: N-Ac-Gly-Gly-NH-Me, N-Ac-Gly-Ala-NH-Me, N-Ac-Gly-Val-NH-Me, N-Ac-Ala-Gly-NH-Me, N-Ac-Ala-Ala-NH-Me, N-Ac-Ala-Val-NH-Me, N-Ac-Val-Gly-NH-Me, N-Ac-Val-Ala-NH-Me and N-Ac-Val-Val-NH-Me; easily summarized as Gly-Gly (GG), Gly-Ala (GA), Gly-Val (GV), Ala-Gly (AG), Ala-Ala (AA), Ala-Val (AV), Val-Gly (VG), Val-Ala (VA) and Val-Val (VV).

For the nine chosen dipeptides (summarized in Fig. 5), the dihedral angles obtained at the RHF/3-21G, RHF/6-31G(d) and B3LYP/6-31G(d) levels of theory are shown in Table 9.

All variations of each of the 20 amino acids can be easily constructed totaling $20 \times 20 = 400$ possible dipeptides, each following the same rules of the sequential numbering system. Using the standardized numeric definition, one would be able to easily automate the construction of all 400 inputs of any and all topologically possible isomers, enantiomers and conformers. Data extraction is also easily automatable.

As an example of the practical use of this method, Fig. 6 presents the numbering of six such dipeptides from the possible 400. All the structures include proline, which was used interchangeably either as the first or second residue, along with three aromatic amino acid residues: Phe, Tyr and Trp.

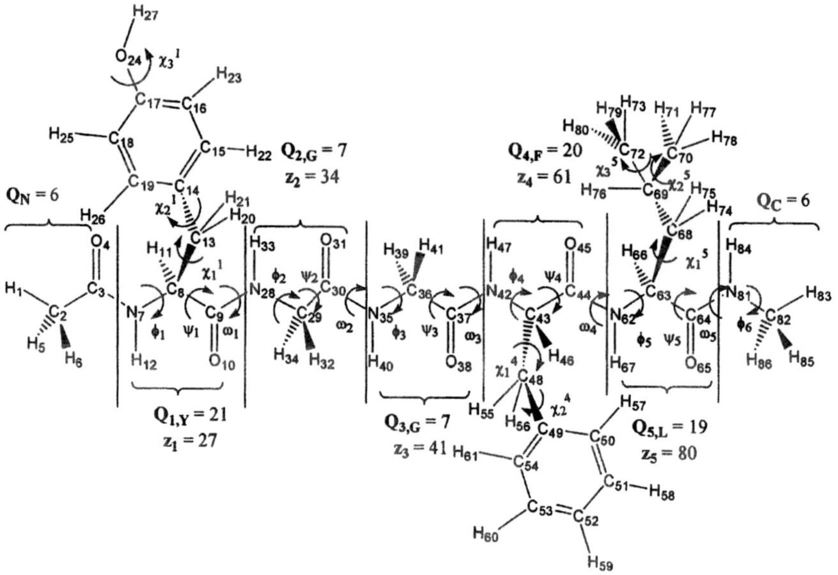

(a)

$$z = Q_N + Q_{1,Y} + Q_{2,G} + Q_{3,G} + Q_{4,F} + Q_{5,L} + Q_C$$
$$= 6 + (z_1 - 6) + (z_2 - z_1) + (z_3 - z_2) + (z_4 - z_3) + 6$$
$$= 6 \quad + 21 + 7 + 7 + 20 + 19 + 6$$
$$= 86$$

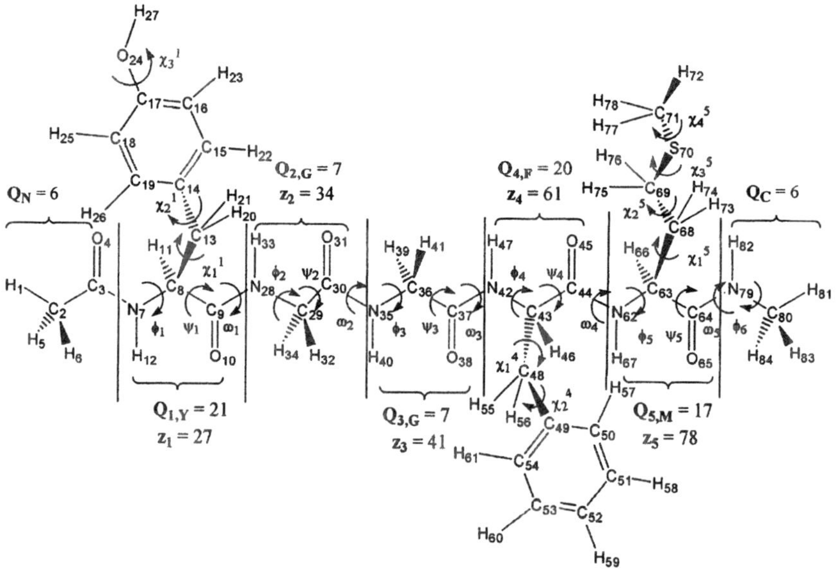

(b)

$$z = Q_N + Q_{1,Y} + Q_{2,G} + Q_{3,G} + Q_{4,F} + Q_{5,M} + Q_C$$
$$= 6 + (z_1 - 6) + (z_2 - z_1) + (z_3 - z_2) + (z_4 - z_3) + 6$$
$$= 6 \quad + 21 + 7 + 7 + 20 + 17 + 6$$
$$= 84$$

Fig. 10. A schematic illustration of two small biologically active pentapeptides (a) Leu-enkaphalin and (b) Met-enkaphalins.

Table 16
A selected pentapeptide: leucine-enkaphalin in its β_L or γ_L backbone conformation geometry optimized at three levels of theory

Level of theory	Energy	ψ_0	ω_0	ϕ_1	ψ_1	ω_1	ϕ_2	ψ_2	ω_2	ϕ_3	ψ_3	ω_3	ϕ_4
RHF/3-21G	−2037.8611800	179.45	178.78	−167.63	171.75	164.34	−81.82	176.40	−40.69	179.29	179.66	178.37	−168.59
RHF/6-31G (d)	−2049.2531600	169.01	175.02	−158.00	167.57	168.06	−71.80	165.48	−61.72	176.07	176.54	175.88	−158.07

ψ_4	ω_4	ϕ_5	ψ_5	ω_5	ϕ_6	χ_1^1	χ_2^1	χ_3^1	χ_1^4	χ_2^4	χ_1^5	χ_2^5	χ_3^5
172.00	168.62	−79.20	84.32	−176.93	−12.76	−144.75	−110.50	−179.74	−150.84	−106.05	−178.73	177.35	58.38
165.01	174.82	−79.35	95.61	−173.14	−171.24	−151.64	−110.86	179.41	−156.43	−106.84	−177.04	176.08	57.39

Table 17
Symbolic internal coordinates for an oligopeptide with a disulfide linkage, N-acetyl-Cys-Ala-Ala-Ala-Ala-Cys-N-methylamide as shown in Fig. 11

P_N	C1 R2
	C2 R31 A3
	O3 R42 A41 D4
	H2 R53 A54 D5
	H2 R63 A64 D6
I	N3 R72 A71 D7
	C7 R83 A82 D8
	C8 R97 A93 D9
	O9 R108 A107 D10
	H8 R117 A113 D11
	H7 R123 A122 D12
	C8 R137 A133 D13
I_{sc}	S13 R148 A147 D14
	H13 R158 A157 D15
	H13 R168 A167 D16
II	N9 R178 A177 D17
	C17 R189 A188 D18
	C18 R191 A199 D19
	O19 R201 A2017 D20
	H18 R211 A219 D21
	H17 R229 A228 D22
	C18 R231 A239 D23
II_{sc}	H23 R241 A2417 D24
	H23 R251 A2517 D25
	H23 R261 A2617 D26
III	N19 R271 A2717 D27
	C27 R281 A2818 D28
	C28 R2927 A2919 D29
	O29 R3028 A3027 D30
	H28 R3127 A3119 D31
	H27 R3219 A3218 D32
	C28 R3327 A3319 D33
III_{sc}	H33 R3428 A3427 D34

Table 17 (continued)

	H33 R3528 A3527 D35
	H33 R3628 A3627 D36
IV	N29 R3728 A3727 D37
	C37 R3829 A3828 D38
	C38 R3937 A3929 D39
	O39 R4038 A4037 D40
	H38 R4137 A4129 D41
	H37 R4229 A4228 D42
	C38 R4337 A4329 D43
IV_{sc}	H43 R4438 A4437 D44
	H43 R4538 A4537 D45
	H43 R4638 A4637 D46
V	N39 R4738 A4737 D47
	C47 R4839 A4838 D48
	C48 R4947 A4939 D49
	O49 R5048 A5047 D50
	H48 R5147 A5139 D51
	H47 R5239 A5238 D52
	C48 R5347 A5339 D53
V_{sc}	N49 R5748 A5747 D57
	C57 R5849 A5848 D58
	C58 R5957 A5949 D59
	O59 R6058 A6057 D60
	H58 R6157 A6149 D61
	H57 R6249 A6248 D62
	C58 R6357 A6349 D63
VI	xS63 R6458 A6457 D64
	H63 R6558 A6557 D65
	H63 R6658 A6657 D66
P_C	N59 R6758 A6757 D67
	C67 R6859 A6858 D68
	H68 R6967 A6959 D69
	H67 R7059 A7058 D70
	H68 R7167 A7159 D71
	H68 R7267 A7259 D72

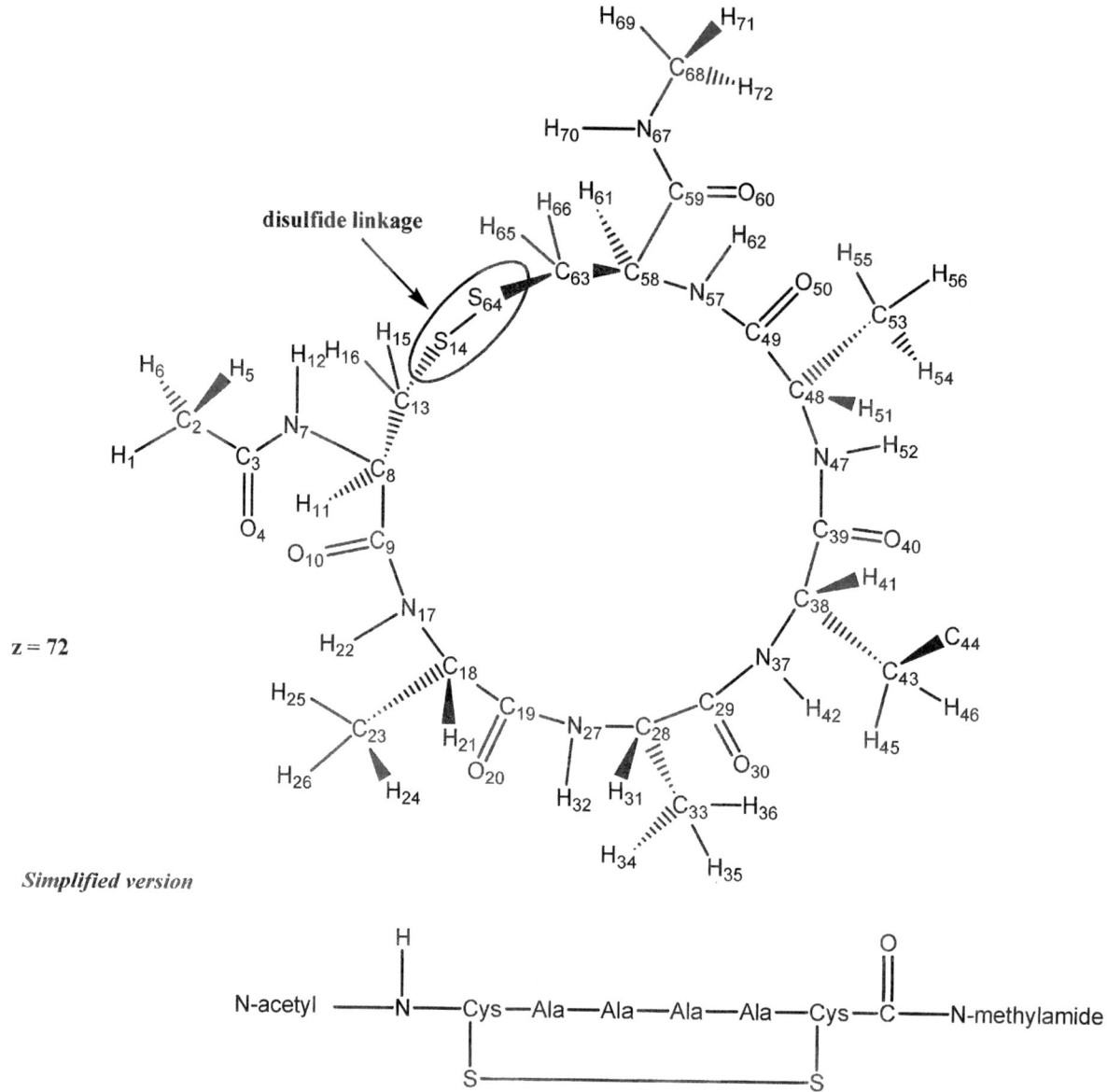

z = 72

Simplified version

Fig. 11. A schematic illustration of an oligopeptide with a disulfide linkage, *N*-acetyl-Cys-Ala-Ala-Ala-Ala-Cys-*N*-methylamide. This particular structure was used as a model for two other oligopeptides with disulfide linkages, oxytocin and vasopression.

Table 10 presents the symbolic internal coordinates of the six dipeptides: N-Ac-Pro-Phe-NH-Me, N-Ac-Pro-Tyr-NH-Me, N-Ac-Pro-Trp-NH-Me, N-Ac-Phe-Pro-NH-Me, N-Ac-Tyr-Pro-NH-Me and N-Ac-Trp-Pro-NH-Me; easily summarized as Pro-Phe (PF), Pro-Tyr (PY), Pro-Trp (PW), Phe-Pro (FP), Tyr-Pro (YP) and Trp-Pro (WP).

4.3. Tripeptides

Table 11 presents the symbolic internal coordinates of the four tripeptides: *N*-acetyl-Gly-Gly-Gly-*N*-methylamide, N-Ac-Gly-Ala-Gly-NH-Me, N-Ac-Ala-Gly-Ala-NH-Me, and N-Ac-Ala-Ala-Ala-NH-Me; easily summarized as Gly-Gly-Gly (GGG),

Table 18
Symbolic internal coordinates for the hormone oxytocin, an oligopeptide with a disulfide linkage, as shown in Fig. 12

P_N	H
I	N1 R2
	C2 R31 A3
	C3 R42 A41 D4
	O4 R53 A52 D5
	H3 R62 A61 D6
	H2 R73 A74 D7
	C3 R82 A81 D8
I_{sc}	S8 R93 A92 D9
	H8 R103 A102 D10
	H8 R113 A112 D11
II	N4 R123 A122 D12
	C12 R134 A133 D13
	C13 R1412 A144 D14
	O14 R1513 A1512 D15
	H13 R1612 A164 D16
	H12 R174 A173 D17
	C13 R1812 A184 D18
II_{sc}	C18 R1913 A1912 D19
	C19 R2018 A2013 D20
	C20 R2119 A2118 D21
	C21 R2220 A2219 D22
	C22 R2321 A2320 D23
	C19 R2418 A2413 D24
	H18 R2513 A2512 D25
	H18 R2613 A2612 D26
	H20 R2719 A2718 D27
	H21 R2820 A2819 D28
	O22 R2921 A2920 D29
	H23 R3022 A3021 D30
	H24 R3119 A3118 D31
	H29 R3222 A3221 D32
III	N14 R3313 A3312 D33
	C33 R3414 A3413 D34
	C34 R3533 A3514 D35
	O35 R3634 A3633 D36
	H34 R3733 A3714 D37
	H33 R3814 A3813 D38
	C34 R3933 A3914 D39
III_{sc}	C39 R4034 A4033 D40
	C40 R4139 A4134 D41
	C41 R4240 A4239 D42
	C39 R4334 A4333 D43
	H43 R4439 A4434 D44
	H39 R4534 A4533 D45
	H40 R4639 A4634 D46
	H40 R4739 A4734 D47
	H41 R4840 A4839 D48
	H41 R4940 A4939 D49
	H43 R5039 A5034 D50
	H43 R5139 A5134 D51

Table 18 (*continued*)

IV	N35 R5234 A5233 D52
	C52 R5335 A5334 D53
	C53 R5452 A5435 D54
	O54 R5553 A5552 D55
	H53 R5652 A5635 D56
	H52 R5735 A5734 D57
	C53 R5852 A5835 D58
IV_{sc}	C58 R5953 A5952 D59
	C59 R6058 A6053 D60
	O60 R6159 A6158 D61
	N60 R6259 A6258 D62
	H62 R6360 A6359 D63
	H58 R6453 A6452 D64
	H58 R6553 A6552 D65
	H59 R6658 A6653 D66
	H59 R6758 A6753 D67
	H62 R6860 A6859 D68
V	N54 R6953 A6952 D69
	C69 R7054 A7053 D70
	C70 R7169 A7154 D71
	O71 R7270 A7269 D72
	H70 R7369 A7354 D73
	H69 R7454 A7453 D74
	C70 R7569 A7554 D75
V_{sc}	C75 R7670 A7669 D76
	O76 R7775 A7770 D77
	N76 R7875 A7870 D78
	H78 R7976 A7975 D79
	H75 R8070 A8069 D80
	H75 R8170 A8169 D81
	H78 R8276 A8275 D82
VI	N71 R8370 A8369 D83
	C83 R8471 A8470 D84
	C84 R8583 A8571 D85
	O85 R8684 A8683 D86
	H84 R8783 A8771 D87
	H83 R8871 A8870 D88
	C84 R8983 A8971 D89
VI_{sc}	S89 R9084 A9083 D90
	H89 R9184 A9183 D91
	H89 R9284 A9283 D92
VII	N85 R9384 A9383 D93
	C93 R9485 A9484 D94
	C94 R9593 A9585 D95
	O95 R9694 A9693 D96
	H94 R9793 A9785 D97
	C94 R9893 A9885 D98
	C98 R9994 A9993 D99
VII_{sc}	C93 R10085 A10084 D100

(continued on next page)

Table 18 (continued)

	H98 R10194 A10193 D101
	H98 R10294 A10293 D102
	H99 R10398 A10394 D103
	H99 R10498 A10494 D104
	H100 R10593 A10585 D105
	H100 R10693 A10685 D106
VIII	N95 R10794 A10793 D107
	C107 R10895 A10894 D108
	C108 R109107 A10995 D109
	O109 R110108 A110107 D110
	H108 R111107 A11195 D111
	H107 R11295 A11294 D112
	C108 R113107 A11395 D113
VIII_sc	C113 R114108 A114107 D114
	C114 R115113 A115108 D115
	H115 R116114 A116113 D116
	C114 R117113 A117108 D117
	H117 R118114 A118113 D118
	H113 R119108 A119107 D119
	H113 R120108 A120107 D120
	H114 R121113 A121108 D121
	H115 R122114 A122113 D122
	H115 R123114 A123113 D123
	H117 R124114 A124113 D124
	H117 R125114 A125113 D125
IX	N109 R126108 A126107 D126
	C126 R127109 A127108 D127
	C127 R128126 A128109 D128
	O128 R129127 A129126 D129
	H127 R130126 A130109 D130
	H126 R131109 A131108 D131
	H127 R132126 A132109 D132
P_C	N128 R133127 A133126 D133
	H133 R134128 A134127 D134
	H133 R135128 A135127 D135

Gly-Ala-Gly (GAG), Ala-Gly-Ala (AGA) and Ala-Ala-Ala (AAA) (Fig. 7).

Selected optimized dihedral angles obtained at the RHF/3-21G, RHF/6-31G(d) and B3LYP/6-31G(d) levels of theory are shown in Table 12. Studies have already been attempted for 3 of these 4 tripeptides [44–46].

As was done with the dipeptides, tripeptide variations of the 20 amino acids can be created, totaling $20 \times 20 \times 20 = 8000$ possible structures, all following the same rules using the sequential numbering system. Again, using the standardized numeric definition, one would be able to easily

automate the construction of all 8000 inputs of any and all topologically possible isomers, enantiomers and conformers. Work is currently under way to extend the automated standard to all possible first order transition structures.

As an example of the practical use of this method, Fig. 8 shows the numbering of three such tripeptides from the possible 8000. The structures include various combinations of the three amino acid residues of Pro, hydroxyproline (Hyp) and Gly.

Table 13 presents the symbolic internal coordinates of these three tripeptides: N-Ac-Hyp-Pro-Gly-NH-Me, N-Ac-Gly-Pro-Hyp-NH-Me and N-Ac-Pro-Hyp-Gly-NH-Me; easily summarized as Hyp-Pro-Gly (P_HPG), Gly-Pro-Hyp (GPP$_H$) and Pro-Hyp-Gly (PP$_H$G).

4.4. Tetrapeptide, PPTP

Ala-Ala-Ala-Ala is the simplest tetrapeptide ever studied using simulations. Clearly the Pro-Pro-Thr-Pro tetrapeptide (Fig. 9) represents a more complicated structural investigation than Ala-Ala-Ala-Ala [38–41], as Pro more frequently exhibits cis- and trans-isomerism in its peptide bonds, as well as gauche($-$) and gauche($+$) ring puckering. Thus, $2 \times 2 = 4$ isomeric forms must be examined for each backbone (BB) conformer. Up to three backbone conformers (γ_L, ϵ_L, α_L) may be expected; thus each proline moiety including the two ring puckerings may have $4 \times 3 = 12$ conformations. Consequently, for the dipeptide N-Ac-Pro-Pro-NH-Me, we may anticipate $12 \times 12 = 144$ conformations. However, even if one disregards ring puckering and considers only trans-peptide bonds, Pro-Pro-Thr-Pro may exhibit $3 \times 3 \times 9 \times 3 = 243$ backbone conformations. Including the $3 \times 3 = 9$ sidechain conformations of the Thr sidechain (SC) moiety, this will lead to a total of 243×9 or 2187 conformers. Thus, the 243 backbone conformations may be best studied on Pro-Pro-Ala-Pro and the Ala sidechain may subsequently be extended to a Thr sidechain.

Fig. 9a presents an arbitrary conformer of the tetrapeptide N-Ac-Pro-Pro-Thr-Pro-NH-Me along with a systematic outline of the numbering according to the formula presented in Fig. 2. With the implementation of this system, input generation is reduced to a fraction of the original time with

Oxytocin

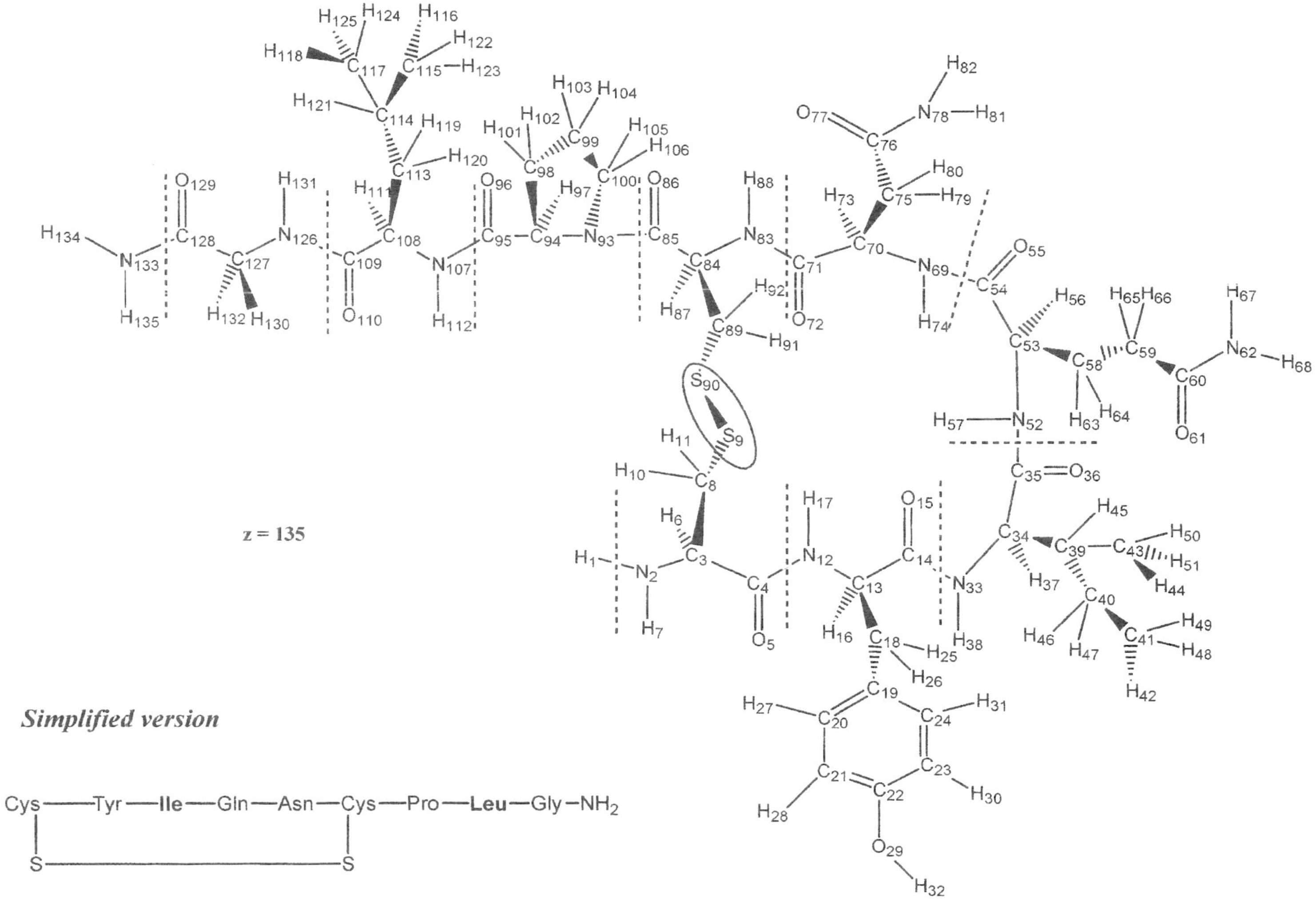

$z = 135$

Simplified version

Cys——Tyr——Ile——Gln——Asn——Cys——Pro——Leu——Gly—NH$_2$

Fig. 12. A schematic illustration of the hormone oxytocin, a small peptide containing nine (9) residues, which is found in the posterior pituitary. Oxytocin stimulates contraction of uterine smooth muscle.

Vasopressin

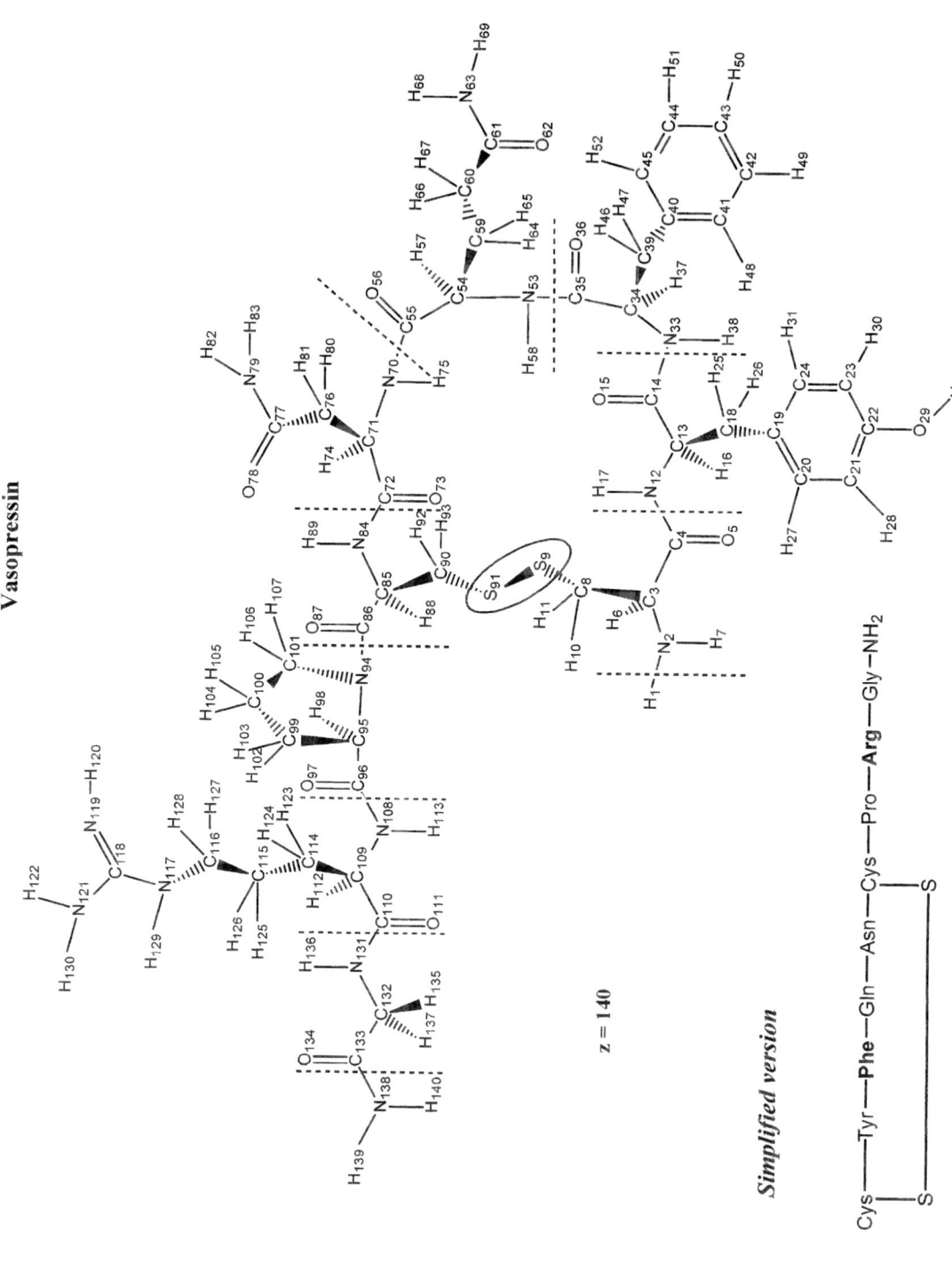

$z = 140$

Simplified version

Cys——Tyr——**Phe**——Gln——Asn——Cys——Pro——**Arg**——Gly—NH$_2$

Fig. 13. A schematic illustration of the antidiuretic hormone (ADH/vasopressin), a small peptide containing nine (9) residues, which is found in the posterior pituitary. Vasopressin controls the resorption of water by the distal tubules of the kidneys and regulates the osmotic content of blood.

Table 19

Symbolic internal coordinates for the hormone vasopressin, an oligopeptide with a disulfide linkage, as shown in Fig. 13

Table 19 (*continued*)

P_N	H
I	N*1* R2
	C2 R3*1* A3
	C3 R42 A4*1* D4
	O4 R53 A52 D5
	H*3* R62 A6*1* D6
	H2 R73 A74 D7
	C3 R82 A8*1* D8
I_{sc}	S8 R93 A92 D9
	H8 R103 A102 D10
	H8 R113 A112 D11
II	N*4* R123 A122 D12
	C*12* R134 A133 D13
	C*13* R14*12* A144 D14
	O*14* R15*13* A15*12* D15
	H*13* R16*12* A164 D16
	H*12* R174 A173 D17
	C*13* R18*12* A184 D18
II_{sc}	C*18* R19*13* A19*12* D19
	C*19* R20*18* A20*13* D20
	C20 R21*19* A21*18* D21
	C*21* R2220 A22*19* D22
	C22 R23*21* A2320 D23
	C*19* R24*18* A24*13* D24
	H*18* R25*13* A25*12* D25
	H*18* R26*13* A26*12* D26
	H20 R27*19* A27*18* D27
	H*21* R2820 A28*19* D28
	O22 R292*1* A2920 D29
	H*23* R3022 A302*1* D30
	H*24* R31*19* A31*18* D31
	H29 R3222 A322*1* D32
III	N*14* R33*13* A33*12* D33
	C*33* R34*14* A34*13* D34
	C*34* R3533 A35*14* D35
	O*35* R3634 A3633 D36
	H*34* R3733 A37*14* D37
	H*33* R38*14* A38*13* D38
	C*34* R3933 A39*14* D39
III_{sc}	C*39* R4034 A4033 D40
	C*40* R4139 A4134 D41
	C*41* R4240 A4239 D42
	C*42* R4341 A4340 D43
	C*43* R4442 A444*1* D44
	C*40* R4539 A4534 D45
	H*39* R4634 A4633 D46
	H*39* R4734 A4733 D47
	H4*1* R4840 A4839 D48
	H*42* R4941 A4940 D49

IV	
	H*43* R5042 A504*1* D50
	H*44* R5143 A5142 D51
	H*45* R5240 A5239 D52
IV	N*35* R5334 A5333 D53
	C*53* R5435 A5434 D54
	C*54* R5553 A5535 D55
	O*55* R5654 A5653 D56
	H*54* R5753 A5735 D57
	H*53* R5835 A5834 D58
	C*54* R5953 A5935 D59
IV_{sc}	C*59* R6054 A6053 D60
	C*60* R6159 A6154 D61
	O6*1* R6260 A6259 D62
	N6*1* R6360 A6359 D63
	H*63* R646*1* A6460 D64
	H*59* R6554 A6553 D65
	H*59* R6654 A6653 D66
	H*60* R6759 A6754 D67
	H*60* R6859 A6854 D68
	H*63* R696*1* A6960 D69
V	N*55* R7054 A7053 D70
	C*70* R7155 A7154 D71
	C*71* R7270 A7255 D72
	O*72* R737*1* A7370 D73
	H*71* R7470 A7455 D74
	H*70* R7555 A7554 D75
	C*71* R7670 A7655 D76
V_{sc}	C*76* R777*1* A7770 D77
	O*77* R7876 A787*1* D78
	N*77* R7976 A797*1* D79
	H*79* R8077 A8076 D80
	H*76* R817*1* A8170 D81
	H*76* R827*1* A8270 D82
	H*79* R8377 A8376 D83
VI	N*72* R847*1* A8470 D84
	C*84* R8572 A857*1* D85
	C*85* R8684 A8672 D86
	O*86* R8785 A8784 D87
	H*85* R8884 A8872 D88
	H*84* R8972 A897*1* D89
	C*85* R9084 A9072 D90
VI_{sc}	S9 R918 A913 D91
	H*90* R9285 A9284 D92
	H*90* R9385 A9384 D93
VII	N*86* R9485 A948*4* D94
	C*94* R9586 A9585 D95
	C*95* R9694 A9686 D96
	O*96* R9795 A9794 D97
	H*95* R9894 A9886 D98
	C*95* R9994 A9986 D99

(*continued on next page*)

Table 19 (*continued*)

VII$_{sc}$	C99R10095A10094D100
	C94R10186A10185D101
	H99R10295A10294D102
	H99R10395A10394D103
	H100R10499A10495D104
	H100R10599A10595D105
	H101R10694A10686D106
	H101R10794A10786D107
VIII	N96R10895A10894D108
	C108R10996A10995D109
	C109R110108A11096D110
	O110R111109A111108D111
	H109R112108A11296D112
	H108R11396A11395D113
	C109R114108A11496D114
VIII$_{sc}$	C114R115109A115108D115
	C115R116114A116109D116
	N116R117115A117114D117
	C117R118116A118115D118
	N118R119117A119116D119
	H119R120118A120117D120
	N118R121117A121116D121
	H121R122118A122117D122
	H114R123109A123108D123
	H114R124109A124108D124
	H115R125114A125109D125
	H115R126114A126109D126
	H116R127115A127114D127
	H116R128115A128114D128
	H117R129116A129115D129
	H121R130118A130117D130
IX	N110R131109A131108D131
	C131R132110A132109D132
	C132R133131A133110D133
	O133R134132A134131D134
	H132R135131A135110D135
	H131R136110A136109D136
	H132R137131A137110D137
P$_C$	N133R138132A138131D138
	H138R139133A139132D139
	H138R140133A140132D140

the help of automation scripts. Table 14 presents the symbolic internal coordinates of N-Ac-Pro-Pro-Thr-Pro-NH-Me.

One of the 17496 topologically possible conformers of Pro-Pro-Thr-Pro, [(3BB × 2SC) × (3BB × 2 SC) × (9BB × 9SC) × (3BB × 2SC)] was optimized in a preliminary study using this standardized numbering system and the structure is presented in

Fig. 9b. The central Pro-Thr structure optimized converged to the desired $\gamma_L \gamma_D$ conformer, previously identified as a new type of β-turn [47].

4.5. Pentapeptides

To illustrate the pentapeptide numbering (Table 15), Leu-enkephalin (Fig. 10a) and Met-enkephalin (Fig. 10b) are used. The enkephalins are relatively weak analgesics, which activate all opioid receptors, but appear to have the highest affinity for the δ receptor [48].

Geometry optimizations were completed for Leu-enkephalin, as an example of the outcome using the standardized numbering system. Due to the size of the Leu-enkaphalin molecule, it was only possible to carry out the calculations at one level of theory using two different basis sets. Optimized dihedral angles for Leu-enkephalin obtained at the RHF/3-21G and RHF/6-31G(d) levels of theory are presented in Table 16.

4.6. Hexapeptides and nonapeptides with disulfide linkages

The modular numbering system was also applied to three other linear peptides: N-Ac-Cys-Ala$_4$-Cys-NH-Me, oxytocin [49] and vasopressin [50]. What makes these peptides unique from the previous examples is that they form cyclic structures as a result of sidechain disulfide linkages.

The hexapeptide N-Ac-Cys-Ala$_4$-Cys-NH-Me is presented in Fig. 11 while Table 17 presents the symbolic internal coordinates.

The nonapeptides oxytocin and vasopressin (Anti-Diuretic Hormone or ADH) are presented in Figs. 12 and 13, respectively.

ADH and oxytocin each have nine (9) amino acids. Both have cysteine residues at amino acid positions $i = 1$ and 6. These cysteine residues form a disulfide bond with one another to create a cyclic six amino acid ring, analogous to the hexapeptide presented before: N-Ac-Cys-Ala$_4$-Cys-NH-Me, with three additional amino acid residues hanging off. ADH and oxytocin share seven amino acids in common and differ only at amino acid positions 3 and 8. Oxytoxin has isoleucine at position 3 and leucine at position 8

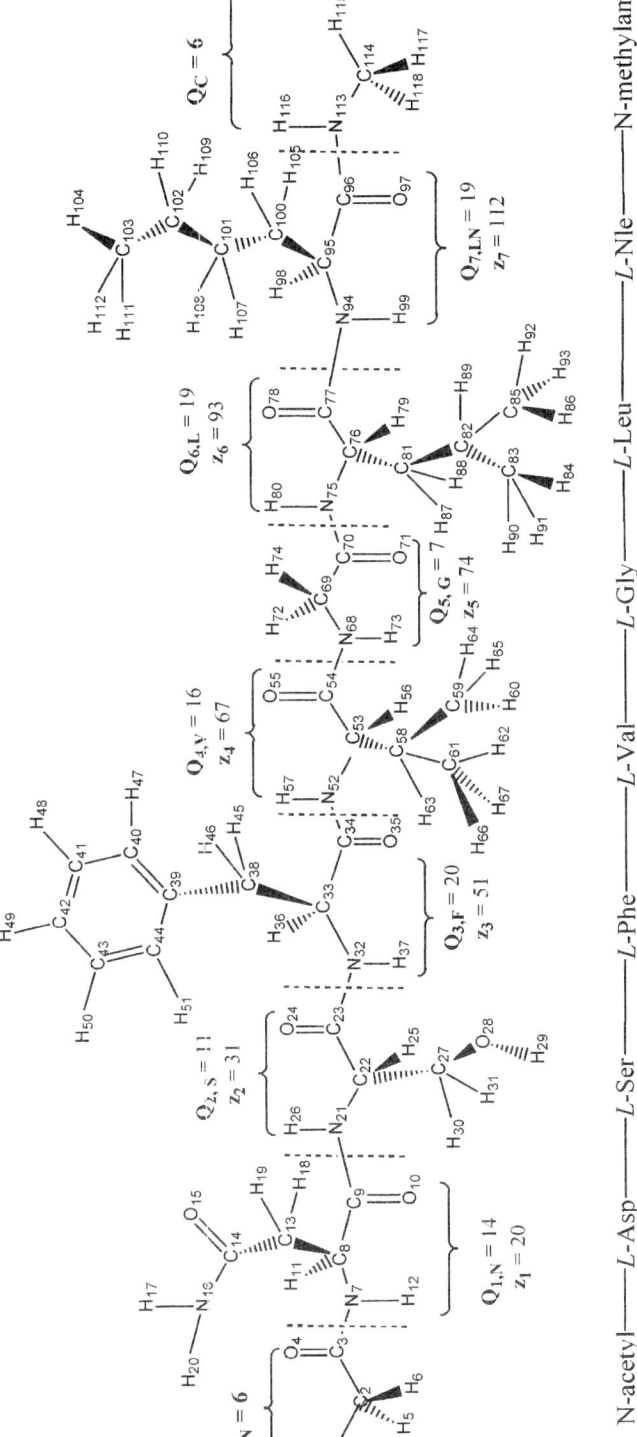

Fig. 14. A schematic illustration of the heptapeptide neurokinin A (NKA) fragment from residues 4–10.

Table 20

Symbolic internal coordinates for a heptapeptide fragment of neurokinin A at residues 4–10 as shown in Fig. 14

P_N	
	H
	C*1*R2
	C2R3*1*A3
	O*3*R42A4*1*D4
	H2R5*3*A54D5
	H2R6*3*A64D6
I	N*3*R72A7*1*D7
	C7R83A82D8
	C8R97A9*3*D9
	O9R108A107D10
	H8R117A113D11
	H7R123A122D12
	C8R137A132D13
I_{sc}	C*13*R148A147D14
	O*14*R15*13*A158D15
	N*14*R16*13*A168D16
	H*16*R17*14*A178D17
	H*13*R188A187D18
	H*13*R198A197D19
	N*16*R20*14*A20*13*D20
II	N9R218A217D21
	C*21*R229A228D22
	C22R23*21*A239D23
	O23R2422A242*1*D24
	H22R25*21*A259D25
	H2*1*R269A268D26
	C22R27*21*A279D27
II_{sc}	O27R2822A282*1*D28
	H28R2927A2922D29
	H27R3022A302*1*D30
	H27R3122A312*1*D31
III	N23R3222A322*1*D32
	C32R3323A3322D33
	C33R3432A342*3*D34
	O*34*R3533A3532D35
	H*33*R3632A362*3*D36
	H32R3723A3722D37
	C33R3832A382*3*D38
III_{sc}	C38R39*33*A39*32*D39
	C39R4038A40*33*D40
	C*40*R4139A4138D41
	C*41*R4240A4239D42
	C42R43*41*A43*40*D43
	C*43*R4442A444*1*D44
	H38R4533A4532D45
	H38R4633A4632D46
	H*40*R4739A4738D47
	H*41*R4840A4839D48
	H*42*R4941A4940D49

Table 20 (*continued*)

	H*43*R5042A504*1*D50
	H*44*R5143A5142D51
IV	N*34*R5233A5232D52
	C52R53*34*A5333D53
	C53R5452A5434D54
	O*54*R5553A5552D55
	H53R5652A56*34*D56
	H52R5734A5733D57
	C*53*R5852A585*1*D58
IV_{sc}	C58R5952A595*1*D59
	H58R6057A6052D60
	C*57*R6152A615*1*D61
	H*60*R6257A6252D62
	H*57*R6352A635*1*D63
	H58R6457A6452D64
	H58R6557A6552D65
	H*60*R6657A6652D66
	H*60*R6757A6752D67
V	N*54*R6853A6852D68
	C68R6954A6953D69
	C*69*R7068A7054D70
	O70R7169A7168D71
	H69R7268A7254D72
	H68R7354A7353D73
	C69R7468A7454D74
V_{sc}	
VI	N*70*R7569A7568D75
	C*75*R7670A7669D76
	C76R7775A7770D77
	O77R7876A7875D78
	H76R7975A7970D79
	H75R8070A8069D80
	C76R8175A8170D81
VI_{sc}	C8*1*R8276A8275D82
	C82R838*1*A8376D83
	H83R8482A848*1*D84
	C82R858*1*A8576D85
	H85R8682A868*1*D86
	H8*1*R8776A8775D87
	H8*1*R8876A8875D88
	H82R898*1*A8976D89
	H*83*R9082A908*1*D90
	H*83*R9182A918*1*D91
	H85R9282A928*1*D92
	H85R9382A938*1*D93
VII	N77R9476A9475D94
	C*94*R9577A9576D95
	C*95*R9694A9677D96
	O*96*R9795A9794D97
	H*95*R9894A9877D98
	H*94*R9977A9976D99

Table 20 (*continued*)

	C95R10094A10077D100
VII$_{sc}$	C*100*R101 95A10194D101
	C*101*R102*100*A10295D102
	C*102*R103*101*A103*100*D103
	H*103*R104*102*A104*101* D104
	H*100*R10595A10594D105
	H*100*R10695A10694D106
	N*101*R107*100*A10795D107
	H*101*R108*100*A10895D108
	H*102*R109*101*A109*100*D109
	H*102*R110*101*A110*100*D110
	H*103*R111*102*A111*101*D111
	H*103*R112*102*A112*101*D112
P$_C$	N96R11395A11394D113
	C*113*R11496A11495D114
	C*114*R115*113*A11596D115
	H*113*R11696A11695D116
	H*114*R117*113*A11796D117
	H*114*R118*113*A11896D118

while ADH has phenylalanine at position 3 and arginine at position 8.

With seven of nine amino acids in common, the three dimensional structure of ADH and oxytocin is important for each to be specifically recognized by their receptor(s). The disulfide bridge, which closes the ring, and intramolecular hydrogen bonds make the structure rigid. Consider the detailed structures of oxytocin (Fig. 12) and ADH (Fig. 13). The essential features for either hormone's activity are the main hexapeptide's backbone ring, a three amino acid 'tail', an $-NH_2$ blocked C-terminal glycine, and a tyrosine (Tyr) at position 2. Note that the Tyr at position 2 is on the opposite face of the backbone ring from the Ile or Phe at position 3. From this information, one may assume that the receptors for ADH and oxytocin may be similar on the face which recognizes the Tyr, but differ on the face which recognizes position 3.

These structures represent an extension of a model peptide, in this case the hexapeptide N-Ac-Cys-Ala$_4$-Cys-NH-Me. The numbering system presented for each in Figs. 12 and 13 would simplify future geometry optimizations, in terms of initial structural estimates as well as data management and extraction.

The system was also used to generate the symbolic internal coordinates of the two nonapeptides, oxytocin

and vasopressin as presented in Tables 18 and 19, respectively.

4.7. [Nle^{10}] NeurokininA (4–10) fragment as a special heptapeptide

The heptapeptide example is the neurokinin A (NKA) fragment from residue 4 to 10 [51]. At residue 6, the Phe is para substituted and residue 10 (Met) is replaced with Nle, therefore, the structure is written as [*p*-F-Phe^6,Nle^{10}]NKA(4–10) or Asp-Ser-(*p*-Phe)-Val-Gly-Leu-Nle.

The numbering system for this structure is outlined in Fig. 14 and the symbolic internal coordinates are presented in Table 20.

4.8. Leuprolide as a special nonapeptide

Leuprolide (*p*Glu-His-Trp-Ser-Tyr-D-Leu-Leu-Arg-Pro) is a synthetic nonapeptide agonist of GnRH [52].

The numbering system for this structure is outlined in Fig. 15 and the symbolic internal coordinates are presented in Table 21.

4.9. GnRH-III as a special decapeptide

GnRH-III a hypothalamic neuroendocrine hormone in the sea lamprey has been found to directly inhibit proliferation of sex hormone-dependent and independent cancer cells without chemical castration [53]. Its sequence is *p*Glu-His-Trp-Ser-His-Asp-Trp-Lys-Pro-Gly.

The numbering system outlined in Fig. 16 lends another example to the ease with which the system can be applied to complicated or synthetic peptides, having biological activity or application. The symbolic internal coordinates are presented in Table 22.

4.10. Comparisons of optimized geometrical dihedrals and relative energies at the different levels of theory

The correlations between the optimized geometrical dihedrals computed at different levels of theory are shown in Fig. 17. As can be seen, all R^2 values were found to be larger than 0.99, indicating that the ϕ_i and

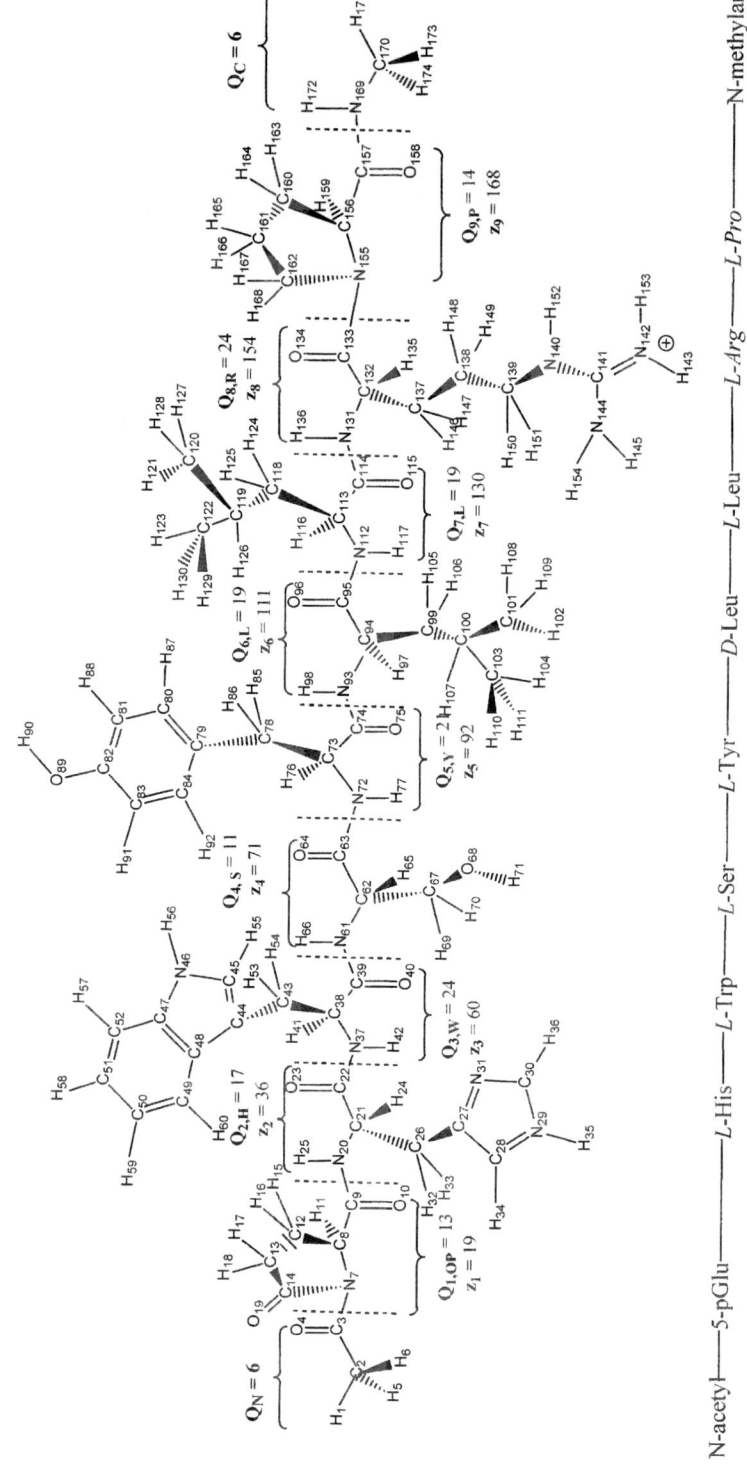

Fig. 15. A schematic illustration of *N*-acetyl-Leuprolide-*N*-methylamide, a synthetic nonapeptide analog of a naturally occurring gonadotropin releasing hormone (GnRH)/luteinizing hormone-releasing hormone (LH-RH) agonist.

Table 21

Symbolic internal coordinates for leuprolide acetate, a synthetic nonapeptide analog of a naturally occurring gonadotropin releasing hormone as shown in Fig. 15

Table 21 (continued)

P_N	H
	H
	C*1*R2
	C2R3*1*A3
	O*3*R42A4*1*D4
	H2R5*3*A54D5
	H2R6*3*A64D6
I	N*3*R72A7*1*D7
	C7R83A82D8
	C8R97A9*3*D9
	O9R108A107D10
	H8R117A11*3*D11
	C8R127A12*3*D12
I_{sc}	C*12*R138A137D13
	C7R143A142D14
	H*12*R158A157D15
	H*12*R168A167D16
	H*13*R17*12*A178D17
	H*13*R18*12*A188D18
	O*14*R197A19*3*D19
II	N9R208A*207*D20
	C20R219A218D21
	C2*1*R22*20*A229D22
	O22R23*21*A23*20*D23
	H2*1*R24*20*A249D24
	H20R259A258D25
	C2*1*R26*20*A269D26
II_{sc}	C26R27*21*A27*20*D27
	C27R28*26*A28*21*D28
	N28R29*27*A29*26*D29
	C29R30*28*A30*27*D30
	N27R31*26*A31*21*D31
	H26R32*21*A32*20*D32
	H26R33*21*A33*20*D33
	H28R34*27*A34*26*D34
	H29R35*28*A35*27*D35
	H*30*R36*29*A36*28*D36
III	N22R37*21*A37*20*D37
	C*37*R38*22*A38*21*D38
	C38R39*37*A39*22*D39
	O*39*R40*38*A40*37*D40
	H38R41*37*A41*22*D41
	H*37*R42*22*A42*21*D42
	C*38*R43*37*A43*22*D43
III_{sc}	C*43*R44*38*A44*37*D44
	C*44*R45*43*A45*38*D45
	N*45*R46*44*A46*43*D46
	C*46*R47*45*A47*44*D47
	C*44*R48*43*A48*38*D48
	C*48*R49*44*A49*43*D49
	C*49*R50*48*A50*44*D50
	C*50*R51*49*A51*48*D51
	C*47*R52*46*A52*45*D52
	H*43*R53*38*A53*37*D53
	H*43*R54*38*A54*37*D54
	H*45*R55*44*A55*43*D55
	H*46*R56*45*A56*44*D56
	H*49*R57*48*A57*44*D57
	H*50*R58*49*A58*48*D58
	H*51*R59*50*A59*49*D59
	H*52*R60*47*A60*46*D60
IV	N*39*R61*38*A61*37*D61
	C*61*R62*39*A62*38*D62
	C62R63*61*A63*39*D63
	O*63*R64*62*A64*61*D64
	H62R65*61*A65*39*D65
	H*61*R66*39*A66*38*D66
	C62R67*61*A67*39*D67
IV_{sc}	O67R68*62*A68*61*D68
	H67R69*62*A69*61*D69
	H67R70*62*A70*61*D70
	H68R71*67*A71*62*D71
V	N63R72*62*A72*61*D72
	C72R73*63*A73*62*D73
	C73R74*72*A74*63*D74
	O*74*R75*73*A75*72*D75
	H73R76*72*A76*63*D76
	H72R77*63*A77*62*D77
	C73R78*72*A78*63*D78
V_{sc}	C78R79*73*A79*72*D79
	C79R80*78*A80*73*D80
	C*80*R81*79*A81*78*D81
	C*81*R82*80*A82*79*D82
	C82R83*81*A83*80*D83
	C79R84*78*A84*73*D84
	H*78*R85*73*A85*72*D85
	H*78*R86*73*A86*72*D86
	H*80*R87*79*A87*78*D87
	H*81*R88*80*A88*79*D88
	O82R89*81*A89*80*D89
	H89R90*82*A90*81*D90
	H83R91*82*A91*81*D91
	H84R92*79*A92*78*D92
VI	N74R93*73*A93*72*D93
	C93R94*74*A94*73*D94
	C94R95*93*A95*74*D95
	O95R96*94*A96*93*D96
	H94R97*93*A97*74*D97
	H93R98*74*A98*73*D98
	C94R99*93*A99*74*D99

(continued on next page)

Table 21 (*continued*)

VI_sc	C99R10094A10093D100
	C100R10199A10194D101
	H101R102100A10299D102
	C100R10399A10394D103
	H103R104100A10499D104
	H99R10594A10593D105
	H99R10694A10693D106
	H100R10799A10794D107
	H101R108100A10899D108
	H101R109100A10999D109
	H103R110100A11099D110
	H103R111100A11199D111
VII	N95R11294A11293D112
	C112R11395A11394D113
	C113R114112A11495D114
	O114R115113A115112D115
	H113R116112A11695D116
	H112R11795A11794D117
	C113R118112A11895D118
VII_sc	C118R119113A119112D119
	C119R120118A120113D120
	H120R121119A121118D121
	C119R122118A122113D122
	H122R123119A123118D123
	H118R124113A124112D124
	H118R125113A125112D125
	H119R126118A126113D126
	H120R127119A127118D127
	H120R128119A128118D128
	H122R129119A129118D129
	H122R130119A130118D130
VIII	N114R131113A131112D131
	C131R132114A132113D132
	C132R133131A133114D133
	O133R134132A134131D134
	H132R135131A135114D135
	H131R136114A136113D136
	C132R137131A137114D137
VIII_sc	C137R138132A138131D138
	C138R139137A139132D139
	N139R140138A140137D140
	C140R141139A141138D141
	N141R142140A142139D142
	H142R143141A143140D143
	N141R144140A144139D144
	H144R145141A145140D145
	H137R146132A146131D146
	H137R147132A147131D147
	H138R148137A148132D148
	H138R149137A149132D149
	H139R150138A150137D150
	H139R151138A151137D151

Table 21 (*continued*)

	H140R152139A152138D152
	H142R153141A153140D153
	H144R154141A154140D154
IX	N133R155132A155131D155
	C155R156133A156132D156
	C156R157155A157133D157
	O157R158156A158155D158
	H156R159155A159133D159
	C156R160155A160133D160
IX_sc	C160R161156A161155D161
	C155R162133A162132D162
	H160R163156A163155D163
	H160R164156A164155D164
	H161R165160A165156D165
	H161R166160A166156D166
	H162R167155A167133D167
	H162R168155A168133D168
P_C	N157R169156A169155D169
	C169R170157A170156D170
	H170R171169A171157D171
	H169R172157A172156D172
	H170R173169A173157D173
	H170R174169A174157D174

ψ_i torsional angles obtained are quite similar irrespective of the level of theory at which the simulations were performed (RHF/3-21G, RHF/6-31G (d) and B3LYP/6-31G (d)). These results are quite significant as they indicate the reliability of a small basis set (i.e. RHF/3-21G) in giving a faithful representation of electron distribution that would be similarly found with a higher basis set (i.e. RHF/6-31G (d) or B3LYP/6-31G (d)), but with no cost of time and the additional computational power.

5. Conclusion

The paper has detailed a stepwise approach, with the aid of a standardized numbering system, to afford the user with an opportunity to efficiently combine pre-optimized monomers into di, tri, penta, and even larger oligopeptides. By understanding the similarities and differences of the amino acid modules, one may see how a standardized definition may be used in applications beyond linear chain peptides. As mentioned, with the advantage of creating these larger

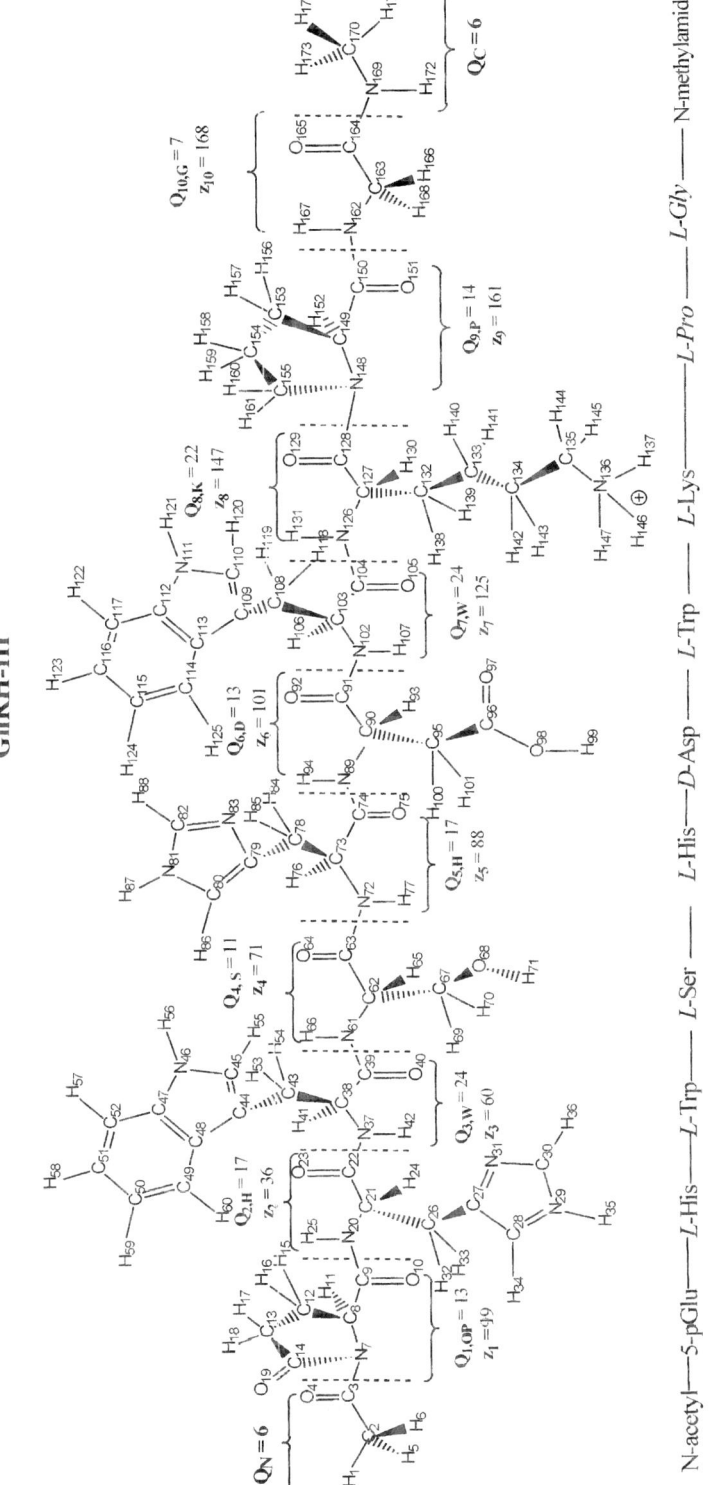

Fig. 16. A schematic illustration of the decapeptide GnRH-III, a hypothalamic neuroendocrine hormone in the sea lamprey.

Table 22

Symbolic internal coordinates for the decapapetide GnRH-III, a hypothalamic neuroendocrine hormone as shown in Fig. 16

P_N	
	H
	C*1* R2
	C2 R3*1* A3
	O3 R42 A4*1* D4
	H2 R53 A54 D5
	H2 R63 A64 D6
I	
	N3 R72 A7*1* D7
	C7 R83 A82 D8
	C8 R97 A93 D9
	O9 R108 A107 D10
	H*8* R117 A113 D11
	C8 R127 A12*3* D12
I_{sc}	
	C*12* R138 A137 D13
	C7 R143 A142 D14
	H*12* R158 A157 D15
	H*12* R168 A167 D16
	H*13* R171*2* A178 D17
	H*13* R181*2* A188 D18
	O*14* R197 A193 D19
II	
	N9 R208 A207 D20
	C20 R219 A218 D21
	C2*1* R2220 A229 D22
	O22 R232*1* A2320 D23
	H2*1* R2420 A249 D24
	H20 R259 A258 D25
	C2*1* R2620 A269 D26
II_{sc}	
	C26 R272*1* A2720 D27
	C27 R2826 A282*1* D28
	N28 R2927 A2926 D29
	C29 R3028 A3027 D30
	N27 R3126 A312*1* D31
	H26 R322*1* A3220 D32
	H26 R332*1* A3320 D33
	H28 R3427 A3426 D34
	H29 R3528 A3527 D35
	H*30* R3629 A3628 D36
III	
	N22 R372*1* A3720 D37
	C37 R3822 A382*1* D38
	C*38* R3937 A3922 D41
	O*39* R4038 A4037 D40
	H*38* R4137 A4122 D41
	H37 R4222 A422*1* D42
	C*38* R4337 A4322 D43
III_{sc}	
	C*43* R4438 A4437 D44
	C*44* R4543 A4538 D45
	N*45* R4644 A464*3* D46

Table 22 (*continued*)

	C*46* R4745 A4744 D47
	C*44* R4843 A4838 D48
	C*48* R4944 A4943 D49
	C*49* R5048 A5044 D50
	C*50* R5149 A5148 D51
	C*47* R5246 A5245 D52
	H*43* R5338 A5337 D53
	H*43* R5438 A5437 D54
	H*45* R5544 A5543 D55
	H*46* R5645 A5644 D56
	H*49* R5748 A5744 D57
	H*50* R5849 A5848 D58
	H*51* R5950 A5949 D59
	H*52* R6047 A6046 D60
IV	
	N*39* R6138 A6137 D61
	C*61* R6239 A6238 D62
	C62 R636*1* A6339 D63
	O*63* R6462 A646*1* D64
	H62 R656*1* A6539 D65
	H6*1* R6639 A663*8* D66
	C62 R676*1* A6739 D67
IV_{sc}	
	O67 R6862 A686*1* D68
	H67 R6962 A696*1* D69
	H67 R7062 A706*1* D70
	H*68* R7167 A7162 D71
V	
	N*63* R7262 A726*1* D72
	C72 R7363 A7362 D73
	C73 R7472 A7463 D74
	O74 R7573 A7572 D75
	H73 R7672 A7663 D76
	H72 R7763 A7762 D77
	C73 R7872 A7863 D78
V_{sc}	
	C78 R7973 A7972 D79
	C79 R8078 A8073 D80
	N*80* R8179 A8178 D81
	C8*1* R8280 A8279 D82
	N82 R838*1* A8380 D83
	H78 R8473 A8472 D84
	H78 R8573 A8572 D85
	H80 R8679 A8673 D86
	H8*1* R8780 A8779 D87
	H82 R888*1* A8880 D88
VI	
	N74 R8973 A8972 D89
	C89 R9074 A9073 D90
	C90 R9189 A9174 D91
	O*91* R9290 A9289 D92
	H90 R9389 A9374 D93
	H89 R9474 A9473 D94
	C90 R9589 A9574 D95

Table 22 (continued)

VI$_{sc}$	C95 R9690 A9689 D96
	O96 R9795 A9790 D97
	O96 R9895 A9890 D98
	H98 R9996 A9995 D99
	H95 R10090 A10089 D100
	H95 R10190 A10189 D101
VII	N91 R10290 A10289 D102
	C102 R10391 A10390 D103
	C103 R104102 A10491 D104
	O104 R105103 A105102 D105
	H103 R106102 A10691 D106
	H102 R10791 A10790 D107
	C103 R108102 A10891 D108
VII$_{sc}$	C108 R109103 A109102 D109
	C109 R110108 A110103 D110
	N110 R111109 A111108 D111
	C111 R112110 A112109 D112
	C112 R113111 A113110 D113
	C113 R114112 A114111 D114
	C114 R115113 A115112 D115
	C115 R116114 A116113 D116
	C116 R117115 A117114 D117
	H108 R118103 A118102 D118
	H108 R119103 A119102 D119
	H110 R120109 A120108 D120
	H111 R121110 A121109 D121
	H117 R122116 A122115 D122
	H116 R123115 A123114 D123
	H115 R124114 A124113 D124
	H114 R125113 A125112 D125
VIII	N104 R126103 A126102 D126
	C126 R127104 A127103 D127
	C127 R128126 A128104 D128
	O128 R129127 A129126 D129
	H127 R130126 A130104 D130
	H126 R131104 A131103 D131
	C127 R132126 A132104 D132
VIII$_{sc}$	C132 R133127 A133126 D133
	C133 R134132 A134127 D134
	C134 R135133 A135132 D135
	N135 R136134 A136133 D136
	H136 R137135 A137134 D137
	H132 R138127 A138126 D138
	H132 R139127 A139126 D139
	N133 R140132 A140127 D140
	H133 R141132 A141127 D141
	H134 R142133 A142132. D142
	H134 R143133 A143132 D143
	H135 R144134 A144133 D144
	H135 R145134 A145133 D145
	H136 R146135 A146134 D146
	H136 R147135 A147134 D147

Table 22 (continued)

IX	N128 R148127 A148126 D148
	C148 R149128 A149127 D149
	C149 R150148 A150128 D150
	O150 R151149 A151148 D151
	H149 R152140 A152128 D152
	H149 R153140 A153128 D153
IX$_{sc}$	C153 R154149 A154148 D154
	C154 R155153 A155149 D155
	H153 R156149 A156148 D156
	H153 R157149 A157148 D157
	H154 R158153 A158149 D158
	H154 R159153 A159149 D159
	H155 R160154 A160153 D160
	H155 R161154 A161153 D161
X	N150 R162149 A162148 D162
	C162 R163150 A163149 D163
	C163 R164162 A164150 D164
	O164 R165163 A165162 D165
	H163 R166162 A166150 D166
	H162 R167150 A167149 D167
	H163 R168162 A168150 D168
P_C	N164 R169163 A169162 D169
	C169 R170164 A170163 D170
	H170 R171169 A171164 D171
	H169 R172164 A172163 D172
	H170 R173169 A173164 D173
	H170 R174169 A174164 D174

peptide structures, one can now begin to model secondary and eventually tertiary structural elements of proteins more quickly and efficiently (in terms of computational resources), perhaps helping to further the understanding of the folding process in proteins.

Any number of combinations or procedures for numbering atomic nuclei can be applied or used. This manuscript seeks only to describe one that is functional, reproducible and automatable, as well as useful in the teaching and education of younger colleagues in the area of quantum chemical computations. Any number of modifications to this methodology is possible and one cannot yet weigh the advantages and disadvantages against the standing system, until the development of the fully functional form is completed.

Nearly 165 years have passed since Gerardus Johannes Mulder coined the term protein [54] and almost a century since Ira Ramsen, the co-discoverer of saccharin, expressed his concerns about peptide and

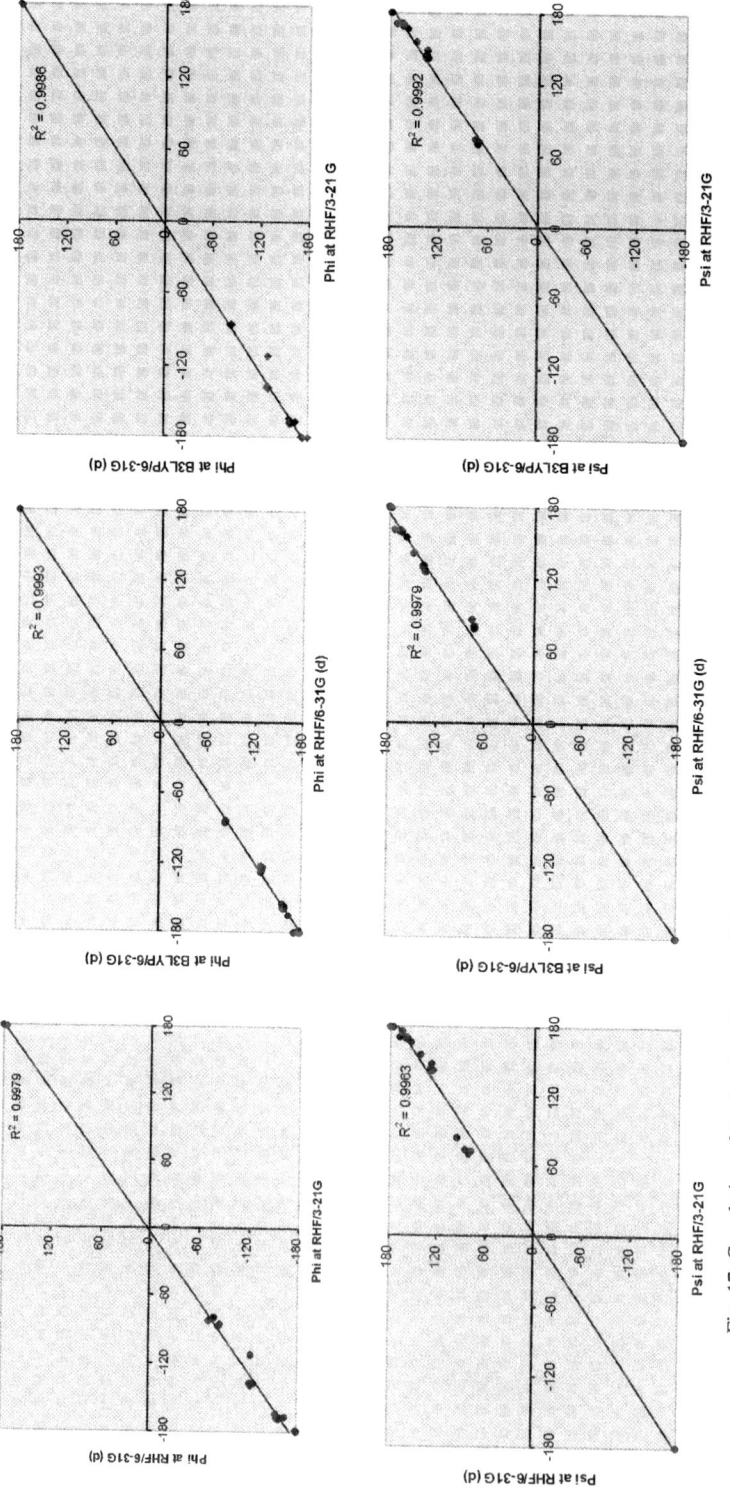

Fig. 17. Correlation of torsional angles (ϕ and ψ) for the oligo-peptides obtained from geometry optimizations at three levels of theory.

protein folding [55]. Even with the dawn of the 21st century, we are still far from fully understanding the folding process in proteins. It is hoped that the relatively flexible, open-ended and modular nature of the system described herein, will allow for future progress.

Acknowledgements

This work was supported by grants from the Global Institute of Computational Molecular and Materials Science (GIOCOMMS), Toronto, Ont., Canada.

One of the authors (IGC) wishes to thank the Ministry of Education of Hungary for a Szent-György Distinguished Visiting Professorship.

The authors thank David H. Setiadi and Tania A. Pecora for helpful discussions during the preparation of this manuscript as well as Graydon Hoare for database management, network support and software and distributive processing development. Special thanks is extended to Andrew M. Chasse, Christopher M. Andrews and Michael R. Sahai for their development of novel scripting and coding techniques as well as data management facilitating a reduction in the number of CPU cycles needed. The pioneering advances of Kenneth P. Chasse, in all composite computer-cluster software and hardware architectures, are also acknowledged.

Finally, GAC would like to extend his gratitude to Emeritus Professors Janos J. Ladik (University of Erlangen), Alex G. Harrison (University of Toronto) and Martin Karplus (Louis Pasteur University, Harvard University) for their continued mentorship and guidance.

The National Science Foundation (EPS-0091900), USA and NIH Grant Number 1 P20 RR16469 from the BRIN Program of the National Center for Research Resources are also acknowledged for allowing some final calculations to be completed.

References

[1] G.A. Chass, M.A. Sahai, J.M.S. Law, S. Lovas, Ö. Farkes, A. Perczel, J.-L. Rivail, I.G. Csizmadia, Int. J. Quantum Chem. 90 (2002) 933–968.

[2] G.A. Chass, A.M. Rodriguez, M.L. Mak, E. Deretey, A. Perczel, C.P. Sosa, R.D. Enriz, I.G. Csizmadia, J. Mol. Struct. (THEOCHEM) 537 (2001) 319–361.

[3] A. Perczel, J.G. Angyan, M. Kajtar, W. Viviani, J.-L. Rivail, J.F. Marcoccia, I.G. Csizmadia, J. Am. Chem. Soc. 113 (1991) 6256–6265.

[4] A. Perczel, W. Viviani, I.G. Csizmadia, Peptide conformational potential energy surfaces and their relevance to protein folding, in: J. Bertran (Ed.), Molecular Aspects of Biotechnology, Computational Models and Theories, Kluwer Academic Publishers, Dordrecht, 1992, pp. 39–82.

[5] W. Viviani, J.-L. Rivail, I.G. Csizmadia, Theor. Chim. Acta 85 (1992) 189–197.

[6] M.A. McAllister, A. Perczel, P. Császár, W. Viviani, J.-L. Rivail, I.G. Csizmadia, J. Mol. Struct. (THEOCHEM) 288 (1993) 161–180.

[7] A. Perczel, Ö. Farkas, I.G. Csizmadia, J. Am. Chem. Soc. 117 (1995) 1653–1654.

[8] H.A. Baldoni, G.N. Zamarbide, R.D. Enriz, E.A. Jauregui, Ö. Farkas, A. Perczel, S.J. Salpietro, I.G. Csizmadia, J. Mol. Struct. (THEOCHEM) 500 (2000) 97–112. (Millennium volume).

[9] A. Perczel, M. Kajtar, J.F. Marcoccia, I.G. Csizmadia, J. Mol. Struct. (THEOCHEM) 232 (1991) 291–319.

[10] G. Endrédi, A. Perczel, Ö. Farkas, M.A. McAllister, G.I. Csonka, J. Ladik, I.G. Csizmadia, J. Mol. Struct. (THEOCHEM) 391 (1997) 15–20.

[11] A.M. Rodriguez, H.A. Baldoni, F. Suvire, R. Nieto-Vasquez, G. Zamarbide, R.D. Enriz, Ö. Farkas, A. Perczel, I.G. Csizmadia, J. Mol. Struct. (THEOCHEM) 455 (1998) 275–302.

[12] W. Viviani, J.-L. Rivail, A. Perczel, I.G. Csizmadia, J. Am. Chem. Soc. 115 (1993) 8321–8329.

[13] A. Perczel, R. Daudel, J.G. Angyan, I.G. Csizmadia, Can. J. Chem. 68 (1990) 1882–1888.

[14] Ö. Farkas, A. Perczel, J.F. Marcoccia, M. Hollósi, I.G. Csizmadia, J. Mol. Struct. (THEOCHEM) 331 (1995) 27–36.

[15] A. Perczel, Ö. Farkas, I.G. Csizmadia, J. Comp. Chem. 17 (1996) 821–834.

[16] A. Perczel, Ö. Farkas, I.G. Csizmadia, J. Am. Chem. Soc. 118 (1996) 7809–7817.

[17] A. Perczel, Ö. Farkas, J.F. Marcoccia, I.G. Csizmadia, Int. J. Quantum Chem. 61 (1997) 797–814.

[18] A. Perczel, Ö. Farkas, I.G. Csizmadia, J. Mol. Struct. (THEOCHEM) 455 (1998) 315.

[19] I. Jakli, A. Perczel, Ö. Farkas, C.P. Sosa, I.G. Csizmadia, J. Comp. Chem. 21 (2000) 626–655.

[20] Ö. Farkas, M.A. McAllister, J.H. Ma, A. Perczel, M. Hollósi, I.G. Csizmadia, J. Mol. Struct. (THEOCHEM) 369 (1996) 105–114.

[21] A. Perczel, Ö. Farkas, I.G. Csizmadia, Can. J. Chem. 75 (1997) 1120–1130.

[22] I. Jakli, A. Perczel, Ö. Farkas, M. Hollosi, I.G. Csizmadia, J. Mol. Struct. (THEOCHEM) 455 (1998) 303–314.

[23] G.A. Chass, S. Lovas, R.F. Murphy, I.G. Csizmadia, Eur. Phys. J. D 20 (2002) 481–497.

[24] M.A. Sahai, S.S. Motiwala, G.A. Chass, E.F. Pai, B. Penke, I.G. Csizmadia, J. Mol. Struct. (THEOCHEM) (2003) in press.

[25] H.A. Baldoni, A.M. Rodriguez, G. Zamarbide, R.D. Enriz, Ö. Farkas, P. Csaszar, L.L. Torday, C.P. Sosa, I. Jakli, A. Perczel,

M. Hollosi, I.G. Csizmadia, J. Mol. Struct. (THEOCHEM) 465 (1999) 79–91.

[26] S.J. Salpietro, A. Perczel, Ö. Farkas, R.D. Enriz, I.G. Csizmadia, J. Mol. Struct. (THEOCHEM) 497 (2000) 39–63.

[27] M. Berg, S.J. Salpietro, A. Perczel, Ö. Farkas, I.G. Csizmadia, J. Mol. Struct. (THEOCHEM) 504 (2000) 127–140.

[28] M.A. Zamora, H.A. Baldoni, J.A. Bombasaro, M.L. Mak, A. Perczel, Ö. Farkas, R.D. Enriz, J. Mol. Struct. (THEOCHEM) 540 (2001) 271–283.

[29] M.A. Zamora, H.A. Baldoni, A.M. Rodriguez, A. Kucsman, R.D. Enriz, C.P. Sosa, A. Perczel, Ö. Farkas, E. Deretey, J.C. Vank, I.G. Csizmadia, Can. J. Chem. 80 (2002) 832–844.

[30] A.R. Sheraly, R.V. Chang, G.A. Chass, J. Mol. Struct. (THEOCHEM) 619 (2002) 21–35.

[31] J.C. Vank, C.P. Sosa, A. Perczel, I.G. Csizmadia, Can. J. Chem. 78 (2000) 395–408.

[32] A. Perczel, I.G. Csizmadia, J. Mol. Struct. (THEOCHEM) 286 (1993) 75–85.

[33] C. Van Alsenoy, M. Cao, S. Newton, B. Teppen, A. Perczel, I.G. Csizmadia, F.A. Monramy, L. Schäfer, J. Mol. Struct. (THEOCHEM) 286 (1993) 149–163.

[34] M.A. McAllister, A. Perczel, P. Császár, I.G. Csizmadia, J. Mol. Struct. (THEOCHEM) 288 (1993) 181–198.

[35] A. Perczel, M.A. McAllister, P. Császár, I.G. Csizmadia, Can. J. Chem. 72 (1994) 2050–2070.

[36] M. Cheung, M.E. McGovern, T. Jin, D.C. Zhao, M.A. McAllister, A. Perczel, P. Császár, I.G. Csizmadia, J. Mol. Struct. (THEOCHEM) 309 (1994) 151–224.

[37] G. Endrédi, C.-M. Liegner, M.A. McAllister, A. Perczel, J. Ladik, I.G. Csizmadia, J. Mol. Struct. (THEOCHEM) 306 (1994) 1–7.

[38] G. Endrédi, M.A. McAllister, A. Perczel, P. Császár, J. Ladik, I.G. Csizmadia, J. Mol. Struct. (THEOCHEM) 331 (1995) 5–10.

[39] M.A. McAllister, G. Endrédi, W. Viviani, A. Perczel, P. Császár, J. Ladik, J.-L. Rivail, I.G. Csizmadia, Can. J. Chem. 73 (1995) 1563–1572.

[40] G. Endrédi, M.A. McAllister, Ö. Farkas, A. Perczel, J. Ladik, I.G. Csizmadia, J. Mol. Struct. (THEOCHEM) 331 (1995) 11–26.

[41] I.A. Topol, S.K. Burt, E. Deratey, T.H. Tang, A. Perczel, A. Rashin, I.G. Csizmadia, J. Am. Chem. Soc. 123 (2001) 6054–6060.

[42] M.A. Berg, G.A. Chasse, E. Deretey, A.K. Füzéry, B.M. Fung, D.Y.K. Fung, H. Henry-Riyad, A.C. Lin, M.L. Mak, A. Mantas, M. Patel, I.V. Repyakh, M. Staikova, S.J. Salpietro, T. Tang, J.C. Vank, A. Perczel, Ö. Farkas, L.L. Torday, Z.

Székely, I.G. Csizmadia, J. Mol. Struct. (THEOCHEM) 500 (2000) 5–58. (Millennium volume).

[43] M.J. Frisch, G.W. Trucks, H.B. Schlegel, G.E. Scuseria, M.A. Robb, J.R. Cheeseman, V.G. Zakrzewski, J.A. Montgomery Jr., R.E. Stratmann, J.C. Burant, S. Dapprich, J.M. Millam, A.D. Daniels, K.N. Kudin, M.C. Strain, Ö. Farkas, J. Tomasi, V. Barone, M. Cossi, R. Cammi, B. Mennucci, C. Pomelli, C. Adamo, S. Clifford, J. Ochterski, G.A. Petersson, P.Y. Ayala, Q. Cui, K. Morokuma, P. Salvador, J.J. Dannenberg, D.K. Malick, A.D. Rabuck, K. Raghavachari, J.B. Foresman, J. Cioslowski, J.V. Ortiz, A.G. Baboul, B.B. Stefanov, G. Liu, A. Liashenko, P. Piskorz, I. Komaromi, R. Gomperts, R.L. Martin, D.J. Fox, T. Keith, M.A. Al-Laham, C.Y. Peng, A. Nanayakkara, M. Challacombe, P.M.W. Gill, B. Johnson, W. Chen, M.W. Wong, J.L. Andres, C. Gonzalez, M. Head-Gordon, E.S. Replogle, J.A. Pople, GAUSSIAN 98, Revision A.9, Gaussian, Inc., Pittsburgh PA, 2001.

[44] A. Mehdizadeh, G.A. Chass, Ö. Farkas, A. Perczel, L.L. Torday, A. Varro, J.G. Papp, J. Mol. Struct. (THEOCHEM) 588 (2002) 187–200.

[45] J.C.C. Liao, J.C. Chua, G.A. Chass, A. Perczel, L.L. Torday, J.G. Papp, J. Mol. Struct. (THEOCHEM) 621 (2003) 163–187.

[46] M.A. Sahai, M.R. Sahai, G.A. Chass, B. Penbe, I.G. Csizmadia. J. Mol. Struct. (THEOCHEM).

[47] A. Perczel, M.A. McAllister, P. Csaszar, I.G. Csizmadia, J. Am. Chem. Soc. 115 (1993) 4849–4858.

[48] C.F. Matta, J. Phys. Chem. A 105 (2001) 11088–11101.

[49] G.K. Toth, K. Bakos, B. Penke, I. Pavo, C. Varga, G. Torok, A. Peter, F. Fulop, Bioorg. Med. Chem. Lett. 9 (1999) 667–672.

[50] F. Laczi, M. Vecsernyes, G.L. Kovacs, B. Penke, I. Jojart, T. Janaky, G. Telegdy, A. Laszlo, Exp. Clin. Endocrinol. 92 (1988) 328–334.

[51] D.S. Gembitsky, M. Murnin, F. Ötvös, J. Allen, R.F. Murphy, S. Lovas, J. Med. Chem. 42 (1999) 3004–3007.

[52] S.C. Hall, M.M. Tan, J.J. Leonard, C.L. Stevenson, J. Pept. Res. 53 (1999) 432–441.

[53] C.R. Watts, M. Mezei, R.F. Murphy, S. Lovas, J. Biomol. Struct. Dyn. 18 (2001) 733–748.

[54] G.J. Mulder, Journal für praktische Chemie 16 (1839) 129 [as translated and excerpted in Mikulás Teich, A documentary history of biochemistry, Fairleigh Dickinson University Press, Rutherford, NJ, 1992, pp. 1770–1940].

[55] I. Remsen, Unsolved problems in chemistry, in Modern inventions and discoveries, J.A. Hill and Co, New York, 1904, Also quoted in H. Neurath, Why protein science? Protein Science 1 (1) (1992) 2.

THEO
CHEM

Journal of Molecular Structure (Theochem) 666–667 (2003) 219–241

www.elsevier.com/locate/theochem

A conformational comparison of N- and C-protected methionine and N- and C-protected homocysteine

András Láng[a], Krisztina György[a], Imre G. Csizmadia[a,b,c,d], András Perczel[a,*]

[a]*Department of Organic Chemistry, Eötvös Loránd University, Pázmány Péter sétány 1/A, H-1117 Budapest, Hungary*
[b]*Department of Medical Chemistry, Szent-Györgyi Medical and Pharmaceutical Center, University of Szeged, Dóm tér 8, H-6720 Szeged, Hungary*
[c]*Department of Chemistry, University of Toronto, 80st George Str., Toronto, Ont., Canada M5S 3HG*
[d]*Global Institute of COmputational Molecular and Materials Science (GIOCOMMS) at 1422 Edenrose St., Mississauga, Ont., Canada L5V 1H3*

Abstract

The conformational space for the five L type backbone forms of N-formyl-L-methioninamide has been explored by ab initio computation. Initially, we expected 135 conformers associated with the five L type backbone forms (right-handed α-helix or α_L, β-pleated sheet or β_L, inverse γ-turn or γ_L, δ_L and poly-Proline II or ε_L). Using computation of RHF/3-21G level of theory, we describe 77 conformers by their five independent variables: φ, ψ, χ_1, χ_2, χ_3. The β_L conformers of N-formyl-L-methioninamide were compared with N-acetyl-L-homocysteine-methylamide and the different backbone forms of N-formyl-L-methioninamide were also compared. Calculation of stabilization energies of the conformers was completed.
© 2003 Elsevier B.V. All rights reserved.

Keywords: Multidimensional conformation analysis (MDCA); Amino acid; Peptide model; Ab initio computations; Molecular conformation; Ramachandran potential energy surface (PES)

Abbreviations: MTHFR, methylenetetrahydrofolate reductase; SAH, S-adenosylhomocysteine; SAM, S-adenosylmethionine; cAMP, cyclic adenosine monophosphate; GRP, glucose regulated protein; NOS, nitric oxide synthase; cGMP, cyclic guanosine monophosphate; HUVEC, human umbilical vascular endothelial cell; BiP, immunoglobulin heavy chain-binding protein; ER, endoplasmic reticulum; CHOP/GADD153, C/EBP (CCAAT/enhancer-binding protein) homologous protein or growth arrest- and DNA damage-inducible gene; HERP, HES (hairy and enhancer of split)-related repressor protein; UPR, unfolded protein response; HepG2, human hepatocarcinoma cell; HASMC, human aortic smooth muscle cells; SREBP-1, sterol regulatory element-binding protein-1; HDL, high density lipoprotein; TCF/LEF, T cell factor/lymphoid enhancer factor; APC, adenomatous polyposis of the colon; Wnt, Wingless-type murine mammary tumour virus integration site; PDB, Protein Data Bank; PEC/PEHS, potential energy curve/potential energy hypersurface; MDCA, multidimensional conformation analysis; Met, methionine; Hcy, homocysteine.

* Corresponding author. Tel.: +36-1-2090-555; fax: +36-1-3722-620.
E-mail address: perczel@para.chem.elte.hu (A. Perczel), http://www.szerves.chem.elte.hu.

1. Introduction

From the three naturally occurring, sulphur atom-containing amino acids, cysteine and methionine are those that can be incorporated into proteins during the process of biological translation.

The third, homocysteine, is the common precursor for the synthesis of methionine and cysteine. The methionine pathway requires a methylation of homocysteine and the path toward the cysteine should possess a two-step *trans*-sulphurational mechanism (Scheme 1).

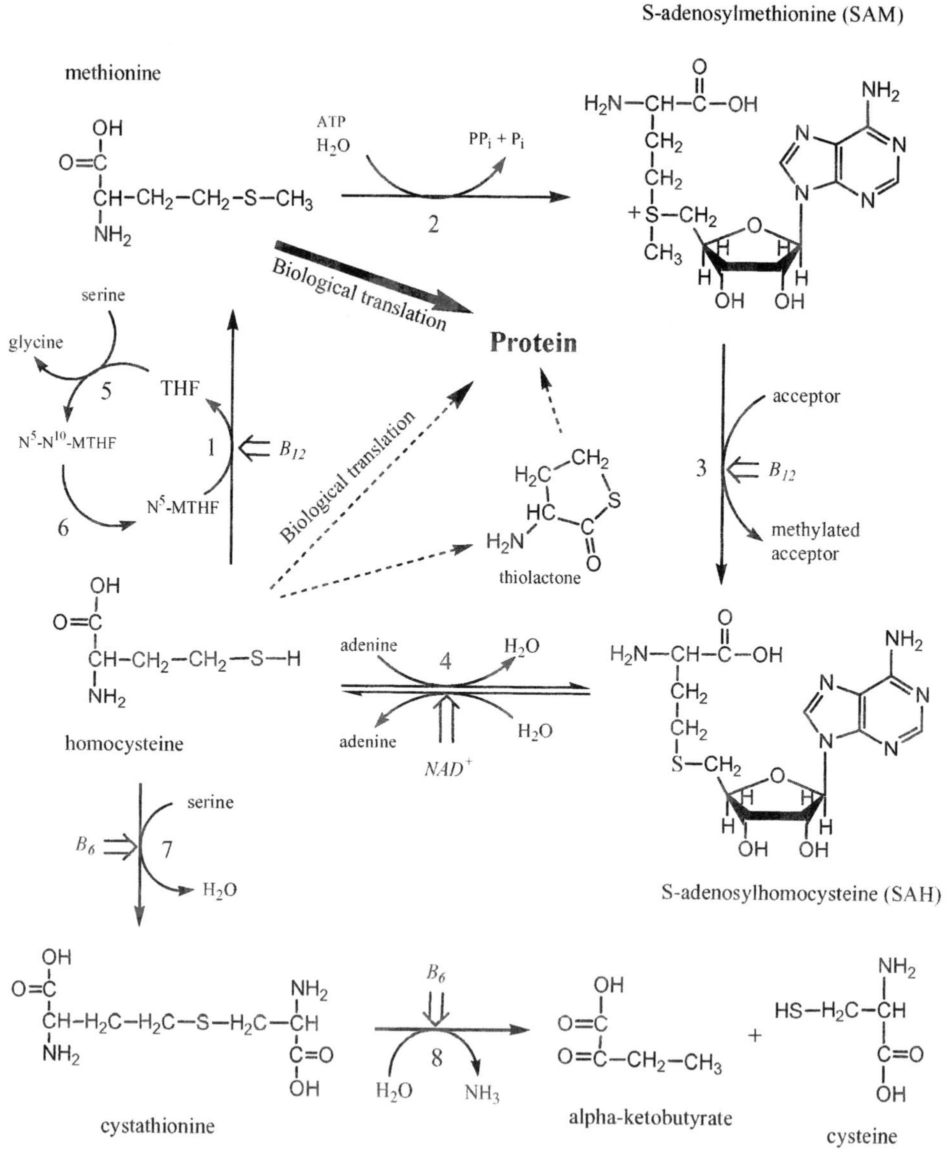

Scheme 1. Interconversion of the sulphur-containing amino acids: the methionine cycle and the *trans*-sulphuration pathway. Numbers stand for the following enzymes: (1) methionine synthase, (2) *S*-adenosylmethionine synthetase, (3) methyltransferases, (4) *S*-adenosylhomocysteinase, (5) serine hydroxymethylase, (6) methylenetetrahydrofolate-reductase, (7) cystathionine-beta-synthase, (8) cystathionine-gamma-lyase. The following abbreviations were introduced: THF, tetrahydrofolate; N^5–N^{10}-MTHF, 5,10-methylenetetrahydrofolate; N^5–MTHF, 5-methyltetrahydrofolate. NAD^+ shows the activation on *S*-adenosylhomocysteinase. Coenzyme B_6 and B_{12} are also indicated.

For the first reaction, it is necessary to have two functional enzymes, namely methylenetetrahydrofolate reductase (MTHFR) that indirectly contributes to the methionine formation, and 5-methyltetrahydrofolate-homocysteine S-methyltransferase, also known as methionine synthase. MTHFR provides the methyl group for the conversion of homocysteine catalyzed by the second enzyme. For methylation, cells also require vitamin B_{12}. This step of the cycle affects the homocysteine/methionine ratio in the cell. In general, the methionine cycle itself does not alter the ratio except when protein synthesis requires methionine.

The other metabolic pathway is via the B_6-vitamin-dependent cystathionine-β-synthase followed by a γ-lysis of the intermediate cystathione. The enzyme joins together a serine and a homocysteine via a thioether link. The second step of the pathway produces cysteine and α-ketobutyrate. This pathway decreases the level of homocysteine, as the latter leaves the methionine cycle.

One enzyme in the methionine cycle, namely the S-adenosylhomocysteinase, that converts S-adenosylhomocysteine (SAH) to homocysteine, was studied intensively and characterized previously [1]. It was described that the enzyme is tetramer and contains nine cysteine residues per subunits, however, does not contain a disulfide bond at all. The tetrameric enzyme binds four NAD^+s. The rate of adenosine binding and enzyme sensitivity to inactivation by adenosine was diminished in the absence of dithiothreitol. Previously, the enzyme was studied in Dictyostelium discoideum and described 25% activation in the enzyme activity by preincubation of SAH hydrolase with NAD^+. Furthermore, these properties of the enzyme were also found in rabbit erythrocytes, the first model for mammalian cells [2]. These results imply that shortage of the reduction power of the cells (low $NADH + H^+$) facilitates the production of homocysteine. This may happen when the glucolysis is stopped at the beginning, as there is no glucose supplement of the cells (e.g. insulin defect), thus there is no motive power of the citrate cycle that produces energy and reduction power. It was described before that glucose plays a role in the expression of many genes. These proteins named as GRPs (glucose regulated proteins). This leads to the increased number of homocysteine and methionine, as the methionine cycle stuck between two reactions: SAH hydrolysis and S-adenosylmethionine synthesis. The ratio of the two compounds is set by methionine synthase, therefore the ammount of 5-methyltetrahydrofolate and B_{12}-vitamin.

Another enzyme that reacts on a free methionine is the methionyl-tRNA synthetase, which binds together a methionine and a corresponding tRNA for protein translation. The reaction mechanism is quite a complicated process involving many amino acids of the enzyme suggested by structural study [3]. The most important side-chains playing role in methionine binding are Leu13, Asp52, Tyr260 and His24 through a water molecule that orient the methionine, while in the structure of the free enzyme the ligand is exchanged by a water molecule. The binding of methionine has an exclusive effect on the enzyme structure, especially on the position of the side-chains of aromatic amino acids, namely Tyr15, Trp253, Phe300, Trp229, Phe304 and Tyr251. This methionine binding favors a conformational rearrangement of these residues compiling a hydrophobic pocket to accommodate the apolar methionine side-chain. The observed methionine conformation regarding the side-chain torsion angles are in the crystal structure: $\chi_1 = -61.51°$, $\chi_2 = -174.64°$, $\chi_3 = 175.9°$. This conformation is $[g^- aa]$ for the side-chain, thus in this structure the sulphur atom is sufficiently separated from the terminal amino and carboxyl groups. It is important that the carboxylic acid moiety should not be in interaction with the sulphur, therefore the carboxyl group can easily react with an ATP molecule forming a high energy linkage.

It is well documented that elevated levels of plasma and intracellular homocysteine caused by enzymatic failures of the two pathways of methionine cycle (Scheme 1) or lack of B_6-B_{12}-vitamin or folic acid, connected to many pathological processes [4,5]. Among others, these are mental retardation, hepatic steatosis and cardiovascular disease. The latter disease is caused by the dysfunction of vascular endothelial cells that contained in the blood vessels, making one of the highest mortality and morbidity rate in the developed countries. This atherosclerotic mechanism of homocysteine is thought to be independent of the mechanism caused by cholesterol [6].

Homocysteine normally can not incorporate into proteins via translation because of the proofreading mechanism of the amino-acyl-tRNA-synthetases.

However, homocysteine derivative can be accumulated in proteins instead of methionine. This derivative is the S-nitroso-homocysteine, that have been proven to incorporate into proteins in E. coli and in rabbit reticulocyte system, recognizing the methionyl-tRNA-synthetase as a methionine, thus translates the methionine codon into S-nitroso-homocysteine [7]. It is known that nitric oxide a highly reactive radical that contribute to vasodilatation is produced in vivo by nitric oxide synthase (NOS). There are three isoforms of this enzyme, namely NOS1, NOS2 and NOS3. The isoforms are expressed in distinct cell types and environments. One of these, NOS3, is expressed constitutively in endothelial cells, thus during hyper-homocysteinaemia S-nitroso-homocysteine may be produced and incorporated into proteins. One experiment shows that not only NO radical is effective on vascular smooth muscle cells but also S-nitroso-cysteine-derivatives of proteins that contain free thiol groups [8]. They suggested that carrier proteins should stabilize and therefore carry the NO radicals from NOS to the active site of guanylate cyclase to produce cGMP. It is known that cGMP is a signaling molecule toward vasodilatation resulting vascular smooth muscle cell relaxation and hypotension.

Moreover, the proofreading mechanism of methionyl-tRNA-synthetase (also valyl-tRNA-synthetase and isoleucyl-tRNA-synthetase) produces thiolactone from homocysteine in E. coli, S. cerevisiae and mammalian cells, thus it can react on the lysin side-chain of proteins, as a posttranslational modification [9–11]. Formyl-methionine and methionine is important in an essential process, starting the translation in prokaryotic and eukaryotic cells toward a properly folded protein. Therefore, it is important to be sure that the first amino acid in a newly synthesized protein is a methionine and not an S-nitroso-homo-cysteine.

In recent years, many experiments focused on the cellular effect of homocysteine. A simpler way to examine the cardiovascular disease affected by vascular endothelial cell dysfunction is examining the vascular endothelial cells in culture (human umbilical vascular endothelial cells (HUVEC)). For this purpose, homocysteine were given to HUVEC and displayed for both the alteration in the transcriptional activity and the differences at the level of translation. The detected genes acted on the endoplasmic reticulum-stress induced response. These are BiP/GRP78 gene, whose product is an endoplasmic reticulum (ER) chaperon known to define the quaternary structure of multimers [12], CHOP/GADD153 for transcriptional activation during ER-stress [13,14] and HERP, whose product is functionally unknown [15]. This unfolded protein response (UPR) is due to the inconvenient ER function. Homocysteine-induced ER-stress altered the biosynthetic pathways of cholesterol and trigly-ceride in HUVEC, human hepatocarcinoma cell line (HepG2) and human aortic smooth muscle cells (HASMC) showing that homocysteine leads to sterol regulatory element-binding protein-1 (SREBP-1) transcriptional activation, therefore contributing to the access of new membrane components [16]. Further, elevated level of homocysteine supplemented in HUVEC medium caused production of thiolactone and post-translational protein homocysteinylation, especially in extracellular proteins those are known to be translated in the endoplasmic reticulum. These processes were affected by homocysteine/methionine ratio and the concentration of high density lipoprotein (HDL) [17].

One explanation of the bulk amount of protein in ER after treatment of HUVEC by homocysteine may be that the elevated level of homocysteine in the cell causes thiolactone formation (Scheme 1) by the proofreading mechanism of many of the amino-acyl-tRNA-synthetases [9–11]. The produced thiolactone can react on the side-chain of the ε amino-group of lysine side-chains of different proteins. Thiolactone has a five-membered ring, however, cysteine if this had a cyclization at all, the produced four-membered ring having a rigid thioamide bond would not be very stable. Once, if an ε amino-group of lysine was modified by producing an isopeptide bond, the protein does not possess any free lysine residue for ubiqui-tylation, thus the protein will not go through a degradation procedure by ubiquitin and thus accumulates in the cell. This is also the case for the non-properly translated polypeptides that remains in the ER lumen. Moreover, it has been described that ubiquitin plays a role in the Wnt signaling (Wnt was named after Wingless-type murine mammary tumour virus integration site, see Ref. [18]), thus regulates the transcriptional activities of many genes. One protein, β-catenin, is known to have important role in this signaling mechanism. The 781 amino acid-long human protein is located mainly in the cytosol but also can be found in the nucleus. β-catenin is known to be inhibited for ubiquitylation and phosphorylation during the Wnt signaling, therefore can enter the nucleus, where it dimerizes with members of TCF/LEF (T cell factor/lymphoid enhancer factor) [19]. Also, the change in the lysine side-chain can affect the dimerization.

Indeed, two structure determinations were done for β-catenin. One, bounded to the intracellular domain of E-cadherin (PDB: 1I7X and 1I7W) [20], the other, bounded to TCF (PDB: 1G3J) [21]. They have suggested and observed two lysine residues in β-catenin that were essential in binding to the partners. β-catenin is also known to interact with other structurally similar proteins, like axin, APC (adenomatous polyposis of the colon). Thus, alteration in β-catenin at lysine residues affects the association rate to these proteins and ruins the degradation by ubiquitin.

These modifications and the supposed incorporation of S-nitroso-homocysteine into proteins, even at the N-terminal end of a protein, make it plausible to suggest that this may ruin the translation or, if the translational mechanism is not effected, disrupt the folded conformation of the newly synthesized protein.

The structures of the other two sulphur-containing amino acids, namely cysteine [22] and homocysteine [23], have been examined at RHF/3-21G level of theory by computational chemistry. In the present study, we aimed to explore the N-formyl-L-methioninamide conformational behaviour in the L-region of the backbone Ramachandran Map, by computational methods and compare the obtained data in β_L backbone conformation with β_L conformers of N- and C-protected homocysteine reported by Sheraly et al.

2. Methods

2.1. Conformational analysis

The five torsion angles (φ, ψ, χ_1, χ_2, χ_3) were specified by using the IUPAC-IUB recommendation [24] (Scheme 2). Torsion angles, traditionally, have been reported in the IUPAC-IUB format i.e. ranging from $-180°$ to $+180°$. In 1991, however, it has been proposed that the format reported by Ramachandran et al. [25] (from $0°$ to $360°$) to be used for graphical representation which we will use through [26]. The five observed backbone conformations studied are: α_L (both φ and ψ are gauche $^-$), β_L (both torsional angles are anti), in γ_L-type backbone fold φ is gauche $^-$ and ψ is gauche $^+$, in δ_L, where φ is anti and ψ is gauche $^+$ and finally in ε_L φ is gauche $^-$ but ψ is anti (Scheme 2).

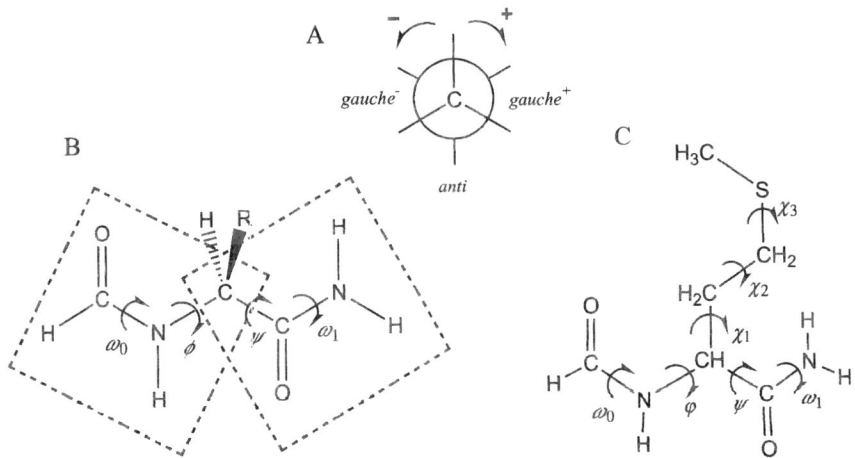

Scheme 2. (A) Conventional representation of a torsion angle illustrating the directions of rotation. (B) Shematic representation of the amide planes in an amino acid emphasizing the limited rotation of ω_0, ω torsional angles. (C) The structural model of methionine indicating all seven examined torsional angles.

2.2. Molecular computations

All ab initio computations were performed on N-formyl-L-methioninamide by the program GAUSSIAN 98 [27] at RHF/3-21G level of theory. Total energies of conformers are given in hartrees, relative and stabilization energies in kcal mol^{-1}. Registration of the potential energy curves (PECs) were carried out through optimized structures obtained by rotating the C–C–S–C torsion angle increased from 0° up to 360° by 15 degrees (reported in Special Issue).

2.3. Stabilization energies

Stabilization energies were calculated at RHF/3-21G level of theory using the isodesmic reaction presented in Scheme 3. The reactant, N-formyl-glycinamide in its β_L conformation. Propyl-methyl-thioether was always in its global minimum corresponded to the $[ag^+]$ or $[ag^-]$ conformation.

3. Results and discussion

3.1. N-formyl-L-methioninamide conformers at RHF/3-21G level of theory

Analyzing structural properties of methionine in peptides and proteins minima of the $E = E(\omega_0, \varphi, \psi, \omega_1, \chi_1, \chi_2, \chi_3, \chi_4)$ hypersurfaces are to be determined. According to multidimensional

conformational analysis (MDCA) each of φ, ψ, χ_1, χ_2, χ_3, χ_4 torsional angles has three expected minima (g^+, a, g^-) and both of the amide bonds (ω_0, ω_1) have two conformations (cis and trans). Thus, the variation of all minima of these conformational variables is $2 \times 3 \times 3 \times 2 \times 3 \times 3 \times 3 \times 3 = 2^2 \times 3^6$. The following restriction or structural features of N-formyl-L-methioninamide peptide models are to be considered.

Both amide bonds of this diamide model are considered to adopt only trans orientation ($\omega_0 \approx \omega_1 \approx \pm 180°$) Scheme 2(B). Therefore, all trans subspace is reduced to 3^6. The terminal torsion angle of the side-chain (χ_4) has three folds potential, but the minima are energetically degenerate, thus they can be considered to be the same. Therefore, the full conformational space has five torsional angles as independent variables, we expect a total of $3^5 = 243$ conformers. The 243 conformers could be separated into nine (3^2) groups according to their backbone conformation (φ and ψ). Our particular interest was to explore the L backbone conformers since they tend to be of lower energy and have more important biological implications. Since there are five of these, we expect $5 \times 3^3 = 135$ stable conformers. The input geometries of the expected conformers were based on the backbone conformations of glycine β_L, γ_L and δ_L [26] and for α_L, ε_L, and for the side-chain torsional angles for g^+, g^- and a —60° ± 30°, +60° ± 30° and 180° ± 30°. After full optimization at the RHF/3-21G level of theory, a total of 77 different conformers

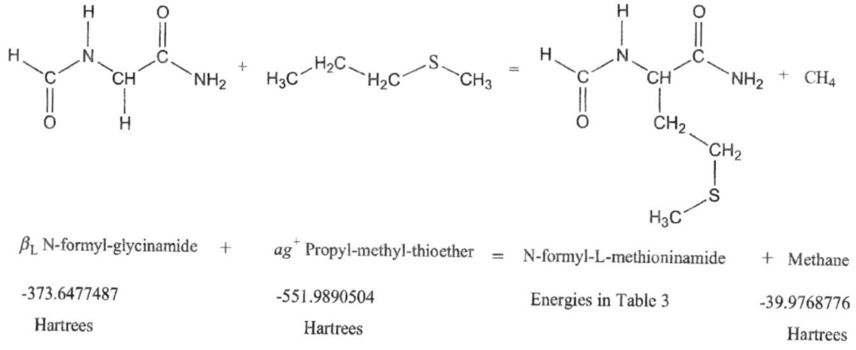

β_L N-formyl-glycinamide	+	ag^+ Propyl-methyl-thioether	=	N-formyl-L-methioninamide	+	Methane
-373.6477487		-551.9890504		Energies in Table 3		-39.9768776
Hartrees		Hartrees				Hartrees

$$\Delta E_{stabilization} = [E_{(N\text{-formyl-L-methioninamide})} + E_{Methane}] - [E_{(N\text{-formyl-glycinamide})} + E_{(Propyl\text{-methyl-thioether})}]$$

Scheme 3. Isodesmic reaction used to calculate stabilization energies while N-formyl-glycinamide was in its β_L conformation, propyl-methyl-thioether was in the conformation corresponding to the side-chain structure of N-formyl-methioninamide. Stabilization energies are presented in Table 1. All the stabilization energy values are given in kcal mol^{-1}.

Table 1
Summary of every conformers of *N*-formyl-L-methioninamide predicted by MDCA and localized after computation at RHF/3-21G level of theory

Final geometry	Final geometry	Final geometry	Final geometry	Final geometry	Final geometry	Final geometry	Final geometry	Final geometry	Final geometry
$\alpha_L(g^+,g^+,g^+)$	Not found	$\beta_L(g^+,g^+,g^+)$	Found	$\gamma_L(g^+,g^+,g^+)$	Found	$\delta_L(g^+,g^+,g^+)$	Found	$\varepsilon_L(g^+,g^+,g^+)$	Not found
$\alpha_L(g^+,g^+,a)$	Not found	$\beta_L(g^+,g^+,a)$	Found	$\gamma_L(g^+,g^+,a)$	Found	$\delta_L(g^+,g^+,a)$	Found	$\varepsilon_L(g^+,g^+,a)$	Not found
$\alpha_L(g^+,g^+,g^-)$	Not found	$\beta_L(g^+,g^+,g^-)$	Not Found	$\gamma_L(g^+,g^+,g^-)$	Not found	$\delta_L(g^+,g^+,g^-)$	Not found	$\varepsilon_L(g^+,g^+,g^-)$	Not found
$\alpha_L(g^+,a,g^+)$	Not found	$\beta_L(g^+,a,g^+)$	Found	$\gamma_L(g^+,a,g^+)$	Found	$\delta_L(g^+,a,g^+)$	Found	$\varepsilon_L(g^+,a,g^+)$	Not found
$\alpha_L(g^+,a,a)$	Not found	$\beta_L(g^+,a,a)$	Found	$\gamma_L(g^+,a,a)$	Found	$\delta_L(g^+,a,a)$	Found	$\varepsilon_L(g^+,a,a)$	Not found
$\alpha_L(g^+,a,g^-)$	Not found	$\beta_L(g^+,a,g^-)$	Found	$\gamma_L(g^+,a,g^-)$	Found	$\delta_L(g^+,a,g^-)$	Found	$\varepsilon_L(g^+,a,g^-)$	Not found
$\alpha_L(g^+,g^-,g^+)$	Not found	$\beta_L(g^+,g^-,g^+)$	Found	$\gamma_L(g^+,g^-,g^+)$	Found	$\delta_L(g^+,g^-,g^+)$	Found	$\varepsilon_L(g^+,g^-,g^+)$	Found
$\alpha_L(g^+,g^-,a)$	Not found	$\beta_L(g^+,g^-,a)$	Found	$\gamma_L(g^+,g^-,a)$	Found	$\delta_L(g^+,g^-,a)$	Found	$\varepsilon_L(g^+,g^-,a)$	Found
$\alpha_L(g^+,g^-,g^-)$	Not found	$\beta_L(g^+,g^-,g^-)$	Found	$\gamma_L(g^+,g^-,g^-)$	Found	$\delta_L(g^+,g^-,g^-)$	Found	$\varepsilon_L(g^+,g^-,g^-)$	Not found
$\alpha_L(a,g^+,g^+)$	Not found	$\beta_L(a,g^+,g^+)$	Found	$\gamma_L(a,g^+,g^+)$	Found	$\delta_L(a,g^+,g^+)$	Found	$\varepsilon_L(a,g^+,g^+)$	Not found
$\alpha_L(a,g^+,a)$	Not found	$\beta_L(a,g^+,a)$	Found	$\gamma_L(a,g^+,a)$	Found	$\delta_L(a,g^+,a)$	Not Found	$\varepsilon_L(a,g^+,a)$	Not found
$\alpha_L(a,g^+,g^-)$	Found	$\beta_L(a,g^+,g^-)$	Found	$\gamma_L(a,g^+,g^-)$	Found	$\delta_L(a,g^+,g^-)$	Found	$\varepsilon_L(a,g^+,g^-)$	Not found
$\alpha_L(a,a,g^+)$	Found	$\beta_L(a,a,g^+)$	Found	$\gamma_L(a,a,g^+)$ 1st	Found	$\delta_L(a,a,g^+)$	Found	$\varepsilon_L(a,a,g^+)$	Not found
$\alpha_L(a,a,a)$	Not found	$\beta_L(a,a,a)$	Found	$\gamma_L(a,a,g^+)$ 2nd	Found	$\delta_L(a,a,a)$	Found	$\varepsilon_L(a,a,a)$	Not found
$\alpha_L(a,a,g^-)$	Not found	$\beta_L(a,a,g^-)$	Found	$\gamma_L(a,a,a)$	Found	$\delta_L(a,a,g^-)$	Found	$\varepsilon_L(a,a,g^-)$	Not found
$\alpha_L(a,g^-,g^+)$	Not found	$\beta_L(a,g^-,g^+)$	Not Found	$\gamma_L(a,a,g^-)$	Found	$\delta_L(a,g^-,g^+)$	Not Found	$\varepsilon_L(a,g^-,g^+)$	Not found
$\alpha_L(a,g^-,a)$	Not found	$\beta_L(a,g^-,a)$	Not Found	$\gamma_L(a,g^-,a)$	Found	$\delta_L(a,g^-,a)$	Not Found	$\varepsilon_L(a,g^-,a)$	Not found
$\alpha_L(a,g^-,g^-)$	Found	$\beta_L(a,g^-,g^-)$	Found	$\gamma_L(a,g^-,g^-)$	Found	$\delta_L(a,g^-,g^-)$	Found	$\varepsilon_L(a,g^-,g^-)$	Not found
$\alpha_L(g^-,g^+,g^+)$	Not found	$\beta_L(g^-,g^+,g^+)$	Found	$\gamma_L(g^-,g^+,g^+)$	Found	$\delta_L(g^-,g^+,g^+)$	Found	$\varepsilon_L(g^-,g^+,g^+)$	Not found
$\alpha_L(g^-,g^+,a)$	Not found	$\beta_L(g^-,g^+,a)$	Found	$\gamma_L(g^-,g^+,a)$	Found	$\delta_L(g^-,g^+,a)$	Not Found	$\varepsilon_L(g^-,g^+,a)$	Not found
$\alpha_L(g^-,g^+,g^-)$	Not found	$\beta_L(g^-,g^+,g^-)$	Found	$\gamma_L(g^-,g^+,g^-)$	Found	$\delta_L(g^-,g^+,g^-)$	Found	$\varepsilon_L(g^-,g^+,g^-)$	Not found
$\alpha_L(g^-,a,g^+)$	Not found	$\beta_L(g^-,a,g^+)$	Found	$\gamma_L(g^-,a,g^+)$	Found	$\delta_L(g^-,a,g^+)$	Found	$\varepsilon_L(g^-,a,g^+)$	Not found
$\alpha_L(g^-,a,a)$	Not found	$\beta_L(g^-,a,a)$	Found	$\gamma_L(g^-,a,a)$	Found	$\delta_L(g^-,a,a)$	Found	$\varepsilon_L(g^-,a,a)$	Not found
$\alpha_L(g^-,a,g^-)$	Not found	$\beta_L(g^-,a,g^-)$	Found	$\gamma_L(g^-,a,g^-)$	Found	$\delta_L(g^-,a,g^-)$	Found	$\varepsilon_L(g^-,a,g^-)$	Not found
$\alpha_L(g^-,g^-,g^+)$	Not found	$\beta_L(g^-,g^-,g^+)$	Found	$\gamma_L(g^-,g^-,g^+)$	Found	$\delta_L(g^-,g^-,g^+)$	Found	$\varepsilon_L(g^-,g^-,g^+)$	Not found
$\alpha_L(g^-,g^-,a)$	Not found	$\beta_L(g^-,g^-,a)$	Found	$\gamma_L(g^-,g^-,a)$	Found	$\delta_L(g^-,g^-,g^-)$ 1st	Found	$\varepsilon_L(g^-,g^-,a)$	Not found
$\alpha_L(g^-,g^-,g^-)$	Not found	$\beta_L(g^-,g^-,g^-)$	Found	$\gamma_L(g^-,g^-,g^-)$	Found	$\delta_L(g^-,g^-,g^-)$ 2nd	Found	$\varepsilon_L(g^-,g^-,g^-)$	Not found

1st and 2nd stands for the two non-equivalent minima localized in same catchment region.

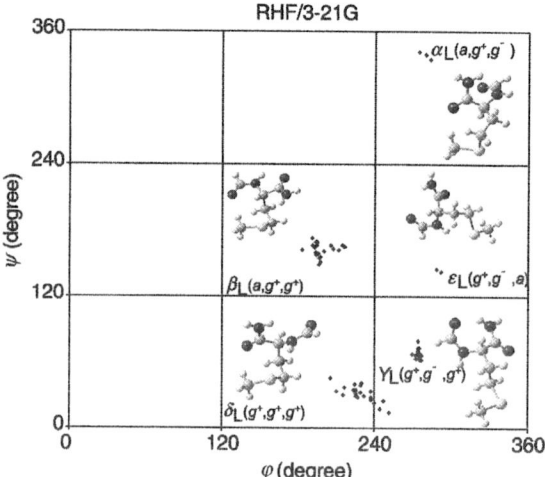

Fig. 1. Ramachandran map of the stable conformers of N-formyl-L-methioninamide optimized at RHF/3-21G level of theory. All conformers are shown on the $[\varphi, \psi]$ projection map of the $E = E(\varphi, \psi, \chi_1, \chi_2, \chi_3)$ hypersurface as dots on the surface. Each main backbone conformer is illustrated by a small structure of the corresponding local minima of the region.

allocated for the five backbone conformers as summarized in Table 1 and Fig. 1. However, in the case of two conformations, namely δ_L [$g^- g^- g^-$] and γ_L [aag^+] the catchment region contained two

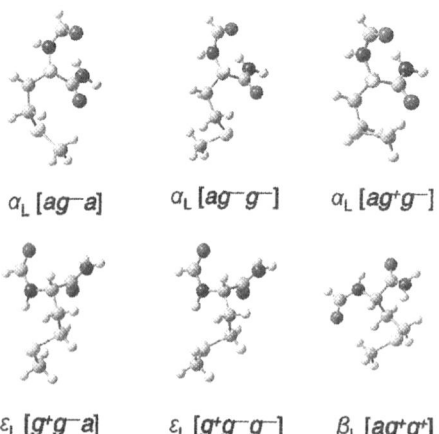

Fig. 2. Structures of all unique stable conformers optimized at the RHF/3-21G level of theory. The upper panel shows the three α_L conformers starting from left to right as follows: α_L [$ag^- a$], α_L [$ag^- g^-$] and α_L [$ag^+ g^-$]. The lower row contains the two stable ε_L conformers that are: ε_L [$g^+ g^- a$] and ε_L [$g^+ g^- g^-$] and the global minimum: β_L [$ag^+ g^+$].

Table 2

(A) Allocation and number of the stable conformers in each region of the Ramachandran map at RHF/3-21G level of theory. (B) Interval of the lowest and highest relative energy values (in kcal mol^{-1}) for each backbone conformers is summarized at RHF/3-21G level of theory. (C) Calculated relative energies at RHF/3-21G level of theory for each region. The first bold values indicate the mean energy for the conformers found in that region, separated by semicolon from the second values that are the deviation from the average. N.e., not explored

A	γ_D N.e.	δ_D N.e.	α_L 3
	ε_D N.e.	β_L 23	ε_L 2
	α_D N.e.	δ_L 23	γ_L 26
B	γ_D N.e.	δ_D N.e.	α_L 6.0–12.3
	ε_D N.e.	β_L 0.0–8.5	ε_L 7.8–8.2
	α_D N.e.	δ_L 3.5–8.0	γ_L 1.3–8.2
C	γ_D N.e.	δ_D N.e.	α_L **9.3**; 3.2
	ε_D N.e.	β_L **5.2**; 2.0	ε_L **8.0**; 0.4
	α_D N.e.	δ_L **6.5**; 1.3	γ_L **3.5**; 1.7

non-equivalent minima. The fact that both α_L and ε_L type backbone conformers were found shows that methionine has a somewhat unusual conformational behaviour although its side-chain is considered to be apolar (Fig. 2). Such backbone conformations were observed in N-formyl-L-histidinamide in different protonation states [28,29]. As can be seen in Table 2, these two backbone conformers represent the highest energy values among the minima, thus they can easily change their conformations toward lower minima. In an amino acid having polar side-chain can obtain a stable conformation corresponding to these regions. Another unique property of N-formyl-L-methioninamide is emphasized by its global minimum localized in β_L [$ag^+ g^+$]. Similar backbone fold has been found for the global minimum of N-acetyl-L-isoleucine-methylamide [30].

The optimized torsion angles, total, relative (local and global), population distributions at 298 K (locally and globally) and stabilization energies of the optimized conformers are summarized in Table 3.

Since, it is impossible to properly represent a 5D potential energy hypersurface (PEHS) selected 2D projections are reported. Fig. 1 shows that the different backbone conformers are localized close to each other gathering into small groups, thus clustering of backbone conformations is straightforward.

Table 3
Summary of the data of the stable conformers found at RHF/3-21G level of theory

Final geometry[a,b]	ω_0	φ	ψ	ω_1	χ_1	χ_2	χ_3	Energy (hartree)	Local ΔE_{rel} (kcal mol⁻¹)	Global ΔE_{rel} (kcal mol⁻¹)	Global population (%)[c]	Local population (%)[c]	Stabilization energy (kcal mol⁻¹)
$\alpha_L(a, g^+, g^-)$	−172.92	−80.34	−20.9	−179.77	−170.79	65.46	−112.77	−885.6651992	0.00	5.98	0.00	99.77	−3.31
$\alpha_L(a, g^-, a)$	−174.81	−76.96	−25.15	−177.79	174.42	−99.11	133.03	−885.6594521	3.61	9.59	0.00	0.23	0.29
$\alpha_L(a, g^-, g^-)$	−174.97	−86.00	−18.44	−178.37	175.17	−96.38	−81.59	−885.6551407	6.31	12.29	0.00	0.00	3.00
$\beta_L(g^+, g^+, g^+)$	177.67	−167.31	165.09	174.05	58.99	89.16	78.45	−885.6646505	6.33	6.33	0.00	0.00	−2.97
$\beta_L(g^+, g^+, a)$	177.72	−168.65	166.98	173.71	61.31	96.47	−172.42	−885.6650776	6.06	6.06	0.00	0.00	−3.23
$\beta_L(g^+, g^+, g^-)$	177.03	−165.44	166.42	175.29	56.10	84.53	−120.54	−885.6682883	4.04	4.04	0.07	0.09	−5.25
$\beta_L(g^+, a, g^+)$	177.35	−168.98	173.85	178.85	60.85	−176.64	75.82	−885.6687961	3.72	3.72	0.11	0.15	−5.57
$\beta_L(g^+, a, a)$	177.77	−165.07	170.63	178.24	59.00	−179.91	179.02	−885.6675639	4.50	4.50	0.03	0.04	−4.80
$\beta_L(g^+, a, g^-)$	177.99	−165.07	168.61	178.44	57.16	−178.98	−77.00	−885.6673418	4.64	4.64	0.02	0.03	−4.66
$\beta_L(g^+, g^-, g^+)$	−178.27	−168.40	166.99	178.33	73.98	−70.56	117.11	−885.6642092	6.60	6.60	0.00	0.00	−2.69
$\beta_L(g^+, g^-, a)$	−178.90	−159.84	160.91	177.05	60.12	−80.47	−161.18	−885.6643708	6.50	6.50	0.00	0.00	−2.79
$\beta_L(g^+, g^-, g^-)$	−176.61	−176.95	162.72	177.09	55.08	−83.88	−53.22	−885.6665459	5.14	5.14	0.01	0.01	−4.16
$\beta_L(a, g^+, g^+)$	178.67	−164.03	159.08	178.63	−172.49	49.10	53.18	−885.6747322	0.00	0.00	61.40	81.26	−9.29
$\beta_L(a, g^+, a)$	178.59	−165.59	160.21	179.89	176.38	56.48	179.06	−885.6684006	3.97	3.97	0.07	0.10	−5.32
$\beta_L(a, g^+, g^-)$	−179.53	−162.55	151.39	177.13	167.67	97.02	−57.37	−885.6638923	6.80	6.80	0.00	0.00	−2.49
$\beta_L(a, a, g^+)$	179.83	−162.96	154.73	178.43	−174.16	176.96	75.67	−885.6645624	6.38	6.38	0.00	0.00	−2.91
$\beta_L(a, a, a)$	179.81	−162.83	155.73	178.28	−173.96	177.35	179.60	−885.6637818	6.87	6.87	0.00	0.00	−2.42
$\beta_L(a, a, g^-)$	179.53	−166.07	158.62	178.07	−172.65	176.19	−74.06	−885.6655266	5.78	5.78	0.00	0.00	−3.52
$\beta_L(a, g^-, g^-)$	179.52	−163.2	149.83	174.08	−177.39	−88.16	−80.79	−885.6612010	8.49	8.49	0.00	0.00	−0.80
$\beta_L(g^-, g^+, g^+)$	176.87	−147.64	162.76	177.03	−58.44	92.53	75.08	−885.6631825	7.25	7.25	0.00	0.00	−2.05
$\beta_L(g^-, g^+, a)$	179.01	−152.82	164.12	177.06	−64.15	92.60	−151.17	−885.6661267	5.40	5.40	0.01	0.01	−3.89
$\beta_L(g^-, a, g^+)$	174.05	−143.55	165.72	176.92	−55.09	−172.68	79.10	−885.6655488	5.76	5.76	0.00	0.00	−3.53
$\beta_L(g^-, a, a)$	173.95	−144.95	166.59	177.00	−58.72	−171.08	−177.67	−885.6659112	5.54	5.54	0.01	0.01	−3.76
$\beta_L(g^-, a, g^-)$	173.79	−144.32	165.84	177.07	−59.31	−171.41	−76.18	−885.6657830	5.62	5.62	0.00	0.01	−3.68
$\beta_L(g^-, g^-, g^+)$	175.17	−154.08	167.09	178.35	−89.32	−66.57	121.67	−885.6732942	0.90	0.90	13.36	17.69	−8.39
$\beta_L(g^-, g^-, g^-)$	176.84	−165.32	169.69	179.78	−108.18	−69.24	−80.06	−885.6700867	2.92	2.92	0.45	0.59	−6.38
$\gamma_L(g^+, g^+, g^+)$	−176.87	−87.30	66.50	−178.29	3.02	65.03	51.63	−885.6699142	1.71	3.02	0.37	1.53	−6.27
$\gamma_L(g^+, g^+, a)$	−178.08	−87.90	63.01	−178.98	21.32	60.68	−175.85	−885.6632110	5.91	7.23	0.00	0.00	−2.06
$\gamma_L(g^+, a, g^+)$	−174.53	−84.48	63.19	−179.74	59.74	177.15	75.60	−885.6677307	3.08	4.39	0.04	0.15	−4.90
$\gamma_L(g^+, a, a)$	−173.71	−84.08	62.60	−179.97	61.02	−178.49	−168.22	−885.6678560	3.00	4.31	0.04	0.17	−4.98
$\gamma_L(g^+, a, g^-)$	−173.92	−84.00	62.14	−179.88	57.64	173.96	−79.13	−885.6694669	1.99	3.30	0.23	0.95	−5.99
$\gamma_L(g^+, g^-, g^+)$	−174.69	−85.31	69.79	−177.94	74.64	−61.01	115.52	−885.6726337	0.00	1.32	6.63	27.42	−7.98
$\gamma_L(g^+, g^-, a)$	−173.67	−84.37	63.39	−179.34	61.34	−82.63	167.58	−885.6711844	0.53	1.85	2.70	11.14	−7.44
$\gamma_L(g^+, g^-, g^-)$	−174.22	−83.36	61.94	−179.47	64.22	−74.62	−68.48	−885.6721391	0.31	1.63	3.93	16.23	−7.67
$\gamma_L(a, g^+, g^+)$	−175.17	−84.97	67.98	−179.15	−173.09	61.09	72.23	−885.6708759	1.10	2.42	1.03	4.25	−6.87

(continued on next page)

Table 3 (continued)

Final geometry[a,b]	ω_0	φ	ψ	ω_1	χ_1	χ_2	χ_3	Energy (hartree)	Local ΔE_{rel} (kcal mol^{-1})	Global ΔE_{rel} (kcal mol^{-1})	Global population (%)[c]	Local population (%)[c]	Stabilization energy (kcal mol^{-1})
$\gamma_L(a,g^+,a)$	-174.55	-84.83	67.87	-178.91	-173.43	64.98	-172.85	-885.6704535	1.37	2.68	0.66	2.72	-6.61
$\gamma_L(a,g^+,g^-)$	-174.75	-84.57	73.32	-177.33	-175.86	59.33	-103.20	-885.6706167	1.27	2.58	0.78	3.23	-6.71
$\gamma_L(a,a,g^+)]1^{st}$	-174.34	-84.40	67.38	-178.67	-169.17	-161.44	80.08	-885.6720016	0.40	1.71	3.39	14.03	-7.58
$\gamma_L(a,a,g^+)2^{nd}$	-174.58	-84.62	67.76	-178.66	-170.55	-179.64	80.51	-885.6710229	1.01	2.33	1.20	4.97	-6.97
$\gamma_L(a,a,a)$	-174.77	-84.95	68.11	-178.43	-172.59	175.00	171.00	-885.6697186	1.83	3.15	0.30	1.25	-6.15
$\gamma_L(a,g^-,g^-)$	-175.20	-84.97	68.16	-178.54	-172.73	177.09	-77.66	-885.6698138	1.77	3.09	0.33	1.38	-6.21
$\gamma_L(a,g^-,g^+)$	-176.89	-85.62	79.27	-175.70	-145.29	-67.92	89.14	-885.6616555	6.89	8.21	0.00	0.00	-1.09
$\gamma_L(a,g^-,a)$	-176.49	-86.02	71.15	-178.93	-147.34	-61.90	170.97	-885.6638002	5.54	6.86	0.00	0.00	-2.43
$\gamma_L(a,g^-,g^-)$	-176.16	-85.82	72.61	-178.37	-145.02	-52.68	-69.63	-885.6706742	1.23	2.55	0.83	3.43	-6.75
$\gamma_L(g^-,g^+,g^+)$	-174.03	-83.86	67.26	-178.50	-53.14	87.35	74.00	-885.6705989	1.28	2.59	0.77	3.17	-6.70
$\gamma_L(g^-,g^+,a)$	-174.25	-83.43	66.55	-178.68	-52.13	94.58	-168.66	-885.6688270	2.39	3.71	0.12	0.48	-5.59
$\gamma_L(g^-,a,g^+)$	-175.57	-85.03	65.27	-179.08	-67.37	174.44	71.02	-885.6690859	2.23	3.54	0.15	0.64	-5.75
$\gamma_L(g^-,a,a)$	-175.96	-85.33	66.61	-178.90	-66.40	176.26	177.84	-885.6681409	2.82	4.14	0.06	0.23	-5.16
$\gamma_L(g^-,a,g^-)$	-176.02	-85.47	67.78	-178.72	-64.03	-177.86	-73.22	-885.6682283	2.76	4.08	0.06	0.26	-5.21
$\gamma_L(g^-,g^-,g^+)$	178.70	-89.78	68.15	-178.89	-65.43	-65.43	103.6	-885.6694049	2.03	3.34	0.22	0.89	-5.95
$\gamma_L(g^-,g^-,a)$	-175.59	-85.05	67.17	-179.11	-64.02	-69.53	175.87	-885.6683125	2.71	4.03	0.07	0.28	-5.26
$\gamma_L(g^-,g^-,g^-)$	-175.20	-84.65	65.41	-179.34	-65.83	-66.50	-73.81	-885.6696841	1.85	3.17	0.29	1.20	-6.13
$\delta_L(g^+,g^+,g^+)$	-175.43	-134.62	29.48	176.66	35.96	51.02	68.77	-885.6691291	0.00	3.52	0.16	65.40	-5.78
$\delta_L(g^+,g^+,a)$	-175.65	-134.60	31.93	177.23	33.83	55.76	-174.09	-885.6620300	4.45	7.97	0.00	0.04	-1.32
$\delta_L(g^+,a,g^+)$	-175.40	-143.78	37.20	176.17	60.52	-179.01	75.03	-885.6633464	3.63	7.14	0.00	0.14	-2.15
$\delta_L(g^+,a,a)$	-174.17	-135.92	34.99	175.82	62.45	-175.88	-171.57	-885.6628159	3.96	7.48	0.00	0.08	-1.82
$\delta_L(g^+,a,g^-)$	-174.26	-135.42	34.88	176.06	60.32	179.98	-75.53	-885.6646201	2.83	6.35	0.00	0.55	-2.95
$\delta_L(g^+,g^-,g^+)$	-172.07	-115.38	17.25	174.24	78.58	-62.20	115.15	-885.6659283	2.01	5.52	0.01	2.20	-3.77
$\delta_L(g^+,g^-,a)$	-173.13	-128.17	29.37	176.12	71.99	-68.65	176.55	-885.6665689	1.61	5.12	0.01	4.33	-4.17
$\delta_L(g^+,g^-,g^-)$	-174.19	-122.43	26.59	176.50	73.75	-62.91	-65.92	-885.6669814	1.35	4.86	0.02	6.71	-4.43
$\delta_L(a,g^+,g^+)$	-176.71	-154.04	46.35	176.22	-166.48	53.86	55.44	-885.6677710	0.85	4.37	0.04	15.49	-4.93
$\delta_L(a,g^+,g^-)$	-173.83	-134.76	35.00	176.07	-156.57	86.77	-74.83	-885.6619150	4.53	8.04	0.00	0.03	-1.25
$\delta_L(a,a,g^+)$	-173.44	-130.17	41.78	176.66	-163.20	-179.72	80.29	-885.6639169	3.27	6.79	0.00	0.26	-2.51
$\delta_L(a,a,a)$	-173.38	-130.89	39.16	176.52	-164.42	176.71	173.04	-885.6628588	3.93	7.45	0.00	0.08	-1.84
$\delta_L(a,a,g^-)$	-174.01	-135.52	40.40	176.59	-165.06	177.47	-76.98	-885.6633341	3.64	7.15	0.00	0.14	-2.14
$\delta_L(g^-,g^+,g^+)$	-176.26	-149.87	33.69	175.10	-104.69	-61.90	79.70	-885.6641808	3.11	6.62	0.00	0.34	-2.67
$\delta_L(g^-,g^+,a)$	-172.98	-133.00	28.83	176.32	-53.17	87.83	-56.23	-885.6636736	3.42	6.94	0.00	0.20	-2.35
$\delta_L(g^-,g^-,a)$	-171.62	-144.31	30.22	176.29	-64.76	86.76	150.31	-885.6652308	2.45	5.96	0.00	1.05	-3.33
$\delta_L(g^-,a,g^+)$	-174.21	-119.86	23.44	176.21	-53.38	178.32	72.71	-885.6639788	3.23	6.75	0.00	0.28	-2.55
$\delta_L(g^-,a,a)$	-174.54	-122.33	26.26	176.09	-65.21	178.70	178.91	-885.6627903	3.98	7.49	0.00	0.08	-1.80
$\delta_L(g^-,a,g^-)$	-174.72	-122.71	28.38	175.91	-63.60	-176.74	-74.66	-885.6624347	4.20	7.72	0.00	0.05	-1.58
$\delta_L(g^-,g^-,g^+)$	-176.55	-131.16	33.21	176.02	-68.71	-87.38	77.54	-885.6620536	4.44	7.96	0.00	0.04	-1.34
$\delta_L(g^-,g^-,a)$	-176.63	-122.96	34.35	175.72	-70.46	-70.87	123.74	-885.6659331	2.01	5.52	0.01	2.21	-3.77

$\delta_L(g^-{\cdot}g^-{\cdot}g^-)1^{st}$	−173.58	−108.69	13.99	175.52	−87.84	−60.35	−50.49	−885.6636459	3.44	6.96	0.00	0.20	−2.34
$\delta_L(g^-{\cdot}g^-{\cdot}g^-)2^{nd}$	−174.02	−112.41	25.53	175.55	−67.66	−67.39	−72.44	−885.6631094	3.78	7.29	0.00	0.11	−2.00
$\varepsilon_L(g^+{\cdot}g^-{\cdot}a)$	177.90	−70.97	144.88	178.42	59.51	−90.03	163.79	−885.6624209	0.00	7.73	0.00	70.49	−1.57
$\varepsilon_L(g^+{\cdot}g^-{\cdot}g^-)$	177.74	−68.03	142.43	178.58	60.37	−85.10	−69.33	−885.6615998	0.52	8.24	0.00	29.51	−1.05

[a] Torsional angles are measured in degree.

[b] 1st and 2nd are conformers of the same catchment region.

[c] Global and local population distribution for the ith conformer is calculated by

$$P_i = \frac{e\left(-\frac{\Delta E_i}{RT}\right)}{\displaystyle\sum_{i=1}^{n} e\left(-\frac{\Delta E_i}{RT}\right)},$$

where R is the universal gase-constant, T is the absolute temperature it is taken here as 298 K. ΔE is the relative energy with respect to the global or local minimum.

3.2. Comparison between stable conformers of the main backbone conformers of N- and C-protected methionine and to N- and C-protected homocysteine β_L conformers computed at RHF/3-21G level of theory

All optimized conformers are shown in Fig. 3. The numbers in the rectangles indicate the relative energy of the conformers compared to the global minimum thus empty rectangles corresponds to lack of low-energy conformations. As can be seen here, the global minimum is in the fully extended backbone region of the Ramachandran plot (β_L [ag^+g^+]). All β_L backbone conformers have been explored by Sheraly et al. [23] for the highly similar N-acetyl-L-homocysteine-methylamide. Their results are displayed in the upper left panel in Fig. 3. As can be seen, there is a slight difference in the unobserved conformers between the two amino acid models in β_L backbone conformation. Undoubtedly, the differences occur because of the different terminal groups of the side-chains in homocysteine (−H) and methionine (−CH$_3$). The β_L backbone conformers can be compared to ε_L or δ_L conformers because these have only one backbone torsional angle with a major difference. Since ε_L conformers are not stable enough most of these conformers have disappeared, thus we have only the δ_L conformers to compare. Both of these backbone types (β_L, δ_L) lack three side-chain conformers namely [$g^-g^+g^-$], [ag^-a] and [ag^-g^+]. Since the major difference between these conformers is the value of the ψ torsional angle, it is reasonable to suggest that the change in the position of the second amide plane disfavors these conformers and that the effect of the first amide on the side-chain conformers does not stabilize them. Furthermore, the latter two side-chain conformers do not exist in β_L homocysteine, either. However, the homocysteine side-chain is stable in the [$g^-g^+g^-$] conformation.

Methionine does not have a stable [$g^-g^-g^+$] side-chain conformation when the backbone is fully extended. Upon changing the second amide plane, a high energy, but stable conformer nevertheless, comes into existence. Change of orientation of this plane allows electrostatically and sterically the side-chain to obtain its relaxed conformation. For homocysteine, the [$g^-g^-g^+$] side-chain conformation indicates that a small and electrostatically favorable group as

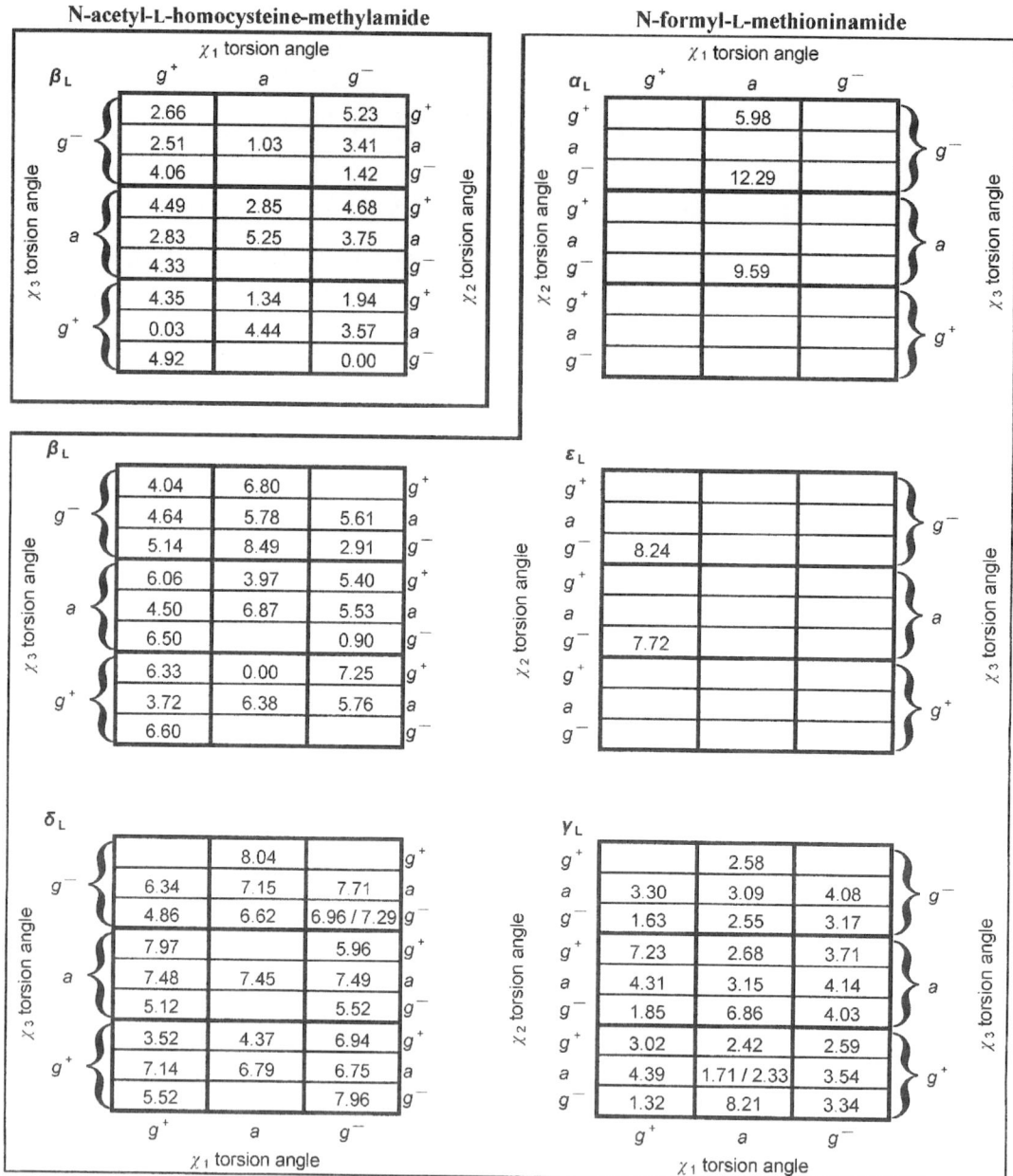

Fig. 3. Illustration of the conformers of *N*-formyl-L-methioninamide found at the RHF/3-21G level of theory by their five torsion angles as independent variables. The five main groups as rectangles show the regions of the Ramachandran map (L-valley) as can be seen in Fig. 1 for backbone conformations. It is important to mention that even δ_L conformers are spreaded just at the border of γ_L backbone conformers, these always considered δ_L because there are no hydrogen bonds observed in these backbones that characteristic of γ_L conformers. Each small rectangle that contains a number represents a stable conformer described by five torsion angles. Therefore, χ_1 is shown at bottom and above, χ_2 is shown between the boxes and χ_3 is indicated on the outer side of the boxes. The conformers that have not been found are illustrated by empty rectangles. The numbers indicate the relative energy values for every conformer compared to the global minimum. In the upper left panel, the β_L backbone conformers of *N*-acetyl-L-homocysteine-methylamide are indicated by their relative energies compared to the minimum found in $\beta_L [g^- g^- g^+]$ at the RHF/3-21G level of theory reported by Sheraly et al. In $\delta_L [g^- g^- g^-]$ and $\gamma_L [aag^+]$, two non-equivalent minima were localized in the same catchment region, thus their relative energy is indicated by two number in these rectangles.

a hydrogen instead of a methyl group allows the conformation to relax. Also, this is the case for the previously mentioned $[g^- g^+ g^-]$ side-chain conformation. In the β_L backbone orientation, methionine has three additional conformers that were not described for homocysteine. These are $[g^- g^- a]$, $[ag^- g^-]$ and $[ag^+ g^-]$. Surprisingly, there is a remarkable difference for $[g^- g^- a]$ and $[g^- g^- g^+]$ in the case of homocysteine and methionine. For homocysteine, the second side-chain conformer represents the minimum but we have not found it for methionine1-on the other hand, for methionine, the first side-chain conformation is the second most stable geometry but such a conformation was not found for homocysteine. The main geometry stabilizing driving force might be due to an interaction between a hydrogen (C–H or S–H) in the side-chain and the $(i - 1)$ carbonyl oxygen in the backbone. In methionine (Met) and in homocysteine (Hcy), the contributing atoms are the same (i.e. oxygen and hydrogen) however two differences exist. First, on the one hand, in methionine, the hydrogen is only indirectly attached to the sulphur atom via a carbon atom; on the other hand, in homocysteine, the hydrogen is directly connected to the sulphur in the thiol group. Thus, the two terminal groups (methyl in Met and hydrogen in Hcy) have different steric attributes as they fill the space differently. The result can be observed in two distinct stable geometries for methionine and homocysteine. Second, the heavy atoms which carry the hydrogen are different in the two molecules. Once, in the case of methionine, it is carbon and in the case of homocysteine, it is sulphur. Furthermore, the α carbon hydrogen and the sulphur atom is in 2.74 Å distance in methionine β_L $[g^- g^- a]$, allowing another effect to consider during stabilization of the conformation. The remaining two side-chain conformers were not very stable in methionine, either ($[ag^- g^-]$: 8.49 kcal mol^{-1}, $[ag^+ g^-]$: 6.80 kcal mol^{-1}).

Comparing δ_L to γ_L backbone types, one can see clearly the differences in Fig. 3: no conformer was found in δ_L when side-chain is $[ag^- g^+]$, $[ag^- a]$ or $[ag^+ a]$. Since δ_L and γ_L are highly similar backbone conformers, the less stable δ_L type of backbone can migrate into γ_L conformation, thus forming a hydrogen bond effectively stabilizes the backbone.

3.3. Behaviour of the side-chain of methionine calculated at RHF/3-21G level of theory. An energetic description

To understand the behaviour of the side-chain of methionine, the C–C–S–C torsional angle of methyl-ethyl-thioether corresponding to χ_3 torsional angle in methionine was varied from 0° to 360° by 15 degrees steps and the energy of each structure was calculated. The resulting potential energy curve (PEC) is very similar to that of butane [31] where a methylene group occupies the position of the sulphur atom. Nevertheless, sulphur and carbon have very similar electronegativity, the difference between the two atoms is that sulphur has two non-bonded electronpairs, hence it does not have bonded protons, and it belongs to the third period, thus is larger than the carbon atom. The non-bonded electron pairs may also require more space than bonded ones. Let us suppose that there is hardly any interaction between the side-chain and the backbone then the side-chain of methionine conformationally behaves as a methyl-ethyl-thioether, hence the behaviour will be very similar to that of an alkane. Thus, we can describe the torsional angles and the rate of the minima compared to each other.

Of methyl-ethyl-thioether, three minima were optimized from which two (g^+ and g^-) were isoenergetic. The third (a) had a slightly higher value. The energy and the torsional angles for the identical (g^+ and g^-) minima are: $E_{min} = -513.169255$ hartree at 72.07° and -72.07°. The *anti* orientation is the second lowest energy structure. The torsional angle is 180° and the relative energy is 0.023 kcal mol^{-1}. Thus, the system exhibits a very slight '*gauche*-effect' [32]. Comparison of these energies and torsional angles with the energies and torsional angles obtained from the methionine minima, can now be made.

Apart from certain cases, the backbone and the side-chain does not interact at all with each other in the methionine optimized conformers. In this ideal case, the terminal torsion angle of the side-chain (χ_3) may behave like that of methyl-ethyl-thioether. Then one may compare the minimum energy conformers, when all other torsional angles are nearly the same, according to the χ_3 torsional angles and the relative energy differences. In Fig. 4, all relative energies for conformers in the five backbone regions are

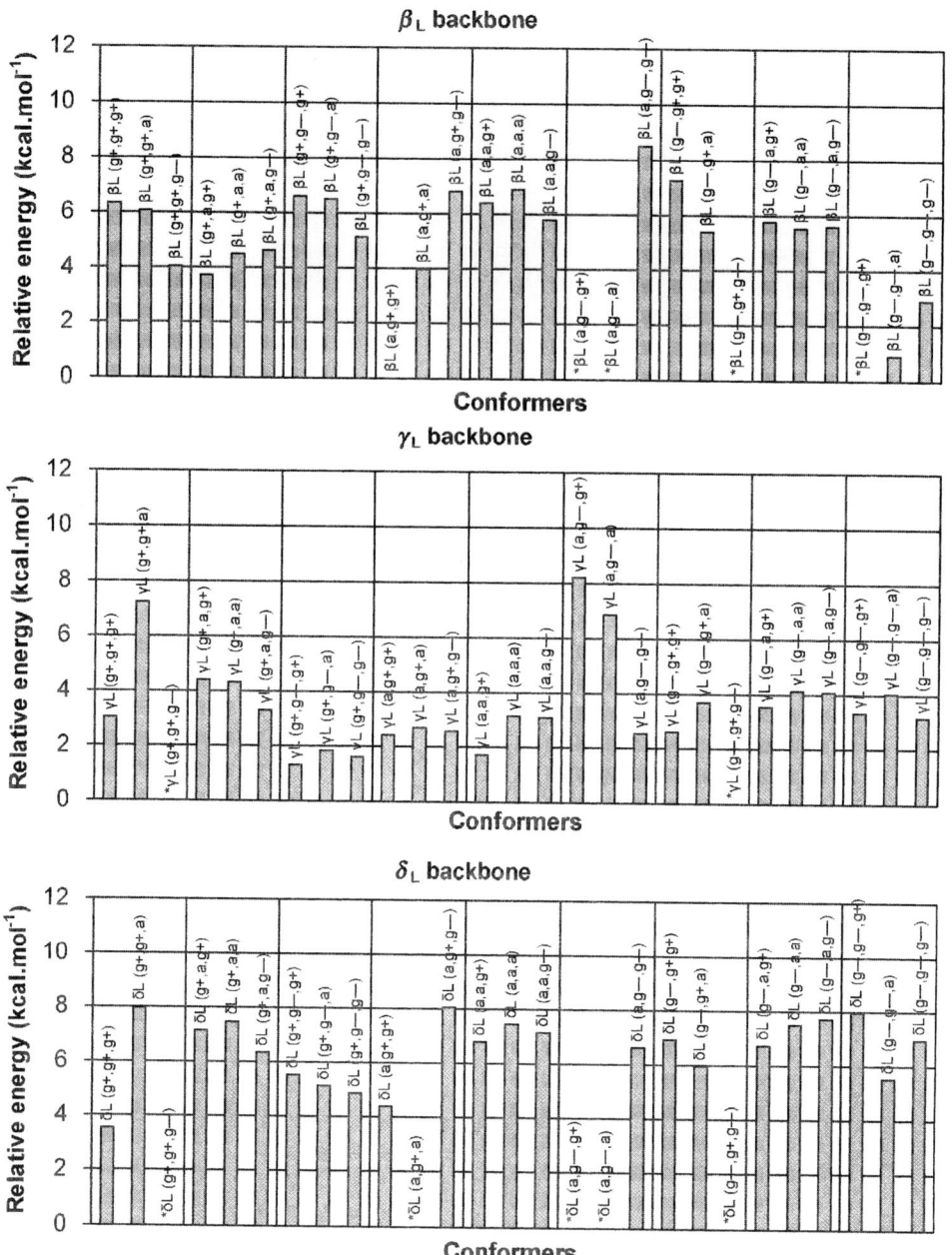

Fig. 4. Relative energies of conformers of *N*-formyl-L-methioninamide are illustrated with three different backbone conformations. All three conformers specified within vertical lines have similar φ, ψ, χ_1, χ_2 torsional angles values and differ only in χ_3. Asterisks indicate the unstable conformations. The α_L and ε_L backbone conformation are not included in this illustration (c.f. Table 3).

illustrated. It can be seen that every three conformers placed beside each other (triplets) have similar descriptions except for the χ_3 torsional angle. A few structures can be found whose relative energy is the highest in the *anti* state, and g^+, g^- conformers whose

energy value is close to the *anti* state and are roughly the same, just like in methyl-ethyl-thioether. In β_L: [*aax*] (and a couple of those that have hardly any difference in the energy; [g^+ax], [g^-ax]), in δ_L these are: [*aax*] ([g^+ax], [g^+g^-x] and [g^-ax]), and in γ_L

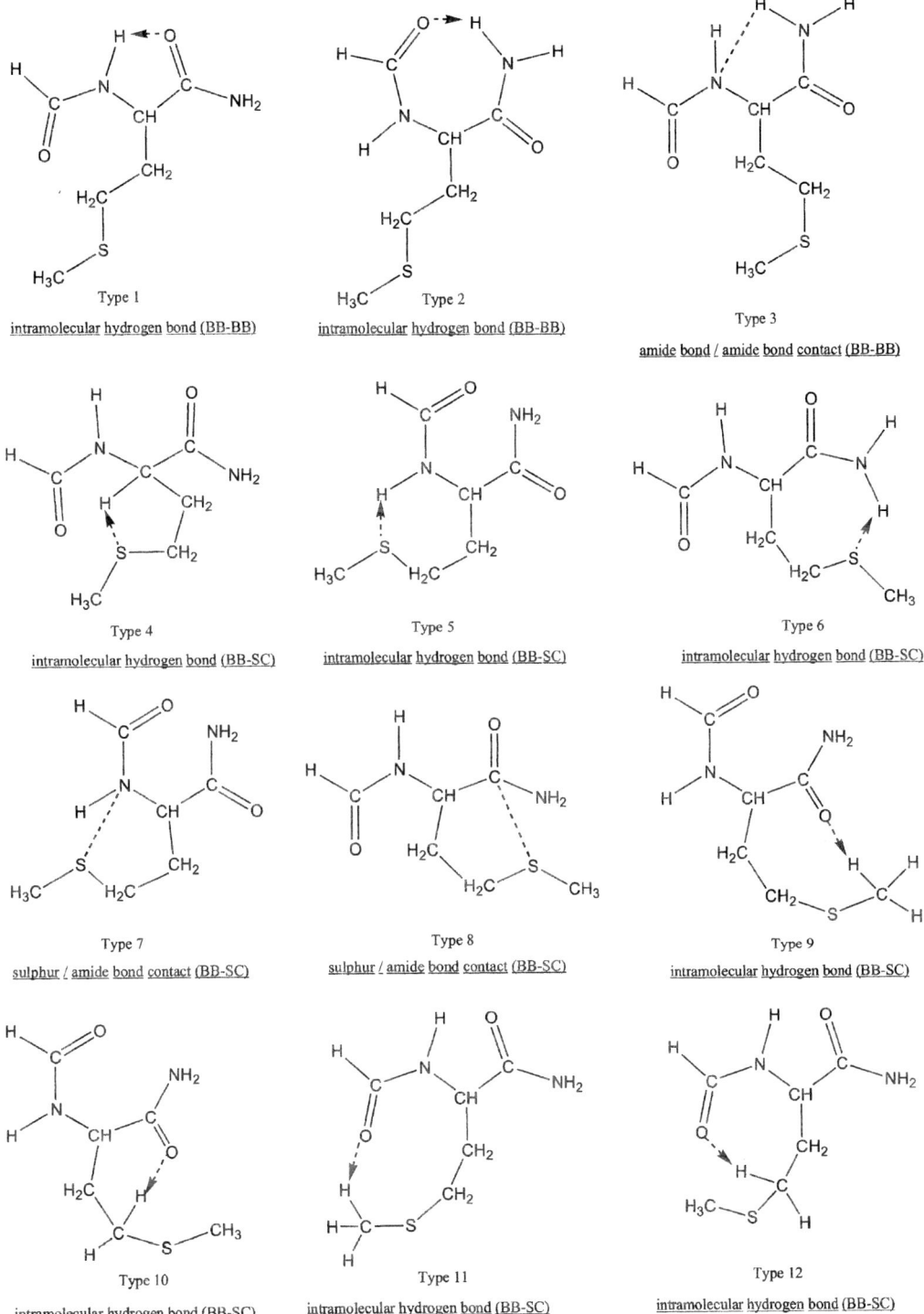

Scheme 4. Illustration of the structures of N-formyl-L-methioninamide. The different types of intramolecular interactions are producing ringlike structures. BB, backbone; SC, side-chain.

Table 4
Illustration of the intramolecular interactions can be described in N-formyl-L-methioninamide at RHF/3-21G level of theory

Final geometry[a]	Backbone/backbone				Backbone/side-chain								ΔE[b]
	Type	Distances		Angle	Participating atoms	Type	Distances		Angle	Type	Distances		
		(H···N)	(N···N)	(N-H···N)			(H···S/O)	(C···S/O) (N–S)	(C–H···S/O) (N–H···S)		(S···N)	(S···C)	
$\alpha_L(a,g^+,g^-)$	3	2.32	2.72	103.19	C–H···S	4	2.85	3.34	108.20				5.98
					C–H···O	9	2.29	3.29	153.38				
					C–H···O	10	2.36	3.05	120.35				
$\alpha_L(a,g^-,a)$	3	2.37	2.74	101.69	C–H···O	9	2.27	3.18	140.85				9.59
					C–H···O	10	2.41	3.01	113.88				
$\alpha_L(a,g^-,g^-)$	3	2.34	2.74	102.81	C–H···O	10	2.35	2.92	111.66				12.29
$\delta_L(g^+,g^+,g^+)$	3	2.35	2.72	101.00	N–H···S	5	2.77	3.36	118.71	8	3.57[c]	3.35	3.52
					C–H···O	9	2.23	3.18	146.09				
$\delta_L(g^+,g^+,a)$	3	2.37	2.73	100.37	N–H···S	5	2.78	3.37	117.72	8	3.43[c]	3.29	7.97
					C–H···O	12	2.51	3.11	114.16				
$\delta_L(g^+,a,g^+)$	3	2.42	2.73	97.69	C–H···O	12	2.51	3.10	114.04				7.14
$\delta_L(g^+,a,a)$	3	2.40	2.74	98.74	C–H···O	10	2.55	3.13	113.03				7.48
$\delta_L(g^+,a,g^-)$	3	2.40	2.74	98.84	N–H···S	5	2.51	3.28	133.60				6.35
$\delta_L(g^+,g^-,g^+)$	3	2.32	2.72	102.97	C–H···O	10	2.44	3.07	116.42				5.52
$\delta_L(g^+,g^-,a)$	3	2.36	2.73	101.10	N–H···S	5	2.55	3.21	123.75				5.12
					C–H···O	10	2.42	3.06	116.60				
$\delta_L(g^+,g^-,g^-)$	3	2.35	2.73	101.58	C–H···S	4	2.71	3.26	111.61				4.86
					C–H···O	11	2.39	3.44	167.36				
$\delta_L(a,g^+,g^+)$	3	2.49	2.75	94.19	C–H···O	10	2.40	3.04	116.26				4.37
$\delta_L(a,g^+,g^-)$	3	2.39	2.73	99.08	C–H···O	11	2.42	3.47	165.04				8.04
					C–H···O	10	2.50	3.07	111.83				
$\delta_L(a,a,g^+)$	3	2.45	2.75	97.12	C–H···O	10	2.36	3.01	117.28				6.79
$\delta_L(a,a,a)$	3	2.42	2.74	98.10	C–H···O	10	2.33	2.99	117.83				7.45
$\delta_L(a,a,g^-)$	3	2.42	2.74	97.60	C–H···O	10	2.33	2.99	117.49				7.15
$\delta_L(a,g^-,g^-)$	3	2.38	2.71	98.68	C–H···S	4	2.76	3.30	110.43				6.62
					C–H···O	9	2.43	3.44	155.05				
$\delta_L(g^-,g^+,g^+)$	3	2.35	2.72	101.17	C–H···O	11	2.54	3.41	137.31	7	3.29	3.57	6.94
$\delta_L(g^-,g^+,a)$	3	2.34	2.70	100.62	C–H···O	12	2.58	3.26	120.42	7	3.33	3.25	5.96
$\delta_L(g^-,a,g^+)$	3	2.35	2.73	101.52									6.75
$\delta_L(g^-,a,a)$	3	2.36	2.72	100.91									7.49
$\delta_L(g^-,a,g^-)$	3	2.37	2.73	100.33									7.72
$\delta_L(g^-,g^-,g^+)$	3	2.39	2.72	98.95	C–H···O	11	2.31	3.34	157.64				7.96
$\delta_L(g^-,g^-,a)$	3	2.42	2.74	98.28	C–H···S	4	2.86	3.37	108.65				5.52
					C–H···O	11	2.29	3.33	162.63				
$\delta_L(g^-,g^-,g^-)$[1st]	3	2.38	2.74	100.64	C–H···S	4	2.87	3.37	108.33				7.29

Conformer	n	(O⋯N)	(O⋯N)	(O⋯H–N)		n	(H⋯S/O)	(C⋯S/O)(N–S)	(C–H⋯S/O)(N–H⋯S)	n	(S⋯N)	(S⋯C)	
$\delta_L(g^-,g^-,g^-)^{2nd}$	3	2.34	2.74	102.94	C–H⋯S / C–H⋯O	4 / 9	2.71 / 2.49	3.30 / 3.52	113.52 / 159.71				6.96
$\varepsilon_L(g^+,g^-,a)$	1				N–H⋯S / C–O⋯H	5 / 10	2.50 / 2.28	3.32 / 3.11	139.45 / 132.00				7.73
$\varepsilon_L(g^+,g^-,g^-)$	1				N–H⋯S / C–O⋯H	5 / 10	2.50 / 2.27	3.33 / 3.09	140.46 / 131.11				8.24
$\beta_L(g^+,g^+,g^+)$	1	2.10	2.60	108.74	C–H⋯O	10	2.95	3.45	108.92	8	3.60[c]	3.37	6.33
$\beta_L(g^+,g^+,a)$	1	2.09	2.59	109.01	C–H⋯O	9	2.30	3.24	145.40	8	3.56	3.42	6.06
$\beta_L(g^+,g^+,g^-)$	1	2.10	2.60	108.72						8	3.60[c]	3.34	4.04
$\beta_L(g^+,a,g^+)$	1	2.10	2.60	108.85									3.72
$\beta_L(g^+,a,a)$	1	2.12	2.61	108.08	C–H⋯O	10	2.90	3.43	110.67				4.50
$\beta_L(g^+,a,g^-)$	1	2.14	2.62	107.65	C–H⋯O	9	2.34	3.35	156.26	7	3.34	3.86	4.64
$\beta_L(g^+,g^-,g^+)$	1	2.19	2.64	105.70	C–H⋯O	10	2.59	3.31	123.98	7	3.17	3.58	6.60
$\beta_L(g^+,g^-,a)$	1	2.23	2.66	103.99	C–H⋯O	10	2.65	3.40	126.19	7	3.32	3.34	6.50
$\beta_L(g^+,g^-,g^-)$	1	2.15	2.63	107.50	N–H⋯S	6	2.79	3.57	136.55				5.14
$\beta_L(a,g^+,g^+)$	1	2.15	2.63	107.06	C–H⋯S	4	2.76	3.26	108.17				0.00
					C–H⋯O	11	2.39	3.45	171.31				
$\beta_L(a,g^+,a)$	1	2.12	2.61	108.31	N–H⋯S	6	2.67	3.47	137.53				3.97
$\beta_L(a,a,g^+)$	1	2.17	2.64	106.40	C–H⋯S	4	2.87	3.28	102.45				6.80
$\beta_L(a,a,a)$	1	2.16	2.63	106.81	C–H⋯O	11	2.32	3.39	171.56				6.38
$\beta_L(a,a,g^-)$	1	2.15	2.63	106.93	C–H⋯O	12	2.35	3.17	132.30	8	3.35	3.46	6.87
$\beta_L(g^-,g^+,g^+)$	1	2.13	2.61	107.74	C–H⋯O	11	2.44	3.32	137.86	7	3.44	3.50	5.78
$\beta_L(g^-,g^+,a)$	1	2.16	2.63	107.11	C–H⋯O	12	2.65	3.20	110.61	7	3.44	3.37	8.49
$\beta_L(g^-,g^+,g^-)$	1	2.22	2.64	103.58	C–H⋯O	12	2.69	3.16	105.95				7.25
$\beta_L(g^-,a,g^+)$	1	2.18	2.62	105.34	C–H⋯O	12	2.76	3.19	103.38				5.40
$\beta_L(g^-,a,a)$	1	2.18	2.63	105.27	C–H⋯S	4	2.75	3.30	111.53				5.76
$\beta_L(g^-,a,g^-)$	1	2.16	2.62	105.88	C–H⋯O	12	2.35	3.19	134.25				5.54
	1	2.17	2.63	105.89	C–H⋯O	11	2.33	3.34	154.19				5.62
	1	2.12	2.62	108.19									0.90
$\beta_L(g^-,a,g^-)$	1	2.12	2.62	108.42	N–H⋯S / C–H⋯S / C–H⋯O	6 / 4 / 12	2.75 / 2.89 / 2.35	3.63 / 3.36 / 3.18	147.11 / 106.43 / 132.60				2.92
$\gamma_L(g^+,g^+,g^+)$	2	2.06	2.91	142.20	N–H⋯S / C–H⋯O	5 / 10	2.46 / 2.28	3.25 / 3.31	136.09 / 159.38				3.02
$\gamma_L(g^+,g^+,a)$	2	2.10	2.95	142.44	N–H⋯S	5	2.60	3.31	127.88				7.23
$\gamma_L(g^+,a,g^+)$	2	2.01	2.87	143.47	C–H⋯O	10	2.29	2.97	119.86				4.39
$\gamma_L(g^+,a,a)$	2	1.99	2.86	144.00	C–H⋯O	10	2.28	2.96	119.44				4.31
$\gamma_L(g^+,a,g^-)$	2	1.99	2.86	144.03	C–H⋯O	10	2.32	2.99	118.57				

(continued on next page)

Table 4 (continued)

Final geometry[a]	Backbone/backbone Type	$(H\cdots N)$	$(N\cdots N)$	Angle $(N-H\cdots N)$	Backbone/side-chain Type	Participating atoms	$(H\cdots S/O)$	$(C\cdots S/O)(N-S)$	Angle $(C-H\cdots S/O)(N-H\cdots S)$	Type	$(S\cdots N)$	$(S\cdots C)$	ΔE^{b}
$\gamma_L(g^+,g^-,g^+)$	2	2.00	2.86	142.18	10	C–H··O	2.22	2.90	119.59				3.30
					9	C–H··O	2.42	3.37	147.64				1.32
					5	N–H··S	2.45	3.25	136.37				
$\gamma_L(g^+,g^-,a)$	2	1.98	2.85	144.22	10	C–H··O	2.29	2.96	118.38				1.85
					5	N–H··S	2.44	3.27	139.71				
$\gamma_L(g^+,g^-,g^-)$	2	1.97	2.85	144.73	10	C–H··O	2.25	2.95	120.68				1.63
					5	N–H··S	2.42	3.27	141.91				
$\gamma_L(a,g^+,g^+)$	2	2.03	2.88	141.29	4	C–H··S	2.85	3.32	106.34				2.42
$\gamma_L(a,g^+,a)$	2	2.02	2.87	141.68	4	C–H··S	2.84	3.29	105.60				2.68
					10	C–H··O	2.42	3.05	116.39				
$\gamma_L(a,g^+,g^-)$	2	2.03	2.87	140.92	4	C–H··S	2.82	3.29	105.89				2.58
					10	C–H··O	2.44	3.06	115.57				
$\gamma_L(a,a,g^-)^{1st}$	2	2.02	2.87	141.94	10	C–H··O	2.46	3.07	114.76				2.33
$\gamma_L(a,a,g^+)^{2nd}$	2	2.01	2.86	142.17	10	C–H··O	2.54	3.08	110.48				1.71
					9	C–H··O	2.45	3.48	159.34				
$\gamma_L(a,a,a)$	2	2.02	2.88	141.88	10	C–H··O	2.43	3.05	115.46				3.15
$\gamma_L(a,a,g^-)$	2	2.03	2.88	141.64	10	C–H··O	2.42	3.05	115.60				3.09
$\gamma_L(a,g^-,g^+)$	2	2.09	2.91	137.50									8.21
$\gamma_L(a,g^-,a)$	2	2.08	2.91	138.78									6.86
$\gamma_L(a,g^-,g^-)$	2	2.08	2.90	137.89	4	C–H··S	3.10	3.26	89.00				2.55
					9	C–H··O	2.24	3.17	144.37				
$\gamma_L(g^-,g^+,g^+)$	2	1.99	2.85	142.67	5	N–H··S	2.60	3.27	125.75				2.59
$\gamma_L(g^-,g^+,a)$	2	1.99	2.85	142.83									3.71
$\gamma_L(g^-,a,g^+)$	2	2.03	2.89	142.46						8	3.78	3.57	3.54
$\gamma_L(g^-,a,a)$	2	2.05	2.90	141.77						8	3.41	3.37	4.14
$\gamma_L(g^-,a,g^-)$	2	2.05	2.90	141.19						8	3.62	3.44	4.08
$\gamma_L(g^-,g^-,g^+)$	2	2.18	3.01	139.25	4	C–H··S	2.82	3.32	108.35				3.34
					11	C–H··O	2.57	3.62	164.64				
$\gamma_L(g^-,g^-,a)$	2	2.04	2.89	141.50	4	C–H··S	2.87	3.33	105.39				4.03
$\gamma_L(g^-,g^-,g^-)$	2	2.02	2.88	142.41	4	C–H··S	2.88	3.36	106.96				3.17

All angles and distances measured in degree and Ångstrom, respectively. Bold letters indicate the δ_L conformers localized in the γ_L region on the Ramachandran map.

[a] Relative to the global minimum in β_L $[a,g^+,g^+]$, $E = -885.6747322$ and measured in kcal mol^{-1}.

[b] 1st and 2nd are conformers of the same catchment region.

[c] Sulphur and amide plane contacts measured between S and O atoms.

backbone state: $[g^+g^-x]$, $[ag^+x]$, $[g^-ax]$ and $[g^-g^-x]$ ($[g^+ax]$, $[aax]$).

Of course, the previous assumption is not true for all methionine conformers. The backbone amides can interact with the side-chain of methionine, in certain conformations thus the angles and the relative energy differences are not exactly the same as those of methyl-ethyl-thioether. To have a more detailed look into the effect of the terminal torsion angle on the energies, this torsional angle was rotated through 360°, and calculated nine PECs, one for every 'triplet' (altogether 27 curves), at the RHF/3-21G level of theory. These PECs are illustrated in a THEOCHEM Special Issue [33]. Only 8 PECs are similar to the PEC of methyl-ethyl-thioether, namely those β_L ($[aax]$, $[g^+ax]$, δ_L ($[aax]$, $[g^+ax]$, $[g^-ax]$), γ_L ($[aax]$, $[g^+ax]$). The similarity would suggest that the non-bonded electronpairs of sulphur are at such a distance from the backbone that the amides have hardly any effect on the conformational behaviour of the side-chain. If there were no interaction at all, the PECs of the pairs ($[g^+ax]$, $[g^-ax]$) would be totally symmetric.

Nevertheless, the sulphur in the thioether functionality can act as an effective nucleophile and that is the reason why of S-adenosylmethionine can be formed. In that case, sulphur atom in the thioether performs an S_{N2} reaction on the phosphate ester moiety of AMP.

3.4. Intramolecular interactions in N-formyl-L-methioninamide at RHF/3-21G level of theory

After full optimization, 77 stable geometry can be described of N-formyl-L-methioninamide at RHF/3-21G level of theory. There are a couple of possible intramolecular interactions in this diamide model and the most determinative contacts in the molecule are illustrated in Scheme 4. There are three main groups of contacts as follows: backbone/backbone, backbone/side-chain and side-chain/side-chain. The main observed intramolecular interactions are illustrated in Table 4. All backbone/backbone contacts (Type 1, 2 and 3 in Scheme 4) are common in other naturally occurring amino acids except proline. However, side-chain/side-chain and backbone/side-chain interactions are more unique than in any other amino acid. Specialty is

given to methionine due to its less reactive sulphur atom placed in the side-chain than that of cysteine. Two contacts of sulphur with backbone atoms can be defined. One of them is a weak hydrogen bond with any of the three hydrogens as can be seen in Type 4, 5 and 6 contacts (Scheme 4). Second one is the interaction of sulphur with the electrons of one of the two amide planes thus the sulphur approaches the amide planes approximately orthogonally (Type 7 and 8 in Scheme 4). Further, the very electronegative oxygen in one of the amide bonds may produce another type of hydrogen bond with hydrogens connected to either γ or ε carbon atoms (Type 9, 10, 11 and 12 in Scheme 4). Both of these hydrogen bonds are plausible from the observed structures and the fact that these carbons connected to the sulphur atom.

3.5. Stabilization energy calculation at RHF/3-21G level of theory

For measuring the stabilization effect of the methionine side-chain, we have calculated the isodesmic reactions as illustrated in Scheme 3. Stabilization energy shows stabilization or destabilization during the production of N-formyl-L-methioninamide, which has ethyl-methyl-tioether as a side-chain, from protected glycine, which has only a hydrogen in the same position, and from propyl-methyl-thioether, having two different torsion angles. The calculated stabilization energies are summarized in Table 3 and shown in Fig. 5. There are two conformers (α_L $[ag^-g^-]$ and α_L $[ag^-a]$) that have positive stabilization energy, therefore in this instance the side-chain has destabilizing effect on the backbone. In general, α_L conformers are not favoured energetically and only those can be observed that have a special positive effect on the conformation. Data of the stabilization energies of the global minimum of diamide amino acids computed before at RHF/3-21G level of theory together with the stabilization energy value of the global minimum of N-formyl-L-methioninamide can be summarized in a diagram (Fig. 6). As can be seen on the diagram the degree of the side-chain polarity of amino acids effectively contributes to the stabilization on the structure by distinct backbone/side-chain interactions. Thus, three main groups can be separated in connection with the stabilization energies. The first group contains amino acids whose side-chain is fully apolar character. This group contains methionine but with the largest stabilizaton

effect on the conformation due to the sulphur atom in the side-chain. This is in good agreement with the above mentioned effects of sulphur on backbone, therefore methionine is fairly in the next group of

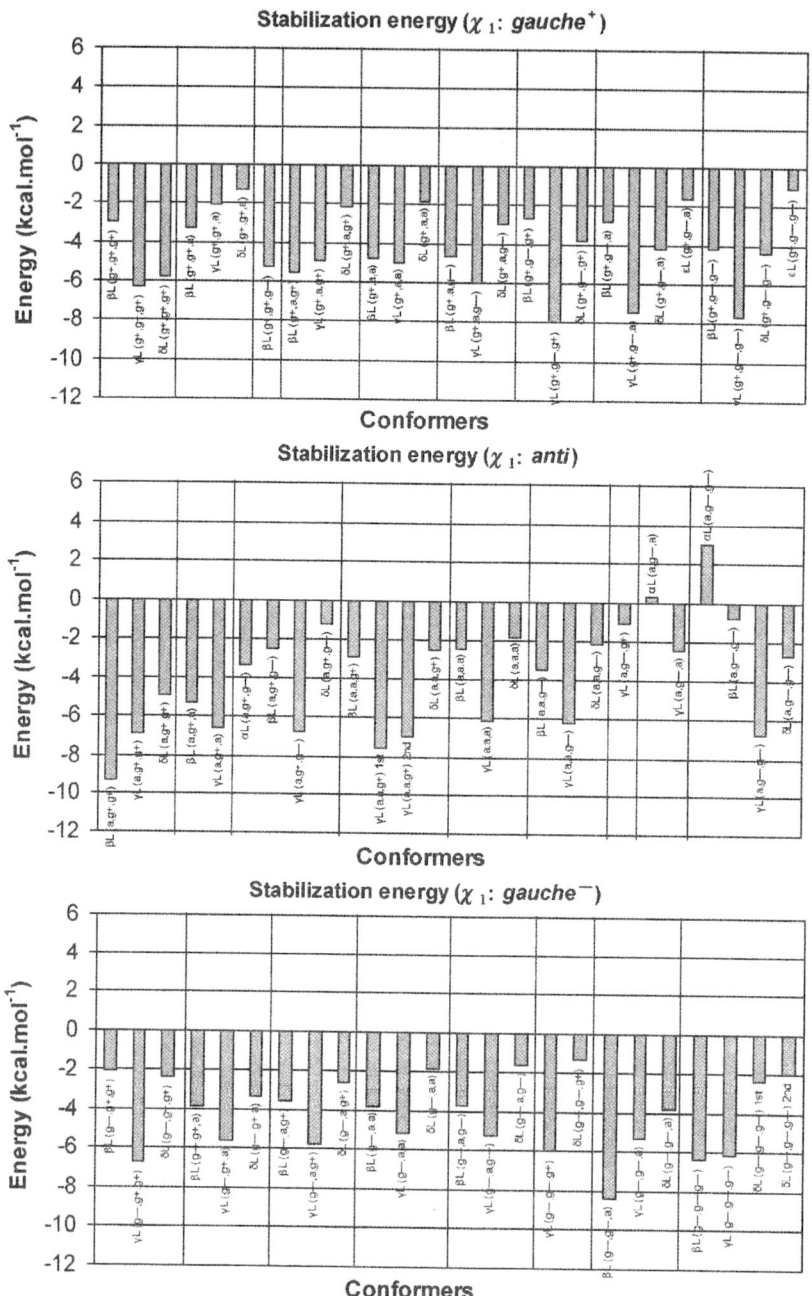

Fig. 5. Stabilization energies of the different conformers calculated as presented in Scheme 3. The stabilization energy was calculated from relative energies.

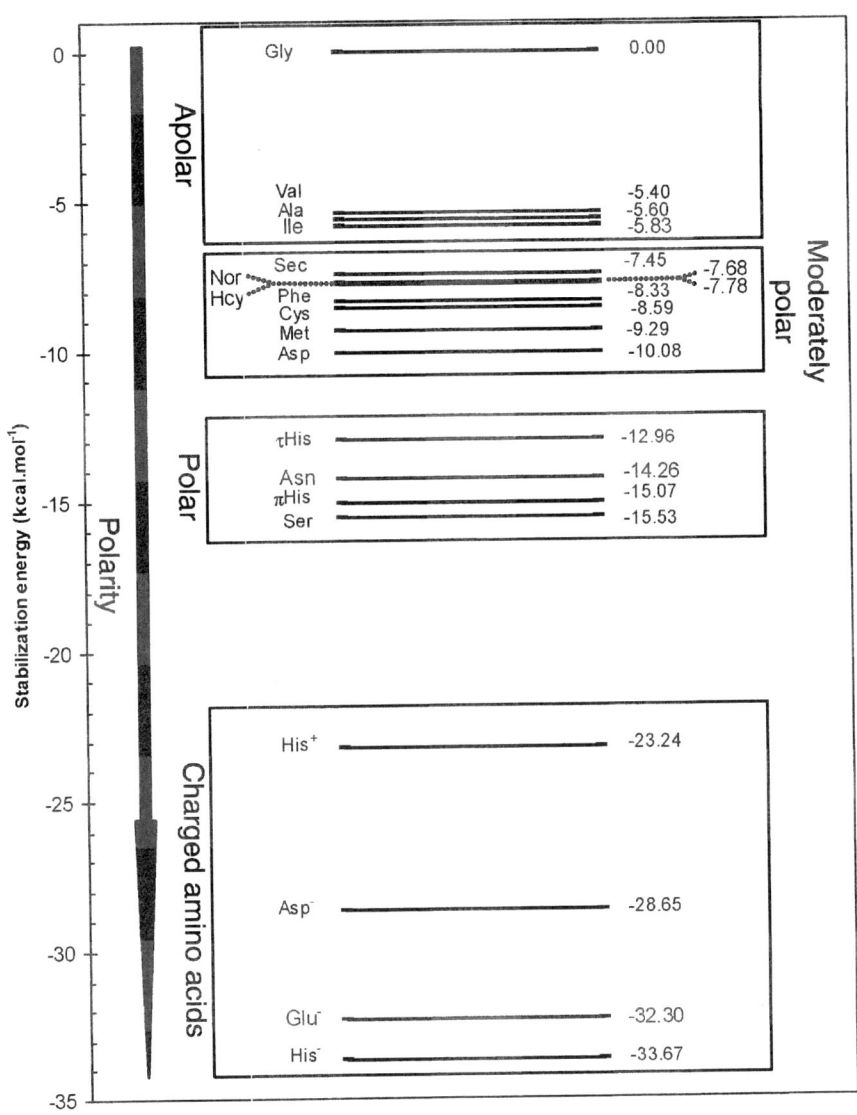

Fig. 6. Stabilization energies of different amino acid diamide models. All stabilization energy values are calculated at RHF/3-21G level of theory. Each amino acid diamide model is presented in its most stable conformation. Glutamate and homocysteine are calculated in N-acetyl-L-aminoacid-N-methylamide forms. For homocysteine, only the full extended backbone forms were computed. The global minimum of N-formyl-L-norleucinamide is localized in γ_L [aaa] (unpublished data). Stabilization energy of norleucine and homocysteine is indicated by a wider line because their stabilization energy is close to each other.

amino acids. This group consists of polar amino acids. The overlapping between the two groups shows that the separation to polar and apolar characters are not that obvious. The last, very discrete group contains amino acids whose have charged side-chain, thus the intramolecular interactions are more expressed allowing much larger stabilization.

4. Conclusion

Our main focus was to explore the multidimensional conformational space of the five main backbone regions on the Ramachandran surface of methionine using RHF/3-21G level of theory because of the high number of the expected minima (135) and describe the conformers and their energies. During the exploration, we obtained an appreciable reduction, altogether 77 minima. We found that *N*-formyl-L-methioninamide has two conformers in ε_L and three in α_L backbone conformation which is unusual for apolar amino acids. In the case of two conformations, the catchment region contained two non-equivalent minimum. Exploration in the β_L region at the RHF/3-21G level of theory allowed us to compare them to the results of homocysteine, a highly similar amino acid, that have been calculated previously [14]. Comparison showed little difference between homocysteine and methionine, however a very stable conformer of methionine was not found in homocysteine, and methionine lacked a stable conformer of homocysteine. Further, comparison of the data of backbone conformers (β_L, δ_L and γ_L) helped to understand the effect of amide planes on the side-chains. The expected δ_L conformers mainly remained in δ_L or in three cases these migrated into a more stable γ_L backbone conformation where the backbone has a hydrogen bond.

The energetic observation of the conformers of *N*-formyl-L-methioninamide and a comparison to the values of the minima of methyl-ethyl-thioether showed a few conformers where the side-chain and the backbone of this methionine model have hardly any interaction. Furthermore, the stabilization energies indicate stabilization expressed on the structures except of two conformers of the three α_L.

Acknowledgements

One of us (IGC) wishes to thank the Ministry of Education of Hungary for a Szent-Györgyi Visiting Professorship. The helpful discussions of Anna K. Füzéry, and the kind instructions by Tamás Beke are gratefully acknowledged.

References

[1] M.S. Hershfield, V.N. Aiyar, R. Premakumar, W.C. Small, Biochem. J. 230 (1985) 43.

[2] R.J. Hohman, M.C. Guitton, M. Veron, Arch. Biochem. Biophys. 233 (1984) 785.

[3] L. Serre, G. Verdon, T. Choinowski, N. Hervouet, J.L. Risler, C. Zelwer, J. Mol. Biol. 306 (2001) 863.

[4] D.E.L. Wilcken, B. Wilcken, J. Clin. Invest. 57 (1976) 1079.

[5] N.A.J. Carson, D.W. Neill, Arch. Dis. Child 37 (1962) 505.

[6] D.W. Jacobsen, Clin. Chem. 44 (1998) 1833.

[7] H. Jakubowski, J. Biol. Chem. 275 (2000) 21813.

[8] J.S. Stamler, D.I. Simon, J.A. Osborne, M.E. Mullins, O. Jaraki, T. Michel, D.J. Singel, J. Loscalzo, Proc. Natl. Acad. Sci. USA 89 (1992) 444.

[9] H. Jakubowski, Proc. Natl. Acad. Sci. USA 87 (1990) 4504.

[10] H. Jakubowski, EMBO J. 10 (1991) 593.

[11] H. Jakubowski, FEBS Lett. 317 (1993) 237.

[12] K. Kokame, H. Kato, T. Miyata, J. Biol. Chem. 271 (1996) 29659.

[13] P.A. Outinen, S.K. Sood, P.C. Liaw, K.D. Sarge, N. Maeda, J. Hirsh, J. Ribau, T.J. Podor, J.I. Weitz, R.C. Austin, Biochem. J. 332 (1998) 213.

[14] S. Althausen, W. Paschen, Brain Res. Mol. Brain Res. 84 (2000) 32.

[15] K. Kokame, K.L. Agarwala, H. Kato, T. Miyata, J. Biol. Chem. 275 (2000) 32846.

[16] G.H. Werstuck, S.R. Lentz, S. Dayal, G.S. Hossain, S.K. Sood, Y.Y. Shi, J. Zhou, N. Maeda, S.K. Krisans, M.R. Malinow, R.C. Austin, J. Clin. Invest. 107 (2001) 1263.

[17] H. Jakubowski, L. Zhang, A. Bardeguez, A. Aviv, Circ. Res. 87 (2000) 45.

[18] R. Nusse, A. Brown, J. Papkoff, P. Scambler, G. Shackleford, A. McMahon, R. Moon, H. Varmus, Cell 64 (1991) 231.

[19] R.C. Conaway, C.S. Brower, J.W. Conaway, Science 296 (2002) 1254.

[20] A.H. Huber, W.I. Weis, Cell 150 (2001) 391.

[21] T.A. Graham, C. Weaver, F. Mao, D. Kimelman, W. Xu, Cell 103 (2000) 885.

[22] M.A. Zamora, H.A. Baldoni, J.A. Bombasaro, M.L. Mak, A. Perczel, Ö. Farkas, R.D. Enriz, J. Mol. Struct. (THEOCHEM) 540 (2001) 271.

[23] A.R. Sheraly, R.V. Chang, G.A. Chass, J. Mol. Struct. (THEOCHEM) 619 (2002) 21.

[24] IUPAC-IUB Commission on Biochemical Nomenclature, Biochemistry 9 (1970) 3471.

[25] G.N. Ramachandran, C. Ramakrishnan, V. Sashisekharan, J. Mol. Biol. 7 (1963) 95.

[26] A. Perczel, J.G. Angyán, M. Kajtár, W. Viviani, J.-L. Rivail, J.-F. Marcoccia, I.G. Csizmadia, J. Am. Chem. Soc. 113 (1991) 6256.

[27] M.J. Frisch, G.W. Trucks, H.B. Schlegel, G.E. Scuseria, M.A. Robb, J.R. Cheeseman, V.G. Zakrzewski, J.A. Montgomery, R.E. Stratmann, Jr., J.C. Burant, S. Dapprich, J.M. Millam, A.D. Daniels, K.N. Kudin, M.C. Strain, Ö. Farkas, J. Tomasi, V. Barone, M. Cossi, R. Cammi, B. Mennucci, C. Pomelli, C. Adamo, S. Clifford, J. Ochterski, G.A. Petersson, P.Y. Ayala, Q. Cui, K. Morokuma, D.K. Malick, A.D. Rabuck, K. Raghavachari, J.B. Foresman, J. Cioslowski, J.V. Ortiz, A.G. Baboul, B.B. Stefanov, G. Liu, A. Liashenko, P. Piskorz, I. Komáromi, R. Gomperts, R.L. Martin, D.J. Fox, T. Keith, M.A. Al-Laham, C.Y. Peng, A. Nanayakkara, M. Challacombe, P.M.W. Gill, B. Johnson, W. Chen, M.W. Wong, J.L.

Andres, C. Gonzalez, M. Head-Gordon, E.S. Replogle, J.A. Pople, GAUSSIAN 98, Revision A.9, Gaussian, Inc., Pittsburgh PA, 1998.

[28] P. Hudáky, T. Beke, A. Perczel, J. Mol. Struc. (THEOCHEM) 583 (2002) 117.

[29] P. Hudáky, I. Hudáky, A. Perczel, J. Mol. Struc. (THEOCHEM) 583 (2002) 199.

[30] F.C. Calaza, M.V. Rigo, A.N. Rinaldoni, M.F. Masman, J.C.P. Koo, A.M. Rodríguez, R.D. Enriz, Comprehensive conformational analysis of N-acetyl-L-isoleucine-N-methy-lamide: an ab initio study, J. Mol. Struc. (THEOCHEM) (2003) in press.

[31] M.R. Peterson, I.G. Csizmadia, J. Am. Chem. Soc. 100 (1978) 6911.

[32] S. Inagaki, S. Ohashi, T. Kawashima, Org. Lett. 1 (1999) 1145.

[33] A. Láng, A.K. Füzéry, T. Beke, P. Hudáky, A. Perczel, Potential Energy Curves, Surfaces and Hypersurfaces. A model to follow and understand the conformational trans-formations in amino acids, J. Mol. Struc. (THEOCHEM) in press.

ELSEVIER

Journal of Molecular Structure (Theochem) 666–667 (2003) 243–249

THEO
CHEM

www.elsevier.com/locate/theochem

The multidimensional conformational analysis for the backbone across the disrotatory axis at selected side-chain conformers of N-Ac-homocysteine-NHMe—an ab initio exploratory study

Aly R. Sheraly[a,b,*], Gregory A. Chass[a,b,c,d], Imre G. Csizmadia[a,b,e]

[a]Global Institute of COmputational Molecular and Materials Science (GIOCOMMS), 1422 Edenrose St, Mississauga, Ont., Canada L5V 1H3
[b]Department of Chemistry, University of Toronto, 80 St George St., Toronto, Ont., Canada M5S 3H6
[c]Institut de Science et d'Ingénierie Supramoléculaires, 8, allée Gaspard Monge, BP 70028, 67083 Strasbourg Cedex, France
[d]Department of Biomedical Sciences, Creighton University, 2500 California plaza, Omaha, NE 68178, USA
[e]Department of Medical Chemistry, Szeged University, H-6720 Szeged, Dom ter 8, Hungary

Abstract

Conformations at 30° intervals along the disrotatory axis of the homocysteine backbone were optimized using three sidechain conformations: ag + a, aaa and ag − a.
© 2003 Elsevier B.V. All rights reserved.

Keywords: Conformational analysis; Disrotatory axis; Conformers; Homocysteine

1. Introduction

Methionine and cysteine are interconvertable through a homocysteine intermediate metabolic pathway [1]. The enzyme methionine synthase, a vitamin B_{12} dependent complex, converts homocysteine to methionine through a methyl transfer process using N^5-methyl-tetrahydrofolate as a methyl donor. N^5-methyl-tetrahydrofolate is generated from its precursor N^5,N^{10}-methylenetetrahydrofolate through the catalytic action of the enzyme methylenetetrahydrofolate reductase (MTHFR) [1]. Homocysteine is then converted to cysteine using the vitamin B_6 dependent complex, cystathionine β-synthase. A genetic defect

or deficiency in any of the key enzymes or a deficiency in folate, B_6 or B_{12} can result in elevated homocysteine levels in the body, a condition known as hyperhomocysteinemia [2].

Hyperhomocysteinemia has been correlated with the development of arteriosclerosis in which the vascular vessels thicken, harden and become less compliant [3,4]. Although the involvement of homocysteine in vascular damage has not be clearly established several mechanisms implicating homocysteine have been proposed [5].

High levels of homocysteine has been proposed to inhibit the enzyme dimethylarginine dimethylaminohydrolase (DDAH) responsible for the degradation of asymmetric dimethylarginine (ADMA) [6]. In endothelium cells, ADMA inhibits NO synthase, the enzyme responsible for the production of the potent vasodilatory, NO. NO diffuses into the vascular smooth muscle activating cytosolic guanlyly cyclase

* Corresponding author.
E-mail addresses: aly.sheraly@utoronto.ca (A.R. Sheraly), gchass@giocomms.og (G.A. Chass), icsizmad@chem.utoronto.ca (I.G. Csizmadia).

resulting in decreased levels of cytosolic Ca^{+2}, thereby causing vasodilation of the vessel. Inhibition of DDAH results in increased concentrations of ADMA, resulting in a decrease in concentration of NO thus preventing NO-mediated vasodilation.

It has also been suggested that homocysteine contributes significantly to endothelium oxidative stress [7]. It has been postulated that homocysteine may inhibit the expression of glutahione peroxidase, an enzyme responsible for the reduction of numerous hydroperoxides [8]. These peroxides contribute to the formation of radical oxygen species responsible for a variety of disruptions in cellular metabolism, including DNA damage [9]. These reactive oxygen species are also created when homocysteine auto-oxidizes to homocystine, in which two homocysteine residues are linked through their sidechains to form a disulfide bridge [9]. The rise in reactive oxidation species leads to the increased expression of endothelium adhesion molecules, which potentially leads to the increased deposition of oxidized LDL particles along the vessel wall [4]. This deposition causes the vascular vessels to become hard and less compliant. Thus some have indicated that increased levels of homocysteine rather than increased levels of cholesterol are the primary cause of vascular damage and arteriosclerosis [4].

High intracellular homocysteine levels have also been implicated with endothelium and hepatocyte endoplasmic reticulum stress resulting in the induction of protein unfolding genes and the increased expression of the sterol regulatory element-binding proteins (SREBP). The upregulation of the SREBP cause an increase in gene expression of the enzymes responsible for cholesterol and triglyceride biosynthesis, and thus is correlated with an accumulation of cholesterol in the liver and secretion of VLDL [10].

Homocysteine, like most other amino acids can exist in many conformational isomers. Nine theoretically probable backbone conformations are topologically probable as stable conformers when rotating the backbone dihedrals ϕ_i and ψ_i from 0 to 360°. The definitions of the backbone dihedral angles are shown for the homocysteine residue in Fig. 1. A traditional plot of dihedral angle rotations from 0 to 360° about ϕ_i and ψ_i for any amino acid diamide yields a Ramachandran map (Fig. 2). Each box corresponds to a region in which one of the nine minima can be located. Each has been assigned a L or D subscripted

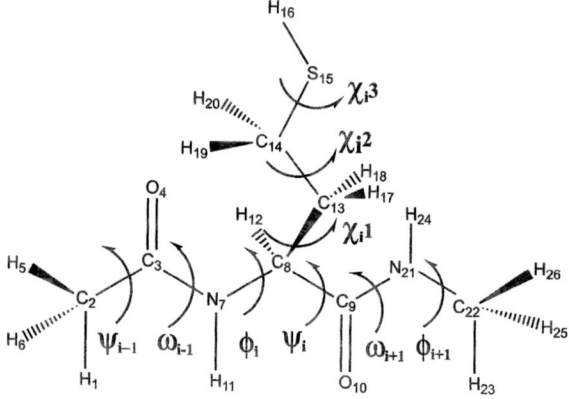

Fig. 1. The model of homocysteine used to conduct the study. Important dihedrals are labeled. The ϕ_i, ψ_i represent the backbone dihedrals that were rotated systematically along the disrotatory axis, while χ_1, χ_2 and χ_3 represent the sidechain dihedrals.

Greek letter for facilitated reference indicating the axis of chirality associated with the conformational twist. L-amino acids tend to be exclusively used by organisms and thus most of the computational research is focused on uncovering the conformational characteristics of these L-enantiomers.

All topologically probable sidechain conformers have been studied for the β_L backbone conformer previously [1]. However, the influence on the potential energy and stability of specific sidechain

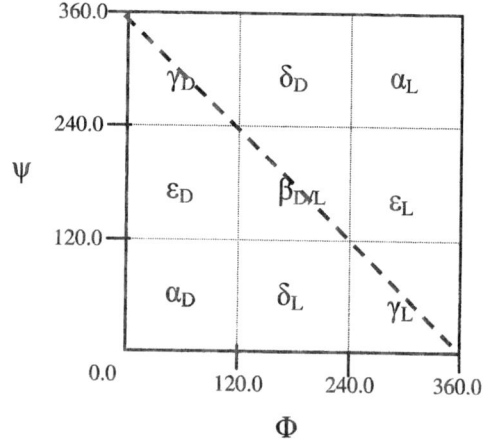

Fig. 2. A plot of the Ramachandran map indicating the nine possible backbone conformation that will yeild potential energy minima. The dashed line represents the disrotatory axis and points at 30.0° intervals along this line were selected as input values for the backbone dihedrals for this study.

conformers on the various backbone conformers has not been examined. Thus in this study, selected sidechain conformers were investigated while the backbone was rotated along the disrotatory axis as shown by the dashed line in the Ramachandran map (Fig. 2). The disrotatory axis was chosen as it represents the γ_D-β_L-γ_L axis. Generally, the backbone conformers that consistently yield the most stable minima are the γ_D, β_L,γ_L backbone conformers. By uncovering the factors that influence the stabilization of the molecule it may become more possible to

understand and predict the mechanisms and biological pathways that generate homocysteine as an intermediate.

2. Method

An acetyl group was attached to the N-terminus and a methylamide group was attached to the C-terminus of the central homocysteine residue (Fig. 1), in order to simulate neighboring residues and to

Table 1

Tabulation of rotation of ϕ_i, ψ_i along disrotatory axis while the homocysteine sidechain is in the selected conformers ag + a, aaa and ag − a

ϕ_i, ψ_i (°)	χ_1 (°)	χ_2 (°)	χ_3 (°)	Energy (Hartrees)	Relative Energy (kcal mol^{-1})
ag + a					
60, −60	180.5215	57.495	178.0963	−924.490066303	0.575306816
90, −90	−178.7154	59.5500	173.6613	−924.482562348	5.284113618
120, −120	−167.7387	73.1171	−168.2837	−924.472078948	11.86255195
150, −150	−153.8825	72.7761	−169.0450	−924.474106152	10.59046117
180, 180	177.8810	51.0967	177.0844	−924.487094697	2.440019297
−150, 150	176.0513	59.3552	−179.5816	−924.490251851	0.45887359
−120, 120	177.0782	61.6782	−179.6640	−924.486443567	2.848609883
−90, 90	−179.9503	63.9463	−173.1741	−924.490983112	0.0
−60, 60	−166.2615	67.291	−165.8767	−924.488240252	1.721172079
aa					
60, −60	−176.4235	173.3731	176.1270	−924.485682468	2.969075488
90, −90	−175.0117	176.405	177.3731	−924.480931530	5.950336592
120, −120	−166.3756	−170.3735	−177.8732	−924.472112728	11.48422304
150, −150	−158.8593	−139.5757	176.2420	−924.476366030	8.815233497
180, 180	−168.4357	168.3507	174.7001	−924.482874490	4.731109763
−150, 150	−174.4277	178.7535	179.6314	−924.486910606	2.198406611
−120, 120	−176.2563	178.0835	177.9445	−924.484777945	3.536672715
−90, 90	−178.7573	174.0817	169.5206	−924.490413987	0.0
−60, 60	−164.0840	176.7146	174.6222	−924.489401597	0.635284849
ag − a					
60, −60	−173.1471	−70.0943	179.0114	−924.477601969	4.716143021
90, −90	−158.8848	−75.9381	−170.0877	−924.475908265	5.778959219
120, −120	−147.8708	−65.5901	−174.2932	−924.468819280	10.2273682
150, −150	−160.0^F	−60.0^F	170.0^F	−924.464388849	13.00750795
180, 180	−160.0^F	−60.0^F	170.0^F	−924.476576664	5.359532162
−150, 150	−155.3312	−65.1981	170.7832	−924.485116887	0.000456827
−120, 120	−160.0^F	−65.0^F	170.0^F	−924.481260762	2.420213826
−90, 90	−155.1814	−65.2575	170.8915	−924.485117615	0.0
−60, 60	−136.1054	−61.4161	168.4527	−924.480552855	2.864432548

For each selected sidechain (ag + a, aaa and ag − a), the optimized sidechain dihedrals χ_1, χ_2 and χ_3 are tabulated for each backbone point (ϕ_i, ψ_i) calculated along the disrotatory axis. For the sidechain ag − a, some of the points failed to optimize when the sidechain were relaxed and thus these calculations were carried out with the sidechain frozen. These cases are noted with the superscript F and thus the sidechain dihedrals represent the input value rather than the optimized value. Potential energy of each conformer was also tabulated in Hartrees and the relative energy was calculated in kcal mol^{-1}. Relative energy was calculated by subtracting the potential energy of each conformer from the most stable conformer for a given sidechain.

conserve the Cα–CO–NH–Cα polypeptide backbone structure. Thus by adding the N- and C-protecting groups, the effects of the neighboring Cα's on the central homocysteine residue is being considered even though homocysteine is not one of the amino acid residues used to synthesize polypeptides.

Input files were constructed using an internal coordinates system. By convention a standardized and modular numbering system was employed to number the homocysteine model, as outlined by a set of rules previously set down [11,12]. The atoms of the acetyl group were numbered first completing the (i − 1) residue, followed by the homocysteine residue numbered along the backbone. The sidechain was then numbered along the sidechain backbone heavy atoms into the peptide backbone, from the Cα to the thiol group. The next module to be numbered was the methyl-amide attached to the homocysteine, which correspond to the (i + 1) residue. For each module all heavy atomic nuclei were numbered prior to the H also present in the module. Fig. 1 shows the model

used for this study and with the inherent atomic numbering system described.

The resultant input file was used as a starting point in two successive and iterative process of GAUSSIAN 98 [13] cycles to bring about a geometry optimization at the ab initio level RHF/3-21G of 'found' AM1 semi-empirical minima. The sidechain dihedrals were relaxed as the backbone dihedrals were systematically rotated along the disrotatory axis. In accordance with the IUPAC-IUB recommendation [14], all dihedral angles were reported within − 180.0 and 180.0° for both backbone and sidechain optimized structures.

3. Results and discussion

Calculations were carried out at the RHF/3-21 level of theory in such a manner that the residue maintained a *trans–trans* (ω_{i-1}, ω_i = 180.0°) conformation. The sidechain conformers selected were ag + a, aaa and ag − a. These conformers were chosen as only χ_2 varied (g + ,a,g −) while χ_1

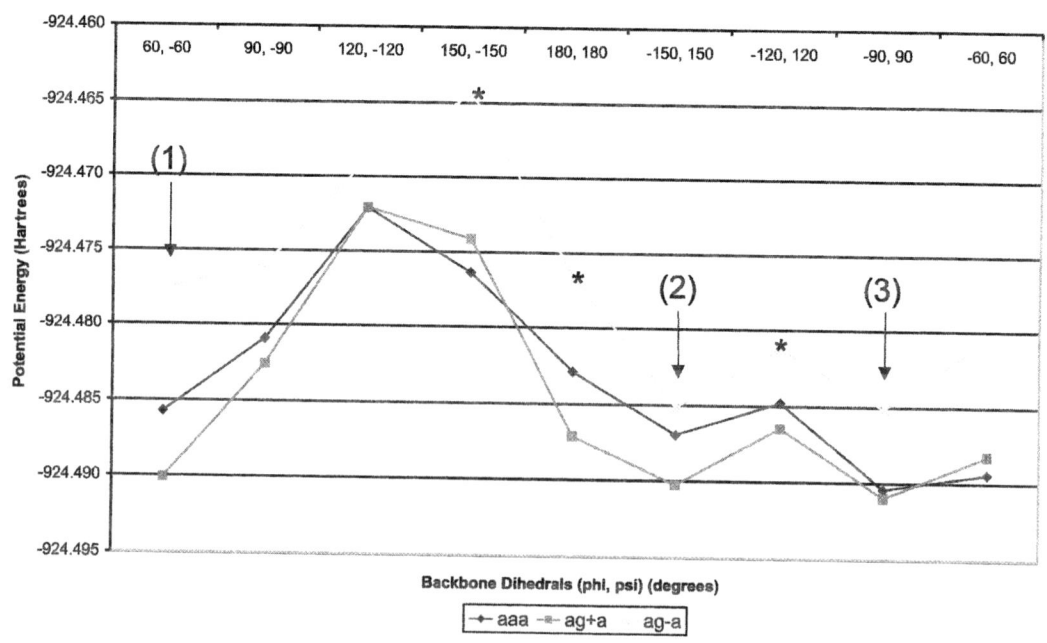

Potential energy curve along disrotatory axis for Homocysteine for the sidechain conformers of ag+a, aaa and ag-a

Fig. 3. A plot of the potential energy for each conformer for a given sidechain using the values tabulated in Table 1. Note that (*) for the ag − a curve represent those conformers that had their sidechain values frozen in order to obtain an energy value. Also the minima γ_D, β_L and γ_L as predicted by the Ramachandran map in Fig. 2 are also labeled in the plot as (1), (2) and (3), respectively.

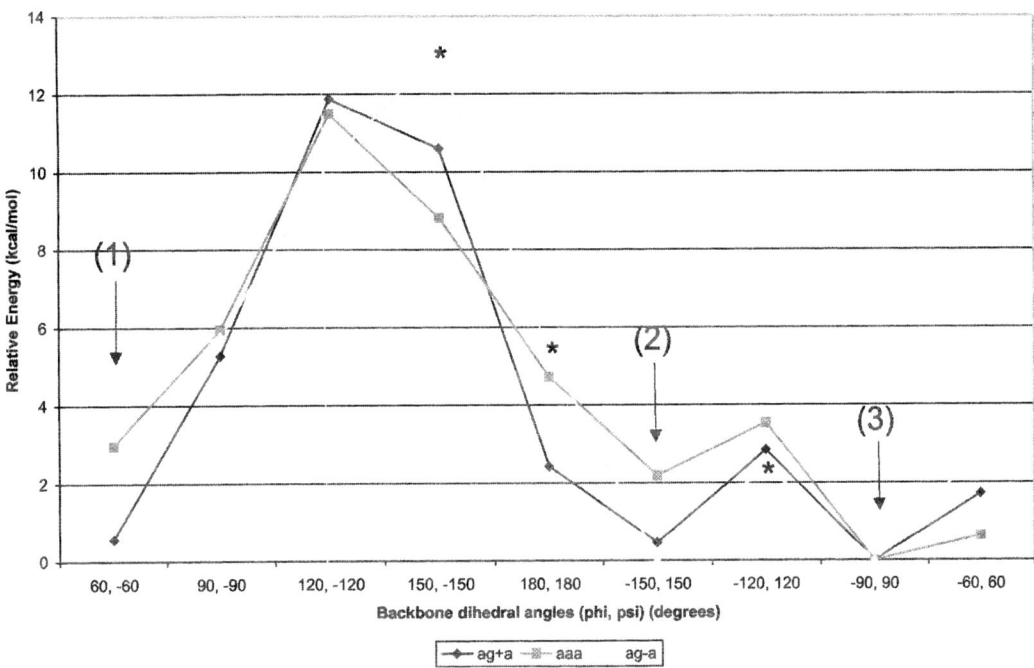

Fig. 4. A plot of the relative potential energy as tabulated in Table 1, for each conformer of a given sidechain with respect to the most stable conformer of that given sidechain. Note that (*) for the ag − a curve represent those conformers that had their sidechain values frozen in order to obtain an energy value. Also the minima γ_D, β_L and γ_L as predicted by the Ramachandran map in Fig. 2 are also labeled in the plot as (1), (2) and (3), respectively.

and χ_3 were maintained at the (a) position. For each sidechain conformation, the backbone dihedrals were systematically varied in increments of 30° intervals along the disrotatory axis to achieve relatively detailed energy plots. The total energy for each conformer was tabulated in Hartrees in Table 1, and converted to kcal/mol using the conversion factor 627.51 kcal/mol = 1 Hartree. When possible the calculation was conducted while the sidechain was fully relaxed. However, several of the ag − a conformers failed to converge when the sidechain was relaxed. Thus, in these cases as denoted in Table 1, the sidechain was also frozen and the computation was completed to obtain a potential energy value. While freezing the sidechain dihedrals will influence the potential energy of the structure, these conformers represented structures that occurred at the potential energy maxima and are generally not considered to exist. Their inclusion was only to better understand

the environment surround the potential energy minima and thus help to generate the shape of the potential energy curve, which is constructed in Fig. 3 The potential energy of the most stable backbone conformer, for a given sidechain, was subtracted from each backbone conformer along the disrotatory axis and yielded the relative potential energy as tabulated in Table 1 For each sidechain conformer of homocysteine, the most stable backbone conformer was the γ_L as indicated by the relative energy plot constructed in Fig. 4.

From the potential energy curve, all three sidechain conformers appear to take a similar shape. For all backbone conformers along the disrotatory axis the most stable sidechain conformer is the ag + a conformer (Fig. 3). Each of the three sidechain conformers yielded three potential energy minima as the backbone was rotated along the disrotatory axis. These minima correspond to the γ_D, β_L, γ_L as indicated in on the plot (Fig. 3). Ideally, as predicted by the Ramachandran map, the ϕ_i and ψ_i backbone dihedrals should correspond to $(60°, -60°)$ for γ_D, $(180°,180°)$ for β_L and $(-60°,60°)$ for γ_L. However, as expected, these minima did not correspond to the ideal backbone dihedral angles but were skewed. Previous studies showed that the β_L backbone dihedrals were skewed [1]. This was expected as the only amino acid that conforms to the ideal values is glycine due to its lack of a bulky sidechain. All other amino acids have a bulky sidechain compared to glycine and thus the backbone dihedrals adjust accordingly to reduce intramolecular strain [15–23].

Interestingly there is a large energy barrier between the γ_D–β_L conformers and a small energy barrier between the β_L–γ_L conformers for all three sidechain conformers (Fig. 4). This may indicate that the β_L–γ_L backbone conformers are easily interconvertable. In each case, the backbone dihedral that showed the most stability was the γ_L followed by the β_L while the γ_D conformer was the least stable (Fig. 4).

The rotation of the backbone dihedrals of a homocysteine residue around the disrotatory axis while relaxing the sidechain conformers in selected conformers yields potential energy plots that allow for comparison. In this study, conformers were selected to observe the effect of varying the sidechain dihedral χ_2 while the remaining two sidechain dihedrals were maintained in the (a) position. It was noted that the ag + a conformer was the most stable conformer throughout the rotation along the disrotatory axis, while the ag − a was the least stable. Also for each sidechain conformer, all three backbone minima were present for each sidechain conformer with the γ_L backbone conformer being the most stable. It was also noted that the energy barrier between the β_L–γ_L backbone conformers was not very large indicating that these backbone conformers may be easily interconverted.

Acknowledgements

This work was supported by grants from the Global Institute of COmputational Molecular and Materials Science (GIOCOMMS), Toronto, Ontario, Canada. The authors thank David H. Setiadi and Tania A. Pecora for helpful discussions during the preparation of this manuscript, as well as Graydon Hoare for database management, network support and software and distributive processing development. Special thanks is extended to Andrew M. Chasse, Christopher M. Andrews and Michael R. Sahai for their development of novel scripting and coding techniques, which facilitate a reduction in the number of CPU cycles needed. The pioneering advances of Kenneth P. Chasse, in all composite computer-cluster software and hardware architectures, are also acknowledged.

One of the authors (IGC) wishes to thank the Ministry of Education for a Szent-Györgyi Visiting Professorship.

References

[1] A.R. Sheraly, R.V. Chang, G.A. Chass, J. Mol. Struct. (THEOCHEM) 619 (2002) 21.

[2] M.J. Baigent, in: H. Kalant, W.H.E. Roschlau (Eds.), Vitamins in Principles of Pharmacology, 6th ed., Oxford University Press, New York, 1998, p. 847.

[3] M. Ward, Int. J. Vitam. Nutr. Res. 71 (3) (2001) 2471.

[4] K.S. McCully, The Homocysteine Revolution, Keats Publishing Inc, New Canaan, CT, 1997.

[5] O. Stanger, M. Weger, W. Renner, R. Konetschny, Clin. Chem. Lab. Med. 8 (2001) 725.

[6] M.C. Stuhlinger, P.S. Tsao, J.H. Her, M. Kimoto, R.F. Balint, J.C. Cooke, Circulation 104 (21) (2001) 2569.

[7] J.C. Chambers, P.M. Ueland, M. Wright, C.J. Dore, H. Refsum, J.S. Kooner, Circ. Res. 89 (2) (2001) 187.

[8] N. Weiss, Y.Y. Zhang, S. Heydrick, C. Bierl, J. Loscalzo, Proc. Nat. Acad. Sci. USA 98 (22) (2001) 12503.

[9] R. Ragone, FASEB J. 16 (3) (2002) 401.

[10] G.H. Werstuck, S.R. Lentz, S. Dayal, G.S. Hossain, S.K. Sood, Y.Y. Shi, J. Zhou, N. Maeda, S.K. Krisans, M.R. Malinow, R.C. Austin, J. Clin. Invest. 107 (10) (2001) 1263.

[11] G.A. Chass, M.A. Sahai, J.M.S. Law, S. Lovas, Ö. Farkas, A. Perczel, I.G. Csizmadia, Int. J. Quantum Chem. 90 (2002) 933.

[12] M.A. Sahai, G.A. Chass, A. Perczel, S. Lovas, A. Varro, J. Papp. A modular numbering system of selected oligopeptides for molecular computations. J. Mol. Struct. (Theochem) (2003) in this issue.

[13] M.J. Frisch, G.W. Trucks, H.B. Schlegel, P.M.W. Gill, B.G. Johnson, M.A. Robb, J.R. Cheeseman, T. Keith, G.A. Petersson, J.A. Montgomery, K. Raghavachari, M.A. Al-Laham, V.G. Zakrzewski, J.V. Ortiz, J.B. Foresman, J. Cioslowski, B.B. Stefanov, A. Nanayakkara, M. Challacombe, C.Y. Peng, P.Y. Ayala, W. Chen, M.W. Wong, J.L. Andres, E.S. Replogle, R. Gomperts, R.L. Martin, D.J. Fox, J.S. Binkley, D.J. Defrees, J. Baker, J.P. Stewart, M. Head-Gordon, C. Gonzalez, and J.A. Pople, Gaussian, Inc., Pittsburgh PA, GAUSSIAN 98. 1998.

[14] IUPAC, Biochem. J. 121 (1971) 577.

[15] H.A. Baldoni, G.N. Zamarbide, R. Enriz, E.A. Jauregui, O. Farkas, A. Perczel, S.J. Salpietro, I.G. Csizmadia, J. Mol. Struct. (THEOCHEM) 500 (2000) 97 (Millenium Volume).

[16] A. Perczel, J. Angyan, M. Kajtar, W. Viviani, J.-L. Rivail, J.F. Marcoccia, I.G. Csizmadia, J. Am. Chem. Soc. 113 (1991) 6256.

[17] W. Viviani, J.-L. Rivail, I.G. Csizmadia, Theoretica Chimica Acta 85 (1992) 189.

[18] M.A. McAllister, A. Perczel, P. Csaszar, W. Viviani, J.-L. Rivail, I.G. Csizmadia, J. Mol. Struct. (THEOCHEM) 288 (1993) 161.

[19] A. Perczel, O. Farkas, I.G. Csizmadia, J. Am. Chem. Soc. 117 (1995) 1653–1654.

[20] A. Perczel, M. Kajtar, J.F. Marcoccia, I.G. Csizmadia, J. Mol. Struct. (THEOCHEM) 232 (1991) 291.

[21] A.M. Rodriguez, H.A. Baldoni, F. Sovire, R. Nieto-Vasquez, G. Zamarbide, R.D. Enriz, O. Farkas, A. Perczel, I.G. Csizmadia, J. Mol. Struct. (THEOCHEM) 455 (1998) 275.

[22] O. Farkas, M.A. McAllister, J.H. Ma, A. Perczel, M. Hollosi, I.G. Csizmadia, J. Mol. Struct. (Theochem) 369 (1996) 105.

[23] A. Perczel, O. Farkas, I.G. Csizmadia, Can. J. Chem. 75 (1997) 1120. (b) O. Farkas, A. Perczel, J.F. Marcoccia, M. Hollosi, I.G. Csizmadia, J. Mol. Struct. (Theochem) 331 (1995) 27.

ELSEVIER

Journal of Molecular Structure (Theochem) 666–667 (2003) 251–267

THEO CHEM

www.elsevier.com/locate/theochem

An ab initio exploratory study of the full conformational space of MeCO-L-threonine-NH-Me

Michelle A. Sahai[a,b,*], Sanober S. Motiwala[a,b], Gregory A. Chass[a,b,c,d*], Emil F. Pai[e], Botond Penke[f,g], Imre G. Csizmadia[a,b,f*]

[a]Global Institute of COmputational Molecular and Materials Science (GIOCOMMS), 1422 Edenrose Street, Mississauga, Ont., Canada L5V 1H3
[b]Department of Chemistry, University of Toronto, 80 St. George Street, Toronto, Ont., Canada M5S 3H6
[c]Institut de Science et d'Ingénierie Supramoléculaires, 8, allée Gaspard Monge, BP 70028, 67083 Strasbourg Cedex, France
[d]Department of Biomedical Sciences, Creighton University, 2500 California plaza, Omaha, NE 68178, USA
[e]Department of Biochemistry, University of Toronto, 1King's College Circle Toronto, Ont., Canada M5S 1A8
[f]Department of Medical Chemistry, University of Szeged, 6720 Szeged, Dóm tér 8, Hungary
[g]Protein Chemistry Research Group, Hungarian Academy of Sciences, University of Szeged, 6720 Szeged, Dóm tér 8, Hungary

Abstract

Ab initio molecular computations were carried out on N- and C-protected L-threonine. Molecular geometry optimizations were conducted on the 81 possible minimum energy conformers of the molecule at the RHF/3-21G level of theory; 39 conformers were located, of which 34 were common with N- and C-protected serine. The relative stabilities of the various conformers have been analyzed in terms of backbone–backbone and backbone–sidechain hydrogen bonding interactions. The stabilization energies exerted by the sidechain of threonine on the backbone are presented as well.
© 2003 Elsevier B.V. All rights reserved.

Keywords: N-Acetyl-L-threonine-N-methylamide; Ab initio geometry optimizations; Backbone and sidechain conformations; Intramolecular hydrogen bonding; Stabilization energies

1. Preamble

Immunoglobulins (Ig) are glycoproteins secreted by plasma cells that function as antibodies and aid in fighting infections. The structure of an Ig molecule is typically represented as being 'Y' shaped and divided into a host-specific 'constant' region (C) and antigen-specific 'variable' regions (V) (Fig. 1) [1].

Ig molecules are produced during immune responses to bind and neutralize invading antigens. Subsequently, the antigen becomes labelled for removal by phagocytosis. However, to combat the recognition mechanism of Ig, antigens have evolved effective deactivation methods, one of which involves a site-specific cleavage of the Ig molecule at the critical hinge region of the molecule. In this way, only the V region remains attached to the antigen. As a result, the body fails to recognize the antigen

* Corresponding authors. Address: Global Institute of Computational Molecular and Materials Science (GIOCOMMS), 1422 Edenrose Street, Mississauga, Ont., Canada L5V 1H3.

E-mail addresses: msahai@giocomms.org (M.A. Sahai), sanober.motiwala@utoronto.ca (S.S. Motiwala), gchass@giocomms.org (G.A. Chass), pai@hera.med.utoronto.ca (E.F. Pai), penke@ovrisc.mdche.szote.u-szeged.hu (B. Penke), icsizmad@chem.utoronto.ca (I.G. Csizmadia).

0166-1280/$ - see front matter © 2003 Elsevier B.V. All rights reserved.
doi:10.1016/j.theochem.2003.08.031

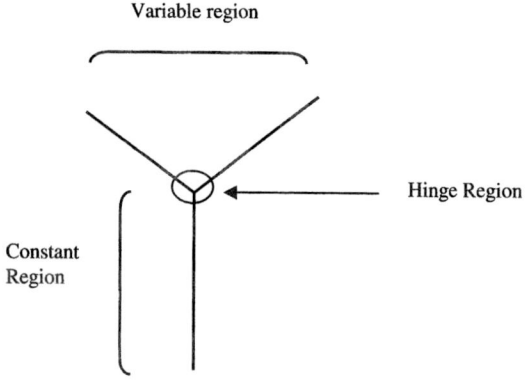

Fig. 1. Simplified pictoral representation of the structure of immunoglobulins (Ig).

as a foreign substance; it mistakenly considers the 'disguised' bacteria to be part of the immune system.

Antibodies are abundantly found in mucous membranes and milk, which are particularly significant for immunity in infants. In these environments, the predominant form of human antibody is IgA1 from the immunoglobulin A (IgA) class, which is regarded as the first line of defence against infection [2]. Therefore, it is not surprising that many human pathogens cleave IgA1 by producing IgA proteases. IgA proteases are extracellular enzymes produced by

I

II

Fig. 2. Structures of For-L-serine-NH$_2$ (I) and N-acetyl-L-threonine-N-methylamide (II), with definitions of backbone and sidechain dihedral angles. The numbering system employed for N-acetyl-L-threonine-N-methylamide is demonstrated in II.

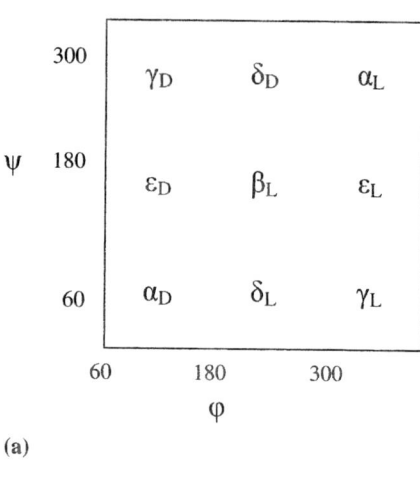

(a)

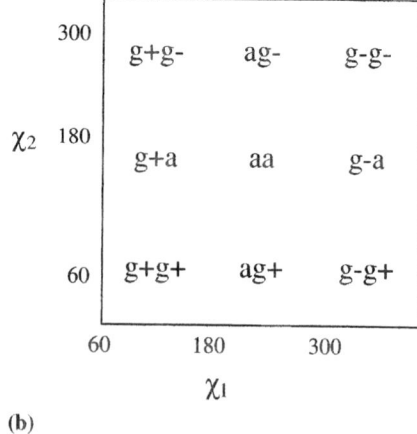

(b)

Fig. 3. Topology of a double-rotor backbone or Ramachandran (a) and sidechain (b) potential energy surface (PES) as cross-section of a potential energy hypersurface (PEHS) of four independent variables: $E = E(\varphi, \psi, \chi_1, \chi_2)$.

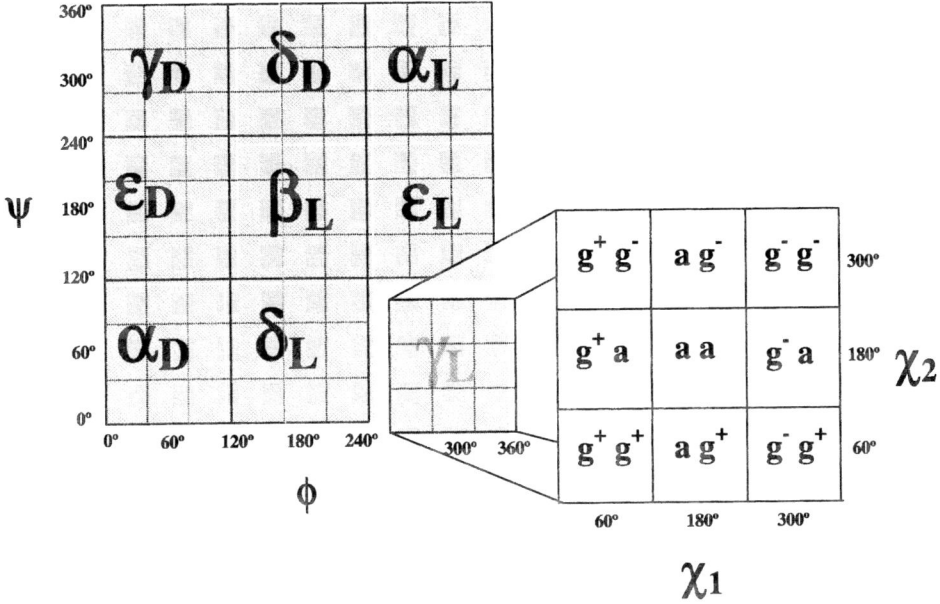

Fig. 4. 2D topology of a Ramachandran PEHS, $E = E(\varphi, \psi)$ of an amino acid residue in a peptide. (Top) conformers are designated by traditional conventions; (bottom) conformers are designated by IUPAC conventions.

a number of Gram-positive and Gram-negative pathogenic bacterial species, including *Streptococcus peneumoniae*, *Bacteroids melaninogenicus*, *Haemophilus influenzae*, *Neisseria gonorrhoeae*, and *N. meningitides* [3,4]. These enzymes cleave IgA1 at specific sites within the proline-rich hinge region, thus dissecting the functional domains of the protein and in turn, impairing its antimicrobial activity [5]. Seven of the 12 known proteases, including proteases from *H. influenzae*, *N. gonorrhoeae* and *S. pneumonia*, have been found to successfully attack and cleave specific sites in the proline-rich hinge region of IgA. More specifically, the Pro-Thr peptide bond is cleaved, rendering IgA inactive.

Medical researchers are increasingly examining diseases at the molecular level and exploring cures at that level. One step towards better understanding of a disease is the conformational analysis of its component structures. In the case of IgA cleavage by proteases, the most obvious place to start is with the tetrapeptide Pro-Pro-Thr-Pro. This proline-containing tetrapeptide is more complicated than some others as the prolines may exhibit *cis* and *trans*-isomerism in their peptide bonds as well as *syn* and *anti*-ring puckering. Even if ring puckering is disregarded and

only *trans*-peptide bonds are considered, Pro-Pro-Thr-Pro may exhibit $3 \times 3 \times 9 \times 3 = 243$ backbone conformations. Incorporating the sidechain conformations for each of the peptides would yield a large number of optimizations. It would be cost-beneficial and time conserving if the tetrapeptide were truncated and optimized geometries were obtained for N- and C-protected mono, di and tri-peptide combinations of the sequence. Since the scission in the body is made at the Pro-Thr peptide bond, it is rational to first obtain conformationally stable structures for proline and threonine diamides. Some preliminary research has been done on the proline diamide [6] and the proline-threonine dipeptide [7]. In this paper, we wish to report the exploration of conformationally

Table 1
Total energy values of the component molecules for isodesmic reactions computed at the RHF/3-21G level of theory

Molecular system	Energy (hartrees)
MeCONH–CH$_2$–CONH–Me (β_L)	-451.29278
Me$_2$–CH–OH	-192.04826
CH$_3$–H	-39.97687

stable structures for one of the diamides, namely *N*-acetyl-L-threonine-*N*-methylamide.

Threonine (Thr) is structurally similar to serine (Ser), another hydroxylic amino acid, differing only by an additional methyl group (Fig. 2). Both amino acid R-groups are short chain –OH residues. Despite their close structural resemblance, serine confers more chemical activity and is generally preferred for certain catalysis reactions that take place in the body. Comparative analysis of the various conformations of Ser and Thr may suggest reasons why enzymes such as proteases often depend on the active site of Ser rather than its Thr counterpart. Intuitively, a single methyl group would not seem considerable enough to cause sufficient disturbance within a molecule. However, it may be the case that steric clash arises from such a moiety in the active site, thereby blocking certain catalytic reactions. In serine, the absence of the extra methyl group may prove to be vital to substrate specificity and thereby the amino acid's mechanistic role in enzymes. Another possible explanation for this is that the extra methyl group on threonine makes it more hydrophobic than serine, leading to an increased tendency to be found on the interior of a protein as opposed to its 'exposed' aqueous environment.

Extensive analysis of N- and C-protected serine [8–14] indicate that out of the 81 possible conformations, 44 optimized structures exist at the ab initio Hartree Fock/3-21G level of theory. Owing to the structural similarity between serine and threonine, results similar to those obtained for For-L-Ser-NH$_2$ are expected for *N*-acetyl-L-Thr-*N*-methylamide.

2. Introduction

The backbone of any amino-acid residue in a peptide or in a protein molecule is a triple rotor. The three dihedral angles ϕ, ψ, and ω measure the rotation about the N–C, the C–CO and the OC–NH (i.e. the peptide) bonds respectively, leading to a potential energy hypersurface: $E = E(\phi, \psi, \omega)$. Since the peptide bond may be either the *cis* or *trans*-isomer, ω usually assumes values in the vicinity of 0° or 180°, respectively. For this reason, it is customary to

consider only a potential energy surface (PES) cross-section of the full potential energy hypersurface (PEHS) such as $E = E(\phi, \psi)$, where ω could be optimized at about 0° or 180°. Most Ramachandran maps of the type $E = E(\phi, \psi)$ found in the literature are associated with the *trans*-peptide bond (where ω is ~180°).

For any given peptide with two optimizable dihedral angles about the peptide bond, ϕ and ψ, the laws of MultiDimensional Conformational Analysis (MDCA) predict nine possible backbone (BB) conformations. These conformations are often depicted on a Ramachandran map (Fig. 3(a)). In threonine, there are two dihedral angles of interest in the sidechain (SC), χ_1 and χ_2. Like the BB dihedrals, each SC dihedral can take three possible conformations, *gauche* + $(g +)$, *anti* (a) and *gauche* − $(g -)$, leading to $3 \times 3 = 9$ possible sidechain conformations (Figs. 3(b) and 4).

Threonine is one of the only two DNA-coded amino acids that demonstrate chirality, with isoleucine being the other chiral amino acid. Here, we have considered threonine in the (S) conformation of the backbone and in the (R) conformation of the sidechain. For the present ab initio study, $3 \times 3 = 9$ conformations for the BB and $3 \times 3 = 9$ for the SC result in a total of 81 possible geometry optimizations, which were carried out on *N*-acetyl-L-Thr-*N*-methylamide at the RHF/3-21G level. Ab initio results obtained for threonine were compared with previous conformational studies performed on *N*-formyl-L-Ser-amide [8–14].

3. Method

3.1. Conformational analysis

A modular numbering system was adopted such that the molecule was divided into three sections, the amino acid residue and the N- and C-protective groups. Each section was numbered separately. This modular system allows the protective groups to be removed and replaced by other amino acids, thus facilitating the use of our results in future studies on oligopeptides [15,16]. The numbering system and definitions of dihedral angles of interest are shown in Fig. 2(II). In accordance with the IUPAC-IUB [17] recommendations, all dihedral angles were described within $-180.0°$ and $180.0°$ for the BB(ϕ, ψ) and SC(χ_1, χ_2) conformations as well as the isomerism of the peptide bond (ω_0, ω_1):

$$-180° \leq \phi \leq 180° \qquad (1a)$$

$$-180° \leq \psi \leq 180° \qquad (1b)$$

$$-180° \leq \chi_1 \leq 180° \qquad (1c)$$

$$-180° \leq \chi_2 \leq 180° \qquad (1d)$$

$$-180° \leq \omega_0 \leq 180° \qquad (1e)$$

$$-180° \leq \omega_1 \leq 180° \qquad (1f)$$

3.2. Molecular computations

Ab initio computations were carried out on N-acetyl-L-Thr-N-methylamide at the RHF/3-21G level of theory [18] using the GAUSSIAN 94 and 98 software [19,20]. The total energies are given in hartrees, the relative energies and stabilization energies are given in kilocalories per mole (using the conversion factor: 1 hartree = 627.5095 kcal mol^{-1}).

3.3. Stabilization energies

The following isodesmic reaction (2), where R = Me–CH$_2$–OH [21], was used to calculate the stabilization energies (3) with respect to the β_L

backbone conformation of N- and C-protected glycine [22]:

$$\text{MeCONH–CH}_2\text{–CONH–Me} + \text{CH}_3\text{–R}$$
$$\text{Reference conformation}(\beta_L)$$

$$\rightarrow \text{MeCONH–CHR–CONH–Me} + \text{CH}_3\text{–H} \qquad (2)$$
$$\text{conformation X}$$

The stabilization energy is defined as follows:

$$\Delta E_{\text{stabilization}} = \{E[\text{MeCONH–CHR–CONH–Me}]_X$$
$$+ E[\text{CH}_3\text{–H}]\} - \{E[\text{MeCONH–CH}_2$$
$$\text{–CONH–Me}]\beta_L + E[\text{CH}_3\text{–R}]\}$$
$$(3)$$

The $\Delta E_{\text{stabilization}}$ value measures the stabilization energy the SC exerts on the BB. The components energy values for Eq. (3) are summarized in Table 1.

4. Results and discussion

4.1. Conformational study

Surprisingly, ab initio computations on L-threonine at the RHF/3-21G level yielded structures in all 9 legitimate backbone conformations, unlike most other amino acids, for which minima are normally not found in α-L and ε-L [6,8–14,22–34]. Calculations conducted at the RHF/3-21G level of theory revealed a total of 39 fully relaxed structures; the remaining 42 conformers migrated

Table 2
Dihedral angles and total energies for optimized structures in alpha-D (α-D)

Sidechain		ψ_0	ω_0	ϕ_1	ψ_1	ω_1	ϕ_2	χ_1	χ_2	χ_3	Energy (hartrees)
$g+$	$g+$	-8.4316	165.5503	47.0243	53.3738	-176.8653	-139.1698	55.9910	59.5836	179.7885	-603.372042451
	a	Not found; migrated to α-D $(g - g +)$									
	$g-$	Not found; migrated to γ-D $(g + g -)$									
a	$g+$	-2.8980	175.5410	50.4148	44.8812	-179.5740	-131.3476	-159.1099	70.9245	-176.1124	-603.354132076
	a	Not found; migrated to α-D $(ag -)$									
	$g-$	-4.0176	173.6084	51.1516	39.5419	179.6803	-132.3476	-167.6824	-65.6599	172.8099	-603.372148414
$g-$	$g+$	-8.6120	174.7621	81.7931	18.0500	179.4537	-122.665	-27.6232	91.3068	-178.1455	-603.368886424
	a	-8.6500	164.9650	63.6478	38.4354	179.8477	109.3541	-42.1546	-177.869	178.8332	-603.370074388
	$g-$	Not found; migrated to α-D $(g - a)$									

Table 3
Dihedral angles and energies for optimized structures in alpha-L (α-L)

Sidechain		ψ_0	ω_0	ϕ_1	ψ_1	ω_1	ϕ_2	χ_1	χ_2	χ_3	Energy (hartrees)
$g+$	$g+$	Not found; migrated to γ-L ($g+g+$)									
	a	Not found; migrated to γ-L ($g+a$)									
	$g-$	Not found; migrated to γ-D ($g+g-$)									
a	$g+$	Not found; migrated to γ-L ($ag-$)									
	a	Not found; migrated to γ-L ($ag-$)									
	$g-$	Not found; migrated to γ-L ($ag-$)									
$g-$	$g+$	Not found; migrated to γ-L ($g-g+$)									
	a	6.5432	−168.6386	−69.7919	−27.4388	−179.4601	126.583	−40.65190	−172.9123	−171.9117	−603.377816741
	$g-$	126.0238	−170.6989	−70.9000	−26.2983	−179.0609	125.8837	−37.9283	−82.9977	−174.2038	−603.372911763

Table 4
Dihedral angles and energies for optimized structures in beta-L (β-L)

Sidechain		ψ_0	ω_0	ϕ_1	ψ_1	ω_1	ϕ_2	χ_1	χ_2	χ_3	Energy (hartrees)
$g+$	$g+$	Not found; migrated to γ-L ($g+g+$)									
	a	−1.8517	173.7719	−140.9507	171.0842	177.4979	−118.6059	70.0047	−177.5233	−171.2762	−603.371747467
	$g-$	−3.8809	170.7811	−144.2401	170.0082	176.2538	−117.7132	69.2297	−62.4293	−175.3136	−603.374072670
a	$g+$	−3.0101	174.16860	−166.6317	−173.8260	−178.3246	127.2238	−172.0540	89.3861	−176.2552	−603.385347220
	a	−3.3022	173.66620	−167.6135	−175.1620	−177.6688	125.9142	−172.2260	164.4791	−175.8533	−603.384775841
	$g-$	Not found; migrated to β-L (aa)									
$g-$	$g+$	−1.7908	176.0353	−168.7497	154.4403	178.4125	−114.1961	−85.5692	45.5974	175.6624	−603.377340013
	a	5.9635	−172.7010	−137.1030	131.9038	−179.9277	−123.7511	−54.9325	178.5837	−178.474	−603.365933169
	$g-$	−113.84000	178.0382	−157.0912	151.2924	178.3555	−120.9468	−49.7189	−27.7094	175.0442	−603.375809615

Table 5
Dihedral angles and energies for optimized structures in delta-D (δ-D)

Sidechain		ψ_0	ω_0	ϕ_1	ψ_1	ω_1	ϕ_2	χ_1	χ_2	χ_3	Energy (hartrees)
$g+$	$g+$	Not found; migrated to α-L $(g + a)$									
	a	Not found; migrated to α-L $(g + a)$									
	$g-$	-6.3269	168.8468	-137.9677	-69.0661	179.6827	-144.3870	46.0411	-70.6205	173.9407	-603.372688457
a	$g+$	-4.8446	172.4059	-177.0448	-37.7448	-177.1699	-137.3458	161.8085	43.7759	176.8927	-603.363573917
	a	-4.9619	172.3379	-169.6216	-47.6180	-179.2521	-133.2230	172.1939	164.7479	-173.4468	-603.359650082
	$g-$	Not found; migrated to γ-D $(ag -)$									
$g-$	$g+$	-7.5642	171.6586	175.1109	-48.3811	-176.8405	-128.2200	-84.3878	57.8143	168.9610	-603.369124241
	a	Not found; migrated to δ-L $(g - a)$									
	$g-$	0.6337	174.5997	-159.1062	-50.2141	-178.1921	-135.1260	-39.2150	-41.1923	-176.0281	-603.367470884

Table 6
Dihedral angles and energies for optimized structures in delta-L (δ-L)

Sidechain		ψ_0	ω_0	ϕ_1	ψ_1	ω_1	ϕ_2	χ_1	χ_2	χ_3	Energy (hartrees)
$g+$	$g+$	Not found; migrated to γ-L $(g + g +)$									
	a	Not found; migrated to γ-L $(g + a)$									
	$g-$	Not found; migrated to γ-L $(g + a)$									
a	$g+$	Not found; migrated to β-L $(ag +)$									
	a	Not found; migrated to β-L (aa)									
	$g-$	4.1846	-172.6190	-127.8158	28.2986	176.3818	122.2607	-170.9935	-57.4592	178.7683	-603.377816356
$g-$	$g+$	Not found; migrated to β-L $(g - g +)$									
	a	9.9056	-158.0731	-140.9566	31.2605	176.7419	119.5061	-35.9368	-157.062	-179.4807	-603.370574497
	$g-$	12.5850	-168.4503	-154.3359	31.9698	177.238	121.3113	-37.9813	-61.4823	177.2315	-603.374531935

Table 7
Dihedral angles and energies for optimized structures in epsion-D (ε-D)

Sidechain		ψ_0	ω_0	ϕ_1	ψ_1	ω_1	ϕ_2	χ_1	χ_2	χ_3	Energy (hartrees)
$g +$	$g +$	Not found; migrated to δ-D ($g + g -$)									
	a	Not found; migrated to δ-D ($g + a$)									
	$g -$	Not found; migrated to δ-D ($g + g -$)									
a	$g +$	131.0341	−160.8408	58.5248	−177.2613	−177.9054	120.2589	−157.6584	80.0303	179.127	−603.36745201
	a	131.2019	−160.616	60.8419	179.096	−177.0975	−121.135	−152.8576	−176.9364	−178.0847	−603.370164213
	$g -$	Not found; migrated to ε-D (aa)									
$g -$	$g +$	Not found; migrated to β-L ($g - g +$)									
	a	Not found; migrated to β-L ($g - g +$)									
	$g -$	5.2641	−172.2096	75.4945	157.1292	179.9159	−122.7317	−51.0655	−65.4833	172.8737	−603.353494970

Table 8
Dihedral angles and energies for optimized structures in epsion-L (ε-L)

Sidechain		ψ_0	ω_0	ϕ_1	ψ_1	ω_1	ϕ_2	χ_1	χ_2	χ_3	Energy (hartrees)
$g +$	$g +$	Not found; migrated to γ-L ($g + g +$)									
	a	Not found; migrated to β-L ($g + a$)									
	$g -$	Not found; migrated to β-L ($g + g -$)									
a	$g +$	Not found; migrated to β-L ($ag +$)									
	a	Not found; migrated to β-L (aa)									
	$g -$	Not found; migrated to β-L (aa)									
$g -$	$g +$	−4.5746	170.4942	−114.945	144.5319	178.3297	−122.5752	−56.6831	50.0768	177.268	−603.370667128
	a	Not found; migrated to γ-L ($g - a$)									
	$g -$	Not found; migrated to γ-L ($g - g -$)									

Table 9
Dihedral angles and energies for optimized structures in gamma-D (γ-D)

Sidechain		ψ_0	ω_0	ϕ_1	ψ_1	ω_1	ϕ_2	χ_1	χ_2	χ_3	Energy (hartrees)
$g+$	$g+$	-4.1025	172.3017	63.2000	-37.1991	-178.0858	66.2008	44.3740	35.854	176.805	-603.368691782
	a	0.1234	167.9171	53.152	-29.5793	-177.7805	71.0815	65.2522	171.1558	-178.8342	-603.361904954
	$g-$	4.4359	-170.3455	89.5521	-118.1964	-179.8876	131.8241	77.0096	-61.6172	68.3662	-603.381685550
a	$g+$	Not found; migrated to γ-D ($ag-$)									
	a	1.9110	177.5553	69.1770	-53.5957	179.0643	121.2115	-153.4655	-160.311	61.1203	-603.362016997
	$g-$	-119.5390	172.3141	56.2990	-21.9984	-178.3028	73.9821	-153.7856	-46.8478	-175.9778	-603.372686143
$g-$	$g+$	-3.0270	175.5404	84.1284	-54.7196	-178.7439	117.3398	-33.2772	88.777	-171.6738	-603.372501346
	a	-8.2307	163.4455	80.7267	-51.6157	-178.8576	117.8253	-44.9057	177.1889	-173.2143	-603.371608327
	$g-$	-8.4459	166.6685	78.4416	-55.0633	-178.8343	121.568	-48.7248	-78.0258	177.9441	-603.369909042

Table 10
Dihedral angles and energies for optimized structures in gamma-L (γ-L)

Sidechain		ψ_0	ω_0	ϕ_1	ψ_1	ω_1	ϕ_2	χ_1	χ_2	χ_3	Energy (hartrees)
$g+$	$g+$	-120.2533	-176.4850	-85.1628	73.6825	-177.5813	-125.9229	50.5993	68.5586	175.5448	-603.390723746
	a	Not found; migrated to γ-L ($g+g+$)									
	$g-$	Not found; migrated to γ-L ($g+g+$)									
a	$g+$	Not found; migrated to γ-L ($ag-$)									
	a	Not found; migrated to γ-L ($ag-$)									
	$g-$	119.1615	-173.3560	-84.9685	60.1216	179.4552	119.5439	-179.9720	-66.0935	173.2008	-603.381686904
$g-$	$g+$	-4.0050	179.8813	-86.8828	73.1510	-178.0267	113.9390	-58.5675	45.3981	174.2631	-603.374397800
	a	4.6509	-167.9740	-80.5953	64.5901	-179.8583	89.3396	-39.4492	-178.2112	-179.5220	-603.379129185
	$g-$	0.5180	-171.1130	-80.2590	65.0445	-179.5855	113.7455	-36.8685	-77.2606	178.2401	-603.379646700

Table 11

Energy, relative energy and stabilization energy associated with the optimized minima at RHF/3-21G

Backbone	Sidechain		Energy (hartrees)	ΔEnergy (kcal mol^{-1})	Stabilization energy
α_L	$g-$	a	−603.377816741	8.099	−8.507
	$g-$	$g-$	−603.372911763	11.177	−5.4291
	$g+$	a	−603.371747467	11.908	−4.6985
	$g+$	$g-$	−603.374072670	10.449	−6.1575
β_L	a	$g+$	−603.385347220	3.374	−13.2324
	a	a	−603.384775841	3.732	−12.8739
	$g-$	$g+$	−603.377340013	8.398	−8.2078
	$g-$	a	−603.365933169	15.556	−1.0499
	$g-$	$g-$	−603.375809615	9.359	−7.2475
	$g+$	$g+$	−603.390723746	0.000	−16.6063
	a	$g-$	−603.381686904	5.671	−10.9356
γ_L	$g-$	$g+$	−603.374397800	10.245	−6.3616
	$g-$	a	−603.379129185	7.276	−9.3306
	$g-$	$g-$	−603.379646700	6.951	−9.6553
	a	$g-$	−603.377816356	8.100	−8.5067
δ_L	$g-$	a	−603.370574497	12.644	−3.9624
	$g-$	$g-$	−603.374531935	10.161	−6.4457
ε_L	$g-$	$g+$	−603.370667128	12.586	−4.0205
	$g+$	$g+$	−603.372042451	11.723	−4.8836
	a	$g+$	−603.354132076	22.962	6.3554
α_D	a	$g-$	−603.372148414	11.656	−4.9501
	$g-$	$g+$	−603.368886424	13.703	−2.9031
	$g-$	a	−603.370074388	12.958	−3.6486
	$g+$	$g+$	−603.368691782	13.825	−2.781
γ_D	$g+$	a	−603.361904954	18.084	1.4778
	$g+$	$g-$	−603.381685550	5.672	−10.9347
	a	a	−603.362016997	18.014	1.4075
	a	$g-$	−603.372686143	11.319	−5.2875
	$g-$	$g+$	−603.372501346	11.435	−5.1715
	$g-$	a	−603.371608327	11.995	−4.6111
	$g-$	$g-$	−603.369909042	13.061	−3.5448
	$g+$	$g-$	−603.372688457	11.317	−5.2889
	a	$g+$	−603.363573917	17.037	0.4305
δ_D	a	a	−603.359650082	19.499	2.8928
	$g-$	$g+$	−603.369124241	13.554	−3.0524
	$g-$	$g-$	−603.367470884	14.591	−2.0149
ε_D	a	$g+$	−603.367454201	14.602	−2.0044
	a	a	−603.370164213	12.901	−3.705
	$g-$	$g-$	−603.353494970	23.361	6.7552

to one of the existing structures. Details of the optimized dihedral angles and energy values obtained for the 9 possible sidechain conformations in each of the backbone conformations are presented in Tables 2–10. Total energies, relative energies and stabilization energies obtained for the 39 structures are listed in Table 11. The number of SC conformers found in each BB conformer is

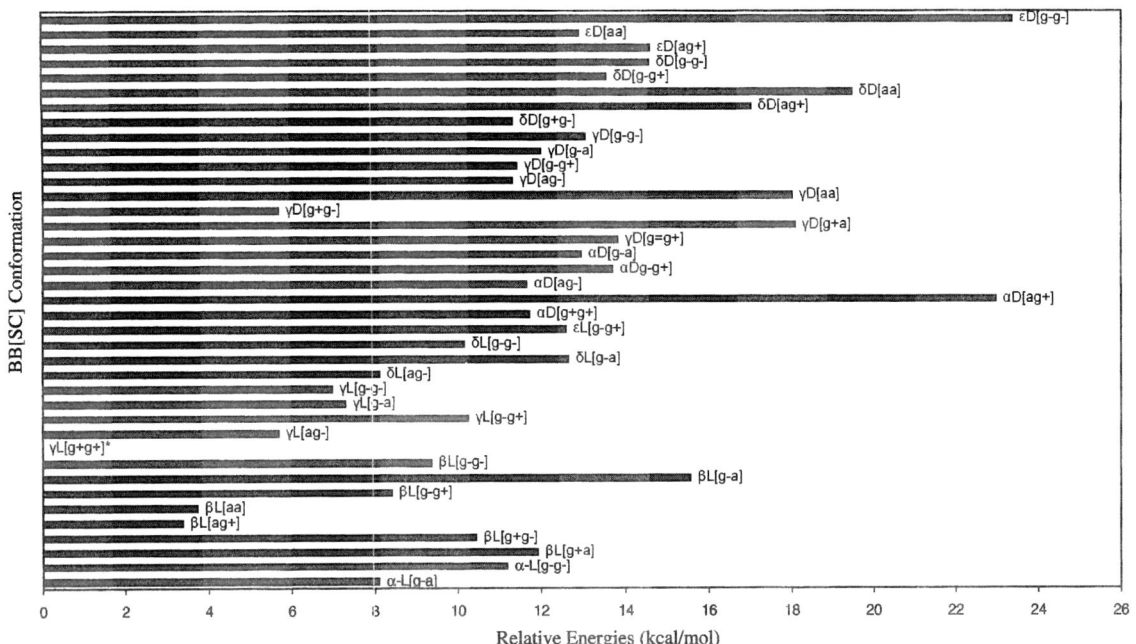

* Global Energy Minimum ($\Delta E_{relative} = 0.000$)

Fig. 5. A graphical presentation of the $\Delta E_{relative}$ values for all existing backbone and sidechain conformations of *N*-acetyl-L-threonine-*N*-methylamide.

noted below:

$$\gamma_D = 8 \quad \delta_D = 5 \quad \alpha_L = 2$$
$$\varepsilon_D = 3 \quad \beta_L = 7 \quad \varepsilon_L = 1 \quad (4)$$
$$\alpha_D = 5 \quad \delta_L = 3 \quad \gamma_L = 5$$

The relative energies for all 39 structures located are depicted graphically in Fig. 5. The range of the lowest and the highest relative energy values (in kcal mol^{-1}) for each of the backbone conformers is summarized below:

$$\gamma_D = 5.672 - 18.084 \quad \delta_D = 11.317 - 19.499 \quad \alpha_L = 8.099 - 11.177$$
$$\varepsilon_D = 12.901 - 23.361 \quad \delta_L = 3.374 - 15.556 \quad \varepsilon_L = 12.586 - 12.586 \quad (5)$$
$$\alpha_D = 11.656 - 22.962 \quad \delta_L = 8.100 - 12.644 \quad \gamma_L = 0.000 - 10.245$$

In L-threonine, conformers with D-subscript (α_D, γ_D, δ_D, δ_L, ε_D) have relatively high energy values, making them energetically less favourable and hence structurally less stable. This finding is consistent with results for other N- and C-protected L-amino acids containing

trans-peptide bonds—Gly [22–28], Ala [23–28], Val [28], Phe [29–31], Ser [8–14], Pro [6], Asp [32], Asn [33] and Sec (selenocysteine) [34]—and can serve as an adequate basis for comparison with the aforementioned amino acids.

4.2. Comparison with serine results

In this computational study, the results for N- and C-protected L-threonine were predicted to parallel those previously obtained for L-Ser-NH$_2$ [9–11]. As a result, approximately 44 out of the 81 possible conformations were expected to exist for L-threonine. Our results indicate 39 existing conformations for L-threonine, a number not far

5.672 g+g- 9.4	11.319 ag- 10.2	13.061 g-g- 12.9	11.317 g+g- 10.5	ag-	14.591 g-g- 15.7	g+g-	ag-	11.177 g-g- 12.5
18.084 g+a 17.1	18.014 aa 12.7	11.995 g-a 12.0	g+a 11.2	19.499 aa 17.2	g-a	g+a	aa 20.6	8.099 g-a 16.6
13.825 g+g+ 14.0	ag+ 12.0	11.435 g-g+ 12.5	g+g+	17.037 ag+ 15.7	13.554 g-g+ 12.3	g+g+	ag+	g-g+
g+g- 4.9	ag-	23.361 g-g- 20.5	10.449 g+g- 9.1	ag-	9.359 g-g-	g+g-	ag-	g-g-
g+a 18.5	12.901 aa 9.4	g-a	11.908 g+a 11.2	3.732 aa 3.8	15.556 g-a 15.4	g+a	aa	g-a
g+g+	14.602 ag+ 10.1	g-g+ 16.3	g+g+	3.374 ag+	8.398 g-g+ 10.5	g+g+	ag+	12.586 g-g+
g+g-	11.656 ag- 9.1	g-g-	g+g-	8.1 ag- 8.3	10.161 g-g- 11.0	g+g-	5.671 ag- 4.8	6.951 g-g- 7.8
g+a	12.958 aa	g-a 12.9	g+a 14.0	aa	12.644 g-a 13.3	g+a	aa	7.276 g-a 0.1
11.723 g+g+ 12.1	22.962 ag+ 20.5	13.703 g-g+	g+g+	ag+	g-g+ 14.0	0.000 g+g+ 0.0	ag+ 12.5	10.245 g-g+ 10.5

Fig. 6. Relative energies represented in a 4D Ramachandran topological matrix, for N-acetyl-L-threonine-N-methylamide (upper value) and For-L-serine-NH_2 (lower value). The conformational assignments (middle) denote the χ_1 and χ_2 conformations, respectively.

from our expectation. The relative energies of the existing conformations for both L-serine and L-threonine are depicted on a 4D Ramachandran map (Fig. 6). A total of 32 optimized structures were found common to both Ac-L-Thr-NH-Me and For-L-Ser-NH_2. Both threonine and serine have the same global minimum, found at γ-L $[g+g+]$. Fig. 7 is a graph depicting the correlation between the BB dihedral angles (ϕ, ψ) of the 32 shared structures of Ac-L-Thr-NH-Me and For-L-Ser-NH_2.

4.3. Stabilization energies

Stabilization energy is a measure of the stabilization ($\Delta E_{stabilization} < 0$) or destabilization ($\Delta E_{stabilization} > 0$) exerted by the SC on the BB with respect to hydrogen, i.e. the SC of glycine. The stabilization energy is calculated according to Eq. (3). Traditionally the global minimum is used for such a calculation. In the case of glycine, the global minimum is found in the γ_L backbone conformation. However, it has recently been demonstrated that the γ_L conformation disappears when the *trans*-peptide

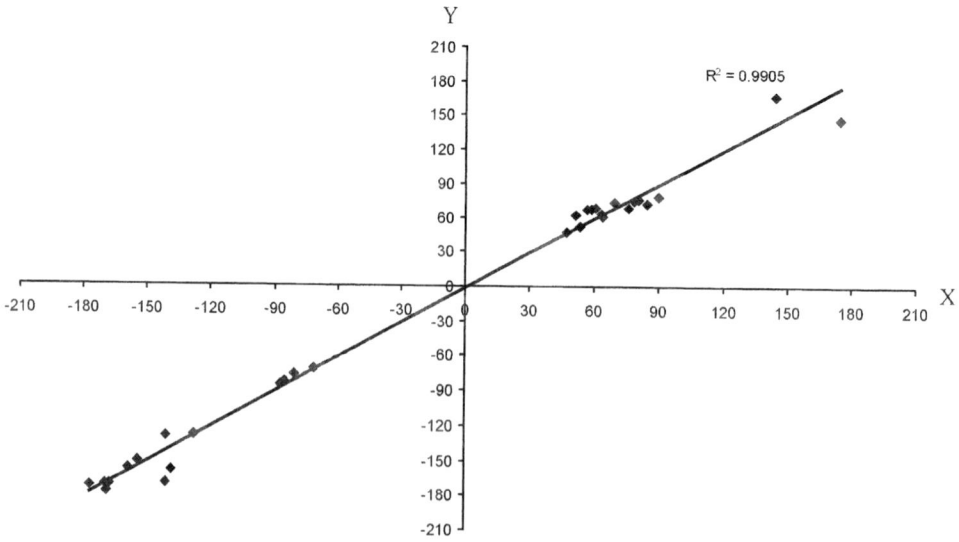

Fig. 7. Correlation of dihedral angles obtained for N-acetyl-L-threonine-N-methylamide with those optimized for For-L-serine-NH_2. Note that the X-axis represents the torsional angles (ϕ), ψ obtained for N-acetyl-L-threonine-N-methylamide and the Y-axis represents the dihedral angle (ϕ, ψ) obtained for For-L-serine-NH_2.

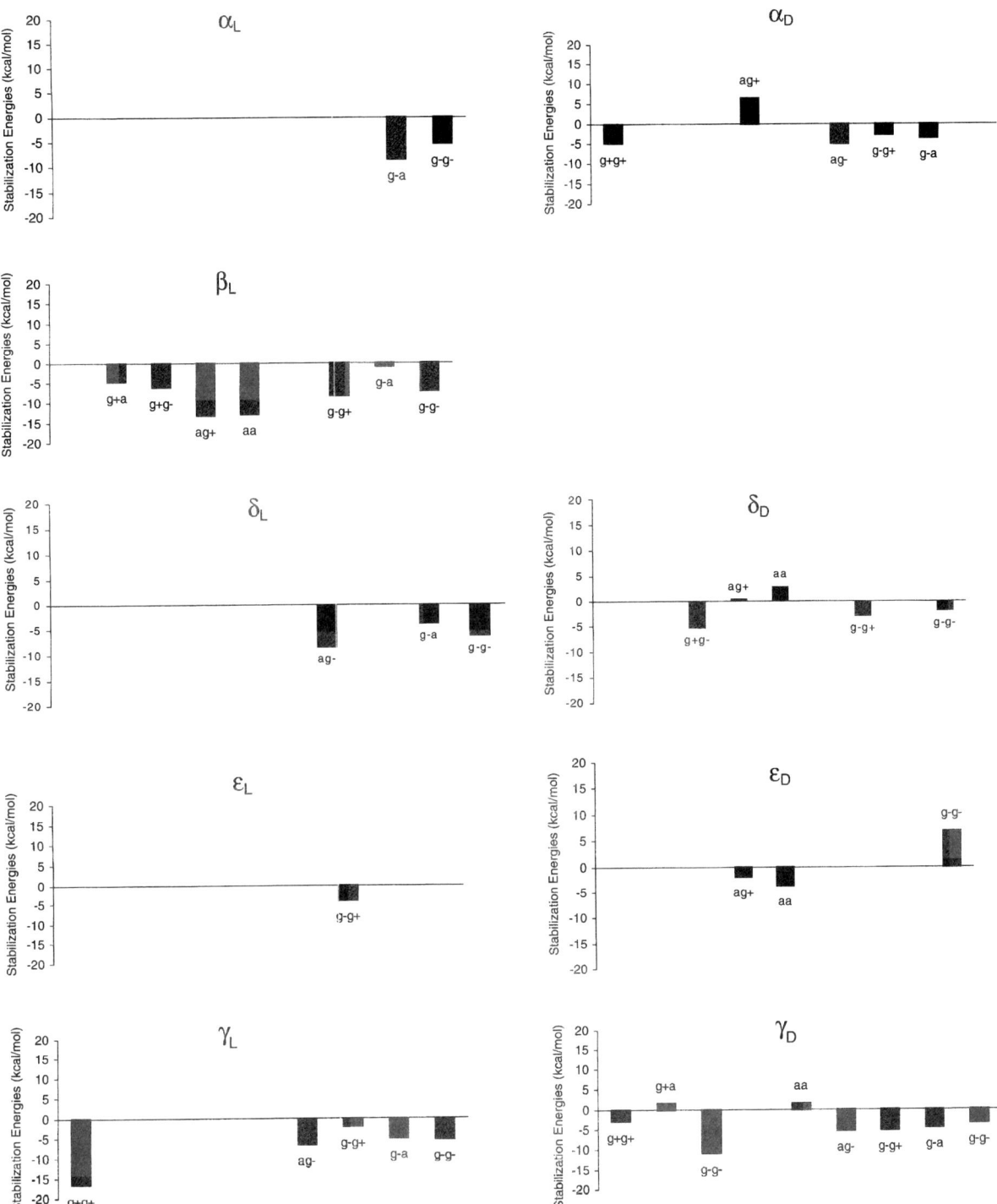

Fig. 8. A graphical presentation of the $\Delta E_{stabilization}$ values for all existing backbone and sidechain conformations of N-acetyl-L-threonine-N-methylamide.

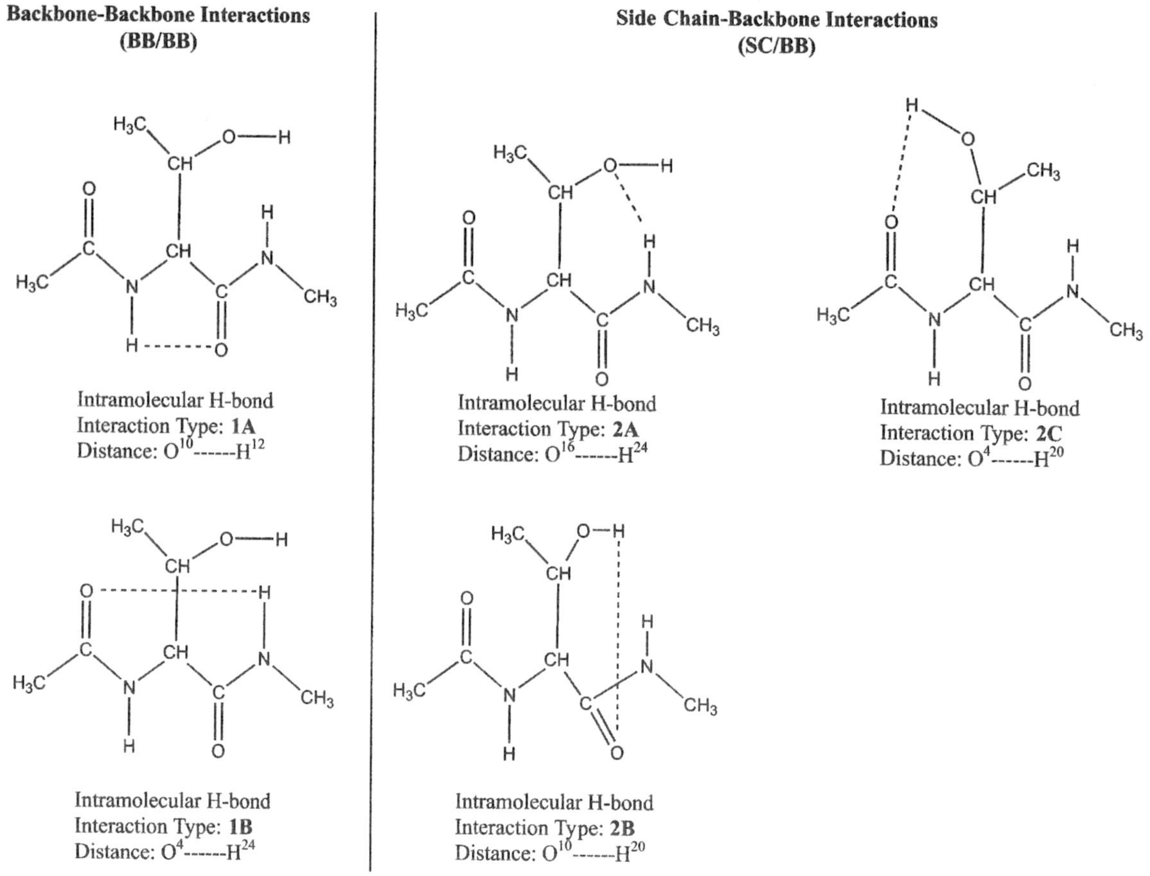

**Backbone-Backbone Interactions
(BB/BB)**

**Side Chain-Backbone Interactions
(SC/BB)**

Intramolecular H-bond
Interaction Type: **1A**
Distance: O^{10}------H^{12}

Intramolecular H-bond
Interaction Type: **2A**
Distance: O^{16}------H^{24}

Intramolecular H-bond
Interaction Type: **2C**
Distance: O^{4}------H^{20}

Intramolecular H-bond
Interaction Type: **1B**
Distance: O^{4}------H^{24}

Intramolecular H-bond
Interaction Type: **2B**
Distance: O^{10}------H^{20}

Fig. 9. Classification of the types of internal hydrogen bonding for *N*-acetyl-L-threonine-*N*-methylamide.

bond is isomerized to its *cis*-form [35]. Owing to this reality, it is speculated that the β_L conformation may become more popular in the future. For the stabilization energies calculated in this paper, the energy associated with the β_L conformation of glycine was used. It should be noted that the difference in total energy values for the β_L and γ_L conformations of glycine is <0.0015 hartrees.

The $\Delta E_{\text{stabilization}}$ values summarized in Table 11 are presented graphically in Fig. 8. All SC conformers of L-threonine in BB conformations with the L-subscript (α_L, γ_L, δ_L, δ_L, ε_L) have a stabilizing effect on their respective BB. Interestingly, most of the existing SC structures in the D-conformers (α_D, γ_D, δ_D, δ_L, ε_D) also have a stabilizing impact on their BB.

4.4. Intramolecular interactions

The existence of hydrogen bonds is largely responsible for the stabilization effects described above and depicted in Fig. 8. There can potentially exist in threonine two backbone–backbone (BB/BB) and three sidechain–backbone (SC/BB) hydrogen bond interactions (Fig. 9). The corresponding distances for these potential hydrogen bonds are reported in Table 12 for the 39 existing structures of L-threonine. Hydrogen bonding is thought to be occurring if the interatomic distance is less than the sum of the Van der Waals radii (<2.6 Å) of the two interacting atoms, oxygen and hydrogen. Each of the five possible interactions exists in at least one of the 39 existing

Table 12

The relative distances of potential hydrogen bonds in N-acetyl-L-threonine-N-methylamide for all its existing conformations computed at RHF/3-21G level of theory

Final conformation	Interaction type	Distance (Å)				
BB(χ_1, χ_2)	BB/BB	SC/BB	$H^{12}-O^{10}$	$H^{24}-O^4$	$H^{24}-O^{16}$	$H^{20}-O^{10}$	$H^{20}-O^4$
α_L Backbone conformation							
$\alpha_L(g-, a)$	–	–	4.370	3.098	4.058	5.090	5.597
$\alpha_L)g-, g-)$	–	–	4.357	3.138	4.088	4.811	4.918
α_D Backbone conformation							
$\alpha_D)g+, g+)$	–	2B, 2C	4.364	3.124	4.569	2.319	2.525
$\alpha_D)a, g+)$	–		4.403	3.159	4.736	3.717	5.300
$\alpha_D)a, g-)$	–	2B	4.409	3.006	4.677	1.976	4.531
$\alpha_D(g-, g+)$	–	2C	4.309	3.335	4.704	4.514	2.194
$\alpha_D(g-, a)$	–	–	4.395	3.169	4.682	4.779	4.177
β_L Backbone conformation							
$\beta_L(g+, a)$	1A	–	2.208	4.898	3.460	4.046	5.274
$\beta_L(g+, g-)$	1A	–	2.083	4.937	3.524	3.153	4.407
$\beta_L(a, g+)$	1A	2A	2.074	5.047	1.920	4.887	3.996
$\beta_L(a, a)$	1A	2A	2.051	5.081	1.901	5.084	4.529
$\beta_L(g-, g+)$	1A	2C	2.063	5.106	3.976	4.803	1.795
$\beta_L(g-, a)$	1A	–	2.475	4.271	4.568	4.872	4.246
$\beta_L(g-, g-)$	1A	2C	2.167	4.847	4.469	4.684	1.918
γ_L Backbone conformation							
$\gamma_L(g+, g+)$	1B	2B	3.455	2.054	4.622	2.041	5.048
$\gamma_L(a, g-)$	1B	2B	3.771	1.978	4.495	1.989	5.273
$\gamma_L(g-, g+)$	1B	–	3.419	2.117	4.818	4.513	3.950
$\gamma_L(g-, a)$	1B	–	3.902	1.880	4.838	4.610	5.509
$\gamma_L(g-, g-)$	1B	–	3.854	1.901	4.905	4.658	4.685
γ_D Backbone conformation							
$\gamma_D(g+, g+)$	1B	2C	4.143	1.833	3.290	4.044	2.011
$\gamma_D(g+, a)$	1B	–	4.227	1.768	2.908	4.278	3.781
$\gamma_D(g+, g-)$	–	2A, 2C	2.762	3.491	1.846	4.399	1.731
$\gamma_D(a, a)$	1B	–	3.904	1.903	4.161	3.962	5.210
$\gamma_D(a, g-)$	1B	2B	4.246	1.847	4.533	1.970	5.016
$\gamma_D(g-, g+)$	1B	2C	3.835	2.055	3.782	5.063	2.211
$\gamma_D(g-, a)$	1B	–	3.998	1.885	4.003	5.253	3.796
$\gamma_D(g-, g-)$	1B	–	3.945	1.880	4.071	4.727	4.322
δ_L Backbone conformation							
$\delta_L(a, g-)$	–	2B	3.838	3.534	4.746	1.923	5.046
$\delta_L(g-, a)$	–	–	3.813	3.624	4.639	4.776	4.056
$\delta_L(g-, g-)$	–·	2C	3.617	3.906	4.795	4.654	2.310
δ_D Backbone conformation							
$\delta_D(g+, g-)$	–	2A	3.382	4.964	1.932	4.732	3.920
$\delta_D(a, g+)$	–	2B	3.578	4.516	3.729	2.476	4.563
$\delta_D(a, a)$	–	–	3.462	4.601	3.660	4.023	4.689
$\delta_D(g-, g+)$	–	2C	3.391	4.324	4.292	4.853	1.766
$\delta_D(g-, g-)$	–	2C	3.558	4.706	3.788	4.835	1.930
ε_L Backbone conformation							
$\varepsilon_L(g-, g+)$	1A	–	2.328	4.280	4.484	4.371	3.356
ε_D Backbone conformation							
$\varepsilon_D(a, g+)$	–	2A	3.119	4.629	1.960	4.839	5.229
$\varepsilon_D(a, a)$	–	2A	3.124	4.721	1.987	5.101	5.158
$\varepsilon_D(g-, g-)$	–	–	3.210	5.072	4.278	5.021	4.083

conformations. The existence of the $(g-, g+)$ conformer in the unstable ε_L backbone conformation may be the result of the reported backbone–backbone hydrogen bond.

5. Conclusions

In summary, from a possible 81 conformations for N-acetyl-L-threonine-N-methylamide, 39 conformers were located, of which 34 were common with N- and C-protected serine. These results demonstrate strong stabilizing effect of intramolecular hydrogen bonding on all the backbone conformations. The comparative study between N-acetyl-L-threonine-N-methylamide and For-L-serine-NH$_2$ suggests correlation between the optimized structures for both compounds, with similar dihedral angles.

Acknowledgements

This work was supported by grants from the Global Institute of Computational Molecular and Materials Science (GIOCOMMS), Toronto, Ontario, Canada. The authors wish to thank the Center for Molecular Design and Information Technology (MDIT), Leslie Dan Faculty of Pharmacy, University of Toronto, for providing computation time. Special thanks goes out to the MDIT support staff. The authors thank David H. Setiadi and Tania A. Pecora for helpful discussions during the preparation of this manuscript, as well as Graydon Hoare for database management, network support and software and distributive processing development. Special thanks is extended to Andrew M. Chasse, Christopher M. Andrews and Michael R. Sahai for their development of novel scripting and coding techniques, which facilitate a reduction in the number of CPU cycles needed. The pioneering advances of Kenneth P. Chasse, in all composite computer-cluster software and hardware architectures, are also acknowledged. One of the authors (IGC) wishes to thank the Ministry of Education for a Szent-Györgyi Visiting Professorship.

References

[1] C.A. Janeway Jr., S. Hunt, M. Walport, Immunobiology: The Immune System in Health and Disease, Third ed., Current Biology Ltd/Garland Publishing, New York, 1997, pp. 3:2–3:4.

[2] B.J. Underdown, J.M. Sciff, Annu. Rev. Immunol. 4 (1986) 389.

[3] M. Kilian, J. Mestecky, R.E. Schrohenloher, Infect. Immun. 26 (1979) 143.

[4] J. Pohlner, R. Halter, K. Bayreuther, T.F. Meyer, Nature 325 (1987) 458.

[5] S.B. Mortensen, M. Kilian, Infect. Immun. 45 (1984) 550.

[6] H.A. Baldoni, A.M. Rodriguez, G. Zamarbide, R.D. Enriz, Ö. Farkas, P. Csaszar, L.L. Torday, C.P. Sosa, I. Jakli, A. Perczel, M. Hollosi, I.G. Csizmadia, J. Mol. Struct. (THEOCHEM) 465 (1999) 79–91.

[7] M.A. Sahai, D.H. Setiadi, G.A. Chass, E.F. Pai, B. Penke, I.G. Csizmadia, J. Mol. Struct. (THEOCHEM) (2003) in this issue.

[8] A. Perczel, R. Daudel, J.G. Angyan, I.G. Csizmadia, Can. J. Chem. 68 (1990) 1882–1888.

[9] Ö. Farkas, A. Perczel, J.F. Marcoccia, M. Hollósi, I.G. Csizmadia, J. Mol. Struct. 331 (1995) 27–36.

[10] A. Perczel, Ö. Farkas, I.G. Csizmadia, J. Comp. Chem. 17 (1996) 821–834.

[11] A. Perczel, Ö. Farkas, I.G. Csizmadia, J. Am. Chem. Soc. 118 (1996) 7809–7817.

[12] A. Perczel, Ö. Farkas, J.F. Marcoccia, I.G. Csizmadia, Int. J. Quantum Chem. 61 (1997) 797–814.

[13] A. Perczel, Ö. Farkas, I. Jákli, I.G. Csizmadia, J. Mol. Struct. (THEOCHEM) 455 (1998) 315.

[14] I. Jakli, A. Perczel, Ö. Farkas, C.P. Sosa, I.G. Csizmadia, J. Comp. Chem. 21 (2000) 626–655.

[15] G.A. Chass, M.A. Sahai, J.M.S. Law, S. Lovas, Ö. Farkes, A. Perczel, J.-L. Rivail, I.G. Csizmadia, Int. J. Quantum Chem. 90 (2002) 933–968.

[16] M.A. Sahai, G.A. Chass, S. Lovas, B. Penke, I.G. Csizmadia, J. Mol. Struct. (THEOCHEM) (in this issue).

[17] IUPAC ± IUB Commission on Biochemical Nomenclature, Biochemistry 9 (1970) 3471.

[18] J.S. Binkley, J.A. Pople, W.J. Hehre, J. Am. Chem. Soc. 102 (1980) 939.

[19] M.J. Frisch, G.W. Trucks, H.B. Schlegel, P.M.W. Gill, B.G. Johnson, M.A. Robb, J.R. Cheeseman, T. Keith, G.A. Petersson, J.A. Montgomery, K. Raghavachari, M.A. Al-Laham, V.G. Zakrzewski, J.V. Ortiz, J.B. Foresman, J. Cioslowski, B.B. Stefanov, A. Nanyakkara, M. Challacombe, C.Y. Peng, P.Y. Ayala, W. Chen, M.W. Wong, J.L. Andres, E.S. Replogle, R. Gomperts, R.L. Martin, D.J. Fox, J.S. Binkley, D.J. Defrees, J. Baker, J.P. Stewart, M. Head-Gordon, C. Gonzalez, J.A. Pople, GAUSSIAN 94, Revision D.2, Gaussian, Inc., Pittsburgh, PA, 1995.

[20] M.J. Frisch, G.W. Trucks, H.B. Schlegel, G.E. Scuseria, M.A. Robb, J.R. Cheeseman, V.G. Zakrzewski, J.A. Montgomery, Jr., R.E. Stratmann, J.C. Burant, S. Dapprich, J.M. Millam, A.D. Daniels, K.N. Kudin, M.C. Strain, O. Farkas, J. Tomasi, V. Barone, M. Cossi, R. Cammi, B. Mennucci, C. Pomelli, C. Adamo, S. Clifford, J. Ochterski,

G.A. Petersson, P.Y. Ayala, Q. Cui, K. Morokuma, P. Salvador, J.J. Dannenberg, D.K. Malick, A.D. Rabuck, K. Raghavachari, J.B. Foresman, J. Cioslowski, J.V. Ortiz, A.G. Baboul, B.B. Stefanov, G. Liu, A. Liashenko, P. Piskorz, I. Komaromi, R. Gomperts, R.L. Martin, D.J. Fox, T. Keith, M.A. Al-Laham, C.Y. Peng, A. Nanayakkara, M. Challacombe, P.M.W. Gill, B. Johnson, W. Chen, M.W. Wong, J.L. Andres, C. Gonzalez, M. Head-Gordon, E.S. Replogle, J.A. Pople, GAUSSIAN 1998 (Revision A.1x), Gaussian, Inc., Pittsburgh PA, 2001.

[21] J. Mestres, M. Duran, J. Bertran, I.G. Csizmadia, J. Mol. Chem. (THEOCHEM) 358 (1995) 229–249.

[22] M.A. McAllister, G. Endredi, W. Viviani, A. Perczel, P. Csaszar, J. Ladik, J.L. Rivail, I.G. Csizmadia, Can. J. Chem. 73 (1995) 563–572.

[23] A. Perczel, J.G. Angyan, M. Kajtar, W. Viviani, J.L. Rivail, J.F. Marcoccia, I.G. Csizmadia, J. Am. Chem. Soc. 113 (1991) 6256–6265.

[24] A. Perczel, W. Viviani, I.G. Csizmadia, Peptide conformational potential energy surfaces and their relevance to protein folding, in: J. Bertran (Ed.), Molecular Aspects of Biotechnology, Computational Models and Theories, Kluwer Academic, Dordrecht, 1992, pp. 39–82.

[25] W. Viviani, J.-L. Rivail, I.G. Csizmadia, Theor. Chim. Acta 85 (1992) 189–197.

[26] M.A. McAllister, A. Perczel, P. Császár, W. Viviani, J.L. Rivail, I.G. Csizmadia, J. Mol. Struct. (THEOCHEM) 288 (1993) 161–180.

[27] A. Perczel, Ö. Farkas, I.G. Csizmadia, J. Am. Chem. Soc. 117 (1995) 1653–1654.

[28] W. Viviani, J.-L. Rivail, A. Perczel, I.G. Csizmadia, J. Am. Chem. Soc. 115 (1993) 8321–8329.

[29] Ö. Farkas, M.A. McAllister, J.H. Ma, A. Perczel, M. Hollósi, I.G. Csizmadia, J. Mol. Struct. (THEOCHEM) 369 (1996) 105–114.

[30] A. Perczel, Ö. Farkas, I.G. Csizmadia, Can. J. Chem. 75 (1997) 1120–1130.

[31] I. Jakli, A. Perczel, Ö. Farkas, M. Hollosi, I.G. Csizmadia, J. Mol. Struct. (THEOCHEM) 455 (1998) 303–314.

[32] S.J. Salpietro, A. Perczel, Ö. Farkas, R.D. Enriz, I.G. Csizmadia, J. Mol. Struct. (THEOCHEM) 497 (2000) 39–63.

[33] M. Berg, S.J. Salpietro, A. Perczel, Ö. Farkas, I.G. Csizmadia, J. Mol. Struct. (THEOCHEM) 504 (2000) 127–140.

[34] J.C. Vank, C.P. Sosa, A. Perczel, I.G. Csizmadia, Can. J. Chem. 78 (2000) 395–408.

[35] H.A. Baldoni, G.N. Zamarbide, R.D. Enriz, E.A. Jauregui, O. Farkas, A. Perczel, S.J. Salpietro, I.G. Csizmadia, J. Mol. Struct. (THEOCHEM). 500 (2000) 97–111.

Journal of Molecular Structure (Theochem) 666–667 (2003) 269–271

www.elsevier.com/locate/theochem

THEO
CHEM

Deciphering the 'biological morse-code': a preliminary ab initio study of phosphoserine

S. Yarligan[a], A.K. Füzery[b,c], C. Öğretir[a], I.G. Csizmadia[b,c,d,*]

[a]Department of Chemistry, Faculty of Arts and Sciences, Osmangazi University, Eskişehir 26020, Turkey
[b]Global Institute of COmputational Molecular and Materials Science (GIOCOMMS), 1422 Edenrose Street, Mississauga, Ont., Canada L5V 1H3
[c]Lash Miller Chemical Lab, Department of Chemistry, University of Toronto, 80 St George Street, Toronto, Ont., Canada M5S 3H6
[d]Department of Medical Chemistry, University of Szeged, Dóm tér 8, 6720 Szeged, Hungary

Abstract

Exploratory ab initio conformational analysis has been carried out on phosphoserine (Pse) residue using formyl-L-serinamide-phopate-esther magnesium salt $[HCO-NHCH(CH_2OPO_3Mg)-CONH_2]$. On the basis of the computed stabilities it appeared that the Mg salt of For-L-Pse-NH$_2$ has greater flexibility than the unphosphorilated counter part: For-L-Ser-NH$_2$. © 2003 Elsevier B.V. All rights reserved.

Keywords: Phosphoserine; Phosphorylation; Conformations

1. Introduction

Protein phosphorylation is one of the major signal transduction mechanisms for controlling and regulating intracellular processes. To fully understand the role of phosphorylation in biological processes, characterization of the site at which phosphorylation occurs is essential. Serine, with its $-CH_2OH$ side chain, is one of the most frequently phosphorylated amino acid residues. While there is clearly a charge difference between serine and phosphoserine, phosphorylation may also bring about a change in the conformational behaviour of the residue even if the phosphate ion is neutralized by Mg^{2+}. Thus, we have investigated and compared the conformational properties of phosphoserine [1] using the For-L-Pse(NH$_2$)$^{2--}$ Mg^{2+} model and compared it with the unphosphorilated serine counter part: For-L-Ser-NH$_2$ [2].

2. Computational methods

Ab initio computations were performed using GAUSSIAN 98 at the RHF/3-21G [3] level of theory. Energies are given in kcal mol^{-1} using the conversion factor 1 hartree = 627.5095 kcal mol^{-1}.

3. Results and discussion

Initial conformational analysis was carried out on the Mg salt of ethyl-phophate (I). The energetics of formyl-serinamide (II) was reported earlier [2]. Four

* Corresponding author. Address: Lash Miller Chemical Lab, Department of Chemistry, University of Toronto, 80 St George Street, Toronto, Ont., Canada M5S 3H6. Tel.: +1-416-978-3564; fax: +1-416-978-3598.
E-mail address: icsizmad@ovrisc.mdche.szote.u-szeged.hu (I.G. Csizmadia).

Table 1
Backbone conformational labelling used for Serine and Serinphosphate residues

Poster labeling scheme	Alternative labeling scheme
α_L	α_R
β_L	C5
δ_L	β_2
ε_L	β
γ_L	$C7_{eq}$
α_D	α_L
δ_D	α'
ε_D	α_D
γ_D	$C7_{ax}$

backbone torsional angles (ω_0, ϕ, ψ and ω_1) and three sidechain dihedral angles (χ^1, χ^2, χ^3) were optimized, together with all other geometrical parameters of [For-L-Pse$(NH_2)^{2-}$] Mg^{2+} (**III**).

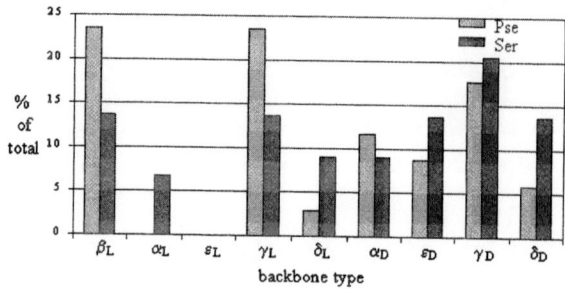

Fig. 2. Percentage of backbone distribution of (For-L-Pse-$NH_2)^{2-} Mg^{2+}$ and For-L-Ser-NH_2.

The optimized conformations were labeled according to the symbols given in Table 1.

The sidechain conformations were classified according to the following scheme.

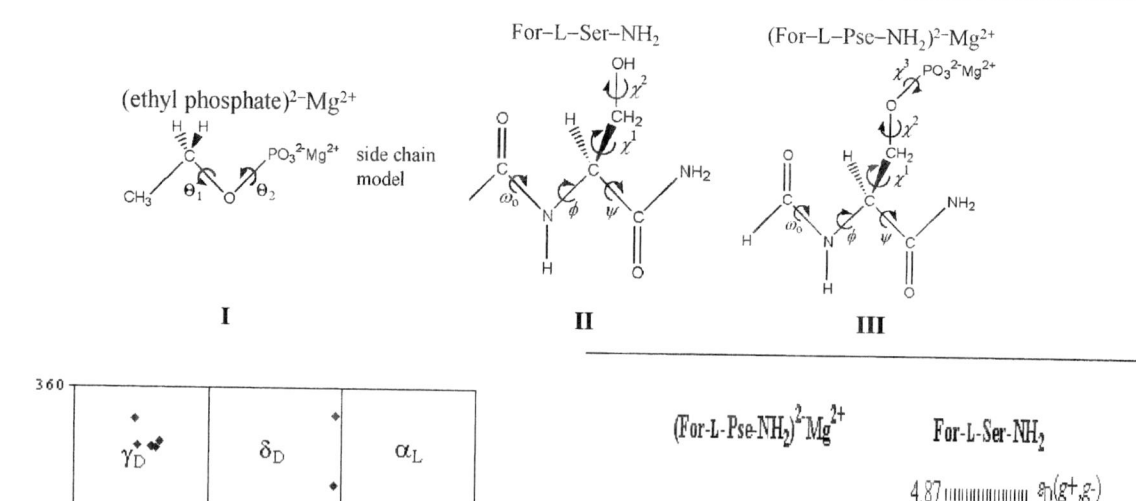

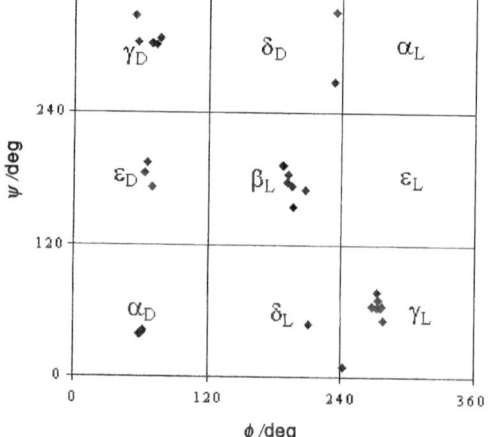

Fig. 1. The location of 32 minima found on the Ramachandran map at the RHF/3-21G level of theory.

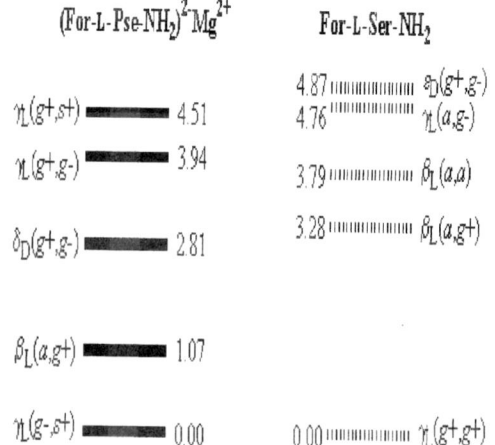

Fig. 3. Comparison of the five lowest energy structures of the Pse and Ser models.

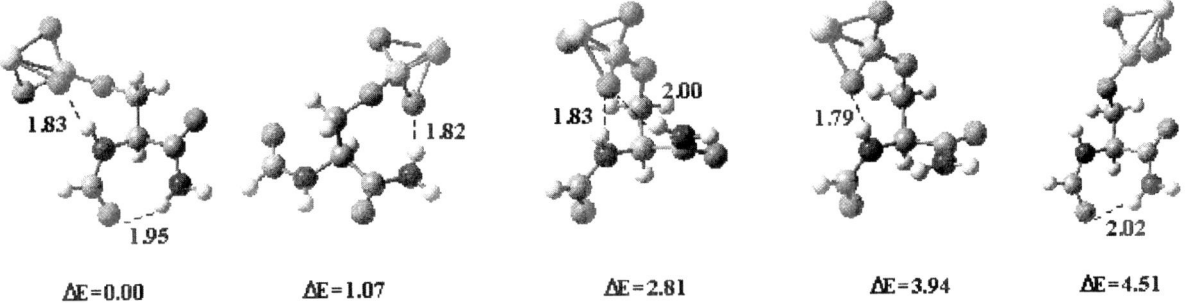

Fig. 4. The five lowest energy structures of $(For\text{-}L\text{-}Pse\text{-}NH_2)^{2-}Mg^{2+}$.

g^+ from $+30°$ to $+90°$
g^- from $-30°$ to $-90°$
a from $-150°$ to $-180°$ and from $+150°$ to $+180°$
s^+ from $+90°$ to $+150°$

All 32 conformers obtained are shown on Ramachandran map in Fig. 1. The pattern clearly shows the predominance of β_L and γ_L backbone types. Fig. 2 shows, in terms of percentage of total number of the conformers, that γ_L, β_L and γ_D dominate the conformational space (17 conformers out of the 32 optimized structures). However, there were more β_L and γ_L conformers in the case of Pse than in Ser.

The computed relative stabilities are illustrated in Fig. 3. Since the conformational energy levels of the Pse residue is denser than that of the Ser residue therefore the Pse residue may be considered to be conformationally more flexible than the Ser residue.

The five lowest energy structures of For-L-$Pse(NH_2)^{2-}Mg^{2+}$ are shown in Fig. 4.

Hydrogen bonds are indicated with dashed lines, distances are given in Å and ΔE values are given in kcal/mol. Note, tin some of the structures, there are certain types of hydrogen bonds ($N-H\cdots O-P$) which would not be possible in For-L-Ser-NH_2.

4. Conclusion

There are marked differences in the conformational behaviour of $(For\text{-}L\text{-}Pse\text{-}NH_2)^{2-}Mg^{2+}$ and

For-L-Ser-NH_2. Thus, phosphorylation introduces not only a change in the charge of the molecule but also in its conformational properties even when the phosphate ion is neutralized by a counter ion such as Mg^{2+}. This could also have an important role in the signalling pathway.

Acknowledgements

The authors would like to express their gratitude for the generous allocation of CPU time provided by the National Cancer Institute (NCI) at the Frederick Biomedical Supercomputing Center. One of the authors (IGC) wishes to thank the Ministry of Education for a Szent-Györgyi Visiting Professorship.

References

[1] Lodish, et al., Molecular Cell Biology, 4th ed.
[2] I. Jakli, A. Perczel, Ö. Farkas, C.P. Sosa, I.G. Csizmadia, J. Comput. Chem 21 (2000) 626.
[3] M.J. Frisch, G.W. trucks, H.B. Schelegel, P.M.W. Gill, B.G. Johnson, M.A. Robb, J.R. Cheeseman, T.A. Keith, G.A. Petersson, J.A. Montgomery, K. Raghavachari, M.A. Al-Laham, V.G. Zakrzewski, J.V. Ortiz, J.B. Foresman, J. Cioslowski, B.B. Stefanov, A. Nanayakkara, M. Challacombe, C.Y. Peng, P.Y. Ayala, W. Chen, M.W. Wong, J.L. Andres, E.S. Replogle, R. Gomperts, R.L. Martin, D.J. Fox, J.S. Binkley, D.J. Defrees, J. Baker, J.P. Stewart, M. Head-Gordon, C. Gonzales, J.A. Pople, GAUSSIAN 94, Gaussian Inc., PA, 1995.

ELSEVIER

Journal of Molecular Structure (Theochem) 666–667 (2003) 273–278

THEO
CHEM

www.elsevier.com/locate/theochem

Asparagine—ab initio structural analyses

Monee Rassolian[a,b,*], Gregory A. Chass[a,b,c,d], David H. Setiadi[b], Imre G. Csizmadia[a,b,e]

[a]Department of Chemistry, University of Toronto, 80 St George St., Toronto, Ont., Canada M5S 3H6
[b]Global Institute of COmputational Molecular and Material Science, 1422 Edenrose Street, Mississauga, Ont., Canada, L5V 1H3
[c]Institut de Science et d'Ingénierie Supramoléculaires, 8, allée Gaspard Monge, BP 70028, 67083, Strasbourg Cedex, France
[d]Department of Biomedical Sciences, Creighton University, 2500 California plaza, Omaha, NE 68178 USA
[e]Department of Medical Chemistry, Szeged University, H-6720 Szeged, Domter 8, Hungary

Abstract

Amino acids can be defined separately by the unique characteristics rendered to each amino acid molecule as a result of the varying reactive abilities of their side chains. Asparagine is among less than a handful of amino acids capable of reacting with saccharides (referring to the biologically active L-asparagine stereoisomer). Recently, it has been found that asparagine in potatoes reacts with glucose through a Maillard reaction to form acrylamide, a neurotoxin and potential carcinogen. Otherwise, its side chain can react to form N-glycosidic bonds with oligosaccharides, establishing a prevalent role in the glycosylation of proteins during the latter phase of protein synthesis. In addition to regulating glycoprotein assembly, asparagine is critical for specific protein functions, such as antibodies, collagen assembly, enzyme function, and cell-to-cell recognition. Asparagine is involved in the metabolic control of cell functions in nerve and brain tissue, and is important as a nitrogen reserve substance. Determination of the conformations adopted in vitro and in vivo is crucial to elucidating the mechanism by which asparagine functions in its various roles. Structural analyses are performed using ab initio calculations in the GAUSSIAN98 program for Linux.
© 2003 Elsevier B.V. All rights reserved.

Keywords: Aspargine; Mailard reaction; Acrylamide

1. Introduction

Amino acids can be defined separately by the unique characteristics rendered to each amino acid molecule as a result of the varying reactive abilities of their side chains. The biologically active L-stereoisomer of asparagine (Fig. 1) is among less than a handful of amino acids capable of reacting with saccharides. Its side chain reacts to form N-glycosidic bonds with oligosaccharides, establishing a prevalent role in the glycosylation

of proteins during protein synthesis. More recently, extended theoretical, computational and experimental analyses have been performed on asparagine in the various instances it is found in nature [1]. Asparagine is known to adopt conformations in the left-handed α-helical region and other partially allowed regions of the Ramachandran plot more readily than any other non-glycyl amino acids. The reason for this preference has not been established, but an examination of the local environments of asparagine and aspartic acid moieties in protein structures with a resolution better than 1.5 Å have revealed that their side chain carbonyls are frequently within 4 Å of their own backbone

---

0166-1280/$ - see front matter © 2003 Elsevier B.V. All rights reserved.
doi:10.1016/j.theochem.2003.08.032

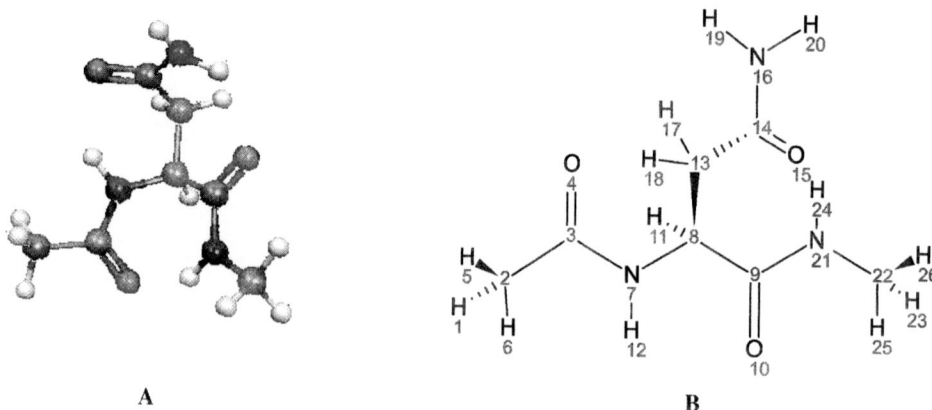

Fig. 1. (a) A pictorial representation of asparagines at the 3rd point of convergence as described in Table 1, (b) The asparagines molecule numbered according to a standardized numbering system.

carbonyl or the backbone carbonyl of the previous residue. This close contact occurs in almost 80% of cases in proteins [2]. In another study, a comparative theoretical analysis of the structure and vibrations of the amino acid L-asparagine was performed using ab initio Hartree–Fock calculations and the Self-Consistent Reaction Field (SCRF) theory. The 3-21G and 6-31 + G basis sets were used to evaluate quadratic force fields of this amino acid in an aqueous solution and in isolation (gas phase) [3]. In a separate study, Raman spectroscopy and SCRF were used to study structural and vibrational features of the amino acid L-asparagine in H_2O and D_2O solutions. Again, ab initio methodology at the RHF/6-31 + G level was used to evaluate the minimum energy structure, quadratic force field, vibrational wavenumbers, and infrared intensities of this molecule in a polar medium [4]. The most relevant study was a side-chain conformational analysis of the asparagine side chain, N-formyl-L-asparaginamide and N-acetyl-L-asparagine N-methylamide, in their γ_l backbone conformation [5]. Finally, in a recent experimental study, researchers discovered that the asparagine in numerous food products (specifically potatoes) could react with glucose monosaccharides in a Maillard reaction to form acrylamide, a neurotoxin and potential carcinogen. For the many reasons listed above, a thorough ab initio computational analysis of N-acetyl-L-asparagine N-methylamide, as it is found in polypeptides,

would be essential in the completion of a full basis set for the asparagine molecule.

2. Computational methods

The structure of the asparagine side-chain is governed by some of the same general principles that affect all molecules built from planar moieties of two-fold symmetry connected to a tetrahedral carbon of three-fold symmetry [5]. The different conformations assumed by N-acetyl-L-asparagine N-methylamide are dictated by the relative values of four different dihedral angles: ϕ (along N7–C8), Ψ (C9–C8), χ_1 (C8–C13), and χ_2 (C14 − C13) (Fig. 1). Since each dihedral has three possible positions (g^+, a, g^-), and rotation about four dihedrals is considered, there exist a total of 81 possible energy minima conformations. The L-asparagine molecule was initially numbered according to a standardized numbering system [6] (Fig. 1) developed for all amino acids, including the flanking protector side groups. This defined arrangement of atomic positions in the molecule was used as the basis for a z-matrix, which was input into the GAUSSIAN 98 computational analysis program [7]. The asparagine molecule was first optimized at the semi-empirical AM1 level of theory, and then advanced to the RHF/3-21G level of theory.

Table 1
Optimized torsional angles and computed energy values for the conformational minima of N-acetyl-l-asparagines N-methylamide, computed at the RHF/3-21G level of theory. Note, that only 13 significant points of convergence have been displayed.

Initial parameters				Optimized parameters				
Φ	Ψ	χ_1	χ_2	Φ	Ψ	χ_1	χ_2	E_{min} (Hartree)
a	a	a	g^+	176.08	−172.46	−158.84	−14.6	−656.973913
a	a	g^-	g^-					
a	g^+	a	g^-					
g^-	a	a	g^+					
a	a	a	a	176.52	−172.35	−138.51	−92.76	−656.977092
a	a	a	g^-					
a	a	g^-	a					
g^-	a	a	g^-					
g^-	a	g^+	a	−171.88	−80.56	49.32	−80.51	−656.978171
g^-	a	g^+	g^-					
g^-	g^+	g^+	g^+					
g^-	g^+	g^+	a					
g^-	g^+	g^+	g^-					
g^-	g^-	g^+	a					
g^-	g^-	g^+	g^-					
g^-	g^-	g^+	g^+					
a	g^-	g^+	g^-					
a	g^+	a	g^+	−176.02	−85.91	−170.59	23.39	−656.964884
g^-	g^+	a	g^+					
g^-	g^+	a	g^-					
g^-	g^-	a	g^+					
g^+	g^+	g^-	g^+	170.05	60.99	−63.35	1.26	−656.964561
a	g^-	g^-	g^-					
g^+	g^+	g^-	a					
g^+	a	a	a	−159.37	68.1	−151.79	−10.25	−656.964129
g^+	a	a	g^+					
g^+	a	a	g^-					
a	g^+	g^+	g^+	−172.92	−127.94	61.34	−23.25	−656.970229
a	g^+	g^+	a					
a	g^+	g^-	g^+	−170.52	−116.94	−139.4	122.4	−656.962348
a	g^+	a	a					
g^-	g^-	a	a					
g^+	g^+	g^+	a	157.86	49.59	50.26	−88.99	−656.961767
g^+	g^+	g^+	g^-					
g^+	g^-	g^+	g^-					
g^+	g^+	a	g^+	175.43	62.26	−160.68	38.52	−656.959612
g^+	g^+	a	g^-					
g^+	g^-	g^-	g^+	170.87	74.19	−65.08	−6.05	−656.963864
g^+	g^-	g^-	a					
g^-	g^-	g^-	a	−159.39	−132.53	−54.5	68.65	−656.960626
g^-	g^-	g^-	g^-					
g^-	g^+	g^-	g^+	−170.66	−82.98	−44.07	62.19	−656.968102
g^-	a	g^-	g^+					

Table 2
Optimized values and results for all 81 conformers

Initial parameters	Optimized parameters						
	ϕ	Ψ	χ_1	χ_2	ω_i	ϕ_{i+1}	E_{min} (hartree)
β_L++	178.65	172.61	51.79	74.62	177	177.56	−656.963522
$\beta_L + a$	−178.85	−156.41	42.83	−142.58	154.21	176.76	−656.961677
$\beta_L + -$	−167.06	−178.87	52.14	−82.45	171.68	−178.73	−656.959063
$\beta_L a +$	176.08	−172.46	−158.84	−14.6	173.82	−179.98	−656.973913
$\beta_L a\, a$	176.52	−172.35	−138.51	−92.76	−175.11	−178.72	−656.977092
$\beta_L a -$	176.5	−172.38	−138.5	−92.67	−175.08	−178.74	−656.977092
$\beta_L - +$	−173.84	−152.13	−53.44	89.12	162.12	176.92	−656.955638
$\beta_L - a$	176.5	−172.38	−138.5	−92.74	−175.11	−178.72	−656.977092
β_L--	176.1	−172.45	−159.16	−14.82	173.47	−179.91	−656.973839
δ_D++	174.99	−170.12	31.52	72.3	−66.6	179.42	−656.965495
$\delta_D + a$	169.3	−160.99	68.13	125.85	−47.8	−177.15	−656.95053
$\delta_D + -$	178.39	−162.65	45.19	23.33	−54.79	179.21	−656.963377
$\delta_D - +$	−170.19	−145.27	−52.54	85.43	−58.61	179.54	−656.947454
$\delta_D - a$	178.41	−162.7	45.25	23.18	−54.71	179.22	−656.963377
δ_D--	169.97	61.08	−63.44	1.69	36.22	179.17	−656.964561
$\delta_D a +$	171.03	176.65	−162.86	26.92	−46.12	−177.15	−656.947757
$\delta_D a\, a$	172.89	−175.87	−163.15	159.62	−41.4	−176.27	−656.952555
$\delta_D a -$	175.35	−167	−132.4	−119.96	−55.08	−177.89	−656.955098
ε_D++	172.56	165.42	62.48	86.58	−138.95	−176.24	−656.967617
$\varepsilon_D + a$	−6.33	−169.91	23.77	−117.51	148.49	176.21	−656.951546
$\varepsilon_D + -$	−173.14	57.44	63.01	−114.58	−164.6	−172.95	−656.94161
$\varepsilon_D a +$	−159.37	68.1	−151.79	−10.25	174.19	−179.89	−656.964129
$\varepsilon_D a\, a$	−159.39	68.09	−151.77	−10.27	174.2	−179.88	−656.964129
$\varepsilon_D a -$	−159.43	68.04	−151.67	−10.48	174.3	−179.9	−656.964128
$\varepsilon_D - +$	−160.71	73.59	−56.83	−72.31	171.41	179.01	−656.958454
$\varepsilon_D - a$	−154.35	62.26	−77.47	−134.8	177.46	−179.43	−656.951756
ε_D--	Not found						
δ_L++	−172.92	−127.94	61.34	−23.25	32.72	177.85	−656.970229
$\delta_L + a$	−172.97	−128.17	61.42	−23.03	32.8	177.84	−656.970229
$\delta_L + -$	−171.62	−81.33	49.29	−80.24	59.03	179.06	−656.978166
$\delta_L - +$	−170.5	−116.97	−139.39	122.39	20.28	176.34	−656.962348
$\delta_L - a$	−176.54	−127.12	−60.45	157.17	26.17	176.44	−656.959955
δ_L--	−177.72	−168.7	−109.32	−29.39	47.08	176.29	−656.95293
$\delta_L a +$	−176.02	−85.91	−170.59	23.39	68.58	−179.9	−656.964884
$\delta_L a\, a$	−170.52	−116.94	−139.4	122.4	20.29	176.35	−656.962348
$\delta_L a -$	176.03	−172.51	−159.3	−15.09	174.38	−179.88	−656.973884
ε_L++	169.51	−164.49	79.69	119.28	−161.56	−177.96	−656.961428
$\varepsilon_L + a$	−171.88	−80.56	49.32	−80.51	58.72	178.77	−656.978171
$\varepsilon_L + -$	−171.63	−82.0	49.3	−80.22	59.03	179.08	−656.978166
$\varepsilon_L a +$	176.07	−172.5	−158.85	−14.58	173.79	−179.99	−656.973913
$\varepsilon_L a\, a$	Not found						
$\varepsilon_L a -$	176.34	−172.28	−138.53	−93.06	−175.14	−178.69	−656.97709
$\varepsilon_L - +$	−170.66	−82.96	−44.07	62.19	67.51	−178.78	−656.968102
$\varepsilon_L - a$	168.63	−120.47	−73.5	−112.92	162.48	177.23	−656.956505
ε_L--	174.73	−108.06	−71.52	−36.01	154.5	177.31	−656.954706
α_D++	172.19	53.41	43.37	104.29	37.06	179.54	−656.939055
$\alpha_D + a$	157.86	49.59	50.26	−88.99	49.66	−178.0	−656.961767
$\alpha_D + -$	157.88	49.59	50.26	−88.99	49.65	−177.99	−656.961767
$\alpha_D a +$	175.43	62.26	−160.68	38.52	37.22	178.6	−656.959612
$\alpha_D a\, a$	171.94	63.29	−143.98	114.79	34.96	179.05	−656.961565

Table 2 (*continued*)

Initial parameters	Optimized parameters						
	ϕ	Ψ	χ_1	χ_2	ω_i	ϕ_{i+1}	E_{min} (hartree)
$\alpha_D\,a\,-$	175.38	62.32	−160.72	38.46	37.2	178.58	−656.959612
$\alpha_{D-}\,+$	170.03	61.03	−63.37	1.32	36.29	179.17	−656.964562
$\alpha_{D}-a$	170.05	60.99	−63.35	1.26	36.32	179.16	−656.964562
$\alpha_{D-}\,-$	Not found						
γ_D++	−167.96	49.2	65.17	86.53	−131.68	−177.4	−656.952064
γ_D+a	174.91	79.41	−51.66	−66.6	−58.65	−179.37	−656.962523
$\gamma_D+\,-$	157.95	49.63	50.25	−88.82	49.66	−177.81	−656.961766
$\gamma_D\,a\,+$	177.08	77.37	−152.82	32.49	−63.43	179.63	−656.956907
$\gamma_D\,a\,a$	172.3	75.1	−150.11	151.38	−49.7	−177.7	−656.962268
$\gamma_D\,a\,-$	177.17	74.63	−175.56	−105.52	−58.44	−176.29	−656.952637
$\gamma_{D-}\,+$	170.87	74.19	−65.08	−6.05	−56.43	−178.97	−656.963864
$\gamma_{D}-a$	170.86	74.21	−65.1	−6.01	−56.48	−178.97	−656.963864
$\gamma_{D-}\,-$	Not found						
γ_L++	−171.63	−81.31	49.28	−80.26	59.02	179.11	−656.978166
γ_L+a	−171.61	−81.33	49.29	−80.25	59.04	179.08	−656.978166
$\gamma_L+\,-$	−171.61	−81.3	49.29	−80.25	59.05	179.1	−656.978166
$\gamma_L\,a\,+$	−175.99	−85.91	−170.55	23.28	68.54	−179.89	−656.964884
$\gamma_L\,a\,a$	−173.36	−85.5	−161.04	107.63	67.84	−178.84	−656.969028
$\gamma_L\,a\,-$	−176.07	−85.9	−170.61	23.17	68.63	−179.89	−656.964884
$\gamma_{L-}\,+$	−170.65	−82.97	−44.10	65.22	67.54	−178.77	−656.968102
$\gamma_{L}-a$	Not found						
$\gamma_{L-}\,-$	−167.62	−88.79	292	−5.63	65.21	−179.06	−656.964169
α_L++	−171.65	−81.30	49.29	−80.19	59.03	179.09	−656.978166
α_L+a	−171.88	−80.56	49.32	−80.5	58.72	178.78	−656.978171
$\alpha_L+\,-$	−171.62	−81.33	49.28	−80.23	59.06	179.08	−656.978166
$\alpha_L\,a\,+$	−176.01	−85.91	−170.59	23.37	68.58	−179.87	−656.964884
$\alpha_L\,a\,a$	−170.53	−116.91	−139.44	122.44	20.35	176.38	−656.962348
$\alpha_L\,a\,-$	−173.51	−136.42	−172.29	−134.63	32.92	176.37	−656.951572
$\alpha_{L-}\,+$	−169.71	−78.37	−53.73	66.07	−15.58	−179.71	−656.96113
$\alpha_{L}-a$	−159.38	−132.54	−54.49	68.7	22.87	176.62	−656.960626

The backbone conformation (ϕ and ψ) have been summarized by one symbol (ie. β_1) which is properly defined by Fig. 2, the Ramachandran plot. ' + ' represents g^+, ' − ' represents g^-, and 'a' represents anti.

3. Results and discussion

The asparagine molecule was optimized at the RHF/3-21G level of theory. Of the 81 tested conformations, 77 ended in the normal termination. Of the 77 conformations, 43 conformations converged to 13 separate points distinguished by 13 unique energy minima (Table 1). In total, approximately 46 separate energy minima conformations of the 81 tested conformations were found for the amino acid asparagine (Table 2, Fig. 2)

The energy minima of the N-acetyl-L-asparagine N-methylamide in the γ_1 backbone conformation had been previously calculated [5]. Interestingly, many of the values for ϕ rested in the a position, providing evidence for a possibly different stable backbone conformation that the already proposed γ_1. The third point of convergence listed in Table 1 reveals a point commonly shared by many different conformers that were initially defined under different parameters. This point is in the δ_d backbone conformation. In fact, many of the conformers initially defined with the γ_1 backbone, converged to δ_d backbone conformation, although there were varying side chain conformations. Berg et al. described similar energy values for all conformers [5]. However, those conformers

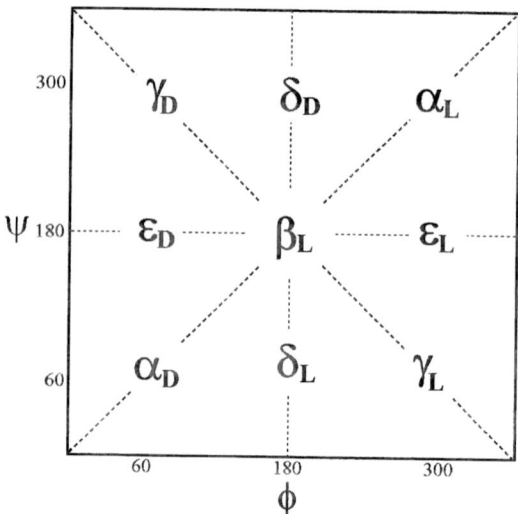

Fig. 2. The Ramachandran plot, which was used to organize the 81 conformers for efficient calculation.

initially set in the γ_l backbone configuration resulted in the most stable conformers (Table 2). On the other hand, the ε_d conformer appeared to be the least stable (specifically $\varepsilon_d\ g^+\ g^-$, see Table 2).

4. Conclusion and future work

Further work is necessary to complete a full analysis of the asparagine molecule at a higher level of theory (RHF/6-31G). As mentioned above, the interaction of asparagine with saccharides such as glucose is prevalent in many biological pathways, the most common being the formation of *N*-glycosidic bonds, and more recently, the formation of acrylamide. Thus, future work towards conducting an ab initio study of asparagines-disaccharide interactions appears to be the next logical step.

Acknowledgements

This work was supported with grants from the Global Institute Of Computational Molecular and Materials Sciences (GIOCOMMS), Toronto, Ontario, Canada. Thanks are extended to Ashton Connor, Reza Mehdizadeh, Jacqueline M.S. Law, and Suzanne K. Lau for helpful discussion.

One of the authors (IGC) wishes to thank the Ministry of Education for a Szent-Györgyi Visiting Professorship.

References

[1] V. Rao, T. Cui, C. Guan, P.V. Roey, Protein Science. 8 (1999) 2338–2346.

[2] C.M. Deane, F.H. Allen, R. Taylor, T.L. Blundell, Protein Engineering. 12 (1999) 1025–1028.

[3] J.T. Lopez, Navarrette, J. Casado, V. Hernandez, F.J. Ramirez. Theor Chem Acc. 98 (1997) 5–15.

[4] J.T. Lopez, Navarrette, J. Casado, V. Hernandez, F.J. Ramirez. Journal of Raman Spectroscopy. 28 (1997) 501–509.

[5] M.A. Berg, S.J. Salpietro, A. Perczel, O. Farkas, I.G. Csizmadia, Theochem. 504 (2000) 127–139.

[6] G.A. Chass, M.A. Sahai, J.M.S. Law, S. Lovas, O. Farkas, A. Perczel, J.L. Rivail, I.G. Csizmadia, International Journal of Quantum Chemistry. 90 (2002) 933–968.

[7] Gaussian 98 (Revision A.9), M.J. Frisch, G.W. Trucks, H.B. Schlegel, G.E. Scuseria, M.A. Robb, J.R. Cheeseman, V.G. Zakrzewski, J.A. Montgomery, Jr., R.E. Stratmann, J.C. Burant, S. Dapprich, J.M. Millam, A.D. Daniels, K.N. Kudin, M.C. Strain, Ö. Farkas, J. Tomasi, V. Barone, M. Cossi, R. Cammi, B. Mennucci, C. Pomelli, C. Adamo, S. Clifford, J. Ochterski, G.A. Petersson, P.Y. Ayala, Q. Cui, K. Morokuma, D.K. Malick, A.D. Rabuck, K. Raghavachari, J.B. Foresman, J. Cioslowski, J.V. Ortiz, A.G. Baboul, B.B. Stefanov, G. Liu, A. Liashenko, P. Piskorz, I. Komaromi, R. Gomperts, R.L. Martin, D.J. Fox, T. Keith, M.A. Al-Laham, C.Y. Peng, A. Nanayakkara, C. Gonzalez, M. Challacombe, P.M.W. Gill, B.G. Johnson, W. Chen, M.W. Wong, J.L. Andres, M. Head-Gordon, E.S. Replogle, J.A. Pople, Gaussian, Inc., Pittsburgh PA, 1998.

ELSEVIER

Journal of Molecular Structure (Theochem) 666–667 (2003) 279–284

THEO
CHEM

www.elsevier.com/locate/theochem

Ramachandran backbone potential energy surfaces of aspartic acid and aspartate residues: implications on allosteric sites in receptor–ligand complexations

Joseph C.P. Koo[a,b,e,*], Janice S.W. Lam[a,b], Gregory A. Chass[a,b,c,d], David H. Setiadi[b], Jacqueline M.S. Law[a,b], Julius Gy. Papp[e], Botond Penke[f,g], Imre G. Csizmadia[a,b,f,*]

[a]Global Institute of Computational Molecular and Materials Science (GIOCOMMS), 1422 Edenrose St., Mississauga, Ont., Canada, L5V 1H3
[b]Department of Chemistry, University of Toronto, 80 St. George St., Toronto, Ont., Canada, M5S 3H6
[c]Institut de Science et d'Ingénierie Supramoléculaires, 8, allée Gaspard Monge, BP 70028, 67083 Strasbourg Cedex, France
[d]Department of Biomedical Sciences, Creighton University, 2500 California plaza, Omaha, NE, 68178, USA
[e]Department of Pharmacology and Pharmacotherapy, Division of Cardiovascular Pharmacology, Hungarian Academy of Sciences and University of Szeged, Dóm tér 12, H-6701, Szeged, Hungary
[f]Department of Medicinal Chemistry, University of Szeged, 8 Dóm tér, H-6720, Szeged, Hungary
[g]Protein Chemistry Research Group, Hungarian Academy of Sciences, University of Szeged, Dóm tér 8, 6720 Szeged, Hungary

Abstract

Ramachandran backbone potential energy surfaces (PES) were generated for *N*-acetyl-L-aspartic acid-*N'*-methylamide and *N*-acetyl-L-aspartate-*N'*-methylamide. Relatively few minima were observed from the Ramachandran PES of the aspartate ion while many existed for both *endo* and *exo* forms of the aspartic acid residue. By comparing the relative stabilization energies as well as the vertical and adiabatic proton affinities of the two forms the aspartic acid residue, it was previously determined that aspartic acid may rather change its backbone conformation before deprotonation into the aspartate ion. Taken together, many stable conformers may exist for both *endo* and *exo* forms of aspartic acid but when deprotonated they lead to the same aspartate ion which has remarkably few conformations. This feature may have significant biological implications, and a receptor–ligand complexation model is proposed where aspartate and aspartic acid could represent, respectively, resting and active states of a putative binding site in a hypothetical protein. More importantly, however, the aspartyl residues may represent important allosteric sites for regulatory ions, directly modulating the specificity of ligand docking in a receptor protein and affecting downstream biological mechanisms.
© 2003 Elsevier B.V. All rights reserved.

Keywords: Ramachandran map; Receptor–ligand complexations; Allosteric actions; Aspartate residue; Aspartic acid residue

1. Introduction

The aspartic acid residue is highly involved in many biological systems. According to the Protein Data Bank, the aspartic acid residue has an average occurrence of approximately 5% in proteins [1].

* Corresponding authors.
  E-mail addresses: joseph.koo@utoronto.ca (J.C.P. Koo), icsizmad@chem.utoronto.ca (I.G. Csizmadia).

0166-1280/$ - see front matter © 2003 Elsevier B.V. All rights reserved.
doi:10.1016/j.theochem.2003.08.055

Undoubtedly, the molecular conformations that an aspartic acid (or aspartate) residue partakes in a peptide chain may ultimately affect the overall three dimensional shape of a protein, altering its structure and function in a biological system. As a result, it is interesting and important to investigate the various geometric implications involving the aspartic acid residue. Various mutational studies have reported that aspartic acid, or aspartate, are important in ligand binding as well as receptor protein function [2–4].

The conformational preferences for N-acetyl-L-aspartic acid-N'-methylamide and its deprotonated form, N-acetyl-L-aspartate-N'-methylamide, were previously explored in great detail (Fig. 1) [5–8]. The sidechain potential energy surfaces (PES) for each of the nine backbones (γ_L, β_L, δ_L, α_L, ε_L, γ_D, δ_D, α_D, ε_D) of the aspartic acid as well as for the aspartate residues were also examined previously [5–8].

In this paper, the Ramachandran backbone PESs for both *endo* and *exo* forms of the aspartic acid residue as well as for the aspartate ion are reported. By comparing the relative minima that occur in these PESs for all three molecular species, biological implications (specifically, receptor–ligand complexations) involving the aspartic acid and aspartate sidechains are analyzed.

2. Computational methods

All geometric minima for both *endo* and *exo* forms of N-acetyl-L-aspartic acid-N'-methylamide as well as for N-acetyl-L-aspartate-N'-methylamide were determined as previously described [5–8]. To generate the respective Ramachandran backbone PESs, $E = E(\phi, \psi)$, normal optimization condition (FOPT = Z-MATRIX) was employed for all three molecular species. In addition, all representative sidechains (CH_3-CH_2-COOH or $CH_3-CH_2-COO^-$), as defined by the variables χ_1 and χ_2 (χ_3 determines whether the aspartic acid residue will be in *endo* or *exo* form), were relaxed while the backbone variables, ϕ and ψ, were rotated with 30.0° increments. Consequently, a total of $12 \times 12 = 144$ geometric points were generated for each Ramachandran backbone surface and all critical points had gradients of less than 4.5×10^{-4} a.u. All calculations were performed using GAUSSIAN98 [9].

3. Results and discussion

The Ramachandran PESs for both *endo* and *exo* forms of the aspartic acid as well as for the aspartate residues are shown in Fig. 2. It is rather astonishing to observe relatively few minima for aspartate while many exist for the aspartic acid residue. Consequently, an interesting trend can be deduced: while

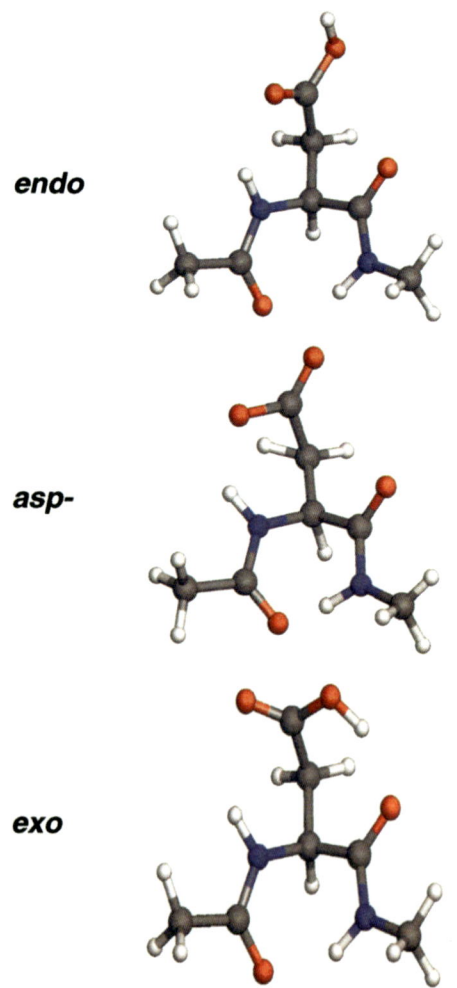

endo

asp-

exo

Fig. 1. Ball-and-stick representations of (top) the *endo* form of N-acetyl-L-aspartic acid-N'-methylamide (denoted as *endo*), (middle) N-acetyl-L-aspartate-N'-methylamide (denoted as *asp-*), and (bottom) the *exo* form of N-acetyl-L-aspartic acid-N'-methylamide (denoted as *exo*).

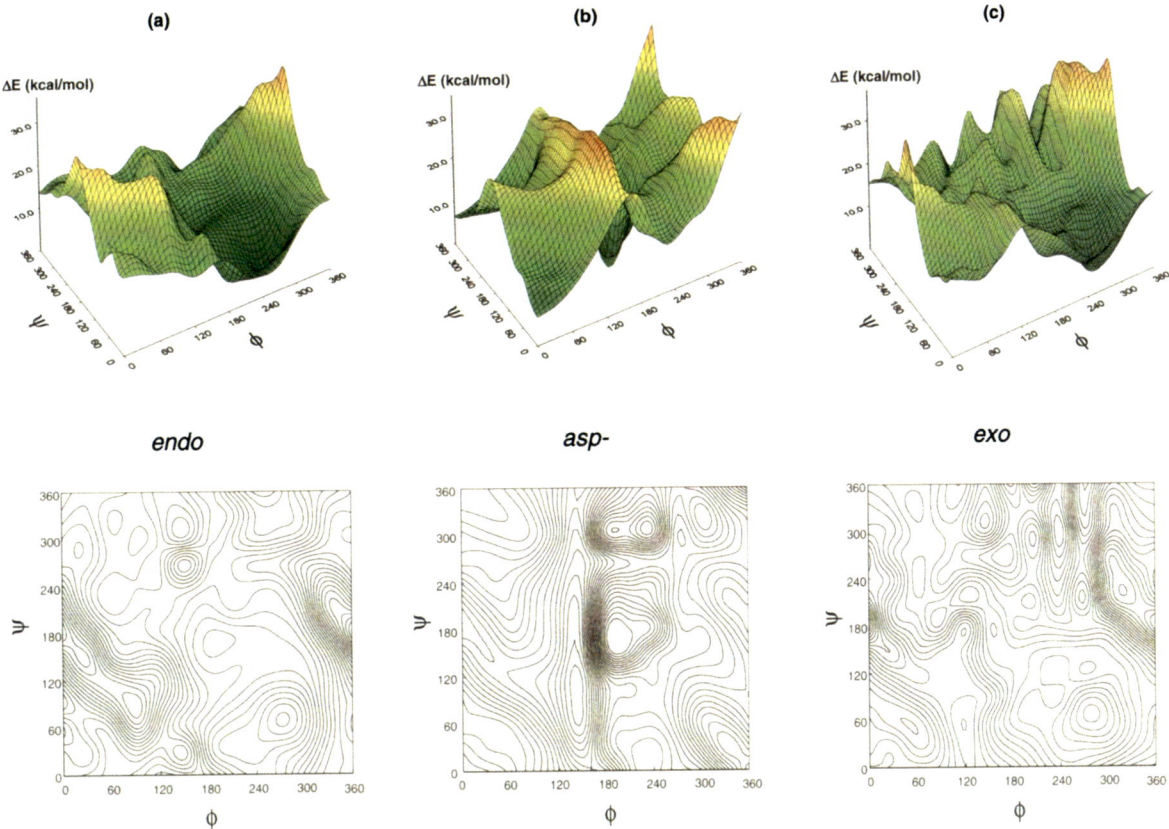

Fig. 2. Ramachandran backbone potential energy surfaces, shown in (top) landscape and (bottom) contour representations, of (a) the *endo* form of *N*-acetyl-L-aspartic acid-*N'*-methylamide (denoted as *endo*), (b) *N*-acetyl-L-aspartate-*N'*-methylamide (denoted as *asp-*), and (c) the *exo* form of *N*-acetyl-L-aspartic acid-*N'*-methylamide (denoted as *exo*). Torsional angles ϕ and ψ are given in degrees.

endo and *exo* forms of the aspartic acid residue may have numerous conformations, when deprotonated, both species lead to the same aspartate ion, which has remarkably few conformations. On a physical-chemical level, such geometric specificities entail the fact that deprotonation of aspartic acid is a highly regulated mechanism where backbone conformations (as presented by the torsions ϕ and ψ in the Ramachandran PESs in Fig. 2) may govern the overall stability of the molecular species. Indeed, there is evidence which indicates that the aspartic acid residue may rather change its backbone conformation in an explicit manner before undergoing deprotonation [8]. Consequently, deprotonation specificities may have important biological implications, such as receptor–ligand complexations.

In the absence of external hindrance, the relative energies for the *endo* form of aspartic acid is lower

[5–7] (and hence producing overall more stable geometric structures) than those of the *exo* form. Consequently, it may be possible that ligand binding is more favourable if the interaction is achieved through the electron pairs that are located at the *endo* positions. As a result, it may be expected that in a protein whose binding domain has rich aspartate regions, receptor–ligand complexation is achieved through the mutual and symmetrical sharing of electron density between a number of aspartate sidechains and the ligand (Fig. 3). After protonation, the aspartic acid residue may still participate in receptor–ligand complexations, although the electron density is now partially distributed to the –OH bond that is formed, resulting in weaker interactions with the ligand. In sum, the aspartate and aspartic acid sidechains may represent, respectively, resting and active forms of a hypothetical receptor binding region

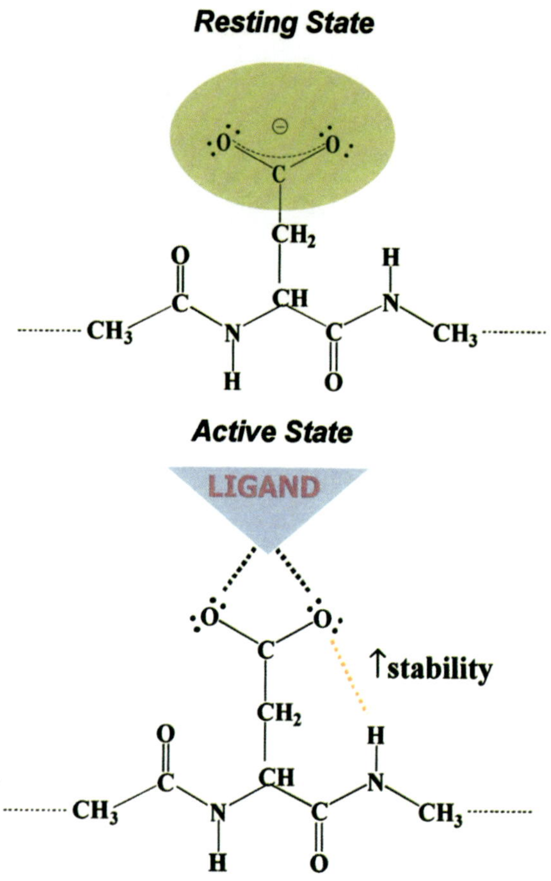

Resting State

Active State

Fig. 3. Receptor–ligand complexation involving aspartate or aspartic acid. (Top) When deprotonated, the aspartate residue has a high electron density in its sidechain that may attract free ligands for binding; represented by the green-shaded area. In this state, the aspartate residue may be considered to assume in a resting mode. (Bottom) When bound by a ligand (such as a proton H^+, a regulatory ion such as Ca^{2+} or Mg^{2+}, or a drug; represented by the blue-shaded triangle), the aspartate residue now behaves more alike an aspartic acid residue. In this state, the aspartate residue may be considered to assume in an active mode. The dotted lines protruded from the N- and C-protective groups represent other amino acids or functional groups in a peptide chain. The red-dotted line represents a sidechain-backbone interaction that may enhance conformational stability during ligand binding.

that has only one sidechain responsible for ligand interaction (Fig. 3).

When not bound by a ligand, the aspartate sidechan represents a resting stage with low stabilization energy. Subsequent to protonation (by H^+, Ca^{2+}, Mg^{2+} or a ligand), however, the overall stabilization energy of the aspartate residue (and hence the protein) increases and the binding domain of the respective

receptor is in an active state, where further interactions or downstream regulatory mechanisms may be triggered to bring forth pharmacological responses. Of course, other intramolecular forces may exist to enhance receptor–ligand complexations when the aspartate sidechain is in an active mode. Here, sidechain-backbone hydrogen bond interactions may help stabilize the backbone geometry of the aspartic acid residue while its sidechain interacts with a ligand (Fig. 3 bottom). Hence, during the resting state of a hypothetical protein, the aspartyl residue may assume in relatively few conformations with no ligand interactions. In the active state, however, the aspartyl residue may adopt many conformations, all of which resulted from the sharing of electron density with different ligands at the *endo* or the *exo* sidechain positions.

Although it is obvious that a receptor protein usually has binding domains composed of different amino acids (rather than just one aspartate or one aspartic acid residue), aspartate or aspartic acid residues themselves may still affect the overall function of the receptor. For instance, single amino acid residues critical in controlling the equilibrium between active and inactive states of receptors have been well documented [10,11]. Also, it is well known that some proteins require regulatory ions in order to facilitate ligand binding [12,13], while downstream biological activity of functional domains may also be regulated by the presence or absence of these ions [14]. In these cases, aspartate or aspartic acid residuse in a receptor protein may act as allosteric sites for the binding of these regulatory ions, modulating the binding or dissociation of ligands on target receptors and altering protein function. Certainly, these interactions must be thoroughly examined, perhaps in solvation studies, in order to fully characterize the pharmacological implications for a receptor that has aspartate-rich regions in its binding domain. Nonetheless, ab initio computations on molecular structures, even for relatively small species such as the aspartate or the aspartic acid residues here examined, allow for the generation of study models that can enhance knowledge of complex biological and chemical systems.

Recently, molecular modelling has become a popular tool for pharmacological studies, especially for ligand screening on receptors as well as for de novo drug discovery. In order to maximize the probability of success and to minimize costs in drug research, there is a great need for efficient and accurate high-throughout screening of lead compounds. Furthermore, with the human genome being completed sequence [15], disease sequences previously unknown or unavailable are now subjected to direct drug targeting. A chemogenomics approach has been described in detail for direct screening of lead compounds against genomic sequences [16]. In such combinatorial informatics settings, virtual libraries that contain thousands of potential leads are crucial components in the screening process. In addition, quantitative structure–activity relationships (QSAR) and molecular descriptors are often used to fully analyze the pharmacological characteristics of all lead compounds constructed in virtual library databases. If a universal numbering module [17] can be utilized for characterizing all molecules and macromolecules studied, then it may be possible to categorize the pharmacological profiles of these molecular structures at the ab initio level.

One possible advantage of utilizing ab initio methods for ligand screening is the higher relative accuracy that could be generated in comparison to conventional molecular modeling techniques. Consequently, ab initio computations may be employed in a drug discovery setting, and genomic drug targeting can be operated at high accuracy and efficiency. Due to the rather burdensome task of recreating a molecule's complete three dimensional arrangement in the form of a z-matrix, it was conventionally thought that ab initio computations can only be used to optimize molecular structures that are composed of approximately 500–600 atoms. However, with the emergence of more powerful computational softwares and hardwares, together with more refined systems of GAUSSIAN scripting, ab initio calculations on large molecules that were previously impossible may now be achievable. In addition, ab initio methods may be useful in the virtual construction of ligands, a feature that is appealingly compatible in a chemogenomics setting.

4. Conclusions

Ramachandran backbone PESs, $E = E(\phi, \psi)$, were generated for the aspartate as well as both *endo* and *exo* forms of the aspartic acid residues. By comparing their respective PESs, it was observed that while *endo* and *exo* forms of the aspartic acid may have numerous conformations, when deprotonated, they lead to the same aspartate ion which has remarkably few conformations. This feature may have important biological implications and a receptor–ligand complexation is proposed, where aspartate and aspartic acid may respectively represent resting and active states of a hypothetical receptor protein with a single amino acid binding site. Such receptor–ligand interactions may account for the relatively few conformations found for the aspartate residue (when no ligand is bound) while many conformations are allowed for the aspartic acid residue (when ligands of different structural and pharmacological characteristics bind to the aspartyl site in a receptor through *endo* or *exo* interactions).

Acknowledgements

The authors are grateful for the generous allocation of CPU time granted by the Frederick Biomedical Super Computering Centre at the National Cancer Institute (NCI). This work was supported by grants (JCPK and IGC) from the Global Institute of Computational Molecular and Materials Science (GIOCOMMS), Toronto, Ontario, Canada. One of the authors (IGC) wishes to thank the Ministry of Education for a Szent-Györgyi Visiting Professorship.

References

[1] http://www.rcsb.org/pdb/
[2] D. Boehning, D.O. Mak, J.K. Foskett, S.K. Joseph, J. Biol. Chem. 276 (2001) 13509.
[3] D.W. Shin, J. Ma, D.H. Kim, FEBS Lett. 486 (2000) 178.
[4] B. Nilius, R. Vennekens, J. Prenen, J.G. Hoenderop, G. Droogmans, R.J. Bindels, J. Biol. Chem. 276 (2001) 1020.
[5] J.C.P. Koo, G.A. Chass, A. Perczel, Ö. Farkas, L.L. Torday, A. Varro, J.G. Papp, I.G. Csizmadia, Eur. Phys. J., D 20 (2002) 499.
[6] J.C.P. Koo, G.A. Chass, A. Perczel, Ö. Farkas, L.L. Torday, A. Varro, J.G. Papp, I.G. Csizmadia, J. Phys. Chem., A 106 (2002) 6999.
[7] J.C.P. Koo, J.S.W. Lam, S.J. Salpietro, G.A. Chass, R.D. Enriz, L.L. Torday, A. Varro, J.G. Papp, J. Mol. Struct., (THEOCHEM) 619 (2002) 143.

[8] J.C.P. Koo, J.S.W. Lam, G.A. Chass, L.L. Torday, A. Varro, J.G. Papp, J. Mol. Struct., (THEOCHEM) 620 (2003) 231.

[9] M.J. Frisch, G.W. Trucks, H.B. Schlegel, G.E. Scuseria, M.A. Robb, J.R. Cheeseman, V.G. Zakrzewski, J.A. Montgomery, Jr., R.E. Stratmann, J.C. Burant, S. Dapprich, J.M. Millam, A.D. Daniels, K.N. Kudin, M.C. Strain, O. Farkas, J. Tomasi, V. Barone, M. Cossi, R. Cammi, B. Mennucci, C. Pomelli, C. Adamo, S. Clifford, J. Ochterski, G.A. Petersson, P.Y. Ayala, Q. Cui, K. Morokuma, D.K. Malick, A.D. Rabuck, K. Raghavachari, J.B. Foresman, J. Cioslowski, J.V. Ortiz, A.G. Baboul, B.B. Stefanov, G. Liu, A. Liashenko, P. Piskorz, I. Komaromi, R. Gomperts, R.L. Martin, D.J. Fox, T. Keith, M.A. Al-Laham, C.Y. Peng, A. Nanayakkara, C. Gonzalez, M. Challacombe, P.M.W. Gill, B.G. Johnson, W. Chen, M.W. Wong, J.L. Andres, M. Head-Gordon, E.S. Replogle, J.A. Pople, GAUSSIAN 98 (Revision A.x), Gaussian, Inc., Pittsburgh PA, 1998.

[10] J. Marie, E. Richard, D. Pruneau, J.L. Paquet, C. Siatka, R. Larguier, C. Ponce, P. Vassault, T. Groblewski, B. Maigret, J.C. Bonnafous, J. Biol. Chem. 276 (2001) 41100.

[11] L. Joubert, S. Claeysen, M. Sebben, A.-S. Bessis, R.D. Clark, R.S. Martin, J. Bockaert, A. Dumuis, J. Biol. Chem. 277 (2002) 25502.

[12] R.I. Saba, E. Goormaghtigh, J.M. Ruysschaert, A. Herchuelz, Biochemistry 40 (2001) 3324.

[13] E.A. Nalefski, M.A. Wisner, J.Z. Chen, S.R. Sprang, M. Fukuda, K. Mikoshiba, J.J. Falke, Biochemistry 40 (2001) 3089.

[14] I.I. Senin, S.A. Vaganova, O.H. Weiergraber, N.S. Ergorov, P.P. Philippov, K.W. Koch, J. Mol. Biol. 330 (2003) 409.

[15] J.D. McPherson, et al., Nature 409 (2001) 934 The number of authors involved in this work far exceeds the limitation of proper citation in this paper, please refer to the research article for a complete list of authors.

[16] D.K. Agrafiotis, V.S. Lobanov, F.R. Salemme, Nat. Rev. Drug Discov. 1 (2002) 337.

[17] G.A. Chass, M.A. Sahai, J.M.S. Law, S. Lovas, O. Farkas, A. Perczel, J. Rivail, I.G. Csizmadia, Int. J. Quantum Chem. 90 (2002) 933.

THEO
CHEM

Journal of Molecular Structure (Theochem) 666–667 (2003) 285–289

www.elsevier.com/locate/theochem

Predicting the conformational preferences of *N*-acetyl-4-hydroxy-L-proline-*N'*-methylamide from the proline residue

Janice S.W. Lam[a,g,*], Joseph C.P. Koo[a,d,g], Ilona Hudáky[b], Andras Varro[c], Julius Gy. Papp[c,d], Botond Penke[e,f], I.G. Csizmadia[a,e,g]

[a]*Department of Chemistry, University of Toronto, Toronto, Ont., Canada M5S 3H6*
[b]*Department of Organic Chemistry, Eötvös Loránd University (ELTE), 1117 Budapest, Hungary*
[c]*Department of Pharmacology and Pharmacotherapy, University of Szeged, Dóm tér 12, H-6701, Szeged, Hungary*
[d]*Division of Cardiovascular Pharmacology, Hungarian Academy of Sciences and University of Szeged, Dóm tér 12, H-6701, Szeged, Hungary*
[e]*Department of Medical Chemistry, University of Szeged, Dóm tér 8, H-6720, Szeged, Hungary*
[f]*Protein Chemistry Research Group, Hungarian Academy of Sciences, University of Szeged, Dóm tér 8, 6720 Szeged, Hungary*
[g]*Global Institute of Computational Molecular and Materials Science (GIOCOMMS), 1422 Edenrose St., Mississauga, Ont., Canada L5V 1H3*

Abstract

Ab initio Hartree–Fock (RHF/3-21G* and RHF/6-31G(d)) and density functional (B3LYP/6-31G(d)) molecular computations were performed on *N*-acetyl-4-hydroxy-L-proline-*N'*-methylamide with varying torsional angles ω_0, ω_1, ψ, χ_1, and χ_5; while the ϕ torsional angle is restricted in the vicinity of $-60°$ (ω_0, ω_1, ψ, χ_1, and χ_5 are 'governing coordinates' and other torsional angles $\chi_0, \chi_2, \chi_3, \chi_4$ change with respect to them). The computed results of *N*-acetyl-4-hydroxy-L-proline-*N'*-methylamide were then compared to those previously reported for the proline residue at the RHF/6-31 + G* level of theory. The calculated torsional angles of the hydroxyproline and proline residues were remarkably similar. When optimized torsional angle values of hydroxyproline were compared to those of proline, a strong correlation was observed ($R^2 = 0.9997$).
© 2003 Elsevier B.V. All rights reserved.

Keywords: Hydroxyproline; Proline; Ab initio; Hartree–Fock; Density functional geometry optimizations; Collagen; Backbone conformation; Conformer; *Cis–trans* isomerism; Ring puckering

1. Introduction

Proline (Pro) is a unique naturally-occurring amino acid due to the presence of both the α-carbon and the nitrogen of the amide group in a ring [1]. This in turn results in the formation of a rigid saturated five-membered cycle with characteristic conformational restrictions [2], and it is these special structural features that play an important role in determining the secondary structure of proteins. In fact, a proline residue in a peptide chain often causes turns [3] and bends [4] in the chain. Hydroxyproline (Hyp) is a hydroxylated derivative of proline and it shares the same features as its parent amino acid. It is formed by a post-translational modification where a proline residue is converted to hydroxyproline by an enzyme, with a ferrous ion at its active site, called prolyl hydroxylase. Both proline and hydroxyproline, along with glycine (Gly), are found in collagen, the most abundant protein in vertebrates. Collagen consists of three helices coiled around each other to form

* Corresponding author.
  E-mail address: jan.lam@utoronto.ca (J.S.W. Lam).

0166-1280/$ - see front matter © 2003 Elsevier B.V. All rights reserved.
doi:10.1016/j.theochem.2003.08.033

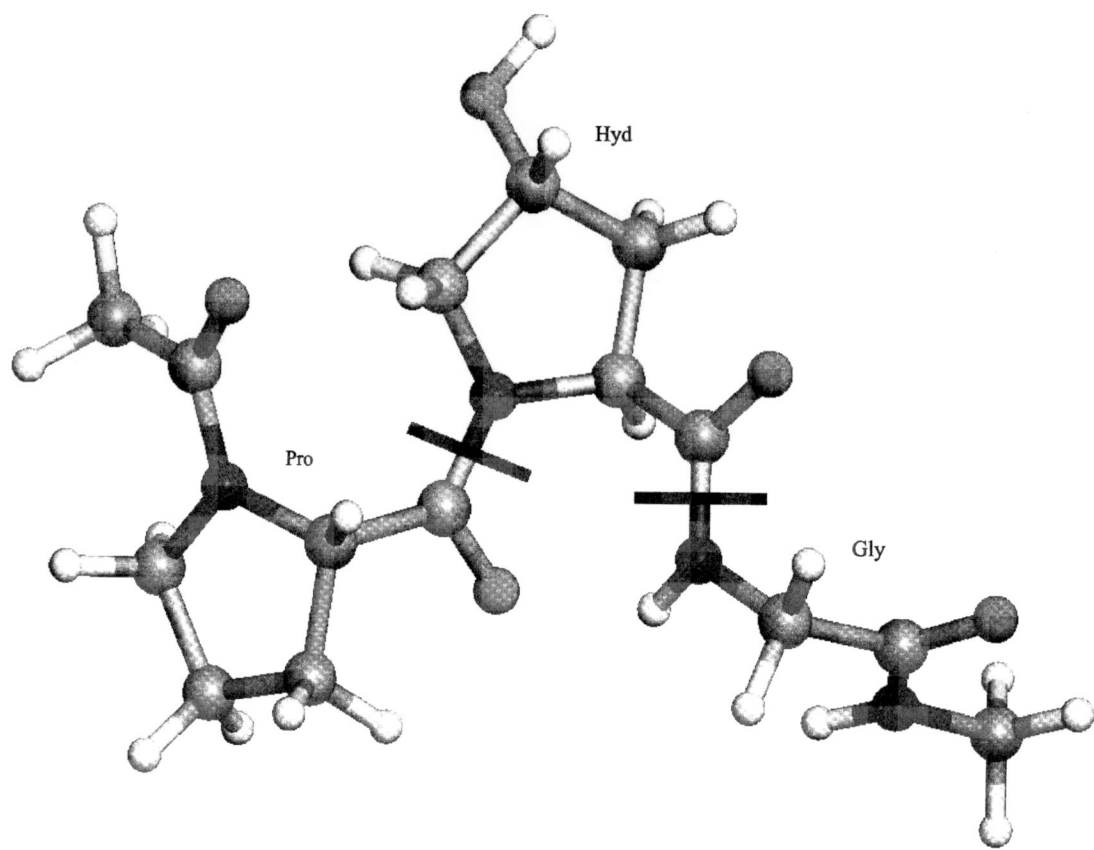

Fig. 1. A view of the Pro-Hyp-Gly triplet. In collagen, this Pro-Hyp-Gly triplet repeats itself for about 300 times to form a helix; and three of these helices again coil together to form a supercoil.

a supercoil. Each helix in this triple helix is generally composed of approximately 300 repeats of the amino acid sequence (Pro-Hyp-Gly) [5]. Therefore it would be of our interest to investigate the part proline and hydroxyproline residues play in the overall stability of the collagen triple helix. Fig. 1 illustrates a Pro-Hyp-Gly triplet found in the collagen helices.

As hydroxyproline and proline share the same structural characteristics, both residues are expected to only take up ϕ values in the vicinity of g^- (i.e. $\phi = -60$ or $300°$) due to the ring structure of its sidechain, resulting in no more than three backbone conformations, namely α_L, ε_L, and λ_L, as shown in Fig. 2. In addition, both residues are capable of forming *cis–trans* isomers, denoted by the ω_0 torsional angle. Ring puckering is another feature unique to proline and hydroxyproline,

denoted by χ_1. However, since hydroxyproline is hydroxylated at the γ-carbon, there exists yet another variable in determining the total number of possible conformers, where the $-OH$ group could be oriented in the g^+, a or g^- position. Hence for each of the three backbone conformations (α_L, ε_L, and γ_L), there is a possible total of 12 conformers, as shown in Fig. 3, giving a grand total of 36 possible conformers in the hydroxyproline residue. A pictorial representation of the definition for various torsional angles in hydroxyproline is shown in Fig. 4. Here we report all the conformers found in N-acetyl-4-hydroxy-L-proline-N'-methylamide and compare their optimized values with those of proline. N-acetyl and N'-methylamide protecting groups are used in order to mimic the local effects of other amino acids in a chain of peptides.

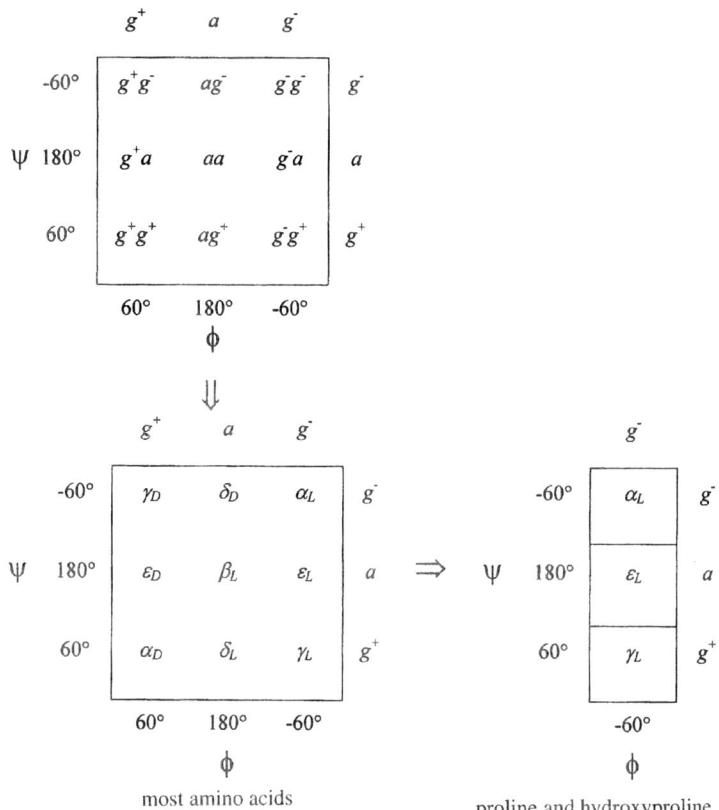

Fig. 2. The Ramachandran map of most common peptides. All nine conformations are possible, with the ϕ and ψ angles ranging from 0 to 360°. Only three discrete conformations are allowed for hydroxyproline and proline due to their special ring structure. The ring limits the ϕ values in the vicinity of the $g-$ (i.e. $\phi = -60$ or 300°) and hence proline and hydroxyproline can only exist in the α_L, ε_L, and γ_L conformations.

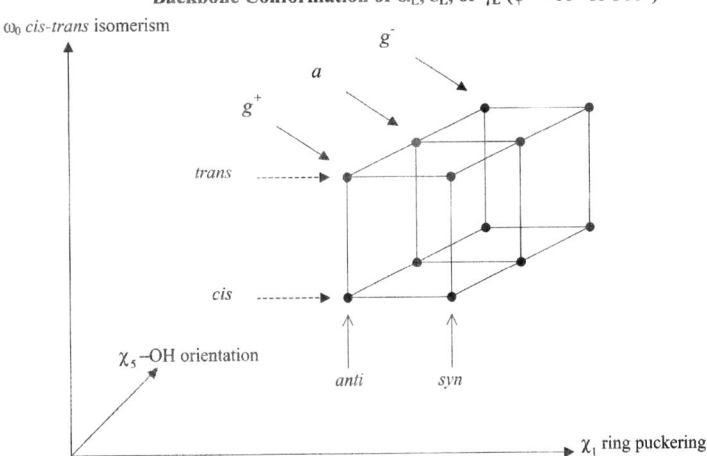

Fig. 3. The conformational space in which a hydroxyproline residue can occupy. It is defined by the three independent variables ω_0, χ_1 and χ_5. In addition, due to the existence of three possible backbone conformations in hydroxyproline (namely α_L, ε_L, and γ_L), there is a total of 36 possible conformers.

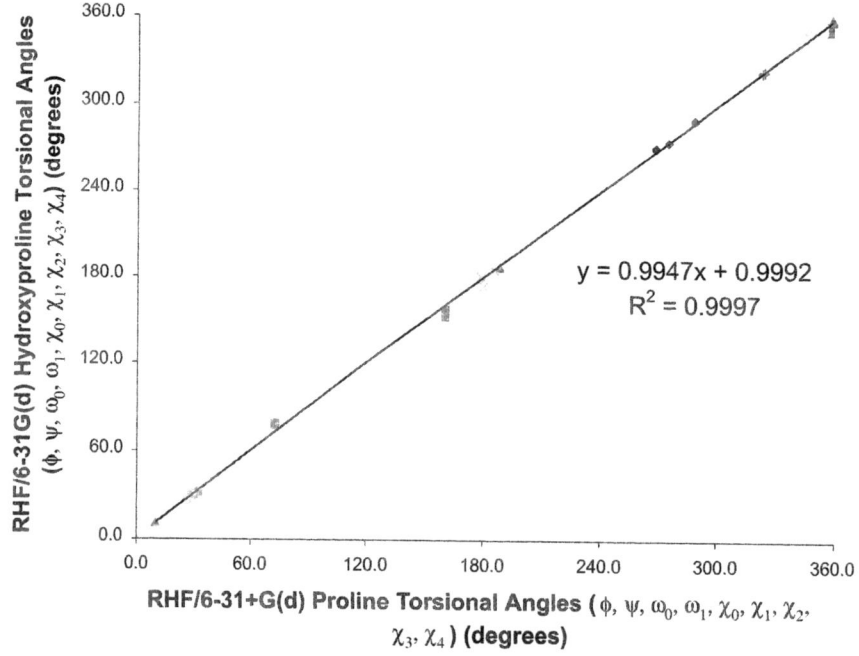

Fig. 4. The molecular structure of *N*-acetyl-4-hydroxy-L-proline-*N'*-methylamide. Definitions of torsional angles are shown by curved arrows.

2. Computational methods

Ab initio Hartree–Fock (HF) and density functional (DFT) geometry optimizations were carried out using the GAUSSIAN 94 [6] and GAUSSIAN 98 [7] program systems under tight optimization conditions for all conformations of *N*-acetyl-4-hydroxy-L-proline-*N'*-methylamide. The basis sets 3-21G* and 6-31(d) were employed at the restricted Hartree–Fock

(RHF) and 6-31(d) at the B3LYP levels of theory using Berny Optimization (FOPT = Z-MATRIX,-TIGHT).

3. Results and discussions

Previously optimized values for the proline residue at the RHF/6-31 + G(d) level [1] were used as initial parameters for corresponding conformers in *N*-acetyl-4-hydroxy-L-proline-*N'*-methylamide (data not shown), including initial parameters not previously found in proline. At B3LYP/6-31G(d), 20 out of 36 possible conformers were found for *N*-acetyl-4-hydroxy-L-proline-*N'*-methylamide (data not shown). Non-existing conformers in proline were not found in hydroxyproline either, suggesting that optimized results from proline in fact provide a helpful initial estimate of existing conformers for its derivative. Moreover, when optimized values of hydroxyproline and proline were compared at the RHF/6-31G(d) level, a

$$y = 0.9947x + 0.9992$$
$$R^2 = 0.9997$$

Fig. 5. A graph showing the correlation of optimized torsional angles at the RHF/6-31G(d) level of theory of proline (previously studied) and *N*-acetyl-4-hydroxy-L-proline-*N'*-methylamide. There exists a strong correlation with $R^2 = 0.9997$, suggesting that optimized torsional angle values of proline could be used reliably to approximate values for hydroxyproline.

strikingly high correlation was seen, with a least square value of $R^2 = 0.9997$, as shown in Fig. 5.

4. Conclusions

The results reported here indicate that optimized torsional angle values of the hydroxyproline residue closely resemble those of the proline residue, giving a significantly high correlation and showing that hydroxyproline structures can be accurately predicted by optimized structures of the proline residue. One could predict, with confidence, the conformational preferences of hydroxyproline given optimized values of its parent amino acid proline. This finding could prove beneficial when geometric optimizations are performed on a modified derivative of a previously studied molecule such as quisqualate, which is a modified compound of glutamate. Less time and effort would be spent on estimating a reasonable initial parameter for optimization to be carried out as previously optimized values of the parent compound could be used as a reliable approximation. Molecular structures could also be predicted in advance. Ab initio studies on single amino acid residues may prove to be highly valued in drug designs, receptor pharmacology and ligand docking.

Acknowledgements

The authors would like to express their gratitude to the National Cancer Institute (NCI) for generously providing computer time. One of the authors (IGC) wishes to thank the Ministry of Education for a Szent-Györgyi Visiting Professorship.

References

[1] I. Hudáky, H.A. Baldoni, A. Perczel, J. Mol. Struct. (THEOCHEM) 582 (2002) 233.
[2] C. Benzi, R. Improta, G. Scalmani, V. Barone, J. Comput. Chem. 23 (2002) 341.
[3] A.M. Sapse, L. Mallah-Levy, S.D. Daniels, B.W. Erickson, J. Am. Chem. Soc. 109 (1987) 3526.
[4] S. Tanaka, H.A. Scheraga, Macromolecules 7 (1974) 698.
[5] R. Improta, C. Benzi, V. Barone, J. Am. Chem. Soc. 123 (2001) 12568.
[6] GAUSSIAN 94, Revision D.2, M.J. Frisch, G.W. Trucks, H.B. Schlegel, P.M.W. Gill, B.G. Johnson, M.A. Robb, J.R. Cheeseman, T. Keith, G.A. Petersson, J.A. Montgomery, K. Raghavachari, M.A. Al-Laham, V.G. Zakrzewski, J.V. Ortiz, J.B. Foresman, J. Cioslowski, B.B. Stefanov, A. Nanayakkara, M. Challacombe, C.Y. Peng, P.Y. Ayala, W. Chen, M.W. Wong, J.L. Andres, E.S. Replogle, R. Gomperts, R.L. Martin, D.J. Fox, J.S. Binkley, D.J. Defrees, J. Baker, J.P. Stewart, M. Head-Gordon, C. Gonzalez, J.A. Pople, Gaussian, Inc., Pittsburgh Pa, 1995.
[7] GAUSSIAN 98 (Revision A.x), M.J. Frisch, G.W. Trucks, H.B. Schlegel, G.E. Scuseria, M.A. Robb, J.R. Cheeseman, V.G. Zakrzewski, J.A. Montgomery, Jr., R.E. Stratmann, J.C. Burant, S. Dapprich, J.M. Millam, A.D. Daniels, K.N. Kudin, M.C. Strain, O. Farkas, J. Tomasi, V. Barone, M. Cossi, R. Cammi, B. Mennucci, C. Pomelli, C. Adamo, S. Clifford, J. Ochterski, G.A. Petersson, P.Y. Ayala, Q. Cui, K. Morokuma, D.K. Malick, A.D. Rabuck, K. Raghavachari, J.B. Foresman, J. Cioslowski, J.V. Ortiz, A.G. Baboul, B.B. Stefanov, G. Liu, A. Liashenko, P. Piskorz, I. Komaromi, R. Gomperts, R.L. Martin, D.J. Fox, T. Keith, M.A. Al-Laham, C.Y. Peng, A. Nanayakkara, C. Gonzalez, M. Challacombe, P.M.W. Gill, B.G. Johnson, W. Chen, M.W. Wong, J.L. Andres, M. Head-Gordon, E.S. Replogle, J.A. Pople, Gaussian, Inc., Pittsburgh PA, 1998.

ELSEVIER

Journal of Molecular Structure (Theochem) 666–667 (2003) 291–301

THEO
CHEM

www.elsevier.com/locate/theochem

An ab initio exploratory study on the conformational features of the dipeptide MeCO-Ala-Ala-NH-Me in its four different configurations: determination of the behaviour of D-enantiomer amino acids within a peptide chain

Sonya U. Brijbassi[a,b,*], Michelle A. Sahai[a,b,*], David H. Setiadi[a],
Gregory A. Chass[a,b,c,d,*], Botond Penke[e,f], Imre G. Csizmadia[a,b,e]

[a]Global Institute of COmputational Molecular and Materials Science (GIOCOMMS), 1422 Edenrose St., Mississiauga, Ont., Canada L5V 1H3
[b]Department of Chemistry, University of Toronto, 80 St. George St., Toronto, Ont., Canada M5S 3H6
[c]Institut de Science et d'Ingénierie Supramoléculaires, 8, allée Gaspard Monge, BP 70028, 67083 Strasbourg Cedex, France
[d]Department of Biomedical Sciences, Creighton University, 2500 California Plaza, Omaha, NE 68178, USA
[e]Department of Medical Chemistry, University of Szeged, H-6720 Szeged, Dóm tér 8, Hungary
[f]Protein Chemistry Research Group, Hungarian Academy of Sciences, University of Szeged, H-6720 Dóm tér 8, Szeged, Hungary

Abstract

Ab initio conformational studies at the RHF/3-21G level of theory were carried out for the dipeptide MeCO-Ala-Ala-NH-Me in its four different configurations (MeCO-L-Ala-L-Ala-NH-Me, MeCO-D-Ala-D-Ala-NH-Me, MeCO-L-Ala-D-Ala-NH-Me and MeCO-D-Ala-L-Ala-NH-Me). From this method the conformations for these dipeptides were found to be $\beta_L\gamma_L$, $\beta_L\gamma_D$, $\beta_L\gamma_D$ and $\beta_L\gamma_L$, respectively. Patterns were investigated on Ramachandran maps to identify annihilated critical points among the four dipeptides to determine whether the D-enantiomer was in fact the 'mirror image' of the L-enantiomer dipeptide in addition to determining the role of the D-isomer in the peptide chain. The differences in energies for each conformation were also compared between the dipeptides.
© 2003 Elsevier B.V. All rights reserved.

Keywords: L-Enantiomer; D-Enantiomer; Protein folding; Ramachandran maps; Dipeptides; Ab initio

1. Introduction

The 20 naturally occurring amino acids can occur both in the L- and D-enantiomeric configuration. In this work both the L and D forms will be used. Fig. 1 shows the mirror imaging of the L- and D-amino acids. These two molecules are not identical and cannot be superimposed.

Although living organisms use L-amino acids for protein synthesis, D-enantiomers have generated a great deal of interest in the drug industry with the advent of drugs consisting of D-isomers, such as the cancer drug leuprolide acetate. Leuprolide acetate is a drug used for the treatment of ovarian cancer

---

* Corresponding authors.
E-mail addresses: s_brijbassi@hotmail.com (S.U. Brijbassi), msahai@giocomms.org (M.A. Sahai), dsetiadi@giocomms.org (D.H. Setiadi), gchass@giocomms.org (G.A. Chass), pbotond@mdche.szote.u-szeged.hu (B. Penke), icsizmad@alchemy.chem.utoronto.ca (I.G. Csizmadia).

0166-1280/$ - see front matter © 2003 Elsevier B.V. All rights reserved.
doi:10.1016/j.theochem.2003.08.034

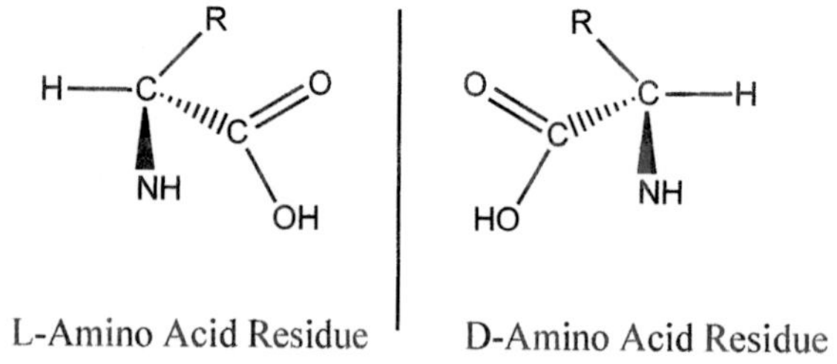

L-Amino Acid Residue　　　D-Amino Acid Residue

Fig. 1. Schematic representation of the mirror imaging of the L- and D- amino acids.

and endometriosis. It consists of the amino acids D-Leucine-L-Leucine within its nonapeptide [1]. Another example is penicillin, a widely used antibiotic, which mimics the dipeptide D-Ala-D-Ala in its structure [2]. These examples show that conformational analysis of D-enantiomers of all amino acids within a peptide sequence would allow for a better understanding of the behaviour of these molecules, such that effective drugs can be synthesized and produced.

In this study, the dipeptide MeCO-Ala-Ala-NH-Me was investigated using molecular orbital computations. The amino acid alanine was chosen because it is the simplest amino acid with a side chain, which implies that it can be used as a template for the study of other amino acids. Computations were carried out for the dipeptide MeCO-Ala-Ala-NH-Me in its four different configurations, MeCO-L-Ala-L-Ala-NH-Me, MeCO-D-Ala-D-Ala-NH-Me, MeCO-L-Ala-D-Ala-NH-Me and MeCO-D-Ala-L-Ala-NH-Me at the Restricted Hartree Fock/3-21G level of theory. (Fig. 2) The basis of this study was to identify the global and local minima of these four dipeptides on the Ramachandran map. These minimas were then compared and contrasted.

2. Methods

N-Acetyl and N-methylamide functionalities were added to the ends of the dipeptides as protecting

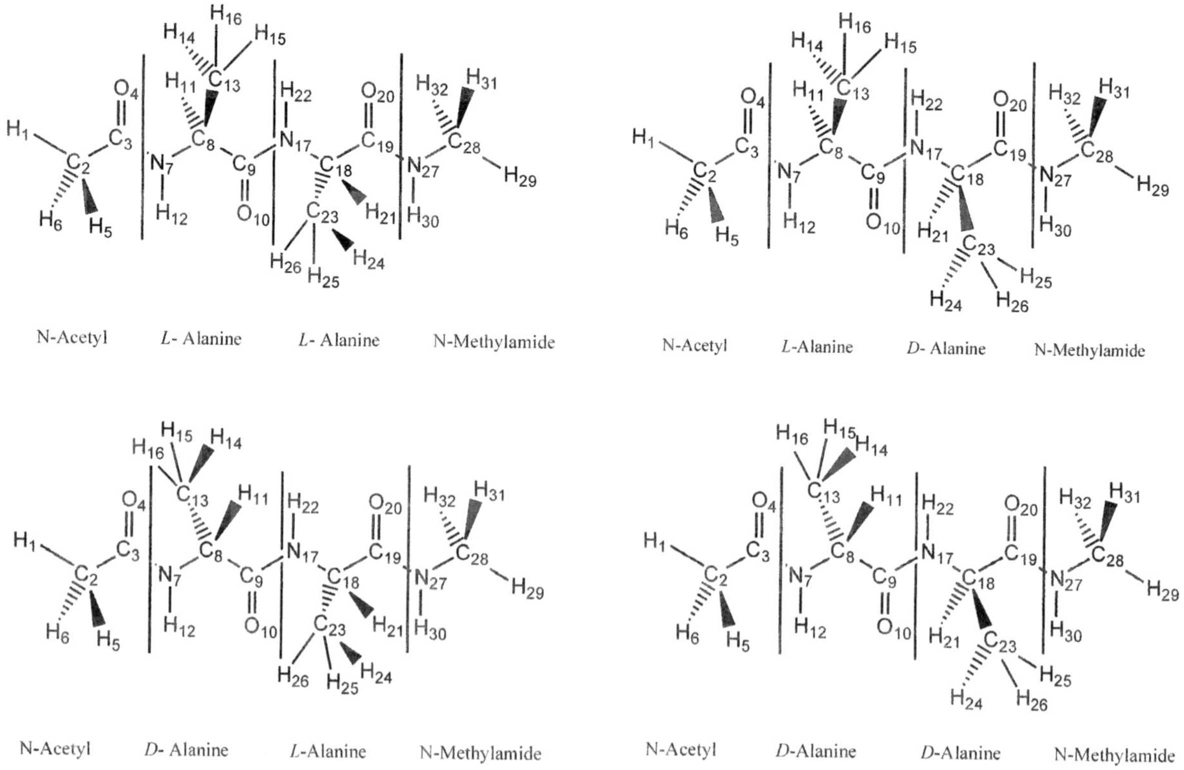

Fig. 2. Schematic illustration of the dipeptide MeCO-Ala-Ala-NH-Me in its four different configurations, MeCO-L-Ala-L-Ala-NH-Me, MeCO-D-Ala-D-Ala-NH-Me, MeCO-L-Ala-D-Ala-NH-Me and MeCO-D-Ala-L-Ala-NH-Me.

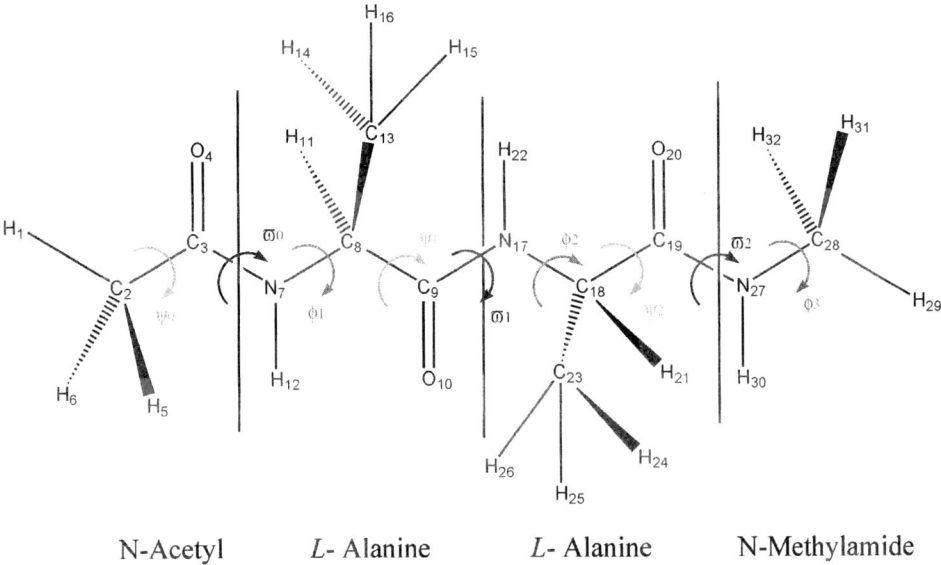

N-Acetyl   L- Alanine   L- Alanine   N-Methylamide

Fig. 3. Schematic representation of the standardized numbering system applied to the dipeptide MeCO-Ala-Ala-NH-Me, showing all backbone torsional angles.

groups to mimic the α-carbon relationship in longer peptide chains. A standardized numbering system was used to define the molecules (Fig. 3) as it was numbered into and along the peptide backbone, for each 'unit' separately so that any of the units comprising it can be truncated at any time to model the structure of a larger polypeptide [1,3].

The internal coordinate system was then created for the molecule with each bond length, angle and dihedral within the molecule fully defined. The dihedrals ϕ_1, ψ_1, ϕ_2 and ψ_2 were the dihedrals of interest (Fig. 3) [1].

The resultant input file was used as a starting point in a successive and iterative process of GAUSSIAN 98 cycles to bring about an optimization at the predicted minima, using the RHF/3-21G level of theory [4].

The different conformations of the dipepetide MeCO-Ala-Ala-NH-Me was optimized by restraining the first alanine residue to the β_L conformation, and varying the second residue for each of the nine optimized conformations (Table 1).

In this paper, we wish to examine 68 possible conformers of the dipeptide MeCO-Ala-Ala-NH-Me. These 68 conformers may be expected to exist, because of the different conformational studies using

the β_L conformation as the deciding factor, whereby, in any particular instant an alanine residue is fixed at the β_L conformation and the second was varied for the nine possible minima present on the Ramachandran Map. Up to 9 minima may be expected in each of these two scans for one dipeptide conformation. This would lead to the $2 \times 9 = 18$ possible conformations. However, the fully extended $\beta_L\beta_L$ conformer would occur two times so only 17 unique conformers are to

Table 1

Optimized ϕ, ψ torsional angle pairs completed at the RHF/3-21G level of theory. The idealized torsional angle pairs, together with their conformational classification, are also shown

Conformer	Conformation classification	Optimized		Ideal	
		ϕ	ψ	ϕ	ψ
α_L	$g - g-$	-66.6	-17.5	-60.0	-60.0
α_D	$g + g+$	61.8	31.9	60.0	60.0
β_L	aa	-167.6	169.9	-180.0	180.0
γ_L	$g - g+$	-84.5	68.7	-60.0	60.0
γ_D	$g + g-$	74.3	-59.5	60.0	-60.0
δ_L	$ag+$	-126.2	26.5	-180.0	60.0
δ_D	$ag-$	-179.6	-43.7	180.0	-60.0
ε_L	$g - a$	-74.7	167.8	-60.0	180.0
ε_D	$g + a$	64.7	-178.6	60.0	-180.0

γ_D γ_D	γ_D δ_D	γ_D α_L	δ_D γ_D	δ_D δ_D	δ_D α_L	α_L γ_D	α_L δ_D	α_L α_L
γ_D ε_D	**γ_D β_L**	γ_D ε_L	δ_D ε_D	**δ_D β_L**	δ_D ε_L	α_L ε_D	**α_L β_L**	α_L ε_L
γ_D α_D	γ_D δ_L	γ_D γ_L	δ_D α_D	δ_D δ_L	δ_D γ_L	α_L α_D	α_L δ_L	α_L γ_L
ε_D γ_D	ε_D δ_D	ε_D α_L	**β_L γ_D**	**β_L δ_D**	**β_L α_L**	ε_L γ_D	ε_L δ_D	ε_L α_L
ε_D ε_D	**ε_D β_L**	ε_D ε_L	**β_L ε_D**	**β_L β_L**	**β_L ε_L**	ε_L ε_D	**ε_L β_L**	ε_L ε_L
ε_D α_D	ε_D δ_L	ε_D γ_L	**β_L α_D**	**β_L δ_L**	**β_L γ_L**	ε_L α_D	ε_L δ_L	ε_L γ_L
α_D γ_D	α_D δ_D	α_D α_L	δ_L γ_D	δ_L δ_D	δ_L α_L	γ_L γ_D	γ_L δ_D	γ_L α_L
α_D ε_D	**α_D β_L**	α_D ε_L	δ_L ε_D	**δ_L β_L**	δ_L ε_L	γ_L ε_D	**γ_L β_L**	γ_L ε_L
α_D α_D	α_D δ_L	α_D γ_L	δ_L α_D	δ_L δ_L	δ_L γ_L	γ_L α_D	γ_L δ_L	γ_L γ_L

Fig. 4. Minima on the 4D-Ramachandran map of a dipeptide diamide. Those conformers seen in bold are the 17 possible existing conformers for each of the four dipeptides under study.

be considered. Therefore, since there are four dipeptides $17 \times 4 = 68$ possible structures can exist. Of course, it is possible that some conformers will be annihilated and fewer than 68 conformers will be found. Schematic representations of the 17 possible conformers to be considered for each of the four dipeptides and their relative proximity to the 81 possible conformers of a dipeptide given in subscripted Greek letters can be seen in bold in Fig. 4.

Since the goal of this study was to identify the global and local minima of these dipeptides on the Ramachandran map, $E(\phi_1, \psi_1)$ and $E(\phi_2, \psi_2)$, the dihedrals in the final optimized conformations were located on the Ramachandran map to determine if they remained in their original conformations (Fig. 5). Those that deviated from their restrained conformer were thought not to exist. The differences in the energy calculated at the RHF/3-21G level of theory was noted between each of the conformations. This was done by converting hartrees to kcal/mol using 1 hartree is equal to 627.5095 kcal/mol. The conformation with the lowest energy minima of all the conformers was identified and compared.

3. Results and discussion

Enantiomers are isomers, which are compounds that have the same molecular formula but differ in

structure and are 'mirror images' of each other. For the D-enantiomer of a dipeptide to actually be the 'mirror image' of the L-isomer dipeptide, it needs to be shown that it is the non-superimposable image of the other, having the opposite configuration at all chiral centers. This can be shown by locating the minima of each conformation of the two dipeptides

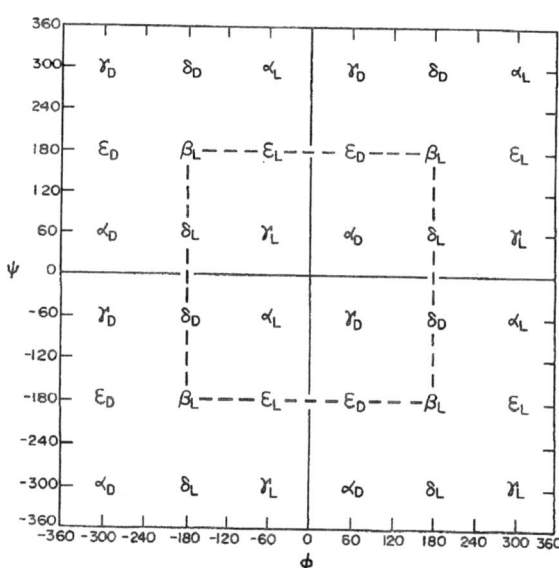

Fig. 5. A schematic representation of the conformation of a peptide (PCONH-CHR-CONHQ). The approximate location of the conformations are symbolized by subscripted Greek letters.

within the Ramachandran map. Ideally, all nine minima on the Ramachandran map are expected to converge; however, with the annihilation of certain critical points, patterns can be found and compared between the two-enantiomer dipeptides, MeCO-L-Ala-L-Ala-NH-Me and MeCO-D-Ala-D-Ala-NH-Me. The other two dipeptides in this study, MeCO-L-Ala-D-Ala-NH-Me and MeCO-D-Ala-L-Ala-NH-Me, could also be described and compared, allowing for a final comparison of all four structures. Among all four configurations of the dipeptide, the behaviour of the D-enantiomer of amino acids could be characterized and its purpose in polypeptide chains determined. This would be useful in the design and synthesis of drugs with D-enantiomer amino acids, and results from these calculations may well be used in the future as templates for further optimizations of polypeptides containing the D isoform of amino acids.

3.1. Description of the potential energy minima of the dipeptides, MeCO-L-Ala-L–Ala-NH-Me and MeCO-D-Ala-D-Ala-NH-Me

The optimization for the energy minima of both these dipeptides at the RHF/3-21G level of theory has revealed some interesting patterns. Results for the optimized conformers of the dipeptide MeCO-L-Ala-L-Ala-NH-Me are summarized in Table 2. The found minima are depicted in Fig. 6. The first noticeable feature observed in this dipeptide was that five of

the found minima were shared between both dipeptides [γD, δD, εD, βL, γL].

The αDβL and αLβL conformers existed in the MeCO-L-Ala$_{[x]}$-L-Ala$_{[βL]}$-NH-Me molecule when the second alanine was restrained to βL, however, the βLαD and βLαL conformations were not observed when the first alanine was restrained to the βL conformation in MeCO-L-Ala$_{[βL]}$-L-Ala$_{[x]}$-NH-Me. Also noted was that the βLδL conformer existed in the dipeptide when the first alanine was restrained to βL but δLβL did not exist when the second alanine was confined to βL. Another interesting feature was that both βLεL and εLβL conformers for each dipeptide did not exist on the Ramachandran map and both migrated to the βLβL conformation. Both βLαL and βLαD migrated to βLδL, and the δLβL conformer migrated to the αLβL position on the Ramachandran map.

Finally, the βLγL conformer was found to be the lowest energy conformer for the MeCO-L-Ala-L-Ala-NH-Me dipeptide (Table 2).

Optimized conformers for the dipeptide MeCO-D-Ala-D-Ala-NH-Me are summarized in Table 3 and the existing energy minima are illustrated in Fig. 7. With this dipeptide 4 of the found minima were shared between both dipeptides [γD, βL, εL, γL].

The βLδD and βLαL conformers existed in the MeCO-D-Ala$_{[βL]}$-D-Ala$_{[x]}$-NH-Me molecule when the first alanine was restrained to βL, however, the δDβL and αLβL conformations were not observed when

Table 2

Optimized dihedral angles of MeCO-L-Ala-L-Ala-NH-Me as well as tabulated energies (hartree and kcal/mole)

Found conformation		RHF/3-21G E (hartree)	ΔE (kcal/mol)	φ0	ω0	φ1	ψ1	ω1	φ2	ψ2	ω2	φ3
βL	δL	−734.6012301	3.10	179.43	178.49	−166.88	171.18	−175.51	−127.21	27.09	179.50	−115.88
βL	δD	−734.5955882	6.64	−179.80	179.09	−166.36	168.91	173.61	−174.36	−46.80	−177.04	−119.48
βL	εD	−734.5946445	7.23	178.10	177.52	−167.56	171.92	−171.39	65.86	−176.35	−179.03	−119.48
βL	γL	−734.6061633	0.00	179.76	178.76	−166.43	169.01	−177.65	−85.72	69.54	−178.58	−115.88
βL	γD	−734.6018734	2.69	179.72	179.04	−166.13	167.99	171.97	75.99	−57.71	−178.46	−119.48
βL	βL	−734.6057512	0.26	179.61	178.83	−167.37	−170.70	177.35	−167.42	169.48	177.91	−115.88
αL	βL	−734.6024479	2.33	178.20	−176.94	−119.83	−16.75	176.48	−169.73	168.72	178.09	−118.74
αD	βL	−734.6003670	3.64	−177.57	176.99	59.53	33.15	−176.40	−175.49	168.56	177.55	−118.61
δD	βL	−734.5971559	5.65	176.83	176.36	−174.42	−44.40	−178.55	−167.27	169.48	177.72	−119.70
εD	βL	−734.5958614	6.46	−170.00	−167.42	63.75	−171.53	179.94	−166.61	168.99	178.17	−119.76
γL	βL	−734.6026994	2.17	179.55	−176.46	−87.63	71.01	179.66	164.42	167.24	177.74	−119.76
γD	βL	−734.5981099	5.05	−178.75	175.47	76.30	−50.38	−178.78	−161.04	168.29	177.59	−120.07

(a)

γD	δD	.
εD	βL	.
.	δL	γL

(b)

γD	δD	αL
εD	βL	.
αD	.	γL

*Where x is one of the nine conformers $(\beta_L, \gamma_L, \gamma_D, \delta_L, \delta_D, \alpha_L, \alpha_D, \varepsilon_L, \varepsilon_D)$

Fig. 6. Depiction of the found minima on the Ramachandran Map for (a) MeCO-L-Ala$_{[\beta L]}$-L-Ala$_{[x]}$-NH-Me and (b) MeCO-L-Ala$_{[x]}$-L-Ala$_{[\beta L]}$-NH-Me.

the second alanine was restrained to the βL conformation in MeCO-D-Ala$_{[x]}$-D-Ala$_{[\beta L]}$-NH-Me. In the molecule MeCO-D-Ala$_{[\beta L]}$-D-Ala$_{[x]}$-NH-Me the conformers $\beta L\alpha D$, $\beta L\delta L$, and $\beta L\varepsilon D$ migrated to the $\beta L\delta D$, $\beta L\alpha L$, and $\beta L\beta L$ conformations, respectively. For the molecule MeCO-D-Ala$_{[x]}$-D-Ala$_{[\beta L]}$-NH-Me the conformers $\alpha L\beta L$, $\alpha D\beta L$, $\delta L\beta L$, $\delta D\beta L$ and $\varepsilon D\beta L$ migrated to the $\varepsilon L\beta L$, $\gamma D\beta L$, $\alpha L\beta L$, $\gamma D\beta L$ and $\beta L\beta L$ conformations, respectively.

Table 3 clearly outlines that $\beta L\gamma D$ was the lowest energy conformer for the dipeptide MeCO-D-Ala-D-Ala-NH-Me.

3.2. Comparision of the enantiomer dipeptides MeCO-L-Ala-L-Ala-NH-Me and MeCO-D-Ala-D-Ala-NH-Me

Studies have already illustrated that for the monopeptide For-L-Ala-NH$_2$ and its enantiomer For-D-Ala-NH$_2$, different pairs of conformations were annihilated upon comparison between the two

structures [5]. A distinct pattern was shown between the two enantiomers; critical points that were missing from one monopeptide were clearly observed in the other, and vice versa. The D-enantiomer was shown to be the 'mirror image' of the L-enantiomer. However, this was not as defined as clearly for our dipeptide of interest in this study.

The minima defined for each of the dipeptides MeCO-L-Ala-L-Ala-NH-Me and MeCO-D-Ala-D-Ala-NH-Me do not display distinct patterns or differences between them, but can still be compared in order to illustrate the role D-enantiomers in a peptide sequence. By comparing the minima of the Ramachandran maps illustrated in Figs. 6(a) and 7(a), it is observed that for the dipeptides where the first alanine in the sequence was restrained to the βL conformer and the other was set to one of the nine conformers, four of the found minima were present in both the D- and L-enantiomer dipeptides [$\beta L\gamma D$, $\beta L\delta D$, $\beta L\beta L$ and $\beta L\gamma L$]. The conformations $\beta L\varepsilon D$ and $\beta L\delta L$ existed exclusively for MeCO-L-Ala-L-Ala-NH-Me

Table 3

Optimized dihedral angles of MeCO-D-Ala-D-Ala-NH-Me as well as tabulated energies (hartree and kcal/mole)

Found Conformation		RHF/3-21G E (hartree)	ΔE (kcal/mol)	$\phi 0$	$\omega 0$	$\phi 1$	$\psi 1$	$\omega 1$	$\phi 2$	$\psi 2$	$\omega 2$	$\phi 3$
βL	αL	-734.5980832	5.07	179.69	-179.19	166.45	-168.17	-172.16	-62.70	-33.95	-178.77	-112.99
βL	δD	-734.6012301	3.10	-179.38	-178.50	166.92	-171.15	175.52	127.25	-27.19	-176.50	-123.17
βL	εL	-734.5946445	7.23	-177.97	-177.48	167.53	-171.93	171.40	-65.38	176.38	179.64	-120.11
βL	γL	-734.6018733	2.69	-179.79	-179.05	166.14	-168.00	-171.97	-75.99	57.66	178.42	-118.83
βL	γD	-734.6061633	0.00	-179.84	-178.80	166.47	-168.96	177.65	85.73	-69.55	178.58	-114.39
βL	βL	-734.6057512	0.26	-179.79	-178.85	167.37	-170.71	-177.33	167.41	-169.47	-177.90	-119.33
εL	βL	-734.5958615	6.46	169.86	167.43	-63.76	171.51	-179.90	166.56	-168.96	-178.17	-119.04
γL	βL	-734.5981097	5.05	178.74	-175.46	-76.29	50.30	178.77	161.04	-168.29	-177.57	-118.89
γD	βL	-734.6026995	2.17	-179.46	176.47	87.64	-71.04	179.63	164.46	-167.22	-177.74	-118.94

(a)

γD	δD	αL
·	βL	εL
·	·	γL

(b)

γD	·	·
	βL	εL
•	·	γL

*Where x is one of the nine conformers *(β_L, γ_L, γ_D, δ_L, δ_D, α_L, α_D, ε_L, ε_D)*

Fig. 7. Depiction of the found minima on the Ramachandran Map for (a) MeCO-D-Ala$_{[\beta L]}$-D-Ala$_{[x]}$-NH-Me and (b) MeCO-D-Ala$_{[x]}$-D-Ala$_{[\beta L]}$-NH-Me.

while the same was true for βLεL and βLαL in MeCO-D-Ala-D-Ala-NH-Me.

When the MeCO-L-Ala-L-Ala-NH-Me and MeCO-D-Ala-D-Ala-NH-Me dipeptides had the second alanine confined to βL, only three of the found minima were present in both the D- and L-enantiomer dipeptides [γDβL, βLβL and γLβL]. This is shown in Figs. 6(b) and 7(b). The conformations δDβL, αLβL, εDβL and αDβL are all present for MeCO-L-Ala-L-Ala-NH-Me, but are annihilated in MeCO-D-Ala-D-Ala-NH-Me, whereas εLβL was exclusive for MeCO-D-Ala-D-Ala-NH-Me.

As seen previously in ab initio studies of For-L-Gly-NH_2 and For-D-Ala-NH_2 [5], γL and γD gave the minimum energy conformations. With the dipeptides MeCO-L-Ala-L-Ala-NH-Me and MeCO-D-Ala-D-Ala-NH-Me, βLγL and βLγD were respectively seen to be the most stable conformations with the lowest potential energy minima. These values show that the lowest energy minima for each dipeptide consisted of the opposite pair of conformers as well as the same minimas are similar to the monopeptide. This revealed some evidence that confirmed the notion of non-superimposable dipeptides.

3.3. Description of the potential energy minima of the dipeptides, MeCO-L-Ala-D-Ala-NH-Me and MeCO-D-Ala-L-Ala-NH-Me

The values for the optimized conformers for the dipeptide MeCO-L-Ala-D-Ala-NH-Me are summarized in Table 4. The minima that were found are illustrated in Fig. 8. The first noticeable feature observed in this dipeptide was that only three of the found minima were shared between both dipeptides [γD, βL and γL].

The βLαL and βLεL conformers existed in the MeCO-L-Ala$_{[\beta L]}$-D-Ala$_{[x]}$-NH-Me molecule when the first alanine was restrained to βL, however, the αLβL and εLβL conformations were not observed when the second alanine was restrained to the βL conformation in MeCO-L-Ala$_{[x]}$-D-Ala$_{[\beta L]}$-NH-Me.

Table 4
Optimized dihedral angles of MeCO-L-Ala-D-Ala-NH-Me as well as tabulated energies (hartree and kcal/mole)

Found conformation		RHF/3-21G E (hartree)	ΔE (kcal/mol)	$\phi0$	$\omega0$	$\phi1$	$\psi1$	$\omega1$	$\phi2$	$\psi2$	$\omega2$	$\phi3$
βL	αL	−734.5981585	5.14	179.42	178.51	−166.91	171.27	−176.55	−61.81	−34.43	−178.69	−113.68
βL	εL	−734.5955867	6.76	−179.39	179.69	−166.47	168.75	162.32	−63.40	178.87	179.49	−119.70
βL	γL	−734.6015675	3.00	179.96	178.92	−166.11	168.45	−177.36	−74.63	58.77	178.42	−119.57
βL	γD	−734.6063541	0.00	179.73	179.02	−166.31	169.24	173.11	86.64	−68.63	178.57	−114.57
βL	βL	−734.6057733	0.36	179.72	178.72	−167.65	170.89	179.15	167.63	−169.68	−178.02	−119.40
δL	βL	−734.6027502	2.26	179.26	−176.75	−120.17	18.35	178.56	166.27	−170.81	−177.96	−119.15
δD	βL	−734.5969202	5.92	177.31	176.40	−173.99	−44.55	−176.49	166.58	−170.54	−178.35	−119.59
εD	βL	−734.5961124	6.43	−169.97	−167.58	63.37	−170.97	−178.33	167.77	−169.74	−177.74	−119.76
γL	βL	−734.6025681	2.38	178.34	−176.17	−89.08	64.01	−178.97	163.20	−168.52	−177.56	−118.87
γD	βL	−734.5979990	5.24	−179.97	176.03	75.58	−61.58	−177.66	163.47	−167.02	−177.71	−118.91

(a)

γD	·	αL
·	βL	εL
·	·	γL

(b)

γD	δD	·
εD	βL	·
·	δL	γL

*Where x is one of the nine conformers (β_L, γ_L, γ_D, δ_L, δ_D, α_L, α_D, ε_L, ε_D)

Fig. 8. Depiction of the found minima on the Ramachandran Map for (a) MeCO-L-Ala$_{[\beta L]}$-D-Ala$_{[x]}$-NH-Me and (b) MeCO-L-Ala$_{[x]}$-D-Ala$_{[\beta L]}$-NH-Me.

Comparatively, the δDβL, εDβL and δLβL conformers existed in the MeCO-L-Ala$_{[x]}$-D-Ala$_{[\beta L]}$-NH-Me molecule when the second alanine was restrained to βL, however, the βLδD, βLεD and βLδL conformations were not observed when the first alanine was restrained to the βL conformation in MeCO-L-Ala$_{[\beta L]}$-D-Ala$_{[x]}$-NH-Me. In the molecule MeCO-L-Ala$_{[\beta L]}$-D-Ala$_{[x]}$-NH-Me the conformers βLαD, βLδL βLδD and βLεD migrated to the βLδD, βLαL, βLαL and βLβL conformations, respectively. For the molecule MeCO-L-Ala$_{[x]}$-D-Ala$_{[\beta L]}$-NH-Me the conformers αLβL, αDβL and εLβL migrated to the δLβL, γDβL and βLβL conformations, respectively.

Finally, it is clearly shown that βLγD is the lowest energy conformation for the MeCO-L-Ala-D-Ala-NH-Me dipeptide (Table 4).

Optimized conformers for the dipeptide MeCO-D-Ala-L-Ala-NH-Me are summarized in Table 5 and the existing energy minima are illustrated in Fig. 9. As with the dipeptide MeCO-L-Ala-D-Ala-NH-Me,

the three conformers consisting of γL and γD, as well as the βLβL conformer, existed for both dipeptides: MeCO-D-Ala$_{[\beta L]}$-L-Ala$_{[x]}$-NH-Me and MeCO-D-Ala$_{[x]}$-L-Ala$_{[\beta L]}$-NH-Me. The only other shared conformer found to exist was δD.

The βLεD and βLδL conformers existed in the MeCO-D-Ala$_{[\beta L]}$-L-Ala$_{[x]}$-NH-Me molecule when the first alanine was restrained to βL, however, the βLεD and βLδL conformations were not observed when the second alanine was restrained to the βL conformation in MeCO-D-Ala$_{[x]}$ L-Ala$_{[\beta L]}$-NH-Me. Comparatively, the εLβL conformer existed in the MeCO-D-Ala$_{[x]}$ L-Ala$_{[\beta L]}$-NH-Me molecule when the second alanine was restrained to βL, however, the βLεL conformation was not observed when the first alanine was restrained to the βL conformation in MeCO-D-Ala$_{[\beta L]}$-L-Ala$_{[x]}$-NH-Me. In the molecule MeCO-D-Ala$_{[\beta L]}$-L-Ala$_{[x]}$-NH-Me the conformers βLαL, βLαD and βLεL migrated to the βLδL, βLβL and βLβL conformations, respectively. For the molecule

Table 5

Optimized dihedral angles of MeCO-D-Ala-L-Ala-NH-Me as well as tabulated energies (hartree and kcal/mole)

Found conformation		RHF/3-21G E (hartree)	ΔE (kcal/mol)	$\phi0$	$\omega0$	$\phi1$	$\psi1$	$\omega1$	$\phi2$	$\psi2$	$\omega2$	$\phi3$
βL	δL	−734.6011738	3.25	179.75	−179.20	166.31	−167.84	−171.18	−126.76	27.91	176.48	−115.70
βL	δD	−734.5957182	6.67	−179.44	−178.53	167.13	−170.50	177.68	−175.76	−45.44	−176.80	−133.53
βL	εD	−734.5955867	6.76	179.43	−179.70	166.46	−168.69	−162.34	63.40	−178.64	−179.49	−119.07
βL	γL	−734.6063540	0.00	−179.69	−179.02	166.31	−169.24	−173.12	−86.64	68.64	−178.58	−124.42
βL	γD	−734.6015675	3.00	−179.97	−178.93	166.08	−168.44	177.36	74.63	−58.82	−178.43	−119.11
βL	βL	−734.6057734	0.36	−179.62	−178.72	167.64	−170.90	−179.15	−167.65	169.61	178.02	−119.37
δD	βL	−734.6027503	2.26	−179.36	176.72	120.26	−18.39	−178.57	−166.25	170.85	177.96	−119.69
εL	βL	−734.5961124	6.43	170.03	167.59	−63.37	170.94	178.33	−167.77	169.77	177.76	−119.22
γL	βL	−734.5979989	5.24	179.83	−176.07	−75.54	61.59	177.65	−163.38	166.99	177.69	−119.78
γD	βL	−734.6025681	2.38	−178.32	176.16	89.07	−63.95	178.96	−163.17	168.52	177.55	−119.74

(a)

γD	δD	.
εD	βL	.
.	δL	γL

(b)

γD	δD	.
.	βL	εL
.	.	γL

*Where x is one of the nine conformers $(\beta_L, \gamma_L, \gamma_D, \delta_L, \delta_D, \alpha_L, \alpha_D, \varepsilon_L, \varepsilon_D)$

Fig. 9. Depiction of the found minima on the Ramachandran Map for (a) MeCO-D-Ala$_{[\beta L]}$-L-Ala$_{[x]}$-NH-Me and (b) MeCO-D-Ala$_{[x]}$-L-Ala$_{[\beta L]}$-NH-Me.

MeCO-L-Ala$_{[x]}$-D-Ala$_{[\beta L]}$-NH-Me the conformers αLβL, αDβL, δLβL and εDβL migrated to the γLβL, δDβL, γLβL and βLβL conformations, respectively.

Table 5 clearly outlines that the conformation βLγL was the lowest energy conformers for the dipeptide MeCO-D-Ala-L-Ala-NH-Me.

3.4. Comparision of the enantiomer dipeptides MeCO-L-Ala-D-Ala-NH-Me and MeCO-D-Ala-L-Ala-NH-Me

To accurately define the behavior of D-alanine in a peptide, the study must include the conformational analysis of the D-enantiomer along with its L-isomer within the dipeptide. Using the dipeptides MeCO-L-Ala-D-Ala-NH-Me and MeCO-D-Ala-L-Ala-NH-Me would be the simplest method to describe this behaviour, and would also create the template for further study. By comparing these dipeptides, it would be possible to outline the contributions of D-stereoisomers to a protein, in addition to investigate the possible differences created due to its presence. This information would be useful in studying more complex amino acids in the D-isomer form in larger polypeptides. Furthermore, it could also play a future role in determining the structure and function of protein folding and the design and synthesis of novel pharmaceutical products.

By comparing the minima of the Ramachandran maps illustrated in Figs. 8(a) and 9(a), it is observed that for the dipeptides where the first alanine in the sequence was restrained to the βL conformer and the other was set to one of the nine conformers, three of the found minima were present in both the D- and L-enantiomer dipeptides [βLγD, βLβL and βLγL]. The conformations βLαL and βLεL existed exclusively for

MeCO-L-Ala-D-Ala-NH-Me while the same was true for βLεD, βLδD and βLδL for MeCO-D-Ala-L-Ala-NH-Me.

When the MeCO-L-Ala-D-Ala-NH-Me and MeCO-D-Ala-L-Ala-NH-Me dipeptides had the second alanine confined to βL, four of the found minima were present in both the D- and L-enantiomer dipeptides [γDβL, δDβL, βLβL and γLβL]. This is shown in Figs. 8(b) and 9(b). The conformations εDβL and δLβL are all present for MeCO-L-Ala-D-Ala-NH-Me, but are annihilated in MeCO-D-Ala-L-Ala-NH-Me, whereas εLβL was exclusive for MeCO-D-Ala-L-Ala-NH-Me.

3.5. Summary of the comparison of the dipeptide MeCO-Ala-Ala-NH-Me in its four different configurations as outlined by the study

All of the four dipeptides under study reveal some interesting properties of the D-isomers of amino acids within a dipeptide chain. Table 6 lists the lowest potential energy minima identified for each dipeptide. From this table, some similarities emerge between the dipeptides. In MeCO-L-Ala-L-Ala-NH-Me when the first alanine in the sequence was confined to βL, the lowest energy conformation was βLγL, which was the same conformer identified for the lowest energy minima of MeCO-D-Ala-L-Ala-NH-Me. Therefore, the behaviour of MeCO-D-Ala-L-Ala-NH-Me could be said to mimic that of MeCO-L-Ala-L-Ala-NH-Me, but only when the first alanine in the chain is restrained to βL. For the dipeptide MeCO-D-Ala-D-Ala-NH-Me, the lowest energy conformer was identified to be βLγD when the first alanine was restrained to βL, which seemed to be the same for MeCO-L-Ala-D-Ala-NH-Me in the same scenario.

Table 6
Summary of the lowest potential energy minima for each dipeptide. The conformation in bold indicates the global minima for the specific dipeptide under study

Dipeptide	Conformation	
	First alanine restrained to βL	Second alanine restrained to βL
MeCO-L-Ala-L-Ala-NH-Me	βLγL	γLβL
MeCO-D-Ala-D-Ala-NH-Me	βLγD	γDβL
MeCO-L-Ala-D-Ala-NH-Me	βLγD	δLβL
MeCO-D-Ala-L-Ala-NH-Me	βLγL	δDβL

In the dipeptides when the second alanine in the sequence was confined to βL, the lowest energy conformation was found to be γLβL, γDβL, δLβL and δDβL for MeCO-L-Ala-L-Ala-NH-Me, MeCO-D-Ala-D-Ala-NH-Me, MeCO-L-Ala-D-Ala-NH-Me and MeCO-D-Ala-L-Ala-NH-Me, respectively.

Another interesting point to discuss concerning the four dipeptides is that the βLβL conformer was found in all structures. Some of the minima that were not found among the dipeptides preferentially migrated towards the βLβL conformation. This shows that some conformations of the dipeptides containing D-amino acids have found a stable conformer in βLβL, even though the βLβL conformations is not the global minima found in this study for this dipeptide. The study here also suggests that the dipeptide is most stable when at least one of the alanine residues is in the βL position.

The computations and tabulations performed here illustrate that the dipeptides MeCO-L-Ala-L-Ala-NH-Me and MeCO-D-Ala-D-Ala-NH-Me demonstrate enantiomeric properties in many cases. Although this distinction is not as clearly seen as it was in ab initio computations carried out on the monopeptide, results for the dipeptides have generated some interesting patterns and have annihilated critical points, which suggest that their configurations might, in fact, be mirror images. This also implies that an increase in the number of D-enantiomer amino acids in a peptide chain may result in the loss of some properties of the non-superimposable 'mirror image'. This could be due to structural stabilization by Van der Waals forces

or hydrogen bonds. With protein folding, the opposite configurations of the D-enantiomer are so altered that the 'mirror image' relationship does not fully exist. It was also shown with certain conformations that the difference in energy was exactly the same between the enantiomers of the dipeptides. With the dipeptides, MeCO-L-Ala-D-Ala-NH-Me and MeCO-D-Ala-L-Ala-NH-Me, it was shown that there were some common properties, which should be further investigated. In addition, the lowest potential energy minima state mimicked the behaviour of the enantiomeric dipeptides.

4. Conclusion

With the synthesis and production of new drugs consisting of D-enantiomeric amino acids, it is essential to determine their role in a peptide chain. By performing computations on the dipeptides MeCO-L-Ala-L-Ala-NH-Me and MeCO-D-Ala-D-Ala-NH-Me, some interesting features concerning the enantiomeric properties between them have been revealed. Current belief holds that MeCO-D-Ala-D-Ala-NH-Me is the true enantiomer of MeCO-L-Ala-L-Ala-NH-Me, but from the results presented here, it is suggested that stereoisomers within a peptide chain are no longer mirror images. Thus, the role of D-enantiomer amino acids within polypeptide chains must be more clearly defined.

From this study 41 of the possible 68 conformers were found. Computations should be undertaken for all possible combinations of the dipeptide structure; that is, 81 conformers per dipeptide x 4 dipeptides (MeCO-L-Ala-L-Ala-NHM, MeCO-D-Ala-D-Ala-NH-Me, MeCO-L-Ala-L-Ala-NH-Me and MeCO-D-Ala-D-Ala-NH-Me) resulting in 324 possible structures. However, since the original 68 conformers are already part of the proposed 324 possible structures, this study is well under way with only 256 more structures to study. It would be useful to see if in the end the structures can be compared in terms of their energies and to determine if there are distinct patterns of minima that can found on the Ramachandran map.

From studying the dipeptides, MeCO-L-Ala-D-Ala-NH-Me and MeCO-D-Ala-L-Ala-NH-Me, distinct patterns can be seen between them, as well as with the dipeptides, MeCO-L-Ala-L-Ala-NH-Me and MeCO-D-Ala-D-Ala-NH-Me. When the four structures are compared, it can be seen that these mimic the enantiomer dipeptides. The D-amino acid appears to depend on its orientation within the peptide sequence in addition to the conformations of other amino acids within the chain. The conclusions here can then be tested to see if in fact this holds true for larger peptide chains and to determine whether the same properties exist. Nonetheless, results shown here lead to an understanding of the conformational role of the D-enantiomer of an amino acid within a peptide sequence. Using this as a starting model for the study, longer peptides that consist of D-enantiomer peptide residues could be further investigated.

Acknowledgements

This work was supported by grants from the Global Institute of Computational Molecular and Materials Science (GIOCOMMS), Toronto, Ont., Canada.One of the authors (IGC) wishes to thank the Ministry of Education for a Szent-Györgyi Visiting Professorship. The authors thank Tania A. Pecora for helpful discussions during the preparation of this manuscript as well as Graydon Hoare for database management, network support and software and distributive processing development. Special thanks is extended to Andrew M. Chasse, Christopher M. Andrews and Michael R. Sahai for their development of novel scripting and coding techniques as well as data management facilitating a reduction in the number of CPU cycles needed. The pioneering advances of Kenneth P. Chasse, in all composite computer-cluster software and hardware architectures, are also acknowledged.

References

[1] M.A. Sahai, S. Lovas, G.A. Chass, B. Penke, I.G. Csizmadia, J. Mol. Struct. (THEOCHEM), in this issue.
[2] R.A. Daniel, J. Errington, Cell 113 6 (2003) 767–776.
[3] G.A. Chass, M.A. Sahai, J.M.S. Law, S. Lovas, Ö. Farkes, A. Perczel, J.-L. Rivail, I.G. Csizmadia, Int. J. Quantum Chem. 90 (2002) 933–968.
[4] M.J. Frisch, G.W. Trucks, H.B. Schlegel, G.E. Scuseria, M.A. Robb, J.R. Cheeseman, V.G. Zakrzewski, J.A. Montgomery, Jr., R.E. Stratmann, J.C. Burant, S. Dapprich, J.M. Millam, A.D. Daniels, K.N. Kudin, M.C. Strain, Ö. Farkas, J. Tomasi, V. Barone, M. Cossi, R. Cammi, B. Mennucci, C. Pomelli, C. Adamo, S. Clifford, J. Ochterski, G.A. Petersson, P.Y. Ayala, Q. Cui, K. Morokuma, P. Salvador, J.J. Dannenberg, D.K. Malick, A.D. Rabuck, K. Raghavachari, J.B. Foresman, J. Cioslowski, J.V. Ortiz, A.G. Baboul, B.B. Stefanov, G. Liu, A. Liashenko, P. Piskorz, I. Komaromi, R. Gomperts, R.L. Martin, D.J. Fox, T. Keith, M.A. Al-Laham, C.Y. Peng, A. Nanayakkara, M. Challacombe, P.M. W. Gill, B. Johnson, W. Chen, M.W. Wong, J.L. Andres, C. Gonzalez, M. Head-Gordon, E.S. Replogle, J.A. Pople, Gaussian 98, Revision A.9, Gaussian, Inc., Pittsburgh PA, 2001.
[5] A. Perczel, J.G. Angyan, M. Kajtar, W. Viviani, J.L. Rivail, J.F. Marcoccia, G. Csizmadia, J. Am. Chem. Soc. 113 (1991) 6256–6265.

THEO
CHEM

Journal of Molecular Structure (Theochem) 666–667 (2003) 303–310

www.elsevier.com/locate/theochem

Conformational effects of the valine sidechain on the $\beta_L\beta_L$ extended and Type I beta turn backbone structures of MeCO-Val-Ala-NHMe and MeCO-Ala-Val-NHMe. An ab initio exploratory conformational study

Szilárd N. Fejér[a,*], Zsuzsanna A. Jenei[a], Gábor Paragi[b]

[a]Department of Medical Chemistry, Szent-Györgyi Medical Center, University of Szeged, Dóm tér 8., Szeged H-6720, Hungary
[b]Protein Chemistry Research Group, Hungarian Academy of Sciences, University of Szeged, Dóm tér 8., H-6720 Szeged, Hungary

Abstract

Ab initio molecular studies were carried out on two dipeptide models, MeCO-Ala-Val-NHMe and MeCO-Val-Ala-NHMe at the RHF/3-21G level of theory. All possible sidechain conformations were investigated for the extended $\beta_L\beta_L$ and Type I beta turn ($\alpha_L\delta_L$) backbone structures. The relative stabilities of the sidechain conformers and also the topologies of the PEC-s obtained by rotating the valine sidechain are discussed. Our exploratory study shows that these curves of the two isomeric dipeptides are almost identical in the case of the $\beta_L\beta_L$ extended backbone conformations, while in the case of the $\alpha_L\delta_L$ backbone conformations they are completely different. The results also show a considerable effect of the rotation of the valine sidechain on the backbone dihedral angles, the magnitude of this effect varies in the different backbone conformations.
© 2003 Elsevier B.V. All rights reserved.

Keywords: Ab initio; Ala-Val; Val-Ala; RHF/3-21G; Dipeptide; Conformational analysis

1. Introduction

Proteins are classified in their primary, secondary and tertiary structures. The genome codes only the primary structure (the amino acid sequence) of the proteins, therefore it is assumed that secondary and tertiary protein structures must be a consequence of the primary structure [1]. The biological activity of a protein is determined by these structures. Protein folding can be defined as a function of backbone torsional angles (φ, ψ, ω). Different interactions between amino acid sidechains modify these torsional angles. According to the theory regarding protein folding, proteins with biological activity exist in conformations with the lowest energy possible (the lowest minima on their potential energy hypersurfaces). Any conformational change from this minimum to another one can result in the loss of this activity and malfunction.

To understand the protein folding it is necessary to build peptide models and study their conformations and the effects of one residue on the others. The conclusions drawn may be used to characterize the potential energy hypersurfaces of

* Corresponding author.
E-mail addresses: fejer.szilard@stud.u-szeged.hu (S.N. Fejér), jenei.zsuzsanna@stud.u-szeged.hu (Z.A. Jenei), paragi@physx.u-szeged.hu (G. Paragi).

0166-1280/$ - see front matter © 2003 Elsevier B.V. All rights reserved.
doi:10.1016/j.theochem.2003.08.035

larger peptides [2]. However, ab initio calculations can only be used at present to characterize simple peptide models.

Protein chemists made the study of protein folding less difficult by separating the problem of backbone conformation from sidechain conformation assuming that the sidechain conformational changes have only minimal effect on the backbone conformation. They also separated the problem of nearest neighbors and long-range interactions.

Alanine and valine are the simplest chiral nonpolar (*hydrophobic*) amino acids. Alanine is the most abundant amino acid; valine is the sixth most abundant one. All possible conformations of the simplest amino acids (glycine, alanine and valine), as well as several di-, tri- and oligopeptides' with no sidechain torsional angles (dialanine, trialanine, tetraalanine, oligoalanine, triglycine, etc.) were widely studied [3–7].

The simplest dipeptides with two chiral centers and one prochiral sidechain torsional angle are Ala-Val and Val-Ala. Since each amino acid has nine conformers and the Valine sidechain can be oriented in g^+, g^- and *anti* conformations, a total of $9 \times 9 \times 3 = 243$ conformers are expected.

This work presents the conformational study of these two isomeric N- and C-protected dipeptides: MeCO-Ala-Val-NHMe and MeCO-Val-Ala-NHMe in their $\beta_L\beta_L$ extended and Type I beta turn ($\alpha_L\delta_L$) backbone structures. These studies were carried out in order to investigate the effect of the sidechain conformational changes on the backbone conformation.

2. Method

Ab initio molecular computations were carried out on two selected backbone conformations of the dipeptide models at the RHF/3-21G level of theory. The Gaussian98 [8] program was used for the geometry optimizations of the g^+, g^- and *anti* conformers of each isomer. The Z-matrices of the dipeptide models were built up using an universal atomic numbering system of amino acids in peptides [2]. The optimizations were performed initially using the semi-empirical AM1 method. These optimizations were followed by ab initio HF computations. One dimensional relaxed PES scans were made on the two isomeric dipeptides in both backbone conformations studied, using the sidechain torsional angle as variable of the energy function. These scans were run at 15° intervals, from 0 to 360°. The corresponding bond angles, bond lengths and dihedral angles were compared among the g^+, g^- and *anti* conformers of the two isomers.

3. Results and discussion

All of the 12 conformers expected (six for each isomer) were found on the potential energy hypersurface. Among these conformations, the Ala-Val Type I beta turn g^- conformer was found to have the lowest energy, as shown in Table 1. Because the Type I beta turn structure has a stabilizing intramolecular H-bond, the energy computed at the RHF/3-21G level

Table 1

Energies in Hartree and relative energies of the conformers found for MeCO-Ala-Val-NHMe and MeCO-Val-Ala-NHMe

| | Energy (Hartrees) | | ΔE (kJ/mol) | |
	β Extended	Type I beta turn	β Extended	Type I beta turn
Ala-Val				
g^+	− 812.2447351	− 812.2440478	2.798	4.603
a	− 812.2426799	− 812.2432270	8.195	6.758
g^-	− 812.2450118	− 812.2458006	2.070	0.000
Val-Ala				
g^+	− 812.2446682	− 812.2446956	2.974	2.901
a	− 812.2424661	− 812.2428104	8.757	7.852
g^+	− 812.2450657	− 812.2448078	1.929	2.607

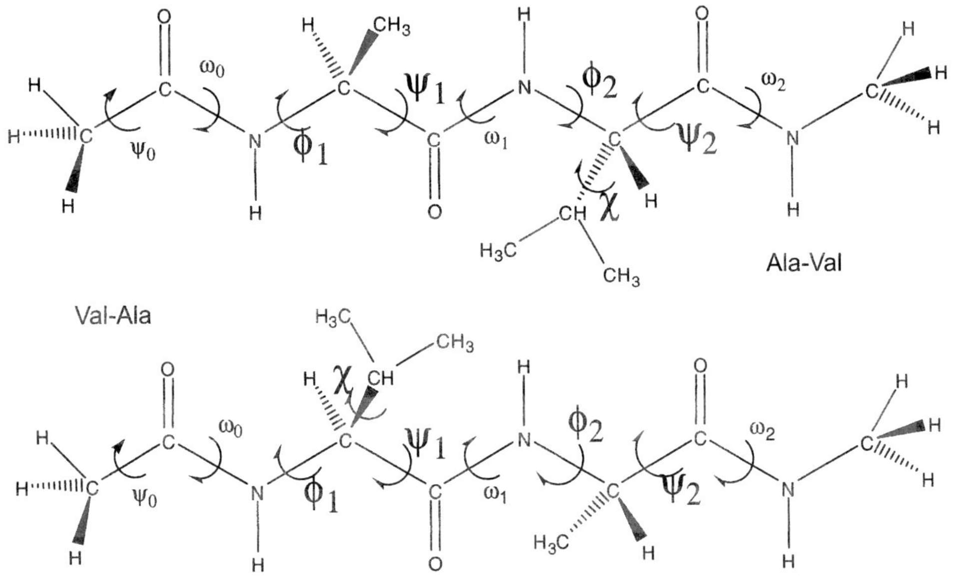

Fig. 1. The MeCO-Ala-Val-NHMe and MeCO-Val-Ala-NHMe model dipeptides with the definition of the dihedral angles.

of theory isn't precise enough. The energies of the conformers should be recomputed using larger basis sets.

The studied model dipeptides are shown in Fig. 1, with the backbone dihedral angles (φ, ψ and ω) and the prochiral sidechain dihedral χ. The backbone dihedral angles are specified by the convention regarding the numeration of dihedral angles in peptides [1]. The optimized dihedral angles are listed in Table 2.

The four optimized backbone conformations (two for each dipeptide model) are illustrated in Fig. 2a–d.

The summary of the relaxed one-dimensional potential energy surface scans is shown in Figs. 3 and 4. The potential energy curves of the extended backbone conformations of the two isomeric dipeptides are almost identical: the maximum deviation in the potential energy curves is only 1 kJ/mol. However, in the case of the Type I beta turn conformations the potential energy profiles are completely different. The deviation in energies is maximum 12 kJ/mol. Although the topology of these curves is different, the minima are close to each other, except the anti-conformations, where the χ dihedral angles' difference is 20°. Moreover, the Ala-Val PEC has two strange regions, in which the rotation of the valine sidechain by 15° causes a sudden change of

the backbone conformation (changes up to 70° in the Valine backbone dihedral angles, and up to 20° in the alanine dihedrals). The PEC in these points looks like an overlap of two curves, each one corresponding

Table 2
Optimized dihedral angles of MeCO-Ala-Val-NHMe and MeCO-Val-Ala-NHMe

	φ_1	ψ_1	ω_1	φ_2	ψ_2	ω_2
$\beta_L\beta_L$ extended						
Ala-Val						
g^+	−167.36	170.64	177.45	−162.11	155.80	178.72
a	−167.17	169.90	174.88	−137.33	142.37	178.90
g^-	−167.09	169.61	172.17	−141.95	162.63	177.17
Val-Ala						
g^+	−162.93	159.69	177.04	−165.42	169.92	178.02
a	−137.62	144.18	177.78	−167.22	169.44	177.87
g^-	−140.86	163.95	176.49	−167.49	169.49	177.88
Type I beta turn						
Ala-Val						
g^+	−69.08	−17.75	177.28	−117.84	26.19	176.39
a	−69.11	−17.60	175.63	−124.93	33.65	176.72
g^-	−69.24	−17.89	176.48	−114.76	24.08	176.52
Val-Ala						
g^+	−68.04	−20.15	177.62	−111.49	21.76	176.51
a	−62.82	−30.48	−179.62	−110.02	25.55	176.02
g^-	−72.78	−12.16	174.60	−111.59	19.22	176.91

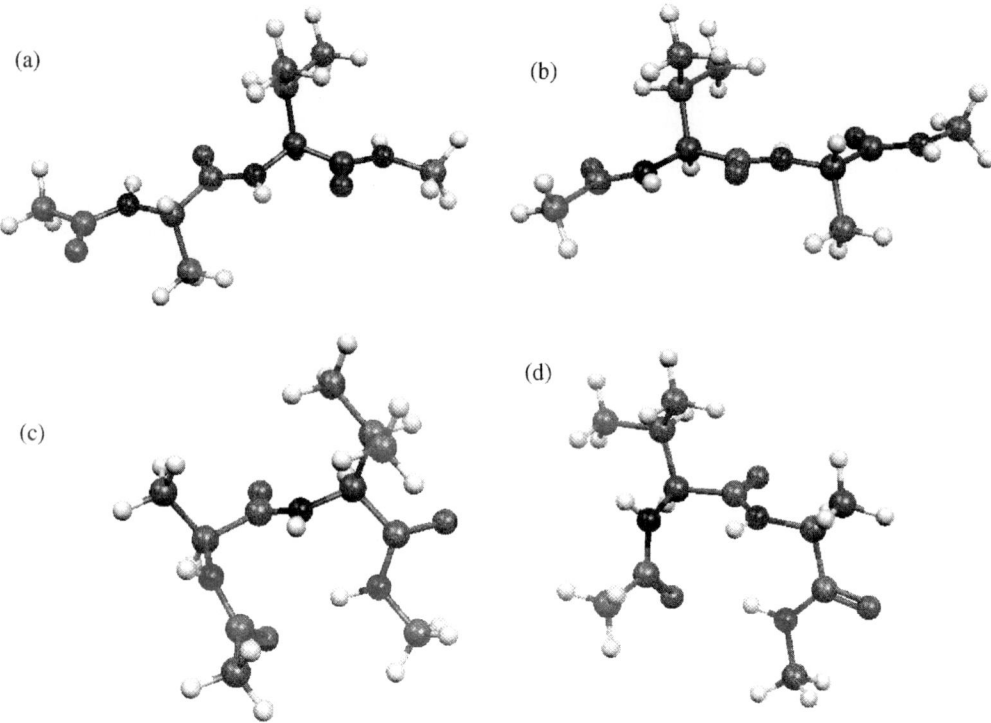

Fig. 2. Optimized structures of MeCO-Ala-Val-NHMe and MeCO-Val-Ala-NHMe in their extended $\beta_L\beta_L$ and Type I beta turn ($\alpha_L\delta_L$) conformations (a) Ala-Val $\beta_L\beta_L$ extended, (b) Val-Ala $\beta_L\beta_L$ extended, (c) Ala-Val Type I beta turn ($\alpha_L\delta_L$), (d) Val-Ala Type I beta turn ($\alpha_L\delta_L$).

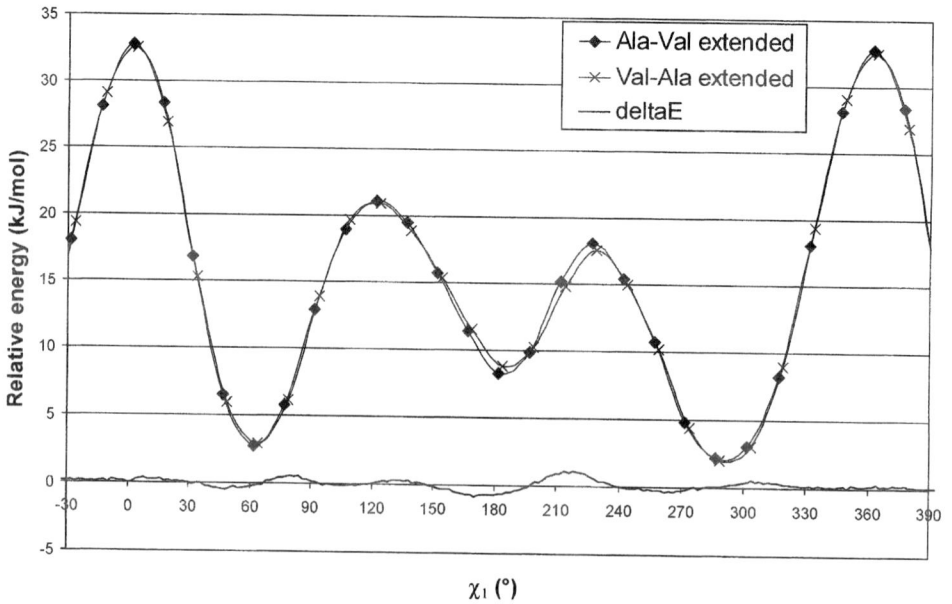

Fig. 3. Potential Energy Curves (PECs) of MeCO-Ala-Val-NHMe and MeCO-Val-Ala-NHMe in their extended $\beta_L\beta_L$ conformations as a function of χ.

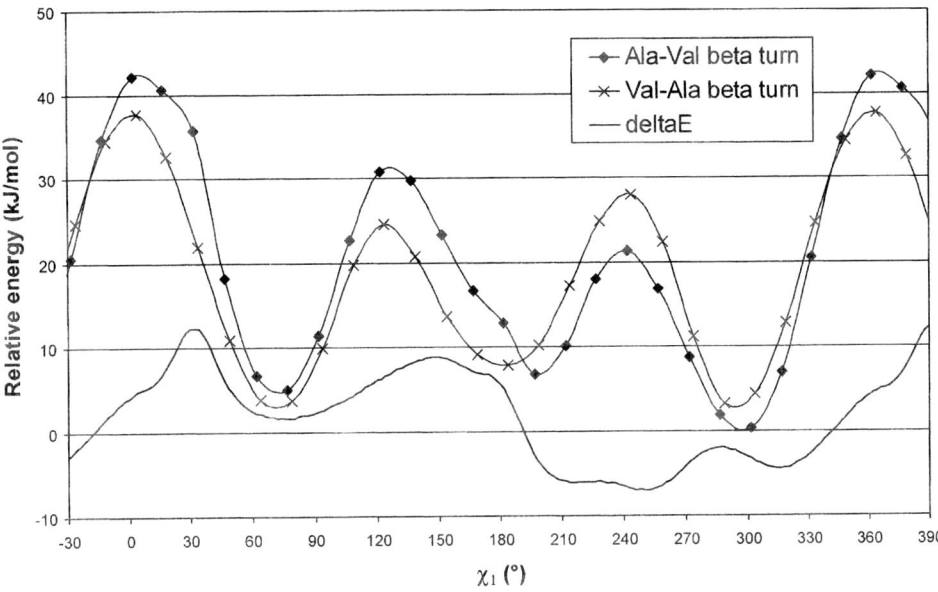

Fig. 4. Potential Energy Curves (PECs) of MeCO-Ala-Val-NHMe and MeCO-Val-Ala-NHMe in their Type I beta turn ($\alpha_L \delta_L$) conformations as a function of χ.

to one backbone conformation. In order to determine the exact shape of the curve between these points, more calculations are necessary. These calculations are in progress.

In Figs. 5–8, the effect of the valine sidechain rotation on the backbone dihedral angles is illustrated. In the case of the extended backbone conformations of the two isomeric dipeptides, the rotation of the valine sidechain has much effect on the φ_2, ψ_2 dihedral angles of the valine residue (changes up to 30°), but

has practically no effect on the neighboring alanine φ_1, ψ_1 dihedral angles. The same phenomenon can be observed in the case of the Val-Ala Type I beta turn conformer: the valine sidechain has much effect on the valine dihedral angles (changes up to 40°), and very little effect on the neighboring alanine dihedral angles. The Ala-Val Type I beta turn conformer's isopropyl (iPr) sidechain influences not only the valine dihedral angles (changes up to 70°), but the alanine dihedrals too (changes up to 30°). The sudden

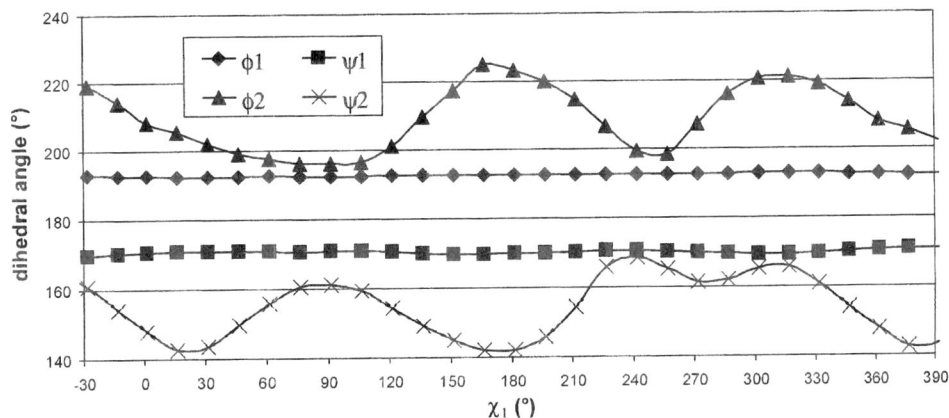

Fig. 5. Backbone torsional angles of MeCO-Ala-Val-NHMe in its extended $\beta_L \beta_L$ conformation as a function of χ.

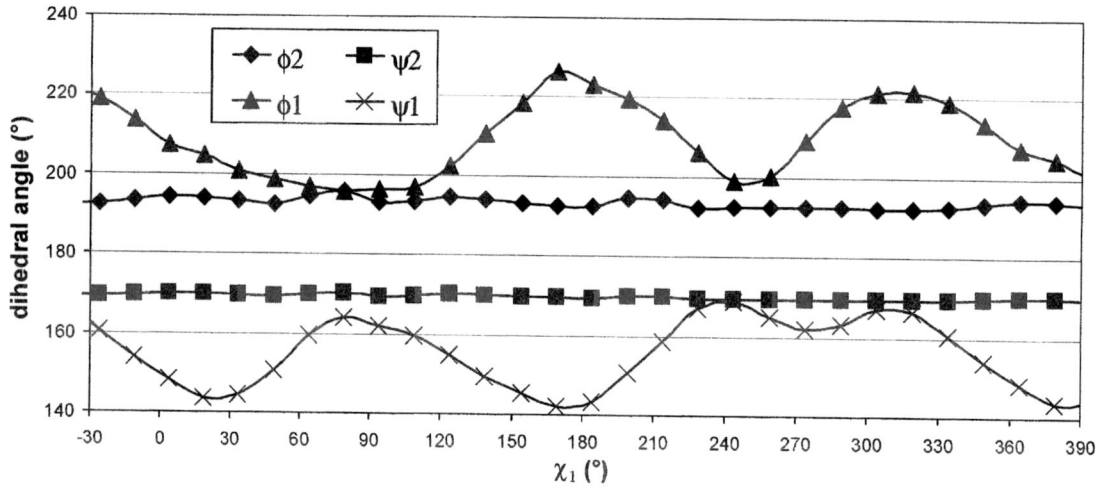

Fig. 6. Backbone torsional angles of MeCO-Val-Ala-NHMe in its extended $\beta_L\beta_L$ conformation as a function of χ.

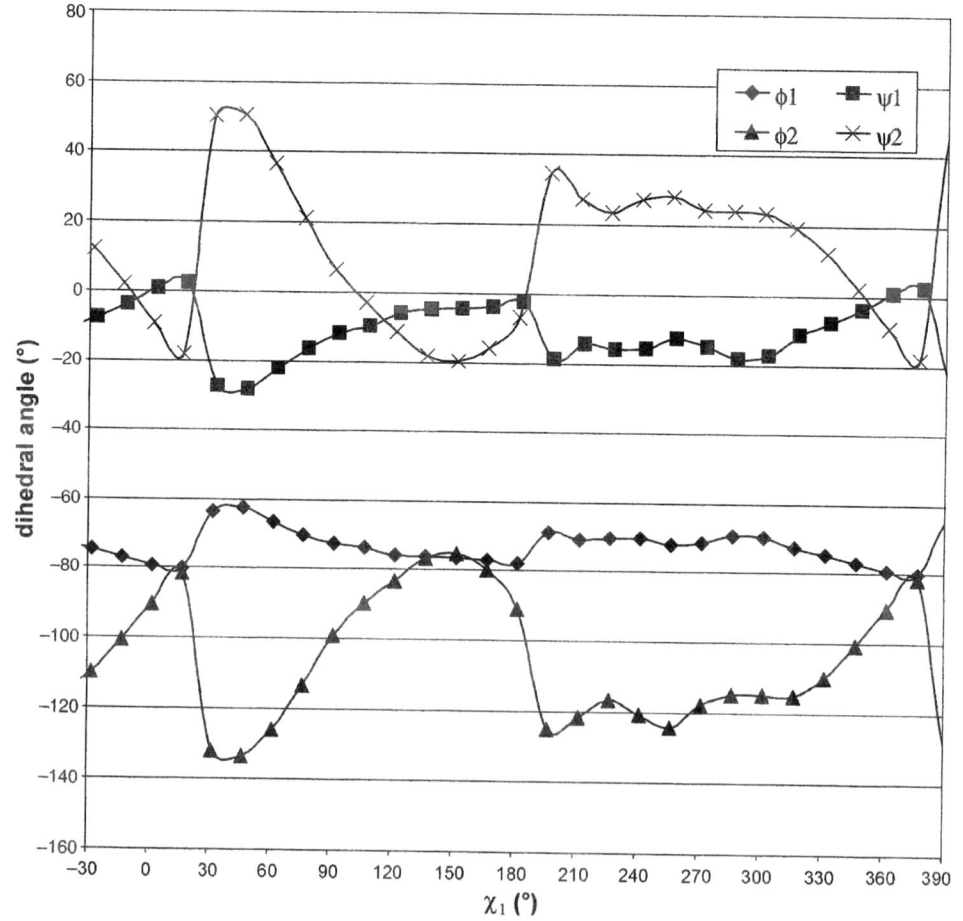

Fig. 7. Backbone torsional angles of MeCO-Ala-Val-NHMe in its Type I beta turn ($\alpha_L\delta_L$) conformation as a function of χ.

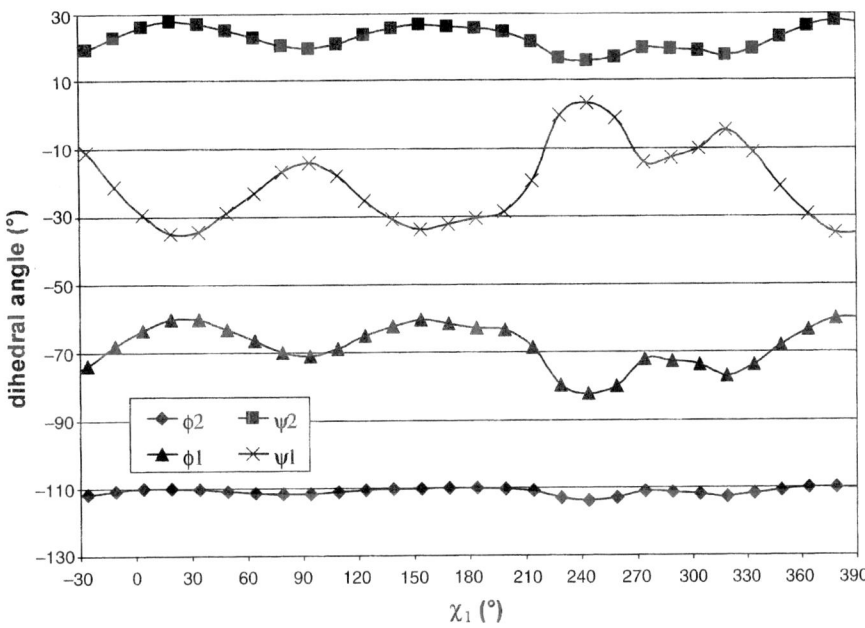

Fig. 8. Backbone torsional angles of MeCO-Val-Ala-NHMe in its Type I beta turn ($\alpha_L \delta_L$) conformation as a function of χ.

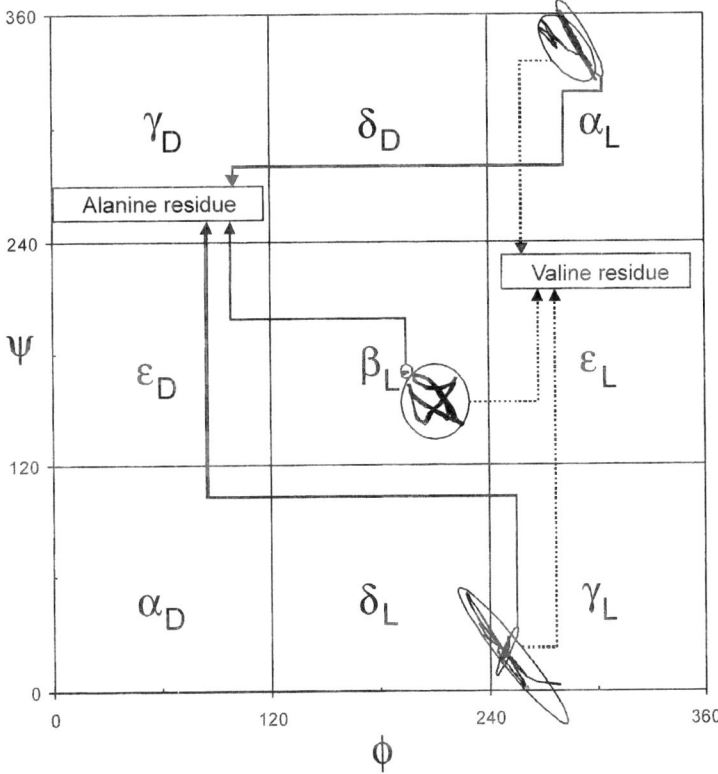

Fig. 9. Ramachadran map showing the variations of the alanine and valine residue dihedrals during the scans made on the two isomeric dipeptides, MeCO-Ala-Val-NHMe and MeCO-Val-Ala-NHMe, without distinguishing the isomers.

Table 3
Backbone torsional angle (φ and ψ) variations as a function of sidechain torsion (χ)

	$\beta_L\beta_L$ extended	Type I beta turn ($\alpha_L\delta_L$)
Self	± 15.2	± 34.9
Neighbor	± 1.8	± 15.2

conformational changes described above can be well observed.

The variations during the relaxed PES scans in the backbone dihedral angles (in both of the isomers) are illustrated on the following Ramachadran map (Fig. 9). Each curve corresponds to the dihedrals of one residue (alanine or valine), without distinguishing the isomers. It can be observed that the dihedral angles of the neighboring alanine residue change significantly only in the α_L and δ_L conformers, while the conformational change of the iPr group has up to 10 times greater effect on the valine backbone conformation.

4. Conclusions

The isomers studied are very close energetically (within 8.7 kJ/mol) in both backbone conformations; there is practically no preference in the amino acid sequence, so the probability of a sequence in a protein like Ala-Val or Val-Ala can be considered equal in all of the conformations studied.

The interconversion of the sidechain conformers has been estimated to have energy barriers between 15 and 45 kJ/mol. In the case of the extended ($\beta_L\beta_L$) conformation the largest barrier is estimated to be in the vicinity of 30 kJ/mol. In the case of the Type I beta turn conformations the highest barrier is considerably higher (45 kJ/mol), because an intramolecular hydrogen bond reduces the flexibility of the molecule.

The rotation of the valine sidechain has great effect on the backbone conformation of the dipeptide, including the neighboring amino acid dihedral angles. This effect is anticipated to be more pronounced in the case of larger sidechain groups.

On the basis of these investigations it appears that sidechain conformational problems cannot be separated completely from backbone conformations (Table 3). For this reason, the sidechain–backbone interactions should be extensively studied, in order to be able to describe their ability to stabilize certain conformations and conformational changes, and consequently their contribution to protein folding.

References

[1] G.A. Chass, A.M. Rodriguez, M.L. Mak, E. Deretey, A. Perczel, C.P. Rosa, R.D. Enriz, I.G. Csizmadi, Peptide and protein folding;, J. Mol. Struct. (Theochem) 500 (2000) 5–58.

[2] G.A. Chass, M.A. Sahai, J.M.S. Law, S. Lovas, Ö Farkas, Toward a computed peptide structure database: the role of a universal atomic numbering system of amino acids in peptides and internal hierarchy of database, Int. J. Quantum Chem. 90 (2002) 933–968.

[3] A. Perczel, M.A. McAllister, P. Császár, I.G. Csizmadia, Peptide models IX. A complete conformational set of for-Ala-Ala-NH$_2$ by ab initio computations, Can. J. Chem. 72 (1994) 2050–2070.

[4] M. Ramek, C.-H. Yu, L. Schäfer, Ab initio conformational analysis of the model tripeptide *N*-formyl-L-alanyl-L-alanine amide, Can. J. Chem./Rev. Can. Chim. 76 (5) (1998) 566–575.

[5] M.A. McAllister, A. Perczel, P. Császár, I.G. Csizmadia, Peptide models V: topological features of molecular mechanics and ab-initio 4D-Ramachandran maps. a conformational database for Ac-L-Ala-L-Ala-NHMe and for For-L-Ala-L-Ala-NH2, J. Mol. Struct. (Theochem) 288 (1993) 181–198.

[6] A. Perczel, M.A. McAllister, P. Császár, I.G. Csizmadia, Peptide models VI: new β-turn conformations from the ab initio calculations confirmed by X-ray data of proteins, J. Am. Chem. Soc. 115 (1993) 4849–4858.

[7] A. Mehdizadeh, G.A. Chass, Ö. Farkas, A. Perczel, L.L. Torday, A. Varro, J.Gy. Papp, Conformational effects of one glycine residue on the other glycine residues in the Ac-Gly-Gly-Gly-NHMe tripeptide motif, J. Mol. Struct. (Theochem) 588 (2002) 187–200.

[8] GAUSSIAN 98 (Revision A.7), M.J. Frisch, G.W. Trucks, H.B. Schlegel, G.E. Scuseria, M.A. Robb, J.R. Cheeseman, V.G. Zakrzewski, J.A. Montgomery Jr., R.E. Stratmann, J.C. Burant, S. Dapprich, J.M. Millam, A.D. Daniels, K.N. Kudin, M.C. Strain, O. Farkas, J. Tomasi, V. Barone, M. Cossi, R. Cammi, B. Mennucci, C. Pomelli, C. Adamo, S. Clifford, J. Ochterski, G.A. Petersson, P.Y. Ayala, Q. Cui, K. Morokuma, D.K. Malick, A.D. Rabuck, K. Raghavachari, J.B. Foresman, J. Cioslowski, J.V. Ortiz, A.G. Baboul, B.B. Stefanov, G. Liu, A. Liashenko, P. Piskorz, I. Komaromi, R. Gomperts, R.L. Martin, D.J. Fox, T. Keith, M.A. Al-Laham, C.Y. Peng, A. Nanayakkara, C. Gonzalez, M. Challacombe, P.M.W. Gill, B. Johnson, W. Chen, M.W. Wong, J.L. Andres, C. Gonzalez, M. Head-Gordon, E.S. Replogle, J.A. Pople, Gaussian, Inc., Pittsburgh PA, 1998.

ELSEVIER

Journal of Molecular Structure (Theochem) 666–667 (2003) 311–319

THEO
CHEM

www.elsevier.com/locate/theochem

A model study of the IgA hinge region: an exploratory study of selected backbone conformations of MeCO-L-Pro-L-Thr-NH-Me

Michelle A. Sahai[a,b,*], David H. Setiadi[a], Gregory A. Chass[a,b,c,d], Emil F. Pai[e], Botond Penke[f,g], Imre G. Csizmadia[a,b,f]

[a]Global Institute of COmputational Molecular and Materials Science (GIOCOMMS), 1422 Edenrose Street, Mississauga, Ont., Canada, L5V 1H3
[b]Department of Chemistry, University of Toronto, 80 St George Street, Toronto, Ont., Canada M5S 3H6
[c]Institut de Science et d'Ingénierie Supramoléculaires, 8, allée Gaspard Monge, BP 70028, 67083 Strasbourg Cedex, France
[d]Department of Biomedical Sciences, Creighton University, 2500 California Plaza, Omaha, NE 68178, USA
[e]Department of Biochemistry, University of Toronto, 1King's College Circle, Toronto, Ont., Canada M5S 1A8
[f]Department of Medical Chemistry, University of Szeged, H-6720 Szeged, Dóm tér 8, Hungary
[g]Protein Chemistry Research Group, Hungarian Academy of Sciences, University of Szeged, H-6720 Szeged, Dóm tér 8, Hungary

Abstract

Using the principles of multidimensional conformational analysis (MDCA), an approximate geometry for selected conformations of MeCO-Pro-Thr-NH-Me, the central dipeptide of the target sequence of immunoglobulin A protease, an enzyme secreted by pathogenic *Neisseria*, was determined at the RHF/3-21G level of theory. Type I ($\alpha_L\delta_L$ and $\alpha_L\gamma_L$) and Type II ($\varepsilon_L\delta_D$, $\varepsilon_L\alpha_D$ and $\varepsilon_L\gamma_D$) β-turns were generated. With nine sidechain combinations 18 different Type I β-turn conformations were studied. Similarly, three Type II β-turns: $\varepsilon_L\delta_D$, $\varepsilon_L\alpha_D$ and $\varepsilon_L\gamma_D$ were investigated each with nine possible sidechain conformations. Therefore, 3 × 9 structures are expected to yield 27 different Type II β-turn conformations. The conformational and energetic consequences of the optimized conformers are discussed in terms of relative stabilities and degree of backbone twisting or foldedness.
© 2003 Elsevier B.V. All rights reserved.

Keywords: MeCO-Pro-Thr-NH-ME; Proline; Threonine; Dipeptide; Ab initio MO computations; Internal hydrogen bonding

1. Introduction

At the beginning of the 20th century, the medical profession had no weapon against bacterial infection.

Antibacterial agents were then developed, the most dramatic of them being antibiotics. In general, antibiotics work by interfering with the life cycle of the bacteria (e.g. protein or bacterial cell wall synthesis). However, due to extensive and indiscriminate use of antibiotics virulent bacterial strains have evolved that are highly resistant to antibiotics. Two notable examples are Group A β-Hemolytic *Streptococcus* and *Mycobacterium tuberculosis*, which give rise to the flesh eating disease and tuberculosis, respectively.

* Corresponding author. Address: Global Institute of Computational Molecular and Materials Science (GIOCOMMS), 1422 Edenrose Street, Mississauga, Ont., Canada L5V 1H3.

E-mail addresses: msahai@giocomms.org (M.A. Sahai), gchass@giocomms.org (G.A. Chass), pbotond@mdche.szote.u-szeged.hu (E.F. Pai), icsizmad@alchemy.chem.utotonto.ca (I.G. Csizmadia).

0166-1280/$ - see front matter © 2003 Elsevier B.V. All rights reserved.
doi:10.1016/j.theochem.2003.08.036

As antibiotic resistance cannot be reversed, new ways are urgently needed to combat microbial pathogens. One such approach is to protect and support the immune system in its attack of foreign intruders.

Antibodies or immunoglobulins (Ig) represent the first line of defense and are crucial in the response to microbial challenge. IgA represents the major antibody in mucosal tissue, the point of entry of many pathogens with *Neisseria* being a prominent one. A schematic representation of the structure of a typical immunoglobulin molecule [1], is shown in Fig. 1. The structure of this protein can be divided into a constant region (C), which is host-specific, and variable regions (V), which are antigen-specific. Based on the constant region, Ig can be categorized into five types, some of these (IgG, IgD and IgA) possess hinges while others (IgM and IgE) do not.

Ig molecules are produced during a humoral immune response to bind and neutralize invading antigens. Subsequently, the antigen is labelled for removal by phagocytosis. The fight, however, is not one-sided. To help neutralize the role of Ig in antigen recognition, bacteria have developed effective deactivation mechanisms. As an example we discuss the site-specific cleavage of an IgA molecule at the hinge region with the aid of the secreted IgA protease produced by *Neisseria*, where the IgA molecule is separated into functional Fab and Fc domains. In this way, only the Fab region remains attached to the antigen and provides a 'camouflage'. The loss of the Fc part deprives the complement system of a proper docking site for the attack of the microbial intruder and provides protection for the pathogen.

As a first line of defense, antibodies are abundantly found in mucous membranes and in milk, which is particularly important for the infection protection in infants. The predominant form of human antibody found in these environments is IgA1. Thus, not surprisingly, many human pathogens have adopted strategies to counter the effects of IgA [2]. Some examples are both Gram-positive and Gram-negative pathogenic bacterial species like *Streptococcus pneumoniae* [3], *Haemophilus influenzae* [3,4], *Bacteroids melaninogenicus* (which has been strongly implicated in caries and periodontal disease [5]), *Ureaplasma urealyticum* strains [6], *Neisseria gonorrhoeae* [7], and *Neisseria meningitidis* [7]. They all secrete proteases that cleave human IgA1 at specific sites within the proline rich hinge region abolishing or at least impairing its antimicrobial activity [8]. Seven of the twelve IgAses (*H. influenzae, N. meningitidis, N. gonorrhoeae, U. urealyticum, S. pneumonia, S. oralis,* and *S. sanguis*) cleave a Pro-Thr peptide bond in the hinge region.

2. Background of peptide conformations

Since the pioneering work of Lothar Schafer [9–14] in the early 1980s, several other groups reported ab initio molecular computations [15–22] on

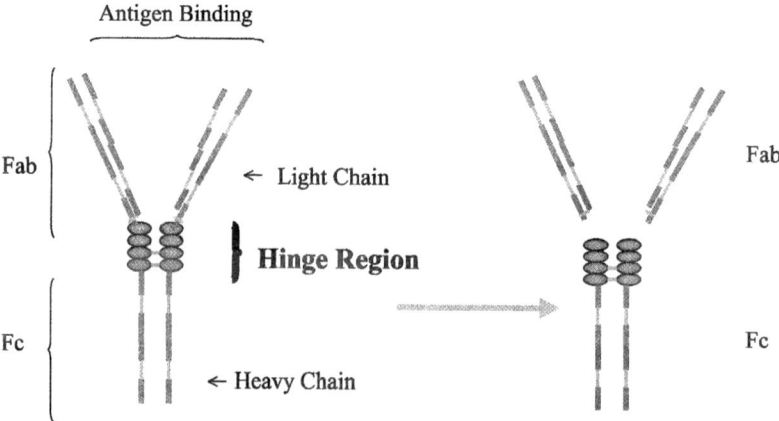

Fig. 1. Symbolic representation of an immunoglobulin molecule showing the Hinge region and the site-specific cleavage of the invading bacteria.

simple peptide models (Eq. (1)) where P, Q and R may be H or $-CH_3$. Research in this area received new momentum in the 1990s when a variety of R groups were studied [23–29] and when the work was extended to oligopeptides [30–35] (Eq. (2)). Since each amino acid residue may assume nine backbone (BB) conformations, the number of possible conformers (N_0) grows exponentially [35] with the number (n) of amino acids; thus we have $N_0 = 9^n = 10^{n \log 9}$. A peptide of modest size generates a conformational labyrinth. Table 1 illustrates the enormity of this exponential growth of the number of possible minima (N_0) with the degree of polymerization (n).

$$PCO-[-NH-CHR-CO-]-NHQ \qquad (1)$$

$$PCO-[-NH-CHR-CO-]_n-NHQ \qquad (2)$$

Clearly, two problems have to be clarified early on:

(i) How can one identify the various conformers in this labyrinth such that each conformer would have a unique name, and therefore, a unique address on the multidimensional conformational potential energy surface (PES)?

(ii) How can one use such unique addresses to generate reliable input for ab initio geometry optimizations?

Although there are numerous notations in naming the conformers, none of them enjoy universal acceptance. Section 2.1 describes the most recent attempt to classify the conformers. This approach is based on the concept that there is chirality in the conformational twist in addition to the chirality associated with the stereocenter. The procedure has been developed based on the principles of multidimensional conformational analysis (MDCA), and is also useful in generating approximate geometries as input for ab initio molecular computations.

2.1. Multidimensional Conformational Analysis (MDCA) of open-chain peptides

The amino acid residue in a peptide or a protein molecule is a triple rotor. The three torsional angles: ϕ, ψ, and ω measure the rotation about the N–C_α, the C_α–CO and the OC–NH (i.e. the peptide bond) bonds, respectively, leading to a potential energy hypersurface (PEHS): $E = E(\phi, \psi, \omega)$. Since the peptide bond is not free to rotate and can either assume a *cis*- or a *trans*- configuration, ω usually assumes values in the vicinity of 0° or 180°, respectively. For this reason, it is customary to consider only a PES cross-section of the full (PEHS), such as: $E = E(\phi, \psi)$ for ω to be optimized at about 180°. In the literature almost every Ramachandran map of the type $E = E(\phi, \psi)$ is associated with the *trans*-peptide bond.

Work has been completed for the following N- and C-protected amino acids containing a *trans*-peptide bond: Gly [22], Ala [22] Val [23], Phe [24,25], Ser [26–29] and Thr [36]. Again, with *trans*-peptide bonds, the following protected amino acid residues are currently under investigation: Arg, Asn, Asp, Cys, His and Lys.

Proline is fundamentally different from all the other 18 chiral amino acids in more than one respect:

(i) the R group forms a five-membered ring with the backbone.

(ii) there is no peptidic N–H group in the residue to be involved in hydrogen bonding, and

(iii) since there are two carbon atoms connected to the nitrogen, there is a higher chance of *cis/trans* isomerization in the peptide bond.

Certain selected conformations of the oligopeptides HCO-[NH-CHR-CO]$_n$-NH$_2$ have been studied by ab initio method, for R = H and $n = 1, 2, 3$ as well as R = CH$_3$ and $n = 1, 2, 3, 4$. Higher members of the series up to $n = 12$ are currently under investigation [37].

Table 1
The variation of the number of possible conformers with degree of polymerization

Degree of polymerization (n)	Number of possible conformers (N_0)
1	$9^1 = 10^{0.954242}$
10	$9^{10} = 10^{9.542425}$
100	$9^{100} = 10^{95.42425}$
1000	$9^{1000} = 10^{954.2425}$

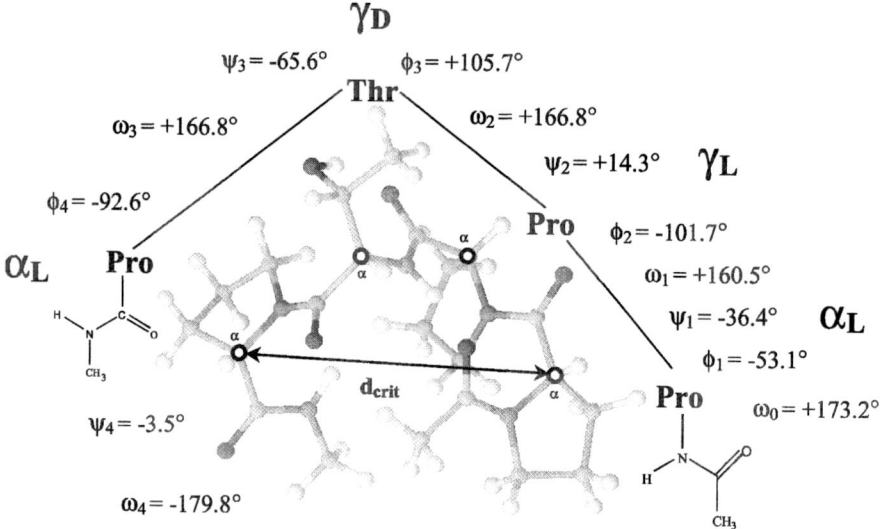

Fig. 2. One of the tetrapeptide structures (Ac-Pro-Pro-Thr-Pro-NH-Me)optimized at the HF/3-21G level of theory. This optimized structure corresponds to an α_L γ_L γ_D α_L conformation. The γ_L γ_D conformer was identified previously as a β-turn [43], with $d_{crit} < 7$ Å and $-90° \leq \tau_{crit} \leq +90°$. The above structure is only marginally β-turn, with $d_{crit} = 7.49$ Å and $\tau_{crit} = -97.77°$. (Note that τ_{crit} is defined as the torsional angle involving the four Cα atoms, denoted by the heavy opened circles above).

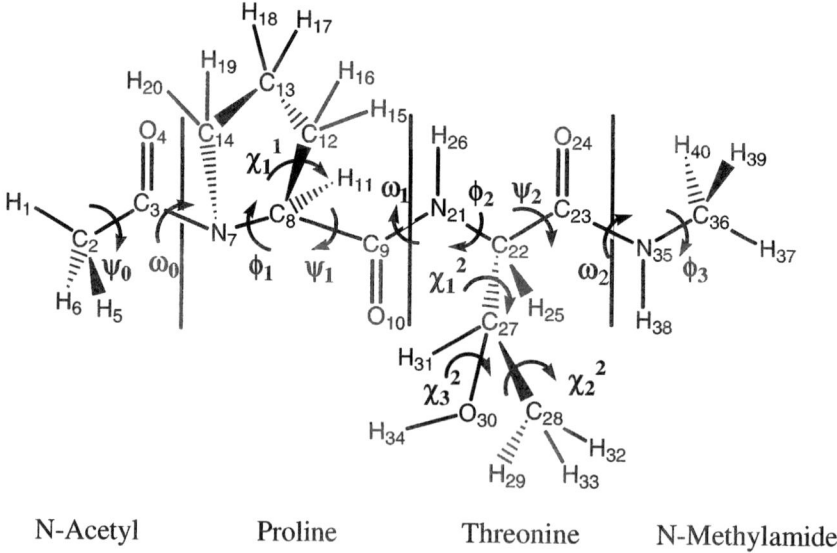

N-Acetyl   Proline   Threonine   N-Methylamide

Dihedral Defintions

χ_1^1 [13-12-8-7]

χ_1^2 [28-27-22-21]

χ_2^2 [29-28-27-22]

χ_3^2 [34-30-27-22]

Fig. 3. Standardized molecular numbering of the dipeptide MeCO-Pro-Thr-NH-Me.

Table 2
Symbolic internal coordinates of MeCO-Pro-Thr-NH-Me as shown in Fig. 3

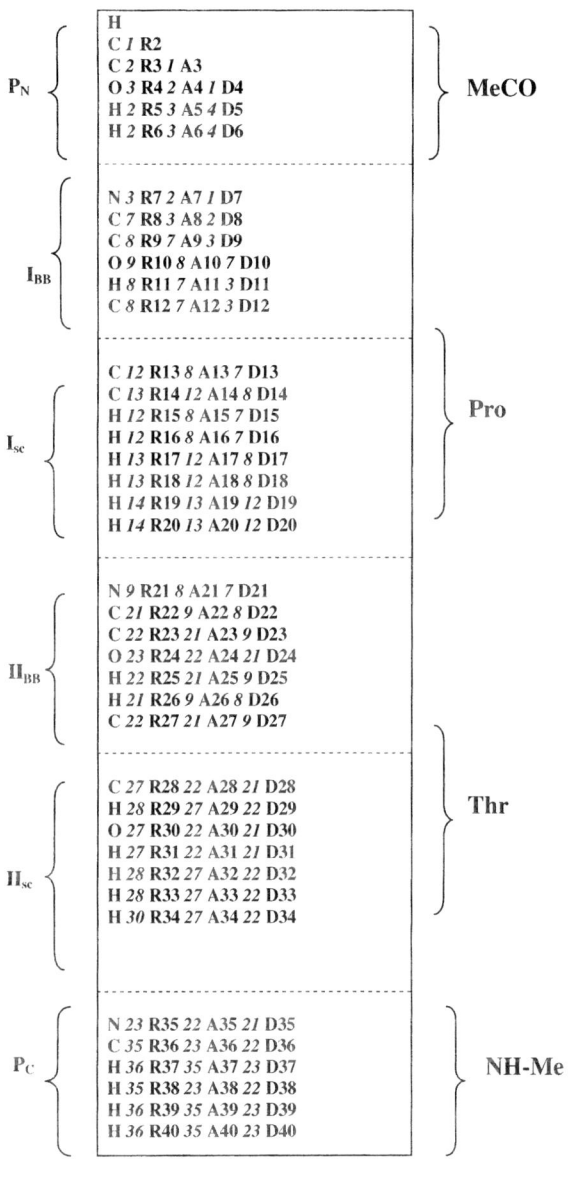

have decided to start a computational characterization of the energy levels of the accessible conformations of the IgAse target sequence: Pro-Pro-Thr-Pro with an emphasis on the central Pro-Thr bond.

Although the conformations of tetrapeptides [30] together with those tripeptides [31] and dipeptides [32–34] have been studied before they involved more simplistic approach as shown in Eq. (2) and discussed above. This Pro-Pro-Thr-Pro tetrapeptide is more complicated than Ala-Ala-Ala-Ala because Pro may exhibit *cis*- and *trans*-isomerism in its peptide bonds as well as *syn*- and *anti*- ring puckering. Thus, $2 \times 2 = 4$ isomeric forms may exist for each BB conformer. According to earlier studies [38], up to three conformers (γ_L, ε_L, α_L) may be expected; thus each proline moiety may have $4 \times 3 = 12$ conformations. Consequently, for the dipeptide MeCO-Pro-Pro-NH-Me, we may anticipate $12 \times 12 = 144$ conformations. However, even if one disregards ring puckering and considers only *trans*-peptide bonds, Pro-Pro-Thr-Pro may exhibit $3 \times 3 \times 9 \times 3 = 243$ BB conformations. Including $3 \times 3 = 9$ sidechain (SC) conformations of the Thr SC moiety [36] will lead to a total of 243×9 or 2187 conformers. Thus,

3. Method

3.1. Structural considerations

As no structural information is available on any IgAse, in complex with substrate or ligand free, we

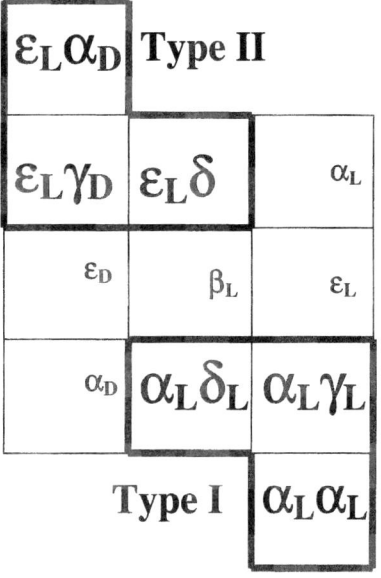

Fig. 4. The topological arrangements of Type I ($\alpha_L x$) and Type II ($\varepsilon_L x$) β-turns of MeCO-Pro-Thr-NH-Me according to the conformations (x) of the threonine residue.

Table 3
Optimized dihedral angles of MeCO-Pro-Thr-NH-Me in its Type I and Type II β-turn conformations

Conformation	β-turn type	Total energy (Hartrees)	Relative energy (kcal/mol)	Dihedral angles												
				ψ_0	ω_0	ϕ_1	ψ_1	ω_1	ϕ_2	ψ_2	ω_2	ϕ_3	χ_1^2	χ_2^2	χ_2^R	χ_3^2
$\alpha_L[g+]\delta_L[g+(a)g-]$	I	−924.345120	0.00	178.17	−169.87	−78.72	−3.21	170.44	−113.97	19.79	177.59	174.83	30.23	65.24	−174.76	−52.92
$\varepsilon_L[g+]\alpha_D[g+(a)g-]$	II	−924.341677	2.16	170.22	179.29	−63.12	131.07	−175.17	52.52	28.94	177.29	−179.89	30.04	72.77	173.68	−62.65
$\varepsilon_L[g+]\gamma_D[a(a)a]$	II	−924.332676	7.81	172.46	177.11	−70.13	157.01	156.98	80.67	−48.50	179.98	116.22	31.61	−164.28	−171.50	−176.39

the 243 BB conformations may be best studied on Pro-Pro-Ala-Pro and the Ala SC may be extended to a Thr SC subsequently. One of the 2187 conformers has been optimized in our preliminary study and the structure is shown in Fig. 2 The central Pro-Thr structure turned out to be γ_L γ_D, which has been previously identified as a new type of β-turn [43]. It is interesting to note that several of the oxygen atoms occupy a close region of space creating an electron-rich domain. Such a display of high electron density may represent a distinguishing factor in substrate enzyme recognition. From previous works, a Type 1 β-turn has been identified at the α_L conformation [39]. At any particular instance, the proline residue was fixed in the [α_L] conformation, including the ring puckering at gauche $+$, while the nine possible SC minima of threonine were optimized with the BB fixed initially at two structures: the [δ_L] and [γ_L] conformations. Consequently two Type I β-turns were generated: $\alpha_L\delta_L$ and $\alpha_L\gamma_L$ with various SC combinations. Therefore, 2×9 structures are expected to yield 18 different Type I β-turn conformations. Similarly, three Type II β-turns: $\varepsilon_L\delta_D$, $\varepsilon_L\alpha_D$ and $\varepsilon_L\gamma_D$ were investigated each with nine possible SC conformations, so 3×9 structures are expected to yield 27 different Type II β-turn conformations.

3.2. Practical considerations

The GAUSSIAN 98 [40] programs were used to carry out ab initio calculations on the 45 possible conformers of the dipeptide in order to determine the locations of all minima on their respective conformational PES.

For automation of input generation and subsequent data extraction from the output, the numbering of the atomic sequence in the peptides was standardized. In this way, scripts and programs may be used to immediately recognize patterns necessary for chemical analysis [41].

The molecule was numbered along the peptide BB (N–C$_\alpha$–CO), for each peptide 'unit' separately (See Fig. 3). The N-acetyl and N-methyl groups were defined as separate units and could be added at any time, as their coordinates do not disturb the definitions of the internal parameters specific to the core. As a result, each amino acid residue as well as, the N-acetyl and N-methylamide end protecting groups were exclusively defined, using the z-matrix internal

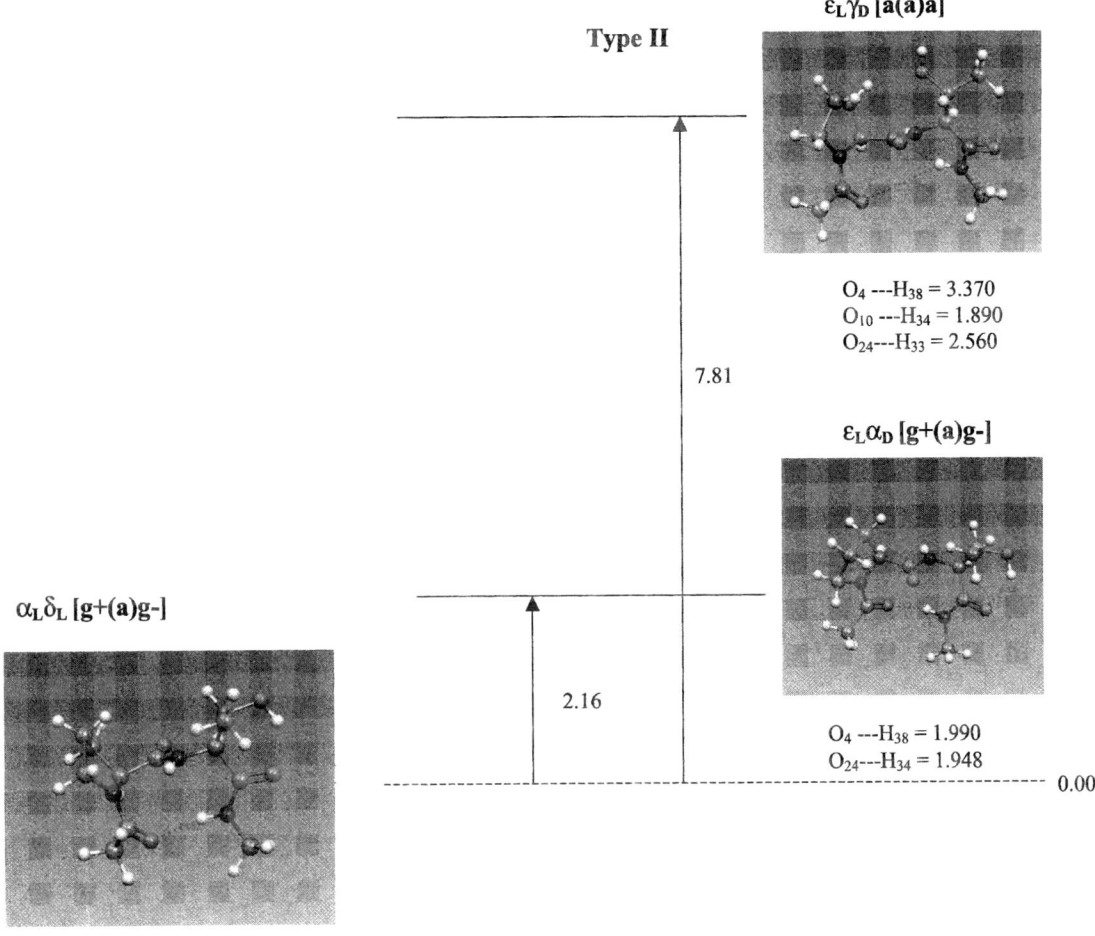

$\varepsilon_L\gamma_D$ [a(a)a]

O_4 --- H_{38} = 3.370
O_{10} --- H_{34} = 1.890
O_{24} --- H_{33} = 2.560

Type II

7.81

$\varepsilon_L\alpha_D$ [g+(a)g-]

O_4 --- H_{38} = 1.990
O_{24} --- H_{34} = 1.948

2.16

0.00

$\alpha_L\delta_L$ [g+(a)g-]

O_4 --- H_{38} = 2.083
O_{24} --- H_{34} = 1.889

Fig. 5. Relative energies and molecular structures of MeCO-Pro-Thr-NH-Me in its Type I and Type II β-turns. The hydrogen bond lengths are indicated under each molecular structure.

coordinate system to characterize molecular structure, stereochemistry and geometry. The resultant input file was used as a starting point in a successive and iterative process of GAUSSIAN 98 cycles to bring about geometry optimizations at the RHF/3-21G level of theory.

Table 2 illustrates the z-matrix that was used for the input file.

4. Results and discussions

It can be seen from the previous analysis of β-turns [39] that there may be three BB conformers for Type I

and another three BB conformers for Type II β-turns. This is illustrated in Fig. 4. Note that the $\alpha_L\alpha_L$ BB conformation listed under Type 1 β-turns could be part of any helical structure. Consequently there were effectively only two BB conformations ($\alpha_L\delta_L$ and $\alpha_L\gamma_L$) of Type I β-turn, which were to be studied while for Type II β-turns all three BB conformations ($\varepsilon_L\alpha_D$, $\varepsilon_L\gamma_D$, $\varepsilon_L\delta_D$) were investigated. Turning to the SC conformations we may note the following. There is only one SC conformation for proline $x_1^1 = g+$ and threonine can exist at $g+$, a and $g-$ conformations for all of x_1^2, x_2^2 and x_3^2. Thus, altogether, the following total number of conformers were considered:

Type I   2(BB) × 9(SC) = 18 conformers
Type II 3(BB) × 9(SC) = 27 conformers

Consequently a grand total of 45 initial structures were subjected to geometry optimization. Surprisingly, only three conformers of the 45 attempted structures were optimized successfully.

The optimized structures listing the key torsional angles are summarized in Table 3. Interestingly enough the Type I and lowest energy Type II β-turns have the same SC conformations [$g + (a) g-$]. In this orientation the SC hydroxyl group was able to hydrogen bond to the carbonyl oxygen of the C-terminal. This corresponds to the dominant SC orientation reported in the threonine diamide [36, 42]. In contrast to the above the least stable Type II β-turn, corresponding to $\varepsilon_L \gamma_D$ conformation, does not have such a SC–BB internal hydrogen bonding (Fig. 5).

Furthermore, the hydrogen bond involving a 10 membered ring so typical of β-turns, are substantially weakened and therefore stretched (from about 2.0 to about 3.4 Å). This was due to the formation of hydrogen bonding involving a seven-membered ring typical of the γ_D or C_7^{ax} conformation.

5. Conclusion

The results of these computations might have no relevance for inhibitor design at this stage. Many of the relevant biological information is not available. For example, there is no indication that the hinge region is locked into a specific conformation or that enzymes bind substrates in their low-energy conformations. So we really cannot claim drug design here at this stage. Consequently, this is a study of basic scientific interest. The relevance of the present results will hopefully be cleared once a structure of a ligand complex of IgAse is available.

However, the results, reported here, suggest that further study of the BB and SC conformers of the dipeptide is necessary, so as to systematically explore all possibilities of β-turn conformers.

Acknowledgements

This work was supported by grants from the Global Institute of Computational Molecular and Materials Science (GIOCOMMS), Toronto, Ontario, Canada. The authors wish to thank the Center for Molecular Design and Information Technology (MDIT), Leslie Dan Faculty of Pharmacy, University of Toronto, for providing computation time. Special thanks goes out to the MDIT support staff. One of the authors (IGC) wishes to thank the Ministry of Education for a Szent-Györgyi Visiting Professorship. The authors also wish to thank Tania A. Pecora for helpful discussions during the preparation of this manuscript, as well as Graydon Hoare for database management, network support and software and distributive processing development. Special thanks is extended to Andrew M. Chasse, Christopher M. Andrews and Michael R. Sahai for their development of novel scripting and coding techniques, which facilitate a reduction in the number of CPU cycles needed. The pioneering advances of Kenneth P. Chasse, in all composite computer-cluster software and hardware architectures, are also acknowledged.

References

[1] C.A. Janeway Jr., P. Travers, S. Hunt, M. Walport, Immunobiology: The Immune System in Health and Disease, third ed., Current Biology Ltd, Garland Publishing, 1997, pp. 3:2–3:4.
[2] B.J. Underdown, J.M. Sciff, Annu. Rev. Immunol. 4 (1986) 389.
[3] M. Kilian, J. Mestecky, R.E. Schrohenloher, Infect. Immun. 26 (1979) 143.
[4] J. Pohlner, R. Halter, K. Bayreuther, T.F. Meyer, Nature 325 (1987) 458.
[5] S.B. Mortensen, M. Kilian, Infect. Immun. 45 (1984) 550.
[6] R.K. Spooner, W.C. Russell, D. Thirkell, Infect. Immun. 60 (1992) 2544.
[7] A.G. Plaut, J.V. Gilbert, M.S. Artenstein, J.D. Capra, Science 190 (1975) 1103.
[8] M. Kilian, J. Mestecky, M.W. Russell, Microbiol. Rev. 52 (1988) 269.
[9] L. Schäfer, C. Van Alsenoy, J.N. Scarsadale, J. Chem. Phys. 76 (1982) 1439.
[10] J.N. Scarsdale, C. Van Alsenoy, V.J. Klimkowski, L. Schäfer, F.A. Momany, J. Am. Chem. Soc. 105 (1983) 3438.
[11] L. Schäfer, V.J. Kimokowski, F.A. Momany, H. Chuman, C. Van Alsenoy, Biopolymers 23 (1984) 2335.

[12] V.J. Kimokowski, L. Schäfer, F.A. Momany, C. Van Alsenoy, J. Mol. Struct. (THEOCHEM) 124 (1985) 143.

[13] M. Ramek, V.K.W. Cheng, R.F. Frey, S.Q. Newton, L. Schäfer, J. Mol. Struct. (THEOCHEM) 235 (1991) 1.

[14] R.F. Frey, J. Coffin, S.Q. Newton, M. Ramek, V.K.W. Cheng, F. Momany, L. Schäfer, J. Am. Chem. Soc. 114 (1992) 5369.

[15] S.J. Wiener, P. Kollman, D.A. Case, U.C. Singh, C. Ghio, G. Alagona, S. Profeta Jr., P. Weiner, J. Am. Chem. Soc. 106 (1984) 765.

[16] S.J. Wiener, U.C. Singh, T.J. O'Donnell, P. Kollman, J. Am. Chem. Soc. 106 (1984) 6243.

[17] T. Head-Gordon, M. Head-Gordon, M.J. Frisch, C.L. Brooks III, J.A. Pople, Int. J. Quantum Chem. Quantum Biol. 16 (1989) 311.

[18] T. Head-Gordon, M. Head-Gordon, M.J. Frisch, C.L. Brooks III, J.A. Pople, J. Am. Chem. Soc. 113 (1991) 5989.

[19] H.J. Böhm, S. Brode, J. Am. Chem. Soc. 113 (1991) 7129.

[20] A. Perczel, J.G. Angyan, M. Kajtar, W. Viviani, J.L. Rivail, J.F. Marcoccia, G. Csizmadia, J. Am. Chem. Soc. 113 (1991) 6256.

[21] G. Endredi, A. Perczel, Ö. Farkas, M.A. McAllister, G.I. Csonka, J. Ladik, G. Csizmadia, J. Mol. Struct. (THEOCHEM) 391 (1997) 15.

[22] M.A. McAllister, A. Perczel, P. Csaszar, W. Viviani, J.L. Rivail, I.G. Csizmadia, J. Mol. Struct. (THEOCHEM) 288 (1993) 161.

[23] W. Viviani, J.L. Rivail, A. Perczel, G. Csizmadia, J. Am. Chem. Soc. 115 (1993) 8321.

[24] Ö. Farkas, M.A. McAllister, J.H. Ma, A. Perczel, M. Hollosi, I.G. Csizmadia, J. Mol. Struct. (THEOCHEM) 369 (1996) 105.

[25] A. Perczel, Ö. Farkas, I.G. Csizmadia, Can. J. Chem. 75 (1997) 1120.

[26] G. Endredi, M.A. McAllister, Ö. Farkas, A. Perczel, J.F. Marcoccia, M. Hollosi, I.G. Csizmadia, J. Mol. Struct. (THEOCHEM) 331 (1995) 27.

[27] A. Perczel, Ö. Farkas, I.G. Csizmadia, J. Comp. Chem. 17 (1996) 821.

[28] A. Perczel, Ö. Farkas, I.G. Csizmadia, J. Am. Chem. Soc. 118 (1996) 7809.

[29] A. Perczel, Ö. Farkas, J.F. Marcoccia, I.G. Csizmadia, Int. J. Quantum Chem. 61 (1997) 797.

[30] G. Endredi, M.A. McAllister, A. Perczel, J. Ladik, G. Csizmadia, J. Mol. Struct. (THEOCHEM) 331 (1995) 5.

[31] M. Cheung, M.A. McAllister, A. Perczel, P. Csaszar, I.G. Csizmadia, J. Mol. Struct. (THEOCHEM) 309 (1994) 151.

[32] A. Perczel, M.A. McAllister, P. Csaszar, G. Csizmadia, Can. J. Chem. 72 (1994) 2050.

[33] M.A. McAllister, A. Perczel, P. Csaszar, I.G. Csizmadia, J. Am. Chem. Soc. 115 (1993) 4849.

[34] M.A. McAllister, A. Perczel, P. Csaszar, W. Viviani, J.L. Rivail, I.G. Csizmadia, J. Mol. Struct. (THEOCHEM) 288 (1993) 181.

[35] C.M. Liegener, G. Endredi, M.A. McAllister, A. Perczel, J. Ladik, I.G. Csizmadia, J. Am. Chem. Soc. 115 (1993) 8275.

[36] M.A. Sahai, S. Motiwala, G.A. Chass, E.F. Pai, B. Penke, I.G. Csizmadia, J. Mol. Struct. (THEOCHEM). (in this issue).

[37] M.A. Sahai, S. Lovas, G.A. Chass, B. Penke, I.G. Csizmadia, J. Mol. Struct. (THEOCHEM). (in this issue).

[38] H.A. Baldoni, A.M. Rodriguez, G. Zamarbide, R.D. Enriz, Ö. Farkas, P. Csaszar, L.L. Torday, C.P. Sosa, I. Jakli, A. Perczel, M. Hollosi, G. Csizmadia, Peptide models XXIV, J. Mol. Struct. (THEOCHEM) 465 (1999) 79–91.

[39] A. Perczel, M.A. McAllister, P. Csaszar, I.G. Csizmadia, J. Am. Chem. Soc. 115 (1993) 4849.

[40] J. Frisch, G.W. Trucks, H.B. Schlegel, G.E. Scuseria, M.A. Robb, J.R. Cheeseman, V.G. Zakrzewski, J.A. Montgomery, Jr., R.E. Stratmann, J.C. Burant, S. Dapprich, J.M. Millam, A.D. Daniels, K.N. Kudin, M.C. Strain, Ö. Farkas, J. Tomasi, V. Barone, M. Cossi, R. Cammi, B. Mennucci, C. Pomelli, C. Adamo, S. Clifford, J. Ochterski, G.A. Petersson, P.Y. Ayala, Q. Cui, K. Morokuma, P. Salvador, J.J. Dannenberg, D.K. Malick, A.D. Rabuck, K. Raghavachari, J.B. Foresman, J. Cioslowski, J.V. Ortiz, A.G. Baboul, B.B. Stefanov, G. Liu, A. Liashenko, P. Piskorz, I. Komaromi, R. Gomperts, R.L. Martin, D.J. Fox, T. Keith, M.A. Al-Laham, C.Y. Peng, A. Nanayakkara, M. Challacombe, P.M.W. Gill, B. Johnson, W. Chen, M.W. Wong, J.L. Andres, C. Gonzalez, M. Head-Gordon, E.S. Replogle, J.A. Pople, GAUSSIAN 98, Revision A.9, Gaussian, Inc., Pittsburgh, PA, 2001.

[41] G.A. Chass, M.A. Sahai, J.M.S. Law, S. Lovas, Ö Farkes, A. Perczel, J.-L. Rivail, I.G. Csizmadia, Int. J. Quantum Chem. 90 (2002) 933–968.

[42] In Ref. [36] the χI torsional angle was defined starting with the sidechain oxygen rather than with the sidechain methyl group as it is done in the present paper. Consequently there is a 120° shift between the two conventions. The "g + " in the present work corresponds to "a" in Ref. [38]..

[43] M.A. Berg, G.A. Chasse, E. Deretey, A.K. Füzéry, B.M. Fung, D.Y.K. Fung, H. Henry-Riyad, A.C. Lin, M.L. Mak, A. Mantas, M. Patel, I.V. Repyakh, M. Staikova, S.J. Salpietro, T. Tang, J.C. Vank, A. Perczel, Ö. Farkas, L.L. Torday, Z. Székely, I.G. Csizmadia, J. Mol. Struct. (THEOCHEM) (Millennium Volume) 500 (2000) 5–58.

ELSEVIER

Journal of Molecular Structure (Theochem) 666–667 (2003) 321–326

THEO
CHEM

www.elsevier.com/locate/theochem

Molecular orbital analysis of the effect of D- and L-alanyl residues on the glycine chirality within the tripeptide N-Ac-Ala-Gly[β]-Ala-NH-Me. An ab initio and DFT study

Jack C.C. Liao[a,b,*], Gregory A. Chass[a,b,c,d], David H. Setiadi[a,b], Imre G. Csizmadia[a,b,e]

[a]Global Institute of Computational Molecular and Materials Science (GIOCOMMS), 1422 Edenrose St, Mississauga, Ont., Canada L5V 1H3
[b]Department of Chemistry, University of Toronto, 80 St George St, Toronto, Ont., Canada M5S 3H6
[c]Institut de Science et d'Ingénierie Supramoléculaires, 8, allée Gaspard Monge, BP 70028, 67083 Strasbourg Cedex, France
[d]Department of Biomedical Sciences, Creighton University, 2500 California plaza, Omaha, NE 68178, USA
[e]Department of Medical Chemistry, Szeged University, Dom ter 8, H-6720 Szeged, Hungary

Abstract

A conformational analysis of the tripeptide models, N-Ac-D-Ala-Gly[β]-L-Ala-NH-Me and N-Ac-L-Ala-Gly[β]-L-Ala-NH-Me, was carried out using ab initio Molecular Orbital (MO) computations at the RHF/6-31G(d) and B3LYP/6-31G(d) levels of theory. At both levels of theory, global and local minima on the Ramachandran potential energy surfaces (PES), $E = f(\phi_1, \psi_1)$ and $E = f(\phi_3, \psi_3)$, of the two systems associated with the possible 18 backbone orientations were geometries optimized. When a glycyl residue was sandwiched between two chiral amino acids such as alanyl residues, its C_α–H_R and C_α–H_S bond lengths were no longer equal but instead perturbed slightly; inducing a chirality on the glycyl residue. Depending on the respective conformations of the adjacent alanyl residues, the magnitude of the conformational impact was investigated in this article.
© 2003 Elsevier B.V. All rights reserved.

Keywords: N-Ac-D-Ala-Gly-L-Ala-NH-Me; N-Ac-L-Ala-Gly-L-Ala-NH-Me; Glycyl residue; Alanyl residue; Multidimensional conformational analysis; Ab initio molecular orbital computations; Conformational potential energy surfaces; Chiral induction on glycine α carbon

1. Introduction

Protein folding can be represented by the following partitioning of the potential energy function: $E(\text{polypeptide}) = f(\psi_0, \omega_0, \phi_1, \psi_1, \chi_1, \omega_1, \dots, \omega_{i-1}, \phi_i, \psi_i, \chi_i, \omega_i, \dots, \omega_{n-1}, \phi_n, \psi_n, \chi_n, \omega_n, \phi_{n+1})$, and

the sidechain interaction is characterized by the dihedral angle vector χ. For the tripeptide systems studied here, only the backbone variables are included and its energy function is shown below:

$E(\text{tripeptide})$

$$= f(\psi_0, \omega_0, \phi_1, \psi_1, \omega_1, \phi_2, \psi_2, \omega_2, \phi_3, \psi_3, \omega_3, \phi_4).$$

In this context, the conformational properties of the N- and C-terminal protected amino acids can mimic the conformational properties of

* Corresponding author. Address: Global Institute of Computational Molecular and Materials Science (GIOCOMMS), 1422 Edenrose St, Mississauga, Ont., Canada L5V 1H3.
E-mail addresses: jack.liao@utoronto.ca (J.C.C. Liao), gchass@fixy.org (G.A. Chass), dsetiadi@lookingout.org (D.H. Setiadi), icsizmad@chem.utoronto.ca (I.G. Csizmadia).

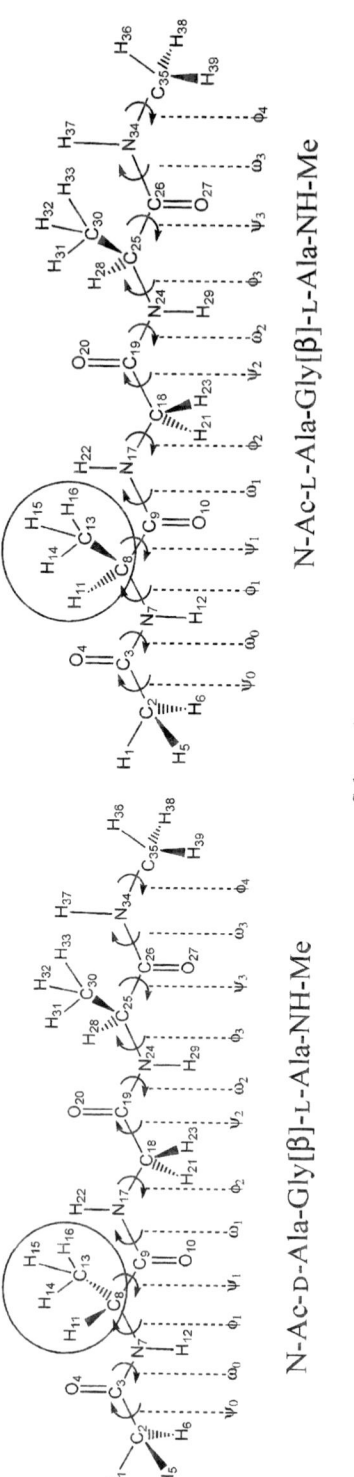

Scheme 1.

polypeptides and protein segments, as these variables can be influenced by the orientation of the neighbouring residues.

2. Computational methods

The GAUSSIAN 98 [1] program was used to carry out ab initio calculations on the 18 possible conformers of all nine backbone conformations ($\beta_{L/D}$, α_L, δ_L, ε_L, γ_L, α_D, δ_D, ε_D, γ_D) of the first and last alanyl residues of N-Ac-D-Ala-Gly[β]-L-Ala-NH-Me and N-Ac-L-Ala-Gly[β]-L-Ala-NH-Me, in order to determine the locations of all minima on their respective conformational PES.

The standard numbering system [2] was utilized such that the molecules were numbered into and along the peptide backbone (N–C_α–CO), for each peptide 'unit' separately and explicitly (Scheme 1). The N-Acetyl or N-Methyl groups are defined as distinctive units and can be added at any time, as their coordinates do not disturb the definitions of the internal parameters specific to the core. Previous geometry optimization at the RHF/3-21G level of theory (Figs. 1 and 2) was used [3] as the starting point in a successive and iterative process of GAUSSIAN 98 cycles to bring about an optimization at the predicted minima, using RHF/6-31(d) and B3LYP/6-31G(d).

3. Results and discussion

Selected optimized parameters of N-Ac-D-Ala-Gly[β]-L-Ala-NH-Me and N-Ac-L-Ala-Gly[β]-L-Ala-NH-Me are tabulated in Tables 1 and 2, respectively. It was demonstrated that, at both RHF/6-31G(d) and B3LYP/6-31G(d) levels of theory (data not shown for B3LYP/6-31G(d)), seven stable minima were found for each system: α_L, β_D, δ_L, δ_D, ε_L, γ_L, γ_D for N-Ac-D-Ala[X_D]-Gly[β]-L-Ala[β_L]-NH-Me, and α_D, β_L, δ_L, δ_D, ε_D, γ_L, γ_D for N-Ac-L-Ala[X_L]-Gly[β]-L-Ala[β_L]-NH-Me.

In addition, β_D-β-β_L and β_L-β-β_L were the lowest energy conformations for N-Ac-D-Ala-Gly[β]-L-Ala-NH-Me and N-Ac-L-Ala-Gly[β]-L-Ala-NH-Me, respectively. All conformations computed at the B3LYP/6-31G(d) level of theory were 6 Hartrees

Table 1
Summary of total energies, relative energies and dihedral angles for N-Ac-D-Ala-Gly[β]-L-Ala-NH-Me, computed at the RHF/6-31G(d) level of theory

Ala	Gly	Ala	Energy (hartree)	ΔE (kcal mol^{-1})	ψ_0	ω_0	ϕ_1	ψ_1	ω_1	ϕ_2	ψ_2	ω_2	ϕ_3	ψ_3	ω_3	ϕ_4	χ_1^1	χ_2^1
β	β	β_L	−945.5325394	0.000	12.33	−179.57	157.93	−160.15	−177.39	179.83	−179.87	178.61	−157.40	160.17	179.23	−169.75	178.36	−178.62
β	β	γ_L	−945.5314930	0.657	13.17	−179.82	157.67	−159.86	−177.90	178.54	−178.20	−178.41	−86.50	82.01	−174.10	−168.93	178.37	−178.29
β	β	γ_D	−945.5268955	3.542	13.62	−179.87	157.53	−159.47	−177.11	178.56	177.73	175.78	76.38	−55.14	−177.39	−175.79	178.38	−174.74
β	β	δ_L	−945.5279434	2.884	108.74	−177.04	157.31	−160.53	−177.95	177.45	−174.82	−168.76	−131.49	22.12	172.54	−70.62	178.31	−179.63
β	β	δ_D	−945.5232876	5.806	5.37	−179.23	157.82	−160.06	−177.40	−179.75	−179.75	171.01	−164.93	−41.51	−172.08	−169.73	178.38	178.32
β	β	ε_L	N/F	N/F	N/F	N/F	N/F	N/F	N/F	N/F	N/F	N/F	N/F	N/F	N/F	N/F	N/F	N/F
β	β	ε_D	−945.5247684	4.876	−7.80	−177.25	157.50	−160.28	−177.96	175.13	−173.16	−165.84	58.57	−145.38	177.86	170.44	178.35	−169.87
β	β	α_L	N/F	N/F	N/F	N/F	N/F	N/F	N/F	N/F	N/F	N/F	N/F	N/F	N/F	N/F	N/F	N/F
β	β	α_D	−945.5245268	5.028	13.70	−179.90	157.63	−159.96	−176.65	−178.39	173.57	166.98	66.35	31.73	−177.16	−169.70	178.31	−175.11
γ_L	β	β_L	−945.5244771	5.059	−100.78	−179.22	−76.89	40.42	178.54	163.41	−178.33	179.76	−157.69	159.54	179.26	−169.24	173.23	−178.50
γ_D	β	β_L	−945.5293508	2.001	7.76	−178.57	88.19	−92.68	176.10	175.19	−174.71	179.85	−157.41	159.44	179.39	−169.41	178.16	−178.55
δ_L	β	β_L	−945.5242820	5.182	84.04	−169.49	164.21	−13.83	173.53	−174.12	−174.18	−179.90	−157.86	160.30	179.14	−169.69	−178.82	−178.60
δ_D	β	β_L	−945.5292447	2.067	45.67	170.99	122.54	147.61	−176.06	−178.13	178.73	177.72	−156.95	159.47	179.54	−168.76	−179.88	−178.64
ε_L	β	β_L	−945.5251314	4.649	−81.82	164.48	−58.33	−32.24	−178.71	175.19	175.06	176.96	−156.80	159.26	179.55	−168.49	168.84	−178.63
ε_D	β	β_L	N/F	N/F	N/F	N/F	N/F	N/F	N/F	N/F	N/F	N/F	N/F	N/F	N/F	N/F	N/F	N/F
α_L	β	β_L	−945.5259531	4.133	−41.93	−169.16	−65.17	32.24	176.43	−170.88	−175.65	−179.75	−157.68	159.36	179.26	−169.12	175.61	−178.52
α_D	β	β_L	N/F	N/F	N/F	N/F	N/F	N/F	N/F	N/F	N/F	N/F	N/F	N/F	N/F	N/F	N/F	N/F

Table 2
Summary of total energies, relative energies and dihedral angles for N-Ac-L-Ala-Gly[β]-L-Ala-NH-Me, computed at the RHF/6-31G(d) level of theory

Ala	Gly	Ala	Energy (hartree)	ΔE (kcal mol^{-1})	ψ_0	ω_0	ϕ_1	ψ_1	ω_1	ϕ_2	ψ_2	ω_2	ϕ_3	ψ_3	ω_3	ϕ_4	χ_1^1	χ_2^1
β	β	β_L	−945.5325456	0.000	−11.63	179.51	−157.88	160.14	177.40	−179.20	179.60	179.60	−157.26	160.10	179.36	−169.12	178.36	−178.63
β	β	γ_L	−945.5315190	0.644	−14.68	179.93	−157.56	159.52	177.12	179.36	−179.44	−178.98	−86.40	82.03	−174.02	−169.18	178.38	−178.26
β	β	γ_D	−945.5268787	3.556	−15.17	−179.91	−157.68	159.71	178.03	−178.06	176.49	175.23	76.47	−55.43	−177.43	−176.73	178.37	−174.80
β	β	δ_L	−945.5279740	2.869	102.27	−179.94	−157.67	159.90	176.81	178.93	−175.18	168.99	−131.16	22.20	172.56	−70.45	178.34	−179.62
β	β	δ_D	−945.5232528	5.831	−12.78	179.68	−157.94	160.04	177.46	−178.41	177.86	170.78	−164.90	−41.62	−172.19	−169.78	178.35	−181.69
β	β	ε_L	N/F	N/F	N/F	N/F	N/F	N/F	N/F	N/F	N/F	N/F	N/F	N/F	N/F	N/F	N/F	N/F
β	β	ε_D	−945.5248250	4.845	−13.41	179.83	−157.78	159.66	176.41	177.21	−174.06	−166.19	58.57	−145.57	177.95	170.85	178.46	−169.82
β	β	α_L	N/F	N/F	N/F	N/F	N/F	N/F	N/F	N/F	N/F	N/F	N/F	N/F	N/F	N/F	N/F	N/F
β	β	α_D	−945.5244805	5.061	−12.42	179.65	−157.92	160.46	178.21	177.05	173.50	166.87	66.45	31.77	−177.09	−169.94	178.24	−175.13
γ_L	β	β_L	−945.5293727	1.991	−24.79	179.91	−88.33	91.79	−176.16	−174.08	173.89	177.02	−156.63	159.32	179.27	−168.88	−178.15	−178.64
γ_D	β	β_L	−945.5245434	5.021	−21.45	179.62	76.86	−41.18	−179.01	−160.53	178.34	177.44	−156.46	158.86	179.23	−168.15	−173.25	−178.56
δ_L	β	β_L	−945.5292036	2.097	72.91	−171.12	−122.29	13.53	176.03	179.01	−179.21	179.37	−157.76	160.05	179.33	−168.71	179.88	−178.55
δ_D	β	β_L	−945.5243039	5.172	35.93	169.55	−164.35	−40.19	−173.53	175.53	173.35	176.79	−157.00	160.13	179.44	−169.14	178.73	−178.66
ε_L	β	β_L	N/F	N/F	N/F	N/F	N/F	N/F	N/F	N/F	N/F	N/F	N/F	N/F	N/F	N/F	N/F	N/F
ε_D	β	β_L	−945.5251108	4.665	82.05	−164.42	58.34	−147.74	178.81	−174.65	−175.80	−179.89	−157.60	159.56	179.41	−169.10	−168.76	−178.59
α_L	β	β_L	N/F	N/F	N/F	N/F	N/F	N/F	N/F	N/F	N/F	N/F	N/F	N/F	N/F	N/F	N/F	N/F
α_D	β	β_L	−945.5259439	4.143	80.50	174.34	64.33	31.90	−175.98	169.74	175.29	177.01	−156.77	158.97	179.75	−168.47	−175.59	−178.65

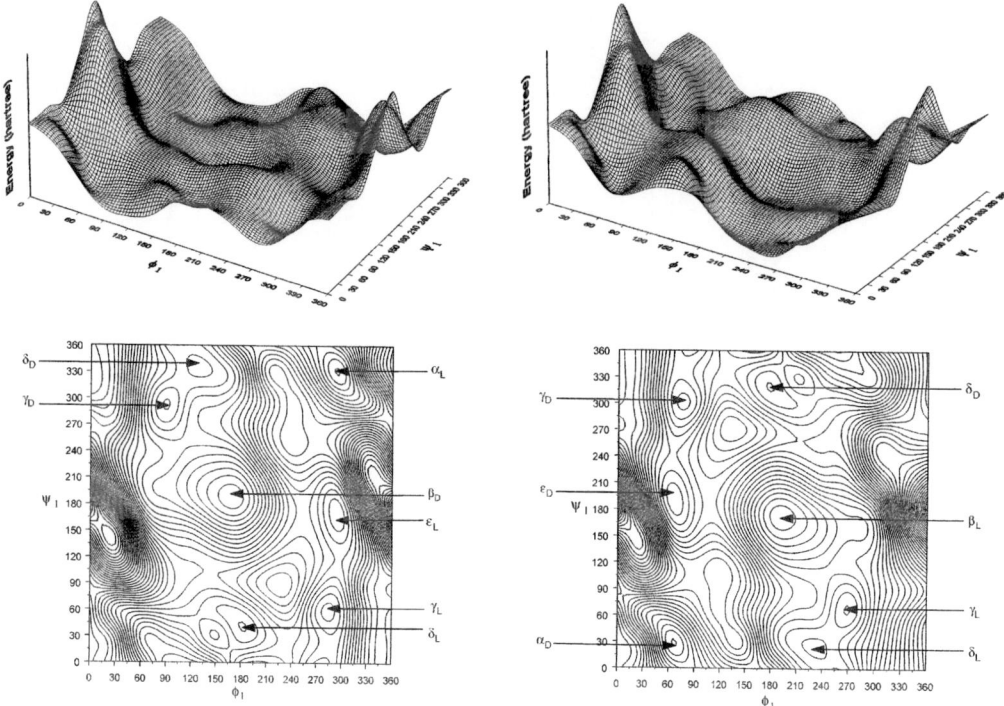

Fig. 1. Ramachandran PES of (Left) *N*-Ac-D-Ala[X$_D$]-Gly[β]-L-Ala[β$_L$]-NH-Me (X$_D$-β-β$_L$) and (Right) *N*-Ac-L-Ala[X$_L$]-Gly[β]-L-Ala[β$_L$]-NH-Me (X$_L$-β-β$_L$) computed at the RHF/3-21G level of theory [3]. (Top) Landscape representation. (Bottom) Contour diagram representation.

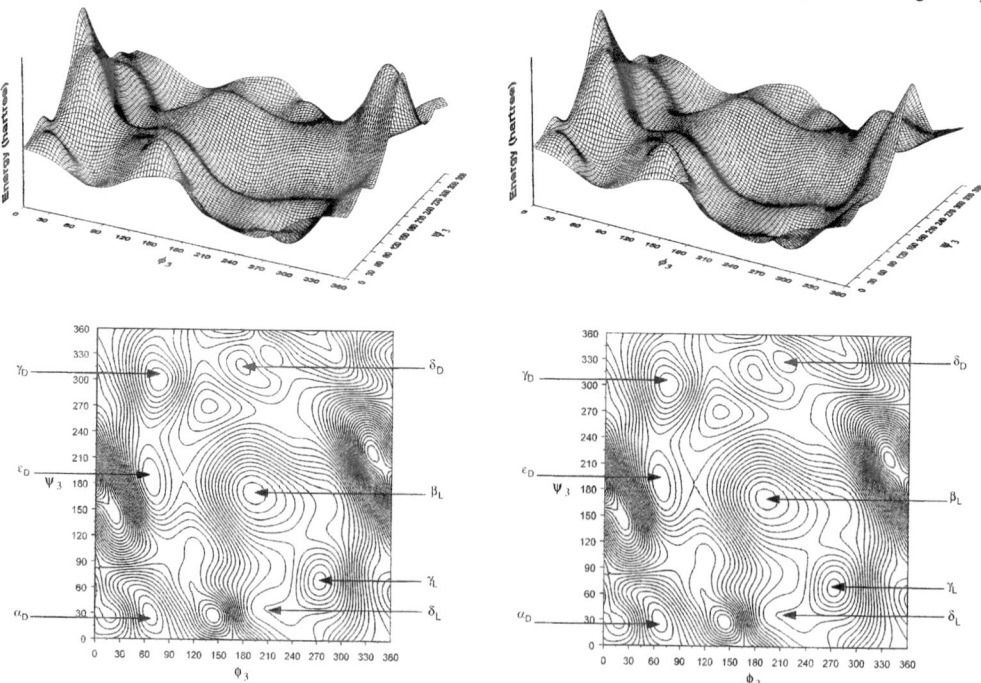

Fig. 2. Ramachandran PES of (Left) *N*-Ac-D-Ala[β$_D$]-Gly[β]-L-Ala[X$_L$]-NH-Me (β$_D$-β-X$_L$) and (Right) *N*-Ac-L-Ala[β$_L$]-Gly[β]-L-Ala[X$_L$]-NH-Me (β$_L$-β-X$_L$) computed at the RHF/3-21G level of theory [3]. (Top) Landscape representation. (Bottom) Contour diagram representation.

Table 3

C–H bond lengths and N–C_α–CO bond angles of N-Ac-D-Ala-Gly[β]-L-Ala-NH-Me, computed at the RHF/6-31G(d) level of theory

Conformer			Bond length of C–H (Å)				Angle of N–C_α–CO (°)		
D-Ala 1	Gly 2	L-Ala 3	D-Ala 1 (C–H_S)	Gly 2		L-Ala 3 (C–H_R)	D-Ala 1	Gly 2	L-Ala 3
				C–H_R	C–H_S				
$β_D$	β	$β_L$	1.082800	1.084800	1.084800	1.083000	107.35	108.91	107.09
$β_D$	β	$γ_L$	1.082770	1.085208	1.084779	1.080668	107.41	109.20	109.48
$β_D$	β	$γ_D$	1.082701	1.084869	1.085219	1.079366	107.42	109.27	114.17
$β_D$	β	$δ_L$	1.082842	1.086153	1.084441	1.081480	107.42	109.16	112.51
$β_D$	β	$δ_D$	1.082759	1.084790	1.085555	1.080731	107.36	109.12	109.99
$β_D$	β	$ε_L$	N/F	N/F	N/F	N/F	N/F	N/F	N/F
$β_D$	β	$ε_D$	1.082834	1.086158	1.083583	1.084581	107.38	109.13	109.22
$β_D$	β	$α_L$	N/F	N/F	N/F	N/F	N/F	N/F	N/F
$β_D$	β	$α_D$	1.082717	1.084559	1.086168	1.085383	107.41	109.13	113.25
$γ_L$	β	$β_L$	1.080471	1.087405	1.082584	1.083032	115.403	110.115	107.30
$γ_D$	β	$β_L$	1.079444	1.086904	1.083333	1.082956	108.99	109.62	107.25
$δ_L$	β	$β_L$	1.081108	1.085874	1.083711	1.082956	110.39	108.91	107.11
$δ_D$	β	$β_L$	1.080975	1.084353	1.085836	1.082966	113.12	109.05	107.17
$ε_L$	β	$β_L$	1.085478	1.083960	1.085605	1.082962	109.53	109.03	107.20
$ε_D$	β	$β_L$	N/F	N/F	N/F	N/F	N/F	N/F	N/F
$α_L$	β	$β_L$	1.086010	1.085630	1.084543	1.082902	112.66	109.10	107.24
$α_D$	β	$β_L$	N/F	N/F	N/F	N/F	N/F	N/F	N/F

Table 4

C–H bond lengths and N–C_α–CO bond angles of N-Ac-L-Ala-Gly[β]-L-Ala-NH-Me, computed at the RHF/6-31G(d) level of theory

Conformer			Bond length of C–H (Å)				Angle of N–C_α–CO (°)		
L-Ala 1	Gly 2	L-Ala 3	L-Ala 1 (C–H_R)	Gly 2		L-Ala 3 (C–H_R)	L-Ala1	Gly 2	L-Ala 3
				C–H_R	C–H_S				
$β_L$	β	$β_L$	1.082772	1.084675	1.084942	1.082960	107.35	108.91	107.09
$β_L$	β	$γ_L$	1.082709	1.084932	1.085079	1.080637	107.41	109.19	106.60
$β_L$	β	$γ_D$	1.082741	1.084566	1.085496	1.079359	107.42	109.28	114.14
$β_L$	β	$δ_L$	1.082719	1.085943	1.084669	1.081494	107.41	109.17	112.52
$β_L$	β	$δ_D$	1.082772	1.084548	1.085806	1.080720	107.39	109.12	109.98
$β_L$	β	$ε_L$	N/F	N/F	N/F	N/F	N/F	N/F	N/F
$β_l$	β	$ε_D$	1.082707	1.085921	1.083864	1.084599	107.37	109.11	109.25
$β_L$	β	$α_L$	N/F	N/F	N/F	N/F	N/F	N/F	N/F
$β_L$	β	$α_D$	1.082829	1.084370	1.086320	1.085389	107.36	109.14	113.25
$γ_l$	β	$β_L$	1.079488	1.083165	1.087028	1.082953	109.03	109.62	107.28
$γ_D$	β	$β_L$	1.080445	1.082369	1.087510	1.082930	115.34	110.18	107.33
$δ_l$	β	$β_L$	1.080945	1.085724	1.084470	1.082997	113.14	109.05	107.17
$δ_D$	β	$β_L$	1.081121	1.083470	1.086107	1.083005	110.39	108.90	107.11
$ε_L$	β	$β_L$	N/F	N/F	N/F	N/F	N/F	N/F	N/F
$ε_D$	β	$β_L$	1.085495	1.085476	1.084099	1.082885	109.53	109.03	107.20
$α_L$	β	$β_L$	N/F	N/F	N/F	N/F	N/F	N/F	N/F
$α_D$	β	$β_L$	1.086011	1.084622	1.085548	1.082984	112.67	109.03	107.24

more stable than those computed at the RHF/6-31G(d) level. The study further confirmed the enantiomeric relationship between pairs of minima when both axial and point chirality were changed, as noted previously at the RHF/3-21G level of theory [3].

Calculated values of C_α–H_R and C_α–H_S bond lengths are given in Tables 3 and 4. In an achiral environment, the glycyl C_α-H bonds are expected to be identical in length. Based on the difference in the glycyl C_α–H bond lengths, the result clearly indicated that a chirality was induced on the glycyl residue due to the introduction of chiral environments (two alanyl residues) surrounding it, allowing the two C_α-H bonds to be in different chemical environments. γ_L-β-β_L and γ_D-β-β_L were the two backbone orientations with the greatest difference in glycyl C_α-H bonds lengths, for N-Ac-D-Ala-Gly[β]-L-Ala-NH-Me and N-Ac-L-Ala-Gly[β]-L-Ala-NH-Me respectively.

This study not only verified and improved the accuracy of the structures obtained earlier at the RHF/3-21G level of theory [3], but also helps us construct rules to further understand the folding of polypeptides containing the tripeptide Ala-Gly-Ala, such as receptors and enzymes. The concept of dynamic chirality may also have great implications in future drug design, as it may have an impact on the molecule's chemical, structural and functional properties.

Acknowledgements

This work was supported by grants from the Global Institute of Computational Molecular and Material Sciences (GIOCOMMS). The help of Kenneth P. Chass (math@velocet.ca) and Graydon Hoare (graydon@pobox.com) are acknowledged for their contribution of spare CPU cycles, network support, database management and for the ongoing development of distributive processing. A special thanks is extended to Andrew M. Chass (fixy@fixy.org) for his continuing and progressive integration of novel software techniques, hardware configurations and well planned scripting and coding, significantly reducing the necessary CPU time for each calculation.

One of the authors (IGC) wishes to thank the Ministry of Education for a Szent-Györgyi Visiting Professorship.

References

[1] M.J. Frisch, G.W. Trucks, H.B. Schlegel, G.E. Scuseria, M.A. Robb, J.R. Cheeseman, V.G. Zakrzewski, J.A. Montgomery, Jr., R.E. Stratmann, J.C. Burant, S. Dapprich, J.M. Millam, A.D. Daniels, K.N. Kudin, M.C. Strain, O. Farkas, J. Tomasi, V. Barone, M. Cossi, R. Cammi, B. Mennucci, C. Pomelli, C. Adamo, S. Clifford, J. Ochterski, G.A. Petersson, P.Y. Ayala, Q. Cui, K. Morokuma, D.K. Malick, A.D. Rabuck, K. Raghavachari, J.B. Foresman, J. Cioslowski, J.V. Ortiz, A.G. Baboul, B.B. Stefanov, G. Liu, A. Liashenko, P. Piskorz, I. Komaromi, R. Gomperts, R.L. Martin, D.J. Fox, T. Keith, M.A. Al-Laham, C.Y. Peng, A. Nanayakkara, M. Challacombe, P.M.W. Gill, B. Johnson, W. Chen, M.W. Wong, J.L. Andres, C. Gonzalez, M. Head-Gordon, E.S. Replogle, J.A. Pople, GAUSSIAN 98, Revision A.9, Gaussian, Inc., Pittsburgh PA, 1998.

[2] G.A. Chass, M.A. Sahai, J.M.S. Law, S. Lovas, Ö. Farkas, A. Perczel, J.-L. Rivail, I.G. Csizmadia, Int. J. Quantum Chem. 90 (2002) 933–968.

[3] J.C.C. Liao, J.C. Chua, G.A. Chass, A. Perczel, A. Varro, J.Gy. Papp, J. Mol. Struct. (THEOCHEM) 621 (2003) 163–187.

ELSEVIER

Journal of Molecular Structure (Theochem) 666–667 (2003) 327–336

THEO
CHEM

www.elsevier.com/locate/theochem

An ab initio exploratory study on selected conformational features of MeCO-L-Ala-L-Ala-L-Ala-NH-Me as a XxxYyyZzz tripeptide motif within a protein structure

Michelle A. Sahai[a,b,*], Michael R. Sahai[a], Gregory A. Chass[a,b,c,d],
Botond Penke[e,f], Imre G. Csizmadia[a,b,e]

[a]Global Institute of COmputational Molecular and Materials Science (GIOCOMMS), 1422 Edenrose St., Mississiauga, Ont., Canada L5V 1H3
[b]Department of Chemistry, University of Toronto, 80 St George St., Toronto, Ont., Canada M5S 3H6
[c]Institut de Science et d'Ingénierie Supramoléculaires, 8, allée Gaspard Monge, BP 70028, 67083 Strasbourg cedex, France
[d]Department of Biomedical Sciences, Creighton University, 2500 California plaza, Omaha, NE 68178, USA
[e]Department of Medical Chemistry, University of Szeged, H-6720 Szeged, Dóm tér 8, Hungary
[f]Protein Chemistry Research Group, Hungarian Academy of Sciences, University of Szeged, H-6720 Szeged, Dóm tér 8, Hungary

Abstract

A conformational study of the tripeptide model MeCO-L-Ala-L-Ala-L-Ala-NH-Me was carried out using ab initio molecular orbital computations in order to investigate the preferred conformations. At any particular instant, two alanine residues were fixed at the $[\beta_L]$ conformation and the third was varied for the nine possible minima present on the Ramachandran map. Subsequently, all minima were optimized. The conformational and energetic consequences of these findings are discussed in terms of relative stabilities and degree of backbone twisting or foldedness.
© 2003 Elsevier B.V. All rights reserved.

Keywords: MeCO-L-Ala-L-Ala-L-Ala-NH-Me; Alanine residue; Alanine motifs; Tripeptide; Ab initio; Molecular orbital computations

1. Preamble

In recent years serious efforts have been made by a large number of university researchers to develop non-empirical or ab initio (from the beginning) methods to compute PEHS from first principles (i.e. from quantum mechanics). Much of this research has been centered on deciphering the rules that govern protein folding. Whereby if we could pronounce the law by which proteins fold, we would be able to proceed with unprecedented speed and efficiency in answering medical questions and remove the many obstacles behind sophisticated drug discovery.

Since protein synthesis occurs sequentially, one amino acid at a time and there are 21 amino acids (Fig. 1), the job of using conformational analysis to determine the most energetically stable and most probable conformations of all 21 amino acids is a monumental task. Research is already taking place to construct a database for all 21 amino acids and their

* Corresponding author. Address: Global Institute of COmputational Molecular and Materials Science (GIOCOMMS), 1422 Edenrose St., Mississiauga, Ont., Canada L5V 1H3.

E-mail addresses: michelle.sahai@utoronto.ca (M.A. Sahai), msahai@giocomms.org (M.R. Sahai), gchass@giocomms.org (G.A. Chass), penke@ovrisc.mdche.szote.u-szeged.hu (B. Penke), icsizmad@chem.utoronto.ca (I.G. Csizmadia).

0166-1280/$ - see front matter © 2003 Elsevier B.V. All rights reserved.
doi:10.1016/j.theochem.2003.08.041

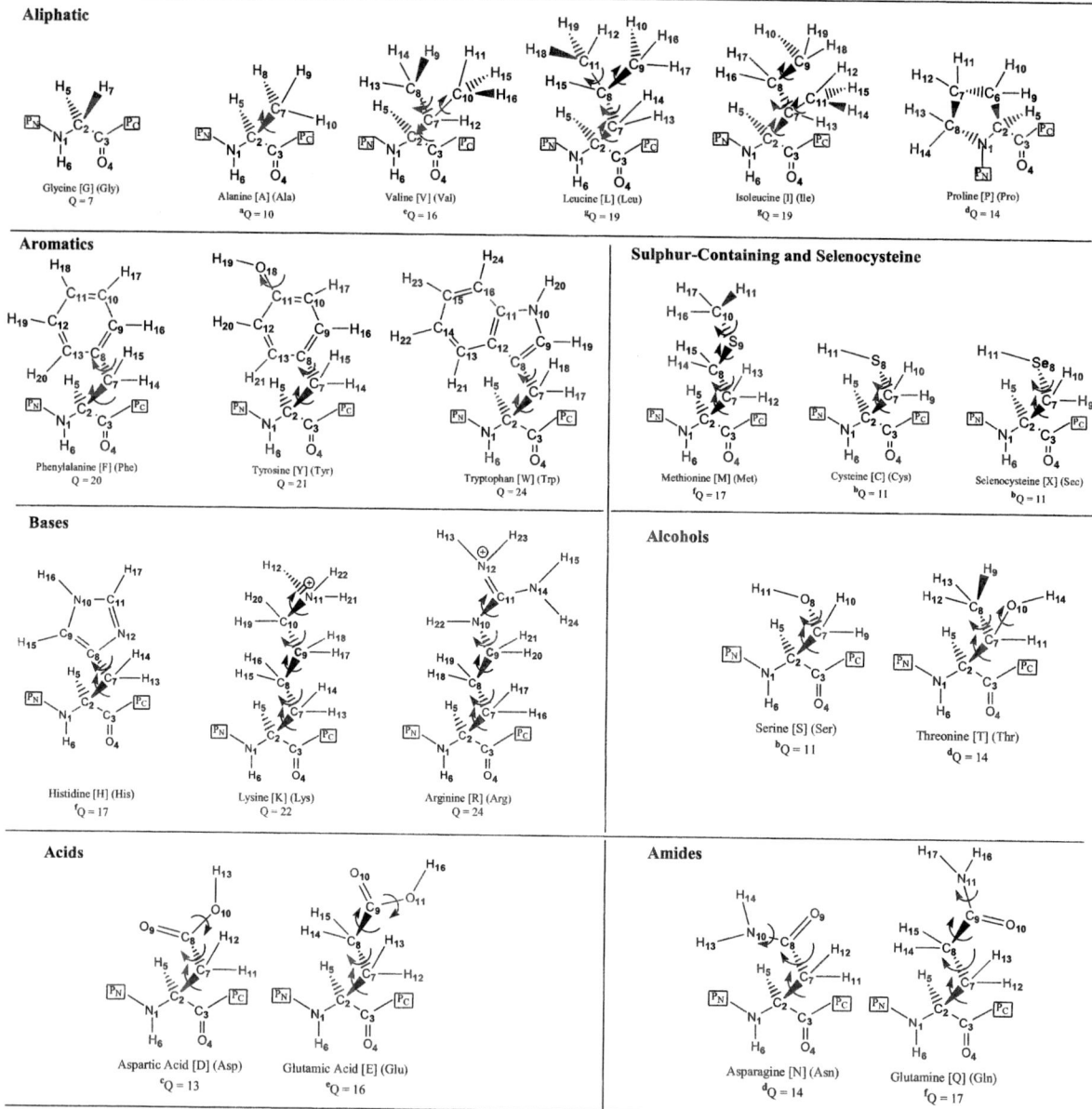

Fig. 1. A schematic illustration of amino acid residues. Note that in addition to the 20 naturally occurring amino acids, which have both DNA and RNA codons, the 21st amino acid, Selenocysteine (Sec), which has only an RNA codon, is also included.

backbone conformations which are 'found' [1–4]. This research aims at using conformational analysis to determine the most energetically stable and most probable conformation of one of these amino acids, thereby contributing to the database already in progress.

2. Introduction

Most of the oligopeptides studied so far at the ab initio level of theory involved alanine. All backbone conformers of dialanine diamides have been studied [5–9]. Analogously trialanine diamides (tripeptide)

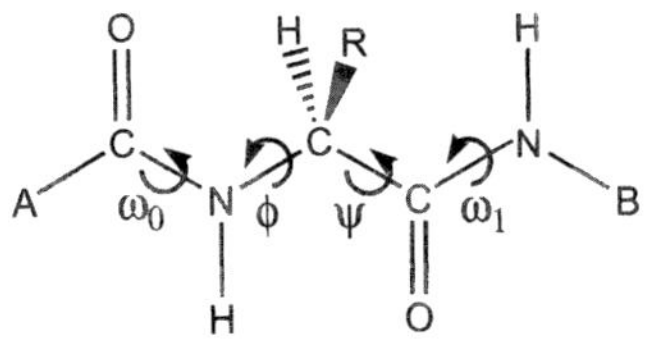

Scheme 1. The conformational structure of diamides of single amino acids.

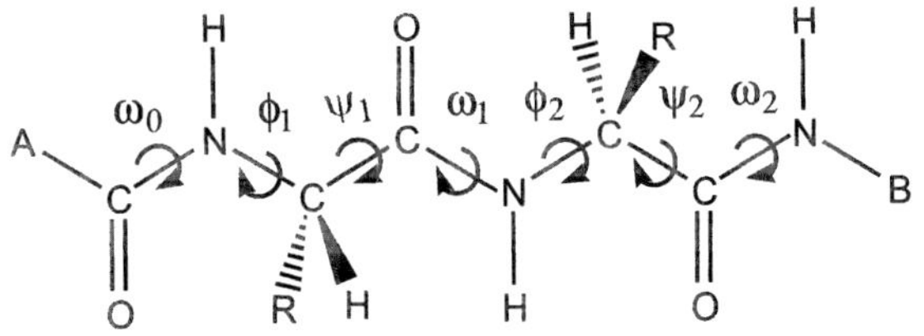

$$E = E(\phi_1, \psi_1, \phi_2, \psi_2)$$

Scheme 3. The conformational structure of diamides of dipeptides.

[10] and tetra-alanine diamides [11–14] have also been investigated at the ab initio level, but only a selected few conformers were studied. In this paper, we wish to illustrate in the case of the tripeptide Ala–Ala–Ala the conformational features as a tripeptide motif Xxx–Yyy–Zzz.

The conformations of diamides of single amino acids (Scheme 1) where A and B may be CH_3 or H is utilized for the 2D Ramachandran map illustrated by Scheme 2. From Scheme 1, in which ϕ and ψ run between 0° and 360°, it can be seen that a monopeptide has nine legitimate minima.

As the chain of the peptide elongates, the complexity increases and the dimensions of the map increases as well. In order to describe diamides of dipeptides (Scheme 3.) where A and B may be CH_3 or H, a 4D map is required which is expected to have $9^2 = 81$ legitimate minima. This is shown in Scheme 4.

In the case of tripeptide derivatives (Scheme 5) where A and B may be CH_3 or H, each of the 81 dipeptide conformations have nine conformers for the third amino acid. This results in $9 \times 9 \times 9 = 729$ conformers. Thus they may be described by a 6D-Ramachandran map (Fig. 2), which contains the 729 legitimate conformations denoted in terms of subscripted Greek letters.

In this paper, we wish to examine 25 possible conformers of the tripeptide MeCO-L-Ala-L-Ala-L-Ala-NH-Me. These 25 conformers may be expected to

exist because of the different conformational studies using the β_L conformation as the deciding factor, whereby, in any particular instant two alanine residues are fixed at the β_L conformation and the third was varied for the nine possible minima present on the Ramachandran map. Up to nine minima may be expected in each of these three scans. This would lead to the $3 \times 9 = 27$ possible conformations for β_L. However, the fully extended $\beta_L \beta_L \beta_L$ conformer would occur three times, so only 25 unique conformers are to be considered. Of course, it is possible that some conformers will be annihilated and less than 25 conformers would be found. Schematic representations of the 25 possible conformers to be considered and their relative proximity to the aforementioned 729 conformers given in subscripted Greek letters can be seen in bold in Fig. 2. The 25 possible conformers to be considered are also listed below for convenience

$$
\begin{array}{lll}
\alpha_L \beta_L \beta_L & \beta_L \beta_L \alpha_L & \beta_L \alpha_L \beta_L \\
\alpha_D \beta_L \beta_L & \beta_L \beta_L \alpha_D & \beta_L \alpha_D \beta_L \\
\beta_L \beta_L \beta_L & \beta_L \beta_L \delta_L & \beta_L \delta_L \beta_L \\
\delta_L \beta_L \beta_L & \beta_L \beta_L \delta_D & \beta_L \delta_D \beta_L \\
\delta_D \beta_L \beta_L & \beta_L \beta_L \varepsilon_L & \beta_L \varepsilon_L \beta_L \\
\varepsilon_L \beta_L \beta_L & \beta_L \beta_L \varepsilon_D & \beta_L \varepsilon_D \beta_L \\
\varepsilon_D \beta_L \beta_L & \beta_L \beta_L \gamma_L & \beta_L \gamma_L \beta_L \\
\gamma_L \beta_L \beta_L & \beta_L \beta_L \gamma_D & \beta_L \gamma_D \beta_L \\
\gamma_D \beta_L \beta_L & &
\end{array}
\tag{1}
$$

The names of the minima are written in Greek letters. The Greek letters originate from earlier nomenclature (involving α, β and γ) while the L and D subscripts originate from the observation that L-amino acids favour L conformations while D-amino acids favour D conformations. The names also suggest

γ_D	δ_D	α_L
ε_D	β_L	ε_L
α_D	δ_L	γ_L

$$E = E(\phi, \psi)$$

Scheme 2. Minima on the 2D-Ramachandran map of diamides of single amino acid.

$\gamma_D\ \gamma_D$	$\gamma_D\ \delta_D$	$\gamma_D\ \alpha_L$	$\delta_D\ \gamma_D$	$\delta_D\ \delta_D$	$\delta_D\ \alpha_L$	$\alpha_L\ \gamma_D$	$\alpha_L\ \delta_D$	$\alpha_L\ \alpha_L$
$\gamma_D\ \varepsilon_D$	$\gamma_D\ \beta_L$	$\gamma_D\ \varepsilon_L$	$\delta_D\ \varepsilon_D$	$\delta_D\ \beta_L$	$\delta_D\ \varepsilon_L$	$\alpha_L\ \varepsilon_D$	$\alpha_L\ \beta_L$	$\alpha_L\ \varepsilon_L$
$\gamma_D\ \alpha_D$	$\gamma_D\ \delta_L$	$\gamma_D\ \gamma_L$	$\delta_D\ \alpha_D$	$\delta_D\ \delta_L$	$\delta_D\ \gamma_L$	$\alpha_L\ \alpha_D$	$\alpha_L\ \delta_L$	$\alpha_L\ \gamma_L$
$\varepsilon_D\ \gamma_D$	$\varepsilon_D\ \delta_D$	$\varepsilon_D\ \alpha_L$	$\beta_L\ \gamma_D$	$\beta_L\ \delta_D$	$\beta_L\ \alpha_L$	$\varepsilon_L\ \gamma_D$	$\varepsilon_L\ \delta_D$	$\varepsilon_L\ \alpha_L$
$\varepsilon_D\ \varepsilon_D$	$\varepsilon_D\ \beta_L$	$\varepsilon_D\ \varepsilon_L$	$\beta_L\ \varepsilon_D$	$\beta_L\ \beta_L$	$\beta_L\ \varepsilon_L$	$\varepsilon_L\ \varepsilon_D$	$\varepsilon_L\ \beta_L$	$\varepsilon_L\ \varepsilon_L$
$\varepsilon_D\ \alpha_D$	$\varepsilon_D\ \delta_L$	$\varepsilon_D\ \gamma_L$	$\beta_L\ \alpha_D$	$\beta_L\ \delta_L$	$\beta_L\ \gamma_L$	$\varepsilon_L\ \alpha_D$	$\varepsilon_L\ \delta_L$	$\varepsilon_L\ \gamma_L$
$\alpha_D\ \gamma_D$	$\alpha_D\ \delta_D$	$\alpha_D\ \alpha_L$	$\delta_L\ \gamma_D$	$\delta_L\ \delta_D$	$\delta_L\ \alpha_L$	$\gamma_L\ \gamma_D$	$\gamma_L\ \delta_D$	$\gamma_L\ \alpha_L$
$\alpha_D\ \varepsilon_D$	$\alpha_D\ \beta_L$	$\alpha_D\ \varepsilon_L$	$\delta_L\ \varepsilon_D$	$\delta_L\ \beta_L$	$\delta_L\ \varepsilon_L$	$\gamma_L\ \varepsilon_D$	$\gamma_L\ \beta_L$	$\gamma_L\ \varepsilon_L$
$\alpha_D\ \alpha_D$	$\alpha_D\ \delta_L$	$\alpha_D\ \gamma_L$	$\delta_L\ \alpha_D$	$\delta_L\ \delta_L$	$\delta_L\ \gamma_L$	$\gamma_L\ \alpha_D$	$\gamma_L\ \delta_L$	$\gamma_L\ \gamma_L$

Scheme 4. Minima on the 4D-Ramachandran map of dipeptide diamides.

the combination of the chirality of a constitutional structure (R and S configuration) and the chirality of the conformational twist or folding.

Tripeptides were chosen as an ideal model because they allow all N-acetyl and N-methyl interactions and resultant conformations possible with a central peptide to be evaluated. With the completion of a tripeptide study, any of the other $20 \times 20 \times 20 = 8000$ tripeptides could be studied as well by binding it to a previously scanned peptide. The optimizations of this 'fused' hexapeptide would be optimized much more quickly and efficiently than a single calculation of a hexapeptide. The reason behind this is due to the fact that the internal structure of each tripeptide would not change radically after binding to a second one [2–4]. This idea is supported by the modular N- and C-protecting groups approach whereby N-acetyl and N-methyl groups mimic the alpha carbons of the neighboring amino acids in a polypeptide chain [2,3]. Thus, these relatively quick 'block' calculations could be used to give further insight quickly and efficiently

into the most stable geometry of much larger polypeptide chains.

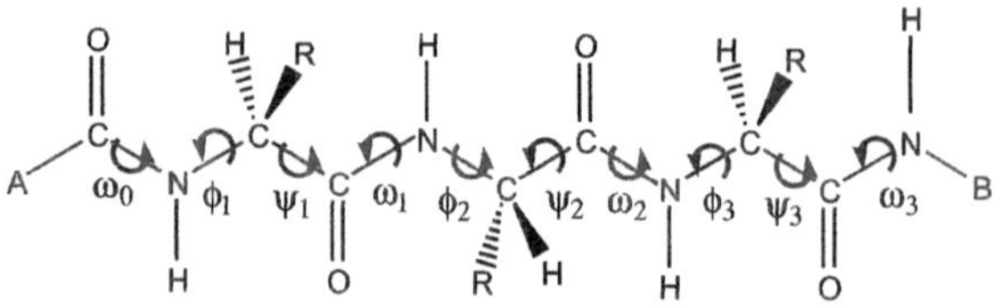

3. Computational methods

3.1. Practical considerations

Peptide and protein conformations are usually studied by empirical methods due to the large dimensionality of the problem. As expected on the basis of qualitative multidimensional conformational analysis, nine minima, associated with the typical backbone conformations (labeled α_L, α_D, β_L, γ_L, γ_D, δ_L, δ_D, ε_L, ε_D), were found. In this work, ab initio molecular orbital calculations were performed using the GAUSSIAN 98 program [15] at the RHF/3-21G basis set level.

For the automation of the input generation and subsequent data extraction from the output, the numbering of the atomic sequence in the peptides was standardized. In this way, scripts and programs may be used to immediately recognize patterns necessary for one's chemical analysis [1–4].

The molecule was numbered into and along the peptide backbone (N–C_α–CO), for each peptide 'unit' separately (Fig. 3) [3]. By this, a 'modular' numbering system was formed that may be truncated at any time, at any of the units comprising it, to model the structure of a larger polypeptide. The N-acetyl or

Scheme 5. The conformational structure of diamides of tripeptides.

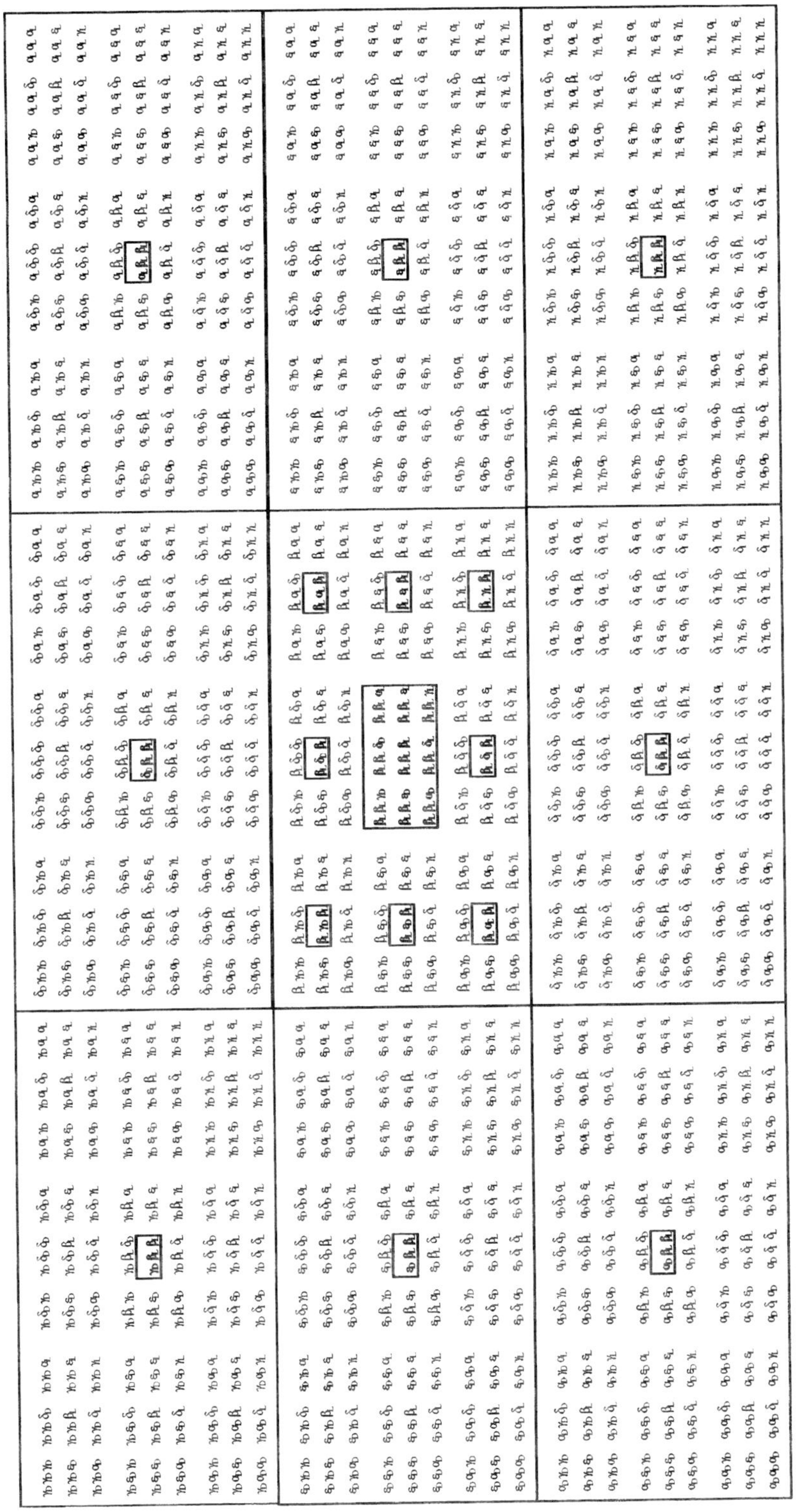

Fig. 2. Names, in terms of subscripted Greek letters, of the 729 legitimate backbone conformers of tripeptides and the relative proximity of the 25 conformers to be considered in this study can be seen in bold.

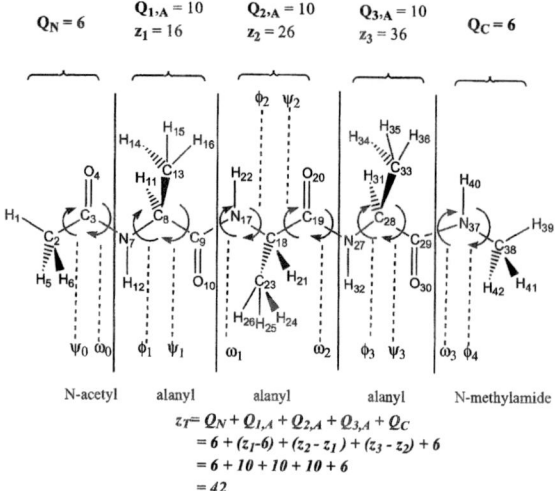

$$z_T = Q_N + Q_{1,A} + Q_{2,A} + Q_{3,A} + Q_C$$
$$= 6 + (z_1 - 6) + (z_2 - z_1) + (z_3 - z_2) + 6$$
$$= 6 + 10 + 10 + 10 + 6$$
$$= 42$$

Dihedral Definitions

$\psi_0 = D7\ (7,3,2,1)$

$\omega_0 = D8\ (8,7,3,2)$

$\phi_1 = D9\ (9,7,3,2)$

$\psi_1 = D17\ (17,9,8,7)$

$\omega_1 = D18\ (18,17,9,8)$

$\phi_2 = D19\ (19,17,9,8)$

$\psi_2 = D27\ (27,19,18,17)$

$\omega_2 = D28\ (28,27,19,18)$

$\phi_3 = D29\ (29,28,27,19)$

$\psi_3 = D37\ (37,29,28,27)$

$\omega_3 = D38\ (38,37,29,28)$

$\psi_4 = D39\ (39,38,37,29)$

Fig. 3. Schematic of the standardized numbering system applied to the tripeptide MeCO-L-Ala-Ala-L-Ala-NH-Me, showing all backbone torsional angles. This figure is taken from Ref. [3].

N-methyl groups are defined as separate units and can be added at any time, as their coordinates do not disturb the definitions of the internal parameters specific to the core. As a result, each amino acid residue, the *N*-acetyl and *N*-methylamide end protecting groups are exclusively defined using the *z*-matrix internal coordinate system to characterize molecular structure, stereochemistry and geometry (see Fig. 4) [2,3]. The resultant input file was used as a starting point in a successive and iterative process of GAUSSIAN 98 cycles to bring about an optimization at the predicted minima, using the RHF/3-21G level of theory.

The initial step for generating PES for MeCO-L-Ala-L-Ala[β_L]-L-Ala-NH-Me was a double scan about ϕ_1 and ψ_1 for the alanine unit being varied, with the other alanine residues maintained in the β_L conformation (Fig. 5). A double scan was performed about the torsional angles ϕ_3 and ψ_2 for the right terminal alanine of MeCO-L-Ala[β_L]-L-Ala[β_L]-L-Ala-NH-Me (β_L-β_L-Scan), while keeping the left terminal alanine at β_L (Fig. 6). This scan was produced using 30.0° increments, resulting in a total of $12 \times 12 = 144$ points. However, if both 0 and 360° are included, defining an entire period of torsion, then a total of $13 \times 13 = 169$ points are available for plotting. The result obtained from 144 points is satisfactory because the 0° and 360° points are degenerate.

3.2. Theoretical considerations

It has been noted that the number of identical bond types is conserved during an isodesmic reaction. To be able to compare the conformational energy of peptides with different sidechains, we have proposed the use of an isodesmic reaction, which can be represented by the following equation:

$$MeCO-(NH-CH_2-CO)_n-NH-Me + n(R-CH_3)$$
$$\rightarrow MeCO-(NH-CHR-CO)_n-NH-Me + n(CH_4)$$
$$(2)$$

Eq. (2) represents an isodesmic reaction that compares the relative effects of changing a substituent (R) on a peptide backbone to that of a simple alkane (R–CH$_3$). If the calculated ΔE_{ID} for this reaction is greater than zero, it would indicate that the substituent prefers to be on the alkane rather than on the peptide. On the other hand, if the ΔE_{ID} is less than zero, then the substituent will prefer to stabilize the peptide relative to the alkane. For the tripeptide model MeCO-L-Ala-Ala-L-Ala-NH-Me the isodesmic reaction is defined by Eq. (3):

$$MeCO-(NH-CH_2-CO)_3-NH-Me + 3(CH_3-CH_3)$$
$$\rightarrow MeCO-(NH-CHR-CO-NH-CHR-CO$$
$$-NH-CHR-CO)-NH-Me + 3(CH_4) \quad (3)$$

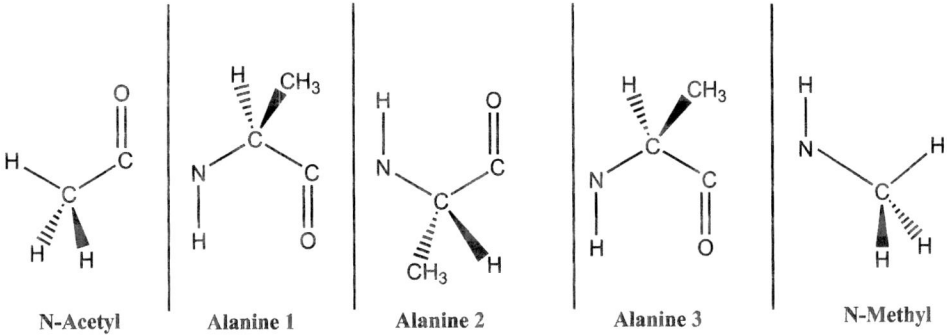

Fig. 4. Modularity of the numbering system allows for easy separation of the different components of the tripeptide.

where $R = CH_3$. The corresponding isodesmic reaction/stabilization energy is shown by Eq. (4):

$$\Delta E_{ID} = E_{products} - E_{reactants} = 3E(CH_4) + E[MeCO$$
$$-(NH-CHR-CO-NH-CHR-CO-NH$$
$$-CHR-CO)-NH-Me] - 3E(R-CH_3)$$
$$- E[MeCO-(NH-CH_2-CO)_3-NH-Me]$$
$$(4)$$

where $R = CH_3$. In Eq. (4), $E(CH_4)$, $E(R-CH_3)$, and $E(MeCO-(NH-CH_2-CO)_3-NH-Me)$ are the energy values in their global minimum energy conformation.

4. Results and discussions

The tabulated torsional angles ω_i, ϕ_i, ψ_i for the 19 legitimate conformers of MeCO-Ala-Ala-Ala-NH-Me are given in Table 1. In general, the α_L and ε_L conformer for any particular alanine residue of

the tripeptide was annihilated, thereby eliminating six conformations of the possible 25 structures.

Based on the study of MeCO-Ala[β_L]-Ala[β_L]-Ala[β_L]-NH-Me, we can predict that all torsional angles should have ideal values of 180°, as well as the two C–H bonds of the central alanine should have equal lengths since the model is symmetrical. Thus it would be interesting to compare the result with the present MeCO-L-Ala[β_L]-L-Ala[β_L]-L-Ala[β_L]-NH-Me to see how the C–H bonds of the central alanine is affected by having terminal alanine residues.

Furthermore, isodesmic reactions will be used to determine the stabilization exerted by the two methyl side chains on the backbone.

Ala–Ala–Ala as prototypical tripeptide for Xxx–Yyy–Zzz. Most amino acids carry a substituent in the $-CH_3$ side chain of alanine. Consequently Ala–Ala–Ala is a prototype of a general tripeptide Xxx–Yyy–Zzz. Glycine has no constituent attached to its α-carbon, yet it is still one of the 20 naturally

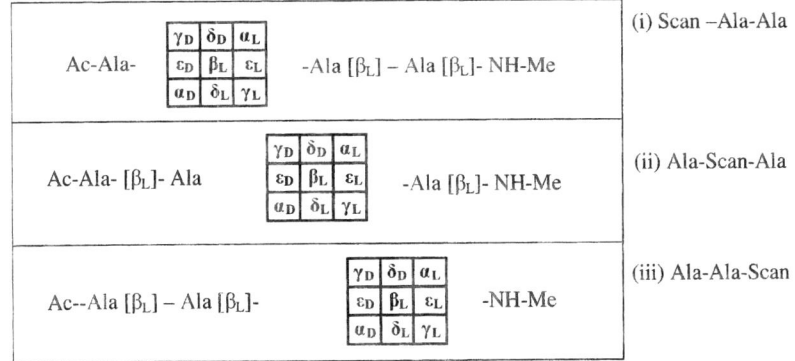

Fig. 5. Schematic of the conformations under study for the tripeptide MeCO-L-Ala-L-Ala-L-Ala-NH-Me.

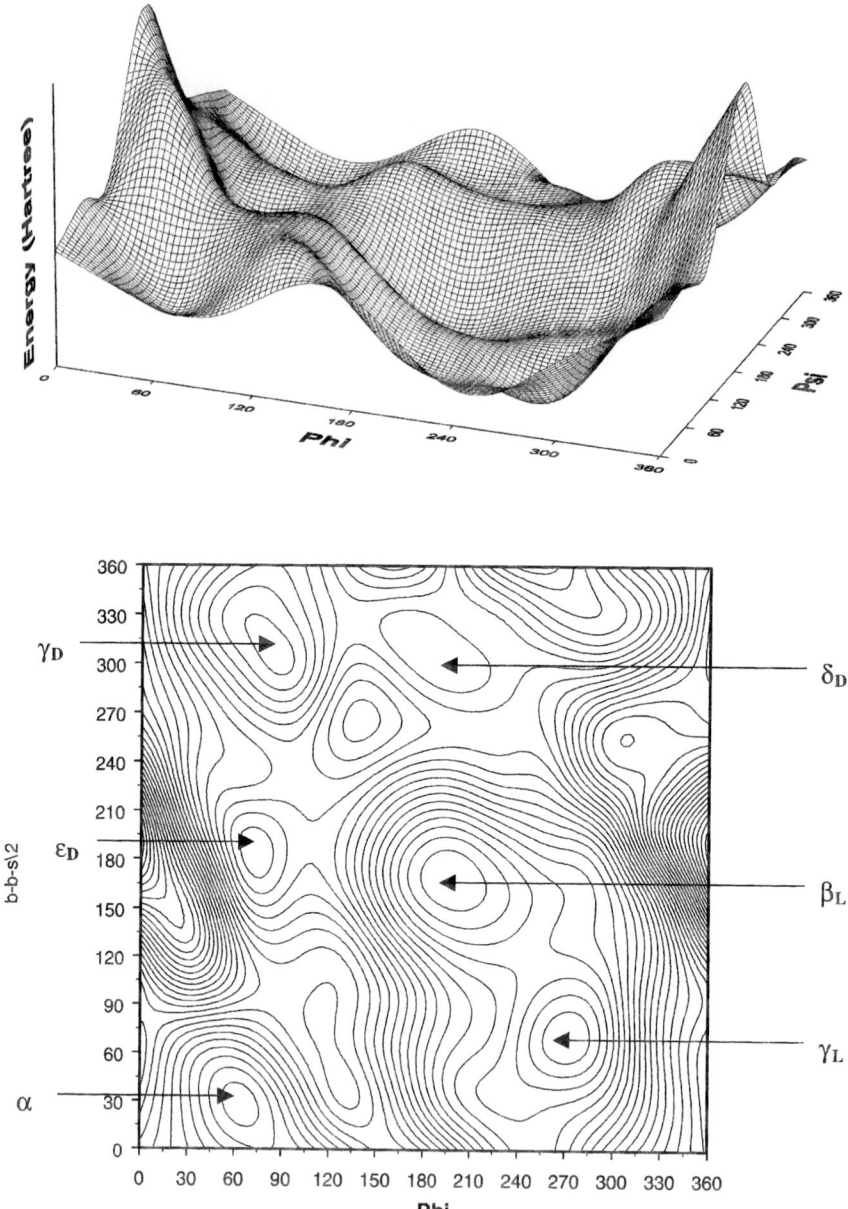

Fig. 6. Ramachandran potential energy hyper-surface landscape (top) and contour map (bottom) of MeCO-L-Ala [β_L]-L-Ala[β_L]-L-Ala-NH-Me.

occurring amino acids. Therefore, there can be 20 possible amino acids for Xxx and another 20 amino acids for Yyy, as well as for Zzz. As a result, Ala–Ala–Ala is one of the $20 \times 20 \times 20 = 8000$ possible tripeptide motifs and therefore it can be the model of any kind of tripeptide. The backbone conformations of these general tripeptides are expected to be analogous to that of Ala–Ala–Ala. As an example, the tripeptide Arg–Gly–Asp (RGD) is discussed in Refs. [1,17].

Stability of tripeptides. The stability of tripeptides is an important question to consider. The computed values in Table 1 are as expected on the basis of previously reported data [18].

Table 1
Optimized dihedral angles of MeCO-Ala-Ala-Ala-NH-Me as well as tabulated energies (Hartree and kcal/mole) and stabilization energies

Found conformation	Dihedral angles															Energy (Hartrees)	Relative energy ΔE (kcal mol^{-1})	Stabilization energy ΔE_{BD} (kcal mol^{-1})
	ψ_0 D7	ω_0 D8	ϕ_1 D9	ψ_1 D17	ω_1 D18	ϕ_2 D19	ψ_2 D27	ω_2 D28	ϕ_3 D29	ψ_3 D37	ω_3 D38	ϕ_4 D39	χ_1^1 D14	χ_1^2 D24	χ_1^3 D34			
$\beta_L\beta_L\beta_L$	179.72	178.82	-167.43	170.70	177.32	-167.78	170.77	177.29	-167.60	169.62	177.92	0.06	179.47	179.25	179.21	-979.093665	0.00	-12.97
$\beta_L\beta_L\gamma_L$	179.75	178.85	-167.03	170.51	177.29	-166.96	169.05	-177.96	-85.75	70.08	-178.67	-5.32	179.50	179.37	-177.31	-979.093402	0.17	-12.80
$\beta_L\beta_L\gamma_D$	179.84	178.88	-167.03	170.60	177.50	-166.46	168.14	172.19	76.03	-58.20	-178.41	0.00	179.53	179.33	-177.54	-979.089028	2.91	-10.06
$\beta_L\beta_L\delta_L$	179.92	178.93	-167.03	170.57	177.08	-167.26	171.22	-175.49	-128.27	28.48	176.39	3.71	179.55	179.93	177.17	-979.088315	3.36	-9.61
$\beta_L\beta_L\delta_D$	-179.91	178.83	-167.30	170.64	177.46	-166.76	168.92	173.53	-173.99	-47.27	-177.01	-15.13	179.51	179.59	178.08	-979.082825	6.80	-6.17
$\beta_L\beta_L\alpha_D$	179.68	178.79	-167.27	170.75	177.58	-166.82	168.17	172.06	62.58	34.08	178.71	-5.84	179.56	179.45	-179.57	-979.085096	5.38	-7.59
$\beta_L\beta_L\varepsilon_D$	-179.95	179.05	-167.22	170.32	176.30	-167.85	172.22	-170.74	65.62	-176.22	-179.11	0.83	179.29	179.16	-170.32	-979.082531	6.99	-5.98
$\beta_L\gamma_L\beta_L$	-178.87	179.14	-166.82	170.49	-179.29	-87.25	69.53	-179.16	-166.31	167.07	177.66	0.38	179.85	-177.47	179.11	-979.090290	2.12	-10.85
$\beta_L\gamma_D\beta_L$	179.87	178.08	-166.31	165.72	173.97	76.12	-51.59	-179.29	-162.18	167.97	177.60	-0.28	179.22	-176.46	179.19	-979.086218	4.67	-8.30
$\beta_L\delta_L\beta_L$	-179.35	-179.92	-164.61	162.64	179.09	-125.54	15.81	178.59	-173.61	169.54	177.98	0.47	176.37	176.61	179.50	-979.089972	2.32	-10.65
$\beta_L\delta_D\beta_L$	179.57	178.79	-167.46	169.84	175.07	-171.10	-47.42	-179.25	-168.25	169.03	177.74	0.06	179.76	178.44	179.19	-979.084203	5.94	-7.03
$\beta_L\alpha_D\beta_L$	178.84	178.42	-167.73	169.82	173.59	60.17	34.61	-179.09	-172.66	168.96	177.69	0.03	179.74	179.84	177.22	-979.087059	4.15	-8.82
$\beta_L\varepsilon_D\beta_L$	178.48	177.73	-167.49	171.94	-172.29	65.34	-175.26	179.99	-167.00	169.25	178.07	-0.19	179.35	-171.81	179.27	-979.083407	6.44	-6.53
$\gamma_L\beta_L\beta_L$	179.18	-176.71	-87.65	71.22	-179.58	-164.44	168.78	177.14	-167.33	169.37	177.84	0.22	-177.29	179.42	179.34	-979.090598	1.92	-11.04
$\gamma_D\beta_L\beta_L$	-178.96	175.54	76.40	-51.50	-178.58	-162.11	169.62	176.73	-167.04	169.12	177.92	0.09	-176.72	179.34	179.31	-979.085819	4.92	-8.04
$\delta_L\beta_L\beta_L$	178.68	-176.72	-121.14	18.02	176.32	-169.53	170.04	177.68	-167.66	169.71	177.85	0.03	176.67	179.42	179.31	-979.090062	2.26	-10.71
$\delta_D\beta_L\beta_L$	176.76	176.24	-174.51	-44.60	-178.49	-167.86	170.74	176.95	-167.42	169.53	177.92	0.25	178.27	179.27	179.09	-979.084795	5.57	-7.40
$\alpha_D\beta_L\beta_L$	-178.15	176.85	60.13	32.70	-177.20	-175.22	169.38	176.82	-167.37	169.29	177.95	-0.04	179.33	176.30	179.28	-979.087617	3.80	-9.17
$\varepsilon_D\beta_L\beta_L$	-169.82	-167.02	63.77	-171.11	-179.82	-166.67	170.72	177.71	-167.48	169.65	177.85	0.10	-170.91	179.31	179.31	-979.083463	6.40	-6.57

5. Conclusion

The results of the present work clearly indicate that at the ab initio RHF/3-21G level of theory, the behaviour of the central alanine is influenced by the backbone conformations of the two adjacent alanines. Following the footsteps of previous works on tripeptides [16,17], it is hoped that the results found herein is possible to continue the study on all 8000 possible tripeptides. With the use of a standardized numbering system [2–4], it would then be possible to expand the number of tripeptide studies in a shorter period of time and at a lower basis set (RHF/3-21G), reducing computational power.

Acknowledgements

This work was supported by grants from the Global Institute of Computational Molecular and Materials Science (GIOCOMMS), Toronto, Ontario, Canada. The authors thank David H. Setiadi and Tania A. Pecora for helpful discussions during the preparation of this manuscript, as well as Graydon Hoare for database management, network support and software and distributive processing development. Special thanks is extended to Andrew M. Chasse and Christopher M. Andrews for their development of novel scripting and coding techniques, which facilitate a reduction in the number of CPU cycles needed. The pioneering advances of Kenneth P. Chasse, in all composite computer-cluster software and hardware architectures, are also acknowledged. One of the authors (IGC) wishes to thank the Ministry of Education for a Szent-Györgyi Visiting Professorship.

References

[1] M.A. Berg, G.A. Chasse, E. Deretey, A.K. Fuezery, B.M. Fung, D.Y.K. Fung, H. Henry-Riyad, A.C. Lin, M.L. Mak, A. Mantas, M. Patel, I.V. Repyakh, M. Staikova, S.J. Salpietro, T.-H. Tang, J.C. Vank, A. Perczel, G.I. Csonka, O. Farkas, L.L. Torday, Z. Szekely, I.G. Csizmadia, J. Mol. Struct. (THEOCHEM) 500 (2000) 5–58.

[2] G.A. Chass, M.A. Sahai, J.M.S. Law, S. Lovas, Ö. Farkes, A. Perczel, J.-L. Rivail, I.G. Csizmadia, Int. J. Quantum Chem. 90 (2002) 933–968.

[3] M.A. Sahai, S. Lovas, G.A. Chass, B. Penke, I.G. Csizmadia, J. Mol. Struct. (THEOCHEM) (in this issue).

[4] G.A. Chass, J. Mol. Struct. (THEOCHEM) (in this issue).

[5] A. Perczel, M. Kajtar, J.F. Marcoccia, G. Csizmadia, J. Mol. Struct. (THEOCHEM) 232 (1991) 291–319.

[6] A. Perczel, G. Csizmadia, J. Mol. Struct. (THEOCHEM) 286 (1993) 75–85.

[7] C. Van Alsenoy, M. Cao, S. Newton, B. Teppen, A. Perczel, I.G. Csizmadia, F.A. Monramy, L. Schäfer, J. Mol. Struct. (THEOCHEM) 286 (1993) 149–163.

[8] M.A. McAllister, A. Perczel, P. Császár, G. Csizmadia, J. Mol. Struct. (THEOCHEM) 288 (1993) 181–198.

[9] A. Perczel, M.A. McAllister, P. Császár, G. Csizmadia, Can. J. Chem. 72 (1994) 2050–2070.

[10] M. Cheung, M.E. McGovern, T. Jin, D.C. Zhao, M.A. McAllister, A. Perczel, P. Császár, G. Csizmadia, J. Mol. Struct. (THEOCHEM) 309 (1994) 151–224.

[11] G. Endrédi, C.-M. Liegner, M.A. McAllister, A. Perczel, J. Ladik, G. Csizmadia, J. Mol. Struct. (THEOCHEM) 306 (1994) 1–7.

[12] G. Endrédi, M.A. McAllister, A. Perczel, P. Császár, J. Ladik, G. Csizmadia, J. Mol. Struct. (THEOCHEM) 331 (1995) 5–10.

[13] M.A. McAllister, G. Endrédi, W. Viviani, A. Perczel, P. Császár, J. Ladik, J.-L. Rivail, G. Csizmadia, Can. J. Chem. 73 (1995) 1563–1572.

[14] G. Endrédi, M.A. McAllister, Ö. Farkas, A. Perczel, J. Ladik, G. Csizmadia, J. Mol. Struct. (THEOCHEM) 331 (1995) 11–26.

[15] M.J. Frisch, G.W. Trucks, H.B. Schlegel, G.E. Scuseria, M.A. Robb, J.R. Cheeseman, V.G. Zakrzewski, J.A. Montgomery, Jr., R.E. Stratmann, J.C. Burant, S. Dapprich, J.M. Millam, A.D. Daniels, K.N. Kudin, M.C. Strain, Ö. Farkas, J. Tomasi, V. Barone, M. Cossi, R. Cammi, B. Mennucci, C. Pomelli, C. Adamo, S. Clifford, J. Ochterski, G.A. Petersson, P.Y. Ayala, Q. Cui, K. Morokuma, P. Salvador, J.J. Dannenberg, D.K. Malick, A.D. Rabuck, K. Raghavachari, J.B. Foresman, J. Cioslowski, J.V. Ortiz, A.G. Baboul, B.B. Stefanov, G. Liu, A. Liashenko, P. Piskorz, I. Komaromi, R. Gomperts, R.L. Martin, D.J. Fox, T. Keith, M.A. Al-Laham, C.Y. Peng, A. Nanayakkara, M. Challacombe, P.M.W. Gill, B. Johnson, W. Chen, M.W. Wong, J.L. Andres, C. Gonzalez, M. Head-Gordon, E.S. Replogle, J.A. Pople, GAUSSIAN 98, Revision A.9, Gaussian, Inc., Pittsburgh, PA, 2001.

[16] A. Mehdizadeh, G.A. Chass, Ö. Farkas, A. Perczel, L.L. Torday, A. Varro, G. Papp, J. Mol. Struct. (THEOCHEM) 588 (2002) 187–200.

[17] J.C.C. Liao, J.C. Chua, G.A. Chass, A. Perczel, L.L. Torday, G. Papp, J. Mol. Struct. (THEOCHEM) 621 (2003) 163–187.

[18] M.A. McAllister, G. Endrédi, W. Viviani, A. Perczel, P. Császár, J. Ladik, J.-L. Rivail, G. Csizmadia, Can. J. Chem. 73 (1995) 1563–1572.

THEO
CHEM

ELSEVIER   Journal of Molecular Structure (Theochem) 666–667 (2003) 337–344

www.elsevier.com/locate/theochem

Exploring the conformational space of the μ-opioid agonists endomorphin-1 and endomorphin-2

Balázs Leitgeb[a,*], András Szekeres[b]

[a]*Institute of Biochemistry, Biological Research Center of the Hungarian Academy of Sciences, Temesvári krt. 62, Szeged H-6726, Hungary*
[b]*Department of Microbiology, University of Szeged, P.O. Box 533, Szeged H-6701, Hungary*

Abstract

Endomorphins (EM1, H-Tyr-Pro-Trp-Phe-NH$_2$ and EM2, H-Tyr-Pro-Phe-Phe-NH$_2$) were isolated from bovine brain, and these flexible tetrapeptides are highly potent and selective ligands for the μ-opioid receptor. We have explored the $\Phi_2 - \Psi_2$ and $\Phi_2 - \Psi_1$ conformational spaces of all four types of EMs (neutral or charged N-terminal amino groups, *cis* or *trans* Tyr^1-Pro^2 peptide bonds) using the simulated annealing. The conformational distributions were illustrated in the three-dimensional Ramachandran plots and three-dimensional pseudo-Ramachandran plots, and the mainly populated $\Phi_2 - \Psi_2$ and $\Phi_2 - \Psi_1$ regions were determined. The main combinations of side-chain rotamers of the Tyr^1, Trp^3 and Phe^4 residues for EM1 and the Tyr^1, Phe^3 and Phe^4 residues for EM2 were identified in χ_1 conformational space.
© 2003 Elsevier B.V. All rights reserved.

Keywords: Endomorphin-1; Endomorphin-2; Simulated annealing; Three-dimensional Ramachandran plot; Three-dimensional pseudo-Ramachandran plot

1. Introduction

Endomorphin-1 (EM1, H-Tyr-Pro-Trp-Phe-NH$_2$) and endomorphin-2 (EM2, H-Tyr-Pro-Phe-Phe-NH$_2$) are endogenous μ-opioid agonists. These tetrapeptides display high affinity and selectivity for the μ-opioid receptor [1]. As the EMs have many significant biological effects, therefore it is important to determine their pharmacophore elements and bioactive conformation.

Different structures were suggested to preferred conformational state and the possible bioactive conformer of EM1 and EM2: a conformation with a bent structure for *trans* isomer [2], a conformer with an extended peptide backbone for both *cis* and *trans* isomers [3,4], a structure with a β-turn similar to type III (1 ← 4) [5], a conformation with a *cis* Tyr^1-Pro^2 peptide bond [6], and a conformer with a *trans* Tyr^1-Pro^2 peptide bond and an inverse γ-turn located in the N-terminal tripeptide fragment [7,8].

In this study, simulated annealing was used to explore the conformational space of EM1 and EM2. The $\Phi - \Psi$ conformational spaces were presented in the three-dimensional Ramachandran (3DR) and pseudo-Ramachandran (3DPR) plots [9,10] for the Pro residue in position 2. The preferred regions of these 3DR and 3DPR plots were identified. The analysis of side-chains of the aromatic amino acid residues was carried out, and the main combinations of rotamers were determined in χ_1 conformational space.

* Corresponding author. Tel.: +36-62-599-600; fax: +36-62-433-506.
  E-mail address: leitgeb@rosi.szbk.u-szeged.hu (B. Leitgeb).

0166-1280/$ - see front matter © 2003 Elsevier B.V. All rights reserved.
doi:10.1016/j.theochem.2003.08.043

2. Methods

Simulated annealing calculations were performed using the AMBER force field [11,12], the distance-dependent dielectric constant ($\varepsilon = 4.5r$), and no cutoff was applied for nonbonding interactions. The calculations were carried out with INSIGHT II DIS-COVER 3 software [13] on an Origin2000 workstation running the Irix 6.5 operation system (Silicon Graphics, Inc.).

For the starting structures with extended peptide backbones were applied energy minimization. The following parameters were used during the SA calculations: heating to 1000 K during 1000 fs, equilibration at 1000 K for 4000 fs, cooling to 50 K during 10,000 fs by an exponential cooling

protocol. After the dynamics stages of SA, a three-step minimization procedure was performed using the steepest descent (SD), conjugated gradient (CG, Polak-Ribiere) and a modified Newton method (BFGS) with convergence criteria of 1000, 10 and 0.001 kcal mol^{-1} Å^{-1}, respectively. The SA cycle and the final minimization were repeated 1000 times, resulting in 1000 conformers for each molecule.

EM1 and EM2 were modeled in the neutral (NH_2-) and cationic (NH_3^+-) forms of the N-terminal amino group. In the latter case, Cl^- was used as counter ion to neutralize the system, with a distance restraint of 3.5 Å measured from the N-terminal N atom.

The NMR investigations showed an equilibrium mixture of *cis* and *trans* Tyr^1-Pro^2 peptide bonds for

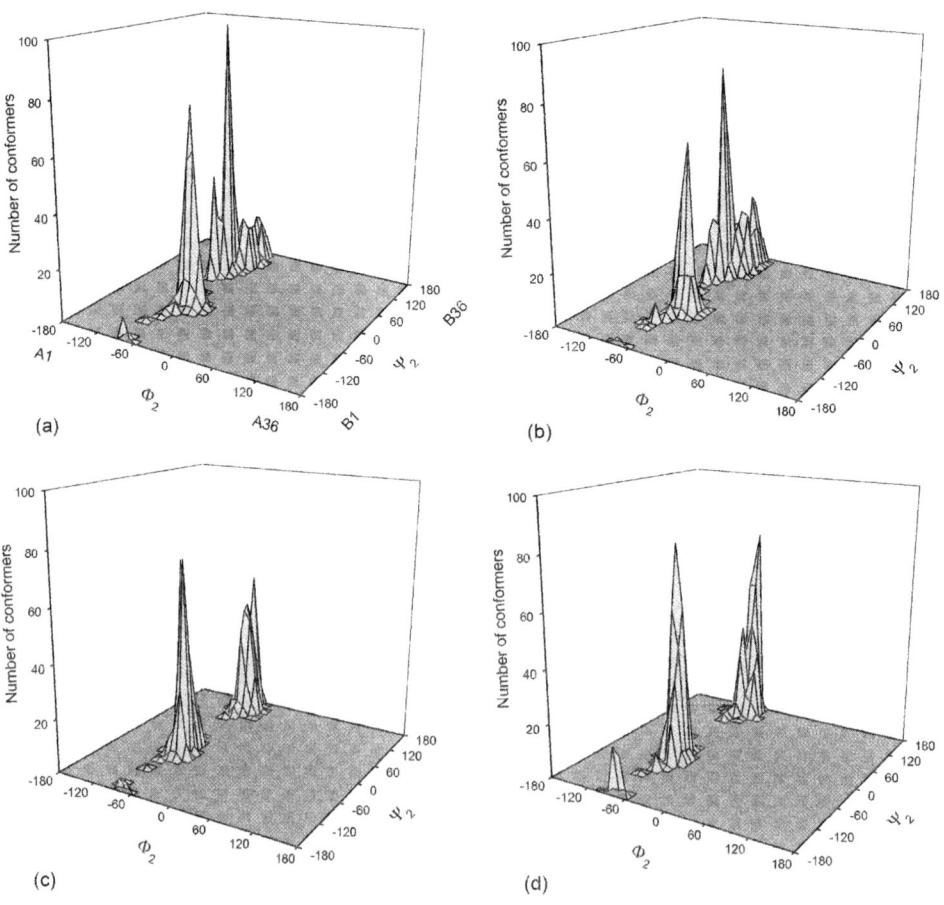

Fig. 1. The three-dimensional Ramachandran (3DR) plots of all four types of EM1 molecules (*cis/trans*, charged/uncharged). (a) *trans*-H$_2$N-EM1, (b) *trans*-H$_3$N$^+$-EM1, (c) *cis*-H$_2$N-EM1, (d) *cis*-H$_3$N$^+$-EM1.

EMs in solution [2–4], therefore simulations were performed with the omega torsion of this peptide bond set at either *cis* ($\omega = 0°$) or *trans* ($\omega = 180°$). All other peptide bonds were kept in the *trans* configuration.

3. Results

3.1. Three-dimensional Ramachandran plots

Conventional $\Phi_2-\Psi_2$, $\Phi_3-\Psi_3$ and $\Phi_4-\Psi_4$ Ramachandran plots of 1000 conformers for Pro[2], Trp[3] and Phe[4] of EM1 and Pro[2], Phe[3] and Phe[4] of EM2 were constructed previously [7,8]. These conformational distributions revealed significant differences only in the $\Phi_2-\Psi_2$ Ramachandran plots of *cis* and *trans*

isomers. In the $\Phi_3-\Psi_3$ and $\Phi_4-\Psi_4$ Ramachandran plots of both isomers similar distributions of the conformers were found. Therefore, the 3DR and 3DPR plots were prepared only for Pro[2].

For 1000 conformers of all four types of EM1 and EM2 molecules (*cis/trans*, charged/uncharged) 3DR plots were constructed using the Φ_2 and Ψ_2 torsion angles of Pro[2] (see Figs. 1 and 2). The ranges from -180 to $180°$ of Φ_2 and Ψ_2 torsion angles were divided up into $10°$ intervals, thus 36×36 regions were obtained in the $\Phi_2-\Psi_2$ conformational space. The numbers of conformers in all $\Phi_2-\Psi_2$ regions were identified, and then the three-dimensional plots of Φ_2 vs. Ψ_2 vs. number of conformers were produced. The 36 intervals of Φ_2 torsion angle were labeled A1–A36 across the range of -180 to $180°$,

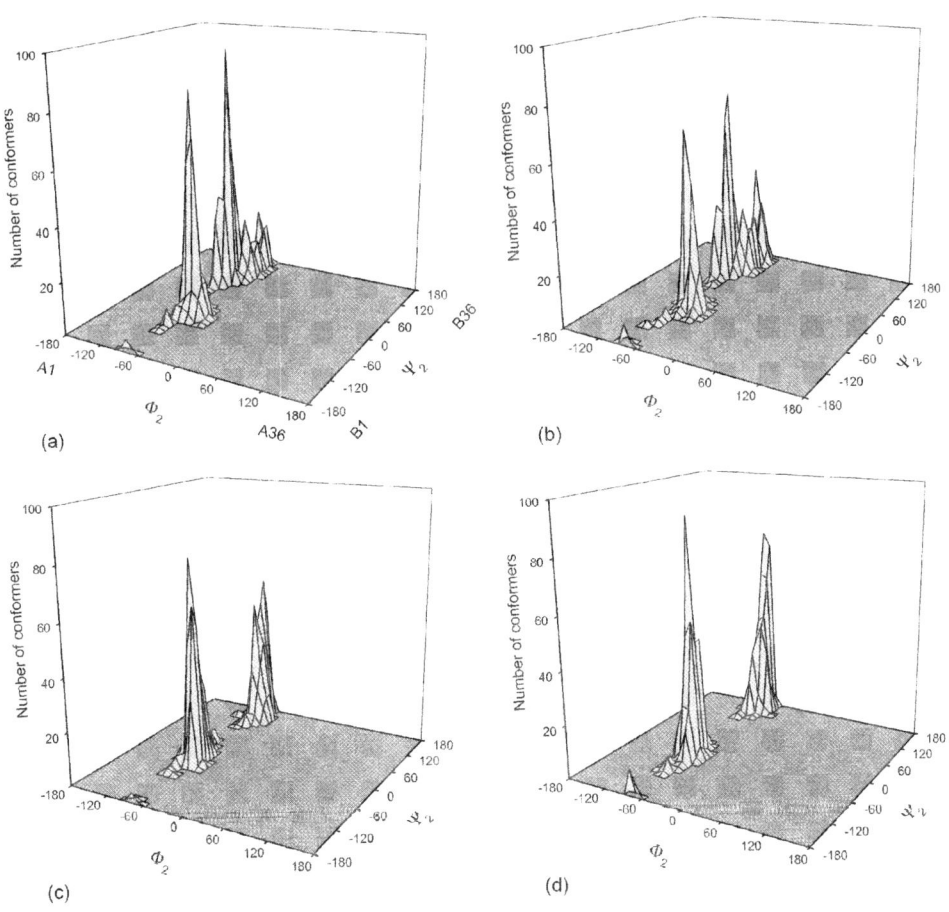

Fig. 2. The three-dimensional Ramachandran (3DR) plots of all four types of EM2 molecules (*cis/trans*, charged/uncharged). (a) *trans*-H$_2$N-EM2, (b) *trans*-H$_3$N$^+$-EM2, (c) *cis*-H$_2$N-EM2, (d) *cis*-H$_3$N$^+$-EM2.

and the intervals of Ψ_2 torsion angle were labeled B1–B36 covering the range from -180 to $180°$.

The main conformer-containing $\Phi_2 - \Psi_2$ regions were determined (see Table 1). The *trans* conformers mainly populated regions are A11–B26, A11–B27, A12–B14, A12–B15, and A13–B14. A significant amounts of *cis* conformers were found in regions A11–B14, A11–B15, A11–B32, A11–B35, A11–B36, A12–B14, A12–B15, A12–B35 and A12–B36. For all types of *trans*-EMs, the major populations of conformers were located in regions A11–B26 and A11–B27, while for neutral and charged forms of *cis*-EMs, the mainly populated regions were A11–B15 and A11–B35.

These preferred conformational regions are related to the possible secondary structural elements of EMs. Previously these secondary structures were determined [7,8]. In the conformers of *trans*-EMs

a significant amounts of β-turns of type III and N-terminal inverse γ-turns were found. In the *cis* conformers of EMs β-turns of type III were also observed, while the inverse γ-turns in the N-terminal tripeptide fragment were absent. The range of β-turns of type III overlap with regions A10-15–B13-18 in the $\Phi_2 - \Psi_2$ 3DR plots. In this range the following mainly populated $\Phi_2 - \Psi_2$ regions were found: A12–B14, A12–B15 and A13–B14 for *trans* isomers, and A11–B14, A11–B15, A12–B14 and A12–B15 for *cis* isomers. To determine the inverse γ-turns the following ranges of torsion angles were used: $-95° < \Phi_{i+2} < -70°$ and $45° < \Psi_{i+2} < 75°$. For the *trans*-EMs, this range containing only one of the preferred conformational regions (A11–B26), while the conformers of *cis*-EMs are completely absent in this range.

Table 1

The preferred $\Phi_2 - \Psi_2$ conformational regions of the three-dimensional Ramachandran plots

	Preferred conformational regions		
trans-H$_2$N-EM1	A11_B26 (100) A11_B27 (69)	A12_B14 (55) A12_B15 (75)	A13_B14 (59)
trans-H$_3$N$^+$-EM1	A11_B26 (85) A11_B27 (65)		A13_B14 (64)
cis-H$_2$N-EM1	A11_B15 (73) A11_B35 (55)	A12_B14 (74)	
cis-H$_3$N$^+$-EM1	A11_B14 (57) A11_B15 (81) A11_B35 (54) A11_B36 (60)	A12_B15 (57) A12_B35 (54) A12_B36 (75)	
trans-H$_2$N-EM2	A11_B26 (94) A11_B27 (57)	A12_B14 (58) A12_B15 (84)	A13_B14 (68)
trans-H$_3$N$^+$-EM2	A11_B26 (64) A11_B27 (74)	A12_B14 (68) A12_B15 (54)	
cis-H$_2$N-EM2	A11_B15 (78) A11_B32 (50) A11_B35 (58)	A12_B14 (61) A12_B15 (53)	
cis-H$_3$N$^+$-EM2	A11_B15 (91) A11_B35 (58) A11_B36 (75)	A12_B14 (53) A12_B35 (57) A12_B36 (70)	

The numbers between brackets are the numbers of conformers in the regions.

3.2. Three-dimensional pseudo-Ramachandran plots

For the *cis* and *trans* conformers of the neutral and charged forms of EMs 3DPR plots were constructed using the Φ_2 torsion angle of Pro[2] and the Ψ_1 torsion angle of Tyr[1] (see Figs. 3 and 4.). The $\Phi_2 - \Psi_1$ conformational space was divided as mentioned above. Axes were labeled in the same way as described above using A1–A36 for Φ_2 and B1–B36 for Ψ_1 torsion angles. In the three-dimensional plots of Φ_2 vs. Ψ_1 vs. number of conformers the mainly populated $\Phi_2 - \Psi_1$ regions were determined (see Table 2).

Significant amounts of the *trans* conformers were found in regions A11–B26, A11–B27, A11–B32,

A11–B33, A11–B34, A12–B33, A12–B34, A13–B27 and A13–B34. The *cis* conformers preferred regions A11–B32, A11–B33, A12–B28, A12–B29, A12–B31 and A12–B32. For neutral and charged forms of *trans*-EMs, the mainly populated regions were A11–B32, A11–B33, A11–B34 and A12–B33, while for all types of *cis*-EMs, the major populations of conformers were located in regions A11–B32, A11–B33 and A12–B28.

The 3DPR plots of *cis* and *trans*-EMs showed that a significant amounts of conformers were found in regions A11–B32 and A11–B33 for both isomers. For *cis* isomers, the greater proportion of conformers exists in these two regions as compared with those of the *trans* isomers. In the 3DPR, for *cis*-EMs greater,

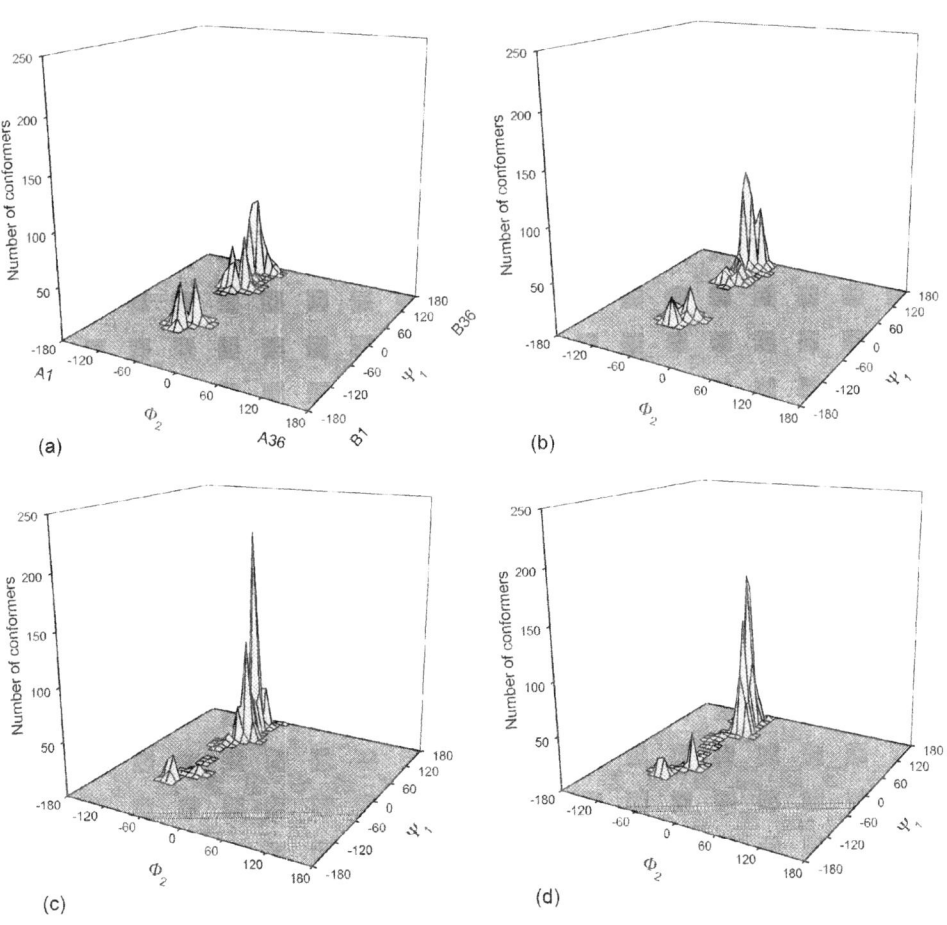

Fig. 3. The three-dimensional pseudo-Ramachandran (3DPR) plots of all four types of EM1 molecules (*cis/trans*, charged/uncharged). (a) *trans*-H_2N-EM1, (b) *trans*-H_3N^+-EM1, (c) *cis*-H_2N-EM1, (d) *cis*-H_3N^+-EM1.

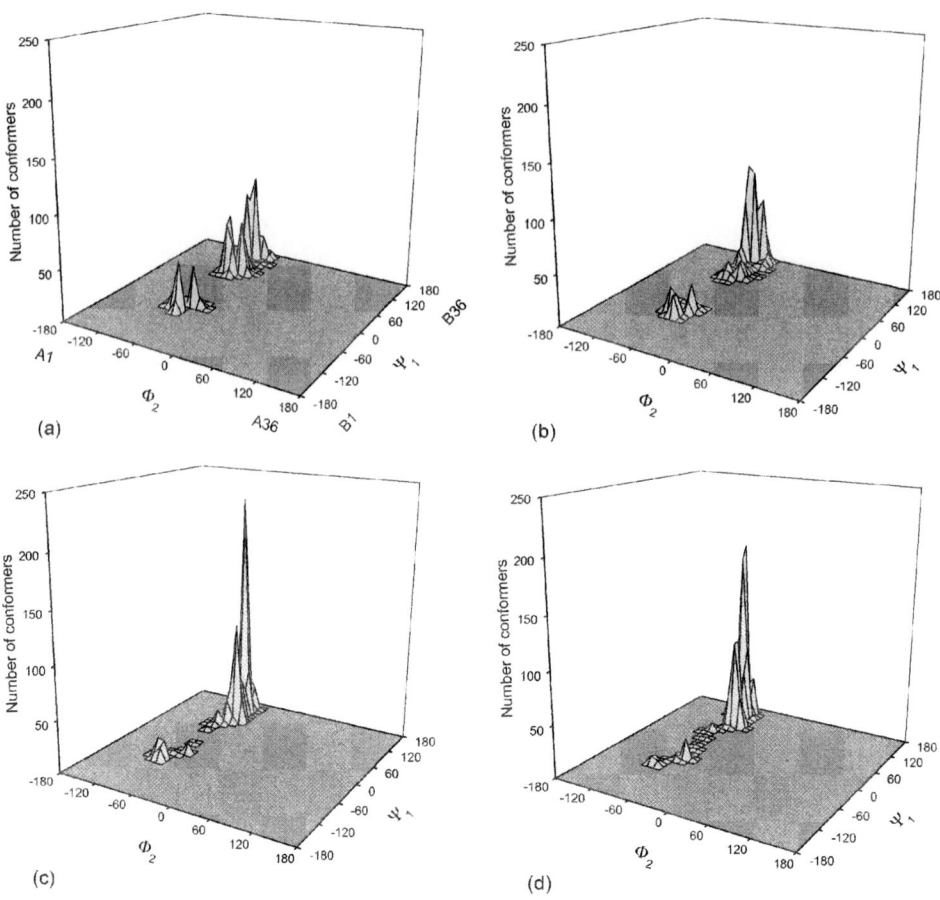

Fig. 4. The three-dimensional pseudo-Ramachandran (3DPR) plots of all four types of EM2 molecules (*cis/trans*, charged/uncharged). (a) *trans*-H_2N-EM2, (b) *trans*-H_3N^+-EM2, (c) *cis*-H_2N-EM2, (d) *cis*-H_3N^+-EM2.

tight peaks were found in these $\Phi_2 - \Psi_1$ regions, while for *trans*-EMs these peaks were smaller and broad.

3.3. Side-chain rotamers

In the course of earlier conformational analyses of EMs we have determined the conformational distributions of the χ^1 torsion angles of the aromatic amino acid residue side-chains. We also investigated the $g(+)$, $g(-)$ and *trans* rotamer populations of the side-chains of the Tyr[1], Trp[3] and Phe[4] for EM1 and the Tyr[1], Phe[3] and Phe[4] for EM2 [7,8]. For all side-chains of EMs, these results revealed the largest populations of the $g(-)$ and *trans* rotamers, while the $g(+)$ rotamers were less preferred.

For *cis* and *trans* conformers of the neutral and charged forms of EMs, the main rotamer combinations of side-chains were determined (see Tables 3 and 4). In these combinations mainly the $g(-)$ and *trans* side-chain conformations were observed. Four types of these combinations exists in the conformers of all types of EMs (*cis/trans*, charged/uncharged): $g(-)$, $g(-)$, $g(-)$; $g(-)$, $g(-)$, *trans*; *trans*, $g(-)$, $g(-)$ and *trans*, *trans*, $g(-)$.

4. Discussion and conclusion

Preferred conformational regions of the $\Phi_2 - \Psi_2$ 3DR and $\Phi_2 - \Psi_1$ 3DPR plots were identified.

Table 2
The preferred $\Phi_2 - \Psi_1$ conformational regions of the three-dimensional pseudo-Ramachandran plots

	Preferred conformational regions		
trans-H$_2$N-EM1	A11_B32 (60)	A12_B33 (78)	A13_B27 (54)
	A11_B33 (74)		
	A11_B34 (57)		
trans-H$_3$N$^+$-EM1	A11_B32 (86)	A12_B33 (79)	A13_B34 (64)
	A11_B33 (103)		
	A11_B34 (91)		
cis-H$_2$N-EM1	A11_B32 (209)	A12_B28 (103)	
	A11_B33 (137)	A12_B29 (53)	
cis-H$_3$N$^+$-EM1	A11_B32 (159)	A12_B28 (63)	
	A11_B33 (144)	A12_B29 (118)	
		A12_B32 (69)	
trans-H$_2$N-EM2	A11_B26 (51)	A12_B33 (87)	A13_B27 (54)
	A11_B27 (58)		
	A11_B32 (71)		
	A11_B33 (63)		
	A11_B34 (53)		
trans-H$_3$N$^+$-EM2	A11_B32 (69)	A12_B33 (98)	A13_B34 (69)
	A11_B33 (105)	A12_B34 (56)	
	A11_B34 (97)		
cis-H$_2$N-EM2	A11_B32 (224)	A12_B28 (102)	
	A11_B33 (149)		
cis-H$_3$N$^+$-EM2	A11_B32 (169)	A12_B28 (89)	
	A11_B33 (182)	A12_B29 (89)	
		A12_B31 (51)	
		A12_B32 (74)	

The numbers between brackets are the numbers of conformers in the regions.

The conformers of trans-H$_2$N-EM1, trans-H$_3$N$^+$-EM1, trans-H$_2$N-EM2 and trans-H$_3$N$^+$-EM2 were found approximately in the same $\Phi - \Psi$ conformational regions in the 3DR and 3DPR plots. For cis-H$_2$N-EM1, cis-H$_3$N$^+$-EM1, cis-H$_2$N-EM2 and cis-H$_3$N$^+$-EM2, the conformers showed broadly similar conformational distributions in the 3DR and 3DPR plots.

Table 3
The main rotamer combinations of the aromatic amino acid residue side-chains for all four types of EM1 molecules (cis/trans, charged/uncharged)

trans-H$_2$N-EM1			trans-H$_3$N$^+$-EM1		
Tyr^1-χ_1	Trp^3-χ_1	Phe^4-χ_1	Tyr^1-χ_1	Trp^3-χ_1	Phe^4-χ_1
g(−)	g(−)	g(−)	g(−)	g(−)	trans
trans	g(−)	g(−)	g(−)	g(−)	g(−)
g(−)	trans	g(−)	trans	g(−)	g(−)
g(−)	trans	trans	trans	trans	g(−)
trans	trans	g(−)	trans	trans	trans
trans	trans	trans	g(−)	trans	trans
g(−)	g(+)	trans			

cis-H$_2$N-EM1			cis-H$_3$N$^+$-EM1		
Tyr^1-χ_1	Trp^3-χ_1	Phe^4-χ_1	Tyr^1-χ_1	Trp^3-χ_1	Phe^4-χ_1
g(−)	trans	g(−)	g(−)	trans	g(−)
g(−)	trans	trans	trans	g(−)	g(−)
g(−)	g(−)	g(−)	trans	trans	g(−)
trans	trans	trans	g(−)	trans	trans
trans	g(−)	g(−)	g(−)	g(−)	trans
g(−)	g(−)	trans	trans	trans	trans
g(−)	g(+)	g(−)	g(−)	g(−)	g(−)
trans	trans	g(−)			

Table 4
The main rotamer combinations of the aromatic amino acid residue side-chains for all four types of EM2 molecules (cis/trans, charged/uncharged)

trans-H$_2$N-EM2			trans-H$_3$N$^+$-EM2		
Tyr^1-χ_1	Phe^3-χ_1	Phe^4-χ_1	Tyr^1-χ_1	Phe^3-χ_1	Phe^4-χ_1
g(−)	g(−)	g(−)	g(−)	g(−)	g(−)
trans	g(−)	g(−)	g(−)	trans	trans
trans	trans	g(−)	trans	trans	g(−)
trans	trans	trans	trans	g(−)	g(−)
g(−)	trans	trans	g(−)	trans	g(−)
g(−)	g(−)	trans	trans	g(−)	trans
g(−)	trans	g(−)	g(−)	g(−)	trans
trans	g(−)	trans			

cis-H$_2$N-EM2			cis-H$_3$N$^+$-EM2		
Tyr^1-χ_1	Phe^3-χ_1	Phe^4-χ_1	Tyr^1-χ_1	Phe^3-χ_1	Phe^4-χ_1
g(−)	g(−)	g(−)	trans	g(−)	g(−)
trans	g(−)	g(−)	g(−)	trans	g(−)
g(−)	trans	g(−)	g(−)	trans	trans
g(−)	trans	trans	g(−)	g(−)	trans
trans	trans	g(−)	trans	trans	g(−)
g(−)	g(−)	trans	g(−)	g(−)	g(−)
			trans	trans	trans
			g(+)	trans	trans

The conformational distributions of *cis* and *trans* conformers of EMs were different in the 3DR plots. The preferred $\Phi_2 - \Psi_2$ conformational regions are related to the secondary structural elements of EMs, which are β-turns of type III and N-terminal inverse γ-turns. In the 3DPR plots, the conformers of *cis* and *trans*-EMs showed approximately similar conformational distribution, but the numbers of conformers are different in the same $\Phi_2 - \Psi_1$ regions.

The main combinations of rotamers of the aromatic amino acid side-chains were determined in χ_1 space. For all types of EMs molecules (*cis/trans*, charged/uncharged), four types of these rotamer combinations were found.

References

[1] J.E. Zadina, L. Hackler, L.J. Ge, A.J. Kastin, Nature 386 (1997) 499–502.

[2] S. Fiori, C. Renner, J. Cramer, S. Pegoraro, L. Moroder, J. Mol. Biol. 291 (1999) 163–175.

[3] B.L. Podlogar, G. Paterlini, D.M. Ferguson, G.C. Leo, D.A. Demeter, F.K. Brown, A.B. Reitz, FEBS Lett. 439 (1998) 13–20.

[4] Y. In, K. Minoura, H. Ohishi, H. Minakata, M. Kamigauchi, M. Sugiura, T. Ishida, J. Pept. Res. 58 (2001) 399–412.

[5] M. Eguchi, R.Y.W. Shen, J.P. Shea, M.S. Lee, M. Kahn, J. Med. Chem. 45 (2002) 1395–1398.

[6] M. Keller, C. Boissard, L. Patiny, N.N. Chung, C. Lemieux, M. Mutter, P.W. Schiller, J. Med. Chem. 44 (2001) 3896–3903.

[7] B. Leitgeb, F. Ötvös, G. Tóth, Biopolymers 68 (4) (2003) 497–511.

[8] B. Leitgeb, A. Szekeres, G. Tóth, J. Pept. Res. 62 (4) (2003) 145–157.

[9] B.M. Grail, J.W. Payne, J. Pept. Sci. 6 (2000) 186–199.

[10] N.J. Marshall, B.M. Grail, J.W. Payne, J. Pept. Sci. 7 (2001) 175–189.

[11] D.A. Pearlman, D.A. Case, J.W. Caldwell, W.S. Ross, T.E. Cheatham III, S. DeBolt, D. Ferguson, G. Seibel, P. Kollman, Comp. Phys. Commun. 91 (1995) 1–41.

[12] W.D. Cornell, P. Cieplak, C.I. Bayly, I.R. Gould, K.M. Merz, D.M. Ferguson, D.C. Spellmeyer, T. Fox, J.W. Caldwell, P.A. Kollman, J. Am. Chem. Soc. 117 (1995) 5179–5197.

[13] INSIGHT/DISCOVER: Biosym Technologies, Scranton Road, San Diego, USA.

THEO
CHEM

Journal of Molecular Structure (Theochem) 666–667 (2003) 345–353

www.elsevier.com/locate/theochem

Structure-activity relationships of endomorphin-1, endomorphin-2 and morphiceptin by molecular dynamics methods

Ferenc Ötvös[a,*], Tamás Körtvélyesi[b], Géza Tóth[a]

[a]*Institute of Biochemistry, Biological Research Center of the Academy of Sciences, Temesvári krt. 62, Szeged H-6726, Hungary*
[b]*Institute of Physical Chemistry, University of Szeged, Rerrich B. tér 1, Szeged H-6720, Hungary*

Abstract

A conformational analysis of endomorphin-1 (Tyr-Pro-Trp-Phe-NH$_2$, EM1), endomorphin-2 (Tyr-Pro-Phe-Phe-NH$_2$, EM2) and morphiceptin (Tyr-Pro-Phe-Pro-NH$_2$, MC), as μ-opioid receptor ligands was performed by the simulated annealing (SA) method and solvated molecular dynamics (MD) calculations. In SA experiments, 1000 conformers were generated for each peptide with both *cis* and *trans* Tyr–Pro peptide bond isomers. The populations of the conformers in the different regions of the Ramachandran plots and the numbers of the various types of turns and the distribution of distances of the main pharmacophore elements (i.e. phenolic OH of Tyr^1, tyramine N and the aromatic side-chain of Phe^3) for the three peptides were compared. EM1 and EM2 seemed to be almost identical in their conformational features, ca. 30% of their conformers possessed turns with a significant ratio of β-turn type III, while in MC a drastic decrease in the number of turns and an increase of the number of extended structures were observed. According to the differences in the $\Phi^3 - \Psi^3$ Ramachandran plots, the Phe^3 pharmacophore was affected in MC. Also, a pairwise comparison of the interaction energies between all fragments of the peptides (i.e. between backbone and side-chain fragments of each amino acid residue) showed that the presence of Pro^4 in MC initiated increased repulsive interactions toward the Pro^2–Phe^3 fragment relative to Phe^4 toward the Pro^2–Phe^3 and Pro^2–Trp^3 fragments in EM2 and EM1, respectively, accounting for its decreased opioid activity. The conformational analysis was repeated by isotherm molecular dynamics calculations in a periodic box with a modified GROMOS96 force field. The results support our previous experience.

© 2003 Elsevier B.V. All rights reserved.

Keywords: Endomorphin-1; Endomorphin-2; Morphiceptin; Simulated annealing; Molecular dynamics; Conformational analysis; Intramolecular H-bonds; β-turns; Regular and inverse γ-turns; Intramolecular interaction energy

1. Introduction

Morphiceptin, isolated from milk, was reported as a prototype of opioid peptide ligands possessing Pro following the N-terminal Tyr residue. It is of moderate activity, highly μ selective ligand [1]. Two novel, structurally related tetrapeptides, endomorphin-1 (EM1, H-Tyr-Pro-Trp-Phe-NH$_2$) and endomorphin-2 (EM2, H-Tyr-Pro-Phe-Phe-NH$_2$) were recently isolated from bovine brain. Both are highly potent and selective endogenous μ-opioid receptor ligands [2]. As the EMs induce significant biological responses in vertebrates, e.g. antinociception, it is important to determine the structural elements responsible for their μ-opioid affinity. The pharmacophores identified so far are the N-terminal amino group, the Tyr^1 residue and a further aromatic

* Corresponding author.
E-mail address: otvos@rosi.szbk.u-szeged.hu (F. Ötvös).

residue (Trp or Phe) in position 3 while Pro^2 functions as a spacer [3,4].

According to the 'message-address concept', the message part of the opioid peptides is possibly the N-terminal tripeptide fragment, while the address part is the remaining C-terminal fragment [5,6].

NMR measurements revealed important structural features of opioid tetrapeptides in solution. For EM1, both extended and bent structures were assumed as bioactive conformations [3,4]. A ratio of ca. 1:3 for the *cis* and *trans* Tyr^1–Pro^2 peptide bond populations was determined. For EM2, a ratio of ca. 1:2 between the *cis* and *trans* isomers was measured and mainly an extended backbone was suggested for both *cis* and *trans* isomers [6,7]. For morphiceptin, *cis/trans* isomerism of both peptidyl–proline bonds was observed, although the population of the *cis* Phe^3–Pro^4 bonds was significantly lower than that of the *cis* Tyr^1–Pro^2 peptide bonds [8]. For EM2, a previous molecular modeling study suggested that, if a turn is required for the bioactive conformation, the *trans* isomer would be preferred because the *cis* isomer is less prone to adopt turns in the N-terminal region [9].

Supporting the steric requirements for the bioactive conformation, different peptide derivatives were probed. Structural determination of a biologically active bicyclic β-turn mimic peptidomimetics by 2D ^{1}H NMR measurements suggested EMs to adopt a β-turn similar to type III (1 ← 4) in the bioactive conformation [10]. EM2 analogs with pseudoproline replacements support the involvement of *cis* Tyr^1–Pro^2 peptide bond in the receptor-bound state [11].

In this work we compared the conformational features of EM1, EM2 and MC, using SA for neutral molecules with restrained *cis* and *trans* Tyr^1–Pro^2 peptide bonds, respectively, and MD methodology for solvated ionized molecules. The populations of conformers with different secondary structure elements (β-turns, N- and C-terminal γ-turns and inverse γ-turns) were calculated.

2. Methods

2.1. Simulated annealing (SA) calculations

Simulated annealing calculations were performed with the AMBER force field [12,13] with the use of the distance-dependent dielectric constant ($\epsilon = 4.5r$) to simulate the effect of water. No cut-off was applied for non-bonding interactions. The calculations were carried out with the Insight II/Discover 3 molecular modeling software package [14] on an Origin2000 workstation running the Irix 6.5 operation system (Silicon Graphics, Inc.).

The initial structure with an extended peptide backbone was optimized by energy minimization and then heated to 1000 K during 100 fs. The following equilibration trajectory at 1000 K was sampled every 4000 fs to gain 1000 starting structures for the SA cycles where the temperature was decreased to 50 K for 10,000 fs by an exponential cooling protocol. After the SA stages, 1000 final minimized conformers were obtained by a three-step minimization procedure, using the steepest descent (SD), conjugated gradient (CG, Polak-Ribiere) and a modified Newton method (BFGS) with convergence criteria of 1000, 10 and 0.001 kcal mol^{-1} Å^{-1}, respectively.

NMR data revealed *cis/trans* isomerism of the Tyr^1–Pro^2 peptide bonds in solution for all the three molecules [6–8]. SA calculations were performed with Tyr^1–Pro^2 peptide bonds restrained in either the *cis* or the *trans* configuration ($\omega = 0°$ and 180°), while all other peptide bonds were kept in the *trans* configuration. The Phe^3–Pro^4 peptide bond was kept in the *trans* configuration in morphiceptin, because its *cis* isomers were present in significantly smaller amounts by solution NMR measurements [8]. The molecules were modeled with neutral N- and C-termini to exclude additional charge effects.

2.2. Isothermal MD calculations

The MD simulations were performed with GRO-MACS program package [15] using a modified GROMOS87 force field. In this force field all the non-polar H atoms are handled together with the atoms to which they connect except the aromatic H atoms to be considered explicitly. The initial structures of the peptides were extended. 1.2 nm of solvent (1224-1378 SPC water molecules) were placed on all sides of the peptides. The dielectric constant was 1.0. The system was energy minimized by the steepest descent method with a gradient criterion of 10^{-3} kJ mol^{-1} nm^{-1}. The charged peptide was neutralized by inserting one Cl-counterion. The structures were again energy minimized positionally restraining the heavy atoms of the peptide. After this minimization a 20 ps NVT positionally restrained MD simulation at 293.16 K was performed, which eliminated the differences in the density of the solvent in the periodic box.

In the 10 ns productive MD simulations with NPT ensemble, the temperature and the pressure was kept constant (293.16 K and 1 bar, respectively) with the relaxation time of 0.1 and 0.5 ps. The compressibility was 4.8×10^{-5} bar^{-1}. The time step was 2 fs. LINCS was used to eliminate the high order vibrational motions. The long-range electrostatic interactions were calculated by Particle Mesh Ewald method (PME). The cut-off distances for the non-bonded interactions was 0.9 nm.

Total and potential energies, root mean square deviations (RMSD) of the backbone (N–C$_\alpha$–C) atoms and the root mean square fluctuation (RMSF) of the backbone atoms related to the structure at the end of equilibration (20 ps) were calculated. Distances between groups in residues and the change in the secondary structures were also evaluated along the trajectories.

3. Results

3.1. Backbone dihedrals

The Ramachandran plots of 1000 conformers obtained by SA simulations for Pro^2, Phe^3 and Phe^4 and the $\Phi^2 - \Psi^1$ correlation Ramachandran plot were calculated for both the *cis*- and *trans* isomers of EM1, EM2 and MC. Conventional regions to navigate on plots were marked as they were established by ECEPP calculations for protected amino acids [16], as seen in Fig. 1.

EM2 and EM1 showed near identical distribution in their conformers (not shown). Only in region A of the $\Phi^2 - \Psi^1$ plot there was noticeable decrease in the number of conformers of EM1 compared to that of EM2. As illustrated in Fig. 2 for the *trans* isomers, Ramachandran plots for MC show variations in the populations of the conformers compared with those obtained for EM2 previously [9]. Table 1 shows the difference in the conformer populations between MC

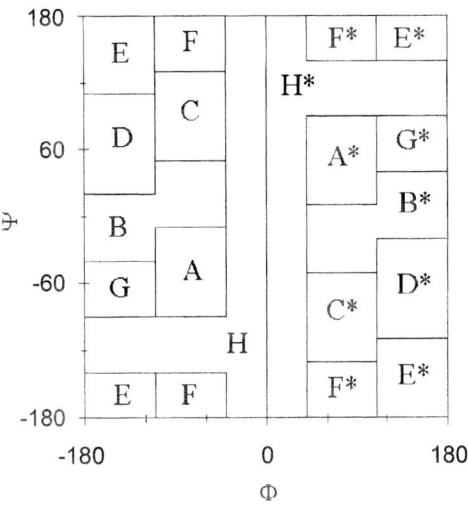

Fig. 1. Conventional regions of the Ramachandran plot.

and EM2 in each region for the *trans* and *cis* isomers, respectively.

For the *trans* isomers, the conformer populations of MC, compared to EM2, decreased in regions A, B, C, G, C*, D* and F*, especially in regions C and C* where region C* became empty. The populations increased in regions D, E, F and A*, particularly in region D. The same tendency was observed for the *cis* isomers too. Further details can be seen on the Ramachandran plots. The conformers of EM2 moved from region C into regions B and F in the $\Phi^2 - \Psi^2$ plot, while in the $\Phi^2 - \Psi^1$ correlation plot, the conformers of all three peptides moved from regions A and B to region F (Table 2).

A comparison of the distributions of the backbone dihedrals revealed that EM1 and EM2 are quite similar to each other. In the distribution of the Ψ^3 values, the populations at -60, $+60$ and $+120°$ show increasing tendency in this order for MC in contrast to the case of EM1 and EM2. The Ψ^4 values for MC show sharper ranges in the distribution plot than do EM1 and EM2. As a consequence of the proline ring, Φ^4 values does not populate around $+60°$ in MC. Comparing the *cis* and *trans* isomers, all three peptide analogs show the same variations in the overall distribution of the backbone dihedrals, i.e. the population of Ψ^2 values around $+60°$ disappeared and the population around $+120°$ became predominant. In the distribution of the Φ^3 values,

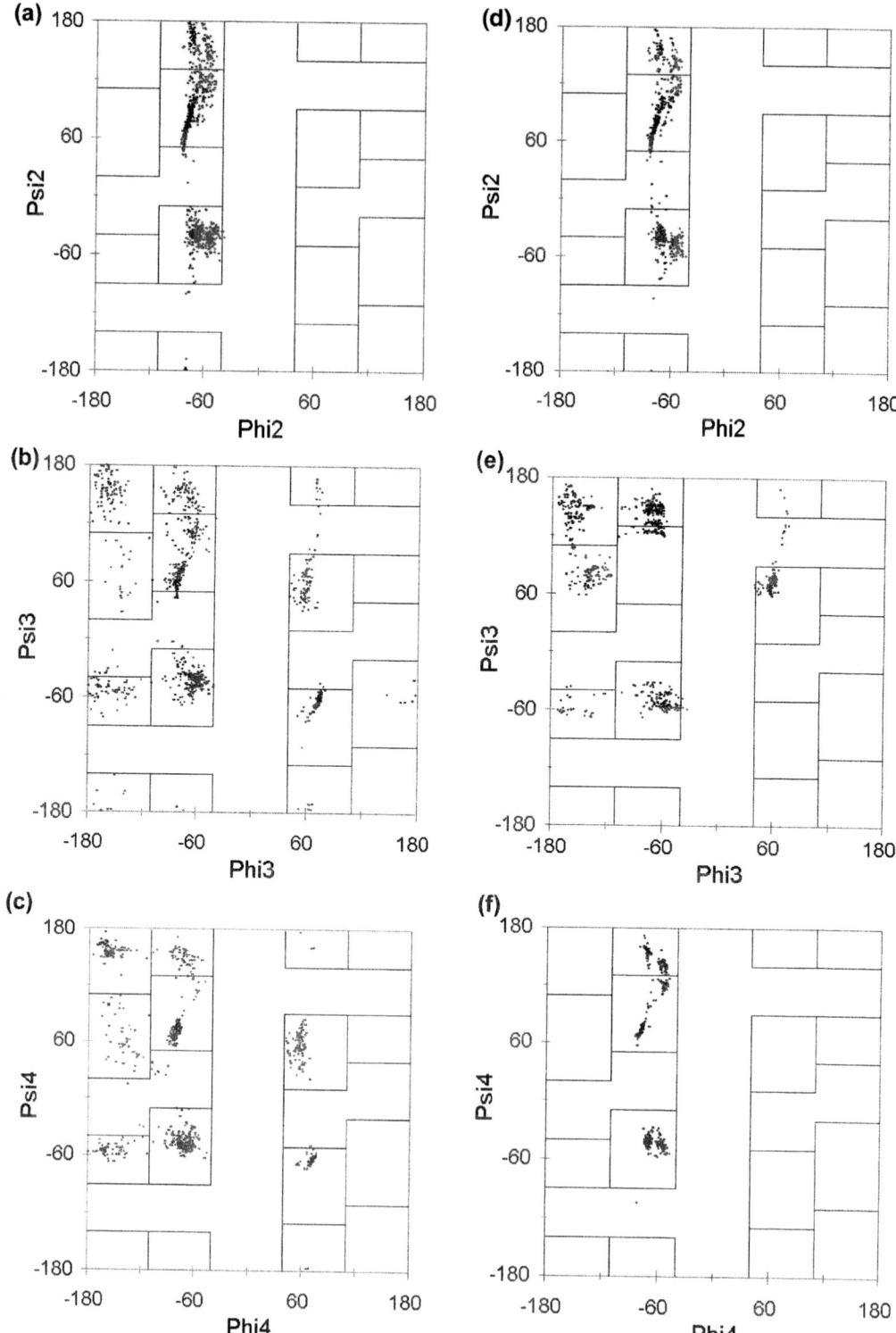

Fig. 2. Ramachandran plots for *trans* EM2 (a, b, c) and *trans* (d, e, f) eMC conformers.

Table 1
Difference of the conformer populations between *cis* and *trans* MC and EM2 in different regions

	tMC–tEM2				cMC–cEM2			
	$\Phi^2-\Psi^2$	$\Phi^3-\Psi^3$	$\Phi^4-\Psi^4$	$\Phi^2-\Psi^1$	$\Phi^2-\Psi^2$	$\Phi^3-\Psi^3$	$\Phi^4-\Psi^4$	$\Phi^2-\Psi^1$
A	−27	−87	136	−11	74	−99	123	−28
B	−1	−18	−11	0	−24	−17	−22	5
C	52	−156	170	−7	−11	−128	206	36
D	0	98	−56	0	0	76	−57	0
E	0	61	−117	0	0	64	−130	0
F	−20	205	152	20	−38	196	126	−13
G	0	−35	−70	0	0	−34	−70	0
H	−4	3	1	−2	−1	−3	0	0
A*	0	45	−113	0	0	78	−83	0
B*	0	−2	−1	0	0	−8	0	0
C*	0	−98	−85	0	0	−119	−86	0
D*	0	−7	0	0	0	0	0	0
E*	0	0	0	0	0	0	0	0
F*	0	−13	−6	0	0	−10	−6	0
G*	0	0	0	0	0	0	0	0
H*	0	4	0	0	0	4	−1	0

the population at −130° increased at the expense of the population at −170° (Fig. 3).

During the 10 ns MD simulation the RMSD of the backbone related to the structure at 20 ps, were 0.08–0.09, 0.09–0.1 and 0.07–0.08 nm for EM1, EM2 and MC, respectively. In the MC simulation there were two peaks at 5 and 7.5 ns with an increase to 0.14–0.15 nm. The initial structure was extended, thus, after 20 ps equilibration the structure did not change significantly. The RMSF were similar but higher in EM1 (0.085 nm) and EM2 (0.08 nm) than in MC (0.07 nm) at the peptide bond between residues 2 and 3. The small peptides are very flexible, the RMSF values are also large at the N- (0.05–0.06 nm) and

Table 2
Difference of the populations in different regions between the *cis* and *trans* peptide isomers

	cEM2–tEM2				cEM1–tEM1				cMC–tMC			
	$\Phi^2-\Psi^2$	$\Phi^3-\Psi^3$	$\Phi^4-\Psi^4$	$\Phi^2-\Psi^1$	$\Phi^2-\Psi^2$	$\Phi^3-\Psi^3$	$\Phi^4-\Psi^4$	$\Phi^2-\Psi^1$	$\Phi^2-\Psi^2$	$\Phi^3-\Psi^3$	$\Phi^4-\Psi^4$	$\Phi^2-\Psi^1$
A	17	28	−9	−104	85	15	−50	−78	118	16	−22	−121
B	59	−2	11	3	49	−8	8	4	36	−1	0	8
C	−375	10	−1	−71	−426	29	14	−41	−438	38	35	−28
D	0	8	1	0	0	16	−3	0	0	−14	0	0
E	0	−27	13	0	0	−14	37	0	0	−24	0	0
F	303	−5	14	174	292	0	−1	118	285	−14	−12	141
G	0	−3	0	0	0	−15	−1	0	0	−2	0	0
H	−4	2	0	−2	0	2	0	−3	−1	−4	−1	0
A*	0	−27	−30	0	0	−28	−17	0	0	6	0	0
B*	0	6	−1	0	0	−9	0	0	0	0	0	0
C*	0	23	1	0	0	13	9	0	0	1	0	0
D*	0	−6	0	0	0	−1	0	0	0	0	0	0
E*	0	0	0	0	0	1	0	0	0	0	0	0
F*	0	−4	0	0	0	−4	2	0	0	−1	0	0
G*	0	0	0	0	0	0	0	0	0	0	0	0
H*	0	−3	1	0	0	3	2	0	0	−3	0	0

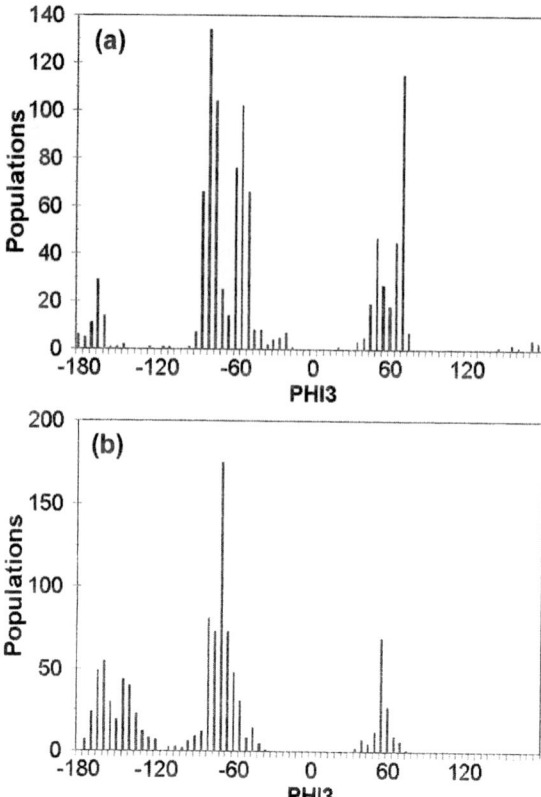

Fig. 3. Distribution of Φ^3 values in *trans* EM2 and in *trans* MC.

C-terminal (0.09 nm). The structure of the Ramachandrans, obtained by sampling the trajectory at every 5 ps, is very similar for the three peptides, but Ψ^4 and Φ^4 values for MC show sharper ranges in the distribution plot, very similarly to the SA results. We also found differences in the range Φ^4 it was 80–150° for EM1 and EM2 and 60–100° for MC. As a consequence of the proline ring, the Φ^4 range is narrow and smaller than for EM1 and EM2.

3.2. Side-chain conformations

In the *trans* isomers of all three peptides, the $g(-)$ and *trans* χ^1 rotamer populations of Tyr[1] were ca. 40% with the predominance of $g(-)$. In MC, this predominance is less enhanced. For Phe[3], the number of $g(+)$ rotamers increased at the expense of the *trans* rotamers in all three peptides. For Phe[4], the $g(-)$ and the *trans* χ^1 rotamer populations were equal in EM1,

while in EM2 the $g(-)$ predominated at the expense of the *trans* rotamers.

In the *cis* isomers, the side-chain rotamer populations of Tyr[1] were similar than in the *trans* isomers. For Phe[3], the $g(+)$ rotamer population increased in all three peptides. In EM1 and MC, the *trans* rotamers became predominant, while in EM2, the $g(-)$ rotamers remained predominant.

3.3. Distances between the putative pharmacophoric elements

The distribution of the pairwise distances between the three known pharmacophoric elements, i.e. the Tyr[1] N, the Tyr[1] phenolic oxygen and the center of the Phe[3] side chain (N1, OPh and PSE3, respectively), were calculated. All three peptides showed similar distributions for both the *trans* and the *cis* isomers. Two distinct populations were found in highly different numbers. The larger population seems to form at least three subpopulations and numerous diffuse points.

EM1, EM2 and MC contains residues with aromatic side chains. These aromatic fragments can interact with each other [17,18]. In EM1, the distance between Tyr[1] and Phe[4] aromatic plane is 0.6–0.8 nm after 2 ns which supports the interaction. For EM2, great fluctuation of the distances was observed between Tyr[1] and Phe[3], Tyr[1] and Phe[4] between the values of 0.5–1.0 and 0.6–0.8 nm, respectively. The distance between Phe[3] and Phe[4] is bigger: 0.8–1.0 nm. In MC, the distance between Tyr[1] and Phe[3] is almost constant, with an average value of 0.5 nm with some fluctuations.

3.4. Energy distribution of the conformers

In the SA experiments Gaussian type energy distribution of the conformers was observed for both the *cis* and *trans* isomers of EM2 and EM1. For MC, the energy of the *cis* conformers showed also a Gaussian type distribution, while the *trans* isomers showed a combination of two Gaussian type distributions. The energy of the *trans* peptide isomers is slightly lower than that of the *cis* peptide isomers. The energy ranges were slightly broader for the *trans* isomers.

3.5. Interaction energies between side chains and backbone fragments in the conformers

The pairwise interaction energies between the side chains and backbone fragments of the residues were calculated for the conformers obtained by SA. The van der Waals, electrostatic and total interaction energies, averaged for the 1000 conformers, were compared between EM1, EM2 and MC to trace back the change of interactions in MC compared to EM1 and EM2 leading to the decreased biological activity of MC. The backbone fragments in each residue involved the alpha N (regardless that it is involved in an amide bond or not) and H atoms, the alpha C and H, and the carbonyl group. The side-chain fragments involved the beta C and H atoms and the further part of the side-chain. As it is expected, EM2 and EM1 showed high similarity, while MC developed repulsive interactions due to the Pro[4] replacement as shown in Fig. 4 for the van der Waals energies. In the *trans* isomers, Pro[4] and its C-terminal amide group caused repulsive interactions with the side chains of Tyr[1], Pro[2] and Phe[3] and with the backbone atoms of Tyr[1]

and Phe[3], compared with EM1 and EM2. This disturbance of the interactions then spread over to develop repulsive interactions between these affected fragments and, additionally, the backbone of Pro[2] was also involved. In the *cis* isomers, the C-terminal Pro[4] had different interactions with the side chains and backbones of Tyr[1], Pro[2] and Phe[3]. The disturbance caused here a slight increase of attractions between the side chains of Tyr[1] and Phe[3], and between their backbones.

3.6. Intramolecular H-bonding

In the MD simulation we did not find significant difference in the number of H-bonds along the trajectory. One H bond was found in a snapshot, very rarely two or three in EM1 and EM2. In the SA studies, resulting energy minimized structures, over 75% of the conformers of *trans* EM1 and EM2 and over 60% of the conformers of *trans* MC contained at least one H-bond. For the *cis* isomers, these numbers decreased by ca. 10%. In the *trans* conformers, the most H-bonds formed between Phe[3]/Trp[3] amide

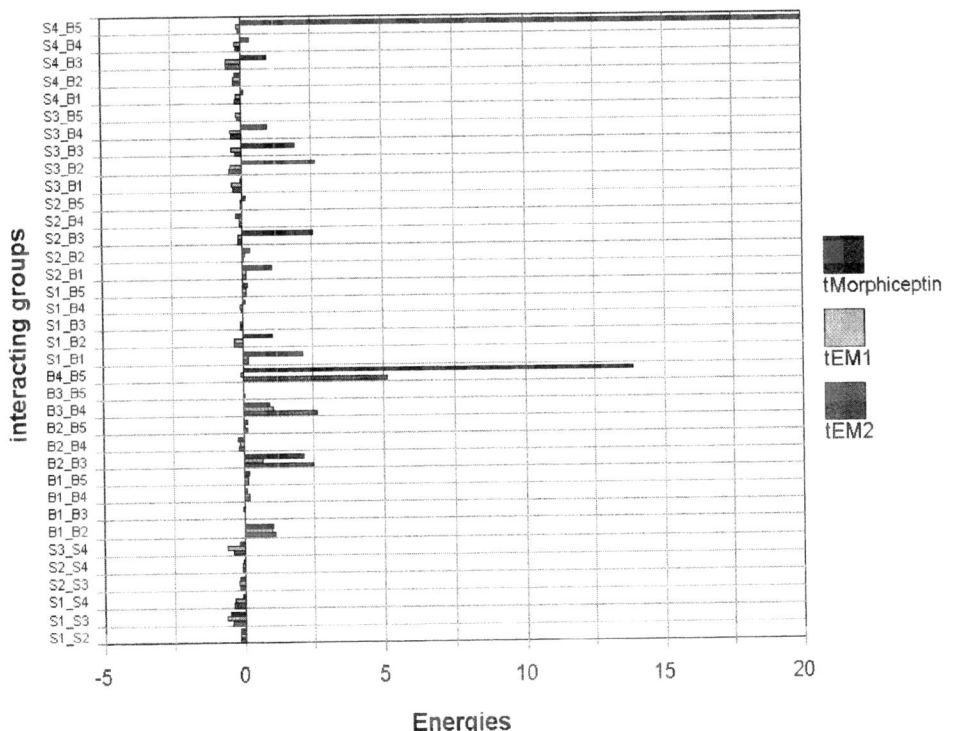

Fig. 4. The averaged pairwise van der Waals interaction energies between the fragments for the *trans* isomers.

protons and the Tyr^1 carbonyl group, and between the C-terminal amide and the carbonyls of Phe^3/Trp^3, Tyr^1 or Pro^2. In the *cis* isomers there were no H-bonds between the Phe^3/Trp^3 amide protons and the Tyr^1 carbonyl, but numerous H-bonds appeared between the Tyr^1 amino protons and the amide N of Phe^3/Trp^3, Phe^4/Pro^4, the C-terminal amide, the carbonyls of Pro^2, and Phe^3/Trp^3. Additionally, both *cis* and *trans* isomers of EM1 and EM2 possessed large number of H-bonds between the Phe^4 amide and the Pro^2 carbonyl.

3.7. Secondary structures

In *trans*-EM2 and EM1, β-turns of types I, III, and V were found and also inverse γ-turns located in the N-terminal fragment and regular and inverse γ-turns were located in the C-terminal fragment. In the *trans*-MC only β-turns of type III was found and inverse γ-turns were located in the N-terminal fragment. The number of β-turns type III in MC was significantly lower than in EM2 and EM1, while the number of the N-terminal inverse γ-turns was slightly higher than in EM2 and EM1.

In the case of the *cis* isomers the N-terminal inverse γ-turns were missing in all three peptides and also the β-turn of type V in EM2 and EM1. Thus, the only secondary structure in *cis*-MC was the β-turn of type III.

In the MD simulations, the result of the dssp [19] analysis of the secondary structure was random coil. No really dominating secondary structure elements were found.

4. Conclusions

All three tetrapeptides are highly flexible shown by both SA and MD results. Unrestrained SA simulations resulted in a large variety of extended and bent structures for both *cis* and *trans*-EM2, EM1 and MC. The bent structures may be categorized into the classic turn types. These secondary structures are stabilized by $1 \leftarrow 4$, $1 \leftarrow 3$ and $2 \leftarrow 4$ H-bonds located in the $Tyr^1–Phe^4$, $Tyr^1–Phe^3$ and $Pro^2–Phe^4$ fragments, respectively. The only secondary structure suggested experimentally so far is the β-turn of type III [10]. According to the SA results, MC is less prone

to adopt this secondary structure than EM1 and EM2, which may explain its decreased affinity to the μ-opioid receptors. However, the 3D distribution of the pharmacophoric elements is quite similar for all three peptides regardless of the energy of the conformers. It may be explained by the high flexibility of the pharmacophoric elements, which may compensate the structural disturbance caused by Pro^4 in MC. The moderate activity of MC compared to the EMs may be explained by intrusion to an exclusion volume inside the receptor, which may be in connection with the restricted range of Φ^4 and the reduced flexibility of Pro^4.

Acknowledgements

This research was supported by the operating grants from the Applied Research and Development Fund, Hungary (OMFB-00656/2003), János Bolyai Foundation (BO/00286/99) and National Research and Development Programs, Hungary (OM0217/2001).

References

[1] K.J. Chang, A. Lillian, E. Hazum, P. Cuatrecasas, J.K. Chang, Science 212 (1981) 75.

[2] J.E. Zadina, L. Hackler, L.J. Ge, A.J. Kastin, Nature 386 (1997) 499.

[3] B.L. Podlogar, G. Paterlini, D.M. Ferguson, G.C. Leo, D.A. Demeter, F.K. Brown, A.B. Reitz, FEBS Lett. 439 (1998) 13.

[4] S. Fiori, C. Renner, J. Cramer, S. Pegoraro, L. Moroder, J. Mol. Biol. 291 (1999) 163.

[5] R. Schwyzer, Ann. NY Acad. Sci. 297 (1977) 3.

[6] Y.K. Minoura, H. Ohishi, H. Minakata, M. Kamigauchi, M. Sugiura, T. Ishida, J. Pept. Res. 58 (2001) 399.

[7] F.L. Ötvös, B. Leitgeb, K.E. Kövér, T. Martinek, T. Körtvélyesi, G. Tóth, Hungarian-German-Italian-Polish Joint Meeting on Medicinal Chemistry, Budapest, Hungary, 2001.

[8] M.A. Castiglione-Morelli, B. Hartrodt, K. Neubert, P.A. Temussi, E. Trivellone, Biochem. Biophys. Res. Commun. 152 (1988) 512.

[9] B. Leitgeb, F.L. Ötvös, G. Tóth, Biopolymers 68 (2003) 497.

[10] M. Eguchi, R.Y.W. Shen, J.P. Shea, M.S. Lee, M. Kahn, J. Med. Chem. 45 (2002) 1395.

[11] M. Keller, C. Boissard, L. Patiny, N.N. Chung, C. Lemieux, M. Mutter, P.W. Schiller, J. Med. Chem. 44 (2001) 3896.

[12] D.A. Pearlman, D.A. Case, J.W. Caldwell, W.S. Ross, T.E. Cheatham III, S. DeBolt, D. Ferguson, G. Seibel, P. Kollman, Comp. Phys. Commun. 91 (1995) 1.

[13] W.D. Cornell, P. Cieplak, C.I. Bayly, I.R. Gould, K.M. Merz, D.M. Ferguson, D.C. Spellmeyer, T. Fox, J.W. Caldwell, P.A. Kollman, J. Am. Chem. Soc. 117 (1995) 5179.

[14] INSIGHT/DISCOVER: Accelrys, Inc., Scranton Road, San Diego, USA.

[15] GROMACS, Version 3.1.4. Groningen, 2003.

[16] S.S. Zimmerman, M.S. Pottle, G. Némethy, H.A. Sheraga, Macromolecules 10 (1977) 1.

[17] S.K. Burley, G.A. Petsko, Science 229 (1985) 23.

[18] T. Körtvélyesi, R.F. Murphy, S. Lovas, J. Biomol. Struct. Dyn. 17 (1999) 393.

[19] W. Kabsch, C. Sanders, Biopolymers 22 (1983) 2577.

Journal of Molecular Structure (Theochem) 666–667 (2003) 355–359

www.elsevier.com/locate/theochem

The benefits of a pre-computed amino acid structure database in quantum chemical geometry optimizations of β-turns of peptides

Attila Borics[a], Gregory A. Chass[a,b,c], Imre G. Csizmadia[b,c,d], Richard F. Murphy[a], Sándor Lovas[a,*]

[a]*Department of Biomedical Sciences, School of Medicine, Creighton University, 2500 California Plaza, Omaha, NE 68178, USA*
[b]*Department of Chemistry, University of Toronto, Toronto, Ont. M5S 1A1, Canada*
[c]*Global Institute of Computational Molecular and Materials Sciences (GIOCOMMS), 1422 Edenrose St, Mississauga, Ont. L5V 1H3, Canada,*
[d]*Department of Medical Chemistry, University of Szeged, Dóm tér 8, Szeged 6720, Hungary*

Abstract

Energetically most favored conformations of individual amino acids can be used as building blocks, using a recently proposed system for logical numbering of atomic nuclei, to enhance the ab initio quantum chemical geometry optimizations of oligopepides. The structures of amino acid residues previously optimized at the RHF/3-21G and the RHF/6-31G(d) levels of theory were used for input generation and the structures of Ac-Asn-Pro-Gly-Gln-NH$_2$ and Ac-Val-Pro-D-Ala-His-NH$_2$, in their experimentally proposed type I and type II β-turn conformations, respectively, were optimized by using the B3LYP and the B3PW91 density functionals on the 6-31G(d) basis set. Input generation, by extracting the internal coordinates of residues from their previously optimized structures, resulted in quicker convergence of optimization showing no significant influence of the basis set on which the pre-optimization was carried out. However, efficiency of the two density functional theory methods differed.
© 2003 Elsevier B.V. All rights reserved.

Keywords: Ab initio; Pre-computed peptides; Database; Density functional theory

1. Introduction

With the aid of rapidly developing computer technology, the capabilities of ab initio quantum mechanical (QM) calculations are increasing so that it is possible to elucidate theoretically the structure of increasingly large molecular systems. Despite this evolution, the size of peptides of which structure can be efficiently optimized on high levels of theory is still limited to approximately to 10 residues [1], but recent

proposals offer an opportunity to extend this limit [1–7].

Scanning the conformational space of peptides involves the optimization of a high number of possible conformers, but this number dramatically increases with the degree of polymerization [7]. Search for minima on the potential energy hypersurface (PEHS) by QM methods seems to be impossible even for short oligopeptides. Therefore, a database which consists of all theoretically possible conformations of N- and C-terminally protected amino acid diamides, previously optimized using ab initio QM methods, could facilitate multidimensional conformational analysis

---

* Corresponding author. Tel.: +1-402-280-5753; fax: +1-402-280-2690.
   E-mail address: slovas@creighton.edu (S. Lovas).

0166-1280/$ - see front matter © 2003 Elsevier B.V. All rights reserved.
doi:10.1016/j.theochem.2003.08.046

(MDCA) [8]. With the established methods [7], non-existing conformations could be eliminated and more settled structures could be used as inputs for optimization of larger systems. Inputs can be built up by extracting the internal coordinates of pre-optimized amino acid residues. This can be done readily using a recently introduced modular numbering of the atomic nuclei [7]. Though this method takes no account of the structure-stabilizing role of long-range interactions and the effect of solvation, it is still possible to obtain a good approximation of the real structure based only on the conformational preference [1, 9–11] of single amino acid residues. Furthermore, the database of pre-optimized structures can be extended to include larger fragments (di-, tri-, even oligopeptides) in which the interactions between adjacent or close residues are taken into account [12].

In this study, internal coordinates of previously optimized amino acid residues of small peptides were used and time efficiency of ab initio QM geometry optimizations was tested. Three different inputs were generated for Ac-Asn-Pro-Gly-Gln-NH$_2$ and for Ac-Val-Pro-D-Ala-His-NH$_2$ the structures of which were previously characterized by spectroscopic methods as β-turns or β-turn-forming cores of longer peptides [13,14]. Comparison of results of full geometry optimizations of different inputs show the benefits of using pre-optimized structures as building modules of larger peptides.

2. Methods

All QM calculations were performed with 1.3 GHz AMD Athlon processor computers by the GAUSSIAN 98 program package [18] using either the restricted Hartree–Fock (RHF) method or the Density Functional Theory (DFT). All geometry optimizations were done on internal coordinates using the Berny algorithm with in Gaussian 98 (max. force = 4.5×10^{-4} a.u., RMS force = 3.0×10^{-4} a.u., max. displacement = 1.8×10^{-3} a.u., RMS

Table 1

Input and output coordinates (in degrees) of the DFT geometry optimization of Ac-Asn-Pro-Gly-Gln-NH$_2$

Residue	Dihedral[a]	Input coordinates[b]			B3LYP/6-31G(d)[c]			B3PW91/6-31G(d)[c]		
		I	3-21G	6-31G(d)	*I*	3-21G	6-31G(d)	*I*[d]	3-21G[e]	6-31G(d)
Asn	ω_1	180.0	176.0	166.3	175.6	176.6	176.6	N/A	N/A	−178.4
	Φ_1	−139.0	−172.5	−179.2	−97.2	174.2	174.4	N/A	N/A	170.8
	ψ_1	135.0	174.8	218.9	109.1	178.3	178.4	N/A	N/A	171.4
	χ_1^1	180.0	201.2	55.5	177.0	54.2	54.2	N/A	N/A	54.9
Pro	ω_2	180.0	184.7	188.8	179.3	−178.4	−178.4	N/A	N/A	−177.2
	Φ_2	−60.0	−66.6	−69.4	−55.8	−61.9	−62.0	N/A	N/A	−65.2
	ψ_2	−30.0	−29.9	−21.3	−29.5	−21.8	−21.4	N/A	N/A	−20.4
	χ_2^1	−9.4	−29.3	−27.4	−25.5	−29.2	−29.1	N/A	N/A	−27.1
Gly	ω_3	180.0	186.1	183.2	178.5	173.9	173.8	N/A	N/A	178.8
	Φ_3	−90.0	−122.0	−85.1	−70.0	−93.2	−93.6	N/A	N/A	−79.3
	ψ_3	0.0	25.3	70.1	−3.9	5.8	5.8	N/A	N/A	3.0
Gln	ω_4	180.0	172.6	168.7	171.3	−172.7	−172.8	N/A	N/A	−179.4
	Φ_4	−139.0	−145.8	−137.9	−108.3	−146.2	−146.1	N/A	N/A	−119.8
	ψ_4	135.0	169.1	160.7	15.6	59.5	61.0	N/A	N/A	16.2
	χ_4^1	−60.0	−59.6	−62.0	−68.6	−56.8	−56.5	N/A	N/A	−60.9

[a] Atomic definition of backbone and side chain dihedral angles can be found in Ref. [15–17].

[b] Coordinates in column *I* are ideal values [15–17]. Coordinates in the other two columns are extracted from previously optimized coordinate sets of single amino acids.

[c] Output coordinates obtained from DFT geometry optimization of the corresponding input coordinate set.

[d] Optimization did not converge in 200 iteration steps.

[e] Optimization crashed repeatedly after 77 iterations due to the breaking up of the Pro ring.

Table 2
Input and output coordinates (in degrees) of the DFT geometry optimization of Ac-Val-Pro-D-Ala-His-NH$_2$

Residue	Dihedral[a]	Input coordinates[b]			B3LYP/6-31G(d)[c]			B3PW91/6-31G(d)[c]		
		I	3-21G	6-31G(d)	I	3-21G	6-31G(d)	I	3-21G	6-31G(d)
Val	ω_1	180.0	176.6	177.1	171.3	174.0	176.3	172.9	173.3	174.2
	Φ_1	− 139.0	− 136.5	− 85.2	− 87.5	− 125.8	− 100.6	− 100.8	− 123.1	− 100.1
	ψ_1	135.0	143.2	88.0	147.4	143.4	106.3	150.1	143.7	111.4
	χ_1^1	180.0	185.6	181.6	177.7	− 175.4	177.7	177.3	− 175.5	− 178.4
Pro	ω_2	180.0	185.1	184.8	172.3	− 168.9	− 172.6	174.6	− 168.5	− 172.8
	Φ_2	− 60.0	− 83.3	− 81.4	− 65.9	− 82.5	− 83.2	− 68.8	− 82.9	− 82.6
	ψ_2	120.0	71.3	86.3	95.8	69.4	68.0	90.3	69.3	69.8
	χ_2^1	− 9.4	− 9.3	− 14.1	− 25.2	− 16.3	− 10.7	− 23.7	− 16.1	− 10.8
D-Ala	ω_3	180.0	172.1	179.7	− 178.6	160.9	− 174.4	− 179.1	160.5	− 179.4
	Φ_3	80.0	128.2	85.1	112.0	129.5	82.4	115.7	128.1	86.1
	ψ_3	0.0	− 29.8	− 78.6	− 1.7	− 19.2	− 70.8	− 1.6	− 16.9	− 55.0
His	ω_4	180.0	176.0	176.0	177.8	177.9	168.8	177.5	177.7	174.2
	Φ_4	− 139.0	− 172.1	− 155.3	− 164.5	− 164.5	− 116.6	− 163.5	− 163.8	− 156.7
	ψ_4	135.0	183.2	150.4	174.7	− 179.5	150.3	178.1	− 178.7	− 173.1
	χ_4^1	180.0	212.3	181.9	− 141.5	− 147.0	− 172.0	− 142.9	− 147.1	− 147.9

[a] Atomic definition of backbone and side chain dihedral angles can be found in Ref. [15–17].

[b] Coordinates in column I are ideal values [15–17]. Coordinates in the other two columns are extracted from previously optimized coordinate sets of single amino acids.

[c] Output coordinates obtained from DFT geometry optimization of the corresponding input coordinate set.

displacement $= 1.2 \times 10^{-3}$ a.u.). In case of tetrapeptides, redundant internal coordinates [19–21] were also introduced to avoid the breaking up of Pro and His rings.

2.1. Database generation

A small database consisting of the structures of the amino acids in the sequences selected for this study was generated. The N- and C-termini of the amino acids were capped with acetyl and amido groups, respectively, to mimic a continuous polypeptide chain. Since the conformation of Ac-Asn-Pro-Gly-Gln-NH$_2$ [13] and Ac-Val-Pro-D-Ala-His-NH$_2$ [14] has been elucidated as type I and type II β-turn structures, respectively, only the desired conformations of individual amino acids were optimized using the RHF method on the 3-21G and the 6-31G(d) basis set. Initial internal coordinates were set to their ideal value [15–17], assuming either type I or type II β-turn for the two central residues, and antiparallel

β-pleated conformation for the terminal residues. Initial and optimized values of ω, Φ, ψ, and χ^1 dihedrals (where applicable) are summarized in Tables 1 and 2.

2.2. Geometry optimization of Ac-Asn-Pro-Gly-Gln-NH$_2$ and Ac-Val-Pro-D-Ala-His-NH$_2$

Three coordinate sets were generated for each sequence using input and output coordinates of the geometry optimizations of the single amino acids. All the internal coordinate variables (bond lengths, bond angles and dihedral angles) were first set to the same ideal values [15–17] and were used for the optimization inputs of individual amino acids. The second input coordinate set was built using RHF/3-21G optimized structures, while the third set was constructed from RHF/6-31G(d) optimized coordinates. All six input geometries (Tables 1 and 2) were then optimized at the B3LYP/6-31G(d) and the B3PW91/6-31G(d) levels of theory [22–24].

Table 3
Time efficiency of optimizations

	Ac-Asn-Pro-Gly-Glu-NH$_2$[a]						Ac-Val-Pro-D-Ala-His-NH$_2$[a]					
	B3LYP/6-31G(d)			B3PW91/6-31G(d)			B3LYP/6-31G(d)			B3PW91/6-31G(d)		
	I	3-21G	6-31G(d)	*I*	3-21G	6-31G(d)	*I*	3-21G	6-31G(d)	*I*	3-21G	6-31G(d)
Number of iterations	83	53	55	>200	N/A	96	83	65	63	98	132	69
CPU time (h)	358.53	241.02	219.16	>827.11	N/A	444.96	374.13	256.01	266.17	425.32	475.61	297.75

[a] Inputs are generated using ideal bond lengths, angles and β-turn dihedral angle values (*I*) or by using preoptimized internal coordinates.

3. Results and discussion

Geometry optimizations of tetrapeptides with previously optimized internal coordinates converged in 20–50% less iteration steps, than those of non-optimized variables (Table 3). When the B3PW91/6-31G(d) level of theory was used for the geometry optimization of Ac-Val-Pro-D-Ala-His-NH$_2$ using the RHF/3-21G pre-optimized coordinates as input, a slight increase was observed in both number of optimization cycles and CPU time. In all cases, when the B3LYP functional was applied, the number of optimization cycles decreased significantly with the use of pre-optimized coordinates. Results obtained for different basis sets were not significantly different (Tables 1–3) from those with the previous optimizations of single amino acids suggesting that the 3-21G basis set is adequate for database generation. This is in agreement with previous results [2–3], where the 3-21G basis set was found to be reasonably accurate in reproducing experimentally observed geometric parameters.

The application of the B3PW91 hybrid functional yielded somewhat different results. Higher efficiency of the optimization was maintained only with the application of coordinates, which were previously optimized using the 6-31G(d) basis set. Moreover, full geometry optimization of Ac-Asn-Pro-Gly-Gln-NH$_2$, could carried out only when the RHF/6-31G(d) pre-optimized coordinate set was used. Since, the input coordinate sets for both DFT methods were the same, the possibility that differences emerged from the use of an inappropriate starting structure can be excluded. Therefore, the difference between the results obtained for different DFT methods can be explained only by the difference between the LYP correlation and the PW91 exchange functionals.

4. Conclusion

This study showed that ab initio QM geometry-optimized amino acids can be used to provide internal coordinates sets as inputs for DFT for geometry optimization of polypeptides. With this input generation, faster convergence was achieved and up to 50% CPU time was saved. The saving of 4–5 days in the geometry optimization of four residue peptides in this study indicates that for larger oligopeptides the efficiency could be substantial. When a complete amino acid database is available, including all possible backbone and even side chain conformers, the data can be used not only to generate input coordinate sets for oligopeptides with known or predicted structure, but actually to predict structure based on conformational preference of single amino acids. The extension of the database to include dipeptides and even oligopeptides, could open a new field of protein structure elucidation.

Acknowledgements

This work was supported with grants from the BRIN Program of the National Center for Research (NIH Grant Number 1 P20 RR16469), from the National Science Foundation (EPS-0091900) and by the Carpenter Endowed Chair in Biochemistry, Creighton University.

One of the authors (IGC) wishes to thank the Ministry of Education for a Szent-Györgyi Visiting Professorship.

References

[1] I.A. Topol, S.K. Burt, E. Deretey, T. Tang, A. Perczel, A. Rashin, I.G. Csizmadia, J. Am. Chem. Soc. 123 (2001) 6054.

[2] A. Perczel, M.A. McAllister, P. Császár, I.G. Csizmadia, Can. J. Chem. 72 (1994) 2050.

[3] G. Endrédi, A. Perczel, Ö. Farkas, M.A. McAllister, G.I. Csonka, J. Ladik, I.G. Csizmadia, J. Mol. Struct. (THEOCHEM) 391 (1997) 15–20.

[4] M.A. Patel, E. Deretey, I.G. Csizmadia, J. Mol. Struct. (THEOCHEM) 492 (1999) 1.

[5] M.A. Berg, G.A. Chasse, E. Deretey, A.K. Füzéry, B.M. Fung, D.Y.K. Fung, H. Henry-Riyad, A.C. Lin, M.L. Mak, A. Mantas, M. Patel, I.V. Repyakh, M. Staikova, S.J. Salpietro, T. Tang, J.C. Vank, A. Perczel, Ö. Farkas, L.L. Torday, Z. Székely, I.G. Csizmadia, J. Mol. Struct. 500 (2000) 5 (Millennium Volume).

[6] G.A. Chasse, A.M. Rodriguez, M.L. Mak, E. Deretey, A. Perczel, C.P. Sosa, R.D. Enriz, I.G. Csizmadia, J. Mol. Struct. (THEOCHEM) 537 (2001) 319–361.

[7] G.A. Chass, M.A. Sahai, J.M.S. Law, S. Lovas, Ö. Farkas, A. Perczel, I.G. Csizmadia, Int. J. Quantum. Chem. 90 (2) (2002) 933 (P.O. Löwdin memorial issue pt. II).

[8] M.A. Sahai, G.A. Chass, S. Lovas, A. Varro, J.Gy. Papp, J. Mol. Struct. (Theochem) (2003) in press.

[9] M. Levitt, Biochemistry 17 (20) (1978) 4277.

[10] S. Lifson, C. Sander, Nature 282 (1979) 109.

[11] K.T. O' Neill, W.F. DeGrado, Science 250 (1990) 646.

[12] G.A. Chass, S. Lovas, R.F. Murphy, I.G. Csizmadia, Eur. Phys. J. D 20 (2002) 481.

[13] F.R. Carbone, S.J. Leach, Int. J. Pept. Protein. Res. 26 (1985) 498.

[14] B. Imperiali, S.L. Fisher, R.A. Moats, T.J. Prins, J. Am. Chem. Soc. 114 (1992) 3182.

[15] G.N. Ramachandran, V. Sasisekharan, Adv. Protein Chem. 23 (1968) 283.

[16] G.N. Ramachandran, et al., Biochim. Biophys. Acta 359 (1974) 298.

[17] IUPAC-IUB, Commission on Biochemical Nomenclature, Biochemistry 9 (1970) 3471.

[18] M.J. Frisch, G.W. Trucks, H.B. Schlegel, G.E. Scuseria, M.A. Robb, J.R. Cheeseman, V.G. Zakrzewski, J.A. Montgomery Jr., R.E. Stratmann, J.C. Burant, S. Dapprich, J.M. Millam, A.D Daniels, K.N. Kudin, M.C. Strain, O. Farkas, J. Tomasi, V. Barone, M. Cossi, R. Cammi, B. Mennucci, C. Pomelli, C. Adamo, S. Clifford, J. Ochterski, G.A. Petersson, P.Y. Ayala, Q. Cui, D. Morokuma, K. Malick, A.D. Rabuck, K. Raghavachari, J.B. Foresman, J. Cioslowski, J.V. Ortiz, A.G. Baboul, B.B. Stefanov, G. Liu, A. Liashenko, P. Piskorz, I. Komaromi, R. Gomperts, R.L. Martin, D.J. Fox, T. Keith, M.A. Al-Laham, C.Y. Peng, A. Nanayakkara, M. Challacombe, P.M.W. Gill, B. Johnson, W. Chen, M.W. Wong, J.L. Andres, C. Gonzalez, M. Head-Gordon, E.S. Replogle, J.A. Pople, Gaussian 98, Revision A.9, Gaussian, Inc., Pittsburgh PA, 1998.

[19] P. Pulay, G. Fogarasi, J. Chem. Phys. 96 (4) (1992) 2856.

[20] J. Baker, J. Comp. Chem. 14 (9) (1993) 1085.

[21] C. Peng, P.Y. Ayala, H.B. Schlegel, M.J. Frisch, J. Comp. Chem. 17 (1) (1996) 49.

[22] A.D. Becke, J. Chem. Phys. 98 (1993) 5648.

[23] C. Lee, W. Yang, R.G. Parr, Phys. Rev. B 37 (1988) 785.

[24] J.P. Perdew, K. Burke, Y. Wang, Phys. Rev. B 54 (1996) 16533.

ELSEVIER

Journal of Molecular Structure (Theochem) 666–667 (2003) 361–371

THEO
CHEM

www.elsevier.com/locate/theochem

The functional benefits of protein disorder

Peter Tompa*

Institute of Enzymology, Biological Research Center, Hungarian Academy of Sciences, H-1518, P.O. Box 7, Budapest, Hungary

Abstract

Intrinsically unstructured proteins, which exist and function without a well-defined folded structure, constitute a newly recognized class of the protein world. Such proteins are common in living organisms and occupy a unique niche in which function is intimately linked with structural disorder. In this paper it is shown that these proteins assume no compact structural state in vivo, despite a significant macromolecular crowding effect that could force them fold under cellular conditions. Further, it is argued that they are not fully unstructured, but contain functionally indispensable residual structure. Their recently suggested functional classification is delineated and extended through novel examples. Finally, the functional benefits of structural disorder encompassing some newly recognized features, such as direct proteasomal degradation and chaperone activity, will be thoroughly discussed. Through all these details, the message is conveyed that our understanding of protein function will not be complete as long as the functional benefits of protein disorder are not fully appreciated.
© 2003 Elsevier B.V. All rights reserved.

Keywords: Intrinsically unstructured protein; Natively unfolded protein; Natively disordered protein; Unstructured in vivo; Functional classification

1. Introduction

Ever since we came to realize that our life is founded on the multifarious functioning of proteins, our basic concepts regarding their action have been dominated by the view that a well-defined 3D structure is essential for their function. A recent surge of reports, however, caution that this view portrays too simple a picture as for many proteins and protein domains the native state is intrinsically unstructured (also termed natively disordered or unfolded) [1–4]. As systemized in these recent reviews, the evidence for the unstructured nature of these IUPs in vitro is manifold and covers data obtained by all possible physicochemical techniques. In general, IUPs are structurally highly flexible and lack a compact, globular fold, in which they resemble the denatured states of globular proteins. In contrast to

Abbreviations: CD, circular dichroism; CREB, cAMP response element binding protein; Cdk, cyclin-dependent kinase; CST, calpastatin; CTD, C-terminal domain; Cyc, cyclin; DLS, dynamic light scattering; 4EBP(1), eukaryotic translation initiation factor 4E binding protein (1); FnBP, fibronectin binding protein; FTIR, Fourier-transformed infrared spectroscopy; IUP, intrinsically unstructured protein; MAP(2), microtubule-associated protein (2); NACP, non-A beta component of Alzheimer's disease amyloid plaque (also termed α-synuclein); NMR, nuclear magnetic resonance; PCNA, proliferating cell nuclear antigen; PEVK, region rich in Pro, Glu, Val and Lys; PKI, protein kinase inhibitor; PP II, polyproline II helix; PRG, Pro-rich glycoprotein; RNAP II, DNA-dependent RNA polymerase II; SAXS, small-angle X-ray scattering; SDS-PAGE, sodium dodecyl sulfate-polyacrylamide gelelectrophoresis; TAD, transactivator domain.

* Tel.: +361-279-3143; fax: +361-466-5465.
E-mail address: tompa@enzim.hu (P. Tompa).

these, however, they fulfill essential functions [1,3,4] and they are rather common in living organisms. By estimations based on their sequence signature [3,5], low compositional complexity [6] and heat-stability [7], about 10–20% of full-length proteins belong to this class and 25–40% of all residues fall into such regions (Table 1 for examples). Protein disorder, thus, is a general phenomenon and the structural and functional exploration of this protein class is one of the most challenging and imperative task various branches of protein science face in the post-genomic era.

2. IUPs are unfolded in vivo

There is ample evidence that IUPs lack a compact, globular fold and are highly flexible in vitro [1–4]. The unusual flexibility of their polypeptide chain is often demonstrated by missing backbone coordinates in X-ray structures or diminished chemical shift dispersion in NMR experiments, whereas their lack of significant ordinary secondary structure is documented by far-UV CD and FTIR spectroscopy. Near-UV CD and UV spectroscopy, as well as hydrodynamic techniques, such as gel-filtration, SAXS and DLS demonstrate their extended, open conformation and lack of a tightly packed hydrophobic core. Complemented by some indirect observations, such as proteolytic sensitivity, heat stability and unusual SDS-PAGE mobility, all relevant data suggest that IUPs assume an extended, highly flexible structure with very limited secondary and practically no tertiary interactions.

Recently, nevertheless, it has been raised that this situation is an artefact, resulting only from studying IUPs in highly diluted solutions in vitro. At the core of the argument is the fact that conditions in vivo are basically different, as the extreme macromolecular concentrations (300 mg/ml) in living cells give rise to a crowding effect that may significantly shift chemical potentials in favor of association and folding reactions [8]. Signs of such an effect, in fact, have been detected with FlgM, the inhibitor of σ^{28} transcription factor. When this protein is ectopically expressed in E. coli, or studied at extreme solute concentrations, its C-terminal half undergoes local ordering, as seen by 2D NMR [9]. This part of the inhibitor is the same, which

preferentially samples locally structured conformations in dilute solutions and folds when the molecule contacts its partner [10]. The issue of whether IUPs in general assume a folded structure in vivo is of paramount importance as our entire novel concept of unstructured functional states hinges on this. In this Section, I will recite several direct observations and indirect considerations, which all point to that the mostly disordered, unstructured state of IUPs does prevail in vivo.

First, the preference of several IUPs or denatured globular proteins for a locally ordered or collapsed state has been probed under crowded conditions. In all the studies practically no [11] or only marginal [12–14] tendency for the formation of ordered structure has been observed. Clearly, the crowding effect cannot promote an overall folding of the IUPs studied; if at all, probably only local structural elements of functional importance form, as seen for FlgM [9].

The second argument is related to extracellular IUPs, for which this issue is fully irrelevant, as they do not experience a crowded environment in vivo. To give a few examples only, one should recall caseins, which occur in the milk, proline-rich glycoproteins (PRGs), abundant in human saliva and fibronectin-binding proteins (FnBPs), protruding from the extracellular side of bacteria. The heat-stability of a large fraction of serum proteins [7] predicts that we are to see many more such examples.

Third, the observed evolutionary rate of IUPs also argues against these proteins generally assuming a compact fold in vivo (also [3]). In general, changes in protein sequences are impeded by constraints on residues directly involved in functional or structural interactions. This has been assessed in the case of the sex-determining transcription factor, SRY, for which the level of synonymous vs. nonsynonymous mutations is low (0.1–0.2) within its globular DNA-binding domain, but much higher (0.4–0.8) for its rapidly-evolving, Gln-rich transactivator domain [15, 16]. A similar observation has been made in the case of casein, for which the mutation rates of translated and non-translated regions have been compared. A significantly higher mutation rate for the translated region has been found which argues for the lack of structural constraints for this IUP ([17]). This issue of evolutionary rates has been comprehensively

Table 1
Functional classification of IUPs. IUPs can be classified in terms of their functional modes into five broad categories; this table represents an extension of the previous classification presented in [4]. For novel examples not covered there, the respective reference is given in square brackets. The binding partner of the protein (when applicable), and its physiological function are shown. An important feature to note is that several proteins/domains belong to more than one categories, which attests to the structural and functional malleability of IUPs

Protein (IUP)	Target	Action/function
Entropic chains		
tau/MAP2 projection domain	Not applicable	Entropic bristle (spacing in cytoskeleton)
Titin PEVK domain	Not applicable	Entropic spring (passive contractile force in muscle)
SNAP-25 linker region	Not applicable	Flexible spacer/linker of binding domains
Neurofilament-H KSP domain [22]	Not applicable	Entropic bristle (spacing in neurofilament lattice)
Nup2p FG repeat region [92]	Karyopherins	Gating in nuclear pore complex
K channel N-terminal region [93]	Not applicable	Entropic clock/inactivation gate
Oct-1 POU linker [73]	Not applicable	Flexible linker of DNA-recognition domains
Grb2 linker [74]	Not applicable	Flexible linker of SH2-SH3 domains
Effectors		
Calpastatin	Ca^{2+}-activated protease (calpain)	Inhibitor of calpain in Ca^{2+} signalling
p21^{Cip1}/27^{Kip1}	Cyclin-dependent kinases	Inhibitors in cell cycle regulation
4EBP1, 2, 3	Eucaryotic translation initiation factor (eIF4E)	Inhibitor of translation initiation
PKI	cAMP-dependent protein kinase	Inhibitor in signal transduction
PP I1, I2; DARPP32	Phosphorylase phosphatase	Inhibitor in signal transduction
Securin	Separase	Inhibitor of chromosome separation
FlgM	sigma28 transcription factor	Inhibitor of flagellin-specific gene expression
Stathmin	Tubulin dimers	Microtubule disassembly, catastrophe
IA₃ [94]	Aspartic proteinase A	Inhibitor of proteinase
α-synuclein (NACP) [63]	Partially unfolded proteins	Chaperone
Caseins [63]	?	Chaperone
Dehydrins [65]	Membrane	Modulation of membrane fluidity (?)
DHPR loop II–III C fragment [66]	Ryanodine receptor	Inhibitor and activator of receptor
Scavengers		
Thymosins (proTα)	Zn^{2+}, histone	Not reported
Caseins	Calcium phosphate	Nanocluster formation, inhibition of precipitation in milk
Chromogranins A, B, C	Ca^{2+}/CaM, catecholamine	Catecholamine sequestration
Salivary proline-rich glycoprotein (PRG)	Tannin	Binding/neutralization of polyphenolic plant compounds
Desiccation stress protein (Dsp) 16	Water	Retaining water to prevent desiccation of plants
NACP	?	Synaptic plasticity

(continued on next page)

Table 1 (*continued*)

Protein (IUP)	Target	Action/function
Assemblers		
tau/MAP2 microtubule-binding domain	Microtubules	Microtubule polymerization, bundling
Caldesmon	Ca^{2+}/CaM, F-actin, myosin, tropomyosin	Actin polymerization, bundling
Bob1	Oct1 transcription factor, Igk promoter, $TAF_{II}105$	B-cell specific expression of Ig genes
Histone-binding Protein N1/N2	Histone	Histone storage, chromatin assembly
L7/L12	rRNA	Ribosome assembly/stability
λphage N protein	mRNA, NusA, RNA Pol II	Translation antitermination
SIBLING proteins	Integrin, complement factor H, CD44, fibronectin	Assembly of bone extracellular matrix
FnBP D_1–D_4	Fibronectin	Adherence to extralellular matrix of host in bacterial invasion
CREB TAD	TATA-box associated factors (TAFs)	Assembly of transcription preinitiation complex
$p21^{Cip1}/27^{Kip1}$ [95]	Cyclin D-Cdk 4	Assembly of functional CycD–Cdk4 complex
Display sites		
CREB TAD	Protein kinases (e.g. PKA, CaMKIV)	Regulation by phosphorylation
MAP2 microtubule-binding domain	Protein kinases (e.g. PKA, MARK)	Regulation by phosphorylation
Bcl-2 linker region	Proteases (e.g. caspase)	In vivo proteolysis site
Caseins [67]	Proteasome	Turnover by proteasome
tau [68]	Proteasome	Turnover by proteasome
NACP [69]	Proteasome	Turnover by proteasome
Cyclin B N-terminal domain [71]	Ubiquitination system	Ubiquitination

addressed in a study carried out by Dunker and colleagues [18]. By comparing the pairwise genetic distances within ordered and disordered regions of 26 protein families, the disordered regions were found to evolve significantly faster in 19 families, and more slowly in 2 families only. For these, the disordered regions are known to be binding sites. All these observations suggest that proteins/regions unstructured in vitro are subject to much less structural constraints in their native state than their structured counterparts, i.e. they are, by all probability, also unstructured in vivo.

The fourth point in favor of the lack of a compact fold for several IUPs in vivo comes from inspecting their mode of binding to their partners. As noted [4], in three out of their five functional categories (i.e. assemblers, effectors and scavengers), IUPs function by molecular recognition and often they can bind to several partners at the same time. In certain cases the 3D structure of an IUP bound to its partner(s) is solved, and is seen to bind in an extended conformation (Fig. 1). For these proteins, it would be rather counterintuitive to assume a folded, compact state prior to binding that would have to unfold for the functional interactions to occur. A similar argument applies to components of certain multiprotein complexes, which cannot be assembled from rigid partners due to steric clashes: an assumption of a compact fold prior to formation of the functional complex does not make much sense [19].

Finally, for the other two functional categories (entropic chains and display sites), function more directly stems from the disordered state and thus in vivo folding is out of question. Display sites are regions that are recognized by enzymes that carry out posttranslational modification preferentially on

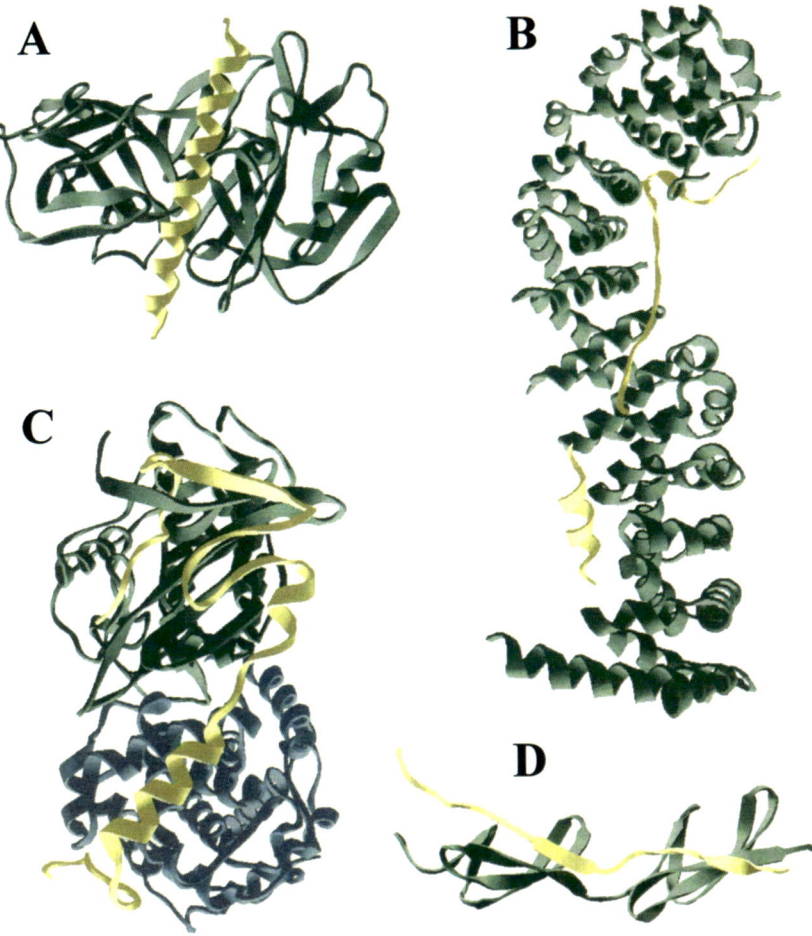

Fig. 1. IUPs bind to their partners in an extended conformation. The structure of some IUPs bound to their partner(s), determined by X-ray crystallography or NMR, is shown (in yellow). The PDB file (identifier in parenthesis) with the 3D coordinates of (A) IA_3 aspartic proteinase inhibitor bound to proteinase A (1DPJ); (B) the transactivator domain CBD of Tcf3 bound to β-catenin (1G3J), (C) the kinase activator domain of p27[Kip1] bound to CycA-Cdk2 (1JSU) and (D) the B3 segment of FnBPA bound to two Fn type 1 repeats (^{1}F1^{2}F1) of fibronectin (1O9A), has been downloaded from the Protein Data Bank, and the structures have been visualized by the Swiss-PDB viewer. It is to be noted that although IUPs attain structural order, they do not fold into a compact, globular state upon binding to their physiological partner.

disordered regions [3]. Proteins or protein domains of entropic chain function simply could not fulfill their function if they had a well-defined, let alone compact, structure; to give the most characteristic examples only, the reader is referred to the PEVK region of titin, an entropic spring that generates passive tension in muscle [20], the projection domain of MAP2 [21] and the side-arm of neurofilament H [22], entropic bristles, which ensure proper spacing of cytoskeletal filaments or the FG repeat region of nucleoporins, which regulate transport through the nuclear pore complex via entropic gating, i.e. spatial exclusion and specific recognition of transport proteins [23].

3. Residual structure in IUPs

Overall, IUPs resemble the denatured state of globular; proteins prior to recognizing their commonness and functional importance there had been no serious reason to pass this structural analogy. As most of the data imply their lack of secondary and tertiary structure, IUPs have been mostly considered random-coil or random-coil-like proteins. Besides the objection that a true random coil does not exist even under strong denaturing conditions [24] and that a featureless conformational state is not compatible with most IUP functions [4], a meticulous perusal of literature reveals that IUPs display signs of structural order so far largely overlooked.

CD, the technique most frequently used for characterizing proteins in terms of secondary structure is rather unreliable at low percentages of structural elements. When given a closer look, however, it often shows a small but significant (10–20%) α and/or β structure, such as for caseins [17], NACP (α-synuclein) [7], tau [25], stathmin [26], p21^{Cip1} [27], 4EBP1 [28], λN [29], Dsp16 [30], CST [31] and CREB TAD [32], for example. As another mark of structural order, the spectrum of the full-length protein sometimes cannot be assembled from the spectra of its fragments due to long-range interactions, as shown for stathmin [26]. FTIR also shows some secondary structure for NACP [33] and tau [25]. In other cases structure can be inferred from the observed shift of the CD spectrum toward the random-coil state upon heating/denaturation, such as for caseins [7,17], CST [31] and FnBPA [34]. Further, the CD spectrum often reveals the presence of PP II conformation. This secondary-structural motif, very effective in molecular recognition [35], occurs frequently in the loop regions of globular proteins [36] and has been observed for tau (25%) [37], casein (23%) [38], stathmin (not quantified) [26], Bob1 (not quantified) [39], the PEVK region of titin [40] and the CTD of RNAP II [41]. PP II structure has also been detected in some IUPs by a related spectroscopic technique, Raman optical activity measurements: the spectra of casein, NACP, tau [42] and some wheat gluten proteins [43] show substantial contributions by PP II conformation. A significant deviation from random-coil parameters, which implies a restriction of flexibility due to medium- or long-range interactions, has been observed by NMR for FnBPA [44], Dsp16 [30] and PRG [45]. In a systematic analysis of residual structure encompassing hydrodynamic and CD spectroscopic data for a bunch of IUPs, Uversky has pointed out that IUPs may be classified into coil-like

and premolten-globule-like classes, with significant residual structure in the latter [2].

Whether such residual structure represents a small amount of stable conformation or an ensemble of rapidly interconverting conformers with some structural preference, cannot be told by traditional spectroscopic approaches. Such information is provided by NMR, which has shown transient secondary structures for FlgM [10] and λN [29]. An attractive possibility is that the protein preferentially samples conformational states which occur upon binding to its target, as explicitly suggested in several cases, such as FlgM [10], CREB TAD [46] and p27^{Kip1} [47], for example.

A deviation from a fully disordered state has also been demonstrated by other techniques. Specific phosphorylation of tau, for example, causes the protein to become extended and stiff [48], as shown by examining paracrystal structure. Such a dramatic conformational change is incompatible with a random-coil state. In another study, an Alzheimer's-disease specific tau epitope only formed upon the orderly action of kinases: it was concluded that certain conformational states have to follow in succession in order to generate the defined conformation characteristic of the epitope [49]. Pertinent to these points is the demonstration that certain residues in FnBP [50] and Sp1 TAD [51], not in contact with the target, contribute to binding; it has been suggested that they are important for presenting the binding site in an optimal conformation [50].

Perhaps most revealing with respect to residual structure subserving function are the results of limited proteolysis. As proteases attack sites which are both sterically accessible and flexible, this technique is traditionally used to probe the tertiary organization of globular proteins and their folding intermediates [52] and is not generally considered suitable for studying IUPs. In certain cases, however, IUPs undergo limited proteolysis at very low protease concentrations, which implies their non-random structural organization. The best examples, again, are tau and MAP2, which have two large functional domains; at low concentrations all the proteases tested cleaved both tau [53] and MAP2 [54] within the same narrow region, which separates the two functional domains. Thus, functional and structural domain organization correlate for these proteins, and similar behavior was reported for stathmin [55], CST [56], nucleoplasmin [57], caldesmon [58] and GC4N TAD [59].

A final point to note is that IUPs sometimes recognize and bind each other: such is the case of multiple Pro-rich region systems, e.g. the interaction between Bob1 and Oct1 TAD [60], RNAP II and TADs of transcription activators, or multiple vesicle-associated proteins [35]. These cases can hardly be reconciled with a fully featureless conformational state.

In conclusion, the evidence is overwhelming that many IUPs have residual structure. In order to understand the mechanistic details of their functioning, it will be crucial to find out if, and how, their limited and possibly transient residual structure is related to their functions.

4. Functional classification of IUPs

The unique structural features of IUPs predisposes them for functional modes in which stable structure is not needed, but function either directly stems from the disordered state or is realized via recognition of, and adaptation to, a partner molecule. In keeping with this point, Dunker and colleagues suggested that the functions of IUPs can be classified into 28 distinct categories [3]. In my subsequent review, I put forward that these functional modes actually segregate into five broad categories [4]. This classification was based on a limited set of some 20 proteins and its validity is worth checking on the broader basis of more than 100 IUPs and unstructured domains we have come to know by now [1–4,61,62]. Although limitations in space do not allow to cover all individual examples, several novel cases are enlisted in Table 1: it appears that the classification suggested is corroborated and somewhat complemented by the new examples.

The first class of entropic chains contains proteins with functions that directly stem from the ability of their polypeptide chain to attain a continuum of structural states and to interfere with the location of other domains or proteins that way. The cases show that functional sub-classes, such as entropic springs, bristles/spacers, linkers, clocks, etc. apply, in accord with [3]. The second class of effectors seems to be more diverse than previously thought. In the first scheme [4], only inhibitors have been enlisted; now

more subtle ways of modulating the activity of their partner have been recognized. For some IUPs, such as NACP [63] and casein [64], chaperone activity has been demonstrated, for others, such as plant dehydrins [65], so far uncharacterized modulation of membrane function (possibly membrane fluidity) can be surmised. An exciting novelty is the observation that the C fragment of dihydropyridine receptor II–III loop can bind to the ryanodine receptor in two distinct conformations, one inhibiting but the other activating it [66]. The third class of scavengers, which store and/ or neutralize small ligands, appears rock solid, with some new examples. In the fourth class of assemblers, which assemble, stabilize and regulate multi-protein complexes such as the ribosome, cytoskeleton, transcription pre-initiation complex, chromatin, or even the extracellular matrix, some novel cases have emerged. Intriguing novel examples in this category are $p21^{Cip1}$ and $p27^{Kip1}$, correctly classified as effectors (inhibitors) of cyclin-dependent kinase action before [4]. The observation that they are also needed for the assembly of the CycD–Cdk4 complex puts them also into this class, underlying the point that any protein or protein domain can belong to several classes within this scheme ([4] for other cases). The fifth category is a special case of molecular recognition by display sites, which undergo regulatory posttranslational modification, most frequently phosphorylation or limited proteolysis. This often requires intrinsic disorder that enables transient but specific interaction with the active site of practically any kind of modifying enzyme. Two novel observations worthy of note are the recent observations of direct recognition and cleavage of some unstructured, non-ubiquitinated proteins, such as casein [67], tau [68], NACP [69] and $p21^{Cip1}$ [70] by the 20S proteasome and ubiquitination of securin and cyclin B, that requires the signal of structural disorder itself [71].

In general, the classification of IUPs into five functional classes appears to be suitable for systemizing all the diverse functional modes of this rapidly growing family of proteins. An important aspect of this classification scheme, as already stated above, is that the various functional modes are not exclusive. Three examples in Table 1 illustrate this point. As shown, $p21^{Cip1}/p27^{Kip1}$ may function both as effectors and assemblers, depending on the actual conditions and physiological context of their action. With MAP2, the projection domain is enlisted as an entropic bristle, whereas the microtubule-binding domain serves both as an assembler for microtubules and a display site for phosphorylation. The situation is similar for the TAD of CREB, which binds TATA-box associated factors in the assembly of the transcription preinitiation complex and also subtly regulated by phosphorylation.

5. The benefits of structural disorder

The major point with respect to understanding IUP function is the realization that for these proteins the lack of a well-defined structure not only is tolerated, but, on the contrary, functionally advantageous. In each of the functional classes (Table 1) the highly flexible, malleable structural state enables unique functional features which are unparalleled by ordered proteins. These features have already been covered in several recent reviews [1–4]; novel observations extend their range in unexpected directions.

The functional imprint of the unique conformational freedom of IUPs is most apparent with entropic chains, for which function stems directly from the ability of their polypeptide chain to wander over a large configurational space. This way, they may exert a long-range, entropic exclusion of other proteins or cellular constituents, which may provide for spacer functions (MAPs and neurofilamtn H), but also for gating (nucleoporins). Another molecular setting where such regions abound is in multidomain proteins, where globular domains are often separated by flexible linkers. These regions regulate distance and enable much freedom in orientational search [72], that permits the recognition of distant and/or discontinuous determinants on the target. Such binding may allow for specificity and adaptability not possible by an overall well-structured protein, as shown in the case of Oct1 POU transcription factor [73] or Grb2 adaptor protein [74].

This unique capacity is also exploited by IUPs in the other classes, which function via molecular recognition. These proteins undergo a significant disorder-to-order transition, i.e. induced local folding upon binding to their target, which is accompanied by a large decrease in conformational entropy. This unfavorable energy term uncouples binding strength from specificity and renders highly specific interactions reversible, which is fundamental in regulation. This effect was explicitly demonstrated by comparing the interaction of a constitutive (c-Myb) and an inducible (CREB) transcription factor with the KIX domain of CREB-binding protein (CBP). Inducibility of KIX binding was shown to be linked with disorder and the negative entropy change upon binding of the TAD of CREB, as opposed to the binding of c-Myb, which is structured in isolation [75]. By definition, such disorder-to-order transition occurs for the proteins shown in Fig. 1, but has also been demonstrated for other IUPs, such as FlgM [76], PKI [77], 4EBP1 [78] and stathmin [26], and may apply to all unstructured proteins which function in molecular recognition [1,4,79].

Another prominent feature of IUPs involved in molecular recognition is that their extended structure enables them to contact their partner(s) over a large binding surface for a protein of the given size. For effectors, the major benefit of this feature is the specificity that derives from multiplicity of contact points. The situation is somewhat different for scavengers, for which multivalent binding is essential for the high-stoichiometry binding of small ligands. The benefits of an extended binding surface are most apparent with assemblers, which by definition function by bringing multiple partners together. In recent proteomic initiatives one major objective is to uncover all the interactions within the proteome, as complicated multiprotein complexes appear to be of great functional significance [80]. In a recent review [81] it has been convincingly argued that disordered proteins, as compared to folded proteins, have the same interaction capacity with smaller protein size, which allows the same functional potential to be encoded by an overall more compact and economical genome. Further, flexibility is directly linked to the assembly process itself, as large complexes cannot be put together from rigid components due to steric hindrance. As shown, for example, for viral capsids and bacterial flagella [19], significant structural rearrangements must occur for these to achieve the final functional state.

A further exquisite functional capacity that results from structural flexibility is the malleability of IUPs, i.e. that they can structurally adapt to different partners. This phenomenon, termed binding diversity/promiscuity [27] or one-to-many signaling [62], enables an exceptionally plastic behavior in response to the needs of the cell. The classical and probably best characterized case is the Cdk inhibitor p21[Cip1], which can interact with CycA-Cdk2, CycE-Cdk2, CycD-Cdk4 [27], the Rho kinase [82] apoptosis signal-regulating kinase 1 [83] and PCNA [84] under different conditions; further examples can be found in [4].

The flexibility of IUPs may also be advantageous due to enabling an increased speed of interaction. Macromolecular association rates are highly enhanced by random search in the conformational space and an initial, relatively nonspecific, association [85]. The extended conformation of IUPs is well suited to accomplish such rate enhancements via this 'fly-casting' mechanism [86]; the rate-enhancement can be of several orders of magnitude, as exemplified by casein, which neutralizes calcium-phosphate in milk by binding small seeds very rapidly as they form. Kinetic data put the binding-site activity of casein to 10^6–$10^7 \, s^{-1}$, paralleling the turnover rate of the fastest enzymes [87].

An additional point to note is that the extreme proteolytic sensitivity of IUPs, in principle, may allow for their effective control via rapid but regulated turnover. Although this mechanism has not been directly tested, it is supported by several recent observations. A comprehensive analysis of the frequency of protein disorder in various protein classes [88] has revealed a significant excess of disorder in signaling, regulatory and cancer-associated proteins, which are usually short-lived proteins. It was also pointed out, that the PEST hypothesis, formulated to account for the structural background of protein instability in vivo [89], may be intimately linked with the characteristic amino-acid composition of IUPs [4]. Furthermore, it appears now that disorder itself constitutes part of the signal recognized by the protein destruction machinery, in two distinct ways. First, IUPs can be destroyed by the 20S proteasome *without* prior ubiquitination; such a fate has been ascertained for casein [67], tau [68], NACP [69] and p21[Cip1] [70]. This mechanism may be important in regulating turnover of these proteins, but may also fulfill a more subtle regulatory role, as demonstrated by the endoproteolytic activity of the proteasome [90]. By attaching globular domains to the two ends of NACP and p21[Cip1], it was demonstrated that the proteasome can process the unstructured middle part, releasing the terminal globular domains. As globular domains in multidomain proteins are often linked by unstructured linker regions, this mechanism may provide for an efficient means of generating constitutively activated functional domains from these. Second, disorder may be recognized by the ubiquitination system itself. As shown for two cell-cycle regulating proteins, securin and cyclin B [71], their regions recognized by the ubiquitination machinery are natively unfolded. The lack of a folded structure for these regions explains prior enigmatic observations, such as polyubiquitination and multiple sites of ubiquitination. As an unexpected twist of the tale, ubiquitination of such unstructured regions may not only affect degradation, but may also directly stimulate activity, as shown for certain transcription factors [91]. Destruction and transcriptional activation signals often overlap on transactivator domains, and it was demonstrated directly in the case of VP16 TAD that its ubiquitination not only marks it for destruction but it is also mandatory for its activation. The generality of this mechanism remains to be seen but the idea of linking activation to destruction via ubiquitination undoubtedly has an appeal.

6. Conclusion

Intrinsically unstructured proteins display functional features not witnessed among globular proteins we have been so used to until now. Their exceptional adaptability and sensitivity to control provide a unique combination for unusual functional features that are just beginning to be appreciated. To achieve their full understanding, the exploration of the molecular details of their ways of function has to be of high priority in our endeavors for many years to come.

Acknowledgements

This work was supported by grants ISRF GR067595 from the Wellcome Trust and T32360 from OTKA.

References

[1] P.E. Wright, H.J. Dyson, J. Mol. Biol. 293 (1999) 321.

[2] V.N. Uversky, Protein Sci. 11 (2002) 739.

[3] A.K. Dunker, C.J. Brown, J.D. Lawson, L.M. Iakoucheva, Z. Obradovic, Biochemistry 41 (2002) 6573.

[4] P. Tompa, Trends Biochem. Sci. 27 (2002) 527.

[5] A.K. Dunker, Z. Obradovic, P. Romero, E.C. Garner, C.J. Brown, Genome Inform. Ser. Workshop Genome Inform. 11 (2000) 161.

[6] J.C. Wootton, Curr. Opin. Struct. Biol. 4 (1994) 413.

[7] T.D. Kim, H.J. Ryu, H.I. Cho, C.H. Yang, J. Kim, Biochemistry 39 (2000) 14839.

[8] R.J. Ellis, Trends Biochem. Sci. 26 (2001) 597.

[9] M.M. Dedmon, C.N. Patel, G.B. Young, G.J. Pielak, Proc. Natl Acad. Sci. USA 99 (2002) 12681.

[10] G.W. Daughdrill, L.J. Hanely, F.W. Dahlquist, Biochemistry 37 (1998) 1076.

[11] S.L. Flaugh, K.J. Lumb, Biomacromolecules 2 (2001) 538.

[12] B. van den Berg, R. Wain, C.M. Dobson, R.J. Ellis, Embo. J. 19 (2000) 3870.

[13] A.S. Morar, A. Olteanu, G.B. Young, G.J. Pielak, Protein Sci. 10 (2001) 2195.

[14] Y. Qu, D.W. Bolen, Biophys. Chem. 101–102 (2002) 155.

[15] P.K. Tucker, B.L. Lundrigan, Nature 364 (1993) 715.

[16] L.S. Whitfield, R. Lovell-Badge, P.N. Goodfellow, Nature 364 (1993) 713.

[17] C. Holt, L. Sawyer, J. Chem. Soc. Faraday Trans. 89 (1993) 2683.

[18] C.J. Brown, S. Takayama, A.M. Campen, P. Vise, T.W. Marshall, C.J. Oldfield, C.J. Williams, A. Keith Dunker, J. Mol. Evol. 55 (2002) 104.

[19] K. Namba, Genes Cells 6 (2001) 1.

[20] K. Trombitas, M. Greaser, S. Labeit, J.P. Jin, M. Kellermayer, M. Helmes, H. Granzier, J. Cell Biol. 140 (1998) 853.

[21] R. Mukhopadhyay, J.H. Hoh, FEBS Lett. 505 (2001) 374.

[22] H.G. Brown, J.H. Hoh, Biochemistry 36 (1997) 15035.

[23] D.P. Denning, S.S. Patel, V. Uversky, A.L. Fink, M. Rexach, Proc. Natl Acad. Sci. USA 100 (2003) 2450.

[24] D. Shortle, Faseb J. 10 (1996) 27.

[25] O. Schweers, E. Schonbrunn-Hanebeck, A. Marx, E. Mandelkow, J. Biol. Chem. 269 (1994) 24290.

[26] G. Wallon, J. Rappsilber, M. Mann, L. Serrano, Embo J. 19 (2000) 213.

[27] R.W. Kriwacki, L. Hengst, L. Tennant, S.I. Reed, P.E. Wright, Proc. Natl. Acad. Sci. USA 93 (1996) 11504.

[28] J. Marcotrigiano, A.C. Gingras, N. Sonenberg, S.K. Burley, Mol. Cell 3 (1999) 707.

[29] M.R. Van Gilst, W.A. Rees, A. Das, P.H. von Hippel, Biochemistry 36 (1997) 1514.

[30] T. Lisse, D. Bartels, H.R. Kalbitzer, R. Jaenicke, Biol. Chem. 377 (1996) 555.

[31] M. Hackel, T. Konno, H. Hinz, Biochim. Biophys. Acta 1479 (2000) 155.

[32] J.P. Richards, H.P. Bachinger, R.H. Goodman, R.G. Brennan, J. Biol. Chem. 271 (1996) 13716.

[33] P.H. Weinreb, W. Zhen, A.W. Poon, K.A. Conway, P.T. Lansbury Jr., Biochemistry 35 (1996) 13709.

[34] K. House-Pompeo, Y. Xu, D. Joh, P. Speziale, M. Hook, J. Biol. Chem. 271 (1996) 1379.

[35] M.P. Williamson, Biochem. J. 297 (1994) 249.

[36] A.A. Adzhubei, M.J. Sternberg, J. Mol. Biol. 229 (1993) 472.

[37] V.N. Uversky, S. Winter, O.V. Galzitskaya, L. Kittler, G. Lober, FEBS Lett. 439 (1998) 21.

[38] A.L. Andrews, D. Atkinson, M.T.A. Evans, E.G. Finer, J.P. Green, M.C. Phillips, R.N. Robertson, Biopolymers 18 (1979) 1105.

[39] J.F. Chang, K. Phillips, T. Lundback, M. Gstaiger, J.E. Ladbury, B. Luisi, J. Mol. Biol. 288 (1999) 941.

[40] K. Ma, L. Kan, K. Wang, Biochemistry 40 (2001) 3427.

[41] E.A. Bienkiewicz, A. Moon Woody, R.W. Woody, J. Mol. Biol. 297 (2000) 119.

[42] C.D. Syme, E.W. Blanch, C. Holt, R. Jakes, M. Goedert, L. Hecht, L.D. Barron, Eur. J. Biochem. 269 (2002) 148.

[43] E.W. Blanch, D.D. Kasarda, L. Hecht, K. Nielsen, L.D. Barron, Biochemistry 42 (2003) 5665.

[44] C.J. Penkett, C. Redfield, J.A. Jones, I. Dodd, J. Hubbard, R.A. Smith, L.J. Smith, C.M. Dobson, Biochemistry 37 (1998) 17054.

[45] J. Muenzer, C. Bildstein, M. Gleason, D.M. Carlson, J. Biol. Chem. 254 (1979) 5629.

[46] Q.X. Hua, W.H. Jia, B.P. Bullock, J.F. Habener, M.A. Weiss, Biochemistry 37 (1998) 5858.

[47] E.A. Bienkiewicz, J.N. Adkins, K.J. Lumb, Biochemistry 41 (2002) 752.

[48] T. Hagestedt, B. Lichtenberg, H. Wille, E.M. Mandelkow, E. Mandelkow, J. Cell Biol. 109 (1989) 1643.

[49] Q. Zheng-Fischhofer, J. Biernat, E.M. Mandelkow, S. Illenberger, R. Godemann, E. Mandelkow, Eur. J. Biochem. 252 (1998) 542.

[50] M.J. McGavin, G. Raucci, S. Gurusiddappa, M. Hook, J. Biol. Chem. 266 (1991) 8343.

[51] G. Gill, E. Pascal, Z.H. Tseng, R. Tjian, Proc. Natl Acad. Sci. USA 91 (1994) 192.

[52] A. Fontana, P. Polverino de Laureto, V. De Filippis, E. Scaramella, M. Zambonin, Fold. Des. 2 (1997) R17.

[53] B. Steiner, E.M. Mandelkow, J. Biernat, N. Gustke, H.E. Meyer, B. Schmidt, G. Mieskes, H.D. Soling, D. Drechsel, M.W. Kirschner, M. Goedert, E. Mandelkow, Embo J. 9 (1990) 3539.

[54] H. Wille, E.M. Mandelkow, E. Mandelkow, J. Biol. Chem. 267 (1992) 10737.

[55] V. Redeker, S. Lachkar, S. Siavoshian, E. Charbaut, J. Rossier, A. Sobel, P.A. Curmi, J. Biol. Chem. 275 (2000) 6841.

[56] R. De Tullio, M. Averna, F. Salamino, S. Pontremoli, E. Melloni, FEBS Lett. 475 (2000) 17.

[57] C. Dingwall, S.M. Dilworth, S.J. Black, S.E. Kearsey, L.S. Cox, R.A. Laskey, Embo J. 6 (1987) 69.

[58] S.B. Marston, C.S. Redwood, Biochem. J. 279 (1991) 1.

[59] I.A. Hope, S. Mahadevan, K. Struhl, Nature 333 (1988) 635.

[60] L. Lee, E. Stollar, J. Chang, J.G. Grossmann, R. O'Brien, J. Ladbury, B. Carpenter, S. Roberts, B. Luisi, Biochemistry 40 (2001) 6580.

[61] V.N. Uversky, J.R. Gillespie, A.L. Fink, Proteins 41 (2000) 415.

[62] A.K. Dunker, J.D. Lawson, C.J. Brown, P. Romero, J.S. Oh, C.J. Oldfield, A.M. Campen, C.M. Ratliff, K.W. Hipps, J. Ausio, M.S. Nissen, R. Reeves, C. Kang, C.R. Kissinger, R.W. Bailey, M.D. Griswold, W. Chiu, E.C. Garner, Z. Obradovic, J. Mol. Graphics Modelling 19 (2001) 26.

[63] S.M. Park, H.Y. Jung, T.D. Kim, J.H. Park, C.H. Yang, J. Kim, J. Biol. Chem. 277 (2002) 28512.

[64] J. Bhattacharyya, K.P. Das, J. Biol. Chem. 274 (1999) 15505.

[65] M.C. Koag, R.D. Fenton, S. Wilkens, T.J. Close, Plant Physiol. 131 (2003) 309.

[66] C.S. Haarmann, D. Green, M.G. Casarotto, D.R. Laver, A.F. Dulhunty, Biochem. J. 372 (2003) 305.

[67] K.J. Davies, Biochimie 83 (2001) 301.

[68] D.C. David, R. Layfield, L. Serpell, Y. Narain, M. Goedert, M.G. Spillantini, J Neurochem. 83 (2002) 176.

[69] G.K. Tofaris, R. Layfield, M.G. Spillantini, FEBS Lett. 509 (2001) 22.

[70] R.J. Sheaff, J.D. Singer, J. Swanger, M. Smitherman, J.M. Roberts, B.E. Clurman, Mol. Cell 5 (2000) 403.

[71] C.J. Cox, K. Dutta, E.T. Petri, W.C. Hwang, Y. Lin, S.M. Pascal, R. Basavappa, FEBS Lett. 527 (2002) 303.

[72] H.X. Zhou, J. Mol. Biol. 329 (2003) 1.

[73] H.C. van Leeuwen, M.J. Strating, M. Rensen, W. de Laat, P.C. van der Vliet, Embo J. 16 (1997) 2043.

[74] S. Yuzawa, M. Yokochi, H. Hatanaka, K. Ogura, M. Kataoka, K. Miura, V. Mandiyan, J. Schlessinger, F. Inagaki, J. Mol. Biol. 306 (2001) 527.

[75] D. Parker, M. Rivera, T. Zor, A. Henrion-Caude, I. Radhakrishnan, A. Kumar, L.H. Shapiro, P.E. Wright, M. Montminy, P.K. Brindle, Mol. Cell Biol. 19 (1999) 5601.

[76] G.W. Daughdrill, M.S. Chadsey, J.E. Karlinsey, K.T. Hughes, F.W. Dahlquist, Nat. Struct. Biol. 4 (1997) 285.

[77] J.A. Hauer, S.S. Taylor, D.A. Johnson, Biochemistry 38 (1999) 6774.

[78] C.M. Fletcher, G. Wagner, Protein Sci. 7 (1998) 1639.

[79] H.J. Dyson, P.E. Wright, Curr. Opin. Struct. Biol. 12 (2002) 54.

[80] G.D. Bader, C.W. Hogue, Nat. Biotechnol. 20 (2002) 991.

[81] K. Gunasekaran, C.J. Tsai, S. Kumar, D. Zanuy, R. Nussinov, Trends Biochem. Sci. 28 (2003) 81.

[82] H. Tanaka, T. Yamashita, M. Asada, S. Mizutani, H. Yoshikawa, M. Tohyama, J. Cell Biol. 158 (2002) 321.

[83] M. Asada, T. Yamada, H. Ichijo, D. Delia, K. Miyazono, K. Fukumuro, S. Mizutani, Embo J. 18 (1999) 1223.

[84] S. Waga, G.J. Hannon, D. Beach, B. Stillman, Nature 369 (1994) 574.

[85] B.W. Pontius, Trends Biochem. Sci. 18 (1993) 181.

[86] B.A. Shoemaker, J.J. Portman, P.G. Wolynes, Proc. Natl Acad. Sci. USA 97 (2000) 8868.

[87] C. Holt, N.M. Wahlgren, T. Drakenberg, Biochem. J. 314 (1996) 1035.

[88] L. Iakoucheva, C. Brown, J. Lawson, Z. Obradovic, A. Dunker, J. Mol. Biol. 323 (2002) 573.

[89] M. Rechsteiner, S.W. Rogers, Trends Biochem. Sci. 21 (1996) 267.

[90] C.W. Liu, M.J. Corboy, G.N. DeMartino, P.J. Thomas, Science 299 (2003) 408.

[91] S.E. Salghetti, A.A. Caudy, J.G. Chenoweth, W.P. Tansey, Science 293 (2001) 1651.

[92] D.P. Denning, V. Uversky, S.S. Patel, A.L. Fink, M. Rexach, J. Biol. Chem. 277 (2002) 33447.

[93] R. Wissmann, T. Baukrowitz, H. Kalbacher, H.R. Kalbitzer, J.P. Ruppersberg, O. Pongs, C. Antz, B. Fakler, J. Biol. Chem. 274 (1999) 35521.

[94] M. Li, L.H. Phylip, W.E. Lees, J.R. Winther, B.M. Dunn, A. Wlodawer, J. Kay, A. Gustchina, Nat. Struct. Biol. 7 (2000) 113.

[95] M. Cheng, P. Olivier, J.A. Diehl, M. Fero, M.F. Roussel, J.M. Roberts, C.J. Sherr, Embo J. 18 (1999) 1571.

ELSEVIER

Journal of Molecular Structure (Theochem) 666–667 (2003) 373–380

THEO
CHEM

www.elsevier.com/locate/theochem

Molecular chaperones, evolution and medicine

Peter Csermely*, Csaba Sőti, Eva Kalmar, Eszter Papp, Balint Pato,
Akos Vermes, Amere S. Sreedhar[1]

Department of Medical Chemistry, Semmelweis University, P.O. Box 260, H-1444 Budapest 8, Hungary

Abstract

Protein folding has numerous steps, which need assistance in vivo. Molecular chaperones are required for many proteins to fold, or re-fold into native structures forming an ancient, primary system for 'intracellular self-defense'. Molecular chaperones participate in the organization of the cytoarchitecture, were necessary for the development of modern enzymes and—by stabilizing the genome—for the development of the first stable cells. They have a profound importance in medical practice. Chaperone induction provides cytoprotection in various pathological conditions, while chaperone inhibition can be an efficient tool to fight against cancer. Chaperones are inefficient enzymes and have low-affinity interactions, therefore their assays require unusual methods, which will be summarized in the concluding part of the paper.
© 2003 Elsevier B.V. All rights reserved.

Keywords: Heat shock proteins; Low affinity; Molecular chaperones; Protein folding; Stress proteins

1. Chaperones and protein folding

Protein folding is characterized by three major steps in vitro (Fig. 1) [1–6]. Under in vitro conditions in the *first few milliseconds* most of the secondary structure is already formed. In most cases folding starts with the formation of alpha-helices, since here the participation of adjacent amino acids is required. Beta-sheet formation establishes H-bonds between amino acids, which are far from each other in the primary sequence, therefore a greater decrease of entropy occurs than in the formation of alpha-helices. In the end of this first step, the hydrophobic segments are segregated by the surrounding water and they form

a hydrophobic core of an intermedier, which is often called as the 'molten globule'. If the protein is larger than 30 kDa, this intermedier can be fairly stable.

The partially folded state of *molten globules* can be characterized by a developed secondary structure, which is mostly un-organized showing almost no tertiary structure [3–5]. Molten globules still have large unburied hydrophobic surfaces, therefore are subjects of extensive aggregation. The volume of molten globules, however, is almost as small as that of the final, folded protein.

The *last steps* of protein folding are the slow, rate-limiting steps [1,2]. Here the inner, hydrophobic core of the protein is re-organized [6]. Parallel with this, unique, high-energy bonds are formed, such as disulfide bridges, ion-pairs, and the isomerization of proline *cis/trans* peptide bonds occurs. The free energy gain of these processes enables the formation of local, thermodynamically unstable, 'high-energy' protein structures, which are stabilized by

---

* Corresponding author. Tel.: +36-1-266-2755x4102; fax: +36-1-266-7480.

E-mail address: csermely@puskin.sote.hu (P. Csermely).

[1] On leave from Centre for Cellular and Molecular Biology, Hyderabad 500 007, India.

0166-1280/$ - see front matter © 2003 Elsevier B.V. All rights reserved.
doi:10.1016/j.theochem.2003.08.048

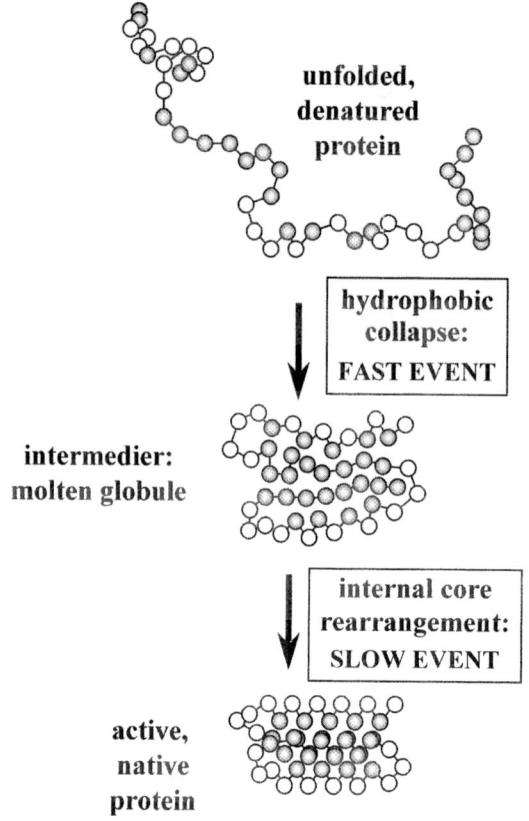

**unfolded,
denatured
protein**

**hydrophobic
collapse:
FAST EVENT**

**intermedier:
molten globule**

**internal core
rearrangement:
SLOW EVENT**

**active,
native
protein**

Fig. 1. Major steps of protein folding in vitro. Adapted from Csermely [6].

thermodynamically favorable conformation of the rest (bulk) of the protein. These high-energy segments of proteins can stabilize themselves by forming complexes with another molecule, thus they often serve as active centers of enzymes or as contact surfaces between various proteins involved, e.g. in signal transduction.

Protein folding is not a straightforward process. Dead-end streets, reverse reactions, futile cycles are all characteristic to it. A minor amount of fully folded, native protein always co-exists with various forms of molten globules and with traces of remaining unfolded specii. This unordered flow of events needs a lot of help. Aggregation of unfolded proteins and of molten globules is a great danger, which would drive the majority of folding intermediates to a nonproductive side-reaction, much before reaching their fully folded, competent state.

Molecular chaperones serve to prevent this. They recognize and cover hydrophobic surfaces successfully competing with the aggregation process. Moreover, molecular chaperones can leave their complex with misfolded proteins, utilizing the energy of ATP-hydrolysis-driven conformational changes.

Unaided protein folding often leads to folding traps. Steric hindrance (and most probably the lack of water in the inner, hydrophobic core of the protein) many times prevent the rearrangement of the hydrophobic core. Unfolding of the core is aided by periodic pulling and water-percolation [6] and the 'traditional' molecular chaperones may also provide better circumstances for ion-pair formation and for the establishment of high-energy protein segments. Disulfide bridge formation and proline *cis/trans* isomerization are promoted by protein disulfide isomerases and by peptidyl-prolyl-*cis/trans* isomerases, respectively [7,8].

Nascent proteins have the unique situation that they have to fold, when they are not even ready yet. The first protein segment, which leaves the ribosome surely has a different energy minimum, than the whole protein. In many cases, in vivo protein folding has to be delayed. Molecular chaperones are attached to the ribosomes 'waiting' for the nascent protein chain. When it appears, the chaperones 'sit on it' preventing premature protein folding before the rest of the protein is synthesized [9,10].

Chaperones also direct proteins inside the cell. Pores of the mitochondria or of the endoplasmic reticulum are too small to accommodate fully folded, globular proteins. Proteins have to unfold to get through, and to re-fold in the lumen of the organelle [11].

Molecular chaperones not only help, but also destroy. Some incorrectly folded proteins—maybe those which have lost not only their tertiary, but also secondary structure leaving their peptide bonds accessible—are presented to the lysosomal protein degradation [12] or to the extralysosomal proteasome [13]. In case of massive protein damage, when the amount of degradable proteins exceeds the capacity of the intracellular proteolytic systems, chaperones help to form inclusion bodies to segregate damaged proteins [14].

2. Nonconventional roles for chaperones: organization of the cytoarchitecture

Heat shock proteins are regarded as molecular chaperones, thus their major cellular function is considered to be established. However, most of the protein folding experiments are conducted in an in vitro environment. When protein folding is studied in vitro, the experimenter has to use rather diluted conditions to prevent unwanted aggregation. Dilution also helps to make the kinetical analysis easier, and spares precious research materials. On the contrary to these usual experimental conditions, the cellular environment is crowded [15]. Molecular crowding promotes protein aggregation thus calls for an enhanced need of chaperone action. On the other hand, bona fide chaperones are not the only cellular solutions for aggregation-protection. Several 'innocent bystanders', such as tubulin [16] or even small molecules (lipids, other amphiphyles, sugars, a class of compounds called as chemical chaperones [17]) may assist folding and prevent aggregation albeit at much higher concentrations than the efficient concentration of heat shock, or other stress-induced proteins. Though we have several important lines of evidence, which undoubtedly show the necessity of chaperones in folding of numerous protein kinases, receptors, actin, tubulin, etc. [18] we do not really know, how big is the segment of the life of an ordinary chaperone, when it 'chaperones' unfolded or misfolded proteins in eukaryotic cells.

To make it clear, with the argumentation above we do not want to question the importance of chaperones in folding-assistance. Nevertheless, we would like to stress, that there is enough room to think about other important functions of chaperones related to, but not equal with their participation in protein folding. One of these possibilities is, that peptide-binding chaperones are the 'dustmen' of the cells. The proteasomal apparatus is most probably linked with oligo- and dipeptidases and therefore the 'leaking' peptide-endproducts of proteasomal degradation [19] are usually cleaved further into single amino acids. However, the coupled protein/peptide degradation can leak especially under stressed conditions, like in oxidative stress. Released peptide segments may often contain elements of important binding sites and thus may efficiently interfere with signaling and, metabolic

processes. If this happened at a massive scale, this would be a disaster for the cell. Peptides need to be eliminated, and safeguarding mechanisms must exist to correct the occasional 'sloppiness' of degradative processes. Chaperones are excellent candidates for this purpose and their role in collection of 'peptide-rubbish' must be considered besides their well-established function in peptide presentation for the immune system [20].

As yet another important, and nonconventional aspect of chaperone action (from the many more possible) lies in their incredible stickiness. Chaperones often form dimers, and tend to associate to tetra-, hexa-, octamers and to even higher oligomers [21–23]. Oligomerization usually affects only a few percent of the total protein; but addition of divalent cations, certain nucleotides, heat treatment enhances oligomer formation. It is important to note that oligomerization studies were usually performed under 'normal', in vitro experimental conditions, using a few µg/ml of purified chaperone. The in vivo concentration of chaperones is estimated to be around a hundred-, or thousand-fold higher. This may significantly enhance the in vivo oligomerization tendencies of these proteins. Oligomer formation of chaperones might be further promoted by the large excluded volume effect of the 'molecularly crowded' cytoplasm [15].

Different chaperones also associate with each other. The Hsp90-organized foldosome may contain almost a dozen independent chaperones, or co-chaperones. The stoichiometry and affinity of these associations dynamically varies, and the variations are affected by the folding state of the actual target (or targets) which associate with these extensive folding machinery [23].

Besides binding to themselves, to their sibling-chaperones, and to their targets, many chaperones bind to actin filaments, tubulin, and other cellular filamentous structures, such as intermediate filaments. There is a chaperone complex associated with the centrosome [24] and several chaperones, especially Hsp90 were considered to be involved in the direction of cytoplasmic traffic [25].

The above model describing chaperones as a highly dynamic 'appendix' of various, and often quite poorly identifiable, cytoplasmic filamentous structures is reminiscent of the early view [26,27] about the microtrabecular lattice of the cytoplasm. Although later studies efficiently questioned the validity of the original electronmicroscopical evidence of the microtrabeculae, pointing out many possibilities for artifact formation during sample preparation, several indirect evidence, such as diffusion anomalies support the existence of a cytoplasmic mesh-like structure [28–31]. The major cytoplasmic chaperones (Hsp90, TCP1/ Hsp60 and their associated proteins) may well form a part of this network in cells [32].

Our experiments showing the acceleration of the efflux of cytoplasmic constituents after the inhibition of the major cytoplasmic chaperone, Hsp90, both in case of Jurkat cells [33] and erythrocytes [34] (Fig. 2) suggest the involvement of the 90 kDa molecular chaperone, Hsp90 in the maintenance of the cytoarchitecture. Interestingly, we did not see an acceleration of cytoplasmic release in *E. coli* [33], which is in agreement with the lower level of cytoplasmic organization of prokaryotes compared to eukaryotes. We cannot ascertain at the moment that the faster release of cytoplasmic proteins after the disruption of Hsp90 complexes by Hsp90

inhibitors [33,34] or anti-Hsp90 ribozyme treatment [33] is a consequence of a disrupted cytoplasmic meshwork or shows the involvement of Hsp90 in the stabilization of the traditional cytoskeleton. However, our ongoing experiments may show the reorganization of Hsp90 in the cytoplasm after these treatments as well as changes in the intracellular diffusion rates.

3. Chaperones and evolution

Chaperones are ancient protein structures, which were highly conserved throughout all the known parts of evolution, and are repeatedly emerging as parts of the minimal genomes of various organisms suggesting their presence in the hypothetical Last Universal Common Ancestor (LUCA) [35]. The increasing size of constituent proteins (a necessity for modern enzyme action, where conformational changes make induced-fit, and allosteric regulation possible) caused more and more folding traps. To prevent this, and to help de novo protein folding an increase of chaperone capacity was probably needed [36,37]. According to the above assumptions, chaperone capacity was likely to grow in parallel with cellular complexity of primordial cells.

As we described in Section 2, chaperones help the organization of the cytoarchitecture [32–34]. Chaperones also stabilize lipid membranes [38]. Both effects may have helped the occurrence of the first stable ancestors of modern cells by worsening the chances for 'membrane-leaks', which would help the exchange of various cell constituents (including genetically coding material) between neighboring organisms.

In the last few years, several experiments were published, which suggested that chaperones behave as 'buffers of evolutionary changes'. Chaperones may correct the conformational changes caused by various mutations, and make the genetical changes phenotypically silent in various organisms studied [39–42]. Thus chaperones were probably not only contributors to the emerging cellular organization of primordial cells, but in parallel, they also increased genetical stability by buffering the phenotypical consequences of mutational events.

After a large stress, the suddenly increased amount of damaged proteins may cause a 'chaper-

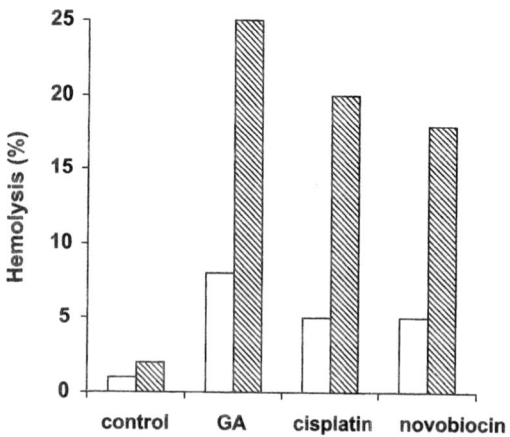

Fig. 2. Hsp90 inhibition induces an accelerated hemolysis. Hemolysis was induced by a 15 min (open bars) or 30 min (filled bars) treatment of 0.002 % of the nonionic detergent, Brij58 at room temperature. Mouse red blood cells were pre-incubated with 1 µM geldanamycin, 100 µM cisplatin or 1 mM novobiocin, inhibitors of Hsp90 at its N-terminus, C-terminus or both termini, respectively [60]. Hemolysis was measured by monitoring the amount of released hemoglobin after centrifugation of lysed cells at 528 nm. Data are representatives of three independent experiments. (The results were published in an abstract form in Ref. [34].)

one overload', and may prevent the conformational repair of misfolded mutants. Therefore, many previously hidden genotypical changes may appear in the phenotype resulting in a 'boom' of genetical variations in the whole population. This may help the selection of a beneficial change, which, in turn, may help the adaptation of the population to changed environmental conditions. However, most of the exposed mutations are disadvantageous, and tend to disappear from the population by natural selection.

Changes in living conditions and the significantly better medical care throughout life in the last 150 years have significantly reduced the occurrence of large physiological stresses that would normally result in significant intracellular proteotoxicity. There is little chaperone overload during reproductive years in the present times. Even major stressful events, such as critical infections and extreme and unexpected changes in the environment, etc. that do cause a massive chaperone overload, can be mitigated by improved medical care, thus saving lives that would otherwise have been lost. Thus, a larger proportion of people harboring deleterious mutations probably survive today and transmit their genes to later generations. Thus improved medical care may have led to a rise in phenotypically silent mutations in the human genome. As a consequence we may be carrying more and more chaperone-buffered, silent mutations from generation to generation [43].

The chance of the phenotypic manifestation of these mutations becomes especially large in aged subjects, where protein damage is abundant, and both chaperone induction and chaperone function are impaired [44]. Here the background of misfolded proteins increases and by competition prevents the chaperone-mediated buffering of silent mutations. Phenotypically exposed mutations may contribute to a more abundant manifestation of multigene-diseases, such as atherosclerosis, autoimmune-type diseases, cancer, diabetes, hypertensive cardiovascular disease and several psychiatric illnesses (Alzheimer disease, schizophrenia, etc.). The chaperone overload hypothesis emphasizes the need for efficient ways to enhance chaperone-capacity in aging subjects [43].

4. Beneficial effects of chaperone induction

Molecular chaperones are responsible for the 'conformational homeostasis' of cellular proteins. When the homeostasis of the host organism is perturbed, an increased capacity of the 'chaperone machines' is highly advantageous. Many of the perturbations (such as alcohol, other poisons, sunburn, anxiety, etc.) may induce the synthesis of these chaperone proteins per se, but in case of bacterial and viral infections the developing fever also helps this process. Ischemia and the consecutive oxidative damage of reperfusion are also common environmental perturbations in higher organisms. Since Currie et al. [45] have shown that the induction of molecular chaperones, most notably Hsp70, may prevent the cardiac muscle from the damage of both ischemia and reperfusion, molecular chaperones are actively investigated as possible tools in the treatment of heart attack or stroke. Their protective role is also used in organ transplantation, where a prior heat treatment induces a more efficient organ-survival and diminishes the occurrence of rejection by the host organism. Several common drugs, such as aspirin [46] promote the induction of the chaperone defense system, however, recently a specific chaperone co-inducer drug family [47,48] interacting with the heat shock factor [49] has been also described.

5. Advantages of chaperone inhibition

The above examples show the advantages of chaperone induction. Chaperones protect our cells—chaperones are good. Not always. When chaperones protect our malignant cells—they are not really beneficial. Still, chaperone inhibition as a pharmacological tool to prevent cancer development seems to be a wild idea. However, if we consider that chaperones are necessary for the folding of numerous cyclin-dependent kinases, which promote the cell cycle [50], and some of the chaperone inhibitors are selectively enriched in tumor cells [51], we begin to believe that chaperone inhibition might be a valid pharmacological intervention against tumors. Indeed, many chaperone inhibitors are currently in clinical trials against various forms of cancer [52].

Since the 90 kDa molecular chaperone (Hsp90) has the most specific and most cell-permeable inhibitors, and since this chaperone is the center of the kinase-related chaperone machinery, in most cases chaperone-based inhibition is achieved by using Hsp90 inhibitors. The first Hsp90 inhibitor drug was geldanamycin, a natural product isolated from *Streptomyces hygroscopicus*. Though the antitumor effects of geldanamycin were initially thought to be due to specific tyrosine kinase inhibition, later studies revealed that the antitumor potential relies on depletion of oncogenic protein kinases via the proteasome [53]. The major regulatory signaling proteins, which are affected by geldanamycin, include the proto-oncogene kinases ErbB2, EGF, v-Src, Raf-1 and Cdk4 [50].

Radicicol, another Hsp90 inhibitor [54], is a macrocyclic antibiotic isolated from *Monosporium bonorden*. However, radicicol lacks antitumor activity in vivo in experimental models because of its instability. The oxime derivatives of radicicol [55] exhibit antitumor activity in vivo as well as in vitro, hence serve as good anticancer drug candidates. Radicicol binds to the N-terminal domain of Hsp90 with much higher affinity than the structurally different drug, geldanamycin [56]. Moreover, radicicol reduces hypoxia-induced VEGF expression, which is an efficient way to decrease hypoxia-induced angiogenesis [57]. As a recent development, PU3, a purine-based Hsp90 inhibitor was designed using X-ray crystallographic data [58]. PU3 behaves like geldanamycin in inhibiting Hsp90 client protein degradation, and possesses a robust antitumor potential [58].

Recently it was shown that Hsp90 contains a second nucleotide binding site at the C-terminal domain [59–61]. Nucleotide binding to this site can be inhibited by the commonly used chemotherapeutic agent, cisplatin [60], which displays a rather selective binding to Hsp90 among proteins [62]. Inhibition of the C-terminal site of Hsp90 by cisplatin results in the differential inhibition of Hsp90-assisted protein folding: that of Raf kinase remains unaffected, however, folding of luciferase is strongly inhibited under these conditions [60]. The C-terminal nucleotide binding site displays a markedly different nucleotide specificity if compared to that of the N-terminal site [63], where all the known inhibitors (geldanamycin,

radicicol and PU3) bind. It is the task of the future to find truly selective and high affinity inhibitors of the C-terminal Hsp90 nucleotide binding site.

6. Assays for chaperone action

Chaperone induction can be monitored by assessing the mRNA or protein levels. However, the paramount importance of molecular chaperones in medicine necessitates the measurement of their activity in patients. It is rather likely that the 'free chaperone capacity' will be a common marker of health in the near future. Unfortunately, until recently we did not have easy methods to determine chaperone-related activities in whole cellular homogenates. However, recent progress in the biochemistry of molecular chaperones enables us to construct and try such methods.

Chaperone activity can be assessed by measuring the chaperone-induced prevention of protein aggregation [64]. This is a measure of passive chaperone function which does not require ATP. The active, ATP-dependent assistance in the refolding of misfolded proteins may be assessed using several test systems such as the luciferase-renaturation assay of whole cell homogenates [64]. To check the chaperone activity of purified chaperones and chaperone-complexes (immune precipitates) several other assays, such as the citrate synthase-assay, are also available [64].

Autophosphorylation in the presence of Ca-ATP is a common feature of almost all molecular chaperones [23]. Since most of protein kinases cannot utilize Ca-ATP, phosphorylation of cellular proteins in the presence of Ca-ATP gives a surprisingly clear pattern. As an example, we have shown earlier, that in streptozotocin-diabetic rats the phosphorylation of the 94 kDa glucose-regulated protein, Grp94, is diminished. Insulin-treatment reversed the effect [65].

Unfortunately, the *ATPase* reaction is not so specific, than the autophosphorylation. Therefore, specific measurement of chaperone/ATPases may only be accomplished by using specific inhibitors of ATPase activity, such as geldanamycin for Hsp90 [23]. Moreover, in many cases the chaperone ATPase activity is too small for efficient measurements.

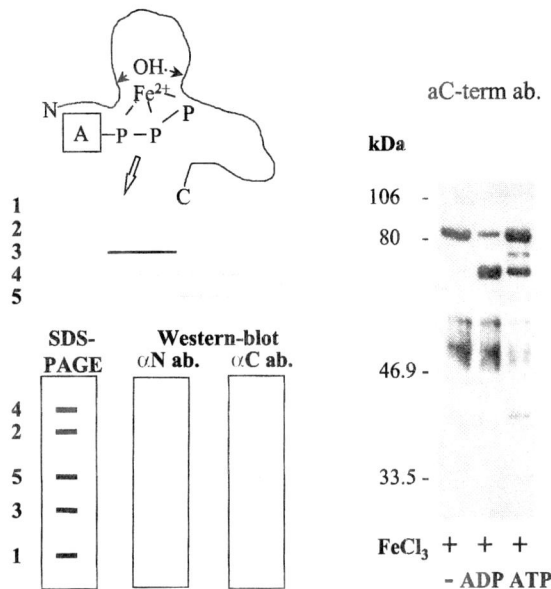

Fig. 3. Nucleotide affinity cleavage: an efficient method to analyze nucleotide binding of molecular chaperones. The Fe-ATP (Fe-nucleotide) complex at the nucleotide-binding site induces a local Fenton reaction producing hydroxyl radicals. These reactive compounds cleave the protein (in this case the 90 kDa heat shock protein, Hsp90) just in the vicinity of the binding site producing a remarkably clean cleavage pattern, which can be detected by antibodies against the N- or C-terminus of the protein (see more details in Soti et al. [60]).

Another efficient method to test the integrity of molecular chaperones is the nucleotide affinity cleavage assay, which has been recently applied to this class of nucleotide binding proteins [60]. This assay is good to detect the ability of the nucleotide binding site to bind the nucleotide, thus to be able to assess the ability of the given chaperone to perform the ATP-dependent, active refolding of various proteins (Fig. 3).

7. Summary and perspectives

In recent years we have learned a lot about the molecular mechanism of protein folding in vitro, and some important features were also revealed of the in vivo formation and repair of protein structure. Many aspects of the molecular mechanism of chaperone action were cleared, and we also recognized the importance of these proteins in the clinical practice. Chaperones participate in the maintenance of the cytoarchitecture and played a prominent role during various important steps of evolution. Both their induction and inhibition have grown to important pharmacological targets and several methods have been established to assess their function both in vitro and in vivo. We hope that with this short review we may increase the courage of some fellow scientists to enter this difficult, but very promising path of multidisciplinary research.

Acknowledgements

Work in the authors' laboratory was supported by research grants from the EU 6th Framework program (FP6506850), from the Hungarian Science Foundation (OTKA-T37357), from the Hungarian Ministry of Social Welfare (ETT-32/03) from the International Centre for Genetic Engineering and Biotechnology (ICGEB, CRP/HUN 99-02). A.S.S. is a recipient of National Overseas Scholarship from Ministry of Social Justice and Empowerment, Government of India.

References

[1] P.S. Kim, R.L. Baldwin, Annu. Rev. Biochem. 59 (1990) 631.
[2] C.R. Matthews, Annu. Rev. Biochem. 62 (1993) 653.
[3] C.M. Dobson, P.A. Evans, S.E. Radford, Trends Biochem. Sci. 19 (1994) 31.
[4] K. Kuwajima, Proteins 6 (1989) 87.
[5] O.B. Ptitsyn, FEBS Lett. 285 (1991) 176.
[6] P. Csermely, BioEssays 21 (1999) 959.
[7] S. Raina, D. Missiakas, Annu. Rev. Microbiol. 51 (1997) 179.
[8] G.S. Hamilton, J.P. Steiner, J. Med. Chem. 41 (1998) 5119.
[9] R.P. Beckmann, L.A. Mizzen, W.J. Welch, Science 248 (1990) 850.
[10] J. Frydman, E. Nimmesgern, K. Ohtsuka, F.U. Hartl, Nature 370 (1994) 111.
[11] W.J. Chirico, M.G. Waters, G. Blobel, Nature 332 (1988) 805.
[12] H.-L. Chiang, S.R. Terlecky, C.P. Plant, J.F. Dice, Science 246 (1989) 382.
[13] D. Voges, P. Zwickl, W. Baumeister, Annu. Rev. Biochem. 68 (1999) 1015.
[14] R.J. Mayer, J. Arnold, L. Laszlo, M. Landon, J. Lowe, Biochim. Biophys. Acta 1089 (1991) 141.
[15] S.B. Zimmerman, A.P. Minton, Annu. Rev. Biophys. Biomol. Struct. 22 (1993) 27.

[16] S. Guha, T.K. Manna, K.P. Das, B. Bhattacharya, J. Biol. Chem. 273 (1998) 30077.

[17] W.J. Welch, C.R. Brown, Cell Stress Chaperones 1 (1996) 109.

[18] F.-U. Hartl, Nature 381 (1996) 571.

[19] A.F. Kisselev, T.N. Akopian, A.L. Goldberg, J. Biol. Chem. 273 (1998) 1982.

[20] P.K. Srivastava, A. Menoret, S. Basu, R.J. Binder, K.L. McQuade, Immunity 8 (1998) 657.

[21] N. Benaroudj, F. Triniolles, M.M. Ladjimi, J. Biol. Chem. 271 (1996) 18471.

[22] J.D. Trent, H.K. Kagawa, T. Yaoi, E. Olle, N.J. Zaluzec, Proc. Natl Acad. Sci. USA 94 (1997) 5383.

[23] P. Csermely, T. Schnaider, Cs. Soti, Z. Prohászka, G. Nardai, Pharmacol. Ther. 79 (1998) 129.

[24] W.C. Wigley, R.P. Fabunmi, M.G. Lee, C.R. Marino, S. Muallem, G.N. DeMartino, P.J. Thomas, J. Cell Biol. 145 (1999) 481.

[25] W.B. Pratt, A.M. Silverstein, M.D. Galigniana, Cell. Signal. 11 (1999) 839.

[26] J.J. Wolosewick, K.R. Porter, J. Cell Biol. 82 (1979) 114.

[27] M. Schliwa, J. van Blerkom, K.R. Porter, Proc. Natl Acad. Sci. USA 78 (1981) 4329.

[28] J.S. Clegg, Am. J. Physiol. 246 (1984) R133.

[29] K. Jacobson, J. Wojcieszyn, Proc. Natl Acad. Sci. USA 81 (1984) 6747.

[30] K. Luby-Phelps, F. Lanni, D.L. Taylor, Annu. Rev. Biophys. Biophys. Chem. 17 (1988) 369.

[31] A.S. Verkman, Trends Biochem. Sci. 27 (2002) 27.

[32] P. Csermely, News Physiol. Sci. 15 (2001) 123.

[33] A.S. Sreedhar, K. Mihály, B. Való, T. Schnaider, A. Steták, K. Kis-Petik, J. Fidy, T. Simonics, A. Maráz, P. Csermely, J. Biol. Chem. 278 (2003).

[34] B. Pato, K. Mihaly, P. Csermely, Eur. J. Biochem. 268 (2001) S107.

[35] E.V. Koonin, M.Y. Galperin, Sequence, evolution, function, Computational Approaches in Comparative Genomics, Kluwer, Dordrecht, 2003.

[36] P. Csermely, Trends Biochem. Sci. 22 (1997) 147.

[37] S. Walter, J. Buchner, Angew. Chem. 41 (2002) 1098.

[38] Z. Torok, I. Horvath, P. Goloubinoff, E. Kovacs, A. Glatz, G. Balogh, L. Vigh, Proc. Natl Acad. Sci. USA 94 (1997) 2192.

[39] S.L. Rutherford, S. Lindquist, Nature 396 (1998) 336.

[40] S.P. Roberts, M. Feder, Oecologia 121 (1999) 323.

[41] M.A. Fares, M.X. Ruiz-Gonzalez, A. Moya, S.F. Elena, E. Barrio, Nature 417 (2002) 398.

[42] C. Queitsch, T.A. Sangster, S. Lindquist, Nature 417 (2002) 618.

[43] P. Csermely, Trends Genet. 17 (2001) 701.

[44] G. Nardai, P. Csermely, Cs. Soti, Exp. Gerontol. 37 (2002) 1255.

[45] R.W. Currie, M. Karmazyn, M. Kolc, K. Mailer, Circulation Res. 63 (1988) 543.

[46] D.A. Jurivich, L. Sistonen, R.A. Kroes, R.I. Morimoto, Science 255 (1992) 1243.

[47] L. Vigh, P. Literati Nagy, I. Horvath, Zs. Torok, G. Balogh, A. Glatz, E. Kovacs, I. Boros, P. Ferdinandy, B. Farkas, L. Jaszlits, A. Jednakovits, L. Koranyi, B. Maresca, Nat. Med. 3 (1997) 1150.

[48] Zs. Török, N.M. Tsvetkova, G. Balogh, I. Horváth, E. Nagy, Z. Pénzes, J. Hargitai, O. Bensaude, P. Csermely, J.H. Crowe, B. Maresca, L. Vígh, Proc. Natl Acad. Sci. USA 100 (2003) 3131.

[49] J. Hargitai, H. Lewis, I. Boros, T. Rácz, A. Fiser, I. Kurucz, I. Benjamin, Z. Pénzes, L. Vígh, P. Csermely, D.S. Latchman, Biochem. Biophys. Res. Commun. 307 (2003) 689.

[50] W.B. Pratt, D.O. Toft, Exp. Biol. Med. 228 (2003) 111.

[51] G. Chiosis, H. Huezo, N. Rosen, E. Mimnaugh, L. Whitesell, L. Neckers, Mol. Cancer Ther. 2 (2003) 123.

[52] L. Neckers, Trends Mol. Med. 8 (2002) S55.

[53] L. Whitesell, E.G. Mimnaugh, B. De Costa, C.E. Myers, L.M. Neckers, Proc. Natl Acad. Sci. USA 91 (1994) 8324.

[54] S. Soga, T. Kozawa, H. Narumi, S. Akinaga, K. Irie, K. Matsumoto, S.V. Sharma, H. Nakano, T. Mizukami, M. Hara, J. Biol. Chem. 273 (1998) 822.

[55] T. Agatsuma, H. Ogawa, K. Akasaka, A. Asai, Y. Yamashita, T. Mizukami, S. Akinaga, Y. Saitoh, Bioorg. Med. Chem. 10 (2002) 3445.

[56] S.M. Roe, C. Prodromou, R. O'Brien, J.E. Ladbury, P.W. Piper, L.H. Pearl, J. Med. Chem. 42 (1999) 260.

[57] E. Hur, H.H. Kim, S.M. Choi, J.H. Kim, S. Yim, H.J. Kwon, Y. Choi, D.K. Kim, M.O. Lee, H. Park, Mol. Pharmacol. 62 (2002) 975.

[58] G. Chiosis, M.N. Timaul, B. Lucas, P.N. Munster, F.F. Zheng, L. Sepp-Lorenzino, N. Rosen, Chem. Biol. 8 (2001) 289.

[59] M.G. Marcu, A. Chadli, I. Bouhouche, M.G. Catelli, L.M. Neckers, J. Biol. Chem. 275 (2000) 37181.

[60] Cs. Soti, A. Racz, P. Csermely, J. Biol. Chem. 277 (2002) 7066.

[61] C. Garnier, D. Lafitte, P.O. Tsvetkov, P. Barbier, J. Leclerc-Devin, J.M. Millot, C. Briand, A.A. Makarov, M.G. Catelli, V. Peyrot, J. Biol. Chem. 277 (2002) 12208.

[62] H. Itoh, M. Ogura, A. Komatsuda, H. Wakui, A.B. Miura, Y. Tashima, J. Biochem. 343 (1999) 697.

[63] Cs. Sóti, A. Vermes, T.A. Haystead, P. Csermely, Eur. J. Biochem. 270 (2003) 2421.

[64] B.C. Freeman, A. Michels, J. Song, H.H. Kampinga, R.I. Morimoto, Meth. Mol. Biol. 88 (2001) 393.

[65] P. Csermely, Cell Biol. Int. 18 (1994) 566.

ELSEVIER

Journal of Molecular Structure (Theochem) 666–667 (2003) 381–385

THEO
CHEM

www.elsevier.com/locate/theochem

Correlation corrected band structures of homopolypeptides IV. BLYP band structures of 19 homopolypeptides

F. Bogar[a,b], C. Van Alsenoy[c], J. Ladik[a,*]

[a]*Institute for Theoretical Chemistry, Laboratory of the National Foundation for Cancer Research, Friedrich-Alexander University Erlangen-Nuremberg, Egerlandstr 3, Erlangen D-91058, Germany*
[b]*Protein Chemistry Research Group of Hungarian Academy of Sciences, University of Szeged, Dóm tér 8, Szeged H-6720, Hungary*
[c]*Department of Chemistry, University of Antwerp, Universiteitsplein 1, Antwerpen B-2610, Belgium*

Abstract

Using the BLYP DFT crystal orbital method for 19 homopolypeptides in their β pleated sheet conformation their band structures were calculated. The position of the conduction and valence bands, their widths and their fundamental gap values are very similar to the previous simple LDA DFT results. This indicates that without the elimination of the self-interaction error one cannot obtain realistic gap values. The B3LYP calculations, which are in progress, most probably will increase the too small gap values because of the inclusion at least a part of the exact HF exchange. However, one expects that a DFT method tailored for the calculation of excited states and gaps of periodic polymers will bring a satisfactory solution of the problem.
© 2003 Elsevier B.V. All rights reserved.

Keywords: Ab initio band structures; Homopolypeptides; BLYP DFT calculation of the homopolypeptides

1. Introduction

As it is well know most density functional (DFT) methods give quite good ground state properties of molecules and polymers (bond lengths, conformation, dipole moments, harmonic vibrational energies etc). On the other hand for the excitation energies of molecules or polymers and the fundamental gap of solids nearly all DFT methods provide too small values.

In the previous papers [1–3] of our systematic investigation of the homopolypeptides we have found the ab initio HF crystal orbital (CO) +MP2 method gives too large gaps. In the case of HF CO the gaps are larger by about 5–6 eV and for the correlation corrected bands at the MP2 level by ∼2.5 eV as compared to the values, which can be estimated on the basis of the experimental exciton spectrum of polyglutamate [4] and polyleucine [5] (estimated experimental gap value 7–8 eV)

On the other hand the LDA gap values are far too small (values between 3.9 and 4.9 eV) in the case of seven homopolypeptides with aliphatic side chains [1]. To investigate further the problem we have calculated the band structures of 19 homopolypeptides using the BLYP method. In the case of polytryptophan we could not get the calculation converged. As we shall see at the results, the introduction of gradient terms in the exchange and correlation parts of the functional hardly changed the gap values. This is understandable, because the introduction of the gradient terms does not influence

* Corresponding author. Tel.: +49-9131-85-28831; fax: +49-9131-85-27736.
E-mail address: janos.ladik@chemie.uni-erlangen.de (J. Ladik).

0166-1280/$ - see front matter © 2003 Elsevier B.V. All rights reserved.
doi:10.1016/j.theochem.2003.08.049

the main reason of the too small level splitting, namely that the diagonal Coulomb and exchange terms J_{ii} and K_{ii} do not cancel in the DFT theory as Perdew et al. [6] have pointed out. In a second paper [7] a method for the correction of this (so-called self-interaction correction (SIC)) was proposed, but this makes the calculation quite tedious and doesn't seem to be suitable in the case of polymers.

In this situation the next step is to use instead of BLYB the B3LYP method which contains a part of the original HF exchange. One expects that in this case the gap will increase somewhat, but will be still too small.

Work is in progress to make the optimal effective potential (OEP) method (which worked reasonable well for the excited states of molecules) suitable also for polymers with larger unit cells. One expects that this method will provide finally good approximate values for the gaps.

Though we know that the different DFT methods which we have applied in these investigations are not suitable to calculate the gap (though they give good total energies) we feel that it is worthwhile to do these studies to learn first of all more about the role of SIC in the too small level splittings (gaps).

2. Methods

In the present calculation instead of the simple exchange, used in the LDA approximation of DFT (applied in the Mintmire program [8] for polymers) we have applied the Becke approximation [9] for the exchange which contains also the gradient of

the density and the Lee, Yang and Parr [10] formula for the correlation functional in which again density gradient terms occur (BLYP DFT). These methods have been built in into our general periodic polymer program instead of the LDA approximation of the exchange. This means that the program can be applied not only in the case of simple translation, but also for the general periodic case (like helical or zig-zag conformations).

In a later paper [11] Becke has shown that one obtains still better total energies if one uses as exchange functional a linear combination of the local spin density exchange of density gradient terms and of the exact (HF) exchange functional [11].

As next step of our comparative investigation of different methods applied to homopolypeptides we are going to perform calculation also with this method (the so-called B3LYP DFT). As one can expect the presence of an exact exchange term will considerably increase the fundamental gap.

We have done the band structure calculations in the β pleated sheet conformation of the 19 homopolypeptides. For the main chain (see Fig. 1) we have used the data given in our previous paper [3] in which the dihedral angles ψ and ϕ were taken to be 140° and −140°, respectively, as a typical value [12]. (See Table 1).

The side chain geometries were optimized using a 5 units long amino acid chain and applying molecular mechanics (for the details see Ref. [3]). The conformation of the main chain was kept fixed in this optimization. As unit cell geometry of the main chain the geometry of the middle unit was taken.

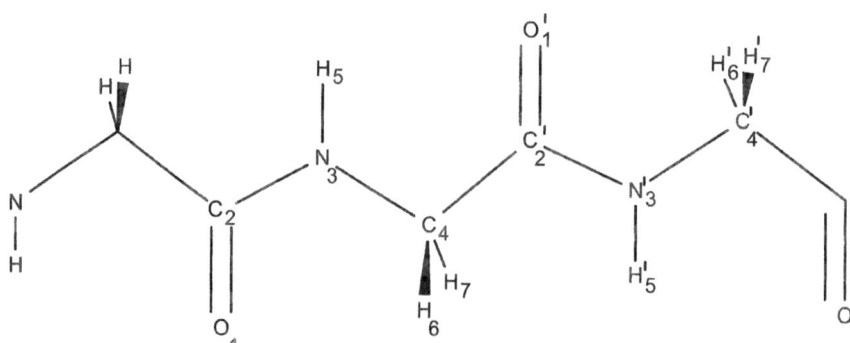

Fig. 1. The numbering of atoms of PolyGly in the β pleated sheet conformation.

Table 1
The bond distances (in Å), bond angles and dihedral angles of PolyGly chain in the β pleated sheet conformation

Bond distances		Bond angles		Dihedral angles	
O_1–C_2	1.23	O_1–C_2–N_3	122.9	O_1–C_2–N_3–H_5	180.0
C_2–N_3	1.34	C_2–N_3–C_4	121.9	O_1–C_2–N_3–C_4	0.0
N_3–C_4	1.45	N_3–C_4–C_2'	110.4	C_2–N_3–C_4–$C_2' = \Phi$	−140.0
C_4–C_2'	1.52	C_4–C_2'–N_3'	116.6	N_3–C_4–C_2'–$N_3' = \Psi$	140.0
N_3–H_5	1.01	C_2–N_3–H_5	119.5	H_6–C_4–N_3–C_2	100.58
C_4–H_6	1.09	N_3–C_4–H_6	109.47	H_7–C_4–N_3–C_2	−19.42
C_4–H_7	1.09	N_3–C_4–H_7	108.85	C_4–C_2'–N_3'–C_4'	180.0

The calculation were done with $2n + 1$ k-points where n was equal to 10. For the computations Clementi's double ζ basis set [13] was applied. In these studies as in Mintmire's program a multipole expansion was used for the long range Coulomb terms [14].

3. Results and their discussion

In Table 2 we present the conduction and valance band edges of the calculated 19 homopolypeptides together with the band widths. Table 3 shows the values of the calculated fundamental gap values.

Comparing the 19 band structures given in Table 2 with those calculated with the aid of the simple LDA DFT approximation again in the β-pleated sheet conformation (only the first 7 polypeptides given in Table 2 were computed in this case [2]) one can notice: (1) the positions of the conduction and valence bands in both cases as well (2) the widths of these bands are very similar. One can find only in the case of PolySer and PolyiLeu somewhat larger differences in the band widths (at PolySer $CB_{width}^{LDA} = 0.38$ eV and $CB_{width}^{BLYP} = 0.23$ eV, at PolyiLeu $CB_{width}^{LDA} = 0.67$ eV and $CB_{width}^{BLYP} = 0.55$ eV). At the same time for PolyThre $CB_{min}^{LDA} = -0.67$ eV and $CB_{min}^{BLYP} = -0.55$ eV. One should mention that in the case of the B3LYP calculations (which are in progress) most probably larger deviations will occur because of the occurrence of the exact exchange term in the total exchange functional [9]. One would expect also larger differences in the band positions and widths in the cases of the homopolypeptides with larger side chains.

One should notice also that in most cases the bands show a monotonic dispersion (the band edges occur at the $k = 0$ and $k = 10$ points, respectively) in accordance with previous HF + MP2 [3] and LDA DFT [1] calculations. This is, however, not in accordance with the results of HF + MP2 calculations done for homopolypeptides with more complicated side chains, where the dispersion of the conduction and valence bands, respectively, are in nearly all cases non-monotonic [3].

Looking at Table 3 one can notice that in the comparable seven cases the fundamental gap values are very similar with the exceptions of PolyThre, PolyLeu and PolyiLeu, where the BLYB gap values are a little larger than the LDA ones. This shows again that the DFT also in this formulation is giving far too small gaps. This situation cannot be expected to be different in the other 12 cases either, because their gaps have very similar values to the first seven homopolypeptides. By performing the B3LYP calculations the gaps very probably will be increased because of the partial reintroduction of the exact exchange, but one would not expect that the gap values will increase enough to reach the estimated values around 7.0–8 eV-s (no gap measurement exist for the homopolypeptides, the measured UV spectra correspond to exciton bands which usually lie below the lower edges of the conduction bands).

To settle the question from the theoretical side one has to calculate the band structures with the help of the optimized effective potential DFT method applied for polymers [15]. The other possibility would be to perform good basis HF + MP2 + MP3 + MP4 computations.

Table 2
The conduction and valance bands with their widths of the essential amino acids with the exceptions of polycysteine and polytriptophan (in eV)-s

	PolyGly(8)	PolyAla(8)[a]	PolySer(10)	PolyThre(8)	PolyLeu(8)	PolyiLeu(8)	PolyVal(8)	PolyAspAc(8)	PolyGlutAc(10)	PolyAsp(8)
CB_{max}	$-0.39(8)$	$-0.30(8)^b$	$-0.36(10)$	$-0.13(8)$	$-0.17(8)$	$-0.29(8)$	$-0.35(8)$	$-2.13(0)$	$-1.39(10)$	$-0.14(8)$
CB_{min}	$-0.85(0)$	$-0.84(0)$	$-0.59(0)$	$-0.58(0)$	$-0.72(0)$	$-0.82(0)$	$-0.87(0)$	$-2.17(6)$	$-1.59(0)$	$-0.35(0)$
CB_{width}	0.46	0.50	0.23	0.45	0.55	0.53	0.52	0.04	0.20	0.21
VB_{max}	$-5.73(0)$	$-5.70(0)$	$-5.40(0)$	$-5.29(0)$	$-5.58(0)$	$-5.63(0)$	$-5.69(0)$	$-6.06(0)$	$-6.08(0)$	$-4.58(8)$
VB_{min}	$-6.02(8)$	$-5.97(6)$	$-5.69(10)$	$-5.46(5)$	$-5.84(8)$	$-5.82(5)$	$-5.88(5)$	$-6.08(5)$	$-6.38(10)^c$	$-4.70(0)$
VB_{width}	0.29	0.27	0.29	0.17	0.26	0.19	0.19	0.02	0.30	0.12

	PolyGlut(10)	PolyHist(10)	PolyProl(10)	PolyCyst(8)	PolyMeth(10)	PolyTir(10)	PolyPheala(10)	PolyArg(10)	PolyLys(10)
CB_{max}	$-0.60(0)$	$-0.55(10)$	$0.07(10)$	$-0.86(8)$	$-0.77(10)$	$-1.21(0)$	$-1.06(3)^d$	$0.13(5)$	$-0.41(10)$
CB_{min}	$-1.07(10)$	$-0.89(0)$	$-0.09(3)$	$-1.04(0)$	$-1.26(0)$	$-1.35(5)$	$-1.22(5)$	$0.02(0)$	$-0.93(0)$
CB_{width}	0.47	0.34	0.16	0.18	0.49	0.14	0.16	0.11	0.52
VB_{max}	$-5.56(8)$	$-5.37(0)$	$-5.06(10)$	$-5.69(0)$	$-5.34(5)$	$-5.46(0)$	$-5.52(0)$	$-4.27(0)$	$-4.5389(0)$
VB_{min}	$-5.58(5)$	$-5.66(7)$	$-5.16(0)$	$-5.70(8)$	$-5.36(10)$	$-5.63(10)$	$-5.84(10)$	$-4.29(5)$	$-4.5443(7)$
VB_{width}	0.02	0.29	0.10	0.01	0.02	0.17	0.32	0.02	0.0054

The homopolypeptides are denoted in the Table by the abbreviation of the name of the amino acid residue which is repeated in them.
The number in parenthesis after the abbreviation of the name of the homopolypeptides is the number of k-points for which the calculation was done.
The number given in parenthesis after the numbers characterizing given the band edges indicate the k value at which the band edge occurs.
In the case of PolyGlutAc the first two valence bands overlap [$VB_{min} = 6.28(10)$ and $VB-1_{min} = -6.24(0)$]. In this case we have taken the two bands as one.
For PolyPheala between the CB_{max} and $CB+1_{min}$ band edges there is no gap ($CB_{max} = 1.12(0) = CB+1_{min}(10)$). We take them also as one band.

[a] The homopolypeptides are denoted in the Table by the abbreviation of the name of the amino acid residue which is repeated in them.
[b] The number in parenthesis after the abbreviation of the name of the homopolypeptides is the number of k-points for which the calculation was done.
[c] The number given in parenthesis after the numbers characterizing given the band edges indicate the k value at which the band edge occurs.
[d] In the case of PolyGlutAc the first two valence bands overlap [$VB_{min} = 6.28(10)$ and $VB-1_{min} = -6.24(0)$]. In this case we have taken the two bands as one.

Table 3
The gaps of the 19 homopolypeptides calculated by the BLYP
method (in eV-s)

PolyGly	4.88	PolyGlut	4.49
PolyAla	4.86	PolyHist	4.48
PolySer	4.81	PolyProl	4.99
PolyThre	4.71	PolyCyst	4.65
PolyLeu	4.86	PolyMeth	4.08
PolyiLeu	4.81	PolyTir	4.11
PolyVal	4.82	PolyPheala	4.30
PolyAspAc	3.89	PolyArg	4.29
PolyGlutAc	4.69	PolyLys	3.61
PolyAsp	4.23		

4. Conclusion

The calculation of the 19 homopolypeptides built
up from the essential 19 amino acids (polycystein
because of its S–S bridge cannot be treated as an 1D
system and in the 2D case a much larger unit cell has to
be taken because in this case the crystal orbital theory
can be formulated only for simple translation and not
for a general periodic system with a combined
symmetry operation). This would make the calculation
even to day prohibitively CPU time consuming
using classical quantum chemical methods (HF +
Moller–Plesset many body perturbation theory [16]).
The same is true for other homopolypeptides with
larger side chains if one wants to use HF + MP2 +
MP3 or MP4 theory or HF + coupled cluster theory
[17]. This is the reason that we try to perform these
large scale band structure calculations with the aid of
different DFT methods.

One hopes that with the further developments of
the DFT methods or of other new less CPU time
consuming methods it will be possible to perform
the band structure calculations for all the homopolypeptides including PolyCys and PolyTry and
obtain realistic gap values for them both in the β
pleated sheet and α-helix conformations. This would
promote very much the study of their transport
properties.

Further with the help of such methods combining
them with procedures applied for disordered systems
it will be possible to contribute to the prediction of the
conformations of proteins for which the sequences are
known.

Acknowledgements

This work was supported by the Flemish-Hungarian Scientific and Technological Joint Fund (TeT B-
2/01 and BIL01/72). One of us (F.B.) is indebted to
the National Scientific Research Fund Hungary
(OTKA M36803) and the Hungarian Ministry of
Education (MU-00094/2002).

References

[1] J. Ladik, F. Bogar, Acta Phys. Chem. Debr. 34 (2002) 51.
[2] F. Bogar, V. Van Doren, J. Ladik, PCCP 3 (2001) 5436.
[3] F. Bogar, J. Ladik, PCCP 5 (2003) 953.
[4] N.C. Johnson Jr., I. Tinoco Jr., J. Am. Chem. Soc. 94 (1972)
    4389.
[5] S. Onari, J. Phys. Soc. Jpn 29 (1970) 528.
[6] J.P. Perdew, R.G. Paer, M. Levy, J. Baldus, Phys. Rev. Lett. 49
    (1982) 169.
[7] J.P. Perdew, M. Levy, Phys. Rev. B56 (1977) 16021.
[8] see for instance:J.W. Mintmire, in: J. Labanowski, J. Andzelm
    (Eds.), Density Functional methods in Chemistry, Springer,
    New York, 1991, p. 125.
[9] A.D. Becke, J. Chem. Phys. 96 (1992) 2155.
[10] Ch. Lee, W. Young, R.G. Parr, Phys. Rev. B37 (1988) 785.
[11] A.D. Becke, J. Chem. Phys. 98 (1993) 5648.
[12] K. Branden, J. Tooze, Introduction to Protein Structure,
     Garland Publishing Co., London, 1991.
[13] L. Gianolo, E. Clementi, Gazz. Chim. Ital. 110 (1980) 179.
[14] See for instance:J. Delhalle, L. Piella, J.-L. Brédas, J.-M.
     André, Phys. Rev. B22 (1980) 6254.
[15] P. Süle, S. Kurth, V. Van Doren, J. Chem. Phys. (2000) 7355.
[16] C. Moeller, M.S. Plesset, Phys. Rev. 46 (1934) 618.
[17] (a) J. Cizek, J. Chem. Phys. 45 (1966) 4256.
     (b) J. Cizek, Adv. Quant. Chem. 3 (1969) 35.
     (c) J. Cizek, J. Paldus, Int. J. Quant. Chem. 5 (1971) 359.

ELSEVIER

Journal of Molecular Structure (Theochem) 666–667 (2003) 387–392

THEO
CHEM

www.elsevier.com/locate/theochem

Oxidative stress and free radicals

Svend J. Knak Jensen*

Department of Chemistry, University of Aarhus, Langelandsgade 140, Aarhus C DK-8000, Denmark

Abstract

Oxidative stress is linked to serious diseases and longevity of aerobic cells, including those in humans, which makes the topic of special interest. In this article the basic chemistry involved in the formation of free radicals and other reactive oxygen and nitrogen species will be reviewed along with an account of the major reaction types they induce. In addition, it will be shown how the technique of computational chemistry may provide detailed information on specific reaction channels of importance in oxidative stress. Specifically, preliminary data for a mechanism for the NO mediated production of HO radicals in the absence of transition metal catalysts will be presented.
© 2003 Elsevier B.V. All rights reserved.

Keywords: Oxidative stress; Aerobic cells; Computational chemistry; Free radicals

1. Introduction

At the core of the topic of oxidative stress are free radicals along with reactive oxygen and nitrogen species, which collectively are called oxidants. All oxidants are small molecules/ions in comparison to, e.g. proteins and very few oxidants have a molecular weight in excess of 100. In the last few years, there has been a considerable interest in the reactions of the oxidants with bio-molecules. For example, the number of research papers dealing with the reactions of a well-known oxidant, peroxynitrite, has been around a thousand annually in the past two years. The small sizes of the oxidants make them attractive to detailed investigations and in the last decade much insight has been added to the understanding of the chemistry of the oxidants. In the first part of this paper the basic chemistry in the production of the oxidants will be reviewed along with an account of the major

types of reactions they engage in. The second part describes a preliminary computational chemistry investigation of a mechanism for nitric oxide mediated production of hydroxyl radicals.

2. Production of free radicals and other reactive species

A free radical is a chemical species that has an odd number of electrons. In the context of oxidative stress the radicals are small molecules/ions that are reactive with small activation energies and short lifetimes. The small size makes it possible for many of them to penetrate cell membranes. The free radicals can be considered as a subset of reactive oxygen or nitrogen species. A major part of reactive oxygen species originates as by-products of the aerobic metabolism in the mitochondria. The superoxide anion, O_2^- is produced in the inner membrane of the mitochondria as part of the mechanism, which reduces O_2 to water. O_2^- is the corresponding base to hydroperoxyl, HO_2,

* Tel.: +45-8942-3862; fax: +45-8619-6199.
E-mail address: kemskj@chem.aau.dk (S.J.K. Jensen).

0166-1280/$ - see front matter © 2003 Elsevier B.V. All rights reserved.
doi:10.1016/j.theochem.2003.08.037

which has a pK_a value of 4.9. O_2^- may enter disproportionation reactions like

$$2O_2^- + 2H_2O \rightarrow H_2O_2 + O_2 + 2HO^- \tag{1}$$

In addition to reaction (1), a major source of H_2O_2 is the oxidative deamination of biogenic amines catalyzed by the mitochondrial outer membrane monoamine oxidase.

Hydrogen peroxide is an oxidant, although it is not a radical. It is relatively long lived and may diffuse long distances before it enters into reactions. It may produce the highly reactive hydroxyl radical, HO, through the Fenton reaction, if a transition metal is available

$$H_2O_2 + Fe^{2+} \rightarrow Fe^{3+} + HO^- + HO \tag{2}$$

$$O_2^- + Fe^{3+} \rightarrow Fe^{2+} + O_2 \tag{3}$$

Hydrogen peroxide may produce other reactive oxygen species like hypochlorous acid, HOCl ($pK_a = 7.53$), by enzymatic [1] (myeloperoxidase) oxidation of chloride ions

$$H_2O_2 + Cl^- \rightarrow HOCl + HO^- \tag{4}$$

HOCl may lead to another powerful oxidant [2]: singlet oxygen, $^1O_2$,

$$H_2O_2 + OCl^- \rightarrow {}^1O_2 + H_2O + Cl^- \tag{5}$$

or to the hydroxyl radical

$$HOCl + O_2^- \rightarrow O_2 + HO + Cl^- \tag{6}$$

Hydrogen peroxide may also oxidize the abundant hydrogen carbonate ion to hydrogen peroxycarbonate [3]

$$H_2O_2 + HCO_3^- \rightarrow H_2O + HCO_4^- \tag{7}$$

The reactive nitrogen species, NO, is produced in biological tissue, e.g. in vascular endothelial cells by nitric oxide synthase. In the context of oxidative stress, the most noted reaction of the radical NO is with the radical O_2^-

$$NO + O_2^- \rightarrow ONOO^- \tag{8}$$

When two radicals react, the product will either be in a spin singlet state or a spin triplet state. In the present case the product, the peroxynitrite anion, is a spin singlet. The reaction (8) is very fast with a rate constant $k = 7 \times 10^9 \text{ M}^{-1}\text{ s}^{-1}$ close to the one for

a diffusion-controlled reaction. $ONOO^-$ is a powerful oxidant, in contrast to NO and O_2^-. Peroxynitrite can react with the abundant CO_2 to form nitrosoperoxycarbonate

$$ONOO^- + CO_2 \rightarrow ONOOCO_2^- \tag{9}$$

$ONOOCO_2^-$ is short-lived and has a dissociation channel that leads to two radicals [4].

$$ONOOCO_2^- \rightarrow NO_2 + CO_3^- \tag{10}$$

The corresponding acid to peroxynitrite is peroxynitrous acid ONOOH ($pK_a = 6.8$), which decomposes within a matter of seconds in aqueous solution. The decomposition pathway for peroxynitrous acid has been a matter of some controversy. However, recent experimental [5] and computational [6] evidence suggests that peroxynitrous acid undergoes homolysis to HO and NO_2 radicals, which may subsequently recombine to nitric acid.

$$ONOOH \rightarrow HO + NO_2 \rightarrow HNO_3 \rightarrow H^+ + NO_3^- \tag{11}$$

2.1. Major reaction types of reactive species

The most reactive among the reactive species is the hydroxyl radical. Its half-life in organisms is about 10^{-9}s, indicating it will do chemistry within a short distance from its place of generation [7]. In fact, it has been estimated that more than half of the damage inflicted by free radicals is attributable to the hydroxyl radical [8]. HO may initiate lipid peroxidation by hydrogen abstraction from an unsaturated fatty acid, XH

$$XH + HO \rightarrow X + H_2O \tag{12}$$

The fragment X may then participate in a radical chain reaction, like

$$X + O_2 \rightarrow XO_2 \tag{13}$$

$$XO_2 + XH \rightarrow XOOH + X \tag{14}$$

The chain can be terminated when two radicals react to form a stable product, e.g.

$$X + O_2^- \rightarrow XO_2^- \tag{15}$$

Other types of reactions involving the hydroxyl radical are hydroxylation, oxidation and cleavage of proteins and nucleic acids. Reactive nitrogen species may induce nitration, nitrosation and deamination of bases. Oxidants—other than the hydroxyl radical—generally exhibit selectivity in the type of biomolecule they will react with. If such reactions had the possibility of proceeding unchecked then the result is structural alteration of proteins, inhibition of enzymatic activity and interference in the regulatory functions and eventual cell death. Luckily, there are defense systems, endogenous antioxidants, which will suppress the deleterious reactions. Antioxidants can be enzymatic, like superoxide dismutase (SOD) or non-enzymatic. The chemical principle behind the protection offered by an effective antioxidant is that it will react with the oxidant before the oxidant will react with important biomolecules. Thus the reaction of SOD with O_2^- to produce H_2O_2 is fast [9] and the only reaction that can compete with it in speed is reaction (8) [10]. Non-enzymatic antioxidants act mainly as scavengers in oxidative chain reactions. They can be divided into lipid phase chain breaking antioxidants such as tocopherols and aqueous phase chain breaking antioxidants such as ascorbate. Moreover, other non-enzymatic antioxidants function by sequestering transition metals, thus reducing the occurrence of Fenton-like reactions (2).

It should be emphasized that free radicals and other reactive species also serve a useful function in the organism. For example, NO is important in regulating blood pressure [11] and as a neural messenger and HOCl fights microbial infection. Under normal (healthy) circumstances free radicals can be detected in cells [12]. In a healthy cell there is a balance between reactive species and antioxidants and oxidative stress can be defined as the imbalance between oxidants and antioxidants in favor of the oxidants. Higher than normal levels of antioxidants can likewise have negative effects on the cell [13,14].

There is a considerable amount of evidence that oxidative stress is associated with many serious diseases [15,16]. The central nervous system is particularly susceptible to oxidative stress [17]. There are several reasons for this [18]: (i) the brain has a high consumption of oxygen and has a low level of endogenous antioxidants; (ii) it has a relative high concentration of polyunsaturated fatty acids and—in some areas—it is relative rich in unbound Fe^{2+} which may lead to chain reaction like (2)–(3). The oxidative damage is made worse due to the fact that neurons normally do not renew themselves. Ref. [18] contains a table listing the neurological diseases and processes, which are believed to be associated with oxidative stress.

The aging process has also been linked to oxidative stress. The free radical theory of aging [19] is based on the hypothesis that since the endogenous antioxidants do not completely suppress the oxidants then there are cumulative effects of oxidative damage over a life span [20,21]. This leads to mutations in the mitochondrial DNA resulting in reduced functionality of the cell and eventual death. The hypothesis is supported by observations of increased concentrations of markers of oxidative damage in aged cells [22,23]. The current opinion seems to be that aging is multifactorial process where oxidative stress is an important but not the only mechanism [20,22,24].

3. NO mediated production of HO. A preliminary computational chemistry study

The Fenton reaction has traditionally been considered to be dominant mechanism leading to the production of the hydroxyl radical [25]. However, recently the reaction of H_2O_2 and NO was considered as an alternative pathway [26] to the hydroxyl radical. The reaction mechanism is not known in detail and all the reaction products have not been identified, but the hydroxyl radical has been established. The reaction will be used as an illustration of how the technique of computational chemistry may be useful in elucidating a specific reaction channel, i.e.

$$H_2O_2 + NO \rightarrow HNO_2 + HO \qquad (16)$$

The method has the following elements:

- The reactants as well as the products form complexes, H_2O_2,NO and HNO_2,HO.
- Quantum mechanical methods are used to find the geometry of the potential energy minima of the complexes (conformers). The thermodynamic data for the conformers in the gas phase are calculated.
- The structure of a transition state connecting a reaction conformer and a product conformer is determined and the thermodynamic data in the gas phase are calculated.

- The energies of the transition state and the two conformers it connects are calculated for an aqueous solution using a self-consistent reaction field method [27].

The structures of the conformers are found using geometry optimization methods. The level of theory is chosen as a density functional method, B3LYP [27] that is widely used. Two basis sets are applied, 6-311 + G(d,p) [28] and AUG-cc-pVTZ [29] to assess the variation of the theoretical data with the basis set. The determination of transition states (TS) is facilitated by using a synchronous transit-guided Quasi-Newton method [30]. Explicit calculations are performed to identify the conformers linked by the transition states.

The thermodynamic properties of the species in aqueous solution are estimated using the

approximate Tomasi theory of polarized continuum simulations (IPCM) [31,32]. That is, we find the energies of the various species in a medium with a dielectric constant of liquid water at 25 °C. Normally this theory leads to good agreement with experiment for systems without hydrogen bonding [27]. However, this shortcoming is likely to be of little importance here as we are

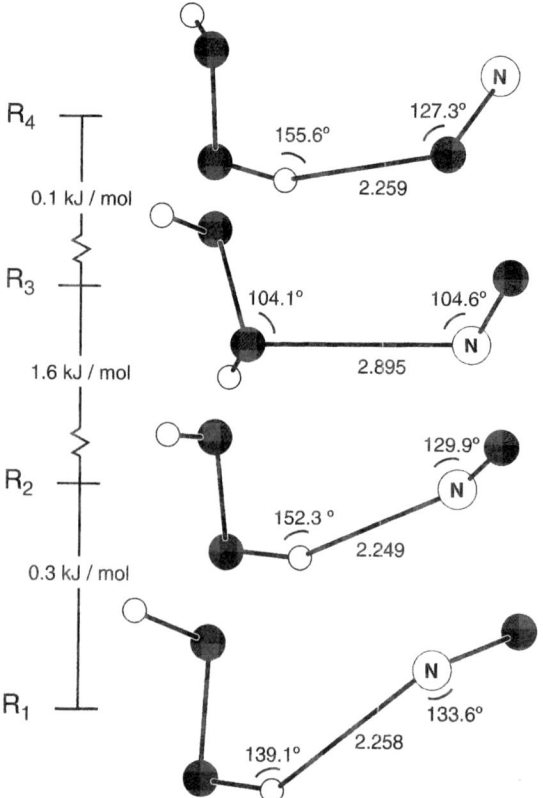

Fig. 1. Structures of complexes between NO and H_2O_2 calculated at the B3LYP/6-311 + G(d,p) level of theory. The distances and angles are in Å and degrees, respectively. ● indicates an oxygen atom and ○ indicates a hydrogen atom.

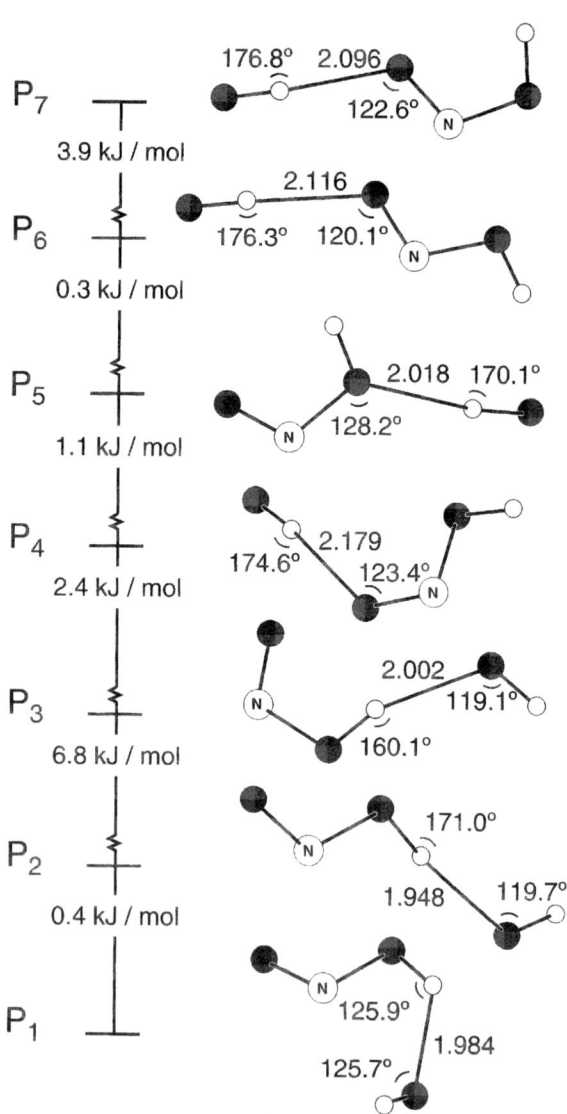

Fig. 2. Structures of complexes between HO and HNO_2 calculated at the B3LYP/6-311 + G(d,p) level of theory. The symbols are explained in Fig. 1.

considering differences between energies of similar molecules.

All the calculations are done using the software package GAUSSIAN 98 [33]. Some indication of the ability of the chosen theoretical methods to predict activation energies may be obtained by considering the reaction

$$trans\text{-}HNO_2 \rightarrow cis\text{-}HNO_2 \qquad (17)$$

where the activation energy in the gas phase has been measured to 48.5 ± 0.8 kJ/mol [34]. The theoretical estimates are 48.3 and 46.3 kJ/mol for B3LYP/6-311 + G(d,p) and B3LYP/AUG-cc-pVTZ, respectively.

In Figs. 1 and 2 we show the reaction complexes, R_i, and product complexes, P_i, respectively. The large number of product complexes is related to the presence of *cis/trans* isomers in nitrous acid. The binding energies of both types of complexes are a few kJ/mol only. In Fig. 3 we show two transition states TSA and TSB. Both of them look like a HO fragment loosely bound to a NO fragment and to another HO fragment. The dissociating mode in the transition states is basically a movement of the central HO fragment between its two neighboring fragments. Detailed studies show that TSA connects R_2 and P_3

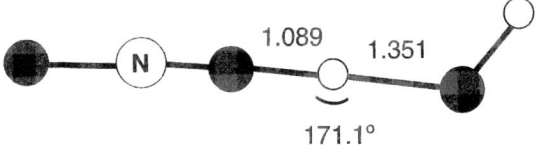

Fig. 4. Structure of a transition state calculated at the B3LYP/6-311 + G(d,p) level of theory. The transition state links P_2 with the NO_2, H_2O complex shown in Fig. 5.

(a *cis* conformer), whereas TSB connects R_3 and P_2 (a *trans* conformer). The activation energy for $R_2 \rightarrow$ TSA is 74.5 and 78.5 kJ/mol at the B3LYP/6-311 + G(d,p) and B3LYP/AUG-cc-pVTZ levels, respectively. The corresponding numbers in the case of TSB are 55.5 and 65.3 kJ/mol. If these numbers can be supported by additional calculations using different theoretical methods then they suggest that the reaction channel (16) is feasible in aqueous solution.

The structures of some of the product complexes in Fig. 2 (e.g. P_2) suggest that they could isomerise to a NO_2, H_2O complex. This is supported by direct calculations. In Fig. 4 we show a transition state which links P_2 and the NO_2, H_2O complex depicted in Fig. 5. In aqueous solution the activation energy from

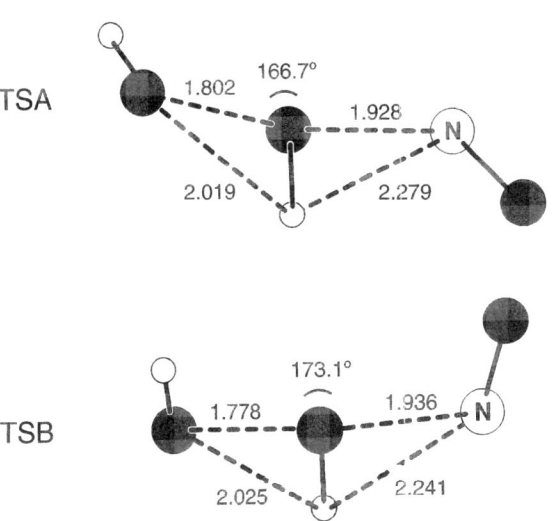

Fig. 3. Structures of two transition states calculated at the B3LYP/6-311 + G(d,p) level of theory. TSA links R_2 and P_3 whereas TSB links R_3 and P_2. The energy of TSA is about 12 kJ/mol higher than that of TSB.

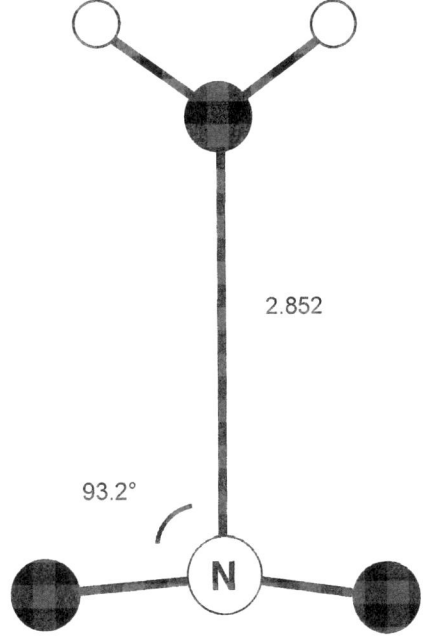

Fig. 5. Structure of a NO_2, H_2O complex calculated at the B3LYP/6-311 + G(d,p) level of theory.

P_2 is 35.4 and 37.9 kJ/mol at the B3LYP/6-311 + G(d,p) and B3LYP/AUG-cc-pVTZ levels, respectively. This low value suggests that the HNO_2, HO complexes may be short-lived intermediates which may dissociate, isomerise or—depending on the composition of the solution—engage in other reactions.

It is should be noted that the reaction

$$H_2O_2 + NO \rightarrow NO_2 + H_2O \tag{18}$$

has been studied in the gas phase [35]. It was found that 90% of the initial NO was found as NO_2 in equilibrium. The identity of the remaining 10% was not established.

Acknowledgements

This work was supported by a grant (HDW-1101-14) from the Danish Center for Scientific Computing.

References

[1] H. Rosen, J.R. Crowley, J.W. Heinecke, J. Biol. Chem. 277 (2002) 30463.

[2] U. Khan, M. Kasha, Proc. Natl Acad. Sci. USA 91 (1994) 12362.

[3] D.E. Richardson, H. Yao, K.M. Frank, D.A. Bennett, J. Am. Chem. Soc. 122 (2000) 1729.

[4] G.L. Squadrito, W.A. Pryor, Chem. Res. Toxicol. 15 (2002) 885.

[5] G. Merényi, J. Lind, G. Czapski, S. Goldstein, Proc. Natl Acad. Sci. USA 97 (2000) 8216.

[6] D.A. Dixon, D. Feller, C.-G. Zhan, J.S. Francisco, J. Phys. Chem. A 106 (2002) 3191.

[7] R.J. Reiter, D. Tan, C. Osuna, E. Gitto, J. Biomed. Sci. 7 (2000) 444.

[8] H.M. Hassan, in: L.B. Clerch, D.J. Massaro (Eds.), Oxygen, Gene Expression and Celluar Function, Marcel Dekker, New York, 1997, p. 27.

[9] H.M. Hassan, Free Radic. Biol. Med. 5 (1988) 377.

[10] W.A. Pryor, G.L. Squadrito, Am. J. Physiol. (Lung Cell. Mol. Physiol. 12) 268 (1995) L699.

[11] A. Moncada, A. Higgins, N. Engl. J. Med. 329 (1993) 2002.

[12] B. Chance, H. Sies, A. Boveris, Physiol. Rev. 59 (1979) 527.

[13] R.J. Reiter, Drug News Perspect 11 (1998) 291.

[14] J.M. McCord, Oxidative Stress and Diseases 1(Oxidative Stress in Cancer, AIDS, and Neurodegenerative Diseases (1998) p. 1.

[15] N. Kaul, H.J. Forman, in: C.J. Rhodes (Ed.), Toxicology of the Human Environment, Taylor and Francis, London, 2000, p. 311.

[16] B.P.F. Rutten, H.W.M. Steinbusch, H. Korr, C. Schmitz, Curr. Opin. Clin. Nutr. Metab. Care 5 (2002) 645.

[17] F. Torreilles, S. Salman-Tabcheh, M-C. Guérin, J. Torreilles, Brain Res. Rev. 30 (1999) 153.

[18] R.J. Reiter, Prog. Neurobiol. 56 (1998) 359.

[19] D. Harman, J. Gerontol. 2 (1957) 298.

[20] B.N. Ames, M.K. Shigenaga, T.M. Hagen, Proc. Natl Acad. Sci. USA 90 (1993) 7915.

[21] W.C. Orr, R.S. Sohal, Science 263 (1994) 532.

[22] T. Finkel, N.J. Holbrook, Nature 408 (2000) 239.

[23] D. Harman, Ann. N.Y. Acad. Sci. 959 (2002) 384.

[24] A.P. Wickens, Respiration Physiol. 128 (2001) 379.

[25] M. Saran, C. Michel, K. Stettmaier, W. Bors, Free Radic. Res. 33 (2000) 567.

[26] J. Nappi, E. Vass, Biochem. Biophys. Acta 1380 (1998) 55.

[27] J.B. Foresman, Æ. Frisch, Exploring Chemistry with Electronic Structure Methods, Pittsburg, USA, Gaussian, Inc., 1996.

[28] W.J. Hehre, L. Radom, P. von Schleyer, J.A. Pople, Ab initio molecular orbital theory, Wiley, New York, USA, 1986.

[29] T.H. Dunning Jr., J. Chem. Phys. 90 (1989) 1007.

[30] C. Peng, P.Y. Ayala, H.B. Schlegel, M.J. Frisch, J. Comput. Chem. 17 (1996) 49.

[31] S. Miertuš, J. Tomasi, Chem. Phys. 65 (1982) 239.

[32] S. Miertuš, E. Scrocco, J. Tomasi, Chem. Phys. 55 (1981) 117.

[33] M.J. Frisch, G.W. Trucks, H.B. Schlegel, G.E. Scuseria, M. Robb, J.R. Cheeseman, V.G. Zakrzewski, J.A. Montgomery Jr., R.E. Stratmann, J.C. Burant, S. Dapprich, J.M. Millam, A.D. Daniels, K.N. Kudin, M.C. Strain, O. Farkas, J. Tomasi, V. Barone, M. Cossi, R. Cammi, B. Mennucci, C. Pomelli, C. Adamo. S. Clifford, J. Ochterski, G.A. Petersson, P.Y. Ayala, Q. Cui, K. Morokuma, D.K. Malick, A.D. Rabuck, K. Raghavachari, J.B. Foresman, J. Cioslowski, J.V. Ortiz, B.B. Stefanov, G. Liu, D.J. Fox, T. Keith, M.A. Al-Laham, C.Y. Peng, A. Nanayakkara, C. Gonzales, M. Challacombe, P.M.W. Gill, B. Johnson, W. Chen, M.W. Wong, J.L. Andres, M. Head-Gordon, E.S. Replogle, J.A. Pople, GAUSSIAN 98, Revision A.6, Gaussian, Inc., Pittsburgh, PA, USA, 1998.

[34] G.E. McGraw, D.L. Bernitt, I.C. Hisatsune, J. Chem. Phys. 45 (1966) 1392.

[35] B.J. Tyler, Nature 195 (1962) 279.

ELSEVIER

Journal of Molecular Structure (Theochem) 666–667 (2003) 393–396

www.elsevier.com/locate/theochem

An exploratory conformational analysis of D and L β-6-deoxyglucose. An ab initio and DFT approach

Gabrille F.C. Yeung[a], David H. Setiadi[a,b], Gregory A. Chass[a,b,c,d], Imre G. Csizmadia[a,b,e,*]

[a]Department of Chemistry, University of Toronto, Toronto, Ont., Canada M5S 3H6
[b]Global Institute of COmputational Molecular and Materials Science (GIOCOMMS), 1422 Edenrose Street, Mississauga, Ont., Canada L5V 1H3
[c]Institut de Science et d'Ingénierie Supramoléculaires (ISIS), Université Louis Pasteur, Strasbourg, France
[d]Department of Biomedical Sciences, Creighton University, 2500 California plaza, Omaha, NE 68178, USA
[e]Department of Medical Chemistry, Szeged University, Dom ter 8, H-6720 Szeged, Hungary

Abstract

6-Deoxyglucose is not only a model compound for glucose, but it is also a hexokinase inhibitor. The molecule may be $^1C_4$ and $^4C_1$ ring structures. It occurs on α and β isomers, each having assigned 81 MDCA-predicted nomenclature to the conformations. An exploratory ab initio conformational analysis was undertaken on β-6-deoxyglucose at each of the RHF/3-21G, RHF/6-31G(d) and B3LYP/6031G(d) levels of theory. Apart from the anomeric hydroxyl group, there are four C–OH rotamers. Full conformational analysis will reveal the global minima of D- and L-β-6-deoxyglucose. Calculations revealed both D and L of $^1C_4$ and $^4C_1$ conformers to have the lowest energy conformation. It was found that the molecule was greatly stabilized by multiple hydrogen bonds. These hydrogen bonds took several forms; unidirect or non-unidirect clockwise and anti-clockwise hydrogen bonding or tunneling motion, which is dependent on the chair configurations. These stable conformations may suggest the effectiveness of this inhibitor and structural mechanisms in forming polymers.
© 2004 Elsevier B.V. All rights reserved.

Keywords: 6-Deoxyglucose; Conformational analysis; Tunneling motion; Ab initio; DFT; Hydrogen bonding

1. Introduction

Polyhydroxylated compounds, such as carbohydrates, play an important role in biochemistry. These hydroxyl groups serve as donors and acceptors in hydrogen bond interactions which function in the molecular recognition of sugar molecules and in the binding of receptor sites [1–6]. Hydroxyl hydrogens complicate the measurement of equilibrium structures by creating tunneling motions that stabilize conformations [2]. As a result, the characterization of conformational behaviour is essential for the understanding of their biological functions and importance.

As a non-metabolizable analogue, 6-deoxyglucose acts as a valuable tool. Since it does not metabolize inside cells, it is used in studies of sugar transport [7,8]. It is also one of the ingredients of glycolipids of the plant origin, where their cytotoxicity against cancer cells, anitimicrobial activity, and

* Corresponding author. Address: Department of Chemistry, University of Toronto, Toronto, Ont., Canada M5S 3H6. Tel.: +1-416-978-3598; fax: +1-416-978-3598.
E-mail address: icsizmad@chem.utoronto.ca (I.G. Csizmadia).

plant growth regulatory effects have been the subject of a series of recent studies [9].

In this study, examinations of energetics and geometries of D,L-β-6-deoxyglucose will be analyzed.

2. Computational methods

The GAUSSIAN 98 program [10] was used to perform ab initio and density functional analysis on both D and L-β-6-deoxyglucose. Molecular orbital computations were carried out at each of the RHF/3-21G, RHF/6-31G(d), and B3LYP/6-31G(d) levels of theory. Final lowest energy conformations were derived from the optimizations from a systematic conformational search. Ramachandran maps (potential energy surfaces) are generated subsequently to determine the co-ordinates of the minima by visual estimation.

3. Results and discussion

3.1. Relative energies

Recent results from conformational searches reveal the possible stable conformers of D and L enantiomers (Tables 1 and 2). From the tables, it is noted that the relative energies from the L enantiomer are relatively smaller than the ones from D enantiomer. However, a trend of possible stable conformers is drawn from both enantiomers, and their chair conformations. The tables are separated by first the different enantiomers, and then chair conformers (Figs. 1 and 2). The stability of the two enantiomers may also be explained qualitatively by using PES analysis of Ramachandran maps (Figs. 3–6).

Results show that the change in energetic order is consistent across the levels of theories studied. In addition, all levels of theory show relative trends.

3.2. Molecular geometry

In the case of $^4C_1$ conformers, the lowest energy structures of both D and L orientations resulted in an intramolecular hydrogen bonding chain in the anti-clockwise direction. Other structures with co-operative intramolecular hydrogen bonds oriented

Table 1
Optimized stabilization energies for D-β-6-deoxyglucose; $^1C_4$ (top portion) and $^4C_1$ (lower portion) chair forms

Total E (Hartree)	Relative E (kcal/mol)	X^1	X^2	X^3	X^4
−605.116469	4.945	+	a	a	−
−605.118674	3.561	+	+	a	−
−605.116831	4.717	+	−	−	a
−605.115740	5.403	+	−	−	+
−605.118880	3.432	+	−	a	−
−605.116990	4.618	+	−	a	a
−605.116056	5.204	+	−	a	+
−605.120493	2.420	−	a	+	−
−605.107951	10.290	−	a	+	−
−605.110540	8.665	−	a	a	−
−605.110731	8.545	−	+	−	+
−605.103739	12.933	−	+	−	a
−605.110331	8.797	−	+	a	−
−605.124349	**0.000**	−	−	+	+
−605.124025	0.204	−	−	+	a
−605.118797	3.484	−	−	+	−
−605.110955	8.405	−	−	a	a
−605.109110	9.563	−	−	a	+
−605.109361	9.405	+	a	a	−
−605.107052	10.854	+	a	a	−
−605.106237	11.366	+	a	+	+
−605.082635	26.176	+	a	+	a
−605.107868	10.342	+	+	a	a
−605.105995	11.518	+	+	a	−
−605.109783	9.140	−	a	a	+
−605.107341	10.673	−	a	+	−
−605.107302	10.697	−	a	+	+
−605.109050	9.600	−	+	a	a
−605.119442	3.079	−	+	−	+

clockwise are often broken, less stable and have higher energy values. These broken chains can be directed clockwise or anti-clockwise. In contrast, the orientations of the hydroxyl groups for $^1C_4$ conformers are directed in left and right combinations. Nonetheless, the most stable conformers, like $^4C_1$ chairs, result in the maximum number of possible hydrogen bonding chain formation. These results show that the orientations of the hydroxyl groups are not independent of each other.

Fixation of the hydroxyl groups at certain positions may lead to the breaking of the ring between C1 and O6. This may be due to their intramolecular interactions and anomeric effect which are the primary source for ring distortion.

Table 2
Optimized stabilization energies for L-β-6-deoxyglucose; $^{1}C_4$ (top portion) and $^{4}C_1$ (lower portion) chair forms

Total E (hartree)	Relative E (kcal/mol)	X^1	X^2	X^3	X^4
−605.123107	3.883	+	a	a	−
−605.124789	2.828	+	+	a	−
−605.122323	4.375	+	−	a	−
−605.122046	4.549	+	−	a	+
−605.122279	4.403	+	−	−	−
−605.122260	4.415	+	−	−	+
−605.114097	9.537	−	a	a	−
−605.110007	12.104	−	a	a	+
−605.127395	1.193	−	a	+	−
−605.113688	9.793	−	+	a	−
−605.125893	2.135	−	+	+	−
−605.111497	11.168	−	+	−	+
−605.106636	14.219	−	+	−	−
−605.114771	9.114	−	−	a	−
−605.112097	10.792	−	−	a	+
−605.129295	**0.000**	−	−	+	+
−605.128809	0.305	−	−	+	a
−605.126699	1.629	−	−	+	−
−605.106212	14.485	+	a	a	a
−605.103169	16.395	+	a	+	+
−605.103154	16.404	+	a	+	+
−605.102762	16.650	+	a	a	−
−605.104972	15.263	+	+	a	a
−605.101835	17.232	+	+	a	−
−605.106356	14.394	−	a	a	a
−605.103861	15.960	−	a	+	+
−605.102620	16.739	−	a	a	−
−605.105703	14.805	−	+	a	a
−605.116113	8.272	−	+	−	+
−605.123107	3.883	+	a	a	−
−605.124789	2.828	+	+	a	−
−605.122323	4.375	+	−	a	−
−605.122046	4.549	+	−	a	+

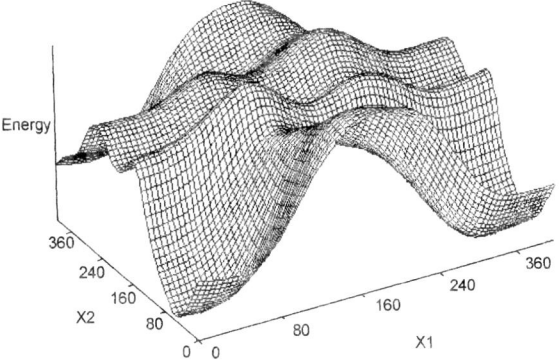

Fig. 3. PES of L-β-6-deoxyglucose- $^{4}C_1$.

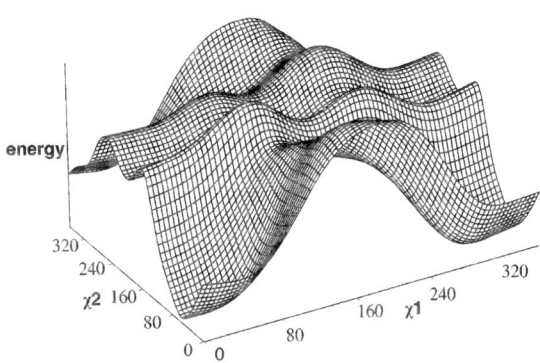

Fig. 4. PES of D-β-6-deoxyglucose- $^{4}C_1$.

Further investigation of the O⋯H interactions, possible bond lengthening effect, and angles will be discussed in a subsequent paper. A more detailed analysis will also be presented with more focus on the performed higher level of theories.

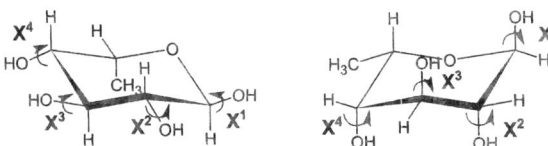

Fig. 1. L-β-6-deoxyglucose, $^{4}C_1$ and $^{1}C_4$ chair forms.

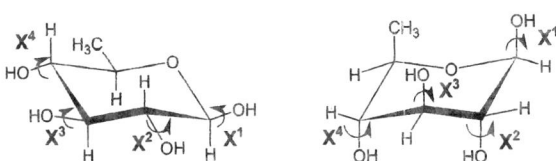

Fig. 2. D-β-6-deoxyglucose, $^{4}C_1$ and $^{1}C_4$ chair forms.

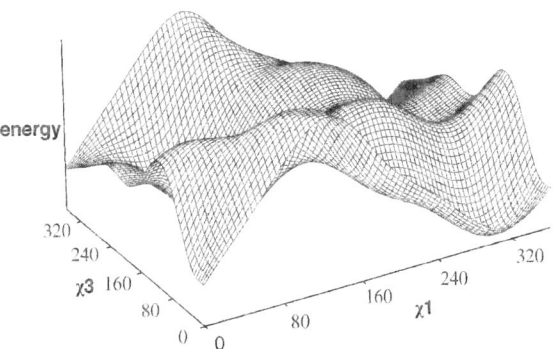

Fig. 5. PES of L-β-6-deoxyglucose- $^{4}C_1$.

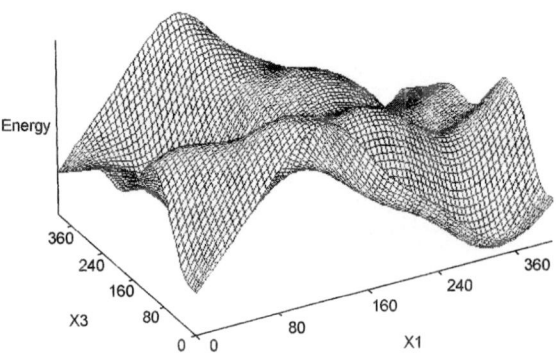

Fig. 6. PES of D-β-6-deoxyglucose- $^1C_4$.

4. Conclusions

From further investigations, the above methodology is believed to be helpful in analyzing 1,4 and 1,2 polysaccharides that might explain possible receptor bindings, and chain linkages in a more biological aspect. Researches of alpha versus beta enantiomers may also be of interest in the aforementioned studies. The following preliminary conclusions can be drawn from the discussion above.

(1) High energy barriers between minima cause conformer stability. Since $^4C_1$ conformers have higher barriers than $^1C_4$ conformers, $^1C_4$ conformers are found to be more reactive and energetically stable in RHF/3-21G calculations.

(2) Due to the increase in unfavourable interactions between hydroxyl groups and distortion due to anomeric effect, fixation of the hydroxyl groups may lead to ring breakage.

(3) Orientations of the hydroxyl groups are dependent on each other. The maximal number of possible hydrogen bonding interaction occurs in the most stable conformers. These hydrogen bond chains form in anti-clockwise or clockwise unidirected, or non-unidirected patterns.

Acknowledgements

One of the authors (IGC) wishes to thank the Ministry of Education for a Szent-Györgyi Visiting Professorship.

References

[1] M.D. Lee, G.A. Ellestad, D.B. Borders, Acc. Chem. Res 24 (1991) 235.

[2] K.C. Nicolaou, W.M. Dai, S.C. Tsay, V.A. Estevez, W. Wrasildo, Science 256 (1992) 1172.

[3] G.I. Csonka, I.G. Csizmadia, Chem. Phys. Lett. 243 (1995) 419.

[4] G.I. Csonka, I. Kolossvary, P. Csaszar, K. Elias, I.G. Csizmadia, J. Mol. Struct. (Theochem) 395 (1997) 29.

[5] G.I. Csonka, K. Elias, I.G. Csizmadia, Chem. Phys. Lett. 257 (1996) 49.

[6] I. Tvaroka, F.R. Taravel, J.P. Utille, J.P. Carver, Carbohydr. Res. 337 (2002) 353.

[7] Y.K. Kim, M. Kitaoka, M. Krishnareddy, Y. Mori, K. Hayashi, J. Biochem. 132 (2002) 197.

[8] Olympus Microscopy Resource Center. http://www.olympusmicro.com/galleries/gluorescence/pages/3t3dapismall.html.

[9] J. Furukawa, N. Sakairi, Trends Glycosci. Glycotech. 13 (69) (2001) 1.

[10] M.J. Frisch, G.W. Trucks, H.B. Schlegel, G.E. Scuseria, M.A. Robb, J.R. Cheeseman, V.G. Zakrzewski, J.A. Montgomery, Jr., R.E. Stratmann, J.C. Burant, S. Dapprich, J.M. Millam, A.D. Daniels, K.N. Kudin, M.C. Strain, O. Farkas, J. Tomasi, V. Barone, M. Cossi, R. Cammi, B. Mennucci, C. Pomelli, C. Adamo, S. Clifford, J. Ochterski, G.A. Petersson, P.Y. Ayala, Q. Cui, K. Morokuma, P. Salvador, J.J. Dannenberg, D.K. Malick, A.D. Rabuck, K. Raghavachari, J.B. Foresman, J. Cioslowski, J.V. Ortiz, A.G. Baboul, B.B. Stefanov, G. Liu, A. Liashenko, P. Piskorz, I. Komaromi, R. Gomperts, R.L. Martin, D.J. Fox, T. Keith, M.A. Al-Laham, C.Y. Peng, A. Nanayakkara, M. Challacombe, P.M.W. Gill, B. Johnson, W. Chen, M.W. Wong, J.L. Andres, C. Gonzalez, M. Head-Gordon, E.S. Replogle, and J.A. Pople, GAUSSIAN 1998 (Revision A.9), Gaussian, Inc., Pittsburg PA, 2001.

ELSEVIER

Journal of Molecular Structure (Theochem) 666–667 (2003) 397–400

THEO
CHEM

www.elsevier.com/locate/theochem

Conformational analysis of oxidized vitamin-C

Viktoria V. Kónya[a,b,*], Peter G. Meszaros[a,b], Bela Viskolcz[b], Imre G. Csizmadia[a,c]

[a]Department of Medical Chemistry, Szent-Györgyi Medical Centrum, University of Szeged, H-6720 Szeged Dóm tér 8, Hungary
[b]Department of Chemistry, JGyTFK University of Szeged, H-6725 Szeged, Boldogasszony sgt., Hungary
[c]Department of Chemistry, University of Toronto, Toronto, Ont., Canada M5S 3H6

Abstract

The oxidized form of vitamin-C has the same side chain as the reduced form of vitamin-C. The side orientation is governed by the torsional angles of the carbon skeleton involving two C–C dihedral angles denoted by Φ and Ψ. The relaxed potential energy surface (PES) were generated at the RHF/3-21G level of theory. Ab inito conformational studies were carried out on all the local and global minima of PES at the MP2/6-311G(d,p) level of theory. The conformational and energetic consequences of these findings are discussed in terms of the mechanism of oxidation.
© 2003 Elsevier B.V. All rights reserved.

Keywords: Structure of oxidized Vitamin C; Ascorbic acid; Ab initio; Potential energy surface

1. Introduction

Vitamin C was first isolated by Professor Albert Szent-Györgyi from the adrenal cortex of cattle, and later from orange juice and from sauerkraut [1]. Vitamin C, although technically it is not a vitamin, is an important antioxidant. Some time ago it was considered as a vitamin because it played an essential role in the protecting mechanism of the organism. As a water-soluble antioxidant, vitamin C is in a unique 'scavenger' of aqueous peroxy (HOO·) and hydroxy (HO·) radicals, working effectively before these destructive substances have had a chance to damage the lipids and other structurally and functionally important molecules in the human body [2–5].

* Corresponding author.
E-mail addresses: konya.viktoria@stud.u-szeged.hu (V.V. Kónya), meszaros-peter@freemail.hu (P.G. Meszaros), viskolcz@jgytf.u-szeged.hu (B. Viskolcz), icsizmad@chem.utoronto.ca (I.G. Csizmadia).

Earlier, the structure of ascorbic acid was studied by both experimental and theoretical methods [6]. Nine stable structures of ascorbic acid were found in ascorbic acid crystals at low temperature study [6]. All the side-chain variables in the nine conformations were optimized, while the ring variables were fixed at the values corresponding to the crystal structure. Recently, full optimizations for 36 conformers of ascorbic acid have been reported [7]. The conformations of protonated and deprotonated ascorbic acid were studied recently [8].

We focused our intention on conformational analysis of the oxidized form of ascorbic acid. Both OH group of five membered ring were dehydrogenated, but the side chain retained the two OH groups as two quasi free rotors. Fig. 1 shows the structure and also the rotors which were analyzed. We have performed ab initio conformational analysis of the isolated molecule, in order to assess the importance of intra-molecular hydrogen bonds in dictating, intrinsically the most stable conformations.

0166-1280/$ - see front matter © 2003 Elsevier B.V. All rights reserved.
doi:10.1016/j.theochem.2003.08.040

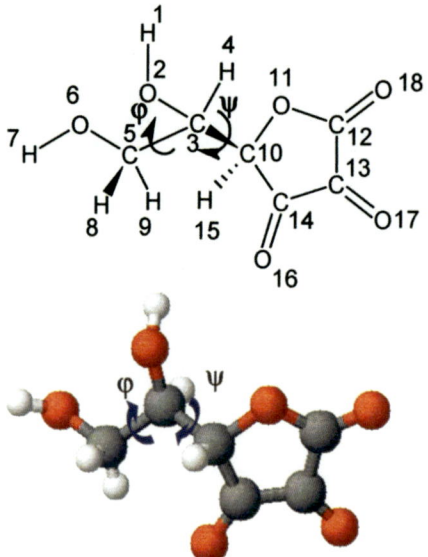

Fig. 1. The structure and numbering of dehydroascorbic acid. Definition of dihedral angels: φ (O6–C5–C3–O2) and ψ (O2–C3–C10–O11).

2. Theoretical methods

The potential energy surface (PES) was obtained by HF/3-21G level of theory. The relative energies of the six minima observed on the PES were obtained by using a variation of Gaussian-3 theory [9] with reduced Møller–Plesset order (G3(MP2) [10]. In this method the geometries were optimized by MP2/6-311G(d,p) level of theory instead of geometries from MP2(FU)/6-31G(d). The harmonic vibrational frequencies were calculated by HF/6-31G* level of theory and zero-point energies were scaled by 0.89 as usual. All ab initio calculations were achieved with GAUSSIAN 98 [11] program package.

3. Results and discussion

3.1. Scan of the surface

The Φ and Ψ dihedral angles (Fig. 1) were rotated by 5 degree intervals using HF/3-21G, so that after

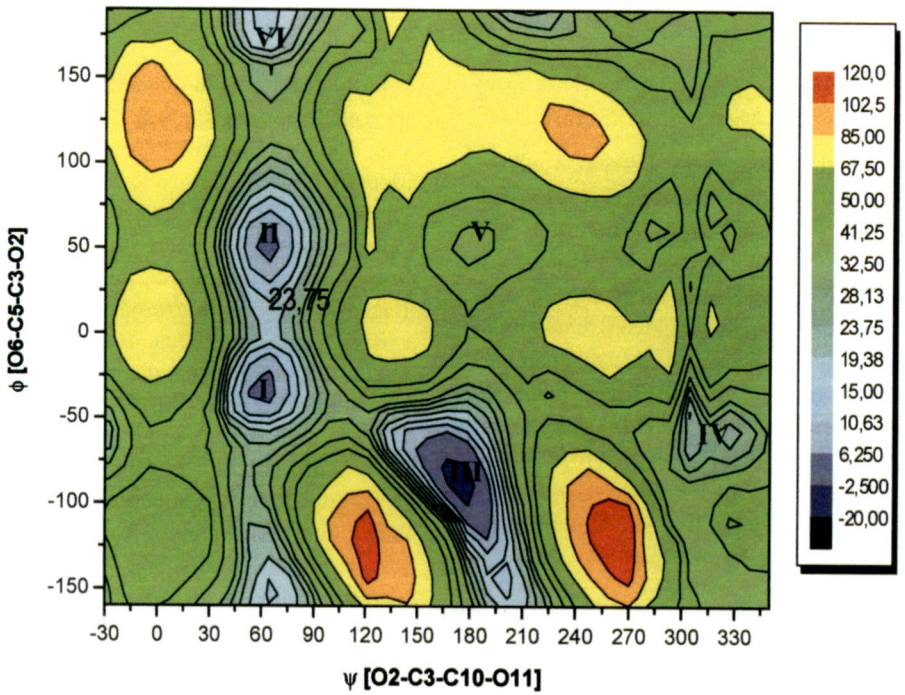

Fig. 2. The PES of dehydroascorbic acid.

every rotation all geometric parameters of the structures were optimized. These involved a total of $72 \times 72 = 5184$ points on the relaxed PES (Fig. 2).

3.2. Properties of PES

The potential energy surface shows the values from -30 to $330°$ along the Φ (O2–C3–C10–O11) dihedral angle and from -170 to $190°$ along the Φ (O6–C5–C3–O2) dihedral angle. To help better see the characteristic points of the surface we have chosen this special cut of the potential surface. On the PES one can see six maxima, and between the minima transition states. The lowest and most characteristic six minima of the relaxed PES are labeled **I–VI**, respectively. Along the Φ dihedral angle rotation minima were found about every 120°. All conformations were assigned by gauche +, anti, and gauche $-$ corresponding to dihedral angles 60, 180 and $-60°$, respectively (Fig. 3).

After the re-optimazation of the six minima we analyzed the H-bond distances. We found that the H-bonds stabilized the structures. On the other hand the stability not only depends on the formation of these secondary bonds. The other cause of the optimal energy level is the planarity of the five-member-ring of the molecule.

The average H-bond distance is between 2.0–2.3 Å. The bond lengths of H-bonds are listed in Table 1. Note that the only minimum is **VI** in which there is no H-bond. Minimum **I** has only one H-bond,

Table 1
The H–O distances in Å for different intramolecular H-bonds

		H1–O6	H1–O16	H7–O2	H7–O11
g^+g^+	**I**	2,161	–	–	–
g^+a	**II**	2,040	–	–	2,151
$a g^-$	**III**	–	2,083	–	2,365
g^-g^-	**IV**	–	2,164	2,285	–
$a g^+$	**V**	–	2,214	2,276	–
g^+g^-	**VI**	–	–	–	–

and all other structures have two different kind of H-bonds. One of them appears to be weaker as the H-bridgehead atom distance is longer. The shortest H-bond was found in the structure of **II**, and the length of the second H-bond was also the shortest one of the second H-bond of all minima, which has two H-bonds. We expected structure **II** to be the most stable conformer, but stability of conformer **I** is approximately is same.

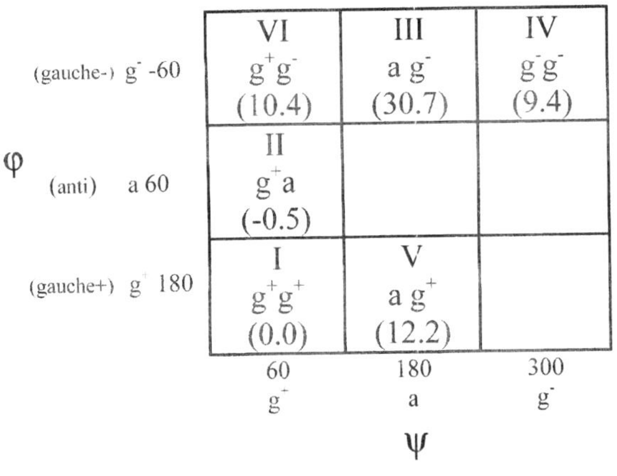

By using HF/3-21G level of theory the minimum **III** appeared to be the lowest minimum, but at higher level of geometry optimization (MP2/6-311G(d,p)) minimum **III** is the highest one of the six minima studied in this work. The G3MP2 level of theory predict also the same trend as the MP2/6-311G(d,p). The agreement in relative energies between the different higher levels of theories was rather good, within 8 kJ mol^{-1}. Minimum **II** seems to be the global one of the PES. One can expect that minima **I** and **II** will have nearly the same stability as these two structures differ only in the orientation of the H-bonds. There is consistency amongst all levels of theory computed as can be see in Table 2.

As we have already mentioned, planarity also has a major influence on the stability of the oxidized

		VI	III	IV
(gauche-) g$^-$ -60		g^+g^- (10.4)	$a\,g^-$ (30.7)	g^-g^- (9.4)
φ (anti) a 60		II g^+a (-0.5)		
(gauche+) g$^-$ 180		I g^+g^+ (0.0)	V $a\,g^+$ (12.2)	
		60 g$^-$	180 a	300 g$^-$

ψ

Fig. 3. The dehydroascorbic acid conformers and their energies in kJ mol^{-1}.

Table 2
The ZPE corrected relative energies in kJ mol^{-1} for local minima of oxidized ascorbic acid

		HF/6-31G*	MP2/6-311G(d,p)	G3MP2
g^+g^+	**I**	0.0	0.0	0.0
g^+a	**II**	0.1	-6.1	-0.5
$a g^-$	**III**	30.7	32.4	29.4
g^-g^-	**IV**	9.4	6.4	7.7
$a g^+$	**V**	12.2	12.6	10.9
g^+g^-	**VI**	10.4	10.9	8.7

ascorbic acid. The five-member-ring of minima **I**, **II**, **IV**, **V** is almost ideally a plane. On the other hand in the ring of minimum **III**, atom C10 was out of plane by about 8°. The atoms of C12, C13, C14, O16, O17 and O18 are in one planar in all structures, except structure **III**. Perhaps the fact that the five member ring in **III** are forced to be non-planar is the underlying cause of it relatively high energy. The planarity of the ring is very important for the delocalization of π-electrons of C=O-bonds.

4. Conclusions

The complete PES of oxidized ascorbic acid was studied using HF/3-21G level of theory. The MP2/6-311G(d,p) level of theory was used to optimize the minima. The relative energy values depend on H-bonds and also the ring planarity has a very important stabilization factor. The expected accuracy of the relative energies is within 8 kJ mol^{-1}.

The study of transition states between the minima is in progress. We expected that by using the activation energies we can obtain more information about the influence on stability of H-bonds and ring planarity.

Acknowledgements

One of the authors (IGC) wishes to thank the Ministry of Education for a Szent-Györgyi Visiting Professorship. One of the authors (BV) thanks the Hungarian Scientific Research Found (OTKA F-037648) for financial support.

References

[1] A. Szent-Gyorgyi, Biochem. J. 22 (1928) 1387.
[2] H. Sapper, S.O. Kang, H.H. Paul, W. Lohmann, Z. Naturfosch. C: Biosci. 37C (1982) 129.
[3] J.R. Woods Jr., M.A. Plessinger, R.K. Miller, Am. J. Obstet. Gynecol. 185 (2001) 5.
[4] P. Evans, B. Halliwell, Brit. J. Nutr. 85 (2001) 67.
[5] F. Shang, M. Lu, E. Dudek, J. Reddan, A. Taylor, Free Rad. Bio. Med. 34 (2003) 521.
[6] M. Milanesio, R. Bianchi, P. Ugliengo, C. Roetti, D. Viterboa, J. Mol. Struct.(Theochem) 419 (1997) 139.
[7] M.A. Mora, F.J. Melendez, J. Mol. Struct.(Theochem) 454 (1998) 175.
[8] J.R. Juhasz, L.F. Pisterzi, D.M. Gasparro, D.R.P. Almeida, I.G. Csizmadia J. Mol. Struct.(Theochem) in press, this issue.
[9] L.A. Curtiss, K. Raghavachari, P.C. Redfern, V. Rassolov, J.A. Pople, J. Chem. Phys. 109 (1998) 7764.
[10] Anwar G. Baboul, Larry A. Curtiss, Paul C. Redfern, K. Raghavachari, J. Chem. Phys. 110 (1999) 7650.
[11] M.J. Frisch, G.W. Trucks, H.B. Schlegel, G.E. Scuseria, M.A. Robb, J.R. Cheeseman, V.G. Zakrzewski, J.A. Montgomery, Jr., R.E. Stratmann, J.C. Burant, S. Dapprich, J.M. Millam, A.D. Daniels, K.N. Kudin, M.C. Strain, Ö. Farkas, J. Tomasi, V. Barone, M. Cossi, R. Cammi, B. Mennucci, C. Pomelli, C. Adamo, S. Clifford, J. Ochterski, G.A. Petersson, P.Y. Ayala, Q. Cui, K. Morokuma, P. Salvador, J.J. Dannenberg, D.K. Malick, A.D. Rabuck, K. Raghavachari, J.B. Foresman, J. Cioslowski, J.V. Ortiz, A.G. Baboul, B.B. Stefanov, G. Liu, A. Liashenko, P. Piskorz, I. Komaromi, R. Gomperts, R.L. Martin, D.J. Fox, T. Keith, M.A. Al-Laham, C.Y. Peng, A. Nanayakkara, M. Challacombe, P.M.W. Gill, B. Johnson, W. Chen, M.W. Wong, J.L. Andres, C. Gonzalez, M. Head-Gordon, E.S. Replogle, and J.A. Pople, GAUSSIAN 98, Revision A.9, Gaussian, Inc., Pittsburgh PA, 2001.

ELSEVIER

Journal of Molecular Structure (Theochem) 666–667 (2003) 401–407

THEO
CHEM

www.elsevier.com/locate/theochem

The effects of conformation on the acidity of ascorbic acid: a density functional study

Jason R. Juhasz[a,*], Luca F. Pisterzi[a], Donna M. Gasparro[a],
David R.P. Almeida[a], Imre G. Csizmadia[a,b,c]

[a]*Department of Chemistry, Lash Miller Laboratories, University of Toronto, 80 St George Street, Toronto, Ont., Canada M5S 3H6*
[b]*Global Institute of Computational Molecular and Materials Science (GIOCOMMS), 1422 Edenrose St., Mississauga, Ont., Canada L5V 1H3*
[c]*Department of Medical Chemistry, University of Szeged, Dóm tér 8, Szeged 6720, Hungary*

Abstract

Ascorbic acid (AsA; ascorbate; vitamin C; $C_6H_8O_6$) is a potent antioxidant in both the cytosolic and membrane components of the human body. AsA serves a protective function as a free radical scavenger and plays a non-direct antioxidant role by aiding in the reduction of the α-tocopheroxy radical back to α-tocopherol. Two hydroxyl moieties (defined as χ_1 and χ_2) are paramount to the conformational profile of AsA. One hydroxyl group, χ_1, is initially deprotonated in AsA's antioxidant mechanism. To explore AsA's acidic character, selected conformations were optimized at the B3LYP/6-31G(d) level of theory and their adiabatic energies of deprotonation calculated. The lowest and highest energies of deprotonation were found to occur in the conformers aa3h-p ($\chi_1 = g^+$, $\chi_2 = g^+$, $\chi_3 = g^-$, $\chi_4 = anti$, $\chi_5 = g^-$, $\chi_6 = g^-$) at 320.92 kcal mol^{-1} and aa11-p ($\chi_1 = g^-$, $\chi_2 = anti$, $\chi_3 = anti$, $\chi_4 = g^+$, $\chi_5 = g^-$, $\chi_6 = anti$) at 343.84 kcal mol^{-1}, respectively. Conformers with minimal intramolecular stabilization had the lowest energies of deprotonation while conformers exhibiting multiple hydrogen bonds had larger energies of deprotonation. In light of these findings, it is hypothesized that higher energy conformations will be favoured in the deprotonation of AsA antioxidant mechanism of action.
© 2003 Elsevier B.V. All rights reserved.

Keywords: Ascorbic acid; Deprotonation; Antioxidant; DFT-Beche 3LYP hybrid functional

1. Introduction

L-Ascorbic acid (AsA; ascorbate; vitamin C; $C_6H_8O_6$) has been a hub of scientific research since its isolation by Hungarian scientist Szent-Györgyi in 1928 [1]. AsA is a potent antioxidant and has an essential role as an enzymatic co-factor for the synthesis of biologically important molecules such as collagen, carnithine, catecholamine, myelin, and neuroendocrine peptides [2,3].

AsA has the ability to protect both cytosolic and membrane components of cells from oxidative stress generated by normal cellular metabolism and exogenous agents [2–12]. Oxidative stress occurs when the production of free radicals such as reactive oxygen species and reactive nitrogen species is exceeded by the rates of removal [2,5,7]. The outcome has the potential to cause various degenerative diseases including: mascular degeneration, cataract, arteriosclerosis and cancer [7].

* Corresponding author.
E-mail addresses: jasonjuhasz@medscape.com (J.R. Juhasz), lpisterzi@medscape.com (L.F. Pisterzi), dgasparro@medscape.com (D.M. Gasparro), dalmeida@medscape.com (D.R.P. Almeida), icsizmad@alchemy.chem.utoronto.ca (I.G. Csizmadia).

0166-1280/$ - see front matter © 2003 Elsevier B.V. All rights reserved.
doi:10.1016/j.theochem.2003.08.042

In the cytosol, AsA is able to combat such oxidative stress by acting as a primary antioxidant and scavenging free radicals. It is also active in cellular membranes, where it is able to recycle the α-tocopheryl radical back to α-tocopherol thereby replenishing the antioxidant potential of the lipid medium [4,5,7,11,13,14].

At physiological pH, AsA is predominantly a delocalized monoanion following a loss of a proton from the carbon-2 hydroxyl moiety (defined as χ_1) (Figs. 1 and 2). Once in its ionized state, AsA is able to donate either one or two electrons in a redox antioxidant mechanism (Fig. 2) [4]. The first step in this reaction requires the dissociation of a proton from the aforementioned hydroxyl group, a phenomenon shown to be dependent on conformation [15]. The present work examines the effects of conformation on the acidity of AsA with molecular orbital computations of adiabatic energies of deprotonation. Elucidation of the conformational-dependent acidity of AsA is paramount to fully deciphering the molecular antioxidant mechanism.

2. Methods

To explore AsA's acidic character, conformations were selected from a previous multidimensional conformational analysis study on the full conformational space of AsA by geometry optimizations of all possible conformational minima of the AsA potential energy hypersurface (PEHS) [16]. The conformers selected were grouped into one of seven conformational assignments based on the conformations of two hydroxyl moieties (defined as χ_1 and χ_2) which have shown to be paramount to the conformational profile of AsA (Table 1) [16]. Both the highest and lowest energy conformers of each family were selected and evaluated in the current study (total of seven low energy and seven high energy conformers belonging to the seven families of AsA) (Table 2). All AsA conformers were defined according to the structure code aaxy-z (aa = ascorbic acid, x = family number, y = high or low relative energy, z = protonated (p) or deprotonated (d)). All conformational assignments were made according to Eq. (1).

$$gauche\ plus\ (g^+) = 60°\ (ideal) \pm 60°$$

$$anti\ (a) = 180°(ideal) \pm 60° \tag{1}$$

$$gauche\ minus\ (g^-) = -60°(ideal) \pm 60°$$

AsA possesses six torsional angles of interest and the conformers of its PEHS can be described by Eq. (2) (Fig. 1). All 14 protonated conformers in this study were subject to full geometry optimizations. Subsequently, the converged structures were deprotonated of proton H14 at torsional angle χ_1 (Fig. 1) and re-optimized. Adiabatic (fully relaxed optimizations) energies of deprotonation were then calculated according to Eq. (3). Energies of deprotonation are quoted as positive values because bond-breaking is always an energy-consuming process.

$$E = f(\chi_1, \chi_2, \chi_3, \chi_4, \chi_5, \chi_6) \tag{2}$$

Energy of deprotonation $= \Delta E_{opt}$

$$= |E_{opt}(\text{protonated}) - E_{opt}(\text{deprotonated})| \tag{3}$$

All computations were performed using the GAUSSIAN98 software program [17]. AsA was exclusively

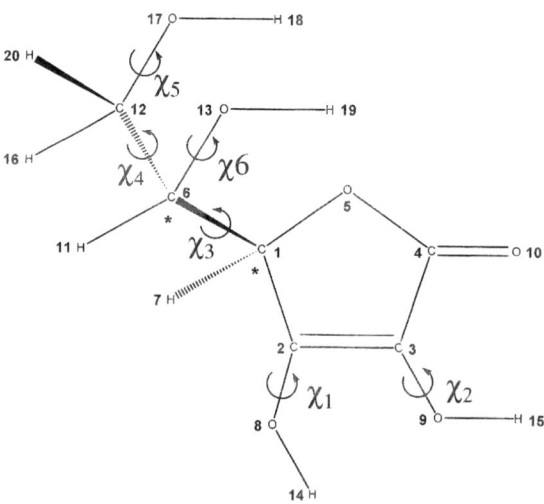

Fig. 1. Molecular structure (top) and torsional angle definitions (bottom) of ascorbic acid (AsA). The antioxidant mechanism of AsA requires an initial deprotonation of the χ_1 proton. Numbers beside atoms indicate input used for GAUSSIAN98. The torsions are defined: χ_1: H14–O8–C2–C1; χ_2: H15–O9–C3–C2; χ_3: O13–C6–C1–O5; χ_4: O17–C12–C6–C1; χ_5: H18–O17–C12–C6; χ_6: H19–O13–C6–C1. Relative conformational energies are dependent on the relation: $E = f(\chi_1, \chi_2, \chi_3, \chi_4, \chi_5, \chi_6)$.

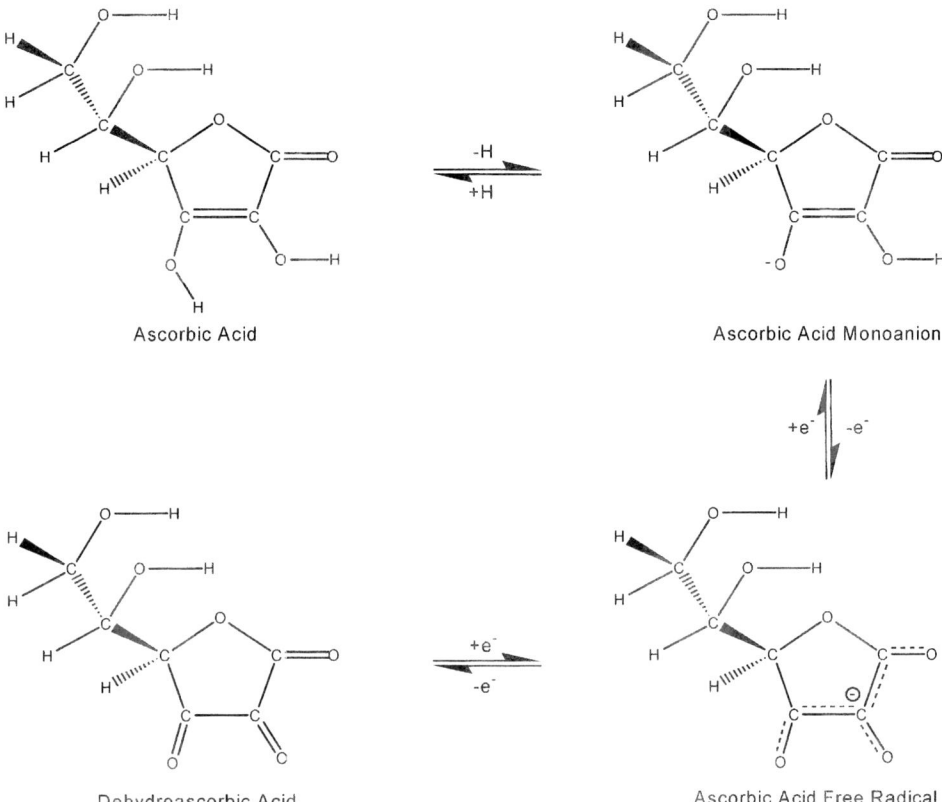

Fig. 2. Antioxidant redox mechanism of ascorbic acid (AsA) adapted from Ref. [4]. At physiologically pH, AsA is initially deprotonated to a resonance stabilized monoanion which can then undergo two separate one-electron transfers to an AsA free radical and dehydroascorbic acid [5]. In the current study, the deprotonation step of this mechanism is investigated.

defined using the GAUSSIAN98 z-matrix internal coordinate and Cartesian internal coordinate systems, to specify molecular structure, stereochemistry, and geometry. All calculations were performed using density functional theory with the Becke 3LYP hybrid exchange-correlation functional [18] at the B3LYP/6-31G(d) level of theory.

3. Results and discussion

Upon completion of all calculations, 14 unique deprotonated conformers successfully optimized to 12 unique deprotonated conformers. All AsA conformations are defined according to the structure code aaxy-z (aa = ascorbic acid, x = family number, y = high or low relative energy, z = protonated (p) or deprotonated (d)). Upon optimization,

Table 1

AsA torsional angle definitions for $\chi 1$ and $\chi 2$. In this study, AsA conformers were grouped into seven different families according to conformations of torsional angles $\chi 1$ and $\chi 2$

Family	Torsional angle definitions	
	χ_1	χ_2
1	g^+	g^+
2	g^+	a
3	g^+	g^-
4	a	g^+
5	a	a
6	a	g^-
7	g^-	g^+
8	g^-	a
9	g^-	g^-

Note that families 4 and 6 had no converged minima, and thus, were not evaluated in this study.

Table 2

Optimized data for AsA conformers at the B3LYP/6-31G(d) level of theory. All conformers are defined according to structure code aaxy-z where aa = AsA, x = family number (Table 1), y = high or low relative energy, and z = protonated (p) or deprotonated (d) (see Section 2)

Code	χ_1	χ_2	χ_3	χ_4	χ_5	χ_6	Energy (hartrees)	Relative energy (kcal mol^{-1})	Energy of deprotonation (kcal mol^{-1})
aa1h-p	−2.06	0.93	77.71	−50.64	−163.40	68.78	−684.743714732	7.45	324.40
aa1h-d	n/a	−1.07	141.34	−80.97	63.92	94.34	−684.226753354	2.47	
aa1l-p	−20.62	−178.89	−173.10	80.37	−58.90	−179.07	−684.750055966	3.47	343.84
aa1l-d	n/a	−178.11	174.06	76.48	−45.23	178.44	−684.202117797	17.93	
aa2h-p	−20.62	−178.89	−173.10	80.37	−58.90	−179.07	−684.750055966	3.47	343.84
aa2h-d	n/a	−178.11	174.06	76.48	−45.23	178.44	−684.202117797	17.93	
aa2l-p	2.24	177.86	72.81	−39.93	−60.37	−56.45	−684.755581553	0.00	336.72
aa2l-d	n/a	−179.86	42.57	−71.79	−36.73	−34.29	−684.218990434	7.34	
aa3h-p	35.17	1.78	−63.24	171.49	−64.97	−63.70	−684.729361358	16.45	320.92
aa3h-d	n/a	−3.10	−99.71	−178.92	41.69	−44.77	−684.217947104	8.00	
aa3l-p	−51.10	0.34	−102.21	−74.42	75.51	88.16	−684.746687507	5.58	332.24
aa3l-d	n/a	−2.17	−65.13	−70.39	48.66	74.66	−684.217227271	8.45	
aa5h-p	179.54	179.82	−66.34	169.26	−66.58	−74.31	−684.741941561	8.56	328.46
aa5h-d	n/a	−172.51	−98.77	−179.10	40.96	−46.69	−684.218501706	7.65	
aa5l-p	179.64	178.78	48.86	−61.96	−52.14	−46.31	−684.753418884	1.36	335.36
aa5l-d	n/a	−179.84	42.57	−71.79	−36.72	−34.26	−684.218990446	7.34	
aa7h-p	−58.33	−0.53	−83.41	63.38	70.81	172.85	684.731969327	14.82	327.10
aa7h-d	n/a	−3.13	−55.63	67.47	48.70	−17.62	684.210702453	12.55	
aa7l-p	−26.32	−0.05	50.39	66.46	73.12	−46.52	684.752090416	2.19	327.18
aa7l-d	n/a	−1.84	49.69	74.32	−51.89	−41.48	684.230694402	0.00	
aa8h-p	−12.60	−179.58	−55.87	179.35	47.06	51.66	684.744399789	7.02	330.82
aa8h-d	n/a	−176.84	−42.90	−169.26	36.69	33.54	684.217199579	8.47	
aa8l-p	−24.79	−179.72	46.66	66.64	−178.75	−46.66	684.754537097	0.66	329.11
aa8l-d	n/a	−178.83	48.09	74.56	−51.72	−41.44	684.230066314	0.39	
aa9h-p	−16.51	−3.77	175.81	49.61	48.10	−80.24	684.739800581	9.90	321.12
aa9h-d	n/a	−1.90	−172.37	72.50	−51.07	35.31	684.228068008	1.65	
aa9l-p	−27.78	−2.03	48.35	67.07	−179.57	−46.41	684.751140482	2.79	326.58
aa9l-d	n/a	−1.84	49.69	74.32	−51.89	−41.48	684.230694402	0.00	

aa1l-p and aa2h-p converged to the same protonated conformation. As such, only one unique conformation was identified from the deprotonation of this conformer (Table 2). Optimization also resulted in aa7l-p and aa9l-p converging to the same deprotonated conformer, aa7l-d. All converged data and energies of deprotonation are presented in Table 2.

Following optimizations of the protonated conformers, the following hydrogen bonds (H-bonds) were found as the dominant interactions: H15···O10, H14···O9, H14···O17, H19···O5, H18···O13, H19···O17, H18···O8, H15···O8, H19···O8, H18···O5, H14···O13 (Fig. 1). While only H-bonding was evident in the protonated conformations, both H-bonding and intramolecular ion–dipole interactions were revealed as the dominant interactions in the deprotonated conformations. The dominant H-bonds that existed were: H15···O10, H19···O5, H18···O13, H19···O17, H18···O5 and the dominant ion–dipole interactions that existed were: H15···O8, H19···O8, H18···O8.

In reviewing all converged protonated and deprotonated conformers it is evident that protonated conformers possessing major intramolecular attractive forces (IMAF) such as extensive H-bonding had the highest energies of deprotonation. Conversely, conformers with minimal IMAF had the lowest energies of deprotonation. An example of the latter is conformer aa3h-p, which possessed the lowest energy of deprotonation (Fig. 3). This conformer possessed a moderate H-bond at H15···O8 and a weak H-bond at H19···O8 (Table 2). Due to these minimal IMAFs, this conformer possessed a high relatively energy at 16.45 kcal mol^{-1}. In addition, this conformer possessed a side-chain that was fully extended

away from H14 at χ_1 such that the deprotonation required a minimal investment of energy. This protonated conformation is therefore best suited for deprotonation because it lacks any major involvement of the H14 proton. The deprotonated and optimized conformer, aa3h-d, was found to possess a novel intramolecular H-bond-ion–dipole–ion–dipole motif in the form of: H18···O13···O8$^-$···H15 (Fig. 3).

The route with the highest energy of deprotonation involved a H14 proton abstraction from aa11-p. This conformer possessed the following significant H-bonds: H14···O13, H15···O10, and H18···O5 (Fig. 4). Given that this conformer involved H14 in a distinct H-bond, it would be expected that this conformer require a substantial investment of energy. Upon deprotonation and re-optimization, conformer aa11-d maintained the H15···O10 and H18···O5 H-bonds. However, the fact that the remaining ionic charge on O8 is in the vicinity of O9 and O13, made for the deprotonated conformer with the highest energy of deprotonation computed in this study (Fig. 4). The large relative energy in aa11-d is indicative of its inability to re-structure itself into a more stable structure.

In mechanisms of deprotonation, as that for AsA in its antioxidant mechanism of action, energies of deprotonation and the acidic character of AsA will be dependent on the protonated molecular

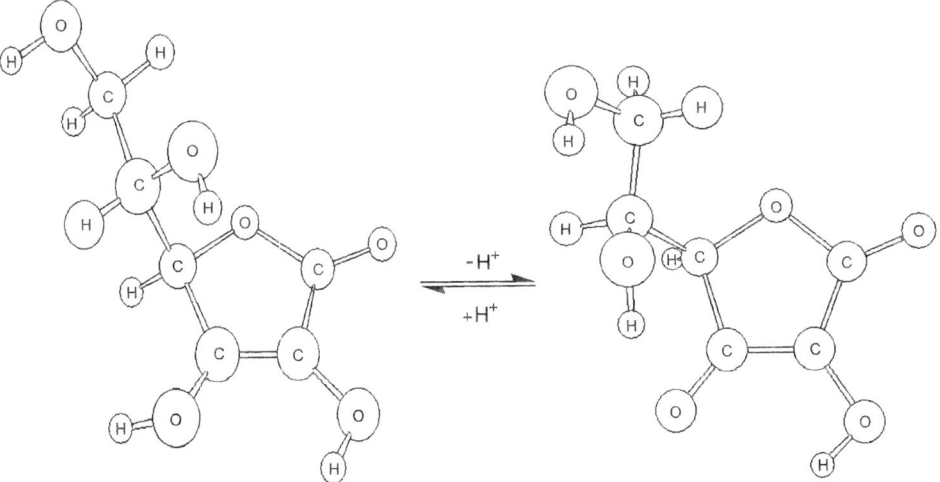

Fig. 3. Route of lowest energy of deprotonation for ascorbic acid involved the deprotonation of conformers aa3h-p to aa3h-d (320.92 kcal mol^{-1} at the B3LYP/6-31G(d) level of theory).

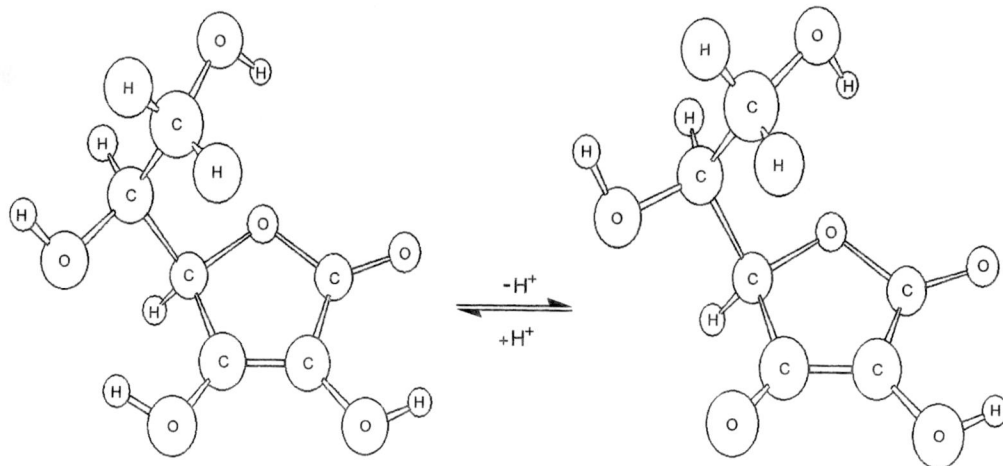

Fig. 4. Route of highest energy of deprotonation for ascorbic acid involved the deprotonation of conformers aa11-p to aa11-d (343.84 kcal mol^{-1} at the B3LYP/6-31G(d) level of theory).

conformations. Thus, when evaluating the AsA PEHS, it is important to evaluate all structures and address their respective mechanisms in question. This work has shown that for the acidic profile of AsA, the ideal structures for deprotonation will be with minimal IMAF. More specifically, structures lacking IMAF with proton H14 will be the best candidate structures for deprotonation schemes.

4. Conclusion

The present study has successfully characterized the role of conformation on the acidity of AsA. It has been shown that conformations possessing strong IMAF will have higher energies of deprotonation than conformations with a lack of IMAF. In particular, structures lacking significant interactions with H14 proton will be favoured in deprotonation schemes. These findings reveal the conformations favoured in the first step of the AsA's redox mechanism and thus provide a crucial stepping stone to fully deciphering the full molecular antioxidant mechanism of AsA.

Acknowledgements

One of the authors (IGC) wishes to thank the Ministry of Education for a Szent-Györgyi Visiting Professorship.

References

[1] M. Milanesio, R. Bianchi, P. Ugliengo, C. Roetti, D. Viterbo, J. Mol. Struct. (Theochem) 419 (1997) 139–154.

[2] J.R. Woods Jr., M.A. Plessinger, R.K. Miller, Am. J. Obstet. Gynecol. 185 (2001) 5–10.

[3] M. Castro, T. Caprile, A. Astuya, C. Millan, K. Reinicke, J.C. Vera, O. Vasquez, L.G. Aguayo, F. Nualart, J. Neurochem. 78 (2001) 815–823.

[4] J.M. May, FASEB J. 13 (1999) 995–1006.

[5] P. Evans, B. Halliwell, Br. J. Nutr. 85 (2001) S67–S74.

[6] S.C. Rumsey, M. Levine, J. Nutr. Biochem. 9 (1998) 116–130.

[7] F. Shang, M. Lu, E. Dudek, J. Reddan, A. Taylor, Free Radical Biol. Med. 34 (2003) 521–530.

[8] N. Nagao, T. Nakayama, T. Etoh, I. Saiki, N. Miwa, J. Cancer Res. Clin. Oncol. 126 (2000) 511–518.

[9] I. Fujita, Y. Akagi, J. Hirano, T. Nakanishi, N. Itoh, N. Muto, K. Tanaka, Res. Commun. Mol. Pathol. Pharmacol. 107 (2000) 219–229.

[10] A. Vojdani, M. Bazargan, E. Vojdani, J. Wright, Cancer Detect. Prevent. 24 (2000) 508–523.

[11] S.J. Padayatty, A. Katz, Y. Wang, P. Eck, O. Kwon, J. Lee, S. Chen, C. Corpe, A. Dutta, S.K. Dutta, M. Levine, J. Am. Coll. Nutr. 22 (1) (2003) 18–35.

[12] V.H. Guaiquil, J.C. Veras, D.W. Golde, J. Biol. Chem. 276 (2001) 40955–40961.

[13] T. Osakai, H. Jensen, H. Nagatani, D.J. Fermin, H.H. Girault, J. Electroanal. Chem. 510 (2001) 43–49.

[14] E. Niki, Ann. N. Y. Acad. Sci. 498 (1987) 186.

[15] D.R.P. Almeida, D.M. Gasparro, L.F. Pisterzi, L.L. Torday, A. Varro, J.G. Papp, B. Penke, J. Mol. Struct. (Theochem) 637 (2003) 251–270.

[16] J.R. Juhasz, L.F. Pisterzi, D.M. Gasparro, D.R.P. Almeida, submitted for publication.

[17] M.J. Frisch, G.W. Trucks, H.B. Schlegel, G.E. Scuseria, M.A. Robb, J.R. Cheeseman, V.G. Zakrzewski, J.A. Montgomery Jr.,

R.E. Stratmann, J.C. Burant, S. Dapprich, J.M. Millam, A.D. Daniels, K.N. Kudin, M.C. Strain, O. Farkas, J. Tomasi, V. Barone, M. Cossi, R. Cammi, B. Mennucci, C. Pomelli, C. Adamo, S. Clifford, J. Ochterski, G.A. Petersson, P.Y. Ayala, Q. Cui, K. Morokuma, D.K. Malick, A.D. Rabuck, K. Raghavachari, J.B. Foresman, J. Cioslowski, J.V. Ortiz, A.G. Baboul, B.B. Stefanov, G. Liu, A. Liashenko, P. Piskorz, I. Komaromi, R. Gomperts, R.L. Martin, D.J. Fox, T. Keith, M.A. Al-Laham, C.Y. Peng, A. Nanayakkara, C. Gonzalez, M. Challacombe, P.M.W. Gill, B.G. Johnson, W. Chen, M.W. Wong, J.L. Andres, M. Head-Gordon, E.S. Replogle, J.A. Pople, GAUSSIAN98 (Revision A.9), Gaussian Inc, Pittsburgh, PA, 1998.

[18] A.D. Becke, J. Chem. Phys. 98 (1993) 5648.

ELSEVIER

Journal of Molecular Structure (Theochem) 666–667 (2003) 409–414

THEO
CHEM

www.elsevier.com/locate/theochem

An isodesmic comparison of the C_1 modified reduced pteridine ring as a folic acid model

Jeremy H. Keller[a,b], Gregory A. Chass[a,b,c,d,*], Imre G. Csizmadia[a,b,e]

[a]*Global Institute of COmputational Molecular and Materials Science (GIOCOMMS),
1422 Edenrose St., Mississauga, Ont., Canada L5V 1H3*
[b]*Department of Chemistry, University of Toronto, 80 St George St., Toronto, Ont., Canada M5S 3H6*
[c]*Institut de Science et d'Ingénierie Supramoléculaires, 8, allée Gaspard Monge, BP 70028, 67083 Strasbourg Cedex, France*
[d]*Department of Biomedical Sciences, Creighton University, 2500 California plaza, Omaha, NE, 68178 USA*
[e]*Department of Medical Chemistry, Szeged University, H-6720 Szeged, Dom ter 8, Hungary*

Abstract

Tetrahydrofolate or pteroyl glutamate (H_4PteGlu) functions as a single carbon (C_1) carrier and a cofactor for intracellular reactions. The reduced pteridine ring of H_4PteGlu was used as a model for the six C_1 substituted forms and the unsubstituted form of H_4PteGlu. An isodesmic reaction was performed using ab initio molecular orbital calculations at the Restricted Hartree–Fock 3-21G level of theory. The isodesmic reaction allows for further characterization of the relationships between the C_1 units beyond the biochemical and provides the foundation for further analysis
© 2003 Published by Elsevier B.V.

Keywords: Ab initio; Tetrahydrofolate; C_1 Unit; Isodesmic

1. Introduction

Tetrahydrofolate or pteroyl glutamate (H_4PteGlu) [1] is the fully reduced bioactive oxidation state of folic acid. The reduction of folate to dihydrofolate and finally to tetrahydrofolate generates a chiral centre at C6 where the absolute *S* configuration is the bioactive enantiomer, *6S*-5,6,7,8-tetrahydrofolate [2]. H_4PteGlu is composed of a pteridine ring (2-amino-4-oxo-6-methyltetrahydropteridine) linked to a *para*-amino benzoic acid (PABA) which is linked to an L-glutamate residue via its α-amino group (Fig. 1a). In

this reduced state the pteridine ring of H_4PteGlu can be modified to act as a single carbon (C_1) carrier that participates as a donor for intracellular reactions that include amino acid and purine synthesis, thymine generation and the source for all intracellular methylation reactions [3,4]. Six C_1 modified structures are present naturally [5] and were examined here in addition to the unsubstituted H_4PteGlu (U). The six C_1 units are: N^5-methyl H_4PteGlu (A), $N^{5,10}$-methylene H_4PteGlu (B), $N^{5,10}$-methenyl H_4PteGlu (C), N^5-formyl H_4PteGlu (D), N^{10}-formyl H_4PteGlu (E) and N^5-formimino H_4PteGlu (F) (Fig. 1b).

The C_1 units are generated through four pathways: the catabolism of L-serine and glycine to generate $N^{5,10}$-methylene H_4PteGlu [3], the catabolism of L-histidine to generate N^5-formimino H_4PteGlu and the incorporation of formate to produce N^{10}-formyl

---

* Corresponding author.
  E-mail addresses: jeremy.keller@utoronto.ca (J.H. Keller), gchass@giocomms.org (G.A. Chass), icsizmad@chem.utoronto.ca (I.G. Csizmadia).

0166-1280/$ - see front matter © 2003 Published by Elsevier B.V.
doi:10.1016/j.theochem.2003.08.045

(a)

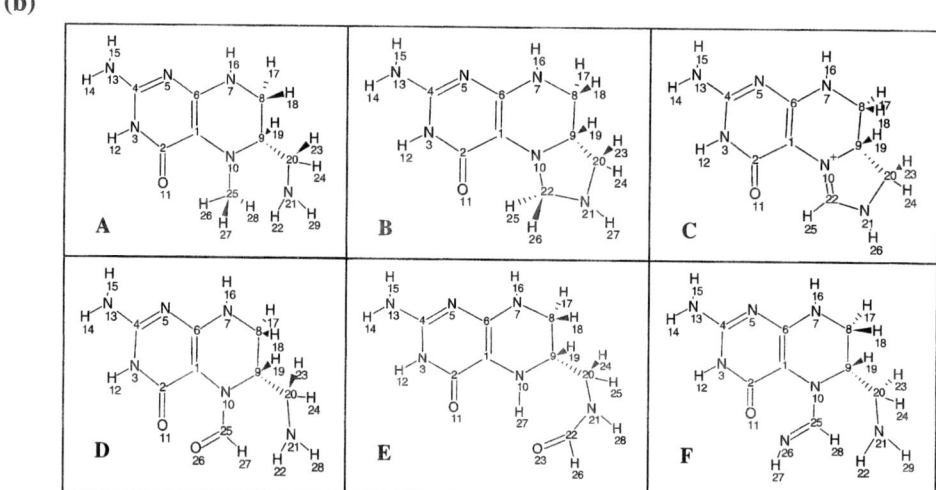

(b)

Fig. 1. The complete H₄PteGlu (U) structure with IUPAC numbering in the pteridine ring (1a). The molecule is truncated for calculations at N^{10}. The truncated C_1 units with z-matrix numbering N^5-methyl H₄PteGlu (A), $N^{5,10}$-methylene H₄PteGlu (B), $N^{5,10}$-methenyl H₄PteGlu (C), N^5-formyl H₄PteGlu (D), N^{10}-formyl H₄PteGlu (E) and N^5-formimino H₄PteGlu (F) (1b).

H₄PteGlu [3,6]. Once a C_1 unit is generated enzymatic interconversion occurs readily to generate the other C_1 units with the exception that N^5-methyl H₄PteGlu generation is irreversible [7] (Fig. 2).

An isodesmic reaction was used to compare the energies of the six C_1 (A–F) units and the unsubstituted H₄PteGlu (U). An isodesmic reaction maintains the number and nature of the bonds while changing the relationship between bonds [8].

Reactions of this type are generated with artificial reactants and products but they ultimately allow comparison of the relative stabilities of each molecule.

2. Methods

To simplify calculations the H₄PteGlu molecule was truncated to include only the pteridine ring and

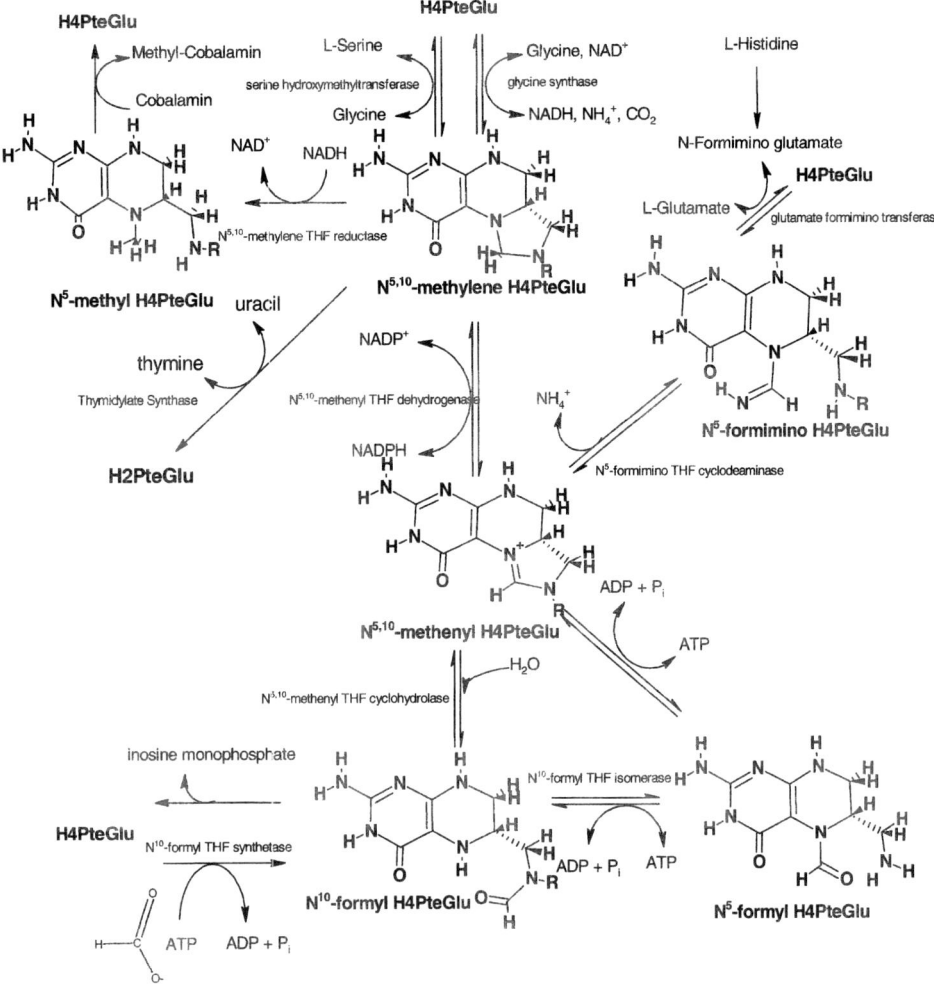

Fig. 2. The complete C_1 H_4PteGlu reaction cycle representing the enzyme mediated intracellular conversion and use of the C_1 modified H_4PteGlu units.

the amino group of PABA while still maintaining the absolute *S* arrangement at C6 with the hydrogen above the plane of the molecule as shown in Fig. 1b. Ab initio molecular orbital calculations were carried out at the Restricted Hartree–Fock 3-21G (RHF/3-21G) level of theory using the GAUSSIAN 98 program package [9]. The molecules were numbered in a modular fashion to explicitly define internal co-ordinates for input into the program (Fig. 1b) [10].

The isodesmic reactants, numbered 1–6, were artificially conceived to best represent the addition of the C_1 units to the unsubstituted H_4PteGlu. Isodesmic

products 7 and 8 represent H_4PteGlu bonds remaining after the addition of the C_1 units (Fig. 3).

Upon investigation each molecule had several degrees of freedom including ring puckering at the sites of reduction in the pteridine ring and single bond rotations at the sites of C_1 modification. Initial RHF/3-21G calculations were performed for each potential conformation of each of the molecules. The most stable conformer's energy was used in the calcu-lations. For the isodesmic reactants scans about the major dihedrals allowed for determination of the minimum energy conformation of each of the reactants

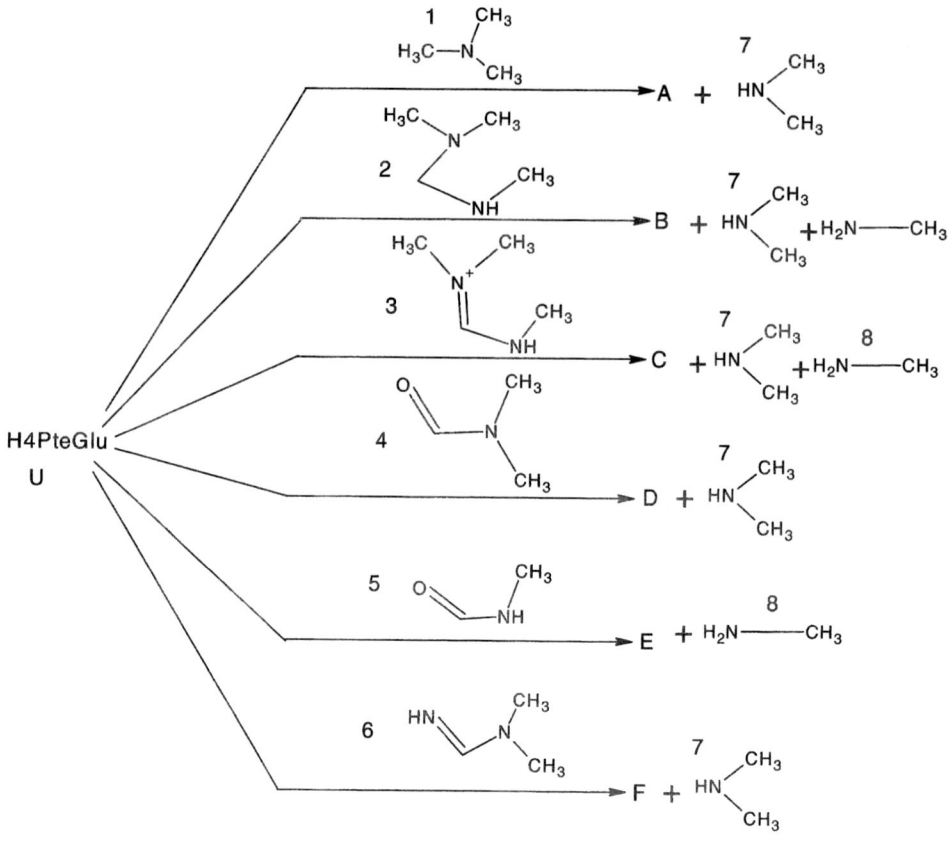

Fig. 3. The structures of the isodesmic reactants (1–6) used to generate each of the C_1 units individually and the isodesmic products (7 and 8) that are generated in their formations.

and products. In all cases the minimum energy conformation was used in the isodesmic calculation (data not shown).

Calculation of the relative energies of each of the C_1 units was completed by summing the isodesmic reactants not required to generate the given C_1 unit and the isodesmic products generated from H_4PteGlu when it becomes the given C_1 unit in the following way.

$$E_U = U + 1 + 2 + 3 + 4 + 5 + 6$$

$$E_A = A + 2 + 3 + 4 + 5 + 6 + 7$$

$$E_B = B + 1 + 3 + 4 + 5 + 6 + 7 + 8$$

$$E_C = C + 1 + 2 + 4 + 5 + 6 + 7 + 8$$

$$E_D = D + 1 + 2 + 3 + 5 + 6 + 7$$

$$E_E = E + 1 + 2 + 3 + 4 + 6 + 8$$

$$E_F = F + 1 + 2 + 3 + 4 + 5 + 7$$

Results were then converted from Hartrees to kcal mol^{-1} using the 627.5095 kcal mol^{-1}?Hartree^{-1} conversion factor.

3. Results

The calculation results (Table 1) all seven molecules are within 10 kcal mol^{-1} of each other with N^{10}-formyl H_4PteGlu (E) being the most stable C_1 structure. $N^{5,10}$-methylene H_4PteGlu (B) is the least stable of the seven. The degradation of L-serine and glycine to generate $N^{5,10}$-methylene H_4PteGlu is the major source of C_1 units [3,6,7]. It is also the C_1 unit that participates in the methylation of uracil to thymine a major target

Table 1
The calculated isodesmic reaction energies for each of the seven molecules in Hartrees calculated at the RHF/3-21G level of theory. Relative energies are recorded in Hartrees and kcal mol^{-1}

Isodesmic summation	E_T (Hartrees)	ΔE (Hartrees)	ΔE (kcal mol^{-1})
U + 1 + 2 + 3 + 4 + 5 + 6	− 2051.414528	0.002940	1.844567
A + 2 + 3 + 4 + 5 + 6 + 7	− 2051.413971	0.003497	2.194394
B + 1 + 3 + 4 + 5 + 6 + 7 + 8	− 2051.400406	0.017062	10.706467
C + 1 + 2 + 4 + 5 + 6 + 7 + 8	− 2051.416098	0.001370	0.859568
D + 1 + 2 + 3 + 5 + 6 + 7	− 2051.416490	0.000978	0.613679
E + 1 + 2 + 3 + 4 + 6 + 8	− 2051.417468	0.000000	0.000000
F + 1 + 2 + 3 + 4 + 5 + 7	− 2051.416763	0.000705	0.442262

for anti-cancer and anti-bacterial therapy through dihydrofolate reductase inhibitors [11–13]. Since this C_1 unit is a major component in the cycle its less stable nature may increase the ease in which this structure is used in any one of its pathways.

By comparing the biochemical conversion pathways to the energetic stabilities (Fig. 4) provides an alternate method of discussion the relationship between the C_1 units. However, the differences in energy between the stable conformations of each of the individual C_1 units can be greater than differences in stability between them. By changing the conformation of a given C_1 unit it can become relatively less stable when compared to the other C_1 units. A potential enzymatic mechanism of conversion would be to shift conformation of a C_1 unit making it

relatively unstable and easing its conversion to another C_1 unit.

4. Conclusion

An isodesmic reaction provides a basis for further characterization of the relationship between the C_1 units by supplying a direct measurement of the energetic stability of each of the C_1 units. It is an alternate method of discussing the energies associated with these biochemically important C_1 units. An examination of the complete $H_4PteGlu$ molecule is required for a true appreciation of the energies of the system, which could yield significant results in the quest for the ultimate understanding of this complex biochemical system.

Acknowledgements

This work was supported with grants from the Global Institute Of COmputational Molecular and Material Sciences (GIOCOMMS), Toronto, Ontario, Canada. The author sends many thanks to Ephraim Tang, Suzanne K. Lau and David H. Setiadi for their input and discussion as well as to the Keller family for their tireless support and motivation.

One of the authors (IGC) wishes to thank the Ministry of Education for a Szent-Györgyi Visiting Professorship.

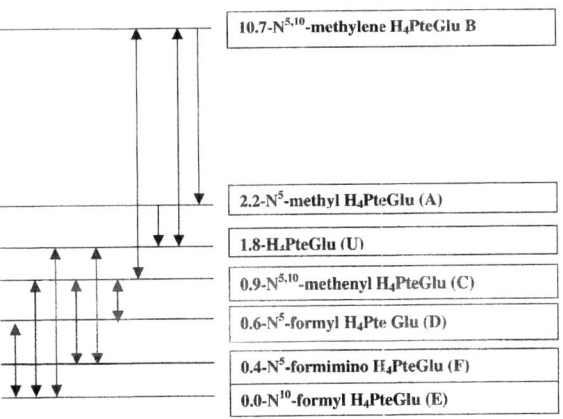

Fig. 4. The relationship between the molecules listed in terms of relative energy. Arrows represent direct conversion pathways between structures from Fig. 1.

References

[1] IUPAC-IUB Commission on Biochemical Nomenclature (CBN).

[2] J.E. Baggott, T. Tamura, H. Baker, British Journal of Nutrition 85 (2001) 653.

[3] B. Fowler, Kidney International 59 (2001) S221.

[4] J.M. Scott, Proceedings of the Nutrition Society 58 (1999) 441.

[5] D. Voet, J.G. Voet, C.W. Pratt, Fundamentals of Biochemistry: Upgrade Edition, Wiley, New York, 2002, p.631.

[6] E.A. Cossins, L. Chen, Phytochemistry 45 (3) (1997) 437.

[7] J. Selhub, Annual Review of Nutrition 19 (1999) 217.

[8] W.J. Hehre, R. Ditchfield, L. Radom, J.A. Pople, Journal of the American Chemical Society 91 (1970) 4796.

[9] Revision A.9, M. J. Frisch, G. W. Trucks, H.B. Schlegel, G. E. Scuseria, M. A. Robb, J.R. Cheeseman, V.G. Zakrzewski, J.A. Montgomery, Jr., R.E. Stratmann, J.C. Burant, S. Dapprich, J.M. Millam, A. D. Daniels, K. N. Kudin, M. C. Strain, O. Farkas, J. Tomasi, V. Barone, M. Cossi, R. Cammi, B. Mennucci, C. Pomelli, C. Adamo, S. Clifford, J. Ochterski, G.A. Petersson, P.Y. Ayala, Q. Cui, K. Morokuma, D.K. Malick, A.D. Rabuck, K. Raghavachari, J. B. Foresman, J. Cioslowski, J.V. Ortiz, A.G. Baboul, B.B. Stefanov, G. Liu, A. Liashenko, P. Piskorz, I. Komaromi, R. Gomperts, R. L. Martin, D.J. Fox, T. Keith, M.A. Al-Laham, C.Y. Peng, A. Nanayakkara, M. Challacombe, P.M.W. Gill, B. Johnson, W. Chen, M.W. Wong, J.L. Andres, C. Gonzalez, M. Head-Gordon, E.S. Replogle, J.A. Pople, Gaussian, Inc., Pittsburgh PA, GAUSSIAN 98, 1998.

[10] G.A. Chass, et al., International Journal of Quantum Chemistry 90 (2002) 933.

[11] I.H. Gilbert, Biochimica et Biophysica Acta 1597 (2002) 249.

[12] A.E. Klon, et al., Journal of Molecular Biology 320 (2002) 677.

[13] J. Bajorath, J. Kraut, Z. Li, D.H. Kitson, A.T. Hagler, PNAS USA 88 (1991) 6423.

ELSEVIER

Journal of Molecular Structure (Theochem) 666–667 (2003) 415–429

THEO CHEM

www.elsevier.com/locate/theochem

An exploratory ab initio conformational analysis of selected fragments of nicotinamide adenine dinucleotide (NAD$^+$). Part I: 5-deoxyribose nicotinamide N-glycoside

Suzanne K. Lau[a,b,*], Gregory A. Chass[a,b,c,d], Sándor Lovas[c], Botond Penke[e,f], Imre G. Csizmadia[a,b,e]

[a]Department of Chemistry, University of Toronto, Toronto, Ont., Canada, M5S 3H6
[b]Global Institute of COmputational Molecular and Materials Science, 1422 Edenrose Street, Mississauga, Ont., Canada, L5V 1H3
[c]Department of Biomedical Sciences, Creighton Medical University, 2500 California Plaza, Omaha, NE 68178, USA
[d]Institut de Science et d'Ingénierie Supramoléculaires, 8, allée Gaspard Monge, BP 70028, 67083 Strasbourg Cedex, France
[e]Department of Medical Chemistry, University of Szeged, H-6720 Szeged, Domter 8, Hungary
[f]Protein Chemistry Research Group, Hungarian Academy of Sciences, University of Szeged, Domter 8, H-6720 Szeged, Hungary

Abstract

Ab initio molecular orbital computations were carried out on selected fragments of Nicotinamide Adenine Dinucleotide (NAD$^+$) at the RHF/3-21G and RHF/6-31G(d) levels of theory. The focus of this work was on the nicotinamide-ribose moiety. In this model study, the ribose was replaced by 5-deoxyribose. Key points of examination included the rotation of the carboxyamide functionality, rotors between the nicotinamide and ribose rings as well as ring puckering in the ribose sugar. The structural definitions of the adenosine fragment and diphosphate chain were defined herein, facilitating future theoretical investigation. A preliminary attempt was made to refine the results at the elevated RHF/6-31G and RHF/6-31g(d) basis sets. The former showed results in agreement with the RHF/3-21G data sets, whereas the inclusion of preliminary polarization showed deviance in the χ_i^1 dihedral of the $a[g^-a]$ conformer.
© 2003 Elsevier B.V. All rights reserved.

Keywords: Ab initio; Coenzyme; Nicotinamide adenine dinucleotide; Nicotinamide-ribose

1. Introduction

Nicotinamide adenine dinucleotide (NAD$^+$/NADH) is a coenzyme and substrate that is responsible for most enzyme-catalyzed dehydrogenation reactions [1]. Over 100 NAD-dependent dehydrogenases are known, catalyzing oxidation reactions of lactate, malate, alcohol, and glyceraldehyde-3-phosphate among others.

The structural characterization of NAD$^+$ (Fig. 1) presents a particular challenge, as it is a flexible molecule with over a dozen rotatable bonds that can adopt a wide variety of environmentally dependent conformations. In addition, NAD$^+$ consists of a diverse set of functional groups including neutral (adenine) and formal positive (nicotinamide ring) heterocyclic ring systems. There are also negative (pyrophosphate) charges, and hydrogen-bonding

* Corresponding author.
  E-mail addresses: suzanne.lau@utoronto.ca (S.K. Lau), gchass@giocomms.org (G.A. Chass).

0166-1280/$ - see front matter © 2003 Elsevier B.V. All rights reserved.
doi:10.1016/j.theochem.2003.08.117

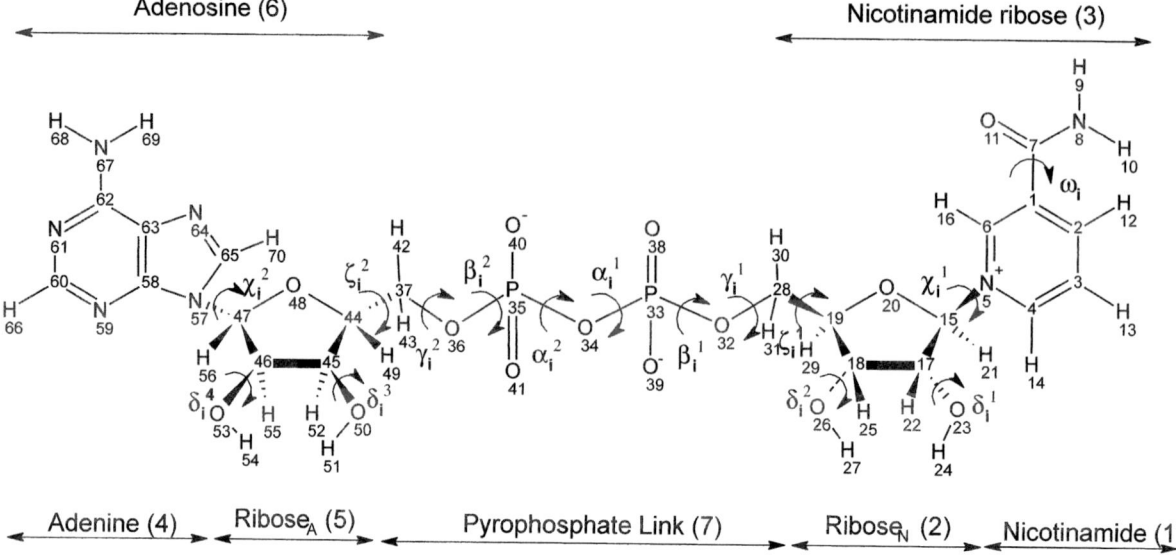

Fig. 1. Systematic modular numbering system of NAD^+ including pertinent dihedrals. The arrows indicate the approximate partitioning of the molecule. (1) Nicotinamide, (2) $Ribose_N$, (3) Nicotinamide ribose (NmR), (4) Adenine, (5) $Ribose_A$, (6) Adenosine and (7) Pyrophosphate bridge. This paper focuses primarily on (1), (2) and (3).

groups (ribose hydroxy groups, carboxamide) in the molecule. In combination, the elastic and chemical nature of NAD^+ allows for it to accommodate a wide variety of active site profiles.

A number of studies [2,3] have been published discussing the merit of using NADH in the treatment of Parkinsons' disease. Parkinsonian patients suffer from a deficiency in dopamine and a loss of tyrosine hydroxylase (TH) [4]. The NAD^+/NADH redox system is coupled with the synthesis of tetrahydrobiopterin (BH_4) by the action of quinoid-dihydropteridine reductase, and BH_4 is a cofactor with TH in the production of levodopa (a dopamine precursor). Applications of NADH in conjunction with TH stimulation may encourage the biosynthesis of levodopa with clinical benefits for patients [5]. Use of NADH has also been investigated as a therapeutic approach in the treatment of Chronic Fatigue Syndrome (a disease of unknown etiology, with debilitating, prolonged fatigue and a host of other symptoms), as it may trigger energy production through ATP generation [6]. In addition, studies have been completed using analogues of NAD^+ as anticancer [7], antibacterial [8] and antitrypanosomal agents [9]. On the other hand, others have questioned the effectiveness of NADH on cognitive improvement

in patients with dementia [10], reveals the constant struggle faced in attempts to develop successful new drug candidates.

A number of studies seeking to characterize the structure of NAD have been undertaken using NMR [11], ab initio structural investigations of the nicotinamide group (3-21G and 6-31G basis sets) using the HF method [12] and classical molecular dynamic simulations [13]. In all, it has been determined that the environmental effect on the conformational character of NAD^+ is significant. It has been shown that NAD^+ adopts a compact, folded conformation in aqueous solution with the nicotinamide and adenine rings approximately 4–5 Å apart [14,15]. In contrast, enzyme-bound NAD^+ is typically observed in an extended conformation, with a distance of 8–18 Å between the rings [16]. The extended conformation is favourable for catalysis as it exposes more surface area for interaction with the enzyme effectively reducing the solvent volume around the nicotinamide to allow for hydride transfer to occur. Examples of the folded and extended forms of NAD^+ are seen in Fig. 2.

While Wu and Houk [12] have already completed an ab initio study on the conformational differences in the nicotinamide group of NAD^+ and NADH in relation to dehydrogenase specificity, their

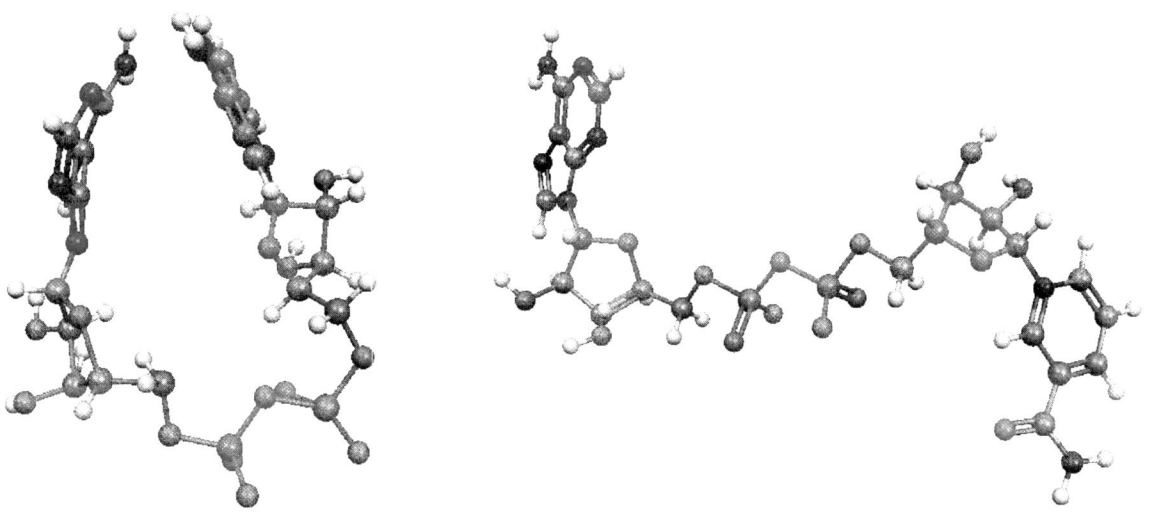

Fig. 2. Representative structures of folded NAD^+ (left) and extended NAD^+, as seen in water, and in a Rossmann fold enzyme, glyceraldehyde-3-phosphate, respectively. The rings in the folded conformation have been observed as close as 5.2 Å [17].

calculations appear to be based primarily on model systems such as N-hydroxymethylpyridinium ion and N-ethylpyridinium ion. They report strong basis-set dependence, with data from the 3-21G and 6-31G* calculations predicting differing minima. This phenomenon has been previously reported for analogue allyl alcohols and ethers [17]. In addition to completing a full study of the nicotinamide-ribose moiety (NmR), a comparison of newly generated data and that presented in [12] can be undertaken. Furthermore, this will be the first in a series of conformational studies on fragments of NAD^+. After each of the individual fragments has been independently analyzed, they will be incorporated into the whole for further computational study.

By beginning a study of NAD^+, data can be easily manipulated and modified for studies of NADH, the $NADP^+$/NADPH complex (Fig. 3), or other NAD analogues. $NADP^+$ also contains a phosphorylated ribose on the adenylate moiety, which would change the specificity of its corresponding ligand and receptor [1]. NADPH is often used as a reducing agent in biosynthetic reactions rather than in energy forming reactions. Insight into the conformation of NAD^+ in different environments will allow for a more complete understanding of the mechanism of its diverse functions at the molecular level, allowing for drug candidates with tailored ligand/receptor binding sites to be developed in the treatment of NAD-dependent pathologies.

2. Method

The GAUSSIAN 98 program package [18] has been used to perform preliminary ab initio calculations on the NmR fragment of NAD^+. The definitions of the relative spatial orientations of all constituent atomic nuclei have been formulated in a standardized and modular fashion so as to allow for portions of the whole to be studied independently (Fig. 1). The focus of this paper is on the NmR group, which was first modelled by the semi-empirical AM1 method (data not shown).

A rotational scan was performed on dihedral angle ω_i^1 of the modified nicotinamide fragment (Fig. 4, on left). A fluorine atom was chosen to represent the ribose fragment since its high electronegative value would most closely mimic the effect of the ribose fragment containing numerous electronegative oxygen atoms on the nicotinamide ring. The result of this has been incorporated into further examination of the moiety. The dihedral χ_i^1 was then examined in conjunction with the hydroxy rotors on the ribose portion of the molecule.

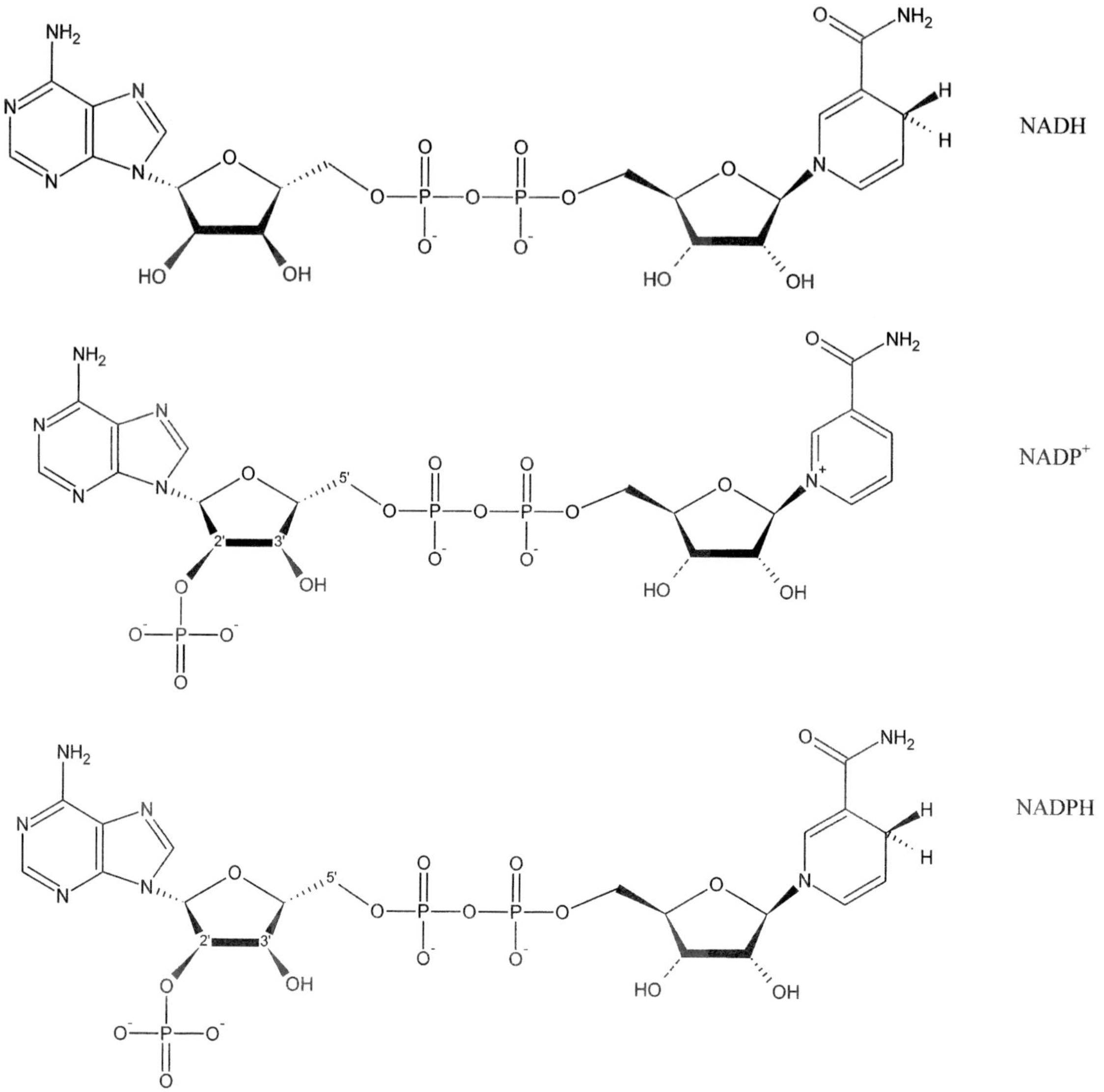

Fig. 3. Three NAD$^+$ analogues. Top: Reduced form of NAD$^+$ (NADH). Middle: Oxidized form of NADP (NADP$^+$), the 2′-hydroxy group of the ribose ring of adenosine is phosphorylated. Bottom: Reduced form of NADP (NADPH).

The two ribose fragments in NAD$^+$ (Fig. 1) appear to be structurally equivalent. Although they are located in different chemical environments and some structural perturbations are expected, results from a single pre-optimization of the ribose fragment should provide satisfactory inputs for further integration.

Before engaging in further studies of the ribose moiety, a model lacking the ring structure was made (2,3-dihydroxybutane) in order to examine the effect

of the hydroxy rotors on the central portion of ribose independent of ring strain among other factors. The structural definitions of selected dihedral angles for both furanose and 2,3-dihydroxybutane are shown in Table 1. Although the two molecules are not stereochemically equivalent at C_3 and C_4 (ribofuranose is RS, while the model is SR), they can be said to be diastereomers. According to Henry-Riyad et al. [19], the dihedral defined as ν_i^2 was shown to be

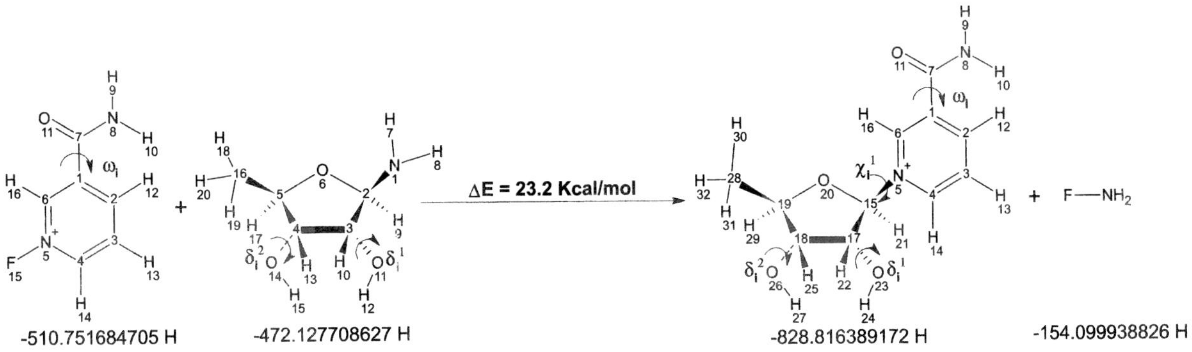

	Dihedral Definitions
ω_1	N_8-C_1-C_6-N_5
χ_1^1	C_{17}-C_{15}-N_5-C_6
δ_1^1	H_{12}-O_{11}-C_3-C_2 and H_{24}-O_{23}-C_{17}-C_{15}
δ_1^2	H_{15}-O_{14}-C_4-C_3 and H_{27}-O_{26}-C_{18}-C_{17}

Fig. 4. Progression from Ribose$_N$ and modified Nicotinamide models (with fluorine replacing ribose moiety) to full Nicotinamide-ribose (NmR) fragment.

$37.372°$ in the optimization of tetrahydrofuran. It can be assumed that ν_i^2 in furanose and the model system should closely correlate to the previous findings. Since ν_i^2 was found in the *gauche*$^+$ (g^+) range $(0-120°)$, it is suspected that it may exist in the *gauche*$^-$ (g^-) region as well $(-120$ to $0°)$. Input files were prepared accordingly for $\nu_i^2 = +40°$ based on the estimate from Henry-Riyad et al. [19], and

Table 1
Structural definitions of selected dihedral angles in the furanose ring and 2,3-dihydroxybutane model system

Dihedral angle	Atoms involved	
	Furanose	2,3-Dihydroxybutane
ν_i^0	C_3–C_2–O_6–C_5	–
ν_i^1	C_4–C_3–C_2–O_6	–
ν_i^2	C_5–C_4–C_3–C_2	C_5–C_4–C_3–C_2
ν_i^3	O_6–C_5–C_4–C_3	–
ν_i^4	C_2–O_6–C_5–C_4	–
δ_i^1	H_{12}–O_{11}–C_3–C_2	H_1–O_9–C_3–C_2
δ_i^2	H_{15}–O_{14}–C_4–C_3	H_{13}–O_{12}–C_4–C_3
Structure	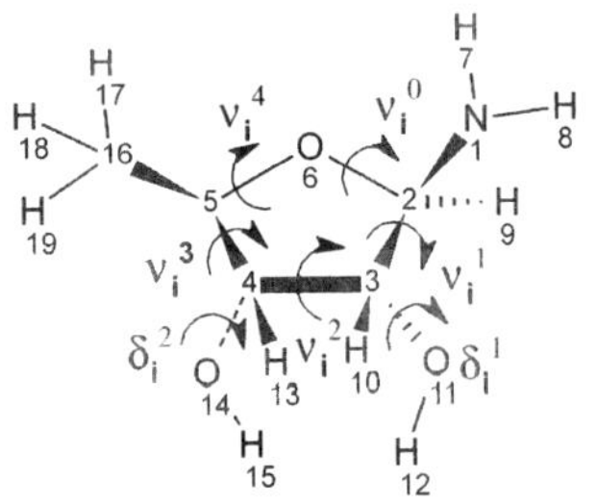	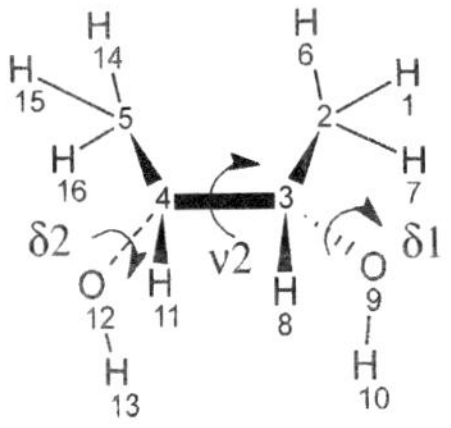

Stereochemistry		
δ_i^1	R	S
δ_i^2	S	R

Table 2
Internal dihedral angles of the modified nicotinamide fragment, computed at the RHF/3-21G level of theory

Dihedral	Input	Optimized value (deg)
$C_4-C_3-C_2-C_1$	0.000	0.000
$C_5-C_4-C_3-C_2$	0.000	0.012
$C_6-C_5-C_4-C_3$	0.000	0.000
$C_1-C_6-C_5-C_4$	0.000	0.000
$C_2-C_1-C_6-C_5$	0.000	0.015
$C_3-C_2-C_1-C_6$	0.000	0.000

$\nu_i^2 = -40°$ with the assumption that the two molecules are diastereomers, and that the relationship between the g^+ and g^- conformers should hold.

In order to compare the structure and energy of ribofuranose analogues, calculations were also made for the RR (arabinofuranose), SS (lyxofuranose) and SR (xylofuranose). Each of these molecules showed a different structural preference for the hydroxyl rotors in relation to the furanose ring; ν_i^2 was varied as described above.

It was anticipated that the dihedral results from the truncated and modified fragments will closely resemble those of the NAD as a whole.

3. Results and discussion

Initial calculations showed that the oxidized nicotinamide ring remained planar, as expected (Table 2). The conjugated bonding pattern in the aromatic moiety restrains the conformational phase space of the ring dihedral angles. Results from a scan of dihedral angle ω_t showed a symmetrical curve, with the global minima at $\omega_i^1 = anti\ (a)$ (Fig. 5). Table 3 summarizes the results of a free rotation with inputs of $\omega_i^1 = g^+, a, g^-$, as well as an extra point at $\omega_i^1 = syn\ (s)$.

The model system for the rotation of the hydroxy groups was expected to closely reflect the results found for the hydroxy spins on a full ribose ring. For the model system at $\nu_i^2 = +40$ and $-40°$, minima were observed at g^+a and aa, respectively. A pictorial comparison of the conformations found for each

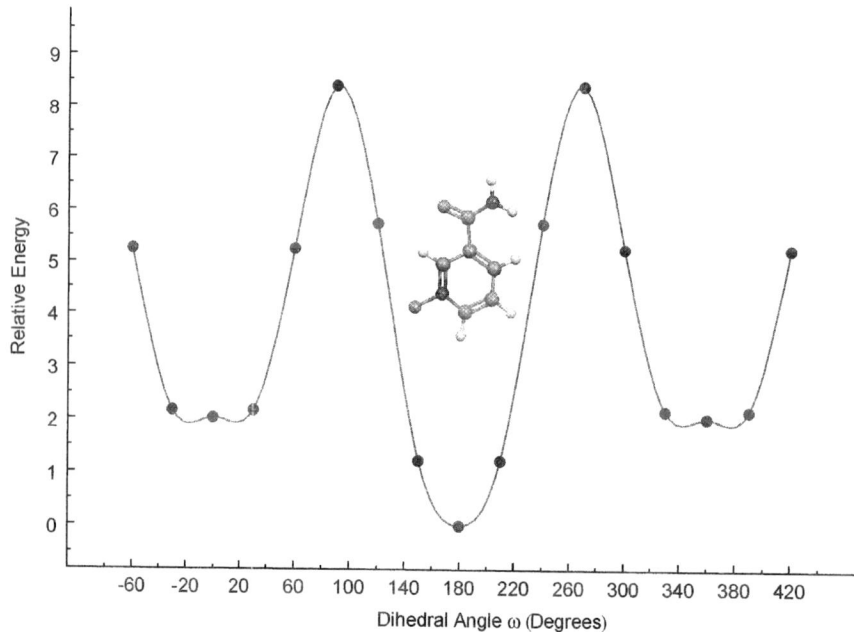

Fig. 5. Potential energy curve associated with the rotation about ω_i of the carboxyamide group of the modified nicotinamide fragment, computed at the RHF/3-21G level of theory.

Table 3
Nicotinamide fragment, computed at the RHF/3-21G level of theory

Conformer ω_i	Optimized values (deg) ω_i	Total energy (Hartrees)	Relative energy (kcal mol^{-1})
0	0.0000	−510.748448042	+2.0310
60	17.4625	−510.748658057	+1.8993
180	180.0000	−510.751684705	0.0000
300	−17.3300	−510.748657782	+1.8868

model is shown in Fig. 6. The data collected (Table 4) also suggested that a negative ν_i^2 value gives a slightly more energetically stable conformer. However, it is expected that more constrained convergence thresholds may show the two to have the same total energies, supporting the enantiomeric and diastereomeric relationships proposed.

The results from the study of the ribose fragment are reported in Table 5. The molecule exhibited behaviour similar to that reported for tetrahydrofuran by Henry-Riyad, et al. [19]; values for the dihedrals $\nu_i^0 - \nu_i^4$ remained comparable. The twist in the furanose ring remains major C_3/minor C_2. All six observed conformers are shown in Fig. 7. Deviances in the structure as compared to tetrahydrofuran are to be expected, as calculations were made on N-glycoside-5-deoxy-D-ribose, a highly substituted ring.

When δ_i^1 was restrained to various values and a rotational scan was performed on δ_i^2, the results were quite different from those found in the 2,3-dihydroxybutane model system. Here, a global minimum was found at $(\delta_i^1, \delta_i^2) = (90°, 30°)$, which was not a minima in either model previously discussed above. Furthermore, the minima in the model systems were the most energetically unfavourable here. It is hypothesized that the model did not accurately reflect the nature of the ribose system, as it lacked the amine group and ring oxygen, which are two strong electron-donating substituents.

The potential energy surfaces (PESs) in landscape and contour format of the ribose fragment are shown in Fig. 8, which also includes a bar distribution graph of ν_i^2 values for varying angles of δ_i^1 and δ_i^2. This last graph confirmed the hypothesis that ν_i^2 can exist in two conformers (positive and negative, or g^+ and g^-, respectively). The concentration of $\nu_i^2 = g^+$ values were found where $\delta_i^1 = (90° - 150°)$, and $(\delta_i^1, \delta_i^2) = (210 - 240°, 240 - 360°)$. The minima were

evenly distributed in the positive and negative regions of ν_i^2. The global minimum of this fragment was located in a region where $\nu_i^2 = g^+$.

Results from the studies on ribofuranose analogues showed that conformers were more stable when $\nu_i^2 = 40°$ (Table 6). In all three analogues studied, many of the $\nu_i^2 = -40°$ conformers ceased to exist, as they migrated to other ν_i^2 conformers. While the RR and SR analogues preferred the $\nu_i^2 = +40°$, the SS analogue preferred $\nu_i^2 = -40°$. It was found that using pre-optimized data constructed by restraining both δ_i^1 and δ_i^2 dihedrals considerably improved the output consistency, in terms of efficient and error-free structural optimizations. The comparison of the relative energies of the ribofuranose models with ribofuranose itself suggests that ribofuranose is not the most energetically favoured of all stereoisomers. In fact, the SR (xylofuranose) isomer with $\nu_i^2 = +40°$ showed an energy of 8.70 kcal mol^{-1}

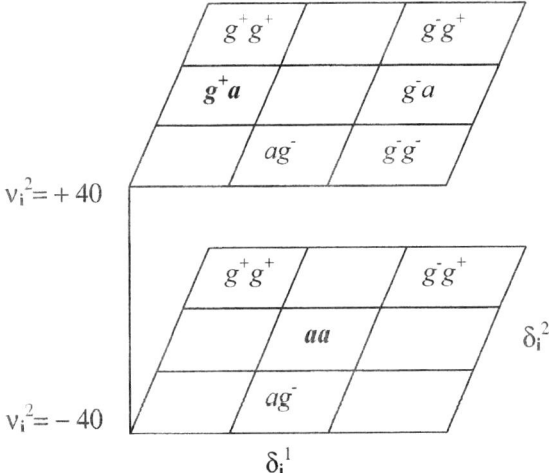

Fig. 6. Pictorial representation of Table 3, showing the possible conformations of δ_i^1 and δ_i^2 in relation to input values for ν_i^2, computed at the RHF/3-21G level of theory.

Table 4

Optimized conformers of the ribose model system (2,3-dihydroxybutane), varying dihedral angles δ_i^1 and δ_i^2, with $\nu_i^2 = +40$ or -40, computed at the RHF/3-21g level of theory

Conformers			Optimized values (deg)		Total energy (Hartrees)	Relative energy (kcal mol^{-1})
ν_i^2	δ_i^1	δ_i^2	δ_i^1	δ_i^2		
+40	g^+	g^+	94.3116	87.8807	−305.304816222	+1.1117
+40	g^+	a	83.8506	−172.5679	−305.306587768	0.0000
+40	g^+	g^-	Not found. Migrated to $g^- g^-$.			
+40	a	g^+	Not found. Migrated to $g^+ g^+$			
+40	a	a	Not found. Migrated to $g^+ a$			
+40	a	g^-	−152.9982	−31.4696	−305.305292098	+0.8130
+40	g^-	g^+	−51.2431	75.6956	−305.298372848	+5.1549
+40	g^-	a	−43.4091	−177.0971	−305.300560432	+3.7822
+40	g^-	g^-	−47.3845	−71.6722	−305.300556401	+3.7847
−40	g^+	g^+	38.1464	31.4045	−305.304816356	+1.2879
−40	g^+	a	Not found. Migrated to aa			
−40	g^+	g^-	Not found. Migrated to aa			
−40	a	g^+	Not found. Migrated to $g^+ g^+$			
−40	a	a	164.0057	178.7630	−305.306868793	0.0000
−40	a	g^-	154.3299	−83.9063	−305.305290336	+0.9905
−40	g^-	g^+	−64.7837	41.4851	−305.306587743	+0.1764
−40	g^-	a	Not found. Migrated to aa			
−40	g^-	g^-	Not found. Migrated to $g^- g^+$			

more favourable than the RS ribofuranose model in NAD$^+$ itself. Thus, it is concluded that there must be an advantage to the RS arrangement of the hydroxy rotors for the structure, stability and function of NAD$^+$. Relative energy appears to be dependent on the distance ν_i^2 migrates from the 40 or $-40°$. The RR and SS conformers were most stable given less deviance in ν_i^2, while the SR conformer was most stable when ν_i^2 had significantly migrated, suggesting that the homogenous isomers (RR and SS) were less likely to perturb the ring system. One may conclude that a coupling exists between the deviance of ν_i^2 from $+40$ or $-40°$ values and the stability of the models. While this phenomenon may not be observable in solvated models, this is clearly shown in the gas phase. As there are two ribofuranose rings contained

Table 5

Optimized conformers of the ribose fragment, varying dihedral angles δ_i^1 and δ_i^2, computed at the RHF/3-21g level of theory

Conformers		Optimized values (deg)		Total energy (Hartrees)	Relative energy (kcal mol^{-1})	Furanose ring dihedral angles (deg)				
δ_i^1	δ_i^2	δ_i^1	δ_i^2			ν_i^0	ν_i^1	ν_i^2	ν_i^3	ν_i^4
g^+	g^+	Not found. Migrated to $g^+ g^-$				13.167	−33.035	39.935	−32.125	12.167
g^+	a	78.3991	169.3875	−472.113198428	+9.1053	11.201	−34.775	44.260	−38.278	17.018
g^+	g^-	72.2591	−36.1972	−472.126425670	+0.8051	13.243	−33.045	39.291	−32.041	12.065
a	g^+	−140.4998	92.6928	−472.125908964	+1.1293	12.965	−33.019	39.897	−32.909	12.752
a	a	−150.0405	176.701	−472.125868708	+1.1546	12.499	−32.980	40.060	−33.283	13.236
a	g^-	Not found. Migrated to $g^+ g^-$				13.275	−33.077	39.300	−32.032	12.038
g^-	g^+	Not found. Migrated to ag^+				12.958	−32.997	39.874	−32.890	12.744
g^-	a	Not found. Migrated to aa				12.776	−33.127	40.029	−33.100	12.949
g^-	g^-	−30.89	−26.3892	−472.127708627	0.0000	37.180	−44.161	34.436	−12.885	−15.133

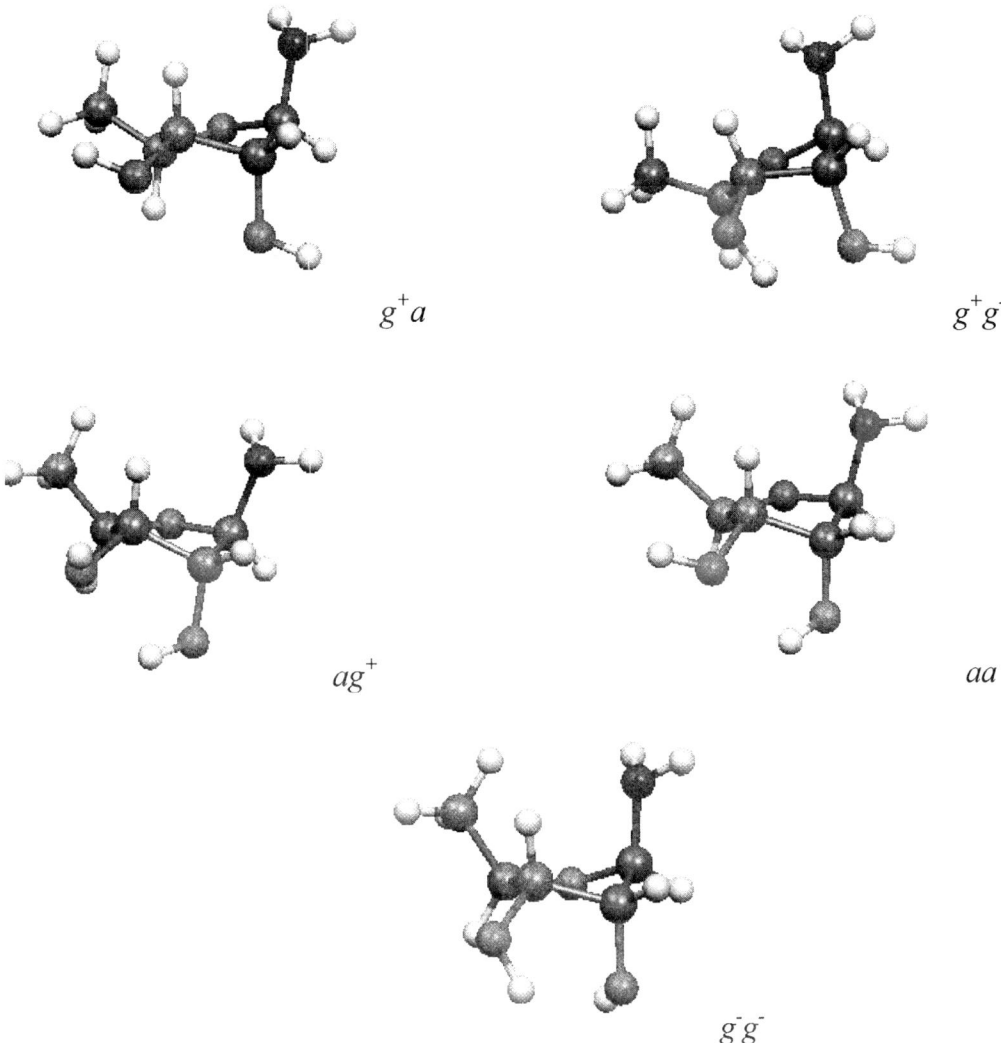

g^+a

g^+g^-

ag^+

aa

g^-g^-

Fig. 7. Optimized geometries of N-glycoside-5-deoxy-D-ribose with variations in δ_i^1 and δ_i^2, computed at the RHF/3-21G level of theory. All show an unsymmetrical twist, major $C_3{}'$-endo and minor $C_2{}'$.

in NAD$^+$, the complete structure must be investigated in order to make conclusive observations for the full NAD$^+$ molecule.

Data emerging from preliminary studies on the rotation between the adenine and ribose moieties (χ_i^2) showed that minima were concentrated around the g^+ and g^- conformers (data not shown). As it was expected that the results for the χ_i^1 rotamer be similar, calculations have been made for the g^+ and g^- conformers. This would allow for an effective and simultaneous investigation of three rotamers at once

(χ_i^1, δ_i^1, δ_i^2) Initial input structures were formed with estimates for δ_i^1 and δ_i^2 developed for the study of the ribose moiety (Table 4). However, to ensure that no conformers were missed, all eight possible conformers of the two hydroxyl groups were attempted, in addition to structures in the $\chi_i^1 = a$ conformation. The results in Table 7 show the lowest energy conformer to be (χ_i^1, δ_i^1, δ_i^2) = (a, g^-, a). Results further indicate that some of the hydroxy group conformations that were not found in the ribose study such as (χ_i^1, δ_i^1, δ_i^2) = (g^+, g^+, g^+) and (g^+, g^-, a) were indeed stable

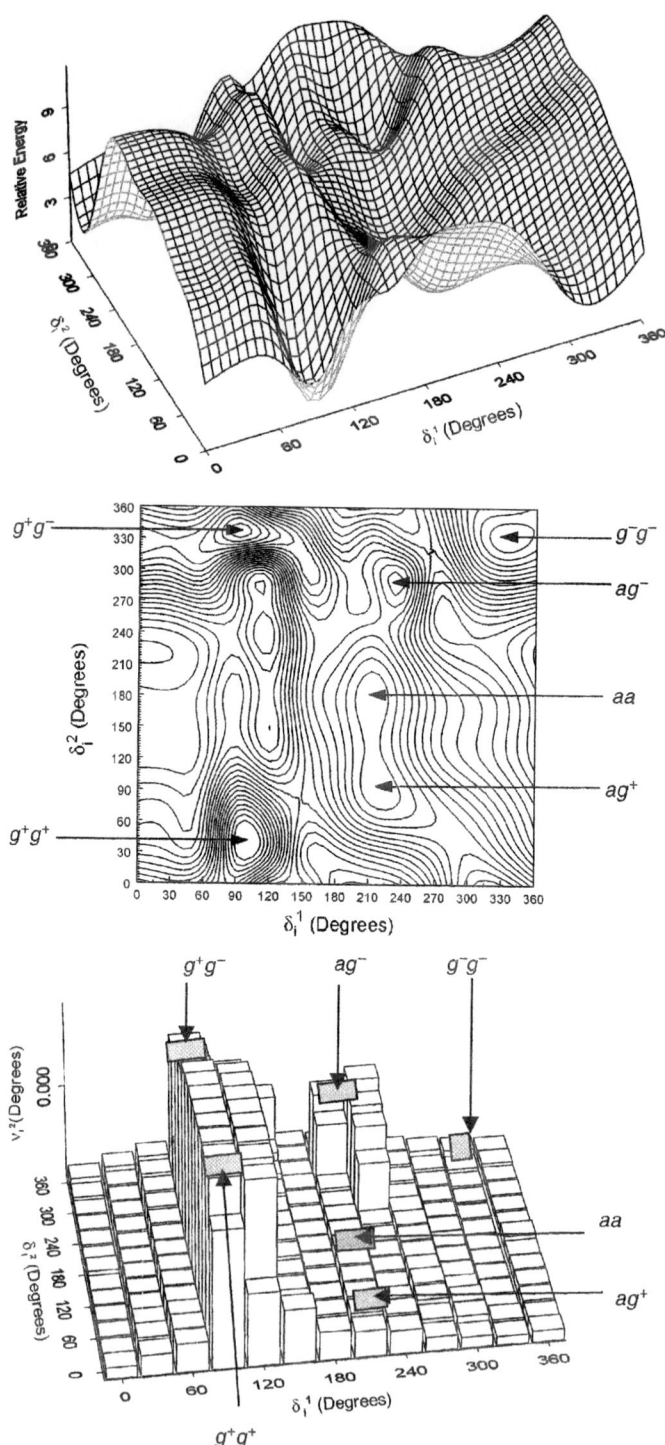

Fig. 8. A potential energy surface, $E = E(\delta_i^1, \delta_i^2)$ of the ribose subgroup, computed at the RHF/3-21G level of theory. Top: Landscape representation. Middle: Contour representation. Bottom: Bar Graph showing the distribution of RHF/3-21G geometry optimized ν_i^2 values while varying δ_i^1 and δ_i^2. Optimized minima are also indicated, with the raised and lowered bars representing the $\nu_i^2 = g^+$ and $\nu_i^2 = g^-$ conformers, respectively.

Table 6

Optimized conformers of the ribose fragment, varying dihedral angles δ_i^1 and δ_i^2, computed at the RHF/3-21G level of theory

Isomer	Conformers			Optimized values (deg)			Energy (Hartrees)	Relative energy[a] (kcal mol−1)	Global relative energy[b] (kcal mol−1)
	v_i^2	δ_I	δ_i^2	v_i^2	δ_I	δ_i^2			
RR Arabinofuranose	+40	g^+	g^+	41.676	68.2051	41.652	−472.131103785	+0.6531	+6.5738
	+40	g^+	a	40.648	64.3822	−172.8681	−472.130499867	+1.0321	+6.9528
	+40	g^+	g^-	28.0226	55.6369	−72.0366	−472.113681120	+11.5860	+17.5067
	+40	a	g^+	41.538	−177.5615	40.7772	−472.132144641	0.0000	+5.9207
	+40	a	a	40.4304	168.4245	−178.5702	−472.128718096	+2.1502	+8.0709
	+40	a	g^-	40.6959	172.767	−77.3584	−472.128062433	+2.5616	+8.4823
	+40	g^-	g^+	39.2379	−40.8099	46.7021	−472.131869356	+0.1727	+6.0934
	+40	g^-	a	38.7615	−30.9583	−165.8539	−472.130633512	+0.9482	+6.8689
	+40	g^-	g^-	Not found. Migrated to +40[g^- a]					
	−40	g^+	g^+	Not found. Migrated to −40[g^+ g^-]					
	−40	g^+	a	Not found. Migrated to −40[g^+ g^-]					
	−40	g^+	g^-	−40.0326	57.2493	−60.4432	−472.126088152	0.0000	+9.7212
	−40	a	g^+	−37.2811	−179.6864	60.2481	−472.117875136	+5.1537	+14.8750
	−40	a	a	Not found. Migrated to +40[g^+ g^-]					
	−40	a	g^-	Not found. Migrated to −40[g^+ g^-]					
	−40	g^-	g^+	−39.989	−60.73	57.216	−472.126087637	+0.0003	+9.7215
	−40	g^-	a	Not found. Migrated to +40[g^- a]					
	−40	g^-	g^-	Not found. Migrated to −40[g^+ g^-]					
SS Lyxofuranose	+40	g^+	g^+	39.2592	15.702	59.159	−472.129954756	0.0000	+7.2949
	+40	g^+	a	Not found. Migrated to +40[g^+ g^+]					
	+40	g^+	g^-	Not found. Migrated to +40[g^+ g^+]					
	+40	a	g^+	Not found. Migrated to +40[aa]					
	+40	a	a	36.4966	173.1691	172.6289	−472.124366979	+3.5064	+10.8013
	+40	a	g^-	35.4075	163.1193	−66.6749	−472.123180815	+4.2507	+11.5456
	+40	g^-	g^+	36.082	−82.9273	70.3062	−472.123157853	+4.2651	+11.5600
	+40	g^-	a	Not found. Migrated to +40[aa]					
	+40	g^-	g^-	32.513	−89.3437	−60.9104	−472.118755333	+7.0277	+14.3226
	−40	g^+	g^+	Not found. Migrated to +40[g^+ g^+]					
	−40	g^+	a	−38.6412	37.8911	178.7361	−472.125233874	+3.1118	+10.2573
	−40	g^+	g^-	Not found. Migrated to +40[g^+ g^+]					
	−40	a	g^+	−37.6105	−163.3564	73.3645	−472.126493674	+2.3213	+9.4667
	−40	a	a	−38.2623	−156.3697	194.4524	−472.127993306	+1.3802	+8.5257
	−40	a	g^-	Not found. Migrated to +40[aa]					
	−40	g^-	g^+	Not found. Migrated to +40[g^+ g^+]					
	−40	g^-	a	−36.0206	−50.042	−172.4543	−472.130527466	0.0000	+7.1455
	−40	g^-	g^-	Not found. Migrated to −40[g^- a]					
SR Xylofuranose	+40	g^+	g^+	37.3843	101.6711	55.0544	−472.138785588	+1.7534	+1.7534
	+40	g^+	a	19.3144	103.4834	−170.4409	−472.127368783	+8.9176	+8.9176
	+40	g^+	g^-	36.6198	2.8595	−49.4095	−472.141579876	0.0000	0.0000
	+40	a	g^+	Not found. Migrated to +40[g^+ g^+]					
	+40	a	a	Not found. Migrated to +40[g^+ a]					
	+40	a	g^-	Not found. Migrated to +40[g^+ g^-]					
	+40	g^-	g^+	Not found. Migrated to +40[g^+ g^+]					
	+40	g^-	a	Not found. Migrated to +40[g^+ a]					
	+40	g^-	g^-	Not found. Migrated to +40[g^+ g^-]					

(continued on next page)

Table 6 (*continued*)

Isomer	Conformers			Optimized values (deg)			Energy (Hartrees)	Relative energy[a] (kcal mol^{-1})	Global relative energy[b] (kcal mol^{-1})
	ν_i^2	δ_l	δ_i^2	ν_i^2	δ_l	δ_i^2			
	-40	g^+	g^+	-30.5793	37.7823	22.4601	-472.129320583	$+3.9098$	$+7.6928$
	-40	g^+	a	Not found. Migrated to $+40[g^+g^-]$					
	-40	g^+	g^-	Not found. Migrated to $+40[g^+g^-]$					
	-40	a	g^+	Not found. Migrated to $-40[g^-g^+]$					
	-40	a	a	Not found. Migrated to $-40[ag^-]$					
	-40	a	g^-	-34.1079	143.9796	-100.6930	-472.129452024	$+3.8274$	$+7.6103$
	-40	g^-	g^+	-26.6569	-21.5228	23.9505	-472.135551298	0.0000	$+3.7830$
	-40	g^-	a	-35.2109	-61.3261	-171.7249	-472.120619437	$+9.3699$	$+13.1529$
	-40	g^-	g^-	Not found. Migrated to $-40[g^-g^+]$					

[a] Relative energy is in reference to the conformers within each isomeric model.
[b] Global relative energy is in reference to the conformers in relation to the global minima, SR $+40[g^+g^-]$.

Table 7

Optimized conformers of the nicotinamide-ribose subgroup, varying dihedral angles χ_i^1, δ_i^1 and δ_i^2, computed at the RHF/3-21G level of theory

Conformer $[\chi_i^1, \delta_i^1, \delta_i^2]$	Optimized values (deg)			Total energy (Hartrees)	Relative energy (kcal mol^{-1})
	χ_i^1	δ_i^1	δ_i^2		
$g^+[g^+g^+]$	Not found. Migrated to $g^+[ag^+]$				
$g^+[g^+a]$	Not found. Migrated to $g^+[g^-a]$				
$g^+[g^+g^-]$	Not found. Migrated to $g^+[ag^+]$				
$g^+[ag^+]$	57.4630	161.0107	54.0260	-828.810301733	$+3.8197$
$g^+[aa]$	Not found. Migrated to $g^+[g^-a]$				
$g^+[ag^-]$	Not found. Migrated to $g^+[g^-a]$				
$g^+[g^-g^+]$	Not found. Migrated to $g^+[g^-a]$				
$g^+[g^-a]$	55.6273	-84.0164	171.9860	-828.814918356	$+0.9929$
$g^+[g^-g^-]$	Not found. Migrated to $g^+[g-a]$				
$a[g^+g^+]$	Not found. Migrated to $g^-[ag^+]$				
$a[g^+a]$	Not found. Migrated to $a[g^-a]$				
$a[g^+g^-]$	Not found. Migrated to $g^-[ag^+]$				
$a[ag^+]$	Not found. Migrated to $g^-[ag^+]$				
$a[aa]$	Not found. Migrated to $a[g^-a]$				
$a[ag^-]$	Not found. Migrated to $a[g^-a]$				
$a[g^-g^+]$	Not found. Migrated to $a[g^-a]$				
$a[g^-a]$	-125.2456	-85.1002	173.6605	-828.816388851	0.0000
$a[g^-g^-]$	Not found. Migrated to $a[g^-a]$				
$g^-[g^+g^+]$	Not found. Migrated to $g^-[ag^+]$				
$g^-[g^+a]$	Not found. Migrated to $a[g^-a]$				
$g^-[g^+g^-]$	Not found. Migrated to $g^-[ag^+]$				
$g^-[ag^+]$	-118.1744	165.6213	55.4595	-828.809205194	$+4.5080$
$g^-[aa]$	Not found. Migrated to $a[g^-a]$				
$g^-[ag^-]$	Not found. Migrated to $a[g^-a]$				
$g^-[g^-g^+]$	Not found. Migrated to $a[g^-a]$				
$g^-[g^-a]$	Not found. Migrated to $a[g^-a]$				
$g^-[g^-g^-]$	Not found. Migrated to $a[g^-a]$				

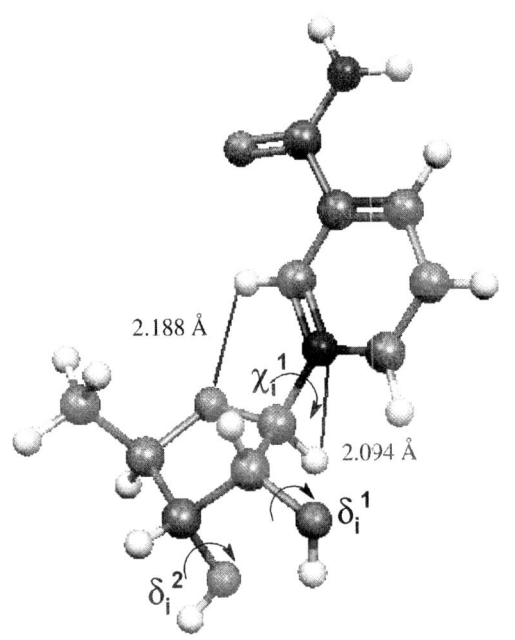

Fig. 9. Optimized geometry of Nicotinamide ribose (NmR) at its minima $(\chi_i^1, \delta_i^1, \delta_i^2) = (a, g^-, a)$, computed at the RHF/3-21G level of theory. Potential hydrogen bonds are shown between N_5/H_{21} and H_{16}/O_{20}, with distances of 2.188 and 2.094 Å, respectively.

2.188 Å

χ_i^1

2.094 Å

δ_i^1

δ_i^2

when combined with the nicotinamide ring. Many of the structures migrated to others conformers, with $(\chi_i^1, \delta_i^1, \delta_i^2) = (a, a, g^-)$ and (g^+, g^-, a) being the most common. These results differ from those presented by Wu and Houk, in which a coplanar structure was predicted to be the most stable using the 3-21G basis set [12]. While the model system chosen by the authors could not completely model the interactions between the nicotinamide and ribose rings, it helped in developing an understanding of the nature of ring systems and adjacent functional groups. As more data is produced, it will be of interest to compare the newly generated results with those from the model system. The optimized minimum is shown in Fig. 9, with two potential hydrogen bonds between N_5/H_{21}–C and C–H_{16}/O_{20}, having distances of 2.188 and 2.094 Å, respectively. It is suggested that these relatively weak hydrogen bonds between the nicotinamide and ribose rings stabilize the conformation.

The five optimized conformers for the NmR subgroup were further optimized in exploratory calculations at the RHF/6-31G and RHF/6-31G(d) basis sets, and the results of the latter are presented in Table 8. It was found that the inclusion of

Table 8
Optimized conformers of the nicotinamide-ribose subgroup, varying dihedral angles χ_i^1, δ_i^1 and δ_i^2, computed at the RHF/6-31G(d) level of theory

Conformer $[\chi_i^1, \delta_i^1, \delta_i^2]$	Optimized values (deg) χ_i^1	δ_i^1	δ_i^2	Total energy (Hartrees)
$g^+[ag^+]$	68.6834	155.8252	57.4827	−833.429700100
$g^+[g^-a]$	70.6516	−87.6016	167.7738	−833.432701035
$a[g^-a]$	−113.7539	−88.6256	168.7907	−833.433765178
$g^-[ag^+]$	−102.0277	160.3673	59.3537	−833.429881068

Table 9
An in-depth examination of the conformer $a[g^-a]$, with particular reference to dihedral angle χ_i^1 of the nicotinamide-ribose subgroup, computed at a increasing levels of theory

Level of theory	Optimized values (deg) χ_i^1	δ_i^1	δ_i^2	Total energy (Hartrees)
RHF/3-21G	−125.2456	−85.1002	173.6605	−828.816388851
RHF/6-31G	−122.603	−90.3071	170.1815	−833.084072847
RHF/6-31G(d)	−113.7539	−88.6256	168.7907	−833.433765178
B3LYP/6-31G	−122.242	−84.7384	166.4259	−838.125289012
B3LYP/6-31G(d)	−116.3917	−86.7735	163.1243	−838.367975464

the preliminary polarization function showed deviance in the χ_i^1 dihedral for the $a[g^- a]$ conformer (as shown in bold). Different basis sets were used in an attempt to reconcile the data and uncover the foundation of this phenomenon, but to no avail. Table 9 briefly summarizes the basis sets attempted. It appears that the preliminary polarization function forces the χ_i^1 dihedral to adopt a g^- conformation that is not found elsewhere. The δ_i^1 and δ_i^2 dihedrals remained stable throughout. Previous experience has shown good correlation between the RHF/3-21G and RHF/6-31G(d) basis sets, and this progression from lower to higher levels of theory has been employed as standard practice by our group to generate the most accurate results. However, peptide work has shown some variability (within 10°) of backbone torsional angles depending on the level of theory [20], which mirrors data presented here. However, the discrepancy between RHF/3-21G and RHF/6-31G(d) data remains a concern, and possible explanations for this are being further examined.

4. Conclusions

Thus, the NmR molecule has one preferred conformation at $(\chi_i^1, \delta_i^1, \delta_i^2) = (a, g^-, a)$. While computations of model systems were helpful, it was found that these did not always accurately predict conformations in integrated molecules. Furthermore, calculations were made more precise by the input of pre-optimized data from initial calculations run with the dihedrals of interest in a frozen conformation. This methodology will be incorporated into future studies.

The conflict between the RHF/3-21G and RHF/6-31G(d) data must be resolved before increasing the level of theory. Once this has been achieved, the next endeavour would be to begin computations with higher levels of theory. In addition, the NmR moiety under examination will be gradually integrated with other fragments of the molecule for computation until the complete structure has been studied. With a better understanding of the oxidized form of NAD, ab initio studies of NADH and its analogues can be easily conducted, allowing for further research encompassing the computational, biochemical and pharmacological fields.

Acknowledgements

This work was supported with grants from the Global Institute Of COmputational Molecular and Materials Sciences (GIOCOMMS), Toronto, Ontario, Canada. S. Lovas would like to acknowledge the support from the BRIN Program of the National Center for Research (NIH Grant Number 1 P20 RR16469) and from the National Science Foundation (EPS-0091900). Thanks are extended to Jacqueline M.S. Law, Christopher N.J. Marai, Jeremy H. Keller, Michelle A. Sahai and David H. Setiadi for helpful discussion.

One of the authors (IGC) wishes to thank the Ministry of Education for a Szent-Györgyi Visiting Professorship.

References

[1] C.R. Bellamacina, FASEB J. 10 (1996) 1257.
[2] W. Kuhn, T. Muller, R. Winkel, S. Danielczik, A. Gerstner, R. Hacker, C. Mattern, H. Przuntek, J. Neural Transm. 103 (1996) 1187.
[3] J.G. Birkmayer, C. Vrecko, D. Volc, W. Birkmayer, Acta Neurologica Scandinavia 87 (Suppl 1) (1993) 32.
[4] R.J. Uitti, D.B. Calne, Eur. Neurol. 33 (Suppl 1) (1993) 6–23.
[5] P.M. Iuvone, J.F. Reinhard Jr., M.M. Abou Donia, O.H. Viveros, C.A. Nichol, Brain Res. 359 (1985) 392.
[6] L.M. Forsyth, H.G. Preuss, A.L. MacDowell, L. Chiazze Jr., G.D. Birkmayer, J.A. Bellanti, Ann. Allergy Asthma Immunol. 82 (1999) 185.
[7] P. Franchetti, L. Cappelacci, P. Perlini, H.N. Jayaram, A. Butler, B.P. Schneider, F.R. Collart, E. Huberman, M. Grifantini, J. Med. Chem. 41 (1998) 1702.
[8] R-G. Zhang, G. Evans, F.J. Rotella, E.M. Westbrook, D. Beno, E. Huberman, A. Joachimiak, F.R. Collart, Biochemistry 38 (1999) 4691.
[9] A.M. Aronov, C.L. Verlinde, W.G. Hol, M.H. Gelb, J. Med. Chem. 41 (1998) 4790.
[10] M. Rainer, E. Kraxberger, M. Haushofer, H.A.M. Mucke, K.A. Jellinger, J. Neural Transm. 107 (2000) 1475.
[11] W.A. Catterall, D.P. Hollis, C.F. Walter, Biochemistry 8 (1969) 4032.
[12] Y-D. Wu, K.N. Houk, J. Am. Chem. Soc. 113 (1991) 58.
[13] P.E. Smith, J.J. Tanner, J. Mol. Recognit. 13 (2000) 27.
[14] A.P. Zens, T.J. Williams, J.C. Wisowaty, R.R. Fisher, R.B. Dunlap, T.A. Bryson, P.D. Ellis, J. Am. Chem. Soc. 97 (1975) 2850.
[15] A.P. Zens, T.A. Bryson, R.B. Dunlap, R.R. Fisher, P.D. Ellis, J. Am. Chem. Soc. 98 (1976) 7559.
[16] C.E. Bell, T.O. Yeates, D. Eisenberg, Protein Sci. 6 (1997) 2084–2096.

[17] S.D. Kahn, W.J. Hehre, J. Am. Chem. Soc. 109 (1987) 666.

[18] M.J. Frisch, G.W. Trucks, H.B. Schlegel, G.E. Scuseria, M.A. Robb, J.R. Cheeseman, V.G. Zakrzewski, J.A. Montgomery, Jr., R.E. Stratmann, J.C. Burant, S. Dapprich, J.M. Millam, A.D. Daniels, K.N. Kudin, M.C. Strain, Ö. Farkas, J. Tomasi, V. Barone, M. Cossi, R. Cammi, B. Mennucci, C. Pomelli, C. Adamo, S. Clifford, J. Ochterski, G.A. Petersson, P.Y. Ayala, Q. Cui, K. Morokuma, D.K. Malick, A.D. Rabuck, K. Raghavachari, J.B. Foresman, J. Cioslowski, J.V. Ortiz, A.G. Baboul, B.B. Stefanov, G. Liu, A. Liashenko, P. Piskorz, I. Komaromi, R. Gomperts, R.L. Martin, D.J. Fox, T. Keith, M.A. Al-Laham, C.Y. Peng, A. Nanayakkara, C. Gonzalez, M. Challacombe, P.M.W. Gill, B.G. Johnson, W. Chen, M.W. Wong, J.L. Andres, M. Head-Gordon, E.S. Replogle, J.A. Pople, Gaussian, Inc., Pittsburgh PA, (Revision A.9), GAUS-SIAN 98, 1998.

[19] H. Henry-Riyad, T.-H. Tang, I.G. Csizmadia, J. Mol. Struct. (Theochem) 492 (1999) 67.

[20] A. Perczel, I.G. Csizmadia, in: A. Greenberg, C.M. Breneman, J.F. Liebman (Eds.), The Amide Linkage: Structural Significance in Chemistry, Biochemistry, and Materials Science, Wiley, New Jersey, 2003, Chapter 13.

ELSEVIER

Journal of Molecular Structure (Theochem) 666–667 (2003) 431–437

THEO CHEM

www.elsevier.com/locate/theochem

An exploratory ab initio conformational analysis of selected fragments of nicotinamide adenine dinucleotide (NAD+). Part II: adenosine

Suzanne K. Lau[a,b,*], Gregory A. Chass[a,b,c,d], Botond Penke[e,f], Imre G. Csizmadia[a,b,f]

[a]Department of Chemistry, University of Toronto, Toronto, Ont., Canada M5S 3H6
[b]Global Institute of Computational Molecular and Materials Science, 1422 Edenrose Street, Mississauga, Ont., Canada L5V 1H3
[c]Institut de Science et d'Ingénierie Supramoléculaires, 8, allée Gaspard Monge, BP 70028, Strasbourg Cedex 67083, France
[d]Department of Biomedical Sciences, Creighton University, 2500 California plaza, Omaha, NE 68178, USA
[e]Protein Chemistry Research Group, Hungarian Academy of Sciences, University of Szeged, Dóm tér 8, Szeged 6720, Hungary
[f]Department of Medical Chemistry, University of Szeged, Dóm tér 8, Szeged H-6720, Hungary

Abstract

Ab initio molecular orbital computations were carried out on selected fragments of nicotinamide adenine dinucleotide (NAD+) at the RHF/3-21G level of theory. The definitions of the relative spatial orientation of all constituent atomic nuclei have been formulated in such a modular fashion so as to allow for portions of the whole to be studied independently. Key points of examination included the rotation of adjacent moieties and ring puckering. This work focuses on the possible conformations of the adenosine moiety. It is anticipated that the structural results from the truncated and modified fragments will closely resemble those of NAD as a whole.
© 2003 Elsevier B.V. All rights reserved.

Keywords: Ab initio; Coenzyme; Nicotinamide adenine dinucleotide; Adenine; Adenosine

1. Introduction

The redox pair formed by the oxidized and reduced forms of nicotinamide adenine dinucleotide (NAD+/NADH) holds great biological importance, most notably in the energy producing process [1]. As depicted in Fig. 1, NAD+ contains an adenylic acid and nicotinamide-5'-ribonucleotide combined by a pyrophosphate bridge. Nicotinamide can carry

a positive charge at the ring nitrogen, while the pyrophosphate link holds two negative charges.

Adenosine is also a significant biological molecule in its own right. 2'-Deoxyadenosine 5'-phosphate forms an integral nucleotide in DNA, while its oxidated counterpart adenosine 5'-phosphate is present in RNA. It is involved in many signalling pathways as cyclic AMP, and has been shown to offer cardioprotection through the regulation of the adenosine receptor [2]. A quick perusal of the literature has shown that little ab initio structure work has been done on adenosine, although some have attempted to examine hydrogen bonding and hydration of the molecule using NMR spectra and molecular

---

* Corresponding author. Department of Chemistry, University of Toronto, Toronto, Ont., Canada M5S 3H6.
E-mail addresses: suzanne.lau@utoronto.ca (S.K. Lau), gchass@giocomms.org (G.A. Chass).

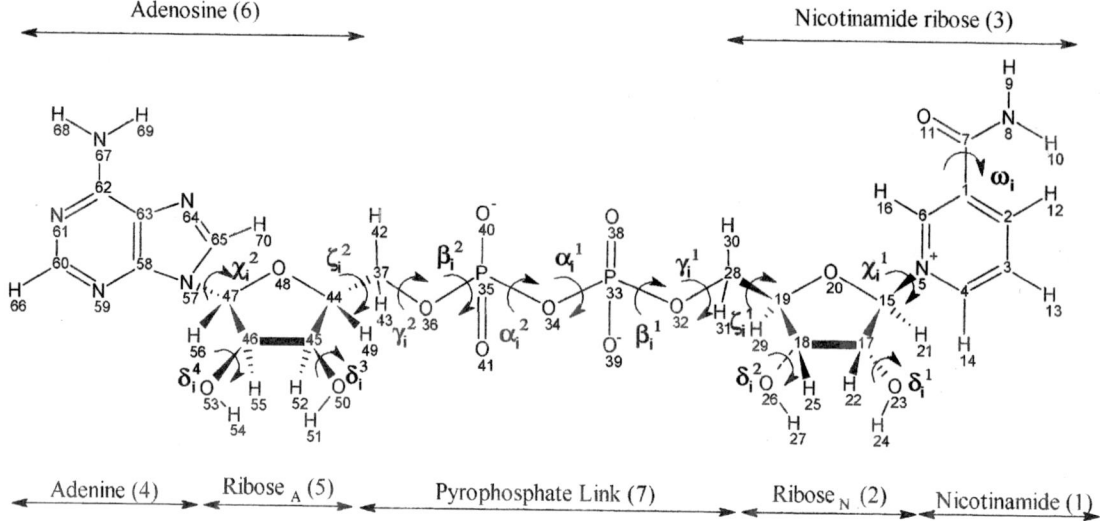

Fig. 1. Systematic modular numbering system of NAD$^+$ including pertinent dihedrals. The arrows indicate the approximate partitioning of the molecule. (1) Nicotinamide, (2) ribose$_N$, (3) nicotinamide ribose (NmR), (4) adenine, (5) ribose$_A$, (6) adenosine and (7) pyrophosphate bridge. This paper focuses primarily on (4)–(6).

modelling [3]. Much of the research interest has been focussed on base pair interactions in DNA and RNA.

The structural characterization of NAD$^+$ presented a particular challenge, as it is a flexible molecule with over a dozen rotatable bonds that can adopt a wide variety of environmentally dependent conformations. Our first work characterized the possible conformations of the nicotinamide-5′-ribonucleotide portion of NAD$^+$ [4]. We now aim to study the adenosine portion in a similar manner. The study of

the adenosine moiety (Fig. 2a) can be considered a precursor for the future examination of adenosine monophosphate, diphosphate or triphosphate. Furthermore, a study of the interactions between the two aromatic ring groups would be suitable. Previous X-ray data has suggested that a tryptophane derivative with similar aromatic rings exists at a perpendicular state [5,6], while molecular dynamic simulations of NAD$^+$ have indicated that ring interactions occur primarily in the stacked conformation [7]. Thus,

(a)   (b)

Fig. 2. (a) Fully numbered adenosine moiety, with dihedrals of interest labelled. (b) Fully numbered positively charged adenosine moiety, with dihedrals of interest labelled.

it would be of interest to complete the conformational analysis of the two ring groups, and then investigate the interaction between them.

Insight into the conformation of NAD^+ in different environments will allow for a more complete understanding of the mechanism of its diverse functions at the molecular level, allowing for drug candidates with tailored ligand/receptor binding sites to be developed in the treatment of NAD-dependent pathologies.

2. Methods

The GAUSSIAN 98 [8] program package has been used to perform preliminary ab initio calculations on the adenosine fragment of NAD^+. The definitions of the relative spatial orientations of all constituent atomic nuclei have been formulated in a standardized and modular fashion so as to allow for portions of the whole to be studied independently (Fig. 1). The focus of this paper is on the adenosine group, which was first modelled by the semi-empirical AM1 method (data not shown).

Previously obtained data for the optimization of the ribose molecule was used for the optimizations [4]. Few problems were encountered in the calculations of this moiety to date, primarily because the techniques developed while computing the conformations of the nicotinamide–ribose moiety were employed. It had been found that freezing the dihedrals of interest and sequentially unfreezing them with pre-optimized data generated the most accurate results. Each dihedral of interest was examined independently, and then incorporated into the larger study with multiple scans of up to three dihedrals. The dihedral χ_i^2 was examined in conjunction with the hydroxy rotors (δ_i^3, δ_i^4) on the ribose portion of the molecule.

Preliminary calculations were also made on a positively charged adenosine molecule found in the acidic environment created by crystallization solutions (Fig. 2b) [9,10]. These calculations would be of use in the event that a comparison between the X-ray crystallography and ab initio structures of NAD^+ was required.

3. Results and discussion

As expected, bicyclic adenine rarely deviated from planarity due to the conjugated bonding pattern of

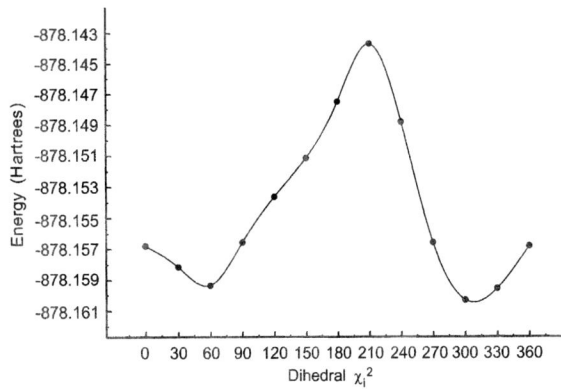

Fig. 3. Potential energy curve associated with the rotation about χ_i^2 of the adenosine moiety, computed at the RHF/3-21G level of theory.

the rings. A scan was performed on the link between the adenosine and ribose rings, labelled as χ_i^2. The minima were located at the g^+ and g^- positions (Fig. 3). This differs from the nicotinamide–ribose dihedral, where conformers were found in all three positions (g^+, a and g^-). This is likely due to the fact that adenine contains a fused bicyclic ring structure rather than a single ring, which would increase the steric hindrance around the rotor. A point optimization of the fragment with χ_i^2 set to the anti (a) position was also made to verify that only two minima existed. Indeed, χ_i^2 converged in the g^+ conformation ($\chi_i^2 = 54.3959$), confirming the scan results. Although two virtually identical minima were found at g^+ and g^-, it appeared that χ_i^2 preferred the g^+ conformation. The examination of all three dihedrals ($\chi_i^2, \delta_i^3, \delta_i^4$) are summarized in Table 1. The fully optimized structure of adenosine is shown in Fig. 4 at its minima ($\chi_i^2, \delta_i^3, \delta_i^4$) = ($g^+, g^+, a$). The g^+ structure must be preferred due to potential hydrogen bond stabilization between O_6/H_{31} and N_{18}/H_{17}, with distances of 2.427 and 2.089 Å, respectively. The most energetically favourable g^- conformer, ($\chi_i^2, \delta_i^3, \delta_i^4$) = ($g^-, a, g^+$), lacked the O_6/H_{31} hydrogen bond (3.736 Å), while N_{18}/H_{17} remained fairly stable at 2.050 Å (Fig. 5). Thus, it appears that the O_6/H_{31} bond between the ribose ring and adenine is critical in generating an energetically favourable conformation of adenosine. Minima obtained were further confirmed by frequency calculations (Table 2).

It was expected that the scan of χ_i^2 of the charged adenosine moiety should closely resemble

Table 1
Conformations obtained for the adenosine subgroup varying the torsional angles χ_i^2, δ_i^3 and δ_i^4, computed at the RHF/3-21G level of theory

Conformer	Optimized values (°)			Energy (hartrees)	ΔEnergy (kcal mol^{-1})
$[\chi_i^2, \delta_i^3 \delta_i^4]$	χ_i^2	δ_i^3	δ_i^4		
$g^+[g^+g^+]$	45.5871	88.5546	78.3654	-878.157847538	$+3.1257$
$g^+[g^+a]$	61.3466	79.2342	-167.1856	-878.162828733	0.0000
$g^+[g^+g^-]$		Not found, migrated to $g^+[g^+a]$			
$g^+[ag^+]$		Not found, migrated to $g^+[ag^-]$			
$g^+[aa]$		Not found, migrated to $g^+[g^+a]$			
$g^+[ag^-]$	52.5532	-157.4757	-35.9853	-878.159931011	$+1.8183$
$g^+[g^-g^+]$		Not found, migrated to $g^+[ag^-]$			
$g^+[g^-a]$		Not found, migrated to $g^+[ag^-]$			
$g^+[g^-g^-]$	54.3876	-76.9195	-46.1692	-878.159444534	$+2.1236$
$g^-[g^+g^+]$	-61.0162	92.5123	71.8611	-878.153416685	$+5.9062$
$g^-[g^+a]$	-58.8578	79.7864	-161.6216	-878.157151647	$+3.5624$
$g^-[g^+g^-]$		Not found, migrated to $g^-[g^+a]$			
$g^-[ag^+]$		Not found, migrated to $g^-[ag^-]$			
$g^-[aa]$		Not found, migrated to $g^-[g^+a]$			
$g^-[ag^-]$	-59.1876	-147.724	-28.4122	-878.160705520	$+1.3323$
$g^-[g^-g^+]$		Not found, migrated to $g^-[g^-g^-]$			
$g^-[g^-a]$		Not found, migrated to $g^-[g^-g^-]$			
$g^-[g^-g^-]$	-61.9397	-83.1485	-36.4621	-878.160322927	$+1.5724$

that of neutral adenosine. However, the potential energy curve produced had a very defined minimum at g^+ (Fig. 5). Furthermore, there was an approximate 3.9117 kcal mol^{-1} difference in energy between the g^+ and g^- conformers. A full examination of all three dihedrals (χ_i^2, δ_i^3, δ_i^4)

demonstrated similar results, with eight minima found between the g^+ and g^- conformers (Table 3). In addition, minima for the positively charged moiety were identical as those in neutral adenosine, although the global minima of each moiety were different (Fig. 6). The fully optimized structure of

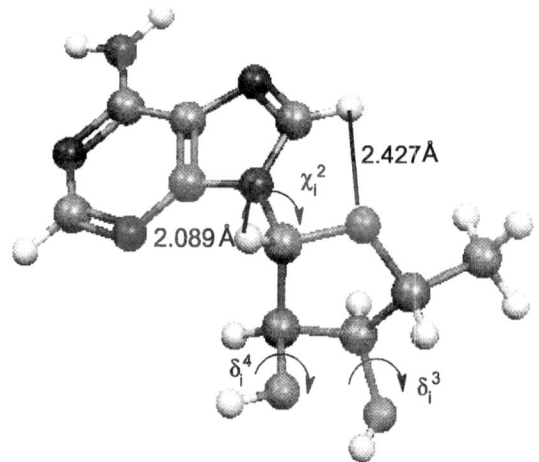

Fig. 4. Optimized geometry of adenosine at its minima (χ_i^2, δ_i^3, δ_i^4) = (g^+, g^+, a), computed at the RHF/3-21G level of theory. Potential hydrogen bonds are shown between O_6/H_{31} and N_{18}/H_{17}, with distances of 2.427 and 2.089 Å, respectively.

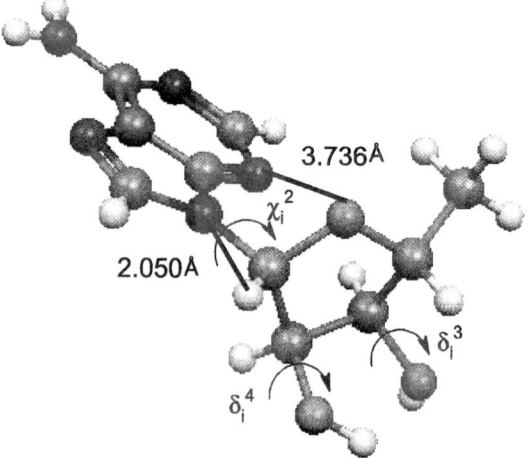

Fig. 5. Optimized geometry of an energetically favourable g^- conformer, (χ_i^2, δ_i^3, δ_i^4) = (g^-, a, g^+) computed at the RHF/3-21G level of theory. Distances between O_6/H_{31} and N_{18}/H_{17} are 3.736 and 2.050 Å, respectively.

Table 2
Confirmation of minima by frequency calculations, computed at the RHF/3-21G level of theory

Conformer $[\chi_i^2, \delta_i^3, \delta_i^4]$	F1	F2	F3
$g^+[g^+g^+]$	26.1958	55.6107	75.8169
$g^+[g^+a]$	28.5352	50.9894	75.4246
$g^+[ag^-]$	28.0533	50.7084	79.894
$g^+[g^-g^-]$	28.621	50.9111	79.6078
$g^-[g^+g^+]$	32.8401	60.4951	86.456
$g^-[g^+a]$	21.0353	53.273	83.8274
$g^-[ag^-]$	33.7452	55.399	86.4975
$g^-[g^-g^-]$	36.9539	58.0549	89.2193

positively-charged adenosine is shown in Fig. 7 at its minima $(\chi_i^2, \delta_i^3, \delta_i^4) = (g^+, g^-, g^-)$ with potential hydrogen bonds at O_6/H_{31} and N_{18}/H_{17} of 2.354 and 2.10 Å, respectively. Similar to neutral adenosine, loss of hydrogen bond stabilization can be observed in the g^-, where distances between O_6/H_{31} and N_{18}/H_{17} are 3.872 and 2.063 Å, respectively (Fig. 8). Minima were confirmed by frequency calculations (Table 4).

The presence of the extra hydrogen on N_{22} in the adenine ring contributes to an approximate 245 kcal mol^{-1} difference in energy between neutral and positively charged adenosine. The positive charge to N_{22} induced by the presence of hydrogen clearly stabilizes the heterocyclic ring system of adenosine, perhaps by increasing the electron-withdrawing effects of nitrogen and increasing a resonance effect. At some point in the future, it may be of interest to determine whether NAD$^+$ in vivo prefers the conformation of adenosine induced by the acidic environment, as the positive charge may influence the interaction between the two ring moieties.

Strong basis-set dependence with data from the 3-21G and 6-31G* calculations predicting differing minima has been shown by Wu and Houk, who studied NAD$^+$ model systems such as N-hydroxymethylpyridinium ion and N-ethylpyridinium ion [11]. Furthermore, this phenomenon has been previously reported for analogue allyl alcohols and ethers [12]. It would therefore be of interest to note the correlation between basis sets as calculations here are brought to increasing levels of theory.

Table 3
Conformations obtained for the positively charged adenosine subgroup varying the torsional angles χ_i^2, δ_i^3 and δ_i^4 computed at the RHF/3-21G level of theory

Conformer	Optimized values (°)			Energy (hartrees)	ΔEnergy (kcal mol^{-1})
$[\chi_i^2, \delta_i^3, \delta_i^4]$	χ_i^2	δ_i^3	δ_i^4		
$g^+[g^+g^+]$	55.1857	87.0527	76.4222	−878.544926258	+5.0444
$g^+[g^+a]$	62.6176	74.9216	−164.2027	−878.551080792	+1.1824
$g^+[g^+g^-]$		Not found, migrated to $g^+[g^+a]$			
$g^+[ag^+]$		Not found, migrated to $g^+[ag^-]$			
$g^+[aa]$		Not found, migrated to $g^+[g^+a]$			
$g^+[ag^-]$	66.7451	−143.0699	−35.5169	−878.552157445	+0.5067
$g^+[g^-g^+]$		Not found, migrated to $g^+[g^-g^-]$			
$g^+[g^-a]$		Not found, migrated to $g^+[g^-g^-]$		−878.552965137	
$g^+[g^-g^-]$	65.3539	−75.9683	−43.206	−878.552965002	0.0000
$g^-[g^+g^+]$	−3.4387	95.4318	67.4728	−878.539101223	+8.7000
$g^-[g^+a]$	−72.4606	76.2185	−156.1714	−878.545247615	+4.8427
$g^-[g^+g^-]$		Not found, migrated to $g^-[g^+a]$			
$g^-[ag^+]$		Not found, migrated to $g^-[ag^-]$			
$g^-[aa]$		Not found, migrated to $g^-[g^+a]$			
$g^-[ag^-]$	−70.2056	−143.1771	−30.5331	−878.548669705	+2.6953
$g^-[g^-g^+]$		Not found, migrated to $g^-[g^-g^-]$			
$g^-[g^-a]$		Not found, migrated to $g^-[g^-g^-]$			
$g^-[g^-g^-]$	−70.8383	−76.8192	−38.718	−878.549121389	+2.4119

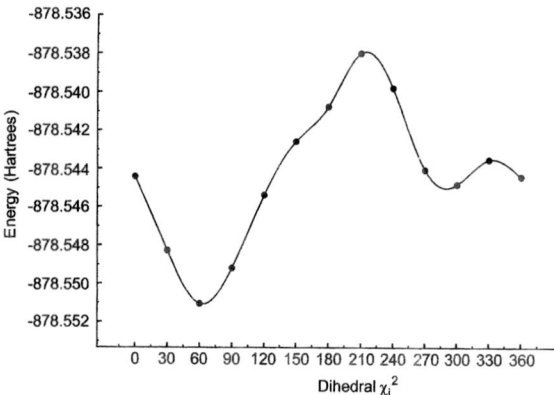

Fig. 6. Potential energy curve associated with the rotation about χ_i^2 of the positively-charged adenosine moiety, computed at the RHF/3-21G level of theory.

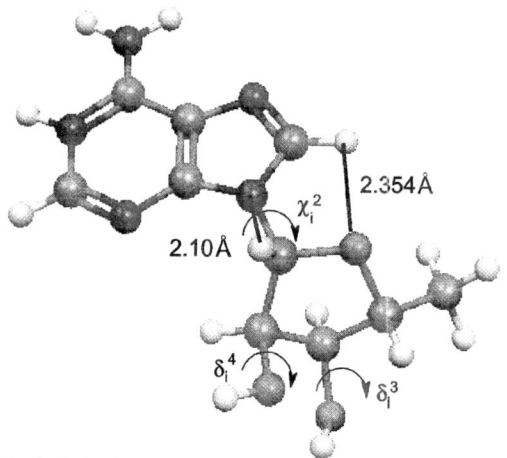

Fig. 7. Optimized geometry of positively-charged adenosine at its minima $(\chi_i^2, \delta_i^3, \delta_i^4) = (g^+, g^-, g^-)$, computed at the RHF/3-21G level of theory. Potential hydrogen bonds are shown between O_6/H_{31} and N_{18}/H_{17}, with distances of 2.354 and 2.10 Å, respectively.

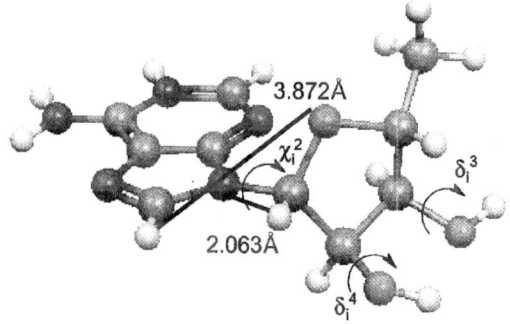

Fig. 8. Optimized geometry of an energetically favourable g^- conformer, $(\chi_i^2, \delta_i^3, \delta_i^4) = (g^-, g^-, g^-)$ computed at the RHF/3-21G level of theory. Distances between O_6/H_{31} and N_{18}/H_{17} are 3.872 and 2.063 Å, respectively.

Table 4
Confirmation of minima by frequency calculations, computed at the RHF/3-21G level of theory

Conformer $[\chi_i^2, \delta_i^3, \delta_i^4]$	$F1$	$F2$	$F3$
$g^+[g^+g^+]$	26.1996	55.0869	76.6393
$g^+[g^+a]$	28.9924	53.1113	77.3137
$g^+[ag^-]$	26.9605	50.9243	74.1477
$g^+[g^-g^-]$	27.7951	51.3511	75.0869
$g^-[g^+g^+]$	33.198	63.2129	77.9338
$g^-[g^+a]$	32.0525	62.0061	79.7404
$g^-[ag^-]$	32.9831	59.6958	76.5724
$g^-[g^-g^-]$	32.0124	60.1255	76.6985

4. Conclusion

The minima for the three dihedrals χ_i^2, δ_i^3 and δ_i^4 in adenosine was found at g^+, g^+ and a using the RHF/3-21G level of theory. The positively charged adenosine moiety was also optimized to $(\chi_i^2, \delta_i^3, \delta_i^4) = (g^+, g^-, g^-)$ at the RHF/3-21G level of theory. All minima were confirmed by frequency calculations. The next step would be to bring the converged conformations to higher levels of theory, and begin combining this fragment with a structurally optimized pyrophosphate bridge.

In this way, we hope to completely characterize each portion of NAD^+ so that we may integrate the segments and fully develop an ab initio model for the molecule. With a better understanding of the oxidized form of NAD, ab initio studies of NADH and its analogues can be easily conducted, allowing for further research encompassing the computational, biochemical and pharmacological fields.

Acknowledgements

This work was supported with grants from the Global Institute Of COmputational Molecular and Materials Sciences (GIOCOMMS), Toronto, Ont., Canada. Thanks are extended to Christopher N.J. Marai, Jeremy H. Keller and David H. Setiadi for helpful discussion. One of the authors (IGC) wishes to thank the Ministry of Education for a Szent-Györgyi Visiting Professorship.

References

[1] L. Stryer, Biochemistry, Freeman, New York, 1988, pp. 320–321.

[2] V. Shneyvays, L.K. Mamedora, D. Leshem, Exp. Clin. Cardiol. 7 (2002) 138–145.

[3] P. Acharya, J. Chattopadhyaya, J. Org. Chem. 67 (2002) 1852.

[4] S.K. Lau, G.A. Chass, S. Lovas, B. Penke. Submitted for publication.

[5] T. Ishida, M. Tarui, Y. In, M. Ogiyama, M. Doi, M. Inoue, FEBS Lett. 333 (1993) 214.

[6] Y. Shimohigashi, I. Maeda, T. Nose, K. Ikesue, H. Sakamoto, T. Ogawa, Y. Ide, M. Kawahara, T. Nezu, Y. Terada, K. Kawano, M. Ohno, J. Chem. Soc., Perkin Lett. 1 2479 (1996).

[7] P.E. Smith, J.J. Tanner, J. Mol. Recognit. 13 (2000) 27.

[8] M.J. Frisch, G.W. Trucks, H.B. Schlegel, G.E. Scuseria, M.A. Robb, J.R. Cheeseman, V.G. Zakrzewski, J.A. Montgomery, Jr., R.E. Stratmann, J.C. Burant, S. Dapprich, J.M. Millam, A.D. Daniels, K.N. Kudin, M.C. Strain, Ö. Farkas, J. Tomasi, V. Barone, M. Cossi, R. Cammi, B. Mennucci, C. Pomelli, C. Adamo, S. Clifford, J. Ochterski, G.A. Petersson, P.Y. Ayala, Q. Cui, K. Morokuma, D.K. Malick, A.D. Rabuck, K. Raghavachari, J.B. Foresman, J. Cioslowski, J.V. Ortiz, A.G. Baboul, B.B. Stefanov, G. Liu, A. Liashenko, P. Piskorz, I. Komaromi, R. Gomperts, R.L. Martin, D.J. Fox, T. Keith, M.A. Al-Laham, C.Y. Peng, A. Nanayakkara, C. Gonzalez, M. Challacombe, P.M.W. Gill, B.G. Johnson, W. Chen, M.W. Wong, J.L. Andres, M. Head-Gordon, E.S. Replogle, J.A. Pople, Gaussian 98 (Revision A.9), Gaussian, Inc., Pittsburgh PA, 1998.

[9] B. Guillot, C. Jelsch, C. Lecomte, Acta Cryst. C56 (2000) 726.

[10] B. Guillot, C. Lecomte, A. Cousson, C. Scherf, C. Jelsch, Acta Cryst. D57 (2001) 981.

[11] Y-D. Wu, K.N. Houk, J. Am. Chem. Soc. 113 (1991) 58.

[12] S.D. Kahn, W.J. Hehre, J. Am. Chem. Soc. 109 (1987) 666.

ELSEVIER

Journal of Molecular Structure (Theochem) 666–667 (2003) 439–443

www.elsevier.com/locate/theochem

Exploratory study on the full conformation space of α-tocopherol and its selected congeners

David H. Setiadi[a,b,*], G.A. Chass[a,b,c], Joseph C.P. Koo[a,b], Botond Penke[d,e],
Imre G. Csizmadia[a,b,d]

[a]*Global Institute of Computational Molecular and Materials Science (GIOCOMMS), 1422 Edenrose Street, Mississauga,
Ont., Canada, L5V 1H3*
[b]*Department of Chemistry, University of Toronto, Lash Miller Chemical Laboratories, 80 St George St., Toronto, Ont., Canada, M5S 3H6*
[c]*Department of Biomedical Sciences, Creighton University, 2500 California Plaza, Omaha, NE 68178, USA*
[d]*Department of Medical Chemistry, University of Szeged, Dóm tér 8, 6720 Szeged, Hungary*
[e]*Protein Chemistry Research Group, Hungarian Academy of Sciences, University of Szeged, Dóm tér 8, 6720 Szeged, Hungary*

Abstract

Preliminary results for a full and comprehensive study of the tocopherol family of compounds. Previous studies have allowed for the full modelling of α-tocopherol as well as its S and Se containing congeners to be subjected to ab initio [RHF/3-21G and RHF/6-31G(d)] and DFT [B3LYP/6-31G(d)] computation. Molecular geometries with full optimized total energies were determined. Initial discussion for trends of the side-chain, which has included computation of helical tails as well as the effect on ring stabilities is also investigated.
© 2003 Elsevier B.V. All rights reserved.

Keywords: Tocopherol; Tocotrienol; Stereo isomers; Ab inito; Congeners; Molecular structure

1. Introduction

Vitamin E [1] is made up of two families of compounds: tocopherols and tocotrienols [2]. Both families consist of a chroman [benzpyran] ring structure and a side-chain. The side-chain has the characteristics of an isoprenoid skeleton, typical of terpenes with some members of the tocopherol families having saturated side-chains. There are also methyl substituted homologues of tocopherol which

can be seen in Fig. 1. There also exists eight different stereo isomers for each homologue of tocopherol with known different activities associated to them.

Past molecular computational studies have concentrated on the fused ring system and the different stabilities associated with hetero-atom substitution of the chroman molecule. It has also been suggested [3] that the selenium congener of α-tocopherol (may be a very effective antioxidant. The ring-closing and ring-opening mechanisms and transition states of ring structure was also investigated.

It has also been suggested that in a biological system where vitamin E is recycled, a single α-tocopherol molecule may convert numerous HOO· radicals to H_2O_2 [4]. This accumulation of peroxide

---

* Corresponding author. Address: Global Institute of Computational Molecular and Materials Science (GIOCOMMS), 1422 Edenrose Street, Mississauga, Ont., Canada, L5V 1H3.

E-mail addresses: dsetiadi@giocomms.org (D.H. Setiadi); gchass@giocomms.org (G.A. Chass).

0166-1280/$ - see front matter © 2003 Elsevier B.V. All rights reserved.
doi:10.1016/j.theochem.2003.08.051

molecules may be referred to as a 'peroxide traffic jam' that could lead to the pro-oxidant effect of Vitamin E.

Conformational analysis on vitamin E models with both the fused ring and the side-chain tail undetermined in the past by ab initio computations is now investigated with the premise of further understanding the role of vitamin E in eliminating 'reactive oxygen species' ROS [5].

2. Methods

As a progression from previous studies where concentration was placed on the fused ring systems, the full model of the tocopherol family of compounds is now studied. Outline of previous models studied is shown in Fig. 1.

Since the previous calculations were modelled in a modular, scalable and reusable manner, it is now possible to add the side-chain tail to the fused ring.

Molecular computations were performed using the Gaussian 98 [6] program package. Preliminary conformers were constructed from initial model A to the full tocopherol model. Energy is initially minimized at the RHF/3-21G level of theory. These minima are brought through to the RHF/6-31G(d) and B3LYP/6-31G(d) levels of theory.

3. Results and discussions

The initial findings of the conformational study of α-tocopherol are shown in Table 1. Considering the nature of this publication, discussion will be kept to a minimum and only focus on the side-chain tail structure. A full analysis of both fused ring, and side-chain will provided in future work.

The full α-tocopherol along with its congeners have geometry optimized stable energy conformers with all dihedrals of the side-chain in the anti position. Fig. 2a shows the fully extended alpha-tocopherol model optimized at the B3LYP/6-31g(d) level of theory with all side-chain dihedrals in the anti position. Fig. 2b shows geometry optimized energy conformer of the hetero-atom substitution of selenium computed at the RHF/6-31g(d) level of theory with all side-chain dihedrals in the anti position.

A comparison is made of the sequence of stabilization for various hetero-atoms. The geometrical

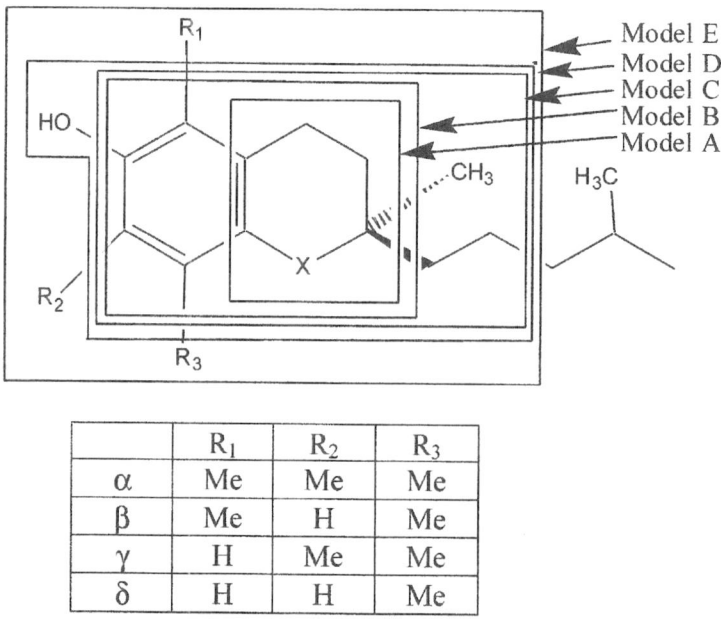

	R_1	R_2	R_3
α	Me	Me	Me
β	Me	H	Me
γ	H	Me	Me
δ	H	H	Me

Fig. 1. Model map used for previous study of tocopherol families and selected congeners.

Table 1
Optimized total energies and side-chain dihedral angles for stable conformers ($\lambda = 0$) of α-tocopherol and its selected congeners

X	Level of theory	Dihedral trend	Selected dihedral angles												Total E (Har)	Rel. E (kcal mol^{-1})
			$DA_{(1)}$	$DB_{(1)}$	$DC_{(1)}$	$DD_{(1)}$	$DA_{(2)}$	$DB_{(2)}$	$DC_{(2)}$	$DD_{(2)}$	$DA_{(3)}$	$DB_{(3)}$	$DC_{(3)}$	$DD_{(3)}$		
O	RHF/3-21G	a	173.35	187.27	183.68	170.94	189.24	175.99	183.97	170.47	189.34	175.92	183.52	171.57	−1269.99688	–
O	RHF/6-31G(d)	a	178.35	187.92	186.73	172.06	189.07	174.86	185.25	171.06	188.77	174.60	184.55	172.26	−1277.04046	–
O	B3LYP/6-31G(d)	a	176.52	185.55	184.65	171.76	189.31	175.40	184.25	170.88	188.58	175.11	183.21	171.70	−1285.69823	–
S	RHF/3-21G	a	175.63	183.03	184.33	171.00	189.35	176.01	183.99	170.54	189.46	175.98	183.56	171.57	−1591.10457	–
S	RHF/6-31G(d)	a	177.44	183.10	186.75	172.00	189.12	174.86	185.24	171.04	188.74	174.59	184.53	172.25	−1599.68471	–
S	B3LYP/6-31G(d)	a	175.53	179.99	184.03	169.90	187.99	174.81	183.40	170.40	188.46	175.63	183.66	172.02	−1608.66284	–
Se	RHF/3-21G	a	175.06	181.41	184.29	171.00	189.45	175.98	183.99	170.54	189.43	175.92	183.54	171.61	−3584.04040	–
Se	RHF/6-31G(d)	a	176.59	183.86	186.47	171.95	189.09	174.90	185.27	171.09	188.78	174.60	184.55	172.26	−3599.75844	–
Se	B3LYP/6-31G(d)	a	Under computation													
O	RHF/3-21g	a	173.35	187.27	183.68	170.94	189.24	175.99	183.97	170.47	189.34	175.92	183.52	171.57	−1269.99688	0.000
O	RHF/3-21g	g^+	48.40	66.68	168.43	56.50	59.95	66.79	173.81	58.42	56.93	60.03	60.32	62.18	−1269.99031	4.122
O	RHF/3-21g	g^+	41.71	65.21	101.24	62.52	56.07	61.30	172.01	57.64	57.32	58.56	58.65	61.42	−1269.98435	3.741
O	RHF/3-21g	g^-	−68.35	−104.44	−68.97	−60.51	−45.90	−69.95	−57.45	−56.49	−63.48	−101.40	−94.10	−67.57	−1269.97133	8.168

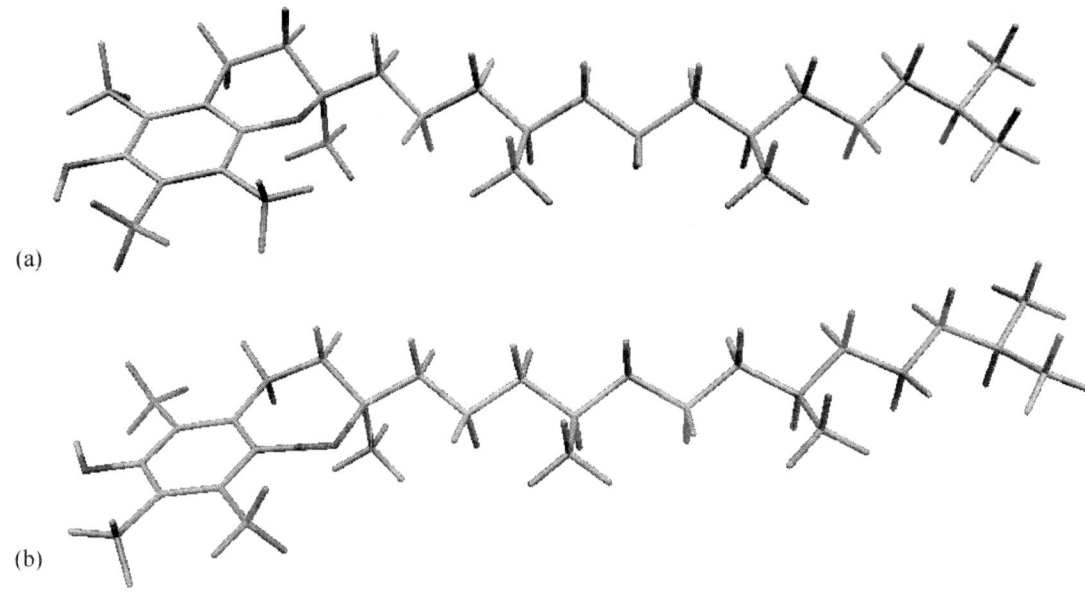

Fig. 2. (a) B3LYP/6-31G(d) optimized minimum energy structure for full extended tail structure of α-tocopherol and (b) B3LYP/6-31G(d) optimized minimum energy of its selenium congener.

parameters also show a non-monotonic change. The R–C–Me bond angles at the ring stereo center varied the following way: 112.2° (O), 111.6° (S), 112.2° (Se). This is very similar to previous reported results of the similar Et–C–Me bond angle for Model E which were also non-monotonic 112.1° (O), 111.6° (S), 112.4° (Se). It is interesting to note that when comparing the dihedral angles positioned further

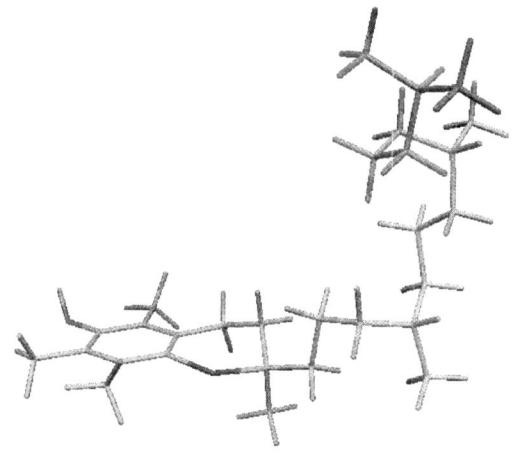

Fig. 3. B3LYP/6-31G(d) optimized minimum energy structure for g^+ trend of the side-chain tail for the selenium congener of α-tocopherol.

away from the fused ring $D[A,B,C,D]_{(2)}$– $D[A,B,C,D]_{(3)}$, there is no significant difference between the dihedrals of any of the congeners while there is a monotonic change for the dihedral closest to the fused ring.

Fig. 3 shows the selenium congener of the α-tocopherol model optimized at B3LYP/6-31g(d) with a g^+ helices where the trend of the atoms in the backbone of the α-tocopherol tail is to conform to the g^+ position.

The model of α-tocopherol with all side-chain tail conformers in the anti position appears to have the least energy of stabilization at this point. While the model of α-tocopherol with all side-chain tail conformers in the g^- position appears to have the largest energy of stabilization at this point 8.168 kcal mol^{-1} from the lowest energy value computer at the RHF/3-21g level of theory.

4. Conclusions

Taking into consideration the results shown is only a partial set of a larger, comprehensive study, further results are expected. The full study will encompass all eight stereo isomers of all homologues and possible congeners of the tocopherol model coupled with previously reported results. This will allow a complete analysis of the full conformational states of tocopherol.

A complete study of ring stability with respect to gradual methylation as well as the trends of biological activity should also be studied. It is important to note that past studies have indicated an inverse relationship between ring stability and biological activity. A comprehensive structure activity analysis of the full model will provide a clearer answer.

With the full conformation space of the Vitamin E models defined, the doors are open for modelling the mechanisms of the antioxidant behaviour of tocopherol. Computations of the full structure of the tocopherol family with respect to exothermic free radical oxidative ring opening and endothermic ionic ring closing can be examined. Computational modelling may also be able to shed light on Vitamin E's role as an oxidative protector of LDL [7] and its relationships with enzymes such as gluthathione peroxidase and catalase, and other bioactive molecules such as Vitamin C.

Acknowledgements

The authors would like to thank the Global Institute of Computational Molecular and Materials Sciences, Aristo Systems and Uniseti Inc. for their support and resources. One of the authors (I.G.C.) wishes to thank the Ministry of Education for a Szent-Györgyi Visiting Professorship. We would especially like to thank all our many friends in many places, the work would not be possible without them. The author would also like to thank Lucha for the constant support.

References

[1] M. Evans, K.S. Bishop, Science 55 (1922) 650.
[2] J.B. Bauernfeind, Vitamin E: a Comprehensive Treatise, Marcel Dekker, New York, 1980, pp. 99–167.
[3] N. Al-Maharik, L. Engman, J. Malmstrom, C. Schiesser, Intramolecular homolytic substitution at selenium: synthesis of novel selenium-containing vitamin E analogues, J. Org. Chem. 66 (2001) 6286–6290.
[4] D.H. Setiadi, G.A. Chass, L.L. Torday, A. Varro, J.Gy. Papp, Vitamin E models. Can the anti-oxidant and pro-oxidant dichotomy of [alpha]-tocopherol be related to ionic ring closing and radical ring opening redox reactions?, J. Mol. Struct.: Theochem. 620 (2003) 93–106.
[5] B.N. Ames, M.K. Shigenaga, in: J.G. Scandalios (Ed.), Molecular Biology of Free Radical Scavenging Systems, Cold Spring Harbor Laboratory Press, New York, 1992, p. 1.
[6] M.J. Frisch, G.W. Trucks, H.B. Schlegel, G.E. Scuseria, M.A. Robb, J.R. Cheeseman, V.G. Zakrzewski, J.A. Montgomery, Jr., R.E. Stratmann, J.C. Burant, S. Dapprich, J.M. Millam, A.D. Daniels, K.N. Kudin, M.C. Strain, Ö. Farkas, J. Tomasi, V. Barone, M. Cossi, R. Cammi, B. Mennucci, C. Pomelli, C. Adamo, S. Clifford, J. Ochterski, G.A. Petersson, P.Y. Ayala, Q. Cui, K. Morokuma, D.K. Malick, A.D. Rabuck, K. Raghavachari, J.B. Foresman, J. Cioslowski, J.V. Ortiz, A.G. Baboul, B.B. Stefanov, G. Liu, A. Liashenko, P. Piskorz, I. Komaromi, R. Gomperts, R.L. Martin, D.J. Fox, T. Keith, M.A. Al-Laham, C.Y. Peng, A. Nanayakkara, M. Challacombe, P.M.W. Gill, B. Johnson, W. Chen, M.W. Wong, J.L. Andres, C. Gonzalez, M. Head-Gordon, E.S. Replogle, J.A. Pople, Gaussian Inc., Pittsburgh, PA, 1998.
[7] M.N. Diaz, B. Frei, J.A. Vita, J.F. Keaney, Antioxidants and atherosclerotic heart disease, New Engl. J. Med. 337 (1997) 408–416.

ELSEVIER

Journal of Molecular Structure (Theochem) 666–667 (2003) 445–449

THEO
CHEM

www.elsevier.com/locate/theochem

Molecular orbital computations on lipids: modular numbering

Jacqueline M.S. Law[a,b,c,*], Joseph C.P. Koo[a,b], David H. Setiadi[a],
Gregory A. Chass[a,b,d,e], Bela Viskolcz[f], Imre G. Csizmadia[a,b,g]

[a]Global Institute Of COmputational Molecular and Materials Science, 1422 Edenrose St., Mississauga, Ont., Canada, L5V 1H3
[b]Department of Chemistry, University of Toronto, Toronto, Ont., Canada M5S 3H6
[c]Simbiosys Inc. 135 Queensplate Drive, Toronto Ont., Canada
[d]Institut de Science et d'Ingénierie Supramoléculaires, 8, allée Gaspard Monge, BP 70028, 67083 Strasbourg Cedex, France
[e]Department of Biomedical Sciences, Creighton University, 2500 California plaza, Omaha, NE 68178, USA
[f]Department of Chemistry, JGyTFK, University of Szeged, P.O. Box 396, Szeged H-6701, Hungary
[g]Department of Medical Chemistry, University of Szeged, Dóm tér 8, 6720 Szeged, Hungary

Abstract

Ab initio molecular orbital calculations have been carried out on selected phospholipids models to explore their structural properties. A modular numbering of constituent molecular components has been employed to effectively model all MDCA-predicted conformers. This system defines a phospholipid into five fragments consisting of the glycerol, the phosphate group (for phospholipids) and three substituents (X-group and two fatty acid chains). Each fragment is converted into distinct section of an internal coordinates matrix. The relative spatial orientation of all atoms in a given fragment will all be defined in an explicit manner. With a multi-lipid system, several of these modules may be used as a macro-module representing one lipid molecule. Subsequently, several macro-modules may be assembled into a lipid bilayer model. All macro-modules may then be combined to create an internal coordinate matrix describing a multi-lipid system without the loss of an explicit definition of every 3N-6 degree of freedom as well as intermolecular coupling and interaction. Molecular orbitals computations are completed on four selected conformations of glycerol. The results reveal that the dihedral values are not ideal as predicted by MDCA, implying that dihedrals properties are coupled to other inter-atomic interactions within the glycerol module.
© 2003 Elsevier B.V. All rights reserved.

Keywords: Computations; Lipids; Modular numbering

1. Introduction

With the rapid development of technology, preliminary biological analyses are performed on computers. Therefore, a systematic method for such studies becomes necessary for to achieve high precision.

The conversion of peptide structural geometry into explicit numbering definitions may be attained through the use of an established systematic modular numbering system [1]. As a result, all atoms from one peptide residue are in an ordered and independent section of the numbering definition [1]. In turn, organized modeling of larger peptide models are made highly efficient in this manner, ensuring effective use of computational resources. Another structural 'building block', larger than an amino acid, is a phospholipid. Unlike the peptide models, phospholipids require no protective end groups for increased accuracy in modeling and it

* Corresponding author.
   E-mail addresses: jlaw@giocomms.org (J.M.S. Law); joseph.koo@utoronto.ca (J.C.P. Koo); dsetiadi@giocomms.org (D.H. Setiadi); gchass@fixy.org (G.A. Chass); icsizmad@fixy.org (I.G. Csizmadia).

0166-1280/$ - see front matter © 2003 Elsevier B.V. All rights reserved.
doi:10.1016/j.theochem.2003.08.053

comprised of 5 structural moieties (glycerol, phosphate fatty acid chains and an X-group). Therefore, the peptide model cannot be transposed to phospholipid systems. In this paper, we will discuss a numerical standard, which may be employed in order to generate an explicit definition of all 3N-6 degrees of intramolecular freedom. The later sections of this paper addresses the application of the standard to modeling glycerol and larger (phospho)lipid systems.

2. Method

2.1. Modular numbering system

This modular system comprises of a set of formulae that converts a chemical structure into groups of explicit coordinates used in ab initio calculations, easily accommodating changes in all moieties on

the backbone. Fig. 1 is a schematic representation of the numbering system including sets of formulations that may be applied to any phospholipid model. The constituent atomic nuclei are separated into five fragments. In fragments III, IV, and V, the atoms are separated into hydrogen and non-hydrogen atoms. In each moiety, all of the non-hydrogen atoms are numbered first followed by the hydrogen atoms. In this way, the hydrogen atoms, which are most likely to be substituted in larger models may be changed without gross perturbation of the remainder of the model [2]. Fragment I and II are numbered first as they are conserved between different phospholipids. Conversely, fragments IV and V could include moieties of varying saturation, thus for sake of convenience are numbered later.

The standard may be used for any type of glycerolipids with any kind of substitutions. It is also applicable to systems with more than one lipid and it

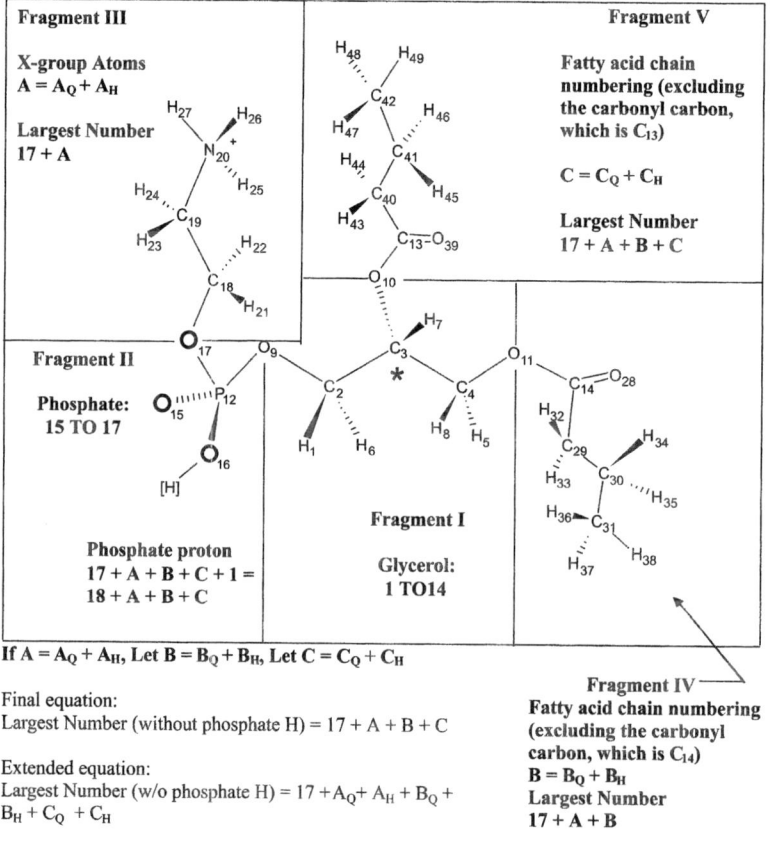

Fig. 1. Schematic drawing of the numbering system for lipids.

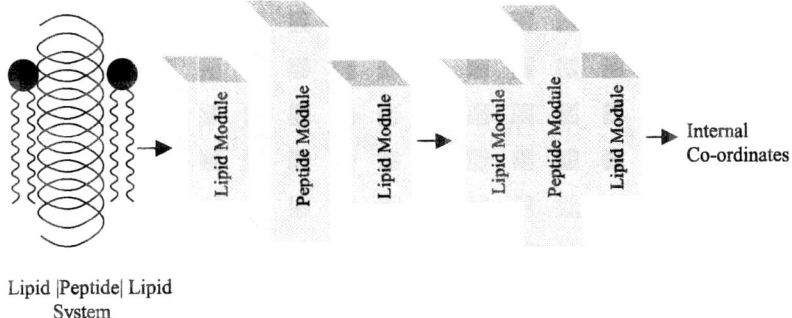

Fig. 2. Combination of different module to convert structural formulae to internal coordinates.

can be used along with other numbering systems (i.e. peptide numbering system when modeling transmembrane peptides). Therefore, when modeling a system with many lipids and peptides, the structural formulae can first be broken down to many separate of lipids and peptides. Consequently, each module of the internal-coordinate matrix will represent a molecular part in the system. Fig. 2 shows how modular numbering can be used to convert a system of structural formulae into a set of coordinates. With such organized design, the process of structure-coordinate as well as the substitution of different fragments to build a molecule can easily be automated. Furthermore, pre-optimized fragments from a database can be used to assemble a layer model system [1]. Due to the explicit definitions, highly-dimensional QSAR (i.e. 3D analyses) become quite practical.

3. Results and discussion

3.1. Glycerol

Glycerol is the backbone structure of many different types of lipids. By itself, it is a molecule that can for various types of inter and intra-molecular hydrogen bonds. The numerical definition of glycerol follows that of Fragment I (Fig. 1). Fig. 3 is a diagram showing how glycerol is numbered along with the defined dihedrals ($\phi, \psi, \chi^1, \chi^2, \chi^3$). Previous studies have been completed on the investigation of the distribution of glycerol conformations through the use of Molecular Dynamics, DFT, and Frequency Calculations [3,4]. For example, ϕ and ψ are the backbone dihedrals of

the lipid molecule; dihdedrals χ^1, χ^2, χ^3 are the first dihedrals of the fatty acid chain. Using this system, all conformations of glycerol predicted by MDCA can be computed and their results tabulated in a systematic manner; without excessive used of computational resources. In turn, trends in such tabulations can help in identification of intramolecular interactions that contribute to stabilizing such conformations.

Four matrices are used to model four conformations. Since all structures are converted into four separate internal coordinate matrices following the same numbering fragment I (Fig. 1), all dihedrals are found in the same segments of matrices. Fig. 4 depicts the sections (squared off sections) of the internal coordinate matrices, which contains the five dihedrals studied. Fully relaxed optimizations were performed on these selected structures at RHF/3-21 g level of theory using GAUSSIAN98 [5]. Table 1 depicts the optimized results of the selected conformations. The results reveal that the dihedral values are not ideal as predicted by MDCA. Therefore,

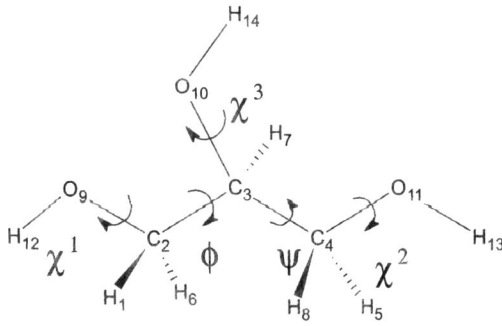

Fig. 3. Schematic drawing of glycerol with numbering.

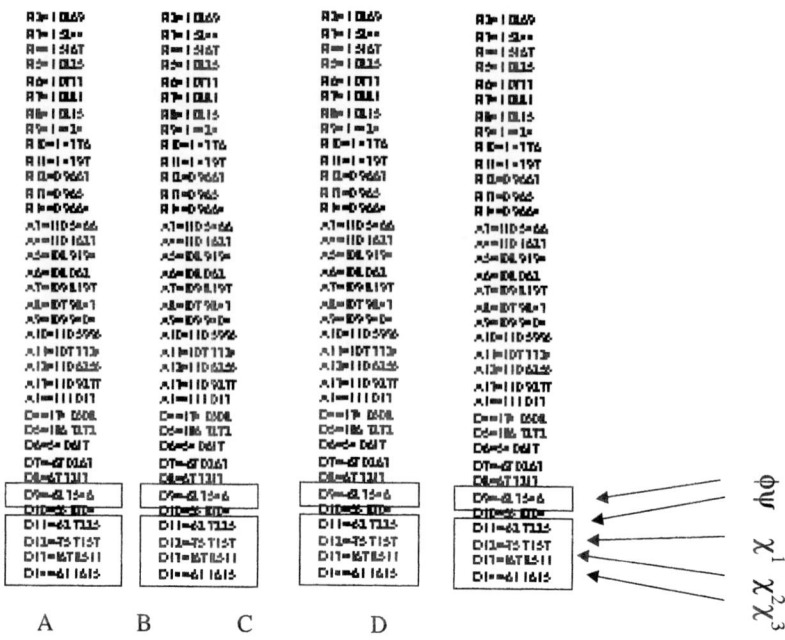

A   B   C   D

Fig. 4. Internal coordinates set up using the numbering system; the five dihedrals in the four structures are identified with boxes.

dihedrals properties are probably coupled to other intramolecular interaction in glycerol.

With such modular definition of internal coordinates, different lipid models can be swapped and put together with minimum amount of modifications to the newly assemble matrix. At any time, all modules may be used in any other lipid module. The structural results and definition from one fully optimized phospholipid may be used to model any other phospholipid. Despite a need to re-compute and re-optimize the new models, bond lengths, bond angles and dihedrals are relatively well conserved within lipid models.

4. Conclusion

Although developed, the phospholipid numbering system may be subjected to changes and modifications as other types of lipids are modeled, which may not necessarily contain the glycerol group as the backbone. It should be noted that there are no phosphate groups in fats and therefore only four groups exists in such lipids. With the development of the phospholipids and peptide modules, molecular orbital investigation of lipid-peptide system can be performed systematically by changing the structures into a combination of these different blocks.

Table 1
Dihedral values for 5 optimized structures

Structures	Dihedrals				
	ϕ	ψ	χ^1	χ^2	χ^3
$g^- g^+ g^- a g^-$	-62.354	-52.723	-75.716	167.851	-61.162
$g^+ g^+ g^+ g^+ a$	64.757	66.457	51.606	75.818	-177.803
$g^+ a a g^- g^-$	47.391	171.632	177.907	-62.906	-64.751
$g^+ a g^+ a a$	61.639	-164.986	49.970	171.785	160.541

5. Further applications

The use of the lipid numbering system has been used for the study of glycerol and glycerol-3-phosphates while the peptide numbering system has been extended to di- and tri-peptides structures [6]. The use of this system not only enhances the effectiveness of backbone analysis, but it is also applicable for side-chain analysis as well as comparative analysis of two blocks/modules [7]. More generally, molecular modeling has become a prominent tool for the pharmaceutical industry to analyze micro- and macromolecules during the drug discovery process. With the human genome being completely sequenced [8], genomic targets, whose biological and pharmacological consequences were previously unknown, become available for automated screening by potential lead compounds. In order to minimize cost and maximize probability of success, a chemogenomics strategy for drug discovery was thoroughly described [9]. In this approach, thousands of lead compounds were synthesized in virtual libraries and were systematically examined for activities against the genomic target of interest. It is therefore important to realize that the universal numbering system described in this paper may allow ab initio computations to be more easily integrated into such chemogenomic approaches for drug discovery. The modular numbering system described here may allow both chemical and physical characteristics, such as sites of hydrogen bond acceptors or donors, electrophilic regions, as well as aromatic and aliphatic hydrophobes, to be thoroughly and consistently described for any specific compound. In turn, a compound's pharmacology can be fully represented in a computational setting, a role that is highly comparable to the so-called molecular descriptors used in conventional chemogenomics methods. Although the integration of ab initio computations to such drug discovery settings may require extensive scripting that allows for automation, the modular numbering system described in this paper may act as a cornerstone for such pioneer works in the near future.

Acknowledgements

The authors thank Christopher N.J. Marai, Jack C.C. Liao and Jeremy Keller for helpful discussion and preparation of tables and figures. The authors would also like to thank the Global Institute Of Computational Molecular and Material Science for support in the production of this work.

One of the authors (IGC) wishes to thank the Ministry of Education for a Szent-Györgyi Visiting Professorship.

References

[1] G.A. Chass, M.A. Sahai, J.M.S. Law, S. Lovas, O. Farkas, A. Perczel, J. Rivail, I.G. Csizmadia, Toward a computed peptide structure database: the role of a universal atomic numbering system of amino acids in peptides and internal hierarchy of database, Int. J. Quant. Chem. 90 (2002) 933–968.

[2] J.M.S. Law, G.A. Chass, L.L. Torday, J.Gy. Papp, Molecular computations on lipids: a numbering system for phospholipids and triglyceride, J. Mol. Struct. (THEOCHEM) 619 (2002) 1–20R.

[3] R. Chelli, P. Procacci, G. Cardini, S. Califano, Glycerol condensed phases part II: a molecular dynamics study of the conformational structure and hydrogen bonding, Phys. Chem. Chem. Phys. 1 (1999) 879–885.

[4] R. Chelli, F.L. Gervasio, C. Gellini, P. Procacci, G. Cardini, V. Schettino, Density functional calculation of structural and vibrational properties of glycerol, J. Phys. Chem. A 104 (2000) 5351–5357.

[5] GAUSSIAN98.

[6] J.C.C. Liao, J.C. Chua, G.A. Chass, A. Perczel, A. Varro, J.Gy. Papp, An assessment of the chiral environment created by adjacent D- and L-alanyl residues on a glycine unit within the tripeptide N-Ac-Ala-Gly-Ala-NHMe: an ab initio exploratory study, J. Mol. Struct. (THEOCHEM) 621 (2003) 163–187.

[7] G.A. Chass, S. Lovas, R.F. Murphy, I.G. Csizmadia, The role of enhanced aromatic π-electron donating aptitude of the tyrosyl sidechain with respect to that of phenylalanyl in intramolecular interactions, Eur. Phys. J. D 20 (2002) 481–497.

[8] J.D. McPherson, et al., A physical map of the human genome, Nature 409 (2001) 934–941.

[9] D.K. Agrafiotis, V.S. Lobanov, F.R. Salemme, Combinatorial informatics in the post-genomics era, Nat. Rev. Drug Dis. 1 (2002) 337–346.

ELSEVIER

Journal of Molecular Structure (Theochem) 666–667 (2003) 451–453

THEO
CHEM

www.elsevier.com/locate/theochem

Modeling copper-containing enzyme mimics

I. Szilágyi[a], G. Nagy[b], K. Hernadi[c], I. Labádi[a], I. Pálinkó[b,*]

[a]Department of Inorganic and Analytical Chemistry, University of Szeged, Dóm tér 7, Szeged H-6720, Hungary
[b]Department of Organic Chemistry, University of Szeged, Dóm tér 8, Szeged H-6720, Hungary
[c]Department of Applied and Environmental Chemistry, University of Szeged, Rerrich B tér 1, Szeged H-6720, Hungary

Abstract

The ligands in the active centre of Cu–Zn superoxide dismutase was replaced with smaller and simpler ones, but still preserving the catalytically most important features of the original ones. The structure of the bare complex and that of the covalently grafted one was modeled by ab initio quantum chemical method. Tetrahedral geometric arrangement was found around the Cu(II) as well as the Zn(II) central ions. The ligands were loosely coupled allowing easy insertion of the incoming reactants.
© 2003 Elsevier B.V. All rights reserved.

Keywords: Cu–Zn superoxide dismutase mimic; Covalent grafting; Molecular modeling; HF 3-21 G ab initio method

1. Introduction

Enzymes are the catalysts of processes occurring in living systems. They work under mild conditions with high turnover rates and extremely high selectivities—the dreams of any catalytic chemist. Obviously, copying the essential features of these materials would lead to these dreams come true. However, choosing the essential features is a challenging task. If one tries to take them into account finds some sure points, like the structure of the prosthetic group, the mode of its anchoring to the host protein, the structural mobility of the host protein in the physiological salt solution and upon the action of the incoming reactants. When trying to mimic the structural characteristics and the actions of enzymes, the stepwise inclusion of the above factors will lead to

model systems of increasing complexity and hopefully to substances that approach in properties the original catalyst.

In this work, the structures of a coordination compound modeling the Cu–Zn superoxide dismutase (SOD) and its silica-anchored derivative are probed computationally. It was found experimentally that Cu–Zn coordination compound and its immobilised versions do have both *catalase* and *SOD* activities [1,2], thus, it is worthwhile to collect information on their structures by the means of computational chemistry as well.

2. Computational methods

Calculations were performed on two models. The Cu–Zn SOD model was the Cu(II)-diethylenetriamino-μ-imidazolato-Zn(II)-*tris*-aminoethyl amine complex (Fig. 1(a)) and its covalently grafted derivative. Grafting was through a propyl group onto a silica cluster closed by OH groups (Fig. 1(b)).

* Corresponding author. Address: Department of Organic and Analytical Chemistry, University of Szeged, Dóm tér 8, Szeged H-6720, Hungary. Tel.: +36-62-544-288; fax: +36-62-544-200.

E-mail address: palinko@chem.u-szeged.hu (I. Pálinkó).

0166-1280/$ - see front matter © 2003 Elsevier B.V. All rights reserved.
doi:10.1016/j.theochem.2003.08.054

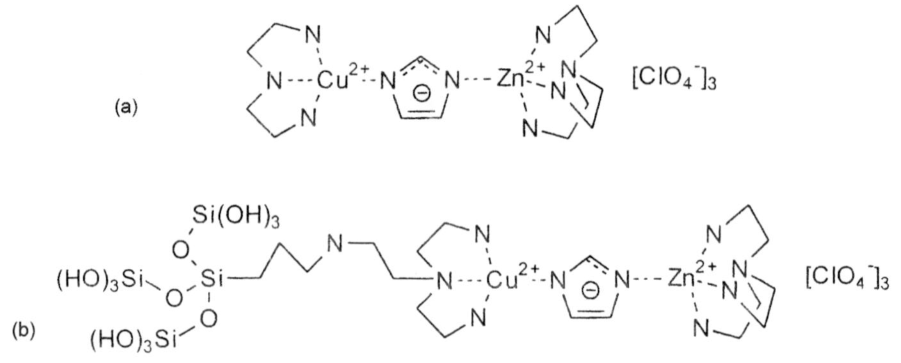

(a)

(b)

Fig. 1. The Cu(II)-diethylenetriamino-μ-imidazolato-Zn(II)-tris-aminoethyl amine complex: the schematic structures of (a) the bare complex and (b) the one grafted onto silica through covalent bonding.

The size of the cluster had to be limited to keep computation tractable.

Two kinds of calculation methods, both built in the HyperChem 7.0 package [3], were applied. First, full geometry optimisation was performed by the MM + force field, then the structures were further optimised at HF level applying the 3-21 G basis set. Terminating criterion was set to gradient smaller than 0.1. Calculating the second derivative gave assurance that minima were obtained indeed.

3. Results and discussion

The structure of the Cu–Zn SOD enzyme is well known. Its active centre is depicted in Fig. 2.

The enzyme transforms the superoxide radical ion to peroxide and oxide ions. The copper ion in the active centre is believed to take part in the redox cycle of the reaction, while the role of the Zn(II) ion is to preserve the structure of the active centre. The dismutase activity can be achieved by applying analogous ligands to the real ones. This was done in the experimental study preceding this work [1,2]. The ligand for making the complex was applied before [4].

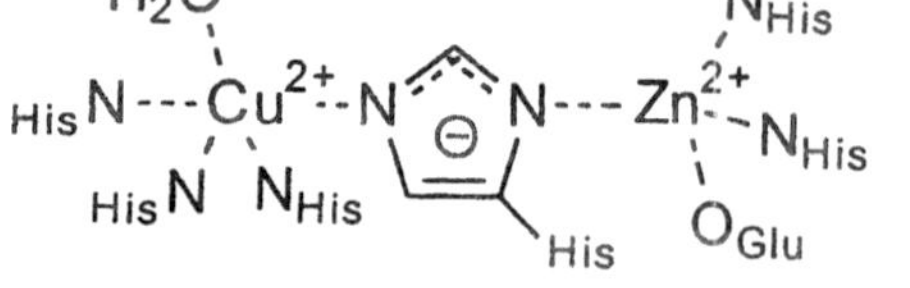

Fig. 2. The active centre of the Cu–Zn superoxide dismutase (SOD) enzyme.

Actually, the nitrogens of the histidine rings were kept and the imidazolium bridge between the copper and zinc ions was also maintained to preserve resemblance to the real complex. During our investigations, the model complex was immobilised on various supports (montmorillonite or MCM-41) and its *catalase* activity was studied. It was found that after certain induction period, it could decompose hydrogen peroxide if it was immobilised in montmorillonite, but it was inactive when the host was MCM-41. Immobilisation occurred via hydrogen bonding in both cases. In further experimental work, the preformed complex was immobilised the same way on silica gel as well [5]. The bare complex had *SOD* activity, but significantly poorer than the enzyme. However, upon immobilisation the activity increased significantly and on silica gel and in/on montmorillonite it became very good and fair, respectively.

The computed structure of the bare complex is shown in Fig. 3.

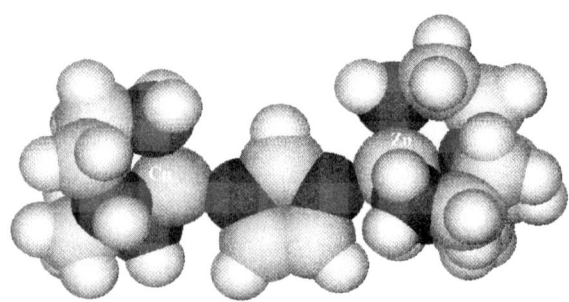

Fig. 3. The structure of the Cu(II)-diethylenetriamino-μ-imidazolato-Zn(II)-tris-aminoethyl amine complex optimised by the HF 3-21 G ab initio method.

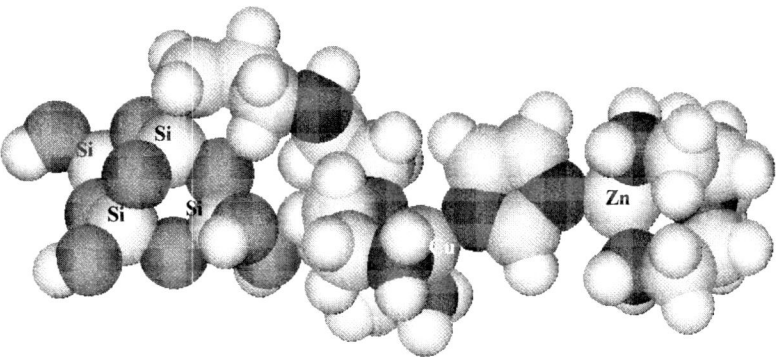

Fig. 4. The structure of the silica-anchored Cu(II)-diethylenetriamino-μ-imidazolato-Zn(II)-tris-aminoethyl amine complex optimised by the HF 3-21 G ab initio method.

It is to be seen that the geometric arrangement is a somewhat distorted tetrahedron around the Cu(II) ion just as around the Zn(II) ion. The coordination sites of the ligands are at 202 pm and that of the imidazolato bridge is 196 pm from the catalytically important central ion (the Cu(II) that is). The connection is loose enough for allowing geometry change around the ion by the incoming reactant.

Covalent grafting through a long enough spacer group (the propyl group that is) is expected to be advantageous compared to both the bare complex and the immobilised one when anchoring was done through adsorption/hydrogen bonding. Immobilisation may increase activity, even if it does not; it definitely makes the recovery of the catalyst easier. Immobilisation through covalent grafting affords more defined system than immobilisation via secondary forces. Moreover, it allows applying spacer group. If it is long enough, it results in significantly more flexibility than the other method(s) of anchoring. It means that the reactant molecules may access easier the active centre, thus, activity may increase significantly. The geometry optimised structure of the covalently anchored complex is displayed in Fig. 4.

It is to be seen that basically the overall geometric arrangement remained, the catalytically important central ion (the Cu(II) ion) is tetrahedrally coordinated, however, the ligands are somewhat closer (187–191 pm) in this complex than in the bare one. The arrangement may allow easy insertion of the incoming molecules to the coordination sphere, which may result in increased activity, however, it has to be checked experimentally.

4. Conclusion

The prosthetic group of the Cu–Zn SOD enzyme can be successfully mimicked by a complex containing the same central ions but simpler ligands than one can find in the real one. The ligands preserved the characteristics of those found in the enzyme, but being smaller and less complex made molecular modelling computationally less expensive, but still reliably pointing at possible reasons of catalytic activity.

Acknowledgements

This work was supported through grants from the National Research Fund of Hungary (T034184 and T034793). The financial help is highly appreciated.

References

[1] K. Hernadi, D. Méhn, I. Labádi, I. Pálinkó, Gy. Bál, E. Sitkei, I. Kiricsi, Stud. Surf. Sci. Catal. 142 (2002) 85.
[2] I. Labádi, I. Szilágyi, N.I. Jakab, K. Hernádi, I. Pálinkó, Mater. Sci. 21 (2003) 235.
[3] HyperChem 7.0, Hypercube, Inc., 2001, Gainesville, FL, USA.
[4] M. Sato, S. Nagae, M. Uehara, J. Nakaya, J. Chem. Soc., Chem. Commun. (1984) 1661.
[5] I. Szilágyi, I. Labádi, K. Hernadi, T. Kiss, I. Pálinkó, J. Biol. Inorg. Submitted for publication.

ELSEVIER

Journal of Molecular Structure (Theochem) 666–667 (2003) 455–467

THEO
CHEM

www.elsevier.com/locate/theochem

Molecular recognition and binding mechanism of *N*-alkyl-benzyltetrahydroisoquinolines to the D$_1$ dopamine receptor. A computational approach

F.D. Suvire[a], N. Cabedo[b], A. Chagraoui[c], M.A. Zamora[a], D. Cortes[b], R.D. Enriz[a,*]

[a]*Departamento de Química, Universidad Nacional de San Luis, Chacabuco 915, 5700 San Luis, Argentina*
[b]*Departamento de Farmacología, Farmacognosia y Farmacodinamia, Facultad de Farmacia,*
Universidad de Valencia, 46100 Burjasot, Valencia, Spain
[c]*Laboratorie de Physiologie, Faculté de Medicine-Pharmacie, Université de Roven, 76183 Roven, France*

Abstract

In order to better understand, at sub-molecular level, the minimal structural requirements for the recognition process in the inhibitory activity, a series of *N*-alkyl-benzyltetrahydroisoquinolines (BTHIQs) were examined as Dopamine D$_1$ receptor antagonist variants. According to the cardinal role of the electrostatic factors during this interaction, ab initio and density functional theory (DFT) calculations were performed for a better understanding of the recognition process at the sub-molecular level. RHF/3-21G, RHF/6-31G(d) and B3LYP/6-31G++(d,p) in the gas phase, plus DFT calculations using the IPCM solvation model were carried out for all the complexes. We simulate the electronic interactions between BTHIQs with its biological receptor in terms of smaller molecules. H$_3$C–COOH was used to mimic the side chain of Aspartic acid and CH$_3$OH mimicked the side chain of Serine; alternative moieties present on the BTHIQ derivatives were used as the different partenaires. Using the above mentioned computational model, we are able to interpret the basic behaviours and predict some additional features of BTHIQ–Dopamine D$_1$ receptor interaction.
© 2003 Elsevier B.V. All rights reserved.

Keywords: Ab initio and density functional theory calculations; Binding mechanism of *N*-alkyl-benzyltetrahydroisoquinolines; D$_1$-Dopamine receptor; Interactions including solvent effect

1. Introduction

The application of gene cloning techniques has allowed the identification of five dopamine receptor subtypes which can be classified into two classes: D$_1$-like dopamine receptors (D$_1$ and D$_5$) and D$_2$-like dopamine receptors (D$_2$, D$_3$ and D$_4$) [1,2]. The D$_2$-like dopamine receptors show high affinities for drugs (antagonist) used for the treatment of schizophrenia (antipsychotics) and those (agonist) used in the treatment of Parkinson's disease [1]. However, at present little is understood concerning the physiological significance of this receptor subtype multiplicity. Future research endeavours should lead to better understanding of these receptor subtypes and eventually to new drugs with more specific actions.

Previously we described the synthesis of (R)-norroefractine [3] a monophenolic unmethylated 1-benzyl-1,2,3,4-tetrahydroisoquinoline (BTHIQ). We

* Corresponding author. Fax: +54-2652-4311301.
E-mail address: denriz@unsl.edu.ar (R.D. Enriz).

0166-1280/$ - see front matter © 2003 Elsevier B.V. All rights reserved.
doi:10.1016/j.theochem.2003.08.070

have also accomplished the synthesis of racemic monophenolic N-alkyl-BTHIQs by a new method incorporating a 'one-pot' cyclization—reduction—alkylation sequence [4]. All those compounds were reported to bind to D_1 dopamine receptor (D_1-d-r) [3,4]. In addition an enantioselective synthesis of dopaminergic (R) and (S) benzyltetrahydroisoquinolines was previously reported [5]. However, in those papers the mechanism of action of BTHIQs at molecular level was not taken into account.

Although there are currently available interesting data about the general characteristics of the D_1-d-r, the molecular mechanism by which dopamine binding to the receptor induces G protein is still unknown. Thus, deduction of the characteristics of the D_1-d-r ligands of relatively small molecular dimensions (i.e. BTHIQs) should be a fruitful addition to the current knowledge of the D_1-d-r. Such a chemical approach to the molecular pharmacology of those compounds could give us a more detailed structural profile for the D_1-d-r which has not yet been entirely elucidated. While helpful empirical rules and correlations are emerging from in vitro activity studies, this is obviously a very complex phenomenon which will undoubtedly require the joint effort of many disciplines to be elucidated (including of course computational approach).

As it was said before, the mechanism by which dopamine bound to the receptor induces G protein activity is unknown but it most likely involves a cascade of intramolecular reactions. In particular, charged and conserved amino acid residues found in transmembrane domains should participate in dopamine recognition.

The molecular recognition and the binding mechanism of dopamine D_1 ligand into its biological receptors might be subdivided into two major steps (A,B) followed by a third phase (C) frequently referred to as an internal interaction:

A: Long range interaction (electrostatic preorientation).
B: Short-range interaction (electronic attachment followed by a 'fine-tune' via aromatic ring orientation) of the BTHIQs compounds.
C: Internal interaction (include fit, conformational readjustment of the protein molecules).

It is assumed that the driving force in the early preorientational phase (A) is electrostatic. This is due to the attraction between Asp 103 in transmembrane III and two Ser (199 and 202) in transmembrane domain V both of which could interact with the amine and phenolic hydroxyl groups of dopamine, respectively, [6]. This model was tested experimentally and proven valid although the two serine residues in transmembrane domain V differentially affect agonist binding [6].

In the present theoretical study, which is exploratory in its nature, we have focused our attention in the first A and B phases. The scope of the present paper does not include the final C binding phase.

Molecular recognition and the converse concept of specificity [7] are explained in mechanistic and reductionistic terms by a stereoelectronic complementarity between the binding molecule and the receptor [8]. In this context it is obvious that the knowledge of the stereoelectronic attributes and properties of BTHIQs compounds will contribute significantly to the elucidation of the mechanism of action at molecular level.

Molecular interactions between molecules are determined fundamentally by molecular size and shape and charge distribution, but when one inspects molecules in these terms one becomes aware that they rarely have a unique description and there may be several different forms or species in equilibrium. In this case an interesting question arises: if one changes drug structure to alter the equilibrium, can one relate the consequences to changes in biological activity? If one can, one may obtain some insight into the mechanism of drug action and provide a method for drug design. The complete molecular structure of the D_1-d-r is not known. How then can theoretical calculations contribute in this area? One way is to seek explanations of differences. We cannot study the absolute, but we can make comparisons, so we will study no single molecular species, but series of chemically closely related compounds and seek explanations of differences between members of an apparently homogeneous series (see Table 1), these may be correlated with differences in the biological activities. Observing the structures and biological activities of representative dopamine D_1 receptor antagonists it is possible to draw a general structural picture for these

Table 1
Structures and biological activities (IC$_{50}$) of representative BTHIQ compounds acting as dopamine D$_1$ receptor inhibitors

Structure	IC$_{50}$ (μ/mol)
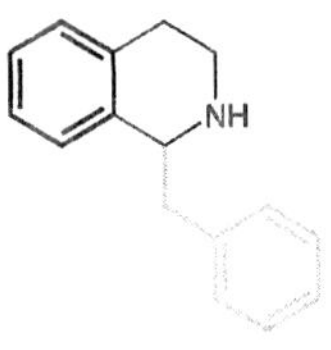	248.80 [9]
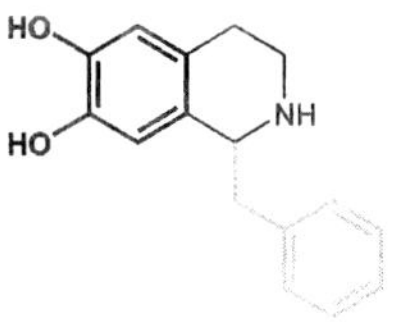	9.27 [9]
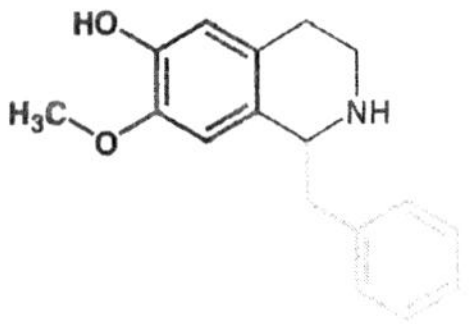	23.22
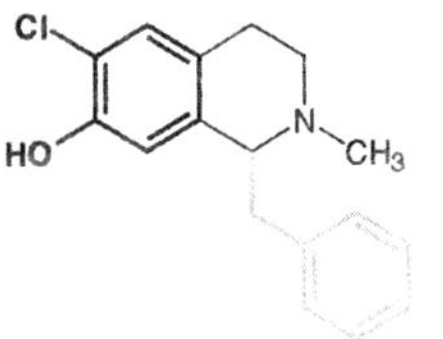	5.75×10^{-2} [10]
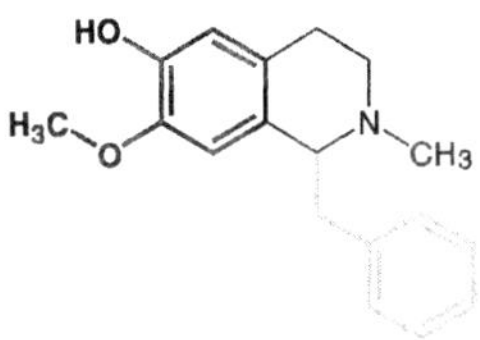	1.84
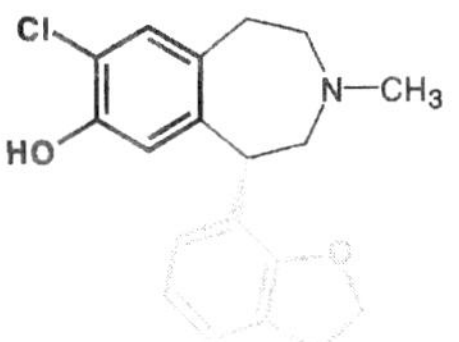	5.80×10^{-3} [11]
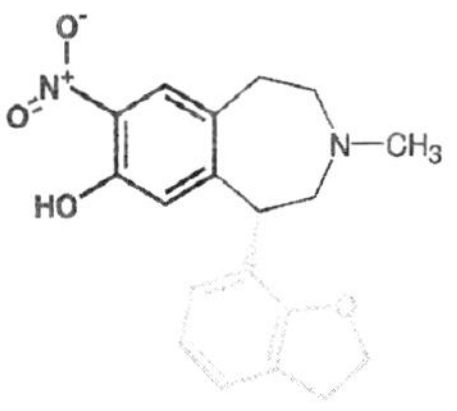	1.80×10^{-4} [11]

compounds. Among the salient molecular features common are:

(a) At least one polar group at C$_7$; specifically a OH group on the benzene ring A at C$_7$ position is present in all potent compounds.
(b) A basic, usually tertiary nitrogen group which is part of a piperidine (or equivalent) ring.
(c) At least one aromatic ring, usually a benzyl group is attached to C$_1$.

The main molecular features apparently required for dopamine D$_1$-receptor inhibitory activity have led to a schematic representation of the pharmacophoric patron, as shown in Fig. 2.

In the present exploratory theoretical study, we simulate the electrostatic interactions between BTHIQ derivatives with its biological receptor in terms of smaller model molecules. With the knowledge of the cardinal role of the electrostatic factors in the molecular interactions, ab initio and density functional theory (DFT) calculations were performed for a better understanding of the recognition process at the sub-molecular level. Thus, we report here the results of a study of BTHIQs with emphasis on the parameters for the construction of a more complete dopamine D$_1$ receptor model: such a model would provide the basis for comparison of various antagonists producing the same response, and for an exploration of the effect that the molecular enviroment in the receptor could have on such a process.

2. Methods

All the calculations reported here were carried out using the GAUSSIAN 98 program [12].

For the molecular interaction (MI) simulations, all the complexes under investigation were initially optimised using RHF/6-31G(d,p) level of theory. Correlation effects were included using DFT with the Becke 3-Lee-Yang-Parr (B3LYP) [13] functional and the 6-31++G(d,p) basis set for all the complexes obtained at the lower level of computation. During the DFT calculations, the RHF/6-31G(d,p) geometries were kept fixed.

While gas phase predictions are appropriate for many purposes, they are inadequate for describing

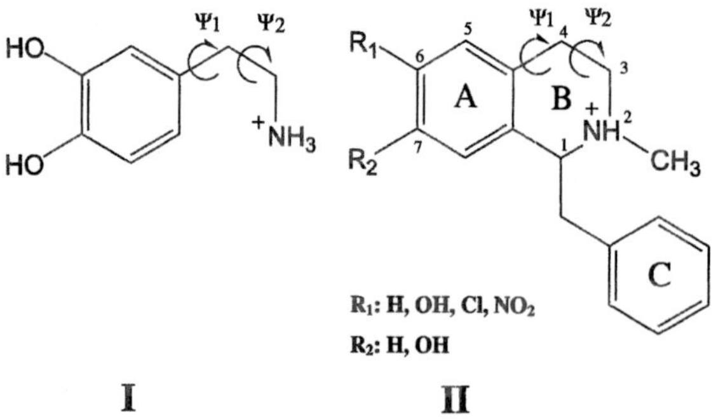

R_1: H, OH, Cl, NO_2

R_2: H, OH

I   **II**

Fig. 1. General structural feature of dopamine and BTHIQs. Definition of the spatial orientation of atomic nuclei used for molecule including the numeric definition of relevant dihedral angles.

the characteristic of polar molecules in solution. Electrostatic effects are often much less important for species placed in a solvent with a high dielectric constant than they are in the gas phase. Recently we have demonstrated the importance of including solvation energies in studying the relative binding free energies of ligand–receptor interactions [14]. Therefore, we have performed DFT (B3LYP/6-31++G(d,p)) calculations using the IPCM solvation model with explicit inclusion of solvent environment to gain deeper insight on the effect of the different molecular interactions reported here. Isodensity polarized continuum model (IPCM) defines the cavity as an isodensity surface of the molecule [15]. It should be emphasised, however, that the evaluation of the solvent effect implies a comparison to the gas phase results. Thus both sets of results (without and with the inclusion of the solvent) are required.

The conformational study of dopamine and its D_1 receptor–antagonists was performed from ab initio and DFT calculations. Rotational energy profiles around torsional angles have been determined by using RHF/3-21G, RHF/6-31G(d) and B3LYP/6-31G(d) calculations. The energy has been calculated at 30° intervals of the dihedral angles.

The electronic study of the compounds was carried out by using molecular electrostatic potentials (MEPs). MEPs have been shown to provide reliable information, both on the interaction sites of molecules with point charges and on the comparative reactivities of these sites [16–19]. These MEPs were calculated by using RHF/6-31 + G(d) wave function from

the SPARTAN program [20]. DFT/6-31G(d) optimised coordinates were imported into PC SPARTAN. To generate the wave functions, HF/6-31 + G(d) single point calculations were performed from PC SPARTAN.

3. Results and discussion

3.1. Stereo-electronic complementarity between dopamine and BTHIQs

The essential conformational problem in dopamine concerns the overall orientation of the side chain with respect to the conjugated ring and is described by the two principal torsion angles ψ_1 and ψ_2 (Fig. 1). The first of these torsion angles defines the position of the plane of the side chain with respect to the plane of the ring (coplanar or perpendicular) and the second, the orientation of the polar head with respect to the ring (*trans* or *gauche*). Dopamine may be considered as a 1,2-disubstituted ethane, therefore they will have *gauche* (*syn*-clinal) and *trans* (*anti*-periplanar) forms, but, where does one form change to the other and how far from an energy minimum? It is clear, therefore, that the information about local and global minima of dopamine is not enough. We need to have at least a good notion of the shape and also some indication about the dynamic behaviour of the internal degree of freedom of dopamine. Probably, the most comprehensive computational method to answer the above questions is to evaluate the potential energy surface (PES) of two independent variables (ψ_1 and ψ_2), associated with dopamine in Fig. 1. Fig. 3 shows the PES obtained for dopamine in the cationic form. In this surface the values obtained for

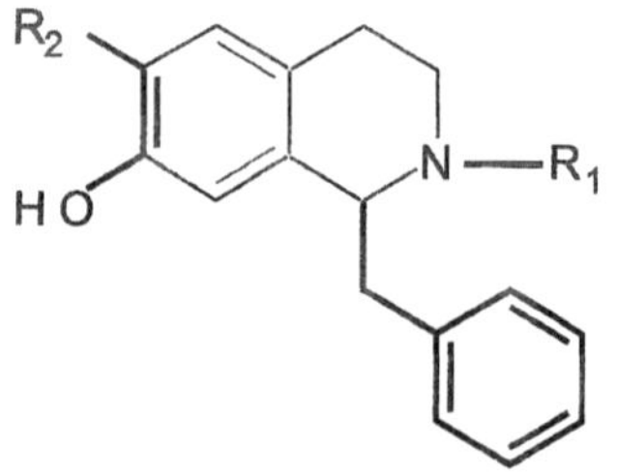

Fig. 2. Pharmacophoric patron of BTHIQ derivatives acting as dopamine-D_1-receptor inhibitors. Schematic representation; the salient structural features are denoted in bold.

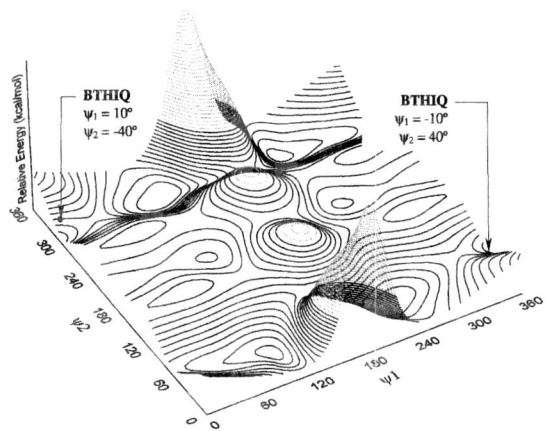

Fig. 3. Conformational PES obtained for dopamine in its cationic form from RHF/3-21G calculations. Full cycle of rotation (from 0 to 360°) is shown for variables ψ_1 and ψ_2.

the torsional angles ψ_1 and ψ_2 of compound II (Fig. 1), a representative BTHIQ molecule have been included for comparison. In Fig. 3 there are six clearly defined potential wells corresponding to regions close to the expected six (ψ_1, ψ_2) rotamers of dopamine. This surface clearly indicates that the g^+ and g^- minima of ψ_2 are noticeably shifting, in the opposite directions, with respect to the *anti* conformer.

Because the 3-21G basis set is a small one, we have attempted a partial verification of the above results by using a more extended basis set (RHF/6-31G(d)) and DFT calculations. For economic reasons we did not recalculate the whole conformational energy surface

but only the essential parts of it. Thus, we have calculated the energies of the perpendicular and antiperpendicular conformers by rotating the ring (varying ψ_1 as the side chain is fixed with $\psi_2 = 180°$ (Fig. 4a). In turn, we have evaluated the g^+, a and g^- conformers by rotating the side chain (varying ψ_2) as the aromatic ring is fixed with $\psi_1 = 90°$ (Fig. 4b). The *gauche* effect (i.e. *gauche* is more stable than *anti*) might be appreciated in this curve.

The barrier of ψ_1 is about 4.3 and 5.0 kcal/mol at DFT and RHF/6-31G(d) levels, respectively (Fig. 4a); while the barrier of ψ_2 is about 8 and 7.6 kcal/mol at DFT and RHF/6-31G(d) levels, respectively (Fig. 4b). Thus, DFT calculations suggest that dopamine possesses a moderate but significant molecular flexibility. Our results are in good agreement with those previously reported for a protonated 2-phenyl-ethyl-amine [21]. However, for dopamine the g^+ and g^- conformers are not equivalent; this is a striking difference which might be attributed to the presence of the OH groups in the aromatic ring.

It is interesting to note that compound II displays the ψ_1 and ψ_2 torsional angles very close to those obtained for the folded conformations of dopamine. Accepting the validity of the DFT calculations and of the obtained results, dopamine seems to possess a significant molecular flexibility, therefore it is reasonable to think that BTHIQs could mimetize the spatial ordering of dopamine in its folded conformation or in a closely related form.

Once the preferred conformations of compounds I and II were obtained, in an attempt to find

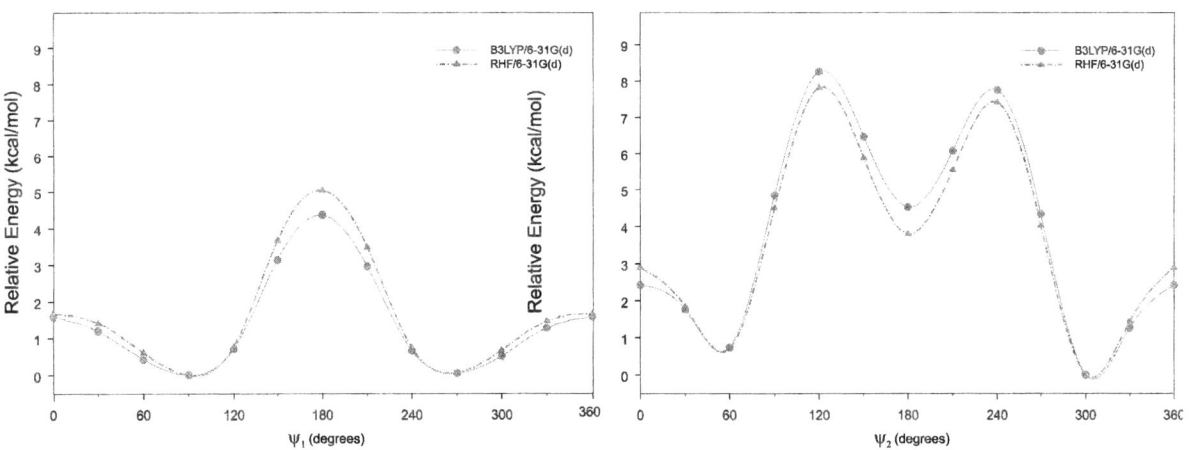

Fig. 4. Rotational energy barrier profiles computed at the RHF/6-31G(d) and DFT levels of theory obtained for dopamine (I).

the potentially reactive sites for dopamine (**I**) and compound (**II**), we have evaluated the electronic aspects of the molecules using MEPs. MEPs are of particular value because they permit visualisation and assessment of the capacity of a molecule to interact electrostatically with a binding site. MEPs can be interpreted in terms of a stereoelectronic pharmacophore condensing all available information on the electrostatic forces underlying affinity and specificity.

Fig. 5 shows the MEPs obtained for compounds **I** and **II**. This figure indicates that the MEP of dopamine in a folded conformation shows a remarkable similarity with that obtained for compound **II**. The close electronic similarity between dopamine and BTHIQ derivatives could account for their common affinity for the D_1-d-r. This assumption is supported by a qualitative comparison of the isopotential maps obtained through the ab initio computer modeling of dopamine and several BTHIQ structures (results not shown).

The analyses of MEPs showed two characteristic regions, one with negative potential and another with positive potential. The negative region is generated by the presence of the OH group at C_7; while the positive region corresponds to the large positive potential around the nitrogen cationic head. In the case of BTHIQs, the electrostatic potential surrounding the aromatic ring C itself does not appear to suggest any characteristic electrostatic interaction (or at least any strong interaction). Therefore, the role of this fragment appears to be one of a 'topological marker' for the proper orientation of the entire molecule at the receptor site allowing optimum interaction of polar substituents with the receptor. Another possibility, however, is that the ring C itself contributes its own interaction with the receptor site through dispersion forces. This problem will be discussed in Section 3.2.3.

3.2. Binding mechanism of BTHIQs to the D_1 dopamine receptors

Reduced model. Several model compounds were used to mimic BTHIQs and related molecules, and to enable quantum mechanical (QM) molecular orbital (MO) calculations. The use of model compounds to calculate MEPs and simulate MI is necessary since BTHIQs and the putative active sites of the D_1-d-r are too large for accurate QM MO calculations and the number of ligand-models to be screened is large. Moreover a model compound representing the modified active BTHIQ moiety may be desirable in order to evaluate the ability of the ligands to interact with the dopamine D_1-receptor pocket. By using a model, one avoids dealing with complexities due to the rest of the BTHIQ molecule. Thus a better understanding of the inherent electronic properties of the active moieties of BTHIQs reflected in the MEPs and MI may be gained. When choosing a model compound, the ability to reproduce electronic properties of the entire molecule (i.e. BTHIQs) was considered. In this pilot study we mainly consider two variations on the BTHIQ structure: (i) our attention was focussed on changes in the phenolic OH at C_7 in BTHIQs with variation on substituents at the adjacent carbon (C_6) (Fig. 6A); (ii) variations which could significantly affect the nature of the potential around the basic nitrogen, for example

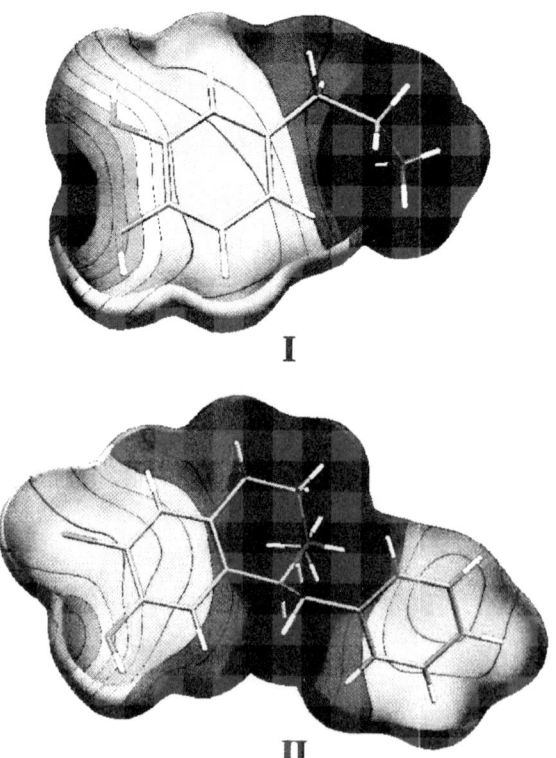

I

II

Fig. 5. MEP energy isosurfaces for compounds **I** and **II** obtained from RHF/6-31G(d,p) calculations.

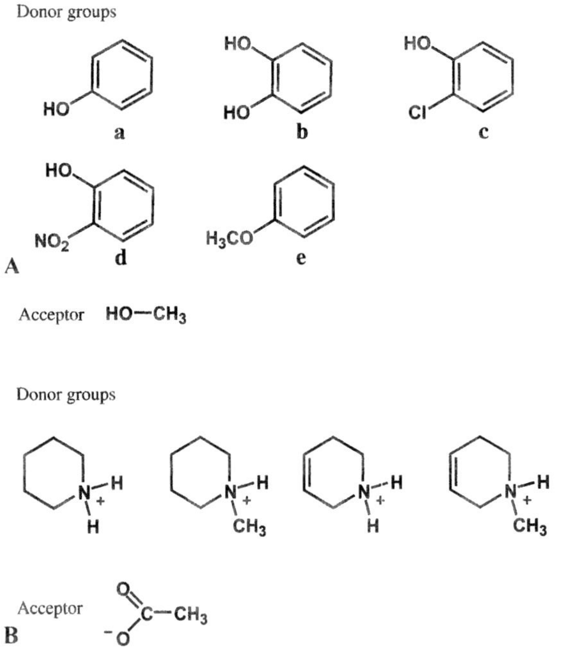

Donor groups

A

Acceptor   HO—CH₃

Donor groups

B   Acceptor

Fig. 6. Reduced models selected to mimic the reactive moieties of BTHIQs. CH₃OH and CH₃COOH were used to mimetize the side chain of Ser and Aspartic acid, respectively.

changes in the substituent on the nitrogen atom as well as in the flexibility of this ring (Fig. 6B).

Methylic alcohol (CH₃OH) was used to mimic the side chain of Serine. Similarly acetic acid (CH₃COOH) mimicked the side chain of Aspartic acid. Alternative moieties present on the BTHIQs (denoted in bold in Table 1 and shown isolated in Fig. 6) were used as the different partenaires (interacting counter parts).

3.2.1. Electrostatic attraction as the basis of preorientation

The preorientation is expected to occur before the ligand can actually bind to the D_1 receptor. The notion of preorientation is based on the idea that in molecular recognition atoms do not see atoms, but the electrostatic field of one of the molecules experiences the electrostatic fields of the other molecule.

Different model systems that mimic alternative moieties present on natural and synthetic BTHIQ molecules (displayed in Fig. 6) were investigated using the MEP approach in an effort to better understand the structural factors accounting for their biological activities. The electrostatic potential has

long been applied as a guide to molecular reactive behaviour [22–24]; for instance, the most negative values of MEP were interpreted as identifying and ranking sites for electrophilic attack, while its overall pattern served as the basis for qualitative analyses of biological recognition interactions.

Fig. 7 shows the MEPs obtained for the reduced models a–d displayed in Fig. 6. It is interesting to note that the maximum values of MEP were obtained for the H atom of the phenolic OH indicating the acid character of this atom (the different values of $V(r)$ are denoted in this figure). The increased electron density at the H position and the result of the electrowithdrawing effect of neighboring chlorine and nitro groups might be also appreciated in this figure. Interestingly compounds possessing electrowithdrawing groups at the adjacent carbon (C_6) of the phenolic group displayed increased biological activity (see Table 1).

Fig. 8 shows the MEPs obtained for the reduced models displayed in Fig. 6B. These MEPs show the minimum values, $V(r)$ min, in the vicinity of the lone pair regions of N, which is in accordance with experimental findings, indicating that this heteroatom is H-bond acceptor. The value of the potential is slightly reduced by the presence of a CH_3 group. Moreover, although the presence of an unsaturated double bond introduce a conformational change, this situation does not introduce appreciable electronic changes.

3.2.2. Electronic attachment

In phase B there is a short range electrostatic interaction, which may result in Brönsted (H-bonded) complexes formation. These are achieved for the Asp 103 and Ser 199, or Ser 204, respectively. We simulate the electrostatic interactions between BTHIQs with the dopamine D_1 receptor in terms of small model molecules (Fig. 6).

The energies of interaction (EI) were calculated with the approximation neglecting the superimposition of error due to the difference between the total energies of the complex with the sum of the total energies of the components:

$$EI = E_{Cx} - (E_{BC} - E_{AC})$$

where EI is the energy interaction, E_{Cx} the complex energy, E_{BC} the energy of proton-donor component

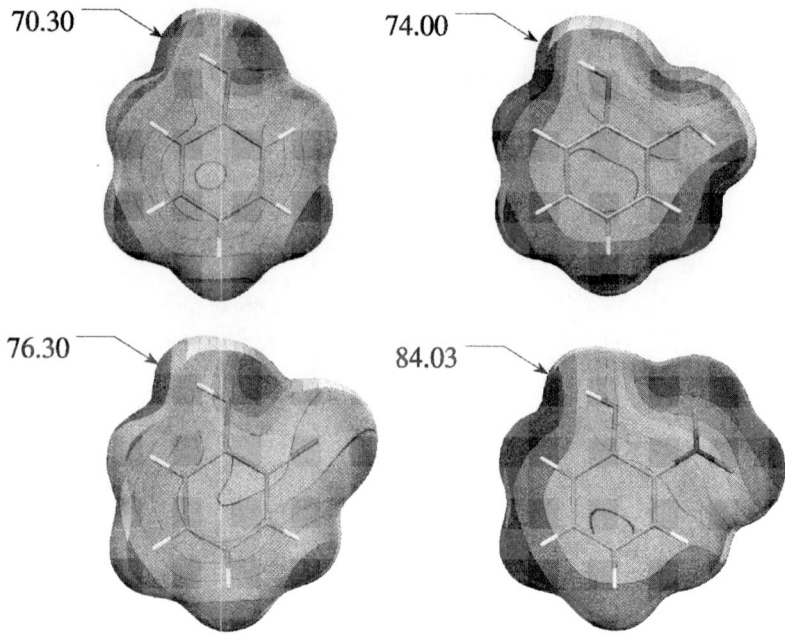

Fig. 7. MEP energy isosurfaces for the reduced models a–d obtained from RHF/6-31G(d,p) calculations.

(i.e. Brönsted acid), and E_{AC} the energy of proton acceptor component (i.e. Brönsted base).

The energies of all the complexes obtained and their components are listed in a summarized way in Table 2; while the interaction energies obtained for the different complexes are shown in Table 3.

Observing the results obtained for complexes I–III (Fig. 9a and Table 3) it is clear that the most favored interaction occurs when the CH_3–OH group is acting as acceptor while the phenol group is the donor counter parts (compare the energies obtained for complexes I and III. This result is in

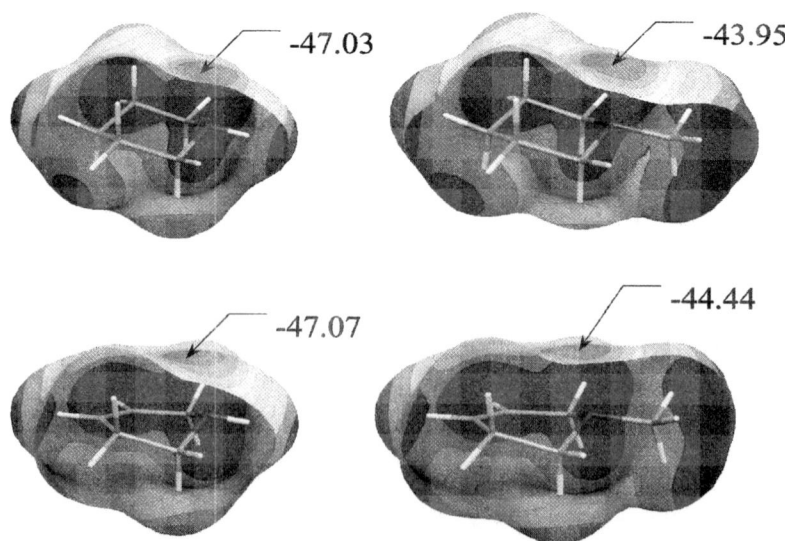

Fig. 8. MEP energy isosurfaces for the reduced models f–i obtained from RHF/6-31G(d,p) calculations.

Table 2
Energies (hartree) obtained at three levels of theory for the complexes and their components

Model compounds	RHF/6-31G**	B3LYP 6-31++G**	
		In vacuo	IPCM
Methanol	− 115.046709586	− 115.733802314	− 115.741803387
a Phenol	− 305.573755062	− 307.491973865	− 307.501889293
b Catechol	− 380.426278834	− 382.710935177	− 382.731344906
c Chloro-phenol	− 764.468134044	− 767.082574279	− 767.094633544
d Nitro-phenol	− 509.032450713	− 511.986385812	− 512.005082855
e Methoxy-benzene	− 344.596958913	− 346.795250672	− 346.800967537
Acetic acid	− 227.822171555	− 229.103261319	− 229.114339899
f Piperidine	− 250.207430902	− 251.929175671	− 251.933020445
g Methyl-piperidine	− 289.238924648	− 291.241193951	− 291.243224610
h Tetrahydro-pyridine	− 249.014553443	− 250.69461484	− 250.699661054
i Methyl-tetrahydro-pyridine	− 288.046307074	− 290.006808341	− 290.010456523
Complexes			
I	− 420.627769654	− 423.232173937	− 423.243579382
II	− 459.650892148	− 462.535433625	− 462.544478111
III	− 420.632071335	− 423.237660588	− 423.249594905
IV	− 495.484534785	− 498.456667964	− 498.474832922
V	− 879.527979234	− 882.829761646	− 882.839592958
VI	− 624.094639926	− 627.735623279	− 627.751690807
VII	− 478.045666056	− 481.049638881	− 481.061547964
VII	− 517.076727520	− 520.361214644	− 520.368948179
IX	− 476.852845990	− 479.815234848	− 479.828844265
X	− 515.884181586	− 519.126868410	− 519.136488717

accordance with our chemical intuition; after all the proton in a phenol group is more acidic than the proton of an alcohol. The replacement of an hydroxyl by a metoxy group gives a decreased interaction (compare the results obtained for complexes **II** and **III**). These results could explain the lack of activity reported for several natural products possessing metoxy substituents at C_7 [25].

On the other hand the results obtained for complexes **IV**–**VI** (Fig. 9b) indicate that the interaction energies obtained for complexes **V** and **VI** are favored over the rest of the complexes. These results illustrate very well the electrowithdrawing effect of chlorine and nitro group at the neighboring C_6. These results are in agreement with experimental data reporting that compounds possessing chlorine and nitro groups at C_6 were the most active in their respective series [9–11].

Fig. 10 gives the geometries obtained for the **VII**–**X** complexes. Starting geometries converged to the same kind of hydrogen–bonding interactions for such

complexes, independent of the level of theory used. It is interesting to note that these complexes are energetically favored with respect to the complexes **I**–**VI**. These results could indicate that the $>NH^+$–CH_3/Asp interactions would be the driving force for the BTHIQ/D_1-d-r complex.

Table 3
Interaction energies (kcal/mol) obtained for the different complexes

Complexes	RHF/6-31G**	B3LYP 6-31++G**	
		In vacuo	IPCM
I	− 4.584	− 4.015	0.071
II	− 4.533	− 4.004	− 1.071
III	− 7.283	− 7.458	− 3.704
IV	− 7.245	− 7.486	− 1.057
V	− 8.243	− 8.399	− 1.980
VI	− 9.714	− 9.686	− 3.015
VII	− 10.080	− 10.794	− 8.903
VII	− 9.781	− 10.517	− 7.143
IX	− 10.116	− 10.893	− 9.314
X	− 9.854	− 10.541	− 7.337

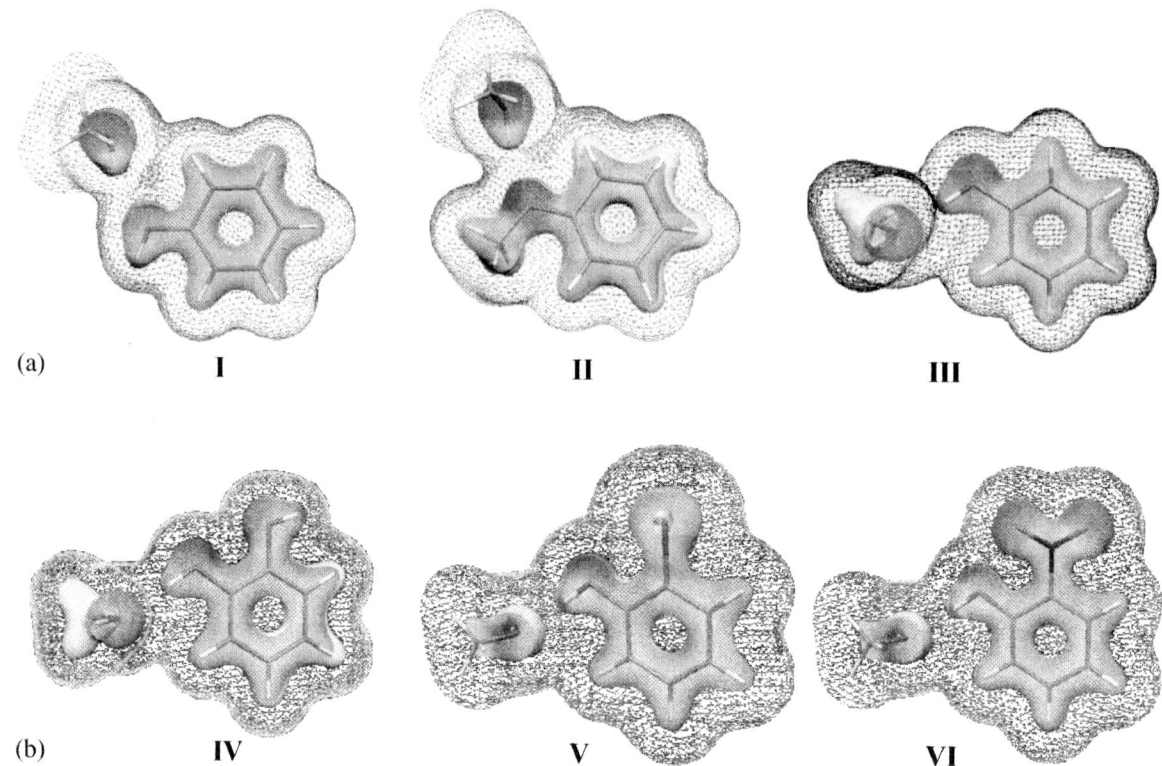

Fig. 9. (a) Spatial view of complexes **I**–**III** obtained from RHF/6-31G(d,p) calculations. The interactions are shown as insolvent surfaces mesh of electronic densities. (b) Spatial view of complexes **VI**–**VI** obtained from RHF/6-31G(d,p) calculations. The interactions are shown as involvent surfaces mesh of electronic densities.

The MI energy of complexes **VII**–**X** was found to change significantly in function of the spatial orientation of the lone pairs of nitrogen atom; being the complexes with the lone pair in axial position the preferred forms. In other words it appears that the complex geometry in this case is highly dependent on the molecular interaction.

3.2.3. The benzyl-group (CH₂–Ph) of BTHIQs (aromatic ring orientation)

To obtain a clear profile of the overall recognition process, it is necessary to underline the role of the benzyl-group (CH₂–Ph), which could play a determinant role in the BTHIQs structure to adopt the proper orientation allowing optimum interactions of ligand with the receptor.

The conformational intricacies of benzyl-group of BTHIQs is itself very complex and therefore it was evaluated in a separate paper. Thus, an exhaustive

conformational study of this flexible moiety of BTHIQs is reported in the companion paper [26]. We take the main conclusions of such study in order to determine the structural role of the benzyl-group in the binding mechanism of BTHIQs.

RHF/6-31G(d,p) indicate [26] that the flexible side chain of BTHIQs display a moderate molecular flexibility; being the extended *trans* conformation the preferred form for this moiety. With respect to the conformational behavior of the ring B of BTHIQs, the energy minima having energies below 2 kcal/mol are separated by a transition barriers of approximately 10.25 kcal/mol and hence, the molecule cannot be expected to move easily from one conformation to the other by gaining this energy from other favorable interactions. In contrast phenyl derivatives display a major molecular flexibility. Thus, our results [26] indicate that the different substituents at C_1 in BTHIQ molecules introduce a significant steric hindrance

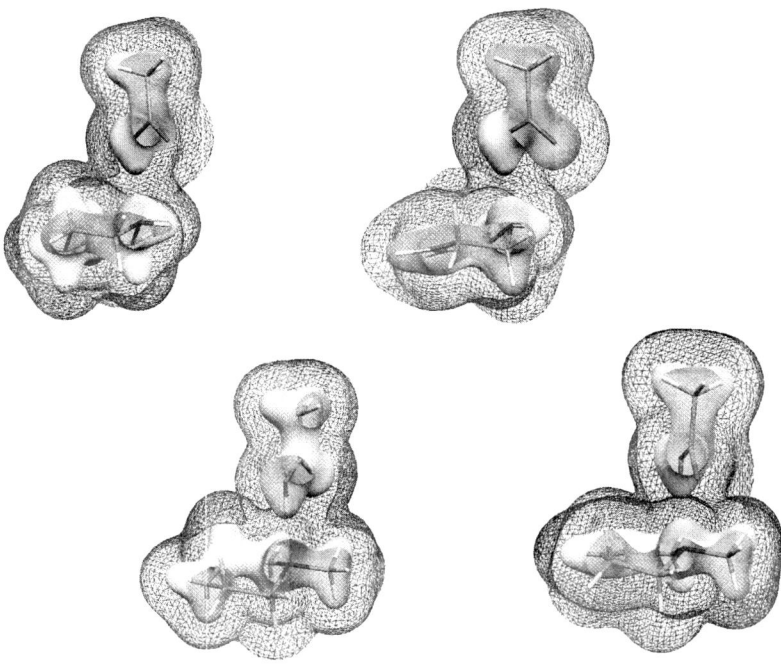

Fig. 10. Spatial view of complexes **VII–X** obtained from RHF/6-31G(d,p) calculations. The interactions are shown as involvent surfaces mesh of electronic densities.

which might, in turn, be responsible for a conformational restriction determining the spatial orientation of the lone pairs of N atom favoring or not the electronic attachment with the side chain of Asp residue. Although the above concepts were formulated independently, it seems that they are interdependent. However, it is clear that on the basis of our calculations it is not possible to discard the possibility of a third binding site which could take place through dispersion forces. Further theoretical and experimental works testing conformationally constrained analogues are necessary to confirm our results.

3.3. A molecular model for the binding mechanism of BTHIQs

Finally we wish to discuss some details about a putative overall recognition process between the BTHIQs molecules and the Dopamine D_1 receptor.

The main molecular features apparently required for the inhibitory activity and the observed effects of changes in structure on activity have led to a reduced representation of the receptor, as shown in Fig. 11,

where the two major complementary features of the receptor site have been inferred. In order to obtain this complex we simulate both interactions (SITES I and II) simultaneously.

Before a ligand can bind, there must be a relatively rigid complementary crevice inside the receptor that complements the ligand in shape in order to be able to accommodate it. Since complementary is rarely

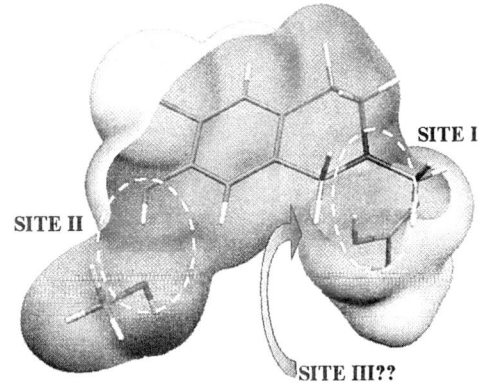

Fig. 11. Schematic representation of a reduced model for the dopamine-D_1-receptor, showing the putative binding sites.

perfect, more than one ligand can fit in the same crevice. Depending on the detailed mechanism of the binding process, the structural characteristics of the small ligand can assume varying degrees of importance. At least, two general cases can be distinguished: (a) one possibility is that binding take place by a stepwise process in which an initial interaction between a single segment of the ligand (in any conformation) and its subsite is followed by a rapid rearrangement of the conformation of the small molecule so as to permit the binding of the remaining segments to their subsites; (b) another possibility is that at any instant, only those molecules having the appropriate single 'biologically relevant' conformation are able to bind to the receptor; each portion of the ligand binds simultaneously to the appropriate subsite of the binding site.

In fact, there are various ways in which BTHIQs may be involved but it is not yet possible to do more than speculate. Accepting the validity of our theoretical calculations, it seems that an intermediate model sharing aspects of both extreme situations (models (a) and (b)) is in particular probable in the case of BTHIQs.

On the basis of our results and using the simple notion of receptor-site occupancy, one may seek chemical features common to dopamine and its competitive antagonist to suggest chemical binding sites. Thus, one may imagine that the amine group of BTHIQ engages the receptor at a specific site (SITE I, Fig. 11). We propose this interaction as the driving one because our results indicate that the molecular interaction between $>NH^+-CH_3$ and Asp 103 would become the major component of the recognition process. In addition the conformational restriction observed in the ring B of BTHIQs appears to play a determinant role on this interaction.

If isoquinoleinic nitrogen engages the receptor at the site which would otherwise accommodate the $-NH_3^+$ of dopamine, then one may envisage that the rest of the BTHIQ molecule contributes additional binding by interacting with at least one accessory region. The molecular structure of dopamine D_1 receptor antagonists appears to be quite critical for activity. The fact that activity is markedly affected by altering the substituents at C_6 and C_7 suggests a co-operative effect between active groups, and one may consider that the HO at C_7 makes a specific contribution to

binding. It should be noted that the molecular interaction at the SITE II is not restricted to a determinate spatial orientation of the OH group; and therefore it is reasonable to think that a rapid arrangement of this portion could takes place. Thus a kind of stepwise binding involving first the (SITE I) followed by the rest of the molecule (SITE II) seems a reasonable possibility and there is some strongly suggestive evidence that this phenomenon takes place. There is not a single and particular substituent at C_1 but different hydrophobic groups which seem to be well tolerated to produce the biological response. It is clear, however, that not any conformation is operative. This is an additional support for an intermediate model. On the basis of our results it is not prudent to discard the possibility of a third binding site, for example a hydrophobic interaction due to the aromatic ring C, and this is still an open question. The proposed molecular model must therefore be refined and tested with the use of molecules known to act as agonists, partial agonists and antagonists on the dopamine D_1-receptor, in order to produce a realistic scale of energies that would be predictive for the activity of other untested compounds. The ability to explain new antagonists structurally related to the BTHIQs reported here on this basis is currently being explored.

4. Conclusions

To better understand how BTHIQs interact with dopamine D_1 receptors, the conformational, electrostatic and physicochemical properties of these agents have been studied using ab initio and DFT calculations.

A putative binding mechanism of BTHIQ and its mimetics to dopamine D_1 receptor has been prepared on the theoretical calculation grounds. The results of calculations presented above show that the molecular mechanism postulated for the recognition process of the dopamine D_1-receptor is energetically feasible. To date, the general picture is that both $>NH^+-CH_3$ and HO at C_7 groups (SITES I and II, respectively) play a key role in the molecular recognition process. Such process is facilitated by the presence of an electro-withdrawing group placed near to the HO at C_7. In addition an appropriate ring orientation of the aromatic ring C of BTHIQs may be operative stabilizing this process.

We were able to explain, on the basis of a combined computational model, several structural characteristics of BTHIQ–D_1-d-r complexes. With this approach we could interpret the conformational and electrostatic factors of BTHIQs, in spite of the missing 3D structure of dopamine D_1 receptor. Our preliminary model might serve as a good basis for rational drug design of the dopamine D_1 binding inhibitors. In this sense, theoretical calculations appear to be of great use for the evaluation of molecular interaction in the molecular recognition process of BTHIQs and its congeners. However, it must be pointed out that high level theory calculations and the inclusion of solvent effects are crucial in order to obtain satisfactory accuracy in the electronic distributions of these compounds. This information is essential to the understanding of the structure-inhibitory activity of BTHIQs congeners from a medicinal chemistry point of view.

Acknowledgements

This work was supported by grants from Fundación Antorchas and Universidad Nacional de San Luis (UNSL)—Argentina. R.D.E. is member of the Consejo Nacional de Investigaciones Científicas y Técnicas de la República Argentina (CONICET).

References

[1] P.G. Strange, Dopamine receptor, Tocris Cooson (1997) 1–5.
[2] D.R. Sibley, F.J. Monsma, Molecular biology of dopamine receptors, Trends Pharmacol. Sci. 13 (1992) 61–69.
[3] N. Cabedo, P. Protais, B.K. Cassels, D. Cortes, J. Nat. Prod. 61 (1998) 709–712.
[4] I. Andreu, D. Cortes, P. Protais, B.K. Cassels, A. Chagraoui, Bioorg. Med. Chem. 8 (2000) 889–895.
[5] N. Cabedo, I. Andreu, M.C. Ramirez de Arellano, A. Chagraoui, A. Serrano, A. Bermejo, P. Protais, D. Cortes, J. Med. Chem. 44 (2001) 1794–1801.
[6] O. Civelli, J.R. Bunzow, D.K. Grandy, Annu. Rev. Pharmacol. Toxicol. 32 (1993) 281–307.
[7] A. Sarai, J. Theor. Biol. 140 (1989) 137.
[8] A.C.T. North, J. Mol. Graphics 7 (1989) 67.
[9] H. Kawai, Y. Kotake, S. Ohta, Biorg. Med. Chem. Lett. 10 (2000) 1669–1671.
[10] B. Hoffman, S.J. Cho, W. Zheng, S. Wyrick, D.E. Nichols, R.B. Mailman, A. Tropsha, J. Med. Chem. 42 (1999) 3217–3226.
[11] P.H. Andersen, F.C. Gronvald, R. Hohlweg, L.B. Hansen, E. Guddal, C. Braestrup, B. Nielsen, Euro. J. Pharm. 219 (1992) 45–52.
[12] GAUSSIAN 98, Revision A.7, M.J. Frisch, G.W. Trucks, H.B. Schlegel, G.E. Scuseria, M.A. Robb, J.R. Cheeseman, V.G. Zakrzewski, J.A. Montgomery, Jr., R.E. Stratmann, J.C. Burant, S. Dapprich, J.M. Millam, A.D. Daniels, K.N. Kudin, M.C. Strain, O. Farkas, J. Tomasi, V. Barone, M. Cossi, R. Cammi, B. Mennucci, C. Pomelli, C. Adamo, S. Clifford, J. Ochterski, G.A. Petersson, P.Y. Ayala, Q. Cui, K. Morokuma, D.K. Malick, A.D. Rabuck, K. Raghavachari, J.B. Foresman, J. Cioslowski, J.V. Ortiz, A.G. Baboul, B.B. Stefanov, G. Liu, A. Liashenko, P. Piskorz, I. Komaromi, R. Gomperts, R.L. Martin, D.J. Fox, T. Keith, M.A. Al-Laham, C.Y. Peng, A. Nanayakkara, C. Gonzalez, M. Challacombe, P.M.W. Gill, B. Johnson, W. Chen, M.W. Wong, J.L. Andres, C. Gonzalez, M. Head-Gordon, E.S. Replogle, and J.A. Pople, Gaussian, Inc., Pittsburgh PA, 1998.
[13] (a) A.D. Becke, Phys. Rev. A. 38 (1998) 3098–3100.
    (b) A.D. Becke, J. Chem. Phys. 98 (1993) 5618–5652.
    (c) C. Lee, W. Tang, R.G. Parr, Phys. Rev. B. 37 (1998) 785–789.
[14] F.D. Suvire, A.M. Rodriguez, M.L. Mak, J. Papp, R.D. Enriz, J. Mol. Struct. (Theochem) 540 (2001) 257–270.
[15] J.B. Foresman, A. Frisch, Modeling System in Solution. Exploring Chemistry with Electronic Structure Methods, Gaussian Inc., Pittsburg, PA, 1996, p. 237–249.
[16] P. Polilzer, D.G. Truhlar, Chemical Applications of Atomic and Molecular Electrostatic Potentials, Plenum, New York, 1991.
[17] P.A. Carrupt, N. El Tayar, A. Karlé, B. Festa, Methods Enzymol. 202 (1991) 638–677.
[18] P. Greeling, W. Langenaeker, F. De Proft, A. Baeten, Molecular Electrostatic Potentials: Concepts and Applications. Theoretical and Computational Chemistry, vol. 3, Elsevier, New York, 1996, p. 587–617.
[19] E.A. Jáuregui, G.M. Ciuffo, R.D. Enriz, Temas Actuales de Química Cuántica; UAM Ediciones 29 (1997) 327–349.
[20] PC SPARTAN PRO Wavefunction, Inc., Pittsburg, PA, 1996–2000.
[21] D.M. Gasparr, D.R.P. Almeida, S.M. Dobo, L.L. Torday, A. Varro, J. Gy Papp, J. Mol. Struct. (Theochem) 585 (2002) 167.
[22] E. Scrocco, J. Tomasi, Topics Curr. Chem. 42 (1973) 95.
[23] P. Politzer, J.S. Murray, in: K.B. Lipkowitz, D.B. Boyd (Eds.), Reviews in Computational Chemistry, vol. 2, VCM, New York, 1991, chapter 7.
[24] G. Naray-Szabo, G.G. Ferenczy, Chem. Rev. 95 (1995) 829.
[25] P. Protais, J. Arbaoui, E.H. Bakkali, A. Bermejo, D. Cortes, J. Nat. Prod. 58 (1995) 1475.
[26] F.D. Suvire, I. Andrew, A. Bermejo, M.A. Zamora, D. Cortes, R.D. Enriz, J. Mol. Struct. (Theochem). (Submited for publication).

ELSEVIER

Journal of Molecular Structure (Theochem) 666–667 (2003) 469–479

THEO
CHEM

www.elsevier.com/locate/theochem

Computational approaches to restriction endonucleases

M. Fuxreiter[a,*], R. Osman[b], I. Simon[a]

[a]*Institute of Enzymology, Biological Research Center, Hungarian Academy of Sciences, H-1113 Budapest, Karolina út 29, Hungary*
[b]*Department of Physiology and Biophysics, Mount Sinai School of Medicine, 1 Gustave L. Levy Place, New York, NY 10029, USA*

Abstract

Type II restriction endonucleases catalyze phosphodiester bond hydrolysis in bacteria to protect the host cell from invading phage DNA. Due to their exquisite sequence selectivity type II restriction endonucleases serve as excellent model systems for studying protein–nucleic acid interactions. Crystal structures of the PD-(D/E)XK superfamily revealed a common α/β core motif and similar active site. In contrast, these enzymes show little sequence similarity and use different strategies to interact with their substrate DNA. Computational approaches have been applied to unify the mechanism of restriction endonucleases and rationalize their diversity.

The first step of type II restriction endonuclease catalysis has been studied on *Bam*HI by semi-microscopic version of the Protein Dipoles Langevin Dipoles method. The substrate-assisted catalysis and the general base mechanism have been concluded as less likely than the metal-catalyzed reaction. A general model for catalysis has been proposed based on the group contributions to the reduction of the activation free energy.

Factors contributing to structural stability of PD-(D/E)XK type II restriction endonucleases have been analyzed to elucidate evolutionary relationship between these enzymes. Residues playing role in catalysis and recognition were highly correlated with those participating in stabilization centers. Thus the main functional motifs were concluded to be evolutionary more conserved than other parts of the structure. This observation is consistent with the proposal that these enzymes have developed from a common ancestor with divergent evolution.
© 2003 Elsevier B.V. All rights reserved.

Keywords: Restriction endonucleases; Phosphodiester hydrolysis; Structural stability; Catalytic mechanism

1. Introduction

Type II restriction endonucleases serve as parts of the restriction–modification system [1] in bacteria by catalyzing phosphodiester bond hydrolysis in 4–8 base pair long DNA sequences [2]. They cleave the invading phage DNA, while the host DNA is protected from hydrolysis by methylation [3]. Recognition of the cognate sequence is therefore of vital importance for the survival of bacterial cells. Hence restriction endonucleases bind their cognate sequence with remarkable specificity [2]. Replacement of a single base pair can decrease the catalytic rate by 6–9 orders of magnitude. Due to their exquisite sequence selectivity type II restriction endonucleases serve as excellent model systems for studying protein–nucleic acid interactions.

Type II restriction endonucleases can be classified as four superfamilies with distinct folds: PD-(D/E)XK, Nuc, GIY-YIG and HNH nucleases [4–6]. The presently available crystal structures correspond to only one, the PD-(D/E)XK superfamily. Although

* Corresponding author. Tel.: +36-279-3138; fax: +36-466-9276.
E-mail address: monika@enzim.hu (M. Fuxreiter).

0166-1280/$ - see front matter © 2003 Elsevier B.V. All rights reserved.
doi:10.1016/j.theochem.2003.08.071

type II restriction endonucleases share only 15–25% sequence similarity [7] the structural and functional characteristics of these enzymes are highly similar. Each enzyme is present in dimer or tetramer form contacting to both DNA strands. The dimerization interface includes a four-helix bundle that carries the recognition residues. Restriction endonucleases can approach DNA from either grooves and an extensive set of direct and water-mediated hydrogen bonded network is formed. Phosphate contacts, although previously believed to be non-specific, can also provide important contribution to the binding of the specific sequence. DNA is often distorted (bent or unwound) in the specific complexes, but this deformation is not present in all cases. Each protein structure contains a common α/β core motif, built up by 5–6 β-strands surrounded by flanking α-helices [8].

PD-(D/E)XK restriction endonucleases also show a similar active site architecture. The active site residues follow a weak consensus sequence (Glu/Asp)-$X_{(9-20)}$-(Glu/Asp/Ser)-X-Lys. Exceptionally Lys can be replaced by Glu, like in *Bam*HI. The catalytic action of restriction endonucleases has the absolute requirement of bivalent metal ion cofactor(s). Due to the controversy in the observed metal ion sites, the role of metals is not fully understood. Three models have been devised for catalysis depending on the number of metal ions involved.

The *substrate-assisted model* requires a single metal ion at the active site. In this mechanism the attacking nucleophile is deprotonated by the $3'$-phosphate group to the scissile bond. The metal ion plays role in the second step, when the nucleophile attacks the phosphorus and a pentavalent transition state (TS) forms. The metal ion helps the migration of the negative charge and contributes to the stabilization of the TS. The metal ion is also proposed to stabilize the OH$^-$, which is formed after protonating the leaving group. The substrate-assisted mechanism conforms to kinetic data on modified substrates. However, there are two main problems with this mechanism. For the phosphate to act as a general base, its pK_a has to increase by ~6 units from 0.76 [9]. The other problem is, that methyl phosphonate or phosphorothioate substitution of the $3'$-phosphate group results in considerable decrease in cleavage activity as far as four nucleotides from the cleaved

bond [10]. These observations suggest a non-specific electrostatic effect of the phosphates rather than a direct involvement in catalysis.

The *two metal mechanism*, that requires the presence of two metal ions at the active site, has been adapted from the exonuclease action of DNA polymerase I [11]. Here the pK_a of the water molecule providing the attacking nucleophile is lowered by ~4 pH units by the neighboring metal ion (metal ion A). The second metal ion (metal ion B) stabilizes the developing negative charge on the pentavalent TS in the nucleophilic attack step. The dominance of electrostatic effects as opposed to entropy factors or strain mechanism have been demonstrated [12]. This mechanism seems to be a plausible explanation for catalysis in *Eco*RV, *Pvu*II *Bgl*I, *Ngo*MIV and *Bam*HI, where two metal ions were observed in the active site. It is not straightforward, however, to extend the *two metal* mechanism to other restriction endonucleases, due to its stringent geometry requirements. Furthermore, to prepare the attacking nucleophile, this mechanism requires a general base at the active site, which is difficult to identify with Lys at the active site.

The *three metal* mechanism recapitulates the two metal pathway and assigns a mainly structural role to the third metal ion [13,14]. Therefore, the problems with this mechanism are similar to those described above for the *two metal* mechanism. Since no direct structural data supports the presence of three metal ions at the active site, this hypothesis has not been studied in detail.

Understanding the role of metal ions is a central theme in developing a unified mechanism for type II restriction endonucleases. In the present work the first step of *Bam*HI action has been investigated and based on energetic considerations a common catalytic mechanism is proposed. This general model requires two positively charged groups at the active site of restriction endonucleases. One is an obligatory metal ion stabilizing the attacking OH$^-$ nucleophile (corresponding to metal A in *Bam*HI), and a second group with variable identity playing role in stabilization of the pentavalent TS (corresponding to metal B in *Bam*HI).

Based on the low sequence and high structural similarity type II restriction endonucleases have been assumed to have evolved from a common ancestor with divergent evolution. This hypothesis is tested by analyzing the structural stabilization factors [15] in PD-(D/E)XK superfamily of type II restriction endonucleases. The sets of cooperative long-range interactions, so-called stabilization centers (SC) that help to prevent structural decay have been shown to be evolutionary more conserved than other parts of the structure. Surprisingly, functionally important residues are more abundant in SCs, than residues belonging to structurally similar motifs. It shows that residues playing role in DNA recognition and catalysis have been designed to contribute not only to the function of restriction endonucleases but also to their structural stability. This result conforms with the divergent evolution hypothesis of PD-(D/E)XK type II restriction endonucleases.

2. Methods

2.1. Calculation of free energy of proton transfer in protein

The first step of *Bam*HI catalysis has been studied by comparing the free energies of proton transfer (PT) in enzyme for the different mechanisms. The energy of deprotonation of the attacking water molecule has been calculated based on the corresponding thermodynamic cycle displayed in Fig. 1 with Eq. (1)

$$\Delta G^p(AH_p + B_p^- \rightarrow A_p^- + BH_p)$$

$$= \Delta G^w(AH_w + B_w^- \rightarrow A_w^- + BH_w)$$

$$- \Delta\Delta G^{w\rightarrow p}(AH + B^-) + \Delta\Delta G^{w\rightarrow p}(A^- + BH) \quad (1)$$

where $\Delta G^p(AH_p + B_p^- \rightarrow A_p^- + BH_p)$ and ΔG^w $(AH_w + B_w^- \rightarrow A_w^- + BH_w)$ is the free energy of PT from water to the general base in protein, and in water, respectively; $\Delta\Delta G^{w\rightarrow p}(AH + B^-)$ and $\Delta\Delta G^{w\rightarrow p}(A^- + BH)$ is the free energy of moving the reactants and products from water to the protein site.

2.2. Calculation of free energy of proton transfer in solution

The free energies of the corresponding reference reactions in water were obtained from the experimental

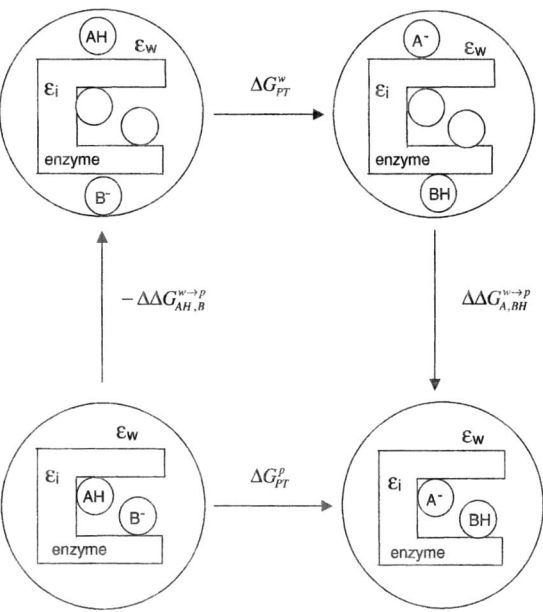

Fig. 1. Thermodynamic cycle used to calculate the free energy of proton transfer in protein.

pK_a values as

$$\Delta G^w(H_2O_w + B_w^- \rightarrow OH_w^- + B - H_w)$$

$$= 2.3RT[pK_a(H_2O)_w - pK_a(B - H)_w] \quad (2)$$

2.3. Calculation of transfer energies

The semi-microscopic version of the Protein Dipoles Langevin Dipoles model (PDLD/S) [16] has been applied to compute the free energies of moving charged groups from water to protein

$$\Delta\Delta G^{w\rightarrow p} = \left[\sum_i - \Delta G_{sol,w}^{i,\infty} + (\Delta G_{sol,w}^p(q=q_0) \right.$$

$$\left. - \Delta G_{sol,w}^p(q=0)) \right] \left(\frac{1}{\varepsilon_{in}} - \frac{1}{\varepsilon_w} \right) + V_{\mu q}\frac{1}{\varepsilon_{in}}$$

$$(3)$$

where $\sum_i - \Delta G_{sol,w}^{i,\infty}$ is the solvation energy of the specific residue in water, when the charged groups are infinitely separated from each other. $\Delta G_{sol,w}^p$ $(q=q_0)$ and $\Delta G_{sol,w}^p$ $(q=0)$ are the solvation energies of protein in water, when the specific group is in charged and uncharged form (with zero residual charges), respectively. $V_{\mu q}$ is the interaction of the reactive groups with the protein charges and polar groups in vacuum. ε_w is the dielectric constant of water, and ε_{in} is a scale factor that accounts for those contributions, which are not considered explicitly in the model. As the protein induced dipoles and the reorganization of permanent dipoles to charge rearrangements were included implicitly, $\varepsilon_{in} = 4$ and was used throughout this work [17].

2.4. Linear response approximation

Structural reorganization of the protein upon transferring charged groups from water to the protein site has been taken into account by the Linear Response Approximation [18]. The configurations for averaging the electrostatic free energies were generated by molecular dynamics simulations with the reacting groups in charged and uncharged form,

$$\Delta\Delta G^{w\rightarrow p} = \frac{1}{2}(\langle\Delta\Delta G^{w\rightarrow p}\rangle_{q=q_0} + \langle\Delta\Delta G^{w\rightarrow p}\rangle_{q=0}) \qquad (4)$$

where the $\langle\Delta\Delta G^{w\rightarrow p}\rangle_{q=q_0}$ and $\langle\Delta\Delta G^{w\rightarrow p}\rangle_{q=0}$ is the average solvation energy of configurations generated when the relevant groups are charged and uncharged, respectively.

2.5. Models

The starting model for our calculations was constructed from the crystal structure of the pre-reactive complex of Bam HI (PDB code: 2bam.pdb) [19]. The two Ca^{2+} ions found at the active site were replaced with catalytically competent Mg^{2+} ions. Residues within 6.5 Å from the scissile phosphate group (Glu-77, Asp-94, Glu-111, Mg^{2+} A, Mg^{2+} B, the susceptible phosphate group and Arg-122) were charged. Four crystallographic water molecules positioned within 5 Å from the scissile phosphate group were treated explicitly, while all other water molecules were replaced by Langevin dipoles. In the simulations the protein/water environment was represented by Surface Constrained All Atom Solvent model [16]. In this approximation the protein/solvent system is divided into four regions: the attacking water molecule and the group that is tested as the general base belong to region I, the rest of

the protein is assigned to region II with all atom treatment, region III is defined by the Langevin grid of the solvent truncated to a sphere with a grid spacing of 1 Å in the proximity of region I, and of 3 Å in the outer part, whereas region IV represents the bulk solvent, which is treated by a macroscopic continuum approximation.

2.6. Evaluation of group contributions

The contribution of an individual residue to the stabilization of the OH^- nucleophile was estimated as the free energy difference between the stabilization of region I when the particular residue is charged and in fully non-polar form (with zero residual charges). Since in the process of the charge annihilation of a given group, the protein structure and the explicit solvent molecules are not allowed to relax, this approach is termed as 'non-relaxed' approximation. For polar, but not ionized residues the interaction between the charges of the reacting system and the residual charges of the protein scaled by the factor ε_{in}, that accounts for implicit dielectric screening ($V_{q\mu}/\varepsilon_{in}$ term in Eq. 3) was found to give the major contribution. Previous studies on analogous systems established the appropriate values for ε to be used in such calculations [17]. Accordingly, $\varepsilon_{in} = 4$ was used for polar groups and $\varepsilon_{in} = 40$ for ionized groups. The group contributions for first-shell residues of the group of interest obtained by the non-relaxed approximation were shown to be in good agreement with the group contributions obtained by the relaxed approach [20].

2.7. Stabilization center calculations

The evolutionary relationship between restriction endonucleases has been explored by analyzing the SCs in the crystal structures of 14 enzymes: Bam HI [19,21–23], Eco RI [24–26], Eco RV [14,27–31], Pvu II [32–35], Bgl I [36], Bgl II [37,38], Fok I [39, 40], Cfr10I [41], Mun I [42], Nae I [43,44], NgoMIV [45], Bso BI [46], Bse634I [47] and Hinc II [48]. These structures represent restriction enzymes in free form, in complex with substrate DNA, and also in catalytically active form with DNA and metal ions. For Bam HI and Eco RV complexes with non-specific DNA were also included. The core motifs were defined to include five β-strands and two α-helices, that are involved in dimerization [49]. For the purpose of the analysis the crystallographic coordinates were not optimized. The SCs were calculated using the original definition [15]. Two residues form an SC element if (i) they are separated by at least 10 residues in sequence and the contact distance of their two closest atoms is less than the sum of their van der Waals radii plus 1 Å, which means that they form a long-range interaction; (ii) two supporting residues can be selected from both of their flanking tetrapeptides, which together with the central residues form at least seven out of the possible nine contacts. SCs calculations can also be performed via the server *SCide* (http://www.enzim.hu/scide).

3. Results

3.1. Free energies of proton transfer

Four reaction mechanisms have been considered as a catalytic model of the first step of *Bam* HI action (Fig. 2). In mechanism I, Glu-113 serves as a general base, that deprotonates a hydrogen bonded water molecule and prepares the attacking nucleophile. Mechanism II corresponds to the substrate-assisted reaction, where the proton from the water is transferred to the 3′-phosphate. Mechanisms III and

IV involve a second water molecule that abstracts proton from the nucleophilic water. These mechanisms are regarded as metal ion catalyzed reactions. The second water molecule is in contact with the bulk phase, but affected by the ionization state of Glu-113, which is charged in mechanism III and neutral in mechanism IV.

Free energies of the PT processes by mechanisms I–IV have been calculated based on the thermodynamic cycle presented on Fig. 1. First the free energies of PT in water ($\Delta G_w(H_2O + B^- \rightarrow OH^- + BH)$) were evaluated using the experimental pK_a values. As it can be expected, the proton abstraction from water by glutamate is the most favorable mechanism in solution with $\Delta G_{PT}^w = 15.7$ kcal/mole (corresponding to mechanism I). PT from water to the phosphate (corresponding to mechanism II) requires by 4.9 kcal/mole more free energy, while the hypothetical process of transferring a proton between two water molecules in the presence of negatively charged glutamate (corresponding to mechanism III) in the same solvent cage requires an additional 3.8 kcal/mole compared to the PT to phosphate.

The free energies of moving the reactants and products from water to the protein site ($\Delta\Delta G^{w\rightarrow p}(H_2O + B^-)$ and $\Delta\Delta G^{w\rightarrow p}(OH^- + BH)$) computed by PDLD/S method are summarized in Table 1. The transfer energies clearly show that in the reactant state the protein environment stabilizes the negatively charged Glu-113 (i), while destabilizes its neutral form (ii) compared to aqueous environment. In product state (iii, iv) the protein environment stabilizes OH^- with Glu-113 in both charged and neutral form, but this effect is more expressed (by 14.4 kcal/mole), when Glu-113 is negatively charged.

Mechanism I

$$Glu - COO^- + H_2O \rightarrow Glu - COOH + OH^-$$

Mechanism II

$$5'O - PO_2^- - O3' + H_2O \rightarrow 5'O - PO(OH) - O3' + OH^-$$

Mechanism III

$$H_2O_M + H_2O_{ext} \xrightarrow{Glu-COO^-} OH^-{}_M + H_3O^+{}_{ext}$$

Mechanism IV

$$H_2O_M + H_2O_{ext} \xrightarrow{Glu-COOH} OH^-{}_M + H_3O^+{}_{ext}$$

Fig. 2. Four mechanisms probed for the first step of *Bam* HI action. Mechanism I corresponds to general base catalysis; mechanism II displays the substrate-assisted catalysis; mechanisms III and IV are metal ion catalyzed reactions in the presence of Glu-113 in charged and neutral forms.

Table 1
Free energy of moving the reactants and products from water to the protein site

	Model	$\Delta\Delta G^{w\rightarrow p}$ (kcal/mole)
(i)	Glu–COO$^-$ + H$_2$O	−9.7
(ii)	Glu–COOH + H$_2$O	2.0
(iii)	Glu–COO$^-$ + OH$^-$	−36.5
(iv)	Glu–COOH + OH$^-$	−22.1
(v)	5′O–(PO$_2$)$^-$–O3$'$ + H$_2$O	2.4
(vi)	5′O–(PO$_2$H)–O3$'$ + OH$^-$	−5.9

This observation can be rationalized by the proximity of metal ion A at the active site of *Bam* HI. Models of the substrate-assisted mechanism (v, vi) indicate that the protein environment prefers the product state (neutral phosphate with deprotonated water) over the reactants (charged phosphate with neutral water) compared to aqueous environment.

Combining the free energies of PT in solution with the transfer energies of the reactants and products according to Eq. (1) gives the free energies of PT in the enzyme by the four possible mechanisms. The results are summarized in Table 2. In the enzyme the most favorable process is, when the proton is transferred from the water, that provides the attacking nucleophile to another water (often denoted as 'spectator' molecule), that is connected to bulk (mechanisms III and IV). These results demonstrate that the pK_a of the attacking water molecule is significantly lowered by the presence of the metal ion. Since the enzyme stabilizes the negatively charged form of Glu-113 in the reactant state, mechanism III is selected as the most preferred pathway. The formation of OH^- by general base catalysis involving Glu-113 (mechanism I) requires an additional 5.8 kcal/mole. This is in agreement with the results of pK_a calculations by Screened Columbic Potentials method [50]. This macroscopic approach, that takes the microenvironment of each ionizable group into account gives pK_a values of 3.2 for Glu-113 and 6.2 for the attacking water molecule. Since the role of Glu-113 as the general base is consistently disfavored by both approaches we conclude that Glu-113 has an electrostatic effect on the *Bam* HI action rather than direct involvement in catalysis. This is also supported by mutational data. When the negative charge of Glu is preserved upon replacement with Asp, the rate is affected only by 10^{-1}. The rate decrease by a factor of

10^{-3}–10^{-4} upon mutating Glu-113 to Gly or Cys can be interpreted as a reduction of shielding of Mg^{2+} A, that stabilizes the nucleophile. The free energy of 12.3 kcal/mole of the substrate-assisted catalysis demonstrates that a pK_a increase of the phosphate group by ~ 6 units is unlikely in the presence of two bivalent metal ions at the active site.

3.2. Group contributions

To better understand the role of the active site groups in the first step of *Bam* HI action, the contributions of individual residues to the stabilization of hydroxide ion have been analyzed. In the framework of the 'non-relaxed' approximation (see Section 2) the contributions were obtained by subtracting the electrostatic interactions of the OH^- with the protein groups from those computed for the H_2O. Those groups, which affect OH^- formation by more than 1 kcal/mole are displayed in Fig. 3.

The two main factors, which enhance the formation of the attacking hydroxide are the two metal ions found in the active site of *Bam* HI. Metal ion A, which coordinates the nucleophilic water molecule

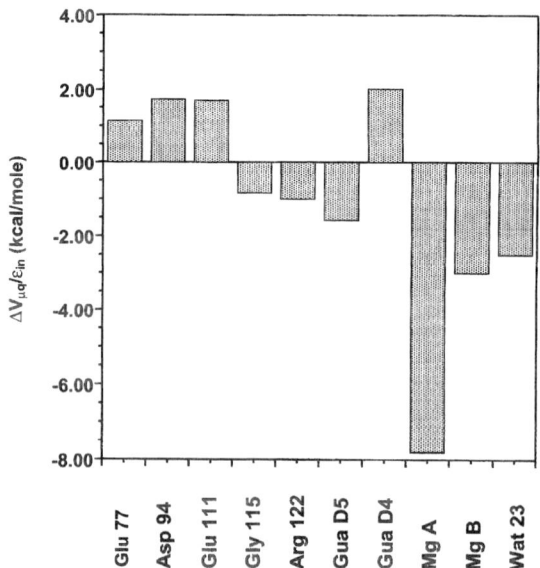

Fig. 3. Group contributions to the reduction of the free energy of proton transfer in the enzyme. Gua D4 is the scissile phosphate, whereas Gua D5 is the neighboring phosphate group 3′ to the scissile phosphate.

Table 2
Free energies of proton transfer in *Bam* HI

Reaction	ΔG_{PT}^p (kcal/mol)
I $Glu-COO^- + H_2O \rightarrow Glu-COOH + OH^-$	3.3
II $5'O-PO_2^- -O3' + H_2O \rightarrow 5'O-POOH-O3' + OH^-$	12.3
III $H_2O + H_2O_{ext} \overset{Glu-COO^-}{\rightarrow} OH^- + H_3O_{ext}^+$	-2.4
IV $H_2O + H_2O_{ext} \overset{Glu-COOH}{\rightarrow} OH^- + H_3O_{ext}^+$	-2.5

contributes to the deprotonation by 7.8 kcal/mole. It suggests that replacement of metal ion A should decrease the rate of the reaction by ~6 orders of magnitude. This is in agreement with the experimental observation that mutation of Glu-113 to Lys, which results in exclusion of metal ion A abolishes the enzymatic activity [51]. Metal ion B, which ligates the O3′ of the scissile phosphate reduces the free energy of OH⁻ formation to less extent, by 3 kcal/mole. Consequently, substitution of a residue, which anchors metal ion B may only decrease the rate by 2 orders of magnitude. Accordingly, the E77K mutant shows a reduced activity by 10^{-2}–10^{-3} [52]. It has to be noted, however, that predictions of overall rates upon metal ion substitutions is problematic due to the different dependence of the two reaction steps on the metal ion. Large decrease in the free energy of the nucleophile can make the nucleophilic attack step more difficult, so ultimately increase the activation energy of the reaction. Nevertheless, our results suggest that metal ion B is of less importance for the catalytic efficiency than metal ion A. This leads to the hypothesis, that metal site B is more variable than metal site A [53].

Besides the metal ions at the active site, a water molecule (water 23 in the crystal structure) also provides considerable stabilization of 2.2 kcal/mole to the formation of the attacking nucleophile. This effect is primary due to dipole–dipole interactions, since the water is perfectly aligned with the OH⁻. This water participates in the last step of the reaction by providing a proton to the O3′ of the leaving group. The fact that this water molecule together with metal ion B provides comparable contribution to the reduction of the free energy of nucleophile formation by metal ion A strengthens the hypothesis that metal site B is more variable than metal site A.

The phosphate–sugar backbone of the neighboring base to the scissile bond stabilizes the OH⁻ by 1.5 kcal/mole. In the absence of direct contacts between this phosphate and the nucleophile, this observation rationalizes why methyl phosphonate or phosphorothioate substitution of this phosphate groups results in considerable decrease in cleavage activity [54]. However, such a conclusion should be considered with considerable caution, because the present study only addresses the energetics

of the formation of the nucleophile, whereas the investigations with the phosphate analogs measure a k_{cat}/K_M for the entire reaction which is affected by many steps including substrate binding and product release.

The catalytic rate is also affected by the negatively charged residues at the active site coordinating the metal ions, which increase the free energy of OH⁻ formation, while the Arg-122 and the backbone dipole of Gly-115 facilitates this reaction. The main catalytic factors, however, identified by the present analysis in the first step of *Bam* HI action are the metal ions and the water molecule ligated to metal ion B.

Bam HI is not the only endonuclease, where general base catalysis is not the most effective way to reduce the reaction barrier. In DNA polymerase I using an external OH⁻ instead of the possible general base Glu-357 for nucleophile preparation has been shown to provide free energies comparable with the experimental results [12]. In equivalent position to Glu-113 in *Bam* HI a Lys can be found in other restriction endonucleases, which has been proposed to deprotonate the attacking water molecule. Although theoretically Lys can act as a general base, but it is very difficult to rationalize the lowering of its pK_a in the presence of a highly ionized active site and in the proximity of the negatively charged backbone of the DNA.

Based on the group contributions and on the conclusion that no general base is involved in the nucleophile formation step a general mechanistic scheme can be proposed for type II restriction endonucleases. According to our hypothesis, the first step in the endonuclease catalysis is carried out with the assistance of an external water molecule. The major energetic factor in this reaction step is the stabilization of the resulted nucleophile by the neighboring metal ion. A second positively charged group with variable identity also contributes to the reduction of the free energy of this reaction step.

A uniform catalytic mechanism of PD-(D/E)XK restriction endonucleases conforms the hypothesis that in spite of their low sequence similarity these enzymes have been developed from a common ancestor. To further investigate the relationship between these enzymes we analyzed those factors that contribute to structural stability of PD-(D/E)XK endonucleases.

3.3. Evolution of PD-(D/E)XK restriction endonucleases

Residues involved in cooperative long-range interactions, defined as SCs have been shown to be evolutionary conserved than average residues. Hence exploring the stability elements in the known structures of PD-(D/E)XK restriction endonucleases identifies the most conserved parts of this enzymes and helps to elucidate their evolutionary relationship. In the SC analysis we focused on in three main parts of the structure: (i) core motif, (ii) active site, and (iii) DNA recognition site.

The number of SCs in the α/β core has been compared to the number of SCs in the whole protein for *Bam*HI, *Eco*RI, *Fok*I, *Bgl*II, *Eco*RV, *Pvu*II, *Cfr*01 and *Bgl*I enzymes. The ratio of SCs in the core vs the whole protein varies in a wide range. SCs are dominant in core of *Eco*RV (~60%), their ratio varies between 30 and 40% in *Bgl*I and *Eco*RI, while it drops to 10–30% in the other enzymes. In most cases the ratio of SCs in the core motif is comparable with the relative size of the core [49]. This observation suggests that contribution of the core to structural stability of the whole enzyme is approximately proportional to the size of the core. Interestingly, *Eco*RV has the highest SC ratio, while the most closely related enzyme, *Pvu*II, has the lowest ratio of core SCs. In *Pvu*II five SCs are sufficient to keep the core motif intact. The core motifs also form few SCs with the rest of the structure. Thus the core does not have a dominant role in determining structural stability of restriction endonucleases. Therefore the α/β core is not the evolutionary most conserved motif in these enzymes.

The active sites of restriction endonucleases include at least two negatively charged sidechains, which primary role is to ligate the catalytically essential metal ion cofactor(s). Those active site residues, which also contribute to structural stability are displayed in Table 3. Interestingly, the active site residues frequently appear in SCs. With the exception of *Eco*RI in complex with DNA (PDB code: 1eri [25]), active site residues are involved in SC formation. In 24 out of the 38 analyzed structures two active site residues participate in stability elements. More than two active site residues contribute to structural stabilization in three

Table 3
Active site residues participating in stabilization centers (in bold)

Enzyme	PDB	SCs including active site residues
*Bam*HI	1bam	**Glu111**-Ile141
	1bhm	**Asp94**-Met110, **Glu111**-Ile141
	2bam	**Asp94**-Met110, **Glu111**-Ile141
	3bam	**Asp94**-Met110, **Glu111**-Ile141
	1esg	**Asp94**-Met110, **Glu111**-Ile141
*Eco*RI	1qc9	**Asp91**-Ala110, **Asp91**-**Glu111**, **Glu111**-Leu167
	1eri	–
	1qps	**Asp91**-Ala110, **Asp91**-**Glu111**, **Glu111**-Leu167
*Bgl*II	1es8	**Asn69**-**Asp84**, **Asn69**-Ile83, **Glu93**-Ile129, **Gln95**-Ile130
	1d2i	**Glu93**-Ile129, **Gln95**-Ile130
	1dfm	**Glu93**-Ile129, **Gln95**-Ile130
*Fok*I	2fok	**Asp450**-Val466, Asp421-**Lys469**
	1fok	**Asp450**-Val466
*Cfr*10I	1cfr	**Asp134**-Leu187, **Asp134**-**Ser188**, **Ser188**-Ala231, **Lys190**-Ala232
*Eco*RV	1rve	**Asp74**-Ile89, **Asp74**-**Asp90**, **Asp90**-Ile133
	4rve	**Asp74**-Ile89, **Asp74**-**Asp90**, **Asp90**-Ile133
	1eoo	**Asp74**-Ile89, **Asp74**-**Asp90**, **Asp90**-Ile133
	1eop	**Asp74**-Ile89, **Asp74**-**Asp90**, **Asp90**-Ile133
	1bgb	**Asp74**-Ile89, **Asp74**-**Asp90**, **Asp90**-Ile133
	1rva	**Asp74**-Ile89, **Asp74**-**Asp90**, **Asp90**-Ile133
	1rvb	**Asp74**-Ile89, **Asp74**-**Asp90**, **Asp90**-Ile133
	1rvc	**Asp74**-Ile89, **Asp74**-**Asp90**, **Asp90**-Ile133
	1az0	**Asp74**-Ile89, **Asp74**-**Asp90**, **Asp90**-Ile133
	1b94	**Asp74**-Ile89, **Asp74**-**Asp90**, **Asp90**-Ile133
	2rve	**Asp74**-**Asp90**
*Pvu*II	1pvu	**Glu68**-Val97, **Glu68**-Pro98, **Glu68**-Trp99
	1eyu	**Glu68**-Pro98, **Glu68**-Trp99
	1pvi	**Glu68**-Pro98, **Glu68**-Trp99
	3pvi	**Glu68**-Val97, **Glu68**-Pro98, **Glu68**-Trp99
	1f0o	**Glu68**-Val97, **Glu68**-Pro98, **Glu68**-Trp99
*Bgl*I	1dmu	**Asp116**-Val141
*Mun*I	1d02	**Asp83**-Gly97, **Asp83**-**Glu98**, Phe84-**Glu98**
*Nae*I	1ev7	**Asp95**-Cys116, **Asp95**-Leu117, Lys97-Leu117
	1iaw	**Asp95**-Ile115, **Asp95**-Cys116, **Asp95**-Leu117
*Ngo*MIV	1fiu	**Asp140**-Ile184,
*Bso*BI	1dc1	**Asp212**-Gly238, **Asp212**-Glu240
*Bse*634I	1knv	**Asp146**-Val195
*Hinc*II	1kc6	**Asp114**-Leu126, **Asp127**-Tyr168, **Asp127**-Leu169, **Lys129**-Glu170

structures, whereas one active site residue is involved in the remaining 11 cases. SCs can be formed between two active site residues like in *Eco*RI, *Eco*RV, *Cfr*10I, *Bgl*II, *Mun*I and *Bso*BI. It is important to note that SC formation within the active site is not general for other types of enzymes (personal communication with É. Tüdős). Results obtained on restriction endonucleases demonstrate that the active site residues provide important contribution to overall structural stability. In other words, although the active site residues belong to different secondary structure elements, they form a structurally stable unit, which has been conserved during evolution.

Type II restriction endonucleases bind 4–8 base pair long palindromic DNA sequences with outstanding specificity. In spite of the wealth of structural information on specific complexes the governing factors of the recognition process have not been fully understood. Binding proteins employ different secondary structure elements to contact with DNA either at the minor or the major groove. Specific DNA binding can be accompanied by conformational changes in the protein or DNA in both. Such diversity explains why no correlation between the cognate bases and the interacting residues has been found so far. We characterized the residues involved in DNA recognition in respect of their contribution to overall structural stability. Those residues, which participate in DNA binding as well as form SCs are summarized in Table 4a and b. Similarly those at the active site, residues contributing to DNA recognition are also involved in SCs. NgoMIV is the only exception, where not Arg-227, that interacts with DNA participates in SC formation, but Ser-40 that orients this binding residue. The presence of SCs reflects an extensive set of long-range interactions and thus a structurally stable unit at the DNA binding site. The stability of the binding site is further increased by the involvement of the neighboring sidechains in SC formation.

More than 200 sequences have been identified as substrates for type II restriction endonucleases [55]. Therefore, the recognition motifs are expected to be highly variable in these enzymes. In contrast, our results indicate that residues involved in specific DNA binding form a stable motif, which is also important for stability of the overall enzyme structure. This contradiction can be rationalized by that only two or three recognition residues are required for SCs, while others are allowed to vary. This structurally stable motif, which is formed at the recognition site has an appropriate conformation to contact with either or both DNA grooves. Since this motif is also important

Table 4a
Residues participating in DNA binding and form stabilization centers (in bold)

Enzyme	PDB	SCs including DNA binding residues
BamHI	1bam	**Asn116-Val156**, Thr114-**Val156**, *Gly115*-**Val156**, Asn55-**Thr153**, Gly56-**Thr153**
	1bhm	**Asn116-Val156**, *Gly115*-**Val156**, *Ile117*-**Val156**, **Asn116**-*Thr157*, Asn53-**Thr153**, Asn55-**Thr153**, Gly56-**Thr153**
	2bam	**Asn116-Val156**, **Asn116**-Thr157, *Ile117*-**Val156**, *Gly115*-**Val156**, Asn55-**Thr153**, Gly56-**Thr153**
	3bam	**Asn116-Val156**, *Ile117*-**Val156**, **Asn116**-*Thr157*, Asn53-**Thr153**, Asn55-**Thr153**, Gly56-**Thr153**
	1esg	Gly115-**Val156**
EcoRI	1qc9	**Asn141-Arg203**
	1eri	*Ile143*-**Arg203**, *Glu144*-**Arg203**, Phe174-**Arg200**
	1qps	*Ile143*-**Arg203**, Phe174-**Arg200**
BglII	1es8	**Gln12-Arg228**, Phe178-**Glu220**
	1d2i	Phe178-**Glu220**
	1dfm	Ser97-*Ser142*
FokI	2fok	**Asn98**-*Ser142*, **Asn98**-Leu143
	1fok	**Asn98**-Leu143
Cfr10I	1cfr	**Thr106**-Gly190, Gly108-**Asn188**, Gly109-**Asn188**, Tyr110-**Asn188**
PvuII	1pvu	**His84-Asn141**, **Thr82-Asn141**, **His83-Asn141**, **Thr82**-*Pro142*, **His84-Asn141**
	1eyu	**Ser81**-*Lys143*, **Thr82**-*Pro142*, **His83-Asn141**, **His84-Asn141**
	1pvi	**Ser81**-*Lys143*, **Thr82**-*Pro142*, **His83-Asn141**
	3pvi	**Ser81**-*Lys143*, **Thr82**-*Pro142*, **His83-Asn141**, **His84-Asn141**
	1f0o	**His83-Asn141**, **His83-Asn140**, **His84-Asn140**, **His84-Asn141**
BglI	1dmu	**Asp154**-*Val280*, **Asp154**-Asp281, **Leu155**-**Arg279**, **Val156**-*Val278*, *Val156*-**Arg279**, Arg263-**Arg279**, Pro264-**Arg279**
MunI	1d02	Lys50-**Gly79**, Asn51-**Gly79**, Leu52-**Gly79**, Tyr53-**Gly79**, **Ile99**-Val151, **Asp103**-Asp155, **Asp103**-Ile156
NaeI	1ev7	**Phe98**-Trp120, **Ser99**-Trp120
	1iaw	**Phe98**-Trp120, **Ser99**-Trp120
NgoMIV	1fiu	Ser40-Thr224
BsoBI	1dc1	Tyr24-**Asp246**, Phe28-**Asp246**
Bse634I	1knv	Ile149-**Ala193**, **Ala193**-Lys235, **Gly194**-Lys235, **Gly194**-Tyr236,
HincII	1kc6	**Asn141**-Ile208, **Asn141**-Gln209, Ala139-Phe210, Pro140-**Gln209**, Asn141-**Gln209**

for the overall stability, it is evolutionary more conserved than other parts of the structure. Conservation of structural stability during evolution in this case leads to conservation of the biological function, as well.

Table 4b
Residues participating in DNA binding and form stabilization centers in *Eco*RV (in bold)

Enzyme	PDB	SCs including DNA binding residues
*Eco*RV	1rve	**Thr106**-Gly190, Gly108-**Asn188**, Gly109-**Asn188**, Tyr110-**Asn188**
	4rve	**Thr106**-*Ile189*, **Thr106**-Gly190, *Leu107*-**Asn188**, Gly108-**Asn188**, Gly109-**Asn188**, Tyr110-**Asn188**
	1eoo	**Thr106**-*Ile189*, **Thr106**-Gly190, *Leu107*-**Asn188**, Gly108-**Asn188**, Gly109-**Asn188**, Tyr110-**Asn188**
	1eop	**Thr106**-*Ile189*, **Thr106**-Gly190, *Leu107*-**Asn188**, Gly108-**Asn188**, Gly109-**Asn188**, Tyr110-**Asn188**
	1bgb	**Thr106**-*Ile189*, **Thr106**-Gly190, *Leu107*-**Asn188**, Gly108-**Asn188**, Gly109-**Asn188**, Tyr110-**Asn188**
	1rva	**Thr106**-*Ile189*, **Thr106**-Gly190, *Leu107*-**Asn188**, Gly108-**Asn188**, Gly109-**Asn188**, Tyr110-**Asn188**
	1rvb	**Thr106**-*Ile189*, **Thr106**-Gly190, Gly108-**Asn188**, Gly109-**Asn188**, Tyr110-**Asn188**
	1rvc	**Thr106**-*Ile189*, *Leu107*-**Asn188**, Gly108-**Asn188**, Gly109-**Asn188**, Tyr110-**Asn188**
	1az0	**Thr106**-*Ile189*, **Thr106**-Gly190, *Leu107*-**Asn188**, Gly108-**Asn188**, Gly109-**Asn188**, Tyr110-**Asn188**
	1b94	**Thr106**-*Ile189*, **Thr106**-Gly190, *Leu107*-**Asn188**, Gly108-**Asn188**, Gly109-**Asn188**, Tyr110-**Asn188**
	2rve	**Thr106**-Gly190, *Leu107*-**Asn188**, Gly108-**Asn188**, Gly109-**Asn188**, Tyr110-**Asn188**

4. Conclusions

The relationship between PD-(D/E)XK type II restriction endonucleases has been studied by computational approaches. The catalytic mechanism of the first step of restriction endonuclease action has been investigated by free energy calculations. Four possible mechanisms have been probed using the PDLD/S method. The substrate-assisted and the general base catalysis have been concluded as not likely pathways for preparing the attacking OH⁻ nucleophile. The PT to a second water molecule, which is connected to the bulk phase was found to be the energetically most favorable reaction. In this process metal ion A plays central role in stabilizing the negative charge of the nucleophile. The contribution of metal ion B is of less importance, but together with a coordinating water molecule it provides comparable stabilization to metal ion A. It leads to the conclusion, that the second metal site is more variable than the first one. Based on the comparison of the different mechanisms and on the analysis of group contributions a uniform model has been proposed for type II restriction endonuclease catalysis. According to our hypothesis restriction endonuclease action requires two positively charged groups at the active site. One is an obligatory metal ion in the proximity of the attacking water molecule (corresponding to metal A in *Bam*HI) and a second group of more variable identity (corresponding to metal B in *Bam*HI), located near the O3′ of the leaving group. The PT step occurs via metal-catalyzed reaction with the assistance of an external water molecule. The role of the metal ion in this step is to stabilize the negative charge of the nucleophile. This model unifies the catalytic mechanism of restriction endonuclease catalysis and is in agreement with structural and kinetic data.

To rationalize the diversity of PD-(D/E)XK endonucleases structural stabilization elements were analyzed. The similar α/β core motif was not found to provide dominant contribution to structure stabilization. Rather, residues involved in specific DNA binding and those found at the active site are persistently involved in SC formation. These stable motifs help to keep the overall structure intact, so they can be predicted to be evolutionary more conserved than other parts of the structure. The structural conservation of the active site and recognition site in PD-(D/E)XK endonucleases leads also to their functional conservation. The variability of these sites is provided by the involvement of generally two residues as central residues in SCs. Based on the SC definition, this is enough to ensure the structural stability of these functional parts, while allows other residues to vary. It also explains the diversity of the substrate DNA sequences.

Both the proposed uniform catalytic model and the correlation found between the functional motifs and structure stabilization factors conform with the divergent evolution hypothesis of PD-(D/E)XK type II restriction endonucleases.

Acknowledgements

The research has been sponsored by the OTKA grant T34131 and Bolyai and OTKA D34572 fellowships.

References

[1] G.G. Wilson, N.E. Murray, Annu. Rev. Genet. 25 (1991) 585.

[2] R.J. Roberts, S.E. Halford, in: S.M. Linn, R.S. Lloyd, R.J. Roberts (Eds.), Nucleases, Cold Spring Harbor, NY, 1993, p. 33.

[3] J. Heitman, Genet. Engng 15 (1993) 57.

[4] J. Bujnicki, M. Radlinska, L. Rychlewski, Trends Biochem. Sci. 26 (2001) 9.

[5] L. Aravind, K. Makarova, E. Koonin, Nucleic Acids Res. 28 (2000) 3417.

[6] R. Sapranauskas, G. Sasnauskas, A. Lagunavicius, G. Vilkaitis, A. Lubys, V. Siksnys, J. Biol. Chem. 275 (2000) 30878.

[7] A. Jeltsch, M. Kroger, A. Pingoud, Gene 160 (1995) 7.

[8] A.K. Aggarwal, Curr. Opin. Struct. Biol. 5 (1995) 11.

[9] J.P. Guthrie, J. Am. Chem. Soc. 99 (1977) 3991.

[10] H. Thorogood, J.A. Grasby, B.A. Connolly, J. Biol. Chem. 271 (1996) 8855.

[11] L.S. Beese, T.A. Steitz, EMBO J. 10 (1991) 25.

[12] M. Fothergill, M.F. Goodman, J. Petruska, A. Warshel, J. Am. Chem. Soc. 117 (1995) 11619.

[13] N.C. Horton, K.J. Newberry, J.J. Perona, Proc. Natl Acad. Sci. USA 95 (1998) 13489.

[14] N.C. Horton, J.J. Perona, J. Biol. Chem. 273 (1998) 21721.

[15] Zs. Dosztanyi, A. Fiser, I. Simon, J. Mol. Biol. 272 (1997) 597.

[16] F.S. Lee, Z.T. Chu, A. Warshel, J. Comput. Chem. 14 (1993) 161.

[17] I. Muegge, T. Schweins, R. Langen, A. Warshel, Structure 4 (1996) 475.

[18] Y.Y. Sham, I. Muegge, A. Warshel, Biophys. J. 74 (1998) 1744.

[19] H. Viadiu, A.K. Aggarwal, Nat. Struct. Biol. 5 (1998) 910.

[20] I. Muegge, H. Tao, A. Warshel, Protein Engng 10 (1997) 1363.

[21] M. Newman, T. Strzelecka, L.F. Dorner, I. Schildkraut, A.K. Aggarwal, Nature 368 (1994) 660.

[22] M. Newman, T. Strzelecka, L.F. Dorner, I. Schildkraut, A.K. Aggarwal, Science 269 (1995) 656.

[23] H. Viadiu, A.K. Aggarwal, Mol. Cell 5 (2000) 889.

[24] Y.C. Kim, J.C. Grable, R. Love, P.J. Greene, J.M. Rosenberg, Science 249 (1990) 1307.

[25] J.A. McClarin, C.A. Frederick, B.C. Wang, P. Greene, H.W. Boyer, J. Grable, J.M. Rosenberg, Science 234 (1986) 1526.

[26] M.M. Horvath, J. Choi, Y. Kim, P. Wilkosz, J.M. Rosenberg, in preparation.

[27] J. Perona, A.J. Martin, Mol. Biol. 273 (1997) 207.

[28] D. Kostrewa, F.K. Winkler, Biochemistry 34 (1995) 683.

[29] M. Thomas, R. Brady, S. Halford, R. Sessions, G. Baldwin, Nucleic Acids Res. 27 (1999) 3438.

[30] F.K. Winkler, D.W. Banner, C. Oefner, D. Tsernoglou, R.S. Brown, S.P. Heathman, R.K. Bryan, P.D. Martin, K. Petratos, K.S. Wilson, EMBO J. 12 (1993) 1781.

[31] N.C. Horton, J.J. Perona, Proc. Natl Acad. Sci. USA 97 (2000) 5729.

[32] J.R. Horton, X. Cheng, J. Mol. Biol. 300 (2000) 1049.

[33] A. Athanasiadis, M. Vlassi, D. Kotsifaki, P.A. Tucker, K.S. Wilson, M. Kokkinidis, Nat. Struct. Biol. 1 (1994) 469.

[34] X. Cheng, K. Balendiran, I. Schildkraut, J.E. Anderson, EMBO J. 13 (1994) 3927.

[35] J.R. Horton, H.G. Nastri, P.D. Riggs, X. Cheng, J. Mol. Biol. 284 (1998) 1491.

[36] M. Newman, K. Lunnen, G. Wilson, J. Greci, I. Schildkraut, S.E. Phillips, EMBO J. 17 (1998) 5466.

[37] C.M. Lukacs, R. Kucera, I. Schildkraut, A.K. Aggarwal, Nat. Struct. Biol. 7 (2000) 134.

[38] C.M. Lukacs, A.K. Aggarwal, Curr. Opin. Struct. Biol. 11 (2001) 14.

[39] D.A. Wah, J.A. Hirsch, L.F. Dorner, I. Schildkraut, A.K. Aggarwal, Nature 388 (1997) 97.

[40] D.A. Wah, J. Bitinaite, I. Schildkraut, A.K. Aggarwal, Proc. Natl Acad. Sci. USA 95 (1998) 10564.

[41] D. Bozic, S. Grazulis, V. Siksnys, R. Huber, J. Mol. Biol. 255 (1996) 176.

[42] M. Deibert, S. Grazulis, J. Arvydas, V. Siksnyis, R. Huber, EMBO J. 18 (1999) 5805.

[43] Q. Huai, J. Colandene, Y. Chen, F. Luo, Y. Zhao, M. Topal, H. Ke, EMBO J. 19 (2000) 3110.

[44] Q. Huai, J. Colandene, M. Topal, H. Ke, Nat. Struct. Biol. 8 (2001) 665.

[45] M. Deibert, S. Grazulis, G. Sasnauskas, V. Siksnys, R. Huber, Nat. Struct. Biol. 7 (2000) 792.

[46] M. van der Woerd, J. Pelletier, S. Xu, A. Friedman, Structure 9 (2001) 133.

[47] S. Grazulis, M. Deibert, R. Rimseliene, R. Skirgaila, G. Sasnauskas, A. Lagunavicius, V. Repin, C. Urbanke, R. Huber, V. Siksnys, Nucleic Acids Res. 30 (2002) 876.

[48] N.C. Horton, L.F. Dorner, J.J. Perona, Nat. Struct. Biol. 9 (2002) 42.

[49] M. Fuxreiter, I. Simon, Protein Sci. 11 (2002) 1978.

[50] E.L. Mehler, F. Guarnieri, Biophys. J. 77 (1999) 3.

[51] H. Viadiu, PhD Thesis, Columbia University, New York, 1999.

[52] S.Y. Xu, I. Schildkraut, J. Biol. Chem. 266 (1991) 4425.

[53] M. Fuxreiter, R. Osman, Biochemistry 40 (2001) 15017.

[54] A. Jeltsch, M. Pleckaityte, U. Selent, H. Wolfes, V. Siksnys, A. Pingoud, Gene 157 (1995) 157.

[55] R.J. Roberts, D. Macelis, Nucleic Acids Res. 29 (2001) 268.

ELSEVIER

Journal of Molecular Structure (Theochem) 666–667 (2003) 481–485

www.elsevier.com/locate/theochem

THEO CHEM

Trypsin: is there anything new under the Sun?

László Gráf*, László Szilágyi

Department of Biochemistry, Eötvös Loránd University, Puskin u. 3, Budapest H-1088, Hungary

Abstract

Though trypsin is the first discovered and probably the best characterized enzyme, recent studies have led to the discovery of new properties and even a new form of this enzyme. The molecular mechanisms of autolysis of both trypsin and chymotrypsin have recently been explored and it has been proposed that the elimination of the major autolytic site by mutation in human cationic trypsin might cause pancreatitis. Other highlights of trypsin research are the discovery, X-ray crystallography and immunohistochemical localization of human brain trypsin.
© 2003 Elsevier B.V. All rights reserved.

Keywords: Trypsin; Autoactivation; Autolysis; Pancreatitis; Human brain trypsin

1. Trypsin is the first discovered and probably the best characterized enzyme

It has been known for more than 130 years that pancreatic juice is able to digest proteins [4]. Kühne suggested that the this property of the juice was due to an 'unorganized ferment' or enzyme that he named 'trypsin'. He also showed that the extracts of fresh pancreas or freshly secreted pancreatic juice had no proteolytic activity, but the activity appeared and were increasing when the pancreas was allowed to stand. After a longer period of standing the proteolytic activity of the pancreatic juice started to decrease (Fig. 1). This was the first description, in 1867, of autoactivation of the inactive (zymogen) form of trypsin and the autolytic inactivation of active trypsin. Since then autoactivation and autolyis of trypsinogen/ trypsin have been favourite topics of biochemical and physiological studies on trypsin.

2. Pancreatitis-associated mutations in trypsin

While the molecular mechanism and biological function of trypsinogen autoactivation are not clear yet, the mechanisms of autolysis of trypsin [6] and chymotrypsin [1] have recently been reported from our laboratory. According to these studies the interdomain loops of both enzymes contain the major autolytic sites, Arg117–Val118 in trypsin and Phe114–Ser115 in chymotrypsin. The cleavages of these peptide bonds lead to the inactivation of the proteinases (Fig. 2). What makes these results particulary interesting is the discovery that an Arg117 to His mutation in human cationic trypsin is associated with hereditary pancreatitis [7]. Since this mutation clearly abolishes the major autolytic site of trypsin, one can speculate that autolyis may play a physiological role in the pancreas also. Autolysis might function as a safety mechanism to eliminate prematurely activated (autoactivated?) trypsin in the pancreas [6,8] (Figs. 3 and 4). The presence and accumulation of an autolysis-resistant trypsin in

* Corresponding author. Tel.: +36-1-2667858; fax: +36-1-2667830.

E-mail address: graf@ludens.elte.hu (L. Gráf).

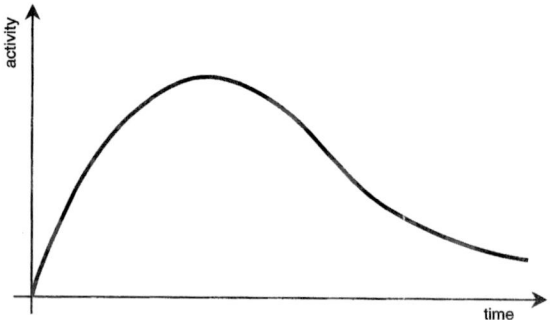

Fig. 1. The time-course of proteolytic activity in extracts from fresh pancreas according to Ref. [4].

the pancreas would lead to the clinical symptoms of pancreatitis (Fig. 4). One weakness of this explanation is that the physiological concentration of human pancreatic secretory trypsin inhibitor (hPSTI) in the pancreas should completely prevent both autoactivation and autolyis of trypsinogen/trypsin. Therefore, in a more recent study of us on another pancreatitis-associated human cationic trypsin mutant, the Asn21Ile one, we proposed that cathepsin B, rather than trypsin, might be the pathological activator of trypsinogen in pancreatitis [5] (Fig. 5). In in vitro experiments, cathepsin B activated the Asn21Ile

mutant 2-3 times as fast as the wild-type zymogen, and the presence of hPSTI did not prevent the activation of the zymogens by cathepsin B.

3. Brain Associated Trypsin (BAT): will exploration of its biochemical properties give a clue to its biological function?

While the involvement of pancreatitis-associated trypsin mutants in the pathomechanism of pancreatitis is evident, the physiological function and possible pathological role of human trypsinogen 4 has not been understood yet. The common gene encoding pancreatic mesotrypsinogen and human trypsinogen 4 (PRSS3) is located on chromosome 9, in contrast to the genes for pancreatic trypsinogens 1 and 2 that are located on chromosome 7. As a result of alternative splicing mesotrypsinogen and human trypsinogen 4 differ (and only differ) in there N-terminal amino sequences: while the former one has a typical signal sequence, human trypsinogen 4, dependent on the translation initiation site, has a 72 or 28 amino acid N-terminal leader sequence (Fig. 6). To our knowledge the mRNA for human trypsinogen 4 was found in

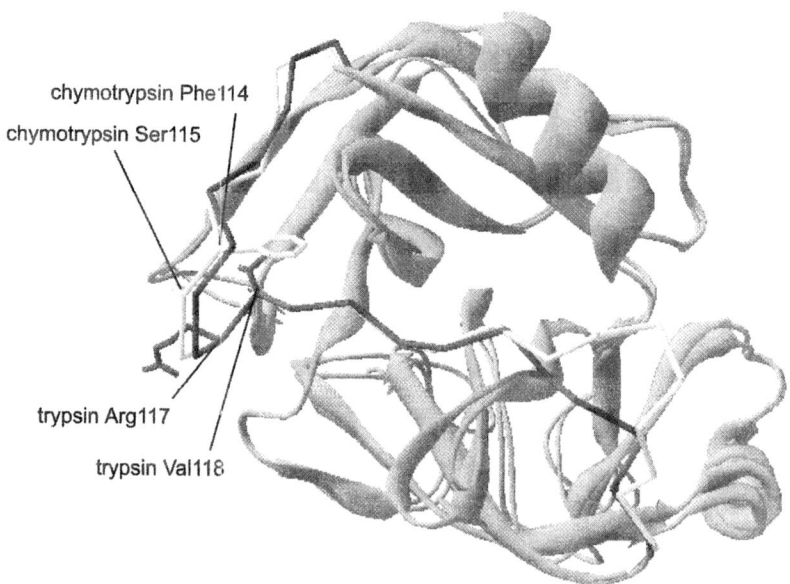

Fig. 2. Superimposed structures of trypsin and chymotrypsin with the major autolytic cleavage sites, Arg117–Val118 for trypsin and Phe114–Ser115 for chymotrypsin [1,6], respectively.

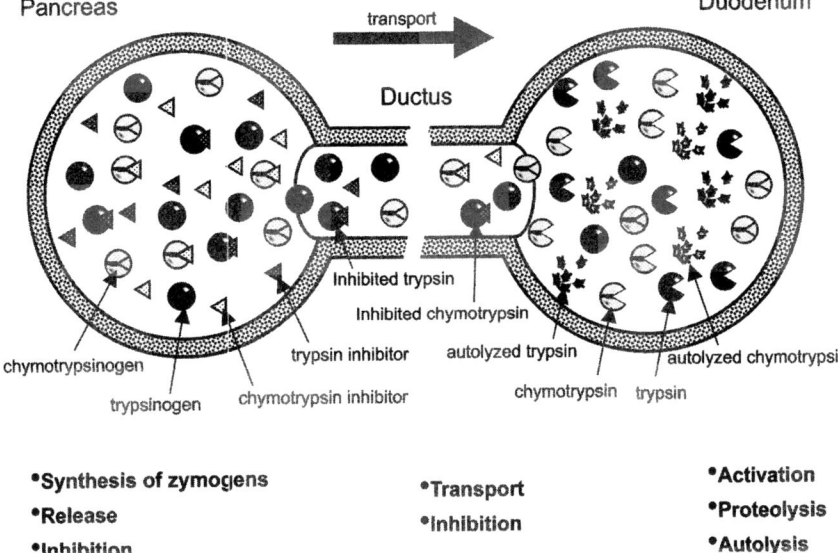

Fig. 3. Schematic representation of the production of trypsinogen, chymotrypsinogen and their inhibitors in the pancreas, their transport through the ductus and their activation/autolysis in the duodenum.

human brain only [9], and for the first time by using two specific monoclonal antibodies raised against the recombinant enzyme and a synthetic fragment of the N-terminal 'leader' peptide we were able to localize human trypsinogen 4-like immunoreactivity in glia cells of human cerebral cortex and spinal cord [2] (Fig. 7). Initiated by these new findings we named human trypsin 4 as Brain Associated Trypsin (BAT). Our preliminary studies on the interaction of some artificial membranes with a synthetic peptide

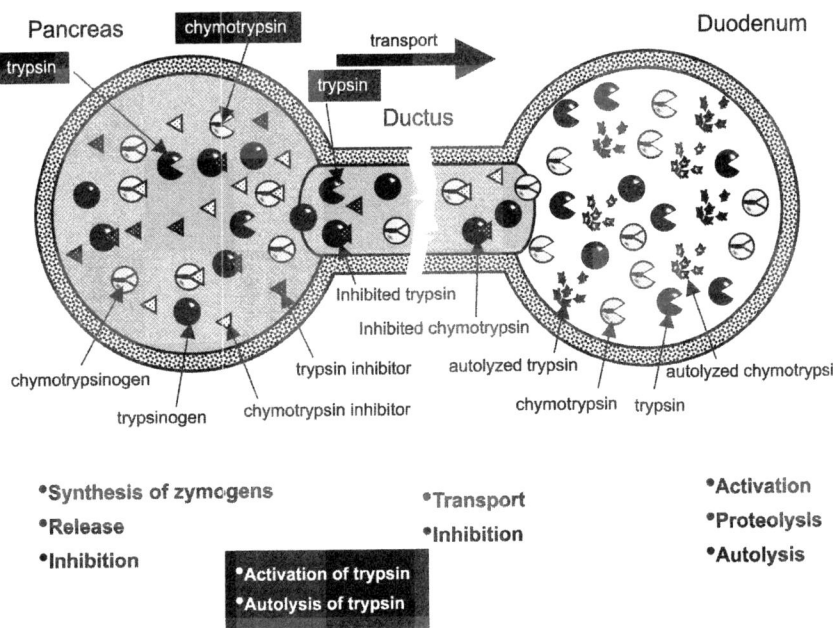

Fig. 4. Schematic representation of the possible consequence of trypsinogen mutation Arg117 to His that prevents autoactivation of trypsin [6].

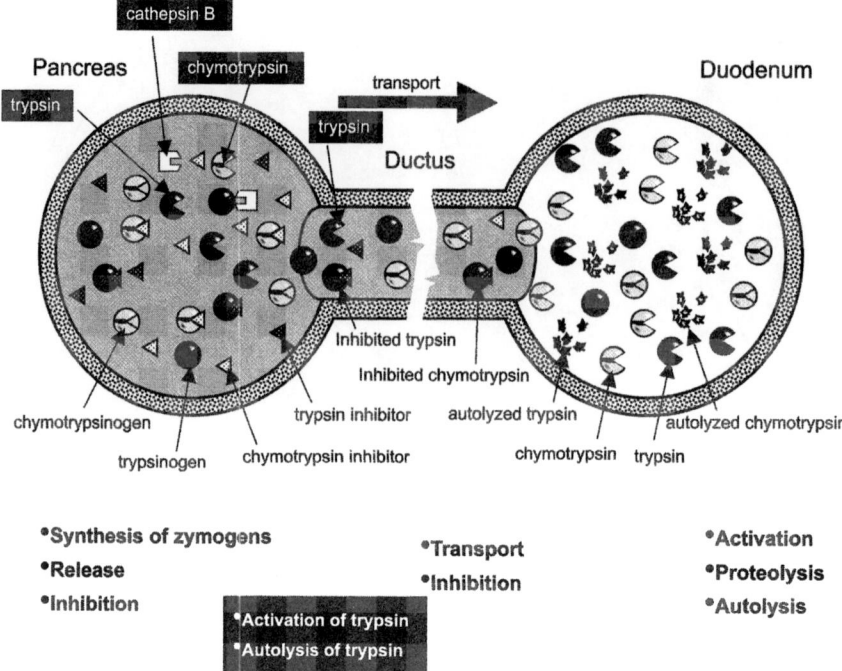

Fig. 5. Cathepsin B might be the pathological activator of trypsinogen in hereditary pancreatitis [5].

Mesotrypsinogen

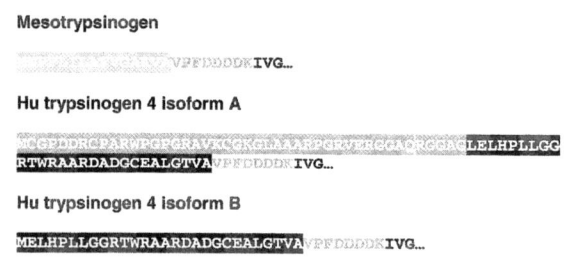

Fig. 6. The *N*-terminal sequences of human mesotrypsinogen and two possible isoforms of human trypsinogen 4.

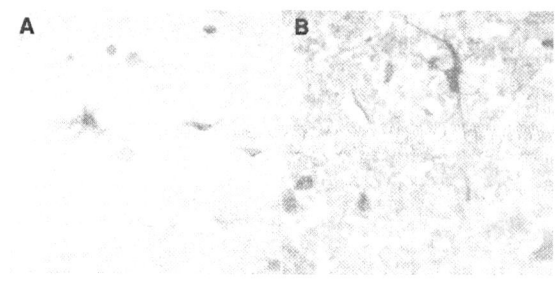

Fig. 7. Immunohistochemical localization of BAT (human trypsin 4) in cortical white matter by staining with monoclonal antibodies raised against the protease domain (A) and the 28-residue synthetic 'anchor' peptide (B).

fragment of the 'leader' sequence of BAT suggest that this region might serve as an anchor to attach the inactive proteinase to the cell membrane.

Mesotrypsin = BAT is a unique isoform of trypsin in which an arginine replaces the conserved glycine at position 193. It has long been thought that this

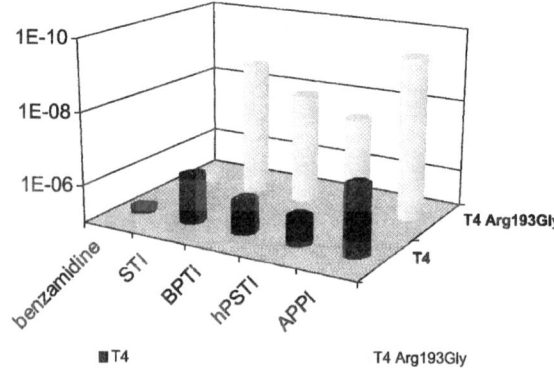

Fig. 8. The effect of mutation Arg193 to Gly in human trypsin 4 (T4). The log K_i inhibitory contants (*M*) for benzamidine, soy-bean trypsin inhibitor (STI), bovine pancreatic trypsin inhibitor (BPTI), human pancreatic secretory trypsin inhibitor (hPSTI) and Alzheimer precursor protein inhibitor (APPI) are shown on the ordinate.

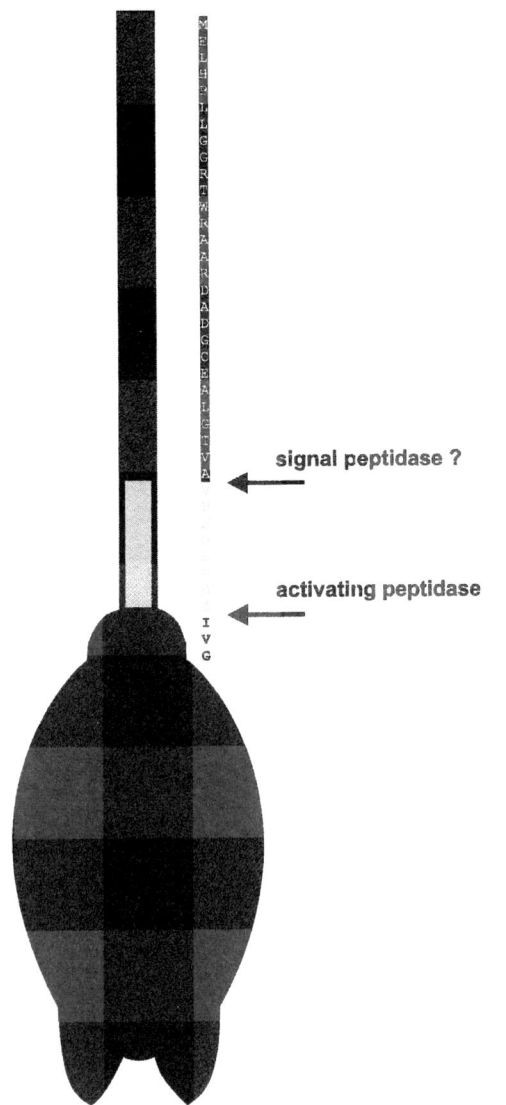

Fig. 9. Brain Associated Trypsinogen (BAT, human trypsinogen 4) might be a membrane associated zymogen. Upon activation with enterokinase or another activating protease the active BAT might be released.

signal peptidase ?

activating peptidase

Arg193 on the enzymatic properties of BAT by changing Arg193 to glycine (Arg193Gly). The results of a comparison of the inhibitor sensitivities of the arginine and glycine containing forms are shown in Fig. 8. The Fig. clearly shows that a single arginine to glycine substitution at position 193 is sufficient to restore the trypsin-like inhibitor sensitivity of BAT towards natural canonical trypsin inhibitors. These data are in agreement with the structural studies that show that in BAT Arg193 occupies the P2′ subsite of the substrate and inhibitor binding surface [3].

Localization of BAT in the glia cells of human brain, the likely association of the zymogen with the cell membrane, the resistance of the active proteinase towards canonical trypsin inhibitors and its limited substrate specificity are properties that make it a unique enzyme of the human brain. Its function, however, does not follow from these properties, and is a mistery at this stage. Our best guess is that the membrane bound BAT gets activated and released from the membrane (Fig. 9) under certain physiological and/or pathological conditions and cleaves specific, so far unknown neuropeptide or neuroprotein substrates in the brain One explanation for its resistance towards the naturally occurring proteinase inhibitors may be that its action is controlled by its release and unique substrate specificity rather than by specific inhibitors.

substitution is responsible for the resistance of mesotrypsin towards naturally occurring protein inhibitors of trypsin. The crystal structure of recombinant mesotrypsin/BAT confirmed this notion revealing the orientation of the side-chain of Arg193. It assumes an extended conformation and fills the S2′ substrate binding subsite of the enzyme [3]. By site-directed mutagenesis also we studied the effect of

References

[1] A. Bódi, G. Kaslik, I. Venekei, L. Gráf, Eur. J. Biochem. 268 (2001) 6238.
[2] K. Gallatz, P. Medveczky, P. Németh, L. Szilágyi, L. Gráf, M. Palkovits, 2003. in preparation.
[3] G. Katona, G.I. Berglund, J. Hajdu, L. Gráf, L. Szilágyi, J. Mol. Biol. 315 (2002) 1209.
[4] W. Kühne, Virchows. Arch. 39 (1867) 130.
[5] L. Szilágyi, E. Kénesi, G. Katona, G. Kaslik, G. Juhász, L. Gráf, J. Biol. Chem. 276 (2001) 24574.
[6] E. Várallay, G. Pál, A. Patthy, L. Szilágyi, L. Gráf, Biochem. Biophys. Res. Commun. 243 (1998) 56.
[7] D.C. Whitcomb, M.C. Gorry, R.A. Preston, W. Furey, M.J. Sossenheimer, C.D. Ulrich, S.P. Martin, L.K. Gates, S.T. Amman, P.P. Toskes, R. Liddle, K. McGrath, G. Uomo, J.C. Post, G.D. Ehrlich, Nat. Genet. 259 (1996) 995.
[8] D.C. Whitcomb, GUT 45 (1999) 317.
[9] U. Wiegand, S. Corbach, A. Minn, J. Kang, B. Muller-Hill, Gene 136 (1993) 167.

ELSEVIER

Journal of Molecular Structure (Theochem) 666–667 (2003) 487–498

THEO CHEM

www.elsevier.com/locate/theochem

α-Amylases of medical and industrial importance

Lili Kandra*

Department of Biochemistry, University of Debrecen, P.O. Box 55, H-4010 Debrecen, Hungary

Abstract

α-Amylases (α-1,4-glucan-4-glucanohydrolases; EC 3.2.1.1) are classical calcium-containing enzymes, which constitute a family of *endo*-amylases catalysing the cleavage of α-D-(1-4) glycosidic bonds in starch and related carbohydrates with retention of the α-anomeric configuration in the products. They can be found in microorganisms, plants and higher organisms where they play a dominant role in carbohydrate metabolism. This study characterizes the substrate binding sites of *Bacillus licheniformis* α-amylase (BLA), human salivary α-amylase (HSA) and its Y151M mutant. It describes the first subsite maps, namely, number of subsites, position of cleavage sites and apparent subsite energies. The product pattern and cleavage frequencies were determined by HPLC, utilising a homologous series of chromophore-substituted maltooligosaccharides of degree of polymerisation (DP) 3–10 as model substrates. 2-Chloro-4-nitrophenyl (CNP) and 4,6-O-benzylidene-modified 4-nitrophenyl (Bnl-NP) β-maltooligosaccharides (DP 4–8) were synthesised from cyclodextrins using a chemical procedure. For the preparation of CNP-maltooligosides of longer chain length a new chemoenzymatic procedure was developed using rabbit skeletal muscle glycogen phosphorylase b. Our results confirmed the presence of eight binding sites in BLA, five glycone sites $(-5, -4, -3, -2, -1)$, three aglycone sites $(+1, +2, +3)$ and the catalytic site is located between subsites $(-1$ and $+1)$. In addition, the subsite map revealed a barrier site at the reducing end of active site which repulses the glucose residue. The binding region of HSA is composed of four glycone and three aglycone-binding sites, while that of Tyr151Met mutant is composed of four glycone and two aglycone-binding sites. The subsite maps show that Y151M has strikingly decreased binding energy at subsite $(+2)$, where the mutation has occurred (-2.6 kJ/mol), compared to the binding energy at subsite $(+2)$ of HSA (-12.0 kJ/mol).

© 2003 Elsevier B.V. All rights reserved.

Keywords: α-Amylase; Maltooligosaccharide; Action pattern; Subsite map; Binding energy

1. Introduction

1.1. α-Amylases

(α-1,4-glucan-4-glucanohydrolases; EC 3.2.1.1) are classical calcium-containing enzymes, which constitute a family of *endo*-amylases catalysing the cleavage of α-D-(1-4) glycosidic bonds in starch and related carbohydrates with retention of the α-anomeric configuration in the products. They can be found in microorganisms, plants and higher organisms where they play a dominant role in carbohydrate metabolism.

1.2. Human amylases—clinical importance

In humans, α-amylase is one of the major secretory products of the pancreas and salivary glands, playing a role in digestion of starch and glycogen. Human amylases of both salivary (HSA) and pancreatic origin (HPA) have been extensively studied from

* Tel.: +36-52-512900x2256; fax: +36-52-512913.
E-mail address: kandra@tigris.klte.hu (L. Kandra).

0166-1280/$ - see front matter © 2003 Elsevier B.V. All rights reserved.
doi:10.1016/j.theochem.2003.08.073

the viewpoint of clinical chemistry, because they are important as indicators of pancreatic and salivary glands disorders (e.g. acute pancreatitis, parotitis).

Our interest was focused on salivary amylase, which is a multifunctional enzyme involved in at least three distinct biological functions. First, the hydrolytic activity is responsible for the initial breakdown of the polymeric starch to short oligomers. Second, amylase in solution binds with high affinity to viridans oral streptococci, a function that may lead to the clearance and/or adherence of these bacteria in the oral cavity. Interestingly, amylase bound to the bacterial surface retains approximately 50% of its enzymatic activity. Thus, bacteria-bound amylase is capable of hydrolysing starch to glucose, which can be used as a food source and then metabolised to lactic acid. Localised acid production by bacteria can lead to the dissolution of tooth enamel, a critical step in dental caries progression. Third, several lines of evidence indicate that salivary amylase bound to tooth enamel or hydroxy-apatite may play a role in dental plaque formation. Taken together, these multiple functions of HSA suggest that it may play a significant role in dental plaque formation and the subsequent process of dental caries formation and progression.

As a result of its role in the oral cavity, we have initiated studies directed towards gaining a detailed understanding of the structural requirements necessary for the breakdown of starch and its oligomers. It should be pointed out that salivary amylase is protected in the gastric environment and therefore it may contribute significantly to duodenal starch hydrolysis in exocrine pancreatic insufficiency.

Furthermore, α-amylases are used as targets for drug design in attempts to treat diabetes, obesity and hyperlipemia. Diabetes mellitus is the most common serious metabolic disease in the world; it effects hundreds of millions.

The widening interest in the treatment of sugar metabolic disorders has stimulated our work to search for new and efficient drugs and apply them as inhibitors of amylolytic enzymes.

We thought that human salivary amylase (HSA), one of the intensively studied α-amylases, should be the enzyme of choice to meet these particular needs. Therefore, we started biochemical studies involving small substrates and substrate-like inhibitors to gain a better understanding of HSA function in obesity,

diabetes and the attempts to manipulate the biological activity of HSA giving rise to ideal microbial ecology in the oral cavity.

1.3. Bacterial amylases—industrial significance

α-Amylases and related amylolytic enzymes are among the most important enzymes and of great significance in the present day biotechnology. They could be potentially useful in the semisynthetic chemistry for the formation of oligosaccharides by transglycosylation.

The spectrum of amylase application has widened in many other fields, such as clinical, medicinal and analytical chemistry; as well as their widespread application in starch saccharification and in the textile, food, brewing and distilling industries. Traditionally, starch hydrolysis was carried out using acid and high temperature. Enzymatic hydrolysis of starch has now replaced acid hydrolysis in over 75% of starch hydrolysing processes due to many advantage, not least its higher yields. Hydrolysis of starch gives rise to small maltooligosaccharides and glucose. Saccharide composition obtained after amylolysis of starch is highly dependent on the effect of temperature, the conditions of hydrolysis and the origin of enzyme. Specificity, thermostability and pH response of the enzymes are critical properties for industrial use.

Specific maltooligosaccharides have a range of potential uses in the food, pharmaceutical and fine chemical industries because of their unique nature and special properties. They are all highly soluble and produce clear viscous solutions which are palatable and superior nutrient foods for infants and aged. Maltopentaose has been used as nutrient food for patients having renal failure and those in a condition of calorie deprivation. Maltotetraose is also being examined as a food additive to improve texture and moisture retention in food. The price of pure maltooligosaccharides is extremely high because of the difficulties encountered in their production.

These facts support that novel α-amylases which can produce desired products in a one-step reaction from starch are needed.

Bacillus licheniformis is a mesophilic bacterium, but produces a highly thermostable α-amylase BLA. It is widely used in alcohol, sugar and brewing industries for the initial hydrolysis of starch to dextrins, which are then converted to glucose by glucoamylases. However, its function on starch and oligosaccharides is poorly understood. Therefore, we were encouraged to study this amylase and it turned out that it was an attractive model enzyme for active centre investigation.

1.4. Enzymological interest—'subsite models-subsite mapping'

X-ray crystallographic analysis, where the proteins in the crystalline state are free or complexed to a substrate-analogue, is a powerful method for mapping the active site of an α-amylase. However, these data can vary according to the crystalline varieties (free enzyme, enzyme-substrate/inhibitor complexes) or are not available at all. Therefore, the use of modified, low-molecular weight substrates could be an effective way to elucidate the number of subsites in the active site area of α-amylases.

In this study we have invoked the popular 'subsite model', which was introduced by Phillips [1], to account for the enzymatic properties of α-amylases such as PPA, BLA and HSA.

The amylase subsite model developed by Robyt and French [2] depicts the substrate binding region of the enzyme to be a tandem array of subsites. Each subsite is complementary to, and interacts with a substrate monomer unit. The subsites are labelled from the catalytic site, with negative numbers for subsites to the left (non-reducing end side) and positive numbers to the right (reducing end side) according to the proposed nomenclature of Davies et al. [3]. There are a number of different ways in which an oligomer substrate can interact with these subsites. A substrate oligomer can bind non-productively so that a susceptible bond does not extend over the catalytic amino acids of the enzyme; alternatively, the substrate can bind productively so that a susceptible bond lies over the catalytic site, in which case the bond is cleaved.

The process of quantifying the subsite model is referred to as *subsite mapping*. To completely map the binding region of α-amylases, we determined the number of subsites, located the position of the catalytic amino acids within the subsites and

determined the binding energies of each subsite-substrate monomer unit.

Subsite mapping is simplified for *exo*-enzymes because there is only one productive binding mode for each substrate. However, endo-acting enzymes form more productive binding modes resulting in a complex product pattern. The relative rate of formation of each product is called bond cleavage frequency (BCF), which gives information about the subsite-binding energy. By using BCFs for a series of oligomeric substrates, it is possible to calculate the subsite binding energy for each subsite on the enzyme binding region, with the exception of the two subsites adjacent to the catalytic site which are occupied by all productive complexes. For subsite map calculation the preferred procedure is that suggested by Allen and Thoma [4]:

- Establish experimental conditions where secondary reactions (transglycosylation, secondary attack) are insignificant.
- Use end-labelled substrates to determine quantitative BCF for chain lengths that are large enough to span the entire binding region.
- Examine bond cleavage frequencies to estimate the number of subsites and the position of the catalytic site.
- Apply a minimisation process to test the differences of measured and calculated BCF data.

A computer program, which was developed by Gyémánt et al. [5] according to the synopsis of Allen and Thoma [4], was used for subsite mapping of α-amylases. Only a few subsite maps have been found in the literature and detailed knowledge about subsite architecture of these well studied enzymes is scarce. Therefore, we hope that our efforts meet a long felt need concerning subsite mapping.

2. Experimental

2.1. Enzymes

α-Amylase (EC 3.2.1.1) from human saliva (Type IXA Sigma) gave a single band on SDS-PAGE and possessed no α- and β-glycosidase activity.

The mutant Y151M enzyme was produced as previously described by Mishra et al. [6].

The Y151M enzyme was purified by a combination of ion exchange chromatography and size exclusion chromatography (data not shown). The SDS-PAGE and western blot analyses showed a very intense band of approximate size of 55 kDa in the culture medium corresponding to the recombinant enzyme. As with the expression of wild type enzyme, hexosamines were not detected in the amino acid analysis indicating that the expressed protein was non-glycosylated. Approximately 5 mg of the protein was finally recovered from a 1 l culture comparable to the amount recovered for HSA. MALDI-TOF MS analysis of the mutant showed a molecular mass of 56.22 kDa corresponding to non-glycosylated enzyme.

α-Amylase (EC 3.2.1.1) from *B. licheniformis* Type XII-A (Sigma) gave a single bond on SDS PAGE and possessed no α- and β-glycosidase activity.

2.2. Substrates

2.2.1. Chemical synthesis of modified maltooligosaccharides

In the course of our studies of convenient substrates for α-amylases, 2-chloro-4-nitrophenyl (CNP), 4-nitrophenyl (NP), and 4,6-O-benzylidene modified 4-nitrophenyl (Bnl-NP) β-maltooligosaccharides (DP 3–8) were synthesised and investigated. For the synthesis of these β-maltooligosaccharide glycosides a classical chemical synthesis was developed, which based on cyclodextrins. α-, β-, and γ-cyclodextrins (CDs), prepared on industrial scale, contain six, seven and eight α-(1-4) bonded glucopyranosyl units, respectively. The conversion of these cyclic oligosaccharides into linear maltooligosaccharides and maltooligosaccharide β-glycosides (DP 3–8) was carried out in our laboratory.

One-pot acetylation and subsequent partial acetolysis of α-, β-, and γ-cyclodextrins resulted in a crystalline peracetylated maltohexaose, -heptaose and -octaose, respectively. Prolonged acetolysis of β-cyclodextrins gave a mixture of acetylated maltooligosaccharides, from which peracetylated maltotriose, -tetraose and -pentaose were isolated. The acetylated oligosaccharides were converted into α-acetobromo

derivatives and then transformed into NP and CNP β-glycosides. From the NP glycosides 4,6-O-benzylidene derivatives were prepared, which were used together with the free glycosides as model substrates to investigate the action pattern and cleavage frequencies of PPA, BLA and HSA. Fig. 1 shows the steps of the synthesis. Detailed procedure, the separation, purification and structural confirmation of the intermediers and the end-products was described by Farkas et al. [7].

2.2.2. Chemoenzymatic syntheses of chromogenic substrates

- Chain shortening via phosphorolysis [8]
- Chain lengthening via transglycosylation [9]

The widening interest in defined maltooligosaccharides and their glycosides has stimulated research for new and efficient syntheses of these compounds. The new synthetic routes may provide us with chemically modified oligosaccharides that posses more desirable physical and biological properties than their natural counterparts. In addition, the homologous maltooligosaccharide substrates are of current interest because of their importance in the investigation of the binding sites and the actions of different depolymerising enzymes. In these studies well-defined, high purity and structurally well-characterised substrates are preferred. Unfortunately, there is not available a really efficient chemical method for carbohydrate chemists to form glycosidic linkages stereospecifically, or to generate higher-molecular-weight oligosaccharide glycosides with chromogenic aglycons, because of the great number of steps necessary for such a synthesis and due to the low yield. Therefore, we decided to focus on the enzymatic synthesis of these compounds. The enzymatic procedures are advantageous alternatives to classical chemical methods since, in many cases, enzymes operate under the mildest reaction conditions (water, pH 6–8, room temperature) and posses a high specificity for the structure they recognise.

In our procedure, chemical synthesis was supplemented with an enzymatic step to gain a series of CNP β-maltooligosides from DP 3–6 and from DP

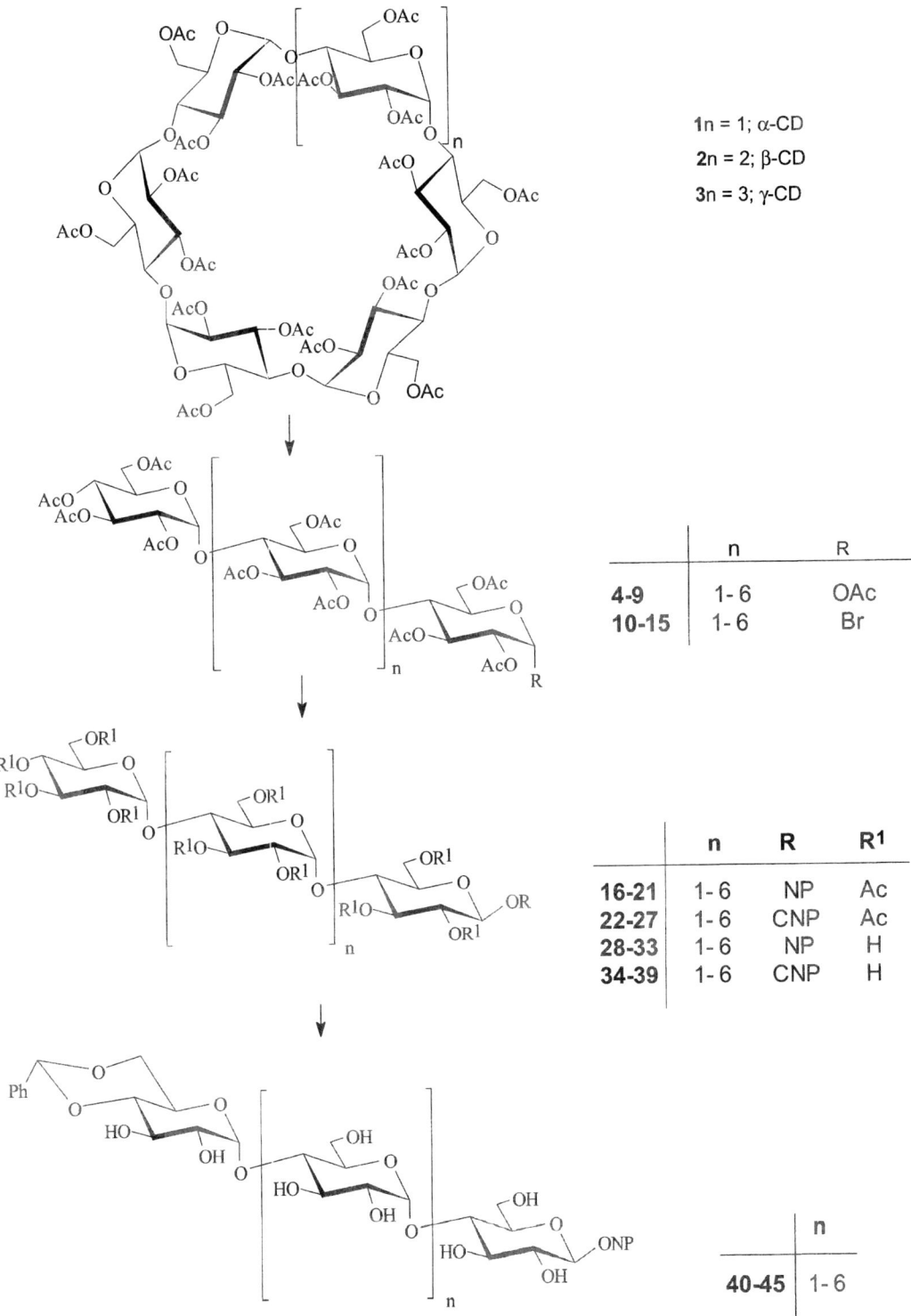

1n = 1; α-CD
2n = 2; β-CD
3n = 3; γ-CD

	n	R
4-9	1- 6	OAc
10-15	1- 6	Br

	n	R	R¹
16-21	1- 6	NP	Ac
22-27	1- 6	CNP	Ac
28-33	1- 6	NP	H
34-39	1- 6	CNP	H

	n
40-45	1- 6

Fig. 1. Synthesis of chromogenic substrates of α-amylases on cyclodextrin basis.

8–12. In the enzymatic conversion, G_7-CNP was used as a starting material and synthesised from β-CD on a preparative scale [7].

We described a new chemoenzymatic procedure of the synthesis of CNP-maltooligosides as a promising alternative to their multistep chemical synthesis, using rabbit skeletal muscle glycogen phosphorylase b [8,9].

Detailed enzymological studies revealed that the conversion of G_7-CNP was highly dependent on the conditions of phosphorolysis and/or transglycosylation. Analysis of reaction products was investigated using an HPLC system. Fig. 2 shows the enzymatic

strategy of chain shortening and the HPLC profile of phosphorolysis products. Fig. 3 represents the transglycosylation reaction for the preparation of longer oligomers and the MALDI-TOF MS of the elongated products.

The preparative scale isolation of the short (DP 3–6) and longer (DP 8–12) oligomers was achieved by size exclusion chromatography and on a semipreparative column, respectively. The productivity of the syntheses was improved by yields up to 75%. The structures of oligomers were confirmed by their chromatographic behaviours and MALDI-TOF MS data.

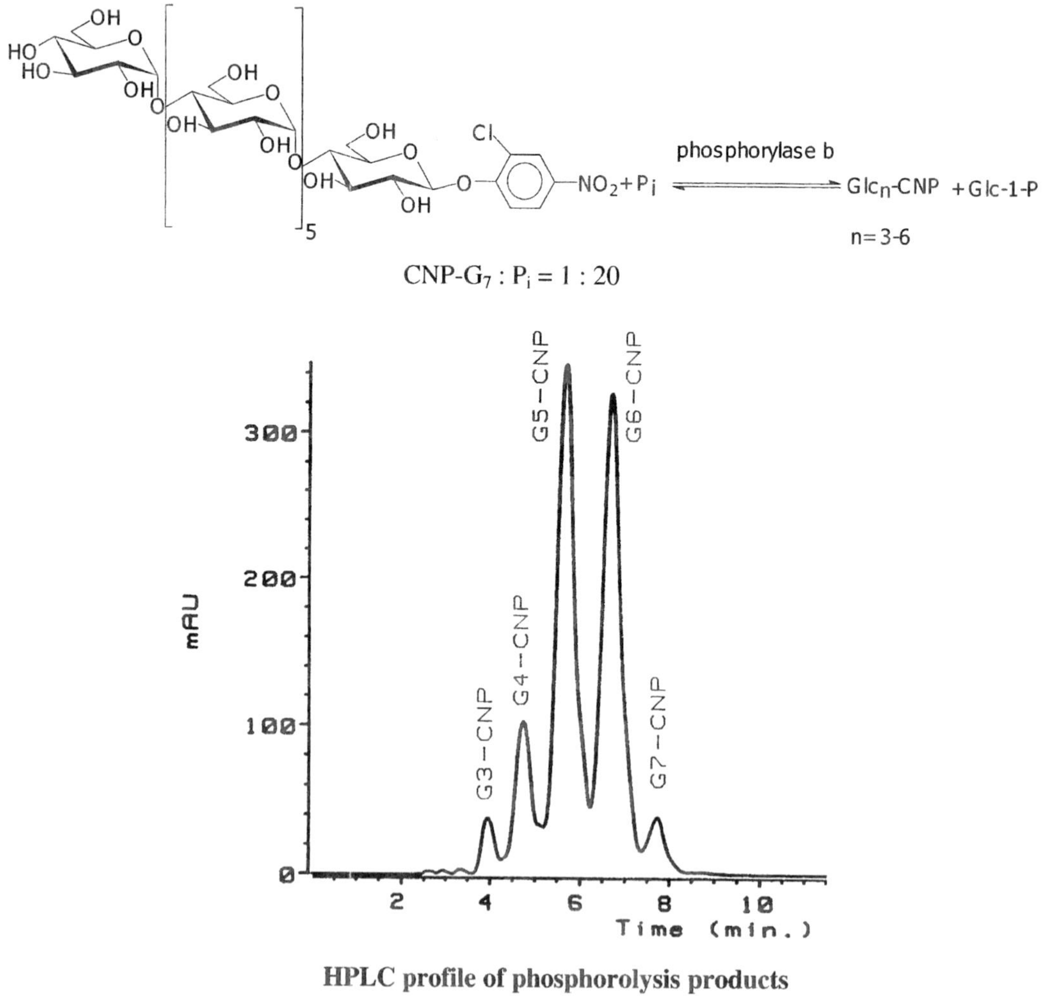

$$CNP\text{-}G_7 : P_i = 1 : 20$$

HPLC profile of phosphorolysis products

Fig. 2. Chemoenzymatic preparation of 2-chloro-4-nitrophenyl β-maltooligosaccharide glycosides using glycogen phosphorylase b (P_i inorganic phosphate) and HPLC profile of phosphorolysis products.

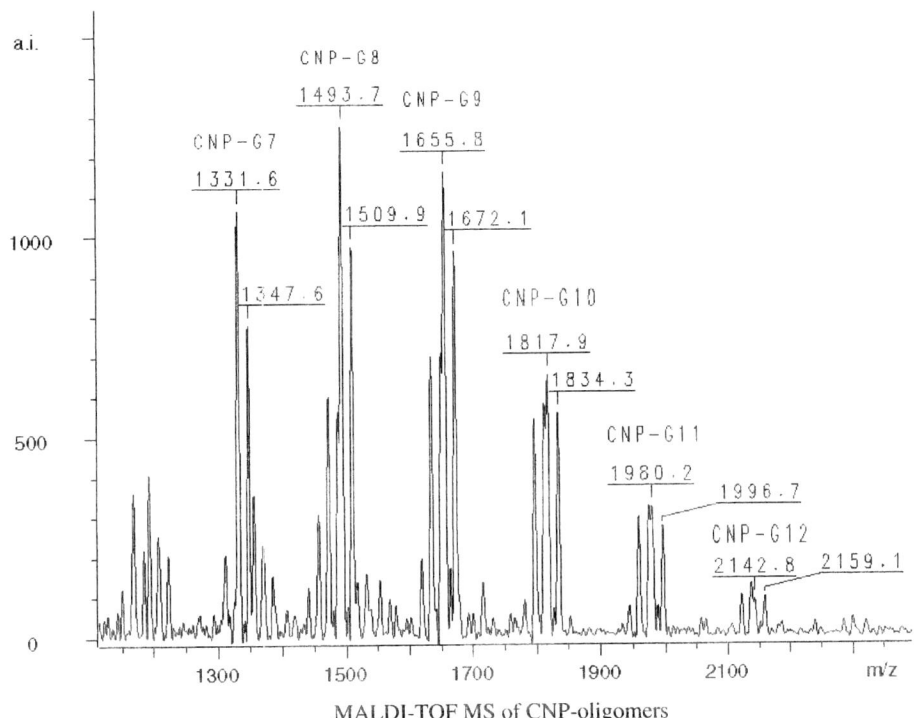

Fig. 3. Chemoenzymatic synthesis of 2-chloro-4-nitrophenyl β-maltoheptaoside acceptor-products glycosides using glycogen phosphorylase b and MALDI-TOF MS of CNP-oligomers. $[M + Na]^+$ are corresponding to the oligosaccharides of DP 7–12.

2.3. Hydrolysis of the maltooligosides

Incubations in 25 mM glycerophosphate buffer (pH 7.0) containing 5 mM $CaCl_2$ and 50 mM NaCl were carried out at 50 °C for BLA and at 37 °C for HSA for 2, 4, 6 and 8 min. The reactions were initiated by the addition of 10 μM of enzyme to solutions containing 1.7 mM of substrate. Samples were taken at the indicated time intervals and the reaction was stopped by the injection of the samples into the chromatographic column. In these studies we have taken care to exclude the secondary attacks on the substrates. The product ratios were always obtained from the early stages of hydrolysis (conversion <10%), before any secondary attack could be detected.

2.4. Chromatographic analysis

For HPLC a Hewlett-Packard 1090 Series II Liquid Chromatograph equipped with a diode array detector, automatic sampler, and ChemStation was used. The samples were separated on Supelco NH_2 5 μm column (20×0.46 cm) and RP-18

10 μm (20 × 0.46 cm^2) with different ratios of acetonitrile–water as the mobile phase flowing at a rate of 1 ml/min at 40 °C. Effluent was monitored for CNP group at 302 nm and the product of the hydrolysis were identified by using relevant standards. The quality of the acetonitrile was gradient grade. The purified water was obtained from a laboratory purification system equipped with both ion-exchange and carbon filters (Millipore, Bedford, MA, USA).

2.5. Mass spectrometry

MALDI-TOF MS analyses of the compounds were performed in positive-ion mode using a Bruker Biflex MALDI-TOF mass spectrometer equipped with delayed-ion extraction. Desorption/ionization of the sample molecules was effected with a 337 nm nitrogen laser with a pulse width of 3 ns. Spectra from multiple (at least 100) laser shots were summarised using 19 kV accelerating and 20 kV reflectron voltage. External calibration was applied using the [M + Na]$^+$ peaks of maltooligosaccharides DP 3–7 and maltoheptaose peracetate, m/z : 527.15, 689.21, 851.26, 1013.31, 1175.36 and 2142.84, respectively. The spectrum was performed in 2,4,6-trihydroxiacetophenon (THAP) matrix by mixing 10 μl of saturated matrix solution with 10 μl of sample dissolved in water, then 0.5 μl was applied to the sample target and it was allowed to dry at room temperature. The identification of compounds was done on the basis of the mass of [M + Na]$^+$ peaks.

2.6. Program for subsite mapping [5]

A computer program has been evaluated for subsite map calculations of depolymerases. The program runs in Windows and uses the experimentally determined BCFs for determination of the number of subsites, the position of the catalytic site and for calculation of subsite binding energies. The apparent free energy values were optimised by minimisation of the differences of the measured and calculated BCF data. The program called *SUMA* (*SU*bsite *M*apping of α-*A*mylases) is freely available for research and educational purposes via the Internet (e-mail: gyemant@tigris.klte.hu).

3. Results and discussion

3.1. 'Action pattern' studies by product-analysis

The application of homologous oligomeric substrates is an effective way to explore the nature of the binding site and the process of catalysis for depolymerising enzymes. Although the overall structures and tertiary foldings of polypeptide chains of α-amylase have been described, less is known about the differences in the mode of actions of these large depolymerases on the homologous maltooligosaccharide series, and the role of the subsites is poorly understood.

Therefore, our substrate series were envisaged as good candidates for further studies of the action pattern of porcine pancreatic α-amylase [10], human salivary α-amylase [11] and *B. licheniformis* α-amylase [12]. *These series are unique since neither their preparation, nor their use in the mapping of the active centre of α-amylases have been reported yet. The β-linkage is stable and not hydrolysed by α-amylases. In addition, the 4,6-O-*benzylidene-modified oligomers are very stable towards hydrolysis by *exo*-glucosidases and are useful to monitor the digestion products modified at their non-reducing end. In this way, the presence or absence of multiple attack can be studied. It is important to note that these compounds are highly water-soluble.

3.2. Action pattern of BLA: eight subsites are suggested [12]

The action pattern and product specificity of *B. licheniformis* amylase BLA was examined by utilising as model substrates CNP β-glycosides of maltooligosaccharides of DP 4–10 prepared by a chemoenzymatical route (Fig. 4).

Bacillus licheniformis α-amylase exhibits a unique pattern of action on CNP-maltooligosaccharides by cleaving maltopentaose units as main products; 68, 84 and 88% from the non-reducing end of CNP-G$_6$, CNP-G$_7$ and CNP-G$_8$, respectively, and leaving CNP-glycosides.

As the chain length increases, the maximum frequency of attack shifts toward the reducing end of the chain and CNP-G$_3$ becomes the major product;

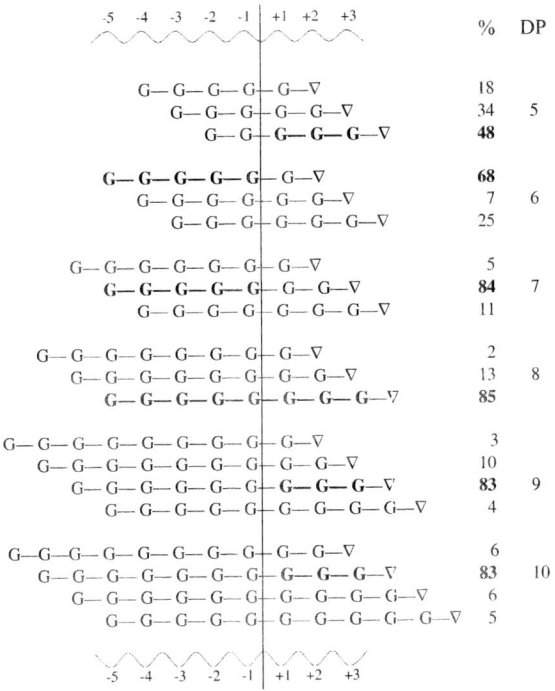

		%	DP
G—G—G—G⊦G—∇		18	
G—G—G—G⊦G—G—∇		34	5
G—G⊦G—G—G—∇		**48**	
G—G—G—G—G⊦G—∇		**68**	
G—G—G—G⊦G—G—∇		7	6
G—G—G⊦G—G—G—∇		25	
G—G—G—G—G—G⊦G—∇		5	
G—G—G—G—G⊦G—G—∇		**84**	7
G—G—G—G⊦G—G—G—∇		11	
G—G—G—G—G—G—G⊦G—∇		2	
G—G—G—G—G—G⊦G—G—∇		13	8
G—G—G—G—G⊦**G—G—G—∇**		**85**	
G—G—G—G—G—G—G—G⊦G—∇		3	
G—G—G—G—G—G—G⊦G—G—∇		10	
G—G—G—G—G—G⊦**G—G—G—∇**		**83**	9
G—G—G—G—G⊦G—G—G—G—∇		4	
G—G—G—G—G—G—G—G⊦G—G—∇		6	
G—G—G—G—G—G—G⊦**G—G—G—∇**		**83**	10
G—G—G—G—G—G⊦G—G—G—G—∇		6	
G—G—G—G—G⊦G—G—G—G—G—∇		5	

Fig. 4. Action pattern of *Bacillus licheniformis* α-amylase (BLA) with modified maltooligosaccharides at 50 °C. G, glucosyl residues; ∇, 2-chloro-4-nitrophenyl groups (CNP is connected to the reducing end in β form); DP, degree of polymerisation.

88, 83 and 83% from CNP-G_8, CNP-G_9 and CNP-G_{10}, respectively. This favourable release of CNP-G_3 was also observed for the pentamer glycoside (CNP-G_5). In this case the substrate was too short to occupy all the subsites and three glycosyl residues from the reducing end were bound in subsites ($+1, +2, +3$) and led to the formation of CNP-G_3 in 48%. The release of CNP-G_3 is much more favourable than the release of CNP-G_2 and CNP-G_1; 34 and 18%, respectively.

Our results strongly suggest the presence of at least eight subsites in BLA, five glycone sites ($-5, -4, -3, -2, -1$) and three aglycone sites ($+1, +2, +3$) and the catalytic site is located between subsites (-1 and $+1$). The subsites are labelled from the catalytic site, with negative numbers for subsites to the left (non-reducing end side) and positive numbers to the right (reducing end side) according to the proposed nomenclature of Davies et al. [3].

For all the substrates the chromogenic aglycon (CNP), which was in β-glycosidic linkages, could interact with subsites ($+2$ and $+3$), but less favourably than a glucose residue. In the ideal arrangement subsite ($+3$) was filled by a glycopyranosyl unit and the aglycone sites ($-5, -4, -3, -2, -1$) were also occupied by glucose residues which resulted in an interesting dual product-specificity for the dominant formation of CNP-G_3 and maltopentaose.

3.3. Subsite mapping of BLA

For subsite mapping we applied the procedure of Allen and Thoma [4]. BCFs were evaluated for chain length of 4–10 of CNP β-maltooligosides and these quantitative data were used to calculate the subsite map for BLA. Primarily calculated binding energies were optimised to give an optimum fit of BCFs to the experimentally measured values. Fig. 5 shows the subsite map and the apparent binding energies of subsites for BLA.

The subsites (-6), ($+5$) of BLA have $+1.0$ and $+1.1$ kJ/mol free-energy, respectively. We can suggest that subsite (-6) and ($+5$) are not real subsites for BLA. Subsite ($+4$) has positive free-energy of binding and will be referred to as 'barrier subsite'. This barrier subsite resulted in the dual product specificity of BLA. Lack of the barrier subsite leads to a more equal distribution between the potential products of the longer substrates as we found in HSA [11]. Results confirm that the eight subsites originally assumed from our experimental data are correct and BCFs are predicted correctly within experimental error.

α-Amylase from *B. licheniformis* was an attractive model enzyme for investigating the active centre. The subsite map of BLA on modified homologous oligosaccharides provided further insight into the structure and function of BLA.

Where naturally occurring enzymes have undesirable characteristics, it is hoped that altered enzymes with improved properties may be made available in the future by protein engineering. A good understanding of the relationship between protein structure and enzyme activity under different conditions is, however, a prerequisite for rational design of modified enzymes.

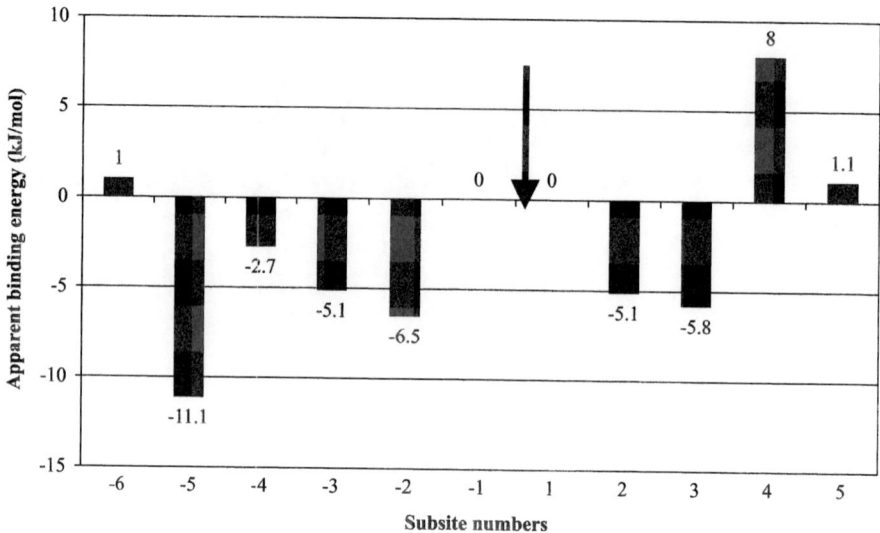

Fig. 5. Subsite map for *Bacillus licheniformis* α-amylase (BLA). The arrow indicates the location of hydrolysis. The reducing end of maltooligomers situated at the right hand of the subsite map. Negative energy values indicate bindings between the enzyme and aligned glucopyranosyl residues, while positive values indicate repulsion.

3.4. Role of the Tyr151 at subsite (+2) in HSA

The active site of human amylases harbors three aromatic residues Trp59, Tyr62 and Tyr151, which provide stacking interactions to the bound glucose moieties (Fig. 6). It has already been shown that Tyr151 occurs at subsite (+2) and may influence the size of the leaving group. To study the role of subsite (+2) in recognition of terminal residue of substrate a mutant Y151M was generated in which the tyrosine at position 151 of HSA was replaced by a methionine [6].

3.5. Product pattern of HSA and Y151M [13]

The product distribution for Y151M mutant, on the same oligosaccharide series, was very interesting and markedly different from that of the HSA. The moiety CNP-G$_1$ is the major product of hydrolysis when

CNP-G$_3$ and CNP-G$_4$ were used as substrates and was significantly released in the hydrolysis of longer oligomers (DP 6–10) as well whereas this monomer glycoside was not recognisable as a product in

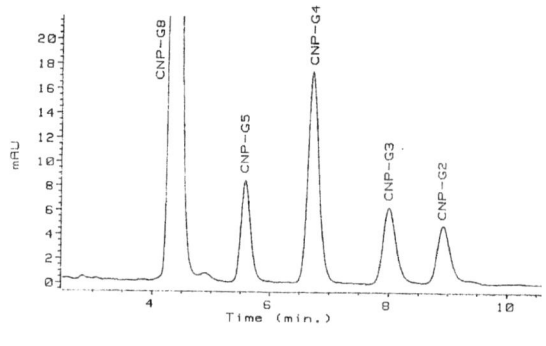

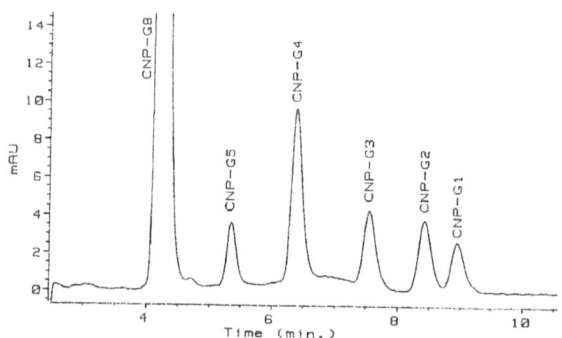

Fig. 7. HPLC chromatogram of CNP-G$_8$ hydrolysis products catalysed by HSA and Y151M mutant. Injected volume: 20 μl.

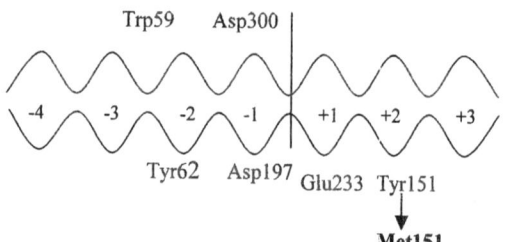

Fig. 6. Subsite model of human α-amylases.

the hydrolysis of the corresponding substrates by HSA. These results can be explained by the presence of methionine at subsite (+2) which is not advantageous to the polar glucose residues. This residue may not provide as strong a stacking and/or hydrophobic interaction as that of the tyrosine present in the wild type enzyme. Fig. 7 shows HPLC chromatograms of CNP-G_8 hydrolysis products catalysed by HSA and Y151M mutant.

3.6. Subsite mapping of HSA and the mutant Y151M

The subsite maps of HSA and Y151M mutant were evaluated by utilising as model substrates 2-chloro-4-nitrophenyl (CNP) β-glycosides of maltooligosaccharides with varying degree of polymerisation DP 3–10 (Fig. 8).

Subsite map shows that four glycone and three aglycone binding sites are present in HSA. Subsite (+3) has a moderate but significant affinity for the glucose residue (−1.5 kJ/mol) and therefore it is reasonable to consider the presence of the third subsite at the aglycone site. This subsite does not exist in Y151M. To clarify the presence of subsite (+3)

further structural and biochemical investigations are necessary. Surprisingly, the calculated binding energy at subsite (+2) (−12.0 kJ/mol) indicates a remarkably good interaction with the bound monomer unit compared with the other subsite energies. This finding is in good agreement with the three-dimensional structural model of HSA described by Ramasubbu et al. [14]. Fig. 9 shows the three-dimensional structure of the active site of HSA and bound inhibitor at subsite (+2).

Although a decreased binding energy was envisaged for subsite (+2) in Y151M, all of the subsites exhibited a lower affinity for a glucose residue compared to the corresponding subsites of HSA. Such a reduction fits well with cooperative binding of the oligosaccharide ligand. The most remarkable decrease in the calculated binding energies can be found at subsite (+2), −2.6 kJ/mol compared to −12.0 kJ/mol when HSA was used. These findings confirm clearly the role of Tyr151 at subsite (+2) and provide evidence that stacking interactions at the reducing end are important in substrate binding and product distribution in the hydrolysis of oligosaccharides catalysed by HSA. The methionine at position

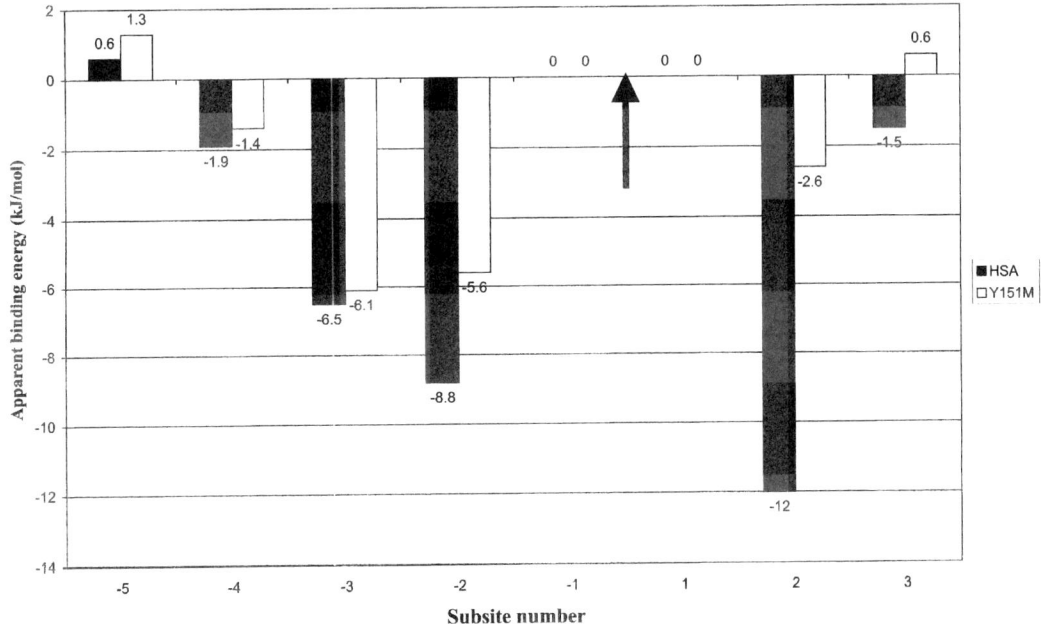

Fig. 8. Subsite maps for HSA (solid bar) and Y151M (open bar). The arrow indicates the scissile bond. The reducing end of maltooligomers situated at the right hand of the subsite map. Negative energy values indicate binding between the enzyme and aligned glucopyranosyl residues, while positive values indicate repulsion.

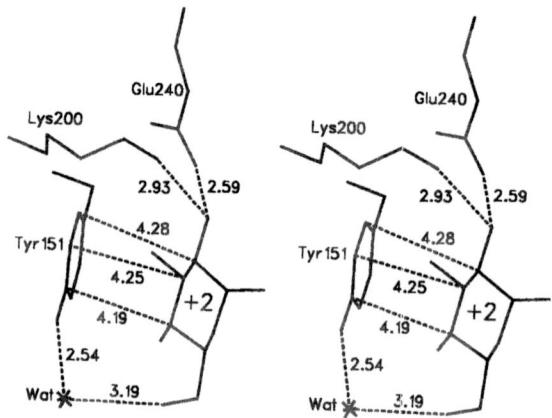

Fig. 9. Stereodiagram of the active centre of HSA. The interaction around subsite (+2) in HSA is shown. The residue Tyr151 interacts with subsite glucose through both a stacking interaction and a water molecule. In addition to its interaction with Tyr151, the glucose also has hydrogen bonding interactions with Lys200 and Glu240.

151 of HSA would not provide the stacking interaction and the water-mediated hydrogen bonding interaction provided by the tyrosine residue of HSA. However, the interactions provided by Glu240 and Lys200 would be expected to be present. The considerable reduction in the affinity at subsite (+2) in the mutant suggests that the stacking interaction should be present. The interactions with Lys200 and Glu240 may be secondary which come into play later. Thus, the disappearance of the affinity at subsite (+3) may be dependent upon the absence of several interactions at subsite (+2). The subsite (+3) appears to be more solvent exposed than (+2) as deduced from the crystal structure of the HSA:acarbose complex [14] which also could account for the relatively smaller reduction in the affinity. Nevertheless, the involvement of Tyr151 may be significant in the hydrolysis mechanism of HSA since the productive binding modes for oligosaccharides include the one in which a glucose unit occupies this site. The results of this study show significant differences in active site architecture of the wild type and the mutant. The reduction in the affinity may arise due to the side of methionine adopting a conformation, which partially occupies the space which would otherwise be occupied by a glucose.

The tyrosine residue at position 151 of the HSA is highly conserved in all mammalian amylases and those which contain a chloride ion near the active site

[15]. A tyrosine residue of similar function exists at position 139 in *Tenebrio molitor* α-amylase, as well [16].

In summary, the tyrosine residue at position 151 of HSA appears to play a major role in the hydrolysis of CNP substrates by not only affecting the ground state binding but also transition state affinities of the substrates.

Acknowledgements

This work was supported by OTKA T032005. I am grateful to Prof. András Lipták member of the Hungarian Academy of Sciences for his guidance and counsel and also to my coworkers Gyöngyi Gyémánt and Judit Remenyik for their support and contribution.

References

[1] D. Phillips, Sci. Am. 215 (1966) 78–90.
[2] J.F. Robyt, D. French, Arch. Biochem. Biophys. 138 (1970) 662–670.
[3] G.J. Davies, K.S. Wilsonand, B. Henrissat, Biochem. J. 321 (1997) 557–559.
[4] J.D. Allen, J.A. Thoma, Biochem. J. 159 (1976) 105–132.
[5] G. Gyémánt, G. Hovánszki, L. Kandra, Eur. J. Biochem. 269 (2002) 5157–5162.
[6] P. Mishra, C. Ragunath, N. Ramasubbu, Biochem. Biophys. Res. Commun. 292 (2002) 468–473.
[7] E. Farkas, L. Jánossy, J. Harangi, L. Kandra, A. Lipták, Carbohydr. Res. 303 (1997) 407–415.
[8] L. Kandra, G. Gyémánt, A. Lipták, Carbohydr. Res. 315 (1999) 180–186.
[9] L. Kandra, G. Gyémánt, M. Pál, M. Petró, J. Remenyik, A. Lipták, Carbohydr. Res. 333 (2001) 129–136.
[10] L. Kandra, G. Gyémánt, E. Farkas, A. Lipták, Carbohydr. Res. 298 (1997) 237–242.
[11] L. Kandra, G. Gyémánt, Carbohydr. Res. 329 (2000) 579–585.
[12] L. Kandra, G. Gyémánt, J. Remenyik, G. Hovánszki, A. Lipták, FEBS Lett. 518 (2002) 79–82.
[13] L. Kandra, G. Gyémán, J. Remenyik, C. Ragunath, N. Ramasubbu, FEBS Lett. 544 (2003) 194–198.
[14] N. Ramasubbu, C. Ragunath, P.J. Mishra, J. Mol. Biol. 325 (2003) 1061–1076.
[15] S. D'Amico, C. Gerday, G. Feller, Gene 253 (2000) 95–105.
[16] S. Strobl, K. Maskos, M. Betz, G. Wiegand, R. Huber, F.X. Gomis-Rüth, R. Glockshuber, J. Mol. Biol. 278 (2000) 617–628.

ELSEVIER

Journal of Molecular Structure (Theochem) 666–667 (2003) 499–505

THEO
CHEM

www.elsevier.com/locate/theochem

Structural biology, ligand binding, metabomics[1]—the changing face of high-field, high-resolution NMR spectroscopy

István Pelczer*

Department of Chemistry, Princeton University, Princeton, NJ 08544, USA

Abstract

NMR spectroscopy has become an indispensable tool for chemistry, molecular biology, biochemistry, and many other disciplines over the last half of century. For the last two decades or so technical developments and investments in high-field, high-resolution NMR were driven dominantly by needs of structural biology. Recently, however, ligand binding and screening, and all kinds of metabolic studies by NMR have joined as major disciplines, and may play increasing role in future developments.
© 2003 Elsevier B.V. All rights reserved.

Keywords: NMR; Structural biology; Ligand screening; Metabomics; Metabolic studies

1. Introduction

NMR spectroscopy went through attractive improvements in the last half of century, such as introduction of PFT methods in the late sixties, affordable isotope labeling of biomolecules, multi-dimensional methods up to 4D in practical terms, for example. Very recent developments include introduction of extremely high-field magnets (currently up to 900 MHz), cryoprobe technology, use of relaxation-optimized methods (TROSY [1]), utilizing residual dipolar couplings (RDC) [2], and measuring microsamples.

Application of NMR spectroscopy to structural biology has been the dominant force behind technology development and investment (primarily in pharmaceutical industry and academic laboratories working on related projects). In recent years, however, additional projects have lined up as important contributors, such as ligand binding/screening by NMR, and the rapidly growing field of metabolic studies.

The following is a brief summary of this changing picture with representative references as the author sees it. It is not intended to be a comprehensive review, rather a quick assessment of the field and these new tendencies, invitation for further analysis and discussion. The scope is reduced to high-resolution liquid-state NMR applications, although it should be noted that solid-state NMR has shown

---

* Tel.: +1-609-258-2342; fax: +1-609-279-6746.

E-mail address: ipelczer@princeton.edu (I. Pelczer).

[1] The term *metabomics* is used here to cover a large family of disciplines and applications in metabolic studies, such as metabonomics, metabolomics, metabolic profiling, metabolic flux analysis, etc. in line with the other sister terms of *genomics* and *proteomics*.

0166-1280/$ - see front matter © 2003 Elsevier B.V. All rights reserved.
doi:10.1016/j.theochem.2003.08.113

exceptional enhancement over the last decade or so, and plays increasing role in biomolecular structural studies.

2. Recent developments of NMR applied to structural biology

NMR spectroscopy and X-ray are siblings, in the same time they are long-term competitors in the field of structural studies, especially those for biomolecules. Both have their advantages and limitations and, certainly in the opinion of the author, they are always better to look at as complementary techniques, which should be applied in an integrated fashion. There are, however, large families of biomolecular structures, which remain difficult targets for either technique, such as membrane proteins, large nucleic acids (RNA at the first place), and highly mobile systems, such as prion proteins and similar. In addition, it remains and interesting conceptual question, how much our view is biased by the technical capabilities in terms of understanding *structure and function*, in other words how extensively we can incorporate dynamics, conformational freedom, and mobility into the picture created by the help of instrumental methods, which rely on relatively stable and rigid models. The discussion of these aspects is out of the scope of this article, however.

NMR has been having a hard time to keep up with X-ray in the last decade, as NMR is a relatively slow method, the result is usually a less well determined structure (or ensemble of structures), and there is an obvious size limitation for NMR, which does not apply to X-ray. In the same time, X-ray analysis has shown a tremendous improvement over this decade with introduction of high throughput, highly automated sample preparation and data acquisition, use of high-energy synchrotron beamlines, and powerful software tools. It reached the point, where the capacity of the system may be larger than the influx of possible reasonable samples, and it may pose difficulty to analyze the tremendous amount of the output data [3].

There are, however, circumstances, where NMR remains a unique tool for structure elucidation. There are some substances, which are difficult or impossible

to crystallize while they make a decent NMR sample. Often dynamics and conformational diversity itself are the subjects of investigation, when NMR is an indispensable tool. It also happens, that packing forces in the crystal lead to different structural behavior than what can be observed in solution [4], which solution structure is more relevant to potential biological function.

NMR in structural biology has reached formerly unbelievable results recently. The largest system characterized by NMR is of 900 kDa size [5]. Molecular species in the 100 kDa regime (and beyond) can be studied successfully [6–8]—although full high-resolution 3D structure determination beyond MW = 30–40 kDa remains an individual challenge. In the same time, sophisticated and robust protocols [9], experimental and software toolkits [10, 11] have been developed for high-throughput (HT) structure determination of smaller proteins. Today a new structure can be determined, usually with relatively low resolution, in a week or two. Roughly a dozen of such HT proteomics centers, about one third of them using NMR, enjoy significant government support in the USA, and similar centers have been established in Japan [12] and Europe also. There is a rapidly increasing number of new entries in NMR [13] and other publicly accessible structural databases [14], let alone proprietary collections in biotech and pharmaceutical industries.

An example of a low-resolution protein structure, determined by NMR (not by HT methods), is shown in Fig. 1 [15]. Skn-1 is a maternally expressed transcription factor with a novel fold for the DNA binding domain. With the help of a ^{13}C-labeled derivative we managed to identify enough number of long-range NOE contacts that establish a tertiary fold for the domain. The NMR solution structure of this domain is highly similar to that determined simultaneously by X-ray spectroscopy in the DNA–Skn-1 complex [16].

3. Ligand binding, ligand screening by NMR

NMR spectroscopy offers unique capabilities when equilibrium systems are concerned. Depending on the kinetics of the binding process and the NMR time scale one can either observe only the complex, two

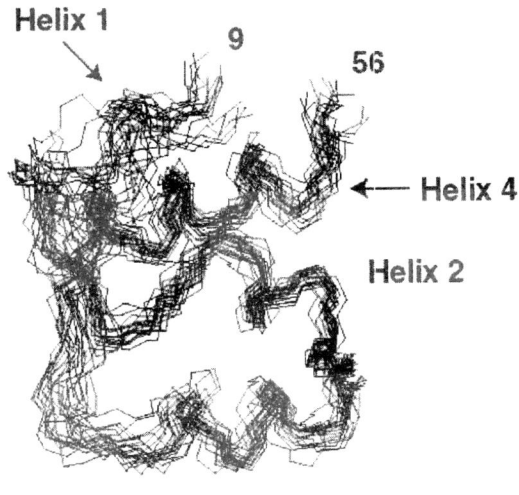

Fig. 1. Low-resolution NMR structure of the residues 9–56 of the Skn-1 domain. Only the backbone is shown for 20 energy-minimized conformers [15].

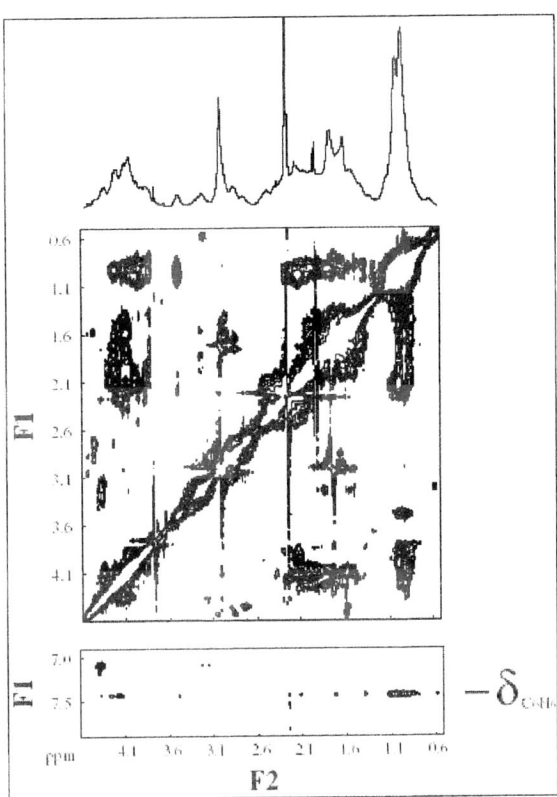

(or more) distinct, but exchanging species, or a weighted average spectrum [17]. The binding process and its kinetics can be followed by observing direct contacts between the interacting species, or by any parameter, which is a function of mobility, conformation, or any other NMR variable. Mobility is reflected in the lineshape, NOE (or ROE) effect, diffusion properties, while conformational features determine spatial contacts (NOE/ROE, RDC), scalar coupling values, chemical shifts, etc. Identification and quantitative characterization of ligand binding (mostly, but not exclusively, a small molecule–large target binding) is a central issue in drug research, and has changed priorities for NMR laboratories in pharmaceutical industry over the last few years. In the same time, it is just as important and interesting in context of all fundamental biochemical, biological research. In the past year or two several excellent reviews [18], books [19], and thematic issues of scientific journals [20] discussed this subject.

The following example is about an engineered three-helix bundle binding small hydrophobic ligands in a cavity [21], as shown in Fig. 2 (insert). A conventional 2D-NOE spectrum clearly shows NOE contact with the benzene dissolved in the same sample (Fig. 2). The system and the binding kinetics (K_d)

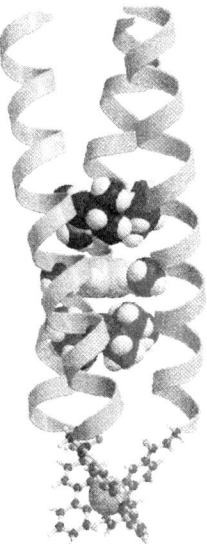

Fig. 2. ^{1}H–^{1}H NOESY spectrum of the engineered three-helix bundle and benzene [21]. Incorporation of benzene into the hydrophobic cavity is proven by NOE contacts between the benzene protons and methyl resonances of selected amino acids.

have been also characterized by chemical shift analysis (also using ^{19}F for perfluoro-benzene), diffusion measurements, and saturation transfer difference (STD) experiments [18,21].

Ligand binding can be characterized basically (but not exclusively) in two ways. One can observe and follow the behavior of the target species (usually a large, slowly tumbling molecule), or that of the ligand (usually a more mobile, small species). The former scenario usually focuses on chemical shifts (chemical shift mapping) or differential dynamics and typically relies on isotope (^{15}N, ^{13}C) labeling of the target molecule. For such applications the target molecule has to be well characterized by NMR in order to identify specific changes. Typical implementations are SAR-by-NMR [22] and SHAPES [23], also SOLVE-NMR [24]. The technique is useful primarily for strong and medium binders.

Observing the ligand, commonly in excess, usually relies on transferred magnetization (tr-NOE, tr-ROE, or saturation transfer) between the bound state and free state. This can be used to quickly screen a large pool of potential ligands, and identify those in a simple and inexpensive fashion, which are medium or weak binders [18]. The sensitivity of the method can be increased tremendously if the initial source of the magnetization is the bulk water (WaterLOGSY [25]). The big advantage of this methodology is, that it needs no isotope labeling, the target system can be very large and be present in very low quantity, even practically invisible on the spectrum. Binding topology can also be identified. Usually the method does not report about the specificity of the binding, however, and works well only in a certain regime of kinetics [18,19].

Fig. 3 presents an example of using saturation transfer difference (STD) method to identify binding. In this case the protein alcohol dehydrogenase was mixed with AMP and the well-known drug ibuprofen. Selective irradiation of the protein resonances at two different frequencies (-0.7 and $+0.3$ ppm, respectively, but otherwise using identical acquisition and data processing parameters) clearly shows binding of both small molecules. The spin-diffusion-based

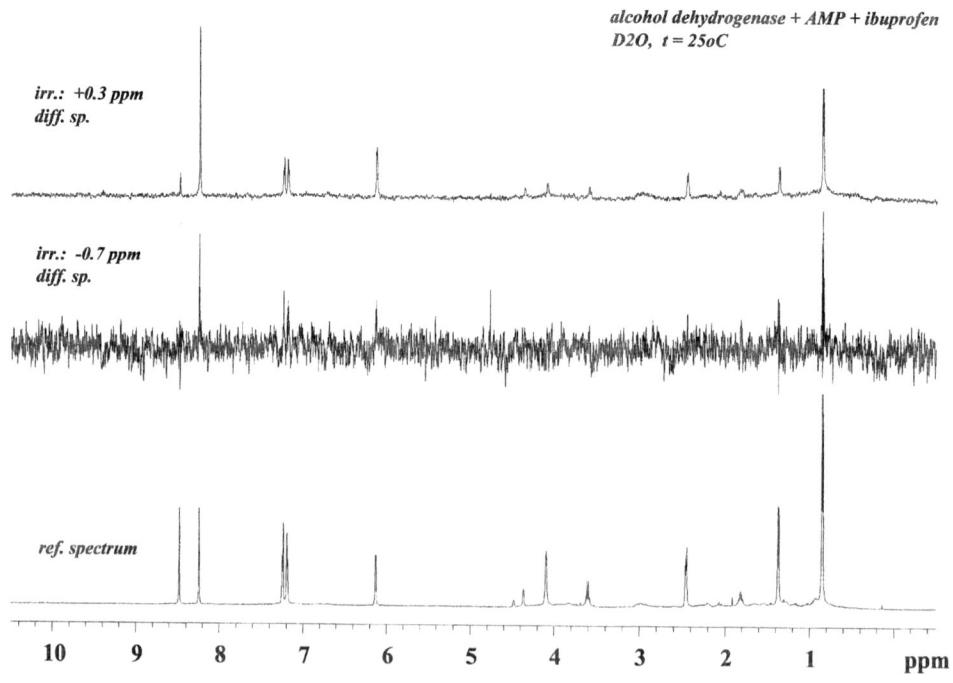

Fig. 3. Saturation transfer difference (STD) spectra of the protein alcohol dehydrogenase, AMP, and ibuprofen in D_2O. On and off irradiations of 2 s each have been applied at -0.7 $+0.3$ ppm, and ca. $+20$ ppm, respectively. The reference spectrum shown at the bottom, while two difference spectra are shown above.

magnetization transfer is much more efficient for the +0.3 ppm irradiation. (The sample was courtesy of Bill Creekmore, FMC Corp., Princeton, NJ, USA.)

Ligand screening by NMR (often in combination with mass spectroscopy, MS) has become essential part of the protocol in drug research and two critical points. Usually a very large pool of potential ligands (molecular fragments, selected scaffolds, or functionalized larger structures) from a random, or a targeted library is screened to find leads. At this point the primary goal is to narrow the pool by a couple of magnitudes or more. NMR may play a critical role again later, when refinement of the few remaining promising structures follows, which have passed the various filters and selection criteria along the earlier part of the process. In this phase structure determination of the ligand–target complex becomes important as it may provide a clue for further modification of the potential drug molecule.

4. NMR in metabomics

Recent developments in genomics and proteomics have revolutionized our understanding of life and living systems. The picture will be complete by learning more about metabolic processes and how they reflect the underlying mechanisms and their kinetics. Metabolic analysis and characterization by NMR, also in combination with MS and ion cyclotron resonance (ICR) spectroscopy, will also serve as indispensable tools for toxicology studies, risk assessment, plant metabolism [26], environmental monitoring, direct medical diagnostics, and for many other purposes. Metabolic flux measurements by NMR provide insight into the mechanism and kinetics of the biochemical processes in vivo [27].

NMR has been used for characterizing metabolites and metabolic processes for a long time [28,29], which field has been pioneered by Jeremy Nicholson and his group at Imperial College (London, UK) [30], also by Metabometrix, a spin-off biotech company [31]. There are several additional new biotech companies, which focus on such studies now, and most pharmaceutical companies have established their research facilities for metabolic analysis. The efficiency of such studies has been increased

dramatically in the last 5 years or so, as the sensitivity of NMR spectroscopy has improved on a large scale. In this context introduction of high field systems, cryoprobes, and microsample NMR [32], as well as high-resolution magic angle spinning (HR-MAS) methods for tissue analysis [33] have made a big difference.

NMR spectra of the metabolome can be analyzed by statistical methods, one can look for individual biomarkers, and full or partial component analysis is also possible—all from the same spectra. As most samples are of small absolute quantity, and low concentration components may carry very important information, sensitivity remains a critical issue. That is one reason why metabolic studies—or metabomics in general terms—present an important factor in technical developments for NMR spectroscopy.

Metabonomics [34] focuses on higher-level organisms, uses biofluids and tissue samples, and is used extensively in toxicology studies and in risk assessment. It is used in combination with sophisticated data analysis [35]. As a result it helps to identify populations of common nature, disease, and individual response to treatment. There is a large-scale joint endeavor between major pharmaceutical companies and the IC group to build a database for toxicology studies (COMET) [36]. Some of the more recent developments include studying and characterizing tropical diseases (parasite infections). Another recent study presented a simple and inexpensive way (NMR analysis of blood plasma) to identify patients with risk for cardiac problems [37]. This simple analysis could even cluster individuals with different number of damaged coronary vessels.

We are only at the doorstep of potential use of metabolic analysis and characterization. Cell and molecular biology, biochemistry, agricultural and environmental studies, direct medical diagnostics will all benefit of such applications. One interesting novel extension is to look at single-cell organisms, such as bacteria. Analysis of their biochemistry, communication (for example, quorum sensing [38]), and response to environmental and other effects will help to enhance our knowledge of fundamental biochemical processes. In the same time, it will provide an avenue to devise better drugs, also against resistant strains, and to manipulate their behavior for novel purposes, such as biological computing [39]. In

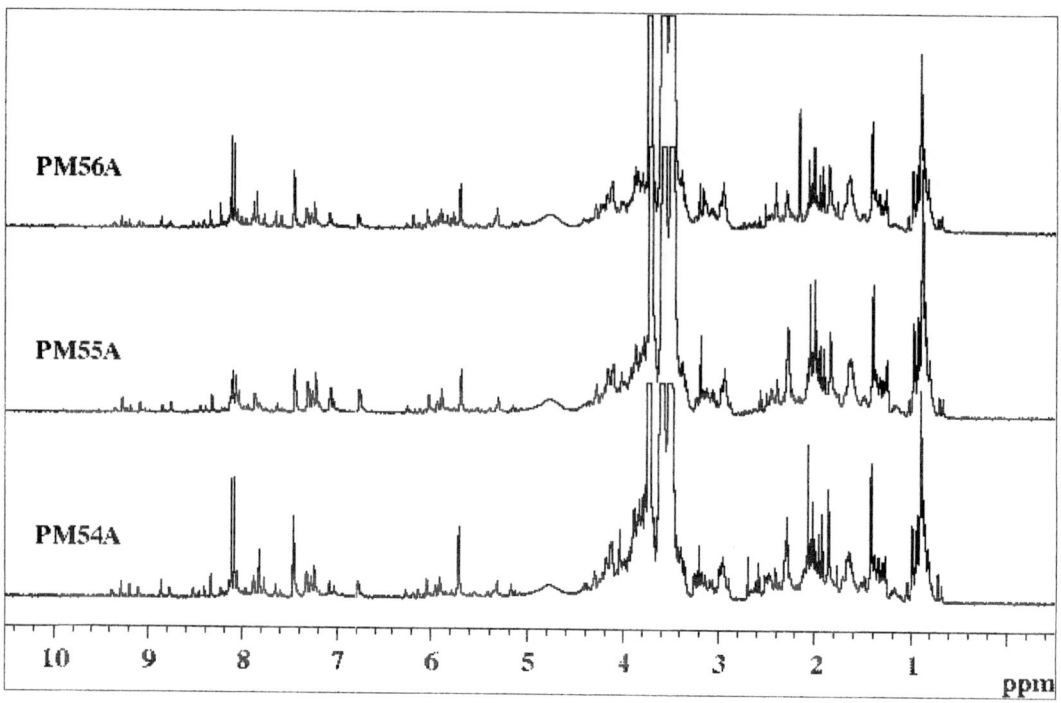

Fig. 4. ¹H NMR spectra of cell-extracts (MW cut-off 3 kDa) of three genetically modified *E.coli* strains with differential strength of the ribosomal binding site (RBS). Ca. 15 μl cell extract was used for each experiment in a capillary flow probe (CapNMR™, Protasis/MRM, Marlboro, MA) at 600 MHz, in a Varian Unity/INOVA NMR spectrometer [40].

Fig. 4 cell-extracts from different genetically modified *E.coli* bacteria are shown. After breaking up the cells, the metabolites were separated from the rest of cell material by simple molecular filtration, therefore no chemical modification or other modulation of the small molecular content was introduced [40]. The NMR spectra show characteristic differences, and will be a subject of detailed analysis at all three levels, mentioned above.

5. Summary

NMR spectroscopy has been tremendously successful supporting structural biology projects and related research over the last couple of decades or so. Such applications have been the driving force for constant technical developments and major investments by pharmaceutical industry at the first place, but also by academic institutions. In recent years additional important applications emerged, such as ligand binding and ligand screening, and a great variety of metabolic studies. Both of them require the highest sensitivity and best instrumental performance possible, and have generated additional interest in improving the hardware, software, and developing new protocols. Ligand binding and screening, and especially metabolic studies by NMR present a growing contribution to the driving force for technical enhancements, and will provide exceptional tools for biological, biochemical, and drug research, as well as for a wide variety of medical applications.

Acknowledgements

Kind invitation to the conference is appreciated to Profs. Imre Csizmadia and Botond Penke.

References

[1] K. Pervushin, R. Riek, G. Wider, K. Wüthrich, Proc. Natl Acad. Sci. USA 94 (1997) 12366.

[2] (a) A. Bax, G. Kontaxis, N. Tjandra, Meth. Enzymol. 339 (2001) 127.
(b) J.H. Prestegard, A.I. Kishore, Curr. Opin. Chem. Biol. 5 (2001) 584.

[3] P. Moore, personal communications, 2003.

[4] V. Gaponenko, A.S. Altieri, J. Li, E.A. Byrd, J. Biomol. NMR 24 (2002) 143.

[5] J. Flaux, E.B. Bertelsen, A.L. Horwich, K. Wüthrich, Nature 418 (2002) 207.

[6] R. Riek, J. Fiaux, E.B. Bertelsen, A.L. Horwich, K. Wüthrich, J. Am. Chem. Soc. 124 (2002) 12144.

[7] V. Tugarinov, R. Muhandiram, A. Ayed, L.E. Kay, J. Am. Chem. Soc. 124 (2002) 10025.

[8] G.M. Clore, C.D. Schwieters, Curr. Opin. Struct. Biol. 12 (2002) 146.

[9] J.L. Markley, E.D. Ulrich, W.M. Westler, B.F. Volkman, in: P.E. Bourne, G. Weissig (Eds.), Structural Bioinformatics, Wiley–Liss, Inc., Hoboken, NJ, USA, 2003, Chapter 5.

[10] see, for example, G.T. Montelione, CABM NMR Software, CABM/Rutgers University, Piscataway, NJ, USA, www-nmr.cabm.rutgers.edu/NMRsoftware/nmr_software.html

[11] S. Kim, T. Szyperski, J. Am. Chem. Soc. 125 (2003) 1385.

[12] S. Yokoyama, Y. Matsuo, H. Hirota, T. Kigawa, M. Shirouzu, Y. Kuroda, H. Kurumizaka, S. Kawaguchi, Y. Ito, T. Shibata, M. Kainosho, Y. Nishimura, Y. Inoue, S. Kuramitsu, Prog. Biophys. Mol. Biol. 73 (2000) 363.

[13] (a) BioMagResBank, www.bmrb.wisc.edu.
(b) F. Doreleijers, S. Mading, D. Maziuk, K. Sojourner, L. Yin, J. Zhu, J.L. Markley, L. Ulrich, J. Biomol. NMR 26 (2003) 139.

[14] Protein Data Bank, www.rcsb.org/pdb.

[15] M.-C. Lo, S. Ha, I. Pelczer, S. Pal, S. Walker, Proc. Natl Acad. Sci. USA 95 (1998) 6455.

[16] P.B. Rupert, G.W. Daughdrill, B. Bowerman, B.W. Matthews, Nat. Struct. Biol. 5 (1998) 484.

[17] E.D. Becker, High Resolution NMR. Theory and Chemical Applications, third ed., Academic Press, San Diego, 2000.

[18] (a) M. Pellecchia, D.S. Sem, K. Wüthrich, Nat. Rev. Drug Discovery 1 (2002) 211.
(b) B.J. Stockman, C. Dalvit, Prog. NMR Spectrosc. 41 (2002) 187.
(c) D.F. Wyss, M.A. McCoy, M.M. Senior, Curr. Opin. Drug Discovery Dev. 5 (2002) 630.
(d) B. Meyer, T. Peters, Angew. Chem. Int. Ed. Engl. 42 (2003) 864.

[19] J. Jiménez-Barbero, T. Peters (Eds.), NMR Spectroscopy of Glycoconjugates, Wiley, New York, 2000.

[20] See, for example, the thematic issues: (a) Comb. Chem. High Throughput Scr. 5 (2002);
(b) Curr. Topics Med. Chem. 3 (2003).

[21] A.J. Doerr, M.A. Case, I. Pelczer, G.L. McLendon, submitted for publication.

[22] S.B. Shuker, P.J. Hajduk, R.P. Meadows, S.W. Fesik, Science 274 (1996) 1531.

[23] J. Fejzo, C.A. Lepre, J.W. Peng, G.W. Bernis, Ajay, M.A. Murcko, J.M. Moore, Chem. Biol. 6 (1999) 755.

[24] M. Pellecchia, D. Meininger, Q. Dong, E. Chang, R. Jack, D.S. Sem, J. Biomol. NMR 22 (2002) 165.

[25] C. Dalvit, G. Fogliatto, A. Stewart, M. Veronesi, B. Stockman, J. Biomol. NMR 21 (2001) 349.

[26] See, for example, the thematic issue: Phytochemistry 62 (2003).

[27] (a) T. Szyperski, Q. Rev. Biophys. 31 (1998) 41.
(b) T. Szyperski, R.W. Glaser, M. Hochuli, J. Fiaux, U. Sauer, J.E. Bailey, K. Wüthrich, Met. Eng. 1 (1999) 189.

[28] J.K. Nicholson, I.D. Wilson, Prog. NMR Spectrosc. 21 (1989) 449.

[29] J.-N. Barbotin, J.-C. Portais (Eds.), NMR in Microbiology. Theory and Applications, Horizon Scientific Press, 2000.

[30] J.K. Nicholson, J. Connelly, J.C. Lindon, E. Holmes, Nat. Rev. Drug Discovery 1 (2002) 153.

[31] Metabometrix Ltd, London, UK, www.metabometrix.com.

[32] (a) M.E. Lacey, R. Subramanian, D.L. Olson, A.G. Webb, J.V. Sweedler, Chem. Rev. 99 (1999) 3133.
(b) Capillary flow NMR probe, CapNMR™, Protasis/MRM www.protasis.com/MRM.
(c) G. Schlotterbeck, A. Ross, R. Hochstrasser, R.H. Senn, T. Kühn, D. Marek, O. Schett, Anal. Chem. 74 (2002) 4464.

[33] M.E. Bollard, S. Garrod, E. Holmes, J.C. Lindon, E. Humpfer, M. Spraul, J.K. Nicholson, Magn. Reson. Med. 44 (2000) 201.

[34] J.K. Nicholson, J.C. Lindon, E. Holmes, Xenobiotica 29 (1999) 1181.

[35] J.C. Lindon, E. Holmes, J.K. Nicholson, Prog. NMR Spectrosc. 39 (2001) 1.

[36] C.M. Henry, C & EN 80 (2002) 66.

[37] J.T. Brindle, H. Antti, E. Holmes, G. Tranter, J.K. Nicholson, H.W.L. Bethell, S. Clarke, P.M. Schofield, E. McKilligin, D.E. Mosedale, D.J. Grainger, Nat. Med. 8 (2002) 1439.

[38] X. Chen, S. Sauder, N. Potier, A. Van Dorsselaer, I. Pelczer, B.L. Bassler, F.M. Hughson, Nature 415 (2002) 545.

[39] R. Weiss, S. Basu, S. Hooshangi, A. Kalmbach, D. Karig, R. Mehreja, I. Netravali, Nat. Comp. 2 (2003) 47.

[40] I. Pelczer, A. Kalmbach, M.A. Case, R. Weiss, submitted for publication.

ELSEVIER

Journal of Molecular Structure (Theochem) 666–667 (2003) 507–513

THEO
CHEM

www.elsevier.com/locate/theochem

Molecular pathomechanisms of Alzheimer's disease

B. Penke[a,b,*], Z. Datki[b], C. Hetényi[b], Z. Molnár[b], I. Lengyel[c],
K. Soós[b], M. Zarándi[b]

[a]*Protein Chemistry Research Group of Hungarian Academic of Science, University of Szeged,
Dóm tér 8, Szeged H-6720, Hungary*
[b]*Department of Medical Chemistry, University of Szeged, Dóm tér 8. Szeged H-6720, Hungary*
[c]*Department of Biochemistry, Biological Research Centre, Szeged H-6726 Hungary*

Abstract

β-Amyloids are neurotoxic compounds produced in the brain and cause Alzheimer-type senile dementia. Their toxicity is in good correlation with their aggregation properties. Short peptide fragments of β-amyloid prevent Aβ-toxicity, probably by binding on the Aβ-aggregate surface.
© 2003 Elsevier B.V. All rights reserved.

Keywords: β-Amyloid; Neurodegeneration; Neuroprotection; Alzheimer's disease

1. Introduction

Alzheimer's disease (AD) is the most common form of dementia in the elderly population. It is clinically characterized by a progressive loss of cognitive abilities, progressive memory and intellectual deficits. The pathological characteristics of AD are as follows:

1. Senile plaques (extracellular insoluble, congophilic peptide and protein aggregates containing β-amyloid peptides of 40–43 amino acids and also τ-proteins).

2. Neurofibrillary tangles, NFTs in the brain cortex (intracellular fibers composed of hyperphosphorylated cytoskeletal tau-proteins).

3. Loss of synapses and neuronal death [1–4].

These hallmarks serve as the major criteria for a definite postmortem diagnosis of AD. Brain capillaries are also damaged. β-amyloid (Aβ) peptides are deriving from amyloid precursor protein (APP), a large 695 amino acid transmembrane molecule expressed in neurons and on 751 or 770 amino acid form in glial cells [5–7]. All Aβ peptides are liberated by the actions of β-secretase (BACE [8,9]) and γ-secretase.

It has been a long-standing debate whether amyloid plaques and NFT-s are essential components of the pathogenic process or only markers of neuredegeneration, so various hypotheses have been suggested to explain the molecular pathogenesis of AD. The two most favoured are the amyloid cascade hypothesis

Abbreviations: MTT, 3-(4,5-Dimethyl-2-thiazolyl)-2,5-diphenyl-2H-tetrazolium bromide; DMSO, dimethyl sulfoxide; TFE, trifluoro ethanol; EGTA, ethylene glycol-bis (β-aminoethyl ether)-N,N,N′,N′-tetraacetic acid; DAMGO, D-Ala[2],N-Me-Phe[4], Gly[5]-ol]-Enkephalin acetate; SA, simulated annealing; Pr, propionyl; Hex, hexanoyl; Oct, octanoyl goup.

* Corresponding author.

E-mail address: penke@ovrisc.mdche.u-szeged.hu (B. Penke).

('baptists': Aβ induce all the subsequent pathology, Hardy et al. [10]) and the tau or tangle hypothesis ('tauists': NFT formation has primacy in the pathological events). Biochemical investigation revealed that all APP-mutations (that were causal for early-onset familial AD) caused increased formation of Aβ peptides and specifically the more fibrillogenic Aβ 1-42, so Aβ peptide is a prime suspect in the pathogenesis of AD. The amyloid cascade theory proposes Aβ as the central trigger of the pathological changes observed in the brains of AD patients including NFT-formation and neuronal death. On the other hand, newly found tau-mutations in other dementias have added weight to the hypothesis as it suggests that tau-pathology is a downstream but essential part of the dementing process.

There are several studies published on the structure of the Aβ peptide and its fragments. It was shown, that equilibrium exists between the helical, random coil and β-sheeted forms of this peptide [11–14]. The plaques consist of mainly the β-sheeted form of Aβ [15–18]. Although, both experimental [19] and theoretical [20] approaches were published, the exact structure of the fibrils and plaques of Aβ is not yet solved at atomic level. Reviews on this topic were published e.g. by Serpell [21] and by Roher et al. [22].

In 1999, two groups independently developed atomic models of Aβ aggregates. Both of them based their models on the data available from the literature: the Aβ molecules adopted antiparallel β-sheet conformation with a type I $G^{25}SNK^{28}$ turn in these studies. While modeled [23] the 12–42 part of the Aβ, and used [24] the $Aβ_{1–43}$ form of the peptide in their study. As the crystallization of Aβ seems to be hopeless this time, such models are of great importance for drug (inhibitor) design.

There are several approaches using Aβ as a therapeutic target [25]. One of these approaches is the discovery of peptide inhibitors of amyloid β-peptide polymerization [26,27] (β-sheet breakers, BSBs). In our laboratory, we use two new terms for neuroprotective peptides: functional antagonists-invers agonists (FA–IA) and amyloid surface binding molecules (ASBIM).

There are evidences that Aβ peptides are generated before tau aggregates. Aβ is neurotoxic to cultured cells and, at least in some conditions, induces tau-phosphorylation. In our work we accept the evidences

of both hypotheses (both Aβ formation and tau-hyperphosphorylation are needed for Alzheimer's dementia), however, we found Aβ peptides as central triggers for further pathological events and neuronal death. The aims of our research work are as follows: (1) To find correlation between amyloid structure, amyloid aggregation and neurotoxicity; (2) To find and design small molecules, peptides or peptidomimetics, which prevent the neurotoxic effect of Aβ peptides, either as BSBs, or by other mechanism; (3) To investigate the effect of Aβ peptides on G-protein activation.

2. Materials and methods

2.1. Molecular docking

The popular docking program: AutoDock [28] has been applied to perform docking simulations. A new approach named blind docking was used in our calculations; details are given in Section 3.

2.2. MTT assay using SH-SY5Y cells

The SH-SY5Y cells used here were obtained from Marcel Ameloot, LUC Diepenbeek. Cells were plated for 24 h at 37 °C on 96-well plates (Nunc, Roskilde, Denmark) at a density of 3×10^4 cells/well, to confluency, with 5% CO_2 in a humidified atmosphere with Dulbecco's modified Eagle's medium (MEM): F-12 (1:1) with phenol red (Sigma). L-Glutamine (4 mM; Gibco), penicillin (200 units/ml; Gibco), streptomycin (200 μg/ml; Gibco), MEM non-essential amino acids (100 ×; Gibco), and 10% fetal bovine serum (FBS; Gibco) were added to the medium. After 24 h the cells were treated with aggregated peptides in a medium containing 2% FBS (100 μl/well). The peptides were ultrasonicated in solution (10 min) and aggregated for 1 or 72 h under gentle shaking at room temperature in the culture medium. The cells were incubated with the aggregated peptides for 24 h, 10 μl of aqueous MTT solution (4 mg/ml) was then added to each well (100 μl), and the mixture was incubated at 37 °C for 3 h. The MTT solution was carefully decanted off, and formazan was extracted from the cells with 100 μl of a 4:1 DMSO–EtOH mixture in each well. Colour was measured with a 96-well ELISA plate reader at 550 nm, with the reference filter set to 620 nm. All MTT assays were repeated three times.

2.3. Preparation of Aβ peptide solution

Aβ peptides were dissolved in TFE in a final concentration of 1×10^{-3} M. For the GTP binding experiments the TFE solutions were diluted $10 \times$ in water and left undisturbed for 1 h prior the addition of the incubation solution (a further $10 \times$ dilution). The final TFE concentration did not exceed 1%.

2.4. Preparation of rat brain membranes

A crude membrane fraction was prepared from Sprague–Dawley rat brains according to an earlier published method [29]. The protein concentration was determined by the Bradford method [30], using bovine serum albumin as standard.

2.5. [^{35}S]GTPγS binding

The assays were carried out according to an earlier published procedure [31,32]. In short, assay tubes containing 10 μg of protein, 30 μM GDP, the appropriate concentration of Aβ peptides, and 0.05 nM [^{35}S]GTPγS, all in 50 mM Tris–HCl buffer containing 1 mM EGTA, 100 mM NaCl and 3 mM MgCl$_2$ in a final volume of 1 ml were incubated for 1 h, at 30 °C. For positive control 10^{-5} M DAMGO was utilised. Non-stimulated activity was measured in the absence of tested compounds, non-specific binding was measured in the presence of 100 μM unlabelled GTPγS. Naloxone was added 10 min before the addition of the protein. The incubation was started by the addition of the [^{35}S]GTPγS and was terminated by filtrating the samples through Whatman GF/B filters in a Millipore filtration instrument. Filters were washed three times with ice-cold 50 mM Tris–HCl buffer (pH 7.4), and then dried. Bound radioactivity was measured in a Wallac 1409 scintillation counter using a toluene based scintillation cocktail. Stimulation is given as percent of the specific binding. Data were calculated from at least three independent experiments performed in triplicates.

3. Results and discussion

3.1. β-Amyloid structure and toxicity

Aggregation of Aβ peptides has been identified as a major risk factor for AD. Monomeric Aβ is always produced during normal metabolism and has no deleterious effects on neurons [28]. As Aβ assembles into oligomeric forms (diffusible aggregates) and eventually into amyloid fibrils, it acquires high neuronal toxicity [33,34].

Aggregation of Aβ peptides was investigated with FT-IR spectroscopy (formation of highly ordered β-structures, characteristic band at 1620 cm^{-1}) which was combined with a cell toxicity assay (formation of formazan from a tetrazolium dye, MTT-test [35]). Different Aβ peptides and fragments (Aβ 1–40, Aβ 1–42, Aβ 25–35, Aβ 31–35) as well as reverse Aβ sequences (Aβ 42–1, Aβ 35–25) and the all-D-Aβ 1–40 were investigated by FT-IR and in MTT cell toxicity test. The results are shown in Fig. 1.

Aβ peptide aggregation and cell toxicity show good correlation. Aβ 1–42 has the most rapid aggregation and the highest toxicity. Even small fragments (Aβ 25–35, Aβ 31–35) proved to be toxic. Interestingly, the Aβ 1–40 containing only D-amino acids shows high toxicity, together with high aggregation grade. Reverse Aβ sequences (Aβ 42–1 and 35–25) can aggregate very slowly and show only moderate toxicity in MTT test.

3.2. BSB peptide design

In our studies, a rational, computational drug design technique was investigated to identify new BSB peptides and active, interacting sequences of

Sequence	Aggregation	Toxicity
1. Aβ 1-40	+++	+++
2. Aβ 1-42	++++	++++
3. Aβ 25-35	+++	+++
4. Aβ 31-35	+++	+++
5. all-D-Aβ 1-40	+++	+++
6. reverse Aβ (42-1)	—	—
7. reverse Aβ (35-25)	—	—

Fig. 1. Correlation between aggregation and toxicity of amyloid peptides.

the Aβ molecule. The above-mentioned model of George and Howlett was applied to represent the amyloid target in our computational studies.

Blind docking. Molecular docking is one of the most important computational drug design techniques [36]. Generally, docking is applied to find the binding sites and exact binding modes of drugs or small molecules [37] in a pre-defined binding pocket of a protein. However, in many cases even the location of the binding pocket on the protein is not known. In our study [38] it was shown, that the detection of the exact binding mode (that is the position of the binding pocket and the coordinates of the ligand molecule in the pocket) is possible for flexible peptide ligand molecules. We named our approach blind docking, as it was possible to find the binding mode of the peptide molecules 'blindly', i.e. without any prior knowledge of their position on the protein target. The popular docking program: AutoDock [28] was applied to perform the blind docking simulations. The blind docking technique was applied for the BSB-amyloid complexes discussed herein.

Mapping of the active sequences of the amyloid molecule. Our first attempt of elucidation of the mechanism of binding of BSB peptides [39] was performed with the aid of the Monte Carlo Simulated Annealing (MC/SA) search technique of the Auto-Dock program. The results were refined and reconsidered in our more recent paper [40] using the Lamarckian Genetic Algorithm (LGA) search technique of the new, 3.0 version of AutoDock. The genetic algorithm-based search methods are more efficient for larger, flexible molecules [38] than the MC/SA methods. In the latter paper a systematic mapping of the binding sequences of two potent BSB peptides was published.

Using our blind docking approach [38], it was possible to find the binding site and mode of ligand molecules with up to 12 rotable sigma bonds on protein targets of moderate size (300 amino acids). In case of the BSB peptides, the number of free torsions is generally higher but at the same time the target Aβ peptide is only 42-43 AA long. Thus, it was realistic to apply the docking technique for the BSB-amyloid systems, as well.

The results of the above-mentioned mapping are in agreement with the experimental findings and the stability of the complexes were checked with

Fig. 2. Schematic representation of part of the results of the mapping of active sequences of the amyloid peptide (grey hairpin) involved in the binding of beta sheet breaker (BSB) peptides Ref. [42]. The active sequences are located at the interface of two amyloid peptide molecules. As these sequences are involved in the interaction with BSB peptides, the interface can not be formed between the blocked amyloid monomers, which may cause the inhibition of the formation of dimers, oligomers and fibrils of the amyloid.

molecular dynamics simulations [40]. On Fig. 2, a schematic picture of a dimer section of the model of the of the amyloid fibrills by George and Howlett is presented.

The marked sequences correspond to the binding regions of the investigated BSB peptides on the monomer amyloid molecule. These regions are at the binding interface of the two amyloid molecules. Therefore, the binding of the BSB peptide molecules may hinder the dimerization and further aggregation of the amyloid peptide into fibrils by blocking the interface residues. In summary, our findings provide an explanation on the molecular mechanism of action of the BSB-like inhibitors of the aggregation of Aβ.

The results of our mapping studies and the experimental research pointed to the importance of the C-terminal sequences of the Aβ peptide in the amyloid–amyloid interactions. In our next paper [41] two active peptide–amides: the GVVIAn and RVVIAn were presented (n denotes the amide terminus) and their modes of action were investigated with experimental and modeling techniques.

Interestingly, the two hydrophobic region of the Aβ showed high affinity to each other (Fig. 3). These 'interface' regions were found by the mapping studies (Fig. 2), as well.

In summary, computer-aided molecular design and modeling proved to be a useful tool in the elucidation

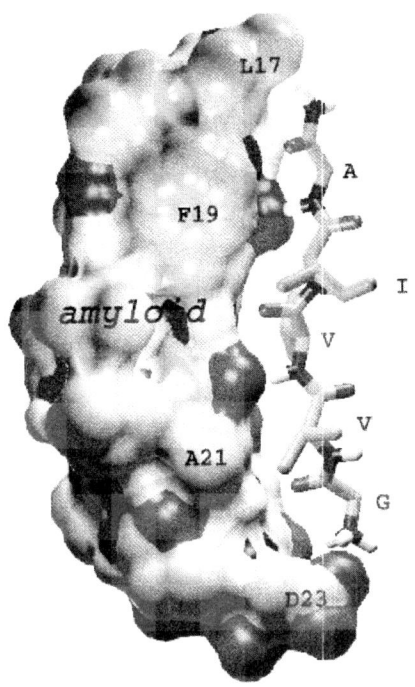

Fig. 3. The docked conformation of the GVVIAn pentapeptide amide (sticks) on the surface of the binding sequences of Aβ (amyloid, SA surface representation). Hydrophobic interactions between Ile and Phe19 (Val and Ala21) and salt bridge between the N-terminal amino group and Asp23 are represented in this view.

of the binding mechanism and the development of new beta sheet breaker (or amyloid surface binding) peptides.

3.3. Biological activity of short Aβ fragments and analogs

Amyloid surface binding molecules. Inhibition of the formation of the toxic Aβ-polymers has emerged as an approach to developing therapeutics for AD [42]. Soto's pentapeptide (LPFFD) is the best known peptidic BSB compound [43]. Some years ago we found a modified Aβ-sequence (propionyl-Ile–Ile–Gly–Leu-amide) which was a potent inhibitor of Aβ neurotocicity, without preventing Aβ aggregation [44]. This compound and its analogs (e.g. Arg–Ile–Ile–Gly–Leu-amide) were called 'functional antagonist—inverse agonist' (FA–IA) because they can antagonize the neurotoxic effect of Aβ peptides, although sometimes they show low neurotoxity, being

also Aβ-sequences. Recently we have found, that GVVIA and RHDS, natural Aβ sequences also prevent Aβ neurotoxicity. Summarizing our findings and the result of Soto and Tjernberg [45], the following Aβ sequences can prevent the neurotoxicity of aggregated Aβ peptides:

1. Aβ 5–8 and related sequences
2. Aβ 17–21 and related sequences
3. Aβ 31–34 and related sequences
4. Aβ 38–42 and related sequences

We do not know yet the exact mechanism of neuroprotection of these short peptides. Spectrophotometric measurements suggest that these peptides can bind to the surface of amyloid aggregates. Computer calculations (docking the above mentioned compounds to monomeric Aβ) also suggest that short Aβ peptides like Aβ 16–20 bind to definite regions of Aβ 1–42.

The protecting mechanism of the short Aβ-fragments is unknown, therefore we introduce a new term for them: Amyloid Surface Binding Molecule, abbreviated as ASBIM. We have synthesized the following ASBIM peptides:

1. Propionyl-Ile–Ile–Gyl–Leu-amide
2. Arg–Ile–Ile–Gly–Leu-amide
3. Phe–Arg–His–Asp–Ser-amide
4. Leu–Pro–Tyr–Phe–Asp-amide
5. Gly–Val–Val–Ile–Ala-amide
6. Arg–Val–Val–Ile–Ala-amide

The neuroprotective effects of different ASBIM peptides were measured by in vitro method (MTT-assay) (Peptides 1 and 2 are Aβ 31–34, 3 is Aβ 5–8, 4 is an Aβ 17–21, 5–6 are Aβ 38–42 derivatives).

Peptide 1, the propionyl-Aβ 31–34 fragment possesses a relatively low protecting capacity. Changing the propionyl-group to arginine (peptide 2) resulted in a peptide, which fully protects neuroblastoma cells against Aβ also in 2-fold molar excess.

Aβ 5–8 sequence analog 3 itself is not toxic in MTT test and protects SH-SY5Y neuroblastoma cells against Aβ 1–42 in 5-fold molar excess. Further experiments show that the short peptides possess very good protective effects also in 2-fold molar excess. On the other hand, Soto's peptide LPFFD has shown no protection in this test. Replacing one Phe to Tyr, however, changed the protecting activity: peptid **4** LPYFD completely prevent Aβ neurotoxicity in this cell toxicity test. Computer simulations also show that the presence of a phenolic OH-group in Tyr increases peptide binding to Aβ.

Similarly, the C-terminal pentapeptide GVVIA and its analog RVVIA **5–6** perfectly potect the SH-SY5Y cells in MTT-test in 2-fold excess.

Summarizing our efforts and results, we have found four neuroprotecting peptide families derived from the original Aβ 1–42 polypeptide. These compounds can bind on the Aβ-aggregate surface ('ASBIM'-peptides) and prevent Aβ-neurotoxicity. The fine details of the protective mechanism are not yet cleared.

G-protein activation. The molecular events underlying the toxic effects is largely unknown, but aggregated Aβ(1–42), and its shorter fragments, have been shown to stimulate GTPase activity [42, 43] in neurons. Although, the connection between G-protein activation and toxicity is yet to be clarified, it was suggested that increasing levels of aggregation of Aβ (1–42) may not only alter the toxic, but also the activation of G-proteins. In this study we determined the effects of Aβ (1–42) on G-protein activation in crude rat brain membranes, and, using synthetic peptide 'antagonists', we were trying to inhibit this activation.

We found that Aβ (1–42) produced a concentration dependent, saturable activation of G-proteins (maximum: 166.0 ± 6.9% and the IC_{50}: 3.9 ± 1.4 uM) in the presence of 1% TFE Fig. 4. Inhibition of the actions of Aβ (1–42) might have therapeutical relevance, the functional agonist propionyl-Arg–Ile–Ile–Gly–Leu-NH2 appeared to be a good inhibitor of Aβ's toxicity (Fig. 4). In our studies this peptide fully inhibited the Aβ (1–42) stimulated GTP binding. In addition it had a profound inhibitory effect on basal G-protein activity. Based on these observations this compound was designated functional-antagonist/inverse-agonist (FAIA). Replacement of propionyl with hexanoyl moiety further potentiated the inhibitory effect of the pentapeptide. (Fig. 5). Given that these compounds inhibit toxicity as well as G-protein

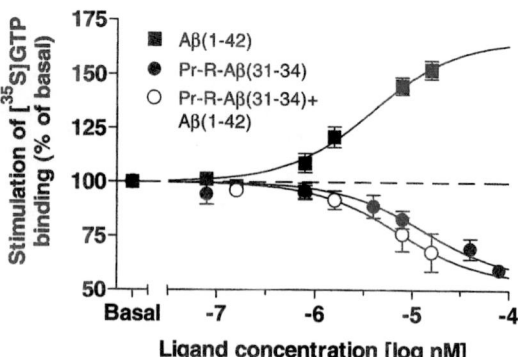

Fig. 4. The Aβ(1–42) [■] induced stimulation was completely inhibited by propionyl-Arg–Ile–Ile–Gly–Leu-NH₂ (Pr–R–Aβ(31–34)) [O]. In fact Pr–R–Aβ(31–34) inhibited the basal G-protein activity [●] by an ED_{50} of 10^{-5} M, acting as an inverse agonist by itself.

activation, FA–IAs might be useful in designing therapeutic agents. The neuroprotective mechanism of FA–IA peptides is not known, however, spectroscopic studies show that these peptides may bind to Aβ aggregates, that is they might form a subclass of ASBIM peptides

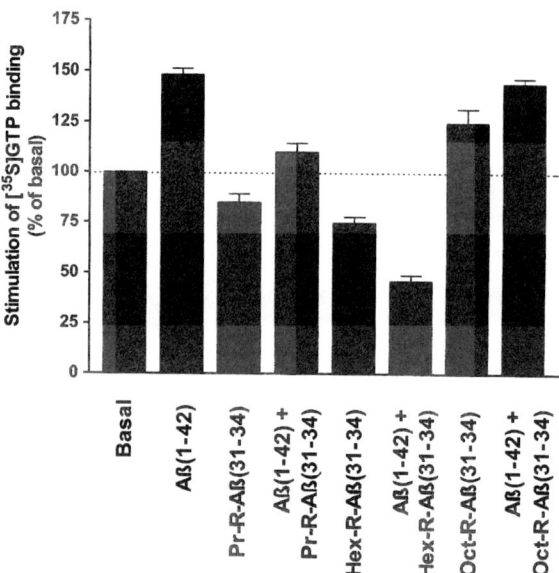

Fig. 5. Increasing the length of carbon chain at the N-terminal affected the inhibition of Aβ((1–42) stimulated G-protein activity by Pr–R–Aβ(31–34). Maximal inhibition was obtained by using Hex–R–Aβ(31–34). A further increase in the length abolished the inhibition of both basal and Aβ (1–42) stimulated G-protein activity.

Acknowledgements

This work was partially supported by the Hungarian research grants OTKA T034895, NKFP 1/010/2001 and 1/027/2001.

References

[1] A. Alzheimer, R.A. Stelzmann, H.N. Schnitzlein, F.R. Murtagh, Clin. Anat. 8 (1995) 429 (English translation of Alzheimer's 1907 paper).

[2] K. Ogomori, T. Kitamoto, J. Tateishi, Y. Sato, M. Suetsugu, M. Aba, Am. J. Pathol. 134 (1989) 243.

[3] V.M. Lee, B.J. Balin, L. Otvos, J. Trojanowski, Q. Sci. 251 (1991) 675.

[4] D.J. Selkoe, Neuron 6 (1991) 487.

[5] P. Ponte, D.P. Gonzalez, J. Schilling, J. Miller, D. Hsu, B. Greenberg, K. Davis, W. Wallace, I. Lieberburg, F. Fuller, Nature 331 (1988) 525.

[6] R.E. Tanzi, E.D. Lamperti, L. Villa-Komaroff, J.F. Gusella, R.L. Neve, Nature 331 (1988) 528.

[7] N. Kitaguchi, Y. Takahashi, Y. Tokushima, S. Shiojiri, H. Ito, Nature 331 (1988) 530.

[8] R. Vassar, B.D. Bennett, F. Collins, J. Treanor, G. Rogers, M. Citron, Science 286 (1999) 735.

[9] S. Sinha, J.P. Anderson, R. Barbour, P. Seubert, S. Wang, D. Walker, V. John, Nature 402 (1999) 537.

[10] J.A. Hardy, C.A. Higgins, Science 256 (1992) 184.

[11] E. Terzi, G. Hölzemann, J. Seelig, Biochemistry 33 (1994) 1345.

[12] S. Zhang, A. Rich, Proc. Natl. Acad. Sci. USA 94 (1997) 23.

[13] J. Jarvet, P. Damberg, K. Bodell, L.E. Eriksson, A. Gräslund, J. Am. Chem. Soc. 122 (2000) 4261.

[14] F. Massi, J.W. Peng, J.P. Lee, J.E. Straub, Biophys. J. 80 (2001) 31.

[15] C. Hilbich, B. Kisters-Woike, J. Reed, C.L. Masters, K. Beyreuther, J. Mol. Biol. 218 (1991) 149.

[16] C.J. Barrow, A. Yashuda, P.T.M. Kenny, G. Zagorski, J. Mol. Biol. 225 (1992) 1075.

[17] J.P. Lee, E.R. Stimson, J.R. Ghilardi, P.W. Mantyh, Y.A. Lu, A.M. Felix, W. Llanos, A. Behbin, M. Cummings, M. Criekinge, W. Timms, J.E. Maggio, Biochemistry 34 (1995) 5191.

[18] T.L.S. Benzinger, D.M. Gregory, T.S. Burkoth, H. Miller-Auger, D.G. Lynn, R.E. Botto, S.C. Meredith, Proc. Natl. Acad. Sci. USA 95 (1998) 13407.

[19] M. Sunde, L.C. Serpell, M. Bartlam, P.E. Fraser, M.B. Pepys, C.F. Blake, J. Mol. Biol. 273 (1997) 729.

[20] F. Massi, J.E. Straub, Proteins 42 (2001) 217.

[21] L.C. Serpell, Biochim. Biophys. Acta 1502 (2000) 16.

[22] A.E. Roher, J. Baudry, M.O. Chaney, Y.M. Kuo, W.B. Stine, M.R. Emmerling, Biochim. Biophys. Acta 1502 (2000) 31.

[23] L. Li, T.A. Darden, L. Bartolotti, D. Kominos, L.G. Pedersen, Biophys. J. 76 (1999) 2871.

[24] A.R. George, D.R. Howlett, Biopolymers 50 (1999) 733.

[25] M.R. D'Andrea, D.H.S. Lee, H.-Y. Wang, R.G. Nagele, Drug Dev. Res. 56 (2002) 194.

[26] M.A. Findeis, Biochim. Biophys. Acta 1502 (2000) 76.

[27] I. Gozes, A.D. Spier, Drug Dev. Res. 56 (2002) 475.

[28] G.M. Morris, D.S. Goodsell, R.S. Halliday, R. Huey, W.E. Hart, R.K. Belew, A.J. Olson, J. Comp. Chem. 19 (1998) 1639.

[29] J. Simon, S. Benyhe, K. Abutidze, A. Borsodi, M. Szucs, G. Toth, M. Wollemann, J. Neurochem. 46 (1986) 695.

[30] M.M. Bradford, Anal. Biochem. 72 (1976) 248.

[31] I. Lengyel, G. Orosz, D. Biyashev, L. Kocsis, M. Al-Khrasani, A. Ronai, C. Tomboly, Z. Furst, G. Toth, A. Borsodi, Biochem. Biophys. Res. Commun. 290 (2002) 153.

[32] I. Szatmari, D. Biyashev, C. Tomboly, G. Toth, M. Macsai, G. Szabo, A. Borsodi, I. Lengyel, Biochem. Biophys. Res. Commun. 284 (2001) 771.

[33] C.J. Pike, A.J. Walencewicz, C.G. Glabe, C.W. Cotman, Brain Res. 563 (1991) 311.

[34] M.P. Lambert, C.E. Finch, G.A. Krafft, W.L. Klein, Proc. Natl. Acad. Sci. USA 95 (1998) 6448.

[35] N. Brooijmans, I.D. Kuntz, Annu. Rev. Biophys. Biomol. Struct. 32 (2003) 335.

[36] C. Loske, A. Neumann, A.M. Cunnigham, K. Nichol, R. Schinzel, R. Riederer, G. Munch, J. Neural. Transm. 105 (1998) 1005.

[37] C. Hetényi, U. Maran, M. Karelson, J. Chem. Inf. Comput. Sci. 43 (2003) 1576.

[38] C. Hetényi, D. van der Spoel, Protein Sci. 11 (2002) 1729.

[39] C. Hetényi, T. Körtvélyesi, B. Penke, J. Mol. Struct. (THEOCHEM) 542 (2001) 25.

[40] C. Hetényi, T. Körtvélyesi, B. Penke, Bioorgan. Med. Chem. 10 (2002) 1587.

[41] C. Hetényi, Z. Szabó, É. Klement, Z. Datki, T. Körtvélyesi, M. Zarándi, B. Penke, Biochem. Biophys. Res. Commun. 292 (2002) 931.

[42] D.B. Schenk, R.E. Rydel, S. Little, J. Panetta, I. Lieberburg, S. Sinha, J. Med. Chem. 38 (1995) 4141.

[43] C. Soto, M.S. Kindy, M. Baumann, B. Frangione, Biochem. Biophys. Res. Commun 226 (1996) 672.

[44] G. Laskay, M. Zarándi, J. Varga, K. Jost, A. Fónagy, C. Torday, L. Latzkovits, B. Penke, Biochem. Biophys. Res. Commun. 235 (1997) 479.

[45] L.O. Tjernberg, J. Näslund, F. Lindqvist, J. Johansson, A.R. Karlström, J. Thyberg, L. Terenius, C. Nordstedt, J. Biol. Chem. 271 (1996) 8545.

Further Reading

M. Shoji, T.E. Golde, J. Ghiso, T.T. Cheung, S. Estus, L.M. Schaffer, X.-D. Cai, D.M. McKay, R. Tintner, B. Frangione, S.G. Younkin, Science 258 (1992) 126.

ELSEVIER

Journal of Molecular Structure (Theochem) 666–667 (2003) 515–520

THEO
CHEM

www.elsevier.com/locate/theochem

The stereochemistry of the chemical expression of darkness

József Csontos[a,*], Miklós Kálmán[a], Gyula Tasi[b]

[a]Bay Zoltán Foundation for Applied Research, Institute for Biotechnology, Derkovits fasor 2., Szeged H-6726, Hungary
[b]Department of Applied and Environmental Chemistry, University of Szeged, Rerrich B. tér 1., Szeged H-6720, Hungary

Abstract

Conformational analysis of melatonin (N-acetyl-5-methoxytryptamine) was performed at ab initio Hartree–Fock level: 128 conformers were obtained using 6-31G* basis set. With the help of equivalence relations, the conformational hypersurface was divided into equivalence classes. It can be concluded that the side chains of melatonin are enantiotopic groups, i.e. melatonin has conformational planar docking chirality, and its conformers are correlated with each other as enantiomer, epimer, or diastereomer pairs. Based on these considerations, it is possible to explain the outcomes of restricted chiral melatonin analog experiments. It is suggested that conformers belonging to different equivalence classes should react with different receptor sites and elements of equivalent classes of 3re and 5si face side of the indole ring prefer the MT1, while 3si face side ones favor the MT2 melatonin receptor site.
© 2003 Elsevier B.V. All rights reserved.

Keywords: Melatonin; Conformational analysis; Chiral analogs; Equivalence relation

1. Introduction

According to their static symmetry, molecules can be placed into various Schönfliess point groups [1]. Molecules classified into either C_n or D_n groups are characterized by the absence of any rotation–reflection axis. Elements of these groups are chiral. 'Any geometrical figure, or group of points, chiral, and it has chirality, if its image in a plane mirror, ideally realized, cannot be brought to coincide with itself' (Lord Kelvin, 1884). Phenomena of chirality might join to the existence of some geometrical elements. In case of a point one can say about central, of an axis axial, of a plane planar and of a spiral helical chirality (Fig. 1). The concepts of axial, planar, and helical chirality are closely

attached to the conformational state. The free torsional angles of a molecule determine its conformation and the conformers are local minima structures on the conformational hypersurface. (In this manner geometrical cis–trans isomer pairs are conformers). In case of a biopolymer possessing α-helical structure (DNA, protein), the helical chirality of chains can be coupled to stable conformations generated by secondary interactions. Similarly, in case of axial chirality there are some rotational barriers, which force the chiral conformation. Chirality due to restricted rotation is called antropisomerism. In this way the stereoisomers having no central chirality generally are antropisomeric conformers [2]. Thus, one can use the concept of chiral conformer or/and conformational chirality.

Planar chirality in space is a direct consequence of two-dimensional (2D) chirality. A geometric figure

---

* Corresponding author. Tel.: +36-62-432-248; fax: +36-62-432-250.

E-mail address: csontos@bio.u-szeged.hu (J. Csontos).

0166-1280/$ - see front matter © 2003 Elsevier B.V. All rights reserved.
doi:10.1016/j.theochem.2003.08.075

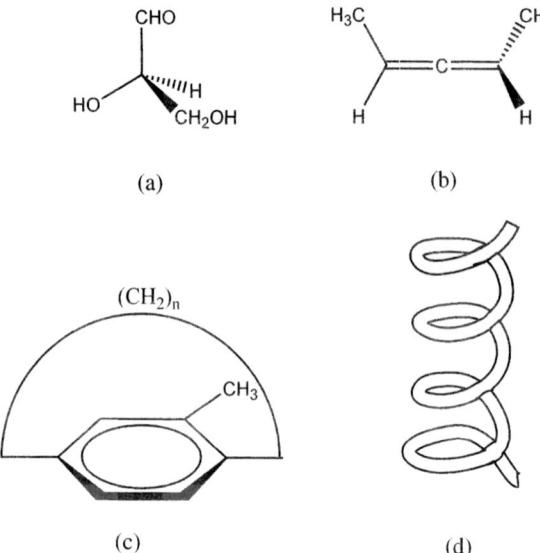

Fig. 1. The types of chirality: central (a), axial (b), planar (c), and helical (d).

possesses chirality in plane, if finite numbers of rotations and translations acting in its plane cannot transform it into its mirror image (Fig. 2a). Of course, applying rotation in space the two figures can be superimposed. The manifestation of 2D chirality in 3D is that space is divided into two distinguishable half spaces. Atoms and groups, which are located within such distinguishable half spaces, have been termed heterotopic atoms and groups by Mislow and Raban [3]. If such a heterotopic object connects to a molecule, which is chiral in plane, then it becomes chiral in space having planar chirality (Fig. 2b).

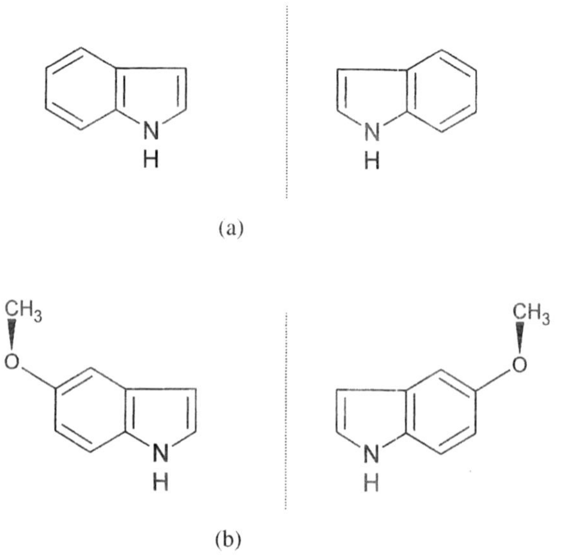

Since indole-ring has 2D chirality, all the conformations of melatonin (N-acetyl-5-methoxy-tryptamine) not locating in the plane are planar chiral conformations. It is noteworthy that the pure mathematical definition of chirality is usually supplemented with some chemical boundary conditions considering lifetime and separability of enantiomers. However, in every case difficulties arise, when the classification of objects depends on a continuously changing parameter like rotational energy. One of the solutions might be the concept of dynamic chirality [4,5]. Recently, a number of articles have focused on the possibility of describing symmetry and chirality as a continuous property instead of a 'yes or no' property [6–8]. The phenomena of conformational chirality can be fitted into this system very well [9]. Up to this point we supposed tacitly in the consideration of conformational chirality that rotational barriers were intrinsic nature of a molecule. In case of melatonin, there are no high inner rotational barriers, the conformers can transform into each other investing a little energy. However, we cannot forget the possibility that enzymes and catalysts might provide such rotational barriers for their substrates and fix that in a chiral conformation. According to Keinan and Avnir [9], one can call these conformations of substrates active site induced chiral conformations. Differentiating this type of rotational barrier from the inner one, one can term the concept of outer rotational barrier and its induced conformational chirality as docking chirality. Since the active conformation of melatonin is probably a folded structure, we can say that melatonin has conformational planar docking chirality [10–15].

2. Results and discussion

Recently, conformational analysis of melatonin has been performed at Hartree–Fock ab initio level [16]. With the help of 6-31G* basis set, 128

Fig. 2. Planar chirality in 2D (a) and in 3D (b).

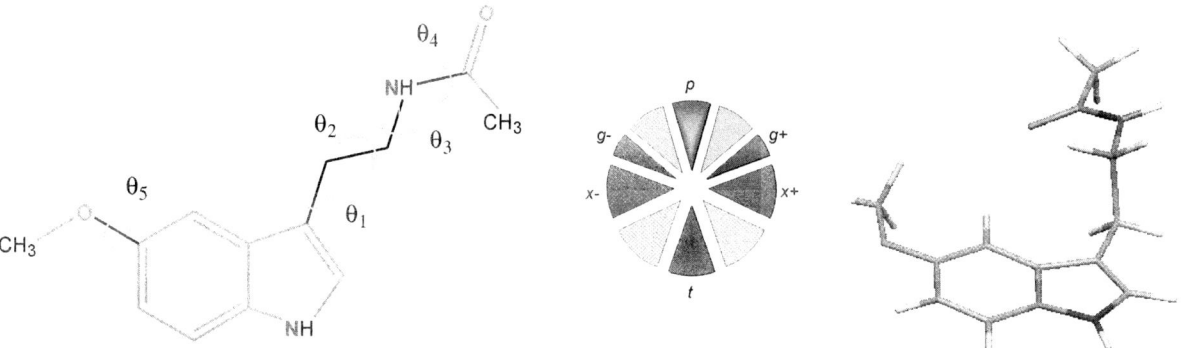

Fig. 3. The free torsional angles of melatonin and our designation system applied to the conformers ($p[-16°, +16°]$, $g^+[+50°, +70°]$, $g^-[-70°, -50°]$, $x^+(+70°, +115°]$, $x^-[-115°, -70°)$, $t[-160°, -180°]U[+160°, +180°]$).

conformers have been obtained. Based on polar histogram, the torsional angles can be classified as follows: synperiplanar or *cis* $p[-16°, +16°]$, synclinal or gauche $g^+[+50°, +70°]$, $g^-[-70°, -50°]$, extended gauche $x^+(+70°, +115°]$, $x^-[-115°, -70°)$, antiperiplanar or *trans* $t[-160°, -180°]$ $U[+160°, +180°]$ ($[+160°, +220°]$). Let us consider the following mapping $\varphi : K \rightarrow L^5$, where K is the set of conformers and L is the set of intervals defined above. For example, x^+tx^+tp sign that conformer, where $70° < \theta_1$, $\theta_3 \leq 115°$, $160° < \theta_2$, $\theta_4 \leq 200°$, $-16° < \theta_5 \leq +16°$ (Fig. 3). One possible classification of the conformational space is plausible. Conformers, which have 3-amido side chains in the 3*re* half space of indole [17] belong to the first class, elements of the second class have amido group in the other half space. The third class consists of planar melatonin conformers. However, one is able to define a more detailed classification in the set of conformers. Let us construct the equivalence relation ρ on K in the following manner: $\forall a, b \in K$, $a\rho b \Leftrightarrow$ if $a(a_1, a_2, a_3, a_4, a_5)$, $b(b_1, b_2, b_3, b_4, b_5)$, a_i, $b_i \in L$ and $|a_i| = |b_i|$ ($i = 1, \ldots, 4$). In this case, conformers belonging to the same class have similar folding pattern in their 3-amido-side chain. If we consider sharper requirements, e.g. $\forall a, b \in K, \forall \sigma \subseteq K \times K$, $a\sigma b \Leftrightarrow$ if $a(a_1, a_2, a_3, a_4, a_5)$, $b(b_1, b_2, b_3, b_4, b_5)$, a_i, $b_i \in L$ and $a_i = b_i$ ($i = 1, \ldots, 4$), then we can divide every equivalence class into two subclasses except two (*tttt, tttp*). In these latter cases, relations ρ and σ are identical, because the elements of these classes and their mirror images are superimposable (Fig. 4a). Thus, one can

use such usual stereo chemical concepts as enantiomer, diastereomer and epimer pairs. Taking our results into consideration, it is possible to explain the outcomes of the restricted chiral melatonin analog experiments. Based on pharmacological profile, melatonin receptors were initially classified into two types: ML_1 and ML_2 [18]. Two mammalian high-affinity melatonin receptor (ML_1) subtypes have been cloned up to now, MT1 and MT2 showing $>80\%$ amino acid identity with each other [19,20]. The MT1 receptor is localized in the suprachiasmatic nucleus and is thought to mediate the circadian and reproductive actions of melatonin, while the MT2 receptor mRNA was detected in the retina. The MT2 receptor is thought to mediate the effects of melatonin on circadian rhythms and is implicated in the regulation of visual function [21]. It has not passed away 10 years yet, since stereo isomers have been applied for melatonin receptors [22]. Optically active analogs showed reversed stereo selectivity for ML_1 and ML_2 receptors, suggesting that melatonin interacts with these sites in two different conformations [10]. The MT1 and MT2 subtypes of the high affinity ML_1 receptor also differ in their receptor pocket [11,15]. In addition, indan derivatives possessing *S*-configuration showed greater affinity for the MT1 receptor than compounds with *R*-stereochemistry [23]. In contrast, melatonin and its bioisosteres without conformational restriction do not have MT1/MT2 subtype selectivity [11]. Taking into account that recognition is not a rigid process, melatonin stereo isomer pairs can oscillate between each other underway the receptor

(a)

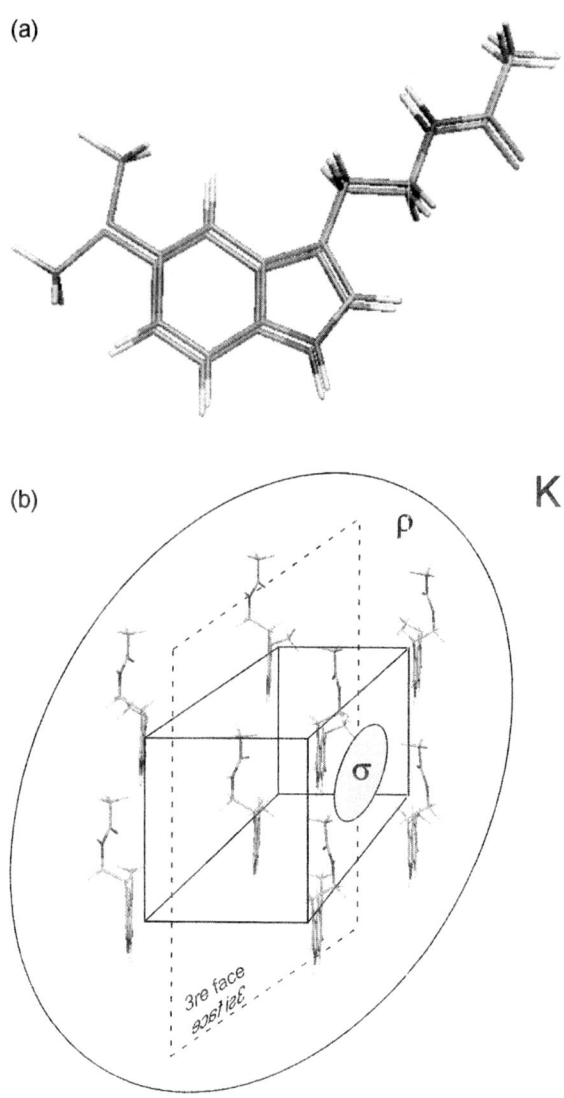

(b)

ρ

K

σ

3re face
3si face

Fig. 4. (a) Two planar conformers of melatonin. (b) One of the 17 ρ equivalence classes, which involves the global minimum; four 3re face side conformers belong to σ and eight (four 3re and four 3si face side) conformers belong to ρ; K is the conformational space of melatonin. It is obvious that all the conformers belong to σ have an enantiomer pair corresponding to the ρ relation.

and the chiral isomer is evolved in the place of docking. However, analogs having central chirality are not able to do such inversions: in this case the receptor–substrate interaction is bias; principally the substrate modulates the binding site.

Table 1
The equivalence classes. Bold italic cells sign the 5si face side conformers.

Enantiomer pairs		Energy (kJ/mol)
3si face	3re face	
1		
x^+tx^+tp	x^-tx^-tp	0.000
$x^+tx^+tx^-$	$x^-tx^-tx^+$	−7.215
$x^+tx^+tx^+$	$x^-tx^-tx^-$	−7.388
x^+tx^+tt	x^-tx^-tt	−10.079
2		
$x^+g^+x^+tp$	$x^-g^-x^-tp$	−2.515
$x^+g^+x^+tx^+$	$x^-g^-x^-tx^-$	−6.018
$x^+g^+x^+tx^-$	$x^-g^-x^-tx^+$	−6.769
$x^+g^+x^+tt$	$x^-g^-x^-tt$	−6.829
3		
$x^+g^-x^-tp$	$x^-g^+x^+tp$	−2.520
$x^+g^-x^-tx^+$	$x^-g^+x^+tx^-$	−6.595
$x^+g^-x^-tt$	$x^-g^+x^+tt$	−8.580
$x^+g^-x^-tx^-$	$x^-g^+x^+tx^+$	−9.804
4		
x^+tx^-tp	x^-tx^+tp	−4.771
$x^+tx^-tx^+$	$x^-tx^+tx^-$	−7.517
$x^+tx^-tx^-$	$x^-tx^+tx^+$	−7.850
x^+tx^-tt	x^-tx^+tt	−7.971
5		
tg^+ttp	tg^-ttp	−8.310
tg^+ttx^-	tg^-ttx^+	−12.159
tg^+ttx^+	tg^-ttx^-	−12.513
tg^+ttt	tg^-ttt	−12.854
6		
ttx^+tp	ttx^-tp	−10.344
ttx^+tx^-	ttx^-tx^+	−13.513
ttx^+tx^+	ttx^-tx^-	−14.259
ttx^+tt	ttx^-tt	−14.658
7		
$x^+g^-x^+tp$	$x^-g^+x^-tp$	−12.182
$x^+g^-x^+tx^+$	$x^-g^+x^-tx^-$	−13.986
$x^+g^-x^+tt$	$x^-g^+x^-tt$	−14.535
$x^+g^-x^+tx^-$	$x^-g^+x^-tx^+$	−15.561
8		
$x^+g^-x^-pp$	$x^-g^+x^+pp$	−13.697
$x^+g^-x^-px^-$	$x^-g^+x^+px^+$	−18.678
$x^+g^-x^-px^+$	$x^-g^+x^+px^-$	−26.118
$x^+g^-x^-pt$	$x^-g^+x^+pt$	−26.287
9		
$x^+g^+x^+pp$	$x^-g^-x^-pp$	−15.267
$x^+g^+x^+px^+$	$x^-g^-x^-px^-$	−18.738
$x^+g^+x^+px^-$	$x^-g^-x^-px^+$	−18.846
$x^+g^+x^+pt$	$x^-g^-x^-pt$	−18.946
10		
$x^+g^+x^-pp$	$x^-g^-x^+pp$	−15.861
$x^+g^+x^-px^+$	$x^-g^-x^+px^-$	−19.116

Table 1 (continued)

Enantiomer pairs		Energy (kJ/mol)
3si face	3re face	
$x^+g^+x^-pt$	$x^-g^-x^+pt$	−19.379
$x^+g^+x^-px^-$	$x^-g^-x^+px^+$	−19.610
11		
x^+tx^-pp	x^-tx^+pp	−17.546
$x^+tx^-px^+$	$x^-tx^+px^-$	−21.114
$x^+tx^-px^-$	$x^-tx^+px^+$	−21.445
x^+tx^-pt	x^-tx^+pt	−21.508
12		
x^+tx^+pp	x^-tx^-pp	−17.635
$x^+tx^-px^+$	$x^-tx^-px^-$	−20.912
x^+tx^+pt	x^-tx^-pt	−21.246
$x^+tx^+px^-$	$x^-tx^-px^+$	−21.450
13		
tg^+tpp	tg^-tpp	−20.513
tg^+tpx^+	tg^-tpx^-	−24.113
tg^+tpx^-	tg^-tpx^+	−24.493
tg^+tpt	tg^-tpt	−24.680
14		
ttx^+pp	ttx^-pp	−21.490
ttx^+px^+	ttx^-px^-	−25.593
ttx^+px^-	ttx^-px^+	−25.596
ttx^+pt	ttx^-pt	−26.208
15		
$x^+x^-x^+pp$	$x^-x^+x^-pp$	−17.688
$x^+x^-x^+px^-$	$x^-x^+x^-px^+$	−21.030
$x^+g^-x^+pt$	$x^-g^+x^-pt$	−28.392
$x^+g^-x^+px^+$	$x^-g^+x^-px^-$	−28.936
16		
$ttttx^+$	$ttttx^-$	−14.860
$ttttp$		−12.176
$ttttt$		−15.186
17		
$tttpx^+$	$tttpx^-$	−26.678
$tttpp$		−22.624
$tttpt$		−27.166

Grey cells sign the 5si face side conformers.

3. Conclusions

It can be concluded that the side chains of melatonin (3-amido, 5-methoxy) are enantiotopic groups; hence conformers are correlated with each other as enantiomer, epimer, or diastereomer pairs. With the help of ρ, the 128 conformers can be divided into 17 equivalence classes and all the classes can be divided into two subclasses applying σ except two $(128 = 15 \times 8 + 2 \times 4$; Fig. 4b, Table 1.). In addition, it is suggested that conformers belonging to different equivalence classes should react with different receptor sites. Elements of equivalent classes of 3re and 5si face side of the indole ring prefer the MT1; while 3si face side ones favor the MT2 receptor (Table 1.). We think that the knowledge of stereochemistry of the hormone of darkness can be the key to understand the interaction between the central chiral melatonin analogs and their receptors.

Acknowledgements

This research was supported by a PhD fellowship grant from the Bay Zoltán Foundation for Applied Research to one of the authors (J. Cs.) and the János Bolyai Research Fund (G. T.).

References

[1] N. Mihály, Introduction to stereochemistry, MŰszaki Könyv-kiadó (1975) 22 in Hungarian.
[2] V. Prelog, G. Helmchen, Angew. Chem. Int. Ed. Engl. 21 (1982) 567.
[3] K. Mislow, M. Raban, Top. Stereochem. (1967) 1.
[4] T. Kawabata, H. Suzuki, N. Yoshikazu, J. Chen, K. Fuji, ICR Annu. Rep. 7 (2000) 36.
[5] T. Kawabata, H. Suzuki, N. Yoshikazu, T. Wirth, K. Yahiro, K. Fuji, Fourth International Electronic Conference on Synthetic Organic Chemistry September 1–30, 2000.
[6] H. Zabrodsky, S. Peleg, D. Avnir, J. Am. Chem. Soc. 114 (1992) 7843.
[7] H. Zabrodsky, D. Avnir, J. Am. Chem. Soc. 117 (1995) 462.
[8] B.K. Lipkowitz, D. Gao, O. Katzenelson, J. Am. Chem. Soc. 121 (1999) 5559.
[9] S. Keinan, D. Avnir, J. Am. Chem. Soc. 120 (1998) 6152.
[10] J.M. Jansen, S. Copinga, G. Gruppen, E.J. Molinari, M.L. Dubocovich, C.J. Grol, Bioorg. Med. Chem. 4 (1996) 1321.
[11] R. Faust, P.J. Garratt, R. Jones, L.K. Yeh, A. Tsotinis, M. Panoussopoulou, T. Calogeropoulou, M.T. Teh, D. Sugden, J. Med. Chem. 43 (2000) 1050.
[12] J.A. Sastre, R.N. Miguel, R.P. Molina, M.C.G. Zarzuelo, C. Romero-Ávila, A. Ramos, J. Mol. Struct. (Theochem) 573 (2001) 271.
[13] P.J. Garratt, S. Vonhoff, S.J. Rowe, D. Sugden, Bioorg. Med. Chem. Lett. 4 (1994) 1559.
[14] S. Sicsic, I. Serraz, J. Andrieux, B. Bremont, M. Mathe-Allainmat, A. Poncet, S. Shen, M. Langlois, J. Med. Chem. 40 (1997) 739.

[15] M. Mor, G. Spadoni, B. Di Giacomo, G. Diamantini, A. Bedini, G. Tarzia, P.V. Plazzi, S. Rivara, R. Nonno, V. Lucini, M. Pannacci, F. Fraschini, B.M. Stankov, Bioorg. Med. Chem. 9 (2001) 1045.

[16] J. Csontos, M. Kálmán, G. Tasi, J. Mol. Struct. (Theochem) (2003) in press.

[17] K.R. Hanson, J. Am. Chem. Soc. 12 (1966) 2731.

[18] M.L. Dubocovich, FASEB J. 2 (1988) 2765.

[19] T. Ebisawa, S. Karne, M.R. Lerner, S.M. Reppert, Proc. Natl Acad. Sci. USA 91 (1994) 6133.

[20] S.M. Reppert, C. Godson, C.D. Mahle, D.R. Weaver, S.A. Slaugenhaupt, J.F. Gusella, Proc. Natl Acad. Sci. USA 92 (1995) 8734.

[21] S.M. Reppert, D.R. Weaver, C. Godson, Trends Pharmacol. Sci. 17 (1996) 100.

[22] D. Sugden, D.J. Davies, P.J. Garratt, R. Jones, S. Vonhoff, Eur. J. Pharmacol. 287 (1995) 239.

[23] K. Fukatsu, O. Uchikawa, M. Kawada, T. Yamano, M. Yamashita, K. Kato, K. Hirai, S. Hinuma, M. Miyamoto, S. Ohkawa, J. Med. Chem. 45 (2002) 4212.

ELSEVIER

Journal of Molecular Structure (Theochem) 666–667 (2003) 521–525

THEO
CHEM

www.elsevier.com/locate/theochem

Endogenous neurotransmitters as anti-amigdaloidic agents: a density functional investigation of the interaction between melatonin and histidine

Luca F. Pisterzi[a,*], David R.P. Almeida[a], Donna M. Gasparro[a], Jason R. Juhasz[a], B. Penke[b,c], G. Tasi[d], Imre G. Csizmadia[a,b,e]

[a]*Department of Chemistry, University of Toronto, Toronto, Ont., Canada M5S 3H6*
[b]*Department of Medical Chemistry, University of Szeged, Dóm tér 8, 6720 Szeged, Hungary*
[c]*Protein Chemistry Research Group, Hungarian Academy of Sciences, University of Szeged, Dóm tér 8, 6720 Szeged, Hungary*
[d]*Department of Applied and Environmental Chemistry, University of Szeged, Rerrich B. tér 1, H-6720 Szeged, Hungary*
[e]*Global Institute of Computational Molecular and Materials Science (GIOCOMMS), 1422 Edenrose Street, Mississauga, Ont., Canada L5V 1H3*

Abstract

Recent insight has implicated the formation of β-amyloid fibril formation as a culpable process for the progression of Alzheimer's disease. Amyloid-β (Aβ) is a peptide naturally secreted by neurons in a random coil structure. It has been deduced that a conformational transition from a random coil to a β-pleated sheet is responsible for Aβ's neurotoxicity. The disruption of a His-Asp salt bridge has been found to inhibit this conformational change to the β-pleated sheet—a phenomenon observed in vitro when the peptides are incubated with melatonin. To further investigate this interaction between melatonin and histidine, the binding was investigated at the B3LYP/6-31g(d) level of theory for the two lowest energy conformers of melatonin: $\chi_2 = g+$, $\chi_3 = anti$, $\chi_4 = g+$, and $\chi_2 = g-$, $\chi_3 = anti$, $\chi_4 = g-$, and imidazolium. The most favorable interaction was found to be an intersecting one; the respective counterpoise corrected binding energies were found to be -17.8 and -19.8 kcal/mol.
© 2003 Elsevier B.V. All rights reserved.

Keywords: Melatonin; Imidazolium; Amyloid beta peptide; Neurofibrillary tangles; Density functional theory

1. Introduction

Alzheimer's disease (AD) is the most common form of dementia in the elderly with a complex etiology not yet fully understood [1]. Two classifications of the disease exist based on the age of onset; when the patient is diagnosed before 65 years of age it is known as early-onset, or presenile AD while a diagnosis in patients over 65 years of age is known as the late-onset, or senile form [1]. The majority of the cases (approximately 95%) of AD are of the late-onset form, which is thought to be a result of idiosyncratic interactions between the individual's environmental surroundings and their genes [1]. Much more is known about the etiology of the early-onset form of

---

* Corresponding author.
E-mail addresses: lpisterzi@medscape.com (L.F. Pisterzi), dalmeida@medscape.com (D.R.P. Almeida), dgasparro@medscape.com (D.M. Gasparro), jasonjuhasz@medscape.com (J.R. Juhasz), penke@ovrisc.mdche.szote.u-szeged.hu (B. Penke), gy.tasi@chem.u-szeged.hu (G. Tasi), icsizmad@ovrisc.mdche.u-szeged.hu (I.G. Csizmadia).

0166-1280/$ - see front matter © 2003 Elsevier B.V. All rights reserved.
doi:10.1016/j.theochem.2003.08.076

AD, with current research suggesting a strong, if not entirely genetic, causal component [1]. To date, mutations in three genes have been identified as causal in the early-onset form, providing further support for a familial etiology. These genes are presenilin 1 (PSEN1) and 2 (PSEN2), and the amyloid precursor protein (APP).

The APP gene was one of the first genetic determinants of AD to be studied based on similar neuropathological features in individuals with Down's syndrome [1]. If individuals with Down's syndrome live beyond 30 years they invariably develop the AD phenotype. The histopathological relation between the two disorders became apparent as the AD phenotype was attributed to the overexpression of APP, which mapped to chromosome 21, present in an extra copy in Down's syndrome [1].

The AD phenotype is an extensive accumulation of amyloid fibrils which are the result of various proteolytic cleavages of the amyloid precursor protein (AβPP), the translation product of APP [2]. The pathologic cleavage products may be 40 or 42 amino acid residues in length, and are known cumulatively as the amyloid-β peptide (Aβ) [3]. The Aβ peptide is normally secreted as a soluble peptide in the α-helix/random coil conformation [4]. Toxicity associated with Aβ is due to a conformational change, forming β-sheets, or amyloid fibrils [4]. Resistance to proteolytic degradation heightens as the concentration of the β-sheet conformer increases, leading to the formation of amyloid fibrils within senile plaques and cerebral and meningeal blood vessels [4].

The conformational change essential to Aβ's pathology has been found to be dependant on the formation of a salt bridge between a histidine residue in position 13 and an aspartate residue in position 23 [5,6]. Melatonin, an endogenous hormone has been found to disrupt this interaction, inhibiting the progressive formation of β-sheets and amyloid fibrils [4]. The protonated imidazole group of histidine is thought to interact with the aromatic indole moiety of melatonin, disturbing the formation of the salt bridge. This study aims to explore this interaction by modeling the binding of melatonin and imidazolium.

2. Computational methods

A conformational analysis of melatonin was performed in order to determine the lowest energy conformer in the gas phase (data to be published). Fully relaxed optimizations were performed with GAUSSIAN 98 [7] at the B3LYP/6-31G(d) level of theory; the torsional angles of interest included the methoxy group and the alkyl amide side chain (Fig. 1). The lowest energy conformers were used as the input structures, in which two classes of binding motifs were instigated: (1) interaction between the protonated imidazole nitrogen and the aromatic carbon in position 4; and (2) interaction between the protonated imidazole nitrogen and the carbon in position 6 (Fig. 1). Note that the numbers describing the position of atoms refer to IUPAC nomenclature, and are outside the ring system in Fig. 1; the numbering in place of the atoms refers to the internal numbering system regarding model assembly, a method used in a previous study [8].

The geometrical parameters of all four binding motifs were optimized at the B3LYP/6-31G(d) level of theory. Harmonic frequencies were also computed to characterize each minimum found at the B3LYP/6-31G(d) level of theory. Full counterpoise corrections were included in this study to account for basis set superposition error (BSSE). In order to explore the effect of solvation on the stability of each structure, self-consistent reaction field (SCRF) solvation calculations were performed; solvent cavities surrounding the structures were defined by Tomasi's polarized continuum model (PCM) [9,10]. Two classes of

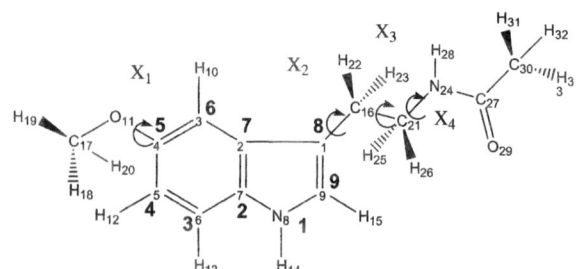

Fig. 1. IUPAC numbering, internal coordinate numbering system, and defined torsional angles of melatonin. Bold numbers outside the ring indicate IUPAC numbering and numbering in place of atoms indicates the internal numbering system. The torsions are defined: $\chi_1 = C_{17}-O_{11}-C_4-C_3$, $\chi_2 = C_{21}-C_{16}-C_1-C_2$, $\chi_3 = N_{24}-C_{21}-C_{16}-C_1$, $\chi_4 = C_{27}-N_{24}-C_{21}-C_{16}$.

solvent were used: polar protic (water and ethanol) and polar aprotic (chloroform and dimethylsulfoxide).

3. Results and discussion

3.1. Reaction model

The four structures modeled in this study may be represented by the following equation:

$$\text{Melatonin} + \text{Imidazole}^{1+} \rightarrow [\text{Melatonin}-\text{Imidazole}]^{1+}$$

The two energetically degenerate minima used in this study ($\chi_1 = syn$, $\chi_2 = g+$, $\chi_3 = anti$, $\chi_4 = g+$, and $\chi_1 = syn$, $\chi_2 = g-$, $\chi_3 = anti$, $\chi_4 = g-$, data to be published) were initially postulated to have two favourable binding motifs with imidazolium, as described in Section 2. Each optimized interaction was found to be most favourable as an intersecting one, as opposed to a 'stacked' interaction. The four input structures were initiated above the aromatic benzene moiety, as opposed to the sp^2 hybridized nitrogen of the pyrrole segment of indole. The rationale behind this strategy was to allow imidazolium to interact with both the indole moiety and the alkyl amide side chain.

3.2. Binding motif 1

The optimized structures for the first binding motif are depicted in Fig. 2(A). The structure designated as

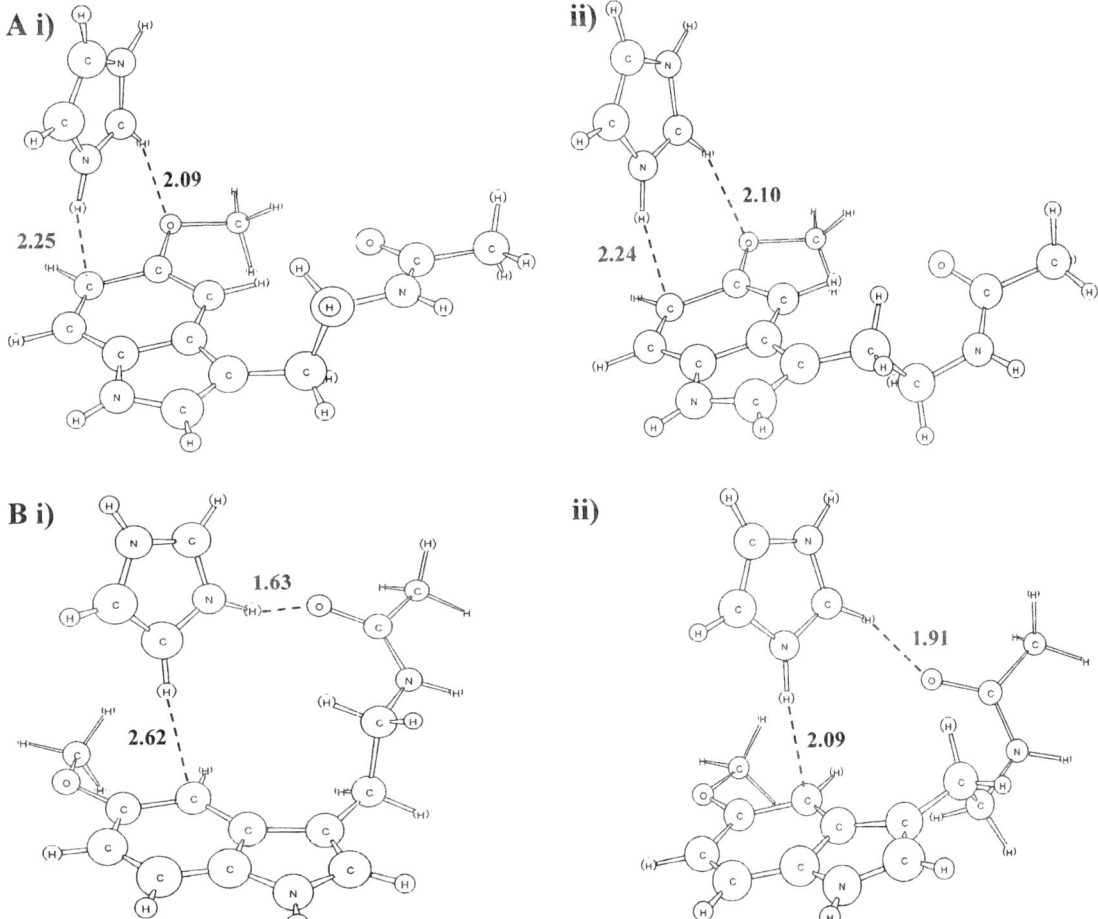

Fig. 2. The four optimized structures observed in this study, and respective bond lengths (in Angstroms). (A) Binding motif 1, (B) Binding motif 2.

Table 1
Solvation energies for the optimized structures in various polar protic and aprotic solvent media

Structure[a]	Gas phase	Chloroform	DMSO	Ethanol	Water
2Ai	−991.609983888	−991.661034061	−991.674704823	−991.693517939	−991.696427361
2Bi	−991.621466793	−991.671630150	−991.684918108	−991.702346138	−991.705397925
2Aii	−991.610225217	−991.666365622	−991.679066070	−991.695585418	−991.698336192
2Bii	−991.617717340	−991.660974228	−991.674493214	−991.693134587	−991.696793227

[a] Refers to the structures as named in Fig. 2.

2Ai shows the interaction between imidazolium and the optimized melatonin conformer $\chi_1 = syn$, $\chi_2 = g+$, $\chi_3 = anti$, $\chi_4 = g+$. The BSSE corrected binding energy was calculated to be −17.80 kcal/mol, with curt hydrogen bond lengths (2.25 and 2.09). The second structure, **2Aii** is structurally similar to **2Ai** with nearly identical bond lengths (2.24 and 2.10). The binding energies were also found to be almost identical (−17.78 kcal/mol).

The interaction between the protonated imidazolium moiety and the aromatic ring is a favourable one as, the methoxy oxygen donates electron density to the ring—the imparted negative charge may be stabilized by resonance. This localized negative charge is then a point of high affinity for hydrogen bonding with the imidazolium ion.

3.3. Binding motif 2

The second binding motif (Fig. 2(B)) proved to be a more favourable interaction between melatonin and imidazolium. Structure **2Bi** was the most exothermic of all calculated structures (−24.60 kcal/mol) with the shortest hydrogen bond distances. However, this interaction was unique among the four structures, it involved a hydrogen bond between pronated nitrogen and the carbonyl group of the alkyl amide side chain. Structure **2Bii** also had a high binding energy (−19.80 kcal/mol), however, the protonated nitrogen did not interact with the carbonyl group; instead, it was found to bind to the aromatic pi system. Although there is some delocalization of density over the peptide bond, the carbonyl oxygen is still a more affinic centre for the imidazolium hydrogen bonding, a phenomenon reflected in the binding energies, and respective bond distances (1.63 vs. 2.09).

3.4. Solvation calculations

Solvation calculations were performed in order to view the additional stability conferred to the structures by solvent interactions. The highest stability was conferred to structure **2Bi** (Fig. 2), by solvation with water and ethanol, respectively (Table 1). This is due to the interaction between the solvents and various polar segments of both melatonin and imidazolium. No significant difference was observed between the converged structures based on surface area available for the formation of solvent spheres, as seen in a previous study [8]. This is most likely because the structures are all quite crowded, and not outstretched. Hydrogen bond stability increased with increasing solvent polarity in each binding motif.

4. Conclusions

The interaction between the two energetically degenerate lowest energy conformers of melatonin and imidazolium were subjected to geometry optimizations in the gas phase. Of the four structures optimized, it was determined that the most stable hydrogen bond motif exists between the protonated imidazolium nitrogen and the carbonyl oxygen of the melatonin's alkyl amide side chain. Solvation of this structure (and all others) conferred stability which increased with solvent polarity, suggesting that this interaction is also stable at the physiological level.

The finding of such stability in the form of an intersecting interaction is not a novel discovery. In a similar study by Carter and Weaver, all interactions were found to be most stable in an intersecting form [11]. The interaction with the alkyl amide side chain found in this study suggests that conformation plays an important role in hydrogen bond formation, one that would be less feasible if the imidazolium were atop the sp^2 nitrogen of the indole moiety.

Acknowledgements

One of the authors (I.G.C.) wishes to thank the Ministry of Education for a Szent-Györgyi Visiting Professorship.

References

[1] A. Rocchi, S. Pellegrini, G. Siciliano, L. Murri, Brain Res. Bull. 61 (2003) 1.

[2] R.L. Neve, N.K. Robakis, Trends Neurosci. 21 (1) (1998) 15.

[3] B. Poeggeler, L. Miravalle, M.G. Zagorski, T. Wisniewski, Y.J. Chyan, Y. Zhang, H. Shao, T. Bryant-Thomas, R. Vidal, B. Fran-gione, J. Ghiso, M.A. Pappolla, Biochemistry 40 (2001) 14995.

[4] M. Pappolla, P. Bozner, C. Soto, H. Shao, N.K. Robakis, M. Zagorski, B. Frangione, J. Ghiso, J. Biol. Chem. 273 (13) (1998) 7185.

[5] P.E. Fraser, D.R. McLachlan, W.K. Surewicz, C.A. Mizzen, A.D. Snow, J.T. Nguyen, D.A. Kirschner, J. Mol. Biol. 244 (1994) 64.

[6] P.E. Fraser, J.T. Nguyen, W.K. Surewicz, D.A. Kirschner, Biophys. J. 60 (1991) 1190.

[7] Gaussian 98, Revision A.11.1, M.J. Frisch, G.W. Trucks, H.B. Schlegel, G.E. Scuseria, M.A. Robb, J.R. Cheeseman, V.G. Zakrzewski, J.A. Montgomery, Jr., R.E. Stratmann, J.C. Burant, S. Dapprich, J.M. Millam, A.D. Daniels, K.N. Kudin, M.C. Strain, O. Farkas, J. Tomasi, V. Barone, M. Cossi, R. Cammi, B. Mennucci, C. Pomelli, C. Adamo, S. Clifford, J. Ochterski, G.A. Petersson, P.Y. Ayala, Q. Cui, K. Morokuma, P. Salvador, J.J. Dannenberg, D.K. Malick, A.D. Rabuck, K. Raghavachari, J.B. Foresman, J. Cioslowski, J.V. Ortiz, A.G. Baboul, B.B. Stefanov, G. Liu, A. Liashenko, P. Piskorz, I. Komaromi, R. Gomperts, R.L. Martin, D.J. Fox, T. Keith, M.A. Al-Laham, C.Y. Peng, A. Nanayakkara, M. Challacombe, P.M.W. Gill, B. Johnson, W. Chen M.W. Wong, J.L. Andres, C. Gonzalez, M. Head-Gordon, E.S. Replogle, J.A. Pople, Gaussian, Inc., Pittsburgh, PA, 2001.

[8] L.F. Pisterzi, D.R.P. Almeida, G.A. Chass, L.L. Torday, J.Gy. Papp, Chem. Phys. Lett. 365 (2002) 542.

[9] S. Miertuš, J. Tomasi, Chem. Phys. 65 (1982) 239.

[10] S. Miertuš, E. Scrocco, J. Tomasi, Chem. Phys. 55 (1981) 117.

[11] M.D. Carter, D.F. Weaver, J. Mol. Struct. (Theochem) 626 (2003) 279.

ELSEVIER

Journal of Molecular Structure (Theochem) 666–667 (2003) 527–536

THEO
CHEM

www.elsevier.com/locate/theochem

Reaction profiling of the MAO-B catalyzed oxidative deamination of amines in Alzheimer's disease

Donna M. Gasparro[a,*], David R.P. Almeida[a], Luca F. Pisterzi[a], Jason R. Juhasz[a], Bela Viskolcz[b], Botond Penke[c,d], Imre G. Csizmadia[a,c,e]

[a]*Department of Chemistry, University of Toronto, 80 St George St, Toronto, Ont., Canada M5S 3H6*
[b]*Department of Chemistry, JGyTFK, University of Szeged, P.O. Box 396, H-6701 Szeged, Hungary*
[c]*Department of Medical Chemistry, University of Szeged, Dóm tér 8, 6720 Szeged, Hungary*
[d]*Protein Chemistry Research Group, Hungarian Academy of Sciences, University of Szeged,*
Dóm tér 8, 6720 Szeged, Hungary
[e]*Global Institute of COmputational Molecular and Materials Science (GIOCOMMS), 1422 Edenrose St,*
Mississauga, Ont., Canada L5V 1H3

Abstract

Monoamine oxidase type B (MAO-B) catalyzes the oxidative deamination of various endogenous and exogenous primary, secondary, and tertiary amines. During this reaction, reactive oxygen species (ROS) are produced which contribute to the oxidative stress in a biological system. Further, research has indicated that increased MAO-B levels and increased ROS production may lead to deposition of β-amyloid (Aβ) and act as contributing factors to the pathogenesis of Alzheimer's disease (AD). As such, irreversible MAO-B inhibitors like selegiline, which decrease the rate of MAO-B catalyzed oxidative deamination, and thus the production of ROS, may have therapeutic potential via neuroprotection. In the current study, molecular orbital calculations were performed using the complete basis set 4m (CBS-4m) theoretical framework employed within the GAUSSIAN 98 software to elucidate the full energetic and thermodynamic profile of the MAO-B catalyzed oxidative deamination reaction. Full geometry optimizations were performed on model compounds to reveal the energies (ΔE) and Gibbs Free Energies (ΔG) of the oxidative deamination reaction as a means to clarify the details of this paramount reaction and how it relates to AD and selegiline's therapeutic potential. Results reveal that the oxidative deamination reaction consists of three successive energy-consuming steps for all amines. Trends indicate that for all amines, as amine substitution increases: (1) energetic investments needed for steps 1 and 3 increase, (2) $\Delta E_{\text{Reaction}}$ and $\Delta G_{\text{Reaction}}$ values for the oxidative deamination reaction as a whole increase, (3) differences between $\Delta G_{\text{Reaction}}$ and $\Delta E_{\text{Reaction}}$ values increase, and (4) ΔG and ΔE values for step 2 of the reaction decrease. The second step of the MAO-B catalyzed reaction is the only step, where increased amine substitution correlates with decreased ΔE and ΔG values. Due to the fact that step two possesses the lowest investment of energy for tertiary and secondary amines but the highest investment of energy for primary amines, it can be postulated that MAO-B has preferred secondary and tertiary substrates. The fact that step 2 is the rate-limiting step relates to the notion that it modulates the potential to increase oxidative stress, and thus, must be tightly regulated to maintain oxidant and anti-oxidant homeostasis. Finally, in disease states such as AD, where increased MAO-B levels have been documented, it is possible that

* Corresponding author.

E-mail addresses: dgasparro@medscape.com (D.M. Gasparro), dalmeida@medscape.com (D.R.P. Almeida), lpisterzi@medscape.com (L.F. Pisterzi), jasonjuhasz@medscape.com (J.R. Juhasz), viskolcz@jgytf.u-szeged.hu (B. Viskolcz), penke@ovrisc.mdche.szote.u-szeged.hu (B. Penke), icsizmad@alchemy.chem.utoronto.ca (I.G. Csizmadia).

MAO-B mediated oxidative stress is a contributing factor to AD. Thus, MAO-B inhibition by drugs like selegiline will provide neuroprotective benefits and help to prevent AD pathogenesis.

Keywords: Monoamine oxidase; Monoamine oxidase type B; Oxidative deamination; Reactive oxygen species; Alzheimer's disease; Complete basis set 4m

1. Introduction

Monoamine oxidase type B (MAO-B) is an integral membrane protein in outer mitochondrial membranes and is found in neuronal and non-neuronal cells in the brain and in peripheral organs [1]. MAO-B is a constitutive enzyme found predominantly in serotonergic neurons of the brain as well as in glial cells and at high concentrations in the thalamus, stratium, cortex and the brainstem [2,3]. MAO-B preferentially breaks down β-phenylethylamine and is selectively and irreversibly inhibited by the drug selegiline (also known as deprenyl) [4]. The half-life of MAO-B is approximately 30 days in the CNS; thus, once selegiline binds to MAO-B, the enzyme remains inactive until it is degraded and new proteins are synthesized [5].

MAO-B exists as a protein homodimer with the C-terminal cysteine of each monomer covalently bound to the 8α position of a flavin adenine dinucleotide (FAD) [6]. As a flavoenzyme, MAO-B catalyzes the oxidative deamination of various neurotransmitters (dopamine, serotonin, noradrenaline) and amines (select primary, secondary, and tertiary amines) [7, 8]. MAO-B is responsible for the oxidative deamination of various amines to an imine followed by the formation of an aldehyde or ketone and a corresponding amine (Fig. 1) [5]. The rate of oxidation of a particular amine by MAO-B is dependent on amine concentration, availability of oxygen (which oxidizes chemically reduced MAO-B via the oxidation of $FADH_2$ to FAD), and the concentration of MAO-B in the mitochondrial outer membranes [3].

MAO-B inhibitors such as selegiline decrease the availability of MAO-B and thus, the rate of oxidative deamination [3]. There are six postulated modes of action for selegiline (1) induces anti-apoptotic activity, (2) increases the amount of *N*-acetylated polyamines hence acting as an antagonist at NMDA receptors, (3) increases nitric oxide synthesis, (4) increases monoamine levels, (5) induces anti-oxidant enzymes such as superoxide dismutase and catalase, and (6) inhibits MAO-B [9]. It is essential to note that the last two modes of action mentioned are anti-oxidant activities of selegiline.

The proposed molecular mechanism for MAO-B catalysis involves the formation of hydrogen peroxide (H_2O_2) concomitant with $FADH_2$ oxidation to FAD (Fig. 1) [10]. If H_2O_2 is not detoxified by glutathione peroxidase, hydroxyl radicals are generated via the Fenton reaction which may propagate free radical damage (Fig. 1) [11]. The hydroxyl radical is cytotoxic due to its high reactivity towards lipids, proteins, and DNA [12,13]. Anti-oxidant enzymes such as catalase and superoxide dismutase are responsible for scavenging radicals, however, these enzymes are found at lowest concentrations in the brain relative to the rest of the body.

The formation of free radicals via MAO-B oxidation is implicated in the neurodegeneration associated with Alzheimer's disease (AD) and in the pathophysiology of aging [14–16]. As levels of MAO-B increase, the brain dopamine system has a corresponding increase in vulnerability to damage via oxidative stress [1]. Various data implicates MAO-B oxidation in AD: postmortem AD patients show an increase in MAO-B levels, MAO-B is expressed in astrocytes of senile plaques in AD patients, and increased binding of radiolabelled selegiline to MAO-B in AD patients [1]. Given the above, selegiline can be viewed as a neuroprotective agent able to delay the progression of neuronal cell death vis-à-vis anti-oxidant modes of action—especially in its inhibition of MAO-B. Consequently, selegiline is currently undergoing human trials to test its efficacy in alleviating AD pathology [13]. Selegiline is also used to treat other neurological diseases such as Parkinson's disease, global ischemia, narcolepsy, and Gilles de la Tourette Syndrome [15,17].

Moreover, since the MAO-B catalyzed oxidative deamination reaction is pathologically linked to AD

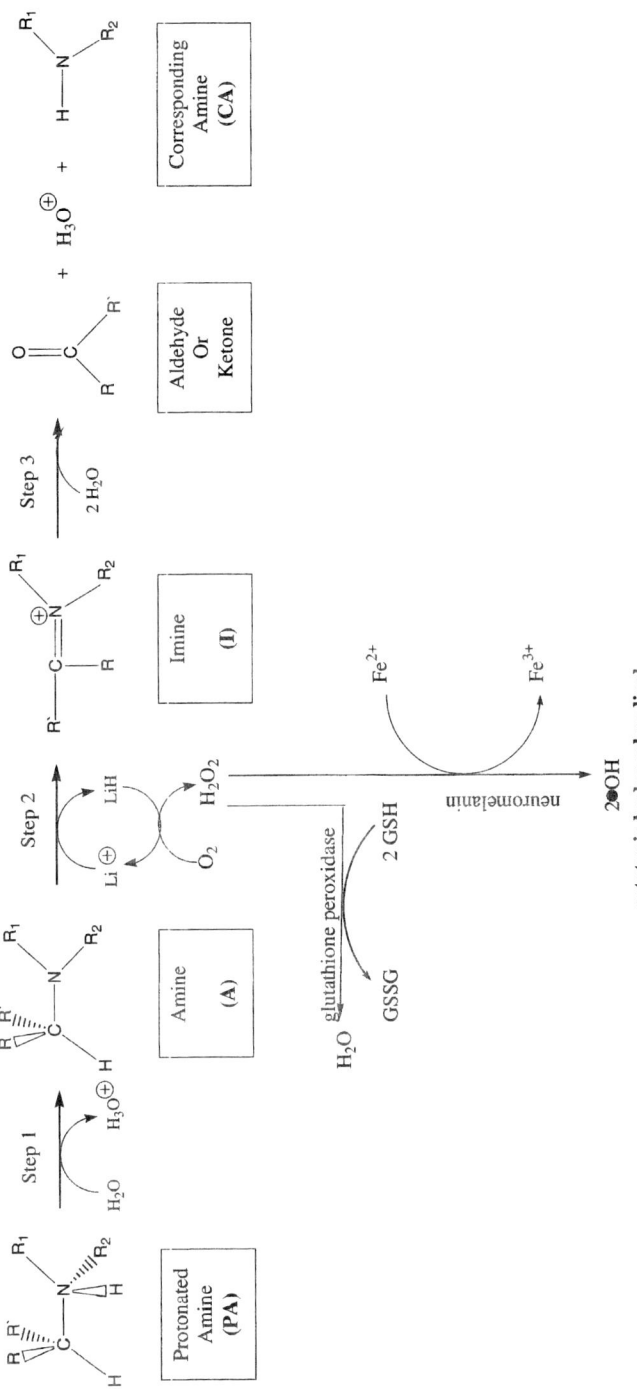

Fig. 1. General reaction path of the MAO-B catalyzed oxidative deamination of primary, secondary, and tertiary amines. Imine formation (step 2) occurs concomitantly with reactive oxygen species (ROS) production, which may contribute to oxidative stress and the neurodegeneration in disease stated like Alzheimer's disease. The irreversible MAO-B inhibitor, selegiline, blocks the oxidative deamination reaction.

and MAO-B inhibition by selegiline is a fundamental basis of selegiline's efficacy towards improving neurological disorders, full characterization of the MAO-B reaction is critical to understanding its role in these disease pathologies. In the current study, the oxidative deamination reaction (presented in Fig. 1) is thermodynamically quantified to elucidate any molecular details or characteristics that may be relevant to AD and selegiline's mode of action in improving neurodegenerative disorders such as AD.

2. Methods

The current study focuses on the thermodynamic assessment and characterization of the MAO-B catalyzed oxidative deamination of primary, secondary, and tertiary amines. The reaction coordinate possesses three distinct steps (and four corresponding reaction plateaus): (1) deprotonation of a protonated amine (primary, secondary, or tertiary) to yield a neutral amine, (2) oxidation of a neutral amine to an imine intermediate, and (3) hydration of an imine to an aldehyde or ketone and a corresponding amine (Fig. 1). In this study, FAD and $FADH_2$ are modeled with Li^+ and LiH, respectively; and the process of FAD oxidation to $FADH_2$ is computed as a one-step hydride reduction of Li^+ to LiH (Fig. 1).

To characterize the full oxidative deamination reaction, model species involved in the reaction coordinate (Table 1) were investigated with molecular orbital (MO) calculations using the GAUSSIAN 98 software [18]. Molecular structure, stereochemistry, and geometry of all species were exclusively defined in terms of GAUSSIAN 98 internal Cartesian coordinates. All calculations were performed at the complete basis set 4m (CBS-4m; m refers to the use of Minimal Population localization) [19] level of theory in the gas phase ($\varepsilon = 0.0$) to extract the best possible energies for this process.

Initially, all species involved in this reaction were subject to full geometry optimizations and their total energy (E) and Gibbs Free Energy (G) calculated (Table 1). Subsequently, E and G values were summated for each respective step of the reaction path according to the molecular species present at that step, while ensuring proper stoichiometry and charge conservation throughout the reaction coordinate. These latter E and G values were used to calculate the energy of reaction (ΔE) and Gibbs Free Energy of reaction (ΔG) between all steps of the reaction path. Finally, energy and Gibbs Free energy of reaction for the complete oxidative deamination reaction (denoted as $\Delta E_{Reaction}$ and $\Delta G_{Reaction}$) were calculated. All graphical data were plotted using Axum 5.0 [20].

Table 1
Model molecules involved in the MAO-B catalyzed oxidative deamination of primary, secondary and tertiary amines

Chemical name	Chemical formula	Energy (SCF) (kcal/mol)	Free energy, G (kcal/mol)
Primary amine	$[NH_3CH_3]$	− 95.246615	− 95.719235
Protonated primary amine	$[NH_4CH_3]^+$	− 95.608019	− 96.060638
Secondary amine	$[NH(CH_3)_2]$	− 134.282784	− 134.950356
Protonated secondary amine	$[NH_2(CH_3)_2]^+$	− 134.656043	− 135.302490
Tertiary amine	$[N(CH_3)_3]$	− 173.320861	− 174.186066
Protonated tertiary amine	$[NH(CH_3)_3]^+$	− 173.701913	− 174.545041
Lithium	$[Li]^+$	− 7.235840	− 7.248587
Lithium hydride	$[LiH]$	− 7.985592	− 8.041151
Water	$[H_2O]$	− 76.053394	− 76.367546
Hydronium	$[H_3O]^+$	− 76.330925	− 76.625856
Imine	$[CH_2NH_2]^+$	− 94.417284	− 94.835524
Methyl imine	$[CH_2NH(CH_3)]^+$	− 133.469364	− 134.083487
Dimethyl imine	$[CH_2N(CH_3)_2]^+$	− 172.518679	− 173.330591
Formaldehyde	$[H_2CO]$	− 113.907358	− 114.381593
Non-substituted amine (ammonia)	$[NH_3]$	− 56.214201	− 56.493922

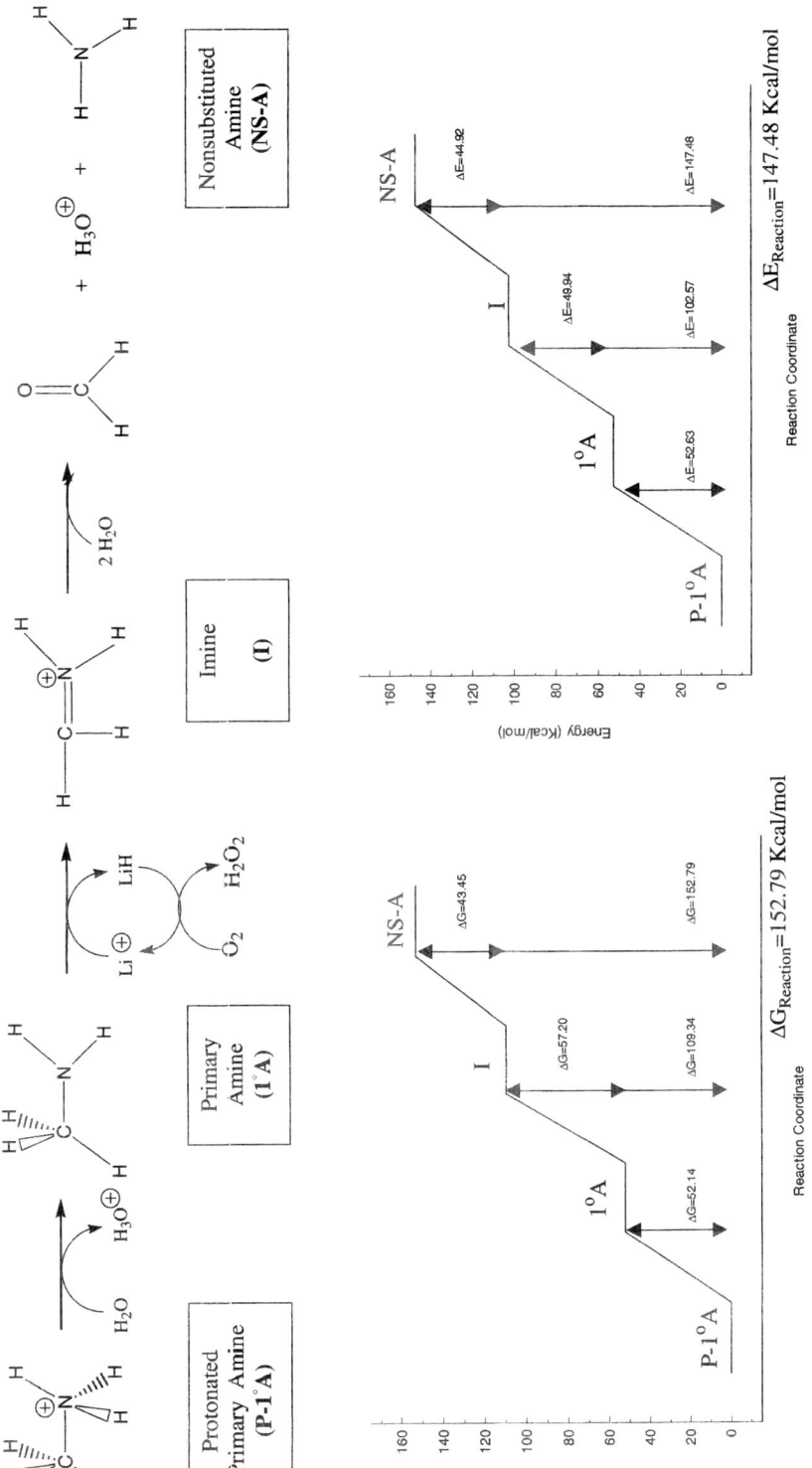

Fig. 2. Thermodynamic and energetic profiles of the MAO-B catalyzed oxidative deamination of primary amines calculated at the CBS-4m level of theory.

3. Results and discussion

Thermodynamic and energetic data revealed that the oxidative deamination reaction catalyzed by MAO-B consists of three successive energy-consuming steps for primary (Fig. 2), secondary (Fig. 3), and tertiary (Fig. 4) amines; all amines computed in this study revealed this same profile. A general decreasing trend in $\Delta E_{Reaction}$ and $\Delta G_{Reaction}$ is evident from tertiary amine ($\Delta G_{Reaction} = 170.34$ kcal/mol, $\Delta E_{Reaction} = 163.37$ kcal/mol), to secondary amine ($\Delta G_{Reaction} = 163.16$ kcal/mol, $\Delta E_{Reaction} = 157.29$ kcal/mol), to primary amine ($\Delta G_{Reaction} = 152.79$ kcal/mol, $\Delta E_{Reaction} = 147.48$ kcal/mol). As amine substitution increases, so does the energetic investment of the oxidative deamination reaction (Eqs. (1) and (2) and Figs. 2–4)

$$\Delta G_{Reaction}(3° \text{ amine}) > \Delta G_{Reaction}(2° \text{ amine})$$

$$> \Delta G_{Reaction}(1° \text{ amine}) \tag{1}$$

$$\Delta E_{Reaction}(3° \text{ amine}) > \Delta E_{Reaction}(2° \text{ amine})$$

$$> \Delta E_{Reaction}(1° \text{ amine}) \tag{2}$$

Moreover, as the substitution of an amine increases, the difference between $\Delta G_{Reaction}$ and $\Delta E_{Reaction}$ for the oxidative deamination of that amine also increases (Fig. 5). Given that ΔG values indicate the amount of work-energy, while ΔE values indicate the total energy of a system, along with the fact that as MAO-B substrates get more complex, the substrates ΔG and ΔE values deviate further from each other, i can be predicted that ΔG values will be the best descriptive terms for the thermodynamics of the oxidative deamination reaction. Thus, for the purpose of this study, the calculated ΔG values will be deemed as the best prognosticators for MAO-B catalyzed oxidative deamination reaction.

As stated above, as substitution increases, so does the $\Delta G_{Reaction}$ and $\Delta E_{Reaction}$ values. This same trend is present for the ΔG and ΔE values for steps 1 and 3 of the oxidative deamination reaction (Eqs. (3) and (4) and Figs. 2–4). Overall, the only exception to this trend of increasing energy consumption with increasing amine substitution occurs in the second step of the reaction, where oxidation of a neutral amine to an imine intermediate is coupled to reactive oxygen species (ROS) formation; here, as substitution increases, ΔG and ΔE values decrease (Eqs. (5) and (6) and Figs. 2–4). The second step of the MAO-B catalyzed reaction is the only step of the reaction that not only deviates, but also is opposite to the trend indicated in Eqs. (1)–(4)

$$\Delta G_{steps\ 1,3}(3° \text{ amine}) > \Delta G_{steps\ 1,3}(2° \text{ amine})$$

$$> \Delta G_{steps\ 1,3}(1° \text{ amine}) \tag{3}$$

$$\Delta E_{steps\ 1,3}(3° \text{ amine}) > \Delta E_{steps\ 1,3}(2° \text{ amine})$$

$$> \Delta E_{steps\ 1,3}(1° \text{ amine}) \tag{4}$$

$$\Delta G_{step\ 2}(1° \text{ amine}) > \Delta G_{step\ 2}(2° \text{ amine})$$

$$> \Delta G_{step\ 2}(3° \text{ amine}) \tag{5}$$

$$\Delta E_{step\ 2}(1° \text{ amine}) > \Delta E_{step\ 2}(2° \text{ amine})$$

$$> \Delta E_{step\ 2}(3° \text{ amine}) \tag{6}$$

Thermodynamically, step two possesses the lowest investment of energy of the entire reaction for tertiary and secondary amines, but possesses the highest investment of energy for primary amines. Thus MAO-B may preferentially bind primary amine substrates (such as β-phenylethylamine), however, once a primary amine is bound to MAO-B a high energetic barrier for the second step is encountered, which is uncharacteristic of secondary and tertiary amines. It can then be postulated that MAO-B has the highest energetic barrier at the second step for primary amines as compared to secondary and tertiary amines, as a means to control ROS formation. This is supported by the fact that both neurotransmitters (which are often primary amines such as dopamine and serotonin), and mitochondria, which contain MAO-B, are abundant in neurons. Thus the high energetic barrier of step two of the MAO-B reaction may serve to decrease ROS formation that would otherwise be excessive since both the substrate (primary amine neurotransmitters) and the enzyme (MAO-B) are in abundance in neurons.

According to kinetic studies, the rate-limiting step in the MAO-B catalyzed reaction is the oxidative cleavage of the C–H bond that is alpha to the amino group (step 2) [1]. Having step 2 as the rate-limiting step is consistent with the idea that this step is

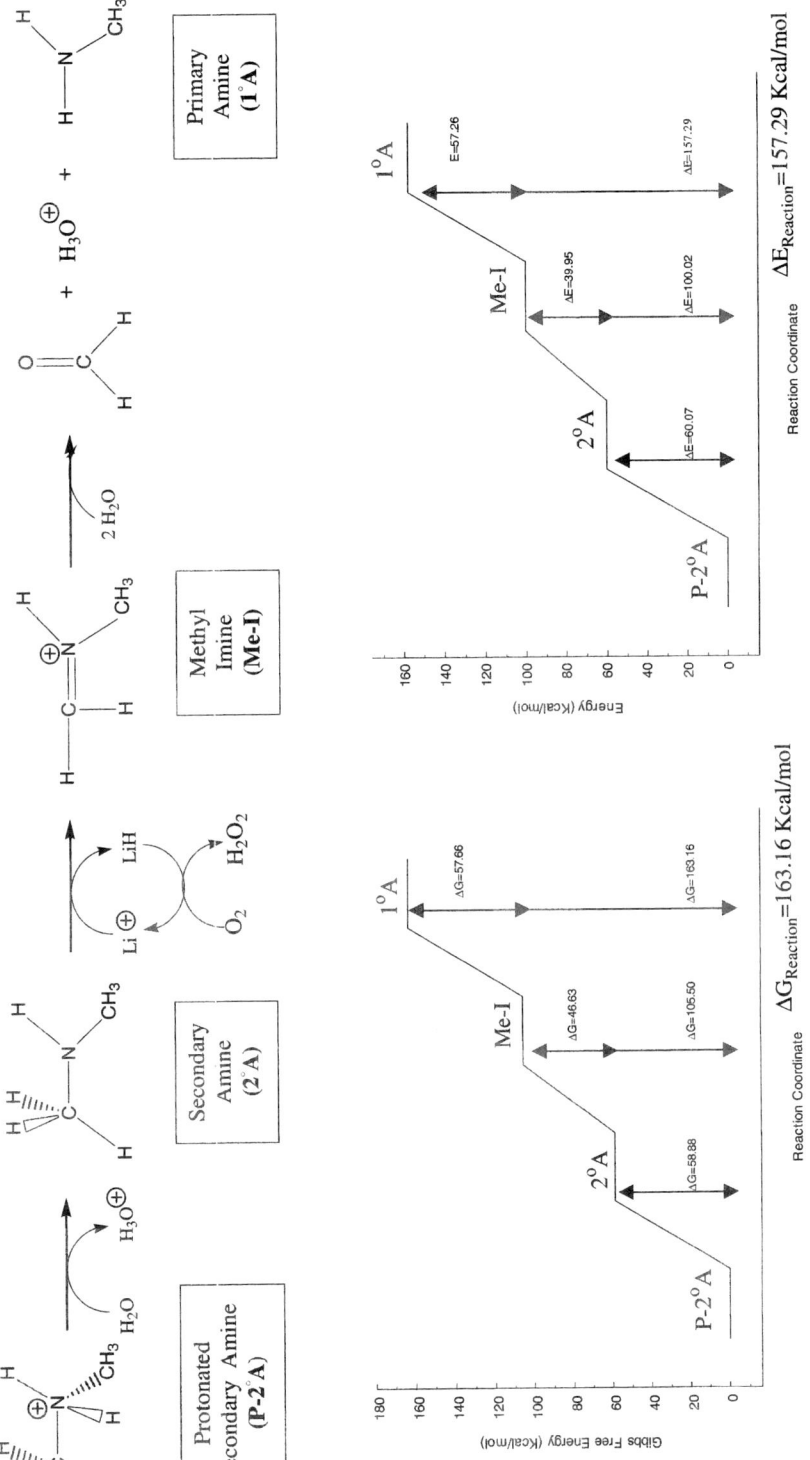

Fig. 3. Thermodynamic and energetic profiles of the MAO-B catalyzed oxidative deamination of secondary amines calculated at the CBS-4m level of theory.

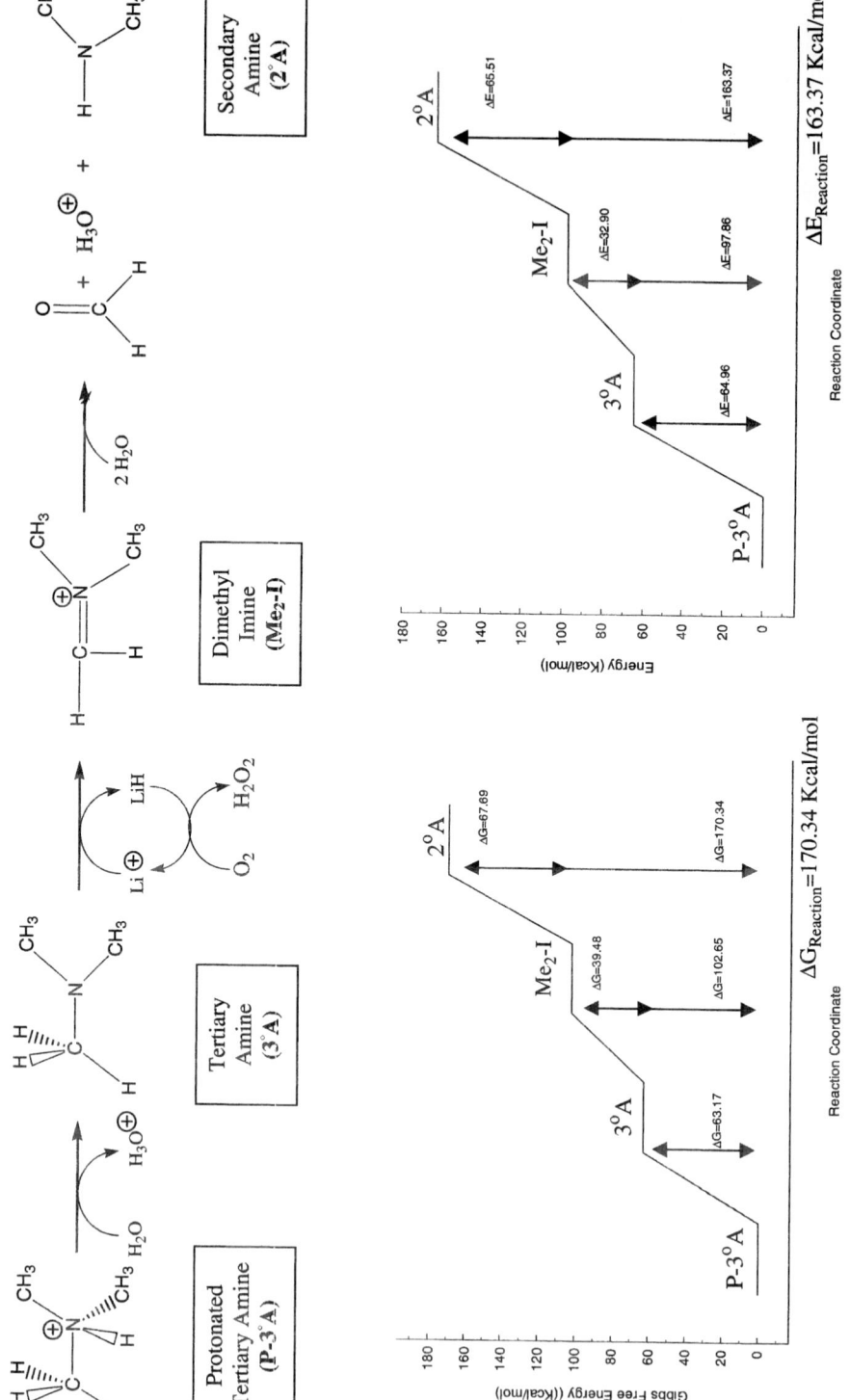

Fig. 4. Thermodynamic and energetic profiles of the MAO-B catalyzed oxidative deamination of tertiary amines calculated at the CBS-4m level of theory.

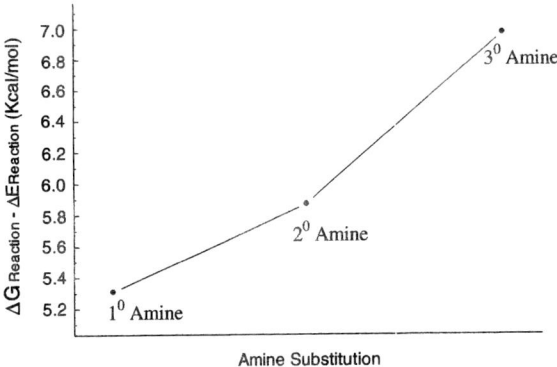

Fig. 5. Graphical representation of the increasing disparity between CBS-4m calculated ΔG and ΔE values as amine substitution increases.

deleterious to living cells because it increases oxidative stress; therefore, it is imperative that this step be kept at a carefully defined and controlled rate. This is to say, the ROS forming step is rate-limiting in order to protect the cell from oxidative stress and to maintain oxidant and anti-oxidant homeostasis.

4. Conclusion

It can be postulated that in disease states such as AD, increased MAO-B levels (as seen in post mortem

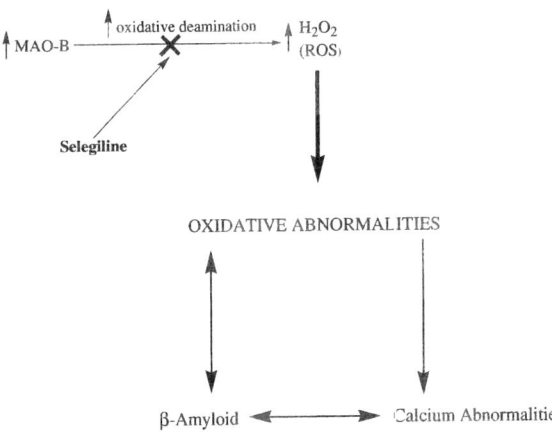

Fig. 6. The MAO-B oxidative deamination reaction is controlled by having the reactive oxygen species forming step as the rate-limiting step. However, it is postulated that increased MAO-B promotes increased ROS production, oxidative and calcium abnormalities, and deposition of β-amyloid (Aβ), which may lead to Alzheimer's disease. Consequently, MAO-B inhibition by selegiline is a neuroprotective reaction via an anti-oxidant mode of action.

AD patients [1]) will lead to oxidative abnormalities. Hence, the therapeutic inhibition of this reaction by a pharmaceutical agent like selegiline will avoid oxidative abnormalities in the cell and subsequently prevent β-amyloid (Aβ) deposition (Fig. 6) [21]. Indirectly, oxidative abnormalities promote calcium abnormalities which promote Aβ formation (Fig. 6) [21]. Since Aβ is a defining feature of AD, and given Aβ deposition is promoted by ROS, it can be inferred that the inhibition of the rate-limiting step (step 2) of the MAO-B catalyzed reaction is neuroprotective via an anti-oxidant mode of action and may help prevent AD pathogenesis.

Acknowledgements

One of the authors (IGC) wishes to thank the Ministry of Education for a Szent-Györgyi Visiting Professorship.

References

[1] J.S. Fowler, J. Logan, N.D. Volkow, G. Wang, R.R. MacGregor, Y. Ding, Methods 27 (2002) 263.
[2] M. Gerlach, M.B.H. Youdim, P. Riederer, Neurology 47 (6) (1996) 137S [S3].
[3] R.R. Ramsay, J. Neural Transm. Suppl. 52 (1998) 139.
[4] D.M. Gasparro, D.R.P. Almeida, S.M. Dobo, L.L. Torday, A. Varro, J.Gy. Papp, J. Mol. Struct. (THEOCHEM) 585 (2002) 167.
[5] S.M. Benedetti, K.F. Tipton, J. Neural Transm. Suppl. 52 (1998) 149.
[6] V. DeRose, J.C.G. Woo, W.P. Hawe, B.M. Hoffman, R.B. Silverman, K. Yelekci, Biochemistry 35 (1996) 11085.
[7] J.M. Rimoldi, Y.X. Wang, S.K. Nimkar, S.H. Kuttab, A.H. Anderson, H. Burch, N.Jr. Castagnoli, Chem. Res. Toxicol. 8 (1995) 703.
[8] Y.X. Wang, S. Mabic, N.Jr. Castagnoli, Biorg. Med. Chem. 2 (1998) 143.
[9] T. Thomas, Neurobiol. Aging 21 (2000) 343.
[10] J.G. Richards, J. Saura, J.M. Luque, A.M. Cesura, J. Gottowik, P. Malherbe, E. Barroni, J. Gray, J. Neural Transm. Suppl. 52 (1998) 173.
[11] P. Dorset, J. Neural Transm. Suppl. 41 (1994) 269.
[12] R. Sotto-Ottero, E. Mendez-Alvares, A. Hermida-Ameijeiras, I. Sanchez-Sellero, A. Cruz-Landeira, M.L.-R. Lamas, Life Sci. 69 (2001) 879.
[13] E.E. Tuppo, L.J. Forman, JAOA 101 (12) (2001) 511.
[14] M. Gerlach, M.B.H. Youdim, P. Riederer, J. Neural Transm. Suppl. 41 (1994) 177.

[15] N.Jr. Castagnoli, K.P. Castagnoli, Natl Inst. Drug Abuse Monogr. Ser. 85 (1997) 173.

[16] K. Kitani, C. Minami, T. Yamamoto, S. Kanai, G.O. Ivy, M. Carrillo, Ann. N. Y. Acad. Sci. 959 (2002) 295.

[17] M. Ebadi, S. Sharma, S. Shavali, H.El. Refaey, J. Neurosci. Res. 67 (2002) 285.

[18] M.J. Frisch, G.W. Trucks, H.B. Schlegel, G.E. Scuseria, M.A. Robb, J.R. Cheeseman, V.G. Zakrzewski, J.A. Montgomery, Jr., R.E. Stratmann, J.C. Burant, S. Dapprich, J.M. Millam, A.D. Daniels, K.N. Kudin, M.C. Strain, O. Farkas, J. Tomasi, V. Barone, M. Cossi, R. Cammi, B. Mennucci, C.Pomelli, C. Adamo, S. Clifford, J. Ochterski, G.A. Petersson, P.Y. Ayala, Q. Cui, K. Morokuma, D.K. Malick, A.D. Rabuck, K. Raghavachari, J.B. Foresman, J. Cioslowski, J.V. Ortiz, A.G. Baboul, B.B. Stefanov, G. Liu, A. Liashenko, P. Piskorz, I. Komaromi, R. Gomperts, R.L. Martin, D.J. Fox, T. Keith, M.A. Al-Laham, C.Y. Peng, A. Nanayakkara, C. Gonzalez, M. Challacombe, P.M.W. Gill, B.G. Johnson, W.Chen, M.W. Wong, J.L. Andres, M. Head-Gordon, E.S. Replogle, J.A. Pople, GAUSSIAN 98 (Revision A.9), Gaussian, Inc., Pittsburgh PA, 1998.

[19] G.A.Petersson,M.A.Al-Laham,J.Chem.Phys.94(1991)6081.

[20] Axum 5.0C for Windows, MathSoft Incorporated, 1996.

[21] G.E. Gibson, Free Radical Biol. Med. 32 (2002) 1061.

ELSEVIER

Journal of Molecular Structure (Theochem) 666–667 (2003) 537–545

THEO
CHEM

www.elsevier.com/locate/theochem

Predicting the conformations of carvedilol based on its pharmacophore fragments: a gas phase and solvation ab initio and density functional study

David R.P. Almeida[a,*], Donna M. Gasparro[a], Luca F. Pisterzi[a], Jason R. Juhasz[a], Ferenc Fülöp[b], Imre G. Csizmadia[a,c,d]

[a]*Lash Miller Laboratories, Department of Chemistry, University of Toronto, 80 St George Street, Toronto, Ont., Canada M5S 3H6*
[b]*Institute of Pharmaceutical Chemistry, Albert Szent-Györgyi Medical University, Szeged University, Eötvös u. 6, Szeged H-6720, Hungary*
[c]*Global Institute of Computational Molecular and Materials Science (GIOCOMMS)@1422, Edenrose St, Mississauga, Ont., Canada L5V 1H3*
[d]*Department of Medical Chemistry, University of Szeged, Dóm tér 8, Szeged 6720, Hungary*

Abstract

Carvedilol is a cardiovascular drug of proven efficacy with multiple modes of action. As a novel anti-fibrillar agent, carvedilol is able to inhibit amyloid-beta (Aβ) fibril formation and may have uses in the treatment of Alzheimer's disease. However, it is currently not known what form of Aβ carvedilol binds to or what type of inhibitory interaction occurs. Previously, we developed a scheme to fragment and study the pharmacophores of carvedilol individually so that the results could be used to predict the dominant features and conformations of whole carvedilol. In the present study, these predictions are tested and this methodology evaluated. Further, the conformational character of carvedilol is investigated to elucidate how it might be related to its anti-fibrillar mechanism. Conformational and structural predictions were tested on protonated S-carvedilol with molecular orbital computations of its potential energy hypersurface using restricted Hartree–Fock (RHF) and density functional theory (DFT with the Becke 3LYP hybrid exchange-correlation functional). Full gas phase optimizations were carried out at the RHF/3-21G and B3LYP/6-31G(d) level of theory and subsequent single point energy (SPE) calculations were performed on B3LYP/6-31G(d) converged conformers at the B3LYP/6-311 + G(2d,p) level of theory. B3LYP/6-31G(d) solvation SPE calculations were performed according to the polarized continuum model (PCM) of Tomasi and coworkers in DMSO and water. Six carvedilol conformations (C1–C6) were evaluated. The lowest energy conformers, C3, C4, and C6, possessed all predicted intramolecular features and their respective conformational assignments were also largely predicted from the carvedilol fragments. Solvation calculations in both aprotic (DMSO) and protic (water) solvents revealed the same trends seen in the gas phase. Generally, gas phase and solvation ab initio and DFT results were in sound agreement with each other. Given these results, the current study gives credence to the idea that large molecular systems can be studied by rational fragmentation of pharmacophores and other structure–activity moieties. Furthermore, the conformations presently evaluated are ideal starting points towards deciphering the molecular basis of the inhibitory interaction between carvedilol and Aβ.
© 2003 Elsevier B.V. All rights reserved.

Keywords: Carvedilol; Molecular fragments; Anti-fibrillar; Restricted Hartree–Fock; DFT-Becke 3LYP hybrid functional

* Corresponding author.
E-mail addresses: dalmeida@medscape.com (D.R.P. Almeida), dgasparro@medscape.com (D.M. Gasparro), lpisterzi@medscape.com (L.F. Pisterzi), jasonjuhasz@medscape.com (J.R. Juhasz), fulop@pharma.szote.u-szeged.hu (F. Fülöp), icsizmad@alchemy.chem.utoronto.ca (I.G. Csizmadia).

0166-1280/$ - see front matter © 2003 Elsevier B.V. All rights reserved.
doi:10.1016/j.theochem.2003.08.078

1. Introduction

Carvedilol, 1-(9H-Carbazol-4-yloxy)-3-[2-(2-methoxy-phenoxy)ethylamino]-2-propanol ($C_{24}H_{26}N_2O_4$), is a cardiovascular drug with multiple modes of action. Its major molecular targets are membrane adrenoceptors (antagonist at α_1, β_1, and β_2), reactive oxygen species (ROS), and ion channels (K^+ and Ca^{2+}) [1]. Carvedilol provides hemodynamic benefits, such as reduction of cardiac work and peripheral vasodilation, from its balanced adrenergic receptor blockage [1,2]. Carvedilol also exerts potent cardiovascular protection (anti-proliferative/anti-atherogenic, anti-hypertrophic, anti-ischemic, and anti-arrhythmic actions) via antioxidant effects, improvement of glucose/lipid metabolism, modulation of neurohormonal factors (e.g. nitric oxides), and beneficial cardiac electro-physiological properties (reviewed in Ref. [1]). Due to its hemodynamic and cardioprotective effects, carvedilol has proven efficacy in the treatment of hypertension, ischemic heart disease (IHD), and congestive heart failure (CHF) [1,3]. A further cardioprotective effect of carvedilol is related to its ability to protect mitochondria from oxidative stress by acting as a mild uncoupler of oxidative phosphorylation by means of a postulated proto-nophoretic mechanism [4].

Work has indicated that carvedilol, and its active hydroxylated analogues, act as novel anti-fibrillar agents [able to inhibit amyloid-beta (Aβ) fibril formation] [5]. According to the amyloid cascade hypothesis of Alzheimer's disease (AD), accumulation of Aβ in extracellular senile plaques (SPs) in brain tissues is the primary influence driving AD pathogenesis [6]. SPs contain aggregated amyloid fibrils formed from 39–43 amino acid Aβ peptides [7]. Aβ peptides consisting of 42 or 43 amino acids (abbreviated as Aβ1-42) are more prone to aggregate than Aβ1-40, and thus, promote aggregation and deposition of fibrillar Aβ [7]. The amyloid hypothesis states that increased Aβ1-42 production and accumulation leads to Aβ1-42 oligomerization and deposition in diffuse plaques [6]. Aβ oligomers cause progressive synaptic and neuritic injury alongside microglial and astrocytic activation (complement factors, cytokines, neuroinflammation, etc.) [6,8]. Neurotoxicity

causes neuronal ionic disturbances and oxidative injury producing neurofibrillary tangles (NFTs), widespread neuronal dysfunction, cell death, and neurotransmitter deficits that all culminate in AD dementia [6]. These observations provide strong rationale that the inhibition and elimination of Aβ fibrils may be viable therapeutic targets for the treatment of AD [7].

It has been shown that Aβ oligomers (dimers, trimers, or higher oligomers), in the absence of monomers and amyloid fibrils, are neurotoxic [8] and there is growing support that Aβ oligomers are the main neurotoxic component of AD [9–11]. Given the above, along with the fact that carvedilol may inhibit oligomer formation [5], it may be postulated that carvedilol may have uses in the prevention or slowing down of AD pathology. The effectiveness of carvedilol's inhibition of Aβ fibril formation is due to three factors: (1) a central basic amino pharmacophore, (2) two cyclic hydrophobic ring centroids, and (3) the molecular flexibility to adopt a specific three-dimensional pharmacophore conformation [5]. Although these three factors are elucidated, it is currently not known if carvedilol binds to Aβ monomers, dimers, or other oligomers [5] or what type of interaction occurs between carvedilol and the Aβ peptide(s).

We have previously [12] developed a scheme to fragment the three pharmacophores of carvedilol vis-à-vis structure–activity (c.f. Fig. 1) as a means to study the conformational profile of each pharmacophore and attempt to predict the conformations and features of the entire drug molecule; such an approach allows for the study of large molecules by rational fragmentation of their pharmacophores. In the current study, these predictions are tested to evaluate such a methodology. Further, since carvedilol's AD effects are due to its pharmacophores and conformational flexibility (factors 1–3 above), which have been studied, the conformational profile of carvedilol is analyzed in the hope of clarifying current and future communications on its anti-fibrillar mechanism of action.

2. Methods

Carvedilol is composed of three distinct pharma-

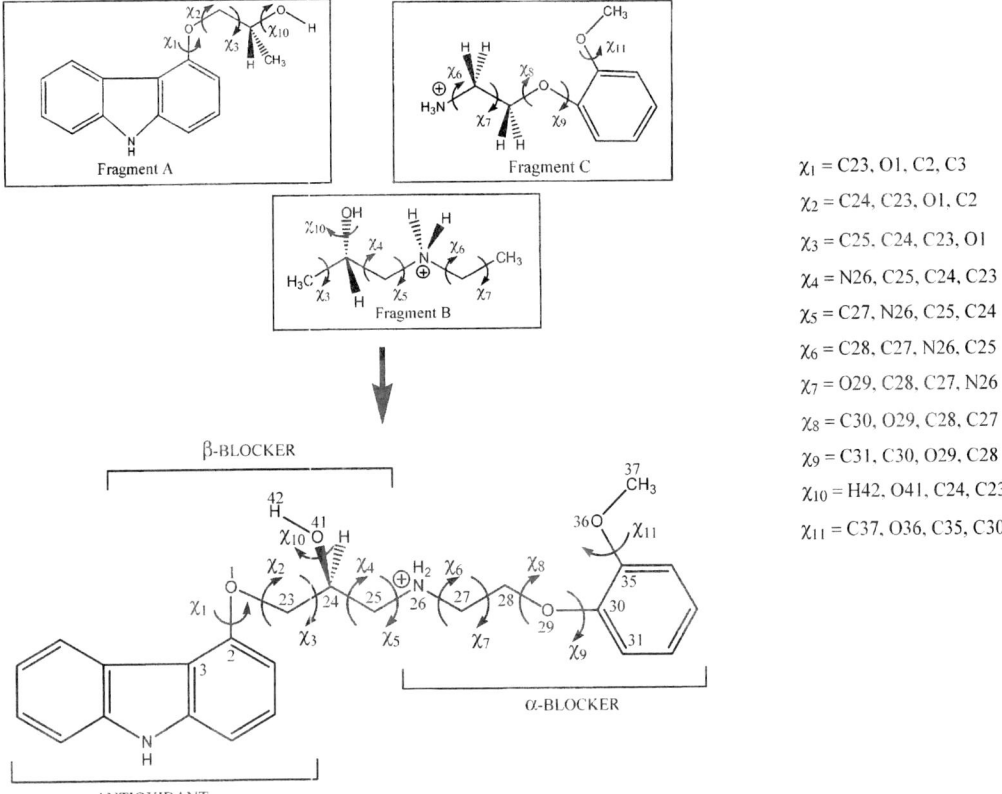

Fig. 1. Carvedilol was divided into three pharmacophoric fragments: *R*- and *S*-4-(2-hydroxypropoxy)carbazol (Fragment A) [12,13], 2(*R* and *S*)-1-(ethylamonium)propane-2-ol (Fragment B) [14], and aminoethoxy-2-methoxy-benzene (Fragment C) [15] (top). These fragments were individually computed and the results were used to predict the conformations and features of *N*-protonated carvedilol (*S*-configuration) (bottom). Numbers placed beside atoms were used to define all torsional angles for carvedilol in the *z*-matrix input for Gaussian 98 (right).

cophores and was thus divided into three molecular fragments: *R*- and *S*-4-(2-hydroxypropoxy)carbazol (Fragment A) possesses the carbazole-related antioxidant effects of carvedilol [12,13], 2(*R* and *S*)-1-(ethylamonium)propane-2-ol (Fragment B) contains the protonophoretic amino group implicated in the cardioprotective uncoupling of mitochondrial oxidative phosphorylation [14], and aminoethoxy-2-methoxy-benzene (Fragment C) is the α_1-adrenergic antagonist pharmacophore of carvedilol [15] (c.f. Fig. 1). Nonselective β-blockage is exerted by both Fragments A and B.

To allow explicit prediction of conformation, a systematic numbering system was used for all structures such that corresponding torsional angles in fragments and carvedilol were defined in the same manner. Further, conformational structural assign-

ments for the conformational minima of a respective potential energy hypersurface (PEHS) were made according to Eq. (1).

gauche plus $(g^+) = 60$ (ideal) $\pm 60°$

anti $(a) = 180$ (ideal) $\pm 60°$ (1)

gauche minus $(g^-) = -60$ (ideal) $\pm 60°$

To fully investigate the predictions, it is necessary to briefly review the individual fragments. The PEHS of Fragment A revealed a planar carbazole ring with intramolecular hydrogen bonding between the ether oxygen and the hydroxyl group proton as the dominant interaction [12,13]. The global minima possessed a conformation with torsional angles χ_1, χ_2, χ_3, and χ_{10} in the *a*, *a*, *a*, and g^- position [*S*-configuration; B3LYP/6-31G(d) results], respectively

[12,13]. The global minima of the Fragment B PEHS had a conformation with torsional angles χ_4, χ_5, χ_6, and χ_{10} in the a, a, a, and g^+ position (S-configuration; RHF/3-21G results), respectively, and the dominant interaction was an intramolecular hydrogen bond between the stereocentre hydroxyl oxygen and an amine proton [14]. Finally, the Fragment C PEHS revealed a pair of axis chiral global minima (see Ref. [13] for explanation) with torsional angles χ_7, χ_8, χ_9, and χ_{11} in the g^+, g^+, g^+, and g^- position, respectively (other axis chiral conformation was g^-, g^-, g^-, and g^+, respectively; RHF/3-21G results) [15]. The dominant interaction was a bifurcated hydrogen bond between an amine proton and the ether and methoxy oxygen atoms on the substituted benzene [15]. Consequently, it is predicted that the lowest energy conformers of carvedilol will possess the above intramolecular features and have a conformation predicted by Eq. (2) (torsional angle positions in brackets [ ] are representative of the possible axis chiral conformations of Fragment C).

$$\chi_1 = a \quad \chi_4 = a \quad \chi_7 = g^+[g^-] \quad \chi_{10} = g^-$$

$$\chi_2 = a \quad \chi_5 = a \quad \chi_8 = g^+[g^-] \quad \chi_{11} = g^-[g^+] \qquad (2)$$

$$\chi_3 = a \quad \chi_6 = a \quad \chi_9 = g^+[g^-]$$

The predictions were tested on protonated S-carvedilol (c.f. Fig. 1) with molecular orbital (MO) optimizations of the PEHS conformational minima. All calculations were carried out using the Gaussian 98 software program [16]. Initially, full optimizations in the gas phase ($\varepsilon = 0.0$) on carvedilol structures

were performed at the restricted Hartree–Fock, RHF/3-21G, level of theory. Converged conformers were subsequently optimized using density functional theory (DFT) with the Becke 3LYP hybrid exchange-correlation functional [17] at the B3LYP/6-31G(d) level of theory; separate vibrational frequency calculations were performed on all converged conformers for full structural characterization. Single point energy (SPE) calculations were computed on B3LYP/6-31G(d) converged conformers at the B3LYP/6-311 + G(2d,p) level of theory. Full optimizations were performed at both RHF and DFT model chemistries for comparison. B3LYP/6-31G(d) solvation SPE computations of the DFT converged minima were computed according to the polarized continuum (overlapping spheres) model (PCM) of Tomasi and coworkers with reaction-field calculations [18–20] in aprotic (dimethyl sulfoxide, DMSO; $\varepsilon = 46.7$) and protic (water; $\varepsilon = 78.39$) solvents. Graphical data was plotted using Axum 5.0 [21].

3. Results and discussion

Six carvedilol conformations (C1–C6) were computed. Conformational assignments of RHF/3-21G and B3LYP/6-31G(d) optimized minima are displayed in Table 1; total energies and relative energies for conformers evaluated at RHF/3-21G, B3LYP/6-31G(d), and B3LYP/6-311 + G(2d,p) are tabulated in Table 2; explicit values of RHF/3-21G and B3LYP/6-31G(d) converged minima are shown in Table 3;

Table 1
Protonated carvedilol structure codes are defined according to optimized molecular conformation

Structure code	Torsional angle conformation										
	χ_1	χ_2	χ_3	χ_4	χ_5	χ_6	χ_7	χ_8	χ_9	χ_{10}	χ_{11}
C1	[a/a]	[a/a]	[a/a]	[a/a]	[a/a]	[a/a]	[a/a]	[a/a]	[a/a]	[a/a]	[a/a]
C2	[a/a]	[a/a]	[a/a]	[a/a]	[a/a]	[a/a]	[a/a]	[a/a]	[a/a]	[g^-/g^-]	[a/a]
C3	[a/a]	[a/a]	[a/a]	[a/a]	[a/a]	[a/a]	[g^+/g^+]	[g^+/g^+]	[g^+/g^+]	[g^-/g^-]	[g^-/a]
C4	[a/a]	[a/a]	[a/a]	[a/a]	[g^-/g^-]	[a/a]	[g^+/g^+]	[g^+/g^+]	[g^+/g^+]	[g^-/g^-]	[g^-/g^-]
C5	[a/a]	[a/a]	[a/a]	[a/a]	[g^-/g^-]	[a/a]	[a/a]	[g^+/g^+]	[g^+/g^+]	[g^-/g^-]	[g^-/g^-]
C6	[a/a]	[a/a]	[a/a]	[g^-/g^-]	[a/a]	[a/a]	[g^+/g^+]	[g^+/g^+]	[g^+/g^+]	[g^-/g^-]	[g^-/g^-]

Molecular conformation is displayed as [RHF/3-21G optimized torsional angle geometry/B3LYP/6-31G(d) optimized torsional angle geometry] for converged conformers (explicit values of torsional angles are found in Table 3).

Table 2
Total and relative energies for fully optimized protonated carvedilol conformers at the RHF/3-21G and B3LYP/6-31G(d) level of theory

Structure code	RHF/3-21G total energy (hartree)	RHF/3-21G relative energy (kcal mol^{-1})	B3LYP/6-31G(d) total energy (hartree)	B3LYP/6-31G(d) Relative energy (kcal mol^{-1})	B3LYP/6-311 + G(2d,p) total energy (hartree)	B3LYP/6-311 + G(2d,p) relative energy (kcal mol^{-1})
C1	− 1325.27429917	42.05	− 1340.93763474	30.53	− 1341.33778950	26.60
C2	− 1325.31609523	15.83	− 1340.96910024	10.78	− 1341.36611515	8.82
C3	− 1325.33656270	2.98	− 1340.98558392	0.44	− 1341.37858877	1.00
C4	− 1325.34028445	0.65	− 1340.98628696	0.00	− 1341.38017747	0.00
C5	− 1325.31708384	15.21	− 1340.96478904	13.49	− 1341.36057420	12.30
C6	− 1325.34131460	0.00	− 1340.98619251	0.06	− 1341.37848989	1.06

Single point energy (SPE) calculations were performed on all B3LYP/6-31G(d) converged conformers at the B3LYP/6-311 + G(2d,p) level of theory.

solvation results for carvedilol conformers are presented in Table 4. Structures of B3LYP/6-31G(d) optimized conformers are shown in Fig. 2.

Conformer C1 was optimized with all torsional angles frozen in the *a* position as a reference structure with no intramolecular interactions (c.f. Fig. 2). B3LYP/6-31G(d) vibrational analysis revealed two negative frequencies associated with this high energy saddle point (c.f. Table 3). C1 was subsequently

optimized in a fully relaxed state to the local minima C2; this structure has an all *a* conformation except torsional angle χ_{10} (*g*$^-$ position) and possesses the intramolecular hydrogen bond motif, O1$\cdots$H42–O41$\cdots$H46, predicted by Fragment A and B (c.f. Fig. 2) with a moderate relative energy of 8.82 kcal mol^{-1} (c.f. Table 2).

The conformational prediction in Eq. (2) was fully tested with C3. Evident in this structure is

Table 3
Optimized values for the converged conformers of the protonated carvedilol surface at the RHF/3-21G and B3LYP/6-31G(d) level of theory

Structure code	χ_1	χ_2	χ_3	χ_4	χ_5	χ_6	χ_7	χ_8	χ_9	χ_{10}	χ_{11}	Relative energy (kcal mol^{-1})
RHF/3-21G optimized torsional angles												
C1	180.00	180.00	180.00	180.00	180.00	180.00	180.00	180.00	180.00	180.00	180.00	42.05
C2	− 174.25	178.62	168.48	− 159.83	160.15	− 154.67	176.54	122.84	124.90	− 39.44	− 171.25	15.83
C3	− 176.23	− 178.75	172.96	− 160.84	− 156.19	156.00	44.94	75.34	74.72	− 44.48	− 111.64	2.98
C4	− 174.64	178.56	169.56	− 156.97	− 80.76	176.72	47.47	66.69	69.46	− 69.03	− 87.84	0.65
C5	− 175.71	177.39	168.39	− 160.43	− 80.19	153.22	− 155.26	98.05	93.12	− 41.56	− 87.84	15.21
C6	179.71	− 173.33	150.96	− 81.36	− 161.72	179.58	48.28	62.13	67.18	− 40.44	− 81.74	0.00
B3LYP/6-31G(d) optimized torsional angles												
C1[a]	180.00	180.00	180.00	180.00	180.00	180.00	180.00	180.00	180.00	180.00	180.00	30.53
C2	168.45	− 168.60	171.48	− 165.94	165.81	− 178.96	175.53	127.97	120.62	− 40.90	− 174.36	10.78
C3	169.54	− 169.80	173.49	− 172.78	− 174.88	166.24	44.73	85.82	80.56	− 43.19	− 166.59	0.44
C4	173.93	− 172.47	172.56	− 162.24	− 74.80	− 179.73	46.80	69.08	68.74	− 40.33	− 95.13	0.00
C5	173.77	− 172.33	172.56	− 163.03	− 77.83	177.01	− 178.32	93.36	89.37	− 42.62	− 92.23	13.49
C6	166.15	− 164.75	166.91	− 79.39	− 158.63	178.63	48.12	66.12	67.09	− 37.87	− 88.28	0.06

[a] Conformer C1 possesses two negative (imaginary) frequencies (− 189.0871 cm^{-1} and − 78.7329 cm^{-1}).

Table 4
Solvation total and relative energies derived from single point energy (SPE) calculations performed on converged conformers of the protonated carvedilol surface with Tomasi's polarized continuum model (PCM) at the B3LYP/6-31G(d) level of theory

Structure code	DMSO total energy (hartree)	DMSO relative energy (kcal mol^{-1})	Water total energy (hartree)	Water relative energy (kcal mol^{-1})
C1	−1341.03014946	15.61	−1341.06581487	8.71
C2	−1341.04332120	7.34	−1341.07123498	5.31
C3	−1341.05501806	0.00	−1341.07970247	0.00
C4	−1341.05205372	1.86	−1341.07538836	2.71
C5	−1341.03866190	10.26	−1340.96478904	72.11
C6	−1341.05181550	2.01	−1341.07479423	3.08

the Fragment A and B hydrogen bond motif, O1···H42–O41···H46, along with a triple hydrogen bond involving an amine proton (H46) and O41, O29, and O36 (c.f. Fig. 2). Fragment C predicted a

bifurcated hydrogen bond, however, it is plausible that this bifurcated hydrogen bond could extend to bond with O41 which was not part of the Fragment C structure. Conformational assignment for C3 was

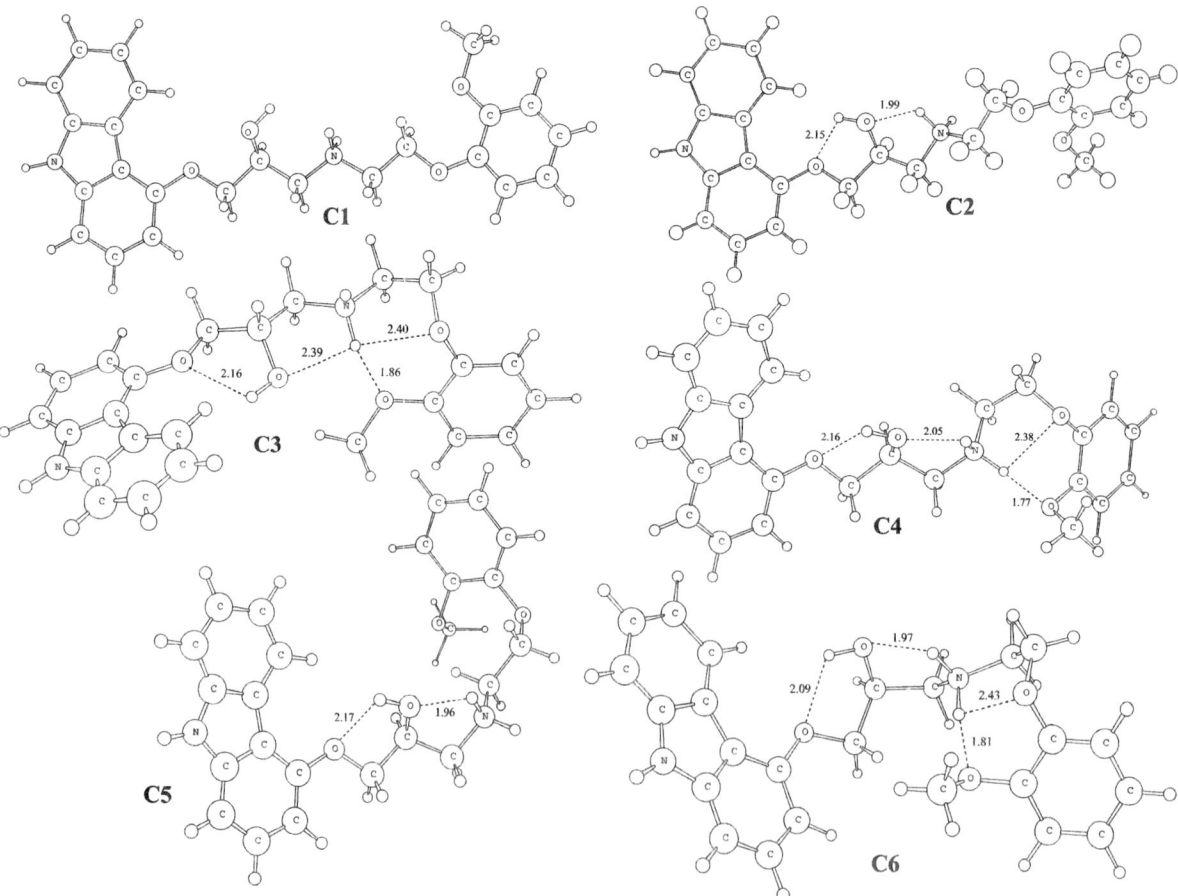

Fig. 2. Molecular structures of fully optimized B3LYP/6-31G(d) converged minima (C1–C6) for the protonated carvedilol potential energy hypersurface (PEHS).

exactly as predicted for the RHF/3-21G results but slightly different for the B3LYP/6-31G(d) results with torsional angle χ_{11} rotating from the g^- to a position (c.f. Tables 1 and 3); this conformer has a relative energy of 1.00 kcal mol^{-1} (c.f. Table 2).

Next, C4 was tested for the sole bifurcated hydrogen bond interaction discovered in Fragment C. C4 revealed all interactions in Fragment A, B, and C; and as predicted, has the lowest DFT relative energy (c.f. Fig. 2 and Table 2). C4 possesses the O1···H42–O41···H57 hydrogen bond motif found in Fragment A and B along the bifurcated intramolecular hydrogen bond, O29···H46···O36, in Fragment C (c.f. Fig. 2). With regards to the conformational assignment, C4 has the conformation predicted by Eq. (2), with the exception of torsional angle χ_5 which optimized in the g^- position to allow both amine protons to be involved in different hydrogen bond networks (c.f. Tables 1 and 3).

The remaining carvedilol conformers, C5 and C6, are miscellaneous converged conformers that display large conformational flexibility (c.f. Fig. 2). C5 is similar to C2 with the same hydrogen bond motif and a relatively high energy (c.f. Tables 2 and 3). C6, like C4, has all the interactions predicted to be in

low energy conformers from the carvedilol fragments (c.f. Fig. 2), and consequently, has a low relative energy of 1.06 kcal mol^{-1} (c.f. Tables 2 and 3). Hence, these results indicate that the intramolecular features of the lowest energy conformers of carvedilol: C3, C4, and C6, were all successfully predicted by the individual fragments of carvedilol. Furthermore, the conformational assignments were, by in large, also predicted from the individual fragments. It is thus evident that this methodology is a viable approach to study large molecular systems explicitly via the rational fragmentation of their pharmacophores.

B3LYP/6-31G(d) SPE solvation calculations performed on conformers C1–C6 using the PCM method revealed that, similar to the gas phase, C3, C4, and C6 had the lowest energies in both DMSO and water (c.f. Table 4). The latter is likely due to a high degree of polarized character inherent in the conformations possessing extensive hydrogen bond networks (c.f. Fig. 2). C5 has a very large relative energy in water (72.11 kcal mol^{-1}; c.f. Table 4) due to the folding of the substituted benzene on itself which makes it difficult to solvate the entire structure and disrupts primary and secondary water molecule ordering surrounding the carvedilol cavity. C2,

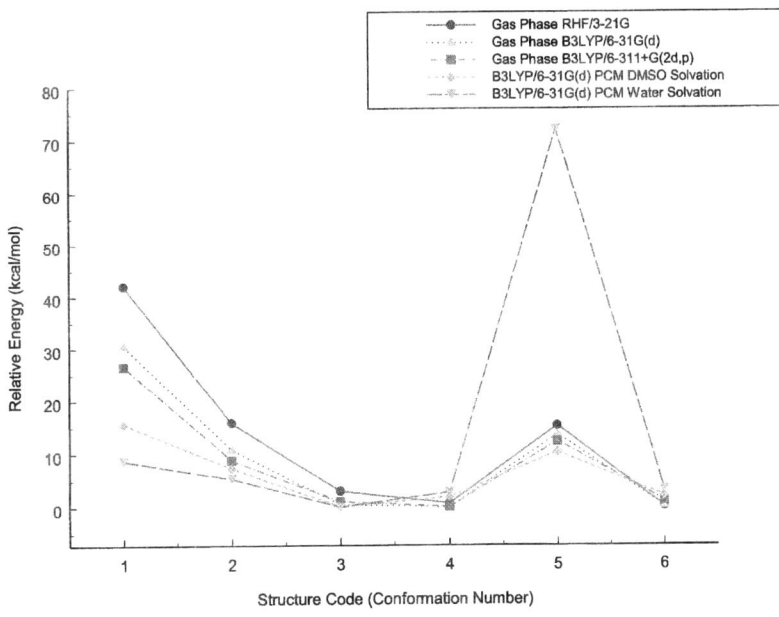

Fig. 3. Graphical plot of the gas phase and solvation ab initio and DFT relative energies of all carvedilol conformers (C1–C6) analyzed in this study.

while possessing the same interactions as C5, has a fully extended conformation, and thus, is more easily solvated by water (c.f. Fig. 2 and Table 4).

In plotting relative energy as a function of conformation number, generally, the trend is similar for all levels of theory in the gas phase and solvation (c.f. Fig. 3). The exception, as described above, is the large DFT water solvation energy of C5. Overall, although DFT energies are regarded as more accurate, ab initio and DFT energies and structural information are in sound agreement with each other. This indicates that the ab initio, RHF/3-21G level of theory, is a good predictor of the conformational character of fragments and molecules at a minimal computational expense.

This study, along with the analysis of the carvedilol fragments, is an ideal starting point for the investigation of the novel carvedilol anti-fibrillar mechanism of action with Aβ because this mechanism is believed to be dependent on the conformational flexibility of the pharmacophores of carvedilol [5]. Thus, it is necessary to decipher bona fide conformations that carvedilol may assume before subsequent testing between carvedilol and Aβ peptide(s) to elucidate the inhibitory interaction between carvedilol and Aβ.

4. Conclusions

Presently, we have predicted all dominant interactions, and to a large extent conformational assignment, of expected low energy carvedilol conformers from the complete analysis of its fragments. The current study validates the idea that large molecular systems with numerous torsional modes and exhaustive conformational possibilities can be studied via the rational fragmentation of structure–activity sections such as pharmacophores. Although only the low energy conformers were evaluated in this study, it is expected that broad analysis of the PEHSs of all fragments will significantly illustrate the trends in carvedilol. This is ideal because one can use the individual fragment PEHSs to identify specific interactions in carvedilol and then solely test these interactions in question without having to fully analyze the entire carvedilol PEHS. Such an approach is useful in deciphering the molecular basis of the

anti-fibrillar action of carvedilol which may have possible uses in alleviating the progression of AD pathology.

Acknowledgements

One of the authors (IGC) wishes to thank the Ministry of Education for a Szent-Györgyi Visiting Professorship.

References

[1] J. Cheng, K. Kamiya, I. Kodama, Cardiovasc. Drug Rev. 19 (2001) 152.
[2] S. Capomolla, O. Febo, M. Gnemmi, G. Riccardi, C. Opasich, A. Carporotondi, A. Mortara, G. Pinna, F. Cobelli, Am. Heart J. 139 (2000) 596.
[3] W. Carlson, K. Oberg, J. Cardiovasc. Pharmacol. Ther. 4 (1999) 205.
[4] P.J. Oliveira, M.P. Marques, L.A.E. Batista de Carvalho, A.J.M. Moreno, Biochem. Biophys. Res. Commun. 276 (2000) 82.
[5] D.R. Howlett, A.R. George, D.E. Owen, R.V. Ward, R.E. Markwell, Biochem. J. 343 (1999) 419.
[6] J. Hardy, J. Selkoe, Science 297 (2002) 353.
[7] V.M.-Y. Lee, Neurobiol. Aging 23 (2002) 1039.
[8] D.M. Walsh, I. Klyubin, J.V. Fadeeva, W.K. Cullen, R. Anwyl, M.S. Wolfe, M.J. Rowan, D.J. Selkoe, Nature 416 (2002) 535.
[9] M.P. Lambert, A.K. Barlow, B.A. Chromy, C. Edwards, R. Freed, M. Liosatos, T.E. Morgan, I. Rozovsky, B. Trommer, K. Viola, P. Wals, C. Zhang, C.E. Finch, G.A. Krafft, W.L. Klein, Proc. Natl Acad. Sci. USA 95 (1998) 6448.
[10] D.M. Walsh, D.M. Hartley, Y. Kusumoto, Y. Fezoui, M.M. Condron, A. Lomakin, G.B. Benedek, D.J. Selkoe, D.B. Teplow, J. Biol. Chem. 274 (1999) 25945.
[11] D.-H. Chui, H. Tanahashi, K. Ozawa, S. Ikeda, F. Checler, O. Ueda, H. Suzuki, W. Araki, H. Inoue, K. Shirotani, K. Takahashi, F. Gallyas, T. Tabira, Nature Med. 5 (1999) 560.
[12] D.R.P. Almeida, L.F. Pisterzi, G.A. Chass, L.L. Torday, A. Varro, J.Gy. Papp, I.G. Csizmadia, J. Phys. Chem. A 106 (2002) 10423.
[13] D.R.P. Almeida, D.M. Gasparro, L.F. Pisterzi, L. Torday, A. Varro, J.Gy. Papp, B. Penke, I.G. Csizmadia, J. Phys. Chem. A107 (2003) 5594.
[14] D.R.P. Almeida, D.M. Gasparro, L.F. Pisterzi, L.L. Torday, A. Varro, J.Gy. Papp, B. Penke, J. Mol. Struct. (Theochem) 631 (2003) 251.
[15] D.R.P. Almeida, D.M. Gasparro, L.F. Pisterzi, J.R. Juhasz, F. Fülöp, I.G. Csizmadia, J. Mol. Struct. (Theochem) (2003) in press.
[16] M.J. Frisch, G.W. Trucks, H.B. Schlegel, G.E. Scuseria, M.A. Robb, J.R. Cheeseman, V.G. Zakrzewski, J.A. Montgomery

Jr., R.E. Stratmann, J.C. Burant, S. Dapprich, J.M. Millam, A.D. Daniels, K.N. Kudin, M.C. Strain, O. Farkas, J. Tomasi, V. Barone, M. Cossi, R. Cammi, B. Mennucci, C.Pomelli, C. Adamo, S. Clifford, J. Ochterski, G.A. Petersson, P.Y. Ayala, Q. Cui, K. Morokuma, D.K. Malick, A.D. Rabuck, K. Raghavachari, J.B. Foresman, J. Cioslowski, J.V. Ortiz, A.G. Baboul, B.B. Stefanov, G. Liu, A. Liashenko, P. Piskorz, I. Komaromi, R. Gomperts, R.L. Martin, D.J. Fox, T. Keith, M.A. Al-Laham, C.Y. Peng, A. Nanayakkara, C. Gonzalez, M. Challacombe, P.M.W. Gill, B.G. Johnson, W. Chen, M.W.

Wong, J.L. Andres, M. Head-Gordon, E.S. Replogle, J.A. Pople, Gaussian 98 (Revision A.9), Gaussian, Inc., Pittsburgh PA, 1998.

[17] A.D. Becke, J. Chem. Phys. 98 (1993) 5648.
[18] S. Miertus, J. Tomasi, Chem. Phys. 65 (1982) 239.
[19] S. Miertus, E. Scrocco, J. Tomasi, Chem. Phys. 55 (1981) 117.
[20] M. Cossi, V. Barone, R. Cammi, J. Tomasi, Chem. Phys. Lett. 255 (1996) 327.
[21] Axum 5.0C for Windows, MathSoft Incorporated, 1996.

ELSEVIER

Journal of Molecular Structure (Theochem) 666–667 (2003) 547–556

THEO
CHEM

www.elsevier.com/locate/theochem

Ab-initio study on the competitive rearrangements of tertiary N-propargylamine-N-oxides

Z. Mucsi*, A. Szabó, I. Hermecz

Chinoin Pharmaceutical and Chemical Works Ltd, P.O. Box 110, Budapest H-1325, Hungary

Abstract

The rearrangement of N-propargyl-morpholine-N-oxide yielding an O-propadienyl-hydroxylamine derivative (Meisenheimer-type Route A) and a formyl-enamin type product in protic solvent (Route B) were investigated by theoretical methods. As the determination of both the exact energy profile and the structures of the transition states and intermediates is rather uncertain for reactions in a protic solvent catalyzed by proton-transfer, a new method was elaborated by applying one or two MeOH molecules associated with the reactant. On this basis the one-step mechanism for the concurrent reactions suggested by experimental results was controlled and corrected in some respect. The competitive rearrangements may proceed through a common intermediate formed in the rate-determining step, which play an important role only in protic environment stabilized by hydrogen bonds.
© 2003 Elsevier B.V. All rights reserved.

Keywords: Propargylamine-N-oxide; Meisenheimer sigmatropic rearrangement; Modelling in protic solvent; Competitive rearrangements; DFT method

1. Introduction

Certain propargyl-substituded tertiary amines are irreversible monoamine oxidase (MAO) inhibitors [1,2]. One major metabolic route is the formation of N-oxide (like **1** in Scheme 1) [3–5]. which may be followed by further transformations [6].

Starting from N-oxide **1**, the well-known Meisenheimer sigmatropic rearrangement [7] (Route A) gives an O-propadienyl-hydroxylamine derivative (**2**) [8–11], which in subsequent steps may transform to other products (**3** and **4**) [12]. We published recently [13] a novel rearrangement of tertiary propargylamine-N-oxides (e.g. **1**) proceeding exclusively in protic medium (Route B)[1], which leads to a new formyl-enamine type product (**5**), competing with the Meisenheimer-transformation through an isoxazoline-type transition state (Route A). For further investigations N-propargyl-morpholine-N-oxide (**1**) [14] was chosen as model compound because Cope elimination, a significant side-reaction (cf. Refs. [15]) can be avoided, and computations can be carried out at a very accurate level, owing to the small and rigid structure of compound **1** with few electrons. The earlier results obtained from preparative experiments and reaction kinetics measurements on N-propargyl-morpholine-N-oxide model (**1**) left several unanswered questions behind.

---

* Corresponding author. Tel.: +36-1-3690-151; fax: +36-1-3690-293.
  E-mail addresses: zoltan.mucsi@sanofi-synthelabo.com (Z. Mucsi), istvan.hermecz@sanofi.com (I. Hermecz).

---

[1] The Routes A and B were designated in previous communications [13,14] as Routes B and A.

0166-1280/$ - see front matter © 2003 Elsevier B.V. All rights reserved.
doi:10.1016/j.theochem.2003.08.079

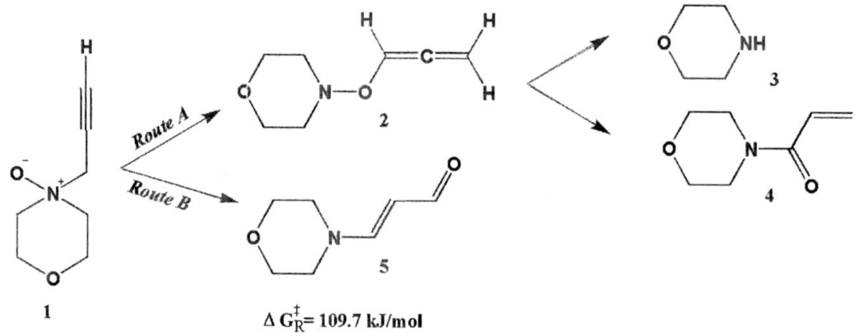

$$\Delta G_R^{\ddagger} = 109.7 \text{ kJ/mol}$$

Scheme 1. The competitive rearrangement of *N*-propargyl-morpholine-*N*-oxide.

For example, it cannot be decided whether the two competitive reactions (Route A and B) proceed through different transition states (TS), or pass through a common rate-determining TS and common intermediate from which different products may form in subsequent fast steps, controlled by different experimental conditions. The characteristic solvent-dependence shows that product-distribution can be regulated by the solvating properties of the protic solvent [13]. To understand the energetic difference between competitive Route A and B, theoretical methods, HF [16], MP2 [17] and DFT [18,19] {B3LYP, [20]} were applied. The mechanism of reactions was studied on an isolated

molecule in vacuo and in presence of one or two solvent (MeOH) molecules, which are required for proton-transfer. A good correlation was found between theoretical and experimental data. On this basis a new energetical profile was drawn which involves a new common isoxazolin-type intermediate formed only in protic media.

Because the new pathway of the rearrangement (Route B) works only in protic media an important theoretical question arises, how the real protic environment and the proton-transfer in a theoretical model can be taken into consideration at a tolerable cost. Unfortunately we are not able to consider a large number of protic solvent molecules (MeOH), which

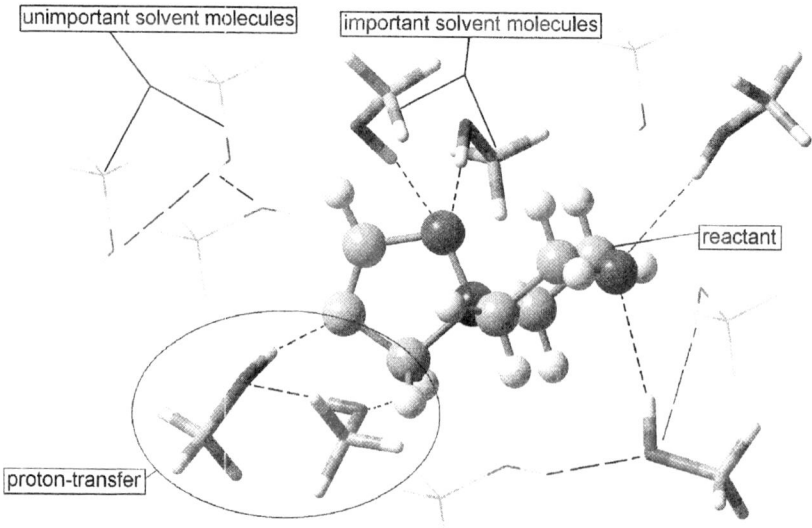

Fig. 1. Solvation of a possible reaction intermediate represented by ball-and-stick model. MeOH molecules playing important and less important role are marked by tube and wire-frame models, respectively.

establish hydrogen bonds in different positions of the reacting molecule (Fig. 1).

Following the reaction pathways involving different reactants, TS and intermediates, hydrogen bonds should play important role in controlling the reaction. We applied one and two methanol (MeOH) molecules simulating the solvent, and these were into different positions about the examined molecule, depending on the different requirements during the reaction.

2. Computational methods

The geometries and vibrational frequencies were calculated by using GAUSSIAN 98W [21] program running on three 1.9 GHz + AMD Athlon processors, with Windows NT operating system. Geometry optimizations were followed by frequency calculations [22,23] at the same level to verify that the structure was a real relative minimum (with no imaginary frequencies) or a transition state (with one imaginary frequency), and to obtain zero point energies (ZPEs) for correcting the frozen nuclei energies (ΔE). The thermodynamical values (ΔH, ΔG, ΔS) were calculated for 298 K. For calculations different methods were used. The preliminary studies were carried out by using the low level RHF/321G calculation [16]. In discussions the B3LYP/631++G(d,p) accurate method were applied [20]. In some cases, when the results were obtained by others methods, it will be mentioned in the text. In one crucial case the optimum was tested

at RHF/631++G(d,p) and MP2(fc)/631++G(d,p) levels [17].

3. Results and discussion

3.1. Conformational studies

Before solving the problems of reaction mechanisms, first of all the stability of N-propargylmorpholine-N-oxide conformers were studied (Scheme 2).

The substituted morpholine ring may assume two chair conformations (**1a** and **1b**), and the conformer having axial N-oxygen atom and equatorial propargyl group (**1a**) (has the lower energy. The propargyl group rotating about the side-chain N2–C3 bond may take up three stable conformations, one anti (180 degree) and two gauche (± 72 degree) positions (**1c**, **1d**, and **1e**). Because the rearrangement of N-propargylmorpholine-N-oxide starts presumably from a gauche conformation about the N2–C3 bond (**1d** or **1e**), the energy of the gauche conformer was chosen as the relative zero point (0 kJ/mol), and the other energy values was compared to this level. The energy of the anti conformer (**1c**) is by $\Delta E = -10.1$ kJ/mol ($\Delta H = -10.3$ kJ/mol $\Delta G = -9.7$ kJ/mol) lower than the standard value. The energy barrier between the gauche and anti positions (**1f** and **1g**, anti-clinal TS) is not too high ($\Delta E = +12.9$ kJ/mol, $\Delta H = +10.8$ kJ/mol, $\Delta G = +16.1$ kJ/mol), allowing a practically free bond rotation. The syn-periplanar position (**1h**) which is the TS between the two gauche positions, is the conformation of highest energy ($\Delta E = +24.4$ kJ/mol, $\Delta H = +22.3$ kJ/mol,

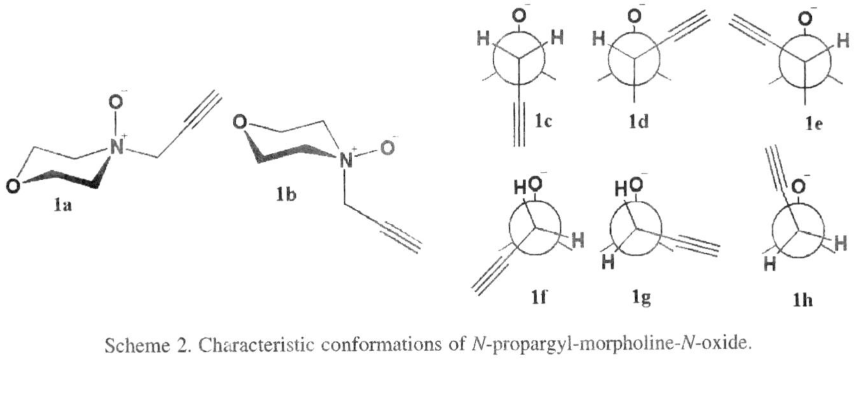

Scheme 2. Characteristic conformations of N-propargyl-morpholine-N-oxide.

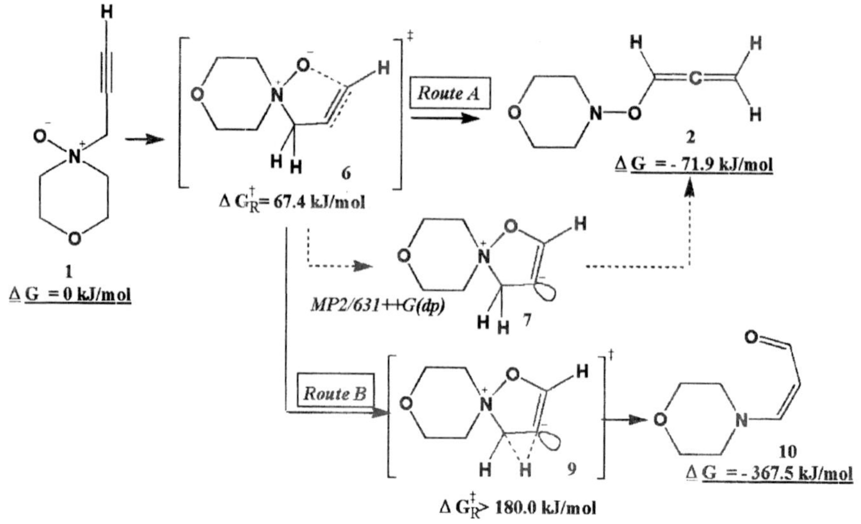

Scheme 3. Reaction pathways (Route A and B) for the rearrangement of *N*-propargyl-morpholine-*N*-oxide in vacuo.

$\Delta G = +27.9$ kJ/mol). The energy difference between the anti-periplanar (**1c**) and syn-periplanar (**1h**) conformations is relatively high ($\Delta E = +34.$ kJ/mol, $\Delta H = +32.6$ kJ/mol, $\Delta G = +39.6$ kJ/mol) but cannot cause a significant restriction of the N2–C3 bond rotation at the reaction temperature (65 °C.).

3.2. Modelling in vacuo and in aprotic solvent:

The rearrangement reaction was studied at first in vacuo then in CH_2Cl_2 aprotic solvent with an isolated molecule (Scheme 3).

Starting from the gauche conformation (**1d** or **1e**), the O1 and C5 atoms approach to each other and the forthcoming reaction passes through only one TS (**6**, $\Delta E = +66.7$ kJ/mol, $\Delta H = 65.4$ kJ/mol, $\Delta G = +67.4$ kJ/mol, $\Delta S = -15.9$ J/mol.K), which leads to only one final product, the *O*-propadienyl-hydroxylamine derivative **3** (Route A). However, the calculated activation energy seems to be very low as compared to the value ($\Delta G = 109$ kJ/mol) obtained experimentally. As will be shown below, the very high difference does not come from the weak efficiency of the theoretical methods but from the inappropriate modelling.

Surprisingly at RHF/321G method/basis set an isoxazoline-type intermediate (**7**) appeared after a TS

(**6**, $\Delta E = +110$ kJ/mol; RHF/321) as shown in Scheme 3 and Fig. 2

However, intermediate **7** seems to be very instable because, when following Route A, only a very low energy barrier (**8** in Fig. 2, $\Delta E = 7.7$ kJ/mol) emerged toward the *O*-propadienyl-hydroxylamine derivative **3**. Because this method leads to an

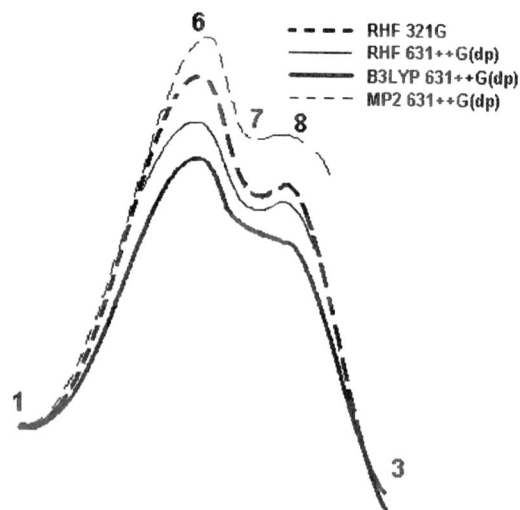

Fig. 2. Potential energy surface for the Meisenheimer-type rearrangement (Route A) of *N*-propargyl-morpholine-*N*-oxide calculated by different methods. The numbers in the Figure refer to structures shown in Scheme 3.

intermediate (**7**) instead of a simple TS (**6**), we tried to find the reason of the different results. At first the effect of the basis set applied was studied by using RHF/631++G(d,p) method where intermediate **7** was also found. Applying a higher correlation method with the same basis set [MP2(fc)/631++G(d,p) level], intermediate **7** remained, but the corresponding minimum appears as a very flat area on the hypersurface (see Fig. 2), and the next lowest energy TS (**8**, related to the cleavage of the N2–C3 bond) is only less than 4 kJ/mol higher in energy. In addition, the distance of the N2–C3 bond in intermediate **7** proved to be very long (1.71 Å), like that in TS **6** obtained by B3LYP/631++G(d,p) method. These date suggest, that the very accurate B3LYP method, which contains several empirical parameters, smoothes out the very flat minimum (see Fig. 2). It should be mentioned, however, that several previous papers consider the Meisenheimer rearrangement (Route A) as an electrocyclic rearrangement, involving only one TS. We suppose that the unstable intermediate **7** appearing on the hypersurface does not play any role in the reaction conducted in vacuo or in aprotic solvent (e.g. in CH_2Cl_2) because intermediate **7** may

practically disappear when the ZPE and the thermal corrections of all of the points on the hypersurface are taken into consideration.

The second, newly explored Route B ($1 \rightarrow 10$ in Scheme 3) can be modelled in vacuo or in aprotic solvent only by hydrogen migration from C3 to C4. However, the corresponding TS **9** represent a very high energy barrier (higher than 180 kJ/mol, the exact value was not determined) as compared to the TS **6** in the Meisenheimer rearrangement. The calculation was started from the TS **6** of the Meisenheimer reaction, with fixed N2–C3 and O1–C5 bonds, and moving the hydrogen atom from C3 to C4. Of course this reaction pathway is not realistic, but in our opinion the energy profile of this step can be predicted reasonably.

3.3. Modelling with one methanol molecule:

The realistic modelling of Route B can be based only on using protic solvent. Therefore our second attempt was to perform a calculation by using one exact methanol molecule, which simulates an aprotic

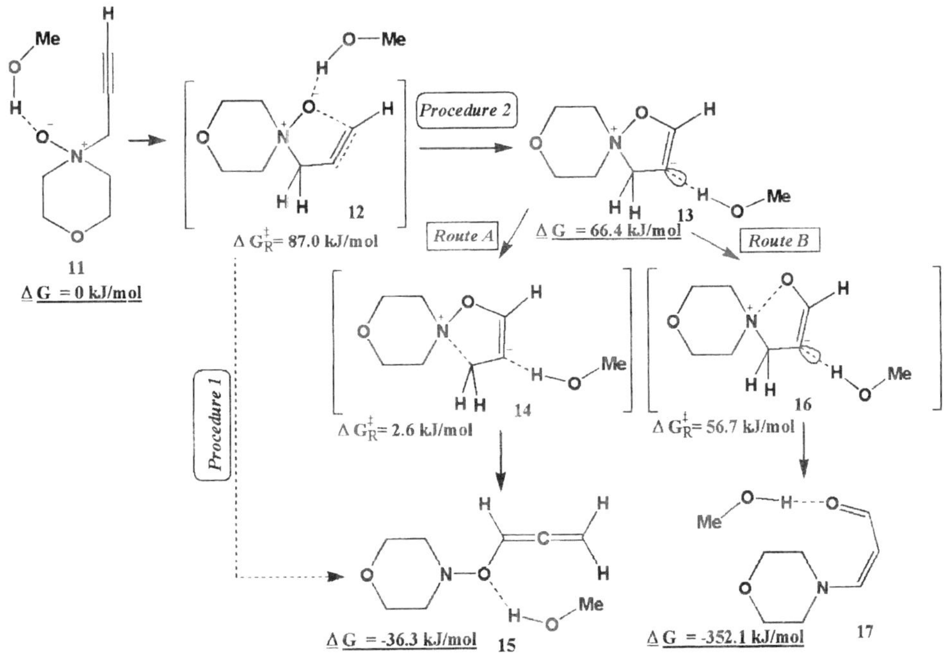

Scheme 4. Reaction pathways (Route A and B) for the rearrangement of *N*-propargyl-morpholine-*N*-oxide in the presence of one MeOH molecule.

solvent containing a protic component in very low concentration (Scheme 4).

The geometry with lowest energy (**11**, cf. **1**) is represented by an arrangement in which the methanol molecule is linked to O1 atom by a strong hydrogen bond (MeOH···O1 distance is 1.73 Å). In this case the reaction passes through TS **12** (cf. **6**), but its energy barrier is higher ($\Delta E = 86.2$ kJ/mol, $\Delta H = 84.8$ kJ/mol, $\Delta G = 87.0$ kJ/mol) than in the reaction without methanol. This result indicates that the hydrogen bond, decreasing the nucleophilic character of O1 atom, decelerate the reaction.

If the MeOH molecule remains in this position (Procedure 1 in Scheme 4), the reaction proceeds directly to the formation of O-propadienyl-hydroxylamine derivative **15** (cf. **3**). However, if the methanol molecule was placed to C4 atom (Procedure 2 in Scheme 4), where the non-bonding electron pair and the negative charge can migrated from the O1 atom, the isoxazoline-type intermediate **13** (cf. **7**) can be formed and stabilized by $C^{-}\cdots$HOMe hydrogen bond (1,96 Å). This intermediate state **13** is favourable than TS **12** by $\Delta E = 22.4$ kJ/mol. If the experimental conditions of the rearrangement are taken into consideration, where the solvent contains a lot of methanol molecules, the 'migration' of one methanol molecule in the theoretical simulation is not

unbelievable. Nevertheless, the 'methanol-stabilized' intermediate **13** seems extremely reactive because the energy barrier leading to the O-propadienyl-hydroxylamine product is almost negligible (TS **14**, ($\Delta E = 2.6$ kJ/mol). The existence of the common intermediate **13** obtained in the simulation indicates that both Route A and B can be realized, although the energy barrier of Route B is too high (TS **16**, ($\Delta E = 56.7$ kJ/mol) as compared to that calculated from the experimental product distribution (3:2 = 1:0.83). In contrast with the experimental findings, the high activation energy of Route B obtained from the simulation suggests that the protonating step is rate-determining, instead of the ring-closure step (Fig. 3.).

3.4. Modelling with two methanol molecules:

As shown above, two parallel rearrangement reactions (Route A and B) can proceed in the presence of one methanol molecule. If two methanol are present, probably two starting structures exist (**18** and **19** in Scheme 5).

In **18** the first methanol molecule (α) is connected to the O1 atom, while the second one (β) is joined to the former by hydrogen bond. In **19** the two methanol molecules are associated with the O1 atom, forming hydrogen bonds. The structure **18** is more favourable by $\Delta E = 6.9$ kJ/mol. One may suppose that in methanol solution containing an excess of methanol molecules, two of them establish hydrogen bonds with O1 as in **19**, and other four methanol molecules join to the former two as in **18**. Therefore it seemed reasonable to look for the TS of the first reaction step by starting from the arrangement **19**. Like the case of using one methanol molecule (Procedure 1 in Schemes 4 and 5) an isoxazoline-type stable intermediate (cf. **7**) cannot be found, but the activation energy represented by TS **20** is $\Delta E = 101.5$ kJ/mol ($\Delta H = 101.7$ kJ/mol, $\Delta G = 102.7$ kJ/mol), which is fortunately almost equal to the value measured experimentally (109 kJ/mol). We believe that the small difference between the measured and calculated values is not a mistake of the calculation method, but can be ascribed to the lack of the necessary amount of methanol molecules. If arrangements **18** and **19** would be considered together, namely four methanol molecules would be

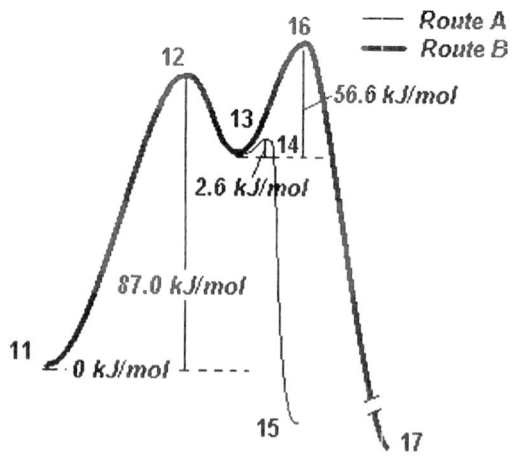

Fig. 3. The comparison of the potential energy surfaces obtained for the rearrangements (Route A and B) of N-propargyl-morpholine-N-oxide calculated by B3LYP-631++G(d,p) method in the presence of one MeOH molecule. The numbers in the Figure refer to structures shown in Scheme 4. Italic numbers indicate the energy barriers in kJ/mol.

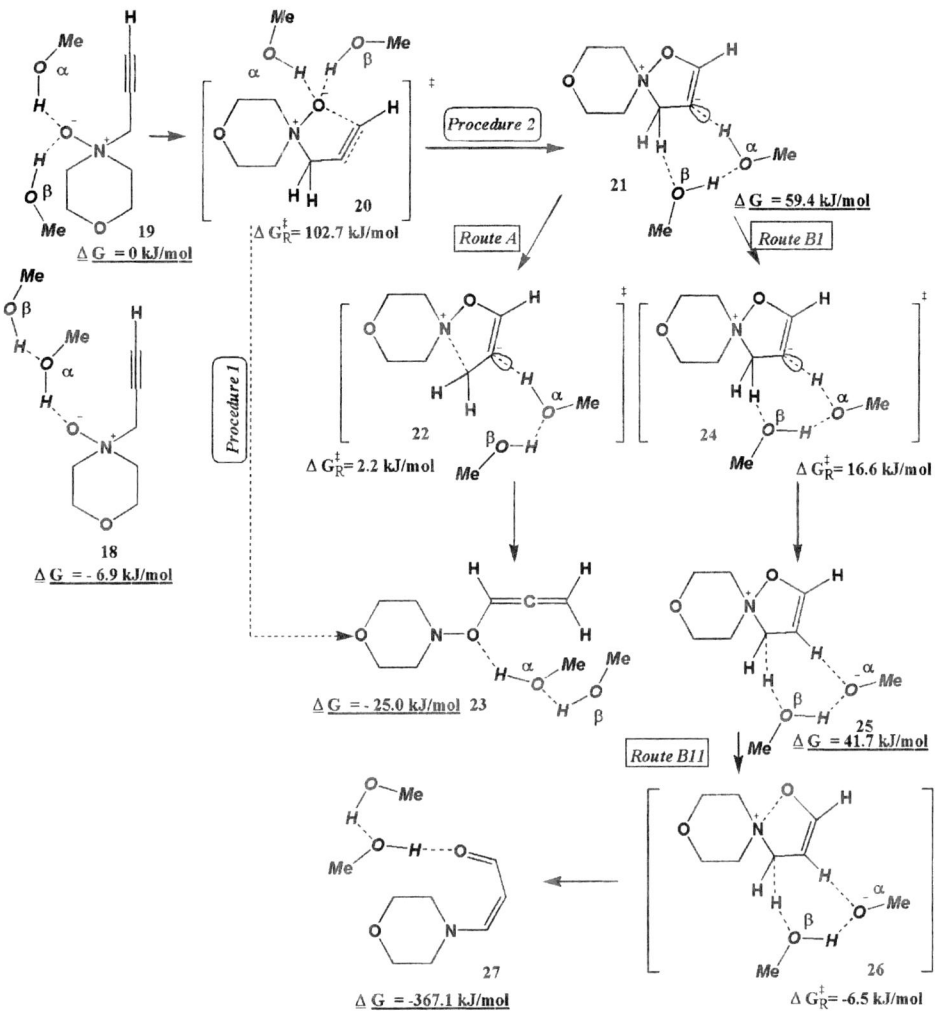

Scheme 5. Reaction pathways (Route A and B) for the rearrangement of *N*-propargyl-morpholine-*N*-oxide in the presence of two MeOH molecules.

used for the calculation of TS, the correct activation energy value should be obtained with a small error coming from the method used.

To obtain the stable intermediate **21**, the two methanol molecules were placed to the C4 atom of the reacting molecule (Procedure 2). This modification seems to be correct, if we consider the experimental conditions with a large number of methanol molecules in the solution, which stabilize immediately the intermediate **21** by establishing hydrogen bond between the negatively charged C4 atom and HOMe (see Fig. 1.). This intermediate **21** seems to be a more stable than **13** in Scheme 4 because one of

the methanol molecules (α-MeOH) establish a strong (1.80 Å), whereas the other (β-MeOH) a weak hydrogen bond (2.4 Å) with the C4 non-bonding electron pair and the C3-hydrogen atom, respectively. For Route A, leading from **21** to *O*-propadienyl-hydroxylamine derivative **23** through the TS **22**, $\Delta E = 3.0$ kJ/mol value of activation energy was computed ($\Delta H = 3.2$ kJ/mol, $\Delta G = 2.2$ kJ/mol). According to reaction coordinates Route B starting from **21** can be realized in four different ways (B1, B2, B3 and B4).

In our opinion the route B1 represent the real mechanism, where the proton of α-MeOH

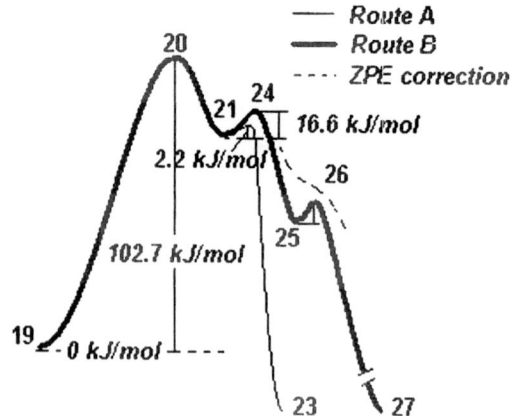

Fig. 4. The comparison of the potential energy surfaces obtained for the rearrangements (Route A and B) of *N*-propargyl-morpholine-*N*-oxide calculated by B3LYP-631++G(d,p) method in the presence of two MeOH molecules. The numbers in the Figure refer to structures shown in Scheme 5. Italic numbers indicate the energy barriers in kJ/mol.

approaches the C4 atom, involving a reasonable and acceptable activation energy ($\Delta E = 9.5$ kJ/mol, $\Delta H = 6.5$ kJ/mol, $\Delta G = 16.6$ kJ/mol). After TS **24** a protonated intermediate **25** was obtained, which does not represent a true minimum because, considering the ZPEs values, this point lies on the downhill of the hypersurface and the minimum disappears. However, we do not know automatically, how to follow the further way of the reaction, starting from this point. Therefore taking the low-entropy structure **25** as a starting point, the final product **27** may form in three different pathways (B11, B12 and B13). Considering ZPEs, the B11 variant proceeds through a proton migration from C3 atom to β-MeOH without developing a real TS (**26**; without ZPEs $\Delta E = +4.1$ kJ/mol, with ZPEs $\Delta E = -5.1$ kJ/mol, $\Delta H = -6.0$ kJ/mol, $\Delta G = -6.5$ kJ/mol), therefore this pathway seems to be true.

Scheme 6. Reaction pathways (Route B12 and B13) for the rearrangement of the protonated intermediate **25** yielding the formyl-enamine product **5** in the presence of two MeOH molecules.

Scheme 7. Reaction pathways (Route B2, B3 and B4) for the rearrangement of the intermediate **21** yielding carbene-type structures **28** and **29** in the presence of two MeOH molecules.

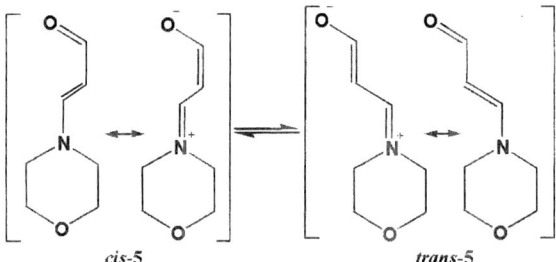

Scheme 8. Resonance structures for the *cis-* and *trans* isomers of formyl-enamine product **5**.

The above route, however, leads to the formation of the *cis-s-cis* stereoisomer of the formyl-enamin product **27** (Fig. 4.), which seems to be in contrast with the experimental findings where the *trans* isomer was obtained after working up the reaction mixture (cf. Ref. [6], and **5** in Scheme 1).

Starting from this contradiction we tried to look for other reaction pathways. The B12 and B13 pathways (Scheme 6) are similar to each others.

In Route B12 an O1–N2 bond stretching occurs, whereas Route B13 involves the torsion of the N2–C3–C4–C5 moiety, both leading to the *trans* isomer **5** found experimentally. However, the activation energies are very high in both cases (ΔE is more than 53 kJ/mol), indicating that the O1–N2 bond is strong enough to hinder the reaction.

Two other pathways (B2 and B3) starting from intermediate **21** were also computed (Scheme 7). Similarly to the previously discussed B12 and B13

mechanisms, the O1–N2 bond stretching and the torsion of the N2–C3–C4–C5 moiety was investigated, yielding a carbene-type intermediate **28**.

In both cases we obtained a relative high activation energy (ΔE is higher than 60 kJ/mol).

In B4 pathway (**21** → **29** in Scheme 7) the one-step deprotonation–protonation (as the reverse of the protonation–deprotonation step in B1–B11 pathways) was taken into account. The C3-hydrogen atom was moved to the oxygen of β-MeOH, and simultaneously proton-transfers occur from β-MeOH to α-MeOH and from α-MeOH to C4 atom. The computed activation energy of this process proved to be very high again (ΔE is more than 100 kJ/mol), therefore this route was also rejected.

The controversy between the computed *cis* structure and the *trans* structure found experimentally for the final product (**27** and **5**, respectively) may be explained by assuming fast *cis-trans* isomerisation at 65 °C, which requires relatively low energy barrier, owing to the conjugated structure of the formyl-enamine product [13] (see resonance structures in Scheme 8).

4. Conclusions

1. The mechanism for the competitive rearrangements of *N*-propargylmorpholine-*N*-oxide (Route A and B) suggested by experimental results was reinvestigated and corrected by applying

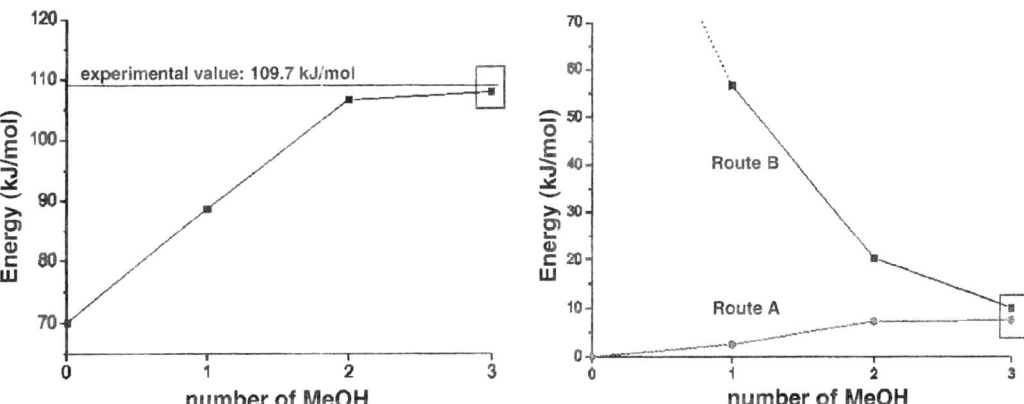

Fig. 5. (A) Calculated activation energies for the rate-determining ring-closure steps **1** → **6**, **11** → **12** and **19** → **20** (see Schemes 3–5) depending on the number of solvent MeOH molecules. (B) Calculated activation energies for rearrangements (see Schemes 3–5) in Route A (**6** → **2**, **13** → **14** and **21** → **22**) and in Route B (**6** → **9**, **13** → **16** and **21** → **24**). Data for the presence of three MeOH molecules come from preliminary calculations.

theoretical methods. The competitive rearrangements in protic solvent do not proceed through two different TS but start from a common intermediate.

2. We have successfully modelled a reaction catalyzed by proton transfer in a protic solvent.

3. We have proved that the energy barrier of the ring closure step depends on the number of hydrogen bonds between MeOH and *N*-oxygen atom (Fig. 5A).

4. We have demonstrated that the common key intermediate in Route A and B exists only in the protic system (MeOH).

5. The activation energy belonging to Route B depends on the number of solvating MeOH molecules, which take part in the stabilization of the common intermediate by hydrogen bonds (Fig. 5B). If a third MeOH molecule in an appropriate arrangement (γ-MeOH is linked to α-MeOH) the activation energy obtained agrees practically with the value calculated from experimental data.

6. The product distribution (Route A and B) is controlled by the number of the stabilizing MeOH molecules. If a model with three MeOH molecules is taken into account the product distributions obtained from calculations and experiments are almost equal (Route A: Route B is 1: 1 and 1: 0.83, respectively).

Acknowledgements

Thanks Prof. I.G. Csizmadia and Prof. Á. Kucsman for discussions.

References

[1] R.H. Abeles, A.L. Maycock, Acc. Chem. Res. 9 (1976) 313.

[2] J. Gaál, I. Hermecz, in: I. Szelenyi (Ed.), Inhibitors of Monoamine Oxidase B, Birkhäuser, Basel, 1999, pp. 75–108.

[3] S.B. Park, P. Jacob, N.L. Benovicz, Chem. Res. Toxicol. 6 (1993) 880.

[4] A.M. Weli, B. Lindeke, Biochem. Pharmacol. 34 (1985) 1993.

[5] M. Katagi, M. Tatsuno, A. Miki, M. Nishikawa, K. Nakajima, H.J. Tsuchihashi, Chromatogr. (B) 759 (2001) 125.

[6] G. Hallström, B. Lindeke, A.H. Khuthier, M.A. Al-Iraqi, Chem. Biol. Interact. 34 (1981) 185.

[7] R.A.W. Johnstone, Mech. Mol. Migr. 2 (1969) 249.

[8] A.H. Khuthier, M.A. Al-Iraqi, Chem. Commun. (1979) 9.

[9] J.C. Craig, N.N. Ekwuribe, L.D. Gruendeke, Tetrahedron Lett. (1979) 4025.

[10] A.H. Khuthier, M.A. Sheat, J. Prakt. Chem. 331 (1989) 187.

[11] A.H. Khuthier, M.A. Sheat, J.M.A. Al-Rawi, Z.S. Salih, J. Chem. Res., Synop. (1987) 184.

[12] G. Hallström, B. Lindeke, A.H. Khuthier, M.A. Al-Iraqi, Tetrahedron Lett. (1980) 667.

[13] A. Szabó, I. Hermecz, J. Org. Chem. 66 (2001) 7219.

[14] A. Szabó, Á. Galambos-Faragó, Z. Mucsi, G. Timái, L. Vasvári-Debreczi, I. Hermecz, Eur. J. Org. Chem. (2003) in press.

[15] C.H. DePuy, R.W. King, Chem. Rev. 60 (1960) 448.

[16] C.C. Roothan, Rev. Mod. Phys. 23 (1951) 69.

[17] C. Moller, M.S. Plesset, Phys. Rev. 46 (1934) 618.

[18] P. Hohenberg, W. Kohn, Phys. Rev. 136 (1964) B864.

[19] W. Kohn, L.J. Shan, Phys. Rev. 140 (1965) A1133.

[20] A.D. Beke, J. Chem. Phys. 98 (1993) 5648.

[21] M. J. Frisch, G. W. Trucks, H. B. Schlegel, G. E. Scuseria, M. A. Robb, J. R. Cheeseman, V. G. Zakrzewski, J. A. Montgomery, Jr, R. E. Stratmann, J. C. Burant, S. Dapprich, J. M. Millam, A. D. Daniels, K. N. Kudin, M. C. Strain, O. Farkas, J. Tomasi, V. Barone, M. Cossi, R. Cammi, B. Mennucci, C. Pomelli, C. Adamo, S. Clifford, J. Ochterski, G. A. Petersson, P. Y. Ayala, Q. Cui, K. Morokuma, D. K. Malick, A. D. Rabuck, K. Raghavachari, J. B. Foresman, J. Cioslowski, J. V. Ortiz, A. G. Baboul, B. B. Stefanov, G. Liu, A. Liashenko, P. Piskorz, I. Komaromi, R. Gomperts, R. L. Martin, D. J. Fox, T. Keith, M. A. Al-Laham, C. Y. Peng, A. Nanayakkara, M. Challacombe, P. M. W. Gill, B. Johnson, W. Chen, M. W. Wong, J. L. Andres, C. Gonzalez, M. Head-Gordon, E. S. Replogle, and J. A. Pople, GAUSSIAN 98W 5. 4, Gaussian, Inc, Pittsburgh PA, 1998.

[22] W.J. Hehre, L. Radom, P. von R. Schleyer, I.A. Pople, Ab Initio Molecular Orbital.

ELSEVIER

Journal of Molecular Structure (Theochem) 666–667 (2003) 557–580

THEO
CHEM

www.elsevier.com/locate/theochem

Conformational-dependent basicity of carvedilol Fragment C: an ab initio study on the primary amine, aminoethoxy-2-methoxy-benzene

David R.P. Almeida[a],[*], Donna M. Gasparro[a], Luca F. Pisterzi[a], Jason R. Juhasz[a], Ferenc Fülöp[b], Imre G. Csizmadia[a],[c],[d]

[a]Lash Miller Laboratories, Department of Chemistry, University of Toronto, 80 St George Street, Toronto, Ont., Canada M5S 3H6
[b]Institute of Pharmaceutical Chemistry, Albert Szent-Györgyi Medical University, Szeged University, Eötvös u. 6., H-6720 Szeged, Hungary
[c]Global Institute of Computational Molecular and Materials Science (GIOCOMMS), 1422 Edenrose Street, Mississauga, Ont., Canada L5V 1H3
[d]Department of Medical Chemistry, University of Szeged, Dóm tér 8, 6720 Szeged, Hungary

Abstract

Carvedilol produces various physiological effects via multiple modes of action. In mitochondria, it is purported that carvedilol is cardioprotective by acting as a mild uncoupler of oxidative phosphorylation; the mechanism is thought to involve the protonable amino group in its side-chain. This uncoupling subsequently leads to a decrease in production of reactive oxygen species and reduced mitochondrial oxidative stress. In the current work, the carvedilol fragment aminoethoxy-2-methoxy-benzene (Fragment C) has been investigated to illustrate the effects of molecular conformation on intrinsic basicity as related to such proton shuttling pathways. It has been previously been shown for carvedilol Fragment B that molecular conformation dictates the energetics of deprotonation. Fragment C is also studied in this context because, as a primary amine which may be deprotonated via three different protons, it provides an ideal structure to elucidate such conformational effects. By calculating the associated energies of deprotonation for each proton, the relative effects of conformation on intrinsic basicity can be determined. The ab initio Hartree–Fock, RHF/3-21G, level of theory was employed for structural analysis and the potential energy hypersurface of Fragment C was computed with geometry optimizations of the conformational minima. Energies of deprotonation were determined with vertical and adiabatic calculations for each proton in each converged minima. Multi-dimensional conformational analysis of the protonated potential energy hypersurface revealed a total of 24 converged minima out of a possible 81 ($\approx 30\%$ convergence). Conformers with the lowest relative energies possessed a motif consisting of bifurcated hydrogen-bonds forming an eight-membered ring. Hydrogen bond networks forming five-membered rings along with intramolecular dipole-type interactions were also evident. In contrast, protonated conformers with large relative energies were devoid of any significant structural features. Geometry optimization of the deprotonated potential energy hypersurface revealed similar structural features; further, optimization of conformational minima belonging to the deprotonated hypersurface revealed a novel amine–aromatic pi electronic interaction currently under study. In analyzing the derived energetics of deprotonation for the primary amine, it was found that conformers lacking significant stabilizing structural motifs were favored and possessed the lowest energies of deprotonation for Fragment C. The route with the lowest energy of deprotonation (optimized) was via the deprotonation of conformer ag^+ag^+ which required 238.34 kcal mol^{-1}. It can thus be concluded that,

---

* Corresponding author.
E-mail addresses: dalmeida@medscape.com (D.R.P. Almeida), dgasparro@medscape.com (D.M. Gasparro), lpisterzi@medscape.com (L.F. Pisterzi), jasonjuhasz@medscape.com (J.R. Juhasz), fulop@pharma.szote.u-szeged.hu (F. Fülöp), icsizmad@alchemy.chem.utoronto.ca (I.G. Csizmadia).

as previously shown for the secondary amine Fragment B, and now for the primary amine Fragment C, proton shuttling mechanisms involving carvedilol, and amines in general, will favor conformations with minimal intramolecular stabilization. As such, molecular conformation and associated structural features will determine, at least with regards to energetics, the intrinsic basicity of compounds and can be used to describe and predict favored substrate conformations for protonophoretic pathways as that postulated for carvedilol in mitochondria.

Keywords: Carvedilol fragment; Aminoethoxy-2-methoxy-benzene; Proton affinity; Basicity; RHF

1. Introduction

1.1. Biological background

Carvedilol, 1-(9*H*-carbazol-4-yloxy)-3-[2-(2-methoxy-phenoxy)ethylamino]-2-propanol ($C_{24}H_{26}N_2O_4$), is a lipophilic multiple-action neuro-hormonal antagonist [1]. As a β-blocker, carvedilol antagonizes noradrenaline non-selectively at $β_1$- (heart muscle, kidney) and $β_2$- (certain blood vessels, smooth muscle of some organs) adrenergic receptors, reducing total cardiac work-load [1–3]. Carvedilol is also a selective $α_1$-adrenergic receptor (found in most sympathetic target tissues) antagonist causing vaso-dilation at peripheral resistance vessels, which decreases pre- and after-load, thereby reducing cardiac work and wall tensions [4]. Therapeutically, carvedilol is a cardiovascular drug of proven efficiency in the treatment of mild to moderate congestive heart failure (CHF), essential hypertension, angina, and in improvement of left ventricular function [1,5]. The US Data and Safety Monitoring Board stopped, for ethical reasons, the clinical investigations of carvedilol before its completion due to greatly lowered mortality rates [6,7].

Aside from its neurohormonal modes of actions, carvedilol possesses direct myocardial-protective antioxidant properties. The carbazole ring gives carvedilol the ability to donate electrons to directly 'scavenge' the activities of oxygen-containing free radicals (or reactive oxygen species; ROS) such as: oxygen superoxide (O_2^-), hydrogen peroxide (H_2O_2), hydroxyl radical (·OH), and peroxynitrite ($ONOO^-$), helping to protect against the deleterious effects of free radical damage [8]. The striking free-radical scavenging ability of carbazole, such as that seen against lipid peroxidation, is enhanced by its relatively high lipid solubility [9].

Further work has indicated that carvedilol may exert indirect cardiac antioxidant effects by protecting mitochondria from oxidative stress by acting as a mild uncoupler of oxidative phosphorylation via a proto-nophoretic pathway involving the amino group of carvedilol's side-chain [10]. Due to the high demand of ATP from working cardiac muscles, a constant mitochondrial input is required and a compromise of mitochondrial bioenergetics leads to a failure to maintain calcium homeostasis which triggers pathways of cell death and suppresses delivery of ATP to heart muscles [11].

Carvedilol may also play a role in the prevention of Alzheimer's disease. It is generally accepted that the neuronal cell loss in Alzheimer's patients is associated with the formation of fibrils from β-amyloid peptide (Aβ) monomers (39-43 residue peptide) and that preventing this aggregation of Aβ might provide a method of slowing the pathology of Alzheimer's disease [12]. Carvedilol acts as a novel anti-fibrillar agent (by inhibiting fibril formation) due to three factors: (1) possession of one central basic amino pharmacophore, (2) possession of two cyclic hydrophobic ring centroids, and (3) the molecular flexibility to adopt a specific three-dimensional pharmacophore conformation [12]. However, it is currently not known if carvedilol binds to Aβ monomers or to small oligomers to prevent fibril formation [12].

Carvedilol contains one stereocentre and is commercially available as a racemic mixture of both its enantiomers (*R*[+] and *S*[−]) (Fig. 1). However, the enantiomers of carvedilol show marked stereoselective properties; both enantiomers have equal $α_1$ blocking activity and antioxidant activity but only the *S*[−] enantiomer contains the non-selective β-adrenergic blocking activity [13]. As such, neither enantiomer alone has the same pharmacologic profile as the racemic mixture of

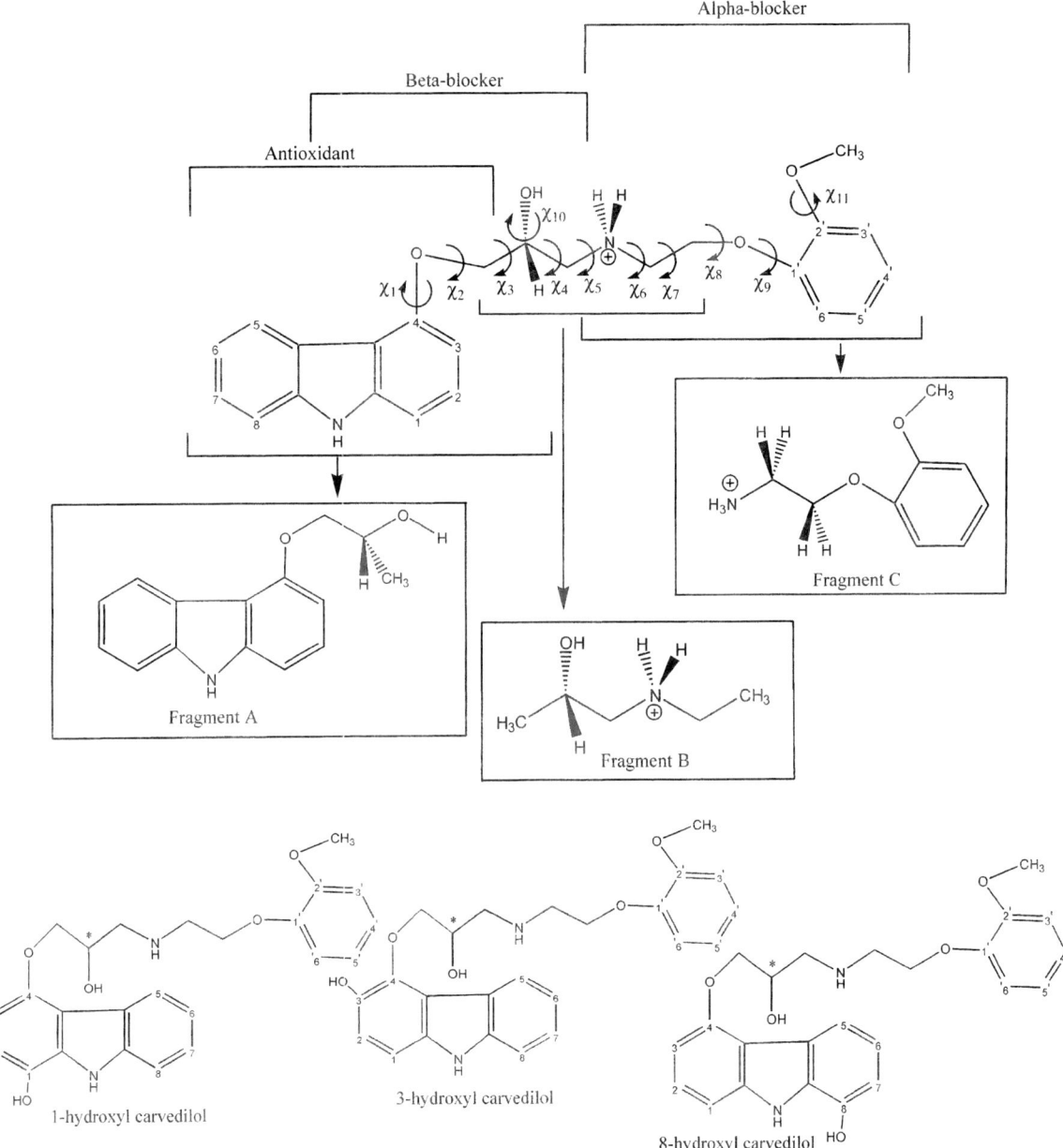

Fig. 1. The complete molecular structure and function of N-protonated carvedilol indicating all eleven torsional angles and its three pharmocophoric fragments: R- and S-4-(2-hydroxypropoxy)carbazol (Fragment A), 2(R and S)-1-(ethylamonium)propane-2-ol (Fragment B), and aminoethoxy-2-methoxy-benzene (Fragment C) (top). Biotransformation of carvedilol produces three hydroxylated metabolites with varying biological activity: 1-hydroxyl, 3-hydroxyl, and 8-hydroxyl carvedilol (bottom).

carvedilol used clinically [14]. The hydroxylated metabolites of carvedilol: 1-, 3-, and 8-hydroxyl carvedilol (Fig. 1), retain the direct antioxidant properties of carvedilol due to the presence of the carbazole moiety [8], the ability to act as uncouplers of oxidative phosphorylation due to the presence of the protonable amino group [10], and to act as anti-fibrillar agents due to possession of the three characteristics described above [12]. As such, it can be postulated that the analysis of the conformational space of carvedilol's molecular fragments can be extrapolated to the active hydroxylated metabolites as well.

1.2. Chemical background

Carvedilol is composed of three distinct pharmacophores (Fig. 1) and was therefore divided into three molecular fragments: R- and S-4-(2-hydroxypropoxy)carbazol (Fragment A) is responsible for the carbazole-related antioxidant effects of carvedilol, 2(R and S)-1-(ethylamonium)propane-2-ol (Fragment B) contains the protonophoretic amino group responsible for the uncoupling of mitochondrial oxidative phosphorylation, and aminoethoxy-2-methoxy-benzene (Fragment C) provides the α_1-antagonist action of carvedilol. (Beta-blockage is exerted by the composite of both Fragments A and B.) While Fragments A [15] and B [16] have been studied, the current objective is the analysis of Fragment C.

Normally, catecholamine (i.e. noradrenaline) binding to an α_1-adrenergic receptor activates phospholipase C (PLC) and inositol 1,4,5-trisphosphate (IP$_3$) leading to the opening of calcium (Ca^{2+}) channels in the cell membrane and/or releasing Ca^{2+} from intracellular stores [17]. The resulting intracellular Ca^{2+} signal leads to muscle contraction. Antagonism at this adrenergic receptor subtype by carvedilol inhibits the cardiac and smooth muscle autonomic nervous system (ANS) effectors responsible for muscle contraction leading to the cardiovascular effect of reduced wall tensions via vasodilation [17]. Since α_1-adrenergic receptor antagonism by carvedilol provides a vital function, the full conformational space of Fragment C will be investigated in this study.

With regards to the uncoupling of oxidative phosphorylation, carvedilol's secondary amino group decreases the mitochondrial electric potential via a weak protonophoretic mechanism [10]. An uncoupler exerts its effects by eliminating the essential mitochondrial proton gradient by freely exchanging protons across the mitochondrial membranes [18]. It is proposed that the amino group of carvedilol (pK$_a$ = 7.9) binds a proton in the low pH intermembrane space, crosses the membrane in the positively charged protonated form into the relatively higher pH mitochondrial matrix (driven by its high lipid solubility and the electric potential which is

negative in the matrix with regards to the intermembrane space), releases the proton in the matrix, and then returns to the cytosolic leaflet in the deprotonated neutral from [10]. The phenomenon known as 'mild uncoupling' occurs when a small decrease in mitochondrial electric potential induces a reduction in the ROS produced by the mitochondrial respiratory chain, thereby producing protective effects [10,19,20].

Classical uncouplers such as salicylanilides [2′,5-dichloro-3-tert-butyl-4′-nitrosalicylanilide (S-13) and 3,4′,5-trichlorosalicylanilide (DCC)], carbonyl cyanide phenylhydrazones [carbonylcyanide p-triflouromethoxyphenylhydrazone (FCCP) and carbonylcyanide m-chlorophenyl hydrazone (CCCP)], along with carvedilol and Fragment B, all possess secondary amino groups with two protons available for deprotonation. However, Fragment C and its primary amino group presents a scenario in which there exists possibilities for deprotonation via three different protons. It has been shown for Fragment B that the conformation adopted will influence the proton abstracted because molecular conformation biases protons differently, thereby resulting in different energies of deprotonation [16]. As such, higher energy structures with lower energies of deprotonation will be better candidates for protonophoretic pathways as that postulated for carvedilol in the mitochondria. Compounding these findings with the high variability in mitochondria uncoupling believed to be due to molecular conformation [10], the intrinsic basicity of Fragment C will also be investigated along with its full conformational space to illustrate if molecular conformation exerts the same influence on the protonophoretics of primary amines. Such a challenge does not appear for tertiary amines because only one proton can be abstracted and neither for quaternary ammonium compounds because of the lack of protons at the nitrogen centre.

2. Computational method

Fragment C possesses four torsional angles of interest and a terminal amino group (Fig. 2). The conformers of its potential energy hypersurface (PEHS) can be described by Eq. (1). It is expected that, as shown for Fragment A [15], the achiral

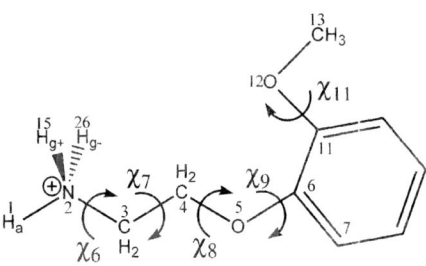

aminoethoxy-2-methoxy-benzene
(**Fragment C**)

$$E = f(\chi_7, \chi_8, \chi_9, \chi_{11})$$

$$\chi_7 = O5\text{-}C4\text{-}C3\text{-}N2$$
$$\chi_8 = C6\text{-}O5\text{-}C4\text{-}C3$$
$$\chi_9 = C7\text{-}C6\text{-}O5\text{-}C4$$
$$\chi_{11} = C13\text{-}O12\text{-}C11\text{-}C6$$

Fig. 2. Numbering and definition of torsional angles for aminoethoxy-2-methoxy-benzene (Fragment C). Numbers placed beside atoms indicates numbering used as z-matrix input for GAUSSIAN 98.

analogues of Fragments A [21], B [16], C will display axis chirality as described by Eq. (2). Eq. (2) states that for an achiral structure with two energetically equivalent minima belonging to the PEHS, plus (P) and minus (M), all torsional angles for those minima are switched from clockwise (CW) to counter-clockwise (CCW) rotation

$$E = f(\chi_7, \chi_8, \chi_9, \chi_{11}) \tag{1}$$

$$E_P = E_M \tag{2}$$

$$f_P(\chi_7, \chi_8, \chi_9, \chi_{11}) = f_M(-\chi_7, -\chi_8, -\chi_9, -\chi_{11})$$

Structural analysis of Fragment C was studied by multi-dimensional conformational analysis (MDCA) and its PEHS was computed with optimizations of the conformational minima. With four torsional angles $(\chi_7, \chi_8, \chi_9, \chi_{11})$ and three possible minima for each torsional angle (*gauche plus*, g^+; *anti*, a; *gauche minus*, g^-), there are expected a grand total of $3^4 = 81$ possible minima for Fragment C.

As mentioned, the structure of Fragment C (Fig. 2) can be deprotonated via the removal of three protons. If any of these protons were replaced with deuterium, one would have deuterium in either of three positions: *anti* position (H1, denoted as H_a), the g^+ position (H15, denoted as H_{g^+}), or the g^- position (H26,

denoted as H_{g^-}). Due to these possibilities, a three prong methodology was applied to analyze the conformational-dependent basicity of Fragment C. (The methodology applied here is similar to the methodology applied to the secondary amine molecular fragment of carvedilol, Fragment B [16].)

For each unique converged conformation of the PEHS, protonated Fragment C (CH^+) was deprotonated of protons H_a, H_{g^+}, and H_{g^-}, independent of each other, and the vertical energies of deprotonation were calculated with single-point energy (SPE) calculations. These energy values are denoted as $\Delta E_{vert}(a)$, $\Delta E_{vert}(g^+)$, and $\Delta E_{vert}(g^-)$ (Eqs. (3)–(5)). The H_a, H_{g^+}, and H_{g^-} deprotonated Fragment C (C) conformers were then geometrically optimized and adiabatic energies of deprotonation were calculated based on fully optimized values. These values are denoted as $\Delta E_{opt}(a)$, $\Delta E_{opt}(g^+)$, and $\Delta E_{opt}(g^-)$ (Eqs. (6)–(8)). A third set of values, denoted as $\Delta\Delta E(a)$, $\Delta\Delta E(g^+)$, and $\Delta\Delta E(g^-)$, represent the differences between the vertical and the optimized adiabatic energies of deprotonation for each conformer (Eqs. (9)–(11)). These latter values can be interpreted as the stabilization experienced by Fragment C conformers as they adopted an optimized conformation after deprotonation. The above methodology is illustrated in Fig. 3. (A positive value for the energy of deprotonation appears because bond-breaking is always an endothermic process.)

$$\Delta E_{vert}(a) = |E_{opt}[CH^+] - E_{SP}[C]| \tag{3}$$

(H_a deprotonation)

$$\Delta E_{vert}(g^+) = |E_{opt}[CH^+] - E_{SP}[C]| \tag{4}$$

(H_{g^+} deprotonation)

$$\Delta E_{vert}(g^-) = |E_{opt}[CH^+] - E_{SP}[C]| \tag{5}$$

(H_{g^-} deprotonation)

$$\Delta E_{opt}(a) = |E_{opt}[CH^+] - E_{opt}[C]| \tag{6}$$

(H_a deprotonation)

$$\Delta E_{opt}(g^+) = |E_{opt}[CH^+] - E_{opt}[C]| \tag{7}$$

(H_{g^+} deprotonation)

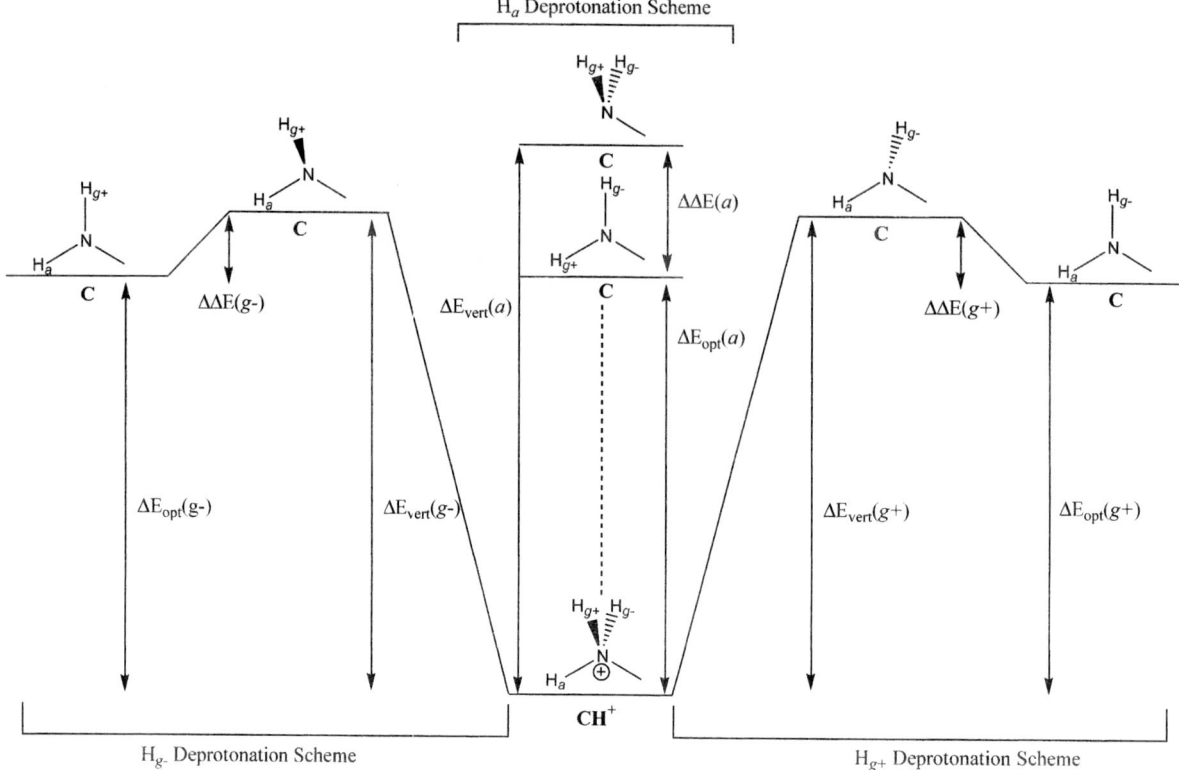

Fig. 3. Methodology employed to analyze the conformational-dependent basicity of the amino group of Fragment C. Each converged minima of the protonated Fragment C PEHS was subject to the deprotonation of H_a (H1), H_{g+} (H15), and H_{g-} (H26) protons.

$$\Delta E_{opt}(g^-) = |E_{opt}[CH^+] - E_{opt}[C]|$$

(8)

(H_{g^-} deprotonation)

$$\Delta\Delta E(a) = \Delta E_{vert}(a) - \Delta E_{opt}(a)$$

(9)

(H_a deprotonation)

$$\Delta\Delta E(g^+) = \Delta E_{vert}(g^+) - \Delta E_{opt}(g^+)$$

(10)

(H_{g^+} deprotonation)

$$\Delta\Delta E(g^-) = \Delta E_{vert}(g^-) - \Delta E_{opt}(g^-)$$

(11)

(H_{g^-} deprotonation)

All computations were performed using the GAUSSIAN 98 software program [22]. Fragment C was exclusively defined using the GAUSSIAN 98 z-matrix internal coordinate system, to specify

molecular structure, stereochemistry, and geometry. All calculations were performed at the Hartree–Fock, RHF/3-21G, level of theory. Graphical data was plotted using Axum 5.0 [23].

3. Results and discussion

3.1. MDCA of the protonated aminoethoxy-2-methoxy-benzene (Fragment C) PEHS

MDCA of the protonated Fragment C PEHS revealed that, instead of the expected 81 conformers, a total of 24 converged minima were found [24/81 ≈ 30% convergence and $(81 - 24)/81 ≈$ 70% annihilated] (Table 1). Conformational structural assignments for the conformational minima were made according to Eq. (12). This is based on the general observation that, if one were to rotate

Table 1

Optimized minima for the PEHS of aminoethoxy-2-methoxy-benzene (Fragment C) at the RHF/3-21G level of theory

Conformational assignment				χ_7	χ_8	χ_9	χ_{11}	E (hartree)	Relative energy (kcal mol^{-1})
χ_7	χ_8	χ_9	χ_{11}						
g^+	g^+	g^+	g^+	\multicolumn{3}{}{Not found moved to}		$g^+g^-g^-g^+$	
g^+	g^+	g^+	a	Not found moved to				$g^+g^+g^+g^-$	
g^+	g^+	g^+	g^-	44.46	71.68	70.76	-98.81	-549.873068247	0.00
g^+	g^+	a	g^+	Not found moved to				g^+aaa	
g^+	g^+	a	A	Not found moved to				g^+aaa	
g^+	g^+	a	g^-	Not found moved to				$g^+g^+g^+g^-$	
g^+	g^+	g^-	g^+	Not found moved to				$g^+ag^-g^{-A}$	
g^+	g^+	g^-	a	Not found moved to				$g^+ag^-g^{-A}$	
g^+	g^+	g^-	g^-	Not found moved to				$g^+g^-g^-g^+$	
g^+	a	g^+	g^+	Not found moved to				$g^+ag^-g^{-B}$	
g^+	a	g^+	a	Not found moved to				g^+aaa	
g^+	a	g^+	g^-	Not found moved to				g^+aaa	
g^+	a	a	g^+	Not found moved to				$g^+ag^-g^{-A}$	
g^+	a	a	a	41.80	122.91	120.56	-179.67	-549.863583899	5.95
g^+	a	a	g^-	Not found moved to				$g^+ag^-g^{-B}$	
g^+	a	g^-	g^+	Not found moved to				$g^+ag^-g^{-C}$	
g^+	a	g^-	a^A	38.30	-137.12	-115.57	-169.54	-549.872527050	0.34
g^+	a	g^-	a^B	59.99	-140.29	-81.29	156.52	-549.870278796	1.75
g^+	a	g^-	a^C	61.03	-135.10	-77.42	144.54	-549.870234030	1.78
g^+	a	g^-	g^-	Not found moved to				$g^+ag^-g^{-B}$	
g^+	g^-	g^+	g^+	Not found moved to				$g^+ag^-g^{-A}$	
g^+	g^-	g^+	a	Not found moved to				$g^+ag^-g^{-B}$	
g^+	g^-	g^+	g^-	Not found moved to				$g^+g^-g^-g^+$	
g^+	g^-	a	g^+	Not found moved to				g^+aaa	
g^+	g^-	a	a	Not found moved to				$g^+ag^-g^{-A}$	
g^+	g^-	a	g^-	95.14	-39.40	135.84	-96.81	-549.861239554	7.42
g^+	g^-	g^-	g^+	62.23	-116.45	-73.64	90.67	-549.871425222	1.03
g^+	g^-	g^-	a	Not found moved to				$g^+g^-g^-g^+$	
g^+	g^-	g^-	g^-	Not found moved to				$g^+g^-g^-g^+$	
a	g^+	g^+	g^+	Not found moved to				$aaaa^A$	
a	g^+	g^+	a	Not found moved to				$aaaa^D$	
a	g^+	g^+	g^-	-175.06	102.35	95.80	-95.49	-549.842004194	19.49
a	g^+	a	g^+	-174.20	77.79	-127.36	93.40	-549.837892957	22.07
a	g^+	a	a	-179.37	85.80	-129.12	152.78	-549.837297717	22.45
a	g^+	a	g^-	Not found moved to				$aaaa^A$	
a	g^+	g^-	g^+	Not found moved to				ag^+ag^+	
a	g^+	g^-	a	Not found moved to				ag^+aa	
a	g^+	g^-	g^-	Not found moved to				$aaaa^D$	
a	a	g^+	g^+	Not found moved to				$ag^-g^-g^+$	
a	a	g^+	a	Not found moved to				$aaaa^D$	
a	a	g^+	g^-	Not found moved to				$aaaa^D$	
a	a	a	g^+	Not found moved to				$aaaa^C$	
a	a	a	a^A	-177.24	-120.95	-125.46	170.49	-549.843128026	18.79
a	a	a	a^B	-173.85	-154.80	136.74	-170.71	-549.839012663	21.37
a	a	a	a^C	173.93	154.89	-136.52	170.67	-549.839012786	21.37
a	a	a	a^D	177.21	120.95	125.47	-170.55	-549.843128037	18.79
a	a	a	g^-	Not found moved to				$aaaa^B$	
a	a	g^-	g^+	Not found moved to				$aaaa^A$	
a	a	g^-	a	Not found moved to				$aaaa^A$	

(continued on next page)

Table 1 (*continued*)

χ_7	χ_8	χ_9	χ_{11}	χ_7	χ_8	χ_9	χ_{11}	E (hartree)	Relative energy (kcal mol^{-1})
a	a	g^-	g^-		Not found moved to			$ag^+g^+g^-$	
a	g^-	g^+	g^+		Not found moved to			$aaaa^A$	
a	g^-	g^+	a		Not found moved to			ag^-aa	
a	g^-	g^+	g^-		Not found moved to			ag^-ag^-	
a	g^-	a	g^+		Not found moved to			$aaaa^D$	
a	g^-	a	a	178.28	-85.73	129.14	-152.73	-549.837297782	22.45
a	g^-	a	g^-	174.22	-77.75	127.31	-93.33	-549.837893074	22.07
a	g^-	g^-	g^+	174.87	-102.32	-95.66	94.96	-549.842004180	19.49
a	g^-	g^-	a		Not found moved to			$aaaa^A$	
a	g^-	g^-	g^-		Not found moved to			$aaaa^D$	
g^-	g^+	g^+	g^+		Not found moved to			$g^-g^+g^+g^-$	
g^-	g^+	g^+	a		Not found moved to			$g^-g^+g^+g^-$	
g^-	g^+	g^+	g^-	-62.22	116.49	73.65	-90.73	-549.871425197	1.03
g^-	g^+	a	g^+	-95.37	38.43	-135.16	96.23	-549.861238102	7.42
g^-	g^+	a	a		Not found moved to			$g^-ag^+g^C$	
g^-	g^+	a	g^-		Not found moved to			g^-aaa	
g^-	g^+	g^-	g^+		Not found moved to			$g^-ag^+g^C$	
g^-	g^+	g^-	a		Not found moved to			$g^-ag^+g^C$	
g^-	g^+	g^-	g^-		Not found moved to			$g^-ag^+g^C$	
g^-	a	g^+	g^+		Not found moved to			$g^-ag^+g^B$	
g^-	a	g^+	a^A	-60.90	135.26	77.29	-144.62	-549.870235279	1.78
g^-	a	g^+	a^B	-60.10	139.67	80.64	-155.33	-549.870277713	1.75
g^-	a	g^+	a^C	-38.27	137.07	115.59	169.53	-549.872526998	0.34
g^-	a	g^+	g^-		Not found moved to			$g^-ag^+g^A$	
g^-	a	a	g^+		Not found moved to			$g^-ag^+g^B$	
g^-	a	a	a	-42.04	-122.90	-120.37	179.24	-549.863582322	5.95
g^-	a	a	g^-		Not found moved to			$g^-ag^+g^C$	
g^-	a	g^-	g^+		Not found moved to			g^-aaa	
g^-	a	g^-	a		Not found moved to			g^-aaa	
g^-	a	g^-	g^-		Not found moved to			$g^-ag^+g^B$	
g^-	g^-	g^+	g^+		Not found moved to			$g^-ag^+g^C$	
g^-	g^-	g^+	a		Not found moved to			$g^-ag^+g^C$	
g^-	g^-	g^+	g^-		Not found moved to			$g^-ag^+g^C$	
g^-	g^-	a	g^+		Not found moved to			$g^-ag^+g^C$	
g^-	g^-	a	a		Not found moved to			g^-aaa	
g^-	g^-	a	g^-		Not found moved to			g^-aaa	
g^-	g^-	g^-	g^+	-44.44	-71.70	-70.74	98.83	-549.873068367	*0.00*
g^-	g^-	g^-	a		Not found moved to			$g^-g^-g^-g^+$	
g^-	g^-	g^-	g^-		Not found moved to			$g^-g^+g^+g^-$	

a tetrahedral carbon against another tetrahedral carbon, the minima would generally fall within the above ranges. Out of the 24 converged minima, only 17 were unique conformational assignments, while the remaining seven conformers (or 29% of the converged minima) were variations of a few conformational assignments (g^+ag^-a, $aaaa$, and g^-ag^+a)

and are denoted with superscripts A, B, C, etc. for a given conformational assignment subset.

gauche plus $(g^+) = 60$ (ideal) $\pm\, 60°$

anti $(a) = 180$ (ideal) $\pm\, 60°$ (12)

gauche minus $(g^-) = -60$ (ideal) $\pm\, 60°$

The axis chirality described by Eq. (2) is evident in the protonated Fragment C PEHS. Axis chirality occurs when the structures adopt conformations with asymmetric electron distribution. Therefore, irrespective of the presence of a point chiral stereocentre with four different substituents, the chirality induced by an asymmetric distribution of electron density is always present whenever the structure adopts asymmetric conformations and, like enantiomers of point chirality, axis chiral conformers come in pairs.

In this case, however, there is a lack of a fully symmetric conformation devoid of axis chirality that separates all minima for a given PEHS. This conformation (usually with all torsional angles in the *anti* position) has a symmetric electron density with respect to a plane of symmetry, and therefore, lacks axis chirality. Fragment C was found to contain a subset of four conformations with all torsional angles in the *anti* position: $aaaa^A$, $aaaa^B$, $aaaa^C$, and $aaaa^D$, all of which were asymmetric and occurred in pairs. To illustrate the axis chirality of these minima, the deviation ($\Delta\chi_T$) of each optimized torsional angle (χ_T) from an ideal value of 180° was calculated based on Eq. (13) (T = torsional angle of interest) (Table 2). The respective $\Delta\chi_T$ values indicate the deviation from an ideal structure with a fully symmetric conformation (this is graphically illustrated in Fig. 4). With regards to Eq. (13), a positive value indicates an optimized torsional angle beyond 180° and a negative value indicates an optimized torsional angle below 180°. From Table 2 and Fig. 4 it is evident that the axis chiral pairs deviated the same amount but in opposite directions (i.e. CW versus CCW) and have the same energetics as expected from Eq. (2). The subset of *aaaa* conformers, like all

converged conformers with torsional angle χ_7 in the *anti* position, did not possess any distinct structural features except the fact that they had fully extended side-chains

$$\Delta\chi_T = \Delta\chi_T - 180° \qquad (13)$$

PEHS conformers were stabilized by internal hydrogen bonding between nitrogen protons (H1, H15, H26) and oxygen atoms (O5 and O12). Strong intramolecular ion–dipole interactions were also present between the protonated positive nitrogen centre and the oxygen atoms. The protonated PEHS global minima, conformation $g^+g^+g^+g^-$ (and it axis chiral pair $g^-g^-g^-g^+$), possesses an eight-membered ring formed by a short hydrogen bond between H1 and O12 and a longer hydrogen bond between H1 and O5 forming a five-membered ring (Fig. 5, left). The bifurcated hydrogen bonding in the global minima allows for the formation of a very stable structure. Further, both oxygen atoms oriented towards the positive nitrogen centre allows dominant intramolecular ion–dipole interactions to be present. The bifurcated hydrogen bond split between both oxygen atoms and forming the eight-membered ring was the most prevalent structural feature of the protonated Fragment C PEHS; conformers with the lowest relative energies, under 2 kcal mol^{-1}, all contained this structural feature in the conformation they adopted (Table 7(I)).

Conformers with moderate relative energies were found in geometries which only possessed one hydrogen bond as opposed to the hydrogen bond split between both oxygen atoms as mentioned above. Conformations g^+aaa and $g^+g^-ag^-$ (and their axis chiral pairs) had relative energies of 5.95 and

Table 2

Optimized geometries (χ_T) and degree deviation ($\Delta\chi_T$) for the converged *aaaa* conformers relative to an ideal symmetric *aaaa* conformation lacking axis chirality ($\chi_T = 180.00$, $\Delta\chi_T = 0$; c.f. Eq. (13)) for aminoethoxy-2-methoxy-benzene (Fragment C) (T = torsional angle of interest)

Converged conformation				$\chi_7(\Delta\chi_7)$	$\chi_8(\Delta\chi_8)$	$\chi_9(\Delta\chi_9)$	$\chi_{11}(\Delta\chi_{11})$	Relative energy (kcal mol^{-1})
χ_7	χ_8	χ_9	χ_{11}					
a	a	a	a^A	−177.24(2.76)	−120.95(59.05)	−125.46(54.54)	170.49(−9.51)	18.79
a	a	a	a^B	−173.85(6.15)	−154.80(25.20)	136.74(−43.26)	−170.71(9.29)	21.37
a	a	a	a^C	173.93(−6.07)	154.89(−25.11)	−136.52(43.48)	170.67(−9.33)	21.37
a	a	a	a^D	177.21(−2.79)	120.95(−59.05)	125.47(−54.53)	−170.55(9.45)	18.79

The geometrical and energetic properties of such axis chiral conformers can be predicted from Eq. (2).

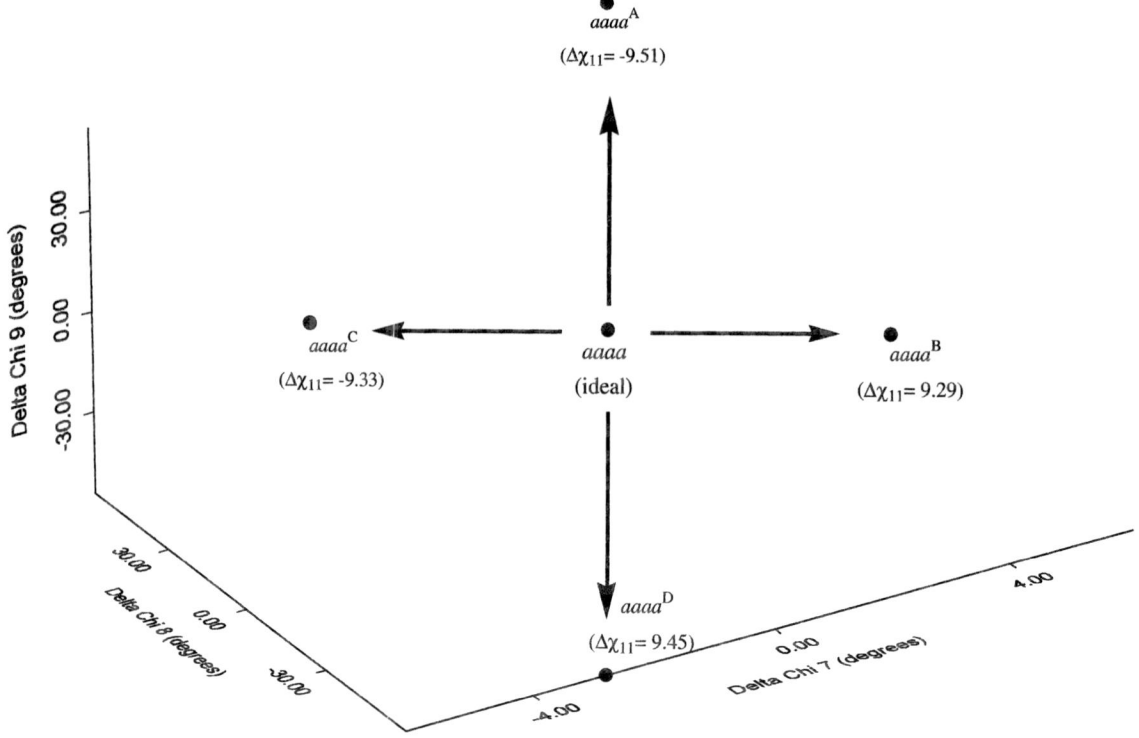

Fig. 4. Graphical illustration of the four converged *aaaa* axis chiral conformers (*aaaa*[A], *aaaa*[B], *aaaa*[C], and *aaaa*[D]) and their deviation from an ideal *aaaa* structure ($\chi_T = 180.00$, $\Delta\chi_T = 0$). Conformation *aaaa* (ideal) would possess symmetrical electron density with regards to a plane of symmetry, and therefore, would be devoid of axis chirality. However, such a structure was not found for the PEHS of Fragment C; instead, two pairs of *aaaa* converged conformers were found and the arrows above represent their geometrical deviation from the *aaaa* (ideal) structure.

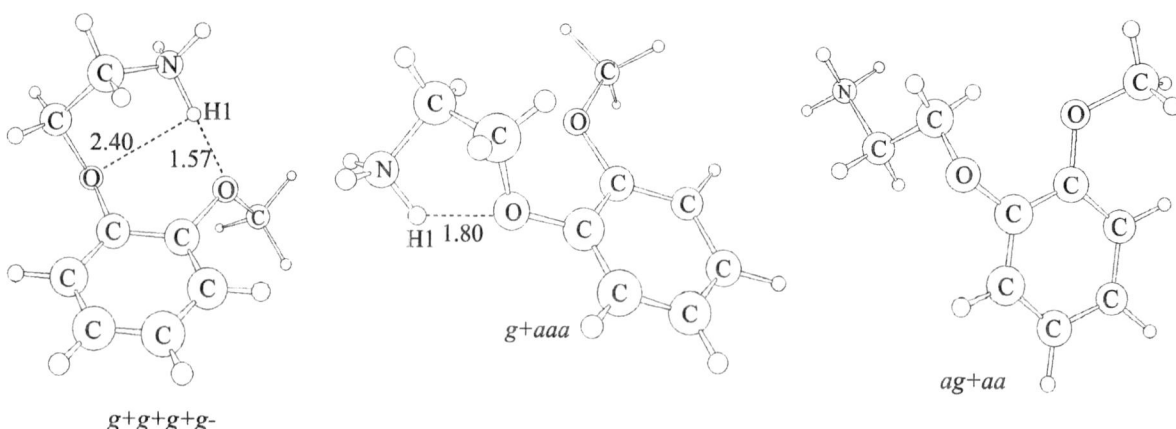

Fig. 5. Optimized global minima of the protonated Fragment C PEHS ($g^+g^+g^+g^-$) exhibits both a five- and eight-membered ring with bifurcated hydrogen bonds (along with the global minima, all protonated conformers with relative energy less than 2 kcal mol^{-1} has this structural feature). Conformer g^+aaa has a moderate relative energy (5.95 kcal mol^{-1}) and exhibits a five-membered ring while conformer ag^+aa had the largest relative energy (22.45 kcal mol^{-1}) of the protonated PEHS and possesses a fully extended conformation.

7.42 kcal mol^{-1}, respectively. Conformation g^+aaa contained a 1.80 Å hydrogen bond forming a five-membered ring with O5 (Fig. 5, middle) while conformation $g^+g^-ag^-$ contained a 1.54 Å hydrogen bond forming an eight-membered ring with O12. The latter eight-membered was similar to that seen in the global minima except that O5 was not involved in any other hydrogen bonding. Finally, structures with torsional angle χ_7 oriented in the *anti* position had the side-chain positioned away from the ring, and therefore, were not able to form any internal hydrogen bonds. The disparity in the relative energies of these conformers, about 20 kcal mol^{-1}, is indicative of their lack of stabilization. Conformation ag^+aa possessed the highest relative energy of the PEHS (Fig. 5, right).

As shown above with the four *aaaa* conformers, the protonated Fragment C PEHS contained a number of different converged structures within a respective conformational assignment (conformations g^+ag^-a, $aaaa$, and g^-ag^+a). Although the majority of subset conformers were very similar to each other, conformer $g^+ag^-a^A$ and $g^+ag^-a^B$ converged to very different geometries. The former conformer was the only structure with two different protons involved in two distinct hydrogen bonds forming separate five- and eight-membered rings (Fig. 6, left). Conformer $g^+ag^-a^B$ and $g^+ag^-a^C$ possessed the more prevalent bifurcated hydrogen bond structural feature (Fig. 6, middle and right).

3.2. Conformational analysis of H_a, H_{g^+}, and H_{g^-} deprotonated aminoethoxy-2-methoxy-benzene (Fragment C) conformers

All converged protonated conformers were deprotonated of H_a, H_{g^+}, and H_{g^-} protons. SPE calculations were initially performed on all deprotonated conformers (Table 3) followed by full geometry optimizations (Tables 4–6). With regards to the H_a (H1) deprotonated conformers (Tables 4 and 7(II)), the global minima $g^+g^+g^+g^{-B}$ possesses a H15···O12 hydrogen bond forming an eight-membered ring similar to that exhibited by protonated conformers (Fig. 7, left). Conformers with bifurcated hydrogen bonds were also evident in the deprotonated state such as that for $g^+ag^-a^A$ (Fig. 7, middle), however, these were longer hydrogen bonds than present in the protonated conformers due to a lack of any intramolecular ion–dipole interactions between the nitrogen and oxygen heteroatoms. In the protonated PEHS, all conformers with torsional angle χ_7 in the *anti* position have fully extended conformations and no significant structural features. In the H_a deprotonated PEHS, converged conformers like $g^+ag^-a^B$ (Fig. 7, right), also contained no structural features and possessed larger relative energies.

H_{g^+} (H15) deprotonated Fragment C conformers converged to different conformations than did H_a

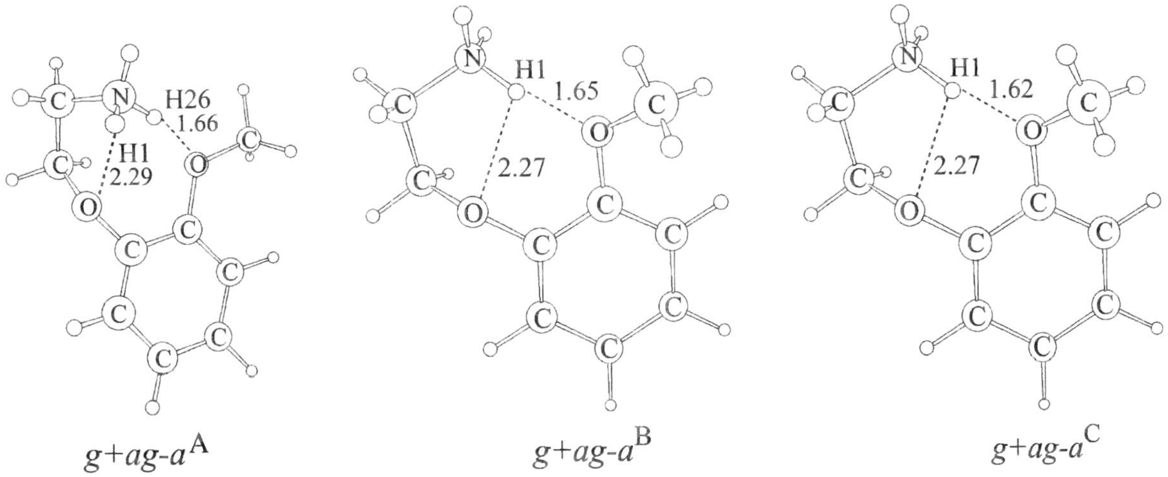

$g+ag\text{-}a^A$ $g+ag\text{-}a^B$ $g+ag\text{-}a^C$

Fig. 6. Converged minima of the protonated Fragment C PEHS g^+ag^-a subset: $g^+ag^-a^A$ (0.34 kcal mol^{-1}), $g^+ag^-a^B$ (1.75 kcal mol^{-1}), and $g^+ag^-a^C$ (1.78 kcal mol^{-1}).

Table 3
Relative energies for the converged minima of the protonated aminoethoxy-2-methoxy-benzene (Fragment C) PEHS computed at the RHF/3-21G level of theory

Conformational assignment				Protonated relative energy (kcal mol⁻¹)	H_a deprotonation SPE (hartree)	$\Delta E_{vert}(a)$ (kcal mol⁻¹)	H_{g^+} deprotonation SPE (hartree)	$\Delta E_{vert}(g^+)$ (kcal mol⁻¹)	H_{g^-} deprotonation SPE (hartree)	$\Delta E_{vert}(g^-)$ (kcal mol⁻¹)
X_7	X_8	X_9	X_{11}							
g^+	g^+	g^+	g^-	0.00	−549.426174969	280.43	−549.442280486	270.32	−549.441928160	270.54
g^+	g^+	a	a	5.95	−549.443611717	263.54	−549.442851892	264.01	−549.434031483	269.55
g^+	a	g^-	a^A	0.34	−549.432586758	276.07	−549.439878833	271.49	−549.430288227	277.51
g^+	a	g^-	a^B	1.75	−549.425684138	278.99	−549.443172996	268.01	−549.441413643	269.12
g^+	a	g^-	a^C	1.78	−549.424896018	279.45	−549.442854134	268.18	−549.440728544	269.52
g^+	a	g^-	a	7.42	−549.437153444	266.12	−549.426047877	273.09	−549.435637888	267.07
g^+	g^-	a	g^+	1.03	−549.427346075	268.66	−549.444060000	268.18	−549.442394356	269.22
g^+	g^+	a	g^-	19.49	−549.445822784	248.61	−549.447282769	247.69	−549.445584545	248.76
a	g^+	a	g^+	22.07	−549.447502744	244.97	−549.448266733	244.49	−549.449480983	243.73
a	g^+	a	a^A	22.45	−549.446176571	245.43	−549.446244896	245.39	−549.447671383	244.49
a	a	a	a^B	18.79	−549.446798634	248.70	−549.448213075	247.81	−549.446483903	248.90
a	a	a	a^C	21.37	−549.448057940	245.33	−549.446928686	246.04	−549.447228542	245.85
a	a	a	a^D	21.37	−549.447239890	245.84	−549.446932064	246.03	−549.448066994	245.32
a	a	a	a	18.79	−549.446488127	246.31	−549.448216115	245.23	−549.446797719	246.12
a	g^-	a	a	22.45	−549.446163539	245.44	−549.447663524	244.50	−549.446240118	245.39
a	g^-	a	g^-	22.07	−549.447513088	244.97	−549.449490572	243.73	−549.448276170	244.49
a	g^-	a	g^+	19.49	−549.445623108	248.73	−549.447324349	247.67	−549.445877477	248.57
g^-	g^-	a	g^+	1.03	−549.427328185	278.68	−549.442367243	269.24	−549.444029127	268.20
g^-	g^-	a	g^+	7.42	−549.426000793	273.12	−549.437205701	266.08	−549.435694955	267.03
g^-	a	g^+	a^A	1.78	−549.424870211	279.47	−549.440701940	269.54	−549.442836385	268.20
g^-	a	g^+	a^B	1.75	−549.425583135	279.05	−549.441348512	269.16	−549.443182477	268.01
g^-	a	g^+	a^C	0.34	−549.430300136	277.50	−549.439888909	271.48	−549.432587532	276.07
g^-	a	a	a	5.95	−549.434112751	269.50	−549.442963271	263.94	−549.443703675	263.48
g^-	g^-	g^+		0.00	−549.426168078	280.43	−549.441929596	270.54	−549.442283555	270.32

Full optimizations were followed by SPE calculations for independent deprotonation of H_a (H1), H_{g^+} (H15), and H_{g^-} (H26) protons.

Table 4
Optimized conformational minima for the H_a (H1) deprotonated PEHS of aminoethoxy-2-methoxy-benzene (Fragment C) at the RHF/3-21G level of theory

Optimized protonated conformation				Optimized deprotonated conformation				χ_7 (deg)	χ_8 (deg)	χ_9 (deg)	χ_{11} (deg)	E (hartree)	Relative energy (kcal mol⁻¹)	$\Delta E_{opt}(g^+)$ (kcal mol⁻¹)	$\Delta\Delta E(g^+)$ (kcal mol⁻¹)
χ_7	χ_8	χ_9	χ_{11}	χ_7	χ_8	χ_9	χ_{11}								
g^+	g^+	g^+	g^-	g^+	g^+	g^+	g^{-A}	70.67	88.77	86.24	−56.91	−549.456031486	3.80	261.69	18.74
g^+	a	a	g^-	g^+	a	a	a	55.44	128.64	122.81	−173.00	−549.457605177	2.82	254.76	8.78
g^+	a	g^-	a^A	g^+	g^-	g^-	a^A	61.11	−153.94	−109.91	173.12	−549.458403320	2.32	259.87	16.20
g^+	a	g^-	a^B	g^+	a	g^-	a^B	71.22	174.11	−119.92	168.63	−549.449098582	8.15	264.29	14.70
g^+	a	g^-	a^C	g^+	a	g^-	a^B					−549.449098582	8.15	264.29	15.16
g^+	g^-	a	g^-	g^+	g^-	g^+	g^{-B}	58.06	59.91	66.85	−69.71	−549.462092585	0.00	250.47	15.65
g^+	g^-	g^-	g^+	g^+	g^-	g^-	g^+	67.80	−115.76	−67.17	70.52	−549.461457783	0.40	257.26	21.40
a	g^-	a	g^-	a	g^+	g^-	g^-	−177.49	93.20	85.38	−68.05	−549.458328292	2.36	240.76	7.85
a	g^-	a	g^+	a	g^+	g^-	g^+	176.07	77.69	−113.64	70.94	−549.458069229	2.52	238.34	6.63
a	a	a	a	a	a	a	a	176.87	78.85	−118.50	173.00	−549.455143308	4.36	239.81	5.62
a	a	a	a^A	a	a	a	a^A	−176.22	−127.35	−123.29	178.77	−549.456320232	3.62	242.73	5.97
a	a	a	a^B	a	a	g^-	a	179.84	179.61	115.49	−171.18	−549.455665083	4.03	240.55	4.78
a	a	a	a^C	a	a	g^-	a	179.78	179.23	−115.19	173.14	−549.456079419	3.77	240.29	5.55
a	a	a	a^D	a	a	a	a^B	177.03	128.42	122.87	−173.47	−549.456075063	3.78	240.30	6.01
a	a	a	a	a	g^-	g^+	a	−176.87	−78.85	118.50	−173.00	−549.455143308	4.36	239.81	5.63
a	g^-	a	a	a	g^-	g^+	a	−176.07	−77.69	113.64	−70.94	−549.458069229	2.52	238.34	6.63
g^-	g^-	a	g^-	a	g^-	g^+	g^-	178.08	−93.71	−87.49	69.23	−549.458374431	2.33	240.73	8.00
g^-	g^+	g^-	g^+	g^-	g^-	g^-	g^+	−67.82	115.72	67.20	−70.49	−549.461457804	0.40	257.26	21.42
g^-	g^+	a	g^+	g^-	g^-	g^-	g^-	−73.56	118.87	−72.23	64.73	−549.455865572	3.91	254.38	18.74
g^-	a	g^+	a^A	g^-	g^+	g^+	a	−71.19	−174.09	119.88	−168.63	−549.449098573	8.15	264.27	15.20
g^-	a	g^+	a^B	g^-	g^+	g^+	a^A					−549.449098573	8.15	264.27	14.78
g^-	a	g^-	a^C	a	a	g^+	a^A	−71.02	154.69	105.64	175.30	−549.458468802	2.27	259.83	17.67
g^-	a	a	a	g^-	a	a	a^B	−60.91	−131.67	−121.62	174.22	−549.457379459	2.96	254.90	14.60
g^-	g^-	g^-	g^+	g^-	g^-	g^-	g^+	−70.67	−88.77	−86.24	56.91	−549.456031485	3.80	261.69	18.74

Table 5

Optimized conformational minima for the H_{g^+} (H15) deprotonated PEHS of aminoethoxy-2-methoxy-benzene (Fragment C) at the RHF/3-21G level of theory

Optimized protonated conformation				Optimized deprotonated conformation				χ_7 (deg)	χ_8 (deg)	χ_9 (deg)	χ_{11} (deg)	E (hartree)	Relative energy (kcal mol^{-1})	$\Delta E_{opt}(g^+)$ (kcal mol^{-1})	$\Delta\Delta E(g^+)$ (kcal mol^{-1})
χ_7	χ_8	χ_9	χ_{11}	χ_7	χ_8	χ_9	χ_{11}								
g^+	g^+	g^+	g^-	g^+	g^+	g^+	g^-	53.60	63.53	67.58	−69.43	−549.461502854	0.37	258.26	12.06
g^+	a	g^-	a	g^+	a	a	a	60.94	131.73	121.62	−174.12	−549.457379474	2.96	254.90	9.11
g^+	a	g^-	a^A	g^+	a	g^-	a	61.12	−153.95	−109.92	173.17	−549.458403331	2.32	259.87	11.62
g^+	a	g^-	a^B	g^+	a	g^-	a					−549.458403331	2.32	258.46	9.55
g^+	a	g^-	a^C	g^+	a	g^-	a					−549.458403331	2.32	258.43	9.75
g^+	g^-	a	g^-	g^+	g^-	a	g^-	73.55	−118.85	72.27	−64.75	−549.455865605	3.91	254.38	18.71
g^+	g^-	g^-	g^+	g^+	g^-	g^-	g^+	63.50	−109.60	−67.91	69.69	−549.461778373	0.20	257.06	11.12
a	g^+	g^+	g^-	a	g^+	g^+	g^-	−178.19	98.38	88.26	−69.43	−549.457609028	2.81	241.21	6.48
a	g^+	g^-	g^+	a	g^+	g^-	g^+	178.07	78.63	−115.16	71.31	−549.458326171	2.36	238.18	6.31
a	a	a	a^A	a	a	a	a^A	178.18	79.54	−119.67	169.09	−549.454953224	4.48	239.92	5.47
a	a	a	a^B	a	a	g^+	a	−176.81	−126.11	−124.42	171.92	−549.456063978	3.78	242.89	4.92
a	a	g^+	a^C	a	a	g^+	a	178.52	177.67	115.38	−174.31	−549.455937215	3.86	240.38	5.66
a	a	g^-	a^D	a	a	g^-	a^B	−178.52	−177.66	−115.39	174.31	−549.455937212	3.86	240.38	5.65
a	a	a	a	a	a	g^-	a	176.81	126.11	124.42	−171.92	−549.456063980	3.78	240.30	4.93
a	g^-	g^+	g^-	a	g^-	g^+	g^+	−177.46	−78.65	118.83	−169.29	−549.454914807	4.50	239.95	4.55
a	g^-	g^+	g^+	a	g^-	g^+	g^+	−177.84	−78.65	114.14	−71.72	−549.458060719	2.53	238.35	5.38
a	g^-	g^+	g^+	a	g^-	g^+	g^+	178.18	−98.38	−88.25	69.42	−549.457609009	2.81	241.21	6.46
g^-	a	a	g^+	g^-	a	g^+	a^A	−67.79	115.72	67.16	−70.49	−549.461457796	0.40	257.26	11.98
g^-	a	a	g^+	g^-	a	g^+	a^A					−549.462092524	0.00	250.47	15.61
g^-	a	g^+	a^A	g^-	a	g^+	a^A	−70.99	154.66	105.61	175.29	−549.458468747	2.27	258.39	11.15
g^-	a	g^+	a^B	g^-	a	g^+	a^B					−549.458468747	2.27	258.41	10.75
g^-	a	g^+	a^C	g^-	a	a	a	−61.12	153.95	109.92	−173.17	−549.458403331	2.32	259.87	11.61
g^-	a	a	a	g^-	a	a	a	−60.94	−131.73	−121.62	174.13	−549.457379476	2.96	254.90	9.04
g^-	g^-	g^-	g^+	g^-	g^-	g^-	g^+	−58.01	−59.89	−66.89	69.63	−549.462092524	0.00	257.89	12.65

Table 6
Optimized conformational minima for the H_{g-} (H26) deprotonated PEHS of aminoethoxy-2-methoxy-benzene (Fragment C) at the RHF/3-21G level of theory

Optimized protonated conformation				Optimized deprotonated conformation				χ_7 (deg)	χ_8 (deg)	χ_9 (deg)	χ_{11} (deg)	E (hartree)	Relative energy (kcal mol^{-1})	$\Delta E_{opt}(g^+)$ (kcal mol^{-1})	$\Delta\Delta E(g^+)$ (kcal mol^{-1})
χ_7	χ_8	χ_9	χ_{11}	χ_7	χ_8	χ_9	χ_{11}								
g^+	g^+	g^+	g^-	g^+	g^+	g^+	g^-	58.06	59.89	66.88	−69.69	−549.462092596	0.00	257.89	12.65
g^+	a	a	a	g^+	a	a	a	60.93	131.73	121.64	−174.18	−549.457379464	2.96	254.90	14.65
g^+	g^-	g^-	a^A	g^+	a	g^-	a	71.03	−154.66	−105.63	−175.26	−549.458468814	2.27	259.83	17.68
g^+	a	g^-	a^B	g^+	a	g^-	a						2.27	258.41	10.71
g^+	g^-	g^-	a^C	g^+	a	g^-	a						2.27	258.39	11.13
g^+	a	g^-	a	g^+	a	g^-	g^-	74.09	−79.55	119.88	−72.92	−549.457178397	3.08	253.55	13.52
g^+	g^-	a	g^-	g^+	g^-	g^-	g^+	67.80	−115.71	−67.18	70.49	−549.461457808	0.40	257.26	11.96
a	g^-	g^-	g^+	a	g^-	g^-	g^+	−178.11	93.55	87.45	−69.22	−549.458374259	2.33	240.73	8.03
a	g^+	a	g^-	a	g^+	a	g^-	177.82	78.70	−114.10	71.72	−549.458060658	2.53	238.35	5.38
a	a	a	g^+	a	a	a	g^+	177.46	78.57	−118.84	169.29	−549.454914788	4.50	239.95	4.54
a	a	a	a^A	a	a	a	a^A	−177.01	−128.27	−123.01	173.58	−549.456074882	3.78	242.88	6.02
a	a	a	a^B	a	a	a	a	−179.80	−179.32	115.11	−173.21	−549.456079310	3.77	240.29	5.56
a	a	a	a^C	a	a	a	a	−179.78	−179.69	−115.48	171.25	−549.455664913	4.03	240.55	4.77
a	a	a	a^D	a	a	a	a	176.24	127.31	123.33	−178.98	−549.456320114	3.62	240.14	5.98
a	a	a	a	a	a	a	a	−178.20	−79.41	119.68	−169.01	−549.454953180	4.48	239.92	5.47
a	g^-	g^+	g^-	a	g^-	g^+	g^-	−178.06	−78.71	115.18	−71.29	−549.458326128	2.36	238.18	6.31
a	g^-	g^-	g^+	a	g^+	g^+	a	177.48	−93.23	−85.36	68.04	−549.458328330	2.36	240.76	7.81
g^-	a	g^+	g^-	g^-	g^+	g^+	g^-	−63.55	109.58	67.97	−69.63	−549.461778325	0.20	257.06	11.14
g^-	g^-	a	g^+	g^-	a	a	a	−74.23	79.52	−119.76	72.89	−549.457178317	3.08	253.55	13.48
g^-	g^-	g^+	a^A	g^-	a	a	a	−61.16	154.09	109.87	−173.16	−549.458403144	2.32	258.43	9.77
g^-	a	a	a^B	g^-	a	g^+	a						2.32	258.46	9.55
g^-	a	g^+	a^C	g^-	a	g^+	a						2.32	259.87	16.20
g^-	a	a	a	g^-	g^-	a	a	−55.45	−128.60	−122.88	173.01	−549.457605176	2.82	254.75	8.73
g^-	g	g	g^+	g^-	g	g^-	g^+	−53.56	−63.60	−67.55	69.45	−549.461502806	0.37	258.26	12.06

Table 7
Structural features of converged conformers of the aminoethoxy-2-methoxy-benzene (Fragment C) PEHS

Conformational assignment				Structural features	Relative energy (kcal mol^{-1})
χ_7	χ_8	χ_9	χ_{11}		
I. Protonated conformers (CH$^+$)					
g^+	g^+	g^+	g^-	1.57 Å H1···O12 H-bond, 2.40 Å H1···O5 H-bond forming an 8-membered ring	*0.00*
g^+	a	a	a	1.80 Å H26···O5 H-bond forming a 5-membered ring	5.95
g^+	a	g^-	a^A	1.66 Å H26···O12 H-bond forming an 8-membered ring, 2.29 Å H1···O5 H-bond forming a 5-membered ring	0.34
g^+	a	g^-	a^B	1.65 Å H1···O12 H-bond, 2.22 Å H1···O5 H-bond forming an 8-membered ring	1.75
g^+	a	g^-	a^C	1.62 Å H1···O12 H-bond, 2.27 Å H1···O5 H-bond forming an 8-membered ring	1.78
g^+	g^-	a	g^-	1.54 Å H15···O12 H-bond forming an 8-membered ring	7.42
g^+	g^-	g^-	g^+	1.56 Å H1···O12 H-bond, 2.47 Å H1···O5 H-bond forming an 8-membered ring	1.03
a	g^+	g^+	g^-	–	19.49
a	g^+	a	g^+	–	22.07
a	g^+	a	a	–	22.45
a	a	a	a^A	–	18.79
a	a	a	a^B	–	21.37
a	a	a	a^C	–	21.37
a	a	a	a^D	–	18.79
a	g^-	a	a	–	22.45
a	g^-	a	g^-	–	22.07
a	g^-	g^-	g^+	–	19.49
g^-	g^+	g^+	g^-	1.56 Å H1···O12 H-bond, 2.47 Å H1···O5 H-bond forming an 8-membered ring	1.03
g^-	g^+	a	g^+	1.53 Å H1···O12 H-bond forming an 8-membered ring	7.42
g^-	a	g^+	a^A	1.63 Å H1···O12 H-bond, 2.27 Å H1···O5 H-bond forming an 8-membered ring	1.78
g^-	a	g^+	a^B	1.65 Å H1···O12 H-bond, 2.22 Å H1···O5 H-bond forming an 8-membered ring	1.75
g^-	a	g^+	a^C	1.66 Å H1···O12 H-bond forming an 8-membered ring, 2.29 Å H26···O5 H-bond forming a 5-membered ring	0.34
g^-	a	a	a	1.80 Å H1···O5 H-bond forming a 5-membered ring	5.95
g^-	g^-	g^-	g^+	1.57 Å H1···O12 H-bond, 2.40 Å H1···O5 H-bond forming an 8-membered ring	*0.00*
II. H$_a$ deprotonated conformers (C)					
g^+	g^+	g^+	g^{-A}	–	3.80
g^+	a	a	a	2.45 Å H26···O5 H-bond forming a 5-membered ring	2.82
g^+	a	g^-	a^A	2.80 Å H26···O12 H-bond, 2.44 Å H26···O5 H-bond forming an 8-membered ring	2.32
g^+	a	g^-	a^B	–	8.15
g^+	g^+	g^+	g^{-B}	2.11 Å H15···O12 H-bond forming an 8-membered ring	*0.00*
g^+	g^-	g^-	g^+	2.10 Å H26···O12 H-bond, 2.60 Å H26···O5 H-bond forming an 8-membered ring	0.40
a	g^+	g^+	g^-	–	2.36
a	g^+	g^-	g^+	–	2.52
a	g^+	g^-	a	–	4.36
a	a	a	a^A	–	3.62
a	a	g^+	a	–	4.03

Table 7 (*continued*)

Conformational assignment				Structural features	Relative energy (kcal mol^{-1})
χ_7	χ_8	χ_9	χ_{11}		
a	a	g^-	a	–	3.77
a	a	a	$a^{\,B}$	–	3.78
a	g^-	g^+	a	–	4.36
a	g^-	g^+	g^-	–	2.52
a	g^-	g^-	g^+	–	2.33
g^-	g^+	g^+	g^-	2.09 Å H15···O12 H-bond, 2.60 Å H15···O5 H-bond forming an 8-membered ring	0.40
g^-	g^+	g^-	g^+	–	3.91
g^-	a	g^+	$a^{\,A}$	–	8.15
g^-	a	g^+	$a^{\,B}$	2.33 Å H15···O12 H-bond, 2.44 Å H15···O5 H-bond forming an 8-membered ring	2.27
g^-	a	a	a	2.35 Å H15···O5 H-bond forming a 5-membered ring	2.96
g^-	g^-	g^-	g^+	–	3.80

III. H$_{g^+}$ deprotonated conformers (C)

g^+	g^+	g^+	g^-	2.12 Å H1···O12 H-bond, 2.73 Å H1···O5 H-bond forming an 8-membered ring	0.37
g^+	a	a	a	2.35 Å H26···O5 H-bond forming a 5-membered ring	2.96
g^+	a	g^-	a	2.80 Å H26···O12 H-bond, 2.43 Å H26···O5 H-bond forming an 8-membered ring	2.32
g^+	g^-	g^+	g^-	–	3.91
g^+	g^-	g^-	g^+	2.09 Å H1···O12 H-bond, 2.72 Å H1···O5 H-bond forming an 8-membered ring	0.20
a	g^+	g^+	g^-	–	2.81
a	g^+	g^-	g^+	–	2.36
a	g^+	g^-	a	–	4.48
a	a	a	$a^{\,A}$	–	3.78
a	a	g^+	a	–	3.86
a	a	g^-	a	–	3.86
a	a	a	$a^{\,B}$	–	3.78
a	g^-	g^+	a	–	4.50
a	g^-	g^+	g^-	–	2.53
a	g^-	g^-	g^+	–	2.81
g^-	g^+	g^+	g^-	2.09 Å H1···O12 H-bond, 2.60 Å H1···O5 H-bond forming an 8-membered ring	0.40
g^-	a	g^+	$a^{\,A}$	2.33 Å H1···O12 H-bond, 2.44 Å H1···O5 H-bond forming an 8-membered ring	2.27
g^-	a	g^+	$a^{\,B}$	2.80 Å H26···O12 H-bond, 2.43 Å H26···O5 H-bond forming an 8-membered ring	2.32
g^-	a	a	a	2.35 Å H1···O5 H-bond forming a 5-membered ring	2.96
g^-	g^-	g^-	g^+	2.11 Å H1···O12 H-bond, 2.58 Å H1···O5 H-bond forming an 8-membered ring	*0.00*

IV. H$_{g^-}$ deprotonated conformers (C)

g^+	g^+	g^+	g^-	2.11 Å H1···O12 H-bond, 2.58 Å H1···O5 H-bond forming an 8-membered ring	*0.00*
g^+	a	a	a	2.35 Å H1···O5 H-bond forming a 5-membered ring	2.96
g^+	a	g^-	a	2.33 Å H1···O12 H-bond, 2.45 Å H1···O5 H-bond forming an 8-membered ring	2.27
g^+	g^-	g^+	g^-	3.11 Å H15···π-electron interaction, 3.96 Å H1···π-electron interaction	3.08

(continued on next page)

Table 7 (*continued*)

Conformational assignment				Structural features	Relative energy (kcal mol^{-1})
χ_7	χ_8	χ_9	χ_{11}		
g^+	g^-	g^-	g^+	2.09 Å H1···O12 H-bond, 2.60 Å H1···O5 H-bond forming an 8-membered ring	0.40
a	g^+	g^+	g^-	–	2.33
a	g^+	g^-	g^+	–	2.53
a	g^+	g^-	a	–	4.50
a	a	a	a^A	–	3.78
a	a	g^+	a	–	3.77
a	a	g^-	a	–	4.03
a	a	a	a^B	–	3.62
a	g^-	g^+	a	–	4.48
a	g^-	g^+	g^-	–	2.36
a	g^-	g^-	g^+	–	2.36
g^-	g^+	g^+	g^-	2.08 Å H1···O12 H-bond, 2.72 Å H1···O5 H-bond forming an 8-membered ring	0.20
g^-	g^+	g^-	g^+	3.11 Å H1···π-electron interaction, 3.96 Å H15···π-electron interaction	3.08
g^-	a	g^+	a	2.81 Å H1···O12 H-bond, 2.43 Å H1···O5 H-bond forming an 8-membered ring	2.32
g^-	a	a	a	2.45 Å H1···O5 H-bond forming a 5-membered ring	2.82
g^-	g^-	g^-	g^+	2.12 Å H1···O12 H-bond, 2.73 Å H1···O5 H-bond forming an 8-membered ring	0.37

All structures presented in this table have been geometrically optimized at the RHF/3-21G level of theory.

deprotonated conformers upon optimization (Tables 5 and 7(III)) further emphasizing that the protonated conformation and the choice of proton abstracted have a large influence on the subsequent converged deprotonated minima. In this case, the global minima structure was conformer $g^-g^-g^-g^+$ which was very similar to the protonated PEHS global minima with bifurcated hydrogen bonds from H1 (Fig. 8, left). Due to the prevalence of this structural motif in the protonated PEHS, combined with the fact that in the majority of protonated conformers the split hydrogen bonds come from H1, it is reasonable to expect that H_{g^+} deprotonated conformers would maintain this bifurcated hydrogen bond intact upon optimization.

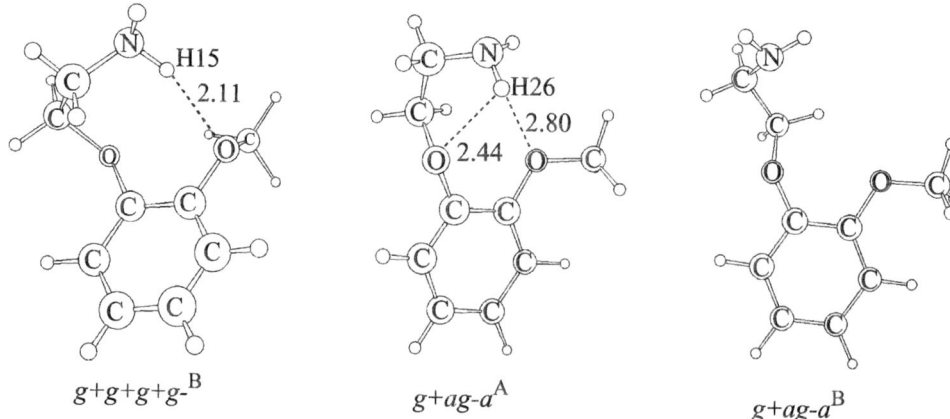

Fig. 7. Converged H_a (H1) deprotonated Fragment C conformers: global minima $g^+g^+g^+g^{-B}$ (0.00 kcal mol^{-1}), $g^+ag^-a^A$ (2.32 kcal mol^{-1}), and $g^+ag^-a^B$ (8.15 kcal mol^{-1}).

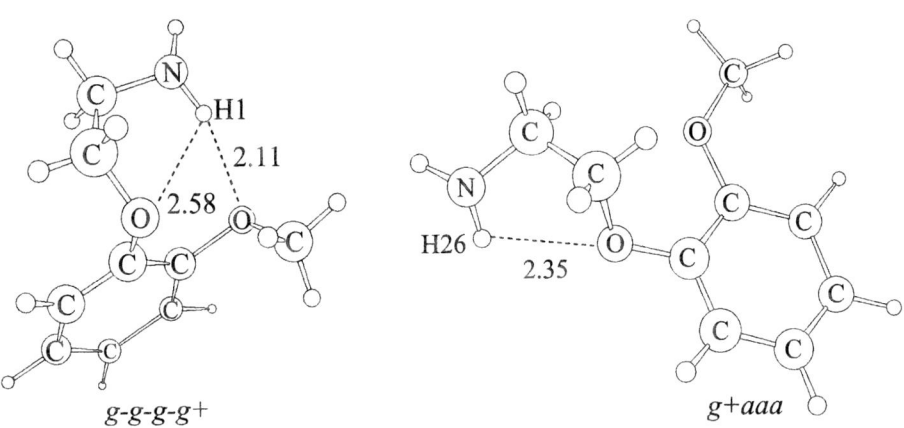

Fig. 8. Converged H_{g^-} (H15) deprotonated Fragment C conformers: global minima $g^-g^-g^-g^+$ (0.00 kcal mol^{-1}) and g^+aaa (2.96 kcal mol^{-1}).

H_{g^+} deprotonated conformer g^+aaa (Fig. 8, right) is very similar to the protonated g^+aaa conformation. Conformers with torsional angle χ_7 in the *anti* position had the side-chain fully extended and lacked any structural features; conformer ag^-g^+a had the largest relative energy of 4.50 kcal mol^{-1}.

The H_{g^-} (H26) deprotonated Fragment C PEHS (Tables 6 and 7(IV)) global minima was conformer $g^+g^+g^+g^-$ (Fig. 9, left) which was similar to the protonated conformation and possessed an intact H1 bifurcated hydrogen bond. Optimization of the H_{g^-} deprotonated conformers revealed a novel interaction in the Fragment C PEHS; conformer $g^+g^-g^+g^-$, and its axis chiral pair $g^-g^+g^-g^+$, possess an internal H1$\cdots$ and H15$\cdots$pi(π)-electron interaction

(Fig. 9, middle) with a relative energy of 3.08 kcal mol^{-1}. This amine–aromatic π interaction is a novel interactions not previously encountered in the literature and confers moderate stability to the deprotonated Fragment C PEHS conformers.

Upon convergence of the deprotonated $g^+g^-g^+g^-$ and $g^-g^+g^-g^+$ conformers, protonated $g^+g^-g^+g^-$ and $g^-g^+g^-g^+$ conformers were re-computed to see if this amine–aromatic π interaction had been missed because it follows intuitively that a protonated positive nitrogen centre would be a more likely structure to form a cation–aromatic π interaction due to its greater electron deficiency. However, these protonated structures were not missed; instead, protonated conformers

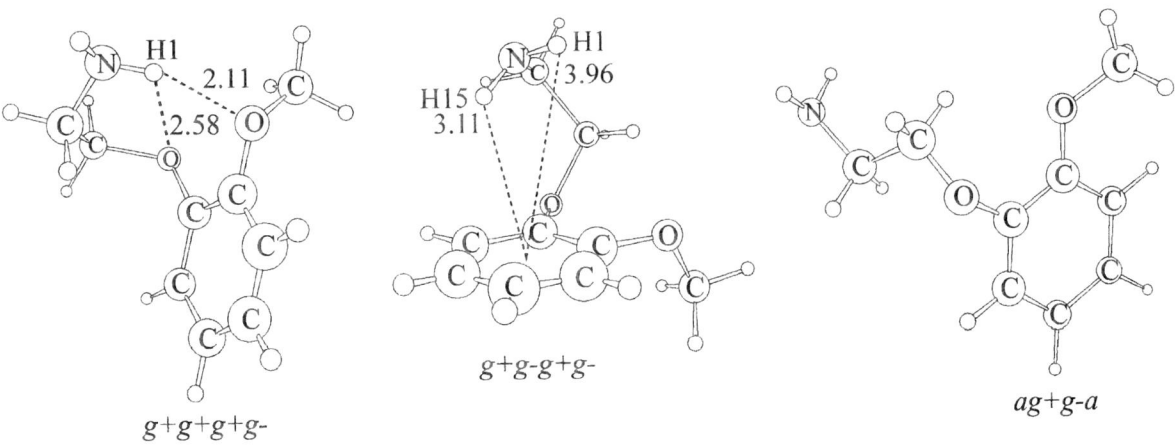

Fig. 9. Converged H_{g^-} (H26) deprotonated Fragment C conformers: global minima $g^+g^+g^+g^-$ (0.00 kcal mol^{-1}), $g^+g^-g^+g^-$ (3.08 kcal mol^{-1}), and ag^+g^-a (4.50 kcal mol^{-1}).

$g^+g^-g^+g^-$ and $g^-g^-g^+g^-g^+$ converged to conformers $g^+g^-g^-g^-g^+$ (1.56 Å H1···O12 and 2.47 Å H1···O5 hydrogen bonds forming an eight-membered ring; 1.03 kcal mol^{-1} relative energy) and $g^-ag^+a^C$ (1.66 Å H1···O12 hydrogen bond forming an eight-membered ring and a 2.29 Å H26···O5 hydrogen bond forming a five-membered ring; 0.34 kcal mol^{-1} relative energy), respectively. This illustrates that the presence of nearby inductively electron-donating oxygen atoms (O5 and O12) favors hydrogen bond and intramolecular ion–dipole formation between the positive nitrogen centre and the oxygen atoms over the cation–aromatic π interaction. Currently, the authors of this work are investigating this novel amine–aromatic π interaction and trying to discern possible confounding activation effects by O5 and O12, which are bound to the benzene ring. Finally, like the protonated PEHS, the only conformers with no structural features were those with torsional angle χ_7 in the *anti* position (conformer ag^+g^-a is shown in Fig. 9, right).

3.3. Intrinsic conformational-dependent basicity of aminoethoxy-2-methoxy-benzene (Fragment C)

Once the protonated PEHS of Fragment C was computed, energies of deprotonation were determined with both SPE calculations and full geometry optimizations for H_a, H_{g+}, and H_{g-} deprotonation events (summary of values in Tables 3 and 8). For vertical energies of deprotonation (i.e. SPE calculations), it was found that conformers showed the greatest difference in their respective $\Delta E_{vert}(a)$, $\Delta E_{vert}(g^+)$, and $\Delta E_{vert}(g^-)$ values when protons involved in internal stabilization were abstracted. In these instances, the proton not involved in intramolecular stabilization (such as hydrogen bond formation) was relatively easily deprotonated. Protons involved in hydrogen bonding required, on average, an additional 5–10 kcal mol^{-1} compared to non-hydrogen bonded protons for deprotonation. Contrasting, high relative energy conformers (i.e. conformations with torsional angle χ_7 in the *anti* position) had energies of deprotonation 25–35 kcal mol^{-1} lower than that of low relative energy conformers. Also, although large variation was found in energies of deprotonation for the same conformation (with regards to the three protons) in low

relative energy conformers, the high relative energy conformers all possessed comparable energies of deprotonation for all three protons for a given conformational assignment because of the lack of intramolecular stabilization directly involving any proton(s).

After all the non-optimized energies of deprotonation were calculated, all H_a (c.f. $\Delta E_{opt}(a)$ in Table 4), H_{g+} (c.f. $\Delta E_{opt}(g^+)$ in Table 5), and H_{g-} (c.f. $\Delta E_{opt}(g^-)$ in Table 6) deprotonated conformers were subject to full geometry optimizations. By majority, unique protonated conformational assignments converged to unique deprotonated conformational assignments with little redundancy. Geometry optimizations reduced most of the energies of deprotonation, however, conformations with internal stabilization still required significantly more energy (20–30 kcal mol^{-1}) for deprotonation. The inherent stability of these conformers made deprotonation a more energetically demanding task; for example, the H_{g-} deprotonation of $g^+g^+g^+g^-$ ($\Delta E_{opt}(g^-) = 257.89$ kcal mol^{-1}) or the H_a deprotonation of $g^+ag^-a^C$ ($\Delta E_{opt}(a) = 264.29$ kcal mol^{-1}) versus the H_{g+} deprotonation of $aaaa^A$ ($\Delta E_{opt}(g^+) = 242.89$ kcal mol^{-1}) or the H_a deprotonation of ag^-aa ($\Delta E_{opt}(a) = 239.81$ kcal mol^{-1}). These dramatic differences in optimized energies of deprotonation are indicative of the fact that the protonated conformation has significant influence on the energetics of protonophoretic pathways and mechanisms.

When analyzing the differences between vertical and optimized energies of deprotonation, the values of $\Delta\Delta E(a)$, $\Delta\Delta E(g^+)$, and $\Delta\Delta E(g^-)$ (Table 8) were used to compare the stabilization gained by a respective deprotonated conformer upon geometry optimization. These values were typically large and in the range of 8–21 kcal mol^{-1} for conformers with intramolecular stabilization present. This can be interpreted as the initial deprotonation event producing a high energy structure that requires major geometrical alterations to reach local minima. On the other hand, conformers with no intramolecular stabilization usually had $\Delta\Delta E$ values of less than 8 kcal mol^{-1} indicating that the protonated conformations were ideal structures for deprotonation and did not require major geometrical changes to rearrange into a local minima.

Carvedilol, with a pK_a of 7.9, is expected to have its amino group involved in a large number of

Table 8
Summary of different energies of deprotonation for each converged conformation of aminoethoxy-2-methoxy-benzene (Fragment C) at the RHF/3-21G level of theory

χ_7	χ_8	χ_9	χ_{11}	$\Delta E_{vert}(a)$ (kcal mol^{-1})	$\Delta E_{opt}(a)$ (kcal mol^{-1})	$\Delta\Delta E(a)$ (kcal mol^{-1})	$\Delta E_{vert}(g^+)$ (kcal mol^{-1})	$\Delta E_{opt}(g^+)$ (kcal mol^{-1})	$\Delta\Delta E(g^+)$ (kcal mol^{-1})	$\Delta E_{vert}(g^-)$ (kcal mol^{-1})	$\Delta E_{opt}(g^-)$ (kcal mol^{-1})	$\Delta\Delta E(g^-)$ (kcal mol^{-1})
g^+	g^+	g^+	g^-	280.43	261.69	18.74	270.32	258.26	12.06	270.54	257.89	12.65
g^+	a	a	a	263.54	254.76	8.78	264.01	254.90	9.11	269.55	254.90	14.65
g^+	a	g^-	a^A	276.07	259.87	16.20	271.49	259.87	11.62	277.51	259.83	17.68
g^+	a	g^-	a^B	278.99	264.29	14.70	268.01	258.46	9.55	269.12	258.41	10.71
g^+	a	g^-	a^C	279.45	264.29	15.16	268.18	258.43	9.75	269.52	258.39	11.13
g^+	g^-	a	g^-	266.12	250.47	15.65	273.09	254.38	18.71	267.07	253.55	13.52
g^+	g^-	g^-	g^+	278.66	257.26	21.40	268.18	257.06	11.12	269.22	257.26	11.96
g^+	g^-	g^+	g^-	248.61	240.76	7.85	247.69	241.21	6.48	248.76	240.73	8.03
a	g^+	a	g^+	244.97	238.34	6.63	244.49	238.18	6.31	243.73	238.35	5.38
a	g^+	a	a	245.43	239.81	5.62	245.39	239.92	5.47	244.49	239.95	4.54
a	a	a	a^A	248.70	242.73	5.97	247.81	242.89	4.92	248.90	242.88	6.02
a	a	a	a^B	245.33	240.55	4.78	246.04	240.38	5.66	245.85	240.29	5.56
a	a	a	a^C	245.84	240.29	5.55	246.03	240.38	5.65	245.32	240.55	4.77
a	a	a	a^D	246.31	240.30	6.01	245.23	240.30	4.93	246.12	240.14	5.98
a	a	a	a	245.44	239.81	5.63	244.50	239.95	4.55	245.39	239.92	5.47
a	g^-	a	g^-	244.97	238.34	6.63	243.73	238.35	5.38	244.49	238.18	6.31
a	g^-	g^-	g^+	248.73	240.73	8.00	247.67	241.21	6.46	248.57	240.76	7.81
g^-	g^-	g^-	g^+	278.68	257.26	21.42	269.24	257.26	11.98	268.20	257.06	11.14
g^-	g^+	g^+	g^+	273.12	254.38	18.74	266.08	250.47	15.61	267.03	253.55	13.48
g^-	a	g^+	a^A	279.47	264.27	15.20	269.54	258.39	11.15	268.20	258.43	9.77
g^-	a	g^+	a^B	279.05	264.27	14.78	269.16	258.41	10.75	268.01	258.46	9.55
g^-	a	g^+	a^C	277.50	259.83	17.67	271.48	259.87	11.61	276.07	259.87	16.20
g^-	a	a	a	269.50	254.90	14.60	263.94	254.90	9.04	263.48	254.75	8.73
g^-	g^-	g^-	g^+	280.43	261.69	18.74	270.54	257.89	12.65	270.32	258.26	12.06

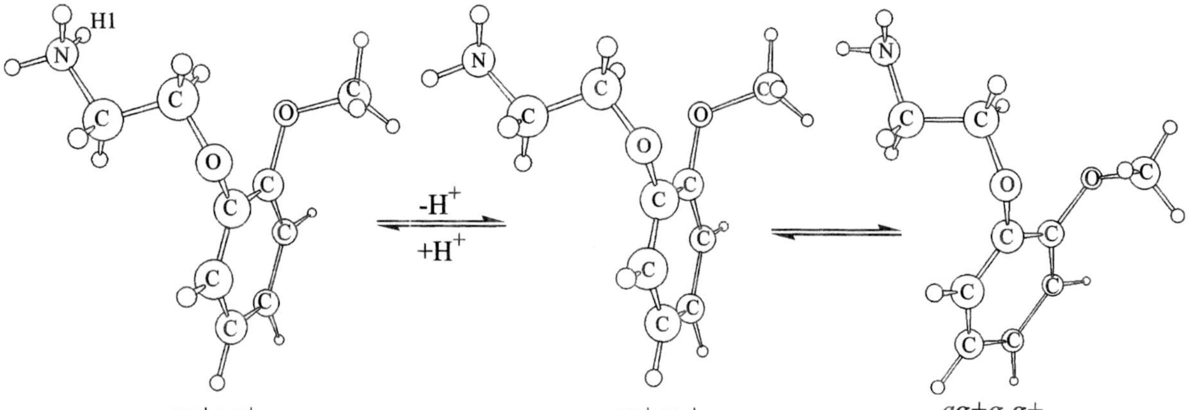

Fig. 10. The route with the lowest energy of deprotonation (optimized) for Fragment C was found to be via the H_a (H1) deprotonation of conformer ag^+ag^+ to the $ag^+g^-g^+$ conformation (238.34 kcal mol^{-1}).

protonophoretic pathways. At a physiological pH of 7.4, about two-thirds of the amino group is in its protonated form. In this light, the pathway with the lowest energy of deprotonation (optimized) for Fragment C was the H_a (H1) deprotonation of conformers ag^+ag^+ (Fig. 10). Conformer ag^+ag^+ has a fully extended conformation with proton H1 oriented away from the core of the structure and as such can be easily deprotonated. The vertical energy for this deprotonation was calculated at 244.97 kcal mol^{-1}. Upon optimization, the converged local minima structure was conformer $ag^+g^-g^+$ with an energy of deprotonation of 238.34 kcal mol^{-1}. Upon deprotonation, only a slight rotation of torsional angle χ_9 from the *anti* to the g^- position was needed for convergence to a nearby local minima on the PEHS. The minor geometry alteration is further emphasized with a low $\Delta\Delta E(a)$ value of 6.63 kcal mol^{-1}.

On the other hand, protonated conformers g^+ag^-a ^{B} and g^+ag^-a ^{C} required 264.29 kcal mol^{-1} for deprotonation of the H1 proton (Fig. 11). The g^+ag^-a ^{B} and g^+ag^-a ^{C} conformations contained a bifurcated H1 hydrogen bond to O5 and O12 forming the preferred eight-membered ring. This deprotonation required that the proton be abstracted from the core of the structure. The SPE for this deprotonation was 278.99 and 279.45 kcal mol^{-1} for g^+ag^-a ^{B} and g^+ag^-a ^{C} conformers, respectively. Upon optimization, both of these conformers converged to

the g^+ag^-a ^{B} with $\Delta\Delta E(a)$ values of 14.70 and 15.16 kcal mol^{-1} for g^+ag^-a ^{B} and g^+ag^-a ^{C} conformers, respectively. Although the conformational assignment remained the same for this deprotonation, the large $\Delta\Delta E$ values indicate the large geometrical alterations that were needed for convergence to a local minima; for example, torsional angle χ_8 rotated about 45° and χ_9 about 40°.

Previous work done on carvedilol has revealed energies of deprotonation of 234 and 238 kcal mol^{-1} for only two conformations, irrespective of the protons deprotonated, at the RHF/6-31G(d) level of theory [10]. Work on Fragment B of carvedilol has revealed energies of deprotonation ranging from 245 to 262 kcal mol^{-1} for the entire PEHS of Fragment B with full optimizations at the RHF/3-21G level of theory [16]. The values obtained here for Fragment C through full optimizations of the entire PEHS range from 238 to 264 kcal mol^{-1} and are comparable with greater selectivity and variation in the values. This indicates that the RHF/3-21G level of theory is adequate to quantify the energetics of carvedilol and its fragments and that the methodology to analyze differences at the proton and conformation level are significant and needed.

With regards to deprotonation schemes, pathways of proton abstraction will vary depending on the molecular conformation of the protonated structure. As such, in assessing the conformers of a particular PEHS, it is important to analyze not only dominant

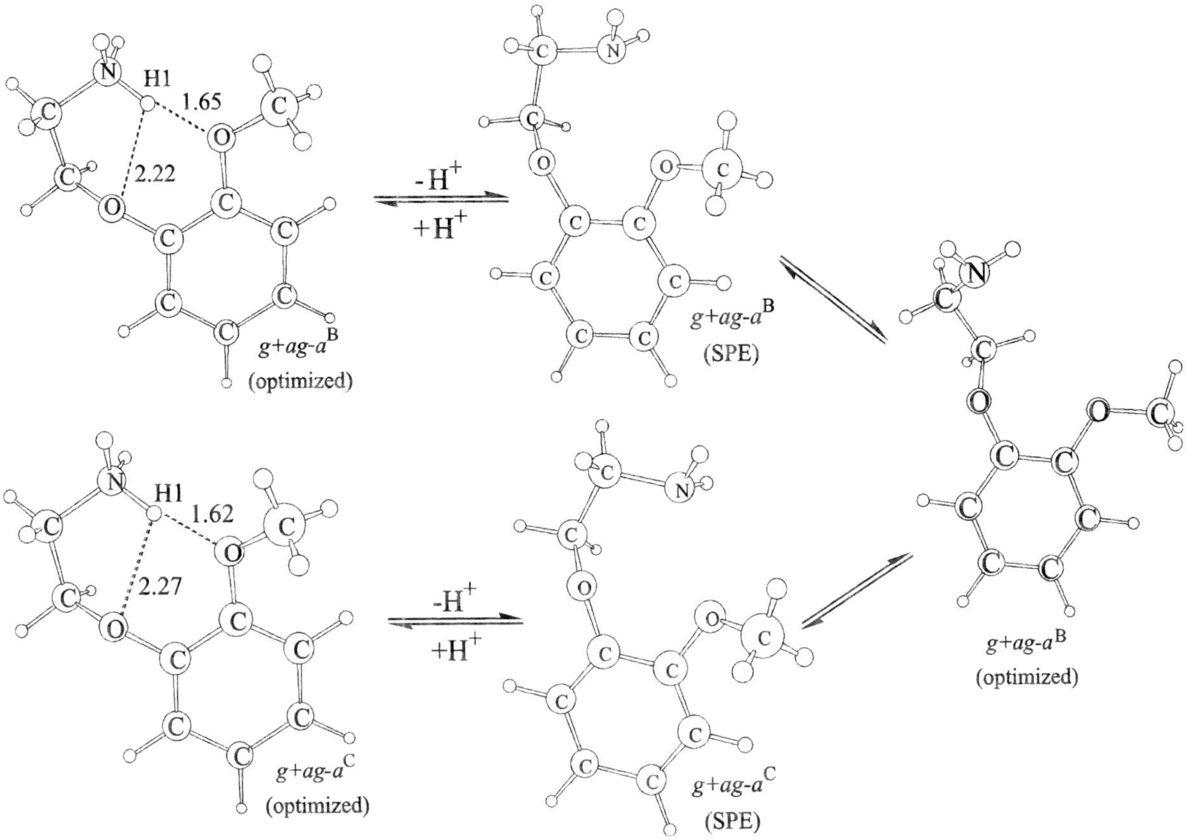

Fig. 11. The route with the largest energy of deprotonation (optimized) for Fragment C was found to be via the H_a (H1) deprotonation of conformers $g^+ag^-a^B$ and $g^+ag^-a^C$ (264.29 kcal mol^{-1}), both of which converged to the $g^+ag^-a^B$ conformation.

structures with low relative energies but also to realize that favored structures will vary widely depending on the mechanism in question. As shown here, and for Fragment B [16], in assessing protonophoretic pathways, the ideal structures for deprotonation will be those of higher relative energies with minimal internal stabilization. Thus, molecular conformation and intrinsic structural motifs will predetermine, at least with regards to energetics, the type of protonophoretic mechanism most utilized or favored by enzymes and substrates in proton shuttling pathways (such as that postulated for carvedilol in the uncoupling of oxidative phosphorylation in the mitochondria). The differences in the energies of deprotonation that are dependent on molecular conformation can help explain why such diverse effects have been seen in mitochondrial uncoupling by carvedilol and attributed to the molecular conformations adopted by carvedilol [10]. To put differently, carvedilol is believed to

possess large conformational flexibility and should possess a PEHS with very diverse conformational minima. Thus, it is expected that different structures will be more likely to participate in proton shuttling across the mitochondrial membrane and be better at the uncoupling effects seen in mitochondria.

4. Conclusions

The data presented here for the primary amine Fragment C agrees with the results obtained for the secondary amine Fragment B and indicates that for carvedilol, and any general protonophoretic pathways involving primary and secondary amines, reactions and events of deprotonation will not only favor higher energy conformations but will also favor protons differentially according to their orientation. It may be expected that the abstracted proton will be one oriented maximally away from the core of the structure devoid of any intermolecular formation for favored energetics.

As is evident from the deprotonation data presented here, it is not the proton per se that determines whether the energy of deprotonation is large or small; rather, it is the molecular conformation adopted by a structure that determines the respective orientation of a given proton, and therefore, predetermines the energetics of the deprotonation event. This is to say that, protons may be indistinguishable one from the other, but depending on the molecular conformations and the protonophoretic mechanisms in question, rarely are two protons the same. Finally, a novel amine–aromatic π interaction was discovered in the deprotonated Fragment C PEHS. Further work is ongoing to further elucidate the characteristics of this interaction.

Acknowledgements

One of the authors (IGC) wishes to thank the Ministry of Education for a Szent-Györgyi Visiting Professorship.

References

[1] W. Carlson, K. Oberg, J. Cardiovasc. Pharmacol. Ther. 4 (1999) 205.

[2] M. Metra, S. Nodari, A. D'Aloia, L. Bontempi, E. Boldi, L. Dei Cas, Am. Heart J. 139 (2000) 511.

[3] G. Feuerstein, R.R. Ruffolo Jr., Adv. Pharmacol. 42 (1998) 611.

[4] O. Saijonmaa, K. Metsarinne, F. Fyhrquist, Blood Press. 6 (1997) 24.

[5] S. Capomolla, O. Febo, M. Gnemmi, G. Riccardi, C. Opasich, A. Carporotondi, A. Mortara, G. Pinna, F. Cobelli, Am. Heart J. 139 (2000) 596.

[6] M. Packer, M.R. Bristow, J.N. Cohn, W.S. Colucci, M.B. Fowler, E.M. Gilbert, N.H. Shusterman, N. Engl. J. Med. 334 (1996) 1349.

[7] M.A. Berg, G.A. Chasse, E. Deretey, A.K. Kuzery, B.M. Fung, D.Y.K. Fung, H. Henry-Riyad, A.C. Lin, M.L. Mak, A. Mantas, M. Patel, I.V. Repyakh, M. Staikova, S.J. Salpietro, T.-H. Tang, J.C. Vank, A. Perczel, G.I. Csonka, O. Farkas,

[8] L.L. Torday, Z. Szekely, I.G. Csizmadia, J. Mol. Struct. (THEOCHEM) 500 (2000) 5.

[8] G. Feuerstein, T.L. Yue, X. Ma, R.R. Ruffolo, Prog. Cardiovasc. Dis. 41 (1 suppl 1) (1998) 17.

[9] A.J.F. Searlee, C. Gree, R.L. Wilson, Ellipticines and carbazoles as antioxidants, in: W. Borns, M. Saran, D. Tait (Eds.), Oxygen Radicals and Biology, Walter de Gruyter, Berlin, 1984, pp. 377–381.

[10] P.J. Oliveira, M.P. Marques, L.A.E. Batista de Carvalho, A.J.M. Moreno, Biochem. Biophys. Res. Commun. 276 (2000) 82.

[11] R. Ferrari, J. Cardiovasc. Pharmacol. 28 (1996) S1.

[12] D.R. Howlett, A.R. George, D.E. Owen, R.V. Ward, R.E. Markwell, Biochem. J. 343 (1999) 419.

[13] H.G. Oldham, S.E. Clarke, Drug Metab. Dispos. 25 (1997) 970.

[14] R.R. Ruffolo Jr., M. Gellai, J.P. Hieble, R.N. Willette, A.J. Nichols, Eur. J. Clin. Pharmacol. 38 (1990) S82.

[15] D.R.P. Almeida, L.F. Pisterzi, G.A. Chass, L.L. Torday, A. Varro, J. Gy. Papp, I.G. Csizmadia, J. Phys. Chem. 106 (2002) 10423.

[16] D.R.P. Almeida, D.M. Gasparro, L.F. Pisterzi, L.L. Torday, A. Varro, J. Gy. Papp, B. Penke, J. Mol. Struct. (Theochem), 631 (2003) 251.

[17] D.U. Silverthorn, W.C. Ober, C.W. Garrison, A.C. Silverthorn, Human Physiology: An Integrated Approach, second ed., Prentice Hall, Upper Saddle River, NJ, 1998.

[18] P.G. Heytler, Uncouplers of Oxidative Phosphorylation, in: Methods in Enzymology, LV, Academic Press, New York, 1979, pp. 462–472.

[19] A. Tzagoloff, Mitochondria, Plenum Press, New York, 1982.

[20] [.2.0.].S.S. Korshunov, V.P. Skulachev, A.A. Starkov, FEBS Lett. 416 (1997) 15.

[21] D.R.P. Almeida, D.M. Gasparro, L.F. Pisterzi, L.L Torday, A. Varro, J. Gy. Papp, B. Penke, I.G. Csizmadia, J. Phys. Chem. A, 107 (2003) 5594.

[22] M.J. Frisch, G.W. Trucks, H.B. Schlegel, G.E. Scuseria, M.A. Robb, J.R. Cheeseman, V.G. Zakrzewski, J.A. Montgomery, Jr., R.E. Stratmann, J.C. Burant, S. Dapprich, J.M. Millam, A.D. Daniels, K.N. Kudin, M.C. Strain, O. Farkas, J. Tomasi, V. Barone, M. Cossi, R. Cammi, B. Mennucci, C. Pomelli, C. Adamo, S. Clifford, J. Ochterski, G.A. Petersson, P.Y. Ayala, Q. Cui, K. Morokuma, D.K. Malick, A.D. Rabuck, K. Raghavachari, J.B. Foresman, J. Cioslowski, J.V. Ortiz, A.G. Baboul, B.B. Stefanov, G. Liu, A. Liashenko, P. Piskorz, I. Komaromi, R. Gomperts, R.L. Martin, D.J. Fox, T. Keith, M.A. Al-Laham, C.Y. Peng, A. Nanayakkara, C. Gonzalez, M. Challacombe, P.M.W. Gill, B.G. Johnson, W. Chen, M.W. Wong, J.L. Andres, M. Head-Gordon, E.S. Replogle, J.A. Pople, GAUSSIAN 98 (Revision A.9), Gaussian Inc., Pittsburgh, PA, 1998.

[23] Axum 5.0C for Windows, MathSoft Incorporated, 1996.

ELSEVIER

Journal of Molecular Structure (Theochem) 666–667 (2003) 581–586

THEO
CHEM

www.elsevier.com/locate/theochem

Hexachlorophene and triclosan—exploratory ab initio structural analyses

Ashton A. Connor[a,b,*], Gregory A. Chasse[a,b,c,d], David H. Setiadi[a], Imre G. Csizmadia[a,b,e]

[a]*Global Institute of COmputational Molecular and Materials Science (GIOCOMMS), 1422 Edenrose St, Mississauga, Ont., Canada L5V 1H3*
[b]*Department of Chemistry, University of Toronto, 80 St George St, Toronto, Ont., Canada M5S 3H6*
[c]*Institut de Science et d'Ingénierie Supramoléculaires, 8 allée Gaspard Monge, BP 70028, 67083 Strasbourg Cedex, France*
[d]*Department of Biomedical Sciences, Creighton University, 2500 California plaza, Omaha, NE 68178, USA*
[e]*Department of Medical Chemistry, University of Szeged, Dóm tér 8, 6720 Szeged, Hungary*

Abstract

Hexachlorophene is primarily used as an antiseptic in germicidal soaps but can be toxic if absorbed through the skin. Rotation of its benzene rings about the two central methyl bonds gives rise to a variety of minima characterized by intramolecular hydrogen bonding. Triclosan is an antibacterial currently used in a wide range of consumer goods, from drugs to hand soaps. With the emerging resistance of bacteria to triclosan, researchers now consider hexachlorophene as a possible option in the design of antibacterial drugs. Clinical studies have demonstrated that these two molecules, though comparable in structure, bind to different adjacent sites in bacterial Enoyl-ACP Reductase (FabI), which is used in fatty acid synthesis. This significant discovery is attributed to subtle structural disparities between the two compounds. Determination of the conformations adopted in vitro and in vivo is crucial in elucidating the binding mechanism of these two compounds. Ab initio structural analyses have been performed on these two molecules using GAUSSIAN98 software.
© 2003 Elsevier B.V. All rights reserved.

Keywords: Hexachlorophene; Triclosan; Dihedral angle

1. Introduction

Only against death can [man] call on no means of escape, but escape from hopeless diseases he has found in the depths of his mind. Sophocles

The study of antibacterial agents is of particular importance for our ever-growing population as new bacteria emerge and old bacteria reinvent themselves to outperform the drugs created to subdue them. Given the complexity of even the simplest bacteria, a myriad of routes exist whereby antibacterials can destroy germ cells. One method that has lately received increased attention is the interruption of the fatty acid biosynthesis pathway [1,2]. This universal pathway builds up fatty acid molecule two carbon units at a time and is catalysed by enzymes encoded by separate genes [3,4]. The enoyl-acyl carrier protein reductase (FabI) catalyses the last step in each cycle and plays a regulatory role in determining the rate of fatty acid synthesis [5,6]. Antibacterial inhibition of FabI is very effective, leading to cell wall lysis and

* Corresponding author. Address: Global Institute of COmputational Molecular and Materials Science (GIOCOMMS), 1422 Edenrose St, Mississauga, Ont., Canada L5V 1H3.
E-mail address: ashton.connor@utoronto.ca (A.A. Connor).

disruption of the cellular membrane [7,8]. Once it was discovered that triclosan (Fig. 1a) acts by inhibiting FabI [9,10], further research identified the binding pocket for triclosan on the enzyme via crystal structures [11,12] and showed that the interaction of triclosan and FabI follows reversible, uncompetitive inhibition [13,14]. Triclosan is a broad-spectrum antibacterial and antifungal agent found in a wide variety of consumer goods; thus, not unexpectedly, bacteria are developing a resistance to it [10,15]. Hexachlorophene (Fig. 1b) has been proposed as an

alternative to triclosan since it also targets FabI [1], although the kinetics markedly vary. Recent research shows that triclosan and hexachlorophene not only bind to different sites on the enzyme, but also bind to different forms of FabI [2]. Although hexachlorophene most likely destroys bacteria via means other than FabI inhibition [10,11], this still exists as a potential mode of action for it and has warranted significant research as of late [1,2]. Hexachlorophene [16–18], triclosan, and related hydroxydiphenylethers contain the hydroxyphenyl moiety [10,11], which is

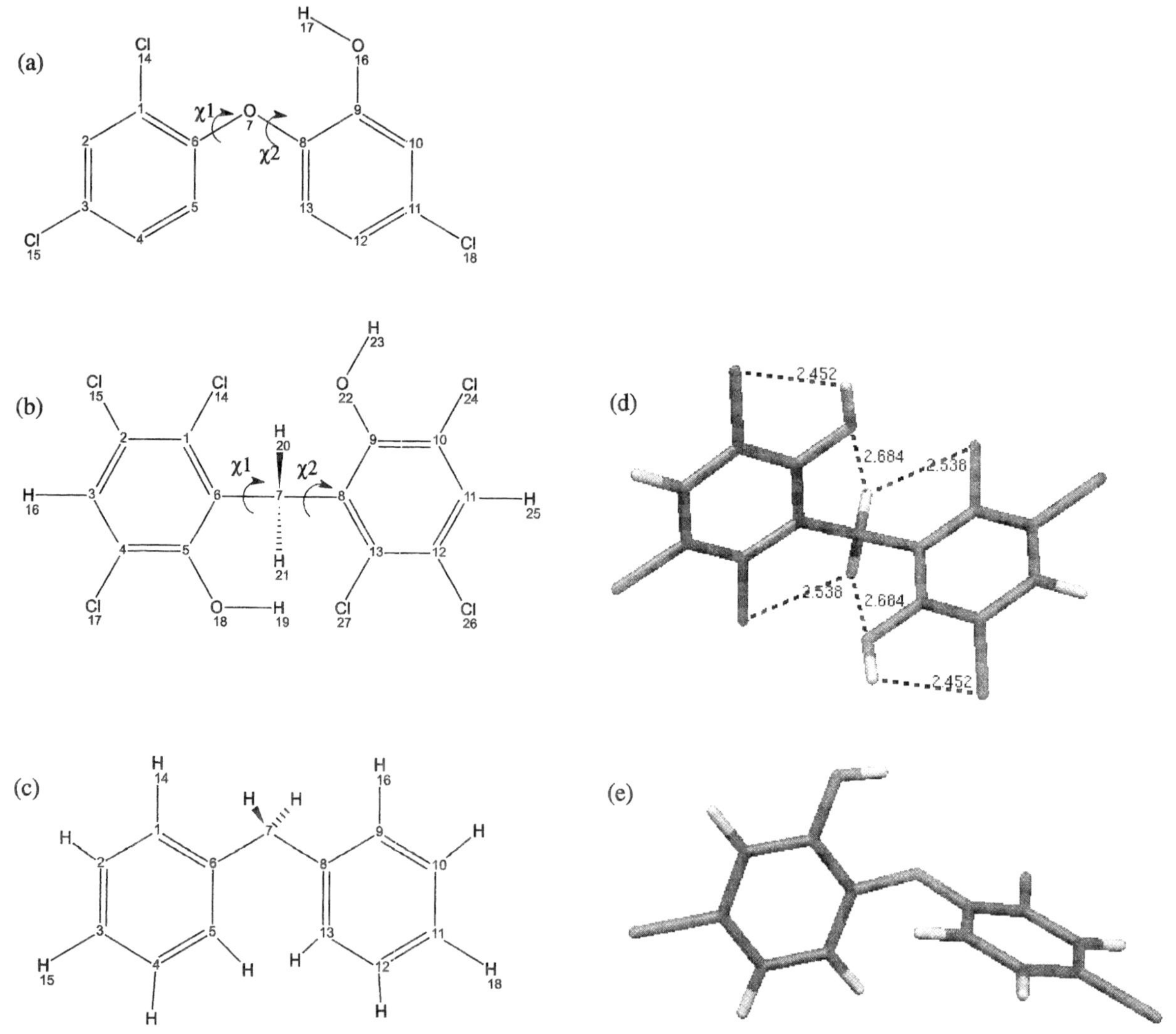

Fig. 1. (a) The structure and numbering of triclosan. (b) The structure and numbering of hexachlorophene. (c) The structure and numbering of diphenyl methane. (d) Hexachlorophene's minimum energy structure, with $X^1 = X^2 = 61.9°$. The hydrogen bonds are depicted in Å. (e) Triclosan's minimum energy structure, with $X^1 = 150°$ and $X^2 = 120°$.

responsible for their similar modes of action. The more subtle differences in their modes of action prompted this ab initio conformational study of their structures. The conformational analyses of hexachlorophene and triclosan are beneficial in determining their docking processes to FabI, which will aid in delineating their behavioural differences. Early results from the study are presented here.

2. Method

Ab initio geometry optimizations were carried out using the GAUSSIAN98 [19] program for Linux. The molecules were numbered in a standardized fashion [20] by first considering the atoms making up the benzene rings and the central methylene or oxygen group, then numbering all substituents extending from the rings. Both hexachlorophene and triclosan were characterized by two principal dihedral angles, named X^1 and X^2, about their central moieties. Z-matrices were constructed to account for each intramolecular bond distance, bond angle, and dihedral. Calculations were aided by a previous study on diphenyl methane [21] (Fig. 1c). The structures were optimized first at the am1 then at the RHF/3-21G level of theory. 144 data points were generated for each molecule to construct potential energy surfaces by holding one principal dihedral constant as the other was rotated at 30° intervals. The distinct minima for hexachlorophene were further optimized at RHF/6-31G(d).

3. Results and discussion

3.1. Structure of hexachlorophene

Previous calculations performed on diphenyl methane show that the two equivalent minima for the molecule occur at $90 \pm 31.7°$ [20]. The symmetry of diphenyl methane is mirrored in hexachlorophene, which is also a symmetric molecule. The true minima exist at $X^1 = X^2 = \pm 61.9°$. A three-dimensional potential energy surface and a two-dimensional contour surface were generated (Fig. 2a). The 18 true and saddle point minima located for hexachlorophene at the RHF/6-31G(d) level of theory have been tabulated in Table 1a. A three-dimensional image of

hexachlorophene's true minimum energy conformer is shown in Fig. 1d. This is likely the structure adopted by hexachlorophene in vitro and in vivo, and it is thus the approximate shape that it adopts when docking to the FabI protein. It is characterized by intramolecular hydrogen bonding between its hydroxyl protons and chlorine atoms (distance of 2.452 Å), between its methylene protons and chlorine atoms (distance of 2.538 Å), and between its methylene protons and oxygen atoms (distance of 2.684 Å) (see Fig. 1d). The latter two hydrogen bonds are unorthodox as they involve methylene protons and deserve further attention in future studies. While these evidently contribute to the overall stability of this conformer, they may not be present in vivo, where surrounding water molecules are likely to disrupt them. A solvation study will be required to determine the behaviour of hexachlorophene in solvent, which will allow for a more precise description of the true minimum energy conformation.

3.2. Structure of triclosan

The triclosan conformers are all of much higher energy than their hexachlorophene counterparts. This lack of stability is likely due to the lack of intramolecular hydrogen bonds present, as there are fewer protons available for this. Triclosan is not a symmetric molecule and its minima reflect this. The true minima was found to be at $X^1 = 150°$ and $X^2 = 120°$, and again at $X^1 = 210°$ and $X^2 = 240°$. Very close saddle points were found at $X^1 = 90°$ and $X^2 = 180°$, again at $X^1 = 270°$ and $X^2 = 180°$, at $X^1 = 120°$ and $X^2 = 150°$, and again at $X^1 = 240°$ and $X^2 = 210°$. These minima are shown graphically in the three-dimensional potential energy surface and the two-dimensional contour surface in Fig. 2b, and Table 1b lists them at the RHF/3-21G level of theory. Since triclosan is better understood and more studied, computations were not carried out as far as those for hexachlorophene. An image of triclosan's lowest energy conformer is given in Fig. 1e. This is likely the structure adopted by triclosan in vitro and in vivo, and approximates the structure it adopts when docking to the FabI protein. Further solvation studies at higher levels of theory should be undertaken for triclosan in order to obtain the best estimate for its lowest energy conformation.

PES of Hexachlorophene: Energy Vs. Dihedrals χ1 and χ2

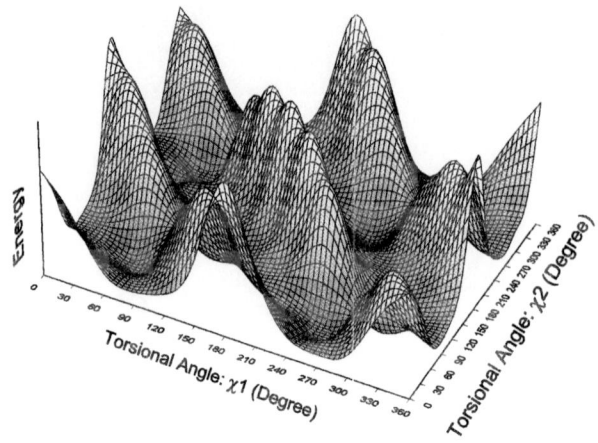

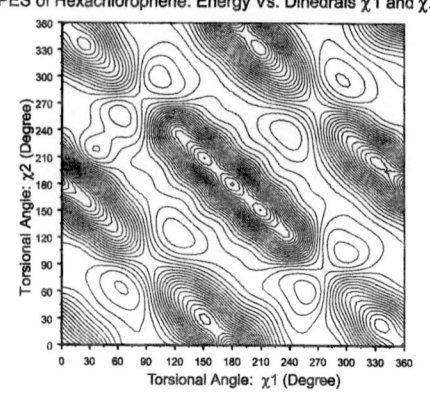

PES of Hexachlorophene: Energy Vs. Dihedrals χ1 and χ2

PES of Triclosan: Energy Vs. Dihedrals χ1 and χ2

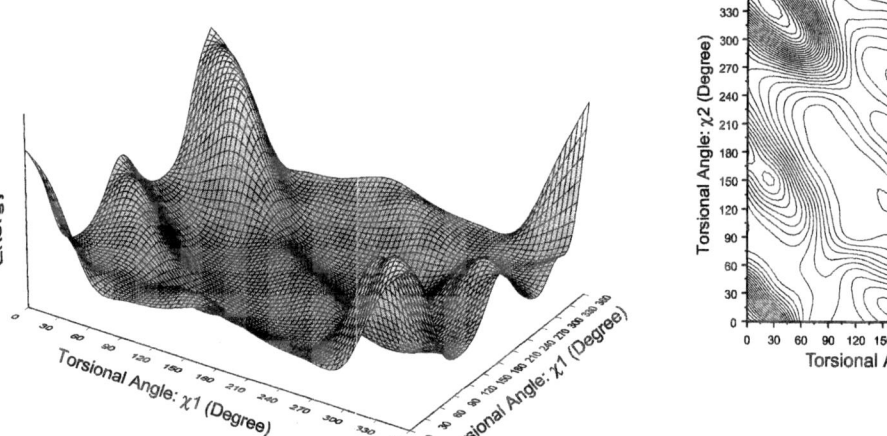

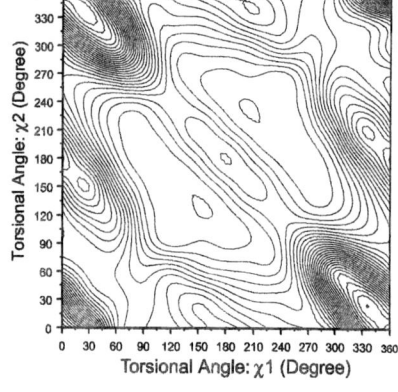

PES of Triclosan: Energy Vs. Dihedral χ1 and χ2

Fig. 2. (a) Potential energy surface $E = f(X^1, X^2)$ in the landscape (left) and contour representation (right), computed at the RHF/3-21G level of theory for hexachlorophene. (b) Potential energy surface $E = f(X^1, X^2)$ in the landscape (left) and contour representation (right), computed at the RHF/3-21G level of theory for triclosan.

3.3. Comparison of hexachlorophene and triclosan

The potential energy surfaces for hexachlorophene and triclosan are markedly different. The energy surface of hexachlorophene is characterized by high peaks, with few possible minima in which to exist. In comparison, triclosan's surface is much more permissible. The true minima for the molecules are in considerably different positions with regards to the relative orientation of their rings. These structural differences could well account for the experimentally observed differences in their binding to FabI. FabI inhibitors must be able to form stable ternary complexes with the protein to have antibacterial activity [1], and triclosan does so with FabI and NAD(H). Hexachlorophene does not require the presence of NAD(H), and its binding mechanism is not well understood [2]. Triclosan's ether oxygen is suspected of being responsible for its ability to form the ternary complex [1], while hexachlorophene's central methylene group would explain its differing activity. Rather than directly accusing the central moieties for the observed activities, it is possible that they are responsible for the different orientations of

Table 1a

Summary of the dihedral angles (in °) and energies (in hartrees) for hexachlorophene's 18 minima located at RHF/6-31g(d). All energies are given to only five decimal places

Conformer	X^1	X^2	Energy
True minima			
1	61.92	61.91	− 3402.36416
2	− 61.94	298.11	− 3402.36416
3	298.09	− 61.91	− 3402.36416
4	− 61.91	− 61.89	− 3402.36416
Saddle points			
5	− 57.18	109.26	− 3402.36271
6	53.72	243.37	− 3402.36299
7	57.20	250.75	− 3402.36271
8	57.03	251.80	− 3402.36185
9	116.63	− 53.70	− 3402.36299
10	123.17	123.02	− 3402.36236
11	116.62	306.30	− 3402.36299
12	109.28	302.79	− 3402.36271
13	108.21	− 56.99	− 3402.36185
14	236.88	236.89	− 3402.36236
15	243.37	53.69	− 3402.36299
16	250.70	57.17	− 3402.36271
17	302.97	108.19	− 3402.36185
18	306.28	116.61	− 3402.36299

the rings in the two molecules, which result in different docking processes. Further steps can be taken to model the actual binding sites on the FabI enzyme with and without NAD(H) present with respect to hexachlorophene and triclosan.

Table 1b

Summary of the dihedral angles (in °) and energies (in hartrees) for triclosan's minima located at the RHF/3-21G level of theory. All energies are given to only five decimal places

Conformer	X^1	X^2	Energy
True minima			
1	150	120	− 1976.70445
2	210	240	− 1976.70445
Saddle points			
3	90	180	− 1976.70402
4	270	180	− 1976.70402
5	120	150	− 1976.70426
6	240	210	− 1976.70426

4. Conclusion and future work

A plausible explanation for the binding differences between hexachlorophene and triclosan to the FabI protein in bacteria has been expounded briefly in this study. More precise structural studies of hexachlorophene and triclosan can decorate the skeleton given here, particularly solvation studies which would remove the supporting intramolecular hydrogen bonds present. This instance is not, likely, an isolated justification for ligand binding involving the hydroxyphenyl moiety. This approach can be extended to other hydroxydiphenylethers and is thus a solid foundation for further interesting studies on the subject. Given the importance this bears towards understanding and designing antibacterials, it should be pursued further. It has not escaped my attention that particular hydroxydiphenylethers can be designed organically to exist in particular conformers which bind preferentially to the desired pocket of FabI and other bacterial proteins with the help of such investigations.

Acknowledgements

I am greatly indebted to Allegra Connor, Monee Rassolian, Reza Mehdizadeh, Jacqueline Law, and Suzanne Lau. This work has been supported by grants from the Global Institute of COmputational Molecular and Materials Sciences.

One of the authors (IGC) wishes to thank the Ministry of Education for a Szent-Györgyi Visiting Professorship.

References

[1] R.J. Heath, J. Li, G.E. Roland, C.O. Rock, J. Biol. Chem. 275 (2000) 4654–4659.

[2] J. Marcinkeviciene, W. Jiang, L.M. Kopcho, G. Locke, Y. Luo, R.A. Copeland, Arch. Biochem. Biophys. 390 (2001) 101–108.

[3] J.E. Cronan Jr, C.O. Rock, *Escherichia coli* and *Salmonella typhimurium*: Cell. Mol. Biol. (1996) 612–636.

[4] C.O. Rock, J.E. Cronan Jr, Biochim. Biophys. Acta 1302 (1996) 1–16.

[5] R.J. Heath, C.O. Rock, J. Biol. Chem. 270 (1995) 26538–26542.

[6] R.J. Heath, C.O. Rock, J. Biol. Chem. 271 (1996) 1833–1836.

[7] F. Turnowsky, K. Fuchs, C. Jeschek, C. Hogenauer, J. Bacteriol. 171 (1989) 6555–6565.

[8] A.F. Egan, R.R.B. Russell, Genet. Res. 21 (1973) 3603–3611.

[9] L.M. McMurry, M. Oethinger, S.B. Levy, Nature 394 (1988) 531–532.

[10] R.J. Heath, Y.T. Yu, M.A. Shapiro, E. Olson, C.O. Rock, J. Biol. Chem. 273 (1998) 30316–30320.

[11] R.J. Heath, J.R. Rubin, D.R. Holland, E. Zhang, M.E. Snow, C.O. Rock, J. Biol. Chem. 274 (1999) 11110–11114.

[12] A. Roujeinikova, C.W. Levy, S. Rowsell, S. Sedelnikova, P.J. Baker, C.A. Minshull, A. Mistry, J.G. Colls, R. Camble, A.R. Stuitje, R. Viner, D.W. Rice, J. Mol. Biol. 294 (1999) 527–535.

[13] W.H.J. Ward, G.A. Holdgate, S. Rowsell, E.G. McLean, R.A. Pauptit, E. Clayton, W.W. Nichols, J.G. Colls, C.A. Minshull, D.A. Jude, A. Mistry, D. Tims, R. Camble, N.J. Hales, C.J. Britton, I.W.F. Taylor, Biochemistry 38 (1999) 12514–12525.

[14] S.L. Parikh, G. Xiao, P.J. Tonge, Biochemistry 39 (2000) 7645–7650.

[15] M.T.E. Suller, A.D. Russell, J. Antimicrob. Chemother. 46 (2000) 11–18.

[16] T.R. Corner, J.L. Joswick, J.N. Silvernale, P. Gerhardt, J. Bacteriol. 108 (1971) 501–507.

[17] H.L. Joswick, T.R. Corner, J.N. Silvernale, P. Gerhardt, J. Bacteriol. 108 (1971) 492–500.

[18] J.N. Silvernale, H.L. Joswick, T.R. Corner, P. Gerhardt, J. Bacteriol. 108 (1971) 482–491.

[19] M.J. Frisch, G.W. Trucks, H.B. Schlegel, G.E. Scuseria, M.A. Robb, J.R. Cheeseman, V.G. Zakrzewski, J.A. Montgomery, Jr., R.E. Stratmann, J.C. Burant, S. Dapprich, J.M. Millam, A.D. Daniels, K.N. Kudin, M.C. Strain, Ö. Farkas, J. Tomasi, V. Barone, M. Cossi, R. Cammi, B. Mennucci, C. Pomelli, C. Adamo, S. Clifford, J. Ochterski, G.A. Petersson, P.Y. Ayala, Q. Cui, K. Morokuma, D.K. Malick, A.D. Rabuck, K. Raghavachari, J.B. Foresman, J. Cioslowski, J.V. Ortiz, A.G. Baboul, B.B. Stefanov, G. Liu, A. Liashenko, P. Piskorz, I. Komaromi, R. Gomperts, R.L. Martin, D.J. Fox, T. Keith, M.A. Al-Laham, C.Y. Peng, A. Nanayakkara, C. Gonzalez, M. Challacombe, P.M.W. Gill, B.G. Johnson, W. Chen, M.W. Wong, J.L. Andres, M. Head-Gordon, E.S. Replogle, J.A. Pople, GAUSSIAN 98 (Revision A.9), Gaussian, Inc., Pittsburgh PA, 1998.

[20] G.A. Chass, M.A. Sahai, J.M.S. Law, S. Lovas, Ö. Farkas, A. Perczel, J.-L. Rivail, I.G. Csizmadia, Int. J. Quantum Chem. 90 (2002) 933–968.

[21] S.E. Villagra, M.B. Santillan, A.M. Rodriguez, G.A. Chasse, M.L. Freile, S. Zacchino, P. Matyus, R.D. Enriz, J. Mol. Struct. (Theochem) 549 (2001) 217–228.

Journal of Molecular Structure (Theochem) 666–667 (2003) 587–598

www.elsevier.com/locate/theochem

Conformational and electronic study of homoallylamines with inhibitory properties against polymers of fungal cell wall

Susana E. Villagra[a], María C. Bernini[a], Ana M. Rodríguez[a], Susana A. Zacchino[b], Vladimir V. Kouznetsov[c], Ricardo D. Enriz[a],*

[a]*Department of Chemistry, National University of San Luis, Chacabuco 915, 5700 San Luis, Argentina*
[b]*Farmacognosia, National University of Rosario, Suipacha 531, 2000 Rosario, Argentina*
[c]*Laboratory of Fine Organic Chemistry, Industrial University of Santander, A.A. 678, Bucaramanga, Colombia*

Abstract

A complete conformational analysis of 4-methyl-2-(3′-pyridyl)quinoline (**I**), 4-methyl-2-(3′-pyridyl)-1,2,3,4-tetrahydroquinoline (**II**) and structurally related compounds with known antifungal properties was carried out using ab initio and DFT calculations. Quasi-planar forms were found as the low-energy conformations for compound **I**. Sixteen conformations, considering the enantiomeric forms, were obtained for compound **II**. The preferred forms are 2R4S and 2R4R displaying the ring B in a half-chair arrangement. Intermediate structures in the conversion of compound **II** into compound **I** were also evaluated.
© 2003 Elsevier B.V. All rights reserved.

Keywords: Ab initio and DFT calculations; Conformation; Tetrahydroquinoline; Antifungal agents

1. Introduction

In the course of our ongoing screening program for new and selective antifungal compounds [1–5], we have previously reported that a series of 4-aryl- or 4-alkyl-*N*-arylamino-1-butenes (Scheme 1) and related tetrahydroquinolines and quinolines [6] display a range of antifungal properties against dermatophytes, fungi that produce most of the dermatomycoses in humans. Regarding their mode of action, active compounds showed in vitro inhibitory activities

against (1,3)-β-D-glucan-synthase and mainly against chitin-synthase, enzymes that catalyze the synthesis of the major fungal cell wall polymers [7].

The interesting antifungal properties displayed by these homoallylamines, tetrahydroquinolines and quinolines prompted us to synthesize new members of this series. One of these new members is 4-*N*-arylamino-1-butene containing the pyridyl moiety at C-4 in order to find new antifungal structures with inhibitory properties of the fungal cell wall synthesis. Thus recently, we reported the synthesis, fungistatic effects, structure–activity relationship and studies of action mechanism of a new series of homoallylamines and related compounds acting against dermatophytes [8]. Among the compounds tested, tetrahydroquinoline (**II**) (Scheme 1) and quinoline derivative (**I**) with a 4-β-pyridyl as ring C, displayed significant

* Corresponding author. Fax: +54-2652-431301.
E-mail addresses: denriz@unsl.edu.ar (R.D. Enriz), villagr@unsl.edu.ar (S.E. Villagra), amrodri@unsl.edu.ar (A.M. Rodríguez), szaabgil@citynet.net.ar (S.A. Zacchino), kouznet@uis.edu.co (V.V. Kouznetsov).

0166-1280/$ - see front matter © 2003 Elsevier B.V. All rights reserved.
doi:10.1016/j.theochem.2003.08.082

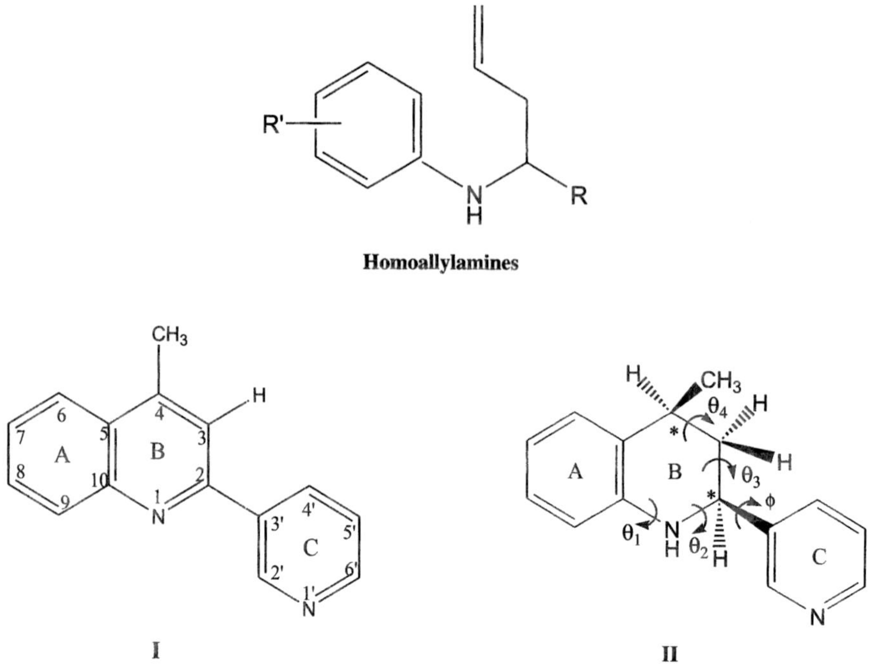

Homoallylamines

I

II

Scheme 1.

antifungal activities as well as interesting aspects from a structural point of view [8].

Since tetrahydroquinoline derivatives have two asymmetric carbons (denoted with asterisks in Scheme 1), they could contain a couple of diastereomers. The GC-MS study of the crude of reaction showed that these tetrahydroquinolines are formed as a diastereoisomeric mixture which differs in the arrangement of the methyl group at C-4 and of pyridyl ring at C-2. We reported that the ratio varies considerably and depends on the chemical nature of *N*-arylamino fragment in the aminobutenes.

Based on previous reports [9,10] and on the NMR data as well as COSY and NOESY techniques, we reported that the major isomers have the *cis* form (*2e,4e* orientation of groups), while minor isomers possess a *2e,4a* disposition [8].

We report here a careful theoretical study of thetrahydroquinoline, quinoline and structurally related compounds in all their possible configurational isomers and respective conformations, at ab initio and DFT levels of theory. The aim of this study is to obtain further and more detailed information on geometrical parameters, molecular symmetries, and on the rotation motion of the pyridyl ring. The results obtained here

are compared with those previously obtained using experimental techniques. In addition, the effects of conjugation and steric interactions on geometry are also discussed.

2. Calculations

The ab initio and DFT calculations, full geometry optimization and calculation of the harmonic vibrational frequencies, were performed using the GAUSSIAN 98 program [11]. Density Functional Theory (DFT) calculations were employed in order to properly account for the electron correlation effects (particularly important in this kind of systems). The widely employed hybrid method denoted by B3LYP [12–14] was used, along with the double-zeta split valence basis set 6-31G(d). This method includes a mixture of HF and DFT exchange terms and the gradient corrected correlation functional of Lee, Yang and Parr [15,16], as proposed and parametrized by Becke [17,18].

Rotational energy profiles around the ϕ torsion angle (Scheme 1) have been determined by using ab initio (3-21G and 6-31G(d) basis set) and DFT

calculations. The energy has been calculated at 30°
intervals of the dihedral angle between the planes of
A–B rings and pyridyl ring. Rotational energy
profiles were calculated using frozen parameters
(bond angles and bond lengths) previously optimized.
All the dihedral angles were optimized during the
scan.

3. Results and discussion

3.1. 4-Methyl-2-(3'-pyridyl)quinoline (I)

Considering that the three rings (A, B and C) in
compound **I** are planar (Scheme 1), this molecule
looks like a simple conformational problem with only
one torsional angle. The overall shape of the molecule
is determined by the orientation of the pyridyl ring
represented by the angle ϕ in Scheme 1.

Looking at the ab initio and DFT data in Fig. 1 and
Table 1, we obtained the barriers of about 4.5 (DFT),
3.9 (RHF/6-31G(d)), and 5.2 kcal mol^{-1} (RHF/3-
21G) for compound **I**, respectively. To understand

the significance of the rotation barrier, it is important
to look not just at the magnitudes of the energy
barriers, but at the complete energy-versus-rotation
angle behaviour.

In Fig. 1 relative energies at different level of
calculations versus torsional angle ϕ for compound **I**
are plotted. For low-level of theory calculation (RHF/
3-21G), energy minima occur at $\phi = 0$ and 180° i.e.
planar conformations, while energy maxima occur at
90 and 270°, i.e. perpendicular conformations. This
profile can be understood by recognizing that the
planar conformations are stabilized by resonance
structures. The resonance stabilization is eliminated
when the $C_{phenyl}-C_{pyridyl}$ bond rotates 90° and there is
an increase of energy. This energy increment due to a
loss of resonance energy is the responsible for the
approximately 5.2 kcal mol^{-1} rotational barrier predicted by RHF/3-21G calculations.

RHF/6-31G(d) and DFT calculations predict the
existence of four conformers instead of the two planar
conformations obtained by RHF/3-21G calculations.
The deviation of planarity is moderate but significant
(approximately $\pm 30°$ at RHF/6-31G(d) and $\pm 20°$ at

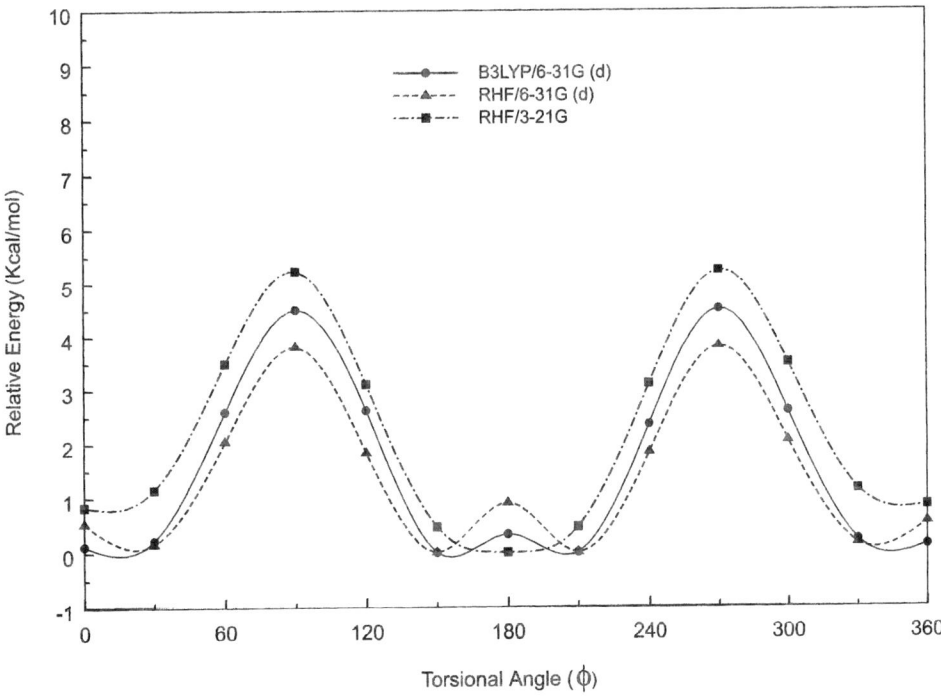

Fig. 1. Comparison of PECs of torsional angle ϕ obtained for compound **I**. The curves were calculated at RHF/3-21G, RHF/6-31G(d),
B3LYP/6-31G(d) levels of theory.

Table 1

Optimized torsional angles, total energy values and relative energies computed at RHF/3-21G, RHF/6-31G(d) and B3LYP/6-31G(d) levels of theory for compound **I**

Torsional angles					Total energy (hartree)	Relative energy (kcal mol^{-1})
θ_1 (9-10-1-2)	θ_2 (10-1-2-3)	θ_3 (1-2-3-4)	θ_4 (2-3-4-5)	ϕ (1-2-3'-2')		
RHF/3-21G						
180.0	0.0	0.0	0.0	0.0	−680.101765	0.85
180.0	0.0	0.0	0.0	180.0	−680.103118	0.00
RHF/6-31G(d)						
−179.8	−0.7	0.9	−0.2	28.2	−683.9341522	0.63
179.7	0.9	−1.0	0.2	153.7	−683.9351501	0.00
−179.7	−0.8	1.0	−0.2	−153.7	−683.9351501	0.00
179.8	0.7	−0.9	0.2	−28.2	−683.9341522	0.63
B3LYP/6-31G(d)						
180.0	−0.7	0.9	−0.2	20.9	−688.3455914	0.70
179.8	0.8	−1.0	0.2	161.5	−688.3467121	0.00
−179.8	−0.8	1.0	−0.2	−161.5	−688.3467121	0.00
180.0	0.7	−0.9	0.2	−20.9	−688.3455914	0.70

DFT level) (Table 1). As we expected, these conformers are isoenergetic. It should be noted that these more accurate calculations predict the planar forms (i.e. $\phi = 0$ and 180°) as transition state structures between their respective minima. However, the negative imaginary frequency for the transition state at $\phi = 0°$ is −42.4158 cm^{-1} indicating a very shallow zone. The existence of four different conformations for compound **I** from RHF/6-31G(d) and DFT calculations is a striking difference with respect to less accurate RHF/3-21G calculations. It is clear, however, that quasi-planar conformations are the preferred forms for compound **I**.

3.2. 4-Methyl-2-(3'-pyridyl)-1,2,3,4-tetrahydroquinoline (II)

Since this compound has two asymmetric carbons (denoted with asterisks in Scheme 1), their conformational intricacies are a little more complex. A complete geometry optimization was carried out for compound **II** in order to obtain the geometries and relative energies of distinct possible configurational isomers as well as their conformations. The effect of several geometrical parameters on the overall stability of the molecule was investigated. From a theoretical point of view, four configurational isomers

2R4R (*cis*-2e-Py/4e-Me), 2S4R (*trans*-2e-Py/4a-Me), 2R4S (*trans*-2a-Py/4e-Me), and 2S4S (*cis*-2a-Py/4a-Me) can be drawn for this molecule (Fig. 2). Thus, one of the structural problem of compound **II** is associated with the saturated ring (ring B) fused to the benzene ring (ring A). Ring B cannot be considered as a cyclohexane or its heterocyclic analogue because it has, at least formally, one carbon–carbon double bond. Our theoretical results indicate that a so-called 'half-chair' conformation is the low-energy form (Fig. 3) which is in agreement with experimental data [8]. Thus, ring B is expected to exist in two interconvertible enantiomeric forms with a transition state for the ring-flip showing some symmetry. We found a transition state (TS) structure connecting the 2R4R minima (up and down) at approximately 4.90 kcal mol^{-1} using RHF/6-31G(d) level of theory. 2R4R-TS structure displays a 'distorted boat' form (Fig. 3). This result is in agreement with those previously reported for other structurally related compounds [19,20]. RHF/3-21G, RHF/6-31G(d) and B3LYP/6-31G(d) calculations predict the existence of four different conformations for each of the four isomers (2R4R, 2S4R, 2R4S and 2S4S). Thus, a total of sixteen different forms were found for compound **II**. However, it should be noted that the eight conformations denoted with an asterisk in Table 2 are

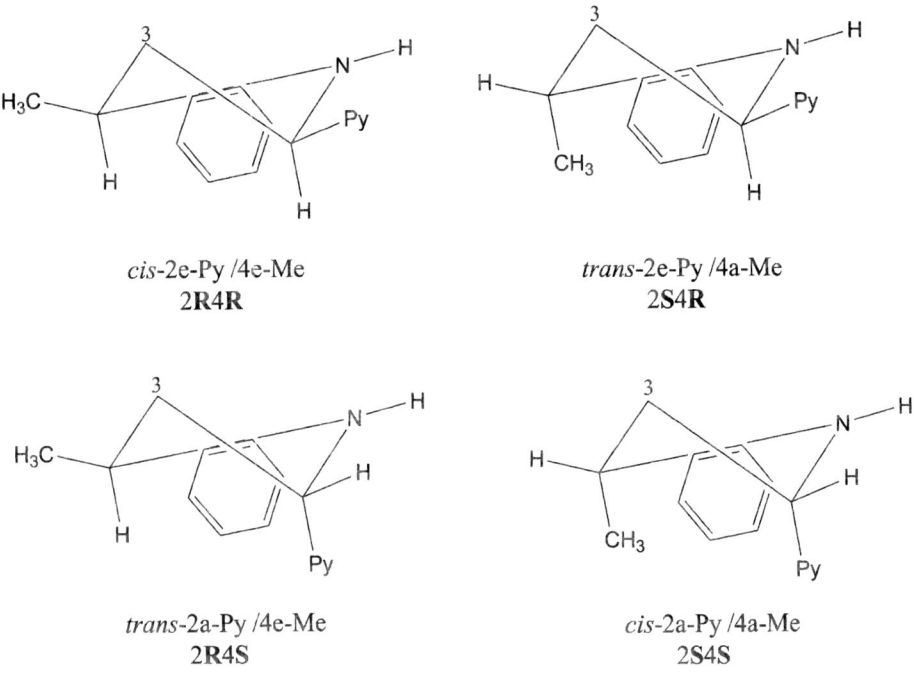

Fig. 2. Schematic spatial representation for the four configurational isomers of compound **II**.

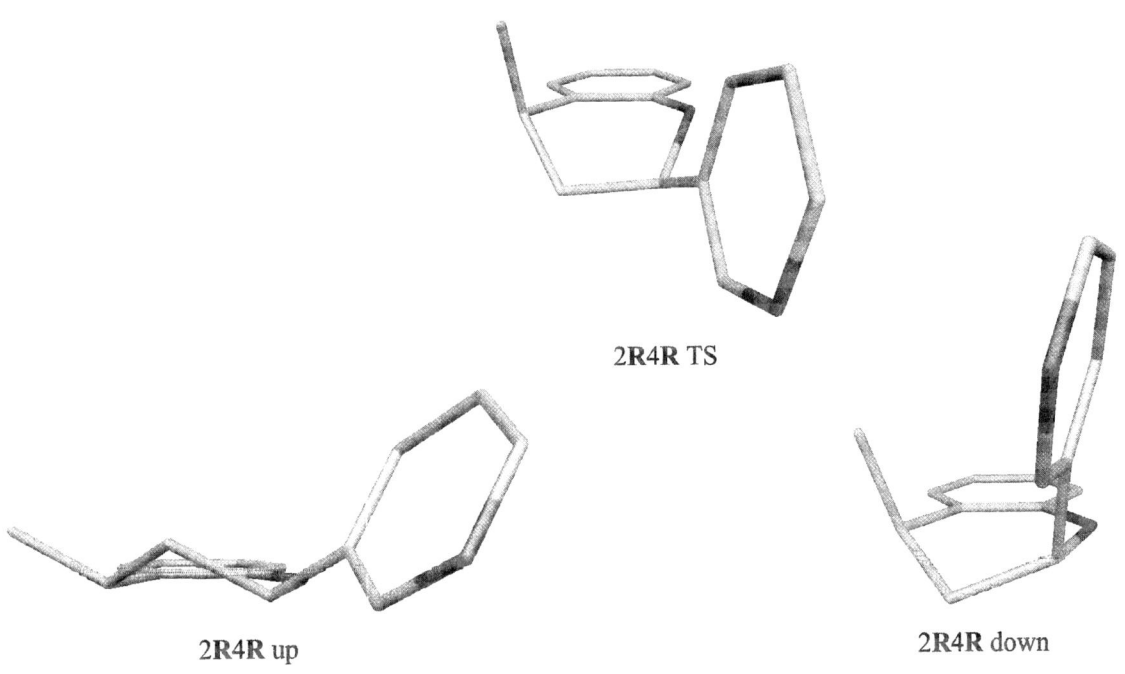

Fig. 3. Spatial view of the low-energy conformations (2*R*4*R*-up and -down) and the transition structure of compound **II** obtained from RHF/6-31G(d) calculations.

Table 2

Optimized torsional angles, total energy values and relative energies computed at RHF/3-21G, RHF/6-31G(d) and B3LYP/6-31G(d) levels of theory for compound **II**

Stereoisomer		Torsional angles					Total energy (hartree)	Relative energy (kcal mol^{-1})
		θ_1 (9-10-1-2)	θ_2 (10-1-2-3)	θ_3 (1-2-3-4)	θ_4 (2-3-4-5)	ϕ (1-2-3'-2')		
RHF/3-21G								
2R4R	Up	− 175.2	− 32.7	57.1	− 54.5	38.3	− 682.4237152	1.94
		− 175.2	− 32.6	57.0	− 54.6	− 142.2	− 682.4245746	1.40
	Down	− 175.7	18.6	− 46.0	50.8	7.7	− 682.4208869	3.71
		− 176.8	20.0	− 46.4	49.9	− 169.1	− 682.4215783	3.28
2R4S	Up	172.1	− 17.6	51.0	− 58.9	160.3	− 682.4244085	1.51
		170.7	− 15.9	50.3	− 59.6	− 18.8	− 682.4234967	2.08
	Down	176.5	30.0	− 54.2	53.5	142.3	− 682.4268107	0.00
		176.9	29.8	− 54.2	53.4	− 38.4	− 682.4259591	0.53
2S4R	Up	− 176.9	− 29.7	54.2	− 53.5	38.4	− 682.4259595	0.53*
		− 176.7	− 29.8	54.2	− 53.6	− 142.7	− 682.4268096	0.00*
	Down	− 170.7	15.9	− 50.3	59.6	18.7	− 682.4234970	2.08*
		− 171.9	17.3	− 50.9	59.0	− 160.2	− 682.4244087	1.51*
2S4S	Up	176.8	− 20.1	46.5	− 49.9	169.1	− 682.4215782	3.28*
		175.8	− 18.7	46.1	− 50.8	− 7.7	− 682.4208871	3.72*
	Down	175.1	32.7	− 57.1	54.5	142.1	− 682.4244966	1.45*
		175.2	32.7	− 57.1	54.8	− 38.4	− 682.4237138	1.94*
RHF/6-31G(d)								
2R4R	Up	− 159.6	− 50.5	61.6	− 43.5	43.2	− 686.2468076	0.84
		− 160.1	− 50.2	61.6	− 43.8	− 136.9	− 686.2477162	0.27
	Down	− 149.8	− 15.5	− 28.5	53.9	14.4	− 686.2396683	5.32
		− 149.4	− 16.2	− 27.9	53.6	− 162.7	− 686.2404449	4.83
2R4S	Up	174.2	− 19.1	50.2	− 55.8	156.0	− 686.2430734	3.18
		− 168.7	− 40.0	57.5	− 49.5	− 22.1	− 686.2429609	3.25
	Down	163.1	46.7	− 60.3	45.9	137.4	− 686.2481389	0.00
		163.0	46.8	− 60.3	45.8	− 42.9	− 686.2472252	0.57
2S4R	Up	− 163.1	− 46.8	60.3	− 45.8	42.8	− 686.2472251	0.57*
		− 163.1	− 46.7	60.3	− 45.9	− 137.2	− 686.2481386	0.00*
	Down	168.6	39.1	57.5	49.4	22.0	− 686.2429613	3.25*
		− 174.0	18.8	− 50.2	55.9	− 156.0	− 686.2430739	3.18*
2S4S	Up	149.4	16.2	27.9	− 53.6	162.7	− 686.2404449	4.83*
		149.9	15.4	28.6	− 53.9	− 14.4	− 686.2396683	5.32*
	Down	159.9	50.2	− 61.6	43.9	137.0	− 686.2477163	0.27*
		159.6	50.5	− 61.6	43.5	− 43.2	− 686.2468079	0.84*
B3LYP/6-31G(d)								
2R4R	Up	− 161.5	− 48.1	60.7	− 45.2	40.9	− 690.7410472	0.83
		− 161.6	− 47.8	60.6	− 45.5	− 138.7	− 690.7418561	0.32
	Down	− 162.1	1.4	− 38.7	54.0	9.9	− 690.7354449	4.34
		− 162.8	2.2	− 39.0	53.6	− 166.9	− 690.7361386	3.91
2R4S	Up	164.9	− 7.8	45.8	− 59.0	156.1	− 690.7387867	2.25
		− 171.5	− 35.8	56.5	− 51.6	− 23.7	− 690.7382181	2.60
	Down	164.1	45.4	− 59.6	46.9	138.6	− 690.7423677	0.00
		164.1	45.6	− 59.8	46.8	− 39.8	− 690.7415819	0.49
2S4R	Up	− 164.5	− 45.0	59.5	− 47.2	40.4	− 690.7414508	0.57*
		− 164.0	− 45.5	59.7	− 46.9	− 138.5	− 690.7423677	0.00*
	Down	171.5	35.8	− 56.5	51.6	23.8	− 690.7382181	2.60*
		− 164.5	7.4	− 45.6	59.1	− 156.1	− 690.7387853	2.25*

Table 2 (continued)

Stereoisomer		Torsional angles					Total energy (hartree)	Relative energy (kcal mol^{-1})
		θ_1 (9-10-1-2)	θ_2 (10-1-2-3)	θ_3 (1-2-3-4)	θ_4 (2-3-4-5)	ϕ (1-2-3'-2')		
2S4S	Up	162.8	−2.2	39.0	−53.6	166.9	−690.7361386	3.91[*]
		162.1	−1.4	38.7	−54.0	−9.9	−690.7354449	4.34[*]
	Down	161.6	47.9	−60.6	45.5	138.7	−690.7418561	0.32[*]
		161.5	48.0	−60.6	45.3	−40.9	−690.7410466	0.83[*]

[*]Enantiomeric forms.

enantiomeric forms. Fig. 4 illustrates this situation very well. The preferred conformations are not planar because pyridyl ring rotates reaching 138.6° for 2R4S form and −138.7° for 2R4R form (B3LYP/6-31G(d) calculations). Ring B is also rotated up or down adopting a half-chair conformation (see Table 2, torsional angles θ_2, θ_3 and θ_4). Therefore, the conformational energy of compound **II** is highly dependent on the geometry of the pyridyl ring. At DFT level, conformations 2R4R (down) and 2R4S (up) possessing the pyridyl ring in axial position displayed 3.91 and 2.25 kcal mol^{-1} over the global minimum, respectively. Thus, spatial orientation of pyridyl ring has a significant influence on the conformation of compound **II**. The preferred conformations for this compound are 2R4S-down (global minimum) and 2R4R-up (with an energy gap of 0.32 kcal mol^{-1} at DFT level). These results are in agreement with our previous experimental data [8].

Fig. 5 gives the energy versus rotation profiles obtained for compound **II** at different levels of theory. Torsional angles for energy minima are at 40.9 and −138.7° (at DFT level), while the maxima are at about 150 and 330°. This is a different behaviour compared with compound **I** reflecting the importance of steric hindrance produced by the different spatial ordering of ring B. Thus, compounds **I** and **II** represent two different situations. In **I**, steric hindrance is of little importance and the molecule becomes planar so as to maximize resonance delocalization. In **II**, steric hindrance is the dominant factor and the molecule twists to about ±139° out of the plane conformation to relieve the strain.

3.3. Compounds A and B

In order to investigate the steric hindrance and the effect of conjugation on the rotation of pyridyl moiety, calculations were extended to compounds **A** and **B**, which display only one asymmetric carbon (Fig. 6).

Fig. 7 and Table 3 show that compound **B** (4R-up) has two low energy conformations at 10.8° and −170.8 out of plane conformation and a rotation barrier of 5.2 kcal mol^{-1} at DFT level. These conformations can be understood as a compromise between the need to minimize steric hindrance and to preserve at the same time as much as possible of the resonance energy associated with a planar molecule. Similarly, the profile obtained for this barrier is also the result of the interplay between resonance stabilization and steric hindrance. It is clear that the conformational behaviour of compound **B** is closely related to that observed for compound **I** (Fig. 1) suggesting a nearly 'aromatic behaviour' for this compound. However, the barrier is higher than that obtained for compound **I** at DFT level. This result indicates that the conformational interconversion is more restricted for compound **B**.

On the other hand, the conformational behaviour obtained for compound A is more related to that observed for compound **II**. The energy minima occur at $\phi = 50.8$ and −129.2° at DFT level. In this case the energy barrier is about 6.6 kcal mol^{-1} which is the highest barrier obtained among the four molecules reported here (Fig. 7).

Comparing the absolute energies obtained for compounds **A** and **B** (note that they are isomers),

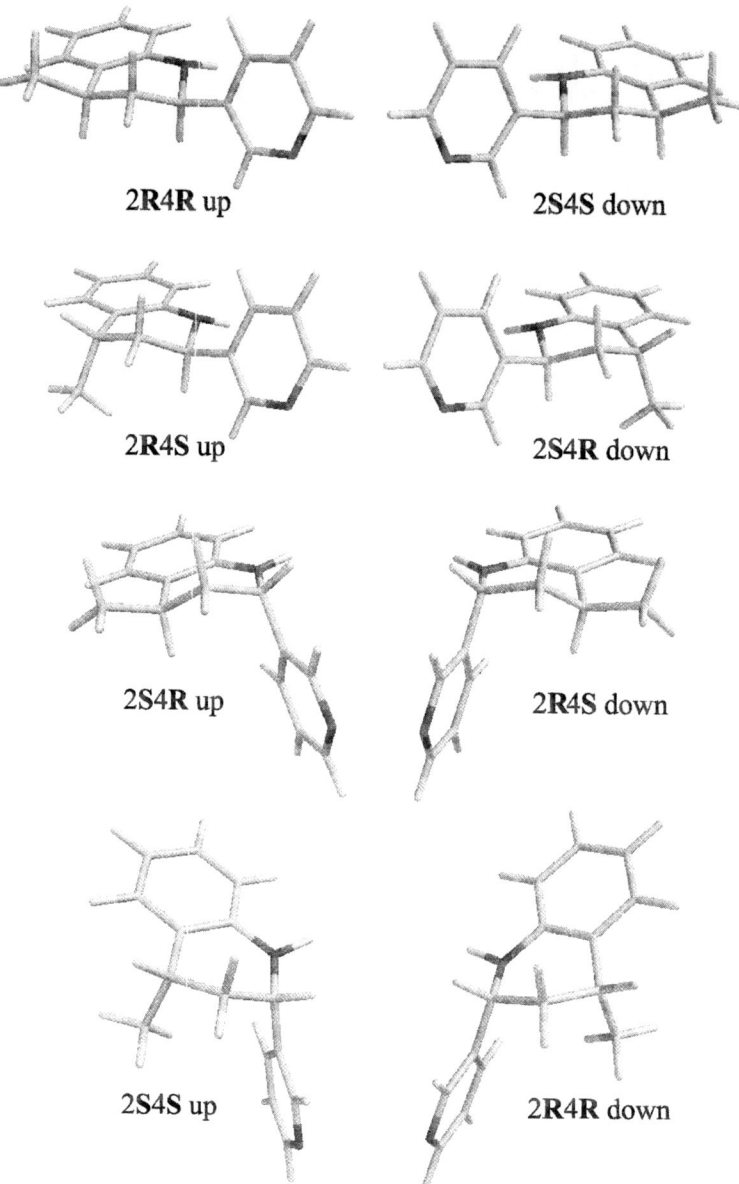

2R4R up

2S4S down

2R4S up

2S4R down

2S4R up

2R4S down

2S4S up

2R4R down

Fig. 4. Spatial view of eight enantiomeric forms of compound **II**.

compound **B** is energetically favoured by more than 9.0 kcal mol⁻¹ at DFT level. Thus from a thermodynamic point of view, molecule **B** seems a more probable intermediate between compounds **II** and **I** (Fig. 6). However, caution is needed in this observation because transition structures interconverting compounds **I**–**A**–**II** and **I**–**B**–**II** were not calculated in the present study. In addition, we have not included here a putative third intermediate structure, i.e. 1,4-dihydro-4-methyl-2-pyridylquinoline which must also be considered. Further theoretical and experimental works are currently being explored in this topic.

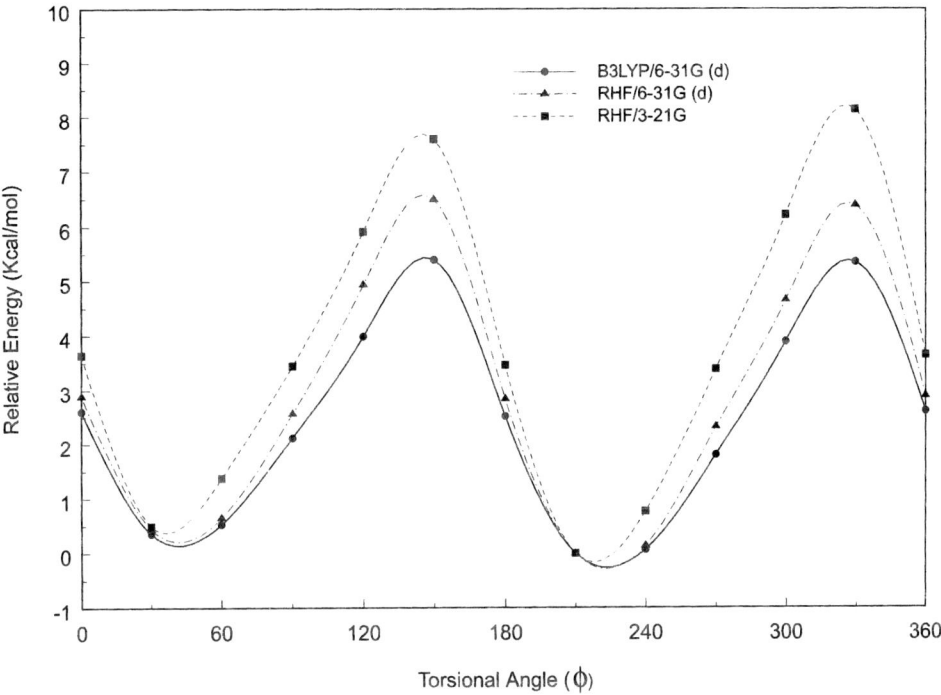

Fig. 5. Comparison of PECs of torsional angle ϕ obtained for compound **II** in the 2R4R-up conformation. The curves were calculated at RHF/3-21G, RHF/6-31G(d), B3LYP/6-31G(d) levels of theory.

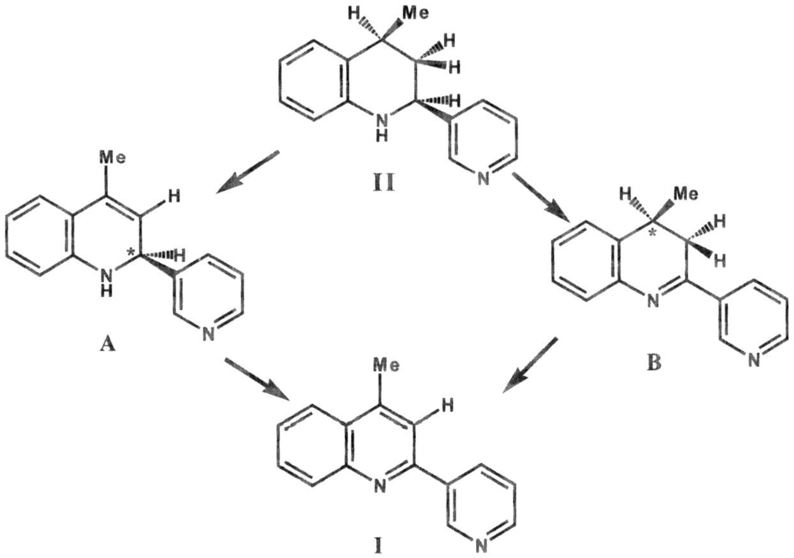

Fig. 6. Schematic representation of compounds **I** and **II** including two putative intermediate structures (compounds **A** and **B**).

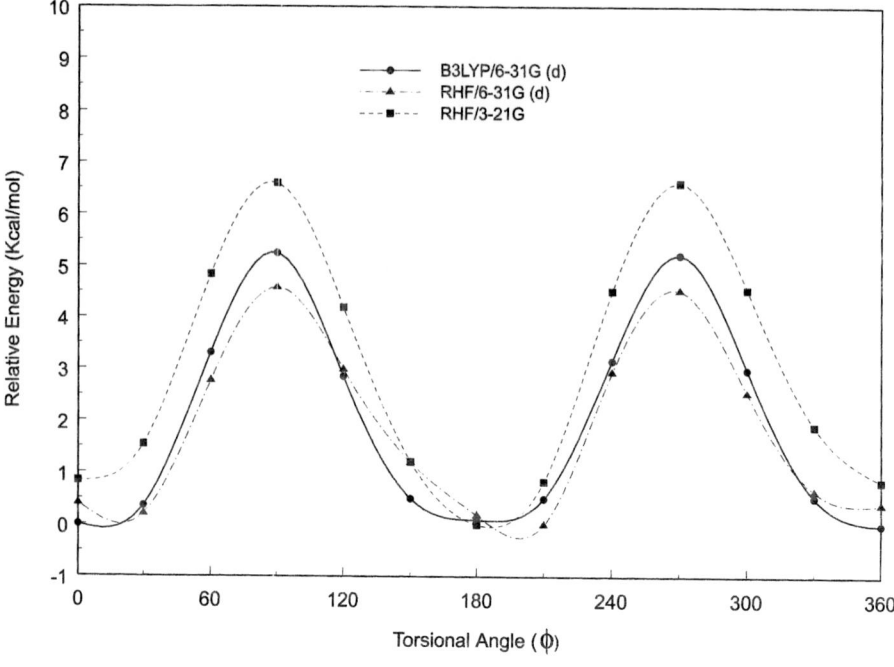

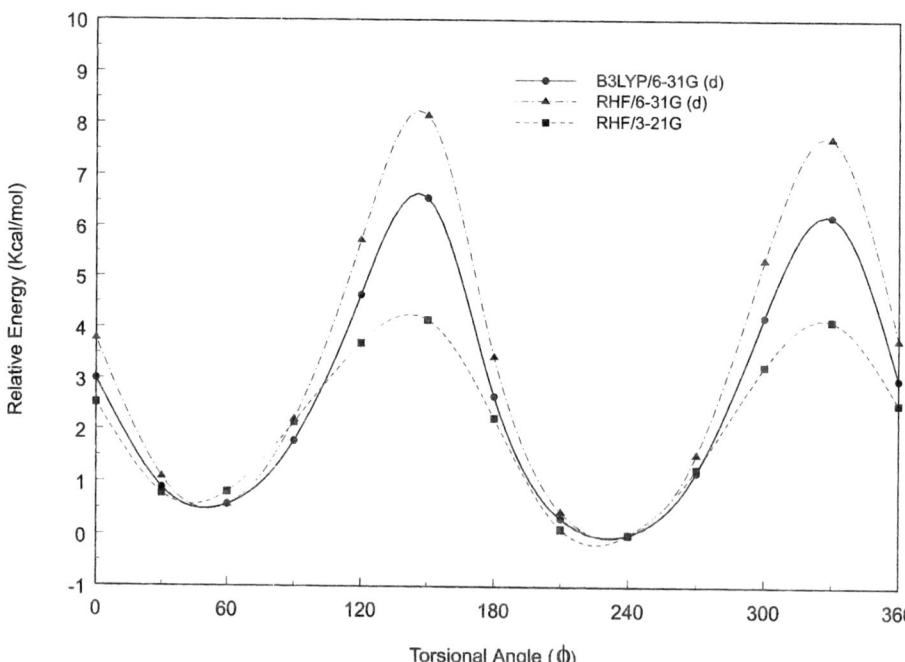

Fig. 7. Potential energy curves of compound **A** from RHF/3-21G, RHF/6-31G(d), B3LYP/6-31G(d) relaxed single scan computations (bottom). The same potential energy curves obtained for compound **B** (top).

Table 3

Optimized torsional angles, total energy values and relative energies computed at RHF/3-21G, RHF/6-31G(d) and B3LYP/6-31G(d) levels of theory for compounds **A** and **B**

Compound	Stereoisomer		Torsional angles					Total energy (hartree)	Relative energy (kcal mol^{-1})
			θ_1 (9-10-1-2)	θ_2 (10-1-2-3)	θ_3 (1-2-3-4)	θ_4 (2-3-4-5)	ϕ (1-2-3'-2')		
RHF/3-21G									
A	2R		−175.2	−7.1	4.3	0.2	44.4	−681.243597	0.73
			−176.1	−5.7	3.2	0.4	−134.0	−681.244772	0.00
	2S		176.2	5.5	−3.1	−0.4	134.1	−681.244772	0.00
			175.4	6.9	−4.2	−0.2	−44.4	−681.243597	0.73
B	4R	Up	−164.8	0.8	−32.2	45.7	8.8	−681.252478	0.82
			−164.7	0.8	−32.2	45.8	−172.2	−681.253789	0.00
		Down	163.1	0.0	33.6	−48.3	171.7	−681.250812	1.86
			163.2	0.0	33.5	−48.3	−9.3	−681.249478	2.70
	4S	Up	164.1	−0.8	32.2	−45.8	172.2	−681.253789	0.00
			164.8	−0.8	32.2	−45.7	−8.7	−681.252478	0.82
		Down	−163.1	0.0	−33.5	48.4	9.4	−681.249478	2.70
			−163.2	0.0	−33.6	48.3	−171.7	−681.250812	1.86
RHF/6-31G(d)									
A	2R		−152.5	−40.8	26.3	−1.6	51.4	−685.0669109	0.69
			−152.6	−40.4	26.0	−1.4	−128.6	−685.0680144	0.00
	2S		152.6	40.4	−25.9	1.4	128.5	−685.0680144	0.00
			152.6	40.6	−26.2	1.5	−51.7	−685.0669113	0.69
B	4R	Up	−164.1	0.6	−32.8	45.6	18.3	−685.0834807	0.71
			−164.0	0.4	−32.6	45.5	−163.6	−685.0846193	0.00
		Down	162.7	0.6	33.2	−47.5	163.4	−685.0835233	0.69
			162.8	0.4	33.4	−47.6	−18.4	−685.0823645	1.41
	4S	Up	164.0	−0.4	32.6	−45.5	163.6	−685.0846193	0.00
			164.1	−0.6	32.8	−45.6	−18.3	−685.0834807	0.71
		Down	−162.8	−0.4	−33.4	47.6	18.3	−685.0834807	1.41
			−162.7	−0.6	−33.1	47.5	−163.4	−685.0835233	0.69
B3LYP/6-31G(d)									
A	R		−156.0	−36.0	22.9	−0.2	−129.2	−689.5210929	0.00
			−156.0	−36.0	22.0	−0.1	50.8	−689.5199894	0.69
B	R	Up	−165.0	1.3	−33.5	46.2	10.8	−689.5348577	0.74
			−165.0	1.3	−33.5	46.2	−170.8	−689.5360372	0.00
		Down	163.3	−0.5	34.6	−48.4	−11.3	−689.5341444	1.19
			163.2	−0.4	34.5	−48.4	170.6	−689.5353673	0.42

4. Conclusions

A conformational study on 4-methyl-2-(3'-pyridyl)quinoline (**I**), 4-methyl-2-(3'-pyridyl)-1,2,3,4-tetrahydroquinoline (**II**) and two structurally related compounds was performed by using DFT and ab initio calculations.

Quasi-planar forms were found as the low-energy conformations for compound **I**.

The conformational analysis carried out for compound **II** in the present work, yielded eight distinct forms (without considering enantiomeric forms), varying in the orientation of both substituents, CH_3 group and pyridyl ring, as well as in the geometry of ring B. The calculated relative energies indicate a clear preference for 2R4S-up and 2R4R-up forms which is in agreement with our previously reported experimental data.

Our theoretical exploratory study suggests the compound **B** as an energetically feasible intermediate structure between **II** and **I**.

This kind of conformational analysis based on theoretical methods may be of significance in future studies aiming at the elucidation of the structure–activity relationships associated with the biological role of these compounds displaying potential anti-fungal activity. These results may supply relevant information at molecular level in order to obtain a better understanding of the mechanism underlying the inhibitory effect of glucan and chitin synthase enzymes displayed by these compounds.

Acknowledgements

This work was supported by grants from Fundación Antorchas, Universidad Nacional de San Luis (UNSL-Argentina). S.A.Z. wish to thank the grant from Agencia de Promociones Científicas y Tecnológicas de la Argentina (PICT nro. 06-06454). This work is part of the Iberoamerican Project X7-PIBEAFUN (Search and Development of New Antifungal Agents) from Red X.A.-RIPRONAMED (Iberoamerican Network on Medicinal Natural Products) of the Iberoamerican Program for the Development of Science and Technology (CYTED). R.D. Enriz is member of Consejo Nacional de Investigaciones Científicas y Técnicas de la República Argentina (CONICET).

References

[1] A.M. Rodríguez, F.A. Giannini, F.D. Suvire, H.A. Baldoni, R. Furlán, S.A. Zacchino, P. Mátyus, R.D. Enriz, I.G. Csizmadia, J. Mol. Struct. (Theochem) 504 (2000) 35.

[2] S.N. López, M.V. Castelli, S.A. Zacchino, J.N. Domínguez, G. Lobo, J. Charris-Charris, J.C. G-Cortés, J.C. Ribas, C. Devia, A.M. Rodríguez, R.D. Enriz, Bioorg. Med. Chem. 9 (2001) 1999.

[3] S.A. Zacchino, G.E. Rodríguez, G. Pezzenati, C.B. Santecchia, F.A. Giannini, R.D. Enriz, J. Ethnopharmacol. 62 (1998) 35.

[4] M.A. Zamora, M.F. Masman, J.A. Bombasaro, M.L. Freile, S. Lopez, V. Cechinel Filho, S. Zacchino, R.D. Enriz, Int. J. Quantum Chem. 93 (1) (2003) 32–46.

[5] L. Karolyhazy, M.L. Freile, M. Anwair, Gy. Beke, F. Giannini, M. Sortino, J.C. Ribas, S. Zacchino, P. Matyus, R.D. Enriz, Arzneimittel Forschung-Drug Res. (2003) In press.

[6] J.M. Urbina, J.C. G-Cortés, A. Palma, S.N. López, S.A. Zacchino, R.D. Enriz, J.C. Ribas, V.V. Kouznetzov, Bioorg. Med. Chem. 9 (2000) 691.

[7] C.P. Selitranvikoff, Antifungal Drugs: (1,3)-β-Glucan-synthase Inhibitors, Springer, Heidelberg, 1995, p. 1.

[8] L.Y. Vargas, M.V. Castelli, V.V. Kouznetsov, J.M. Urbina, S.N. López, M. Sortino, R.D. Enriz, J.C. Ribas, S. Zacchino, Bioorg. Med. Chem. 11 (2003) 1531.

[9] (a) V.V. Kouznetsov, A.E. Aliev, N.S. Prostakov, Khim. Geterotsikl. Soed. 1 (1994) 73. (b) V.V. Kouznetsov, A.E. Aliev, N.S. Prostakov, Chem. Abstr. 121 (1994) 300.738.

[10] L.Y. Vargas Méndez, V.V. Kouznetsov, E. Stashenko, A. Bahsas, J. Amaro-Luis, Heterocycl. Commun. 7 (2002) 323.

[11] M.J. Frisch, G.W. Trucks, H.B. Schlegel, G.E. Scuseria, M.A. Robb, J.R. Cheeseman, V.G. Zakrzewski, J.A. Montgomery Jr., R.E. Stratmann, J.C. Burant, S. Dapprich, J.M. Millam, A.D. Daniels, K.N. Kudin, M.C. Strain, O. Farkas, J. Tomasi, V. Barone, M. Cossi, R. Cammi, B. Mennucci, C. Pomelli, C. Adamo, S. Clifford, J. Ochterski, G.A. Petersson, P.Y. Ayala, Q. Cui, K. Morokuma, D.K. Malick, A.D. Rabuck, K. Raghavachari, J.B. Foresman, J. Cioslowski, J.V. Ortiz, A.G. Baboul, B.B. Stefanov, G. Liu, A. Liashenko, P. Piskorz, I. Komaromi, R. Gomperts, R.L. Martin, D.J. Fox, T. Keith, M.A. Al-Laham, C.Y. Peng, A. Nanayakkara, C. Gonzalez, M. Challacombe, P.M.W. Gill, B. Johnson, W. Chen, M.W. Wong, J.L. Andres, C. Gonzalez, M. Head-Gordon, E.S. Replogle, J.A. Pople, GAUSSIAN 98, Revision A.7, Gaussian, Inc., Pittsburgh PA, 1998.

[12] T.V. Russo, R.L. Martin, P.J. Hay, J. Phys. Chem. 99 (1995) 17085.

[13] A. Ignaczak, J.A.N.F. Gomez, Chem. Phys. Lett. 257 (1996) 609.

[14] F.A. Cotton, X. Feng, J. Am. Chem. Soc. 119 (1997) 7514.

[15] C. Lee, W. Yang, R.G. Parr, Phys. Rev. 337 (1988) 785.

[16] B. Miehlich, A. Savin, H. Stoll, M. Preuss, Chem. Phys. Lett. 157 (1989) 200.

[17] A. Becke, Phys. Rev. A38 (1988) 3098.

[18] A. Becke, J. Chem. Phys. 98 (1993) 5648.

[19] D.H. Setiadi, G.A. Chass, L.L. Torday, A. Varro, J.G. Papp, J. Mol. Struct. (Theochem) 594 (2002) 161.

[20] F.D. Suvire, I. Andreu, A. Bermejo, M.A. Zamora, D. Cortés, R.D. Enriz, J. Mol. Struct. (Theochem) (2003) in press.

Journal of Molecular Structure (Theochem) 666–667 (2003) 599–608

www.elsevier.com/locate/theochem

An ab initio conformational study on captopril

Graciela N. Zamarbide[a,*], Mario R. Estrada[a], Miguel A. Zamora[a], Ladislaus L. Torday[b],
Ricardo D. Enriz[a], Francisco Tomás Vert[c], Imre G. Csizmadia[b,d,e]

[a]*Departamento de Química, Universidad Nacional de San Luis, San Luis 5700, Argentina*
[b]*Department of Medical Chemistry, University of Szeged, Dóm tér 8, Szeged 6720, Hungary*
[c]*Departament de Química-Física, Universitat de València, BURJASSOT, Valencia 46100, Spain*
[d]*Global Institute of Computational Molecular and Materials Science (GIOCOMMS) @ 1422 Edenrose,
St, Mississauga, Ont. L5V 1H3, Canada*
[e]*Department of Chemistry, University of Toronto, 80 St George St, Toronto, Ont. M5S3H6, Canada*

Abstract

Captopril can interact regio- and stereo-specifically with various functional groups present at the active site of angiotensin converting enzyme (ACE). Since no X-ray structure of ACE is available, Captopril, as an ACE inhibitor may be used as a 'molecular caliper', to estimate upper and lower bound values for separation d, where the mercaptidic terminal group of the molecule is linked to the enzyme Zn^{2+} cofactor, while the carboxylate links via an hydrogen bond to the guanidine moiety of an arginine side chain. As the results of this Ab Initio study, the conformations of the dianionic form of the full captopril molecule are reported here.
© 2003 Elsevier B.V. All rights reserved.

Keywords: Captopril; Antihypertensive drug; Role of Zn^{2+} cofactor; Molecular caliper; Organic sulphur

1. Introduction

1.1. Biomedical background

Captopril, 1-(D-3-mercapto-2-methylpropionyl)L-proline, an effective anti-hypertensive drug, is a relatively small molecule [1], which may be in a neutral [**1a**] or biologically active dianion form [**1b**].

[1a]

[1b]

It has several functional groups, and the molecule may assume numerous stable conformations. Consequently, captopril [**1**] can interact regio- and stereo-specifically with various functional groups present at the active site of angiotensin converting enzyme (ACE) [2], as illustrated schematically by Scheme 1.

Since no X-ray structure of ACE is available, one does not know the stereochemical separations of the various sites of the enzyme involved in the reaction mechanism. As an ACE inhibitor, captopril may be

---

* Corresponding author.

E-mail addresses: gzama@unsl.edu.ar (G.N. Zamarbide), mzamora@unsl.edu.ar (M.A. Zamora).

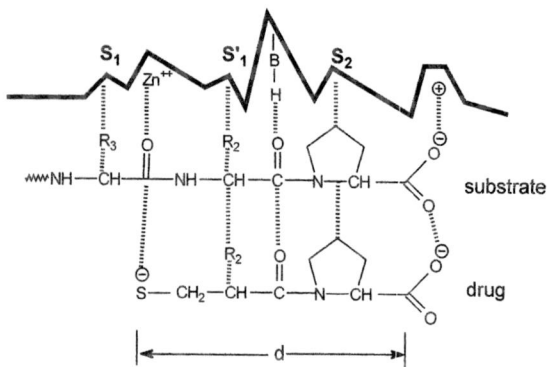

Scheme 1. Schematic illustration of the natural substrate and captopril docking to angitensin converting enzyme (ACE).

used as a 'molecular caliper' to estimate the upper and lower bound values for the internuclear separation (d) where the mercaptidic terminal group of the molecule is linked to the enzyme Zn^{2+} cofactor, while the carboxylate links via an hydrogen bond to the guanidine moiety of an arginine side chain.

Obviously, the internuclear separation between sulfur (S) and the carboxylate (C) cannot exceed a certain limit. In addition to regular conformational change, captopril can go through a *cis-trans* isomerization process [2]:

It is very important to study in considerable detail the conformations of both the *cis* and *trans* isomers to obtain an upper and lower value of d. Such detailed conformational study is necessary to make an estimate of the enzyme active site geometry.

1.2. Computational background

Captopril has been studied [3] by a 'classical potential energy' force field type method as early as 1985. Subsequently, Hillier et al. [4] carried out

AM1 computations on Captopril in 1991. This was followed by Luke [5] in 1994. Luke also reported [6] AM1 calculations on the demethylated analogue of Captopril in 1995. To the best of our knowledge no ab initio computation has been reported as yet on captopril.

Captopril poses a non-trivial multidimensional conformational problem. For this reason, it seemed prudent to study some fragments of captopril first [1], or preferably the fragments of its biologically active dianionic form [1b].

In the present paper, our examination of the *N*-acetyl prolinate fragment and the results of an exploratory study on the conformations of the dianionic form of the full captopril molecule [1] will be reported here.

1.3. Conformational analysis of captopril fragments

1.3.1. One-dimensional conformational analysis

Rotation about a single bond that joins two tetrahedral (sp^3) atoms, leads to clockwise *gauche* or *gauche*$^+(g^+)$, *anti(a)* and counter clockwise *gauche* or *gauche*$^-(g^-)$ conformers [3]. Note that the same anti conformation may be reached by both clockwise and counter clockwise rotations.

This conformational pattern is frequently exemplified by *n*-butane. Here rotations about the $HS–CH_2$ bond and about the $H_2C–CHMe$ bond of captopril are governed by such conformational characteristics.

Rotation about a single bond that joins a tetrahedral (sp^3) and a trigonal planar (sp^2) atoms is not so obvious. Minimum energy conformations may involve perpendicular arrangement or they may occur at an eclipsed/staggered arrangement.

[4]

perpendicular
90°

eclipsed / staggered
0° / 180°

This conformational pattern may be exemplified by ethyl benzene [7], where $1 = CH_3$ and the heavy line is the benzene ring. In that case, the perpendicular structure turned out to be the global minimum while eclipsed staggered turned out to be the local minimum. In the case of the propionate ion [8], where $1 = CH_3$ and the heavy line is $COO(-)$, the 90° is the high TS and 0° (as well as 180°) is almost the global minimum. As a slight variation, the 0 and 180° are low energy local maximal (0.015 kcal/mol) and the actual global minima (0.00 kcal/mol) is displaced by $\pm 15°$. Thus, the potential energy curve for Et–COO(−) is the mirror image of Et–Ph.

The situation is different in the case of propionic acid [8] since only eclipsed/perpendicular structures turned out to be critical point.

[5]

0°
TS

60°
local min

120°
TS

180°
global min

Propionic acidamide [9] is completely analogous to propionic acid.

[6]

0°
TS

60°
local min

120°
TS

180°
global min

The potential energy curves for the four molecular systems are shown in Figs. 7 and 8 of Ref. [9].

The rotation of the $-CONH_2$ functionality in N-formyl-L-prolinamide [7] can be compared to [6].

[7]

Since the $-CONH_2$ in [7] is attached to a stereo centre (the α-carbon of proline) the potential energy curves obtained for the *cis* and *trans* peptide bond (HCO–N <) in [7] is heavily distorted [10] as shown in Fig. 1.

1.3.2. Multi-dimensional conformational analysis

Beyond the proline residue, the other conformational variations of captopril are residing in the flexible open chain which may be represented, at least in a preliminary fashion, as 3-mercapto propanamide [8a] or its S-deprotonated conjugate base [8b].

[8a]

[8b]

In these structures, the five member ring is omitted with respect to captopril and hydrogen atoms are attached to the amide nitrogen.

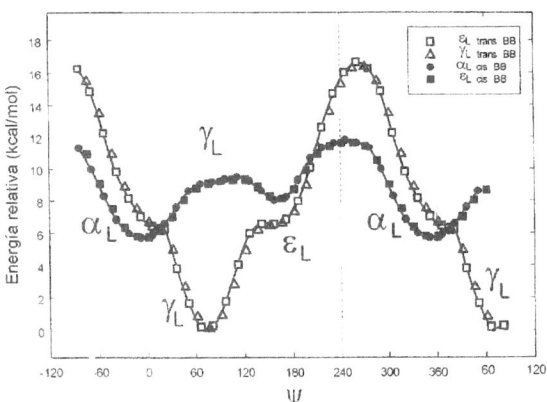

Fig. 1. Conformational potential energy curve, $E = E(\psi)$ for N-formyl-L-prolinamide containing either *trans*- and *cis*- peptide bond. The scan for the *trans*- isomer was started at both the ϵ_L and γ_L conformations while the scan for the cis isomer was started at both the α_L and the ϵ_L conformations.

Of course compounds [8a] and [8b] are missing the methyl group at the α-carbon (i.e. at C^2) which is present in captopril. If that methyl group is included then two enantiomers may be generated (S)-2-methyl-3-mercapto-propanamid [9a] and its conjugate base [9b] as well as (R)-2-methyl-3-mercapto-propanamid [10a] and its conjugate base [10b].

[9a]-S

[9b]-S

[10a]-R

[10b]-R

The S-deprotonated conjugate bases, i.e. [8b], [9b] and [10b], represent a simple conformational problem than their neutral counterparts, i.e. [8a], [9a] and [10a], because there are one fewer torsional angles in the conjugate bases than in the corresponding neutral compounds.

Clearly, in any study the conjugate bases (b-series) should be given the first attention before the neutral counterparts (a-series) are discussed. Such an approach is prudent because only two torsional (φ and ψ) are to be studied in the b-series while three torsional angles (θ, φ and ψ) need to be studied in the a-series. The acidamide moiety is always planar. Consequently, ω may be ignored, as a variable, in both series. The value of ω, however, may be recorded as an optimized value being in the vicinity of 180 or 0° depending on which hydrogen atom of the –NH₂ moiety is considered. The results reported earlier [11] are summarized in Fig. 2.

2. Scope

One of the conformational intricacies is associated with the five member ring on the proline moiety.

The –CO–N < moiety may be of cis or trans configurations. Also, the five-member ring may be in a number of non-planar conformation. Finally, the –COOH moiety of [1a] could assume three conformations (g^+, a, g^-) leading to γ_L, ϵ_L and α_L peptide folding. Consequently, it seemed prudent to study the fragments of captopril or preferably the fragments of its biologically active dianionic form [1b].

[1b]

[9b]

[11]

The conformational intricacies of compounds [8a], [9a] and [10a] as well as their conjugate bases [8b], [9b] and [10b] have been reported previously [11]. Clearly, the R and S isomeric forms represent crucial spatial requirement for docking of the two enantiomers. We also wish to report our finding on the N-acetyl prolinate fragment [11] and its stereochemical similarities and/or differences to the previously studied formylproliamide [7]. Also, as the results of this exploratory study, the conformations of the dianionic form of the full captopril molecule will be reported here.

3. Methods

3.1. Molecular computations

Using the GAUSSIAN 98 program system [12], ab initio computations have been carried out at the HF/3-21G level of theory, with full geometry optimization. HF/6-31 + G(d) and HF/6-311++G(d) calculations were carried out for the minimal energy structures obtained with the minor base set.

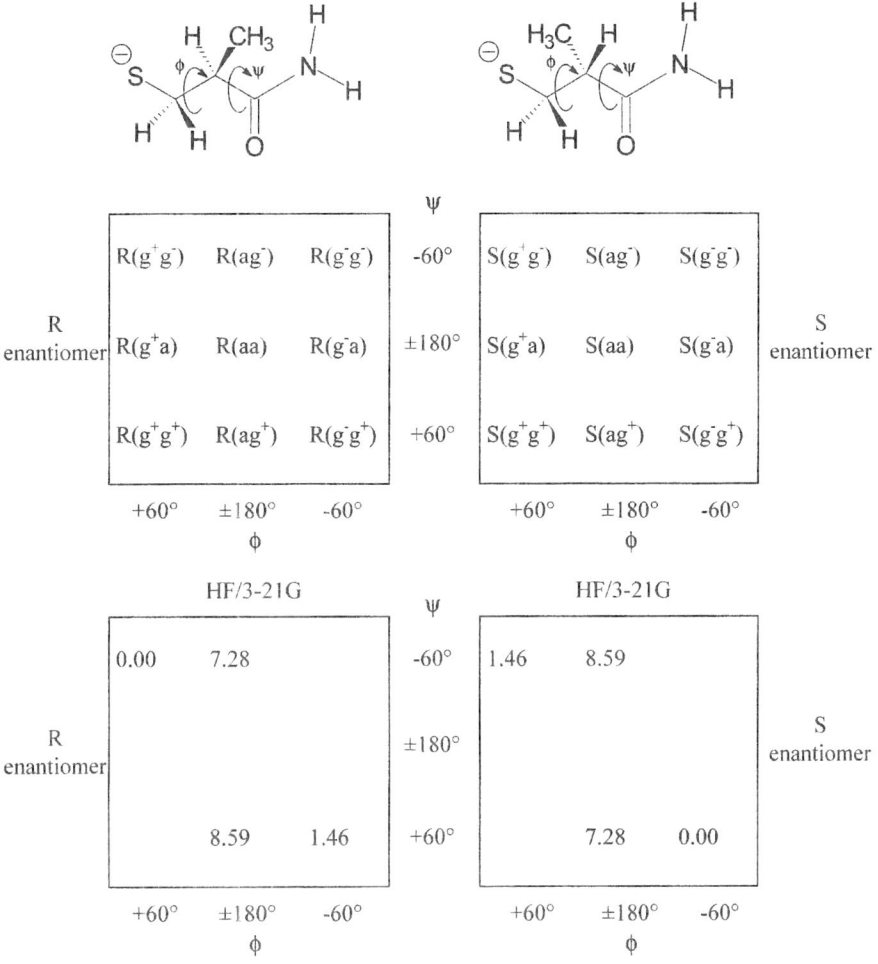

Fig. 2. Optimized (HF/3-21G) conformers (top) and computed relative energies (bottom) of *R*- and *S*-enantiomers of S-deprotonated 2-methyl-3-mercapto-propanamide.

Potential energy curves were plotted using the AXUM 5.0 software.

3.2. Internal co-ordinates

Considering all torsional angles, a complete conformational study on captopril essentially means that the geometric optimization of 324 structures for *R*-Captopril is required, and as many for *S*-Captopril. The predicted 324 structures are the result of the assumed topological periodicities of the seven torsional angles shown below [**1a′**]. Half of the structures (i.e. 162) are for the *trans* and the other half are for the *cis* isomers.

$3 \times 3 \times 3 \times 2 \times 1 \times 3 \times 2 = 324$

[**1a′**]

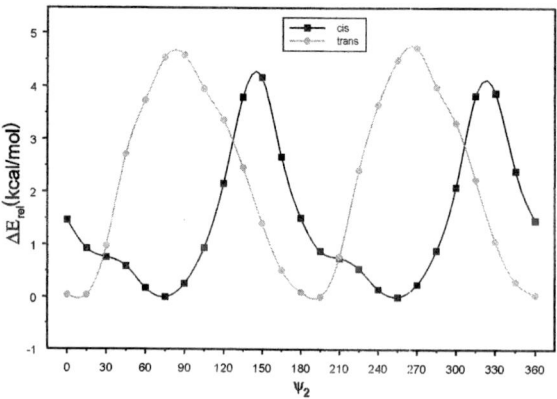

Fig. 3. Conformational potential energy curves (HF/3-21G), $E = E(\psi_2)$ for N-acetyl prolinate fragment [11] for molecule with *cis* and *trans* peptidic bond.

Similarly, for the double anions, the number of structures to be studied may be estimated from the topological periodicities of the five torsional angles involved. The number (54) includes 27 *cis* and 27 *trans* isomers as shown in [**1b'**].

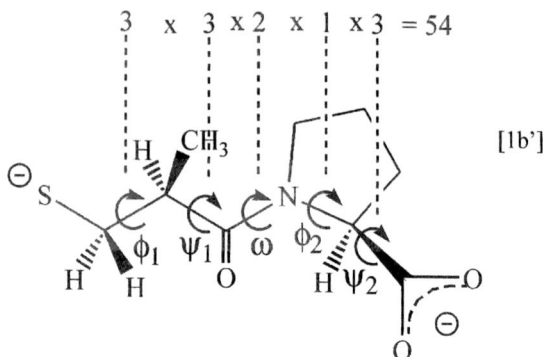

Table 1
Conformations of *trans*-captopril dianion optimized at the HF/3-21G level of theory

Conformations	φ_1	ψ_1	ω	φ_2	ψ_2	d1-12	E (hartree)	ΔE (kcal/mol)
$\alpha_L(g^+g^+)$	No found							
$\alpha_L(g^+a)$	No found							
$\alpha_L(g^+g^-)$	No found							
$\alpha_L(ag^+)$	No found							
$\alpha_L(aa)$	179.565	158.415	−6.011	−109.125	−15.493	6.537	−1019.462417	0.00
$\alpha_L(ag^-)$	No found							
$\alpha_L(g^-g^+)$	No found							
$\alpha_L(g^-a)$	No found							
$\alpha_L(g^-g^-)$	No found							
$\epsilon_L(g^+g^+)$	85.490	78.807	12.856	−129.771	169.819	5.871	−1019.4456347	10.53
$\epsilon_L(g^+a)$	No found							
$\epsilon_L(g^+g^-)$	No found							
$\epsilon_L(ag^+)$	No found							
$\epsilon_L(aa)$	−179.571	158.418	−5.994	−109.138	171.136	6.537	−1019.462417	0.00
$\epsilon_L(ag^-)$	−157.342	−65.316	−5.897	−99.473	124.315	6.233	−1019.4603437	1.30
$\epsilon_L(g^-g^+)$	No found							
$\epsilon_L(g^-a)$	No found							
$\epsilon_L(g^-g^-)$	No found							
$\gamma_L(g^+g^+)$	No found							
$\gamma_L(g^+a)$	No found							
$\gamma_L(g^+g^-)$	No found							
$\gamma_L(ag^+)$	No found							
$\gamma_L(aa)$	174.134	160.675	−4.561	−89.527	68.110	6.200	−1019.460301	1.32
$\gamma_L(ag^-)$	No found							
$\gamma_L(g^-g^+)$	No found							
$\gamma_L(g^-a)$	No found							
$\gamma_L(g^-g^-)$	−49.449	−58.267	10.128	−108.936	75.115	5.528	−1019.4525700	6.18

Table 2
Conformations of *cis*-captopril dianion optimized at the HF/3-21G level of theory

Conformations	φ_1	ψ_1	ω	φ_2	ψ_2	d1-12	E (hartree)	ΔE (kcal/mol)
$\alpha_L(g^+g^+)$	81.096	75.018	176.288	−96.610	−14.378	6.957	−1019.4532255	9.43
$\alpha_L(g^+a)$	No found							
$\alpha_L(g^+g^-)$	No found							
$\alpha_L(ag^+)$	−171.826	87.795	−178.862	−102.625	−15.986	7.393	−1019.4682589	0.00
$\alpha_L(aa)$	No found							
$\alpha_L(ag^-)$	−166.996	−56.039	−172.119	−107.206	−16.944	7.182	−1019.4559158	7.74
$\alpha_L(g^-g^+)$	−83.658	115.111	175.485	−107.551	−15.221	6.520	−1019.4602905	5.00
$\alpha_L(g^-a)$	No found							
$\alpha_L(g^-g^-)$	No found							
	No found							
$\epsilon_L(g^+g^+)$	No found							
$\epsilon_L(g^+a)$	No found							
$\epsilon_L(g^+g^-)$	No found							
$\epsilon_L(ag^+)$	−171.832	87.877	182.004	−102.864	170.242	7.394	−1019.468258	0.00
$\epsilon_L(aa)$	No found							
$\epsilon_L(ag^-)$	No found							
$\epsilon_L(g^-g^+)$	No found							
$\epsilon_L(g^-a)$	No found							
$\epsilon_L(g^-g^-)$	No found							
$\gamma_L(g^+g^+)$	No found							
$\gamma_L(g^+a)$	No found							
$\gamma_L(g^+g^-)$	No found							
$\gamma_L(ag^+)$	No found							
$\gamma_L(aa)$	No found							
$\gamma_L(ag^-)$	No found							
$\gamma_L(g^-g^+)$	No found							
$\gamma_L(g^-a)$	No found							
$\gamma_L(g^-g^-)$	No found							

In the present study, we have investigated structure [1b′] in which the following torsional angles were considered: ϕ_1, ψ_1, ω, ϕ_2 and ψ_2. All $3N − 6 = 3 \times 27 − 6 = 75$ geometrical parameters were optimized for the 116 electron-containing species.

4. Results and discussion

The conformational intricacies of compound [9b] have been reported previously [11] and are shown in Fig. 2. Clearly, the R and S isomeric forms contain crucial spatial requirements for docking of the two enantiomers.

Results obtained on the N-acetyl prolinate fragment [11] are pictorially represented in the potential energy curve $E = E(\psi_2)$ shown in Fig. 3 for molecule with *cis* and *trans* peptidic bond.

Tables 1 and 2 display the geometrical parameters and energies obtained for *cis*- and *trans*- dianionic captopril, respectively. As it can be seem in Fig. 4, only four α_L conformers were found for *trans*-dianionic captopril, while for *cis*- dianionic captopril most and the better structures were found and ϵ_L conformers are prevailing. In Fig. 5, the PEC [$E = E(\psi_2)$] for the *cis*- and *trans*- conformers can be compared. As the PEC obtained for the N-acetyl prolinate fragment, it shows the conformer with *cis*-peptide bond as the most stable, but the corresponding ΔE are much minor indicating that all *cis* structures are possible to be found.

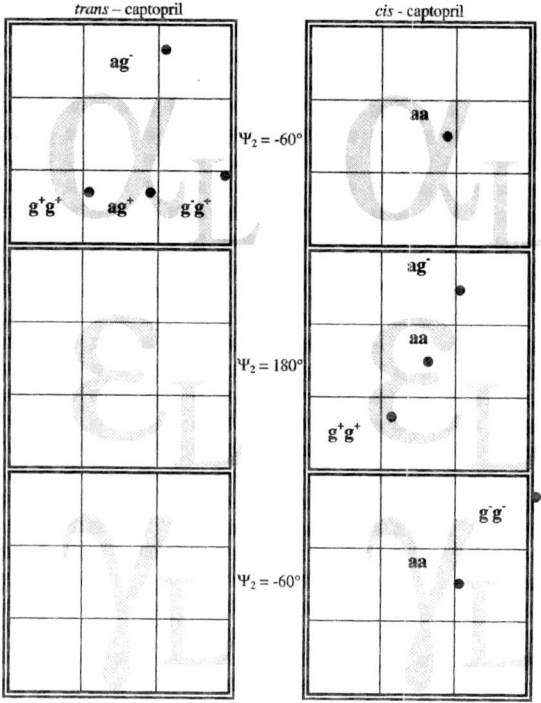

Fig. 4. Topological representation of the HF/3-21G optimized conformers of the *S* and *O* deprotonated Captopril double conjugate base containing either a *trans-* or a *cis-* peptide bond.

The conformational potential energy surfaces (at HF/3-21G level of theory), $E = E(\varphi_1, \psi_1)$ for Captopril in its dianionic form containing either a *trans-* or a *cis-* peptide bond are shown in Figs. 6 and 7, respectively.

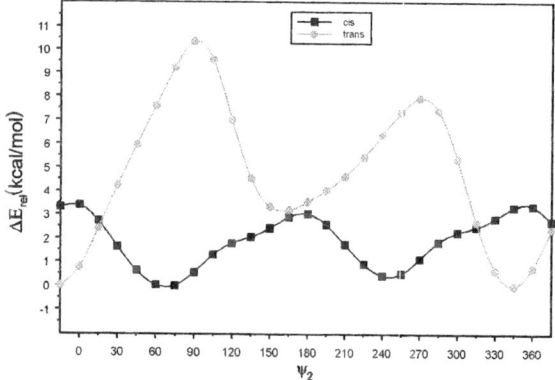

Fig. 5. Conformational potential energy curves (HF/3-21G), $E = E(\psi_2)$ for Captopril in its dianionic form containing either a *trans-* or a *cis-* peptide bond.

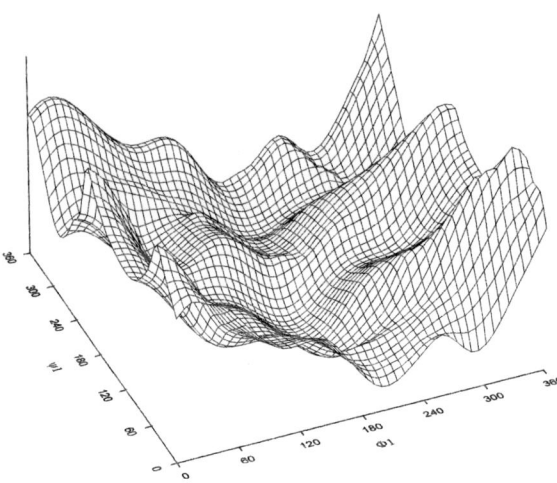

Fig. 6. Conformational potential energy surface (HF/3-21G), $E = E(\varphi_1, \psi_1)$ for Captopril in its dianionic form with *cis-* peptide bond.

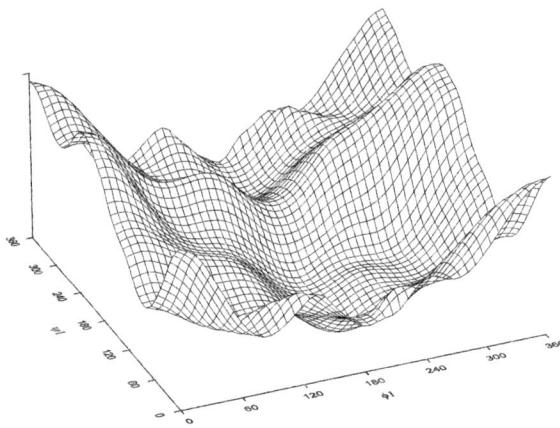

Fig. 7. Conformational potential energy surface (HF/3-21G), $E = E(\varphi_1, \psi_1)$ for Captopril in its dianionic form with *trans-* peptide bond.

5. Conclusion

It is believed [13,14], that the *trans* structure is the active form of captopril. In Tables 3 and 4 the geometrical parameters and energies for the best structures of *cis-* and *trans-* captopril at three levels of theory: HF/3-21G, HF/6-31 + G(d) and HF/6-11++G(d) are shown. The ΔE at the highest level of theory is 4.43 kcal/mol suggesting that

Table 3

Geometrical parameters and energies for the best structures of *cis*-captopril at three levels of theory (HF/3-21G, HF/6-31 + G(d) and HF/6-311++G(d))

Method	Φ_1	Ψ_1	ω	Φ_2	Ψ_2	d	Energy
HF/3-21G	179.565	158.415	− 6.011	− 109.125	− 15.493	6.536762	− 1019.462417
HF/6-31G*	181.762	152.069	− 2.437	− 101.872	170.220	6.248863	− 1024.946834
HF/6-311++G*	− 178.014	151.817	− 1.947	− 102.848	− 12.352	6.252749	− 1025.1175346

Table 4

Geometrical parameters and energies for the best structures of *trans*- captopril at three levels of theory (HF/3-21G, HF/6-31 + G(d) and HF/6-311++G(d))

Method	Φ_1	Ψ_1	ω	Φ_2	Ψ_2	d	Energy
HF/3-21G	188.171	87.756	181.939	− 102.549	− 16.003	7.392116	− 1019.4682589
HF/6-31G*	192.788	92.793	183.177	− 96.660	− 15.262	7.389190	− 1024.9539993
HF/6-311++G*	192.973	92.690	183.396	− 97.004	− 14.158	7.387116	− 1025.1245956

both, the *cis* and *trans* forms may contribute to the activity.

In (Fig. 8), the minima energy structures obtained at the higher level of theory are represented, d_{trans} being more than 1 Å longer that d_{cis}, where d is the separation between the negative charges or docking points in the biologically active captopril molecule.

So far, it can be assumed that distance between the two positive points in the active site of ACE (d_{ACE}), which is the distance between the Zn^{2+} cofactor and the guanidine moiety of an arginine side chain, must obey the relationship:

$$d_{trans} < d_{ACE} < d_{cis} \Rightarrow 7.38 \text{ Å} < d_{ACE} < 6.25 \text{ Å}$$

Acknowledgements

The authors thank the Departament de Química-Física, Universitat de València, Spain for providing computational facilities and acknowledge the financial support of the Universitat de València.

One of the authors (IGC) wishes to thank the Ministry of Education for a Szent-Györgyi Visiting Professorship.

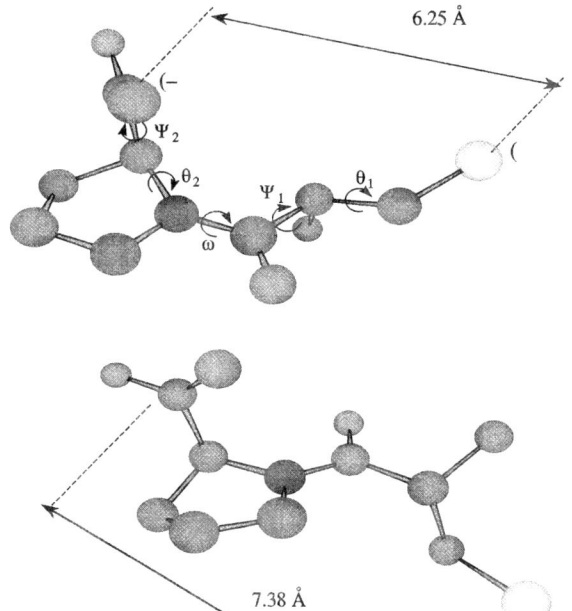

Fig. 8. Lowest energy conformers of *trans*- and *cis*-captopril double anion at HF/6-311G(d,p) level of theory.

G.N. Zamarbide et al. / Journal of Molecular Structure (Theochem) 666–667 (2003) 599–608

References

[1] D.W. Cushman, H.S. Cheung, E.F. Sabo, M.A. Ondetti, Biochemistry 16 (1977) 9484.

[2] (a) S.H. Ferreira, Br. J. Pharmacol. Chemother. 24 (1965) 163.
(b) Y.S. Bakhle, Nature (London) 222 (1969) 956.
(c) Y.S. Bakhle, A.M. Reynard, J.R. Vane, Nature (London) 222 (1969) 956.

[3] P.R. Andrews, J.M. Carson, A. Cassali, M.J. Spark, R. Woods, J. Med. Chem. 28 (1985) 393.

[4] D.V.S. Green, I.H. Hillier, G.A. Morris, N. Gensmantel, D.W. Payling, D.H. Robinson, J. Mol. Struct. (THEOCHEM) 251 (1991) 173.

[5] B.T. Luke, J.Mol.Struct.(THEOCHEM) 309 (1994) 1.

[6] B.T. Luke, J. Mol. Struct. (THEOCHEM) 332 (1995) 25.

[7] O. Farkas, S.J. Salpietro, P. Csaszar, I.G. Csizmadia, J. Mol. Struct. (THEOCHEM) 367 (1996) 25–31.

[8] S.J. Salpietro, A. Perczel, O. Farkas, R.D. Enriz, I.G. Csizmadia, J. Mol. Struct. (THEOCHEM) 497 (2000) 39–63.

[9] M. Berg, S.J. Salpietro, A. Perczel, O. Farkas, I.G. Csizmadia, J. Mol. Struct. (THEOCHEM) 504 (2000) 127–140.

[10] A.M. Rodríguez, H.A. Baldoni, L.L. Torday, I. Jakli, G. Zamarbide, R.D. Enriz, O. Farkas, A. Perczel, C.P. Sosa, I.G. Csizmadia, Proline: the maveric amino acid. An Ab Initio study on the cis–trans isomerization of peptide bonds: N-Formyl-L-Proline, Fifth World Congress of Theoretically Oriented Chemists (WATOC 99), (1–6 August 1999), Book of Abstracts, Imperial College, London, 1999, p. 212.

[11] L.L. Torday, B. Penke, G. Zamarbide, R.D. Enriz, J. Gy, J. Mol. Struct. (THEOCHEM) 528 (2000) 307–317.

[12] GAUSSIAN 98, Revision A.7, M.J. Frisch, G.W. Trucks, H.B. Schlegel, G.E. Scuseria, M.A. Robb, J.R. Cheeseman, V.G. Zakrzewski, J.A. Montgomery, Jr., E. Stratmann, J.C. Burant, S. Dapprich, J.M. Millam, A.D. Daniels, K.N. Kudin, M.C. Strain, O. Farkas, J. Tomasi, V. Barone, M. Cossi, R. Cammi, B. Mennucci, C. Pomelli, C. Adamo, S. Clifford, J. Ochterski, G.A. Petersson, P.Y. Ayala, Q. Cui, K. Morokuma, D.K. Malick, A.D. Rabuck, K. Raghavachari, J.B. Foresman, J. Cioslowski, J.V. Ortiz, A.G. Baboul, B.B. Stefanov, G. Liu, A. Liashenko, P. Piskorz, I. Komaromi, R. Gomperts, R.L. Martin, D.J. Fox, T. Keith, M.A. Al-Laham, C.Y. Peng, A. Nanayakkara, C. Gonzalez, M. Challacombe, P.M.W. Gill, B. Johnson, W. Chen, M.W. Wong, J.L. Andres, C. Gonzalez, M. Head-Gordon, E.S. Replogle, J.A. Pople, Gaussian, Inc., Pittsburgh PA, 1998.

[13] P.R. Andrews, J.M. Carson, A. Casseli, M.J. Spark, R. Woods, J. Med. Chem. 28 (1985) 393.

[14] D.V.S. Green, I.H. Hiller, G.A. Morris, N. Gensmantel, D.W. Payling, D.H. Robinson, J. Mol. Struct. (THEOCHEM) 251 (1991) 173.

ELSEVIER

Journal of Molecular Structure (Theochem) 666–667 (2003) 609–615

THEO
CHEM

www.elsevier.com/locate/theochem

A theoretical approach to acidity–basicity behaviour of some biologically active 6-phenyl-4,5-dihydro-3(2H)-pyridazinone derivatives

Cemil Öğretir[a,*], Selma Yarlıgan[a], Şeref Demirayak[b], Taner Arslan[a]

[a]Department of Chemistry, Faculty of Arts and Sciences, Osmangazi University, 26040 Eskişehir, Turkey
[b]Faculty of Pharmacy, Anadolu University, 26470 Eskişehir, Turkey

Abstract

The acid dissociation constants, pK_a values, of 12 biologically active 6-[p- or p- and m-substituted] phenyl-4,5-dihydro-3(2H)-pyridazinone derivatives were computed using semi-empirical methods. Correlations between the experimental and computed acid dissociations constants were found to be satisfactory.
© 2003 Published by Elsevier B.V.

Keywords: Semi-empirical calculations; Acidity–basicity behaviour; Pyridazinones; Tautomerism; Theoretical calculations; Computed and dissociation constants

1. Introduction

It is well-known that biological processes are based on chemical reactions. The main factors of life such as providing energy, transmission of pulses, metabolism and transfer of genetic information are all chemical reactions in which heteroaromatic molecules take part. Therefore, knowledge about the structure of heteroaromatic is invaluable in understanding their reactivity. The acid dissociation constants have been used in various areas of research, such as stereochemical and conformational structure determinations [1,2], the directions of nucleophilic and electrophilic attack, the stabilities of intermediates, the size of activation energies in organic reactions [3], and determination of the active centres of enzymes in biochemistry [4].

In the present work we report on the acid–base behaviour of some 6-[p- or p- and m-substituted] phenyl-4,5-dihydro-3(2H)-pyridazinone derivatives with potential antihypertensive effects [5–8].

2. Method of calculation

Theoretical calculations were carried out at the Restricted Hartree-Fock level (RHF) using AM1, PM3 semi-empirical SCF-MO method in the MOPAC 7.0 program, implemented in an Pentium Pro 200 MHz computer. The solvent effect was included in the geometry optimisations following the 'COnductor-like Screening Model' (COSMO) implemented in MOPAC 7.0 [9]. The initial estimates of the geometry of all the structures were obtained by a molecular mechanics program of Cs Chem Office

* Corresponding author. Tel.: +90-222-229-04-33; fax: +90-222-239-35-78.
E-mail address: cogretir@ogu.edu.tr (C. Öğretir).

0166-1280/$ - see front matter © 2003 Published by Elsevier B.V.
doi:10.1016/j.theochem.2003.08.085

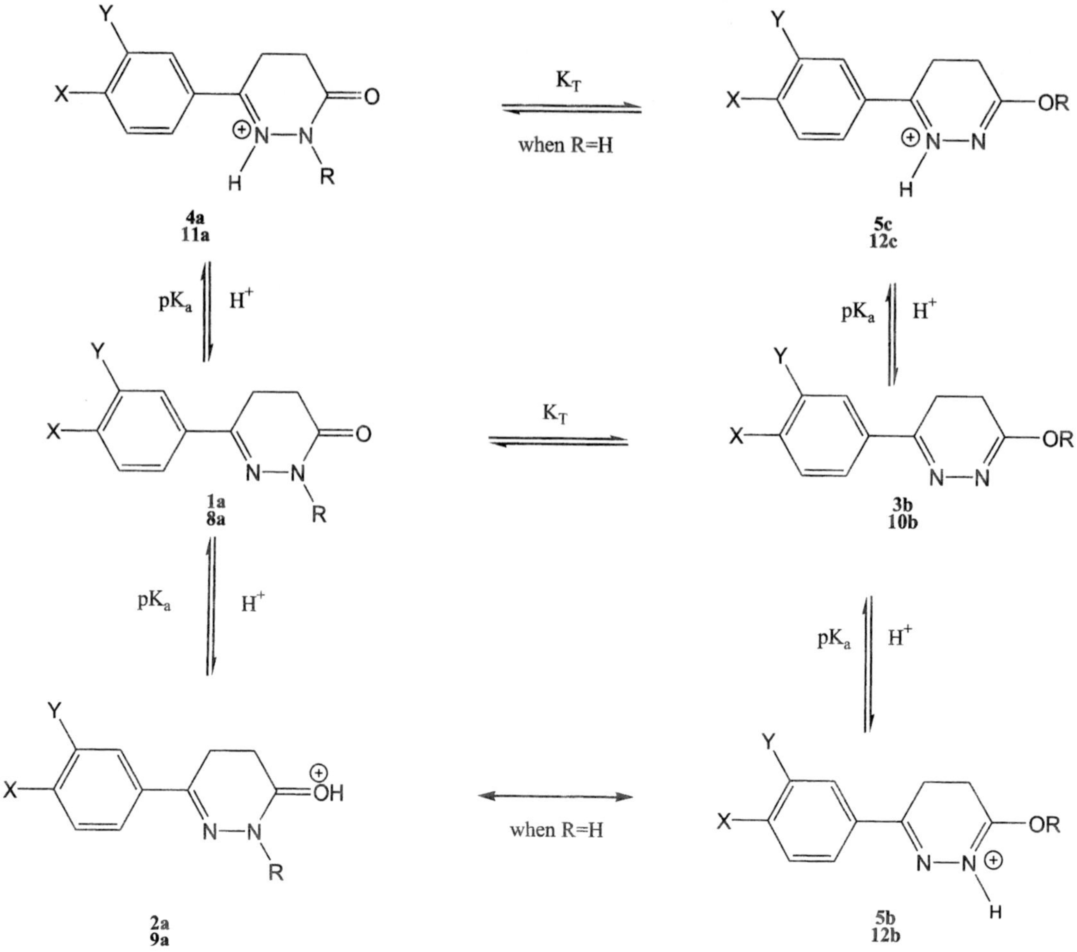

R=H (for compounds, 1-6); CH$_3$ (for model molecules, 7-12)

Scheme 1. Suggested protonation pathways for studied molecules (1–12).

Pro for Windows [10], followed by full optimisation of all geometrical variables (bond lengths, bond angles and dihedral angles) without any symmetry constraint, using semi-empirical AM1, PM3 quantum-chemical methods in the MOPAC 7.0 program.

2.1. Methods for pK$_a$ calculations

Taking the protonation into account we can write the following equation.

$$B + H_3O^+ \underset{K_b \text{ (backward)}}{\overset{K_f \text{ (forward)}}{\rightleftharpoons}} BH^+ + H_2O \qquad (1)$$

from which we can derive the equilibrium constant

and standard free energy change as follows:

$$k_a = \frac{[BH^+][H_2O]}{[B][H_3O^+]} = \frac{k_f}{k_b} \qquad (2)$$

and we know that thermodynamically $\Delta G = -2.303RT \log K_a = \Delta H - T\Delta S$ therefore we can write the following equation

$$\delta\Delta G = \Sigma\Delta G_{(\text{reactant})} - \Sigma\Delta G_{(\text{product})} \qquad (3)$$

$$\delta\Delta G^+_{(\text{BH})} = [\Delta G_{(\text{B})} + \Delta G_{(\text{H}_3\text{O})}] - [\Delta G^+_{(\text{BH})} + \Delta G_{(\text{H2O})}]$$

$$pK_a = \frac{\delta\Delta G}{2.303RT} \qquad (4)$$

where $R = 1.98$ cal mol^{-1}, $T = 298$ K

Table 1
Nomenclature of the studied compounds (**1–12**)

Compound	Name	R	X	Y
1	6-phenyl-4,5-dihydro-3(2H)-pyridazinone	H	H	H
2	6-(4-methylphenyl)-4,5-dihydro-3(2H)-pyridazinone	H	H	CH_3
3	6-(4-methoxyphenyl)-4,5-dihydro-3(2H)-pyridazinone	H	H	OCH_3
4	6-(4-chlorophenyl)-4,5-dihydro-3(2H)-pyridazinone	H	H	Cl
5	6-(3,4-dichlorophenyl)-4,5-dihydro-3(2H)-pyridazinone	H	Cl	Cl
6	6-(4-bromophenyl)-4,5-dihydro-3(2H)-pyridazinone	H	H	Br
7	2-methyl-6-phenyl-4,5-dihydro-3(2H)-pyridazinone	CH_3	H	H
8	2-methyl-6-(4-methylphenyl)-4,5-dihydro-3(2H)-pyridazinone	CH_3	H	CH_3
9	2-methyl-6-(4-methoxyphenyl)-4,5-dihydro-3(2H)-pyridazinone	CH_3	H	OCH_3
10	2-methyl-6-(4-chlorophenyl)-4,5-dihydro-3(2H)-pyridazinone	CH_3	H	Cl
11	2-methyl-6-(3,4-dichlorophenyl)-4,5-dihydro-3(2H)-pyridazinone	CH_3	Cl	Cl
12	2-methyl-6-(4-bromophenyl)-4,5-dihydro-3(2H)-pyridazinone	CH_3	H	Br

Table 2
Calculated thermodynamic values for studied molecules in aqueous phase ($\epsilon = 78.4$)

	AM1			PM3		
	ΔH_f (kcal mol^{-1})	ΔS (cal mol^{-1} K^{-1})	$\Delta G_f{}^a$ (kcal mol^{-1})	ΔH_f (kcal mol^{-1})	ΔS (cal mol^{-1} K^{-1})	$\Delta G_f{}^a$ (kcal mol^{-1})
$X = H; Y = H; R = H$						
1a	−8.606	117.48	−43.62	−16.05	116.91	−50.59
2a	118.21	100.78	88.20	118.31	103.38	87.50
3b	8.27	102.78	−22.36	6.92	104.00	−24.08
4a	109.43	118.06	74.45	114.33	102.96	83.65
5b	120.05	102.65	89.46	113.24	101.13	83.10
5c	140.83	97.56	111.92	133.608	101.21	103.44
$X = H; Y = H; R = CH_3$						
8a	15.39	112.73	−18.20	0.87	111.79	−32.47
9a	129.98	113.72	96.09	121.42	111.80	88.10
10b	19.01	110.37	−13.88	15.55	115.29	−13.81
11a	131.86	112.17	98.43	118.96	110.56	83.33
12b	127.15	111.29	93.99	125.73	115.34	91.36
12c	149.56	106.05	117.97	144.41	110.11	111.63
$X = CH_3; Y = H; R = H$						
1a	−4.33	114.72	−38.52	−8.11	113.19	−41.84
2a	111.49	113.40	77.70	108.79	111.58	75.53
3b	1.03	114.00	−32.94	−2.05	112.04	−35.44
4a	111.24	111.20	78.10	103.25	112.49	69.73
5b	112.99	111.56	79.76	103.28	113.34	69.50
5c	133.29	107.73	101.40	124.11	111.03	91.03
$X = CH_3; Y = H; R = CH_3$						
8a	7.42	119.06	−28.06	−5.82	123.84	−5.10
9a	121.80	123.83	84.90	113.03	126.26	75.40
10b	12.84	128.52	−25.46	4.67	125.31	4.09
11a	121.91	124.45	84.82	106.08	122.22	93.12
12b	120.54	125.81	83.05	115.85	123.29	101.57
12c	141.58	118.16	106.41	134.77	120.86	99.01

(continued on next page)

Table 2 (*continued*)

	AM1			PM3		
	ΔH_f (kcal mol^{-1})	ΔS (cal mol^{-1} K^{-1})	$\Delta G_f{}^a$ (kcal mol^{-1})	ΔH_f (kcal mol^{-1})	ΔS (cal mol^{-1} K^{-1})	$\Delta G_f{}^a$ (kcal mol^{-1})
$X = OCH_3$; $Y = H$; $R = H$						
1a	−38.88	115.28	−73.23	−40.38	117.60	−75.42
2a	89.80	117.75	54.71	92.54	121.21	56.42
3b	−29.83	116.86	−64.45	−30.43	117.77	−65.53
4a	79.19	117.77	44.09	74.30	119.28	38.75
5b	79.49	114.34	45.45	77.21	118.43	41.92
5c	99.76	113.47	66.08	92.79	118.55	57.62
$X = OCH_3$; $Y = H$; $R = CH_3$						
8a	−28.28	125.15	−65.58	−39.63	125.41	−34.66
9a	97.55	122.42	61.07	94.82	128.77	56.45
10b	−22.89	123.29	−59.64	−22.83	124.29	19.99
11a	88.64	123.48	51.84	76.47	125.67	66.86
12b	85.15	123.80	48.26	83.70	126.17	73.14
12c	108.55	118.40	73.38	103.67	124.23	66.72
$X = Cl$; $Y = H$; $R = H$						
1a	−4.04	108.37	−36.33	−6.22	110.67	−39.20
2a	124.92	113.86	90.99	138.41	112.69	104.83
3b	2.60	111.43	−30.61	−0.94	113.37	−34.72
4a	113.27	109.68	80.59	107.46	111.62	74.20
5b	116.00	111.60	82.74	106.00	112.01	72.62
5c	133.78	104.31	102.79	127.07	108.55	94.88
$X = Cl$; $Y = H$; $R = CH_3$						
8a	7.44	120.05	−28.34	−2.08	119.85	−37.80
9a	124.92	113.86	90.99	135.72	124.07	98.75
10b	11.15	117.92	−23.99	7.32	120.76	−28.67
11a	120.88	121.44	84.70	109.93	116.94	75.08
12b	121.36	117.90	86.23	119.70	120.07	83.92
12c	142.524	113.54	108.85	137.68	116.22	103.11
$X = Cl$; $Y = Cl$; $R = H$						
1a	−10.63	116.42	−45.32	−10.79	117.84	−45.91
2a	106.07	114.30	30.06	105.05	117.86	70.38
3b	−3.98	115.31	−38.34	−5.07	115.61	−39.52
4a	108.44	113.38	74.65	102.80	114.98	68.54
5b	110.98	112.78	99.37	102.01	117.50	67.00
5c	128.85	114.64	94.89	121.89	111.79	88.81
$X = Cl$; $Y = Cl$; $R = CH_3$						
8a	0.21	125.48	−37.18	−10.58	129.97	−49.31
9a	128.16	125.98	90.62	107.58	127.79	69.50
10b	7.13	124.54	−29.98	2.02	124.90	−35.20
11a	118.72	123.56	81.90	104.37	123.54	67.55
12b	114.35	123.96	77.41	113.89	127.26	75.97
12c	137.59	119.48	102.13	133.89	121.85	97.83
$X = Br$; $Y = H$; $R = H$						
1a	7.68	113.74	−26.21	8.18	114.81	−26.03
2a	124.89	110.81	91.87	122.31	114.60	88.15
3b	12.26	113.83	−21.66	13.64	112.94	−20.02
4a	124.22	112.65	−90.65	127.87	111.09	94.77

Table 2 (continued)

	AM1			PM3		
	ΔH_f (kcal mol^{-1})	ΔS (cal mol^{-1} K^{-1})	$\Delta G_f{}^a$ (kcal mol^{-1})	ΔH_f (kcal mol^{-1})	ΔS (cal mol^{-1} K^{-1})	$\Delta G_f{}^a$ (kcal mol^{-1})
5b	129.90	111.62	96.64	120.36	113.40	86.57
5c	143.33	107.17	111.44	140.47	112.04	107.09
$X = Br;\ Y = H;\ R = CH_3$						
8a	16.73	121.74	−19.55	8.17	122.49	7.17
9a	133.64	125.66	96.19	119.43	120.85	83.42
10b	21.25	117.10	−13.65	23.90	122.82	20.97
11a	135.53	122.38	99.46	125.42	122.50	110.06
12b	131.31	119.02	95.83	136.74	121.68	120.01
12c	153.97	119.09	118.51	151.79	118.32	116.62

^{a} $\Delta G_f = \Delta H_f - T\Delta S$.

Table 3
The measured and computed acid dissociation constants, pK_a, of studied molecules

Compound	AM1		PM3		Experimental results					
	$\delta\Delta G_f{}^a$ (kcal mol^{-1})	p$K_a{}^b$	$\delta\Delta G_f{}^a$ (kcal mol^{-1})	p$K_a{}^b$	$H^{1/2c}$	m^d	p$K_{a1}{}^e$ (exp)	$H^{1/2c}$	m^d	p$K_{a2}{}^e$ (exp)
$X = H;\ Y = H;\ R = H$										
1a ⇌ 2a	20.25	−14.84	−11.36	−8.33	1.37 ± 0.01	1.03	1.41	−4.65 ± 0.03	0.99	−4.63
1a ⇌ 4a	−6.39	−4.68	−7.37	−5.41						
3b ⇌ 5b	−0.47	−0.35	14.52	10.65						
3b ⇌ 5c	17.79	13.04	−0.79	−0.57						
$X = H;\ Y = H;\ R = CH_3$										
8a ⇌ 9a	−2.94	−2.16	0.71	0.52	−2.48 ± 0.03	1.86	−4.62	−8.45 ± 0.03	0.80	−6.74
8a ⇌ 11a	−5.42	−3.98	3.47	2.55						
10b ⇌ 12b	3.50	2.57	11.79	8.64						
10b ⇌ 12c	20.22	14.82	1.29	0.94						
$X = CH_3;\ Y = H;\ R = H$										
1a ⇌ 2a	−5.08	−3.72	4.32	3.17	−4.65 ± 0.03	0.77	1.45	−1.96 ± 0.03	0.47	−0.92
1a ⇌ 4a	−5.27	−3.86	10.13	7.43						
3b ⇌ 5b	−1.35	−0.98	16.75	12.28						
3b ⇌ 5c	17.73	13	0.26	0.19						
$X = CH_3;\ Y = H;\ R = CH_3$										
8a ⇌ 9a	−1.54	−1.13	3.44	2.52	6.09 ± 0.07	0.82	5.01	−3.18 ± 0.05	0.47	−1.51
8a ⇌ 11a	−1.65	−1.21	9.25	4.75						
10b ⇌ 12b	2.75	2.01	9.93	7.28						
10b ⇌ 12c	20.2	14.81	31.81	23						
$X = OCH_3;\ Y = H;\ R = H$										
1a ⇌ 2a	−16.78	−12.30	−10.33	−7.57	−0.51 ± 0.05	0.77	−0.39	−1.45 ± 0.04	1.34	1.94
1a ⇌ 4a	−6.17	−4.52	7.31	5.36						
3b ⇌ 5b	1.28	0.94	14.26	10.46						
3b ⇌ 5c	21.54	15.79	3,58	2.62						

(continued on next page)

Table 3 (*continued*)

Compound	AM1		PM3		Experimental results					
	$\delta\Delta G_f{}^a$ (kcal mol^{-1})	$pK_a{}^b$	$\delta\Delta G_f{}^a$ (kcal mol^{-1})	$pK_a{}^b$	$H^{1/2c}$	m^d	$pK_{a1}{}^e$ (exp)	$H^{1/2c}$	m^d	$pK_{a2}{}^e$ (exp)
$X = OCH_3; Y = H; R = CH_3$										
8a ⇌ 9a	−15.37	−11.27	−11.86	−8.69	−0.83 ± 0.03	1.07	−0.90	−6.36 ± 0.03	0.85	−5.39
8a ⇌ 11a	−5.57	−4.03	4.19	3.07						
10b ⇌ 12b	3.65	2.67	15.76	11.56						
10b ⇌ 12c	19.05	13.97	80	58.6						
$X = Cl; Y = H; R = H$										
1a ⇌ 2a	−18.81	−13.79	−22.34	−16.38	−3.29 ± 0.04	1.50	−4.92	−6.88 ± 0.07	0.79	−5.46
1a ⇌ 4a	−8.11	−5.94	8.01	5.87						
3b ⇌ 5b	−2.00	−1.47	14.36	10.53						
3b ⇌ 5c	18.67	13.69	−2.87	−2.01						
$X = Cl; Y = H; R = CH_3$										
8a ⇌ 9a	−15.96	−11.70	−15.12	11.09	−3.16 ± 0.03	1.82	−5.78	−7.66 ± 0.01	1.96	−15.06
8a ⇌ 11a	−5.06	−3.71	8.79	6.45						
10b ⇌ 12b	1.14	0.84	9.02	6.61						
10b ⇌ 12c	19.23	14.01	−5,05	−3.07						
$X = Cl; Y = Cl; R = H$										
1a 2a	−6.97	−5.11	5.86	4.30	−3.44 ± 0.02	2.06	−7.09	−8.05 ± 0.02	1.96	−15.81
1a ⇌ 4a	−8.63	−6.33	7.25	5.31						
3b ⇌ 5b	−4.37	−3.21	15.18	11.34						
3b ⇌ 5c	18.84		−1.6							
$X = Cl; Y = Cl; R = CH_3$										
8a ⇌ 9a	−17.00	−12.46	−11.54	−8.47	1.49 ± 0.06	1.45	2.16	1.36 ± 0.01	0.30	−1.86
8a ⇌ 11a	−7.76	−5.69	5.90	4.32						
10b ⇌ 12b	3.82	2.80	10.43	7.65						
10b ⇌ 12c	19.96	13.81	−6.3	−4.42						
$X = Br; Y = H; R = H$										
1a ⇌ 2a	−6.76	−4.95	7.57	5.55	−3.46 ± 0.03	1.45	−5.00	−6.68 ± 0.03	1.05	−7.01
1a ⇌ 4a	−5.79	−4.25	0.82	0.60						
3b ⇌ 5b	−6.90	−5.06	14.98	10.98						
3b ⇌ 5c	18.97	13.27	−0,38	−0.27						
$X = Br; Y = H; R = CH_3$										
8a ⇌ 9a	−6.46	−4.73	1.92	1.41	−3.64 ± 0.03	1.01	−3.70	−8.24 ± 0.02	1.16	−8.33
8a ⇌ 11a	−7.76	−5.69	−4.75	3.49						
10b ⇌ 12b	1.89	1.38	8.56	6.27						
10b ⇌ 12c	19.91	14.6	31.08	22.79						

[a] Calculated using Eq. (3).

[b] Calculated using Eq. (4).

[c] Half protonation values used in $pK_a = m H_x^{1/2}$.

[d] Slopes of log I–pH (or H-) graphs.

[e] Taken from Ref. [11].

3. Results and discussion

The structures of studied molecules (Scheme 1, Table 1) indicate that the first six molecules 1–6 (i.e parent molecules) and their N-Me derivatives 7–12 (i.e model molecules) have two proton gain centres. A close values of acid dissociation constants, pK_a, values of model molecules can let

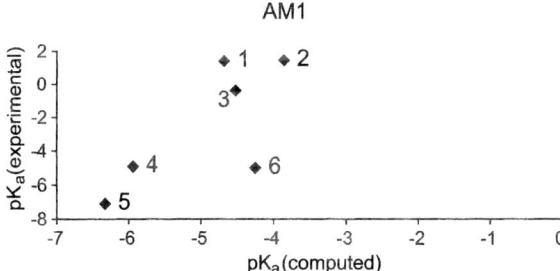

Fig. 1. A graph of experimental acid dissociation constants, pK_a, against AM1 aqueous phase computed acid dissociation constants, pK_a, for oxo form and 1N protonation for parent compounds 1–6.

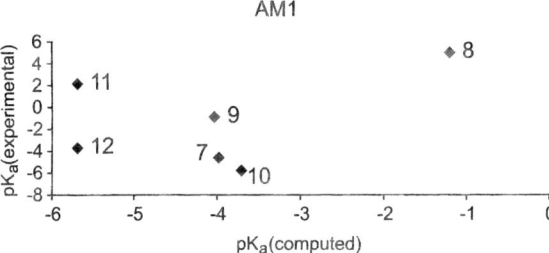

Fig. 2. A graph of experimental acid dissociation constants, pK_a, against AM1 aqueous phase computed acid dissociation constants, pK_a, for oxo form and 1N protonation for model (N-Me) compounds 7–12.

us to evaluate the mechanism of proton gain process (Scheme 1) by way of comparing the absolute values of computed pK_a values, which were obtained using the computed thermodynamic values of related physical properties (Table 2), with that of experimentally obtained pK_a values (Table 3). A graphical search for a correlation between experimental pK_a values and the AM1, which gives better correlation coefficient values, calculated pK_a values elucidate that there is a good parallelism between the computed and experimental

values (Figs. 1 and 2). The very same trend of the first protonation to be applicable for the parent molecules 2 and 4 similar to the model molecules 7, 8, 9 and 11 for the rest of the molecules protonations seems to take place at pyridine type of nitrogen via tautomeric rearrangement as indicated in the literature as an unusual protonation behaviour of pyridazin-3-ones [12].

References

[1] H.C. Brown, O.H. Mc Daniel, O. Haflinger, Determination of Organic Structure by Physical Methods, Academic Press, New York, 1985.

[2] J. Chilton, J.B. Stanlake, J. Pharm. Pharmacol. 14 (1962) 367.

[3] C.D. Johnson, The Hammet Equation, Cambridge University Press, 1973.

[4] P.A. Frey, F.O. Kokesh, F.H. Westheimer, J. Am. Chem. Soc. 93 (1971) 7266–7270.

[5] M. Tishler, B. Stonovnik, Pyridazines, Advances in Heterocyclic Chemistry, Academic Press, New York, 1968, 1979, 1990.

[6] G. Szasz, Z. Budvavi, Bavauy Pharmaceutical Chemistry of Antyhypertensive Agents, CRC Press, Boca Raton, FL, 1991.

[7] G. Henisch, H. Kopelent-Frank, Pharmacalogically active pyridazine derivatives, Prog. Med. Chem. 29 (1992) 141–183.

[8] P.A.V. Zwieten, W.J. Greenlee, Antihypertensive Drugs, Harwood Academic Publishers, New York, 1997.

[9] J.J.P. Stewart, MOPAC 7.0 QCPE, University of Indiana, Bloomington, IN, USA.

[10] CS Chemoffice Pro for Microsoft Windows, Cambridge Scientific Computing, Inc., 875 Massachusets Avaneu, Suite 61, Cambridge MA, 02139, USA.

[11] C. Ogretir, S. Yarlıgan, Ş. Demirayak, J. Chem. Engin. Data 47 (2002) 1396–1400.

[12] J. Elguero, C. Marzin, A.R. Katritzky, P. Linda, In: Advances in Heterocyclic Chemistry, A.R Katritzky, A.J. Boulton (Eds.), Academic Press, 1976.

ELSEVIER

Journal of Molecular Structure (Theochem) 666–667 (2003) 617–623

THEO
CHEM

www.elsevier.com/locate/theochem

Theoretical study of a hydration mechanism in an enaminone pro-drug prototype

Juan C. Garro[a,*], Graciela N. Zamarbide[a], Mario R. Estrada[a],
Francisco Tomás Vert[b], Carlos A. Ponce[a]

[a]*Departamento de Química, Universidad Nacional de San Luis, San Luis 5700 Argentina*
[b]*Departament de Química-Física, Universitat de València, Burjassot (Valencia) 46100, Spain*

Abstract

Enaminones may act as pro-drugs releasing via proton-catalyzed hydrolysis a primary amine, which may be an actual drug. A hydration mechanism of prototype enaminone (2-propenal-3-amine) has been subjected to quantum chemical studies. All involved compounds were investigated in a search for the most likely reactive form. Results revealed that the proposed reaction pathway is thermodynamically possible.
© 2003 Elsevier B.V. All rights reserved.

Keywords: Ab initio computations; Hydration mechanism; Enaminone pro-drug prototype

1. Introduction

Enaminones are a group of organic compounds containing the conjugated system N–C=C–C=O formed between a primary amine and a 1,3-dicarbonyl compound. It may act as a pro-drug, releasing a primary amine that may be an actual drug [1], via proton-catalyzed hydrolysis. A hydration mechanism of prototype enaminone (2-propenal-3-amine) has been subjected to quantum chemical studies.

The present paper aims to determine the thermodynamic possibility of the mechanism proposed by previous experimental reports [2,3] that suggested that the protonation of enaminone followed by the addition of water, forming enol-carbinolamine, which decomposes in the corresponding aldhedide and the primary active amine.

Spectroscopic [4,5] and theoretical [6,7] studies revealed oxo-enaminone as the most stable tautomeric structure, and consequently, the most likely reactive form. Thus, this study includes the conformational study of hydration reaction intermediates (enolcarbinolamine and β-hydroxialdehide), assuming that the most favorable site of protonation in 2-propenal-3-amine is O [7], allowing the first pre-equilibrium protonation proceed through the following reaction (Fig. 1).

2. Molecular computations

Using the GAUSSIAN 98 [8] program package, ab initio computations have been carried out at the HF/3-21G level of theory for all the molecules studied. Higher level calculations, such as HF/6-31 + G* and B3LYP(DFT)/6-31 + G*, were performed on selected molecules.

* Corresponding author.
E-mail address: jcgarro@unsl.edu.ar (J.C. Garro).

0166-1280/$ - see front matter © 2003 Elsevier B.V. All rights reserved.
doi:10.1016/j.theochem.2003.08.086

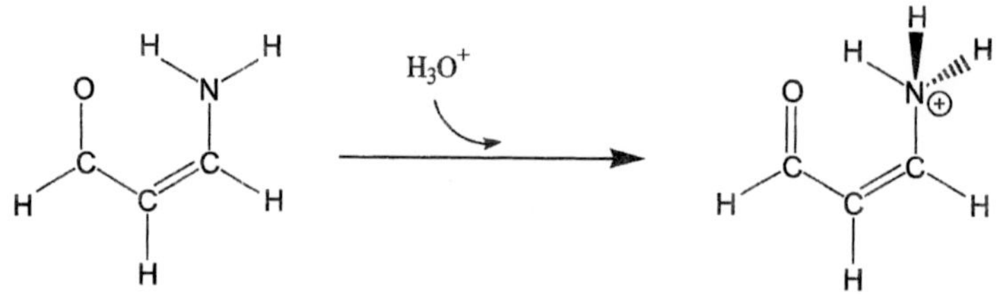

Fig. 1. First pre-equilibrium protonation of 2-propenal-3-amine.

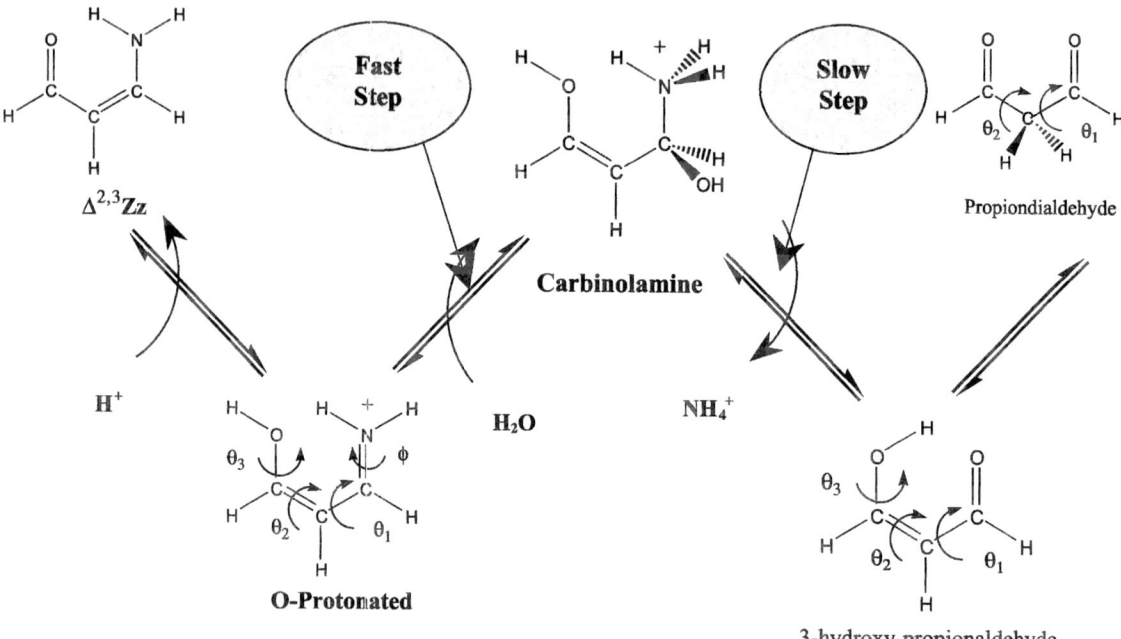

Fig. 2. Proposed mechanism for hydration of 2-propenal-3-amine.

3. Results and discussion

The proposed mechanism for hydration is schemed in Fig. 2, where the slow and fast steps are indicated in agreement with experimental evidence [9–11].

A tetrahedral intermediary compound, propyleno-1,3-diol-3-amino ion (type carbinolamine), is the result of the addition of a water molecule on the O-protonated enaminone (Fig. 3).

The molecule presents several possible conformers and a potential energy hypersurface,

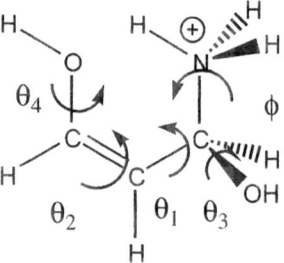

Fig. 3. Tetrahedral intermediary compound, propyleno-1,3-diol-3-amino ion (type carbinolamine).

$E(\text{Carbinolamine}) = E(\theta_1, \theta_2, \theta_3, \theta_4, \phi)$, will be generated.

In the first step, using the restriction that double bonds have for a free rotation, eight PES [$E_{(\phi, \theta_2 \theta_4)} = E(\theta_1, \theta_3)$] were calculated starting on structures with fixed ϕ, θ_2 and θ_4 angles. Results are shown in Fig. 4a and b, while Table 1 sums up the conformational parameters for the minimum energy structures obtained.

When a full optimization was performing at the HF/3-21G level of theory (Table 2), the ϕ angle remained

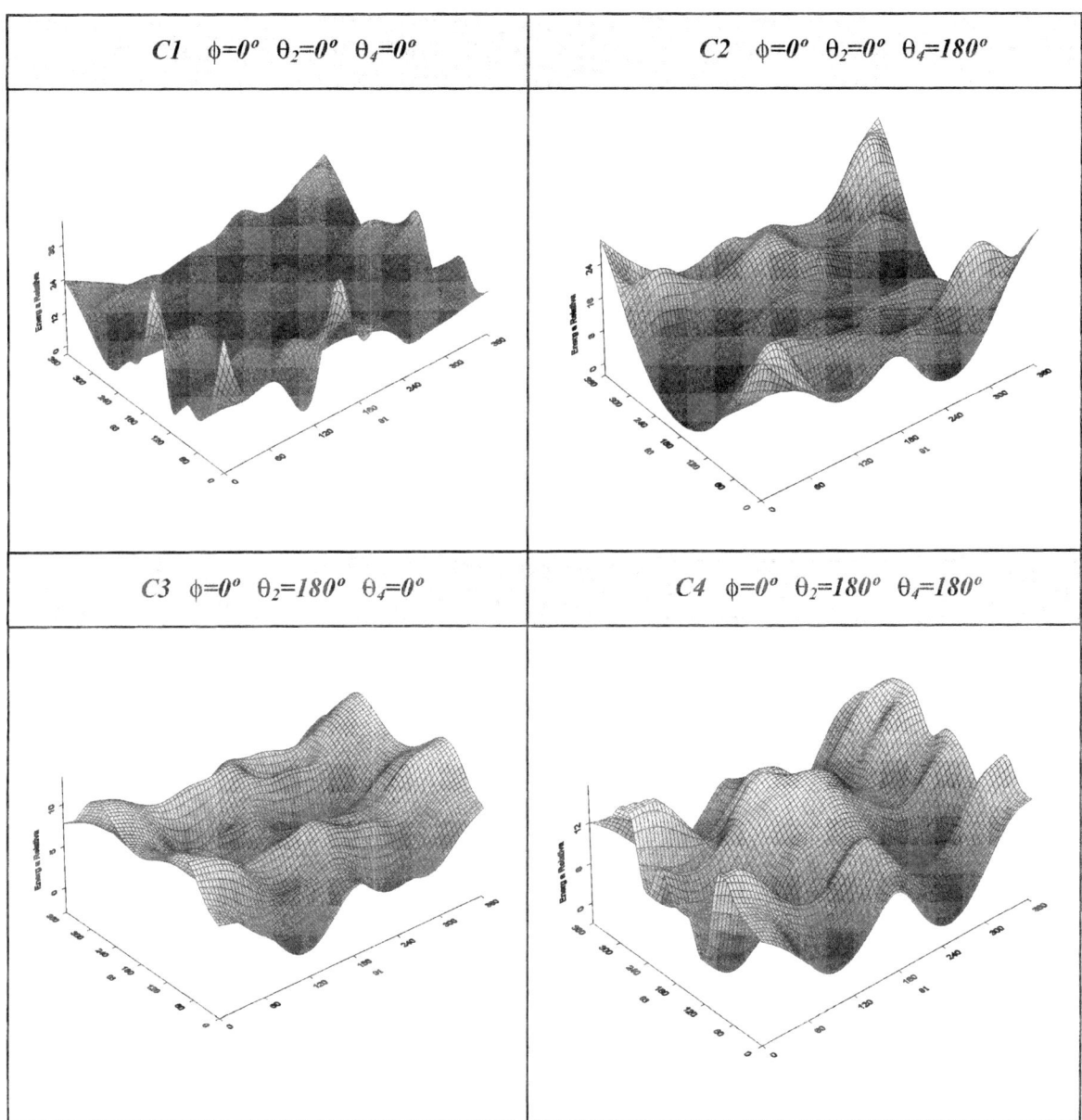

Fig. 4. Conformational potencial energy surface (HF/3-21G) $E_{(\phi, \theta_2, \theta_4)} = E(\theta_1, \theta_3)$ for propyleno-1,3-diol-3-amino ion.

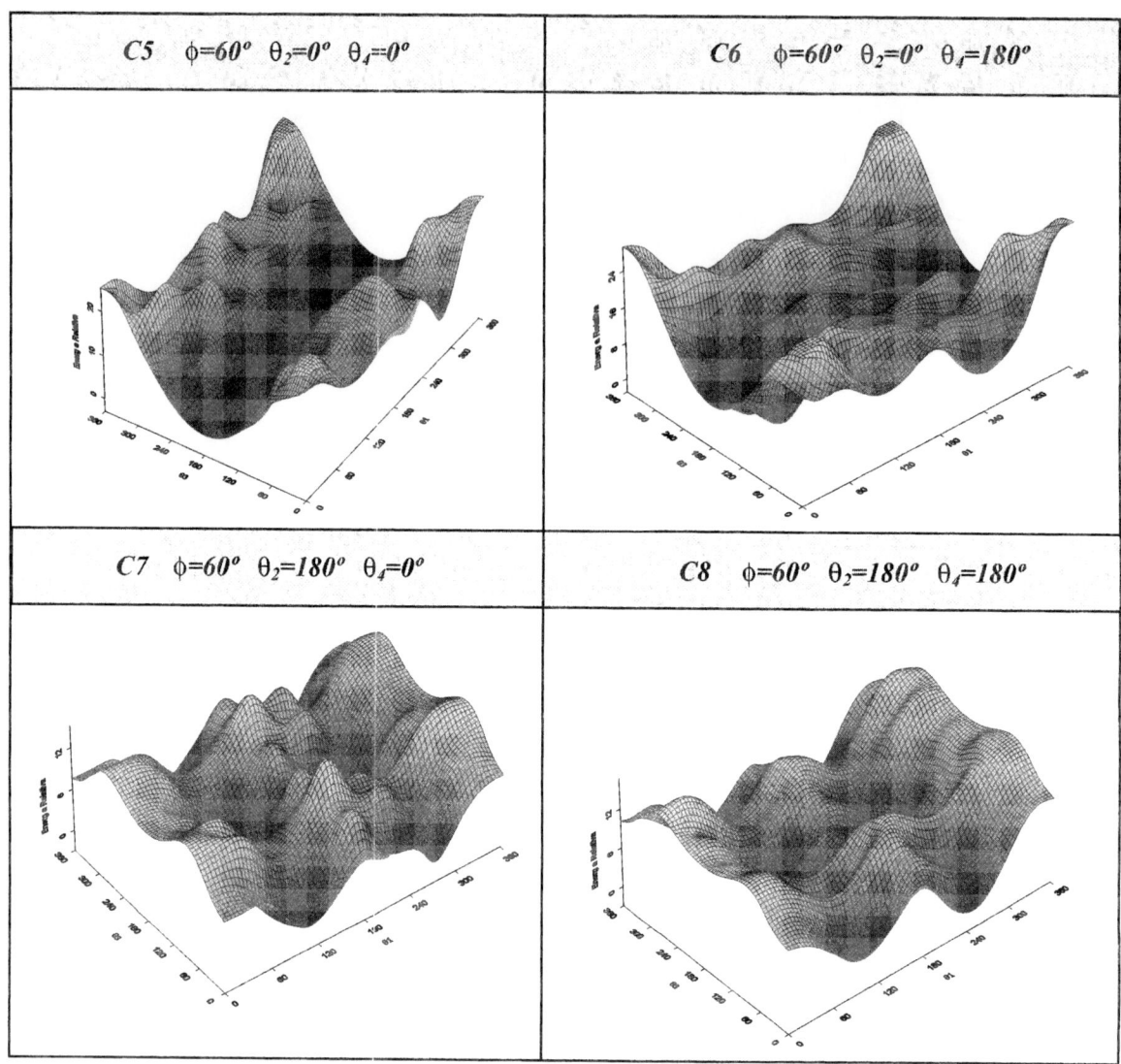

Fig. 4 (*continued*)

Table 1

Conformational parameters for the minimum energy structures (HF/3-21G) obtained for propyleno-1,3-diol-3-amino ion using rotational restriction

RHF/3-21G	θ_1	θ_2	θ_3	θ_4	ϕ	E (Hartrees)
C1	0.087	−0.35	−118.71	0.401	0.037	−320.340551
C2	0.001	0.57	−180.1	180.36	−0.121	−320.40244
C3	124.03	187.369	0.088	−0.004	0.065	−320.39342
C4	−101.99	179.90	0.083	175.24	0.023	−320.38784
C5	−59.98	1.032	179.23	−2.036	−65.27	−320.41342
C6	61.23	0.023	180.056	182.326	62.35	−320.407568
C7	−90.74	180.39	−0.958	0.265	62.35	−320.392472
C8	−120.35	−186.36	0.036	−179.54	65.25	−320.388490

Table 2
Conformational parameters for the minimum energy structures (HF/3-21G) obtained for propyleno-1,3-diol-3-amino ion using full optimization

RHF/3-21G	θ_1	θ_2	θ_3	θ_4	ϕ	E (Hartrees)
C1	50.02	− 5.65	− 23.25	168.141	− 53.02	− 320.41197
C2	56.025	− 5.23	80.309	167.509	− 53.93	− 320.41317
C3	111.947	179.17	52.49	− 1.462	60.9516	− 320.39934
C4	− 101.90	179.90	91.25	175.36	62.80	− 320.39885
C5	− 105.87	1.49	134.11	− 5.90	− 53.60	− 320.42654
C6	52.80	− 5.089	135.68	167.71	67.78	− 320.41360
C7	− 94.34	181.25	80.75	1.29	62.21	− 320.400312
C8	− 146.5	− 178.31	− 33.06	− 184.17	49.32	− 320.392610

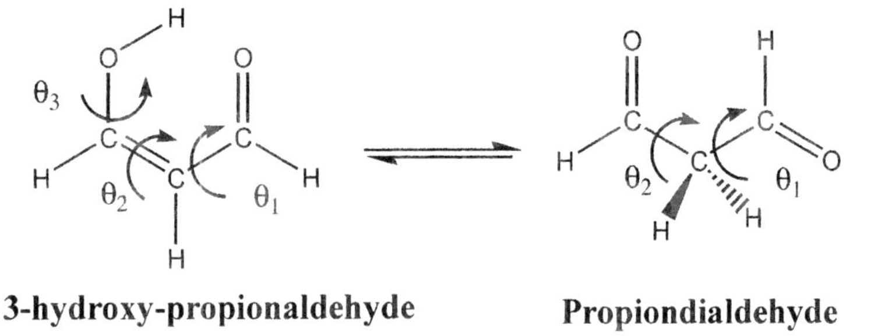

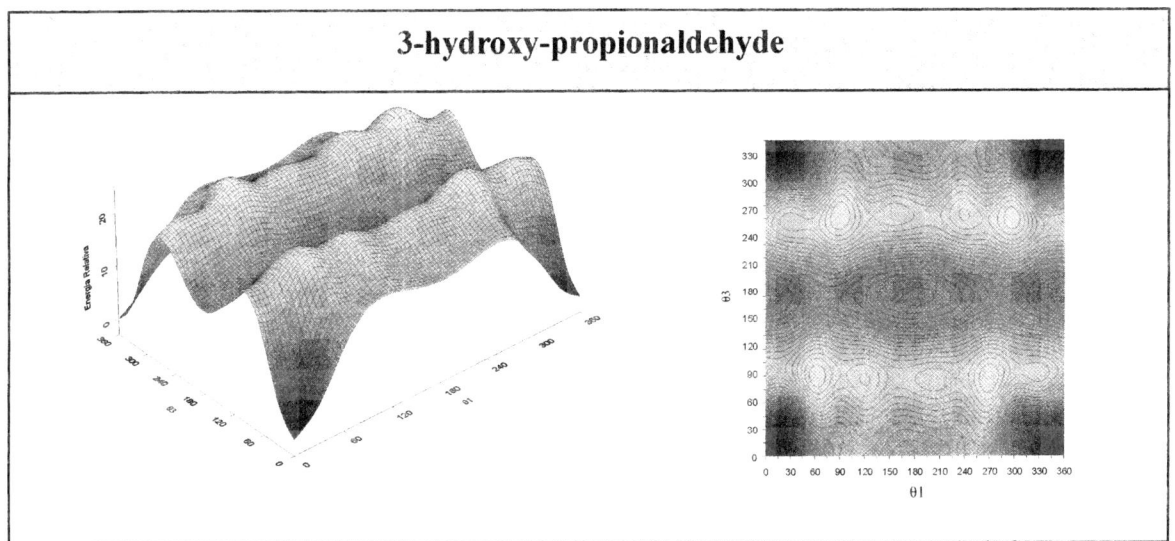

3-hydroxy-propionaldehyde   **Propiondialdehyde**

Fig. 5. Keto−enol tautomerism for 3-hydroxy-propionaldehyde.

Fig. 6. Conformational potencial energy surface (HF/3-21G) for the tautomers of keto−enol equilibrium.

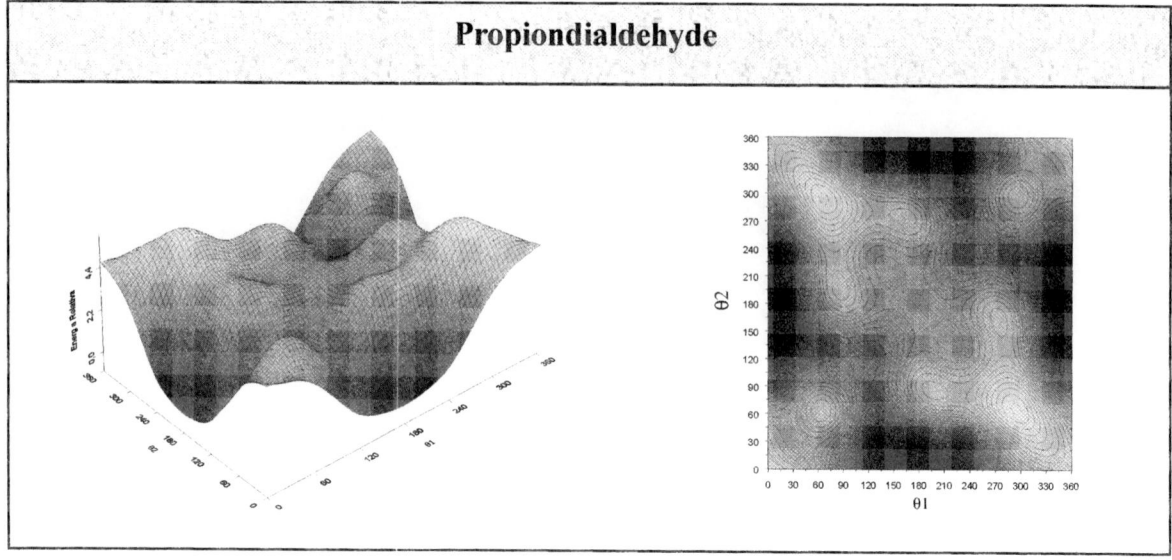

Fig. 6 (*continued*)

Table 3
Conformational parameters for the minimum energy structures obtained for both tautomers of 3-hydroxy-propionaldehyde using full optimization at three levels of theory

	θ_1	θ_2	θ_3	E_{min} (Hartree)	ΔE (kcal/mol)
RHF/3-21G					
3-Hydroxy-propionaldehyde	−0.0243	0.14377	−0.1316	−264.14021	0.00441
Propiondialdehyde	179.9404	0.2439	−	−264.1358	
RHF/6-31 + G*					
3-Hydroxy-propionaldehyde	−0.0090	−0.00071	0.0039	−265.64092	0.00188
Propiondialdehyde	−179.9498	−0.0069	−	−265.6428	
B3LYP/6-31 + G*					
3-Hydroxy-propionaldehyde	−0.00231	0.00138	0.0016	−267.16352	0.0018
Propiondialdehyde	−179.6954	0.18675	−	−267.1646	

very close to 60° for all the minimum energy conformers, so it can be assumed that the nitrogen atom is in its sp³ hybridization state. As in previous calculations, the most stable structure corresponds to $\theta_2 = 1.49°$, $\theta_3 = 134.11°$, $\theta_4 = -5.90°$ and $\phi = -53.60°$, but dihedral θ_1 changes from $\theta_1 = -59.98$ to $-105.87°$.

The last step in the hydrolysis reaction is the release of the amine group and the formation of 3-hydroxy-propionaldehyde, which may be in equilibrium with its keto form (Fig. 5).

The study was carried out at HF/3-21G level of theory and Fig. 6 shows the PES for both tautomers. In Table 3, parameters for minimum energy structures are displayed. It can be noted that the ΔE for both compounds is only 10^{-3} kcal/mol, which indicates that the compounds shown in equilibrium in Fig. 5 are present in equal proportions.

In conclusion, it can be said that the hydrolysis of a simple enaminone will proceed by a similar mechanism that is shown in Fig. 7.

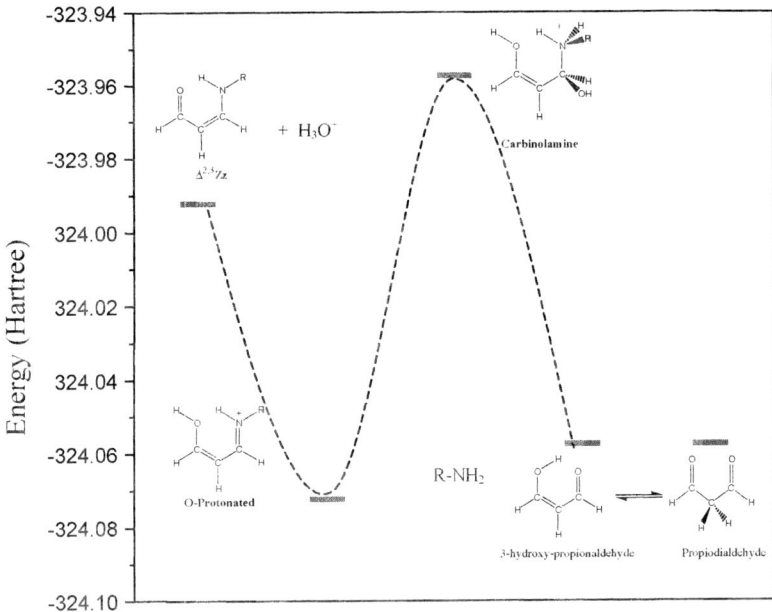

Fig. 7. Reaction way for the hydrolysis of a simple enaminone.

References

[1] (a) in: Z. Rappoport (Ed.), The Chemistry of Enamines, Wiley, New York, 1994. (b) J.V. Greenhill, Chem. Soc. Rev., 6, 1977, pp. 277. (c) H. Bundgaard, in: H. Bundgraad (Ed.), Design of Prodrugs, Elsevier, Amsterdam, 1985, pp. 1–92.

[2] (a) V.H. Naringrekar, V.J. Stella, J. Pharm. Sci. 79 (1990) 138–147. (b) S.J. Kesten, M.J. Degnan, J. Hung, D.J. McNamara, D.F. Ortwine, S.E. Uhlendrofy, L.M. Werbel, J. Med. Chem. 35 (1992) 3429. (c) I.O. Edafiogho, Pharm. World J. 7 (1990) 20.

[3] J.D. Larsen, H. Bundgaard, Arch. Pharm. Chem. Sci. Ed. 14 (1986) 52–63. S.F. Dyke, The Chemistry of Enaminomes, S.F. Dyke, Syndics of the Cambridge University Press Cambridge, UK (1973).

[4] J.P. Guthrie, F. Jordan, J. Am. Chem. Soc. 94 (26) (1972) 9132.

[5] L. Kozerski, J. Dabrowski, Org. Magn. Reson. 4 (1972) 256.

[6] M. Noguiera Eberlin, Y. Takahata, C. Kascheres, J. Mol. Struct. (Theochem) 207 (1990) 143.

[7] J.C. Garro, G.D. Manzanares, G.N. Zamarbide, C.A. Ponce, M.R. Estrada, E.A. Jáuregui, J. Mol. Struct. (Theochem) 545 (2001) 17–27.

[8] M.J. Frisch, G.W. Trucks, H.B. Schlegel, G.E. Scuseria, M.A. Robb, J.R. Cheeseman, V.G. Zakrzewski, J.A. Montgomery, Jr., R.E. Stratmann, J.C. Burant, S. Dapprich, J.M. Millam, A.D. Daniels, K.N. Kudin, M.C. Strain, O. Farkas, J. Tomasi, V. Barone, M. Cossi, R. Cammi, B. Mennucci, C. Pomelli, C. Adamo, S. Clifford, J. Ochterski, G.A. Petersson, P.Y. Ayala, Q. Cui, K. Morokuma, D.K. Malick, A.D. Rabuck, K. Raghavachari, J.B. Foresman, J. Cioslowski, J.V. Ortiz, A.G. Baboul, B.B. Stefanov, G. Liu, A. Liashenko, P. Piskorz, I. Komaromi, R. Gomperts, R.L. Martin, D.J. Fox, T. Keith, M.A. Al-Laham, C.Y. Peng, A. Nanayakkara, C. Gonzalez, M. Challacombe, P.M.W. Gill, B. Johnson, W. Chen, M.W. Wong, J.L. Andres, C. Gonzalez, M. Head-Gordon, E.S. Replogle, J.A. Pople, GAUSSIAN 98, Revision A.7, Gaussian, Inc., Pittsburgh, PA, 1998. A.C.

[9] V.H. Naringrekar, V.J. Stella, J. Pharm. Sci. 79 (1990) 138–147. S.J. Kesten, M.J. Degnan, J. Hung, D.J. McNamara, D.F. Ortwine, S.E. Uhlendrofy, L.M. Werbel, J. Med. Chem. 35 (1992) 3429. I.O. Edafiogho, Pharm. World J. 7 (1990) 20.

[10] J.D. Larsen, H. Bundgaard, Arch. Pharm. Chem. Sci. Ed. 14 (1986) 52–63.

[11] Mc. Murric, Química Orgánica, Ed Latinoamericana, 5° edición, pp. 697–701.

ELSEVIER

Journal of Molecular Structure (Theochem) 666–667 (2003) 625–636

THEO
CHEM

www.elsevier.com/locate/theochem

Novel *ortho*- and *peri*-fused pyridazine ring systems

Olivér Éliás[a], László Károlyházy[a], Gyula Horváth[b], Veronika Harmat[c], Péter Mátyus[a],*

[a]*Department of Organic Chemistry, Semmelweis University, Hőgyes Endre utca 7, Budapest 1092, Hungary*
[b]*Ivax Drug Research Institute, Berlini út 47/49, Budapest 1045, Hungary*
[c]*Department of Theoretical Chemistry, Eötvös Loránd University, Pázmány Péter sétány 1/A, Budapest 1117, Hungary*

Abstract

By utilizing nucleophilic and electrophilic substitution reactions of 3(2*H*)-pyridazinones, and application of various types of ring forming reactions, some novel polycyclic fused compounds containing pyrrolo[*b*]-, pyrido[*b*]-, pyrrolooxazino[*b*]- or pyrrolopyrazino[*b*]pyridazine skeletons were synthesized. From mechanistic point of view, structure–reactivity relationships were analyzed. In most cases, the cyclization reaction was strongly dependent on the substituents of the pyridazine ring.
© 2003 Elsevier B.V. All rights reserved.

Keywords: Pyridazines; *Ortho*- and *peri*-condensed pyridazine ring systems; Ring contraction reaction; Ring closure reaction; *Tert*-amino effect

1. Introduction

Pyridazines including mono- and bicyclic derivatives have attracted much attention due to their potential biological activities; e.g. we have recently described several series of pyridazinones with α-adrenergic blocking properties, antiarrhythmic, CNS and antifungal activities [1,2]. One of the key issues of the synthetic approaches to these types of compounds is selection of the ring closure method and the consideration of the possible choice of substituents of the pyridazine derivative. We have successfully employed several types of carbon–heteroatom and carbon–carbon bond formation reactions for ring closure [3,4]. In particular, it was shown that 2-methyl-6-nitropyridazin-3(2*H*)-one with a hydroxyalkylamino group at 4- or 5-position, easily available from 4,5-dichloro-2-methyl-6-nitropyridazin-3(2*H*)-one (**1**), could be a very useful building block for the syntheses of pyridazino[4,5-*d*]oxazepines, pyridazino[4,5-*d*]thiazepines, and pyridazino[4,5-*d*]diazepines. Theoretical studies on the substitution and cyclization reactions indicated that the regiochemistry might be strongly dependent on both the reaction conditions and substituents. As a continuation of these studies, we now report on the transformation of pyridazino[3,4-*b*]oxazines and oxazepines into *ortho*- and *peri*-fused ring systems by electrophilic substitution followed by a ring closure reaction, and the synthesis of an angularly fused pyrrolo ring system. Furthermore, we also describe an unexpected ring contraction reaction of a pyridazine into a pyrazole derivative with a mechanistic proposal.

2. Experimental section

All melting points were determined on a Boetius type Kofler hot-stage microscope, and are uncorrected. The IR spectra were recorded on

* Corresponding author. Tel./fax: +36-1-2170851.
E-mail address: matypet@szerves.sote.hu (P. Mátyus).

0166-1280/$ - see front matter © 2003 Elsevier B.V. All rights reserved.
doi:10.1016/j.theochem.2003.08.087

a Perkin–Elmer 1600 FTIR instrument in potassium bromide pellets. The ^{1}H NMR spectra were recorded at ambient temperature in the solvent indicated, using the ^{2}H signal of the solvent as the lock and tetramethylsilane as the internal standard. Chemical shifts (δ) are given in ppm and coupling constants (J) in Hz. The signals are designated as follow: s, singlet; d, doublet; t, triplet; m, multiplet; b, broad. Bruker AM at 200 MHz and Bruker AC spectrometer at 400 MHz were used. ^{13}C NMR spectra were recorded on the same spectrometers at 50 and 100 MHz, respectively. The assignments of ^{13}C NMR spectra were supported by DEPT-135 spectra. Mass spectra were taken by a Finnigan MAT 8430 spectrometer (resolution:1250, ion accelerating voltage: 3 kV, ion source temperature: 250 °C/Electron Ionization Mass Spectra (EIMS) and Chemical Ionization Mass Spectra (CIMS), 25 °C/Fast Atom Bombardment Mass Spectra (FABMS).

All new compounds gave satisfactory elementary analytical data (C, H, N); these analyses were performed on a Carlo Erba Elemental Analyzer Model 1012 apparatus.

X-ray crystallography: data were collected on a Rigaku RAXIS-II imaging plate detector using graphite-monochromated Mo K$_\alpha$ radiation at 293 K. For all data sets, data processing was carried out using the software supplied with the diffractometer. Structure solutions with direct methods were carried out with the teXsan Crystal Structure Analysis Package [5]. The refinements were carried out using the SHELXL-93 program [6] with full matrix least squares method on F^2. All non-hydrogen atoms were refined anisotropically. Hydrogen atoms were generated based upon geometric evidence and their positions were refined by the riding model.

For column chromatography Kieselgel 60 (Aldrich, 0.040–0.063 mm silica gel) was used; for TLC analysis Silica gel 60 F$_{254}$ (Merck) plates were applied. Solvent mixtures used for chromatography are always given in a vol/vol ratio. The reagents were purchased from commercial suppliers and used as received, solvents were dried and distilled prior to use.

Starting compounds were prepared according to the literature procedures cited: **1** [7], and its derivatives **4c** and **4d** [3].

2.1. Preparation of 5-chloro-4-[2-(hydroxymethyl)pyrrolidin-1-yl]-2-methyl-6-nitropyridazin-3(2H)-one (2) and 4-chloro-5-[2-(hydroxymethyl)pyrrolidin-1-yl]-2-methyl-6-nitropyridazin-3(2H)-one (3)

To a solution of 4,5-dichloro-2-methyl-6-nitropyridazin-3(2H)-one (**1**) (15 g, 0.067 mol) in dry ethanol (200 ml), a pyrrolidin-2-ylmethanol (9.0 g, 0.089 mol) and triethylamin (3.3 ml, 0.024 mol) were added, and the mixture was heated under reflux for 15 h (monitored by TLC). The reaction mixture was evaporated to dryness in vacuo. Then water (100 ml) was added to the residue, and the mixture was acidified with aqueous 2 M hydrochloric acid. The solution was extracted with chloroform (2 × 200 ml), and the combined organic phases were dried over anhydrous magnesium sulfate. The solvent was evaporated in vacuo, and the 4- and 5-regioisomers thus formed were separated by column chromatography with a mixture of toluene:methanol (4:1) as the eluent.

Compound **2** was obtained as yellow crystals (yield: 36%); mp 98–99 °C; $R_f = 0.83$ ethyl acetate: chloroform (9:1); IR (potassium bromide): ν 3388, 2922, 1641, 1525, 1459, 1402, 1303, 1117, 1041 cm^{-1}; ^{1}H NMR (deuteriochloroform): δ 1.60–2.27 (m, 4H, CH_2), 3.34–3.49 (m, 2H, NCH_{2a} and OCH_{2a}, $J = 6.1$), 3.59–3.67 (dd, 1H, OCH_{2b}, $J_1 = 4.2$, $J_2 = 11.4$), 3.70 (s, 3H, NCH_3), 4.05–4.18 (m, 1H, NCH_{2b}), 5.33–5.45 (m, 1H, NCH); ^{13}C NMR (deuteriochloroform): δ 25.6 and 27.9 (CH$_2$), 40.5 (NCH$_3$), 54.5 (NCH$_2$), 61.6 (NCH), 66.0 (OCH$_2$), 106.0 (C-5), 144.2 (C-4), 149.8 (C-6), 157.8 (C-3). Anal. calcd for C$_{10}$H$_{13}$ClN$_4$O$_4$: C, 41.61; H, 4.54; N, 19.41. Found: C, 41.93; H, 4.53; N, 19.23.

Compound **3** was obtained as orange crystals (yield: 56%); mp 185–186 °C; $R_f = 0.81$ ethyl acetate:chloroform (9:1); IR (potassium bromide): ν 3429, 2925, 1658, 1591, 1539, 1499, 1404, 1360, 1322, 1042, 876 cm^{-1}; ^{1}H NMR (deuteriochloroform): δ 1.77–2.19 (m, 4H, CH_2), 3.13–3.21 (m, 1H, NCH_{2a}), 3.43–3.65 (m, 3H, OCH_2 and NCH_{2b}), 3.78 (s, 3H, NCH_3), 4.31–4.35 (m, 1H, NCH); ^{13}C NMR (deuteriochloroform): δ 25.6 and 27.6 (CH$_2$), 40.8 (NCH$_3$), 52.1 (NCH$_2$), 61.5 (NCH), 64.6 (OCH$_2$), 122.9 (C-4), 139.8 (C-5), 145.9 (C-6), 158.2 (C-3).

Anal. calcd for $C_{10}H_{13}ClN_4O_4$: C, 41.61; H, 4.54; N, 19.41. Found: C, 41.56; H, 4.53; N, 19.48.

2.2. Preparation of 1-chloro-3-methyl-6,6a,7,8,9,10-hexahydropyridazino[3,4-b]pyrido[1,2-d]-[1,4]oxazin-2(3H)-one (4b) and [1-(1-methyl-3-nitro-1H-pyrazol-4-yl)piperidin-2-yl]methanol (5)

To a solution of 4,5-dichloro-2-methyl-6-nitropyridazin-3(2H)-one (1) (0.5 g, 2.23 mol) in a mixture of ethanol and water (6 and 25 ml, respectively), a piperidin-2-ylmethanol (1.25 g, 10.87 mol) was added, and the mixture was heated under reflux for 2 h (monitored by TLC). The reaction mixture was evaporated to dryness in vacuo. Then water (20 ml) was added to the residue, and the solution was extracted with dichloromethane (4 × 20 ml), and the combined organic phases were dried over anhydrous magnesium sulfate. The solvent was evaporated in vacuo, and the title compounds were separated by column chromatography with a mixture of ethyl acetate:chloroform (9:1) as the eluent.

Compound 4b was obtained as white crystals (yield: 20%); mp 178–180 °C; $R_f = 0.17$ ethyl acetate:chloroform (1:9); IR (potassium bromide): ν 2944, 2926, 2854, 1628, 1598, 1530, 1476, 1440, 1400, 1362, 1274, 1232, 1134, 1060 cm^{-1}; ^{1}H NMR (deuteriochloroform): δ 1.22–1.98 (m, 6H, CH_2), 2.71–2.84 (m, 1H, NCH_{2a}), 3.34–3.44 (m, 1H, NCH) 3.59 (s, 3H, NCH_3), 3.84 (dd, 1H, OCH_{2a}, $J_1 = 8.6$, $J_2 = 11.6$), 4.27 (dd, 1H, OCH_{2b}, $J_1 = 4.0$, $J_2 = 11.6$), 4.80 (bd, 1H, NCH_{2b}); ^{13}C NMR (deuteriochloroform): δ 23.4, 25.2 and 27.7 (CH_2), 39.4 (NCH_3), 48.3 (NCH_2), 56.9 (NCH), 69.1 (OCH_2), 108.4 (C-1), 135.7 (C-11a), 142.2 (C-4a), 158.2 (C-2). Anal. calcd for $C_{11}H_{14}ClN_3O_2$: C, 51.67; H, 5.52; N, 16.43. Found: C, 51.95; H, 5.60; N, 16.06.

Compound 5 was obtained as red crystals (yield: 28%); mp 103–105 °C; $R_f = 0.06$ ethyl acetate:chloroform (1:9); IR (potassium bromide): ν 3438, 2944, 2848, 1628, 1566, 1500, 1472, 1424, 1356, 1332, 1280, 1262, 1208, 1164, 1148, 1120, 1068, 1042, 1024, 992 cm^{-1}; ^{1}H NMR (400 MHz, deuteriochloroform): δ 1.50–2.00 (m, 6H, CH_2), 2.89–2.95 (m, 1H, NCH_{2a}), 3.18–3.24 (m, 1H, NCH_{2b}), 3.29–3.34 (m, 1H, NCH), 3.59 (dd, 1H, OCH_{2a}, $J_1 = 5.1$, $J_2 = 11.2$), 3.66 (dd, 1H, OCH_{2b}, $J_1 = 5.6$, $J_2 = 11.2$), 3.94 (s, 3H, NCH_3), 7.32 (s, 1H, CH);

^{13}C NMR (100 MHz, deuteriochloroform): δ 21.4, 24.7 and 26.4 (CH_2), 40.8 (NCH_3), 51.0 (NCH_2), 59.7 (NCH), 69.7 (OCH_2), 125.7 (C-5), 132.2 (C-4), 148.1 (C-3). Anal. calcd for $C_{10}H_{16}N_4O_3$: C, 49.99; H, 6.71; N, 23.32. Found: C, 50.40; H, 6.76; N, 22.96.

2.3. General procedure for preparation of bicyclic derivatives 4a and 6

To a solution of 2 or 3 (3 mol) in dry ethanol (25 ml), a 1 M sodium ethoxide in ethanol was added (3 ml), and the mixture was heated under reflux until the starting material had been consumed (monitored by TLC). The solvent was evaporated in vacuo, and the solid residue was dissolved in water, and extracted with ethyl acetate. The combined organic phases were dried over anhydrous magnesium sulfate, and evaporated in vacuo. The crude product was purified by column chromatography (6) or by crystallization from ethanol (4a).

1-Chloro-3-methyl-6a,7,8,9-tetrahydro-6H-pyridazino[3,4-b]pyrrolo[1,2-d][1,4]oxazin-2(3H)-one (4a). This compound was obtained as white crystals from ethanol (yield: 98%); mp 220–221 °C; $R_f = 0.49$ ethyl acetate:chloroform (9:1); IR (potassium bromide): ν 1626, 1600, 1537, 1481, 1444, 1386, 1332, 1279, 1155, 1075 cm^{-1}; ^{1}H NMR (deuteriochloroform): δ 1.46–2.18 (m, 4H, CH_2), 3.63 (s, 3H, NCH_3), 3.66–3.79 (m, 2H, OCH_{2a} and NCH), 3.92–4.06 (m, 1H, NCH_{2a}), 4.28–4.56 (m, 2H, NCH_{2b} and OCH_{2b}); ^{13}C NMR (deuteriochloroform): δ 24.0 and 26.8 (CH_2), 39.5 (NCH_3), 50.6 (NCH_2), 55.9 (NCH), 68.5 (OCH_2), 108.4 (C-1), 135.7 (C-10a), 142.2 (C-4a), 158.2 (C-2). Anal. calcd for $C_{10}H_{12}ClN_3O_2$: C, 49.70; H, 5.00; N, 17.39. Found: C, 49.57; H, 4.73; N, 17.50.

2-Methyl-4-nitro-6a,7,8,9-tetrahydro-6H-pyridazino[4,5-b]pyrrolo[1,2-d][1,4]oxazin-1(2H)-one (6). This compound was obtained as red crystals, col. chrom., chloroform:methanol (95:5) (yield: 39%); mp 152–153 °C; $R_f = 0.79$ ethyl acetate:chloroform (9:1); IR (potassium bromide): ν 3428, 2925, 1654, 1590, 1523, 1337, 1107 cm^{-1}; ^{1}H NMR (deuteriochloroform): δ 1.41–2.28 (m, 4H, CH_2), 3.32 (dd, 1H, OCH_{2a}, $J_1 = 9.2$, $J_2 = 10.2$), 3.52–3.71 (m, 2H, NCH_{2a} and NCH), 3.74 (s, 3H, NCH_3), 4.28–4.56 (m, 2H, OCH_{2b} and NCH_{2b}); ^{13}C NMR (deuteriochloroform): δ 23.8 and 27.0 (CH_2), 40.3 (NCH_3),

50.5 (NCH_2), 55.7 (NCH), 67.9 (OCH_2), 130.0 and 130.2 (*C*-10a and *C*-4a), 142.9 (*C*-4), 157.5 (*C*-1). Anal. calcd for $C_{10}H_{12}N_4O_4$: C, 47.62; H, 4.80; N, 22.21. Found: C, 48.06; H, 4.81; N, 22.34.

2.4. Preparation of 5-chloro-4-[2-(chloromethyl)pyrrolidin-1-yl]-2-methyl-6-nitropyridazin-3(2H)-one (7)

Compound **2** (1.0 g, 3.47 mol) was added to thionyl chloride (20 ml) with stirring at 0–5 °C and the mixture was refluxed for 1 h. Then, the excess of thionyl chloride was removed in vacuo and the residue was treated with toluene which was evaporated to drive off excess thionyl chloride. Chloroform (20 ml) was added to the residue, and washed with water (2 × 20 ml). After evaporation of the solvent ethanol (20 ml) and charcoal were added to the crude product, and the suspension was refluxed. Charcoal was filtered off, and ethanol was evaporated. Hexane was added to the oily residue, and the yellow crystals thus formed were filtered off. Yield: 79%; mp 72–73 °C; $R_f = 0.92$ ethyl acetate:chloroform (9:1); IR (potassium bromide): ν 3428, 2937, 1651, 1522, 1316, 1115, 870 cm^{-1}; ^{1}H NMR (deuteriochloroform): δ 1.67–2.26 (m, 4H, CH_2), 3.38–3.59 (m, 3H, OCH_2 and NCH_{2a}), 3.71 (s, 3H, CH_3), 4.12–4.25 (m, 1H, NCH_{2b}), 5.55 (m, 1H, NCH); ^{13}C NMR (deuteriochloroform): δ 25.6 and 29.0 (CH_2), 40.5 (NCH_3), 48.2 (ClCH_2), 55.0 (NCH_2), 60.3 (NCH), 106.8 (*C*-5), 143.1 (*C*-4), 149.6 (*C*-6), 157.3 (*C*-3). Anal. calcd for $C_{10}H_{12}Cl_2N_4O_3$: C, 39.11; H, 3.94; N, 18.24. Found: C, 38.95; H, 3.86; N, 18.03.

2.5. Preparation of 2-methyl-4-nitro-5,6,6a,7,8,9-hexahydropyrrolo[1′,2′:1,6]pyrazino[2,3-d]pyridazin-1(2H)-one (8)

Compound **7** (1.0 g, 3.26 mol) was added to saturated aqueous solution of ammonia (20 ml) and the solution was refluxed for 24 h (TLC). Then, water (20 ml) was added to the mixture, and the solution was extracted with chloroform (4 × 50 ml). The combined organic layers were dried over anhydrous magnesium sulfate. After evaporation of the solvent, a red oily residue was obtained, which was then chromatographed on silica gel using a mixture of ethyl acetate:chloroform (9:1) as the eluent, to give

yellow crystals (yield: 31%); mp 179–182 °C; $R_f = 0.26$ ethyl acetate:chloroform (1:9); IR (potassium bromide): ν 3286, 2950, 2865, 1616, 1542, 1391, 1322, 1149, 1023, 873, 755 cm^{-1}; ^{1}H NMR (400 MHz, deuteriochloroform): δ 1.68–1.79 (m, 2H, 8-*H*), 1.88–1.91 (m, 1H, 7-H_a), 2.21–2.26 (m, 1H, 7-H_b), 2.70 (dd, 1H, 6-H_a, $J_1 = 9.0$, $J_2 = 11.2$), 2.83–2.89 (m, 1H, 9-H_a), 3.32–3.41 (m, 2H, NCH and 9-H_b), 3.62 (dd, 1H, 6-H_b, $J_1 = 3.2$, $J_2 = 11.4$), 3.74 (s, 3H, NCH_3), 5.90 (bs, 1H, NH); ^{13}C NMR (100 MHz, deuteriochloroform): δ 22.5 and 28.4 (CH_2), 39.7 (NCH_3), 42.6 (*C*-9), 50.7 (*C*-6), 53.7 (NCH), 119.0 (*C*-10a), 129.0 (*C*-4a), 143.6 (*C*-4), 154.9 (*C*-1). Anal. calcd for $C_{10}H_{13}N_5O_3$: C, 47.81; H, 5.22; N, 27.87. Found: C, 47.99; H, 5.24; N, 28.09.

2.6. General procedure for preparation of compounds 9

To a solution of the appropriate chloro compound **4** (1.0 mol) in dry ethanol (20 ml) ammonium formate (4.8 mol) and 10 percentage Pd–C catalyst (0.1 g) were added. The suspension was stirred at 60 °C for 1 h. The catalyst was removed by filtration, and the filtrate was concentrated in vacuo to give white solid. The residue was dissolved in dichloromethane and the solution was extracted with water. The organic phase was dried over anhydrous magnesium sulfate, and the solvent was evaporated in vacuo to give the crude product which was crystallized from ethanol.

3-Methyl-6a,7,8,9-tetrahydro-6*H*-pyridazino[3,4-*b*]pyrrolo[1,2-*d*][1,4]oxazin-2(3*H*)-one (**9a**). This compound was obtained as white crystals (yield: 97%); mp 200–201 °C; $R_f = 0.19$ ethyl acetate:chloroform (9:1); IR (potassium bromide): ν 2878, 1639, 1598, 1560, 1486, 1354, 1268, 1063, 992, 802 cm^{-1}; ^{1}H NMR (deuteriochloroform): δ 1.42–2.30 (m, 4H, CH_2), 3.19–3.45 (m, 2H, NCH_2), 3.57 (s, 3H, NCH_3), 3.63–3.77 (m, 2H, OCH_{2a} and NCH), 4.54–4.66 (m, 1H, OCH_{2b}) 5.62 (s, 1H, 1-*H*); ^{13}C NMR (deuteriochloroform): δ 23.6 and 28.1 (CH_2), 38.3 (NCH_3), 45.9 (NCH_2), 54.7 (NCH), 69.5 (OCH_2), 97.9 (*C*-1), 138.1 (*C*-10a), 142.9 (*C*-4a), 161.6 (*C*-2). Anal. calcd for $C_{10}H_{13}N_3O_2$: C, 57.96; H, 6.32; N, 20.28. Found: C, 57.95; H, 6.29; N, 20.37.

3-Methyl-6,6a,7,8,9,10-hexahydropyridazino[3,4-*b*]pyrido[1,2-*d*][1,4]oxazin-2(3*H*)-one (**9b**). This compound was obtained as white crystals (yield:

88%); mp 144–150 °C; $R_f = 0.07$ ethyl acetate: chloroform (1:9); IR (potassium bromide): ν 3934, 2856, 1644, 1584, 1538, 1480, 1462, 1444, 1384, 1368, 1352, 1316, 1246, 1188, 1166, 1128, 1094, 1074, 1060, 1016, 832, 818 cm^{-1}; ^{1}H NMR (deuteriochloroform): δ 1.06 (m, 6H, CH_2), 2.59–2.73 (m, 1H, NCH_{2a}) 2.66 (m, 1H, NCH), 3.49 (s, 3H, NCH_3), 3.54–3.60 (m, 1H, NCH_{2b}) 3.91 (dd, 1H, OCH_{2a}, $J_1 = 8.4$, $J_2 = 11.4$), 4.21 (dd, 1H, OCH_{2b}, $J_1 = 3.2$, $J_2 = 11.2$), 5.86 (s, 1H, 1-H); ^{13}C NMR (deuteriochloroform): δ 22.2, 23.8 and 26.7 (CH_2), 38.1 (NCH_3), 44.8 (NCH_2), 51.8 (NCH), 69.2 (OCH_2), 101.4 (C-1), 140.6 (C-11a), 143.5 (C-4a), 161.3 (C-2). Anal. calcd for C$_{11}$H$_{15}$N$_3$O$_2$: C, 59.71; H, 6.83; N, 18.99. Found: C, 59.45; H, 6.95; N, 18.89.

5-Benzyl-2-methyl-6,7-dihydro-5H-pyrida-zino[3,4-b][1,4]oxazin-3(2H)-one (**9c**). This compound was obtained as white crystals (yield: 96%); mp 105–106 °C; $R_f = 0.42$ toluene:methanol (8:2); IR (potassium bromide): ν 3500, 2913, 1645, 1601, 1550, 1490, 1443, 1346, 1276, 1237, 790, 703 cm^{-1}; ^{1}H NMR (deuteriochloroform): δ 3.46 (t, 2H, NCH_2, $J = 4.6$), 3.56 (s, 3H, NCH_3), 4.34 (t, 2H, OCH_2, 4.45 (s, 2H, benzyl CH_2), 5.87 (s, 1H, 4-H), 7.15–7.4 (m, 5H, aromatic of benzyl); ^{13}C NMR (deuterio-chloroform): δ 38.2 (NCH_3), 45.7 (NCH_2), 53.4 (benzyl CH_2), 64.4 (OCH_2), 99.7 (C-4), 126.7 (C-2$'$ and C-6$'$), 127.9 (C-4$'$), 128.9 (C-3$'$ and C-5$'$), 134.4 and 139.6 (C-1$'$ and C-4a), 143.1 (C-8a), 161.4 (C-3). Anal. calcd for C$_{14}$H$_{15}$N$_3$O$_2$: C, 65.35; H, 5.88; N, 16.33. Found: C, 65.46; H, 5.92; N, 16.19.

5-Benzyl-2-methyl-5,6,7,8-tetrahydropyrida-zino[3,4-b][1,4]oxazepin-3(2H)-one (**9d**). This compound was obtained as white crystals (yield: 91%); mp 163–164 °C; $R_f = 0.50$ toluene:methanol (8:2); IR (potassium bromide): ν 3437, 2930, 1631, 1523, 1445, 1325, 1239 cm^{-1}; ^{1}H NMR (deuteriochloro-form): δ 2.09 (qui, 2H, CH_2, $J = 6.6$), 3.57 (s, 3H, NCH_3), 3.60 (t, 2H, NCH_2), 4.31 (t, 2H, OCH_2), 4.43 (s, 2H, benzyl CH_2), 5.86 (s, 1H, 4-H), 7.15–7.4 (m, 5H, aromatic of benzyl); ^{13}C NMR (deuteriochloro-form): δ 28.5 (CH_2), 38.5 (NCH_3), 48.7 (NCH_2), 56.2 (benzyl CH_2), 69.1 (OCH_2), 104.8 (C-4), 126.9 (C-3$'$ and C-5$'$), 127.8 (C-4$'$), 129.0 (C-2$'$ and C-6$'$), 135.4 and 138.1 (C-1$'$ and C-4a), 147.1 (C-9a), 161.6 (C-3). Anal. calcd for C$_{15}$H$_{16}$N$_3$O$_2$: C, 66.40; H, 6.32; N, 15.49. Found: C, 66.10; H, 6.31; N, 15.51.

2.7. General procedure for preparation of aldehydes 10 by Vilsmeier–Haack reaction

A solution of **9** (3.69 mol) in dry dimethylforma-mide (3.12 ml) was cooled by ice-water bath. A solution of POCl$_3$ (0.76 ml) in dry dimethylforma-mide (1.76 ml), was added dropwise under 6 °C. The reaction mixture was allowed to warm to room temperature, and was heated at 70 °C for 75 min. After evaporation of the solvent (under 60 °C in vacuo), ice (25 g) was added to the black oily residue and the mixture was allowed to warm to room temperature. Then it was made alkaline with aqueous 40% sodium hydroxyde (pH 8) and was extracted with ethyl acetate. The combined organic phases were dried over anhydrous magnesium sulfate. The solvent was evaporated in vacuo to give the product in pure form. In case of compound **10b**, the crude product was chromatographed on silica gel (ethyl acetate:chloro-form (1:9)).

3-Methyl-2-oxo-2,3,6a,7,8,9-hexahydro-6H-pyri-dazino[3,4-b]pyrrolo[1,2-d][1,4]oxazine-1-carbalde-hyde (**10a**). This compound was obtained as white crystals (yield: 58%); mp 155–156 °C; $R_f = 0.45$ ethyl acetate:chloroform (9:1, v/v); IR (potassium bromide): ν 2860, 1666, 1618, 1590, 1532, 1455, 1385, 1323, 1295, 1252, 1070, 975 cm^{-1}; ^{1}H NMR (deuteriochloroform): δ 1.58–2.29 (m, 4H, CH_2), 3.23–3.37 (m, 1H, NCH_{2a}), 3.54–3.68 (m, 1H, NCH_{2b}), 3.58 (s, 3H, NCH_3), 3.80–3.98 (m, 2H, OCH_{2a} and NCH), 4.49–4.62 (m, 1H, OCH_{2b}) 10.31 (s, 1H, CHO); ^{13}C NMR (deuteriochloroform): δ 24.0 and 26.7 (CH_2), 38.1 (NCH_3), 54.2 (NCH_2), 56.6 (NCH), 68.0 (OCH_2), 106.7 (C-1), 138.1 (C-10a), 142.2 (C-4a), 161.7 (C-2), 190.2 (CHO). Anal. calcd for C$_{11}$H$_{13}$N$_3$O$_3$: C, 56.16; H, 5.57; N, 17.86. Found: C, 56.06; H, 5.54; N, 17.83.

3-Methyl-2-oxo-2,3,6,6a,7,8,9,10-octahydropyri-dazino[3,4-b]pyrido[1,2-d][1,4]oxazine-1-carbalde-hyde (**10b**). This compound was obtained as yellow crystals (yield: 71%), mp 129–130 °C; $R_f = 0.20$ ethyl acetate:chloroform (1:9); IR (potassium bro-mide): ν 3906, 3552, 3428, 3060, 2930, 2852, 1670, 1624, 1588, 1530, 1478, 1434, 1398, 1350, 1306, 1272, 1248, 1160, 1142, 1080, 1026, 1008, 958, 920 cm^{-1}; ^{1}H NMR (deuteriochloroform): δ 1.18–1.96 (m, 6H, CH_2), 3.04–3.18 (m, 1H, NCH), 3.50–3.57 (m, 5H, NCH_2 and NCH_3), 3.54–3.68 (dd, 1H,

OCH_{2a}, $J_1 = 7.0$, $J_2 = 11.8$), 3.80–3.98 (dd, 1H, OCH_{2b}, $J = 3.8$), 10.10 (s, 1H, CHO); ^{13}C NMR (deuteriochloroform): δ 22.7, 25.1 and 28.0 (CH_2), 38.0 (NCH_3), 54.7 (NCH_2), 57.2 (NCH), 68.2 (OCH_2), 106.7 (C-1), 139.7 (C-11a), 144.9 (C-4a), 162.4 (C-2), 188.2 (CHO). Anal. calcd for $C_{12}H_{15}N_3O_3$: C, 57.82; H, 6.07; N, 16.86. Found: C, 57.35; H, 6.18; N, 16.69.

5-Benzyl-2-methyl-3-oxo-2,3,6,7-tetrahydro-5H-pyridazino[3,4-b][1,4]oxazine-4-carbaldehyde (**10c**). This compound was obtained as yellow crystals (yield: 81%), mp 188–189 °C; $R_f = 0.42$ ethyl acetate:chloroform (9:1); IR (potassium bromide): ν 3439, 2857, 1658, 1615, 1532, 1472, 1354, 1284 cm^{-1}; ^1H NMR (deuteriochloroform): δ 3.44 (t, 2H, NCH_2, $J = 4.9$), 3.58 (s, 3H, NCH_3), 4.28 (t, 2H, OCH_2), 4.57 (s, 2H, benzyl CH_2), 7.13–7.37 (m, 5H, aromatic of benzyl), 10.20 (s, 1H, CHO); ^{13}C NMR (deuteriochloroform): δ 37.9 (NCH_3), 48.1 (NCH_2), 59.7 (benzyl CH_2), 63.6 (OCH_2), 107.4 (C-4), 128.5 and 128.9 (C-2',3',4',5',6'), 133.6 (C-1'), 139.9 (C-4a), 144.3 (C-8a), 161.9 (C-3), 189.2 (CHO). Anal. calcd for $C_{15}H_{15}N_3O_3$: C, 63.15; H, 5.30; N, 14.73. Found: C, 63.35; H, 5.42; N, 14.69.

5-Benzyl-2-methyl-3-oxo-2,3,5,6,7,8-hexahydro-pyridazino[3,4-b][1,4]oxazepine-4-carbaldehyde (**10d**). This compound was obtained as yellow crystals (yield: 82%); mp 140–141 °C; $R_f = 0.48$ toluene: methanol (4:1); IR (potassium bromide): ν 2919, 1689, 1630, 1515, 1447, 1358, 1250, 1086 cm^{-1}; ^1H NMR (deuteriochloroform): δ 2.00 (qui, 2H, CH_2, $J = 5.2$), 3.50 (t, 2H, NCH_2), 3.64 (s, 3H, NCH_3), 4.14 (t, 2H, OCH_2), 4.31 (s, 2H, benzyl CH_2), 7.20–7.36 (m, 5H, aromatic of benzyl), 10.24 (s, 1H, CHO); ^{13}C NMR (deuteriochloroform): δ 27.5 (CH_2), 38.5 (NCH_3), 49.7 (NCH_2), 62.8 (benzyl CH_2), 69.9 (OCH_2), 112.0 (C-4), 128.4 (C-4'), 128.6 and 128.9 (C-2',3',5',6'), 135.9 (C-1'), 146.8 (C-4a and C-9a), 162.9 (C-3), 188.7 (CHO). Anal. calcd for $C_{16}H_{17}N_3O_3$: C, 64.20; H, 5.70; N, 14.00. Found: C, 63.84; H, 5.74; N, 13.82.

2.8. General procedure for the synthesis of 11a and 11c

To a solution of a compound **10** (1.0 mol) in ethanol, Meldrum's acid (1.0 mol) was added. The mixture was stirred at ambient temperature until the starting material had been consumed (TLC). The precipitated product was filtered off, then washed with ethanol to give analytically pure product.

5-(3'-Methyl-2'-oxo-2',3',6a',7',8',9'-hexahydro-6'H-pyridazino[3,4-b]pyrrolo[1,2-d][1,4]oxazine-1-ylmethylene)-2,2-dimethyl-1,3-dioxane-4,6-dione (**11a**). This compound was obtained as yellow crystals (yield: 79%); mp 210–211 °C; $R_f = 0.27$ ethyl acetate:chloroform (9:1); IR (potassium bromide): ν 1758, 1732, 1634, 1598, 1580, 1530, 1454, 1392, 1358, 1330, 1310, 1282, 1230, 1198, 1122, 1096, 1032, 1008, 914 cm^{-1}; ^1H NMR (deuteriochloroform): δ 1.22–2.26 (m, 4H, CH_2), 1.78 and 1.94 (bs, 3H, CH_3 and bs, 3H, CH_3), 3.49 (t, 2H, NCH_2, $J = 5.8$), 3.58 (s, 3H, NCH_3), 3.67–3.91 (m, 2H, NCH and OCH_{2a}), 4.56 (bd, 1H, OCH_{2b}) 8.30 (s, 1H, CH); ^{13}C NMR (deuteriochloroform): δ 24.1 and 26.3 (CH_2), 27.1 and 27.3 (CH_3), 38.9 (NCH_3), 53.0 (NCH_2), 57.0 (NCH), 67.8 (OCH_2), 104.9 (C-2), 106.9 (C-1'), 116.8 (C-5), 139.6 (C-10a'), 141.9 (C-4a'), 147.8 (CH), 157.8 (C-2'), 159.8 and 162.7 (C-4,6). Anal. calcd for $C_{17}H_{19}N_3O_6$: C, 56.51; H, 5.30; N, 11.63. Found: C, 56.29; H, 5.25; N, 11.57.

5-(5'-Benzyl-2'-methyl-3'-oxo-2',3',6',7'-tetrahydro-5'H-pyridazino[3,4-b][1,4]oxazin-4-ylmethylene)-2,2-dimethyl-1,3-dioxane-4,6-dione (**11c**). This compound was obtained as yellow crystals (yield: 94%); mp 175–176 °C; $R_f = 0.49$ ethyl acetate:chloroform (9:1); IR (potassium bromide): ν 2927, 1759, 1723, 1642, 1572, 1493, 1444, 1355, 1277, 1192, 1023, 929 cm^{-1}; ^1H NMR (deuteriochloroform): δ 1.78 and 1.90 (bs, 3H, CH_3 and bs, 3H, CH_3), 3.48 (t, 2H, NCH_2, $J = 4.6$), 3.61 (s, 3H, NCH_3), 4.23–4.57 (m, 4H, benzyl CH_2 and OCH_2), 7.16–7.39 (m, 5H, aromatic of benzyl), 8.39 (s, 1H, CH); ^{13}C NMR (deuteriochloroform): δ 27.5 ($2\times$ CH_3), 39.1 (NCH_3), 48.0 (NCH_2), 58.4 (benzyl CH_2), 63.9 (OCH_2), 105.1 (C-2), 107.3 (C-4'), 116.0 (C-5), 128.1 and 129.3 (C-2'',3'',5'',6''), 128.9 (C-4''), 134.0 (C-1''), 141.9 (C-4a'), 144.0 (C-8a'), 149.3 (CH), 158.7 (C-3'), 159.9 and 162.4 (C-4,6). Anal. calcd for $C_{21}H_{21}N_3O_6$: C, 61.31; H, 5.14; N, 10.21. Found: C, 61.21; H, 5.15; N, 10.20.

2.9. Preparation of 2,2′,2′-trimethyl-6-phenylspiro[2,4,5,6,8,9-hexahydro-3H,7H-10-oxa-1,2,6a-triazacyclohepta[d,e]naphthalene-5,5′-(dihydro-4′H-[1,3]dioxane)(-3,4′,6′-trione (12)

A solution of **10d** (0.3 g, 1.0 mol) and Meldrum's acid (0.15 g, 1.0 mol) in dry ethanol (15 ml), was stirred at room temperature for 4 h. After evaporation of the solvent, residue thus obtained was triturated with petroleum ether and the precipitate was filtered off to give the crude product, which was then recrystallized from toluene. Yield: 59%; mp 201–202 °C; $R_f = 0.37$ ethyl acetate:chloroform (9:1); IR (potassium bromide): 2942, 1771, 1739, 1629, 1598, 1536, 1382 cm^{-1}; ^{1}H NMR (deuteriochloroform): δ 0.91 (s, 3H, CH_3), 1.63 (s, 3H, CH_3), 1.67–2.05 (m, 2H, 4-H_2), 3.10–3.68 (m, 4H, 8-H_2 and NCH_2), 3.64 (s, 3H, NCH_3), 4.47 (m, 2H, OCH_2), 4.81 (s, 1H, NCH), 7.29–7.40 (m, 5H, phenyl); ^{13}C NMR (deuteriochloroform): δ 27.9 and 29.4 (CH_3), 28.6 and 30.8 (C-4 and C-8), 38.8 (NCH_3), 47.2 (NCH_2), 51.9 (C-5), 66.6 (NCH), 68.2 (OCH_2), 105.7 (C-2′), 112.5 (C-3a), 128.8, 129.2 and 129.8 (C-2″,3″,4″,5″,6″), 134.4 (C-1″), 142.6 (C-10b), 146.3 (C-10a), 159.4 (C-3), 163.6 and 168.1 (C-6′ and C-4′). Anal. calcd for $C_{22}H_{23}N_3O_6$: C, 62.11; H, 5.45; N, 9.88. Found: C, 61.75; H, 5.44; N, 9.88.

Crystal data of compound **12**: monoclinic, space group $P2_1/c$, $a = 10.459(7)$ Å, $b = 10.962(6)$ Å, $c = 17.364(14)$ Å, $\beta = 96.13(7)°$, $V = 1979(2)$ Å^3, $Z = 4$, $D_{calc} = 1.428$ g cm^{-3}, $F(000) = 896$, μ(Mo K$_\alpha$) = 0.105 mm^{-1}, $\lambda = 0.71069$ Å. 2406 independent reflections were collected to theta range 1.96–22.60°. Index ranges $-10 \leq h \leq 10$, $-11 \leq k \leq 11$, $-18 \leq l \leq 18$. Full matrix least-squares refinement was carried out on F^2. Data/restraints/parameters ratio: 2406/0/286. Goodness-of-fit on F^2: 1.060. Final R indices for reflections with $I > 2\sigma(I)$: $R = 0.0766$, $wR_2 = 0.2173$. Final R indices for all reflections: $R = 0.0989$, $wR_2 = 0.2377$. The extinction coefficient is 0.115(14). The largest residual peak and hole in the final difference electron density map are 0.300 and -0.333, respectively [8].

2.10. Preparation of 2-methyl-3-oxo-6-phenyl-2,4,5,6,8,9-hexahydro-3H,7H-10-oxa-1,2,6a-triazacyclohepta[1,2,3-de]naphthalene-5-carboxylic acid (13)

A solution of compound **12** (0.1 g, 0.24 mol) in concentrated hydrochloric acid (10 ml) was refluxed for 3 h. Water (10 ml) was added and the pH was adjusted to 5–6 with 5 M sodium hydroxide. After extraction with chloroform (3 × 20 ml), the organic phase was dried over anhydrous magnesium sulfate, then the solvent was evaporated in vacuo to give a yellow solid which was subjected to column chromatography by using a mixture of ethyl acetate:acetone: water (4:4:1) as the eluent, to give white crystals (yield: 86%); mp 172–174 °C; $R_f = 0.16$ ethyl acetate:ace-tone:water (4:4:1); IR (potassium bromide): ν 3430, 2926, 2856, 2574, 2382, 1728, 1604, 1568, 1528, 1484, 1446, 1382, 1358, 1312, 1266, 1240, 1196, 1158, 1088, 954, 752, 702 cm^{-1}; ^{1}H NMR (deuteriochloroform): δ 1.80–2.20 (m, 2H, 8-H), 2.43 (dd, 1H, 4-H_a, $J_1 = 13.8$, $J_2 = 19.4$), 2.94–3.08 (m, 2H, 5-H and 4-H_b), 3.30–3.70 (m, 5H, NCH_2 and NCH_3), 4.12–4.40 (m, 2H, OCH_2), 4.77–4.87 (m, 1H, NCH), 7.07–7.32 (m, 5H, phenyl); ^{13}C NMR (deuteriochloroform): δ 19.1 and 20.4 (C-8), 27.9 and 28.2 (C-4), 39.2 (NCH_3), 42.6 and 43.3 (C-5), 48.6 and 48.8 (NCH_2), 64.0 and 64.1 (C-6), 69.1 and 69.3 (OCH_2), 111.6 and 112.1 (C-3a), 126.4, 127.4, 128.1, 128.4, 128.6 and 128.9 (C-2′,3′,4′,5′ and 6′), 138.1 and 140.2 (C-1′), 141.9 and 142.2 (C-10b), 146.6 and 146.9 (C-10a), 160.3 (C-3), 172.9 and 174.2 (COOH). EIMS spectrum (electron energy: 70 eV, electron current: 500 μA, evaporation temperature: 165 °C) characteristic spectral data: m/z 296, 100%, [M−COOH]$^+$; 341, 47%, [M]$^{+\cdot}$. CIMS spectrum (i-butane as reagent gas, evaporation temperature: 150 °C) characteristic spectral data: m/z 342, 100%, [M + H]$^+$. FAB spectra (matrix: m-nitrobenzylic alcohol, FAB gas: xenon, FAB gun accelerating voltage: 9 kV) characteristic spectral data (negative FAB): m/z 340, 100%, [M−H]$^-$; 296, 21%, [M−H−CO$_2$]$^-$.

2.11. General procedure for the synthesis of pyrrolo fused pyridazines 14a and 14b

To a suspension of NaH (0.03 g) in dimethylformamide (4 ml), a solution of aldehyde **10c** or **10d**

(0.67 mol) in dry dimethylformamide (2 ml) was dropped at 5 °C. The reaction mixture was stirred in an ice-water bath for 15 min, then it was heated at 120 °C for 2 h. After cooling, the solvent was evaporated in vacuo, the residue was taken up in dry toluene (5 ml), and acidified with concentrated hydrochloric acid (pH 2). The mixture was heated at 100 °C for 2 h, then the solvent was evaporated in vacuo. The residue was taken up in water (10 ml), the solution was neutralized by the addition of 2 M sodium hydroxide and extracted with chloroform. The combined organic phases were dried over anhydrous magnesium sulfate and the solvent was evaporated in vacuo. The solid residue was recrystallized from ethyl acetate to give the pure product.

7-Methyl-2-phenyl-3,4-dihydro-5-oxa-2a,6,7-triazaacenaphthylen-8(7H)-one (**14a**). This compound was obtained as white crystals (yield: 20%); mp 196–197 °C; $R_f = 0.33$ ethyl acetate:chloroform (9:1); IR (potassium bromide): ν 3427, 2925, 2844, 1743, 1641, 1582, 1510, 1031 cm^{-1}; ^{1}H NMR (deuteriochloroform): δ 3.77 (s, 1H, NCH_3), 4.36 (dd, 2H, NCH_2, $J_1 = 3.8, J_2 = 4.4$), 4.64 (dd, 2H, OCH_2), 6.94 (s, 1H, CH), 7.43–7.48 (m, 5H, aromatic of phenyl) ^{13}C

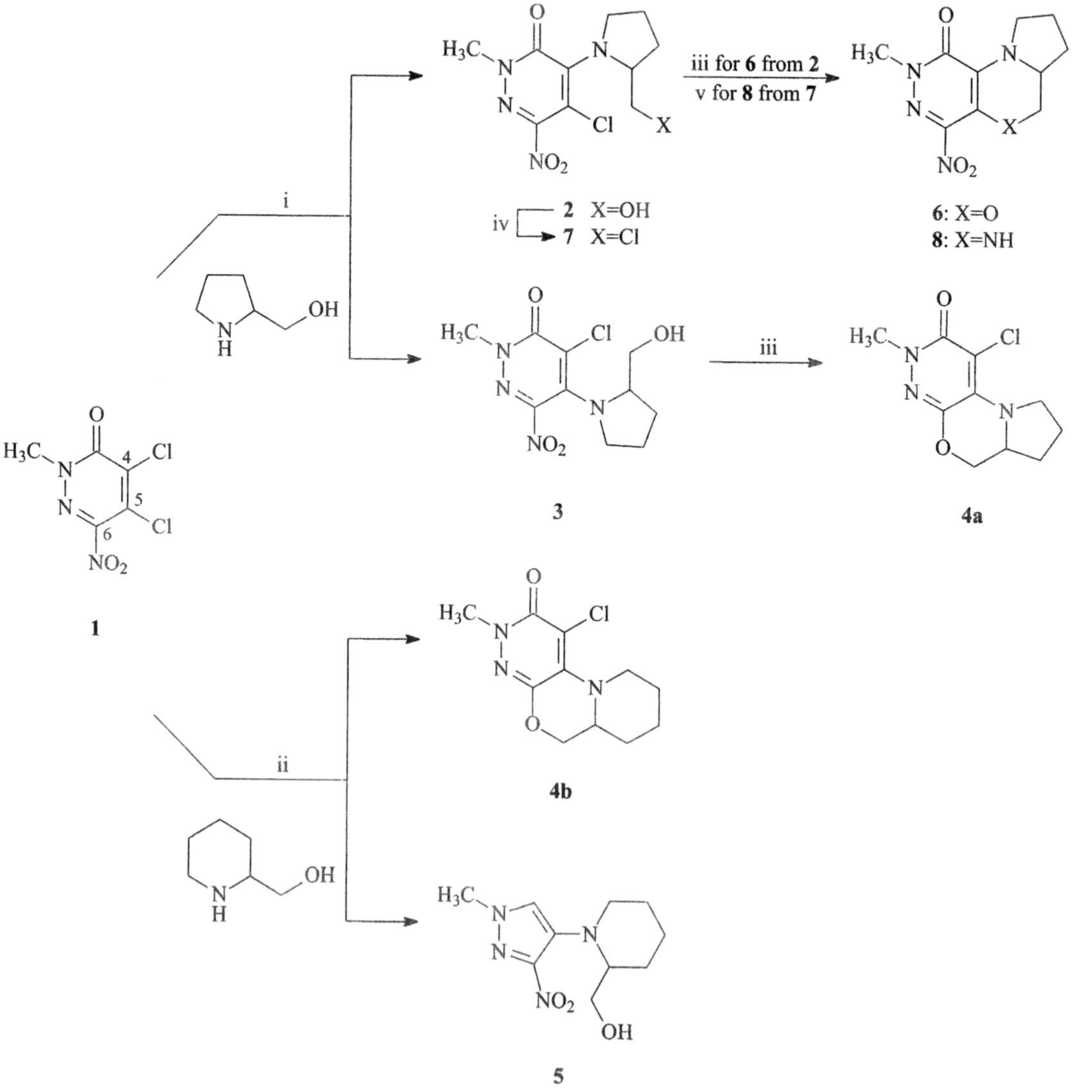

Scheme 1. Reaction conditions: (i) EtOH/Et$_3$N, Δ; (ii) H$_2$O/EtOH (80:20), Δ; (iii) NaOEt/EtOH, Δ; (iv) SOCl$_2$; (v) ccNH$_4$OH, Δ.

NMR (deuteriochloroform): δ 39.0 (NCH$_3$), 43.3 (NCH$_2$), 67.0 (OCH$_2$), 105.1 (CH), 119.9, (C-8a), 123.7 (C-2), 128.1 and 129.2 (C-2′,3′,5′,6′), 129.0 (C-4′), 130.1 (C-1′), 140.4 (C-8b), 144.2 (C-5a) 158.7 (C-8); INAPT OCH$_2$ (C-5a), NCH$_2$ (C-2, C-8b); Anal. calcd for C$_{15}$H$_{13}$N$_3$O$_2$: C, 67.41; H, 4.90; N, 15.72. Found: C, 67.14; H, 4.88; N, 15.70.

4-Methyl-1-phenyl-8,9-dihydro-7H-6-oxa-4,5,9a-triazabenzo[cd]azulen-3(4H)-one (**14b**). This compound was obtained as pale yellow crystals (yield: 46%); mp 127–128 °C; R_f = 0.28 toluene:methanol (4:1); IR (potassium bromide): ν 2925, 1745, 1640, 1550, 1519, 1476, 1369, 1254, 1139 cm^{-1}; ^{1}H NMR (deuteriochloroform): δ 2.42 (qui, 2H, CH$_2$, J = 5.6), 3.76 (s, 3H, NCH$_3$), 4.21 (t, 2H, NCH$_2$), 4.50 (t, 2H, OCH$_2$), 6.94 (s, 1H, CH), 7.40–7.50 (m, 5H, aromatic of benzyl); ^{13}C NMR (deuteriochloroform): δ 20.8 (CH$_2$), 38.5 (NCH$_3$), 48.8 (NCH$_2$), 70.3 (OCH$_2$), 107.1 (CH), 123.7 (C-2a), 126.1 (C-1), 128.9 and 129.6 (C-2′,3′,5′,6′), 129.1 (C-4′), 131.1 (C-1′), 142.7(C-9b), 144.3(C-5a), 157.9 (C-3); INAPTCH (C-1, C-2a), OCH$_2$ (C-5a, NCH$_2$, CH$_2$), NCH$_2$ (C-9b, C-1, CH$_2$). Anal. calcd for C$_{16}$H$_{15}$N$_3$O$_2$: C, 68.31; H, 5.37; N, 14.94. Found: C, 68.34; H, 5.73; N, 14.83.

3. Results and discussion

Reaction of the 4,5-dichloro-2-methyl-6-nitropyridazin-3(2H)-one (**1**) with two cyclic aminoalcohols, 2-pyrrolidine- and 2-piperidinemethanol. In the first case, the expected 4- and 5-hydroxyalkylamino derivatives **2** and **3**, were formed which

upon treatment with sodium ethoxide resulted in the corresponding tricyclic derivatives **6** and **4a**, respectively. Compound **7** possessing a chloroalkyl side chain underwent cyclization with concentrated ammonia to afford the tricyclic compound **8**. Interestingly, treatment of **1** with the piperidine analogue of prolinol under similar conditions, the corresponding 4- and 5-regioisomers could not be isolated. In a more polar medium (ethanol and water, 1/4 vol/vol ratio), after isolation, we observed that the two major components were not the desired regioisomers. Instead, the fused system **4b**, and a pyrazole derivative **5** were obtained. The structures of these compounds were elucidated by ^{1}H and ^{13}C NMR spectral data; in case of **5**, it was confirmed by HMBC measurements. Formation of the tricyclic pyridazino[3,4-b]oxazine **4b** could be explained by a nucleophilic displacement of the 5-chloro atom of **1** followed by a ring closure reaction to C-6. On the other hand, formation of **5** may be a result of a ring contraction reaction with a similar mechanism suggested for some pyridazinone derivatives [9] (Scheme 1).

Next, reactivity of bicyclic pyridazines **9** was investigated. It could be expected that these compounds possessing an enamino-carbonyl moiety might undergo formylation under Vilsmeier–Haack conditions. Indeed, we found that the reaction proceeded smoothly to afford aldehydes **10**. Versatility of the aldehydes could also be demonstrated. Namely, in two synthetic pathways, the aldehydes were utilized for the synthesis of *ortho*- and *peri*-fused ring systems (Scheme 2).

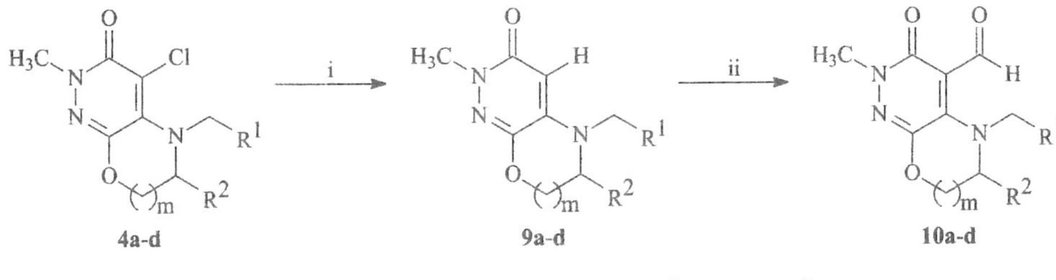

	R^1	R^2	m
a:	—CH$_2$CH$_2$—		1
b:	—CH$_2$CH$_2$CH$_2$—		1
c:	Ph	H	1
d:	Ph	H	2

Scheme 2. Reaction conditions: (i) Pd/C, HCOONH$_4$, EtOH, Δ; (ii) POCl$_3$/DMF, Δ.

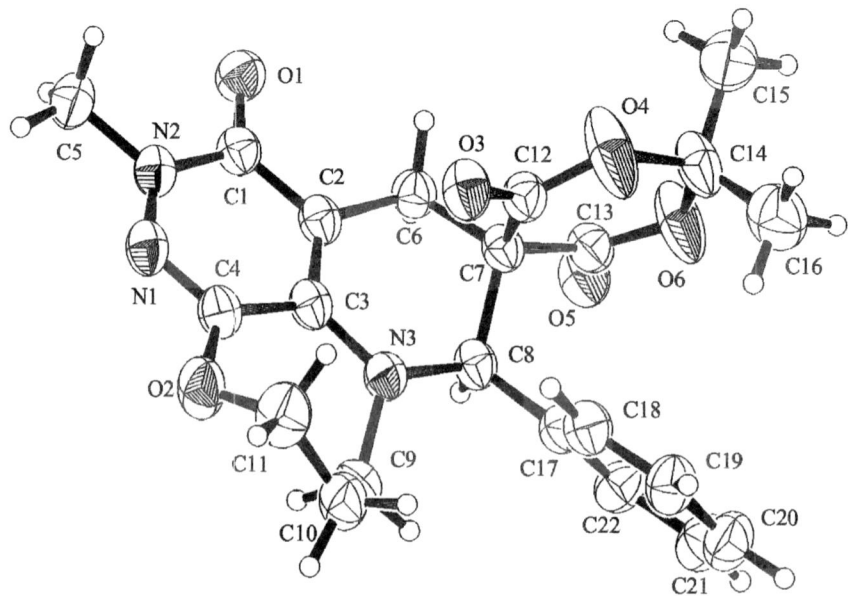

for **a, c, d** see Scheme 2.

Scheme 3. Reaction conditions: (i) Meldrum's acid, EtOH, rt; (ii) cc HCl/rfx.

Fig. 1. ORTEP drawing and numbering of **12** with displacement ellipsoids drawn at 30% probability level.

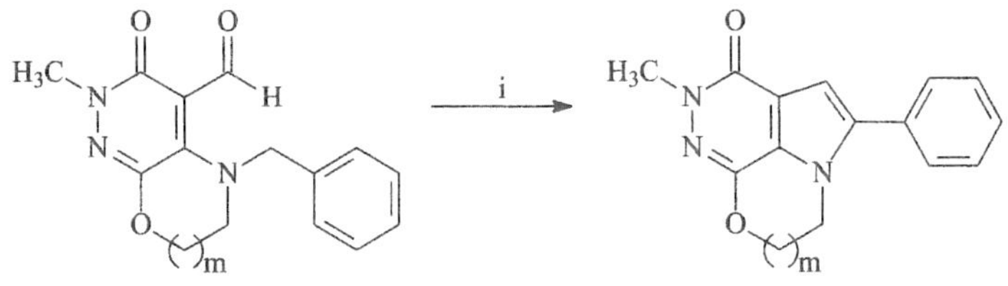

Scheme 4. Reaction conditions: (i): I: NaH/DMF, 120 °C; II: toluene/cc HCl, Δ.

The *tert*-amino effect has been demonstrated to be a powerful approach to obtain ring systems containing a spiro moiety [10]. Application of this methodology to compounds **10**, **11**, fairly unexpectedly, different results were obtained (Scheme 3). The vinyl derivatives **11a, c**, isolated smoothly from the reaction of the respective aldehydes with Meldrum's acid, could not be cyclized to pyridopyridazines under the usual conditions [10]. On the contrary, in case of the aldehyde **10d**, which possesses an oxazepine ring in place of an oxazine, the spirocyclic **12** was formed under very mild conditions. Structure of **12** was also confirmed by X-ray investigation (Fig. 1). The synthetic value of compounds **12**, and the synthetic route of **10–13** as well could be demonstrated by the hydrolytic cleavage of the dioxane part of **12** to form a carboxylic group, and by this way to obtain the otherwise hardly accessible pyridopyridazinecarboxylic acid **13**.

The other pathway for the utilization of aldehydes was shown in Scheme 4. In these cases an intramolecular condensation between the aldehyde carbonyl and benzyl methylene (cf. [11]) led to the formation of the novel ring system **14**.

4. Conclusion

4,5-Dichloro-2-methyl-6-nitropyridazin-3(2*H*)-one (**1**) is a particularly versatile building block for the synthesis of polycyclic pyridazines due to its generally well-predictable functionalization. In this work, starting form compound **1**, several novel pyridazino ring systems were prepared by combination of:

(i) inter- and intra-molecular nucleophilic substitution reactions;
(ii) nucleophilic and electrophilic substitution reactions, and the *tert*-amino effect;
(iii) nucleophilic and electrophilic substitution reactions, and an intramolecular condensation reaction.

It was also demonstrated that the *tert*-amino effect with properly selected aldehyde and active methylene compound might be of high synthetic value for the functionalization of polycyclic compounds.

Acknowledgements

Financial support of this work by National Research Foundation (T-31910) and Ministry of Health (ETT/121/2003) are acknowledged. The authors also express their gratitude to Prof. J. Nyitrai for his help in naming of new structures, B. Podányi for NMR measurements, Á. Puhr for microanalyses, and M. Mogyorósi for the preparative assistance.

References

[1] P. Mátyus, J. Heterocycl. Chem. 35 (1998) 1075.
[2] L. Károlyházy, M. Freile, M.N.S. Anwair, Gy. Beke, F. Giannini, M.V. Castelli, M. Sortino, J.C. Ribas, S. Zacchino, P. Mátyus, R.D. Enriz, Arzneimittel-Forsch. (2003) In press.
[3] O. Éliás, L. Károlyházy, G. Stájer, F. Fülöp, K. Czakó, V. Harmat, O. Barabás, K. Keserű, P. Mátyus, J. Mol. Struct. (Theochem) 545 (2001) 75.
[4] P. Tapolcsányi, G. Krajsovszky, R. Andó, P. Lipcsey, Gy. Horváth, P. Mátyus, Zs. Riedl, Gy. Hajós, B.U.W. Maes, G.L.F. Lemière, Tetrahedron 58 (2002) 10137.
[5] Molecular Structure Co., Houston, Texas, USA, 1992.

[6] G.M. Sheldrick, University of Gottingen, Germany, 1993.

[7] T. Terai, H. Azuma, R. Hattori, Jap. Pat. Appl. (1967); Chem. Abstr., 66 (1967) 65497z; CAS: 13645-28-8.

[8] Atomic coordinates, bond lengths, angles and atomic displacement parameters have been deposited at the Cambridge Crystallographic Data Centre under deposition code: CCDC214424.

[9] Y. Maki, G.B. Beardsley, M. Takaya, Tetrahedron Lett. 19 (1971) 1507. (b) Y. Maki, M. Takaya, Chem. Pharm. Bull. 20 (1972) 747. (c) G. Heinisch, B. Matuszczak, K. Mereiter, Heterocycles 38 (1994) 2081.

[10] A. Schwartz, G. Beke, Z. Kovári, Z. Böcskey, Ö. Farkas, P. Mátyus, J. Mol. Struct. (Theochem) 528 (2000) 49.

[11] G.H. Walizei, E. Breitmaier, Synthesis (1989) 337.

ELSEVIER

Journal of Molecular Structure (Theochem) 666–667 (2003) 637–644

THEO
CHEM

www.elsevier.com/locate/theochem

Computer modelling of enzyme reactions

Gábor Náray-Szabó[a,*], Imre Berente[b]

[a]*Protein Modelling Group, Eötvös Loránd University—Hungarian Academy of Sciences, H-1117 Budapest, Pázmány Péter st. 1A, Hungary*
[b]*Department of Theoretical Chemistry, Eötvös Loránd University, 1117 Budapest, Pázmány Péter st. 1A, Hungary*

Abstract

We give an overview on modern methods and atomistic models for the computation of enzyme mechanisms. It has to be stressed that models and methods must be at the same level of sophistication. If essential components (H-bonding side chains, metal ion, cofactor, etc.) of the active site are absent from the model, even the largest basis set may provide false results. On the other hand, the largest model with dozens of explicit atoms may lead to artefacts if the accuracy of the computational method (minimal basis set ab initio, semi-empirical or molecular force field) is unsatisfactory. Enzyme reactions are especially adequate for the application of an embedding scheme. While the active site should be treated by a high-level quantum mechanical method, distant protein atoms may be considered at a lower level of accuracy, e.g. as fixed point charges. The biophase is best treated by the DelPhi method, which is based on the solution of the linearised Poisson–Boltzmann equation and considers important counter-ion shielding effects on charged surface side chains. An integrated version of the embedding scheme is the combination of quantum mechanical and molecular mechanical methods (QM/MM) allowing treating all protein atoms in a calculation, however, at different levels of sophistication. Here we may have problems for atoms at the boundary, for which we present our approach offering a natural partition of the active site and protein core by applying strictly localised molecular orbitals. Two enzyme mechanisms, those of HIV integrase and xylose isomerase will be discussed in more detail in order to illustrate the principles outlined above.
© 2003 Elsevier B.V. All rights reserved.

Keywords: Quantum mechanical methods; Molecular mechanical methods; Enzyme

1. Introduction

Computational treatment of enzymatic processes developed spectacularly in the past two decades [1]. Though we are far from being able to perform quantitative calculations on precisely defined, molecular models, like in case of small molecules in the gas phase, molecular graphics, molecular dynamics and quantum mechanics provide essential new information not available at present by even the most sophisticated experiments. Development of faster computers that are within the reach of the widest scientific community as well as efficient computational methods allows investigating systems between 50–100 atoms in the frame of quantum mechanics and up to 50,000 atoms with molecular dynamics. Since the models become increasingly realistic, direct comparison with experimental data becomes possible. Besides X-ray crystallography and nuclear magnetic resonance spectroscopy, molecular

---

* Corresponding author. Address: Department of Theoretical Chemistry, Lorand Eotvos University, Pazmany, Peter St. 2, Budapest 1117, Hungary. Tel.: +36-1-209-0555; fax: +36-1-209-0602.

E-mail address: naraysza@para.chem.elte.hu (G. Náray-Szabó).

0166-1280/$ - see front matter © 2003 Elsevier B.V. All rights reserved.
doi:10.1016/j.theochem.2003.08.088

modelling became the third most powerful and generally applicable method for the elucidation of enzyme mechanisms.

In spite of the considerable advancement, none of the above three methods can be applied routinely and exclusively if we wish to obtain appropriate results. Before drawing final conclusions, a careful analysis of their performance should be carried out in order to avoid artefacts and misinterpretations. The best way to get a full and adequate picture on structural aspects of enzymatic processes is to combine various experimental and computational methods. While the primary information is provided by kinetic data, a well-defined molecular model can be obtained by X-ray crystallography or NMR spectroscopy. Since these methods do not provide direct information on energy aspects quantum mechanical or molecular dynamics calculations are needed to complete the picture and to understand the detailed mechanism. The growing importance of this trend has already been stressed more than twenty years ago [2].

In the following, we give an overview on structural models and computational methods applied to the study of enzyme reactions and present two examples, HIV integrase and xylose isomerase, for illustration. Our goal is to help the reader in finding the appropriate method for his/her studies and in critical analysis of published results. For further reading beyond Ref. [1], we offer our book dedicated to computational approaches to biochemical reactivity [3].

2. Models and methods

Most aspects of enzyme reactions can be appropriately understood in terms of layered models, since the crucial chemical event, breaking or forming a bond, can be located to a few atoms at the active site. Accordingly, the majority of applied models are composed of three major regions: (1) atoms of the active site, embedded in the (2) protein core and the (3) biophase with bulk water and dissolved counter ions shielding ionised surface side chains. The non-replaceable pool of protein structures with atomic resolution is the Protein Data Bank [4] containing more than 21,000 entries referring to a wide variety of

wild type and mutant proteins, as well as nucleic acids and carbohydrates. Explicit active site and protein core models can be deduced using this source with some modifications on the substrate or mutated side chain, if necessary. In most cases, the structure of the relatively unstable protein-substrate complex is not available, instead the co-ordinate set for an enzyme-inhibitor complex can be used in molecular modelling in order to provide the desired model. It is very important to consider explicit (structural) water molecules that may be seen by X-ray diffraction; in this case the positions of their oxygen atoms are well defined. Hydrogen atoms should be oriented along potential hydrogen-bonding directions to neighbouring proton acceptor atoms. Sometimes loosely bound water molecules, coming from the bulk, may also get fixed at the active site and change even the qualitative features of the reaction profile. We will provide an example for such a case in Section 3 for the HIV integrase catalytic reaction.

Once, we have the appropriate model, an adequate computational method has to be selected. Owing to the layered nature of enzyme structures, some variant of the available quantum mechanical/molecular mechanical approach seems to be most appropriate for a calculation on enzyme mechanisms. Essentially, two types of QM/MM methods are applied, which differ in the partition scheme of the active site and environment connected to this by one or more covalent bonds. One way of partition is to saturate the dangling bond, formed after dissection of the quantum region and the environment, by hydrogen or a specially defined dummy atom [5]. Maybe the simplest application of this model is a variant, the ONIOM method [6], defining the total energy of the system as follows:

$$E(\text{QM} - \text{MM}) \approx E_{\text{QM}}(\text{QM} - \text{D}) + E_{\text{MM}}(\text{QM} - \text{MM}) - E_{\text{MM}}(\text{QM} - \text{D}) + E_{\text{NB}}(\text{QM}, \text{MM}) \quad (1)$$

Here QM, MM and D refer to the quantum mechanical and molecular mechanical regions, as well as the link atoms, respectively. The subscripts QM, MM and NB stand for energy expressions obtained from the quantum mechanical and molecular mechanical approximations and the non-bonding interaction between MM and QM regions, respectively. In general, this latter term is calculated by including protein atomic monopoles in the Hamiltonian of the quantum region.

The major drawback of this treatment is the potential steric conflict of the dummy atoms with their neighbours in the molecular mechanical region during geometry optimisation, which may lead to convergence problems, and provide even sometimes artefacts. Breaking bonds in the quantum region are also improperly treated by molecular mechanics, thus the correction term, $E_{MM}(QM - D)$, is also a potential source of error. There are several propositions to overcome the above difficulties and, in fact, concrete calculations are also performed following this philosophy, however, we do not believe that a long-term solution will be found for the problem.

More promising is the use of localised orbitals at the boundary of the QM and MM regions [7–11]. In this case, the quantum and classical systems fall apart without overlap (which means that there is no charge transfer between them) and the total energy is written as

$$E(QM - MM) \approx E_{QM}(QM - hybr) + E_{MM}(MM)$$
$$+ E_{NB}(QM, MM) \qquad (2)$$

'hybr' means here a hybrid orbital pointing towards the MM region (Fig. 1). It is also possible to consider other partitions of the four hybrids between the QM and MM regions. In this version of the QM/MM method, combinations with at least one atom in MM are included in the force field and the appropriate parameters can be fitted to the specific environment near the boundary, as done, e.g. by Philipp and Friesner [12].

A further reduction of computational work is possible if we use a locally dense basis set [13].

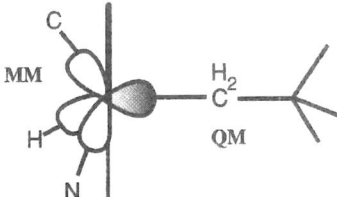

Fig. 1. Partition of a molecule into quantum (QM) and molecular mechanics (MM) regions by using strictly localised hybrids at the boundary.

In this approach, the QM region is further partitioned into a core, i.e. atoms directly involved in the reaction for which more basis functions are used, and surrounding shells to which basis sets of decreasing complexity are assigned. Though this treatment may lead to a certain accumulation of charges on atoms with more basis functions and deprivation of charges from those with smaller or minimal basis sets, this electronic rearrangement has only a minor effect on the reaction if it occurs in relatively distant regions.

The development of ab initio molecular dynamics in the mid-eighties [14] made possible to treat dynamic aspects of enzyme reactions concerning the core of the active site. Conventional molecular dynamics is not suitable for this purpose since the empirical description of bond breaking and bond formation is problematic, an appropriate parametrization is available only for limited bond types. The central concept of ab initio molecular dynamics is the treatment of both nucleic positions and electronic degrees of freedom in the time evolution of a fictitious classical system as dynamic variables. Though the computational procedure is very time consuming, it is possible to take crucial thermal effects into account. A QM/MM hybrid version of this method is available and has been applied to the galactose oxidase catalytic reaction [15].

Recently, a method has been proposed to treat direct quantum effects, like tunnelling and zero-point vibration, in enzyme reactions [16]. The discrete nature of quantum mechanical vibrational energies is incorporated in the treatment of nuclear motions, furthermore, multi-dimensional tunnelling contributions as well as dynamic re-crossing transmission coefficients are included.

Crucial electrostatic effects on the active site by the protein core and the surrounding biophase are best treated by the linearized Poisson–Boltzmann method, which is based on the following equation [17]

$$\nabla \varepsilon(\mathbf{r}) \nabla \phi(\mathbf{r}) - \varepsilon(\mathbf{r}) \lambda \kappa^2 \phi(\mathbf{r}) = -4\pi \rho(\mathbf{r}) \qquad (3)$$

where ε, ϕ and ρ denote the dielectric constant, the electrostatic potential and the charge density, respectively. Instead of having a uniform value as in earlier approaches, $\varepsilon(\mathbf{r})$ is continuously changing in space, thus it is possible to treat protein core, bulk water and the cloud of counter ions on a common and sound theoretical basis. $\lambda = 1$ in the solvent, while it is zero in the solute, while κ^2 is proportional to the ionic strength. It is supposed that counter ions have a Boltzmann distribution and their contribution is considered in the total charge density. Eq. (3) can be solved numerically, by finite difference methods that may become quite complicated. A computer code called DelPhi [18], is commercially available and allows even non-skilled users to perform calculations and estimate electrostatic effects of the environment. In the layered structure of enzyme models, the molecular mechanics region may be dropped and replaced by a continuum as described by the linearized Poisson–Boltzmann method, which makes calculations much simpler, yet quite precise, however, does not allow treating dynamic effects.

It must be stressed that models and methods to be used for the adequate description of a certain enzyme reaction must be at the same level of sophistication. If the model does not contain essential components (H-bonding side chains, metal ion, cofactor, etc.) of the active site, or the protein and solvent environment is not considered, even the largest basis set may provide false results. On the other hand, an extended model with a low quality force field or inadequate semi-empirical quantum mechanical method may also lead to artefacts or unfounded statements. Thus, before beginning a calculation, we should consider this aspect, and harmonise our model with the method applied.

3. Xylose isomerase

D-xylose isomerase catalyses the reversible conversion of D-xylose into D-xylulose and it is also capable to convert other sugars from aldose to ketose [19]. The enzyme requires a bivalent metal ion (Mg^{2+}, Co^{2+} or Mn^{2+}) for activation, while other metals (Zn^{2+} and Ca^{2+}) inhibit catalysis. The xylose–xylulose conversion takes place in two steps, first the α-pyranose ring opens, then a proton shuttle becomes activated and the acyclic intermediate transforms via hydride shift, this latter is the rate-limiting step [20,21]. A minimum type electrostatic potential pattern,

provided by the protein core and catalytic metal, stabilizes the $(- + -)$ charge distribution of the transition state in the ring opening step and thus accelerates it to an extent, which makes the hydride shift rate determining (Fig. 2).

The rate-limiting hydride shift step is also influenced by electrostatics, mainly by the effect of the metal ions located near the reaction center. Semi-empirical quantum mechanical calculations on the active site, surrounded by a polarisable region and fixed atomic point charges in two subsequent layers of enzyme atoms, indicate that the charge transfer to the catalytic metal is an important factor in determining its activation ability [21]. Our calculations indicate that a significant component of the catalytic effect is of local origin, i.e. it emerges from the electrostatic stabilization of the transition state by the close environment of the shifting hydrogen atom. This finding can be transferred to another case, the catalytic reaction by HIV integrase, where the transition-state complex is considerably stabilized by hydrogen bonds to the close-lying side chains as well as by the catalytic metal ion. Remote regions of the protein core have only a minor effect on the amount of electrostatic stabilization [22].

As it is seen from Table 1, larger calculated activation energies correspond to inhibition and they are connected to the amount of charge transfer to the metal. If the charge transfer is too large, the metal becomes inhibiting. It is also seen that the net charge on the O^2 atom, participating in the hydride shift (cf. Fig. 2) plays a role in determining the activation energy. The more negative is the oxygen atom, the higher is the activation barrier.

As we have mentioned earlier, Truhlar and co-workers elaborated a procedure for the estimation of kinetic isotope effects in various enzyme reactions [16]. We summarize their results for the hydride transfer step in xylose isomerase catalysis in Table 2. The experimental value is roughly reproduced, thus two important conclusions can be drawn from the calculations, that seem to be justified. First, the quasi-classical and tunnelling contributions are about of the same importance and second, distant atoms also play a role in the tunnelling contribution increasing from 2.08 to 2.68 if the quantum system is extended from 19 to 79 atoms.

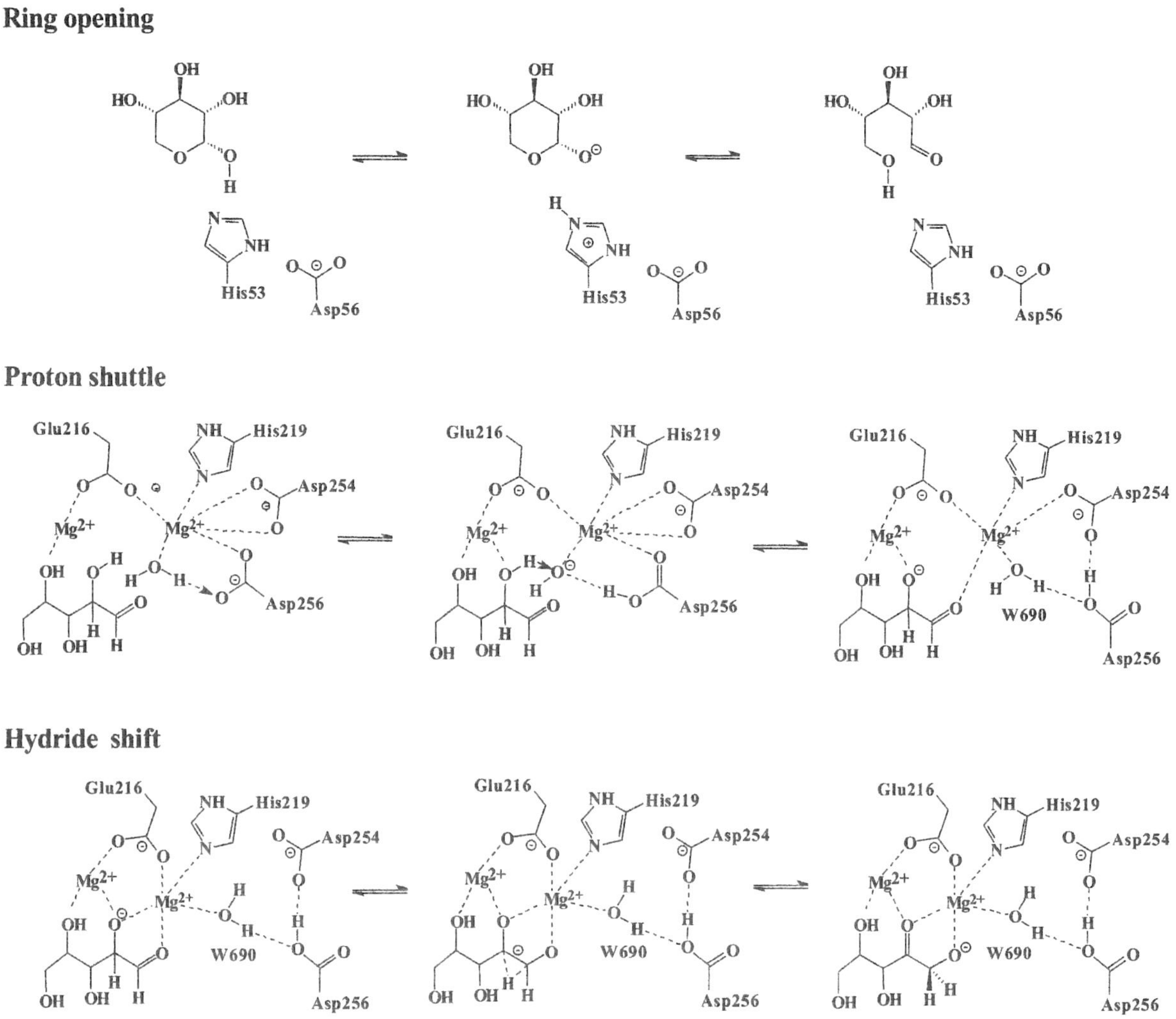

Ring opening

Proton shuttle

Hydride shift

Fig. 2. Catalytic mechanism of the xylose isomerase reaction. Note the formation of the $(- + -)$ charge distribution in the transition state of the ring opening step.

4. HIV integrase

Human immunodeficiency virus (HIV) integrase is a member of the group of retroviral integrases, a class of virus encoded enzymes belonging to the superfamily of nucleases and polymerases. They catalyse the insertion of viral DNA into the host cell genome, which is a crucial step in the virus life cycle. HIV integrase catalyses the hydrolysis of phosphodiester bonds making use of a metal bound at the active site [24]. A probable mechanism for the reaction is depicted in Fig. 3. A structural water

Table 1
Calculated metal properties for the hydride shift reaction in xylose isomerase

Cation	Catalytic role	Activation energy (kcal/mol)	Charge transfer (milielectron)	Net charge on the O^2 atom
Mg^{2+}	Activating	24.8	1568	-154
Al^{3+}	Inhibiting	33.0	2117	-268
Zn^{2+}	Inhibiting	38.7	1738	-310

Table 2
Composition of the kinetic isotope effect in the hydride transfer step of the xylose isomerase catalytic reaction [23]

	Quasi-classical contribution	Tunnelling contribution	Total kinetic isotope effect
19 QM atoms	1.80	2.08	3.75
79 QM atoms	1.93	2.62	5.01
Exp.	–	–	2.7–4.0

molecule, bound to the active site is activated by the nearby Asp-116 side chain and attacks the phosphorus atom to form a trigonal bipyramidal transition state, which decomposes into monophosphate and alcohol as products. A divalent metal cation and other side chains at the active site plays an essential role in catalysis, therefore it is crucial to include them into an adequate model to be used for calculations.

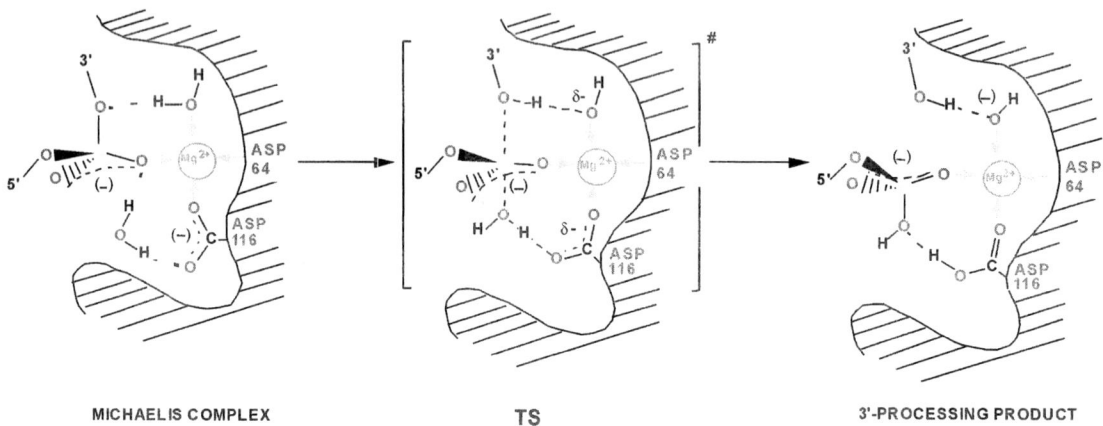

Fig. 3. Mechanism of phosphodiester hydrolysis by HIV integrase.

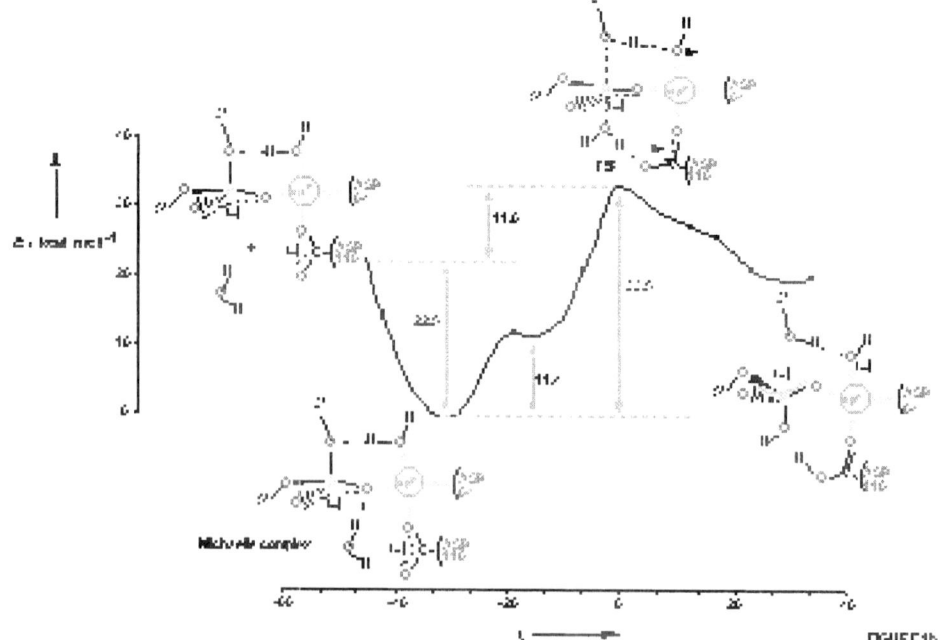

Fig. 4. Reaction energy profile for the hydroxyl attack at the HIV integrase active site.

Our model was based on DNA polymerase with single-stranded DNA as substrate [25] and we considered 50 explicit atoms and 220 electrons in the quantum region [22]. Dangling bonds were saturated with hydrogen atoms, protein atoms were represented by monopoles. We did density functional calculations with the B3LYP functional and locally dense basis sets, namely 6-311G*, 6-31G*, 3-21G* and STO-3G for the Mg cation, the reaction centre, close environment (structural water molecules, carboxylates and the 5′-oxygen), and external methyl groups, respectively. The distant environment was treated by the Poisson–Boltzmann method and after intrinsic reaction co-ordinate optimisation, we obtained a reaction curve depicted in Fig. 4. A Michaelis complex is first formed, which sits in a deep energy minimum, 22.5 kcal/mol lower than the initial state. Then, approaching the transition state a very shallow energy minimum appears that might provide some evidence for the presence of a stable intermediate in the reaction, which is subject of a long debate [26]. However, a refined model may include an additional water molecule (cf. Fig. 5) since there is still free space here. Though this molecule cannot be seen by protein crystallography, indicating that it is only loosely bound, it is quite possible that it may assist proton transfer in the initial phase of

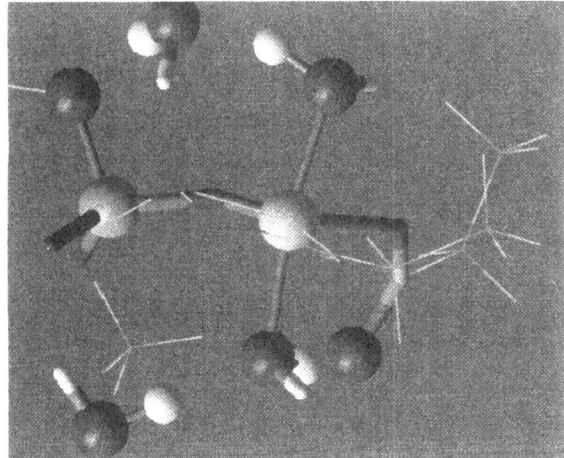

Fig. 5. Refined model of the active site of HIV integrase. Note that addition of an extra water molecule in the lower part of the model (cf. Figs. 3 and 4). Large light grey, dark grey and black spheres represent magnesium, phosphorus and oxygen atoms, respectively, small white spheres stand for hydrogen atoms.

the reaction. In fact, if we include this water molecule in our model, the shallow minimum on the reaction profile in Fig. 4 disappears, as confirmed by our preliminary studies. Both unrestricted Hartree-Fock calculations with a 3-21G basis set and a density functional study with a B3LYP functional and a 6-31^{++}G basis set indicate that there is no local minimum in this region of the reaction energy profile.

The activation energy of a model reaction in the gas phase is 70.3 kcal/mol, that is reduced to 43.1 kcal/mol for our active-site model without the metal ion. Its importance in catalysis can be recognised, if we calculate the activation energy for the 50-atom active-site model including Mg^{2+}, a decrease by 9.6, to 33.5 kcal/mol is observed. At last, the distant environment (protein atoms and bulk water) do not play an important role in catalytic rate acceleration, since incorporating them in a full model decreases the barrier by only 0.4 kcal/mol.

5. Conclusions

Molecular modelling of enzyme reactions became quite reliable by now, which is partly due to the fact that the chemical process is localised on a few atoms, the full system, including explicit protein atoms, bulk water and counter ions has a layered structure. This is, why QM/MM methods provide an appropriate tool for the quantitative analysis of enzymatic processes. We stress that to find the right structural model is crucial, because there is a danger that spurious effects arise for inappropriate models. The influence of closer and distant environment may be crucial and is mainly electrostatic, however, in some cases quantum effects are also important.

Acknowledgements

This work was supported by the National Scientific Research Fund (OTKA) under grant number T034994.

References

[1] L.A. Eriksson (Eds.), Theoretical biochemistry, Processes and Properties of Biological Systems, Elsevier, Amsterdam, 2001.

[2] G. Náray-Szabó (Eds.), Steric effects in biomolecules, Proceedings of International Symposium, Eger, Akadémiai Kiadó/Elsevier, Budapest/Amsterdam, 1981. p. 1982.

[3] G. Náray-Szabó, A. Warshel, Computational Approaches to Biochemical Reactivity, Kluwer, Dordrecht, 1997.

[4] H.M. Berman, T. Battistuz, T.N. Bhat, W.F. Bluhm, P.E. Bourne, K. Burkhardt, Z. Feng, G.L. Gilliland, L. Iype, S. Jain, P. Fagan, J. Marvin, D. Padilla, V. Ravichandran, B. Schneider, N. Thanki, H. Weissig, J.D. Westbrook, C. Zardecki, Acta Cryst. D58 (2002) 899.

[5] D. Bakowies, W. Thiel, J. Phys. Chem. 100 (1996) 10580.

[6] M. Svensson, S. Humbel, R. Froese, T. Matsubara, S. Sieber, K. Morokuma, J. Phys. Chem. 100 (1996) 1200.

[7] A. Warshel, M. Levitt, J. Mol. Biol. 103 (1976) 227.

[8] G. Náray-Szabó, P.R. Surján, Chem. Phys. Lett. 96 (1983) 499.

[9] G.G. Ferenczy, J.L. Rivail, P.R. Surján, G. Náray-Szabó, J. Comput. Chem. 13 (1992) 830.

[10] X. Assfeld, J.L. Rivail, Chem. Phys. Lett. 263 (1996) 100.

[11] J. Gao, P. Amara, C. Alhambra, M.J. Field, J. Phys. Chem. 102 (1998) 4714.

[12] D.M. Philipp, R.A. Friesner, J. Comput. Chem. 20 (1999) 1468.

[13] J.S. Wright, E.R. Johnson, G.A. DiLabio, J. Am. Chem. Soc. 123 (2001) 1173.

[14] R. Car, M. Parrinello, Phys. Rev. Lett. 55 (1985) 2471.

[15] P. Carloni, U. Röthlisberger, Theoretical biochemistry, in: L.A. Eriksson (Ed.), Processes and Properties of Biological Systems, Elsevier, Amsterdam, 2001, p. 215.

[16] D.G. Truhlar, J. Gao, C. Alhambra, M. Garcia-Viloca, J. Corchado, M.L. Sánchez, J. Villà, Acc. Chem. Res. 35 (2002) 341.

[17] K.A. Sharp, B. Honig, Annu. Rev. Biophys. Chem. 19 (1990) 301.

[18] A. Nicholls, B. Honig, J. Comput. Chem. 12 (1991) 435.

[19] B. Asbóth, G. Náray-Szabó, Curr. Prot. Peptide Sci. 1 (2000) 237.

[20] M. Rangarajan, B.S. Hartley, Biochem. J. 283 (1992) 223.

[21] M. Fuxreiter, Ö. Farkas, G. Náray-Szabó, Prot. Eng. 8 (1995) 925.

[22] F. Bernardi, A. Bottoni, M. De Vivo, M. Garavelli, G.M. Keserű, G. Náray-Szabó, Chem. Phys. Lett. 362 (2002) 1.

[23] M. Garcia-Viloca, C. Alhambra, D.G. Truhlar, J. Gao, J. Comput. Chem. 24 (2003) 177.

[24] R.A. Katz, A.M. Skalka, Annu. Rev. Biochem. 63 (1994) 133.

[25] C.A. Brautigam, T.A. Steitz, J. Mol. Biol. 277 (1998) 363 Protein Data Bank File 1 KFS.

[26] M. Fothergill, M.F. Goodman, J. Petruska, A. Warshel, J. Am. Chem. Soc. 17 (1995) 11619.

ELSEVIER

Journal of Molecular Structure (Theochem) 666–667 (2003) 645–649

THEO
CHEM

www.elsevier.com/locate/theochem

Lead conformer prediction based on a library of flexible molecules

A. Kalászi, Ö. Farkas*

Department of Organic Chemistry, Eötvös Loránd University, 1/A Pázmány Péter St, Budapest H-1117, Hungary

Abstract

Flexible molecules are not widely welcome as lead compounds, due to the unlikelihood of their specific activity. However, they may show *some activity* more easily. A library of flexible molecules can provide a wide range of different activities. We use the different activity and conformational behavior of the individual molecules to choose the *lead conformers* which are responsible for the binding or biological activity. The lead conformers then may help us to find less flexible *lead structures* and then drug molecules. Important to note, that lead conformers can also be obtained in situations when no *a priori* spatial information but quantitative binding or biological activity data are available. Our present method automatically evaluates the conformational distribution for each member of a selected library of flexible molecules. The probability of the 'active' conformers, the *lead conformers*, is assumed to correlate with the collected experimental binding affinity or biological activity. © 2003 Elsevier B.V. All rights reserved.

Keywords: Conformational search; Peptide library; Correlation; Flexible molecules; QSAR

1. Introduction

Drug discovery has transformed from a simple trial-and-error process into a much more systematic work including multiple levels of optimizations due to the availability of parallel chemical synthesis, combinatorial chemistry and high throughput biological tests carried out by robots. When an acceptable starting molecule, a lead compound, is available then the systematic machinery can execute the optimization processes to find an acceptable candidate to be a drug. However, finding a lead molecule is almost a hopeless venture in cases when no *a priori* spatial information is available for the target. The number of these kinds of cases will increase, since genetic information become more and more widely available leading to preparations of target proteins without providing any structural information.

The correlation between the conformational flexibility and bioactivity has already been examined before [1–3]. In these studies the common segment of conformational spaces or the overlap with the active conformational space of the different flexible molecules are stated as effective conformers. It has, however, been assumed that a good enough candidate, providing initial spatial arrangement, is present in the set of studied molecules.

The purpose of the present method is to ease the search for lead compounds in cases of drug discovery when no acceptable candidate for lead compound is available yet. Scheme 1 outlines the proposed role of the present method in finding lead compounds.

---

* Corresponding author.
E-mail address: farkas@chem.elte.hu (Ö. Farkas).

0166-1280/$ - see front matter © 2003 Elsevier B.V. All rights reserved.
doi:10.1016/j.theochem.2003.08.089

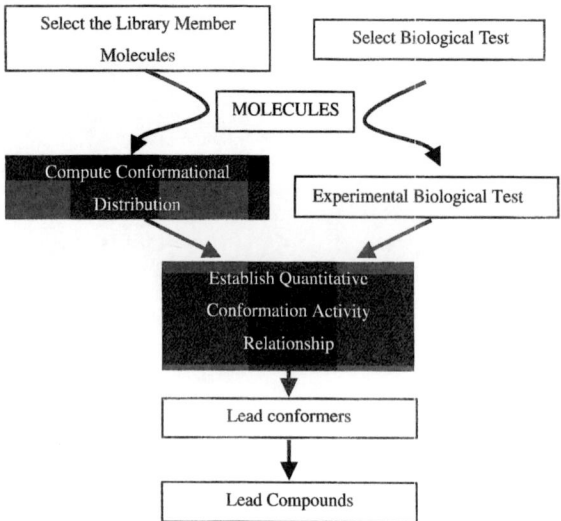

Scheme 1. Outlines of the role of the proposed method in obtaining lead compounds via lead conformers. The parts denoted with black are discussed in detail.

The distinct functional groups of the substrate should be in the correct orientation for the effective binding. Molecules with low flexibility can be described by a discrete structure and well-defined properties in QSAR based analysis. Flexible molecules, on the other hand, may provide detectable binding affinity easier due to the wider range of spatial arrangement of their functional groups. The wider range of spatial arrangements may prevent applying regular QSAR methods on flexible molecules. Our proposed method uses a library of flexible molecules and describes their conformational space in an almost continuous fashion. In practice, this means that each single molecule of our set, the library, acts like another library of their conformations. The conformational analysis, based on long trajectories, obtained by molecular dynamics simulations in the appropriate solvent, provides the relative probability of the conformations. The relative probabilities of the active conformers assumed to correlate with the experimental activities of the corresponding molecules. This correlation is used for choosing active conformers of the members of the library with detectable activity resulting in the description of the lead conformer. The description then provides the necessary spatial information for

further QSAR studies to find an acceptable lead compound.

2. Method

The present method implies a number of assumptions and requirements. For the sake of efficient correlation studies the conformational space of the library members or at least the relative position of their similar functional groups should be described in a uniform fashion. Quantitative results from a reliable biological test for the question of interest (binding, inhibition activity, etc.) should be available for the members of the library. We also assume that the active members of the library have a common conformational motif required for binding. The lead conformer consists of these conformational motifs. The relative frequency, the probability, of the active conformations in the simulation trajectories of the library members assumed to correlate with their experimental activity. If the measured activity data do not show a theoretical linear connection to the concentration of the active species then modifications on the original data for the purpose of linear correlation should be considered. The quality of conformational distribution obtained for the library members by computations greatly affects the reliability of the present method.

2.1. Method for computing the conformational distribution

Distribution of conformers in solvent at room temperature for each member of the library is obtained in the present case by executing molecular dynamic (MD) simulations using the GROMACS force field [4,5]. Parallel MD simulations are started from several low energy conformers of each distinct library member resulting in the distribution. The conformational space of a flexible molecule can be more efficiently sampled by using multiple simulations rather than a single long MD. In order to gain reasonable low energy starting conformers for MDs our automated conformational search (CS) algorithm was applied.

The general CS scheme builds up from the following steps

1. The iterative process starts from an optimized extended structure of the actual molecule.
2. Execute high-energy dynamics simulation in vacuum.
3. Select the structures to optimize.
4. Store the new conformers.
5. Continue with Step 2 again if new conformers have been found. (The low-energy structures are collected until the stop criteria are reached.)
6. Select a suitable set of low-energy conformers.
7. Execute the room temperature simulations in solvent started from the set of conformers.

Numerous methods, like systematic search, random Cartesian search, random dihedral search, distance geometry methods and molecular dynamics are available for conformational search [6]. However, finding conformers of larger flexible molecules is still a challenge because of the dimensionality of the molecular space. Our CS scheme uses high-energy molecular dynamic simulations in vacuum to generate starting ensemble of conformers. Geometry optimized low energy conformers are collected intelligently from that ensemble. The collection is based on differentiating the conformers using the deviations of their internal coordinates. The cycle of ensemble generation and conformer collection is continued until, under a given timescale, no new low energy conformer is found. This CS scheme can also be used for global optimization. The advantage of using high-energy dynamics simulations is, that the search is a priori samples the structures in the valid conformational space, which is usually only a small fraction of the theoretical conformational space. On the other hand, both cyclic and linear molecules can be studied by using MD. We assume that using long enough high-energy MD simulations, molecules can surmount the conformational energy barriers and sample the whole conformational space. Our conformational search method was originally tested on the conformational analysis of oligotuftsin analogues [7,8].

2.2. Method to establish quantitative conformation activity relationship

Once the conformational distribution of each library member molecule is obtained the total interval

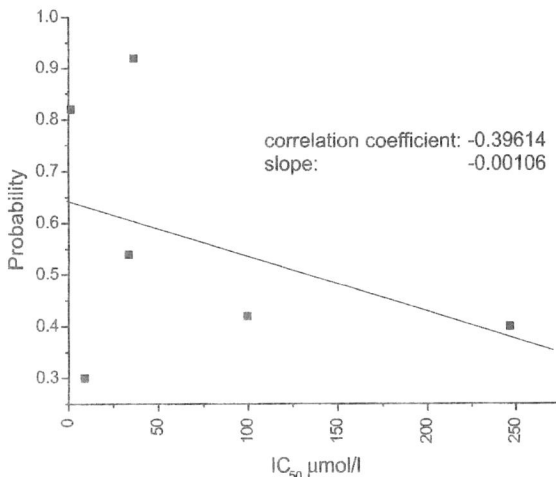

Fig. 1. Example of using linear regression for testing the correlation between biological activity of the molecules (x-axis) and the probability of their common conformational motif (y-axis).

of each conformational descriptor is partitioned to bins. All different conformations with respect to these bins are collected. The correlation between the distribution of these conformations and the biological activity is evaluated. The ones having the correlation coefficient and slope better than a criterion are denoted as lead conformers. Fig. 1 shows the biological activity vs. the relative conformational distribution of one distinct conformation.

3. Results and discussion

For testing our method, we have chosen the active members of the TQTXT (X = all amino acids except C) pentapeptide library [9]. The members of the peptide library represent epitope candidates to prospective tumor marker antibody mAb 994. The IC50 values of the members were obtained from competitive ELISA experiments. The amino acids Thr_1 Thr_3, and Thr_5 were identified as groups essential for binding. Six members had considerable inhibition activity: X = A, F, P, S, Y, W. Among these TQTPT was the most effective candidate.

To gain the lowest energy starting conformers for parallel MD simulations we executed our CS method. Three different values for simulation temperatures were tested for CS. After 200 ns of simulation time on 1200 K the molecule exploded. On 600 K the process

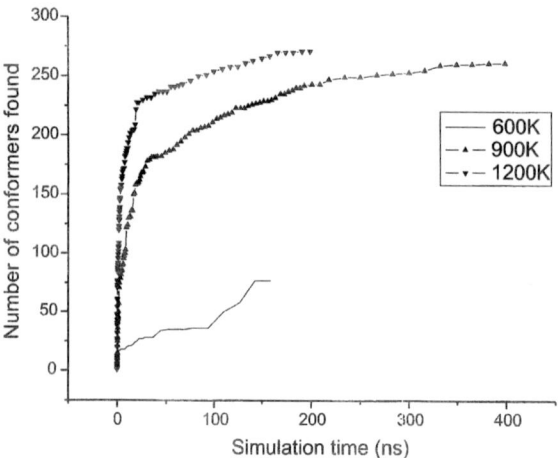

Fig. 2. Testing the effect of the kinetic energy on the number of founded conformers at CS. The example molecule is the TQTFT pentapeptide.

of finding conformers was not effective enough and the convergence criterion was reached too early. On 900 K the effectiveness of the search was acceptable and also the molecule remained stable. Fig. 2 shows the number of found conformers vs. the simulation time for the selected simulation temperatures. Fig. 3 shows the number of conformers vs. simulation time curves for the six active members of the library.

In case of pentapeptides, a uniform description can be accomplished by using their backbone conformation. However, other generally useful description, the matrix of selected intramolecular atomic distances

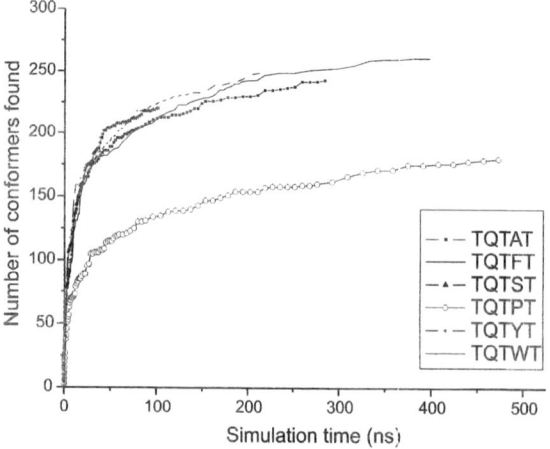

Fig. 3. Number of low energy conformers founded at 900 K by CS vs. the simulation time for the six effectively binding pentapeptides. TQTPT, the best binding candidate, is the less flexible.

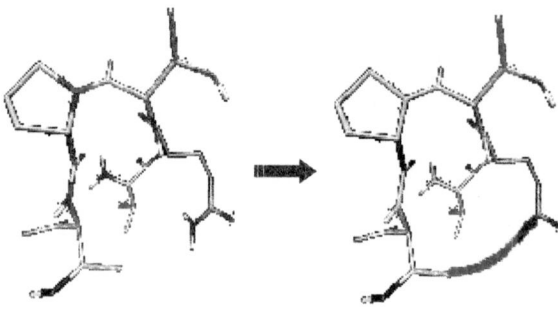

Fig. 4. The lead conformer for the most effectively binding TQTPT. A chemical modification on the lead conformer may lead to a less flexible lead compound still providing the required spatial arrangement.

is also feasible. The correlation of the conformational distribution and binding data resulted in a set of bins for the backbone dihedral angles. This set of intervals can be mapped on any of the library members to obtain the lead conformers. However, it is advisable to map them on the most active candidate, on the TQTPT pentapeptide.

The lead molecule can be obtained from lead conformer either by reducing its flexibility by performing chemical modifications on the molecule or carrying out similarity search on databases containing 3D structures. A schematic process for freezing the most active species in its active conformer is shown on Fig. 4.

4. Conclusion

The present method is a promising way for finding acceptable lead compounds in cases when no a priori spatial information, but a reliable quantitative biological test is available. The construction of a suitable library of flexible molecules with common conformational descriptors, like oligopeptides, allows one to examine the conformational behavior of the members of the library on the same ground. The assumed correlation between the relative frequency of the active conformers, obtained by MD simulations, and the experimental activity of the corresponding molecules is used for choosing the active, lead conformers. The lead conformers then provide the necessary spatial information for finding less flexible lead compounds and gradually drug molecules. The method was tested on the TQTXT pentapeptide library with available experimental binding activities. The resulted lead conformer will be used in further studies for finding less flexible candidates.

Acknowledgements

This work has been supported by Gaussian, Inc., and the Hungarian Research Foundation (OTKA D29446). The authors would also like to thank NSF, MTA and OTKA for an international collaboration grant. The computations have been performed in part using the facilities of the Hungarian Supercomputer Center and the High Performance Computing Center of the Eötvös Loránd University.

References

[1] O.M. Becker, Y. Levy, O. Ravitz, J. Phys. Chem. B 104 (2000) 2123.

[2] J.W. Payne, N.J. Marshall, B.M. Grail, S. Gupta, Curr. Org. Chem. 6 (14) (2002) 1221.

[3] N.J. Marshall, R. Andruszkiewicz, S. Gupta, S. Milewski, J.W. Payne, J. Antimicrob. Chemother. 51 (4) (2003) 821.

[4] H.J.C. Berendsen, D. van der Spoel, R. van Drunen, Comp. Phys. Comm. 91 (1995) 43.

[5] E. Lindahl, B. Hess, D. van der Spoel, J. Mol. Mod. 7 (2001) 306.

[6] M. Saunders, K.N. Houk, Y.-D. Wu, W.C. Still, M. Lipton, W.C. Guida, J. Am. Chem. Soc. 112 (1990) 1419–1427.

[7] A. Kalászi, G. Mező, F. Hudecz, Ö. Farkas, in: E. Benedetti, C. Pedone (Eds.), Peptides 2002, 2002.

[8] A. Kalászi, G. Mező, F. Hudecz, Ö. Farkas, J. Peptide Sci. 8 (2002) S194–P D28.

[9] E. Windberg, F. Hudecz, A. Marquardt, F. Sebestyen, A. Kiss, S. Bosze, H. Medzihradszky-Schweiger, M. Przybylski, Rapid Commun. Mass Spectrom. 16 (9) (2002) 834.

ELSEVIER

Journal of Molecular Structure (Theochem) 666–667 (2003) 651–657

THEO
CHEM

www.elsevier.com/locate/theochem

Validation of the SPROUT de novo design program

Jacqueline M.S. Law[a,b], David Y.K. Fung[c], Zsolt Zsoldos[c,*], Aniko Simon[c],
Zsolt Szabo[c], Imre G. Csizmadia[a,b,d], A. Peter Johnson[e]

[a]Department of Chemistry, University of Toronto, Toronto, Ont., Canada M5S 3H6
[b]Global Institute of Computational Molecular and Materials Science (GIOCOMMS), 1422 Edenrose St., Mississauga, Ont., Canada L5V 1H3
[c]Simulated Biomolecular Systems Inc., 135 Queen's Plate Dr, Unit 355, Toronto, Ont., Canada M9W 6V1
[d]Department of Medical Chemistry, University of Szeged, Dóm tér 8, Szeged 6720, Hungary
[e]Department of Chemistry, Institute for Computer Applications in Molecular Sciences (ICAMS), University of Leeds, Leeds LS2 9JT, UK

Abstract

The validation of SPROUT was carried out on four receptor–ligand complexes: thrombin–NAPAP, calmodulin (CAM)–AAA, Ras P-21–GDP and dihydrofolate reductase (DHFR)–methotrexate (MTX). These complexes were downloaded from the Brookhaven Protein Data Bank (PDB). For the thrombin–NAPAP complex, two structures very similar to NAPAP were generated. These two structures were similar in 3D structure to NAPAP but contained an extra hexane ring. For CAM–AAA and Ras P-21–GDP, the ligands generated were essentially identical to their original ligands. For DHFR, two ligands, one most similar in 2D structure and one most similar in 3D conformation were found. The successful regeneration of the ligands for each case proves the ability and applicability of SPROUT for designing strongly binding, successful drug candidates. When the program is executed with less restricted constraints, it generates a large number of novel structures that are structurally diverse, making it an ideal tool for de novo design.
© 2003 Elsevier B.V. All rights reserved.

Keywords: De novo design; Drug design; Fragment-based exhaustive search; Structure-based; Validation

1. Introduction

The goal of structure-based drug design is to build novel molecular structures or ligands that bind to the receptor of these proteins. Computational methods have been used in the design of novel ligands. Software for structure-based de novo drug design can be divided into three categories: (1) database search techniques, (2) atom based methods that build structures one atom at a time, and (3) fragment joining techniques that build structures from a library of common molecular fragments.

Hook/MCSS [1] falls into the category of database search techniques. Hook/MCSS exhaustively searches potential binding pockets and docks functional groups to these sites. The functional groups are then linked with templates from a database.

LEGEND [2] is an atom based structure generation program which builds structures one atom at a time. The atom by atom building proceeds as follows. The first atom is placed in a specified distance about a potential hydrogen bonding receptor atom. The subsequent atoms are placed by randomly choosing an atom on the existing ligand, and then placing

* Corresponding author. Tel.: +1-416-741-4263; fax: +1-416-741-5084.

E-mail address: zsolt@simbiosys.ca (Z. Zsoldos).

the new atom at a random point on the circle of all possible dihedral angles with fixed bond length and angle. This new atom is assigned an atom type and hybridization. If the new atom occupies a forbidden position on the grid then it is rejected. This procedure terminates when the structure reaches the user defined size.

The Allegro [3] program falls into the third category of de novo drug design software. Allegro starts with a pre-computed grid map that classifies the binding zones within the active site. A root atom is chosen and new atoms and fragments grown from this. Growth points are determined and one is randomly selected to form a connection with another randomly chosen atom or functional group. The new atom or group is placed and a complementary score evaluated from the binding zone classification. A Monte Carlo (MC) sampling criteria is used to decide whether this atom or group is to be accepted. If accepted, new growth points are determined and structure generation is repeated from any one of the potential growth points of the molecule. If the distance and bond angle of the growing structure are appropriate, the MC procedure can also spontaneously perform ring closure. This is also called an undirected random search.

SPROUT [4] also constructs structures using a fragment joining technique. SPROUT carries out five main functions: (i) locate binding pockets in a receptor, (ii) identify potential interaction sites, (iii) dock molecular fragments to target sites, (iv) generate novel chemical structures by incremental construction from templates and (v) score, sort and cluster the solutions for an efficient means of evaluating the results. In this paper, we present results of validation experiments by using SPROUT to regenerate known, strongly binding inhibitors co-crystallized in their target protein. The idea of these examples is to suggest that if SPROUT can regenerate known inhibitors to a number of targets, this proves the program's ability to generate strongly binding inhibitors. Therefore, any other 'novel' structure generated will also likely be strongly binding. The validation examples presented in Section 3 includes inhibitors of thrombin, calmodulin (CAM), Ras P-21 and dihydrofolate reductase (DHFR), all of which are important for mediating and regulating cell cycles.

2. Method

An overview of each of the modules will be given below. However, for a more detailed description of each module, we refer you to Ref. [4].

2.1. Locating binding pockets

This module offers an automatic tool for finding potential binding sites within protein structures. In this way, the program is able to define the receptor site and the cavity (binding pocket) of a target molecule, e.g. an enzyme. This module also contains interactive graphical tools which enable the user to separate co-crystallized protein ligand complexes and then generate a cavity and receptor file, respectively, based on the position of the existing ligand.

2.2. Identification of potential interaction sites

This module identifies potential interaction sites within a cavity that can be used in de novo design. These interaction sites define the positions for potential ligand atoms during structure generation. This module can detect potential hydrogen bonding, covalent bonding, metal ions and hydrophobic interaction sites. An important novel feature of the method is that it deals with multicentered and bifurcated hydrogen bonding possibilities.

2.3. Docking of molecular fragments to target site

Before structure generation can begin a set of target sites, steric boundary, and a set of start templates are required. This module is designed to select and orient start templates representing functional groups at the target sites. The functional group selection is based on the information provided by the target site identification and takes into account the distances, relative positions and directions of those target sites, which are close to each other.

2.4. Generation of structures

This module generates structures that satisfy the input constraints of the binding pocket. The input is a set of steric constraints that describe the 3D shape of the receptor in which the structure must fit. This is done by joining spacer fragments to the template fragments docked at the target sites and then connecting the resulting partial structures together.

The structure generation phase in SPROUT involves two steps: (1) generating the 3D molecular graphs or template structures that are consistent with the steric constraints of a receptor site; and (2) converting a template structure into a molecular structure with different atom and bond types consistent with the desired electrostatic complementarity.

2.5. Clustering and sorting of results

This module enables the user to cluster/sort the structures generated in SPROUT, into sets using a list of parameters specified by the user. In this way, unwanted structures can be identified and discarded leaving only structures that meet the requirements of the user.

The solutions are ranked using a scoring function that assesses hydrogen bonds, van der Waals interactions, rotatable bonds and hydrophobic surface contact area. The sorting of results can be done by various criteria such as number of gauche interactions, stereo centers and ring fusions. An additional function of this module is hetero-atom substitution in which generic atoms are replaced by specific atoms to fulfill interaction type criteria (e.g. hydrogen bond donor) at target sites, leading to final solutions which may have better binding score or stability score.

3. Results and discussion

3.1. Thrombin

Thrombin is a serine protease that mediates the blood clotting cascade and converts fibrinogen into fibrin through hydrolysis. The crystal structure of thrombin complexed with NAPAP from the Protein Data Bank (PDB code 1ets) was used for the first validation of SPROUT. Thrombin has three binding sites, one of which has an aspartic acid (Asp) residue that interacts with a positively charged part of a ligand. The other two sites are hydrophobic sites, one of which allows the binding of an aromatic moiety of a ligand. Studies have shown that the ligand NAPAP

exhibits quite a strong binding affinity towards this pocket [5].

The hydrogen bond donor site OD1 ASP H189, hydrogen bond acceptor site N GLY H216 and manually added hydrophobic spheric sites were chosen on the receptor for the interactions sites in the generation of a ligand. These sites are shown in Fig. 1. The spheric sites were very important because they helped in the regeneration of a structure similar to NAPAP. In addition, they also model the hydrophobic interactions between protein and ligand. Without the spheric target sites, SPROUT would generate a large diverse set of structures, many of them very different from NAPAP.

Starting fragments similar in structure to NAPAP were docked into the target sites with different positions and conformations as shown in Fig. 1. Note that for the fragment with the $NHSO_2$ group, two sites were used to dock the fragment. These two sites anchored the fragment such that the methyl end and the $NHSO_2$ would be oriented properly.

Two structures found to closely resemble NAPAP are shown in Fig. 2. However, they both have an extra six-member ring, hexane, which may result in this new structure being more bulky. However, it seems to

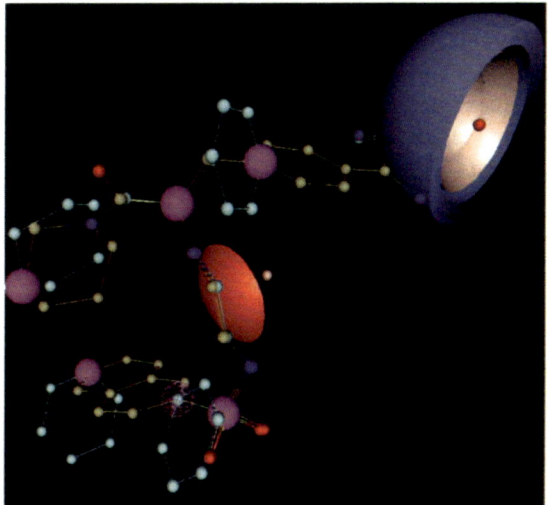

Fig. 1. The target sites chosen for regeneration of NAPAP. The generic spheric sites are in pink and the hydrogen bond donor and acceptor sites are the blue and red semi-spheres, respectively. A comparison of the fragments docked at the target sites (light blue) and NAPAP (brown) are also shown.

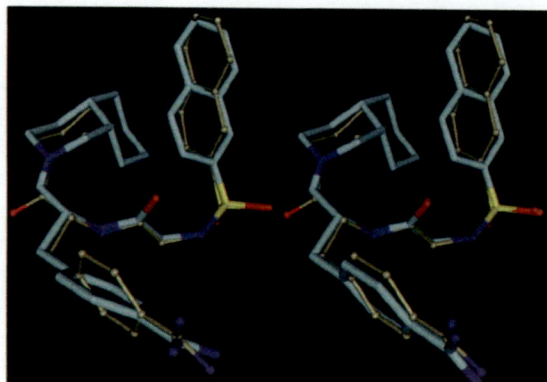

Fig. 2. Two SPROUT generated structures (blue) resembling the original NAPAP ligand (brown).

form additional hydrophobic interactions between the ligand and the receptor.

3.2. Calmodulin

CaM is known to control processes such as cell division and gene expression [6]. To activate different proteins, CaM complexes with Ca^{2+}, then binds onto the polypeptides of different receptor proteins and induces conformational change [7].

Many studies have been carry out on ligands such as N-(3,3-diphenylpropyl)-N'-[1-R-(3,4-bis-butoxyphenyl)ethyl]-propylendiamine (AAA) that mimic the receptor peptides that bind to CaM. The CaM–AAA complex (PDB code 1qiw) was used because AAA has a high affinity for CaM in the presence of Ca^{2+} atoms. Without Ca^{2+}, AAA would not bind to CaM at all.

In total, nine sites were selected. One site is a hydrogen bond donor site OE1 GLU-B11 and eight of them were hydrophobic sites represented by the spheric sites. The sites that were chosen are shown in Fig. 3. Since the biphenyl groups can interact with the hydrophobic cleft, the sites must be put near those clefts to ensure that the phenyl groups can be grown. Fig. 3 also shows the fragments docked onto the target sites. The blue structures are the fragments docked and the brown structure is AAA.

In the structure generation module, these fragments were joined using template fragments similar to those found in AAA. There were many solutions found, however, one solution had a very close resemblance to the original ligand shown in Fig. 4.

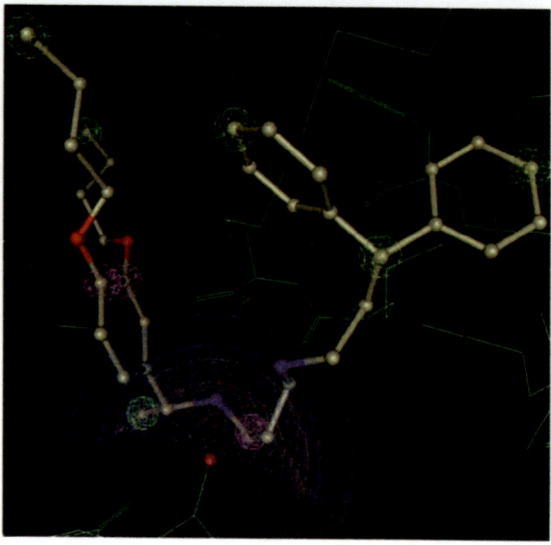

Fig. 3. Fragments docked to target sites of calmodulin. The hydrophobic sites are the green spheres, generic spheric sites in pink and the hydrogen bond donor sites is the blue semi-sphere.

3.3. Ras P-21 and GDP complex

P21 is an oncogene found in the Ras family known to be related to different human tumors [8]. Under normal circumstances, this oncogene is inactivated when GTP is hydrolyzed to GDP by a GAP protein [9]. However, a point mutation in a base would result in GAP being unable to bind to and hydrolyze the P21–GTP complex to P21–GDP [10]. In turn, the Ras protein would always be activated and would become an oncogene [11].

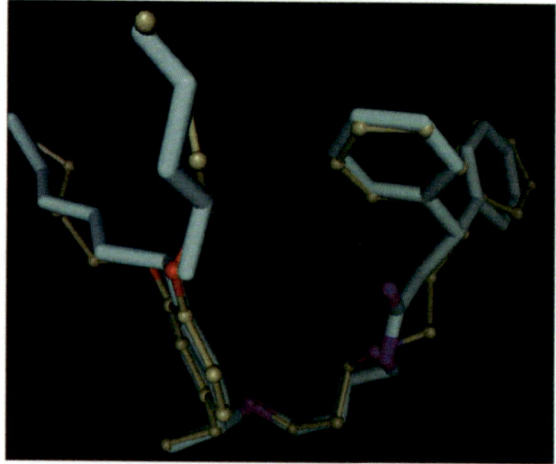

Fig. 4. The final structure (light blue) that resembles AAA (brown).

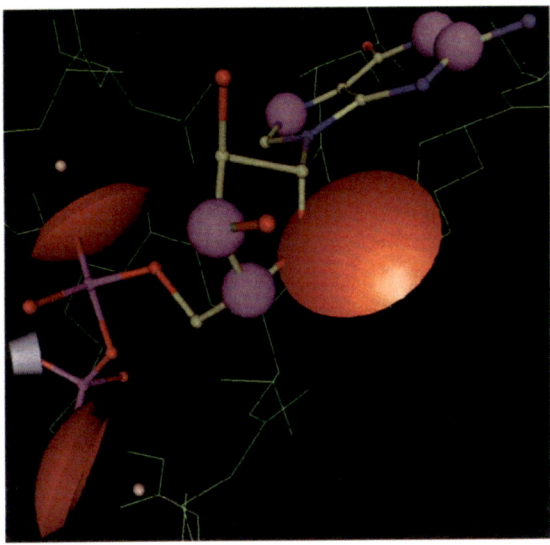

Fig. 5. Sites in Ras P-21 chosen for the docking of fragments.

The Ras P-21 protein complexed with GDP (PDB code 1q21) was studied. The target sites satisfying the ligand include a hydrogen bond acceptor site on the pentose moiety of guanidine (N Gly-13), two hydrogen bond acceptor sites (NZ Lys-117 and N ALA-18), each located at one of the phosphate groups of the ligand, and a divalent metal ion site (Mg Mg-173) at the terminal phosphate group shown as a silver cylinder in Fig. 5. The metal ion is an important binding site for the GDP molecule. Five spheric sites were also chosen to guide the extension and growth of

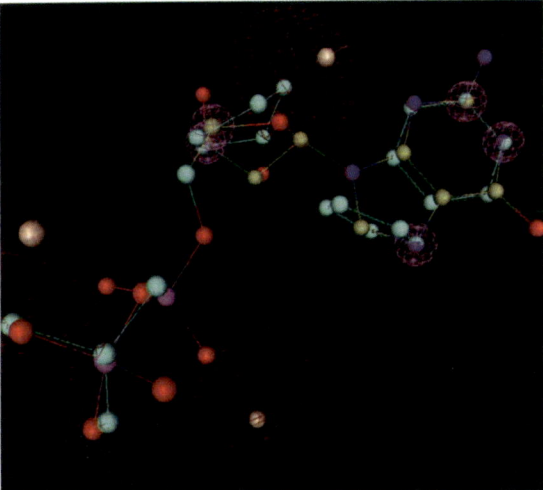

Fig. 6. Fragments chosen for docking at the target sites of Ras P-21.

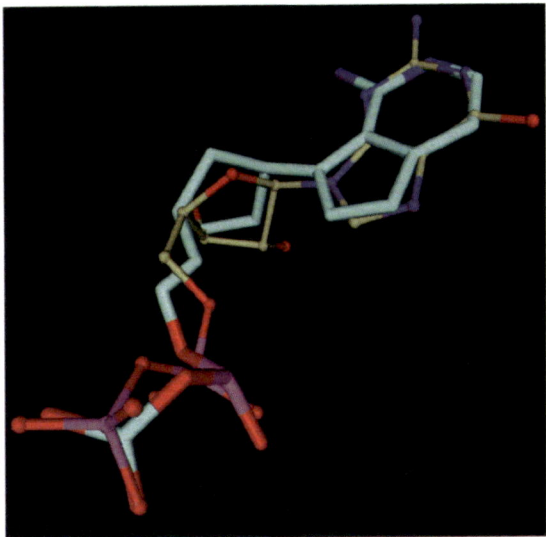

Fig. 7. The solution (light blue) mimicking the GDP ligand (brown).

this molecule. Fig. 5 shows the target sites selected for this molecule. The fragments chosen as starting points are shown in Fig. 6. The three sites used for the pentose sugar ensured that the five-member ring would have the correct geometry. This method also ensures that the geometry of this molecule would also stay unchanged when the joining fragments are added.

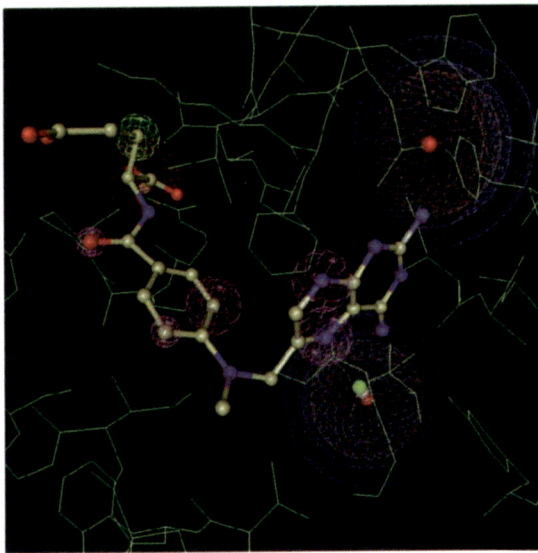

Fig. 8. The target sites chosen in DHFR. The hydrophobic site is the green sphere, generic spheric sites in pink and the hydrogen bond donor and acceptor sites are the blue and red semi-spheres, respectively. The original ligand MTX is in brown.

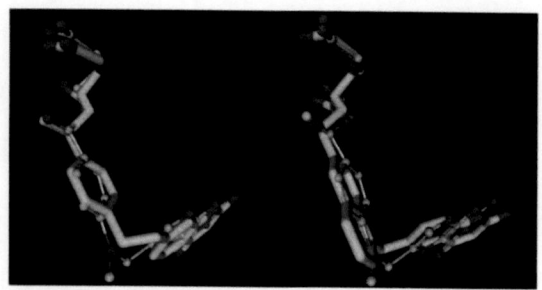

Fig. 9. Structures most similar to MTX (brown). The ligand on the left is most similar in 2D structure and the ligand on the right is most similar in 3D structure.

Fig. 7 shows the final ligand structures mimicking GDP.

3.4. Dihydrofolate reductase and methotrexate

DHFR is essential for producing tetrahydrofolate by the NADPH facilitated reduction of dihydrofolate [12]. Tetrahydrofolate is essential in the synthesis of metabolites in cells. In cancerous cells, the high levels of DHFR are essential for the proliferation of cells. The methotrexate (MTX) ligand can inhibit DHFR and can thus serve as both an anticancer and antitumour drug.

The crystal structure of DHFR complexed with MTX (PDB code 4dfr) was used as the next model for validation of SPROUT. In the DHFR receptor, a number of hydrogen bonding sites were chosen. They included the hydrogen bond donor sites OD1 ASP-A27 and O Ile-A94, the hydrogen bond acceptor site NH2 Arg-A57, as well as a number of hydrophobic sites on the two aromatic rings and at the end of the chain. The hydrogen bonding sites were used such that the newly designed ligand would have a relatively good hydrogen bonding interaction with the active sites of the receptor cleft. Fig. 8 shows the target sites chosen.

In this model, two structures similar to MTX were found. Although only two results are shown, other hetero-atom substitutions on atoms at other positions are possible. An image of the two most similar molecules are shown in Fig. 9.

4. Conclusions

SPROUT provides an exhaustive, deterministic solution for de novo ligand design. This approach can guarantee to find the best solution with the defined constraints. This is a very valuable and unique feature, differentiating SPROUT from the numerous de novo tools that apply random techniques or artificial limitations imposed by database searches.

The SPROUT de novo design software was used to regenerate known, strongly binding ligands for thrombin, CAM, Ras P-21 and DHFR. The ligands structurally and conformationally resembling the co-crystallized X-ray ligand were successfully regenerated for all these receptors. The success of SPROUT in this validation demonstrates its ability to regenerate known strongly binding inhibitors suggesting that it has great potential in deriving unknown, novel ligands for other target receptors.

Acknowledgements

This study was made possible by the Scientific Research and Experimental Development grant from the National Research Council of Canada and the Global Institute of Computational Molecular and Materials Science (GIOCOMMS), Toronto, Ont., Canada. One of the authors (IGC) wishes to thank the Ministry of Education for a Szent-Györgyi Visiting Professorship.

References

[1] (a) A. Miranker, M. Karplus, Protein Struct. Func. Gen. 11 (1991) 29.
(b) A. Caflish, A. Miranker, M. Karplus, J. Med. Chem. 36 (1993) 2142.
(c) M.B. Eisen, D.C. Wiley, M. Karplus, R.E. Hubbard, Protein Struct. Func. Gen. 19 (1994) 199.
[2] (a) Y. Nishibata, A. Itai, Tetrahedron 47 (1991) 8985.
(b) Y. Nishibata, A. Itai, J. Med. Chem. 36 (1993) 2921.
[3] R.S. Bohacek, C. McMartin, J. Am. Chem. Soc. 116 (1994) 5560.
[4] (a) V.J. Gillet, A.P. Johnson, P. Mata, S. Sike, P. Williams, J. Comput.- Aided Mol. Des. 7 (1993) 127.
(b) V.J. Gillet, W. Newell, P. Mata, G.J. Myatt, S. Sike, Z. Zsoldos, A.P. Johnson, J. Chem. Inf. Comp. Sci. 34 (1994) 207.
(c) P. Mata, J. V, A.P. Gillet, J. Johnson, G.J. Lampreia, S. Myatt, A.L. Sike, Stebbings, J. Chem. Inf. Comp. Sci. 35 (1995) 479.

[5] D.W. Banner, P. Hadvar, J. Biol. Chem. 266 (1991) 20085.

[6] A. Crivici, M. Ikura, Ann. Rev. Biophys. Biomol. Struct. 24 (1995) 85.

[7] N. Takuwa, W. Zhou, Y. Takuwa, Cell. Signal. 7 (1995) 93.

[8] A.M. deVos, L. Tong, M.V. Milburn, P.M. Matias, J. Jancarik, S. Noguchi, S. Nishimura, K. Muira, D. Ohtsuka, S. Kim, Science 239 (1988) 888.

[9] A. Wolfman, I.G. Macara, Science 248 (1990) 67.

[10] J. Downward, R. Riehl, L. Wu, R.A. Weinberg, Proc. Natl Acad. Sci. USA 87 (1990) 5998.

[11] T. Schweins, K. Scheffzek, R. Auheuer, A. Wittinghofer, J. Mol. Biol. 266 (1997) 847.

[12] V.M. Reyes, M.R. Sawaya, K.A. Brown, J. Kraut, Biochemistry 34 (1995) 2710.

ELSEVIER

Journal of Molecular Structure (Theochem) 666–667 (2003) 659–665

www.elsevier.com/locate/theochem

Software tools for structure based rational drug design

Zsolt Zsoldos[a,*], Irina Szabo[a], Zsolt Szabo[a], A. Peter Johnson[b,1]

[a]*SimBioSys Inc., 135 Queens Plate Dr, Unit 355, Toronto, Ont., M9W 6V1, Canada*
[b]*ICAMS, University of Leeds, Leeds, LS2 9JT, UK*

Abstract

Scientific advancements during the past two decades have altered the way pharmaceutical research produces new bio-active molecules. Traditional 'trial and error' drug discovery efforts are gradually being replaced by structure based rational drug design. A key technique is the use of methods including X-ray crystallography and NMR for the determination of the 3-dimensional structure of a target protein followed by various modelling techniques for the design of small molecule ligands that could interact with the target structure.

The first generation of the software tools were limited to energy calculations and molecular dynamics simulations based on simple force field models. More automated methods, like flexible ligand docking and de novo ligand design programs have emerged in the 90s. However, many of these software systems relied on stochastic algorithms to perform the search, effectively performing a computerised 'trial and error' search.

SimBioSys provides ligand design and docking tools that are systematic, exhaustive, rely on rational rules, heuristics and dynamic knowledge bases. The novel algorithms designed and optimised to solve the specific problems greatly outperform the general random methods. The presentation gives a brief overview of the efficient algorithms behind the SPROUT (de novo ligand design) and eHiTS (flexible ligand docking and virtual high throughput screening) software tools. Validation test results are presented to demonstrate the effectiveness of these tools for the solution of practical drug design problems.
© 2003 Elsevier B.V. All rights reserved.

Keywords: Rational drug design; Ligand docking; de novo Design; VHTS; Scoring function

1. Introduction

The problem of structure based rational drug design can be seen as a challenge of finding small molecular structures that fulfil both of the following criteria:

- The molecular structure should bind strongly to a given protein due to steric and electrostatic complementarity
- It should be readily available or easy to synthesize

It is difficult to keep equal focus on both of these criteria. Two main approaches have emerged to solve the problem, each prioritizing one of the criteria over the other:

- Flexible docking of potential ligands from a database of 3-D structures of known compounds.

* Corresponding author. Tel.: +1-416-741-4263; fax: +1-416-741-5084.
E-mail address: zsolt@simbiosys.ca (Z. Zsoldos).
[1] http://www.simbiosys.ca/

0166-1280/$ - see front matter © 2003 Elsevier B.V. All rights reserved.
doi:10.1016/j.theochem.2003.08.105

In this approach, the availability criterion is a given, since the considered structures are all available (or readily synthesisable).

- Build up complementary structures from scratch—de novo design focused on generating the best possible fit to the constraints.

Each approach have distinct advantages and disadvantages. The docking approach has no problem with synthetic accessibility, because it typically works with databases of available materials. Alternatively, it may be applied to virtual combinatorial libraries, but any structure in such libraries would, in principle, still be easy to synthesize using automated combinatorial chemistry. However, the docking approach suffers the following major shortcomings:

- It is limited to solutions present in a database or virtual library, typically containing 10^6–10^{12} structures. In comparison, it is estimated that 10^{200}–10^{300} molecules are potential drug candidates that fulfil the Lipinski rules [2]. This means that only a tiny fraction of the possibilities is considered, making it highly probable that molecules with much higher binding affinity could be constructed for any target receptor than the best solution found by flexible docking. On the other hand, the drug discovery process is not concerned with finding the 'perfect' ligand, just one that is good enough for the job!
- Abagyan and Totov [1] concludes that, in a realistic case, an average reputable docking algorithm would dock only about 30–50% of the genuine binders to the receptor pocket with RMSD less than 2–3 Å in about 1–3 min per molecule on a single processor. This is a more of a practical technological limit due to the state of the art of docking methods at the time of that article.

As demonstrated in this paper, the eHiTS method, due to its unique exhaustive systematic search technique, is capable of much higher accuracy and speed than the cited [1].

The de novo design approach has a better chance of exploring the vast possibilities mentioned above to find the best possible fit to a particular receptor target. The difficulties of the method are:

- Synthetic accessibility, i.e. the chemical structures suggested by the method may prove very difficult to synthesize.
- Finding the best solution in the vast search space requires very sophisticated algorithms. Brute force systematic searches are not feasible. Most available tools are stochastic, and are thus limited to search a tiny fraction of the space by arbitrary random tests.

The SPROUT software [3] is able to explore the possibilities in a systematic way while avoiding a combinatorial explosion by using efficient pruning techniques and multi directional search to dramatically reduce the computational costs.

2. Methods

2.1. The eHiTS flexible docking method

The main logic of the docking algorithm is described with the following steps:

(1) The 3D structure of the ligand is divided into rigid fragments and flexible chains connecting them. The ligand is represented by a graph, where the nodes are the rigid fragments and the edges are the flexible chains.
(2) The rigid fragments are docked independently into the receptor site. All suitable poses are enumerated that differ sufficiently from each other. There is an adjustable granularity in terms of RMSD of the atom poses in addition to qualitative topological criteria, i.e. forming interactions between a different pair of atoms constitutes a different docking mode for the fragment.
(3) Rigid fragment poses are stored in DockTable, an SQL database that increases the speed of docking large databases of ligands by the re-use of poses of common molecular fragments (e.g. functional groups) that occur in many ligands.
(4) A novel graph matching algorithm (superior to hyper-graph clique detection) rapidly enumerates all fragment pose combinations where the distance and spatial positioning of the rigid fragments is compatible with the possible length range of the chain that should connect them.

(5) In each of the graph maps, flexible chains are fitted between the rigid fragment poses to satisfy steric criteria imposed by the fragments and the receptor site. It is important to note that this flexing of the chain is a very limited problem in scope since both end points are already fixed. Furthermore, it is sufficient to find a single local minimum energy conformation that fits the steric constraints for this sub-problem.

(6) The ligand structure is then reconstructed from the rigid fragment poses selected by the graph map and the chain conformations tweaked to connect the fragments.

(7) Local energy minimisation is performed on the complete (reconstructed) conformation to optimise the scoring function value (taking into account internal strain energy as well as interaction energy contributions).

An empirical scoring function is used several times during the algorithm. First, for the evaluation of rigid poses, then at the selection of the best graph matching solutions, followed by the flexible chain fitting, and finally during the local optimisation of the reconstructed conformation pose.

The scoring function consists of the following components:

- Hydrogen bonding: distance and angle dependent energy function
- Hydrophobicity: contact surface area + pi stacking energy
- Electrostatic potential (Coulomb formula)
- Van der Waals contact energy
- Metal ion interactions: distance and angle dependent (similar to H-bond)
- Penalty for incompatible contacts (e.g. polar-hydrophobic)
- Exposed surface scored against solvent properties
- Intra-molecular interactions (bumps, torsion penalties and internal interactions, like H-bond and hydrophobic contacts)

2.2. The de novo design method employed in SPROUT

The search space to be explored for de novo design is greater by several orders of magnitude than the conformation and pose space of flexible ligand docking. The reason is very simple: de novo design can be considered as a flexible docking problem, where members of a ligand library of infinite size are considered as candidates. In addition to dealing with all the same problems that docking presents, de novo design also has to deal with the construction of candidate structures.

SPROUT contains a very efficient exhaustive multi-directional tree search with a set of constraints limiting the search space to manageable size. The constraints consist of steric conditions imposed by the protein structure, geometric limits for a selected set of interaction sites, size limitations for the candidate structures (maximum number of heavy atoms, maximum number of rings of each specific size etc.). The structure generation is also limited by a set of joining rules that have been defined based on frequently occurring motifs in small molecule drug databases.

The program offers automated cleft detection and target interaction site exploration based on a configurable knowledge base of residues. The atoms in the knowledge base are labelled to identify typical hydrogen bonding, hydrophobic, electrostatic and metal binding potentials. The user is presented with 3D visualisation of the analysis results and is allowed to choose a specific subset of the detected interaction sites to be target sites, where the program will focus the structure generation effort.

A set of small (rigid) molecular fragments, including functional groups are docked at each target site independently. The program enumerates all possible docking modes (combinations of interacting atom pairs) and selects the most suitable orientation of the fragment considering the steric constraints.

Structure generation is initiated from all these docked seed fragments. Partial structures are grown from each seed by incremental joining of spacer template fragments from a small library of generic 3D structures that allows the building of a very wide variety of chemically sensible and stable structures according to rules in the joining knowledge base. A rapid rigid body docking is applied after each joining step to ensure that the steric and target site geometry constraints are still satisfied. The partial structures are connected when they reach sufficient size and the unified structure is docked again. These docking steps are fast because the interacting atom pairs are forced to maintain their binding with the same receptor atoms as their parent structures.

During this multi directional hierarchical tree search every possible successor is examined to see whether it can fit the binding site satisfying all the steric, target site geometric and user defined constraints. When a sub-structure fails for any of the constraints, the whole branch of its potential successors is eliminated from the search tree. Each search tree is limited in depth based on the distances of the target sites to be spanned by the ligand structure.

3. Results

The described methods are intended to be used as research tools for new problems for which the solution is not yet known. However, their suitability to produce reliable results is best proved by running them for input cases for which the expected output is available as experimental data, e.g. for protein-ligand pairs where X-ray structure of the co-crystalised complex is available. These test runs are typically referred to as 'validation test'.

There is another paper entirely focused on validation test runs of SPROUT by [8].

3.1. Validation results of eHiTS

The eHiTS docking system was validated on a set of 100 protein ligand complex files taken from the PDB database[2]. Water was removed from the PDB files, protonation states and partial charges were assigned automatically by eHiTS during the run. The standard parameter set was used, which generates up to 200 docking poses for each case. The RMS deviation of the heavy atom coordinates of the best pose from the original X-ray ligand pose was calculated. Some statistics on the results are shown in Table 1. Some examples are shown on Figs. 1–3.

[2] 1a28 1a4g 1a6w 1abe 1abf 1acj 1acm 1aco 1ai5 1aoe 1azm 1b59 1b9v 1byb 1byg 1c1e 1c5x 1c83 1cbx 1cil 1ckp 1coy 1dbb 1dbj 1dg5 1dmp 1dog 1dr1 1dwb 1ebg 1eoc 1f3d 1fgi 1fki 1flr 1hsb 1hsl 1hyt 1ibg 1lah 1lcp 1ldm 1lna 1lst 1lyl 1mbi 1mdr 1mrg 1mrk 1mup 1ngp 1nis 1okl 1okm 1pbd 1phd 1ptv 1qcf 1qpe 1qpq 1rnt 1srj 1tng 1tnh 1tni 1tnl 1trk 1tyl 1ukz 1ulb 1wap 1xie 1ydr 2aad 2ack 2cht 2cmd 2cpp 2ctc 2dbl 2gbp 2h4n 2lgs 2mcp 2pcp 2phh 2pk4 2qwk 2ypi 3cpa 3erd 3gpb 3hvt 3tpi 4cts 4fbp 4lbd 5abp 5cpp 7tim.

Table 1
Validation result statistics for eHiTS

Number of cases	100
Cases when RMSD < 2.0 Å (%)	100
Cases when RMSD < 1.5 Å (%)	93
Cases when RMSD < 1.0 Å (%)	56
Cases when RMSD < 0.5 Å (%)	13
Average RMSD	0.93

The X-ray ligand is displayed with stick and ball representation, while the eHiTS solution poses are displayed in cylinder bond representation.

3.2. Speed test results of eHiTS

eHiTS is suitable for virtual high throughput screening of large ligand databases against a receptor target. To measure the speed of the docking, a test was carried out to dock 10000 ligands from the Maybridge database into the 1hri receptor taken from the Protein Data Bank using the '-fast' option. This mode enforces limits on the number of poses tested and evaluated to speed up mass processing at the expense of docking pose accuracy.

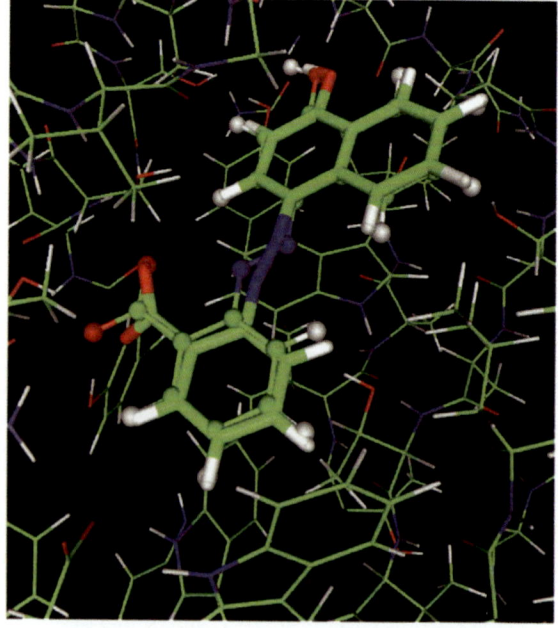

Fig. 1. PDB code 1SRJ, eHiTS pose 0.5 Å RMS.

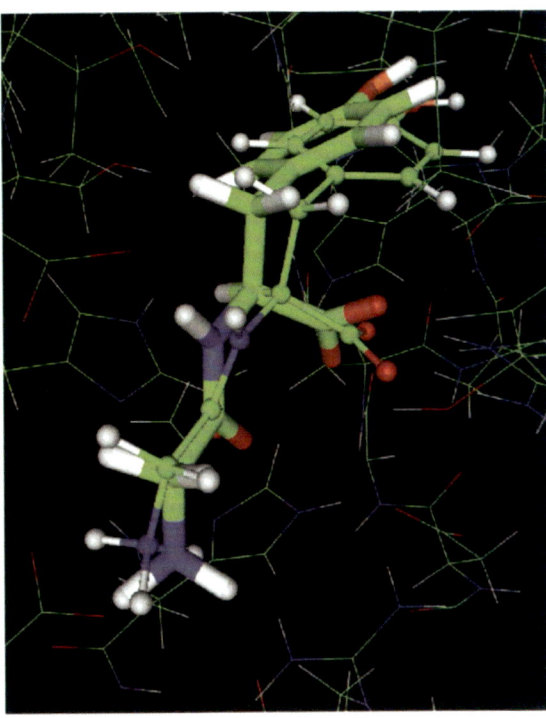

Fig. 2. PDB code 3CPA, eHiTS pose 0.9 Å RMS.

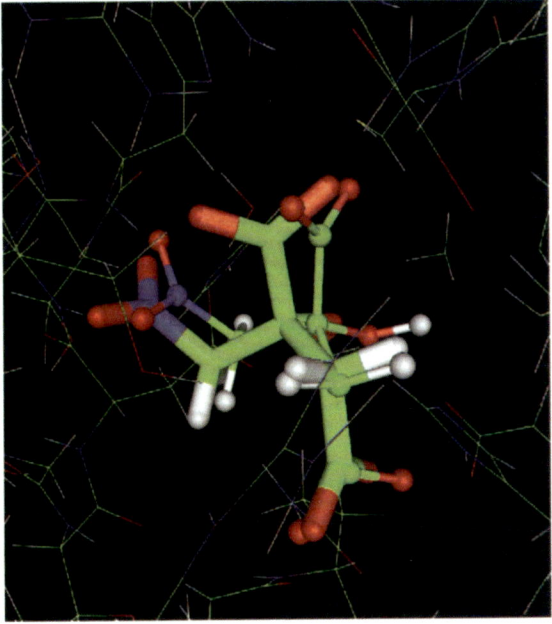

Fig. 3. PDB code 1NIS, eHiTS pose 0.6 Å RMS.

Table 2
Speed test results of eHiTS

Number of ligands	Total time (s)	Time (s)/ligand
100	500.12	5.00
1000	2382.53	2.39
10000	16458.63	1.65

Table 2 contains some statistics about the run time. The time required to dock each ligand decreases as the number of ligands increases. The DockTable feature of eHiTS greatly reduces the computation time especially when large databases are used. This feature is especially advantageous in industry since many virtual libraries of compounds are of the order of millions of ligands.

4. Discussion

Generic stochastic optimisation methods (e.g. Monte Carlo simulated annealing) and evolutionary techniques (e.g. genetic algorithms) can be applied to a wide range of scientific problems. They often provide acceptable results with little research, design and development effort. On the other hand, it would not make sense to use such an algorithm for sorting numbers, because efficient systematic methods are available already for that task. Similarly, we have developed efficient, systematic methods for flexible ligand docking and de novo ligand design problems. These methods offer exhaustive search with better speed performance and accuracy than the random methods. [4] compared stochastic methods with systematic method and concluded that stochastic methods have inferior performance.

There are a number of systematic methods for flexible docking in the literature, e.g. [6,7], however, they are all limited to discrete dihedral angle sampling (e.g. 30°) and heuristic pruning, therefore they are not covering the search space exhaustively. A single dihedral angle twist of 5° does not alter significantly the strain energy of the ligand, but it can make a very significant 1.5 Å displacement of an atom 6 Å away from the axis of the rotated bond as demonstrated on the drug-like molecule on Fig. 4 taken from an actual receptor–ligand co-crystal structure. Such an atom

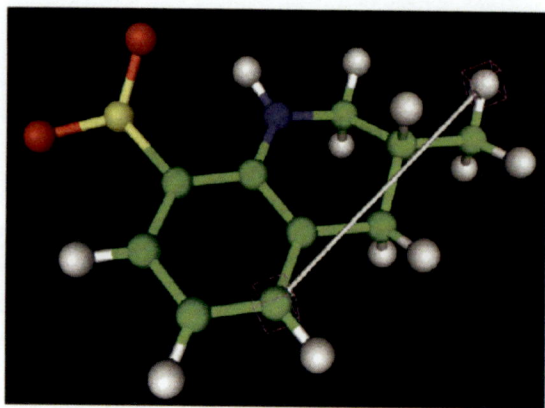

Fig. 4. Amplified effect of a small dihedral angle change.

movement can alter a very severe unacceptable steric clash into a perfect surface fit. Therefore, continuous variation of dihedral angles could significantly improve the chance of finding a good fit.

Another important observation can be made on the flexible ligand conformation in X-ray structures of co-crystallized protein–ligand complexes. The majority of flexible ligands bind in a conformation so that one or more dihedral angles substantially deviate from the gas-phase lowest energy conformer of the same ligand. Experimental evidence also supports the need to consider a wider variety of conformations, including unfavourable dihedral angles. Methods that only use low energy conformers are bound to miss the bio-active binding conformation of many ligands.

Let us consider a typical drug-like molecule, and estimate the size of the conformational search space that need to be explored if a sufficiently fine sampling would be used. Assume, the molecule has 6 rotatable bonds, each of which should be sampled at least every $5°$ as we have seen above. This requires $360/5 = 72$ samples for each independently, giving rise to 72^6 combinations. The orientational freedom of the whole ligand can be expressed as rotations about 3 coordinate axes, with the same granularity requirements, that is 72^3 orientations altogether. The binding study would be performed in a receptor pocket that fits inside a 10Å cube. The translational freedom should be sampled at least with a 0.5 Å precision, resulting in 20^3 grid points to test. Altogether, $20^3 \times 72^3 \times 72^6 \simeq 2 \times 10^{20}$ conformation poses should be considered for a brute force evaluation. A fast, empirical scoring

function could evaluate about 2000 poses per second, which means a brute force exhaustive flexible docking would take about 3 billion years. Stochastic methods explore a tiny fraction of this same vast search space with random selection.

Therefore, the innovative docking method of eHiTS is a unique solution that provides systematic and exhaustive coverage of the search space for flexible ligand docking. Other methods are either unreliable due to the use of random techniques, or limited to search an arbitrary discretised sub-space, which is proven to miss the ideal solution in practical cases.

The validation test results demonstrate the superior accuracy of eHiTS in reproducing the X-ray crystal structure pose of numerous protein–ligand complexes. The speed test results demonstrate that the method is fast enough for screening of large ligand databases in practical time frames. Another key criterion for database screening is that the system must be able to operate in fully automated mode, i.e. without any human intervention, seed fragment selection or any other parameter specification. eHiTS is capable of assigning partial charges and considering all alternative protonation states (tauto-merisms) in a single run.

The other presented software tool, SPROUT, provides a de novo design solution that is also systematic and exhaustively covers the search space within the problem constraints and user defined limits. The efficient pruning of the search trees and depth limitations for partial structures allow SPROUT to perform the exhaustive search in a practical time frame, especially when run on a distributed computing environment, e.g. a Linux cluster.

The search space for this problem is so vast that it would take billions of years to cover all possibilities by brute force evaluation even if the fastest super-computers available are used. Therefore, many de novo tools that are available use stochastic or evolutionary techniques to find some random solutions. However, there are reports of surprisingly good results using random techniques even against the astronomical odds.

There are systematic methods that search bridge fragments from a library of known compounds to connect binding functional groups. Without incremental building process, these semi de novo methods are somewhat similar to the flexible docking of programs.

The introduction of our new, fast and truly exhaustive systematic search methods solves half of both the docking and de novo design problems: the enumeration of suitable conformations, poses and structures. However, another key component remains subject to intense research: construction of a fast and reliable scoring function which would be able to identify the best solution from all the candidates for any protein family. Although scoring functions have been the subject of active research and debates for several years [5], the perfect solutions has still not yet emerged.

The availability of these ultimate search techniques further emphasizes the need for better scoring functions. As we demonstrated, eHiTS and SPROUT are both capable of precisely reproducing X-ray structure poses of known ligands. However, they are also capable of generating hundreds of solutions that score better than the crystal structure ligand for any particular scoring function tested so far. These false 'better' solutions are typically different for the various scoring functions, providing a sharp focus on the weakness of the particular function.

5. Conclusions

Flexible ligand docking and de novo design are two computational approaches to structure based rational drug design. Both of them consist of two major components: search algorithm and scoring function. Many software tools use generic random search techniques relying heavily on the scoring function to drive the search towards the solution. Other tools use systematic methods on arbitrary sub-sampling of the vast search space.

A new continuously exhaustive systematic flexible docking search method has been presented that is available in the eHiTS software. This search engine provides an excellent solution for quickly generating all reasonable candidate conformations and poses for binding the ligand to the target receptor, including a pose that strongly resembles the bound conformation found in the known co-crystallized ligand.

Identifying the pose closest to the X-ray structure remains an elusive challenge for the scoring functions.

In fact, the exhaustive search can produce alternative poses for all tested scoring functions that are judged to be better than the known X-ray pose. Thus, the excellence of the search engine emphasizes the weakness of the scoring functions.

The SPROUT software delivers a versatile search engine for de novo ligand design. It also uses a systematic search engine with no random elements and is capable of exhaustive coverage of the possibilities within the receptor imposed constraints and user defined limits.

Both software tools are available from SimBioSys Inc. (SimBioSys)

Acknowledgements

We would like to thank the National Research Council of the Government of Canada for the financial support of our research work under the following grants: NRC-IRAP project #411461, GeoDock SR and ED project (2001), NRC-IRAP project #468906, EnerMin SR and ED project (2002)

References

[1] R. Abagyan, M. Totrov, High-throughput docking for lead generation, Curr. Op. in Chem. Biol. 5 (2001) 375–382.
[2] R. Carr, M. Hann, Modern Drug Discov. 5 (2002) 45–48.
[3] V.J. Gillet, G. Myatt, Z. Zsoldos, A.P. Johnson, SPROUT, HIPPO and CAESA: tools for de novo structure generation and estimation of synthetic accessibility, Perspect. Drug Discovery Des. 3 (1995) 34–50.
[4] A critical evaluation of several global optimization algorithms for the purpose of molecular docking, J. Comput. Chem., 20, 1999, 1740–1751
[5] C. Bissantz, G. Folkers, D. Rognan, Protein-Based Virtual Screening of Chemical Databases. 1. Evaluation of Different Docking/Scoring Combinations, J. Med. Chem. 43 (2000) 4759–4767.
[6] M. Rarey, B. Kramer, T. Lengauer, G. Klebe, A fast flexible docking method using an incremental construction algorithm, J. Mol. Biol 261 (1996) 470–489.
[7] S. Makino, I.D. Kuntz, Automated flexible ligand docking method and its application for database search, J. Comp. Chem. 18 (1997) 1812–1825.
[8] J.M.S. Law, D.Y.K. Fung, Z. Zsoldos, A. Simon, Z. Szabo, I.G. Csizmadia, A.P. Johnson, Validation of the SPROUT de novo design program, J. Mol. Struct. Theochem (2003).

ELSEVIER

Journal of Molecular Structure (Theochem) 666–667 (2003) 667–680

THEO
CHEM

www.elsevier.com/locate/theochem

Thermochemical study on the ring closure reaction of 5-morpholino-4-vinylpyridazinones by *tert*-amino effect

László Károlyházy[a], Géza Regdon Jr.[b], Olivér Éliás[a], Gyula Beke[a], Tamás Tábi[a], Klára Hódi[b], István Erős[b], Péter Mátyus[a],*

[a]*Department of Organic Chemistry, Semmelweis University, Hőgyes Endre utca 7., 1092 Budapest, Hungary*
[b]*Department of Pharmaceutical Technology, University of Szeged, Eötvös utca 6., 6720 Szeged, Hungary*

Abstract

Cyclization of *tert*-anilines with a properly substituted vinyl moiety by the *tert*-amino effect affords fused pyridines. This type of ring closure reaction of 5-morpholino-4-vinylpyridazinones and a benzene analogue was investigated by differential scanning calorimetry (DSC) measurements. The structure of products was confirmed by spectroscopic and microanalytical methods. The enthalpy values and heats of reactions were obtained from the thermograms by integrating the peak area corresponding to the ring-closure reaction, and by semiempirical (PM3) and DFT (density function theory) calculations, respectively.
© 2003 Elsevier B.V. All rights reserved.

Keywords: Differential scanning calorimetry; *Tert*-amino effect; Vinylpyridazines; Pyridopyridazines

1. Introduction

Application of the *tert*-amino effect for preparation of pyrido-fused ring systems has been well documented [1–4]. In this type of ring closure reaction cyclization of compound **A** occurs between the β-carbon of a vinyl group and the α-carbon of the *ortho-tert*-amino group to form compound **B** (Scheme 1). The presence of strongly electron-withdrawing substituents at the β-carbon seems to be a prerequisite for the ring closure. If the electron-withdrawing substituents are parts of a cyclic moiety (e.g. a pyrimidinetrione ring), spirocyclic ring systems (**C**) may be obtained. In this way, we recently prepared some new racemic, spirocyclic pyridazinone derivatives including dispiro compounds exhibiting central and axial types of chirality [5,6].

As a part of this program, we now report on the differential scanning calorimetry (DSC) study of the ring closure reactions of 5-morpholinopyridazinones possessing cyclic and acyclic vinyl substituents; to study the role of the aryl ring, a phenylmorpholine analogue was also investigated. The aim was to determine the enthalpies of cyclization reactions and to investigate structural changes possibly occurring upon heating. Calculations by semiempirical (PM3) and DFT (density functional theory) methods were also undertaken to gain heats of reactions.

2. Experimental section

2.1. Synthesis

All melting points were determined in a Kofler apparatus, and are uncorrected; the values are given

* Corresponding author. Tel./fax: +36-1-2170851.
 E-mail address: matypet@szerves.sote.hu (P. Mátyus).

0166-1280/$ - see front matter © 2003 Elsevier B.V. All rights reserved.
doi:10.1016/j.theochem.2003.08.090

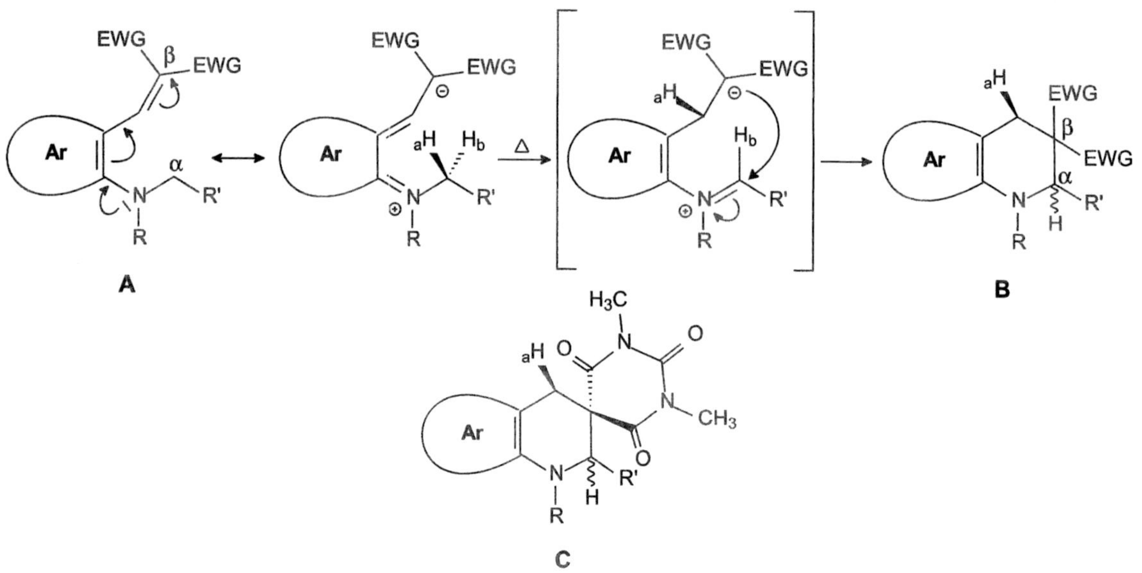

Scheme 1.

in °C. The IR spectra were recorded on a Perkin–Elmer 1600 FTIR instrument in potassium bromide pellets. The NMR spectra were recorded on a Bruker AM spectrometer at 200 MHz (in case of ^{1}H NMR) or at 50 MHz (in case of ^{13}C NMR) in the solvent indicated at room temperature, using the ^{2}H signal of the solvent as the lock, and tetramethylsilane as the internal standard. The assignments of ^{13}C NMR spectra were supported by DEPT-135 spectra. Coupling constants (J) are given in Hz. The elementary analyses (C, H, N) were performed on a Carlo Erba Elemental Analyzer Model 1012 apparatus. For column chromatography, Kieselgel 60 (Aldrich, 0.063–0.2 mm silica gel) was used; for tlc analysis Silica gel 60 F_{254} (Merck) plates were applied. Solvent mixtures used for chromatography are always given in a vol/vol ratio. All reagents, as well as *2-fluorobenzaldehyde, N,N'-dimethylbarbituric acid* and *malononitrile* were purchased from Aldrich (Sigma-Aldrich Kft, Hungary) and used as received. Solvents were dried and distilled prior to use. Organic extracts were dried over anhydrous sodium sulphate.

The synthetic route of the compounds investigated is shown in Scheme 2.

The following compounds were prepared according to the procedures cited: **1** [7], **2** [3], **4** [5], **6** [6], **7** [3], **9** [5], **11** [6], **12** [3].

Names and numbering of new compounds are included in Appendices A and B.

2.1.1. Preparation of 2-benzyl-2-methyl-1,3-dioxane-4,6-dione (13)

Malonic acid (10.40 g, 0.1 mol) was dissolved in a mixture of acetic anhydride (45 ml) and sulfuric acid (96%, 0.3 ml). Benzyl-methyl-ketone (14.7 ml, 0.1 mol) was added dropwise to the reaction mixture, while the inner temperature was kept between 20–25 °C. After the reaction mixture was kept at 5 °C for two days, it was poured onto ice (500 g) and extracted with chloroform (300 ml). The chloroform layer was washed with water (3 × 100 ml). The organic layer was dried and evaporated in vacuo. Hexane (80 ml) was added to the oily residue and the suspension was stirred for 0.5 h. After decantation, hexane was added in several portions (7 × 80 ml), and was subsequently decanted. The solid thus formed was collected by filtration, washed with hexane and recrystallized from 2-propanol. The white product was filtered off, washed with 2-propanol and ether, then dried (3.71 g, 17%; mp 66–67; R_f 0.42 in toluene–methanol, 4:1). $C_{12}H_{12}O_4$ (220.22) Calcd: C 65.45; H 5.49. Found: C 65.76; H 5.53. IR (ν) 3080, 2925, 1732, 1290, 978, 910, 784 cm^{-1}. ^{1}H NMR (CDCl$_3$)

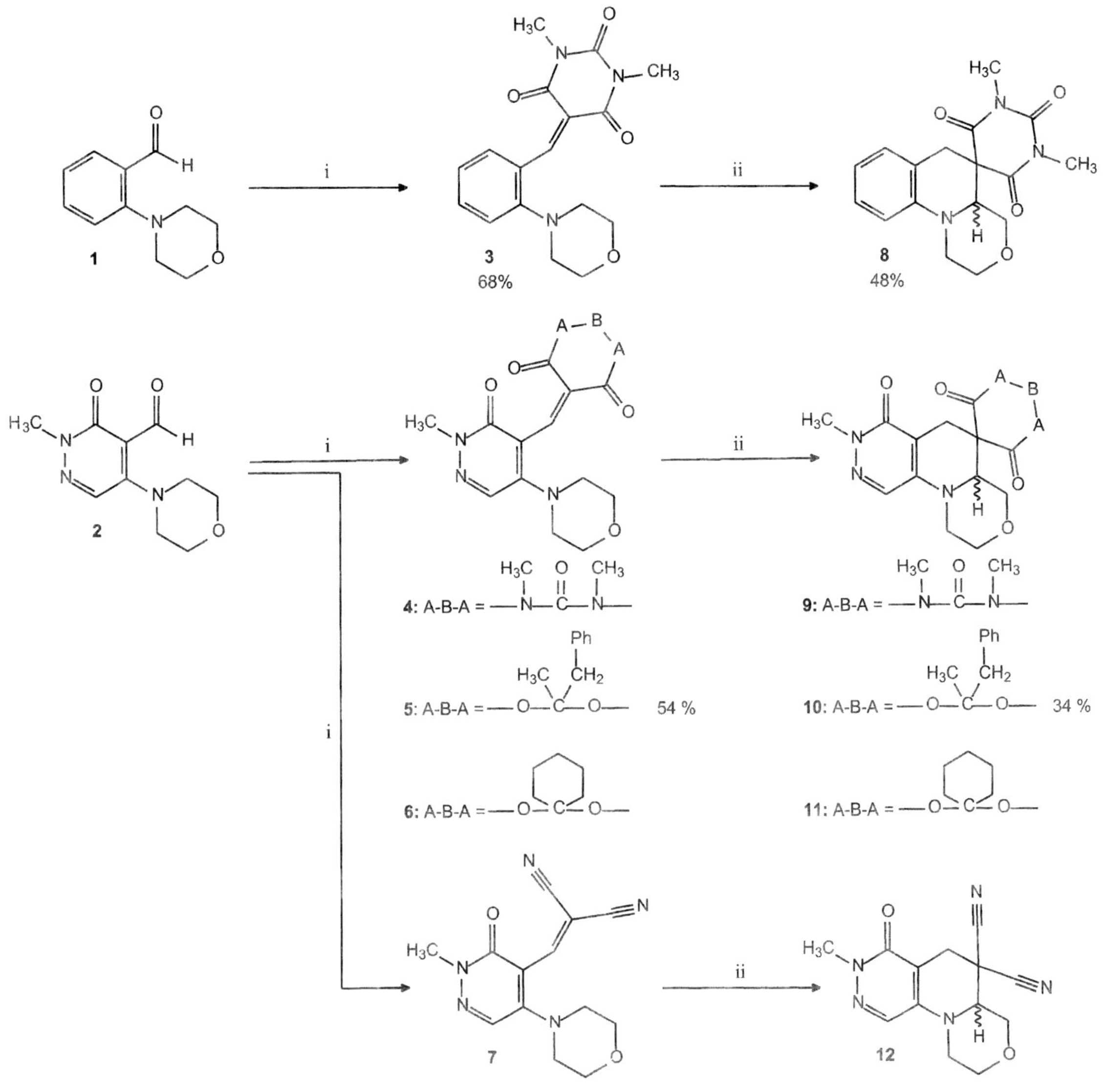

Scheme 2. *Reagents and conditions*: (i) for **3**: *N,N*'-dimethylbarbituric acid, ethanol, rt, 15 min; for **5**: 2-benzyl-2-methyl-1,3-dioxane-4,6-dione (**13**), piperidine, ethanol, rt, 60 min. (ii) for **8**: toluene, 10% AlCl₃, 70 °C, 3 h; for **10**: DMF, 120 °C, 30 h. Compounds **1** [7], **2** [3], **4** [5], **6** [6], **7** [3], **9** [5], **11** [6] and **12** [3] were prepared in the described ways.

δ 7.29–7.21 (5H, m, phenyl protons), 3.26 (2H, s, C2–CH_2), 3.20 (1H, d, 5-H$_a$, $J = 21.2$ Hz), 2.31 (1H, d, 5-H$_b$), 1.81 (3H, s, C2–CH_3). ^{13}C NMR (CDCl₃) δ 162.7 (C-4,6), 131.9 (C-1'), 131.4 and 129.1 (C-2',3',5',6'), 128.4 (C-4'), 107.0 (C-2), 46.9 (C2–CH_2), 35.2 (C-5), 27.3 (C2–CH_3).

2.1.2. Preparation of the vinyl compounds 3, 5

Compound **3**: To a solution of **1** (0.38 g, 2 mmol) in ethanol (3 ml), compound **2** (0.31 g, 2 mmol)

was added. The reaction mixture was stirred at room temperature for 15 min. The yellow solid precipitated was filtered off, washed with ethanol and recrystallized from a mixture of acetone and hexane at -18 °C (0.45 g, 68%; mp 165–166; R_f 0.46 in dichloromethane–ethyl acetate, 9:1). C₁₇H₁₉N₃O₄ (329.36) Calcd: C 62.00; H 5.81; N 12.76. Found: C 62.12; H 5.87; N 12.62. IR (ν) 2944, 2818, 1732, 1669, 1589, 1458, 1368, 1260, 1112, 934, 754 cm^{-1}. ^{1}H NMR (CDCl₃) δ 8.43 (1H, s, –CH=), 7.98 (1H, m, 6-H), 7.49 (1H, m, 4-H), 7.0–7.2 (2H, m, 3,5-H), 3.88 (4H, t, $2 \times O-CH_2$), 3.43, 3.37 ($2 \times 3H$, s, N1–CH_3 and N3–CH_3), 3.02 (4H, t, $2 \times N-CH_2$). ^{13}C NMR (CDCl$_3$) δ 162.7, 160.7, 151.4 (C-2,4,6; pyrimidine), 156.9 (–CH=), 133.7, 132.3 (C-4,6), 126.5 (C-1), 121.7 (C-3), 117.6 (C-5), 116.0 (C-5; pyrimidine), 67.2 ($2 \times O-CH_2$), 53.9 ($2 \times N-CH_2$), 28.4, 28.3 (N1–CH_3 and N3–CH_3).

Compound 5: To a solution of 2 (0.45 g, 2 mmol) in ethanol (3 ml), compound 13 (0.44 g, 2 mmol) was added. The reaction mixture was stirred at room temperature for 60 min. The yellow solid precipitated was filtered and washed with 2-propanol, then with diisopropylether, and recrystallized from a mixture of dichloromethane and petrolether at $-18\,°C$ (0.46 g, 54%; mp 184–185; R_f 0.51 in toluene–methanol, 4:1). C$_{22}$H$_{23}$N$_3$O$_6$ (425.45) Calcd: C 62.11; H 5.45; N 9.88. Found: C 61.85; H 5.67; N 9.55. IR (ν) 3453, 2856, 1728, 1648, 1575, 1148, 1356, 1242, 1207, 1112, 959 cm^{-1}. ^{1}H NMR (DMSO-d_6) δ 8.03 (1H, s, 6-H), 7.79 (1H, s, –CH=), 7.34 (5H, s, phenyl), 3.71 (4H, s, $2 \times O-CH_2$), 3.53 (3H, s, N2–CH_3), 3.50–3.25 (6H, m, $2 \times N-CH_2$ and C2–CH_2), 1.81–1.51 (3H, br, C2–CH_3). ^{13}C NMR (CDCl$_3$) δ 162.5, 159.4 (C-4,6; dioxane), 157.4 (C-3; pyridazine), 151.6 (C-5; pyridazine), 146.2 (–CH=), 134.1, 130.8, 130.1, 128.2, 127.2 (C-4,6; pyridazine and phenyl carbons), 109.5 (C-5; dioxane), 66.3 ($2 \times O-CH_2$), 50.8 (N–CH_2), 45.0 (C2–CH_2), 39.4 (N2–CH_3), 25.1 (C2–CH_3).

2.1.3. Preparation of pyrido fused compounds 8, 10

Compound 8: A solution of 3 (0.175 g, 0.53 mmol) in toluene (2 ml), in the presence of catalytic amount of AlCl$_3$ (0.01 g, 0.07 mmol), was heated at 70 °C for 3 h. After the solvent was removed in vacuo, the residue was taken up in water, the solution was acidified, and decanted. Crystallization of the residue from ethanol gave a white solid (0.08 g, 48%; mp 221–222; R_f 0.48 in dichloromethane). C$_{17}$H$_{19}$N$_3$O$_4$ (329.36) Calcd: C 62.00; H 5.81; N 12.76. Found: C 61.74; H 5.81; N 12.46. IR (ν) 2957, 2846, 1744, 1688, 1669, 1603, 1499, 1450, 1378, 1260, 1127, 753 cm^{-1}. ^{1}H NMR (DMSO-d_6) δ 7.10–6.85 (3H, m, 7,9,10-H), 6.70 (1H, m, 8-H), 3.19, 3.10 ($2 \times 3H$, s, N1′–CH_3 and N3′–CH_3), 3.90–3.10 (8H, m, 1-H$_{eq}$, 2-H$_2$, 4-H$_2$, 4a-H, 6-H$_2$), 2.87 (1H, m, 1-H$_{ax}$). ^{13}C NMR (DMSO-d_6) δ 169.7, 167.2, (C-4′,6′), 150.8 (C-2′), 144.2 (C-10a), 128.3, 126.4 (C-7,9), 121.6 (C-6a),

118.2 (C-10), 112.8 (C-8), 65.8, 65.6 (C-2,4), 59.4 (C-4a), 49.5 (C-5), 46.6 (C-1), 33.6 (C-6), 28.7, 28.2 (N1′–CH_3 and N3′–CH_3).

Compound 10: A solution of 5 (0.26 g, 0.61 mmol) in N,N-dimethylformamide (2 ml) was heated at 120 °C for 30 h. After the solvent was removed in vacuo, the residue was triturated with water, the solid was filtered off and crystallized from 2-propanol to give white crystals (0.09 g, 34%; mp 192–193; R_f 0.34 in toluene–methanol, 4:1). C$_{22}$H$_{23}$N$_3$O$_6$ (425.44) Calcd: C 62.11; H 5.45; N 9.88. Found: C 62.20; H 5.55; N 9.74. IR (ν) 3440, 2958, 1779, 1735, 1728, 1628, 1622, 1446, 1402, 1260, 1125 cm^{-1}. ^{1}H NMR (CDCl$_3$) δ 7.61–7.16 (6H, m, 1′-H and phenyl), 4.00–3.05 (9H, m, 6a′-H, 7′-H$_2$, 9′-H$_2$, 10′-H$_2$, C2–CH_2), 3.69 (3H, s, N3′–CH_3), 3.00–2.80 (1H, m, 5′-H$_{eq}$), 2.50–2.35 (1H, m, 5′-H$_{ax}$), 1.87 (3H, s, C2–CH_3). ^{13}C NMR (CDCl$_3$) δ 167.3, 167.2, 163.2, 162.9 (C-4,6), 159.9 (C-4′), 143.4 (C-11a′), 143.8, 129.4, 128.4, 126.2 (phenyl carbons), 126.1 (C-1′), 111.8, 108.5 (C-4a′), 106.6 (C-2), 66.7, 66.0 (C-7′,9′), 57.5 (C-6a′), 47.9 (C-6′), 46.9 (C2–CH_2), 45.4 (C-10′), 39.5 (N3′–CH_3), 29.2 (C2–CH_3), 28.9 (C-5′).

2.2. DSC measurements

The thermoanalytical examinations [8,9] were carried out with a Mettler–Toledo DSC 821^e (Mettler–Toledo GmbH, Switzerland) instrument. DSC curves were evaluated with STAR^e Software. The starting and final temperatures were 25 °C and 300 or 400 °C, respectively. Heating rate was 2 °C/min. Argon atmosphere was always used. Samples of 4–12 mg were used. Three parallel examinations were made in every case. The instrument was calibrated using indium. For NMR spectroscopy, from the DSC samples, the compound was gained by treatment of the aluminium DSC-cup with DMSO-d_6. The solutions were filtered and used for the measurement.

2.3. Computational chemistry

Semiempirical PM3 and DFT (Density Functional Theory) calculations were carried out by using Spartan program package (Spartan SGI Version 5.1.1., Wavefunction Inc. 1998) on Silicon Graphics (INDY R4400). For DFT (LSDA/pBP86/DN model

[10]) calculations the optimized structures obtained at PM3 level were used as starting geometries.

3. Results and discussion

For the thermochemical investigation of the ring closure reaction by *tert*-amino effect vinylpyridazinones incorporating a pyrimidinetrione (**4**), an 1,3-dioxane (**5**, **6**) and a malonodinitrile moiety (**7**) as well as a benzene analogue of **4**, i.e. compound **3** were prepared by a Knoevenagel reaction of *ortho*-morpholinoaldehydes with the appropriate active methylene compound at room temperature. Thermal ring closure reactions were next carried out to give fused ring systems **8–12**, respectively (Scheme 2). Of these compounds, **3**, **5**, **8**, and **10** have not been reported so far. All compounds gave satisfactory analytical data (Section 2).

The calorimetric investigations of vinyl compounds **3–7** and their ring closed derivatives **8–12** were carried out between room temperature and 300–400 °C.

For the fused systems **8–12**, the DSC thermograms are very simple (Figs. 1–5), and peaks corresponding to their melting points could be easily identify as endothermic peaks: 221.0 °C for **8**, 258.9 °C for **9**, 196.2 °C for **10**, 194.3 °C for **11**, 231.9 °C for **12**.

In the cases of the precursor vinyl compounds **3–7**, each thermogram has more than one peaks (Figs. 6–10).

On the DSC heating curves of the two N,N'-dimethylbarbituric acid derivatives **3** and **4**, three and two, respectively, exothermic peaks, as well as one endothermic peak appeared (Figs. 6 and 7). The endothermic peaks (218.3 °C for **3** and 257.0 °C for **4**) were overlapping with the endothermic peaks of the corresponding ring closed derivatives (221.0 °C for **8**, Fig. 1; and 258.9 °C for **9**, Fig. 2). Evidently, upon heating in the DSC apparatus, the ring closure reactions took place to give the fused systems that then melted to produce the respective endothermic peaks. In order to prove that the cylization indeed

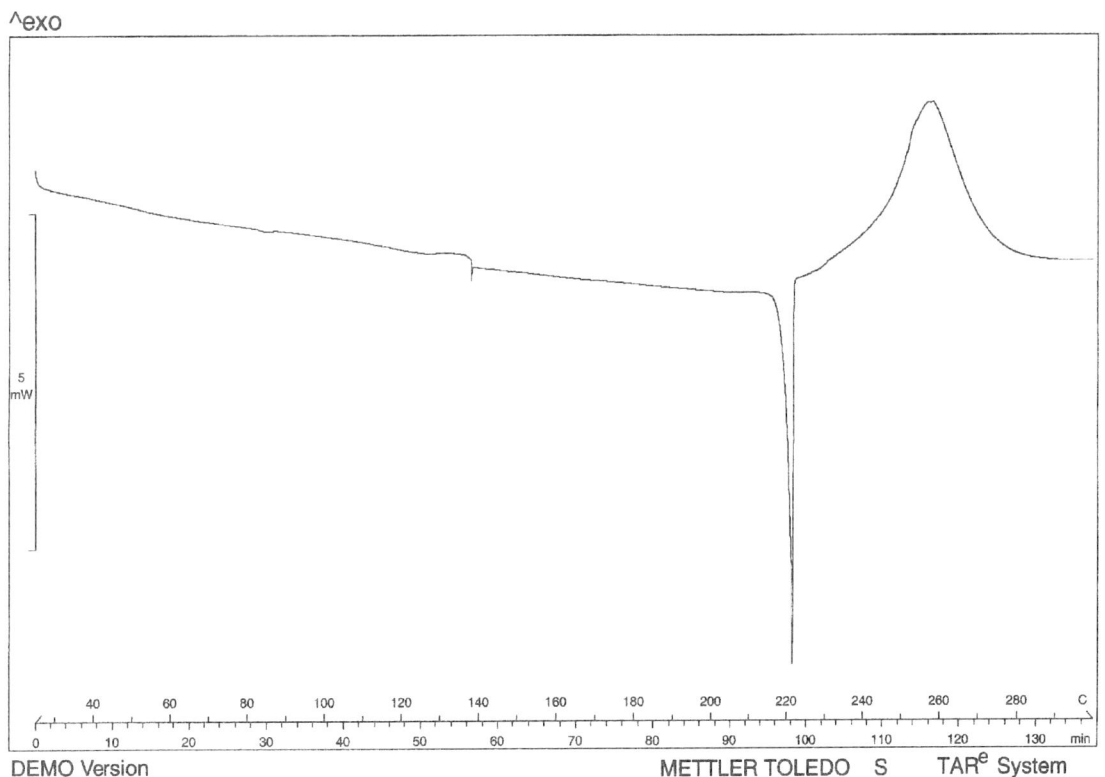

Fig. 1. The DSC curve of compound **8**.

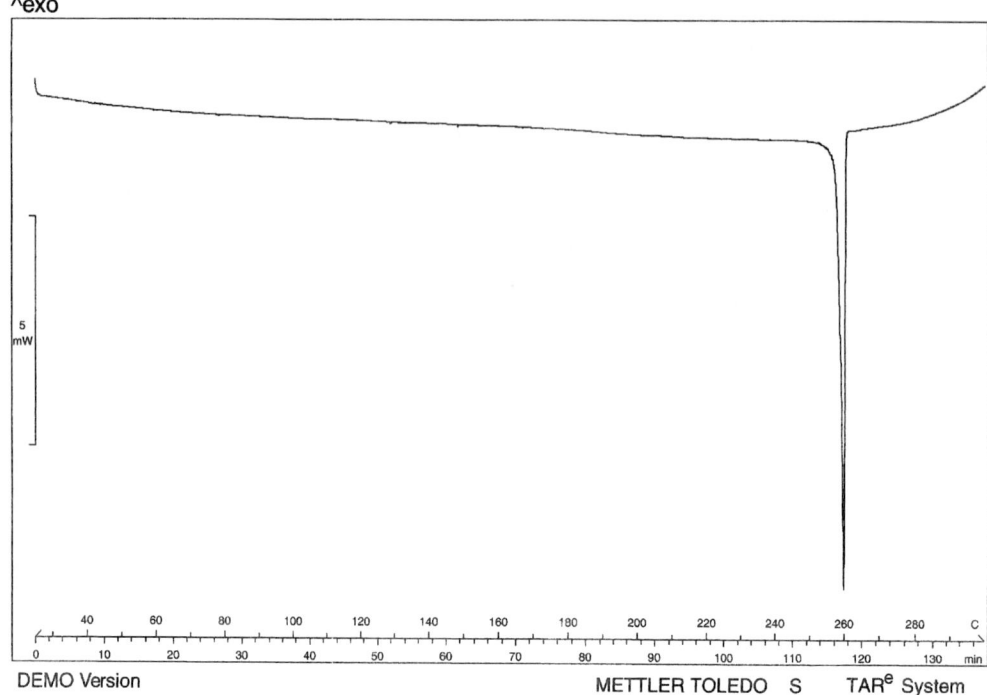

Fig. 2. The DSC curve of compound **9**.

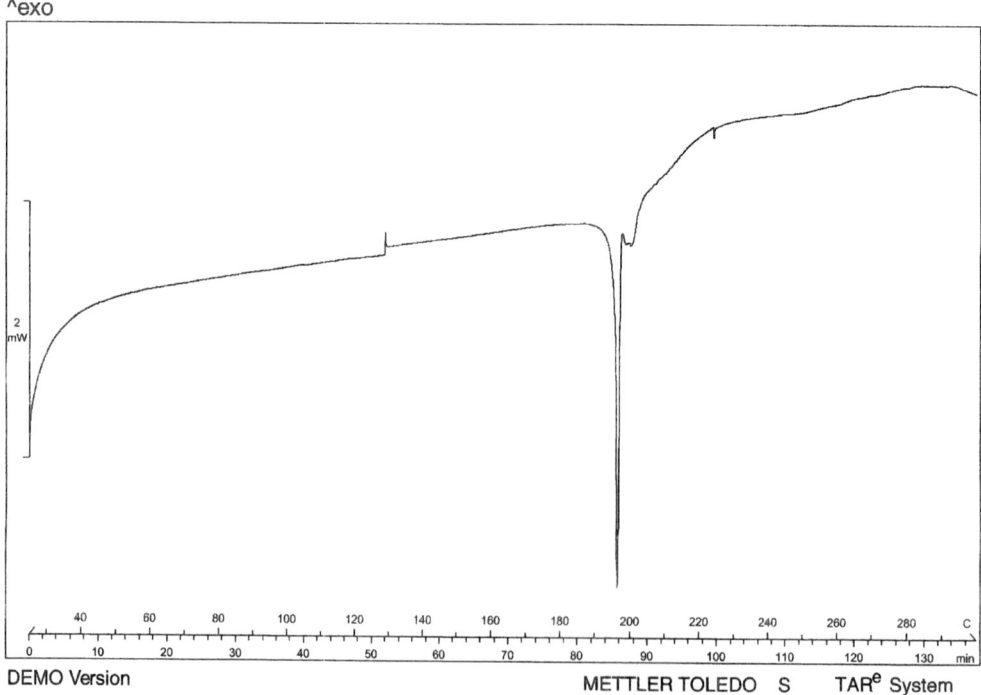

Fig. 3. The DSC curve of compound **10**.

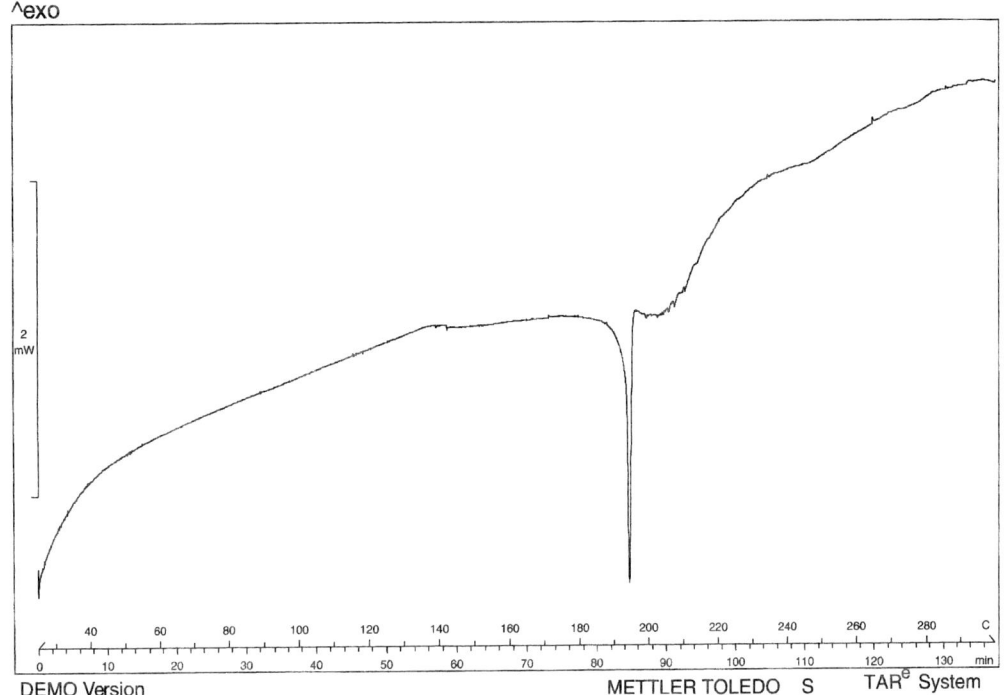

Fig. 4. The DSC curve of compound **11**.

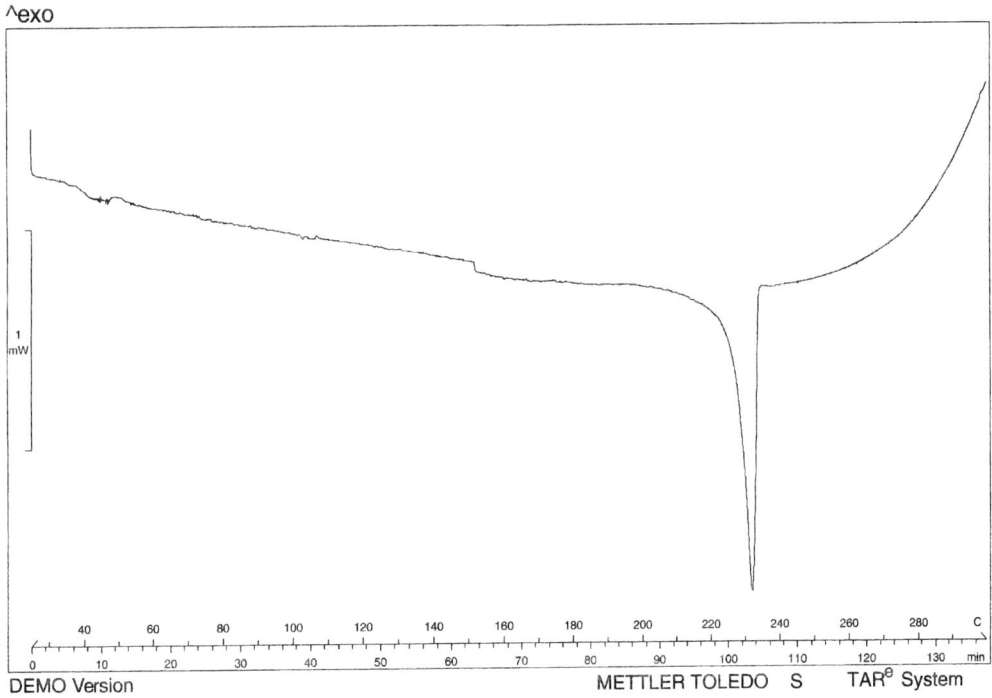

Fig. 5. The DSC curve of compound **12**.

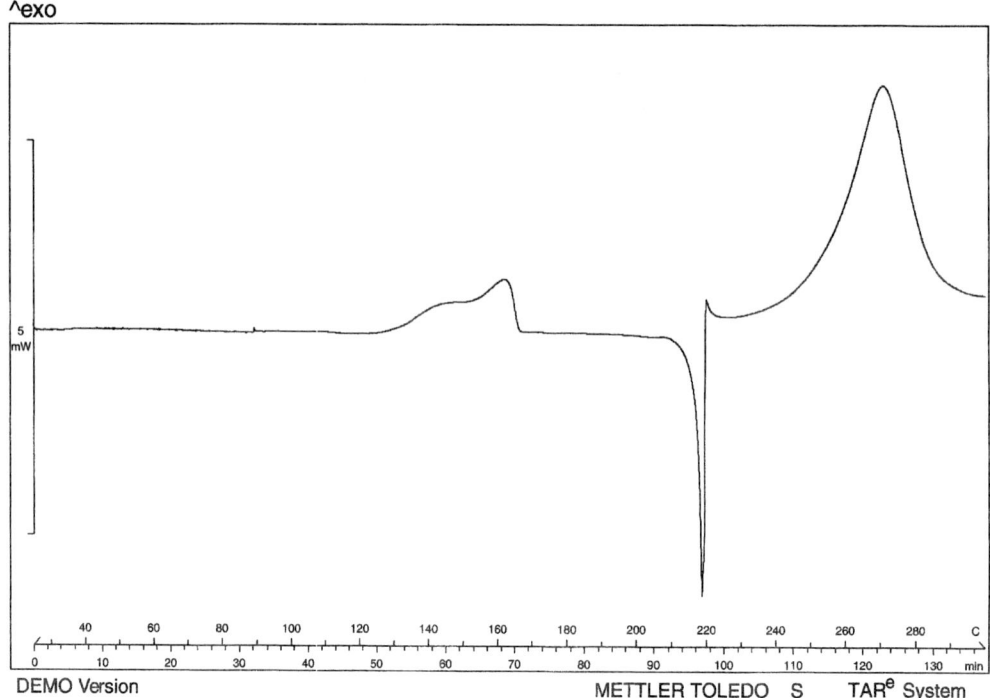

Fig. 6. The DSC curve of compound **3**.

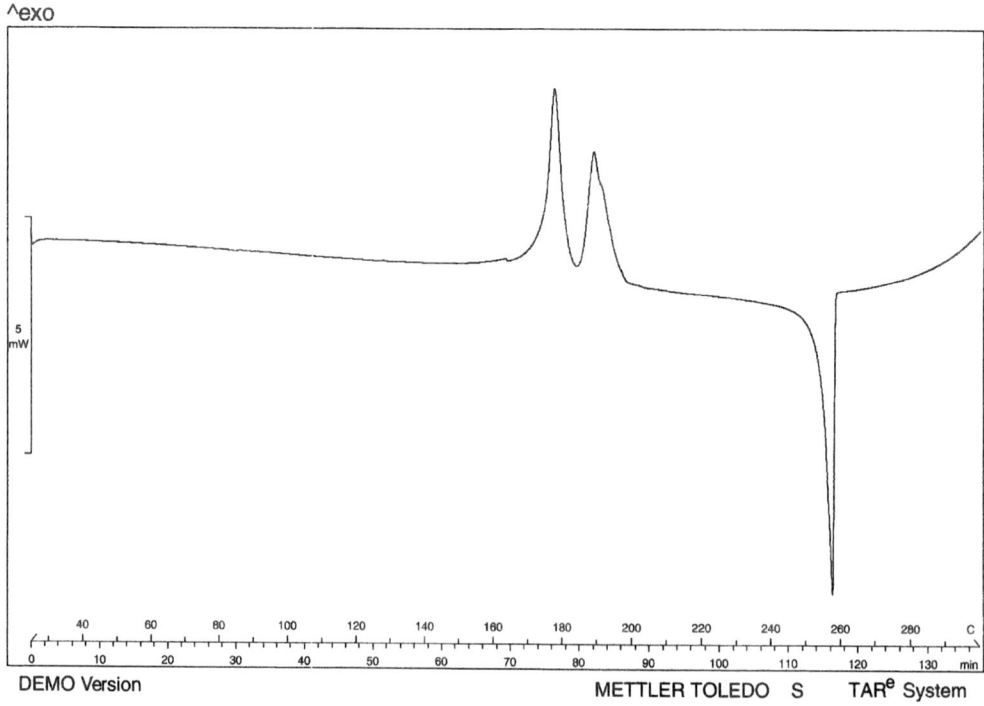

Fig. 7. The DSC curve of compound **4**.

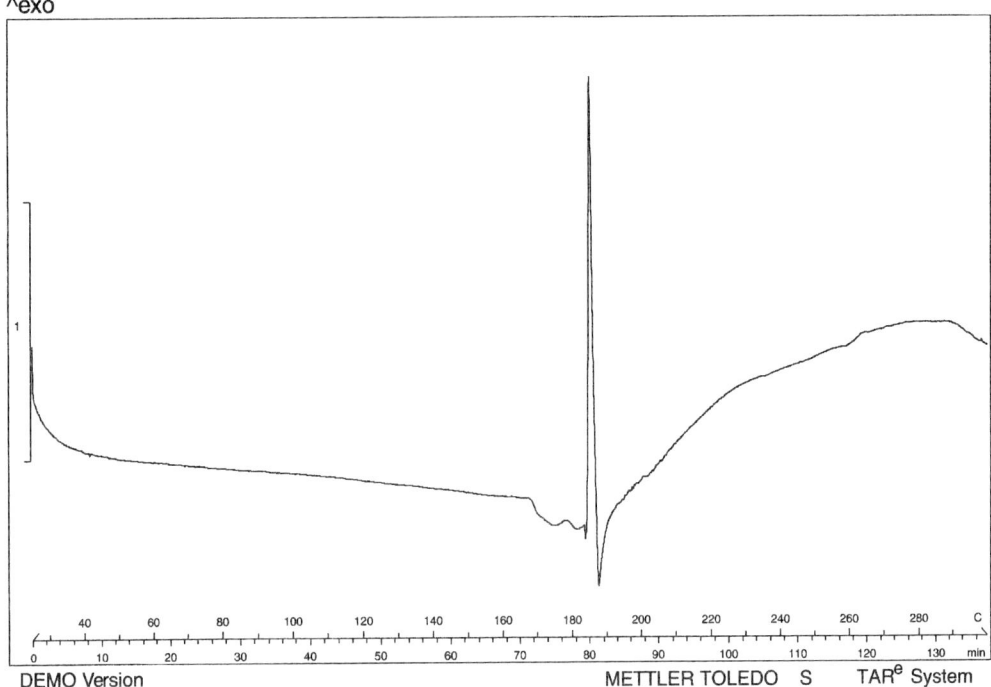

Fig. 8. The DSC curve of compound **5**.

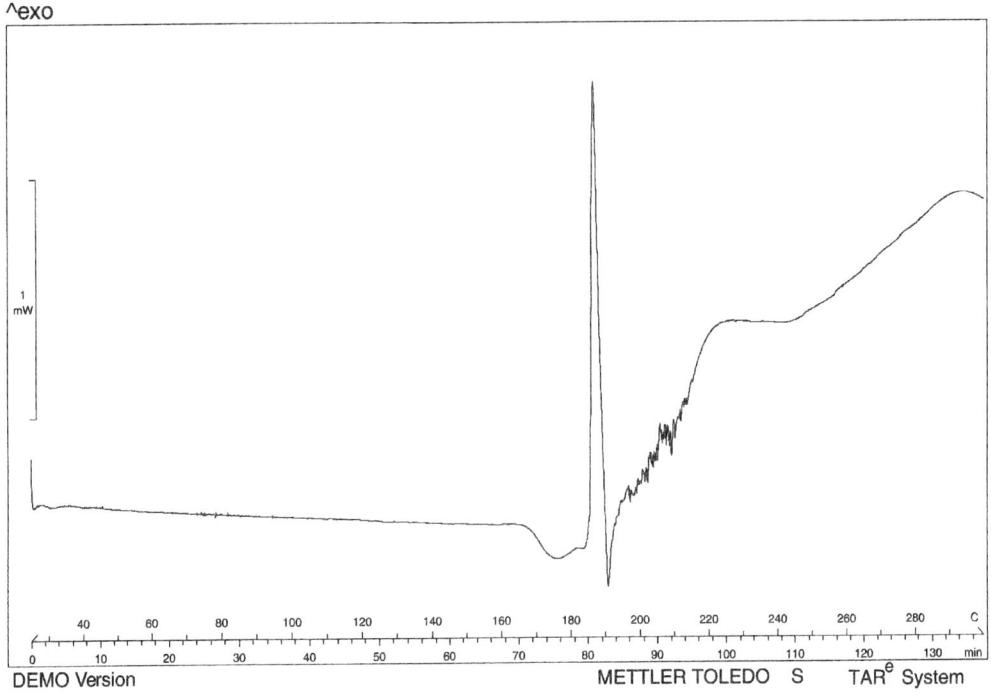

Fig. 9. The DSC curve of compound **6**.

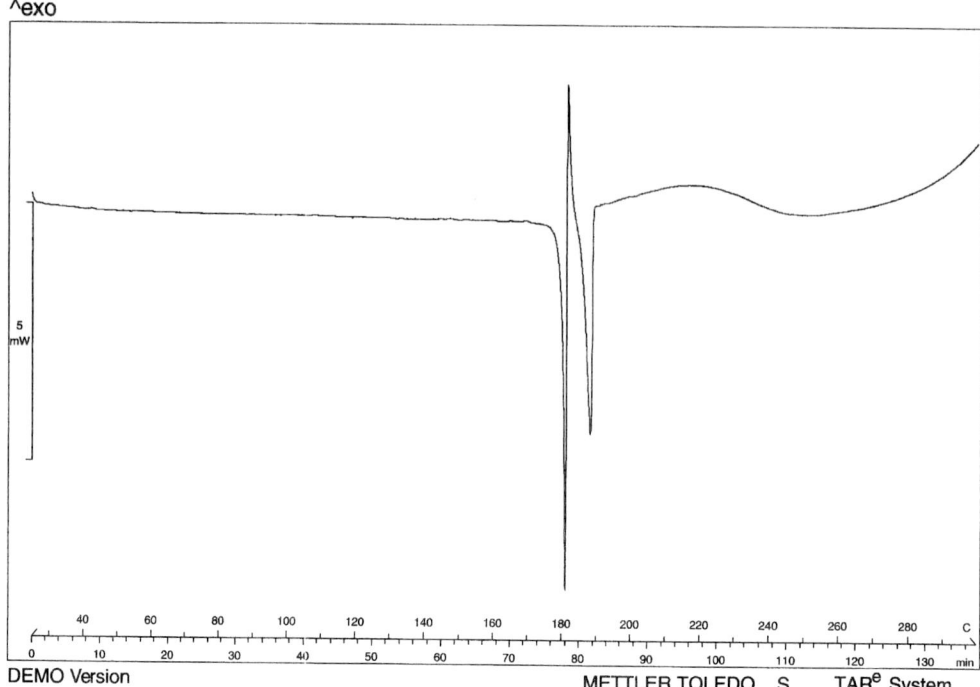

Fig. 10. The DSC curve of compound 7.

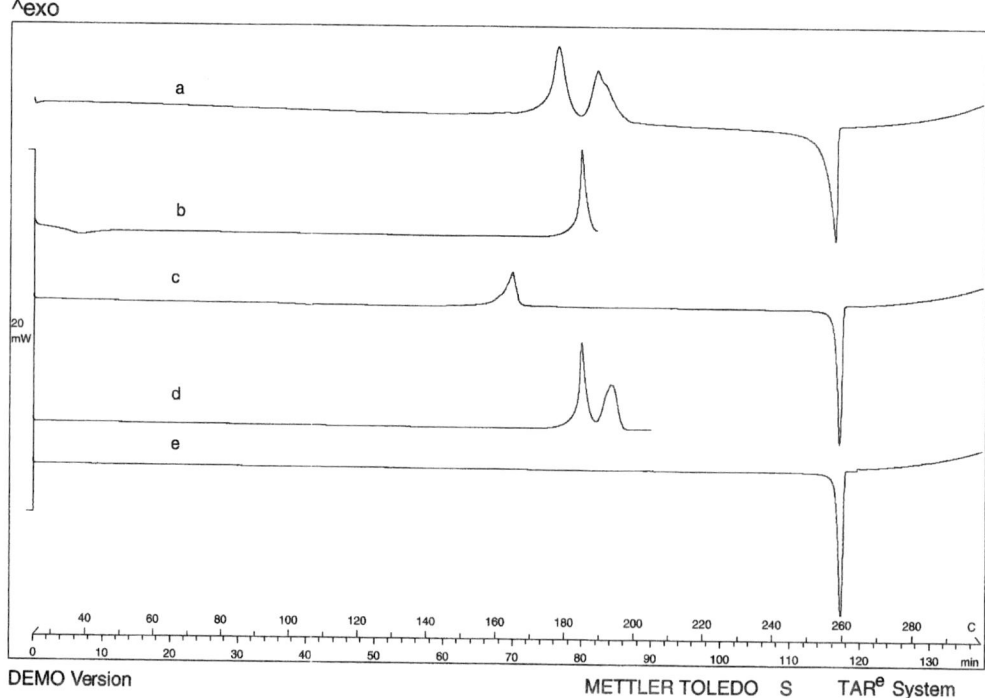

Fig. 11. (a): the DSC curve of compound 4; (b) heating was stopped at 189 °C; (c) sample (b) was cooled and re-heated to 300 °C; (d) heating was stopped at 205 °C; (e) sample (d) was cooled and re-heated to 300 °C.

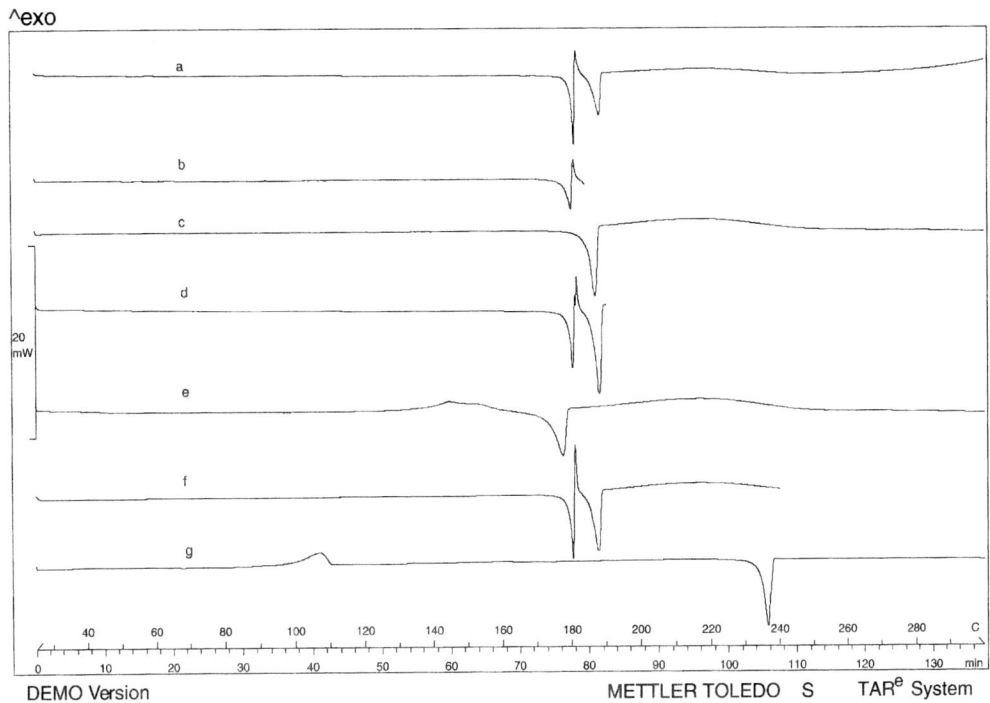

^exo

DEMO Version   METTLER TOLEDO S   TAR^e System

Fig. 12. (a) The DSC curve of compound **7**; (b) the heating was stopped at 184 °C; (c) sample (b) cooled and re-heated to 300 °C; (d) the heating was stopped at 191 °C; (e) sample (d) cooled and re-heated to 300 °C; (f) the heating was stopped at 240 °C; (g) sample (f) cooled and re-heated to 300 °C.

occurred, DSC heating curve of compound **4** was taken in various ways (Fig. 11): (a) corresponds to the thermogram shown in Fig. 7; (b) heating was interrupted at 189 °C, where the first exothermic peak appeared; (c) sample (b) was cooled then re-heated; (d) heating was interrupted at 205 °C, where the second exothermic peak appeared; (e) sample (d) was cooled and then re-heated (this thermogram was overlapping with that of shown in Fig. 2). In all cases, ^{1}H NMR spectra were also recorded from the samples. From these data could be concluded that the first exothermic peak was the temperature of cyclization. The second exothermic peak was due probably to a polymorphic rearrangement or (and this is more likely) crystallization of the amorphous product formed at the temperature of the first exothermic

Table 1
Enthalpies of reaction. Determined by DSC (ΔH_r), and by calculation ($\Delta\Delta H_f = \Delta H_{f\,pr} - \Delta H_{f\,st}$)

Starting compound	Product (T_m, °C)[a]	T_r[b] (°C)	ΔH_r[c] (kcal/mol)	$\Delta\Delta H_f$[d] (kcal/mol)
3	**8**(221.0)	148.7	− 4.4	− 9.6
4	**9**(258.9)	177.7	− 7.7	− 12.2(− 16.8)[e]
5	**10**(196.2)	184.9	− 3.3	− 11.5
6	**11**(194.3)	186.7	− 3.9	− 12.3
7	**12**(231.9)	218.4	− 9.4	− 17.5

[a] T_m corresponds to the melting temperature.
[b] T_r corresponds to the temperature of cyclization.
[c] ΔH_r is the enthalpy of reaction.
[d] $\Delta\Delta H_f$ corresponds to the difference of calculated heats of formation as obtained by PM3 method.
[e] $\Delta\Delta H_f$ corresponds to the difference of calculated heats of formation as obtained by DFT method.

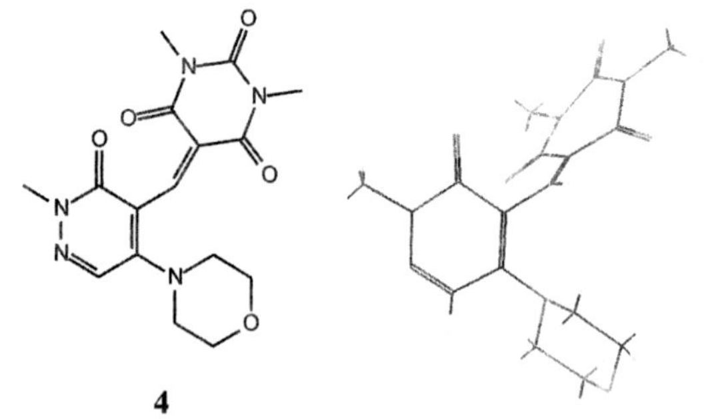

Fig. 13. Optimized structures of **4** and **9**. Calculations were carried out by DFT method at LSDA/pBP86/DN level [10].

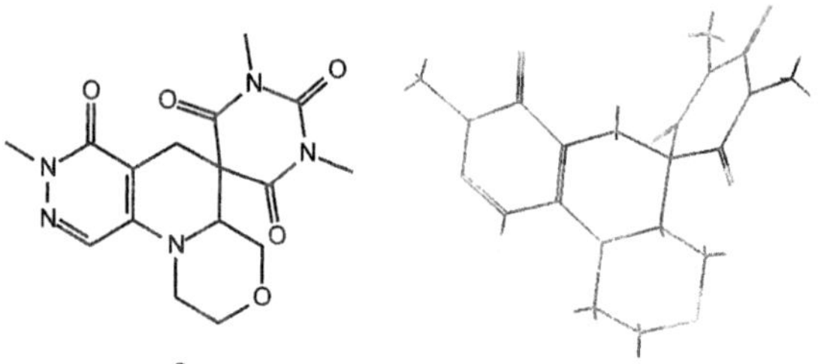

peak. The IR spectra, which are generally suitable for the establishment of polymorphism, were almost fully identical, therefore they were of no diagnostic value in this respect.

On the DSC curves of the Meldrum's acid derivatives **5** and **6** there were two peaks, one exothermic and one endothermic peak (Figs. 8 and 9). The endothermic peaks (187.5 °C for **5** and 190.7 °C for **6**) correspond to the melting temperatures of their ring closed derivatives (196.2 °C for **10**; Fig. 3 and 194.3 °C for **11**; Fig. 4), whereas the exothermic peaks are the temperatures of ring closure reactions.

The DSC heating curve of the malonitrile derivative **7** showed a different pattern. There appeared two exothermic and two endothermic peaks (Fig. 10). Interestingly, the melting temperature could be well identified. The first endothermic peak at 180.6 °C corresponds to the melting point of the vinyl compound. This endothermic peak was followed by an exothermic peak at 180.7 °C due to a crystallization. The second endothermic peak, at 187.9 °C, was assigned as the melting point of the crystalline form. The broad exothermic peak at 218.4 °C was identified as the temperature of ring closure. This assignment was confirmed via interrupted measurements at 184, 191 and 240 °C (Fig. 12, curves b, d and

f), which were followed by cooling and re-heating periods (Fig. 12, curves c, e and g).

Integrals of the exothermic peaks areas related to the ring closure reactions provided the enthalpy changes of the reactions (Table 1; $\Delta\Delta H_r$). All reactions studied were exothermic.

The heats of reaction (Table 1, $\Delta\Delta H_f$) were also calculated. Calculations were carried out with full geometrical optimization by semiempirical PM3 and, for one pair of compounds, by DFT method for their optimized structures (Fig. 13). The heat of reaction values were determined as the differences of heat of formation of fused products and that of the starting vinyl compounds.

For all reactions, the heat of reaction values indicate that the cyclization is an exothermic process, as it has been found by DSC measurements. The reason for the difference between the enthalpy and heat of reaction values, lies among others, in the different conditions of the ring closure: calculations are related to gas phase, whereas DSC measurements take place in a solid/melted phase.

4. Conclusions

The *tert*-amino effect based thermal ring-closure reaction of 5-morpholino-4-vinylpyridazinones and a benzene analogue was studied by differential scanning calorimetry. The enthalpies of reaction were determined experimentally by DSC, and heats of reaction were calculated by semiempirical (PM3) and DFT methods.

It can be concluded that

(i) upon heating of vinyl compounds in the DSC apparatus cyclization occurs and incorporation of the β,β-vinyl carbon into a heterocycle significantly decreases the temperature of cyclization;

(ii) both measurements and calculation indicates that the cyclization is an exothermic process;

(iii) DSC is a valuable method for studying thermal isomerization reactions.

Acknowledgements

Financial supports of this work by National Research Foundation (T-31910); and the ministry of

Health (ETT-121/2003). The authors are grateful to Prof. József Nyitrai for his help in naming of polycyclic systems, and to Benjamin Podányi for valuable discussion on NMR data. The authors are also indebted to Ágnes Puhr for microanalyses, and for the technical assistance to Noémi Bús.

Appendix A. Names of new compounds

3:   1,3-Dimethyl-5-(2-morpholinobenzylidine)pyrimidine-2,4,6(1H,3H,5H)-trione.

5:   2-Benzyl-2-methyl-5-[(2-methyl-5-morpholino-3-oxo-2,3-dihydropyridazin-4-yl)methylene]-1,3-dioxane-4,6-dione.

8:   1′,3′-Dimethyl-1,2,4,4a-tetrahydro-2′H,6H-spiro[1,4-oxazino[4,3-a]quinoline-5,5′-pyrimidine]-2′,4′,6′(1′H,3′H)-trione.

10:   2-Benzyl-2,3′-dimethyl-3′,5′,6a′,7′,9′,10′-hexahydro-4′H-spiro[1,3-dioxane-5,6′-pyridazino[4′,5′:5,6]pyrido[2,1-c][1,4]oxazine]-4,4′,6-trione.

13:   2-Benzyl-2-methyl-1,3-dioxane-4,6-dione.

Appendix B. Numbering of new compounds

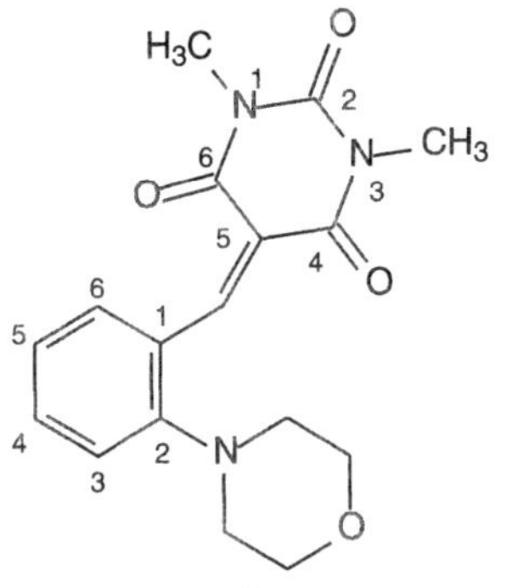

3

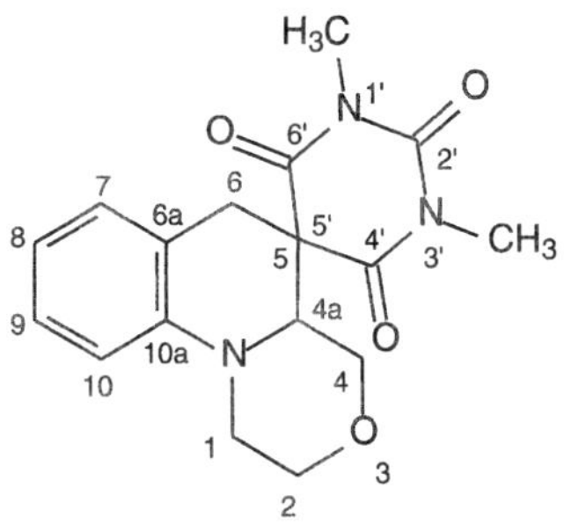

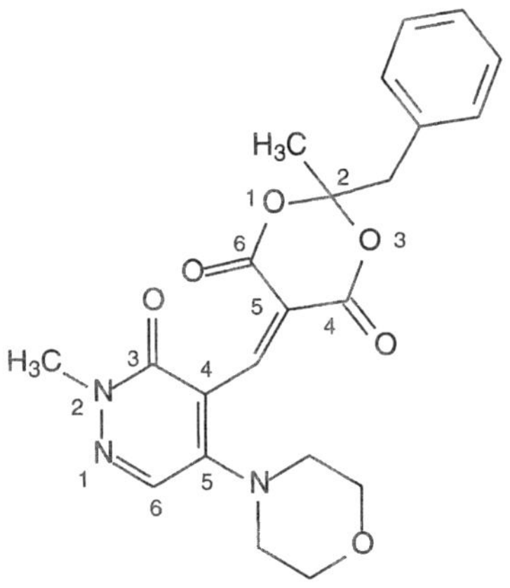

5

8

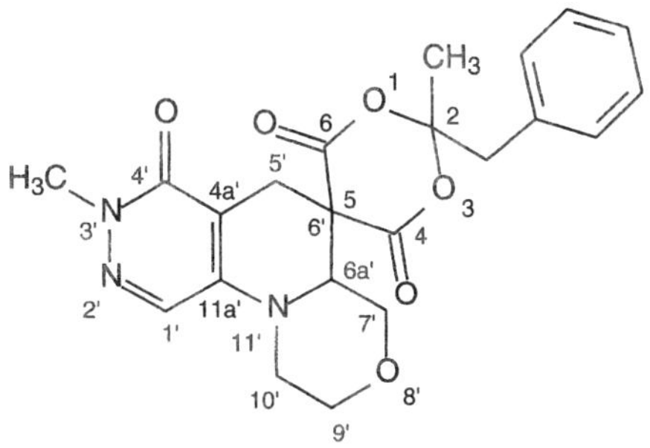

10

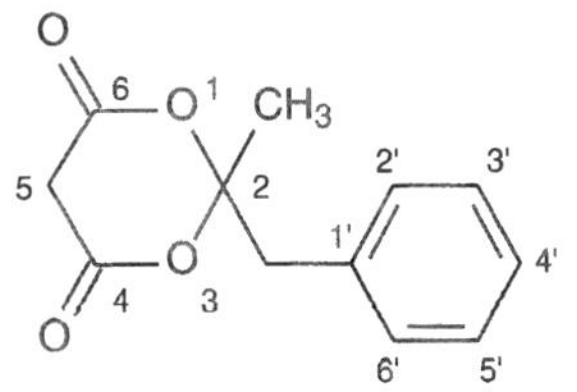

13

References

[1] For a recent review, see:O. Meth-Cohn, The *t*-amino effect: heterocycles formed by ring closure of *ortho*-substituted *t*-anilines, in: A.R. Katritzky, A.J. Boulton (Eds.), Advances in Heterocyclic Chemistry, vol. 65, Academic Press, London, 1996, pp. 1–37.

[2] L.C. Groenen, W. Verboom, W.H.N. Nijhuis, D.N. Reinhoudt, G.J. VanHummel, D. Feil, Tetrahedron 44 (1988) 4637.

[3] P. Mátyus, K. Fuji, K. Tanaka, Heterocycles 37 (1994) 171.

[4] H. Wamhoff, V. Kramer-Hoss, Liebigs Ann. Recl. 7 (1997) 1619.

[5] A. Schwartz, G. Beke, Z. Kovári, Z. Böcskey, Ö. Farkas, P. Mátyus, J. Mol. Struct. (Theochem) 528 (2000) 49.

[6] G. Beke, A. Gergely, G. Szász, A. Szentesi, J. Nyitrai, O. Barabás, V. Harmat, P. Mátyus, Chirality 14 (2002) 365.

[7] K.B. Niewiadomski, H. Suschitzky, J. Chem. Soc., Perkin Trans. 1 (1975) 1679.

[8] G. Widmann, R. Riesen, Thermoanalyse, Anwendungen, Begriffe, Methoden, Hüthig Buch, Heidelberg, 1990.

[9] G.W.H. Höhne, W. Hemminger, H.-J. Flammersheim, Differential Scanning Calorimetry, Springer, Berlin, 1996.

[10] W.J. Hehre, L. Lou, A Guide to Density Functional Calculations in Spartan, Wawefunction Inc., Irvine, 1997.

ELSEVIER

Journal of Molecular Structure (Theochem) 666–667 (2003) 681–686

THEO
CHEM

www.elsevier.com/locate/theochem

Hunt for genetic susceptibility in a complex disease

Ágnes Czibula[a,*], Mónika Mórocz[a], Csanád Z. Bachrati[a], Ágnes Csiszár[a],
László Szappanos[b], Erika B. Szabó[a], Edit Tóth[a], Ferenc Szeszák[b], Éva Morava[c],
Tamás Illés[c], István Raskó[a]

[a]Research Centre of Hungarian Academy of Sciences, Institute of Genetics, BRC, P.O. Box 521, Szeged H-6701, Hungary
[b]Medical Center, University of Debrecen, Debrecen, Hungary
[c]Medical School, University of Pecs, Pecs, Hungary

Abstract

One of the most frequent spine abnormality in early adulthud is adolescent idiopathic scoliosis. It has great significance, since it represents 80% of all types of scoliosis. It is ten times more frequent in girls, than in boys. The mode of inheritance is contraversial, there were data for autosomal dominant, X-linked dominant, or non-mendelian inheritance. So far no candidate genetic locus for IS has been identified. By comparing the gene expression levels in paraventral muscle between the two sides of the spine in IS patients 11 candidate genes were identified for further investigation. With restriction fragment length polymorphism analysis of these genes a polymorphism in the gene of the Bromodomain PHD finger Transcription Factor (BPTF) has been revealed. A deficiency of the sequence appears more frequently in IS ($P = 0.6$) than in control ($P = 0.21$) cases. The molecular basis of the polymorphism is a 4.5 kb deletion in the last intron of the BPTF gene. The BPTF is the human ortholog of Drosophila Nurf301, which is the largest subunit of a chromatin remodeling NURF complex. Malfunction of BPTF in early ontogenesis, together with intense growth rate during adolescence and other environmental factors could influence the development of IS.
© 2003 Elsevier B.V. All rights reserved.

Keywords: Complex disease; Idiopathic scoliosis; Bromodomain containing transcription factor

1. Introduction

Major life threatening diseases of our day, including heart disease, cancer, hypertension, obesity, diabetes, and disorders associated with aging are not due to single gene defects. Their development involves complex mixtures of nature, nurture influences. An individual's risk of disease is combined from three sources: *genetic susceptibility to disease,* *environmental exposure and chance.* Biomedical sciences have now reached the point when large-scale population-based studies based on genetics, environment, and health are feasible.

In a general sort of way, it has been known for a long time that such diseases sometimes run in families, but rather obviously they do not show anything like Mendelian inheritance. It is generally hoped, though, that molecular understanding of those aspects of these diseases that have genetic components will improve their therapy, and allow for their prevention.

* Corresponding author. Tel.: +36-62-432232; fax: +36-62-433503.

E-mail address: czagi@nucleus.szbk.u-szeged.hu (Á. Czibula).

0166-1280/$ - see front matter © 2003 Elsevier B.V. All rights reserved.
doi:10.1016/j.theochem.2003.08.092

The genetic complexity of these diseases is believed to be due partly to their being controlled by polygenic systems, in which the products of more than one gene interact to produce a phenotype [1,2]. An equally confusing factor is the strong environmental contribution in all these diseases; or, from the point of view of a geneticist, weak penetrance. This is most irrefutably seen in studies of identical (monozygotic) twins, in which the appearance of a disease in one is a probable but far from certain indicator of its eventual appearance in the other. A degree of concordance between the appearances of a disease that is higher in monozygotic twins than in dizygotic twins is the best indication of a truly hereditary component in cases where there is a strong environmental component.

The contributing factors include several genes, environmental conditions or exposures, the age, nutritional status, life style and other minor predisposing factors. The diseases appear to result from interactions of the contributing environmental factors with multiple susceptibility genes that confer low or moderate disease risk, or those trigger the expression of disadvantageous genetic traits. In order to achieve the full understanding of the molecular pathology of complex diseases both the genetic and environmental contributions have to be elucidated. Gene-environment research is still in its infancy but promises to have many beneficial results. The extent to which genetic and environmental factors interact is hard to establish. In principle it is possible to calculate the heredibility of a factor of the phenotype: that is, the proportion of the observed variation in a population, which is caused, by genetic variation. However, this calculation is only valid for a population exposed to a range of environments with the same degree of variation that applied to the population studied. Genes that have evaluative and predictive potential for use in screening and diagnostics could be identified. Until recently the progress of the research was hampered and limited because of the lack of human gene sequences and high through-put technological limitations. Due to the advent of the completion of the Human Genome Project and the wide-spread use of DNA chip technology nowadays there have been several projects initiated world-wide. This research could be used to target public health strategies that help individuals avoid or reduce adverse exposures and the professionals to design better treatment personalized regimes, including drug and gene therapy.

Epidemiological analysis of the relative importance of suspected risk factors is most informative if it is conducted in a population whose a priory risk of the diseases is similar. Individuals who carry identical germ-line mutations represent such a population. Identifying carriers of such founding mutations and variants of risk modifiers from a population-based series of incident disease cases with matched healthy control cases will enable the influence of various reproductive, environmental and behavioural factors to be evaluated.

Nutrition as a major contributing factor in the development of complex diseases is one of the beneficiaries of recent developments in genomics. With the new tools in hand nutrition is positioned to finally understand metabolic regulation and it will be possible in the future to improve health by the diet. It has been known for a relatively long time that nutrients influence gene expression, but little is known on the interaction of these induced genes with susceptibility gene networks participating in complex diseases. New genomic technologies can measure simultaneously the response of families of genes in any cell at any time allowing studies on nutrient–gene interactions. Among many others the clearest example of diet-susceptibility gene interactions were discovered in diet-related cancers.

But apart from determining the nature and relevance of environmental factors that affect complex diseases, medical science is mounting a sustained assault upon their genetic components. The strategy is to identify, wherever possible, single genetic factors that contribute to disease susceptibility, even if they may not cause it in the simple manner of monogenic diseases.

Genetic studies, regarding complex diseases are like looking for a needle in a haystack or fishing for the gold fish in a sea of many candidates. There are several strategies to find major contributing, or susceptibility genes. For the use of the classical linkage studies large families are needed, where the disease co-segregates with some observable phenotype, which could be either a genetic trait or a DNA sequence polymorphism. The most abundant differences between individuals are single nucleotide polymorphisms. There are around 10 million predicted SNPs with well above 1% allele frequency in the population, but it is assumed that those variants which have any relevance to disease are rare. Another strategy is the association analysis, where the co-occurrence of alleles or phenotypes is established. In this case one allele is more often associated with a disease than their frequencies in the population. Therefore associations are dependent on the genetic history of the given population. While linkage is usually genome-wide, associations are limited to certain candidate regions. As we have seen previously linkage analysis is carried out in pedigrees, while associations could be found in case (patient with disease)/control relations. Another potential tool for finding disease predisposing genes in complex traits is the use of microarray analysis. This technique is able to distinguish disease-related changes in gene expression or obtain global whole-genome picture of gene expression profiles.

One of the most frequent spine abnormality in early adulthud is adolescent idiopathic scoliosis (IS), with 0.15–4% incidence among schoolchildren [3]. It has great practical and clinical significance, since it represents 80% of all types of scoliosis. It is 10 times more frequent in girls, than in boys, with a 13.6% prevalence [4]. The picture of lateral dominance in the disorder is quite conservative;the upper thoracic curve is convex to the right in 90%, whereas the lumbar curve is convex to the left in 85% of all cases [5]. The development of scoliosis could be due to different growth rate in the anterior and posterior part of the vertebral column.

The mode of inheritance is contraversial, there were data for autosomal dominant, X-linked dominant, or non-Mendelian inheritance [6–8]. Genetic abnormalities in bone, vertebral disc, muscle and collagene development have been previously excluded [9]. The strong genetic influence in the aethiology of the disorder is proven by the twin studies, it shows 73% concordance in monozygotic and 36% in dizygotic twins [10]. The appearance of the disease could be either familial or sporadic. According to our experience in 19 out of 54 families, members of the same family possessed the symptoms [11].

Mouse mutations have proven to be informative in defining genes that are important in development of complex diseases. Giampietro et al. identified synteny-defined genes for IS. These candidate genes were classified as being involved in intracellular signalling, intracellular signal transduction, encoding extracellular matrix proteins, and involved in extra-cellular matrix metabolism [12].

2. Materials and methods

2.1. cDNA expression array

Total RNA from vertebrate muscle biopsy of IS patient was prepared by SV total RNA Isolation kit (Promega) and was labeled and hybridized to membrane containing 588 cDNA arrayed into six regions (Clontech's Atlas™ cDNA Expression Array) according to the instruction of the manufacturer.

2.2. Southern hybridization

Genomic DNA was extracted from peripheral blood leukocytes obtained from 52 IS and 68 control individuals using standard protocol [11,13]. For Southern hybridizations 10-10 μg of total genomic DNA were digested separately with HindIII, PvuII and XbaI restriction endonucleases and were loaded onto a 0.6% agarose gel and allowed to run overnight at 30 V in 1xTBE. DNA was transferred to a nylon support membrane (Amersham, Hybond N +). The P^{32}-labeled probes were prepared from EST clones by using Multiprime Labeling Kit (Amersham) and [λ-^{32}P] dCTP. Prehybridization, hybridization and washes were as described by Church and Gilbert [14]. Radioactivity of membranes was determined by PhosphorImager™ 445 SI (Molecular Dynamics, Inc.).

2.3. Long range PCR amplification and sequencing

PCR primers for the last intron of BPTF were: E-2F, 5'-TAGACCCTAATGATGCACCAG-3' (AC005829:17528-17548), E-1R: 5'-GTCAGGA-TAGCCTCACTAAAAG-3' (AC005829:25605-25585). Amplification reactions were performed using a MJ Research PTC-200 PCR System and contained 200 ng genomic DNA, 0.5 μM each of primers, 1 U TaKaRa Ex Taq in 1 × TaKaRa buffer (TaKaRa). The template DNA was initially

Table 1
Genes with expression differences on the cDNA array

Clontech's Atlas™ cDNA Expression Array	Symbols	Gene map locus
Oncogenes		
Cell cycle regulator	p^{55}CDC	9q13-q21
Stress response		
Intracellular signal transduction modulators	CCK4	6p21.1-p12.2
	JAK1	1p31.3
Apoptosis		
Transcription factors	TFIID	6q27
Receptors		
Growth factors receptor	FGFR2	10q26
Interleukins receptor	IL-4R	16p12.1-p11.2
Cell-surfaces antigenes	CD44	11pter-p13
Cell adhesion molecules	SQM1	19p13.12
Cell–cell communication		
Growth factors	NGF 2	12p13
	BMP-4	14q22-q23
	FPRL1	19

denatured at 93 °C for 3 min followed by 30 cycles of denaturation at 93 °C for 1 min, annealing 65 °C for 1 min and extension at 72 °C for 6 min, the final extension step was 10 min at 72 °C. The PCR products were examined on 0.8% agarose gels. The polymorph, 3.6 kb long PCR product was directly sequenced with an ABI Prism 310 sequencer (Perkin Elmer) using the ABI PRISM BigDye™ Terminator v3.0 Cycle Sequencing Ready Reaction Kit. The primers for the sequencing were E-2F, E-1R (the same as those used for amplification),

Seq3R, 5′-TGGAATCAGAACTGGATGTTG-3′ (AC005829:24402-24423) and Seq3F, 5′-GCCA-CAGTATGAGACCTTGTCT-3′ (AC005829:18260-18282).

3. Results

By the means of a modern gene expression array (Atlas cDNA Expression Array, Clontech) the mRNA expression profile of the muscle biopsies originating from the left and right side of the spine of patients has been compared. Eleven genes were identified which have shown expression differences (Table 1).

The respective EST clones have been purchased and it was investigated whether the clones with expression differences possess any differences in their RFLP profiles in Southern blots on DNA from controls and patients with scoliosis. One clone was found which have exhibited altered RFLP pattern. Luckily that EST proved to be a mixture of two, one from the interleukin 4 gene and the other from the bromodomain transcription factor, BPTF. After purification it was proved that the altered Southern pattern could be linked to the BPTF (Fig. 1). By the Southern blot analysis of DNA from 52 IS patients and 68 controls significant alterations were found between the two groups of people. In the case of IS 28 patients showed the altered RFLP pattern and only 19 individuals from controls.

On the basis of the available GenBank sequences appropriate PCR primers were designed by which

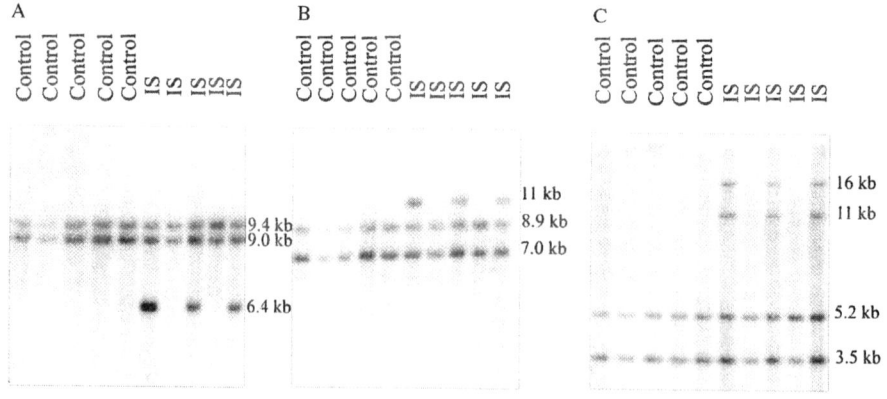

Fig. 1. Southern hybridizations of genomic DNA digested by HindIII (A), by PvuII (B) and by XbaI (C) from healthy control individuals (Control Lanes) and from idiopathic scoliosis patients (IS Lanes) with BPTF EST.

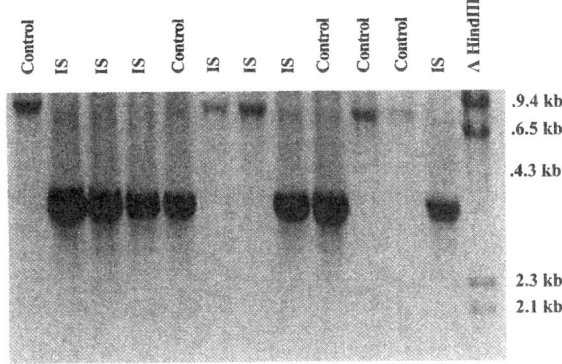

Fig. 2. Long range PCR amplification of genomic DNA from healthy control individuals (control) and idiopathic scoliosis patients (IS) analysed by agarose gel electrophoresis.

the BPTF region responsible for the altered Southern pattern could be amplified. By PCR amplifications two fragments were generated, an 8 kb and a 3.6 kb long (Fig. 2). By means of direct sequencing it was revealed, that the reason for the altered RFLP is a 4.4 kb long intronic deletion in the last intron of BPTF.

The frequency of the deletion was determined by PCR reaction in DNA obtained from peripheral leukocytes of 46 control individuals and 47 patients with IS. It was found that the frequency of deletion in the patients is 0.6, while in the control individuals 0.21, respectively. This proved to be significant, according to the Fisher test ($P = 0.00019$).

4. Discussion

The most important lesson has been over the past decades from the genetic characterization of more than 1200 human diseases that both in diseases with Mendelian inheritance or in complex diseases, clinical variability of the phenotypes are associated with mutations, genotypic heterogeneity, modifier genes and environmental contributors.

In previous studies a strong hereditary component was suggested in IS and the polygenic nature of the disease was proposed, however, no major genetic locus has been identified.

In a linkage study in four multiplex families three loci were identified on chromosome 6p, 10q

and 18q which showed allele sharing in two of the four families [15]. In an other linkage analysis after examining seven, unrelated multiplex families with adolescent IS no proof was found for linkage on chromosomes 6p, 10q and 18q in two families, but a critical region was identified on chromosome 19p13.3 [16].

The region is 5.2 cM and it is 1.9Mb long and harbors ~71 possible gene sequences. Two possible candidate genes were suspected in the region, which might participate in the scoliosis pathogenesis; the leucine-rich a-2-glycoprotein and the GRB2-like 1 gene, which is similar to the endophilin 2 gene. Since the publication of these data, however, no conformations were reported from other laboratories yet.

The success of the strategy of finding susceptibility genes for complex diseases and in our case the success of an association study depends on the detection of an increased frequency of specific disease alleles in affected individuals. In this paper we have described a lucky coincidence (which could also be considered as serendipity: finding interesting or valuable things by chance) by which a possible susceptibility factor for IS has been identified.

The bromodomain motive of BPTF is a 110 amino-acid conserved structural region, which is involved in protein–protein interaction, and it can be found in proteins that regulate signal-dependent, non-basal transcription. BPTF was originally identified by Jones et al. [17], it has a Drosophila homologue, Nurf301, which has been characterized as a factor of nucleosome sliding [18]. The N-terminal 2200 nucleotides of BPTF are identical to the cDNA of FAC1, or FALZ, a developmentally regulated protein, which participates in protein–protein interactions. Increased expression of the FAC1 gene was observed in early stages of Alzheimer's disease [19].

All these data, concerning the function of BPTF support our hypothesis that this factor might play some role in vertebrate development in some stage of ontogenesis, which if malfunctions might have consequences in later life, particularly in adolescence, when intensive growth could bring to the surface the early developmental inadequacies. The identification of this new player in the pathomechanism of adolescent idiopathic scoliosis, might lead to better prevention and early identification of the disease, which affects so many adolescents.

The functional role of the deletion in the pathomechanism has to be elucidated.

References

[1] I. Raskó, C.S. Downes, Genes in Medicine, Chapman & Hall, London, 1994.

[2] A.P.R.T. Strachan, Human Molecular Genetics, BIOS Scientif. Publ. Ltd, USA, 1996.

[3] B.V. Reamy, J.B. Slakey, Am. Fam. Physician 64 (2001) 111.

[4] D.R. Benson, Orthopedics 10 (1987) 1691.

[5] L.A. Rinsky, J.G. Gamble, West. J. Med. 148 (1988) 182.

[6] A. Czeizel, A. Bellyei, O. Barta, T. Magda, L. Molnar, J. Med. Genet. 15 (1978) 424.

[7] T.G. Lowe, M. Edgar, J.Y. Margulies, N.H. Miller, V.J. Raso, K.A. Reinker, C.H. Rivard, J. Bone Joint Surg. Am. 82 (2000) 1157.

[8] R. Wynne-Davies, J. Bone Joint Surg. Br. 50 (1968) 24.

[9] J.A. Byrd III, Clin. Orthop. (1988) 114.

[10] K.L. Kesling, K.A. Reinker, Spine 22 (1997) 2009.

[11] L. Szappanos, É. Oláh, E. Balogh, I. Raskó, F. Szeszák, K. Szepesi, Is there any regulatory gene defect behind idiopathic scoliosis?, in: I.A.F. Stokes (Ed.), Research into Spinal Deformities, IOS Press, Amsterdam, 1999.

[12] P.F. Giampietro, C.L. Raggio, R.D. Blank, Am. J. Med. Genet. 83 (1999) 164.

[13] Short Protocols in Molecular Biology, Greene Publishing Associates/John Wiley and Sons, 1992.

[14] G.M. Church, W. Gilbert, Proc. Natl Acad. Sci. USA 81 (1984) 1991.

[15] C.A. Wise, R. Barnes, J. Gillum, J.A. Herring, A.M. Bowcock, M. Lovett, Spine 25 (2000) 2372.

[16] V. Chan, G.C. Fong, K.D. Luk, B. Yip, M.K. Lee, M.S. Wong, D.D. Lu, T.K. Chan, Am. J. Hum. Genet. 71 (2002) 401.

[17] M.H. Jones, N. Hamana, M. Shimane, Genomics 63 (2000) 35.

[18] H. Xiao, R. Sandaltzopoulos, H.M. Wang, A. Hamiche, R. Ranallo, K.M. Lee, D. Fu, C. Wu, Mol. Cell 8 (2001) 531.

[19] R. Bowser, A. Giambrone, P. Davies, Dev. Neurosci. (1995) 17.

ELSEVIER

Journal of Molecular Structure (Theochem) 666–667 (2003) 687–691

THEO
CHEM

www.elsevier.com/locate/theochem

Albert Szent-Györgyi and his life

István Hannus*

Department of Applied and Environmental Chemistry, Jozsef Attila University, University of Szeged, Rerrich B. tér 1, Szeged H-6720, Hungary

Abstract

Internationally, Albert Szent-Györgyi is one of the best-known Hungarians. His fame is mainly due to his winning the Nobel prize in 1937 'for his discoveries in connection with the biological combustion processes, with special references to vitamin C and the catalysis of fumaric acid'. This led to his becoming a living legend in Hungary, but much less information is publicly available as concerns the second half of his life, which he spent in the US.

As part of an attempt to fill this considerable hiatus, the author has collected all those articles published in one of the most respected American newspapers, *The New York Times*, which involve the personality, opinions or activities of Szent-Györgyi during his life in the US (and earlier). The nearly one hundred newspaper pieces written by or about him provide an excellent picture of his scientific activities and his contributions to the well-being of the wider community.
© 2003 Elsevier B.V. All rights reserved.

Keywords: Nobel prize; Biological combustion; Vitamin C; The New York Times

1. Introduction

In the last century Albert Szent-Györgyi was one of the 12 Nobel laureates of Hungarian origin who was awarded the medal while still a Hungarian citizen [1]. In this way one Nobel medal came to Hungary: the original gold medal (207 g) today remains in the Hungarian National Museum in Budapest. (In 2002 Imre Kertész got the Nobel prize in literature.)

His scientific activities in Hungary and in Szeged, in particular are well known and well-documented [2], but much less information is publicly available as concerns the second half of his life, which he spent in the US. This is so inspite of an interview-based detailed biography entitled *Free Radical* (published in 1988) [3]. It is even so in Hungary, since this book, the title of which refers to his independent personality and

also to his increasingly acknowledged view of the decisive role of radicals in causing cancer, has not been translated into Hungarian.

2. A potted biography of Albert Szent-Györgyi

As the Table 1 shows he was born in 1893 in Budapest. His medical studies, started in 1911, were interrupted by World War I, during which he served in Ukraine. He was wounded and invalided to Budapest. During his convalescence, he completed his medical studies, and was awarded his diploma in 1917. In the same year, he married. The end of the war saw him again at the front, this time in Italy.

The next few years were a hectic period for Hungary because of the war and the ensuing Trianon peace treaty. And it was just as hectic for Szent-Györgyi. He worked successively in Pozsony (today Bratislava, capital of Slovakia), Budapest, Prague and

* Tel.: + 36-62-425-523; fax: + 36-62-544-619.
E-mail address: hannus@chem.u-szeged.hu (I. Hannus).

0166-1280/$ - see front matter © 2003 Elsevier B.V. All rights reserved.
doi:10.1016/j.theochem.2003.08.094

Table 1
Biographical details of Albert Szent-Györgyi

1893	Born in Budapest
1911	Medical studies in Budapest
1914	Military service in World War I
1917	Medical graduation, marriage
1918–1920	Works in Pozsony, Budapest, Prague and Berlin
1921–1922	Hamburg
1922–1923	Leiden
1923–1926	Groningen, resolved dispute of Warburg and Wieland on biological combustion
1926–1930	Cambridge, PhD in chemistry hexuronic acid (later ascorbic acid, vitamin C)
1930	Professor in Szeged
1932	Vitamin C from paprika
1937	Nobel prize
1939	Muscle research
1942	Activity against Nazis
1944	Avoided Gestapo, lived incognito in Budapest
1945	Professor in Budapest
1946–1947	Social activity
1948	Move to USA, Director of Laboratory of Institute for Muscle Research, Woods Hole, Massachusetts
1955	American citizenship
1960	Thymus gland research, cancer research
1970-	Against the Vietnam War
1973	Visit to Hungary after 26 years opening ceremony of Biological Research Center in Szeged
1986	Died in Woods Hole

Berlin, just a few months in each. Then came a longer period in Hamburg. Later, he went to The Netherlands, where in Leiden and especially in Groningen he worked successfully on biological combustion. At that time, the main question in biochemistry was the subject of controversy between Heinrich Wieland and Otto Warburg and their followers.

Nomen est omen. H. Wieland held that the dehydrogenation of substrates was the basic process involved in biological oxidation and that oxygen reacts directly with such activated hydrogen atoms. O. Warburg considered activated oxygen was essential for biological oxidation.

Albert Szent-Györgyi recounted, "I got interested in the controversy and found a very simple way to show that both were right, that active oxygen oxidized active hydrogen. I wrote a very nice paper about it." [3] This later became a landmark of 20th century biochemistry.

Cambridge in England, his next workplace, became his scientific home, where he received his PhD in chemistry. In Groningen he had detected a strong reducing (anti-oxidant) substance in the adrenal gland. A similar material was found in lemon and orange juice and watery extract of cabbages. He extracted and purified it in Cambridge (about 1 g, good crystals), and it proved to be a carbohydrate, probably a sugar acid, with chemical formula $C_6H_8O_6$. At that time, 1 g was not enough for a detailed structural analysis.

His findings immediately aroused interest among biochemists worldwide and he submitted a manuscript to the Biochemical Journal, in this he finally had to decide on a name for what had simply been called 'Szent-Györgyi's substance'. He proposed calling it 'Ignose', from the Latin 'Ignosco' ('I don't know'), while the ending '-ose' means a sugar compound. With this name, the manuscript was not accepted. Szent-Györgyi resubmitted it, renaming his compound 'Godnose'('God knows sugar'). Finally, the editor suggested the much more prosaic 'hexuronic acid' and the paper was published with this nomenclature.

Szent-Györgyi received a professorship in Szeged, Hungary in September in 1928, but he got two sabbatical years to finish his work in Cambridge. He moved in Szeged in 1930. In the next year it was proved that hexuronic acid and the vitamin C are one and the same, and in 1932 Szent-Györgyi found that paprika is 'a regular mine of vitamin C'. Since paprika is not particularly a sweet fruit, there was less difficulty in separating vitamin C from other sugar compounds, than in the case of oranges, for example. 1.5 kg pure, crystalline vitamin C was produced from paprika during a week. This was ample for detailed chemical analysis. Some was sent to N. Haworth (Nobel prize in chemistry in 1937) in Birmingham, who quickly established its chemical nature (see Fig. 1).

The first thing to do was to give 'hexuronic acid' a new name. Szent-Györgyi and Haworth decided to call it 'a-scorbic acid', meaning that it prevented the scorbutic disease, scurvy. Figure shows the reversible oxidation and reduction steps, which comprise the basis of the role of vitamin C as an antioxidant in the living cells. The products of strong, irreversible

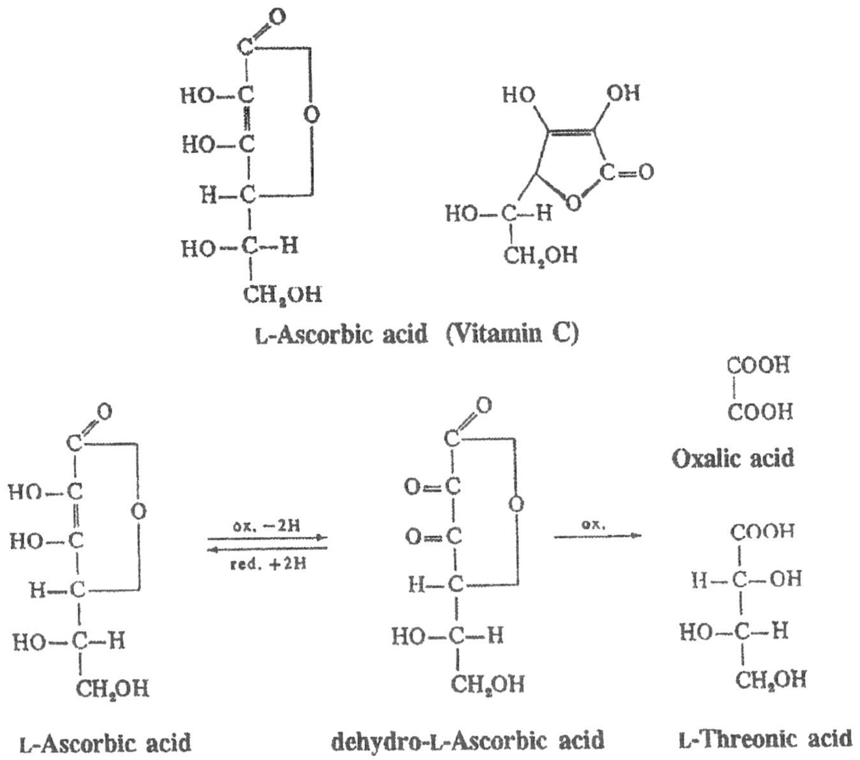

Fig. 1. Two formulations of the same structure, and the reversible or irreversible oxidation reactions involving Ascorbic acid (Vitamin C).

oxidation of ascorbic acid are oxalic and threonic acids.

Szent-Györgyi's idea was that the medically recommended daily adult intake of vitamin C, 60 mg, is sufficient to combat scurvy, but a higher amount will make the body more resistant to other diseases.

The current megavitamin therapy or orthomolecular medicine, based on this concept, was developed by Linus Pauling [4] (1901–1994), twice Nobel Prize winner, in 1954 for chemistry and in 1962 for peace.

These results and his role in resolving the Wieland–Warburg dispute over biological oxidation earned Szent-Györgyi the Nobel Prize in 1937.

3. Albert Szent-Györgyi in The New York Times

The journal was founded in 1851, and in the 20th century, especially after World War II, it became one of the best-known and most-respected American newspaper.

The first report in connection with Szent-Györgyi was published in 1933, in the 'Week of science' section of the Sunday issue, entitled 'The power of sound'. "Astonishing things happened. Small animals and bacteria were killed by the vibrations. Liquids boiled and decomposed. Waves were set up in them of a strange order."... "... physicists and chemists began to explore the new field. The latest of them is Professor A. Szent-Györgyi, known over the world for his work in trying to find out of what vitamins are composed. In fact, Szent-Györgyi is one of the leaders in organic chemistry" [5].

The second piece of news came just a week later from the Budapest correspondent of The New York Times, under the title 'Vitamin C may be obtained from paprika, chemist finds'.

General scientific interest has been aroused by the claim of a prominent Hungarian chemist, Dr Albert SzentGyoergyi, professor at Szegedin University, to have discovered after ten years research a method of producing vitamin C artifically.

Professor Szentgyoergyi says he has established that the vitamin is abundantly present in the Hungarian paprika, or sweet pepper, which he holds contains at least four times as much vitamin C as an orange or a lemon. In his experiments he has used 10 000 paprikas and claims that he has now extracted the vitamin, which can be administered in the form of powder or pills even to tiny babies.

Professor Szentgyoergyi's research work has been financed by wealthy American friends. He has already been invited to lecture on his discoveries in Berlin, Stockholm and Copenhagen [6].

The third article was the news of his winning the Nobel Prize from Stockholm in 1937: "NOBEL PRIZE GOES TO SZENT-GYORGYI; Hungarian wins the award in medicine for discoveries in biological combustion; He isolated vitamin C" [7].

In the following years The New York Times published a number of articles on his vitamin research, and from 1945 on new results concerning muscle contraction, his new research field.

In 1954, a further news item was published in connection with him, illustrated with his photo, indicating the importance of the news and the person: "The 1954 Albert Lasker Award of the American Heart Association will go to Dr Albert Szent-Gyorgyi, it was announced yesterday. The 60-year-old bio-chemist was cited for his 'distinguished research achievements in the field of cardiovascular diseases which have led to new understanding of the basic physiology of the heart'" [8].

A very important event in Szent-Györgyi's life, again making the pages of The New York Times in 1955, was when he received American citizenship after spending 8 years in the US. "The winner of the 1937 Nobel Prize in medicine became a United States citizen today. Dr Albert Szent-Gyorgyi, a native of Hungary, and his wife, Marta, took the oath" [9].

After this, news about him and his work regularly appeared in The New York Times each year, as shown in Table 2. The items included scientific reports about his new research results (9 news from his cancer research), minor news announcements of different events, letters to the editor, philosophical articles of Szent-Györgyi about the future and interviews with him. Two maxima are revealed by the statistics. The first was in the early sixties, when, besides the scientific reports, several 'Letters

Table 2
Statistics of the Szent-Györgyi's articles in The New York Times

Year	No.	Year	No.
1933	2	1964	2
1937	1	1965	3
1939	3	1966	3
1940	1	1967	4
1942	1	1968	1
1944	1	1969	2
1945	2	1970	7
1947	2	1971	6
1951	2	1972	5
1954	2	1973	2
1955	3	1974	2
1956	3	1975	3
1957	2	1976	1
1958	2	1978	1
1961	7	1981	1
1962	9	1986	2
1963	4		
		Total	92

to The Times' were published that warned against the danger of atomic war. The second maximum occurred in the early seventies, when Albert Szent-Györgyi was very active against the Vietnam War. For example, following the publication of his famous booklet 'The Crazy Ape', he participated in a lengthy interview entitled "Biologist Doubts Man's Survival in a World Run by 'Idiots' Too Old to Change" [10] and he wrote an article '15 Minutes to Zero' [11] at that time.

Later, his activity decreased and in 1986 The New York Times published his necrology [12]. He was 93 years old. It is symbolic that, one week after his death, the Patents section of the journal included a short news item: "Patent No. 4,620,014 was granted for compounds to help stimulate the body's immune system. The work was done by the late Albert Szent-Györgyi, a Nobel winner, and Gabor B. Fodor of the University of West Virginia." [13]

References

[1] F. Nagy (Eds.), Nobel prize winners from Hungary for humanity, Federal Chamber of Technical and Scientific Societies, Budapest, 1994.

[2] T. Szabó, A. Zallár, Albert Szent-Györgyi in Szeged and the Szent-Györgyi Collection (published in Hungarian), Csongrád Megyei Levéltár, 1989.

[3] R.W. Moss, Free Radical—Albert Szent-Györgyi and the Battle over Vitamin C, Paragon House Publishers, New York, 1988.

[4] L. Pauling, How to Live Longer and Feel Better, Avon Books, New York, 1987.

[5] The New York Times, March 26, 1933.

[6] The New York Times, April 2, 1933.

[7] The New York Times, October 29, 1937.

[8] The New York Times, March 15, 1954.

[9] The New York Times, February 22, 1955.

[10] The New York Times, February 20, 1970.

[11] The New York Times, September 25, 1970.

[12] The New York Times, October 25, 1986.

[13] The New York Times, November 1, 1986.

ELSEVIER

Journal of Molecular Structure (Theochem) 666–667 (2003) 693–698

THEO
CHEM

www.elsevier.com/locate/theochem

From peptide structure to proteomics structural studies on polypeptides in Hungary

Ilona Laczkó[a],*, Miklós Hollósi[b]

[a]*Institute of Biophysics, Biological Research Center, Temesvári krt. 62, Szeged H-6726, Hungary*
[b]*Department of Organic Chemistry, Eötvös Lóránd University, Pázmány Péter st. 1/A, H-1117 Budapest, Hungary*

Abstract

This paper is devoted to structural studies on peptides and proteins in Hungary from the foundation of peptide chemistry by Viktor Bruckner after 1951.
© 2003 Elsevier B.V. All rights reserved.

Keywords: Anthrax polypeptide; Chiroptical spectroscopy; Conformational studies; Peptide-based drugs; Peptide synthesis

It was Viktor Bruckner who founded peptide chemistry in Hungary (for the most important dates of his scientific career, see Table 1) [1]. In 1926 he joined the Institute of Organic Chemistry and Pharmaceutical Chemistry of the University of Szeged. In 1938 he was appointed assistant professor to Albert Szent-Györgyi. They have a joint paper about the chemistry of citrin which was published in Nature in 1936 [2]. This date shows that their collaboration started earlier than 1938. The scientific career of Bruckner was greatly inspired by the collaboration with Szent-Györgyi. Bruckner was the chemist who synthesized new molecules for Szent-Györgyi to support his biochemical research.

After the war in 1951 he was invited to the Eötvös Loránd University (ELU) in Budapest and he remained the leader of the Department of Organic Chemistry for 20 years. In 1946 he was elected as a corresponding member and three years later an ordinary member of the Hungarian Academy of Sciences (HAS). His scientific work was honoured by the Scheele medal

of the Chemistry Society of Stockholm, Sweden (1948), several prizes and orders [1].

Bruckner worked in various fields of organic chemistry and published many important papers. But his career was highlighted by researches on native polyglutamic acids [1]. In collaboration with the microbiologist György Ivánovics, Bruckner isolated polypeptides from the capsular substance of virulent *Bacillus anthracis* and from the culture medium of the serologically related *B. subtilis*. The structure of the polymers was found to be unique: the anthrax polypeptide consisted exclusively of D-(−)-glutamic acid residues connected by γ-amide bonds. The subtilis polypeptide contained, besides D-(−)-glutamic acid, more or less of the L-Glu form.

There was a scientific debate about the structure of anthrax polypeptide after the war. Some scientists abroad had the view that the polypeptide is built up with α-rather than γ-peptide bonds. Bruckner and coworker József Kovács performed degradation studies on the subtilis and anthrax polypeptide and on synthetic α-polyglutamic acid (Table 2). They proved that the bacterial polypeptides are built up predominantly or exclusively from γ-glutamyl moieties [3].

* Corresponding author. Fax: +36-62-433133.
  E-mail address: laczko@nucleus.szbk.u-szeged.hu (I. Laczkó).

0166-1280/$ - see front matter © 2003 Elsevier B.V. All rights reserved.
doi:10.1016/j.theochem.2003.08.096

Table 1
The scientific career of Győző (Victor) Bruckner (1900–1980) [1]

1925	Chemical engineering diploma (Technical University of Budapest)
1928	Ph.D. in Chemistry
1926	Institute of Organic and Pharmaceutical Chemistry, University of Szeged (under T. Széki)
1927–1928	Institute of Organic Chemistry, Technical University Berlin-Charlottenburg (under the supervision of A. Schönberg)
1929	Institute of Medical Chemistry, University of Graz (in the laboratory of the Nobel Prize winner F. Pregl)
1938–1941	Assistant Professor to Albert Szent-Györgyi
1941–1951	University Professor, Head of the Department in Szeged
1946	Corresponding member of the Hungarian Academy of Sciences (HAS)
1949	Ordinary member of HAS
1951–1970	Head of the Department of Organic Chemistry, Eötvös Lóránd University (ELU)
1949, 1955	National (Kossuth) Prize
1963, 1970	Hungarian State Orders

The next step was the synthesis of the bacterial polyglutamic acids. At the beginning of the 1950s, peptide chemistry was still in its infancy. The main problem was that activation of the γ-carboxyl group leads to the formation of pyrrolidone derivatives instead of γ-polyglutamic acid. By using a dimer (startdipeptide), Bruckner and co-workers succeeded in synthesizing the γ-polymer. They also synthesized

Table 2
Studies on natural polyglutamic acids [1]

1936	Gy. Ivánovics and V. Bruckner. *Isolation* of the anthrax polypeptide (from the capsular substance of *B. anthracis*) and the subtilis polypeptide (from the culture medium of *B. subtilis*) *Structure elucidation*: D-glutamic acid residues, no biuret reaction, $M = 6400–7100$ (or higher)
1950	Bruckner together with J. Kovács and I. Kandel. *Chemical degradation studies* α-glutamyl polymer → α,γ-diamino butyric acid γ-glutamyl polymer → β-formyl propionic acid (the only product of degradation)
1953	Bruckner and his co-workers (J. Kovács, H. Nagy, J. Wein, K. Kovács, M. Kajtár, A. Kótai, M. Szekerke, M. Hollósi) *Synthesis of polyglutamic acids* (α, γ; L,D)
1958	Gy. Ivánovices. Serological identity of the natural polypeptide and γ-poly-D-glutamic acid
1965	The optical rotatory dispersion (ORD) of γ-poly- and γ-oligo-D-glutamic acids

polyglutamic acids with α- and γ-peptide bonds and alternating L- and D-Glu residues. The synthetic work was completed with the synthesis of stereoisomeric γ-polyglutamic acids and γ-oligoglutamic acids via polybenzyl and poly-*tert*-butyl esters [4,5]. The polyglutamic acid research was a unique scientific performance at that time in Hungary. It was an example of the route of modern biomolecular research: from biology (isolation) through chemistry (structure elucidation and synthesis) to biology or life sciences (functional studies).

Ten years after the appointment of Bruckner in Budapest, another great Hungarian scientific adventure of peptide chemistry was started: the ACTH project [1,6]. Work was performed in the laboratories of the second generation of Hungarian peptide chemists: Kálmán Medzihradszky (a co-worker of Bruckner), Sándor Bajusz (a student of Miklós Bodánszky) and Lajos Kisfaludy (a student of Géza Zemplén, Technical University in Budapest) (Table 3). The project was aimed at synthesizing a pure corticotropin drug for the pharmaceutical industry because the natural preparations often contained harmful proteins. The success of the ACTH project [7–9] opened the way to the establishment of several peptide chemistry schools in Hungary (Table 4).

Table 3
Synthesis of biologically active peptides and peptide-based drugs in the Department of Organic Chemistry of ELU and the Peptide Research Group of HAS. Structure/function studies (1960–1980) [6,45]

1960	V. Bruckner (coordinator) K. Medzihradszky (ELU) S. Bajusz (Pharmaceutical Research Institute) L. Kisfaludy (Chemical Works Gedeon Richter Ltd) *ACTH project*: porcine ACTH 1-28 carrying biological activity
1966	*Human ACTH project*: human ACTH 1-32 (Gedeon Richter) National Prize, 1970
1960	Bruckner, M. Szekerke and J. Császár. *Synthesis of peptide cytostatics*
1965	K. Medzihradszky, H. Medzihradszky-Schweiger, H. Süli-Vargha. *Synthesis of peptide hormones (corticotropin and melanotropin fragments)*
1970	A. Kótai, Gy. Szókán, M. Szekerke, F. Hudecz. *Synthesis of polypeptide drug carriers*
1970	M. Szekerke, F. Hudecz. *Synthesis of polylysine-based carrier molecules*
1980	F. Hudecz, G. Mező. *Synthesis of T and B cell epitopic peptides*

Table 4
The establishment of schools of peptide chemistry in Hungary [6]

1960	K. Medzihradszky, M. Szekerke, M. Kajtár, H. Medzihradszky-Schweiger (ELU) Peptide Research Group of HAS (1974)
1960	S. Bajusz. A. Turán (Institute for Pharmaceutical Research, Budapest)
1960	L. Kisfaludy, M. Lőw, I. Schőn (Gedeon Richter Ltd)
1970	K. Kovács, B. Penke, L. Baláspiri (1970–1977 József Attila University, Szeged; 1977 Medical School of Szeged)
1970	I. Teplán, I. Mező, A. Seprődi, Gy. Kéri (Isotope Institute of HAS, Semmelweiss Medical School, Budapest)

Peptide chemists early recognized that peptides and peptide derivatives may carry cytotoxic groups and thus, they may be used as cytostatic drugs (Table 3) [1]. Bruckner and Mária Szekerke prepared the O,O′-dimesyl derivative of serylserine. Some phosphoric acid-based mustard derivatives were also synthesized. András Kótai and Gyula Szókán synthesized the first polypeptide drug carriers. Szekerke and Ferenc Hudecz developed the synthesis of a series of poly(lysine)-based drug carrier molecules. From the 1980s one of the main interest of the group headed by Hudecz was the synthesis and study of T and B cell epitopic peptides (Table 5) [6].

After 1970, the Hungarian peptide schools concentrated on the synthesis of peptide hormones [corticotropin (ACTH), melanotropin (MSH), gonadotropin hormone releasing hormone (GnRH), luteinizing hormone releasing hormone (LH-RH), growth hormone-releasing hormone (GH-RH), thyrotropin-releasing hormone (TRH), cholecystokinin (CCK), gastrins, enkephalins, endorphins and other opiate peptides, ^{3}H-labelled hormones including angiotensins, melanotropins, vasopressins, etc. and their analogs [6]. Bajusz and co-workers synthesized tripeptide aldehydes with selective anticoagulant activity. The synthesis of peptides with potential antitumor activity was continued. The chemistry of opiate binding sites were also studied [6].

This paper is dedicated to structural studies on polypeptides with special emphasis upon conformational research performed at ELU. For peptide studies in other Hungarian laboratories, see Ref. [6] and the references cited therein.

Table 5
Conformational studies in the Department of Organic Chemistry of ELU and in collaboration with other laboratories

1965	ORD spectroscopic studies on γ-linked oligo- and poly-glutamic acids [10]
1967	ORD and IR studies on sequence peptide polymers [38]
1970	CD of cyclo-γ-oligoglutamic acids [12]
1970	Structure-function studies on trypsin enzyme. Investigation of cation and magnetic field effects [6,46]
1975	Conformation of β-endorphin and its constituent fragments. Conformation of enkephalins [11,39]
1975	CD, FTIR, X-ray crystallographic, ^{1}H and ^{13}C NMR studies on antamanid-analogue cyclic peptides [13–15]
1985	Spectroscopic and X-ray crystallographic detection and characterization of β- and γ-turns in polypeptides [16–20,40]
1985	CD studies on T and B cell epitopic peptides [6,21,22]
1985	Convex constraint decomposition of CD curves of proteins [41]
1990	Conformational (CD) studies on glycosylated and phosphorylated peptides [22]
1990	Conformational (CD and FTIR) studies on peptides covering the intersubunit region of influenza virus hemagglutinin [22–24]
1990	Development of the CD/FTIR approach [22]
1990	Conformational (CD and FTIR) studies on peptides related to the structural biology of Alzheimer's disease [25–32]
1990	Effect of micelles and liposomes on the conformation of HIV-1 and hemagglutinin peptides [23,44]
1995	Conformational (CD and FTIR) studies on deletion and multiple antigenic (comb) and phosphorylated peptides [6,22]
1995	CD, FTIR and NMR studies on glycosylated cyclic peptides [42]
1995	CD and FTIR spectroscopic studies on peptide cation interactions [43]
1995	Ab initio calculations of peptide conformation [37]
2000	NMR studies on the conformation of proteins [36]

It was the late Márton Kajtár (another Bruckner student) who used optical rotatory dispersion (ORD) spectroscopy for studying the chiroptical properties and conformation of γ-oligo- and γ-polypeptides [10]. Rydon suggested in a paper which appeared in 1964 that γ-poly-D-glutamic acid in acidic solution adopts left-handed helical conformations. He proposed two structures, both stabilized by intramolecular H-bonds parallel with the helix axis. Márton Kajtár compared the ORD spectrum of the polymer with the ORD spectra of the oligomers both in acidic and basic pH values. It was shown that there is a continuous change of the spectra from the dimer to the heptamer. More importantly, there is a good agreement between the ORD of the polymers and the extrapolated spectra of the oligomers representing the ORD of a γ-glutamyl residue in mid-chain position. This proves that the polymer can adopt an ordered helical conformation neither in basic nor in acidic solution [10].

It was László Gráf (Institute for Pharmaceutical Research) who suggested circular dichroism (CD) studies on opiate peptides after the first CD spectrometer, a Jobin-Yvone Mark III was installed in the Department of Organic Chemistry of ELU [11].

After 1975 Miklós Hollósi (a student of Kajtár) became interested in the CD of cyclic peptides [12–14]. He synthesized a series of antamanid analogue cyclic peptides in Heidelberg, in the laboratory of Theodor Wieland in 1971–1972. After returning to Budapest he succeeded in growing crystals from the hexamer. The X-ray crystallographic structure was determined by Mátyás Czugler and Kálmán Sasvári in the Central Research Institute of Chemistry (CRIC) [15]. This was the first case that the X-ray crystallographic structure of a peptide was determined in Hungary. Lajos Radics (CRIC) and Hollósi performed detailed ^{1}H and ^{13}C NMR studies on the cyclic hexamer and octamer [14]. (The instrument they used was a 100 MHz Varian NMR spectrometer.) To help in assignment, ^{13}C- and ^{15}N-labelled cyclic peptides were also synthesized. The cyclic hexapeptide shows a unique turn conformation both in solid state and in solution [14,15].

Intensive work aimed at the spectroscopic characterization of β- and γ-turn structures was started after the study trip of Hollósi in Boston at the Brandeis University in the laboratory of Gerald D. Fasman [16–20]. In collaboration with Hudecz and researchers in the Wistar Institute in Philadelphia, he started CD studies on a great number and variety of epitopic as well as glycosylated and phosphorylated peptides (Table 5).

Some comments should be made here on the problems associated with the structure determination of polypeptides. There are four levels of polypeptide structure. Structure-function studies in the 1960s were aimed at the modification of the *primary structure*. But let us think of the enormously great number of possible peptide variants. All possible variants differing only in one position of a peptide mean 20 peptides. Replacement of all amino acids of a decapeptide

(one replacement in each peptide only) requires the preparation of 200 peptides. The number of variants increases exponentially if we want to make more than one substitution in a peptide. This is why early peptide chemistry was empirical and highly intuitive. If we analyze the reason of the successful replacements, that is the syntheses of derivatives with improved or modified biological activity, we necessarily come to the conclusion that side-chains in critical positions may be more important than secondary structure. This idea is strongly supported by the X-ray crystallographic structure of the immunologically important major histocompatibility complexes. It was proved that T cell epitopic peptides adopt extended conformation (β-strand or PPII) in the binding groove [21]. The consequence of this is that the amino acid side-chains point upward and downward (β-strand backbone conformation) or spiral round the central axis (PPII backbone) that underlines the importance of the side chain functional groups. We know less about the binding preferences of peptide hormones and other peptides of biological importance but it is reasonable to suppose that they also adopt an extended rather than folded conformation at their receptors.

Concerning *secondary structure (conformation)*, the problem is the flexibility of the structure. This is why cyclic peptides of more rigid conformation became very popular from the 1970s. Small and mid-size (10–30 residues) linear peptides are not easy to crystallize. (The exceptions are N- and C-protected linear peptides.) The characterization of the in solution structure of mid-size peptides by NMR methods is also dubious. NMR data give information about the average structure rather than the individual components of a mixture of two or more rapidly interconverting conformers. This prompted the co-workers of the Chiroptical Spectroscopic Laboratory at ELU to develop the CD/FTIR combined method for the conformational characterization of mid-size peptides of biological importance [22]. Infrared spectroscopy has roughly the same time scale as CD. While CD gives information about the backbone conformation, IR indicates the strength of H-bondings [20]. CD and FTIR also tell about the conformational effect of amino acid substitutions.

The CD/FTIR method was applied for many biologically active peptide synthesized by colleagues in the Department of Organic Chemistry or in the Peptide Research Group affiliated to the Department (Ferenc Hudecz, Helga Süli-Vargha, Gábor Mező, György Orosz). Many peptides were prepared in the Department of Medical Chemistry in Szeged by Botond Penke and Gábor K. Tóth. A great number of 'exotic' peptides (multiple antigenic peptides, phosphorylated peptides, etc.) were prepared by Tóth and co-workers (Table 5). Immunological studies were performed by Éva Rajnavölgyi and co-workers at ELU [21–24].

After 1985 peptide chemists in Hungary became interested in the structural biology of Alzheimer's disease. Research was partly performed in collaboration with groups in the Wistar Institute and at Brandeis University (Table 5) [6,25–32]. Alzheimer research was continued by Penke in Szeged [6].

Árpád Furka at ELU laid down the principles of combinatorial chemistry in the 1980s [33]. Today nobody questions the importance of the work of the Furka group. It is a real invention: a typical peptide chemical idea which, however, has opened new routes for pharmaceutical research.

Tertiary structure is the relative steric arrangement of the secondary structural (conformational) elements of proteins. *Quaternary structure* earlier meant the spacial configuration (relative position) of non-covalently bound subunits of a complex protein, e.g. hemoglobin. Today it is more appropriate to speak about *domain or modular structure* which denotes an ensemble of covalently bound or not bound protein subunits.

Protein tertiary structure determination in Hungary was initiated in the Institute of Enzymology (after 1970 part of the Biological Resarch Center (BRC) in Szeged). Pál Elődi and Péter Závodszky used first ORD and other biophysical methods for the characterization of enzymes and proteins. The first CD spectrometer in Hungary was installed in Szeged in BRC but it was used by András Garay and co-workers for studies on the origin of homochirality on Earth. From the 1970s structure/function studies were supported by protein X-ray crystallography and NMR measurements performed abroad. István Simon started statistical analysis of protein data banks and elaborated methods for 3D structural prediction of proteins. IR studies were performed on bacteriorhodopsin, CD studies on peptides and proteins in the presence of micelles or liposomes [34].

Structural biology is based upon the *three-dimensional (3D) structure* of proteins. The experimantal tools of 3D structure elucidation of proteins are protein X-ray crystallography and NMR spectroscopy. After 1990 Gábor Náray-Szabó and Zsolt Böcskei built up a protein X-ray crystallographic laboratory at ELU (for a recent paper, see Ref. [35]), András Perczel was the pioneer of protein NMR in Hungary [36]. Perczel and his group also started modeling and ab initio calculations of peptide conformation [37].

We came to the year of 2000. The success of the genom projects, in the first place the deciphering of the human genom, is a real revolution of biology and life sciences and it is also a milestone in the history of mankind. Genom research is followed (or accompanied) by *proteomics*, the mapping of the structure and biological function of all the proteins encoded by the genom. But proteome research is also problematic. There are billions of living organisms on Earth, the proteome of which can be chosen as the object of a new project. Which one to start studying, how to compare sequence and 3D structure information and finally, how to use the information for medical or pharmaceutical purposes?

We are only at the beginning of this new scientific game of mankind. Victor Bruckner, the founder of peptide chemistry was an assistant to Albert Szent-Györgyi. Bruno F. Straub, the founder of BRC was a student of Szent-Györgyi. Peptide chemistry and protein chemistry unified in the mid-1990s to respond to the challenges of modern biology and life sciences. The result of this marriage is todays *structural biology*. We live in the century of genomics and proteomics. The perspectives are both frightening and marvellous. Is it necessary to decode the genom of more and more living organisms? Or should we look for principles of simplifications and generalizations? We should learn from Albert Szent-Györgyi who always looked for the basic, most important and relevant issues of scientific research.

Acknowledgements

The authors thank for the support of the Hungarian National Research Foundation, OTKA T029983 and T034866.

References

[1] K. Medzihradszky, Acta Chim. Acad. Sci. Hung. 107 (1981) 287.

[2] V. Bruckner, A. Szent-Györgyi, Nature 138 (1936) 1057.

[3] V. Bruckner, J. Kovács, G. Dénes, Nature 172 (1953) 508.

[4] V. Bruckner, J. Wein, H. Nagy, M. Kajtár, J. Kovács, Naturwiss 42 (1955) 210.

[5] M. Kajtár, V. Bruckner, Zs. Rihmer, Acta Chim. Acad. Sci. Hung. 43 (1965) 161.

[6] K. Medzihradszky, Magyar Kémiai Folyóirat 100 (1994) 235 (in Hungarian, with a list of references).

[7] Gy. Bruckner, K. Medzihradszky, S. Bajusz, L. Kisfaludy, 151.214 Sz. Magyar Szabadalmi Leírás 214 (1962).

[8] S. Bajusz, K. Medzihradszky, Z. Paulay, Zs. Láng, Acta Chim. Acad. Sci. Hung. 52 (1967) 335.

[9] L. Kisfaludy, M. Löw, I. Schön, T. Szirtes, M. Sárközi, S. Bajusz, A. Turán, R. Beke, A. Juhász, L. Gráf, K. Medzihradszky, in: J. Meinhofer (Ed.), Chemistry and Biology of Peptides, 299, 1972.

[10] M. Kajtár, V. Bruckner, Tetrahedron Lett. (1966) 4813.

[11] M. Hollósi, M. Kajtár, L. Gráf, FEBS Lett. 74 (1977) 185.

[12] M. Kajtár, M. Hollósi, G. Snatzke, Tetrahedron 27 (1971) 5659.

[13] M. Hollósi, Th. Wieland, Int. J. Peptide Protein Res. 10 (1977) 329.

[14] M. Hollósi, L. Radics, Th. Wieland, Int. J. Peptide Protein Res. 10 (1977) 286.

[15] M. Czugler, K. Sasvári, M. Hollósi, J. Am. Chem. Soc. 104 (1982) 4465.

[16] M. Hollósi, M. Kawai, G.D. Fasman, Biopolymers 24 (1985) 211.

[17] M. Hollósi, K.E. Kövér, S. Holly, G.D. Fasman, Biopolymers 26 (1987) 1527.

[18] M. Hollósi, K.E. Kövér, S. Holly, L. Radics, G.D. Fasman, Biopolymers 26 (1987) 1555.

[19] A. Perczel, M. Hollósi, B.M. Foxman, G.D. Fasman, J. Am. Chem. Soc. 113 (1991) 9772.

[20] H.H. Mantsch, A. Perczel, M. Hollósi, G.D. Fasman, Biopolymers 33 (1993) 201.

[21] M. Hollósi, in: É. Rajnavölgyi (Ed.), Synthetic Peptides in the Search for B- and T-Cell Epitopes, R.G. Landes Co, Austin, 1994, p. 67.

[22] A. Perczel, M. Hollósi, in: G.D. Fasman (Ed.), Circular Dichroism and the Conformational Analysis of Biomolecules, Plenum Press, New York, 1996, pp. 285, Ch. 9.,.

[23] M. Hollósi, A.A. Ismail, H.H. Mantsch, B. Penke, I.G. Váradi, G.K. Tóth, I. Laczkó, I. Kurucz, Z. Nagy, G.D. Fasman, É. Rajnavölgyi, Eur. J. Biochem. 206 (1992) 421.

[24] Zs. Majer, S. Holly, G.K. Tóth, Gy. Váradi, Z. Nagy, A. Horváth, É. Rajnavölgyi, I. Laczkó, M. Hollósi, Arch. Biochem. Biophys. 322 (1995) 112.

[25] L. Ötvös Jr., J. Thurin, E. Kollát, L. Ürge, M. Hollósi, Int. J. Peptide Protein Res. 38 (1991) 476.

[26] I. Laczkó-Hollósi, M. Hollósi, V.M.-Y. Lee, H.H. Mantsch, Eur. Biophys. J. 21 (1992) 345.

[27] M. Hollósi, L. Ürge, A. Perczel, J. Kajtár, I. Teplán, L. Ötvös Jr., G.D. Fasman, J. Mol. Biol. 223 (1992) 673.

[28] A. Perczel, E. Kollát, M. Hollósi, G.D. Fasman, Biopolymers 33 (1993) 665.

[29] M. Hollósi, Z.M. Shen, A. Perczel, G.D. Fasman, Proc. Natl Acad. Sci. USA 91 (1994) 4902.

[30] Z.M. Shen, A. Perczel, M. Hollósi, I. Nagypál, G.D. Fasman, Biochemistry 33 (1994) 9627.

[31] M. Hollósi, L. Ötvös Jr., J. Kajtár, A. Perczel, V.M.-Y. Lee, Peptide Res. 2 (1989) 109.

[32] I. Laczkó, E. Vass, K. Soós, J.L. Varga, S. Száraz, M. Hollósi, B. Penke, Arch. Biochem. Biophys. 335 (1996) 381.

[33] A. Furka, Drug Discov. Today 7 (2002) 1.

[34] For papers published by researchers of Biological Research Center, see the "blue-book" at the homepage, address: http://www.szbk.u-szeged-hu

[35] G.B. Vértessy, V. Harmath, Zs. Böcskei, G. Náray-Szabó, F. Orosz, J. Ovádi, Biochemistry 37 (1998) 15300.

[36] A. Wiles, G. Shaw, J. Bright, A. Perczel, I.D. Campbell, P.N. Barlow, J. Mol. Biol. 272 (1997) 253.

[37] A. Perczel, I.G. Csizmadia, Int. Rev. Phys. Chem. 14 (1995) 127.

[38] D.F. DeTar, T. Vajda, J. Am. Chem. Soc. 89 (1967) 998.

[39] M. Hollósi, Z. Dobolyi, S. Bajusz, FEBS Lett. 110 (1980) 136.

[40] M. Hollósi, Zs. Majer, A.Z. Rónai, A. Magyar, K. Medzihradszky, S. Holly, A. Perczel, G.D. Fasman, Biopolymers 34 (1994) 177.

[41] A. Perczel, M. Hollósi, G. Tusnády, G.D. Fasman, Protein Engng 4 (1991) 669.

[42] E. Láng, B. Hargittai, Zs. Majer, A. Perczel, M. Mák, M. Kajtár-Peredy, L. Radics, G.D. Fasman, M. Hollósi, Protein Peptide Lett. 3 (1996) 9.

[43] Zs. Likó, J. Botyánszki, J. Bódi, E. Vass, Zs. Majer, M. Hollósi, H. Süli-Vargha, Biochem. Biophys. Res. Commun. 227 (1996) 351.

[44] I. Laczkó, M. Hollósi, L. Ürge, H.H. Mantsch, J. Thurin, L. Ötvös Jr., Biochemistry 31 (1992) 4282.

[45] K. Medzihradszky, H. Medzihradszky-Schweiger, FEBS Lett. 67 (1976) 45.

[46] T. Vajda, A. Garai, J. Inorg. Biochem. 15 (1981) 307.

ELSEVIER

Journal of Molecular Structure (Theochem) 666–667 (2003) 699–706

THEO CHEM

www.elsevier.com/locate/theochem

Bridging the gap between pure science and the general public: comparison of the informational exchange for these extremities in scientific awareness

Tania A. Pecora[a], Michael C. Owen[b], Christopher N.J. Marai[a,b,c], David H. Setiad[a,b,c], Gregory A. Chass[a,b,c,d,*]

[a]GIOCOMMS, 1422 Edenrose St, Mississauga, Ont., Canada L5V 1H3
[b]Department of Biomedical Sciences, 2500 California Plaza, Omaha, NE, USA 68178
[c]Department of Chemistry, 80 St George St, University of Toronto, Toronto, Ont., Canada M5S 3H6
[d]Laboratoire Francis Perrin, CEA/DSM/DRECAM/SPAM-CNRS URA 2453 Cea-Saclay, 91191, Gif-sur-Yvette, France 68178

Abstract

The overwhelming volume of scientific data presently available is still not of much use to the layperson, due to the inefficiency of informational translation and antiquated nomenclature. Despite an enormous evolution of search and indexing mechanisms over the past decade, there presently exists no matching acceleration of the scientific awareness and comprehension in the average citizen. The exponential growth and global establishment of the World Wide Web has had a dichotomic effect on bridging the gap between pure science and the general public. A wealth of information, both too complex/simplistic and off topic is often the result of current database searches. Too often pertinent knowledge is delayed from reaching one extremity or the other. The sobering experience of the drug Thalidamide attests to this inefficiency, as the hints of a potential difference in enantiospecific bioactivity did not reach the proper regulatory channels in time. Similarly, Linus Pauling's announcement of the antioxidant and 'fountain of youth' properties of Albert Szent-Györgyi's Vitamin C resulted in decades of overdosing.

It is proposed that with this new age of pure science, technology, search and indexing mechanisms, databases and an increased desire for knowledge that there now exists a need to more effectively bridge this gap. Science itself must afford a more generalized description of resulting data sets and facilitate opportunities for individuals to more fully saturate disciplines falling into the bridging areas. Lycopene, Vitamin E and Vitamin C are used as examples of the successes and failures of informational exchange.
© 2004 Published by Elsevier B.V.

Keywords: Vitamin-C; Vitamin-E; Lycopene

1. Introduction

With every age comes a lag time for the transfer of knowledge, through the proper informational and educational channels, to bring about a saturation of understanding in the population. Concepts still in their infancy at the turn of the century now find themselves expressed in language, literature, art and daily routine. In order of current saturation one may list the following in decreasing order: mass production, commodity trading, environmentalism, digital technology, nano-technology, quantum mechanics. Many of these ideals,

* Corresponding author. Address: GIOCOMMS, 1422 Edenrose St, Mississauga, Ont., Canada L5V 1H3.
E-mail address: gchass@giocomms.org (G.A. Chass).

0166-1280/$ - see front matter © 2004 Published by Elsevier B.V.
doi:10.1016/j.theochem.2003.08.097

born of academic exploration, are, however, delayed from reaching the populace and subsequently saturating society. Conflicting views find equal support in the debate of whether art and literature or science and technology are responsible for nucleating preliminary informational transfers and reducing delay. This argument is not addressed in this work and a focus is made instead on the informational exchange between pure science and the layperson.

Delays take the form of active and passive elements effectively generating bottle necks in the streams of communication. Elements may include reduced funding for educational programs, socio-economic restrictions limiting access, saturation of employment and educational positions, government and religious restrictions, embargos, disease, disaster, economic depression, conflict and war. All of these are acknowledged as being vastly influential in the delay or acceleration of informational transfer. This work is restricted to one area, specifically the retarding mechanisms of the transfer of information between pure science and the layperson.

Currently, many individuals perform preliminary searches for information using the World Wide Web, or the Internet, often resulting in a mixture of successes and failures. A large number of web sites amorally make use of particular concepts, techniques, methods and even keywords to market and sell their products. A search on a particular vitamin for example, may result in information indirectly related to the compound or to sites promoting a product for sale. Some sites make use of the infamy of the names and keywords used and have absolutely no relation to the actual compound. Vitamin-C and Professor Pauling's name are especially prone to this sort of 'name dropping'.

Nomenclature and terminology therefore prove themselves to be vastly influential in informational transfer in pure science. Many years of scientific research may result in a few published data tables or mechanistic descriptions that are only detectably novel to individuals familiar within that specific area, of a particular discipline. Often, the results are of little use outside the area of focus, even to experienced professionals working under the same discipline, using identical analytical techniques. Examination of these sorts of tabulated or mechanistic data immediately demonstrates this, even for studies of interest to the general public; as in the case of Vitamin-E,

Vitamin-C and Lycopene. Unfortunately, novel discoveries and evolved understanding of these systems remain in 'scientific limbo', awaiting 'informational cascade' to periodicals more saturated with common language, themselves saturating more of the populace.

Citation and 'journal cascades' aid in moving concepts and reproducible results from studies of specific focus, downwards in complexity, through to more generalized ones and their associated literature sources. Often a 'novel discovery' may be decomposed into several established and well proven 'older findings' themselves to be found in forgotten periodical stacks; each idea based backwards in time upon its predecessor.

Terms used to define new mechanisms; trends and phenomena have also aged with these manuscripts, entering common language soon afterwards. Using the biomedically important definitions 'anti-oxidant', 'free radicals', 'radical-scavenging' and 'oxidative stress' as example, one is hard-pressed to find a lack of awareness of these expressions in a growing majority of the populace. The lifeworks of individuals such as Professor Szent-Györgi, Albert and Professor Pauling, Linus, on Vitamin-C may attest to the efficacy of societal saturation. One may imagine that there is a direct coupling between the time for selected expressions to enter common language from publication and the current receptiveness of the populace. This may be quantified, for example, as the time for a novel finding published in an internationally peer-reviewed journal to appear in daily newsprint.

A quantitative exercise of tracing citation lists backwards through publications, may indeed be able to help track the critical points of delay and acceleration of informational transfer, as each article is dated. The results and qualitative understanding gained, would both be informative and demonstrative of a need to refine current methods of informational transfer from pure science to the general public.

2. Methods

An attempt was made to conduct an unbiased search using popular World Wide Web search engines, on selected molecular systems of interest. *Vitamin-E*, *Vitamin-C* and *Lycopene* were chosen as test systems in searches, using three popular internet search engines Google, Lycos and Excite; www.google.com, www.lycos.com and www.excite.com, respectively. Two searches were conducted, the first using only the name of the molecular systems, the second using the names in addition to the term 'molecular structure' (Table 1). The first 20 web sites resulting for each molecular system were investigated for their content, making a total of 180 web sites visited, for the three systems; two searches for each system using three search engines. Any web page that failed to connect or open was discounted and a new site was chosen from the list.

Three categories were used to characterize the results of the searches conducted. A 'hit' with supporting links or references, a 'hit' without support, and finally 'sales' whereby the keywords were used to advertise, sell or announce a product for sale. The labels appear as *Link*, *Non-* and *Sales*, respectively. Sites completely outside the three categories were not considered nor counted.

The first search involved 'hit' criteria of the following: generalized description of the compound, sources of and recommended daily allowances (RDA), general information of the benefits and properties. The second search followed a 'hit' criteria of providing 'basic physical properties', chemical and physical information, chemical structure, or material from/for scientific publication. An attempt was made to make a threshold of acceptance at a first year undergraduate level of information (current Canadian average), with no upper threshold.

Some rules employed during the searches include the following: pages too complex for simple searches were afforded a 'hit' as they were at least representative conduits of reproducible information, multiple 'hits' to same site were each counted since these are multiple the potentials (chances) to arrive at same site and available information, pages promoting products for sale were deemed acceptable 'hit' if informative and correct. It is mentioned here the numerous sites arising in Vitamin-C searches promoting Professor Pauling, Linus's lifetime achievements were not considered as 'hits'.

Secondly, the journal Nature was selected to conduct a search for the most recent published article on Vitamin-E, Vitamin-C and Lycopene. Search criteria were limited to the title containing either the name of the compound or a general title, comprehensible by the average citizen. Albeit a subjective measure, it is deemed likely that the average citizen is familiar with the terms of 'oxidation', 'oxidative stress', 'serum levels' and terms able to be found in national newsprint.

An unsuccessful search was treated by repeating the search in subsequent journals, in the following order, until a successful article was located: Nature, Science, Cell, Journal of the American Medical Association, Cancer, Cancer Research. The latter two journals were chosen due to the term 'Cancer' being well saturated in the general public's vocabulary.

Once a suitable article was located it was used in a successive search to map out the journal cascade attributed to that publication as well as the 'cascade' it was based upon. By definition a journal cascade is the succession of references sited and articles citing the article in question. In both 'directions of cascade'

Table 1
World Wide Web searches of selected molecular systems using differing keywords and search engines (Number of 'Link-Hits', 'Non-Link-Hits' and 'Sales' sites)

Search category		World Wide Web search engine								
		Google			Lycos			Excite		
		Link	Non-	Sales	Link	Non-	Sales	Link	Non-	Sales
1	Vit-E	10(1)	3	6	10(1)	2	7	10	2	8
	Vit-C	13(2)	2	2	5	2	7	4	1	13
	Lyc	7(1)	2	10	6	1	12	3	1	16
2	Vit-E	3	2	7	4	0	3	6	0	7
	Vit-C	2	0	6	2	0	8	4	0	12
	Lyc	11	0	8	8(4)	1	5	6	2	9

(1) KWs = Vitamin-E, Vitamin-C, Lycopene → general information, sources of, benefits, dosage (RDA). (2) KWs = Vitamin-E, Vitamin-C, Lycopene + molecular structure → Molecular Structure, physical properties, basis of activity. 'Sales' makes use of the results for categories (1) and (2), tabulating the number of sites using the molecular system of interest as a commodity in sales or marketing.

Table 2

AUTHORS	JOURNAL	VOL.	PAGES	DATE D/M/Y	TITLE	REFS. CITED	TIMES CITED
KARLBERG EOL, ANDERSSON SGE	NATURE REVIEWS GENETICS	4	391-397	MAY 2003	MITOCHONDRIAL GENE HISTORY AND MRNA LOCALIZATION: IS THERE A CORRELATION? KARLBERG EOL, ANDERSSON SGE NATURE REVIEWS GENETICS	49	0
MARGEOT A, BLUGEON C, SYLVESTRE J, VIALETTE S, JACQ C, CORRAL-DEBRINSKI M	EMBO JOURNAL	21	6893-6904	16/12/2002	IN SACCHAROMYCES CEREVISIAE, ATP2 MRNA SORTING TO THE VICINITY OF MITOCHONDRIA IS ESSENTIAL FOR RESPIRATORY FUNCTION	39	1
RICART J, IZQUERDO JM, DI LIEGRO CM, CUEZVA JM	BIOCHEMICAL JOURNAL	365	417-428	15/07/2002	ASSEMBLY OF THE RIBONUCLEOPROTEIN COMPLEX CONTAINING THE MRNA OF THE BETA-SUBUNIT OF THE MITOCHONDRIAL H+-ATP SYNTHASE REQUIRES THE PARTICIPATION OF TWO DISTAL CIS-ACTING ELEMENTS AND A COMPLEX SET OF CELLULAR TRANS-ACTING PROTEINS	58	1
MARC P, MARGEOT A, DEVAUX F, BLUGEON C, CORRAL-DEBRINSKI M, JACQ C	EMBO REPORTS	3	159-164	FEB. 2002	GENOME-WIDE ANALYSIS OF MRNAS TARGETED TO YEAST MITOCHONDRIA	32	11
BELOGRUDOV GI, LEE PT, JONASSEN T, HSU AY, GIN P, CLARKE CF	ARCHIVES OF BIOCHEMISTRY AND BIOPHYSICS	392	48-58	01/08/2001	YEAST COQ4 ENCODES A MITOCHONDRIAL PROTEIN REQUIRED FOR COENZYME Q SYNTHESIS	53	3
MIYADERA H, AMINO H, HIRAISHI A, TAKA H, MURAYAMA K, MIYOSHI H, SAKAMOTO K, ISHII N, HEKIMI S, KITA K	JOURNAL OF BIOLOGICAL CHEMISTRY	276	7713-7716	16/03/2001	ALTERED QUINONE BIOSYNTHESIS IN THE LONG-LIVED CLK-1 MUTANTS OF CAENORHABDITIS ELEGANS	31	34
FINKEL T, HOLBROOK NJ	NATURE	408	239-247	09/11/2000	OXIDANTS, OXIDATIVE STRESS AND THE BIOLOGY OF AGEING	106	260
HARSHMAN LG, HABERER BA	JOURNALS OF GERONTOLOGY SERIES A-BIOLOGICAL SCIENCES AND MEDICAL SCIENCES	55	B415-B417	SEP. 2000	OXIDATIVE STRESS RESISTANCE: A ROBUST CORRELATED RESPONSE TO SELECTION IN EXTENDED LONGEVITY LINES OF DROSOPHILA MELANOGASTER?	19	7
HARSHMAN LG, HOFFMANN AA	TRENDS IN ECOLOGY & EVOLUTION	15	32-36	JAN 2000	LABORATORY SELECTION EXPERIMENTS USING DROSOPHILA: WHAT DO THEY REALLY TELL US?	49	46
GIBBS AG	JOURNAL OF EXPERIMENTAL BIOLOGY	202	2709-2718	OCT. 1999	LABORATORY SELECTION FOR THE COMPARATIVE PHYSIOLOGIST	67	32
MONGOLD JA, BENNETT AF, LENSKI RE	EVOLUTION	53	386-394	APR. 1999	EVOLUTIONARY ADAPTATION TO TEMPERATURE. VII. EXTENSION OF THE UPPER THERMAL LIMIT OF ESCHERICHIA COLI	27	5
BENNETT AF, LENSKI RE	EVOLUTION	51	36-44	FEB. 1997	EVOLUTIONARY ADAPTATION TO TEMPERATURE. 6. PHENOTYPIC ACCLIMATION AND ITS EVOLUTION IN ESCHERICHIA COLI	30	20
BENNETT AF, LENSKI RE	EVOLUTION	50	493-503	APR. 1996	EVOLUTIONARY ADAPTATION TO TEMPERATURE. 5. ADAPTIVE MECHANISMS AND CORRELATED RESPONSES IN EXPERIMENTAL LINES OF ESCHERICHIA COLI	37	10
TRAVISANO M, VASI F, LENSKI RE	EVOLUTION	49	189-200	FEB. 1995	LONG-TERM EXPERIMENTAL EVOLUTION IN ESCHERICHIA-COLI 3. VARIATION AMONG REPLICATE POPULATIONS IN CORRELATED RESPONSES TO NOVEL ENVIRONMENTS	76	29
TRAVISANO M, MONGOLD JA, BENNETT AF, LENSKI RE	SCIENCE	267	87-90	JAN. 6 1995	EXPERIMENTAL TESTS OF THE ROLES OF ADAPTATION, CHANCE, AND HISTORY IN EVOLUTION	24	67
VASI F, TRAVISANO M, LENSKI RE	AMERICAN NATURALIST	144	432-456	SEP. 1994	LONG-TERM EXPERIMENTAL EVOLUTION IN ESCHERICHIA-COLI .2. CHANGES IN LIFE-HISTORY TRAITS DURING ADAPTATION TO A SEASONAL ENVIRONMENT	56	47
MESTERTON-GIBBONS M	ECOLOGY	74	2467-2468	DEC. 1993	WHY DEMOGRAPHIC ELASTICITIES SUM TO ONE - A POSTSCRIPT TO DEKROON ET-AL	3	16
DEKROON H, PLAISIER A, VANGROENENDAEL J, CASWELL H	ECOLOGY	67	1427-1431	OCT. 1986	ELASTICITY - THE RELATIVE CONTRIBUTION OF DEMOGRAPHIC PARAMETERS TO POPULATION-GROWTH RATE	22	199
TORNQVIST L, VARTIA P, VARTIA YO	AMERICAN STATISTICIAN	39	43-46	1985	HOW SHOULD RELATIVE CHANGES BE MEASURED	22	37
VARTIA P	SCANDINAVIAN JOURNAL OF ECONOMICS	79	485-487	1977	NOTE ON CALCULATION OF ARC-ELASTICITIES	8	1
SATO K	REVIEW OF ECONOMICS AND STATISTICS	56	549-552	1974	IDEAL INDEX NUMBERS THAT ALMOST SATISFY FACTOR REVERSAL TEST	1	5
THEIL H	REVIEW OF ECONOMICS AND STATISTICS	55	498-502	1973	NEW INDEX NUMBER FORMULA	9	9
THEIL H	ECONOMIST	116	677-689	1968	GEOMETRY AND NUMERICAL APPROXIMATION OF COST OF LIVING AND REAL INCOME INDICES	5	7

Table 3

AUTHORS	JOURNAL	VOL.	PAGES	DATE D/M/Y	TITLE	REFS. CITED	TIMES CITED
SCHNEIDER JE, PRICE S, MAIDT L, GUTTERIDGE JMC, FLOYD RA	NUCLEIC ACIDS RESEARCH	18	631-635	11/02/1990	METHYLENE-BLUE PLUS LIGHT MEDIATES 8-HYDROXY 2'-DEOXYGUANOSINE FORMATION IN DNA PREFERENTIALLY OVER STRAND BREAKAGE	22	144
FLOYD RA, WEST MS, ENEFF KL, SCHNEIDER JE	ARCHIVES OF BIOCHEMISTRY AND BIOPHYSICS	273	106-111	15/08/1989	METHYLENE-BLUE PLUS LIGHT MEDIATES 8-HYDROXYGUANINE FORMATION IN DNA	29	159
VANDERPUTTEN WJM, KELLY JM	PHOTOCHEMISTRY AND PHOTOBIOLOGY	49	145-151	FEB. 1989	LASER FLASH SPECTROSCOPY OF METHYLENE-BLUE WITH NUCLEIC-ACIDS - EFFECTS OF IONIC-STRENGTH AND PH	15	13
OHUIGIN C, MCCONNELL DJ, KELLY JM, VANDERPUTTEN WJM	NUCLEIC ACIDS RESEARCH	15	7411-7427	25/09/1987	METHYLENE-BLUE PHOTOSENSITIZED STRAND CLEAVAGE OF DNA - EFFECTS OF DYE BINDING AND OXYGEN	33	65
LEE PCC, RODGERS MAJ	PHOTOCHEMISTRY AND PHOTOBIOLOGY	45	79-86	JAN. 1987	LASER FLASH PHOTOKINETIC STUDIES OF ROSE-BENGAL SENSITIZED PHOTODYNAMIC INTERACTIONS OF NUCLEOTIDES	29	155
RODGERS MAJ	JOURNAL OF THE AMERICAN CHEMICAL SOCIETY	105	6201-6205	1983	SOLVENT-INDUCED DEACTIVATION OF SINGLET OXYGEN - ADDITIVITY RELATIONSHIPS IN NON-AROMATIC SOLVENTS	29	230
MATHESON IBC, RODGERS MAJ	JOURNAL OF PHYSICAL CHEMISTRY	86	884-887	1982	OXYGEN-TRANSFER RATES ACROSS WATER MICELLE INTERFACES DERIVED FROM MEASUREMENTS OF NI-2+ QUENCHING OF SINGLET MOLECULAR-OXYGEN IN AEROSOL-OT HEPTANE REVERSE MICELLES	31	18
FOYT DC	COMPUTERS & CHEMISTRY	5	49-54	1981	A MINICOMPUTER-BASED SYSTEM FOR PROCESS-CONTROL, DATA ACQUISITION, AND DATA-ANALYSIS IN A DIVERSE FAST KINETICS FACILITY	19	121
RODGERS MAJ, BECKER JC	JOURNAL OF PHYSICAL CHEMISTRY	84	2762-2768	1980	ELECTRON-TRANSFER QUENCHING OF THE LUMINESCENT STATE OF THE TRIS(BIPYRIDYL)RUTHENIUM(II) COMPLEX IN MICELLAR MEDIA	32	94
SCHMEHL RH, WHITTEN DG	JOURNAL OF THE AMERICAN CHEMICAL SOCIETY	102	1938-1941	1980	INTRAMICELLAR ELECTRON-TRANSFER QUENCHING OF EXCITED-STATES - DETERMINATION OF THE BINDING CONSTANT AND EXCHANGE-RATES FOR DIMETHYLVIOLOGEN	30	89
YEKTA A, AIKAWA M, TURRO NJ	CHEMICAL PHYSICS LETTERS	63	543-548	1979	PHOTO-LUMINESCENCE METHODS FOR EVALUATION OF SOLUBILIZATION PARAMETERS AND DYNAMICS OF MICELLAR AGGREGATES - LIMITING CASES WHICH ALLOW ESTIMATION OF PARTITION-COEFFICIENTS, AGGREGATION NUMBERS.	28	193
ALMGREN M, GRIESER F, THOMAS JK	JOURNAL OF THE AMERICAN CHEMICAL SOCIETY	101	279-291	1979	DYNAMIC AND STATIC ASPECTS OF SOLUBILIZATION OF NEUTRAL ARENES IN IONIC MICELLAR SOLUTIONS	41	550
ULMIUS J, LINDMAN B, LINDBLOM G, DRAKENBERG T	JOURNAL OF COLLOID AND INTERFACE SCIENCE	65	88-97	1978	H-1,C-13,CL-35, AND BR-81 NMR OF AQUEOUS HEXADECYLTRIMETHYLAMMONIUM SALT-SOLUTIONS - SOLUBILIZATION, VISCOELASTICITY, AND COUNTERION SPECIFICITY	43	111
KAMENKA N, FABRE H, CHORRO M, LINDMAN B	JOURNAL DE CHIMIE PHYSIQUE ET DE PHYSICO-CHIMIE BIOLOGIQUE	74	510-512	1977	INFLUENCE OF SOLUBILIZATION ON COUNTERION BINDING TO SURFACTANT MICELLES FROM SELF-DIFFUSION COEFFICIENTS	12	17
KAMENKA N, LINDMAN B, BRUN B	COLLOID AND POLYMER SCIENCE	252	144-152	1974	TRANSLATIONAL MOTION AND ASSOCIATION IN AQUEOUS SODIUM DODECYL-SULFATE SOLUTIONS	51	55
LINDMAN B, BRUN B	JOURNAL OF COLLOID AND INTERFACE SCIENCE	42	388-399	1973	TRANSLATIONAL MOTION IN AQUEOUS SODIUM OCTANOATE SOLUTIONS	46	99
LINDMAN B, DANIELSS.I	JOURNAL OF COLLOID AND INTERFACE SCIENCE	39	349	1972	NUCLEAR MAGNETIC-RELAXATION OF RB-85 IN SOME AQUEOUS SOAP SOLUTIONS	30	30

Table 3 (continued)

AUTHORS	JOURNAL	VOL.	PAGES	DATE D/M/Y	TITLE	REFS. CITED	TIMES CITED
SCHNEIDER JE, PRICE S, MAIDT L, GUTTERIDGE JMC, FLOYD RA	NUCLEIC ACIDS RESEARCH	18	631-635	11/02/1990	METHYLENE-BLUE PLUS LIGHT MEDIATES 8-HYDROXY 2'-DEOXYGUANOSINE FORMATION IN DNA PREFERENTIALLY OVER STRAND BREAKAGE	22	144
FLOYD RA, WEST MS, ENEFF KL, SCHNEIDER JE	ARCHIVES OF BIOCHEMISTRY AND BIOPHYSICS	273	106-111	15/08/1989	METHYLENE-BLUE PLUS LIGHT MEDIATES 8-HYDROXYGUANINE FORMATION IN DNA	29	159
VANDERPUTTEN WJM, KELLY JM	PHOTOCHEMISTRY AND PHOTOBIOLOGY	49	145-151	FEB. 1989	LASER FLASH SPECTROSCOPY OF METHYLENE-BLUE WITH NUCLEIC-ACIDS - EFFECTS OF IONIC-STRENGTH AND PH	15	13
OHUIGIN C, MCCONNELL DJ, KELLY JM, VANDERPUTTEN WJM	NUCLEIC ACIDS RESEARCH	15	7411-7427	25/09/1987	METHYLENE-BLUE PHOTOSENSITIZED STRAND CLEAVAGE OF DNA - EFFECTS OF DYE BINDING AND OXYGEN	33	65
LEE PCC, RODGERS MAJ	PHOTOCHEMISTRY AND PHOTOBIOLOGY	45	79-86	JAN. 1987	LASER FLASH PHOTOKINETIC STUDIES OF ROSE-BENGAL SENSITIZED PHOTODYNAMIC INTERACTIONS OF NUCLEOTIDES	29	155
RODGERS MAJ	JOURNAL OF THE AMERICAN CHEMICAL SOCIETY	105	6201-6205	1983	SOLVENT-INDUCED DEACTIVATION OF SINGLET OXYGEN - ADDITIVITY RELATIONSHIPS IN NON-AROMATIC SOLVENTS	29	230
MATHESON IBC, RODGERS MAJ	JOURNAL OF PHYSICAL CHEMISTRY	86	884-887	1982	OXYGEN-TRANSFER RATES ACROSS WATER MICELLE INTERFACES DERIVED FROM MEASUREMENTS OF NI-2+ QUENCHING OF SINGLET MOLECULAR-OXYGEN IN AEROSOL-OT HEPTANE REVERSE MICELLES	31	18
FOYT DC	COMPUTERS & CHEMISTRY	5	49-54	1981	A MINICOMPUTER-BASED SYSTEM FOR PROCESS-CONTROL, DATA ACQUISITION, AND DATA-ANALYSIS IN A DIVERSE FAST KINETICS FACILITY	19	121
RODGERS MAJ, BECKER JC	JOURNAL OF PHYSICAL CHEMISTRY	84	2762-2768	1980	ELECTRON-TRANSFER QUENCHING OF THE LUMINESCENT STATE OF THE TRIS(BIPYRIDYL)RUTHENIUM(II) COMPLEX IN MICELLAR MEDIA	32	94
SCHMEHL RH, WHITTEN DG	JOURNAL OF THE AMERICAN CHEMICAL SOCIETY	102	1938-1941	1980	INTRAMICELLAR ELECTRON-TRANSFER QUENCHING OF EXCITED-STATES - DETERMINATION OF THE BINDING CONSTANT AND EXCHANGE-RATES FOR DIMETHYLVIOLOGEN	30	89
YEKTA A, AIKAWA M, TURRO NJ	CHEMICAL PHYSICS LETTERS	63	543-548	1979	PHOTO-LUMINESCENCE METHODS FOR EVALUATION OF SOLUBILIZATION PARAMETERS AND DYNAMICS OF MICELLAR AGGREGATES - LIMITING CASES WHICH ALLOW ESTIMATION OF PARTITION-COEFFICIENTS, AGGREGATION NUMBERS	28	193
ALMGREN M, GRIESER F, THOMAS JK	JOURNAL OF THE AMERICAN CHEMICAL SOCIETY	101	279-291	1979	DYNAMIC AND STATIC A SPECTS OF SOLUBILIZATION OF NEUTRAL ARENES IN IONIC MICELLAR SOLUTIONS	41	550
ULMIUS J, LINDMAN B, LINDBLOM G, DRAKENBERG T	JOURNAL OF COLLOID AND INTERFACE SCIENCE	65	88-97	1978	H-1,C-13,CL-35, AND BR-81 NMR OF AQUEOUS HEXADECYLTRIMETHYLAMMONIUM SALT-SOLUTIONS - SOLUBILIZATION, VISCOELASTICITY, AND COUNTERION SPECIFICITY	43	111
KAMENKA N, FABRE H, CHORRO M, LINDMAN B	JOURNAL DE CHIMIE PHYSIQUE ET DE PHYSICO-CHIMIE BIOLOGIQUE	74	510-512	1977	INFLUENCE OF SOLUBILIZATION ON COUNTERION BINDING TO SURFACTANT MICELLES FROM SELF-DIFFUSION COEFFICIENTS	12	17
KAMENKA N, LINDMAN B, BRUN B	COLLOID AND POLYMER SCIENCE	252	144-152	1974	TRANSLATIONAL MOTION AND ASSOCIATION IN AQUEOUS SODIUM DODECYL-SULFATE SOLUTIONS	51	55
LINDMAN B, BRUN B	JOURNAL OF COLLOID AND INTERFACE SCIENCE	42	388-399	1973	TRANSLATIONAL MOTION IN AQUEOUS SODIUM OCTANOATE SOLUTIONS	46	99
LINDMAN B, DANIELSS I	JOURNAL OF COLLOID AND INTERFACE SCIENCE	39	349	1972	NUCLEAR MAGNETIC-RELAXATION OF RB-85 IN SOME AQUEOUS SOAP SOLUTIONS	30	30

(citing and being cited), a simple criteria of the articles closest in date to each step was used. Two or more articles bearing a similar year of publication and no other information, were to be treated by choosing the article closest to the original topic; albeit a result, it is reported here that this aspect did not arise.

The chosen article is highlighted with hash, also marked by arrow in the left margin, of the 3 data tables.

No physical searches were conducted and only those journals or articles electronically accessible were included in the exercise. Searches were conducted using the proxy server supported by the library system of the University of Toronto, Toronto, Canada. Journals outside the institutions' subscription base were discounted and the next closest article, accessible using the proxy server, was chosen.

When no further articles were electronically accessible, the search was considered complete, with an ending defined to that 'cascade'. This was assumed to be the expected 'end of the road' for 'cascades' moving backwards to articles published more than three to four decades ago. In the forward direction (articles citing the key article), terminus was defined as an article with no further articles citing it.

3. Results and discussion

The search of Internet sites on the World Wide Web was both time consuming and the result of some subjective decision. In many cases the sites accessed were easily categorized as a 'Link-hit', 'Non-Link-Hit' or 'Sales'. However, in a small number of cases a subjective decision was made on how to label the site. The study may be improved in the future by decision being made by a more extensive set of 'hit' criteria rules or else by the decision of a greater number of individuals. It is felt that a relatively good qualitative determination has been achieved, presented in Table 1.

The number of sites requiring a detailed description of the criteria allowing them to be counted as a 'hit', is too large and the specifics themselves too detailed to be included in this work. Readers are encouraged to contact the authors for specific discussion of this aspect of the results. However, the reader may easily conclude that majority of sites may be easily categorized using similar criteria through a simple attempt at these searches. The authors

encourage the readers to investigate this, perhaps in their own areas of expertise.

It should be noted that the results of this work will change over time as the ranking of sites by each search engine changes with time due to a number of factors and elements not covered in this work. The basis and search criteria of the search engines themselves are highly dynamic and change over relatively short periods of time.

Many conclusions may be drawn from Table 1, however, the authors leave the majority of this to the reader, due to the highly subjective nature of the searches. The trend for Vitamin-E and Vitamin-C to far surpass Lycopene in 'Link' for category 1 shows that the general public has been made much more aware by these compounds' publicity or 'societal saturation'. Similarly, the less known Lycopene is more prone to have more 'Sales' hits as companies compete with educational and informative websites to bring the compound to the masses. Institutions established for the scientific investigation of Vitamin-E and Vitamin-C allow for the educational element to have a robust backing against marketing and sales. The inverse relationship is observed in category 2, as Lycopene's relative anonymity in the general public allowed sites bearing scientifically valid information to outrank those that have used it in marketing or sales.

The results of the journal cascades for Vitamin-E, Vitamin-C and Lycopene are listed in Tables 2–4, respectively. The results of this exercise are self-explanatory and investigation of the tables shows that the publicity of the lifeworks of Professor Szent-Gyorgi and Professor Pauling have an overwhelming impact on the transfer of information from publication to the general public. Lycopene shows a limited publication record, with no articles containing the compounds name in their titles in any of the following journals: Nature, Science, Cell, Journal of the American Chemical Society, and Cancer. Only Cancer Research showed an article with the word Lycopene in the title, itself being a proof of the limited amount of transfer of information, cited only one time by an article itself never cited.

In Tables 2–4 cells were shaded anytime an author appeared in adjacent publications or the same journal is a neighbouring source of citing or cited reference. Cells in title column were shaded for articles with

Table 4

AUTHORS	JOURNAL	VOL.	PAGES	DATE D/M/Y	TITLE	REFS. CITED	TIMES CITED
GIOVANNUCCI E	JOURNAL OF THE NATIONAL CANCER INSTITUTE	91	1331-1331	04/08/1999	TOMATOES, TOMATO-BASED PRODUCTS, LYCOPENE, AND PROSTATE CANCER: REVIEW OF THE EPIDEMIOLOGIC LITERATURE - RESPONSE	7	0
GANN PH, MA J, GIOVANNUCCI E, WILLETT W, SACKS FM, HENNEKENS CH, STAMPFER MJ	CANCER RESEARCH	59	1225-	15/03/1999	LOWER PROSTATE CANCER RISK IN MEN WITH ELEVATED PLASMA LYCOPENE LEVELS: RESULTS OF A PROSPECTIVE ANALYSIS	50	113
YOSHIZAWA K, WILLETT WC, MORRIS SJ, STAMPFER MJ, SPIEGELMAN D, RIMM EB, GIOVANNUCCI E	JOURNAL OF THE NATIONAL CANCER INSTITUTE	90	1219-1224	19/08/1998	STUDY OF PREDIAGNOSTIC SELENIUM LEVEL IN TOENAILS AND THE RISK OF ADVANCED PROSTATE CANCER	36	111
CLARK LC, DALKIN B, KRONGRAD A, COMBS GF, TURNBULL BW, SLATE EH, WITHERINGTON R, HERLONG JH, JANOSKO E, CARPENTER D, BOROSSO C, FALK S, ROUNDER J	BRITISH JOURNAL OF UROLOGY	81	730-734	MAY 1998	DECREASED INCIDENCE OF PROSTATE CANCER WITH SELENIUM SUPPLEMENTATION: RESULTS OF A DOUBLE-BLIND CANCER PREVENTION TRIAL	21	126
PARKER SL, TONG T, BOLDEN S, WINGO PA	CA-A CANCER JOURNAL FOR CLINICIANS	46	5-27	JAN.-FEB. 1996	CANCER STATISTICS, 1996	11	978
FEUER EJ, WUN LM, BORING CC, FLANDERS WD, TIMMEL MJ, TONG T	JOURNAL OF THE NATIONAL CANCER INSTITUTE	85	892-897	02/06/1993	THE LIFETIME RISK OF DEVELOPING BREAST-CANCER	20	155
FREY CM, MCMILLEN MM, COWAN CD, HORM JW, KESSLER LG	JOURNAL OF THE NATIONAL CANCER INSTITUTE	84	872-877	03/06/1992	REPRESENTATIVENESS OF THE SURVEILLANCE, EPIDEMIOLOGY, AND END RESULTS PROGRAM DATA - RECENT TRENDS IN CANCER MORTALITY-RATES	12	48
DEVESA SS, SILVERMAN DT, YOUNG JL, POLLACK ES, BROWN CC, HORM JW, PERCY CL, MYERS MH, MCKAY FW, FRAUMENI JF	JOURNAL OF THE NATIONAL CANCER INSTITUTE	79	791-770	OCT. 1987	CANCER INCIDENCE AND MORTALITY TRENDS AMONG WHITES IN THE UNITED-STATES, 1947-84	184	294
WHITE E	AMERICAN JOURNAL OF PUBLIC HEALTH	77	495-497	APR. 1987	PROJECTED CHANGES IN BREAST-CANCER INCIDENCE DUE TO THE TREND TOWARD DELAYED CHILDBEARING	36	54
RUSSO IH, ALRAYESS M, SABHARWAL S	PROCEEDINGS OF THE AMERICAN ASSOCIATION FOR CANCER RESEARCH	26	116-116	26/03/1985	EFFECT OF CONTRACEPTIVE AGENTS ON MAMMARY-GLAND STRUCTURE AND SUSCEPTIBILITY TO CARCINOGENESIS	0	1

the compounds' common or scientific name (i.e. ascorbic acid) appearing in the title. Finally, the titles: 'Refs. Cited' and 'Times Cited' were highlighted anytime the latter had a greater count.

Again, analysis is left in the reader's hands as the amount of data collected is insufficient for the authors to make non-subjective claims of any trends observed. A larger data set is necessary for this task and will be included in future works.

4. Conclusions

Albeit a qualitative study at best, this work perhaps represents a preliminary attempt to characterize the mechanisms retarding informational transfer from pure science to the general public. The results signal the need for this type of investigation to both continue and to evolve to accommodate other influential variables. Relationships between works citing and cited by a particular article must be validated and confirmed as being genuine.

Acknowledgements

The Global Institute Of COmputational Molecular and Materials Sciences (GIOCOMMS) is thanked for generous support in making possible the gathering, compiling and presentation of this work.

The authors wish to thank Lucha Alforque, for helpful discussions, searches and tabulation. A special thanks to Elena Luzi for making this works possible in an international setting, allowing work to be completed across any border.

The authors thank GIOCOMMS for allowing this and all other related works to have been pursued and completed in all of the following settings: Toronto, Ottawa and Montreal, Canada; Budapest and Szeged, hungary; Paris and Strasbourg France; Milano and Napoli, Italy; Girona, Spain; New York City and Omaha, USA. The 'travelling nature' of all of GIOCOMMS work truly attest to the versatility of the World Wide Web and the future need to 'sheppard' informational transfer from the internet to the general public.

ELSEVIER

Journal of Molecular Structure (Theochem) 666–667 (2003) 707–711

THEO
CHEM

www.elsevier.com/locate/theochem

Towards a true community of scholars: undergraduate research in the modern university

Kenneth Bartlett*

Department of History, Victoria College, University of Toronto, Toronto, Ont., Canada M5S 3J3

Abstract

A research intensive university should not divorce the function of teaching undergraduates from the imperative to do innovative research. The two responsibilities are in many ways complementary and the ability of the researcher to integrate his or her students into the collective enterprise of research adds an important dynamic to the best schools.

The University of Toronto has for 7 years practiced this principle through its Research Opportunity Program in the Faculty of Arts and Science. Each year up to 180 carefully selected, gifted second year students are permitted to work in the research projects of professors for course credit. The teams vary according to discipline but all recognize that the role of the undergraduates is to contribute fundamentally to the progress of the project.

The results have been remarkable, as is evident from the posters displayed at this conference. Students have co-authored papers, participated in important international meetings and continued to do outstanding graduate work later in their academic careers.

Undergraduate research, then, provides an instrument for mentoring and modelling; it also permits the best among our students to sample the vivid excitement of cutting edge investigation, to travel into the scientific unknown. Undergraduate research can constitute one of the glues that cements the academy together, by linking in a great intellectual chain of knowledge every element of the human potential of the institution. It is an opportunity, which can help define a great research school and can simultaneously add lustre to the enterprise of research by involving every constituency.

© 2003 Elsevier B.V. All rights reserved.

A research intensive university should not divorce the function of teaching undergraduates from the imperative to do innovative research. The two responsibilities are in many ways complementary and the ability of the researcher to integrate his or her students into the collective enterprise of research adds an important dynamic to the best schools.

The University of Toronto has practised this principle since 1995, the year the Research Opportunity program was introduced into the Faculty of Arts and Science. The ROP (299Y) is designed to permit students in their second year of undergraduate studies to work in the research program of a professor for course credit. Students must be in their second year to ensure that more senior candidates do not exclude the target population of the program. Second year was chosen so that students who had enjoyed the benefits of a First Year Seminar (199Y) could continue in their studies through a program unique in Canada. The First Year Seminars introduced newly admitted students to the idea of curiosity driven investigation by encouraging senior members of the academic community to offer small (no more than 24 students) interactive seminars on subjects of their own choice, with no or little connection with the usual introductory

* Tel.: +1-416-946-8534; fax: +1-416-946-8588.

E-mail address: kenneth.bartlett@utoronto.ca (K. Bartlett).

0166-1280/$ - see front matter © 2003 Elsevier B.V. All rights reserved.
doi:10.1016/j.theochem.2003.08.106

curriculum. Thus, these first year courses could not be counted towards a student's major, minor or specialist certification and this flexibility was advertised by giving the seminars a generic (Humanities, Social Science, Science or Interdisciplinary) rather than a departmental designator (such as Chemistry, Mathematics or Classics). In this way, our colleagues could design courses which suited their personal interests, courses with titles such as 'The Physics of Poetry and the Poetry of Physics', 'The Riddle of Light', 'Sport as Culture, Culture as Sport', and 'Science and Social Choice'. There are now 91 of these courses available and all introduce students new to university life to the fundamental idea that learning can be exciting, stimulating and meaningful. To have a professor of French teach a wonderful course on science fiction or a professor of Weimar German culture contribute to a course in environmental issues reveals most effectively to young minds about to begin the rigorous process of academic studies that it is possible to pursue intellectual interests beyond the confines of a prescriptive curriculum.

The ROP, then, does enjoy a reasonable platform on which to build. It is, however, a much smaller program, offering an average of 75 projects with about 160 places of students. It is a very competitive program and designed to attract an elite of second year students who see their future in research, graduate studies or some form of independent study. The process by which the students are chosen illustrates the self-selecting and demanding admission instruments governing the program. By mid-February of the student's first year of study the binder containing all of the project descriptions is available in the ROP Office, registrarial and departmental offices and elsewhere on campus. These projects are short descriptions of the research plan of a faculty member or research team. They contain not only the general purpose and methodology of the project but the requirements for the ROP students, including the kinds of skills needed and the nature of the student's contribution. Each project entry must be signed by the department chair to guarantee that necessary resources, such as lab space, equipment, materials and time will be provided to the students in order to fulfil their responsibilities. In general, it is expected that each student will invest the same amount of time in a 299 course as with any other second year

course, that is, about 8–10 h per week. The reality is, though, that the students become so involved in the project that the problem is to get them away from their projects and out of their labs so that they can meet their other obligations.

After reviewing the project descriptions and students' responsibilities in the binder, students then apply for admission into a ROP course. Students can apply for up to five 299Y places but can only enrol in one. Because the first year grades are not yet available to applicants, mid-term assessments, letters of reference and personal testimonials constitute the addenda to many applications, freely offered by students to increase the chances of their being summoned to the next step, the interview. As with all research teams it is important that the principal investigators be convinced that they have chosen the right candidates from the many applicants. This is especially true where there has seldom been an opportunity to observe the students in senior work at close quarters, and where reliable grades are not available. The interview, then, is the mechanism by which an informed choice is made.

All applications from students are sent to the project supervisor who then reads through the material and selects a short list of candidates to interview. These are called for appointments and a final selection is made based on the usual criteria for admission to challenging programs: preparation, enthusiasm, perceived ability and collegiality. It can be said that the process has worked extremely well until now, as this personal contact has resulted in very successful student/supervisor collaboration. In the eight years of the program there has been only a small number of complaints about either students or supervisors. The teams have been well constituted and students have learned early about the value of cooperation and collegiality.

With a final list of selected candidates the supervisor then has each chosen student sign a contract, which my office sees as binding on both parties. The contract specifies the students' role in the project, the projected time commitment and the means by which the student is assessed. As a grade is to be assigned to the 299Y, which operates like any other full year credit course and constitutes one of the 20 courses required for a degree. The signed contracts are then copied, with one copy given to the student, one kept by the supervisor, and one delivered to the departmental office and one to the ROP Office. These files are required in case there should be a dispute about any aspect of the course. 299Ys are, unlike First Year Seminars, eligible to be used for major, minor or specialist certification in a degree program and the grade will be averaged in with those of other courses to determine the student's grade point average. The point is bluntly made by my office that ROP students must be given work appropriate to a second year course. This means that no low level assignments, such as washing test tubes, returning library books, light housekeeping or purely clerical functions can represent all of a student's work. We all do some measure of this kind of daily drudgery; but the point of the program is that the students should participate in cutting edge research and understand the nature of the research model. With all of this recorded, the student has become a member of the professor's research team.

I am often asked whether the supervisor is given teaching credit for supervising 299 students. The answer is no, because the role is that of research team leader, despite the need to assign a grade. The students in the ROP represent very capable research assistance in the project; consequently, the reward is not a hour or two on the professor's worksheet but the value of the research done for the project. Rather than pay a research assistant, then, the professor receives work from a keen young person whose reward is the experience, acquisition of skills and a credit. The bargain is in fact a good one, as witnessed by so many of our most eminent scholars and scientists. Another benefit is that the lab or project team supervisor, often a graduate or post doctoral student, gets experience in teaching and managing undergraduates thereby learning the necessary skill of making complex problems clear through effective description and demonstration. The 299s have become instruments to research teaching styles and methods as much as extending knowledge. The diversity of the many constituencies that often constitute ROP teams serve almost as microcosms of the university itself, and the culture revealed in the group a reflection of the larger culture of the academy. It is in part for these reasons that the experiment has been so successful and that some of our supervisors return again and again for new students to train and engage. Equally, a fortuitous secondary effect of the program has been that many

students continue to study with their supervisors, work as paid assistants in their labs or during the summer in the libraries or archives; they continue on to do graduate work and use in their own lives the lessons acquired in their ROP experience.

Another vehicle for the intellectual development and acquisition of communication skills of the ROP students is the opportunity to participate in the Annual ROP Research Fair held each March. This event is altogether a student centred event. Although my office makes all of the logistical arrangements from the rental of boards for posters, computer links, refreshments and room bookings, the content is determined by the ROP students themselves, with, of course, assistance from their supervisors. The Research Fair is an important occasion in the university's calendar, attracting a large audience of invited guests, from the students' family and friends to the president and/or provost of the University of Toronto. First Year Seminar students are specifically encouraged to come as are senior secondary school students who participate in the university's mentoring program. The purpose of these invitations is to advertise the ROP to excellent potential students interested in pursuing research careers and encouraging them to apply to the University of Toronto. It becomes clear to these target students as a result that the university might be a large but that opportunities are available that they will find no where else in Canada.

The Research Fair is a large and complex event with about 35 projects participating each year. The students design posters or demonstrations to illustrate their participation in the research project. They are present to respond to questions and to guide visitors through their material. Those supervisors who choose to attend observe how well the students explain their work, providing another means of assessment and the evidence to reinforce letters of reference and other instruments to help mentor ROP students through their careers. Often there has been media coverage of the event as well, broadening the reach of the ROP and attracting even more highly motivated students to the University of Toronto and to the ROP. Without doubt, the Research Fair is an annual highlight of the program, as it focuses the students' enthusiasm and knowledge and prepares them for the demanding task of linking their research to teaching by requiring each of them to explain often very complex experimental design and results to a general audience.

What has emerged from the ROP experience over the past eight years is a recognition that a research intensive university must use every mechanism possible to link closely its research and teaching functions. And, given the very large number of carefully selected undergraduates in comprehensive publicly supported universities, this function must be extended to this population as well. The experience our best students enjoy during their work for their first degree will determine to a great extent their desire to continue on to graduate studies and research careers. Consequently, it is both prudent and reasonable that some form of undergraduate research program be established to ensure an adequate pool of highly motivated, exceptional candidates. Moreover, the undergraduate research experience helps students themselves decide whether a career in research is suited to them. Rather than engage in the wasteful and extremely expensive experiment of admitting students to doctoral studies with the assumption that some significant percentage will not complete their degrees, an undergraduate research program serves as a reliable pre-screening for potential talent. These programs are very much in the interests of both universities and the governments that support them, as they help in ensuring that graduate research funds are more securely invested.

In addition, the ROP functions extremely well as a mentoring system for highly talented students. Many candidates for ROP positions were first encouraged in First Year Seminars. This occurred not necessarily in the discipline the student subsequently identified for a ROP application; but the recognition of talent, the degree of self-confidence and the reinforcement of what are often inchoate student ambitions all foster the belief that a life as a research scientist or scholar is possible. Students involved in undergraduate research are usually in the first instance self-selecting: this is reasonable and as it should be. Nevertheless, we must step back and ask what elements of their personality and experience conditioned them to engage in this act of self-selection, self-promotion. Often it was mentoring by a senior faculty member or even a youthful energetic junior instructor who served as both a validator of the student's dreams and as a source of practical and useful advice as to how to achieve those dreams.

It should be apparent, then, how much universities and their research faculties benefit from introducing undergraduate research courses or internships into their curricula. The contribution of skilled, enthusiastic labour and fresh perspectives can only benefit the research enterprise. To be sure the training of undergraduates for research positions is a longer process and more complex than required for graduate students or post-doctoral fellows; and much of this training has to be front loaded at the beginning of the academic term, the busiest time for our colleagues. Also, the need to provide assessment based on some form of objective criteria takes time and thought; and the contributions of undergraduate students must be structured according to the academic year, so that grades can be assigned at the appointed times and progress to advanced levels of the student's program allowed. For these reasons some universities, especially in the USA, offer non-credit undergraduate research programs. These function much like the model at the U of T but students cannot count the experience towards their degrees, receiving at most an annotation on their transcripts or a letter describing their contributions. These are useful in gaining volunteer experience often required for entry into a professional faculty, such as medicine or law; and the other benefits of involvement in the research team-such as mentoring and skill acquisition-obtain equally. However, the commitment and recognition that derives from earning a credit towards a degree is missing, and it was for that reason that the University of Toronto decided on the credit model for its ROP.

Another question is why limit this opportunity to the second year. In addition to the need for there to be a continuation of the First Year Seminars already mentioned, there is the observation that students should have this experience at a time when they are still fashioning their academic programs. If this opportunity is so significant and can have such a profound effect on a student's choice of career and course of study, then it is imperative that it be enjoyed early enough in the student's education to permit flexibility. By third year a student's course of study is largely fixed; after second there is still a chance to develop in new directions. Also, the skills, knowledge and attitudes derived from undergraduate research can be applied to subsequent classes, including traditional lecture or seminar courses. To fulfil best a student's potential, it is obviously most efficacious to offer an ROP sufficiently early to have an effect on their senior years.

Have there been problems with individual students, supervisors or projects? Of course. No program involving about two hundred professors and students per year for eight years cannot be without difficulties. Some of these should be obvious: students who do not fulfil their obligations and drop out of the ROP in mid year, leaving the project supervisor to scramble for a paid replacement or a redeployment of human resources. There have been supervisors who have not adequately trained their students or who did not assess them appropriately or who simply did not supervise their undergraduates at all, treating second year students like post docs who could work independently. Grades have been challenged and personality disputes revealed. But, among the more than a thousand students who enrolled in the program the number of such problems has been extremely small. Remember that both supervisors and students are self-selecting: they choose to do this and undergo a rigorous process to match students with research projects. Professors have too much invested in their research to behave in a cavalier manner with any member of their team; and the kind of students who apply for 299Y courses are too ambitious and committed to be irresponsible. Indeed, most of the failure arose from circumstances beyond the control of the student or supervisor: illness, changing conditions of employment, or faults in communication. In such instances my office intervenes to help work out the best arrangement for both parties. Finally, when students sign their contracts they are informed in writing that this constitutes a serious commitment and should not be taken lightly, as the success of a professor's research and their fellow students' learning depends on mutual obligation and responsibility. Consequently, failures are few.

When troubles arise they seldom reach my office, however, as the program enjoys a peer counselling system in which ROP students from one year serve as peer mentors for the next generation as they enter the program. My office matches the students' interests and course selections to ensure that there will always be another student outside the program who can offer advice and share the experience of what it is to be a ROP student. This is a necessary function, as ROP students often work in isolation or one-on-one with their supervisors or through the mediacy of a graduate student or post doc who manages the experiment or team. Since there is no class culture to draw upon, and since some students are nervous discussing unsuccessful experiments, difficult assignments or inappropriate assessment with the research supervisor who will assign the grade for the ROP and serve as the referee and advocate of the student for future studies or employment, the university found it necessary to institute this peer advisory system. Through this peer mentoring function individual ROP projects and the program as a whole have developed a kind of culture of shared experience and firm friendships. In a large public university where classes can number literally in the thousands, this is an achievement in itself.

In conclusion, I cannot recommend the implementation of an undergraduate research program more highly. Its success at the University of Toronto has been illustrated over and over and its contribution to the research of our colleagues and the learning experience of our students remarkable. It presupposes a large and sophisticated research establishment, if it is to operate well and with a sufficient critical mass to attract the best students. It must be monitored and celebrated and participants should feel that they are an elite group making special contributions to the university and receiving special benefits. In short, it should be perceived as an elite program for elite students. This will aid in the process of self-identification and attract the very best candidates. Equally, project supervisors should be honoured both for their participation and for the confidence they show in their students. As a privileged group the students of the ROP will emerge as a model for others and exhibit a career path open to any others with equal energy and ability. The ROP has become a signature program of the University of Toronto, and the message it sends is exactly the message we want heard among all of our constituencies: teaching and research are integrally related, two elements of the same enterprise pursued through different means. It is the definition of a modern university dedicated to excellence in teaching and research, and that is how the University of Toronto defines itself.

ELSEVIER

Journal of Molecular Structure (Theochem) 666–667 (2003) 713–714

THEO
CHEM

www.elsevier.com/locate/theochem

Author Index

Almeida, D.R.P., 401, 521, 527, 537, 557
Andor, J., 159
Andreu, I., 109
Arslan, T., 609

Bachrati, C.Z., 681
Bartha, F., 117
Bartlett, K., 707
Beke, G., 667
Berente, I., 637
Bermejo, A., 109
Bernini, M.C., 587
Bogár, F., 41, 143
Bogar, F., 381
Borics, A., 355
Brijbassi, S.U., 291

Cabedo, N., 455
Chagraoui, A., 455
Chass, G.A., 61, 79, 99, 169, 243, 251, 273, 279, 291, 311, 321, 327, 355, 393, 409, 415, 431, 439, 445, 699
Chasse, G.A., 581
Choi, C., 99
Connor, A.A., 581
Cortes, D., 109, 455
Csermely, P., 373
Csiszár, Á., 681
Csizmadia, I.G., 11, 79, 89, 99, 131, 135, 143, 169, 219, 243, 251, 269, 273, 279, 285, 291, 311, 321, 327, 355, 393, 397, 401, 409, 415, 431, 439, 445, 521, 527, 537, 557, 581, 599, 651
Csontos, J., 515
Czibula, Á., 681

Datki, Z., 507
Demirayak, Ş., 609
Dörnyei, Á., 135

Éliás, O., 625, 667
Enriz, R.D., 109, 455, 587, 599
Erős, I., 667
Estrada, M.R., 599, 617

Farkas, Ö., 31, 51, 645
Fejér, S.N., 303
Fekete, Z.A., 159
Fülöp, F., 537, 557
Fung, D.Y.K., 651
Füsti-Molnár, L., 25
Fuxreiter, M., 469
Füzery, A.K., 269

Garro, J.C., 617
Gasparro, D.M., 401, 521, 527, 537, 557
Gráf, L., 481
György, K., 219

Hannus, I., 69, 687
Harmat, V., 625
Hermecz, I., 547
Hernadi, K., 451
Hetényi, C., 507
Hódi, K., 667
Hollósi, M., 693
Hon, A., 99
Horváth, G., 625
Hudáky, I., 285

Illés, T., 681
Imre, G., 51

Jenei, Z.A., 303
Jensen, S.J.K., 387
Johnson, A.P., 651, 659
Juhasz, J.R., 401, 521, 527, 537, 557

Kalászi, A., 645
Kálmán, M., 515
Kalmar, E., 373
Kandra, L., 487
Kapuy, O., 117
Károlyházy, L., 625, 667
Kehoe, T.A.K., 79
Keller, J.H., 409
Kiss, J.T., 163
Kónya, V.V., 397
Koo, J.C.P., 279, 285, 439, 445
Körtvélyesi, T., 159, 345
Kouznetsov, V.V., 587
Kozmutza, C., 95, 117

Labádi, I., 451
Laczkó, I., 693
Ladik, J., 1, 381
Lam, J.S.W., 279, 285
Láng, A., 219
Lau, S.K., 415, 431
Law, J.M.S., 279, 445, 651
Leitgeb, B., 337
Lengyel, I., 507
Liao, J.C.C., 321
Lovas, S., 169, 355, 415

Magyar, A., 163
Mandity, I.M., 143
Marai, C.N.J., 699
Mátyus, P., 625, 667
Meszaros, P.G., 397

doi:10.1016/S0166-1280(03)00969-2

ELSEVIER

Journal of Molecular Structure (Theochem) 666–667 (2003) 715–726

THEO CHEM

www.elsevier.com/locate/theochem

Subject Index

doi:10.1016/S0166-1280(03)00970-9

Appendix

Albert Szent-Györgyi
Introduction to a Submolecular Biology

Introduction
to a
Submolecular
Biology

Introduction
to a
Submolecular
Biology

ALBERT SZENT-GYÖRGYI

The Institute for Muscle Research
at the Marine Biological Laboratory
Woods Hole, Massachusetts

1960

ACADEMIC PRESS
New York and London

ACADEMIC PRESS INC.
111 Fifth Avenue, New York, New York 10003

United Kingdom Edition published by
ACADEMIC PRESS INC. (LONDON) LTD.
Berkeley Square House, London W.1

Library of Congress Catalog Card Number: *60-14265*

Third Printing, 1968

PRINTED IN THE UNITED STATES OF AMERICA

Acknowledgments

WHEN SETTLING DOWN TO WRITE THIS BOOK my thoughts turn gratefully to my associates who have given me part of their lives, sharing my work, hopes and disappointments.

My thanks go to those who gave me their confidence and help, making this work possible. In the first place, I am thinking of the Commonwealth Fund which helped me from the beginning in spite of my warning that it was supporting a gamble. I think of the Muscular Dystrophy Associations, the American Heart Association, the Association for the Aid of Crippled Children, and the Cerebral Palsy Association. My thoughts also wander back to those who helped me earlier, in my more troubled days, notably Armour and Company.

Last but not least, I am thinking of the people of the USA who gave me a new home, accepting me as one of their own, supporting my work through the National Institutes of Health, Bethesda, Maryland, and the National Science Foundation, Washington, D. C.

Abbreviations

~	High-energy bond (mostly phosphate bond).
A	Acceptor
ADP	Adenosine diphosphate
AMP	Adenosine monophosphate
ATP	Adenosine triphosphate
D	Donor
DPN	Diphosphopyridine nucleotide
EA	Electron affinity
EV	Electron-volt
ESR	Electron spin resonance
FAD	Flavine-adenine-dinucleotide
FMN	Flavine-mono-nucleotide
IP	Ionization potential
IR	Infrared
P	Phosphate
TPN	Triphosphopyridinenucleotide
UV	Ultraviolet

Contents

Part One

General Considerations

I

Introduction

THE BASIC TEXTURE OF RESEARCH CONSISTS OF dreams into which the threads of reasoning, measurement, and calculation are woven.

This booklet is the reincarnation of my "Bioenergetics"[1] which was hardly more than a dream. So, it was a surprise to me to find it translated into Russian by the Academia Nauk USSR, and to find that the introduction was written by A. Terenin, a leading figure of Soviet science. Two years later, in the fall of 1959, the Atomic Energy Committee organized a meeting, in Brookhaven, on "Bioenergetics," which gave to the problem the status of a more or less well-defined field of inquiry.

Since 1957 my thoughts have assumed a somewhat more definite form and even if I am unable to present final solutions, I am capable, at least, of asking a few questions more intelligently, weaving in a few threads of measurement, reasoning, and calculation. All the same, it is not without a great deal of anxiety that I

[1] "Bioenergetics," Academic Press, New York, 1957.

1

publish this booklet which will be the last instance of the repetitive pattern of my being driven into fields in which I was a stranger. I started my research in histology. Unsatisfied by the information cellular morphology could give me about life, I turned to physiology. Finding physiology too complex I took up pharmacology, in which one of the partners, the drug, is of simple nature. Still finding the situation too complicated I turned to bacteriology. Finding bacteria too complex I descended to the molecular level, studying chemistry and physical chemistry. Armed with this experience I undertook the study of muscle. After twenty years' work, I was led to conclude that to understand muscle we have to descend to the electronic level, the rules of which are governed by wave mechanics. So here again, I was driven into a dimension of which I had no knowledge. In earlier phases I always hoped, when embarking on a new line, to master my subject. This is not the case with quantum mechanics. Hence my anxiety.

I have not referred to my personal history as if, in itself, it would be of any importance. I have referred to it because a most important question hinges on it: should biologists allow themselves to be steered away from this electronic dimension because of their being unfamiliar with the intricacies of quantum mechanics? At present, the number of those who master both sciences, biology and quantum mechanics, is very small. Maybe it will never be very great owing to the limited nature of human life and brain. Both sciences claim a whole mind and lifetime. So, at least for the present, developments depend on some sort of hybridization.

2

In my opinion, at least temporarily, the best solution does not lie in the biologists crossing over into physics, and *vice versa,* but in the collaboration of biologist and physicist. For this it is not necessary for the biologist to acquaint himself with the intricacies of wave mechanics. It is sufficient to develop a common language with the physicist, get an intuitive grasp of the basic ideas and limitations of quantum mechanics, to be able to isolate problems for the physicist and understand the meaning of his answer. Similarly, the physicist had better stay on his side of the fence rather than become, perhaps, a second-rate biologist. If, for example, as a biologist, I am interested in energy levels of a substance, and am told that the highest orbital of a substance has a k value of, say, 0.5, I can start from this point. It is sufficient for me to know what $k = 0.5$ means, and there is no need for me to know exactly how the value was arrived at. In exchange, I can bring substances to the notice of the physicist, the k value of which might be of special importance.

There is only one warning I would like to give to the biologists who venture into physical problems. There is a basic difference between physics and biology. Physics is the science of probabilities. If a process goes 999 times one way, and only once another way, the physicist will not hesitate to call the first *the* way. Biology is the science of the improbable and I think it is on principle that the body works only with reactions which are statistically improbable. If metabolism were built of a series of probable and thermodynamically spontaneous reactions, then we would burn up and the machine would run down as a watch does if deprived of its regulators. The reactions are kept in hand by being statistically improbable and made possible by specific tricks which may then be used for regulation. So, for the living organism, reactions are possible which may seem impossible, or, at least, improbable to the physicist. When Tutankhamen's grave was opened, his breakfast was found unoxidized after three thousand years. This represents the physical probability. Had His Majesty risen and consumed his meal this would have been burned in no time. This is biological probability. His Majesty, himself, must have been a very complex and highly ordered structure of nuclei and electrons with a statistical probability of next to zero. I do not mean to say that biological reactions do not obey physics. In the last instance it is physics which has to explain them, only over a detour which may seem entirely improbable on first sight. If Nature wants to do something, she will find a way to do it if there is no contradiction to basic rules of Nature. She has time to do so.*

All this makes the relationship of physicist and biologist rather touchy. The biologist depends on the judgment of the physicist, but must be rather cautious when told that this or that is improbable. Had I always accepted the physicist's verdict as the last word, I would have given up this line of research. I am glad I did not. One can know a good theory from a bad one by the former's leading to new vistas and exciting experiments, while the latter mostly gives birth only to new theories made in order to save their parents.

* Living Nature also often works with more complex systems than the physicist uses for testing his theories.

4

Since I am working on my present line everything seems more colorful and I am even more eager to get to my laboratory in the morning than ever before.

The biologist, embarking on this line, is neither without help nor without encouraging examples. C. A. Coulson, in the first volume of the "Advances in Cancer Research" (Academic Press, New York, 1953) has written an admirably clear unmathematical review of the basic concepts of quantum mechanics. In the third volume of the same reviews (1955) A. and B. Pullman wrote equally clearly about complex indices. Those who like to read French may find a more comprehensive review by the same authors in "Cancérisation par les substances chimiques et structure moléculaire" (Masson, Paris, 1955). For details and some mathematics one may apply to B. and A. Pullman's "Les théories électroniques de la chimie organique" (Masson, Paris, 1952). The other extreme, a very brief and popular summary, is found in B. Pullman's "La structure moléculaire" (Presses Universitaires France, 1957). Naturally, L. Pauling's "The Nature of the Chemical Bonds" (Cornell University Press, 1948) should not be missing from one's desk, nor Th. Förster's "Fluoreszenz Organischer Verbindungen" (Göttingen 1951). Its next (English) edition is eagerly awaited.*

As encouraging examples of the fruitfulness of this field I would like to mention the bold pioneering stud-

---

* In spite of these splendid contributions it is difficult to deny that an up-to-date comprehensive treatment, written especially for the biologist, in a possibly unmathematical language, is badly needed.

ies of A. and B. Pullman which started with the study of the electronic structure of carcinogenic hydrocarbons, in which correlations of electronic structure and carcinogenic power were established, which may be one of the first major steps toward understanding cancer. If this booklet does not contain a chapter on this subject this is because the Pullmans have, themselves, given an account of their work which could not be equaled in clarity. The same authors have broken ground also in various other fields of quantum mechanical biology, establishing electronic indices for many biologically important catalysts,[2-4] and venturing even into the field of enzymes[5] and high-energy phosphate bonds.[6]

While the quoted examples may encourage physicists interested in biology, B. Commoner[7-10] and his associates may be quoted for their pioneering studies in electron spin resonance to cheer biologists interested in submolecular phenomena. It was this work

[2] B. Pullman and A. Pullman, *Proc. Natl. Acad. Sci. U.S.* **44**, 1197, 1958.

[3] B. Pullman and A. Pullman, *Proc. Natl. Acad. Sci. U.S.* **45**, 136, 1959.

[4] B. Pullman and A. M. Perault, *Proc. Natl. Acad. Sci. U.S.* **45**, 1476, 1959.

[5] A. Pullman and B. Pullman, *Proc. Natl. Acad. Sci. U.S.* **45**, 1572, 1959.

[6] B. Pullman, *Radiation Research*, 1960.

[7] B. Commoner, J. Townsend, and G. E. Pake, *Nature* **174**, 689, 1954.

[8] B. Commoner, J. J. Heise, B. B. Lippincott, R. E. Norberg, J. V. Passoneau, and J. Townsend, *Science* **126**, 3263, 1957.

[9] B. Commoner, B. B. Lippincott, and Janet V. Passoneau, *Proc. Natl. Acad. Sci. U.S.* **44**, 1099, 1958.

[10] B. Commoner and B. B. Lippincott, *Proc. Natl. Acad. Sci. U.S.* **44**, 1110, 1958.

which led to the first direct experimental evidence for the participation of free radicals in the one-electron enzymic electron transfer, as outlined in the classic studies of L. Michaelis.[11]

[11] L. Michaelis, Fundamentals of Oxidation and Reduction, *In* "Currents in Biochemical Research." D. E. Green (Ed.), Interscience Publ., New York, 1946.

7

II

Why Submolecular Biology? The Problem is Stated

LOOKED AT FROM A DISTANCE, THE HISTORY OF biochemistry seems to be but a series of astounding successes, a blaze of glory. The rate of progress shows no decrement and it looks as if soon we could strike out "don't know" from our vocabulary, altogether. Why, then, talk about "submolecular biology" until molecular biochemistry has run its full course?

There is no doubt about these successes. All the same, if one does not allow oneself to be blinded by them and approaches biochemistry with a dark-adapted eye, big gaps in our knowledge become evident. Let us consider some of the main problems of chemical biology, starting with metabolism. Biochemistry has unraveled the complex cycles of intermediary metabolism and has shown that the main object of this metabolism is to prepare the foodstuffs for their final oxidation in which their energy is used to couple one molecule of phosphate to ADP, producing, thereby, ATP (Fig. 19). In this process the energy of the food-stuff is translated into the energy of the terminal "high-energy phosphate bond," $\sim$, of a very specific

9

molecule. It is only in this form that the energy of the foodstuffs can serve as fuel for the living machinery and drive it. This "oxidative phosphorylation" is thus the central event of metabolism. Its mechanism is completely unknown.

We are equally ignorant about the reversal of this process, the release of the energy of the $\sim$ of ATP. How these $\sim$'s drive life, how their energy is translated into various forms of work, w, be they mechanical, electrical, or osmotic, we do not know, although this transformation may be the most central problem of biology. We know life only by its symptoms and what we call "life" is, to a great extent, but the orderly interplay of these various w's; since the dawn of mankind, death has been diagnosed, mostly, by the absence of one of these w's, expressing itself in motion. We do not know how motion is generated, how chemical energy is transformed into mechanical work.

Physiology has shown that the various functions of our body are regulated and coordinated by hormones and the biochemist will proudly show the row of vials containing these mysterious hormones mostly in the form of nice, crystalline powders, some of which might have been prepared synthetically. The same is true for the various vitamins, the catalogue of which seems near completion. The biochemist will be able to give us the structural formula of most of these substances. The really intriguing problem, however, is not what these substances *are*, but what they *do*, how they act on the molecular level, how they produce their actions. There is no answer to this question. The same holds true also for the majority of drugs.

As to the living machinery itself, the biochemist will

10

tell you that its central parts are proteins, nucleic acids, and nucleoproteins. He will point out the great progress made in the structural analysis of these substances, show their building blocks, amino acids and nucleotides, their links and relative position, will speak about bond angles and distances and the various helices formed. But, if we ask why Nature has put together that very great number of atoms in that very specific way, what property did she want to achieve, our biochemist will become silent. One of the basic principles of life is "organization" by which we mean that if two things are put together something new is born, the qualities of which are not additive and cannot be expressed in terms of the qualities of the constituents. This is true for the whole gamut of organization, for putting electrons and nuclei together to form atoms, atoms to molecules, amino acids to peptides, peptides to proteins, proteins and nucleic acid to nucleoproteins, etc. What Nature had in mind when doing this we cannot even guess at present. So here, too, we find the door to the central problem locked.

There are various circumstances which make this situation rather disturbing. First, these unanswered questions are the central and most intriguing problems of biology. Another rather disturbing fact is that corresponding to lacunas in our basic knowledge, there are lacunas in medical science and a great number of "endogenous" or "degenerative" diseases still rampage freely, causing endless suffering. But the most disturbing fact is that while biochemistry is still progressing in the fields where it has already been successful, it makes practically no progress in solving the problems mentioned. It looks as if the problems of biology could

11

be divided into two classes: those which current biochemistry can solve and those which it cannot. It looks as if something very important, a whole dimension, might be missing from our present thinking without which these problems cannot be approached.

There is no doubt in the author's mind what this missing dimension is. The story is simple and logical. Biochemistry came into bloom at the end of the last century. At that time, matter was thought to be built of very small, indivisible units, atoms. Molecules were the aggregates of these atoms. There were about 90 different sorts of atoms which were symbolized by various letters, while their links were denoted by dashes. No doubt, this letter-and-dash language ranks among the greatest achievements of the human mind and is responsible for all the amazing successes of biochemistry. If we go through the list of problems enumerated above, we will find that the ones with which biochemistry was successful were problems of structure, or changes of structure taking place in simpler reactions which could be duplicated mostly in homogeneous solutions, and could be expressed and answered in terms of letters and dashes, while the problems which remained unanswered were problems of function of complex systems which cannot be expressed in this language. How could a reaction such as muscle contraction, the main product of which is not a substance, but work, w, be expressed in these terms?

The language of current biochemistry is still that of letters and dashes which means that this science is still moving in the same molecular dimension as it was moving at its birth in the last century. But since that

12

time its parent science, chemistry, allying itself with physics and mathematics, made a dive into a new dimension, that of the submolecular or subatomic dimension of electrons, a dimension the happenings of which can no longer be described in the terms of classic chemistry, the rules of which are dominated by quantum, or wave, mechanics. Looked at through the glasses of this new science the atom is no more an indivisible unit but consists of a nucleus surrounded by a cloud of electrons with varying and fantastic shapes, and it seems likely that the subtler phenomena of life consist of the changing shapes and distributions of these clouds.

Biochemistry did not follow its parent science, chemistry, into this new subatomic dimension, which may hold the key to the understanding of the subtle biological functions. An example may illustrate the point. On the left side of Fig. 1 stands the classic

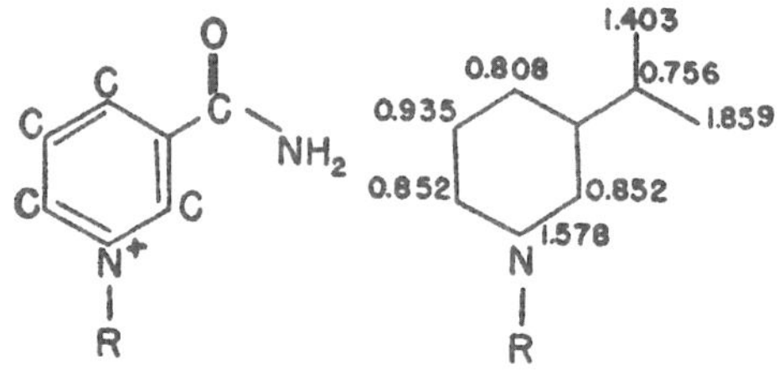

FIG. 1. Classic formula and molecular diagram of the pyridine end of DPN.

formula of the pyridine end of DPN expressed in classic symbols. It tells us that the pyridine ring is built of five equal C atoms and an N which has a positive charge. On the right side of the same figure is the "molecular diagram" of the same substance, as found

13

in a recent publication of the Pullmans.* The numbers coordinated to each atom indicate the electric charge. They tell us that each atom has a different charge and the molecule is thus surrounded by an electronic cloud of very complex structure. The positive charge is divided unequally over the one N and five C atoms of the ring while the negative charges are relegated to the side chain. This figure should be completed by three more sets of numbers, one set giving information about what is called the "free valency" of the single atoms of the ring, the other describing the "bond-order" of the links, and the third giving the "localization energies." While the classic formula attributed to the whole molecule but an overall shape and a dipole moment, in the molecular diagram every atom of the ring assumes a personality, a profile, a high degree of specificity and the whole structure begins to assume that subtlety which we can expect from any structure taking part in biological reactions.

While atoms and molecules were revealed to be complex little universes, their strict individuality has been broken down by "solid state physics." If many atoms form a regular and closely packed system, they may develop new properties. If, for instance, a great number of copper or iron atoms get together in a specific order they may develop electric conductivity, which is a collective property due to the interaction of the wave mechanical properties of the single units. Even macromolecules may develop solid state properties. So, in order to approach the central problems of biology we have to extend our thinking in two opposite directions, into both the sub- and supramolecu-

* Ref. 3, page 6.

14

lar. The two, in a way, are identical, the supramolecular qualities being but the collective action of the submolecular factors, supplying a new example of "organization." Similarly, we can expect entirely new properties to develop also when these molecules or molecular aggregates interact with the general matrix of life, water, forming with it a new and unique system. The elucidation of all these interrelations may eventually lend to our thinking the plasticity which may be necessary to approach life and the meaning of that unique system called the "cell."

The approach to these new dimensions may be a difficult one, and many of the ideas to be presented here may seem hazy and doubtful. The unknown offers an insecure foothold. What admits no doubt in my mind is that the Creator must have known a great deal of wave mechanics and solid state physics, and must have applied them. Certainly, He did not limit himself to the molecular level when shaping life just to make it simpler for the biochemist.

III

The Energy Cycle of Life

IN OUR FIRST APPROACH WE WILL DO WELL TO take a broad view. The broadest view we can take of energetics of life consists of considering the whole living world, trying to see how energy drives it. It is common knowledge that the ultimate source of this energy is the radiation of the sun. If a photon, ejected by the sun, interacts with a material particle on our globe, it lifts an electron from an electron pair in the ground state to a higher empty orbital, as symbolized by the upward pointing arrow in my Fig. 2,A. As a rule, the electron drops back within a very short time to its ground level, as symbolized by the downward pointing arrow. Life has shoved itself between these two processes and makes the electron drop back within its own machinery, utilizing its energy, as symbolized by the semicircle in Fig. 2,B. In order to do this efficiently it must meet the electron with a specially built substance (mostly chlorophyll) and couple this substance to a system which converts the very labile electronic excitation energy into a more stable chemical potential, into chemical energy, that is the energy of

17

a system of electrons of a stable substance. According to our present knowledge this is done, to a great extent, by using the energy of the excited electron to separate

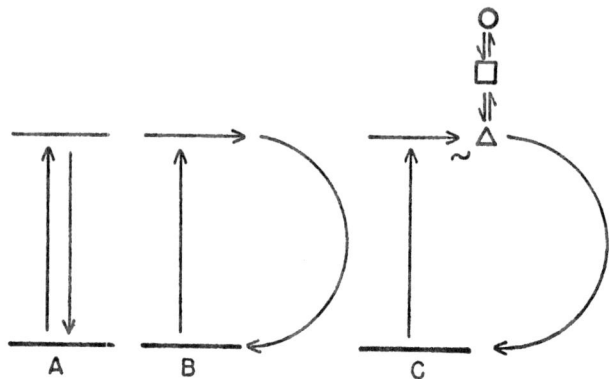

Fig. 2. Symbol of photosynthesis. (See text.)

the elements of water, H_2O.* The O is sent as O_2 into the atmosphere while the H becomes coupled to a pyridine nucleotide, TPN+ or DPN+, which thus becomes reduced into TPNH or DPNH, as shown by the work emanating from the laboratory of D. Arnon[12] and others. Simultaneously, also, ATP is formed from ADP and P, converting part of the energy into that of the terminal "~" of ATP. The pyridine nucleotide is symbolized in my Fig. 2,C, by a triangle, while the square and circle above it express the fact that pyridine nucleotides and ATP are unfit to store energy in quantity and so their energy may be converted into other forms more fit for storage. This is done by absorbing

* I will consider here only the classic "open" cycle of photosynthesis. For the simpler and "closed" forms see D. I. Arnon.[12]

[12] D. I. Arnon, *Nature*, **184**, 10, 1959.

18

CO₂ from the atmosphere and reducing it to carbohydrate and water, as symbolized by the left side of
my Fig. 3, the square standing for carbohydrate, the
circle for fat.

All this requires much time and a bulky apparatus,
and so we let the plants do it and then eat the plant,
or eat the cow which has eaten the plant. This, our
eating, is symbolized in Fig. 3 by the horizontal arrows,
while on the right side of the same figure I tried to
symbolize, in a most sketchy and symbolic manner,
what we do with these substances. We transfer the H
atoms of the carbohydrate mostly back onto a pyridine
nucleotide, releasing the C in the form of CO_2. Vennesland, Westheimer, and their associates,[13,14] by the isotope technique, have actually identified the H de-
tached from metabolites with the H coupled onto the
pyridine nucleotide. The DPNH or TPNH, in their
turn, reduce flavine-nucleotides (FMN) but the H's
found on $FMNH_2$ could no longer be identified with
those of DPNH or TPNH and so it seems likely that
what is transferred from these to FMN are not H's but
are electrons, and the H's found on $FMNH_2$ in test
tube experiments are derived from the universal solvent, water, from which the negatively charged FMN^-
captured protons. Kosower et al.[15] have shown DPN
to be a good electron donor.

DPN⁺ needs for its reduction to DPNH one H and
one electron. It can thus take up (and give off) both

[13] H. F. Fisher, E. E. Conn, B. Vennesland, and F. H. Westheimer,
    J. Biol. Chem. 202, 687, 1953.
[14] F. A. Loewus, F. H. Westheimer, and B. Vennesland, J. Am.
    Chem. Soc. 75, 5018, 1953.
[15] E. M. Kosower and P. E. Klinedinst, J. Am. Chem. Soc. 78,
    3493 and 3497, 1956.

19

H's and electrons; it is, so to speak, an exchange counter at which the H's of the foodstuffs can be exchanged
for electrons, which then are sent down the oxidative
chain over FMN. From FMN the electrons go from
substance to substance, of which I quoted, in Fig. 3
as an example, cytochrome b, c, and a.* Eventually,

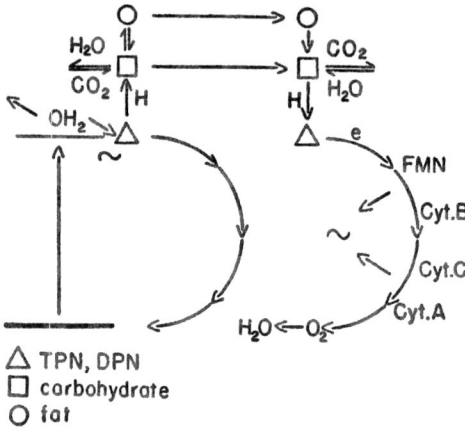

Fig. 3. Symbol of photosynthesis. (See text.)

the electron is taken up by the O_2 which then binds H
ions and is thus reduced to H_2O. In H_2O the electron
reaches its lowest energy level having given up its
energy stepwise. The energy thus given up is converted
with some loss into the energy of the $\sim$ of ATP.

---

*Electrons are sent down the chain also via succinate which
mediates between this chain and the citric acid cycle. For simplicity's sake this has been omitted from the figures, as well as other
intermediary catalysts, as the Q coenzyme and the free Fe which,
according to D. E. Green[16] also may play an important role in
electron transmission.

[16] D. E. Green, *Adv. Enzymol.* **21**, 73, 1959.

20

The point which I want to bring out is that apart from the process of excitation (vertical arrow), the right-hand side and left-hand side of my Fig. 3 are essentially identical. Being separated only by horizontal arrows which represent our (theoretically) uninteresting eating, we can pull the two sides together, as I have done in Fig. 4. Since carbohydrate and fat are

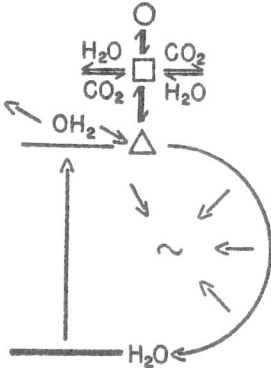

Fig. 4. Symbol of photosynthesis. (See text.)

but side lines, we can leave them off too, and complete the sketch with the main actor of this drama, the good old sun, as has been done in Fig. 5. In this figure we have thus the essentials of the energy cycle of life, which consists of electrons being boosted up by photons, and then dropping back to their ground level through the living systems, giving up gradually their excess energy which then drives the living machinery.

This Fig. 5 contains nothing beyond common knowledge. All the same, it helped me to clarify my mind about three essential points which form the cornerstones of my thinking. First, it made me see that life

21

is driven by nothing else but electrons, by the energy given off by these electrons while cascading down from the high level to which they have been boosted

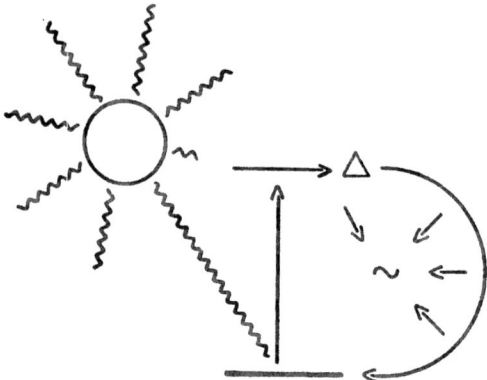

Fig. 5. Symbol of photosynthesis. (See text.)

up by photons. An electron going around is a little current. What drives life is thus a little electric current, kept up by the sunshine. All the complexities of intermediary metabolism are but lacework around this basic fact. The second point is that what is left behind by the electrons is but ATP and DPNH or TPNH. So these substances have to be the real fuel of life. My third point is that in this cycle the electrons go it alone. They are boosted up, one by one, and go through the cytochrome series, one by one, the central Fe atom being capable of a monovalent change only. It seems likely to me that they go alone throughout the whole cycle. I will finish this chapter by telling why these points are so important for me.

I have to go back to my school days when I was taught that life is driven by burning processes and the

22

oxidative energy is released in reactions taking place between colliding molecules. If molecule X collides with molecule Y and oxidizes it, the free energy of the system is decreased by a ΔF which was supposed to drive life. X and Y, as other molecules, are closed systems and so is XY formed during their interaction. I could never see how a change taking place in a closed system can drive anything outside. It is like energy released in a closed box. So the ΔF remained for me but an item for my thermodynamic bookkeeping and not something with which one can buy something.[*] The dropping electron suggested by Fig. 5 is some-

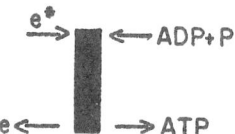

Fig. 6. Symbol of oxidative phosphorylation. (See text.)

thing different: it is a little current by which I could drive anything, a dynamo or a muscle machine, or produce "~'s" by it. In order to produce ~ with it I would simply have to put a "black box"[†] (Fig. 6) between its two levels in question, into which box I would feed high-energy electrons, ADP, and P on top

---

[*] The thermodynamic bookkeeping is most important because it keeps us from talking nonsense. Our "debit" and "credit" accounts must tally and the final figures must conform to the two basic laws, the first of which tells that you cannot get something for nothing, and you cannot overdraw your banking account, while the second warns you that there is an overhead charge for any transaction. So thermodynamics only tells us whether a reaction is possible or not but tells us nothing about its nature or mechanism.

[†] By "black box" is meant, according to current usage, some unknown reaction which one hopes to clarify later.

to pull out ATP and a tired electron at the bottom. In order to drive life later with my ATP I would only have to turn the box upside down, feed in ATP at one end to pull out a high-energy electron (and ADP + P) at the other. So the question, how ATP can drive life, can produce contraction in muscle, narrowed down the problem for me, for instance, to how the energy of its $\sim$ can be transformed into electronic energy (see Chapter X).

The electrons "going alone" also need some additional explanation. At school I was taught that organic reducing agents are substances which can give off two H's or two electrons while oxidizing agents can take up two H's or electrons, have thus two stable states which differ from one another by two H's or two electrons.[*] To quote Michaelis and Schubert[17]: "Usually, in organic compounds, reduction is bivalent, that is it involves taking up two electrons. This is because almost without exception, the compounds which we have considered as stable organic molecules contain an even number of electrons."

What matters here is that in such a bivalent redox process the two molecules, oxidant and reductant, meet, assume a new stable configuration after having lost or gained two electrons, then part. I was unable to see how the ΔF of such an encounter with the subsequent internal rearrangement of molecular structure

[*] H's and electrons are equivalent since a negative charge can easily be exchanged for an H by capturing a proton, and vice versa: $A^{2-} + 2H^+ \rightleftharpoons AH_2$.

[17] L. Michaelis and M. P. Schubert, *Chem. Rev.* 22, 437, 1938. See also ref. 11, on page 7.

24

could drive anything. This situation was not changed by the discovery of monovalent oxidation by Michaelis. To use his words: "all oxidations of organic molecules, *although they are bivalent,* proceed in two successive univalent steps, the intermediate state being a free radical.[18,19] In Michaelis' idea the situation is thus analogous to that in Noah's ark where the animals had to go two-by-two. Michaelis allows them only to pass the bridge one-by-one.

The distinction is important. A bivalent oxidoreduction is a classic chemical reaction which involves a rearrangement of molecular structure while a monovalent electron transfer, an electron going it alone, is a little current which does not necessarily involve such rearrangement. The sideline in Fig. 4 is what is usually summed up as "intermediary metabolism" and is composed of a host of classic chemical reactions, and will not occupy us further; but what goes around in the semicircle and is driving life is a current, single electrons cascading down and giving up their energy piecemeal. A current can do anything but cannot be expressed in classic chemical terms. A wandering electron belongs to the world of changing shapes and

[18] L. Michaelis, The Theory of Oxidation and Reduction, *in* "The Enzymes," J. B. Sumner and K. Myrbäck (Eds.), Vol. II, Part I, p. 1. Academic Press, New York, 1951. See also ref. 11 on page 7.

[19] W. H. Westheimer (in the "Mechanism of Enzyme Action," W. D. McElroy and B. Glass (Eds.), page 321, Johns Hopkins Press, Baltimore, Md. 1954) has challenged the universal applicability of this statement and has given examples in which oxidations may occur without passing through the free radical intermediate. So, according to his views, the word "all" would have to be replaced by "many" in the above quotation.

distributions of those electron clouds which belong to the submolecular, dominated by quantum mechanics.*

* This distinction between classic chemical reactions and monovalent electron transfer is not invalidated by the fact that in the last analysis also a classic chemical reaction is but the end product of a series of quantum changes. Nor is the distinction invalidated by the fact ATP is also formed in the intermediary metabolism.

26

IV

Units and Measures

IN ORDER TO DISCUSS ENERGY CHANGES, CORRELATE facts, make numerical statements or predictions, we need units and measures. This chapter will be devoted to a brief discussion of the main possibilities available at present.

REDOX POTENTIALS

A transmission of electrons from one substance to another means the oxidation of the one and the reduction of the other. We can also place the solution of reductant and oxidant into two different beakers and make the electrons pass from the one to the other through a wire. The "stronger" a reducing agent, the more it will tend to give off electrons and charge up the electrode, while the opposite will hold true for the oxidant. The potential difference between the two solutions will give us information about the free energy change of the oxidoreduction which would take place between the two substances if we were to mix their

27

solutions, one EV (electron volt) being equal to about 23 kcal.*

The redox potentials can give us a great deal of most useful information. They created order among the host of oxidizing and reducing agents, allowing us to arrange them in a row according to their oxidizing or reducing power. They also allow us to make numerical statements about the energy changes taking place in the oxidation cycle. The potential difference between DPNH and O_2, which corresponds to the semicircle in Figs. 2–5, is about 1.1 EV, 25 kcal, a rather modest amount. Life, from the energy point of view, is a very modest phenomenon indeed. But even this small amount of 25 kcal seems to be turned in at the first step for small change, the energy of the ~ of ATP which is about 10 kcal.† Maybe, the subtle structures which make the living machinery cannot be exposed to the destructive action of higher quanta.

These values put limitations on our speculations about bioenergetics. There are many reasons to tempt one to involve excited states of molecules in biological oxidation, but the energies needed to raise an electron from the ground state into the first excited level in one and the same molecule are mostly considerably higher than 25 kcal. For example, to lift an electron

* It is customary to measure the potential of the two substances in question one by one against some standard electrode, as the normal H electrode. Since limiting conditions are not well defined, one usually does not measure the potential of a pure oxidizing or reducing agent but measures the potential of an isomolar mixture of its oxidized and reduced form.

† The smaller values found experimentally 6000–8000 cal, relate to standard conditions, equimolar mixtures of ATP and ADP. The ADP concentration in tissues is negligible and so the free energy of ATP is correspondingly higher.

28

in DPN from the ground state into the first excited level of the same molecule, we would need a quantum of about 85 kcal (these substances having their absorption in the UV). No one has yet demonstrated the production of such a high-energy quanta in oxidative metabolism. Even a substance absorbing at the red end of the spectrum would demand 40 kcal for its excitation. Such transitions can be produced by absorbed photons. So we have to distinguish between processes induced by photons, like photosynthesis or vision, and processes occurring in the normal metabolism. This book will be concerned only with the latter.

It is true, the firefly may object to such a distinction, demonstrating by its greenish light that it is capable of producing quanta of 60 kcal, and one could think that bioluminescence may not be an isolated phenomenon but merely a leak in a metabolic process. In fact, there are even reports about light emitted by germinating seeds.[20] All the same, this does not necessarily mean that electrons have been excited from the ground state to the excited level within one and the same molecule. To this point I will come back later; meanwhile, I will maintain the distinction.

Like any method, the measurement of redox potentials also has its shortcomings. First, it gives us only thermodynamic data which is to say that the potential measured will not tell us how fast a reaction will go. There are strong reducing agents, e.g. agents containing SH groups, which will react very sluggishly with the electrode, or not react at all. Second, the standard conditions applied in our experiment are not identical

[20] L. Colli, U. Facchini, G. Guidotti, R. Dugnani, M. Orsenigo, and O. Sommariva, *Experientia*, **11**, 479, 1955.

29

with the conditions in the living cell. Third, the energy changes may be different according to the dielectric constants of the medium. In our measurements *in vitro* we usually use water as a solvent. D. E. Green[21] has always strongly emphasized that the matrix of the oxidation cycle within mitochondria consists of lipid material, is hydrophobic, and energy values may be quite different in such a medium.

Another more serious limitation is due to the fact that only the potential of substances can be measured which can freely give up electrons, that is, have two stable states which differ from one another, as a rule, by two electrons. As will be shown later, there are reasons to believe that there are organic substances which play an important role in biology and are capable of giving off one electron only and have no stable state which would correspond to such a one-electron transfer. These substances will give no potential at all. This shortcoming is not relieved by the fact that free radicals, which can give or take up one electron, might give a potential and Michaelis and his associates* were capable of measuring the potential of various free radicals formed on the way to bivalent oxidations at unphysiological pH's.

If we measure the free energy change of a redox process then we measure not only the energy change involved in transferring one or two electrons from the one substance to the other, but measure the total energy change which includes also the energy changes taking place in the rearrangements within the two interacting molecules, rearrangements which have to

* Ref. 17 on page 24 and ref. 18 on page 25.

[21] D. E. Green and R. L. Lester, *Fed. Proc.* **18**, 987, 1959.

accompany the transfer of two electrons and cannot do external work. Very often one would like to know the free energy change which corresponds solely to the transfer of an electron, uncomplicated by consecutive structural changes.

IONIZATION POTENTIALS AND ELECTRON AFFINITIES

If an electron passes from one molecule to another, from a donor (D) to an acceptor (A), energy may be gained or lost. In order to evaluate the change in our thoughts (and in our thoughts only) we can make the electron pass over a detour, in two moves. In the first imaginary move we take the electron from D, altogether removing it into infinity, while in the second move we drop it from infinity onto A. This procedure has the advantage that we have a fixed point: the energy of the electron in infinity. In infinity all electrons can be supposed to have the same energy.

This mental experiment is symbolized in Fig. 7A in which we take the electron from the molecule D and then drop it onto one of the A's. In this figure the thick lines stand for the highest filled energy level, the "ground state," occupied by an electron pair. If we remove an electron from an atom or molecule it is, as a rule, this level from which we remove it. In order to remove it we have to impart to it sufficient energy to be able to move through all the empty levels (symbolized by thinner lines), beyond the last one of which lies infinity. This energy, symbolized by the upward arrow, has close relations to what is called "ionization potential."[22] The higher the ground state lies, the shorter the arrow and the less energy will be needed

[22] R. S. Mulliken, *Phys. Rev.* **74**, 736, 1948.

to remove the electron, the smaller its ionization potential, the greater the tendency of the molecule or atom to "donate" an electron, the stronger donor it will be.

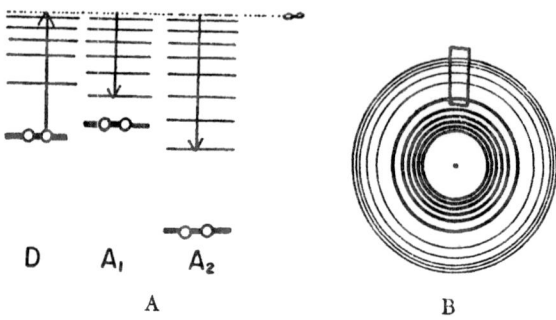

FIG. 7A. Symbol of the ionization potential and electron affinity. (See text.)

FIG. 7B. Schematic representation of energy levels of an atom. Thick lines: occupied levels. Thin lines: empty levels. (See text.)

To make such symbols more intelligible we could suppose that the atom is surrounded by a number of spherical filled and unfilled orbitals. In Fig. 7B the filled ones are symbolized by thick, the empty ones with thin, lines. It is the lower levels which are filled and the higher ones which are empty. The lines in Fig. 7A would correspond to the section of the orbitals enclosed by the oblong. In most other figures, similar to 7A, only the highest filled and lowest empty orbitals will be given. The symbols will also hold for the π electrons of more complex molecules, which π electrons are non-localized and belong to the whole molecule, or at least to its system of conjugated double bonds.

The situation will be more complicated in our second move in which we drop the electron onto the acceptor in which the electron will have to occupy the lowest empty energy level. The energy released in this act is symbolized by the downward arrow and is called the "electron affinity" (EA). What complicates the situation here is the fact that by adding one electron we disturb the postion of all energy levels, and so the energy of the orbital which this added electron eventually occupies will not be identical with the energy of the lowest energy level which existed prior to our manipulation. The lower the new level lies, the greater the EA (the longer the downward arrow), the more energy will be gained in this move and the more the atom or molecule will tend to accept an electron, the stronger acceptor it will be.

If there were no other factors involved, we could predict the energy change accompanying the electron transfer from the difference in length of the upward and downward arrows, EA–IP. More energy would thus be gained in the transfer of an electron from D to A_2 than to A_1, the EA of A_2 being greater, its arrow longer. That the change in energy will not be equal to EA–IP is due to the fact that there are additional complications. Such a complication is introduced by the fact that the transfer of the electron may alter also the interrelation of the D and A molecules, as well as their relation to the solvent. As will be discussed later, D and A must be very close to one another if an electron has to pass between the two. This implies that there can hardly be any solvent molecules between them, there will be solvent molecules only outside the DA complex. It is easy to see that polar solvent molecules, like those of water, with their strong dipole character, will exert a strong attraction on the electron transferred, will tune down the electropolar attraction between the two interacting molecules and may even enable the two molecules to part, each with its unpaired

33

electron, as a "free radical." While the two molecules are in close proximity the transferred electron may also resonate between the two, contributing with its resonance energy to the balance of forces. Summing up, we can thus say that the change in energy will not be equal to EA–IP. If we lump all the other factors as Δ, we could say that the change in energy accompanying the transfer of an electron will be equal to EA–IP + Δ.

In spite of all these complications IP and EA are most useful parameters. In the experiment we can often simplify the complex situation. We can, for instance, as we will discuss later, couple one and the same acceptor to a number of various donors, using always similar solvents. In this case, the only variable will be the quality of the donors, and so the situation will be dominated solely by their IP. We can also reverse the situation and use the same donor with different acceptors to make the EA of the acceptors come to the fore.

Unfortunately, the ionization potentials are but poorly known, and the electron affinities are hardly known at all. There are various methods for the measurement of the ionization potentials, which give fairly consistent results in themselves, but the results obtained by the different methods may differ from one another by as much as one EV, which is about as much as the total energy released in the whole biological oxidation.

Moreover, distinction should be made between ionization potentials measured by fast and slow methods. The electron can, for instance, be torn off by a light quantum within 10^{-15} seconds, which time is too short to allow any changes in the molecular configuration and would thus reveal purely the energy used in this act of removing an electron. Other methods are slower than this optical method and may thus yield complex quantities composed of the energy of tearing the electron off *and* the energy changes due to the changes in molecular configuration accompanying the removal of the electron. The ionization potentials given in tables relate mostly to measurements in the gas phase. In biology we are concerned with condensed phases, mostly watery solutions in which the ionization energies may be quite different.

But even if ionization potentials and electron affinities give us, at present, but modest help in our work, they can provide a backbone for our thinking, presenting a clean-cut case of the removal and addition of an electron, into which we can, at least in thoughts, decompose the transfer of an electron from one substance to another.

ORBITAL ENERGIES

Two methods are commonly used for the calculation of energy levels of molecules. One is the molecular orbital method in the LCAO approximation (Linear Combination of Atomic Orbitals), the other the valence bond method. For calculations involving but the simplest molecules one is restricted to the LCAO method which has been applied extensively by B. and A. Pullman and G. Karreman to calculate the energy levels in a great number of molecules taking part in different biological reactions.* These calculations are very involved and demand not only a heavy mathe-

* See quotations in Table I.

35

matical armory, but also much experience and intuition. The calculations are restricted to the case of molecular orbitals, that is, to cases where the molecule has a system of delocalized electrons, as is the case, for instance, in systems of conjugated double bonds where the electron in question does not belong to any single atom but to the whole molecule, is a π electron belonging to the "π electron pool." This may give the impression that only molecules with extensive conjugated systems with an extensive π pool are important for electron transfer in biology. Certainly, most of the catalysts of biological oxidations have such systems, but we must not allow ourselves to be deceived into believing that only such molecules can be important as electron donors or acceptors. If we talk mostly about them, this is only because we know most about them.

It follows from what has been said before, that we are interested chiefly in the energies of the highest filled and lowest empty orbital.

The energy of an orbital, that is, the energy of the electron on that orbital $E = \alpha + k\beta$, where α is the "Coulomb integral" of the method, and β the "exchange integral" between two C's. For substances belonging to the same chemical series α and β are fairly constant, and so the energy depends on k. Where $k = 0$, $E = \alpha$. This is the zero. As a rule the filled "bonding" levels lie below 0 and as a convention their k has a positive sign. The empty "antibonding" levels lie above 0, their k being denoted with a negative sign. Figure 8 shows a number of k values. The values of k of a greater number of substances, as calculated by the Pullmans and G. Karreman, are given in Table I.

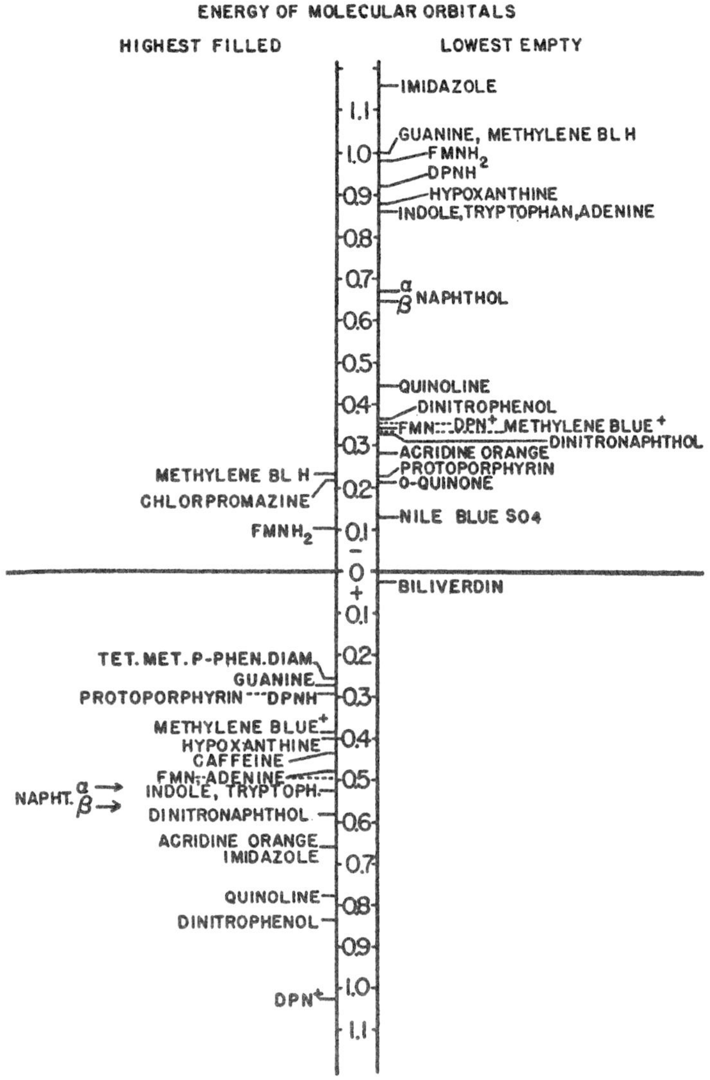

ENERGY OF MOLECULAR ORBITALS

HIGHEST FILLED   LOWEST EMPTY

- IMIDAZOLE
- 1.1
- GUANINE, METHYLENE BL H
- 1.0 — FMNH$_2$
- DPNH
- 0.9 — HYPOXANTHINE
- INDOLE, TRYPTOPHAN, ADENINE
- 0.8
- 0.7 α
- β NAPHTHOL
- 0.6
- 0.5
- QUINOLINE
- 0.4 — DINITROPHENOL
- FMN····DPN$^+$ METHYLENE BLUE$^+$
- 0.3 — DINITRONAPHTHOL
- ACRIDINE ORANGE
- PROTOPORPHYRIN
METHYLENE BL H — 0.2 — O-QUINONE
CHLORPROMAZINE —
- NILE BLUE SO4
FMNH$_2$ — 0.1
—
0
—
+
- 0.1 — BILIVERDIN

TET. MET. P-PHEN.DIAM. — 0.2
GUANINE —
PROTOPORPHYRIN ··· DPNH — 0.3
METHYLENE BLUE$^+$
HYPOXANTHINE — 0.4
CAFFEINE
FMN··ADENINE — 0.5
NAPHT. $\alpha \rightarrow$ INDOLE, TRYPTOPH.
$\beta \rightarrow$ DINITRONAPHTHOL — 0.6
ACRIDINE ORANGE
IMIDAZOLE — 0.7
QUINOLINE — 0.8
DINITROPHENOL —
- 0.9
DPN$^+$ — 1.0
- 1.1

FIG. 8. k Values of the highest filled and lowest empty molecular orbitals.

37

TABLE I

k VALUE OF THE HIGHEST OCCUPIED (LEFT), AND LOWEST EMPTY (RIGHT) MOLECULAR ORBITAL

Substance	k Values		Reference [a]
Acridine	0.494	−0.342	(9)
Acridine orange	0.657	−0.278	(9)
Adenine	0.486	−0.865	(3)
Alloxane	1.033	−1.295	(3)
6-Aminonictonamide	0.735	−0.471	(8)
2-Amino-4-nitrophenol	0.469	−0.354	(10)
4-Amino-2-nitrophenol	0.451	−0.355	(10)
Aniline	0.544	−1.000	(10)
Anthracene	0.414	−0.414	(2)
Antipyrine	0.248	−0.956	(10)
Ascorbic acid	0.529	−0.899	(10)
Atabrine	0.311	−0.486	(10)
Barbituric acid	1.033	−1.295	(3)
Benzanthracene	0.452	−0.452	(2)
1-Benzyl-2-methoxy-(N,N-dimethyl) tryptamine	0.427	−0.866	(6)
Benzpyrene	0.372		(1)
Biliverdine	0.455	+0.021	(5)
Bromanil	0.646	−0.266	(10)
Catechol (Adrenaline)	0.666	−1.049	(10)
Chloranil	0.753	−0.275	(10)
Chlorpromazine	−0.217	−1.000	(6)
Chloroquine	0.478	−0.654	(10)
Chrysene	0.521	−0.520	(2)
Colchicine	0.355	−0.499	(10)
Cytosine	0.595	−0.795	(3)
5-Methylcytosine	0.530	−0.796	(3)
Dibenzanthracene	0.494	−0.501	(2)
2,4-Dichlorophenol	0.698	−1.016	(10)
3,5-Dichlorophenol	0.749	−1.032	(10)
5,7-Dimethyl-3,4-benzacridine radical	−0.277	−0.716	(10)
5,7-Dimethyl-1,2-benzacridine radical	−0.299	−0.682	(10)
9,10-Dimethyl-1,2-benzanthracene	0.387	−0.475	(10)

[a] References for this table are listed at the end of the table.

38

TABLE I (*Continued*)

Substance	k Values		Reference
p-Dinitrobenzene	1.000	−0.232	(10)
m-Dinitrobenzene	1.015	−0.317	(10)
2,4-Dinitronaphthol	0.580	−0.329	(10)
2,4-Dinitrophenol	0.841	−0.352	(9)
2,5-Dinitrophenol	0.809	−0.243	(10)
DPN+	1.032	−0.356	(2)
DPNH	0.298	−1.032	(2)
Duroquinone	0.757	−0.273	(10)
Flavinemononucleotide	0.496	−0.343	(4)
FMNH$_2$	−0.105	−0.979	(4)
Fluoranil	0.960	−0.200	(10)
Fluoropromazine	−0.207	−0.987	(10)
Formilhydrazine	0.192	−1.710	(10)
Guanine	0.307	−1.050	(3)
1-Methylguanine	0.303	−1.064	(3)
9-Methylguanine	0.302	−1.074	(3)
Histidine	0.660	−1.160	(3)
p-Hydroquinone	1.000	−1.175	(10)
Hypoxanthine	0.402	−0.882	(3)
Indole	0.534	−0.863	(2)
Indole acetic acid	0.479	−0.863	(10)
Indophenole	0.516	−0.122	(10)
D-Lysergic acid diethyl amide	0.218	−0.726	(6)
Lumichrome	0.581	−0.434	(10)
Lumiflavine	0.482	−0.342	(10)
20-Methylcholanthrene	0.388	−0.475	(10)
Methylene blue+	0.398	−0.354	(7)
Methylene blue H	−0.232	−1.000	(7)
L-Methylmedmaine	0.348	−0.869	(6)
Methylphloroglucinol	0.649	−1.123	(10)
Naphthacene	0.295	−0.337	(2)
Naphthalene	0.617	−0.618	(2)
α-Naphthol	0.519	−0.671	(9)
β-Naphthol	0.569	−0.637	(9)
1,4-Naphthoquinone	1.000	−0.325	(10)
1,2-Naphthoquinone	1.000	−0.332	(10)
Nile blue sulfate+	0.591	−0.133	(10)
Nile blue sulfate H	0.091	−0.741	(10)
Nitrobenzene	1.000	−0.334	(10)
o-Nitrophenol	0.806	−0.352	(10)

TABLE I (*Continued*)

Substance	*k* Values		Reference
m-Nitrophenol	0.797	−0.334	(10)
p-Nitrophenol	0.827	−0.354	(10)
Phenanthrene	0.607	−0.606	(2)
Phenanthrenequinone	0.721	−0.442	(10)
o-Phenanthroline	0.649	−0.563	(10)
Phenazine	0.555	−0.251	(10)
Phenothiazine	−0.210	−1.000	(10)
Phenylalanine	0.908	−0.993	(3)
p-Phenylene diamine	0.321	−1.000	(10)
Phlorizine	0.731	−0.901	(10)
γ-Picoline	1.000	−0.872	(10)
Picramic acid	0.469	−0.354	(10)
Proflavine	0.657	−0.278	(10)
Protoporphyrin	0.293	−0.233	(5)
Pteroilglutamic acid (Folic acid)	0.52	−0.64	(3)
2,4-Dimethyl-pteridine	0.544	−0.508	(3)
2,3-Dihydroxypteridine	0.653	−0.663	(3)
2-Amino-4-hydroxy-pteridine	0.489	−0.650	(3)
Pteridine	0.864	−0.386	(3)
Pyocyanine (alkal)	0.286	−0.272	(10)
Pyrene	0.445	−0.444	(2)
Pyridine	1.000	−0.871	(9)
Pyrimidine	1.063	−0.820	(10)
Quinine	0.584	−0.563	(10)
Quinoline	0.77	−0.44	(9)
o-Quinone	0.694	−0.211	(10)
p-Quinone	1.000	−0.225	(10)
Reserpine (Indole part)	0.464	−0.892	(10)
Reserpine	0.627	−0.872	(10)
Rhodamine 5G	0.688	−0.184	(10)
Serotonin	0.461	−0.870	(10)
Stilbene	0.503	−0.503	(2)
Stilbestrol	0.369	−0.553	(10)
p-Terphenyl	0.594	−0.592	(2)
Tetramethyl-para-phenylenediamine	0.266	−1.000	(10)
Thionine+	0.398	−0.354	(10)
Thionine H	−0.208	−1.000	(10)
Thymine	0.510	−0.958	(3)
Toluidine blue+	0.391	−0.359	(10)

TABLE I (*Continued*)

Substance	k Values		Reference
Toluidine blue H	−0.217	−0.997	(10)
Trinitrobenzene	1.045	−0.317	(10)
Trinitrophenol	0.859	−0.317	(10)
Triphenylene	0.684	−0.655	(2)
Tryptophan	0.534	−0.863	(3)
Tyrosine	0.792	−1.000	(3)
Uracil	0.597	−0.960	(3)
Uric acid	0.172	−1.194	(3)
1-Methyl uric acid	0.172	−1.202	(3)
3-Methyl uric acid	0.153	−1.204	(3)
7-Methyl uric acid	0.133	−1.120	(3)
9-Methyl uric acid	0.161	−1.204	(3)
Xanthine	0.442	−1.005	(3)
Xanthine H on N9	0.397	−1.097	(3)
1-Methylxanthine	0.397	−1.198	(3)
3-Methylxanthine	0.345	−1.197	(3)
9-Methylxanthine	0.394	−1.213	(3)
Vitamin K_3 (2 methyl-1,4-naphthoquinone)	0.915	−0.340	(10)
Vitamin K_5 (4 amino-2 methyl-1-naphthoquinone)	0.283	−0.738	(10)

REFERENCES FOR TABLE I

[1] A. and B. Pullman, Cancerization par les substances chimiques et structure molecularies. Masson Edit. Paris, 1955.

[2] B. Pullman and A. Pullman, Les théories électroniques de la chimie organique. Masson Edit. Paris, 1952. In this book m values are given instead of k. The k values have been calculated by means of the following formulas: for filled orbitals $k = m/(1 − 0.25m)$ for empty orbitals $k = −m/(1 + 0.25m)$.

[3] B. and A. Pullman, *Proc. Natl. Acad. Sci. U.S.* **44**, 1197, 1958.

[4] B. and A. Pullman, *Proc. Natl. Acad. Sci. U.S.* **45**, 136, 1959.

[5] B. Pullman and A. M. Perault, *Proc. Natl. Acad. Sci. U.S.* **45**, 1476, 1959.

[6] G. Karreman, I. Isenberg, and A. Szent-Györgyi, *Science* **130**, 1191, 1959.

[7] B. and A. Pullman, *Biochim. et Biophys. Acta* **35**, 535, 1959.

[8] B. and A. Pullman, *Cancer Research* **19**, 337, 1959.

[9] B. Pullman, Personal communication.

[10] G. Karreman, Personal communication.

41

We need not concern ourselves here with the exact meaning of α and β. What concerns us here is that the k for the highest filled orbital is a linear function of the ionization potential.[*] The smaller its value the less energy will be needed to take an electron off, the easier the molecule in question will give up this electron and act as an electron donor. Similarly—as experience indicates—the smaller the $-k$ of the lowest empty orbital, the easier the molecule will accept an electron. A $k = 0.5$ for the highest filled orbital means, for instance, that the substance is a "fair" electron donor, the same $-k$ for the lowest empty orbital means that it is a "fair" electron acceptor; 0.2 means "very good." In exceptional cases the k of the highest filled orbital can be even smaller than 0, and have a negative sign, an "antibonding character." This means an excessively good donor. Such substances, of which few examples are known, are mostly unstable and readily auto-oxidize in air, as leucomethylene blue or $FMNH_2$. *Mutatis mutandis*, the same holds for positive values of the lowest empty orbital. The numerical value of β is taken as 1–3 EV, so a difference of 0.5 between two k values means 0.5–1.5 EV difference in the ionization potential for substances belonging to the same series of substances and having, thus, a similar α and β.

We must be very careful in relating the k's to other physical constants. While Mulliken has shown that the position of the highest filled orbital (given by the $+k$'s) actually gives the ionization potential of the molecule, the position of the lowest empty orbital (given by the $-k$'s) does not have such simple relationship to the electron affinity. The reason for this is

[*] Ref. 22 on page 31.

42

that, as pointed out before, upon adding an electron to the molecule, all the energy levels tend to change. This change, however, may be small in big molecules with an extensive system of π electrons, and most biological catalysts have such systems. What we may expect in any case is that, on the whole, the lower the position of the empty orbital (and the lower the $-k$'s) the greater will be the electron affinity, and so even the $-k$'s give a very useful parameter.

There are various ways to convince oneself of the approximate usefulness and reliability of these k's. We can, for instance, compare in various substances, the k's with their tendency to give off or accept electrons. The experience of my laboratory on this line supports, on the whole, the reliability of the k values. Another crude way of checking consists of measuring the absorption spectra. The longest wavelength of the light absorbed by a substance can be expected to lift an electron from the highest filled orbital to the lowest empty one, so there should be some relationship between the wavelength of the absorption and the distance between the $+$ and $-k$ values.[23] Fujimori's curve (see Fig. 11) also contributes to the mass of evidence pleading for the reliability of these k's.

The meaning of the k's has given me an answer to a problem which occupied my mind in earlier days and lured me into this field: why so many biological substances, like flavines, flavones, pteridines, and cyto-

[23] I. Isenberg and the author actually demonstrated such an interrelation (*Proc. Natl. Acad. U.S.* **45**, 519, 1959). It should be noted that in their table two absorptions were quoted erroneously (tyrosine and phenylalanine). These two amino acids would not fit if quoted correctly.

43

chromes, apparently involved in energetics, are colored? What is the meaning of color in cells not exposed to light? The answer is simple. In terms of k, color means that the distance between the two k values, that of the highest filled and lowest empty orbital, are close to one another, making even the energy-poor visible light capable of lifting an electron from the former to the latter. This light being thus absorbed will make the substance colored. In order to be able to act as a catalytic electron transmitter the substance in question has to be both a good electron donor and acceptor, that is, the k values for both the highest filled and lowest empty orbitals must be small, close to 0 and thus close also to one another. The k values of FMN, for instance, are 0.496 and −0.343 respectively, which makes FMN colored* and makes it into a good donor and acceptor. As I have shown in my earlier days,[24] one can knock out the whole respiratory chain by cyanide and then restore oxygen uptake by adding methylene blue which takes the whole electron transport over between dehydrogenases and O_2. This it can do, as pointed out by B. and A. Pullman,[25] because its k values are 0.398 and −0.354 respectively, almost ideally placed to make the dye into a good donor *and* acceptor.*

These examples may suffice to show the k values of

---

* One may object that DPN has no color and all the same it is one of the main mediators of oxidation. But the contradiction is only an apparent one because DPN is not a pure electronic mediator. It takes up H's and gives off electrons. Its k values are 1.032 and −0.356 which indicates that DPN+ is a very good acceptor. The donor will be not DPN+ but DPNH, the highest filled orbital of which has a k of 0.248 making it into a good donor.

[24] A. Szent-Györgyi, *Biochem. Z.* **150**, 195, 1924.
[25] B. and A. Pullman, *Biochim. et. Biophys. Acta*, **35**, 535, 1959.

44

the LCAO approximation to be most valuable tools of research and understanding. They also have their shortcomings. First, the LCAO method is but an "approximation" and the k values obtained depend on the parameters used in the calculations which may be subject to change. As mentioned before, experience and intuition are needed in this line which also means that there is no real numerical rigidity. Another, even more serious, shortcoming is that the k values of the substances belonging to different chemical series cannot be strictly compared, since both α and β may vary in different chemical series by as much as 25%. If the numerical value of α is around 6 EV, 25% means 1.5 EV, more than the total energy of biological oxidation.

To sum up: all three methods of measurement and expressions, redox potentials, ionization potentials and electron affinities, as well as orbital energies, have their merits and shortcomings and at present there is no universally applicable method available. What would be needed is to bring the three methods to a common denominator which has all the merits and none of the shortcomings. This, at present, is but a desire. The relations between redox potentials and the k values of the LCAO approximation are very complex, and there is no theory available to bring the two together. The relation of the ionization potentials and the k values are clearer, but the two cannot be identified since the α and β values are not constant for different chemical series. What would serve our purpose best would be a knowledge of the ionization potentials and electron affinities of all molecules in biological media, but this information is not available. So we have to wait for further development and, meanwhile, get along as well as we can.

45

V

Electronic Mobility

IN ORDER TO GET AROUND IN THE OXIDATIVE CYCLE, the electron must wander from substance to substance, and thus have a certain mobility. We do not know how this mobility is acquired. What we do know is that several members of the oxidation cycle are bound to structure, thus fixed in space, so that they cannot reach one another by diffusion. They are rather bulky and have relatively small active centers so that it seems unlikely that they could be arranged in such a fashion that their active centers touch, even if a small degree of freedom of motion (e.g. free rotation) is granted. All the same, there must be some connection between them. There are different ways known in which energy quanta or electrons can acquire a long-range mobility. But, tentatively, we could also suppose that in biological oxidation the electrons are carried from one structure-bound substance to another by small diffusible molecules which shuttle between fixed members of the oxidation chain, getting alternately oxidized and reduced. In certain cases we can exclude diffusion, as, for instance, between chlorophyll

47

and cytochrome in photosynthetic bacteria where Chance and Nishimura[26] found electron transmission even at liquid N_2 temperature where everything is frozen stiff and no diffusion is possible. All the same, the fact stands that the oxidation system contains diffusible small molecules, as the coenzyme Q or members of vitamin K group,[27] or Fe atoms to which D. E. Green calls attention.[28] These substances could transport electrons between fixed members of the chain. Whether they do so, we do not know.

Being unable to decide between the various possibilities, I will give brief consideration to the most likely modes of long-range energy and electron transfer and in the end I will consider the question, how electrons can be transmitted from one substance to another at short range, in a direct contact, without wasting their energy.

ELECTROMAGNETIC COUPLING. RESONANCE TRANSFER OF ENERGY

If a molecule X is excited and there is a molecule Y not too far away, which molecule is capable of a similar excitation, then there is a chance that the excitation may suddenly die out in X and appear in Y. Something analogous can be demonstrated in two pendulums connected with a weak spring. The motion imparted to the one dies off after a time and is taken over by the other. An electronic excitation can be

[26] B. Chance and M. Nishimura, *Proc. Natl. Acad. Sci. U.S.* **46**, 19, 1960.

[27] A. I. Arnon, *Nature* **184**, 10, 1959.

[28] D. E. Green, *Radiation Research* Suppl. 1960 (In press). Brookhaven Conf. Bioenergetics.

48

looked upon as an oscillation, and the role of the spring may be fulfilled by the electromagnetic field.

Th. Förster[29] has formulated the rules of such an energy transmission and his formula allows us to calculate the distance at which the energy may be transmitted in this way. Karreman and Steele,[30] for instance, have calculated the critical distance through which excitation can be transmitted between aromatic amino acids in a protein and found it to be 17 Å, a distance big enough to allow energy transmission within one molecule or closely adjacent molecules. Förster has also shown that the transmission becomes the more probable the greater the overlap between the absorption spectra of the two molecules and is favored by a somewhat longer wavelength of absorption of the molecule which has to take the energy over.

As to the mechanism of this energy transfer, we have two ways to picture it. Förster, as well as earlier investigators (J. and F. Perrin), picture it as a sort of jump from X to Y. However—as pointed out by Z. Bay at the symposium held in Woods Hole in the summer of 1959—quantum electrodynamics can give a different explanation. If molecule X is surrounded by several Y's, the disappearing excitation in X may bring these Y's into a state which is neither a ground nor an excited state. Then, after a very short time the energy will collect again and suddenly declare itself in the excitation of one of the Y's.* In many cases, the cal-

* Ordinary language is most inadequate in describing such phenomena.

[29] Th. Förster, *Disc. Faraday Soc.* **27**, 1959. See also Th. Förster, "Fluoreszenz Organischer Verbindungen," Göttingen, 1951.

[30] G. Karreman and R. H. Steele, *Biochim. et Biophys. Acta* **25**, 280, 1957.

culation according to the two theories may lead to the same result, but in certain cases, if there is a "coherence," a cooperative action between the Y's, the chances of the transmission may be greater in the second theory, which may have a role in photosynthesis, where several hundred chlorophyll molecules collaborate in leading a photon to its site of action.

SEMICONDUCTION

In 1941 I published an article with the ambitious title "Towards a New Biochemistry." This article grew out of discussions with my young pupil K. Laki and suggested that energy, in living systems, may be transmitted by conduction bands. But, biochemistry did not move; the theory remained a dead duck. Theories which bear no fruit are useless. All the same, a few words may be said about semiconduction. The basic idea is simple. Atoms have single energy levels. If many atoms are close to one another and are placed in a regular array, their energy levels disturb one another and the disturbed levels join to form a continuous band which contains many levels and extends over the whole system. The conductance depends, in this case, on the number of electrons within the band. According to the Pauli principle only two electrons can have the same energy within the same level. So, if the number of electrons occupying the band is double the number of levels, i.e. atoms building the system, then the band is filled and no conduction can take place.

Metals owe their conductivity to their unfilled bands. There may be energy bands also in dielectrics which, then, owe their insulating property to the fact that the band corresponding to the ground state is completely filled and the next higher band is completely empty and much energy is needed to raise electrons into it.

In intrinsic semiconductors, the distance to the first empty band is small so that even heat agitation may raise an electron into it from the nearest filled band, rendering it conductant.

The calculation of Evans and Gergely[31] strongly suggested that proteins actually have conduction bands. Later, Eley[32,33] and his associates demonstrated experimentally the existence of such conduction bands, but also showed that the distance between the filled and the first empty band is about 2–3 EV, a distance too great to be covered by the small biological quanta.

Conduction bands certainly exist in diverse biological systems. The experiments of Arnold and Sherwood,[34] for instance, leave little doubt that such bands do exist in dried chloroplasts which can store light which can be hunted out as such later, by heat, "electron traps" being present under the conduction bands.

That energy can move through proteins was suggested long ago by Bucher and Kaspers[35] who produced dissociation in the heme part of CO-myoglobin by quanta absorbed by the protein. Shore and Pardee[36]

[31] M. S. Evans and J. Gergely, *Biochim. et. Biophys. Acta.* **3**, 188, 1949.

[32] D. D. Eley, S. D. Parfitt, M. J. Perry, and D. H. Taysum, *Trans. Faraday Soc.* **49**, 79, 1953.

[33] M. H. Cardew and D. D. Eley, *Ibid.*, 1959.

[34] W. Arnold and H. K. Sherwood, *Proc. Natl. Acad. Sci. U.S.* **43**, 104, 1957.

[35] T. Bucher and J. Kaspers, *Biochim. et. Biophys. Acta* **1**, 21, 1947.

[36] V. G. Shore and A. B. Pardee, *Arch. Biochem. Biophys.* **62**, 355, 1956.

51

have seen light emitted by the dye in chromoproteins after excitation of the protein. Schubert's[37] experiments cast some doubt on Bucher and Kasper's results but at the same time supplied fresh evidence for such energy transfer. None of these experiments is quite conclusive in the sense that they do not allow us to decide definitely between resonance energy transfer and semiconduction. One of the most important results of electron microscopy is the demonstration of the wide occurrence of membranes and layered structures, as found in mitochondria, chloroplasts, visual rods, and in the protoplasm (Sjöstrand[38]). Even the particles found in extracts, such as microsomes, may be but fragments of such membranes, broken up in the process of extraction. Membranes, especially double membranes (as most biological membranes are), suggest by their ordered molecular structure semiconduction. As has been demonstrated lately by Tollin[39] in Calvin's laboratory, such layered structures acquire a high conductivity if the one layer donates electrons to the other.

There is also the possibility that filled energy bands may become conductant by donating electrons to some outside substance and thus become unsaturated and conductant. Also, the opposite may be true and an empty electron band may become conductant by accepting electrons from some outside substance. Accepted and/or donated electrons may thus enable a

[37] J. Schubert, *Arkiv Kemi.* 15, 97, 1959.
[38] F. S. Sjöstrand, Fine Structure of Cytoplasm: the Organization of Membranous Layers, *in* "Biophysical Science." J. Wiley & Sons, New York, 1959.
[39] G. Tollin, *Radiation Research*, 1960. (AEC conference on Bioenergetics, Brookhaven, 1959). See also "Semiconductors," Chapter 15 by C. G. B. Garrett, Reinhold Press, 1959.

biological system to fulfill its biological role by semiconduction, or else may inhibit or paralyze function by tapping off the conducting electrons or filling up the holes which rendered the system unsaturated. Some of the hormones or drugs might act this way (see later). Accepted and donated electrons may thus enable a system to work, may regulate its function, or else cause profound disturbance. Even a small change along this line may have far reaching consequences, as the damage done to the insulation of an electric wire may paralyze a whole electric network. Mason[40] suggested relations between such changes and malignant growth.

Electromagnetic coupling transfers energy, semiconduction transfers electrons. We must clearly distinguish between the transfer of energy, and that of electrons, even if the electrons carry energy with them, and this for several reasons. In order to transfer energy, we would have to excite an electron first from the ground level of a substance to its first excited level. The energy needed for this excitation corresponds to the distance between the ground levels and the first excited levels in Fig. 7. As mentioned before, this energy is of the order of 40–100 kcal, an energy quantum which may be carried by a photon, but which, probably, cannot be produced in the normal course of oxidation. Hence we must distinguish between processes started up by photons, as is the case in vision and photosynthesis, and processes occurring in metabolism. This book is concerned only with these latter.

Another reason for distinguishing between transfer of energy and electrons will become evident by casting

[40] R. Mason, *Nature*, **181**, 820, 1958.

another glance at Fig. 7, in which three molecules were symbolized which are supposed to belong to different substance groups and have different 0 levels. What is drawn here, at the same level, is infinity, ∞, since all electrons, in infinity, have the same energy. Infinity, in this case, lies just beyond the highest empty orbital. Judging from the length of the arrows (ionization potential in D and electron affinity in the A's) we could transfer no electron without external help from the ground level of D to the first excited level of A_1, but could transfer an electron from D to A_2. The reverse is true for energy. We could transfer excitation energy by resonance from D to A_1 since the distance from the ground level to the first excited level is smaller in A_1 than in D. However, we could not transmit excitation energy from D to A_2, since the distance between the two levels is greater in A_2.

CHARGE TRANSFER[41]

Evidence started accumulating more than thirty years ago suggesting that in certain complexes electrons may trespass between the borderlines of two complexing molecules. It was J. Weiss[42-44] who, in 1942, gave the first clear formulation of the idea that within a complex an electron of one of the two complexing molecules may be transferred to an orbital of the other. This "charge transfer" has been extensively

[41] A very clear, though not very recent, review of charge transfer, is found in the: *Quarterly Reviews of the Chemical Society*, London 8, 422, 1954 by L. E. Orgel.

[42] J. Weiss, *J. Chem. Soc.* 245, 1942.

[43] J. Weiss, *Nature* 147, 512, 1941.

[44] J. Weiss, *Trans. Faraday Soc.* 37, 78, 1941.

studied since by Mulliken[45-47] and his associates, who applied quantum mechanics to it and classified its various forms.

The transfer of electrons from one substance to another is usually termed as an oxidoreduction. However, we must clearly distinguish between "oxidoreduction" and "charge transfer." As a rule, in organic substances, electrons occupy orbitals in pairs and in oxidoreduction an electron pair is transferred from one molecule to the other, two new closed shell molecules being formed. The two molecules then part, the one having become richer, the other poorer, by two electrons. After this, in principle, they have nothing more to do with one another. So both molecules must have two stable states differing by two electrons. The whole redox process consists of the transfer of these two electrons and the consecutive rearrangement in structure. This situation is not changed by the fact that Michaelis[*,48-50] has shown that electrons can pass from one substance to the other, one-by-one, and that most[†] biological oxidations are actually built of two one-electron steps. The basic ideas of charge transfer not having been cleared yet, Michaelis, himself, regarded the one-electron step as an intermediary of the bi-

* Chapter I, ref. 11, page 7; Chapter III, refs. 17, page 24 and 18, page 25.
† Footnote 19, Chapter III, page 25.

[45] R. S. Mulliken, *J. Am. Chem. Soc.* **72**, 600, 1950.
[46] R. S. Mulliken, *J. Am. Chem. Soc.* **74**, 811, 1952.
[47] R. S. Mulliken, *J. Phys. Chem.* **56**, 801, 1952.
[48] L. Michaelis, *Chem. Rev.* **16**, 243, 1935.
[49] L. Michaelis, M. P. Schubert, R. K. Reber, J. A. Kuck, and S. Granick, *J. Am. Chem. Soc.* **60**, 1678, 1938.
[50] L. Michaelis and S. Granick, *J. Am. Chem. Soc.* **66**, 1023, 1944.

valent oxidation even if the free radicals formed in the one-electron step were stable at extreme pH's. In "charge transfer" one electron only is transferred. The two molecules, the "donor" and the "acceptor," usually stay together, and if they part they do not part as closed shell molecules, but as free radicals with an unpaired electron.

The establishment of the idea of charge transfer means a broad extension of our previous ideas. First, it breaks down the rigidity of our thinking about the individuality of molecules. Charge transfer means that the electrons of a molecule D (D for "donor") are capable of using, under certain conditions, orbitals of molecule A (A for "acceptor"). Second, it brings into the realm of electron transfer a host of substances capable of giving up one electron only, substances which do not have two stable states differing by two electrons, which do not affect the electrode, and are thus not regarded as oxidation or reduction agents. The transfer of one electron does not necessarily involve any rearrangement within the molecule. A charge transfer could be symbolized, schematically, by Figs. 9 and 10, as a simple transfer of one electron from the highest filled orbital of D to the lowest empty orbital of A without any further rearrangement.

Several important points are evident. In order to allow a passage of an electron from an orbital of D to an orbital of A the two electron clouds must overlap. This means that the two molecules must come very close to one another. Close fit and approachability become decisive factors. This may explain why many charge transfer reactions are very slow and take hours, the chances of coming together in the right way being remote. Naturally, this does not exclude that charge transfer may, under conditions, be achieved also in a short encounter, in a so-called "contact transfer." Another point which is made clear by Figs. 9 and 10 is

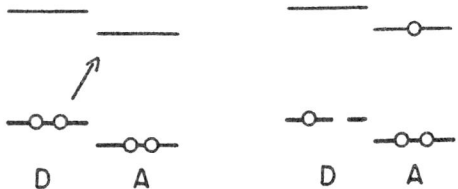

FIGS. 9 and 10. Symbol of the ground state and first excited level of two molecules, before (9) and after (10) charge transfer.

that the relative positions of the two orbitals on the energy scale, that is, the ionization potential of the donor and the electron affinity of the acceptor, have to become dominating factors. Schematically speaking, we could distinguish between two extreme cases: (1) the donating orbital lies considerably lower as compared with the accepting one, as is the case in Figs. 9 and 10, and, consequently, a considerable amount of energy is needed to lift the electron from the former to the latter, $IP \gg EA$; (2) the accepting orbital is lower than the donating one and so the electron can be transferred without outside help, $EA > IP$, as was the case between D and A_2 in Fig. 7.

Considering the first of these two cases, we would have to suppose that the two molecules approach one another and form a complex, held together by the classic forces of complex formation (dispersion forces, polarization, dipole moments). No considerable charge transfer could be expected. All the same, we know that in complex formation a dipole moment may be developed and Mulliken attributes this to a small degree of charge transfer. The transferred electrons resonate between the two molecules contributing thus their "resonance energy" to the binding forces. As single molecules have excited states leading to an absorption, so these complexes also have an excited state, characteristic for the complex. In this case it is the energy of the absorbed photon which lifts the electron of the donor molecule to the excited level of the acceptor. In the excited complex the major part of the electron cloud is now on the acceptor molecule. It is for this reason that Mulliken calls such a spectrum a "charge transfer spectrum" which is characteristic for the complex. The energy needed for the charge transfer would be indicated by the wavelength of the absorbed light and would depend, in part, on the difference between the two levels, the donating and accepting one, that is, on EA–IP. If the difference is great the absorption spectrum falls into the UV. If the difference is small it may fall into the visible and mean a strongly colored complex. If the transfer would demand even less energy the spectrum would fall into the IR. In any case, the spectrum will be characteristic for the complex, and not for its components, as has been shown by Brackman[51]; it will be a "charge transfer spectrum."

Naturally, the IP of D and EA of A will not be the only factors in play because the transfer of an electron may alter the relation of the two molecules to one another, or may alter their relation to the solvent. As will be shown later, there are also other still unknown factors involved. All the same, if the IP of the donor

[51] W. Brackman, *Rec. Trav. Chim.* **68**, 147, 1949.

dominates behavior and all other factors are constant, then we can expect that the energy (that is the frequency) of the light absorbed will depend linearly on this IP. It follows that if we take a series of donors and couple them, one by one, to the same acceptor, then the frequency of the absorbed light, if plotted against the IP of the single donors, must give a straight line. This prediction was first made and verified in the experiment by McConnell, Ham, and Platt,[52] and simultaneously by Hastings, Franklin, Schiller, and Madsen.[53] The lines obtained in this plot were surprisingly straight although the donors used belonged to different substance groups. This straightness indicates that the main factor in charge transfer was actually the ionization potential of the donor. Later Briegleb and Czekalla[54,55] found ways to calculate the charge spectrum in advance and found the same linear relation, their experiments being soon corroborated by Foster.[56] The lines obtained in these plots are so smooth that it is permissible to turn the argument around and read the ionization potential of a donor by plotting the charge transfer spectrum of its complex on the curve of the corresponding acceptor.

If the energy relations greatly favor charge transfer, then the electron could pass spontaneously from D to A. This would correspond to the complexing of a

[52] H. McConnell, J. S. Ham, J. R. Platt, J. Chem. Phys. 21, 66, 1953.

[53] S. H. Hastings, J. L. Franklin, J. C. Schiller, and F. A. Madsen, J. Am. Chem. Soc. 75, 2900, 1953.

[54] G. Briegleb and J. Czekalla, Z. Electrochem. 63, 6, 1959.

[55] J. Czekalla, G. Briegleb, W. Herre, and R. Grier, Z. Electrochem. 61, 537, 1957.

[56] R. Foster, Nature 183, 1253, 1959.

very strong donor with a very strong acceptor, that is, a donor of low IP with an acceptor of high EA. Such cases have been studied by Kainer, Bijl, and Rose-Innes[57-59] as well as Miller and Wynne-Jones.[60] In such a case we can expect the electron to pass spontaneously from D to A, and to resonate between the two molecules, contributing with its resonance energy to the binding forces. The two parted electrons can also impart a dipole moment to the molecule, and, may no more compensate each other's magnetic moment, thus rendering the complex paramagnetic. In the extreme case, the two molecules may even part altogether forming two independent free radicals. Kainer, Bijl, and Rose-Innes have shown that as the difference EA–IP decreases, the molecule complex becomes more and more paramagnetic. Kainer and Überle[61] have shown that such complexes can reach a paramagnetism which corresponds to 40% of the electrons being completely uncoupled. Kainer and Ottig[62] could detect such uncoupling also by IR spectroscopy.

The two cases discussed: $IP \gg EA$ and $IP < EA$ were but extremes. There will be many intermediate cases and as Mulliken has shown, even in one and the same complex, the transfer by photons and spontaneous transfer may be mixed. As the difference IP–EA

[57] H. Kainer, D. Bijl, and A. C. Rose-Innes, *Naturwiss.* **41**, 303, 1951.

[58] H. Kainer, D. Bijl, and A. C. Rose-Innes, *Nature* **178**, 1462, 1956.

[59] D. Bijl, H. Kainer and A. C. Rose-Innes, *J. Chem. Phys.* **30**, 756, 1959.

[60] R. E. Miller and W. F. K. Wynne-Jones, *J. Chem. Soc.* 1959, p. 2378.

[61] H. Kainer and A. Überle, *Chem. Ber.* **88**, 1147, 1955.

[62] H. Kainer and W. Ottig, *Chem. Ber.* **88**, 1921, 1955.

becomes less and less, spontaneous transition will become more and more important, till, eventually, spontaneous transfer may take over altogether and practically a whole electron may be transferred, being lodged more or less permanently on A. The dissociation into two free radicals may make the situation final. Strongly polar solvents will promote the spontaneous transfer and a subsequent dissociation by depolarizing the strong coulombic attractions generated between the two molecules by the transfer of the electron. Weak transfer of the first type will be favored by homoiopolar solvents.[*]

The aforesaid makes it clear also that we have to distinguish between two different events in charge transfer: the bringing of the two molecules together in the desired proximity, and the charge transfer proper. If IP and EA are not favorable, no charge transfer can take place, but there may be cases in which the values of IP and EA are favorable, but no charge transfer can take place because the two molecules do not complex or the desired proximity cannot be achieved owing to steric hindrances. So there may be cases in which there is charge transfer *in vivo*, while there is none *in vitro*. The cell may, for instance, hold two molecules in close proximity, two molecules which would not complex spontaneously. This it could do by binding them to the same enzymic surface or by linking them together by a covalent bond. Nature may also render IP–EA or steric relations favorable by distorting, activating D and A.

To sum up: we could distinguish, schematically, between two types of charge transfer. In the first, the

[*] Refs. 54, 55, page 59.

energy of a photon is needed to transfer the electron, when IP $\gg$ EA. The sequence of events would be:

$$D + A \leftrightarrow DA, \quad DA + h\nu \rightarrow D^+ A^-.$$

In the second type spontaneous charge transfer takes place, we could symbolize events by writing

$$D + A \leftrightarrow DA \leftrightarrow D^+A^- \leftrightarrow D^+ + A^-.$$

As stated, the transition between the two types is gradual. Starting with IP $\gg$ EA we would need light of high energy to work the transfer. As the difference between IP and EA becomes less, light of increasingly longer wavelength would be needed and spontaneous transfer would become more and more important. So we can expect the plot of frequency versus IP to yield a straight line where much energy is needed for the transfer. However, as we come to longer wavelength, and spontaneous transfer gradually takes over, we can expect the curve to deviate from the straight line and become eventually asymptotic. A slight curvature is noticeable even in Briegleb and Czekalla's curves and becomes evident in the plot of Fujimori[63] (Fig. 11). In this curve, instead of the IP of the various donors their k value was plotted against the frequency of the light absorbed. (The k being a linear function of IP, this should make no difference, except for making the scatter somewhat greater since k is somewhat more uncertain than IP.) The four different curves in Fig. 11 were obtained with four different acceptors. The elevation of these curves corresponds to the difference in the EA of the acceptors used. The curvature becomes the stronger the lower we go and the curve

[63] E. Fujimori, unpublished.

62

suggests that spontaneous transfer becomes an important factor where we approach the infrared, the curve tending to become asymptotic.[*] Unfortunately, there

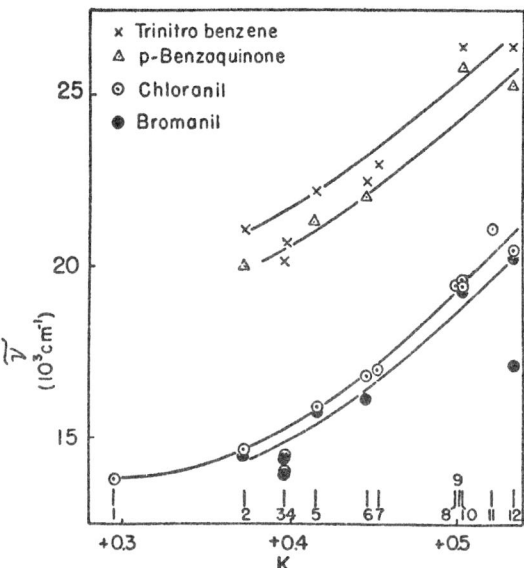

FIG. 11. Frequency of the maximal light absorption as plotted against the k of various donors in complexes formed with four different acceptors. *Key:* 1 = Naphthacene; 2 = Benzpyrene; 3,4 = 9, 10-Methyl-1,2-benzanthracene and 20-methylcholenthrene; 5 = Anthracene; 6 = Pyrene; 7 = Benzanthracene; 8 = Dibenzanthracene; 9 = Picene; 10 = Stilbene; 11 = Chrysene; 12 = Indole.

are too few points in this region because the choice of substances and the technical difficulties, caused by insolubility or color, become rather great. McConnell,

[*] It might be worth noting that two of the strongest carcinogens, 9,10-dimethylbenzanthracene and 20-methylcholanthrene do not fit into the curves, their points falling somewhat under it. Benzpyrene fits the curve. All of these three carcinogens lie very low.

Ham, and Platt[*] have already felt the need to extend research in this region.

In the author's opinion, charge transfer may be one of the most important, frequent, and fundamental biological reactions, a possibility, clearly recognized by Mulliken.[†] It may be objected that there being no light in our body, light-induced charge transfer can have no major biological importance, while substances giving a strong spontaneous charge transfer are rather rare. Most charge transfers which have been studied so far belong to the first type. But, here again, I have to point to the differences between physics and biology. The physicist, when studying charge transfer, will study the interaction of substances which he finds on his shelf, while the biologist works with substances which Nature has developed through millions of years for specific purposes. So the question is not: is charge transfer a common, weak, or strong phenomenon? The question is: does charge transfer belong to Nature's ways, or not? If so, then we can trust that Nature has developed specific substances to fit her purpose. On the subsequent pages, examples of strong charge transfer between biological substances will be given.

The reader may think that the author has dropped into the common error of overemphasizing the importance of a factor which took his fancy. However, he would like to remind the reader of G. N. Lewis' definition of an acid and base. This great pioneer of science defined acids or bases as substances capable of giving up or taking up electrons. So, the acceptor-donor relation of charge transfer merges with the broader acid-

[*] Ref. 52 on page 59.
[†] Refs. 45, 46, 47 on page 55 and ref. 22 on page 32.

base concept which is, undoubtedly, one of the foundations of chemistry. The donor-acceptor relation is also related to oxidoreduction, which is one of the cornerstones of biology. So charge transfer makes part of a most basic triad.

VI

Problems of Charge Transfer

WHAT DO WE GAIN BY INTRODUCING CHARGE TRANS-
fer into the complex texture of biochemistry and biophysics?

Charge transfer allows us to transfer an electron
from one substance to another without major loss in
energy, since it does not involve a rearrangement in
molecular structure.

It brings into the realm of biological oxidation the
great host of substances which are capable of giving
off but one electron, which do not affect an electrode
and were hitherto not regarded as redox agents at all.

Charge transfer brings into play the excited levels
which may not have been available before because the
energies needed to raise an electron into the excited
level of the same molecule are, as a rule too great.

By charge transfer relatively inactive molecules may
acquire a high reactivity. The donor, having developed a "hole" in its low-lying ground level, becomes a
good acceptor, while the acceptor having acquired an

electron on its high-lying excited orbital, becomes a good donor. A charge transfer complex is something between a regular closed shell molecule and a free radical, and the great reactivity of free radicals need not be emphasized. If the complex dissociates, as happens in extreme cases, two real free radicals are formed.

An acceptor may accept an electron also from a saturated energy band, thus creating a hole and making the band conductant. Conversely, a donor may donate an electron to an empty energy band, rendering this band conductant. So charge transfer opens the way for semiconduction into biology. As shown by Tollin,* sheets of donor molecules layered over sheets of acceptor molecules become conductant and show a strong photoelectric effect.

Akamatu et al.[63] found that perylene, violanthrene, and other higher aromatics undergo marked increases in electrical conductivity when complexed with I_2 or Br_2. These complexes show a marked paramagnetic component (Matsunaga[64]). This work demonstrates that an insulator or poor semiconductor can be transformed into a good one and that I_2 may also give strong charge transfer complexes with the spontaneous passage of almost a whole electron.

How can we distinguish between "strong," spontaneous, and "weak" light-induced charge transfer?

---

* Ref. 39 on page 52.

[63] H. Akamatu, H. Ivokuchi, and Y. Matsunaga, *Bull. Chem. Soc. Japan* 29, 213, 1956; *Nature,* 173, 168, 1954.
[64] Y. Matsunaga, *Bull. Chem. Soc. Japan* 28, 475, 1955; *J. Chem. Phys.* 30, 855, 1959.

Having no light in our body this distinction may be important. How can we prove charge transfer at all? There are various methods.

OPTICAL METHODS. Most charge transfer complexes are strongly colored and their spectrum can give information about the nature of the underlying reaction. If the absorption shows the relation between frequency and ionization potential, described in the previous chapter, then, evidently, we are dealing with "weak charge transfer," induced by the light absorbed. If, on the contrary, there is no such relation and the spectrum shows some relation to the (perhaps somewhat shifted or distorted) spectrum of the free radical of one of the complexing substances, then, evidently, we have strong charge transfer in hand with a spontaneous transition of electrons and a trend of going into the ionic state or into free radicals. Many complexes will show a mixed behavior.

MAGNETIC METHODS. In valency saturated compounds electrons occupy orbitals in pairs. Electrons have their spin, and a spinning electron is a tiny magnet. Since the two electrons occupying the same orbital always spin in opposite directions, they cancel out each other's magnetic moments. However, if the two electrons become separated, as may be the case in a strong charge transfer, they may no longer compensate one another, rendering the substance paramagnetic which can be expected to declare itself in magnetic measurements. There is also another possibility which may render the complex paramagnetic: as soon as the two electrons are no longer strongly coupled, and occupy different orbitals, they become relieved of the limitation of the Pauli principle according to which two electrons, forming a pair on the same orbital must spin in opposite directions. If one of the two uncoupled electrons reverts its spin, the complex goes into the triplet state, adding a paramagnetic component. McGlynn and Boggus[65,66] actually found in photolytic experiments such complexes to give a triplet emission.

There are two methods for detecting paramagnetic behavior: the magnetic balance and the electron spin resonance (ESR).[67-70] The former is rather crude as compared to the latter. The signal given in ESR depends on circumstances. Free radicals give a sharp and high ESR signal extending over 10–50 gauss. If the electron is close to other unpaired electrons, it will be perturbed and may give a broad signal. Exposed to the magnetic influence of different nuclei it may split up giving a "hyperfine" structure. Although the ESR method is rich in pitfalls, it is one of the most powerful tools of research finding access, now, to biological laboratories. How far a charge transfer complex will give an ESR signal depends on circumstances. In a weak transfer the two electrons remain strongly coupled and so give no ESR signal, or, give a signal only under strong illumination.

The situation will be different in strong charge transfer, depending on how strongly the transferred elec-

[65] S. P. McGlynn and J. D. Boggus, *J. Am. Chem. Soc.* **80**, 509, 1959.

[66] S. P. McGlynn, *Chem. Rev.* **58**, 1113, 1958.

[67] For technique and theory see: G. E. Pake, S. I. Weissmann, and J. Townsend, *Disc. Faraday Soc.* **19**, 147, 1955.

[68] W. Gordy, W. V. Smith, and R. F. Trambarülo, "Microwave Spectroscopy," Wiley, New York, 1953.

[69] P. B. Sogo and B. M. Tolbert, *Biol. and Med. Phys.* **5**, 1, 1957.

[70] D. J. E. Ingram, "Free Radicals," Academic Press, New York, 1958.

tron is still coupled to its partner, on how "strong" this charge transfer is (that is, how big a portion of the electron is transferred), what proportion of its time it spends on the acceptor,[*] and how far the complex tends to split up, going into the ionic state in which the dissociated partners are free radicals.

The ESR is a wonderful instrument which gives unique information. Its handling requires specific knowledge and experience, complicated (and expensive) machinery, and a great deal of caution in interpretations. One can make a discovery with it in one day, but, then, it takes six months to disprove it. Watery solutions, which might interest the biochemist most, are difficult to study, because only tiny amounts of this solvent will not blur the spectrum. Impurities, present in traces, may mislead the researcher. Special caution is required when dealing with solid state. Distortions of lattices or their physical damage may lead to signals or free radicals. The theory of ESR is still in its infancy; most of the substances which have been studied are simple in nature and surprises may be expected. There appears to be no theory for the explanation for the broad signals reported by Blumenfeld, *et al.*[71,72] in nucleoproteins.[†]

---

[*] Such expressions as "transfer of part of the electron" or "the electron spending part of its time on the new orbital" must be taken with a grain of salt. They do not represent reality, only the shadow of it. These partial reactions, in any case, try to describe a state between the original and the final one in which an electron may be definitely lodged on its new orbital.

† See footnote (†), page 72.
[71] L. A. Blumenfeld, A. E. Kalmanson, and Shen-Pei, *Gen. Doklady Akad. Nauk, USSR,* **124,** 1144, 1959.
[72] L. A. Blumenfeld, *Biophysika* **4,** 515, 1959.

71

Kainer, Bijl, and Rose-Innes,‡ who were the first to apply ESR to charge transfer complexes, have shown that when going from weak charge transfer to an increasingly stronger one, by taking as donor stronger reductants, and as acceptor stronger oxidants, the ESR signal appeared and became gradually stronger.

DIPOLE MOMENT measurements can also support charge transfer, and have been measured by Weiss,* Briegleb, and Czekalla,** Kainer and Überle,# and others. Unfortunately, the measurement of dipole moments is not simple, especially not in polar solvents which favor strong charge transfer (while weak charge transfer is more favored by homoiopolar solvents). Also, the conclusions deduced from dipole moments may be less straightforward than those deduced from optical and magnetic measurements. Crystallography of charge transfer complexes may also give additional evidence. (Planar aromatic molecules can be expected to lie parallel with their centers shifted to allow charge transfer otherwise excluded for symmetry reasons.§)

Does charge transfer open new possibilities for the understanding of oxidation or other biological phenomena?

‡ It may be added that I. Isenberg and the author found in preliminary studies most tested steroids (Δ^4-androstene-3, 17-dione; Δ^5-androstene-3, 17-dione; Δ^5-androstene 3β, 17β-diol, cholesterol; progesterone; stigmasterol, and deoxycorticosterone) to give a broad and low signal while testosterone was negative. Further studies are needed to decide whether this signal was due to impurities, or other disturbing factors, or was a property of these steroids perhaps complexed with O_2.

‡ Refs. 57, 58 on page 60.
* Ref. 42, page 54.
** Ref. 54, page 59.
# Ref. 61, page 61.
§ Ref. 46, page 55.

Let us suppose that D, A₁, A₂, and A₃ in Fig. 12 represent four molecules in close proximity, and suppose that D transfers an electron from its ground level to the excited level of A₁. This electron could not move on being held in place by the electrostatic attraction between D and A₁. This situation would change if, from an outside molecule, we would transfer one elec-

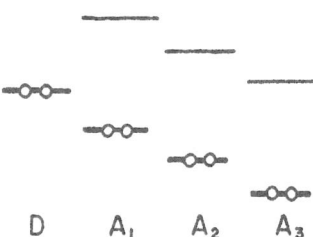

Fig. 12. Symbol of a possible quantum-mechanical framework of electron transport. Thick lines: ground states; thin lines: first excited levels. See text.

tron to D, filling the "hole" in its ground level. The electron transferred earlier to A₁ would become released from its electrostatic bondage and could cascade down over the excited levels of A₂ and A₃. It is not impossible that something like this really happens in biological oxidation. It would be equally possible to remove one electron from the ground state of A₃, whereupon an electron of A₂ could fill the hole, while A₂ would take an electron from A₁, from D, all the electrons cascading one step down.

We could also suppose, as a fascinating possibility, that we transferred an electron from the ground state of D to the excited level of A₃ and then transfer an electron from the ground state of A_3 to an outside molecule, say O_2. In this case the electron could fall down from the excited level of A_3 into its ground level, filling the empty hole on it, and emitting its excess energy in the form of a photon. Some such possibility may underlie bioluminescence. All this is very amusing because it makes a biochemistry without chemistry. There being no rearrangements in molecular structure or bonds, there are no "chemical changes" and the molecules involved serve merely as a quantum-mechanical framework on which the electrons can move. Maybe the cycle of biological oxidation represents some such scaffolding.

Summing up we can thus state that charge transfer opens diverse new, important, and intriguing possibilities.

VII

Three Examples of Charge Transfer

AS MENTIONED BEFORE, IN THE FORMATION OF A charge transfer complex we must distinguish between two events: bringing the molecules together and transferring charge. So, if we want to study charge transfer proper, we must find methods for holding the two molecules close to one another.

There are various possibilities. We may build the two substances into a crystal, in which case the lattice forces will hold the two molecules together. We may simply evaporate the solvent in which we dissolved both substances in question. The evaporating solvent will leave the two substances behind in close proximity. Another method consists of freezing the watery solution of the mixture. The water will, in this case, crystallize out and leave the dissolved molecules in intimate touch. In this case complex formation will be favored also by the decreased temperature. In the living cell, also, association and charge transfer may be promoted by the trend of the matrix, water, to form structures which limit the degrees of its freedom (see next chapter). On the subsequent pages I will

75

give examples of charge transfer produced by these methods.

QUINONE-HYDROQUINONE

A classic material for the study of weak charge transfer is quinhydrone, formed by mixing quinone and hydroquinone. Both of these molecules are flat and have an extensive system of π electrons. So, if they come to lie close enough to one another with their faces parallel, then the orbitals of their π electrons may overlap, forming what Dewar[73] calls a "π-π-bond" in a "π complex." Osaki and Matsuda* studied the crystal structure of quinhydrone by X-rays and found that the two rings, that of the hydroquinone and quinone, lie very close to one another. While the distance usually found between associating aromatic compounds is 3.5–5.7 Å, in quinhydrone it was found to be 3.16 Å, a very close proximity, indeed. In the crystal, the quinone and hydroquinone molecules lie alternating, on top of one another, with their centers slightly shifted, their plane being inclined by 34° toward the long axis of the needle-shaped crystals. So the electrons, excited by light, may oscillate parallel to the plane of the molecules or at right angles thereto, oscillating between the neighboring quinone and hydroquinone molecules which thus form a charge transfer complex. According to the light vector, whether more parallel or at right angles to the plane of the molecules, the quinhydrone will thus absorb light of different wavelength, will be dichroic, as shown by Nakamoto.[74] This can easily be demonstrated by drying down an acetone solution of

* Quoted from Nakamoto, ref. 74 on page 76.

[73] M. J. S. Dewar, *Nature* **156**, 784, 1945. "The Electronic Theory of Organic Chemistry," p. 62, Clarendon Press, London, 1949.
[74] K. Nakamoto, *J. Am. Chem. Soc.* **74**, 1739, 1952.

quinhydrone on an object slide. Under the microscope (Fig. 13) one finds featherlike crystals running in different directions. If these are observed in polarized light (Fig. 14) part of the crystals will be found to be dark blue, while others, running at right angles to the former are light yellow. If the preparation is turned by 90° the colors interchange. (In Fig. 14 the blue ones appear black.) The absorption of quinhydrone shows two maxima,[75] one at 550 and one at 380, corresponding to the two different oscillations.

A 10^{-2} M p-hydroquinone solution is colorless and a 10^{-2} M p-quinone solution is light yellow. On mixing the two, the color hardly changes, though 10^{-2} M quinhydrone can be expected to be red-brown. Our mixture remains colorless (first tube on the left in Fig. 15) because, at this dilution, quinhydrone dissociates, or rather, is not formed at all.

Amusing experiments can be made with this 10^{-2} M mixture. If one suddenly freezes this solution,* it turns into dark ultramarine blue (last tube on the right in Fig. 15). This shows that we have, by freezing, forced the molecules to form a complex, and forced them into an unusual relation which favors the oscillation between quinone and hydroquinone molecules at right angles to their plane.† In order to trap the quinhydrone in its red-brown modification, we have to add to the

---

* As a freezing mixture I usually use powdered dry ice suspended in the monomethyl ester of ethylene glycol.

† If somewhat more concentrated solutions are frozen, then on melting, the ice leaves behind dark blue, almost black, fine needle-shaped crystals which show no dichroism, contrary to the regular quinhydrone crystallized at room temperature, which is red-brown and dichroic.

[75] Keisuke Suzuki, *Busseiron Kenkyu*. Research on the Structure of Matter, No. 102, page 5, Japan.

818

Fig. 14. Quinhydrone in polarized light.

Fig. 13. Quinhydrone crystals in plain light.

mixture of one part of $10^{-2} M$ quinone and quinhydrone, one part of 10% glucose. By sudden freezing we can then trap the quinhydrone molecules in their redbrown form which is metastable in our frozen system (second tube from left). That this is so, can be demonstrated by dipping the frozen tube, for a few seconds, into water of room temperature. The carrot-red color shoots over into ultramarine blue (third tube in Fig. 15). One can amuse oneself (and one's children) with these colorful experiments. All these changes are reversible and the highly colored ice melts into a practically colorless liquid.

What holds the quinhydrone molecule so tightly together in the quinhydrone crystal is, in the first place, not the H-bonds but the π-π interaction.[76] In the blue quinhydrone this interaction seems to be favored. Simply drying down a mixed solution of quinone and hydroquinone will result only in the formation of brown quinhydrone, as can easily be demonstrated by

[76] That the essential point in quinhydrone interaction is not the H-bonding has been shown by Michaelis and Granick (*J. Am. Chem. Soc.* **66**, 1023, 1944), who demonstrated the formation of quinhydrone also when the H's in hydroquinone were exchanged for methyl groups. That there is no obligatory exchange of H's between the two molecules in the crystal has been demonstrated by I. P. Gragerov and G. P. Mikluhin (*Doklady Akad. Nauk. USSR* **62**, 79, 1948) and A. Bothner-By (*J. Am. Chem. Soc.* **73**, 4228, 1951) who showed that if isotopically marked quinone formed quinhydrone with unmarked hydroquinone (or vice versa), then from the crystals the marked or unmarked components could be recovered without having taken over H from, or given H's to, their partner. Quinhydrone formation was thus, in these cases, purely an interaction of the π electron systems. In these experiments the H-atoms of the benzene ring were substituted to prevent a rearrangement of double bonds.

pouring a few drops of a methanoli solution of quin-hydrone on filter paper: the evaporating solution leaves a brown patch behind. If this patch is wetted with water and frozen by placing some powdered dry ice on top, the color changes into a vivid blue.

Quinhydrone is a classic example of a weak charge transfer. Quinone is a good acceptor but hydroquinone is a poor donor (see Table I) and so there is no spontaneous transfer of electrons. The orbitals of the very closely packed quinone and hydroquinone molecules overlap and the light makes the electrons oscillate between the neighboring molecules if the light vector moves them in this direction. Neither the blue, nor the red-brown quinhydrone gives an ESR signal. As mentioned before, charge transfer is very sensitive to distance. The flat benzene rings fit very closely and if no bulky side chains are introduced there is no steric hindrance to disturb close proximity.

Riboflavin (FMN) and Serotonin

As is generally known, FMN is intensely yellow and its solutions show a strong green fluorescence which, on freezing, changes into an orange phosphorescence.[77] Serotonin (5-hydroxytryptamine) is a close relative of tryptophan. It has no color. It is not a "reducing agent" having no two stable states which would correspond to the loss of two H's or two electrons. However, it has a fairly high filled orbital ($k = 0.486$) and can be expected to be a fair monovalent electron donor while FMN with its lowest empty orbital ($k = 0.343$) is a good acceptor. So, if the two molecules are kept close together a charge transfer could occur, an electron

[77] A. Szent-Györgyi, "Bioenergetics," pages 28, 29. Academic Press, New York, 1957.

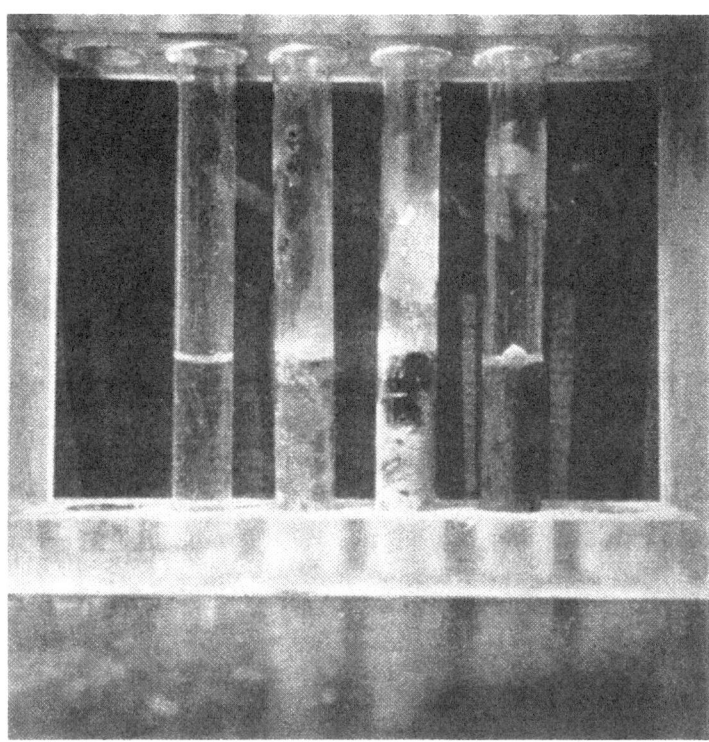

FIG. 15. From left to right: First—mixture of 0.01 M p-quinone and hydroquinone; second—mixture of 0.01 M quinone, hydroquinone, and 10% glucose, frozen; third—same after slight warming; fourth—0.01 M quinone and 0.01 M hydroquinone, frozen. (Details—see text.)

FIG. 16. From left to right: First—5.10⁻⁴ M riboflavin phosphate Na (FMN); second—mixture of 10^{-4} M FMN and serotonin; third—5.10⁻⁴ M FMN, frozen; fourth—FMN—serotonin mixture, frozen; fifth—same at acid reaction. (Details—see text.)

passing from the indole to the vitamin. Figure 16 shows the color changes taking place on freezing of an aqueous solution containing a mixture of the two substances. The first tube on the left contained a 5.10^{-4} M solution of FMN, while the next tube contained, in addition, an equivalent amount of serotonin. There is no striking difference in color between the two although the mixture may be a trace darker. The third tube contained pure FMN frozen to show that freezing, as such, makes no difference in color: the ice is intense yellow. The fourth tube shows the FMN-serotonin mixture frozen. Its color is mahogany brown.[78] The last tube contains the same[79] mixture frozen, containing also 1% strong HCl. It is black. Most striking changes indeed. All of them are reversible: on melting, the original color returns. Freezing promotes a reaction occurring to a smaller extent also at room temperature, as can be shown by the noticeable darkening of the FMN solution obtained on mixing FMN and serotonin of higher concentration.

With the method of Strehler and McGee[80] the absorption spectrum of the frozen solution can be measured. The brown solution shows an absorption maximum around 500 mμ. What makes this mahogany color and absorption interesting is its similarity to the color of the free radical of FMN. This radical, Michaelis' rhubroflavin, in free condition, is stable only at a strongly acid reaction and can readily be produced by reducing FMN by dithionite in presence of strong HCl. This free radical, produced by the monovalent

[78] I. Isenberg and A. Szent-Györgyi, *Proc. Natl. Acad. Sci. U.S.* **44**, 857, 1958.
[79] I. Isenberg and A. Szent-Györgyi, *Ibid.* **45**, 1229, 1959.
[80] B. Strehler and M. McGee (unpublished), quoted from ref. 79.

reduction of FMN, was discovered a quarter of a century ago by Kuhn and Wagner-Jauregg,[81] and has been studied since, extensively, by Michaelis and his associates,[82-84] and Beinert.[85,86] It has its absorption maximum, depending on pH, between 475 and 570 mμ. In partially reduced yellow enzymes Beinert found a strong absorption at 500 mμ. This means that we can actually identify, with great probability, the molecular species which is responsible for the absorption in our neutral frozen mixtures as being related to the free radical of FMN.

An absorption, similar to that of the free radical, can be observed even at room temperature in stronger solutions of pure FMN without an added donor. FMN being both a good acceptor and a fair donor (k of the highest filled orbital $= 0.496$) we can expect a charge transfer from one molecule to the other in the FMN dimers.[*] As pointed out by Mulliken, transfer is possible not only between different, but also between identical, molecules and Michaelis and Granick[87],[†] as well as Beinert,[85] demonstrated the dimerization.

E. Haas,[88] Beinert,[85] and Ehrenberg and Ludwig[89]

[*] The donor properties of FMN can also be demonstrated by the color change produced on freezing of FMN solutions containing some I_2. Molecular iodine with its unsatisfied electronegativity is a good acceptor. On freezing, the solution turns black.

[†] Refs. 48, 49, 50, on page 55.

[81] R. Kuhn and T. Wagner-Jauregg, *Ber.* **67**, 361, 1954.

[82] L. Michaelis, M. P. Schubert, and C. V. Smythe, *J. Biol.* **116**, 15, 1936.

[83] L. Michaelis and G. Schwarzenbach, *J. Biol. Chem.* **123**, 521, 1938.

[84] L. Michaelis, Theory of Oxidation and Reduction, *in* "The Enzymes," Vol. **II**, Part I, J. B. Sumner and K. Myrbäck (Eds.), Academic Press, New York, 1951.

[85] H. Beinert, *Biochim. et Biophys. Acta.* **20**, 588, 1956.

[86] H. Beinert, *J. Biol. Chem.* **225**, 465, 1957.

[87] L. Michaelis and S. Granick, *J. Am. Chem. Soc.* **66**, 1020, 1944.

[88] E. Haas, *Biochem. Z.* **290**, 291, 1937.

[89] A. Ehrenberg and J. D. Ludwig, *Science* **127**, 1177, 1958.

demonstrated the formation of such red radicals also in enzyme-linked FMN and FAD, under influence of metabolic H-donors or reducing agents, while Commoner[*] and his associates, as well as Ehrenberg and Ludwig,[†] showed FMN to give ESR signal on partial reduction, and Ball[90] noted the brown color on xanthin oxidase, a flavoprotein.

The black FMN-serotonin mixture, frozen at acid reaction, shows in the spectroscope[‡] a broad absorption stretching over the whole visible and having a maximum in the near ultraviolet, with its tail in the visible, similar to the absorption of the green FMN radical, Michaelis' "verdoflavin." Another maximum is located at 570 mμ with a shoulder at 620 mμ. Evidently, various spectra become confluent here, maybe that of FMN[-], and FMN[-] with one or two protons bound (as suggested by calculations of Karreman[§]), possibly with some charge transfer spectrum mixed in. In any case, there can be little doubt that this is a very strong charge transfer with nearly a whole electron transferred. In frozen mixtures containing equivalent amounts of FMN and serotonin, the lack of any light emission indicates that the whole FMN has been involved in the formation of the charge transfer complex, at both neutral and acid reactions, one whole electron having been transferred to it.

The transferred electron may have come from the π

[*] Refs. 7, 8 on page 6.
[†] Ref. 89 on page 82.
[‡] Ref. 79 on page 82.
[§] G. Karreman, Personal communication.

[90] E. Ball, *Cold Spring Harbor Symp. Quant. Biol.* **7**, 100, 1939.

pool of the serotonin or from the lone electron pair of its N, which also makes part of the π pool of the molecule.[91]

The assumption that the color changes are due to the absorption of the FMN⁻ formed and not to a charge transfer spectrum, can be supported by freezing FMN in presence of KI. The iodide anion is a fair electron donor. Although its ionization potential is very different from that of indoles,* the color changes obtained on freezing are the same as those produced by serotonin, since the absorbing molecular species, FMN⁻ are the same. If the spectra had been charge transfer spectra and not the spectra of the radical one would have expected marked changes in absorption when using different donors. With the FMN complexes the spectra were practically unchanged.

There is one point about the interaction of FMN and serotonin which demands attention: its great intensity. Although serotonin is a somewhat stronger donor than indole, or some of its derivatives such as tryptamine, this extraordinary reactivity is characteristic for the indole formation in general and is in no way explained by the energy values of the highest filled molecular orbital. According to its k (0.534) indole is only a "fair" donor, not a really good one. Its charge transfer spectra correspond to this k, as shown by the curves of Fujimori (Fig. 11). However, this curve

---

* The electron affinity of an atom is equal to the ionization potential of its monovalent ion. The electron affinity of I and thus the ionization potential of I⁻ is 3, 14 EV. That of indoles is probably considerably higher.

[91] As shown by Fujimori, part of the emission of tryptophan disappears in presence of strong acid indicating that the excited electron came from a lone electron pair (*Biochim. et Biophys. Acta* **40**, 257, 1960).

shows that with bromanil, indole gives a second absorption peak which corresponds to a considerably lower frequency. Such a double peak, according to Kainer, Bijl, and Rose-Innes,[*] is characteristic for very strong charge transfer, the interaction of strong oxidizing and reducing agents. But even the k, corresponding to this second absorption peak[†] does not explain the observed strong donor property. There seem to be other hitherto unknown molecular indices which entail this extraordinary reactivity. There is but one general statement we can safely make about all this, and this is that if Nature developed a substance for electron transfer, then she may have given to this substance extraordinary qualities as an electron transmitter. Indole has such qualities which make it likely that Nature developed this ring actually for services of this kind. This seems to be noteworthy because physiologically rather active and important substances are found among indole derivatives. The plant hormone indole acetic acid, and serotonin, may be quoted, but the most remarkable fact to me is that protein also has an indole side chain contained in its tryptophan. This may throw light, some day, on the real meaning of protein. It is tempting to think that Nature introduced this amino acid into the protein molecule to mediate its charge transfer reactions.[‡,92]

[*] Refs. 57, 58 on page 60.

[†] This absorption would correspond to a k of 0.46.

[‡] Myosin and actin, which transform chemical energy into motion, are rich in tryptophan while paramyosin, which has a passive role, supplying a catch mechanism with its crystallization, has none. The highly crystalline tropomyosin, which seems to have some similar function, is also devoid of tryptophan.

[92] W. H. Johnson and Andrew G. Szent-Györgyi, *Science* **130**, 160, 1959.

FMN too has but a fairly low k for its lowest empty molecular orbital (-0.344) and its reactions as an electron acceptor are stronger than this k allows us to expect. So what has been said for indole holds, also, for FMN of which we do know that Nature made it to serve as an electron transmitter.[*]

According to its k, FMN is not an especially good donor (k of the highest filled orbital $= 0.500$), but it should be remembered that the H-donor in oxidation is not FMN, but FMN[-], which has an electron accepted on an antibonding orbital and can be expected to be an especially good donor. This is the case as has been shown for $FMNH_2$ by the Pullmans,[†] the k value for the highest filled orbital having a negative value, which was the first instance observed of a filled orbital with an antibonding character (see Table I).

Other donor-acceptor pairs of biological substances have been described by Cilento and Giusti,[93] and Isenberg and the author.[‡] The electron transmission between serotonin and DPN[+] can be beautifully demonstrated by the freezing technique: on freezing, the mixture assumes the yellow color of DPNH. Fujimori[94] studied charge transfer, using pteridines as an acceptor.[95,96] The first to point out the possible role of

[*] Karreman (personal communication) showed that the FMN and indole molecule can be laid on top of one another in a way that positively charged C atoms come to lie opposite negatively charged ones, which may contribute to attractions between the two molecules and favor charge transfer. The strong interactions observed, however, cannot be ascribed solely to this factor since indole acts as a strong donor, also, with other acceptors.

[†] Ref. 25 on page 44.

[‡] Ref. 79 on page 81.

[93] G. Cilento and P. Giusti, *J. Am. Chem. Soc.* **81**, 3901, 1959.

[94] E. Fujimori, *Proc. Natl. Acad. Sci. U.S.* **45**, 133, 1959.

[95, 96] H. A. Harbury and K. A. Foley, *Proc. Natl. Acad. Sci. U.S.* **44**, 662, 1958; *Ibid.* **45**, 178, 1959. These authors measured the dissociation constants of complexes of isoallaxizins in various solvents and leave it open as to how far charge transfer was involved.

charge transfer in the function of DPN, and hence in biological oxidation, was Kosower.[*]

CORTISONE I₂

The interaction of hydroquinone and quinone was a classic π–π interaction. The interaction of serotonin and FMN also involved, in all probability, the π pool of these substances, though the transferred electron might have been derived from the lone pair of N. In these cases the planarity of the interacting molecules greatly favored complex formation, the two flat surfaces providing approachability and a free play of the attractive forces leading to intimate union.

Mulliken[†] has pointed out that charge transfer complexes can be formed not only by an extensive planar system of conjugated double bonds with their π pools, but also by local donors or acceptors, atoms or small atomic groups making part of bigger molecules. The lone pair of N or O molecules can serve as local "onium donors." Miller and Wynne-Jones[‡] have shown the NH_2 groups to be strong donors.

Reid and Mulliken[97] studied such a case: the charge transfer between pyridine and I_2. As they point out, the various solvents dissolve I_2 with two different colors. The homoiopolar ones, such as chloroform, dissolve it with a violet color, which is also the color of the iodine vapor. The heteropolar solvents, as alcohols, dissolve it with a brown color (which in higher dilution looks rather yellow). This color change they ascribe to a charge transfer which shifts the absorption

[*] Ref. 15, page 19.
[†] Ref. 47, page 55.
[‡] Ref. 60, page 60.

[97] C. Reid and R. S. Mulliken, *J. Am. Chem. Soc.* **76**, 3869, 1954.

of I$_2$ from the violet 520 mμ to 410–370. Pyridine also dissolves I$_2$ with a yellow color, and at the same time a "charge transfer spectrum" appears in the UV. The donor, in this case, is the lone pair of the N atom of pyridine. According to Mulliken and Reid, first an "outer complex" is formed which then goes over into the "inner one"; in the cortisone complex, also, some such rearrangement may occur.

As mentioned earlier, charge transfer is very sensitive to distance and a very close proximity of acceptor and donor are a *sine qua non*. It is easy to see that two flat aromatic structures can easily establish such proximity with their faces parallel, and conditions will favor charge transfer if the donor is the π electron system of the one while the acceptor is the π system of the other. The situation will be different if we are dealing with a localized donor, as was the case with pyridine in which the N atom donated one of the electrons of its lone pair. In this case the desired proximity can be established much easier if the acceptor itself is but a small molecule such as that of I$_2$ which can easily find the desired steric position in relation to the N atom. Charge transfer will be more difficult to achieve between such a local donor and the π system of a complex molecule.

Ketosteroids are big flat molecules with CO groups in which the O, with its lone pair of electrons, may act as a strongly localized donor. So in order to find out whether it really will do so, we do well to choose a small molecule, such as I$_2$, as acceptor. The question is only how to bring the O and I$_2$ together. Both I$_2$ and steroids are soluble in chloroform, but nonpolar solvents disfavor strong charge transfer. Strong charge transfer is favored by polar solvents. However, most polar solvents, like alcohol, which dissolve both steroids and I_2, are fair local electron donors themselves and so can be expected to compete with the steroid for I_2, with which they form charge transfer complexes themselves. Water is a strongly electropolar solvent and is a very poor donor, so, in principle it would suit our purpose, but it dissolves neither steroids nor I_2 very well. So, evidently, some trickery is needed. We can try, for instance, to dissolve both steroids and I_2 in chloroform, pour some of the solution on filter paper and let the chloroform evaporate, expecting the I_2 and steroid to form a complex. Then we can wet the paper, thus exposing the complex to the desired heteropolar solvent, expecting a charge transfer to take place and declare itself by a change in spectral properties.

If 0.01 M I_2 is dissolved in chloroform and 0.01 M cortisone is added, there is no change in color. If the solution is poured on (Watman 1) filter paper and the chloroform is allowed to evaporate, a yellow complex is left behind (without cortisone the I_2 evaporates too, leaving a colorless filter paper behind).

If the paper is dipped, now, into water, its yellow color turns deep blue. The color will be even more intense if instead of water the paper is dipped into a Lugol solution,* diluted with 25 volumes of water.†

---

* Lugol's solution contains 5 grams iodine and 10 grams KI in 100 ml water. A Lugol diluted 1/25 colors the paper itself but faintly brown.

† Acridine, which is not only a good acceptor ($k = -0.343$) but also a good donor ($k = +0.494$) gives a similar color under similar conditions.

The spectroscope shows a broad absorption with a maximum at 740 mμ, where neither I₂ nor cortisone absorbed. The absorption is structureless, has thus all the earmarks of a charge transfer spectrum, lying also at a longer wavelength than the absorption of the reactants. Kendall's compound S shows a similar, though weaker color change. The color developed by the cortisone-iodine compound is rather unstable and disappears in a few minutes. (Mulliken noted that many charge transfer complexes have a strong tendency to give secondary reactions, as going from "outer" to "inner" complexes, or giving irreversible changes.) Ten per cent methanol completely inhibits the development of the blue color (the alcohol being a not too poor electron donor).

The blue color, with the long wavelength absorption on the border of the infrared, indicates that relatively little energy was needed to transfer an electron from cortisone to I₂. So, with a stronger electron acceptor a charge transfer in the ground state could be expected.

The other ketosteroids studied, give, under similar conditions, an orange color which is more stable than the blue one given by cortisone. Here too the peaks are broad and structureless. The maximum lies with testosterone at 410 mμ, with deoxycorticosterone at 430 mμ, with Δ⁵-androstene-3,17-dione at 420 mμ, with Δ⁴-androstene-3,17-dione at 450 mμ, and with progesterone at 405 mμ. Cholesterol, stigmasterol, Δ⁵-androstene-3β,17β-diol, or estradiol, having no CO, do not show such color changes. Estradiol having a system of conjugated double bonds with their π pool, gives a charge transfer with trinitrobenzene, as shown by Williams-Ashman, while the guest of my laboratory.

90

The I$_2$ compounds can be produced preparatively.*

The color changes, given by steroids with I$_2$, have been described a decade ago by Zaffaroni[98,99] and his associates, who used them for the detection of steroids in chromatography. At that time these reactions were demonstrated to me by Dr. Oscar Hechter of the Worcester Institute for Experimental Biology and Medicine. Charge transfer being yet unknown to me, I could make no sense of these color changes. It was, evidently, the memory of this early demonstration which led me back to these reactions.†

Although there is no final proof for it, the probabilities are that these reactions of steroids and I$_2$ are charge transfer reactions indicating a strong donor ability of ketosteroids. The experiments of Talalay, Williams-Ashman, and Hurlock, to which I will come back later (Chapter XII) actually show that steroids can act as electron donors and acceptors, mediating the electron transfer between pyridine nucleotides in the presence of a catalyst.

---

* Dissolve 0.5% cortisone in hot water, then add 1/20 vol. of Lugol's solution. On cooling, the deep blue iodine complex of the steroid separates, can be spun off, washed with little water and pressed out between filter papers.

† For the generous supply of DOC and cortisone, I am grateful to Merck; for the progesterone and compound S and estradiol, to Dr. O. Hechter and the Worcester Foundation for Experimental Biology and Medicine; for the androstenes and other highly purified sterols, to Prof. Charles B. Huggins.

[98] A. Zaffaroni, R. B. Barton, and E. H. Keitman, *Science* **111**, 6, 1950.

[99] R. B. Barton, A. Zaffaroni, and E. H. Keitman, *J. Biol. Chem.* **188**, 763, 1951.

VIII

Miscellaneous Remarks

Water

IN MY LITTLE BOOK "BIOENERGETICS" TWO CHAP-ters were devoted to water. In the present book hardly any mention has been made of it. This is by no means because the author has changed his mind about structured water which, so to say, is half of the living machinery, and not merely a medium or space-filler. He still thinks that it is a mistake to talk about proteins, nucleic acids or nucleoproteins *and* water, as if they were two different systems. They form one single system which cannot be separated into its constituents without destroying their essence. I am more convinced than ever that half of the contractile matter of muscle is water, contraction the collapse of its structure, induced by actomyosin. Biology has forgotten water as a deep sea fish may forget about it. If water was hitherto passed by in silence in the present booklet this was because the author had no new experimental material to present, although there has been some progress, and also criticism in different quarters.

93

836

In my first little booklet I talked about "ice" formed around various molecules. J. D. Bernal[*] justly objected to the term "ice" which means something very definite for a crystallographer. Instead of ice one should rather speak about structured water, or water with restricted freedoms. All the same, "ice" is a nice and short word. It was probably for this reason that it has been used by outstanding earlier investigators like Frank and Evans.[100]

Speaking about progress made in the study of water in its relation to biological structures, the work of I. Klotz[101-103] should not remain unmentioned. Klotz showed that certain reactions of proteins could be explained by supposing the existence of a strongly bound sheet of water around the particles which greatly modified their accessibility,[101] reactivity,[102] and even mediate energy transmission.[103] Eigen's work[104] on the high proton conductivity of ice could bear on such a possibility.[105] Bernal's[106,107] latest work shows that even random liquids have their geometry.

Water may be important for biological happening both by its presence and its absence. Green[†] stresses

[*] Personal communication.
[†] Ref. 21 on page 30; ref. 28 on page 48.

[100] H. S. Frank and M. W. Evans, *J. Chem. Phys.* **13**, 507, 1945.
[101] I. M. Klotz, *Science* **128**, 815, 1958.
[102] I. M. Klotz and R. G. Heiney, *Proc. Natl. Acad. Sci. U.S.* **43**, 74, 1957.
[103] I. M. Klotz, *J. Am. Chem. Soc.* **80**, 2123, 1958.
[104] M. Eigen and L. de Maeyer, *Proc. Roy. Soc.*, A. **247**, 505, 1958.
[105] Review on the possible biological role of water structures: See P. L. Privalov, *Biofizika*, **3**, 738, 1958. Translation: *In* Biophysics **3**, 691, 1958. "The state and role of water in biological systems."
[106] J. D. Bernal, *Nature*, **183**, 141, 1959.
[107] J. D. Bernal, *Nature*, **185**, 68, 1960.

that the process of oxidative phosphorylation in mitochondria takes place in an anhydrous, lipid matrix. This anhydrous nature of the medium may alter happenings and energy relations. It has been shown by B. Pullman,[†] for instance, that the hydrolytic free energy of ATP depends, to a great extent, on the dielectric properties of the medium, and, as pointed out by Isenberg,[‡] can be expected to be considerably higher in an anhydrous than in an aqueous milieu. There may be processes, such as electron transfer, which simply might not work in a watery medium, as the spark plugs of our car refuse to serve if wetted.

In my little booklet on bioenergetics emphasis was laid on the excited states, singlet as well as triplet. This I did because I thought that till we really know what's what we had better look at everything. As mentioned in my present Chapter IV, a direct transition from the ground state into the excited state within one and the same molecule seems rather unlikely for processes which are not induced by photons. So, in a way, I lost faith in excited states and levels. Charge transfer helped me to recover it.

Ions

The donor-acceptor relations may also lead to a better understanding of ionic activity. Hofmeister has arranged the various ions according to their colloidal activity into the so-called "lyotropic series." This colloidal activity has been connected since with various parameters, as, ionic radius, the intensity of the surrounding electrostatic field, and hydration. For anions the series is mostly quoted as: citrate > tartrate > SO_4

[†] Ref. 6, page 6.
[‡] Woods Hole Conference on Molecular Excitation, 1959.

> acetate > ClO_3 > NO_3 > Br > I > CNS. For cations: Li > Na > K > RB > Cs. The colloidal effects are opposite on the two ends, and accordingly the initial members of the anion series are called "salting out" and the last members the "salting in" anions. My own laboratory has often made use of these activities, "salting out," precipitating proteins by sulfates, or "salting in," bringing into solution and depolymerizing actin, or dissociating actomyosin by elevated concentration of iodides. Rhodanides had similar effects. One may wonder why these ions "salt in," dissolve, dissociate, and whether some other parameter has not been overlooked? Such a parameter may be the ability to act as electron donors or acceptors. This assumption is supported by the ionization potentials and the parallel electron donor properties of the anions. SCN^- and I^- are at the end of the series and are relatively good electron donors. Br^- and $(NO_3)^-$ are less active but even $(ClO_3)^-$ is known as a fair donor.[*] The cations can be expected to act as electron acceptors, though they are less dominating and there is less difference among them, as expressed by their oxidation potentials (Table II). According to the Table, lithium will be the best acceptor, Na will be poorer than K, but there will be very little difference between them. All the same, when studying the action of a salt, we must bear in mind that what we measure in most cases will not be the donor action of the anion nor the acceptor action of the cation, but both, acting simultaneously. So, the solubilizing action of KI on a protein might be due to the prevailing donor activity of the I^-, the solubilizing indicating only the overall result of the dona-

[*] Ref. 47 on page 55.

tion and acceptance of electrons, and will not mean
that the cation did not act at all.

TABLE II
OXIDATION POTENTIALS OF ELEMENTS

Element	Oxidation Potential[a]	Element	Oxidation Potential[a]
Li	+3.045	F_2	−2.85
Na	+2.714	Cl_2	−1.36
K	+2.925	Br_2	−1.066
Rb	+2.925	I_2	−0.536
Cs	+2.923		

[a] Values, in volts, referred to the hydrogen-hydrogen ion couple
as zero, for unit activities at 25°C. For cations, as Li the reaction
is: $Li \rightarrow Li^+ + 1e$, for anions like F the reaction is $2F^- \rightarrow F_2$
$+ 2e$. (Quoted from the Handbook of Chemistry and Physics, 38
Ed., Chemical Rubber Publ. Co.)

The donor activity of iodide can easily be demonstrated by freezing a 10^{-3} M FMN in presence of 10^{-2}
or 10^{-3} M KI. On freezing,* the solution assumes a
brownish color, indicating the charge transfer with the
formation of FMN⁻. If the mixture is frozen in the
presence of 1% HCl, the final color reached within a
few hours is black. If time is given and the mixture is
stored at the low temperature over night, even the
isomolar 10^{-3} M KI will make the FMN turn completely black. Such a slow reaction is typical for many
charge transfer reactions. The phosphorescence of the
frozen FMN is completely wiped out, indicating the
transfer of a whole electron, that is a strong charge
transfer reaction in the ground state. KSCN also turns
the FMN solution brownish, but seems to be less active

* See first footnote on page 77.

than KI, which difference may be due to the poorer approachability of SCN⁻ and the greater "softness" of the I⁻ ion. That KSCN actually is a stronger donor than KI can be shown by using a different acceptor, and freezing a 10^{-3} solution of 1, 2-naphthoquinone-4-sulfate Na in presence of 0.1–0.01 M KI or KSCN. The naphthoquinone radical is brown, and on freezing the solutions turn brown. The tube containing KSCN will be darker brown than the one with KI, all these reactions being reverted on melting.

Charge transfer complexes of FMN are split by 0.1 M "neutral" salts. This holds true for many other charge transfer complexes. This dissociating action was, for a long time, a mystery to me, for heteropolar properties should rather promote than inhibit charge transfer. However, if anions and cations act as electron donors and acceptors, then this dissociating action can be ascribed to a competition of the anion and cation of the salt for the acceptor, respectively the donor of the complex.

Metachromasia

Metachromasia is a fascinating phenomenon, in all probability also connected with charge transfer. By metachromasia, following Ehrlich,[108] is meant the ability of certain dyes to stain different material in different color. Metachromasia has been studied by several investigators, most intensely by Michaelis and Granick.[109] As they showed, very dilute solutions of metachromatic dyes (over $10^{-6} M$) have a sharp ab-

[108] P. Ehrlich, *Arch. mikroskop. Anat. u. Entwicklungsmech.* **13**, 1877.

[109] L. Michaelis and S. Granick, *J. Am. Chem. Soc.* **67**, 1212, 1945.

sorption peak, given by the monomer. If the concentration of the dye is increased another lower peak appears at a somewhat shorter wavelength, which is
generally accepted to be that of the dimer of the
dye[110,111] (Fig. 17). With increasing concentration the

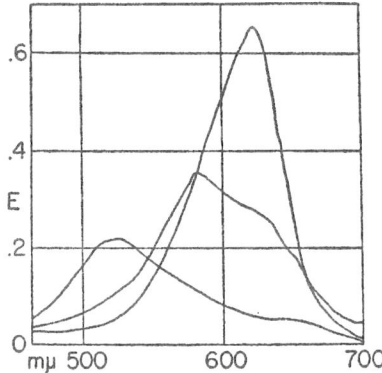

FIG. 17. Highest curve: Absorption of $10^{-6} M$ toluidine blue, 10
mm path. Lower curve: $10^{-5} M$ toluidine blue, path 1 mm. Lowest
curve: same plus equivalent polyethylenesulfonate Na.

monomer peak becomes lower, the dimer peak higher.
The latter always remains considerably lower than the
monomer, even if dimerization is nearly complete. If
a substance is added which is stained metachromatically, a third, still lower and broader structureless peak
appears at a shorter wavelength and both the monomer and dimer peaks disappear (Fig. 17). The classic
metachromatic dyes are toluidine blue, methylene
blue, and thionine.

[110] There is also a smaller peak and shorter wavelength. Th. Förster,
Naturwiss. **33**, 166, 1946.
[111] J. Lavorell, *J. Phys. Chem.* **61**, 1600, 1957.

Lison[112] observed that most naturally occurring substances which color metachromatically such as agaragar or chondroitin, contain SO₃ groups and if these groups are eliminated the metachromasia disappears. Evidently, the metachromatic coloration had to be due to an interaction of the SO₃ group with the dye. Since the absorption shifts to shorter wavelength and becomes smaller on dimerization, it was natural to believe that metachromasia, a further shift and decrease, was due to a further association of the dye, induced, somehow, by the SO₃ groups of the stained substance. The theory of polymerization (suggested with a question mark by Michaelis) was lately revived by Bradley and Wolf.[113]

The naturally occurring metachromatically dyeing substances, such as agar-agar or chondroitin, owing to their complex and only partially known structure, are unfit for quantitative study. The same holds for polyphosphate, polymetaphosphates, or silicates, which also give metachromasia. Thanks to the courtesy of the Upjohn Company, Kalamazoo, I am in possession of a polyethylsulfate, which is an almost ideal material. It consists of repeat units of CH_2—$CHSO_3Na$, has a molecular weight of 12,900 and is soluble in water.

If one of the metachromatic dyes, say toluidine blue, is gradually added to a stronger solution of the polyethylsulfate ($10^{-2} M$ in relation to the ethyl sulfate unit which weighs 134 grams) the first minute quantity of the dye which produces a visible color or meas-

---

[112] L. Lison, Histochimie Animale, Gauther, Paris, 1936; *Arch Biol.* **46**, 599, 1935.

[113] D. E. Bradley and M. Wolff, *Proc. Natl. Acad. Sci. U.S.* **45**, 944, 1959.

urable spectrum is metachromatic. This simple observation strongly speaks against the polymerization theory of metachromasia because the few molecules of the dye added can be expected to disperse rather than aggregate, there being less than one dye molecule per macromolecule.

The energy levels of the three classic metachromatic dyes mentioned show a peculiarity which I did not find in the several scores of dyes studied in collaboration with G. Karreman. The k's, both those of the highest filled and those of the lowest empty orbital, are very small (Table III) which makes the gap between

TABLE III
(condensed from Table I)
k VALUES FOR THE HIGHEST FILLED (hf) AND LOWEST EMPTY (le)
MOLECULAR ORBITALS

	hfmo	lemo
Toluidine blue	0.391	−0.359
Methylene blue	0.398	−0.354
Thionine	0.398	−0.354
Nile blue sulfate	0.591	−0.133

the two very narrow—hence the blue color with its long wave absorption which lies partly in the infrared. These values make the dyes into both good donors and good acceptors. It follows that they should actually be able to form a self-charge transfer complex, one molecule of the dye donating an electron to another, if we bring two molecules into close proximity, as can be done by freezing their dilute aqueous solution. On freezing of their solution all three dyes actually change their color from the "normal" blue to the metachro-

101

matic purple. All this, taken together supports the idea that the metachromatic coloration is due to the formation of a charge transfer complex.* SO_3 groups, according to the tabulation of Mulliken,† are "ketoid" acceptors while the O's with their lone pairs of electrons could also act as donors.

If one molecule of toluidine blue is added pro SO_3 a metachromatic complex is formed which precipitates. An excess of the colloid, in the relation 2:1, keeps the complex in solution. If more than one molecule of dye is added pro SO_3, a 1:1 metachromatic complex precipitates leaving the excess of the dye in the dark blue solution behind.

There are three points which make metachromasia rather fascinating. Lison‡ observed that the SO_3 group gives metachromasia only if linked to a macromolecule. Low-molecular sulfo compounds give no metachromasia. I can fully corroborate this statement. My three polyethylsulfate preparations of molecular weight 6000, 12,900, and 27,000 grams reacted equally strongly. However, when going under 5000 grams molecular weight, taking other big molecules such as heparin or chondroitinsulfuric acid, the metachromasia became poorer. The limit seems to lie around 5,000 grams molecular weight. The size of the molecule seems to

* This does not entail that the metachromatically shifted absorption of the dye corresponds to a charge transfer spectrum. It is, probably, analogous to the shift in the spectrum of iodine in charge transfer reactions. In charge transfer complexes formed with alkaloids or pyridine the spectrum of iodine shifts isosbetically from 520 mμ to 410 mμ while the charge transfer spectrum proper is in the UV.

† Ref. 47 on page 55.

‡ Ref. 112 on page 100.

mean something for charge transfer and it may mean something most important for biology, but what, I am unable to say. The 5,000 gram limit is fascinating because, as judged by the experience on myosin, this is the smallest size unit of which this protein is built.

This relation of metachromasia to molecular weight may have something in common with the experience of Kainer, Bijl, and Rose-Innes, who found that the bulkiness of the reactant molecules favors charge transfer.[*]

The other point which fascinates me is connected with Strugger's[114] observation, according to which live tissues are colored green by acridine orange, and dead tissues, red. According to my experience the situation is not this simple. All the same, metachromatic coloration may open a way to study donor and acceptor properties of live tissues, and the living state or death may have intimate relations to these properties.

The cartilage and synovial membrane is rich in SO_3 groups and one may wonder what their function should be. Two antimalarial agents, atabrine and chloroquine, both of which are good electron donors, have been found to alleviate symptoms of arthritis, alleviated also by cortisone, another donor. These relations might suggest new ideas about this disease which is crippling a considerable percentage of our kind.

The third point which made metachromasia interesting for me is that it may shed light also on the nature of ionic activity. The experiments of B. Kaminer (unpublished) show that salts, containing the two last anions of the "salting in" end of the Hofmeister series,

[*] Ref. 57 on page 60.

[114] S. Strugger, *Jenaische Z. Naturwiss.* **37**, 1941.

103

induce a metachromatic coloration in all three dyes discussed. This pleads for the correctness of both assumptions: that the salting in effect is due to electron donation and that metachromasia is due to charge transfer. The comparison of the dyes discussed supplies further evidence. As the k values indicate (Table III) Nile blue sulfate is a better acceptor than the other three dyes. Accordingly, we can expect that it is also more sensitive to the action of anions, an expectation borne out by the experiment.

The fact that Nile blue sulfate shows a metachromatic change with nitrate, bromide, iodide, and rhodanide is rendered especially fascinating by the observation of Kahn and Sandow,[115] according to which these four anions are capable of increasing the maximal twitch tension of muscle by their action on the membrane. It seems likely that this action is due to the electron donating properties of these anions, which may open the way to the closer correlation of function, charge transfer, and the understanding of ionic activity and drug action.

An apparent inactivity of a salt, like KCl, does not necessarily mean that its ions did not act as donors or acceptors. It may also mean that these just compensate each other's overall action. So even the ions of an apparently inactive salt may render empty or saturated energy bands conductant by donating or accepting electrons.

[115] A. J. Kahn and A. Sandow, *Ann. New York Acad. Sci.* **62**, 137, 1955.

Part Two

Problems and Approaches

IX

On the Mechanism of Drug Action

IT IS NOT THE INTENTION OF THE AUTHOR TO attempt to give an answer to the unsolved problems outlined at the outset. All he hopes to do in this part of his booklet is to show that some of these problems, if looked at through the glasses of the submolecular, may appear in fresh light, suggesting tentative theories which may lead to useful experimentation.

The first example I want to take is that of the mechanism of drug action. A number of drug actions have found an explanation in competitive inhibition. As a false key may fit into a slot without being able to open it, so a drug may fit into some biological binding site, and displace some natural substance without being capable of fulfilling its function. A few biologically active substances were found to act by blocking some important atomic group, as the SH. In spite of these successes, the action mechanism of the great majority of drugs remained unexplained. This pertains for instance, to most of the alkaloids, and pertains also to the hormones, drugs produced by our own body free of charge.

Drug action represents a wide field of inquiry and

107

our approach can be but a very limited one. We may ask, for instance, whether charge transfer is not involved in drug action, some of the drugs acting as electron donors or acceptors. It is difficult to predict the biological effect of such a charge transfer, for the action will not only depend on the question, whether electrons are donated or tapped off, but will also depend on the site of action. The cellular membrane, for instance, which dominates many functions of the cell, has mostly a negative charge inside and a positive one outside. So electrons donated at the inside should increase the charge and lead to hyperpolarization, and with it to inhibition, while electrons donated at the outside will decrease the potential and can be expected to cause excitation. The opposite will hold for electron acceptors. We can be prepared also to meet paradoxical results. Let us suppose that a biological substance acts by donating electrons. A drug, acting as electron donor, may compete with a natural substance for its acceptor. Interfering thus with the normal course of electron transmission this drug, though itself a donor, may produce an effect which corresponds to the inhibition of electron donation. In spite of all these incertitudes we may expect one definite interrelation: If a substance exerts its biological activity by accepting or donating electrons, then it should have exceptional acceptor or donor properties.

An indication that acceptor-donor properties may underlie pharmacological action was given on the preceding pages. It was shown that the indoles are exceptionally good donors, and a great number of biologically active substances contain an indole ring (serotonin, lysergic acid, bufontine, indole acetic acid). Attempts were made earlier by Popov, Castellani-Bisi, and M. Craft[116] to connect pharmacological activity, like convulsant effect, with electron donation. That indoles may actually act in their biological reactions as electron donors is further suggested by the fact that the OH group, induced into the serotonin molecule at position 5 equally increases pharmacological activity and the electron donating property. Serotonin is one of the strongest donors I ever met though the k of the highest filled orbital is but moderately lower than that of indole (0.461). So there are still unidentified molecular parameters which greatly influence donor-acceptor properties. That charge transfer may be involved in pharmacological activity is suggested also by the fact that the various alkaloids, such as quinine, nicotine, atropine, physostigmine, morphine, strychnine, aconitine, etc., all act as good donors toward iodine in an anhydron solution,* and so do the ketosteroids, while estrogens, as mentioned before, prove to be donors toward trinitrobenzene. An additional evidence for the possible pharmacological importance of acceptor-donor properties was given by Fujimori† who showed that various pteridine derivatives are the stronger antimetabolites of folic acid the stronger they act as acceptors toward tryptophan.

---

* The I_2-spectrum absorption (in chloroform solution) shifts in presence of these alkaloids with an isosbestic point from 520 mμ to a shorter wavelength around 400 mμ, while in the UV a broad charge transfer is developed. With nicotine the charge transfer spectrum develops slowly, within an hour or so. This tardiness of the reaction is characteristic for many charge transfer reactions.

† Ref. 94 on page 86.

[116] A. I. Popov, C. Castellani-Bisi, and M. Craft, *J. Am. Chem. Soc.* **80**, 6513, 1958.

As a further example we may quote nitrophenols, in the first place 2,4-dinitrophenol and halophenols. Dinitrophenol is known to have a strong uncoupling activity on oxidative phosphorylation, one of the greatest puzzles of current biochemistry. Polynitrophenols, on the whole, are very good electron acceptors, owing to the unsatisfied electronegativity of their nitro group, and it is believable that they capture the high-energy electrons which should go into the "black box" discussed on page 23, forming charge transfer complexes with their donor. The same holds for halophenols, the strong activity of which, on phosphorylation, has been discovered by Clowes and Krahl.[117-119] Their action is less specific than that of dinitrophenol (Middlebrook et al.[120]). It should also be remembered that one of the most important regulators of cellular oxidation, the hormone thyroxine, is also a halophenol, containing the most electronegative halogen, iodine. It uncouples oxidative phosphorylation. It develops its activity very slowly, in 48 hours or so—a striking sluggishness, characteristic of many charge transfer reactions, one example of which has been analyzed by R. H. Steele, i.e. the very slow complex formation between rhodamine B and dichlorophenol.* Nitro compounds also promote the transition of the attracted electron into

---

* See "Bioenergetics," (A. Szent-Györgyi, Academic Press, New York, 1957), page 54, notes about the depressant action of dichlorophenoxyacetic acid on oxygen uptake; see Ibid. p. 122.)

[117] G. H. A. Clowes and M. E. Krahl, J. Gen. Physiol. 20, 145, 1936.

[118] M. E. Krahl and G. H. A. Clowes, Ibid. 20, 173, 1930.

[119] M. E. Krahl and G. H. A. Clowes, J. Cellular Comp. Physiol. 11, 1, 21, 1938.

[120] M. Middlebrook and A. Szent-Györgyi, Biochim. et Biophys. Acta. 18, 407, 1955.

the triplet state.† This can be expected also from thyroxine with its very heavy iodine atoms.

Another train of thought was followed by B. Kaminer (unpublished). The proboscis (a smooth muscle preparation of *Phascolosoma* (a marine worm), if denervated, shows no spontaneous rhythmicity. Serotonin, a donor, is inactive. However, if the preparation is treated first with good acceptors, serotonin induces rhythmic contractions. The reverse, also, holds: after treatment with serotonin, acceptors induce rhythmic contraction.

The obvious way to find additional evidence for a connection between electron transfer and pharmacological activity could be found by picking drugs with extraordinary pharmacological activity and then asking whether they have extraordinary qualities as electron donors or acceptors. If the extraordinary biological activity of the drug in question is due to an electron transfer then we can also expect that drug to have exceptional qualities as electron donor or acceptor.

A drug with a unique biological activity is the tranquilizer chlorpromazine* (Fig. 18) which is widely used in treating the symptoms of schizophrenia. Since its introduction a great number of hospital beds have become empty, because among all diseases it was schizophrenia which had permanently occupied the most hospital space. With this end in view, Karreman *et al.*[121] calculated the *k*-values of this substance and

† Ref. 52 on page 59.
* The author is grateful for a supply of this substance "Thorazine" to Smith, Kline and French, Philadelphia, Pa.

[121] G. Karreman, I. Isenberg, and A. Szent-Györgyi, *Science* 130, 1191, 1959.

111

found that of the highest filled orbital quite exceptionally high. In fact, it is smaller than zero, has a negative sign, being equal to −0.210, the orbital having an antibonding character. Such values had been found earlier by B. and A. Pullman in leucomethylene

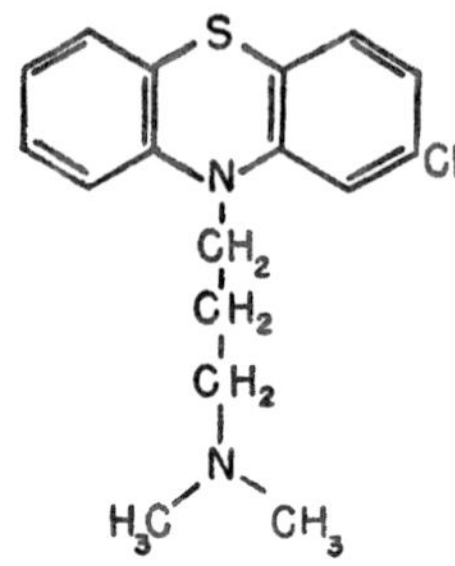

Fig. 18. Chlorpromazine.

blue or reduced riboflavine, but both these substances are unstable, autooxidize rapidly, two electrons having been forced upon them. Chlorpromazine is the first substance found which has an antibonding highest filled orbital in its normal, stable state. It can be expected to be an exceedingly good monovalent electron donor capable of forming stable charge transfer complexes, the complex formation being supported by the planar nature of the extensive system of conjugated double bonds with its pool of π electrons and the lone pairs of electrons on the N and S atoms.

These ideas may become the starting point of various trains of thought. If the action of this drug is due to its qualities as electron donor and these qualities can be calculated, then, maybe, new and even more potent drugs may be found by calculation, which may be still better electron donors and may have an even more favorable action than chlorpromazine. And, if the symptoms of schizophrenia can be influenced favorably by electron donation, then, perhaps, a lack of electrons may be involved in the genesis of this disease, and if so, the question comes up: what has induced it? If the disturbance can be corrected to some extent by electron donation, could it not have been caused by the presence of an electron acceptor? It is most fascinating to note, in this connection, that B. Pullman and A. M. Perault* find hematoporphyrins both good donors and good acceptors, and find the metabolic product of protoporphyrins, biliverdin, a quite exceptionally good electron acceptor. This substance is, in a way, the counterpart of chlorpromazine, the k of its lowest empty orbital having a plus sign (0.021), a bonding character. To my knowledge this is the first substance found to have such a value. Urobilin, a further metabolite has still strong donating properties. What makes these data exciting is the fact that we metabolize a great quantity of hem daily, 0.7% of our total blood, and Klüver[122-125] has for many years called attention to a close relation of hematoporphyrin and mental disturbance. In newborns a hyperbilirubinemia is often found to be accompanied by grave anatomical damage to the brain. Bilirubin inhibits the

* Ref. 4 on page 6.

[122] H. Klüver, *Science* **99**, 482, 1944.
[123] H. Klüver, *J. Psychology* **17**, 209, 1944.
[124] H. Klüver, Functional difference between the occipital and temporal lobes, *in* "Cerebral Mechanisms and Behavior," L. A. Jeffres (Ed.), Wiley and Son, New York, 1951, pp. 172–178.
[125] H. Klüver and E. Barrera, *J. Psychol,* **37**, 199, 1954.

production of hems,[126] and similarly to several other electron acceptors, inhibits oxidative phosphorylation.[127] Should schizophrenia be due to the presence of a strong acceptor in the blood, then, maybe, this substance could be inactivated by a strong donor with which the acceptor could form a charge transfer complex, relieving the patient of all his troubles. I am not proposing here a new theory of schizophrenia. What I am trying to show is merely that quantum mechanics may suggest unexpected new approaches to important problems which have been stagnant such a long time. The examples quoted also show that the distance between those abstruse quantum mechanical calculations and the patient's bed may not be as great as believed. Starting on these lines, schizophrenia, with all its mysterious psychic aberrations, may turn out, some day, to be but the side effect of some, in itself harmless, metabolic disturbance which, once recognized, can perhaps easily be corrected.

[126] R. F. Labbe, R. Zaske, and R. A. Aldrich, *Science* **129**, 1741, 1959.

[127] L. Ernster, L. Herlin, and L. Zellerstrom, *Pediatrics* **20**, 647, 1947.

X

On ATP

ATP IS THE MAIN FUEL OF LIFE PRODUCED IN photosynthesis and oxidative phosphorylation. In both cases it is produced by an electric current, that is, the energy released by a "dropping" electron. So the question arises whether the energy of the $\sim$ is not transformed into a current back again when it has to drive life and produce work, w. I have confessed earlier my inability to make sense of the transformation of chemical energy into w, except by supposing that, somehow, the chemical potential is translated into electronic energy.

A transformation demands a transformer. Looking at the structure of the ATP molecule (Fig. 19) one wonders whether the adenine group is not the transformer. Why should Nature have attached it to the phosphate chain if she only wanted the energy of $\sim$'s? Pyrophosphate or inorganic triphosphate should have done just as well! I have also shown, in my little "Bioenergetics,"[*] that the phosphate chain and the adenine are linked together by the pentose in such a way that

[*] "Bioenergetics," page 64.

115

the former can fold back on the latter making the two terminal phosphates touch upon the two N's at positions 6 and 7 and form a tetradentate chelate with them, using Mg as a link. This possibility has found some, though not conclusive, evidence in the infrared studies of Epp, Ramasarma, and Wetter.[128] The phosphates may fold back onto the adenine also in such a way as to lie flat on the face of the purine. In this case the two terminal phosphates could be in touch with the ring but not the third one. The folding could take place in the pentose and so need not bend the phosphate chain, which bending would be resisted by the negative charges on the chain and would also interfere with the intimate contact with the purine. Adenine is a good donor and the P atom has unoccupied d orbitals which could make it into an acceptor.* So the possibility exists that within the ATP molecule an electron is transmitted from adenine to phosphate, either from the former's π-pool, or from the lone pair of one of its N's.† The rotatory dispersion of ATP solutions also indicates a folded molecule.

* It seems possible to give reasons why Nature may have selected the P atom for its central role. Elements on the left side of the periodic table are metallic and tend to give off electrons and not take them up. The middle ones, C and Si have no special affinity, while the ones farther to the right of P may have a too great electron affinity to serve as reversible electron transmitters. Of the elements in the same column as P, N would not do since it has no d orbitals. The next heavier element, As, seems to be incompatible with life for other reasons. So P may be the only element which could fill the role of a reversible electron transmitter. I am indebted to Dr. M. Calvin for discussions and ideas on this line.

† The metachromasia of polyphosphates supports the assumption of their being good electron acceptors.

[128] A. Epp, T. Ramasarma and L. R. Wetter, *J. Am. Chem. Soc.* **80**, 724, 1958.

It seems worthwhile to reflect about this point since this formation, adenine-pentose-phosphate, returns in almost all important biological catalysts of energy production or transmission: DPN, TPN, FAD, and coenzyme A. The first three have an analogous formation also on the other side of the phosphate chain, the sequence being pyridine-pentose-pyrophosphate-pentose-adenine, or isoalloxazine-pentose-pyrophosphate-pentose-adenine, so that not only the adenine could fold back onto the phosphates but also the pyridine and isoalloxazine. The phosphate thus sandwiched between adenine and pyridine, respectively adenine and iso-

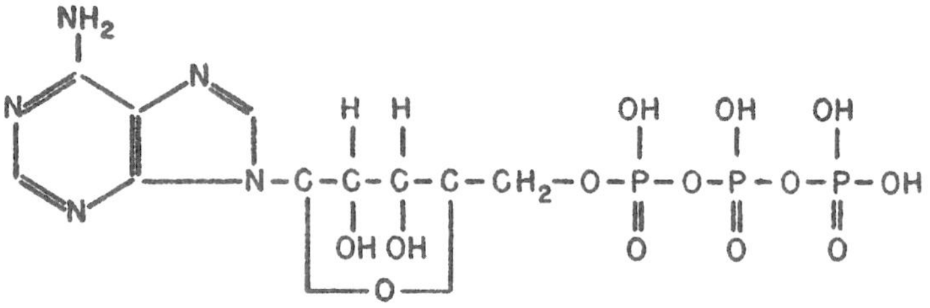

FIG. 19. Adenosinetriphosphate (ATP).

alloxazine could mediate the electron transfer between the two cyclic structures on its two sides. Weber actually showed a strong interaction between adenine and isoalloxazine[129] respectively pyridine.[130] In coenzyme A there is no pentose between the phosphates and pantothenic acid, but this latter has enough freely rotating links to be able to fold back on the phosphate. It may be also worth mentioning that only triphosphate or diphosphate can fold back onto the adenine

[129] G. Weber, *Biochem. J.* **47**, 114, 1950.
[130] G. Weber, *Nature* **180**, 1409, 1957.

117

in a way to allow a sufficient overlap and charge transfer. Monophosphate would not do, as is easy to show by means of an atomic model. All the catalysts mentioned contain triphosphate or diphosphate (pyrophosphate), but not monophosphate.

A charge transfer involves the uncoupling of an electron pair, and represents thus, in a way, an embryo of a current, generating a negatively and a positively charged molecule. So we may turn our phantasy loose for an instant, and ask whether such a charge transfer could help us to explain the function of these nucleotides? Let us consider first the synthesis of ATP from ADP and P in photosynthesis. The Pullmans[*] have shown lately that, almost without exception, the two atoms disconnected in enzymic hydrolysis carry a positive formal charge. This is true also for the P—O bonds in ATP. These charges can be expected to predispose the "dipositive bond" for hydrolytic splitting by their mutual coulombic repulsion. But, if these charges predispose the P—O bond for splitting, they must also resist the formation of this bond, and so an electron, transferred onto the P by charge transfer must turn the repulsion into an attraction promoting synthesis. The empty place on adenine may be filled by an electron of the oxidative cycle, releasing thus the electron donated to the phosphate chain from its electrostatic bondage.

For consideration of the reverse process, the splitting of the P—O—P bond and the transformation of its energy into a current, we could take as an example muscle contraction in which the energy of the $\sim$ is transformed into motion—mechanical work. Let us

[*] Ref. 5 on page 6.

118

start with an ATP molecule in which an electron has been transferred from adenine to the phosphate chain.* By this transfer the adenine has to become a good acceptor, the phosphate a good donor, and so (as a *jeux d'esprit*, a play of the mind) we could suppose that adenine accepts an electron from myosin while phosphate donates one to actin. This would make the positive and negative charges become widely separated. Now let us imagine that we take the myosin molecule between the two fingers of our right hand, the other end (actin) between the two fingers of our left, and pull the chain apart. If we do this slowly the transferred electron can be expected to slip back to its original position. If we would tear the chain in two very fast, allowing no time for the electrons to slip back, we would have to do work to overcome the coulombic attraction between positive and negative charges. We could tear up the complex without resistance only in one case: if, previously, we have hidden in the link to be torn a sufficient amount of free energy to pay for the process. In this case the chain could be broken without external energy. The P—O—P bond actually has about 10 kcal worth of energy hidden in it and so this bond could simply break and thus separate the positive and negative charges in actin and myosin, generating thus an electric potential, a current. There is no difficulty in making theories of how such a current could produce contraction.

I can see here the outlines of an extremely cunning

* Owing to the unshared pairs of electrons of the O atoms the phosphate chain can be looked upon as conjugated with mobile π electrons.

scheme. Possibly, Nature created in ATP a substance which could take up an electron on one end and give one off on the other, while the terminal $\sim$ has enough energy hidden to make it possible for a hydrolytic enzyme to split the bond, thus trapping the positive and negative charge out in the contractile protein, thus transforming chemical energy into a current.[*]

I am conscious of speculation which, at the moment, has no experimental support. All the same, it is this speculation which allows me to make for the first time, after twenty years of muscle research, an intelligible picture of how ATP could drive muscle. This is a most fundamental problem, close to the core of life.

It can be hoped that ESR will give the answer to these problems. It is impossible to predict whether a charge transfer between adenine and phosphate would give a signal or not, depending on the degree at which the two electrons become uncoupled. All the same, there is no reason for not putting ATP into the ESR machine. This has been done by Isenberg et al.[134] with

[*] It seems possible that some such tricks, as represented by the hypothetical function of ATP in muscle, has been applied by Nature in other fields too. So, in vision, the photon absorbed by retinene may cause a charge transfer onto opsin. The cis-trans transformation which follows the photochemical acts, splits the link between dye and protein, or at least may interfere with the overlap of wave functions, trapping, thus, the transferred electron on the protein, starting up "excitation."[131-133]

[131] G. Wald, Scientific American, 201, 92, 1959.
[132] V. J. Wolff, R. S. Adams, H. Linschitz, and D. Kennedy, A.M.A. Archives of Ophthalmology 60, 695, 1958.
[133] V. J. Wolff, R. S. Adams, H. Linschitz, and E. W. Abramson, Ann. New York Acad. Sci. 74, 281, 1958.
[134] I. Isenberg and A. Szent-Györgyi, Proc. Natl. Acad. Sci. U.S. 45, 1232, 1959.

the help of Dr. H. Kallmann of New York University, who kindly offered his equipment. A signal was obtained, consisting of a broad peak, stretching over several hundred gauss and carrying smaller secondary peaks. Since that time the author's laboratory has become the happy owner of an ESR machine. On this machine ATP is now subjected to a more detailed study, allowing us to experience all the difficulties connected with ESR studies. We hope to learn soon to what extent the observed spectra were artifacts, or due to impurities, or were due to the electronic distribution in the ATP molecule itself, giving information about the nature and function of this molecule.

XI

On the Chemistry of the Thymus Gland

RESEARCH IS RARELY GUIDED BY LOGIC. IT IS GUIDED mostly by hunches, guesses, and intuition. All the same, once we get somewhere and present our results, we like to present them as a logical sequence. If the course of our work had to be plotted, we would plot it as the straight line in Fig. 20, where A follows B, B

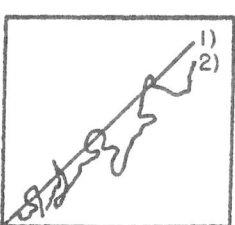

FIG. 20. Curve 1: course of research as presented. Curve 2: actual course. (See text.)

follows C, etc. The real curve would be more like curve 2, at least for my studies on the thymus.

In my book on "Bioenergetics" I dragged in the thymus without any binding reason and tried to show

123

how logically it fitted into my story, though it was but a hunch that this gland had something to do with energy transmission because its extracts were colored more yellow than corresponded to their FMN content. I only guessed that myotonia was due to a defective energization of the membrane and that the absence of the "thymus substance" was responsible for it. However loose my reasoning, I could show that the thymus extracts actually cured the myotonic symptoms in goats. As is generally known, 2,4-dichlorophenoxy-acetic acid produces myotonic symptoms in animals which might be due to the drugs complexing with my hypothetical "thymus substance." So I was encouraged to find that the active agent which benefited my goats, could be shaken out from the watery thymus extracts by 2,4-dichlorophenol.

I am still unable to state what this active substance is, or tell whether it is a specific product of the thymus. If I bring this gland in again, it is for two reasons. In my "Bioenergetics" I left the story unfinished. Although I am still unable to finish it, I feel the obligation toward my earlier readers to tell them that I am still trying to finish it. My second reason is that my observations brought to light the existence of a second substance which is antagonistic to the first, and it is not impossible that the two represent some basic biological balance connected with the acceptor-donor relation. So, even if it does not clear up the physiology of the thymus, my work can be quoted at least as an example of "serendipity," the experience of looking for one thing and finding another.

To be correct, this work on the thymus is not my work at all, but a joint work, started in association with Jane McLaughlin, and pursued at present in association with Dr. Andrew Hegyeli, while the animal tests are conducted by Dr. Hedda Rév.

The difficulty in isolating the active substance lay in the fact that we possessed no suitable biological test for it.* Looking around for a bioassay we also tried its influence on malignant growth. Our extracts seemed to slow it down, or stop it altogether.

I will forego the discussion of the difficulties of using malignant growth as a test object. Eventually the difficulties were solved by using one and the same animal for the test and its control, measuring growth rate while under the influence of extracts, and after. Deviation from logarithmic growth was found to be a reliable indication of an activity.

With this test in hand, isolation was begun. For a little while things went well. Then everything got mixed up and activity decreased or was lost where it had to increase. Eventually the puzzle was solved by the demonstration that our extracts contained two active substances, the one promoting, the other inhibiting malignant growth and the result depended on their balance. It is a common experience that two unknown variables make results messy.

The problem is still vigorously pursued and it is hoped that the next edition of this booklet (if any) will contain the chemical formulas of the two active agents.

* Myotonic goats, as a test object, are very unsatisfactory.

125

XII

The Living State

THIS, TO MY MIND, IS A MOST INTRIGUING PROB-lem. It is also the most obscure one, as difficult to define as life itself. Although we cannot define life, we know life from death and can distinguish between a dead cat and a live one, which corresponds to the two basic states of biological systems. The problem is, perhaps, not quite as abstruse as it appears on first sight because we can produce the transition by simple experimental means from the one state into the other, at least "one way." To stay with the cat, we can, for instance, wipe out consciousness in one blow by clamping the carotid artery, that is, cutting the O_2 supply to the brain. Consciousness being the main product of the brain, this means the cessation of biological activity. It could be objected that this is not death, the change being reversible. However, the reversibility has very narrow limits. If we keep the carotids clamped for a few minutes, the change becomes irreversible. This means that the living system was in a metastable state which needed a permanent supply of energy for its maintenance, the half-lifetime of the living state being

127

of the order of minutes. No doubt, similar changes can be produced also in other organs though the half-lifetime may vary to some extent. The same point can be demonstrated by cyanide in an even more dramatic fashion. If we inject cyanide into one of our veins the first thing we notice is that we are dead. Cyanide, as we know from Warburg's classic work, cuts out the activation of O_2, combining with some catalytic metal present.

That living systems are in a metastable state which demands the permanent supply of energy for its maintenance, is no surprise. The second law of thermodynamics could predict this. What is unexpected is the brevity of the half-lifetime. This is a surprise because most organs have a considerable store of ATP, backed up by creatine phosphate, while limited amounts of ATP can also be produced anaerobically. This ATP should be able to tide us over much longer periods of lack of O_2, if it could supply the energy needed to maintain the metastable living state.

There is no doubt in my mind that the energy for muscular contraction is derived from ATP. When muscle enters into activity to perform its public function, contraction, it suddenly demands great amounts of energy. O_2 supply is slow and continuous. So muscle could not depend on O_2 for the energy of its contraction, directly, and has to use as energy source a substance like ATP, available any time in moderate amounts. But, if ATP can cover the sudden great energy demands of a public function, why can it not cover the modest and continuous demands made by the private life of the cell, the maintenance of its metastable state? Is it not that there are two inde-

128

pendent systems of energy production, both using O_2 as their final electron acceptor, the one located in mitochondria and responsible for the production of ATP, while the other is located in the basic cellular structures themselves which have to be maintained in their peculiar state?

A glance at Fig. 4 makes such an assumption seem reasonable. This figure intends to show that a considerable portion of the electrons, which are sent down the chain of oxidative phosphorylation to produce ATP by their energy, are derived from DPNH or TPNH. But why should the organism take this long detour over mitochondrial oxidative phosphorylation and ATP to satisfy a slow and continuous demand? Why could the high-energy electrons of DPNH or TPNH not be placed more directly on the living structure which eventually could couple them to O_2 using their energy more directly?

All this may be speculation, but once we recognize the existence of a metastable living state we have to ask questions and speculate to find a reasonable working hypothesis. To say that it is ATP which maintains the living state is no more than a speculation either, a speculation burdened with major difficulties. The hypothesis I am proposing is this: the cell derives the energy necessary for the maintenance of its living state directly from DPNH or TPNH. As a first move in this direction, we may ask questions about the two ends of the hypothetical process: how are the electrons of DPNH or TPNH transmitted to the protein edifice, and how could electrons be transmitted from this edifice to O_2? There are tentative answers possible to both questions.

129

As to the first: Talalay, Williams-Ashman, and Hurlock[135-138] have shown that steroids can mediate the electron transfer between pyridine nucleotides in the presence of protein (both specific and unspecific). Why then could they not mediate the electron transfer between TPNH or DPNH and the protein edifice itself? That steroids can, with their $=O$ act as donors, has been shown to be probable in Chapter VII, and a steroid which has donated an electron must be a good acceptor, enabling the molecule to act as an electron transmitter.

As to the other end, O_2, Debye and Edwards[139] have found that proteins, illuminated by shorter UV, emit, at low temperatures, a long-lasting afterglow, which has been studied also in my laboratory[140] using, by preference, proteins of the lens which contain no hem dyes. This light emission is completely quenched by O_2 which, evidently, picks up the excited electrons. The direct transmission of electrons from the protein to O_2 may thus be demonstrated by simple means.*

* It is noteworthy that the long-lived afterglow of tryptophan is not quenched by O_2 while that of protein is, though the emission of protein is, in all probability, an emission of its tryptophans. Though less probable, the possibility should not be disregarded that the quenching by O_2 is due to its paramagnetism, being a "paramagnetic quenching."

[135] P. Talalay and H. G. Williams-Ashman, *Proc. Natl. Acad. Sci. U.S.* **44**, 15, 1958.

[136] P. Talalay, B. Hurlock, and H. G. Williams-Ashman, *Ibid.* **44**, 862, 1958.

[137] B. Hurlock and P. Talalay, *J. Biol. Chem.* **234**, 886, 1958.

[138] B. Hurlock and P. Talalay, *Arch. Biochem. Biophys.* **80**, 468, 1959.

[139] P. Debye and J. O. Edwards, *Science* **116**, 143, 1952.

[140] A. Szent-Györgyi, *Biochim. et. Biophys. Acta.* **16**, 167, 1955.

All the same, the experience with cyanide indicates that some metal-containing catalyst is involved in coupling the electrons to O_2.[141]

The hypothesis presented would also give an explanation for the hitherto unknown mechanism of action of steroids. What we know about it at present is hardly more than that steroids are indispensable for life, and life simply fizzles out in their absence.

What happens between the hypothetical steroid and O_2 end is but a question mark, the whole problem having never been stated before. The fate of the electron within the protein or nucleoprotein structures would depend to a great extent on the nature of these macromolecular aggregates and their hydrate shells, whether they are semiconductors, proton conductors, etc. Our approach to these problems will depend to a great extent, also, on the question of what we mean by energy. Whatever this word may mean, the energy of an electron can conveniently be expressed in terms of ionization potential.* It seems likely to me that as

---

* One could also talk about an "electron pressure" (in analogy to pH) which is inversely proportional to IP. The negative potential of the "inside" of most cells, as compared with the "outside" may have relations to the higher electron pressure inside.

[141] The cyan-sensitive enzymes may be peroxidases. H. G. Williams-Ashman, M. Cassman, and M. Klavins have shown (*Nature* **184**, 427, 1959) peroxidases to oxidize DPNH or TPNH under catalytic influence of estrogens. This is the more remarkable since estrogens, which induce a hypertrophy of the uterus muscle, also cause peroxidases to appear in this organ in high concentration (F. V. Lucas, H. A. Neufeld, J. G. Utterback, A. P. Martin, and E. Stotz, *J. Biol. Chem.* **214**, 775, 1955). These peroxidases could also utilize the H_2O_2 formed in the reduction of O_2. That estrogens, with their aromatic structure can act also as π donors, was demonstrated by Williams-Ashman (see page 90).

131

pH is equalized within the single cells by buffers, so the ionization potentials also are equalized by electron donors and acceptors. Ascorbic acid, with dehydro-ascorbic acid, may be one of these "IP-buffers." Ions, acting as donors or acceptors may also be instrumental in this equalization. As electron acceptors and donors are, in the extended ideas of G. N. Lewis, acids and bases, so "IP-buffer" means but an extension of the idea of an acid-base buffer. The NH groups of the protein backbone can be expected to be strong donors, while the CO groups may act as "ketoid acceptors" or lone-pair donors, both located strategically in the continuous chain of H-bonds which conjugate the whole protein molecule and may create continuous energy bands (see Evans and Gergely[*] and Eley *et al.*).[†]

Certainly, the living state involves many factors and at present we can hardly do more than to try to collect single items in the hope of fitting them together later.

One of the characteristics of the living state is the accumulation of ions against a gradient, concentrations becoming equalized in death. We still have no final answer to the question of how ions are accumulated. We only have theories. Is it a redox pump, as suggested by Conway,[142] or is it some different juggling with carriers, or is it a bulk property,[143] a consequence of structure? We do not know. The fact that ion trans-

[*] Ref. 31 on page 51.
[†] Ref. 32 on page 51.

[142] E. J. Conway, *Science* **713**, 270, 1951.
[143] S. L. Baird, Jr., G. Karreman, H. Müller, and A. Szent-Györgyi, *Proc. Natl. Acad. Sci. U.S.* **43**, 705, 1957.

port against a gradient is paralyzed by cyanide but not by 2,4-dinitrophenol (which stops ATP production by oxidative phosphorylation) also suggests that the energy used is derived from some other source than ATP.

One important characteristic property of the living state seems to be its paramagnetic behavior, as suggested by the work of Commoner[*] and his associates who found the intensity of the ESR signal given by various animal or vegetable tissue proportional to the intensity of metabolism. One is reminded of the Latin saying: "the faster a motion the more it is."[†] The intenser life and metabolism, the more life it is, and the intenser the paramagnetic behavior, while in death diamagnetic susceptibility increases.[144] The ESR signals of live tissue seem to come chiefly from the free radicals formed in metabolism. Commoner's groups, as well as Ehrenberg and Ludwig,[‡] found the free radical formed on partial reduction of FMN and its protein complexes to give a signal. Other free radicals may have contributed too, and so might have contributed other charge transfer complexes going into their ionic state. In any case, the ESR signal seems to be a signal of life, though given also by stable substances such as melanin or various resins.[**] Anything alive seems to give a signal, "your finger or the mouse's tail."[§]

[*] Refs. 7, 8, 9, 10 on page 6.
[†] "Quo celerior motus eo magis motus."
[‡] Ref. 89 on page 82.
[**] Ref. 7 on page 6.
[§] Conversation with M. Calvin.

[144] E. Bauer, *Nature* **138**, 801, 1936.

These observations give the impression that the paramagnetic behavior is due to the free radicals formed in the one-electron redox processes, but we should not forget that a not inconsiderable part of the living protein structure must actually be present in the form of a charge transfer complex, is thus, in a way, a free radical. To convince oneself of this, one has to look only at the intense brown coloration of the liver. In spite of the undoubtedly numerous efforts, nobody ever has isolated the dye which could have been made responsible for this color. I have isolated a considerable quantity of a chocolate brown substance, which on treatment with HCl, fell into a colorless protein and a golden yellow flavine. Evidently, this substance was a charge transfer complex of protein with FMN and the brown coloration of the liver has to be ascribed to the formation of a charge transfer complex with isoalloxazins. The kidney and the adrenal cortex are brown too. The brain may be white because of the great bulk of dielectrics it has to contain, although the cortex, which is richer in cells, has a brownish coloration, and where cells aggregate, as in the "red nucleus," the brown color becomes quite evident. It is not impossible that flavines are not the only substances forming charge transfer complexes with the proteins. Steroids, as donors, might do likewise, giving charge transfer complexes with spectra in the infrared or ultraviolet.

As the last item on my list of the peculiarities of the "living state" I would like to touch upon one of the most intriguing observations of the last years, that of the very broad signal given by nucleoproteins (Blumenfeld, Kalmanson, and Shen-Pei*). This signal (if really due to the protein-nucleic acid complex) indicates a density of unpaired electrons which is almost comparable to that found in metals, to which metals owe their conductivity. This finding, if corroborated, may lift a veil which now obscures the real nature and meaning of protein, nucleic acid, and nucleoprotein.

All these factors are parts of the magnificent edifice of life, as bricks, lying on the roadside may have been parts of a Greek temple.

In an earlier chapter I emphasized the biological importance of "organization," by which I meant that if Nature puts two things together a new structure is born which can no more be described in terms of the qualities of its components. The same holds also for functions. In living systems the various functions, too, seem to integrate into higher units. We will really approach the understanding of life when all structures and functions, all levels, from the electronic to the supramolecular, will merge into one single unit. Until then our distinguishing between structure and function, classic chemical reactions and quantum mechanics, or the sub- and supramolecular, only shows the limited nature of our approach and understanding.

* Refs. 71, 72 on page 71.

135